Laser Devices
and Applications

Laser Devices and Applications

Edited by

Ivan P. Kaminow

Member, Technical Staff
Bell Telephone Laboratories

Anthony E. Siegman

Professor of Electrical Engineering
Stanford University

A Book of Selected Reprints, prepared under the
sponsorship of the IEEE Electron Devices Group
and the IEEE Microwave Theory and Techniques
Group, and consisting principally of papers reprinted
from the PROCEEDINGS OF THE IEEE.

The Institute of Electrical and Electronics Engineers, Inc. New York

International Standard Book Numbers:
Clothbound: 0-87942-025-1
Paperbound: 0-87942-026-X

Library of Congress Catalog Card Number 73-77998

PRINTED IN THE UNITED STATES OF AMERICA

Preface

During the past decade many excellent invited tutorial and review papers on lasers have been published in the PROCEEDINGS OF THE IEEE and IEEE SPECTRUM. Special Issues of the PROCEEDINGS on lasers and laser applications appeared in 1963, 1966, and 1970. This volume gathers together a selected group of these papers, primarily from the PROCEEDINGS, with a few additional papers from SPECTRUM. The resulting collection is designed to serve as a convenient introduction and source of reference material on lasers and their applications. The applications- and device-oriented articles included here will complement the more basic articles on laser theory and laser physics in the IEEE PRESS Selected Reprint Volume, *Laser Theory*, edited by Frank S. Barnes.

Because this volume has been limited to reprints from the PROCEEDINGS and SPECTRUM, a few important areas are not fully covered, and a few articles may need some updating. We have attempted to indicate these deficiencies in the Introduction to each section. A comprehensive book list on lasers has been appended to this volume as a guide to other sources of information on lasers.

Additional laser-related review articles of similar caliber are planned for future issues of the PROCEEDINGS. As they appear, they may serve as the basis for another reprint collection to supplement this one.

Contents

Part I: Introductory Survey

As an introductory overall survey of laser devices, we present only one paper, which in turn gives a large number of references to the literature. Although the title of this paper refers specifically to communications, its scope is, in fact, considerably broader.

Coherent Optical Sources for Communications

JOSEPH E. GEUSIC, WILLIAM B. BRIDGES, FELLOW, IEEE, AND
JACQUES I. PANKOVE, FELLOW, IEEE

Abstract—The development of coherent optical sources, producing usable amounts of power, has provided a stimulus for communications research. Coherent sources in the form of lasers and parametric oscillators are available at wavelengths which span the entire optical spectrum. This paper reviews the state of the art of coherent optical sources with major emphasis on the most highly developed sources.

I. INTRODUCTION

THE introduction of the laser in 1960 and the subsequent work on intense coherent optical sources within the past decade has stimulated research in optical communications. An imposing number of coherent sources have been demonstrated with useful power outputs at wavelengths which span the optical spectrum. Lasers of all types (solid-state ion, semiconductor, liquid, and gaseous), both pulsed and continuous, have been utilized to investigate specific optical communication problems relating to modulation, detection, terminal design, and propagation. Since the discrete nature of laser transitions permits only limited frequency tuning of such sources, the laser has been combined with nonlinear optical materials to produce optical parametric oscillators with wide frequency tuning capabilities. These latter sources have not yet been used in optical communications research but are certain to play an important role in its future.

The purpose of this paper is to review those coherent sources and related techniques which are believed to be important to current and future optical communication research and system planning. In general the discussion will emphasize continuously operating sources at room temperature, since they are the most suitable for communication experiments and systems. Section II is a brief discussion of the principles of operation of coherent optical sources, and Section III covers representative methods of controlling the spatial and temporal behavior of optical oscillators. The state of the art of gas lasers, semiconductor lasers, solid-state ion lasers, and optical parametric oscillators is reviewed in Sections IV–VII.

II. TYPES OF COHERENT OPTICAL SOURCES

A brief description of the principles of operation of the basic optical source types (lasers, parametric oscillators, and second-harmonic generators) is given in this section.

A. Lasers

The maser principle, which was originally developed in the microwave region, is the basis for the laser. Application

Manuscript received in final form August 21, 1970.
J. E. Geusic is with Bell Telephone Laboratories, Murray Hill, N. J.
W. B. Bridges is with Hughes Research Laboratories, Malibu, Calif.
J. I. Pankove is with RCA Laboratories, Princeton, N. J.

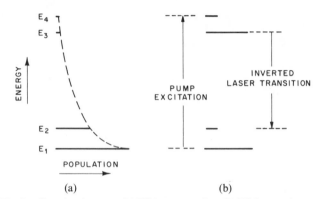

Fig. 1. Four-level system. (a) Without pumping. (b) With pumping excitation applied between levels E_1 and E_4 to produce inverted population between levels E_3 and E_2.

of this principle to optical transitions provided the first method of achieving net gain at optical frequencies. Extension of the maser principle to optics was first proposed by Schawlow and Townes in 1958 [1] and applied by Maiman in 1960 [2] to optical transitions in ruby to produce the pulsed ruby laser. Less than a year later Javan *et al.* [3] reported operation of the first continuous-wave (CW) laser, which used a He–Ne discharge. Since then laser action has been demonstrated in numerous systems.

Strictly speaking, a laser is an amplifier of light. In all the lasers which will be discussed, optical gain is achieved by utilizing a medium which has three or more energy levels. The four-level system shown in Fig. 1 is illustrative. As depicted in Fig. 1(a), in the absence of pumping the population of the lower levels exceeds that of higher levels and all transitions are absorptive. If pumping excitation (the pumping may be optical, electrical, or chemical depending on the system) is applied selectively between levels E_1 and E_4, the level populations are modified. Under some conditions, for example, the case in which particles in levels E_4 and E_2 decay rapidly and predominantly to levels E_3 and E_1, respectively, the population of level E_3 will exceed that of E_2, as is shown in Fig. 1(b) and the system will exhibit optical gain at frequency ν_{32}.

A laser oscillator is constructed by utilizing the laser gain medium inside of an optical cavity as is shown in Fig. 2(a). Optical regenerative gain occurs for light traveling along the cavity axis. The cavity length l is typically 10^3 to 10^6 times larger than the laser wavelength, and typically more than one axial or longitudinal cavity resonance will fall within the laser gain profile. Oscillation occurs [4] at those cavity resonances lying within the inhomogeneous width of the laser transition for which the laser gain exceeds the cavity losses. This is depicted in Fig. 2(b) for a situation where the laser peak gain exceeds the single-pass cavity

Reprinted from *Proc. IEEE*, vol. 58, pp. 1419–1439, Oct. 1970.

2

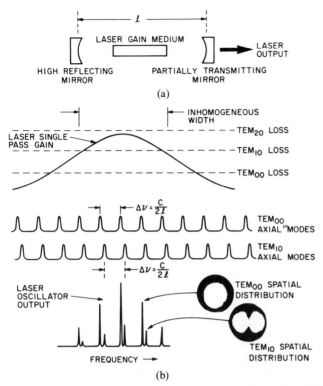

(a)

(b)

Fig. 2. Laser oscillator. (a) Regenerative optical oscillator. (b) Multimode operation of laser showing gain curve, cavity resonance, and cavity modes on which oscillation occurs.

losses for the two lowest order transverse modes. The spatial and temporal coherence of a laser source arises from the regenerative character of the combined laser gain medium and the optical cavity.

Review articles on lasers and on modes in optical resonators are to be found in [5]–[12]. Discussions of the laser oscillator and the statistical properties of its output are found in [13]–[17].

B. Parametric Conversion Sources

Parametric interactions were first considered by Lord Rayleigh [18], and they have been extensively utilized at microwaves [19]. There are many types of optical parametric interactions [20]–[23]; however, we shall discuss only the case involving three optical waves and the nonlinear polarizability of a noncentrosymmetric crystalline material. In such materials the second-order nonlinearity is of the form

$$P_i = \sum_j \sum_k \chi_{ijk} E_j E_k \tag{1}$$

where the χ are the second-rank tensor components of the nonlinear susceptibility, the P are vector components of the generated optical polarization, and the E are the vector components of the applied electric field. For any two applied optical fields at frequencies v_1 and v_2 fields will be generated in the material at both the difference frequency $|v_1 - v_2|$ and the sum frequency $(v_1 + v_2)$.

1) Difference Mixing (Parametric Amplification and Oscillation): Consider an intense optical wave v_p (the pump) and a second wave of lower frequency v_s (the signal) incident on a nonlinear material as depicted in Fig. 3(a). A wave v_I (the idler) will be generated in the nonlinear medium at

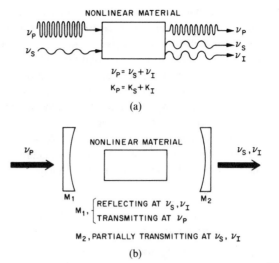

(a)

(b)

Fig. 3. Parametric down-conversion. (a) Schematic diagram of basic process, (b) Its use in doubly resonant (signal and idler frequencies resonated) optical parametric oscillator.

the difference frequency $v_p - v_s$. The three coupled waves satisfy the condition

$$v_p = v_s + v_I \tag{2}$$

which is the condition for conservation of energy. In order for the three waves to interact strongly, allowing an efficient power transfer between each other, momentum conservation or phase matching of the propagating wave vectors

$$\mathbf{k}_p = \mathbf{k}_s + \mathbf{k}_I \tag{3}$$

is required. For the collinear case depicted in Fig. 3(a), (3) becomes

$$n_p v_p = n_s v_s + n_I v_I \tag{4}$$

where the n are the refractive indices of the material at the respective frequencies. In Fig. 3(a) the signal wave beats with the pump wave through the nonlinear polarizability to produce the idler wave. As the idler travels through the nonlinear material it also beats with the pump wave to produce a polarization wave at the signal with just the proper phase to amplify the original signal wave. It follows that the signal and idler waves experience exponential gain at the expense of the pump wave. Since there is gain at the signal and idler frequencies, an oscillator can be built, as is shown in Fig. 3(b). Oscillation occurs when the pump power provided produces sufficient gain to overcome cavity losses at the signal and idler frequencies. With presently existing nonlinear materials, optical parametric oscillation is possible only by using a laser pump. For a given pump frequency and nonlinear material, oscillation occurs, in the collinear case, at those v_s and v_I which satisfy (2) and (4). Tuning of v_s and v_I can be achieved by utilizing means such as temperature or an electric field to change the refractive indices of the crystal. This type of oscillator does not use discrete energy levels and, since the nonlinear χ are approximately constant at optical frequencies, it is inherently broadly tunable. Optical parametric oscillators have recently been reviewed by Harris [24].

2) Sum Mixing (Second-Harmonic Generation): In sum frequency generation, two intense optical waves of frequencies v_1 and v_2 interact in the nonlinear medium to produce a third wave of frequency $v_3 = v_2 + v_1$. As in our previous discussion strong interaction occurs only if there is momentum conservation, $\mathbf{k}_3 = \mathbf{k}_1 + \mathbf{k}_2$. In contrast to difference mixing, there is no possibility for achieving electronic gain and only power conversion from waves v_1 and v_2 to v_3 will occur. Sum mixing of lasers in a nonlinear material is another way of generating new source frequencies. The most common example is second-harmonic generation for which $v_1 = v_2$ and $v_3 = 2v_1$. A detailed discussion of nonlinear optical sum frequency generation can be found in [25], [26] and its practical usage with CW lasers in [27]–[29].

III. CONTROL OF THE SPATIAL AND TEMPORAL CHARACTERISTICS OF OPTICAL OSCILLATORS

The spatial and temporal characteristics of an optical oscillator are important in communication systems. For efficient optical heterodyne detection, the source and local oscillator should have the same transverse mode distribution, preferably the lowest order TEM_{00} mode. Single-frequency operation (i.e., operation on a single axial mode) is required in some cases. In a pulse-code modulation (PCM) system [30], an oscillator which emits repetitive pulses is most useful. In this section some representative techniques which have been developed to control the spatial and temporal behavior of optical oscillators are discussed.

A. Transverse Mode Selection

In the laser oscillator in Fig. 2, transverse mode selection can be achieved by selectively increasing the loss of one transverse mode relative to another by introducing an appropriate obstacle into the cavity. For example, the TEM_{00} can be suppressed if a small absorbing wire is placed at one of the mirrors in a direction coincident with the field null of the TEM_{10} mode. More typically, operation in the TEM_{00} mode and suppression of all other modes including the TEM_{10} mode is desired. Calculations show that the lowest order modes always have lower diffraction losses than higher order modes. As a result, if the diameter of the resonator is sufficiently constricted at some point with an aperture stop, only the lowest order transverse mode can oscillate, all others being prohibited by excessive diffraction loss. Other methods of achieving single transverse mode operation of optical oscillators are mentioned by Kogelnik [31]. The inclusion of the effect of transverse modes on the saturation of the gain medium has been theoretically treated by Bloom [32] and Fox and Li [33].

B. Axial Mode Selection and Stabilization

Operation on a single axial mode can be realized in a number of different ways utilizing a variety of multiple-mirror optical resonator configurations to achieve the appropriate limitations of resonances [34]. One representative method which has been employed utilizes an intracavity Fabry–Perot etalon, as is depicted in Fig. 4. The etalon is inserted at the position of the beam waist in the cavity with a tilt angle that provides sufficient reflection loss to prevent oscillation of all the unwanted axial modes of the long cavity. Oscillation occurs at a transmission maxi-

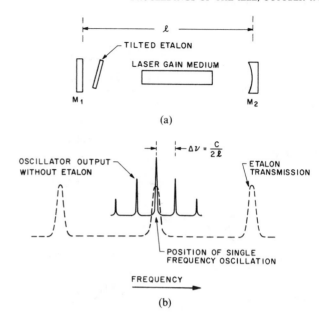

Fig. 4. Single frequency operation laser using tilted intracavity etalon. (a) Schematic diagram. (b) Effect of etalon transmission on selection of single axial resonance.

mum of the etalon. This maximum is made to coincide with a laser cavity mode near the center of the gain profile. The optimum design of the tilted etalon mode selector and its use for stabilizing the output of an optical oscillator has been described by Danielmeyer [35]. The frequency stabilization of gas lasers has been reviewed by Polanyi and Tobias [36].

C. Cavity Modulation Techniques

A variety of methods utilizing intracavity modulation techniques have been developed which permit continuously pumped optical oscillators to be repetitively pulsed. Three such techniques, which have been historically referred to as mode locking, Q switching, and cavity dumping, will be discussed.

In addition, intracavity modulation has been employed by Harris [37] to produce FM or single-frequency outputs. Since a discussion of this work will not be given in this paper, the reader is directed to [37].

1) Mode Locking: A single transverse, multiaxial mode laser can be repetitively pulsed by introducing an intracavity loss or phase modulator which is driven at a frequency equal to the frequency separation, $\Delta v = c/2l$, between axial modes. The effect of the perturbation is to couple the axial modes in an AM manner. For a large number N of locked modes the oscillator has an oscillation envelope which is a pulse train with period $T = 2l/c$, pulsewidth $\tau = T/N$, and a peak power N times the average power. For an oscillator with $l = 30$ cm and $N = 20$, $T = 2 \times 10^{-9}$ second and $\tau = 10^{-10}$ second. Mode-locked pulsewidths of 500 ps, 200 ps, and 50 ps have been obtained, respectively, from He–Ne [38], Ar [39], and Nd:YAlG [40] lasers. Such mode-locked sources are useful in high-data-rate optical PCM systems [30]. The experimental and theoretical aspects of mode-locked lasers have been studied extensively and are discussed in [37]–[46].

2) Q Switching and Cavity Dumping: Although mode locking can be extended to low repetition rates by using

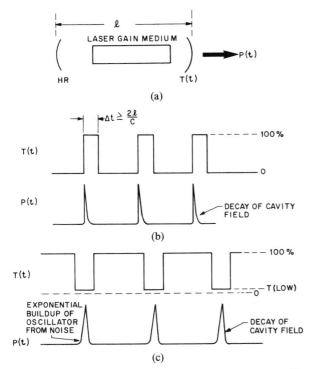

Fig. 5. Repetitive cavity dumping and Q switching of laser oscillator. (a) Laser oscillator which has time-dependent mirror transmission; in practice, this is effectively accomplished using an intracavity modulator and a mirror whose transmission is time-independent. (b) Cavity-dumping operation cycle. (c) Q-switched operation cycle.

long optical cavities, it is somewhat impractical at repetition rates below 30 MHz. Up to 100 kHz, Q switching is effective with a laser such as Nd: YAlG; from 100 kHz to 30 MHz cavity dumping is preferred. The latter two techniques can be described with the aid of Fig. 5 where the transmission of the output mirror of the oscillator is assumed to be time dependent in response to a drive signal. In practice, this can be effectively accomplished using an intracavity modulator and a mirror whose transmission is time independent.

In cavity-dumped operation [Fig. 5(b)] the oscillator stores energy in the optical cavity during the interpulse period when the output mirror transmission is zero. For the case in which maximum peak output power is desired, the transmission is switched to 100 percent (for a time interval $\geq 2l/c$) at a time when the optical cavity has maximum stored energy. As a result, all the stored energy of the cavity is dumped and the output is a narrow pulse whose peak power is equal to the maximum internal circulating power. Restoration of the mirror transmission to zero permits a repetition of the cycle. In general, to maximize the average output power for a given repetition rate the mirror transmission in the high transmission state is less than 100 percent and is held for times longer than $2l/c$. The technique of cavity dumping was first proposed by Vuylsteke [47] for pulse pumped lasers and applied to CW gas lasers by Steier [48]. A theoretical discussion of repetitive cavity dumping and, in particular, its application to the Nd: YAlG laser has recently been given by Chesler and Maydan [49].

The operation cycle for Q switching is shown in Fig. 5(c). Here the mirror transmission is 100 percent during most of the cycle, so as to prevent oscillation and to allow

the energy to be stored in the laser medium (in contrast to the intercavity flux storage used in the cavity-dumping technique). During this period, the single-pass laser gain builds up to a value which exceeds the mirror transmission in the low transmitting state. The mirror is switched to its low transmission state for a period sufficient to allow the oscillator to build up from noise and discharge the inversion energy stored in the gain medium. This period depends on the round-trip transit time of light in the cavity and the maximum single-pass unsaturated gain achieved. Q switching was first proposed by Hellwarth [50] and applied to pulsed solid-state lasers by McClung and Hellwarth [51]. Theoretical and experimental investigations of repetitively Q switched, continuously pumped lasers are reported in [52]–[55].

IV. GAS LASERS

Gas lasers are, perhaps, more difficult to discuss as a group than either semiconductor or solid-state lasers because they possess more diverse properties. Their wavelengths span the range from the shortest (Ne IV, 0.2358 μ [56]) to the longest (ICN) 774 μ [57]) among lasers. Their CW or average power outputs also range from the highest (>9 kW from CO_2 [58]) to the "lowest" (that is, the lowest of practical interest, the stabilized He–Ne laser used for metrological applications [59]). They utilize an amazing variety of pumping methods, e.g., electrical excitation in the form of glow discharges, both RF and dc excited (He–Ne and CO_2); arc discharges, RF, dc, and pulsed (ion lasers); special discharges (the ultraviolet N_2 laser [60] and recent types of CO_2 lasers [61]); optical excitation (He lamp pumping of Ce vapor [62] and CO flame pumping of CO_2 [63]); chemical excitation (the HF laser [64] and the DF pumped CO_2 laser [65]); thermal excitation (expansion-inverted CO_2 [66]); and various combinations (photo-dissociated CH_3I [67], electrically formed HBr [68], etc.). Many of these gas lasers could be used for communication, but because of the wide selection of operational characteristics available to the user, only the most useful few have received the serious attention and development necessary to realize performances approaching their optimum as practical optical communication sources. Even the few selected for development exhibit too wide a variety of characteristics to receive a complete discussion here. We will have to be content with examples of the most familiar types used for communication and restrict ourselves to considering the characteristics most important for communication, i.e., power output, life, efficiency, mode control, frequency control, and growth potential.

The three examples chosen are the three most popular gas lasers now in use, i.e., the He–Ne laser, the Ar ion laser, and the CO_2 laser. Table I lists the important characteristics of these lasers: the "age," the principal wavelength(s), the nature of the active medium (for the most common excitation methods; the (E) or (H) after radio frequency (RF) indicates the principal RF coupling method used), and the usual range of average or CW output powers and efficiencies. A more detailed description of each laser

TABLE I
CHARACTERISTICS OF GAS LASERS

	He–Ne	Ar II	CO_2
Years since "Discovery"	8.5	6.5	5.5
Wavelength (μ)	0.633	0.488	10.6
	1.15	0.515	9.6
	3.39	(0.45–0.53)	(9.6–10.8)
Medium	Glow discharge in He–Ne mix	Arc discharge in pure Ar	Glow discharge in CO_2–N_2–He
Current density (A/cm²)	0.05–0.5	100–2000	0.01–0.1
Excitation	dc, RF (E)	dc, RF (E, H)	dc, RF (E)
Power (at wavelength)	(0.633)	(multicolor)	(10.6)
Best laboratory (watts)	1	100	9000
Best commercial (mW)	100	20	1000
Lowest commercial (μW)	100	1	1
Efficiency (percent)	0.001–0.1	0.01–0.2	1–20

type including the technical problems most relevant to their use as optical communications sources is given in the following subsections.

A. He–Ne Lasers

It is reasonable to say that the He–Ne laser has reached its peak performance because of its age and the relatively "nice" nature of its active medium, a noble gas glow discharge. The principal wavelengths of interest for communication are the red line at 0.6333 μ and the 3.39-μ infrared line; some applications have also been found for the 1.15-μ line. The commercially available output power of 100 mW for the red line listed in Table I is not likely to increase for purely practical reasons. Almost 1 watt has been obtained in a laboratory type He–Ne tube [69]; however, the tube was over 5 meters long, while a 1 watt Ar ion laser is typically less than 1 meter long. The "lowest commercial power" listed for the He–Ne laser is ~100 μW obtainable from small single-frequency stabilized units [59]. The efficiency figures shown correspond to the range of output power and are also not likely to improve.

The power output and efficiency figures for the 3.39-μ line are roughly 25 percent of those given for the 0.633-μ line in any given tube size. This is simply the ratio of photon energies for the two lines and reflects the fact that they share a common upper level. The gain coefficient for the 3.39-μ line is much higher than for the 0.633-μ line (all other things being equal, the gain coefficient being proportional to λ^3). As a consequence, the 3.39-μ laser discharge tube can be used practically as a single-pass amplifier while the 0.633-μ laser is confined to oscillator service by its very low gain coefficient, typically a few percent per meter.

He–Ne lasers exhibit a sharp optimum value in power output as a function of discharge current; a further increase in input power does not yield a corresponding increase in output power. This optimum operating point occurs for quite fundamental reasons, i.e., the neon upper laser level is populated by resonant collisions with excited helium atoms

$$He(2s^1S_0) + Ne \rightarrow He + Ne(5s'[\tfrac{1}{2}]_1^0) \qquad (5)$$

for the 0.633- and 3.39-μ lines. Because the excited helium state involved is metastable (i.e., it cannot decay by the spontaneous emission of a photon), it is both created and destroyed by electron or wall collisions. Thus the helium metastable population at first increases with discharge current (roughly proportional to plasma electron density), but soon saturates when the creation and destruction rates by electron collision are equal. The neon upper laser level follows the helium metastable population closely through process (5) and thus also saturates with current. On the other hand, the lower neon laser levels Ne($3p'[3/2]_2$) for the 0.633-μ line and Ne($4p'[3/2]_2$) for the 3.39-μ line, have allowed transitions to levels below them and are thus primarily destroyed by spontaneous emission (independent of electron density). However, these lower levels are populated by electron collision. For this reason their populations continue to increase with increasing current. Since the laser gain is proportional to the population difference, a maximum gain will occur at a somewhat lower current than that necessary to saturate the helium metastable population [70]. This optimum discharge current density is relatively modest (0.05–0.5 A/cm²) in typical He–Ne lasers, so that the associated technology has, in large measure, also "saturated" with simple glass envelope tubes and hot or cold cathodes with small current loading. The modest technology requirements imposed by the neon-sign-like active medium allowed a fairly complete mapping of the optimum operating condition very early in the game. Subsequent engineering efforts have been devoted primarily to extending tube life by reducing gas cleanup (largely through the proper choice of cathode type) and improving the optical elements within the cavity (mirrors, windows). Electrodeless RF discharges were used early in the research stage, but they have almost disappeared because of their relatively severe gas cleanup rates. Fig. 6 shows a typical small He–Ne laser discharge tube. The glass discharge capillary extends completely through the hollow cold cathode. The anode (weld ring) is evident at the left end of the tube and a getter is suspended from the cathode pins. Brewster's angle optical windows are sealed on each end

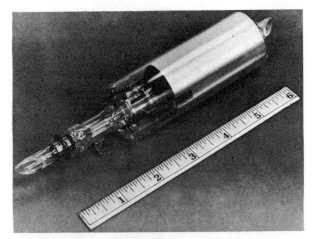

Fig. 6. Typical small (0.5 mW) external-mirror He–Ne laser.

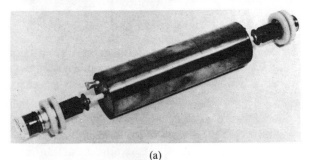

(a)

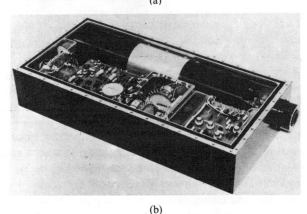

(b)

Fig. 7. (a) Metal–ceramic internal-mirror He–Ne lasers for space-qual-fied systems. (b) Integrated space-qualified He–Ne laser package (5 mW) with all solid-state power conditioning system. Prime power requirements are 30 watts at 24–32 volt dc.

with low vapor pressure epoxy. While this sealing method may seem undesirable to those experienced in the vacuum-tube art, it remains the best compromise to date when window cleanliness and optical quality, vacuum properties and fabrication cost are balanced. Tubes using these construction techniques have given operation on life tests in excess of 22 000 hours and show a statistical mean time between failures (MTBF) in excess of 28 000 hours [71].

A somewhat more challenging engineering development is shown in Fig. 7, a metal–ceramic He–Ne tube designed for use in space [72]. The discharge tube shown in Fig. 7(a) has mirrors internal to the vacuum envelope and thus requires no intracavity sealed windows, although an internal Brewster's angle window (no vacuum seal) is used in the version shown in order to provide a plane-polarized output.

The flexible bellows on each end allow mirror motion and also serve as anodes. Only small portions of the ceramic capillaries are visible at each end of the tube, but they extend into the center of the large hollow cold cathode. Because no epoxy is used in this tube, a high temperature bake-out schedule can be used to ensure complete outgassing of all elements and assure long operating life. The space-packaged laser is shown in Fig. 7(b). The discharge tube is contained within the beryllium cylinder, located at the top of the package; this cylinder maintains the optical alignment of the laser mirrors. The remainder of the packages contains the solid-state power conditioning and telemetry electronics. The specification on this tube is 5-mW output power at 0.633 μ for a minimum of 10 000 hours in a spacecraft environment.

Because of its high state of development, special purpose He–Ne lasers such as the space-qualified unit previously described, are now straightforward engineering projects. The various techniques for controlling the spatial and temporal characteristics referred to in Section III, with the exception of Q switching, have all been used successfully on He–Ne lasers; in fact, these techniques were all originally demonstrated with He–Ne lasers. The remaining challenges for He–Ne laser development is primarily improvement in manufacturing economy. Fully automated production should be able to reduce the price to the level of a conventional glass electron tube, provided large volume applications develop. The availability of such cheap lasers should open the way for new communications uses.

B. Ar–Ion Lasers

If He–Ne lasers are the "glass receiving tubes" of the laser world, CW ion lasers are the klystrons and TWTs in terms of complexity. Despite intensive research and development efforts, the peak in ion laser performance has not yet been reached for several reasons. First, the laser medium is much harder to handle than the He–Ne glow discharge. The ion laser medium is typically a low-pressure (less than 1 torr), small capillary arc (1 to 10 mm in diameter), with current densities ranging from 100 to more than 10 000 A/cm^2 (the latter figure refers to pulsed operation). This is two to four orders of magnitude larger than the densities required by the He–Ne laser. Second, the output power continues to increase with increasing current density; at present this value is limited by discharge technology (2000 A/cm^2 for CW operation). Thus the technology has not yet "saturated;" the solution of a problem which allows an increase in performance merely leads to new technological problems at the new power level.

This property of the ion laser can be understood from first principles in much the same way as the He–Ne laser [73]. In the ion laser, the upper laser levels (the $4p$ ionic states) are populated by a combination of direct electron collision and radiative decay from higher lying levels populated by electron collision. Although the details are not known completely quantitatively, it is certain that at least two successive collisions are required, and that the last collisional step involves the impact of an electron and an ion.

As a consequence, the upper level population varies as the product of electron density and ion density, or approximately as electron density squared, since the plasma has charge neutrality. Over the range of plasma parameters typical of the ion laser the electron density is approximately proportional to the discharge current density, so that the upper level population should vary as the square of the discharge current density. This has been confirmed experimentally by spontaneous emission measurements. The lower laser levels (the $4s$ ionic states) are populated in much the same way, and thus the population difference varies as the square of the current density.

There are processes which can upset this simple picture, such as depletion of the neutral gas from the discharge region of the tube by plasma processes of heating, or trapping of the spontaneous ultraviolet emission which depopulates the lower laser level. However, thermal depletion of the neutral gas in the discharge active region is only now being observed at the highest current densities (>1000 A/cm^2), and radiation trapping is apparently only a problem in pulsed ion lasers, where ion heating does not have time to occur. Thus for CW ion lasers designed for the 0.1 to 100 watt output range, the output continues to increase with increasing current limited only by the tube construction techniques. An empirically derived relation for the power output per unit volume of active discharge is [73]

$$P/V = 10^{-5}J^2 \qquad (6)$$

when P is in watts, V in cm^3, and J in A/cm^2. In this expression, P is the multimode multicolor output of the laser. If single-color operation at 0.488 μ or 0.515 μ is selected by using a dispersive intracavity element, 30–40 percent of this power can be obtained. If TEM$_{00}$ operation is required, some sacrifice in power is usually experienced because of the additional diffraction losses that are usually introduced to select TEM$_{00}$ operation; however, the degradation depends on the size of the laser and the cleverness of the technique used. Usually, 70–80 percent of the multimode power can be obtained in TEM$_{00}$ operation. Selection of a single longitudinal mode can also be accomplished by any one of the several techniques referred to in Section III-B. Because of a strong tendency to behave as a homogeneously broadened line, especially at high power levels, almost all of the multimode power can be obtained in a single longitudinal mode except that part lost due to additional intracavity elements.

Ion laser technology has progressed through several stages with corresponding increases in power output and tube life. Early discharge tubes were made of fused silica, similar to the present He–Ne tubes, except that water cooling was required. Actually, the very first CW ion laser [74] was not water cooled but was operated with incandescent quartz walls. The need for cooling was painfully evident after a few minutes of operation. Fig. 8 shows the cathode throat region of a typical quartz discharge tube after several hours of operation. Localized sputtering and decomposition of the capillary wall has occurred, especially at the entrance

Fig. 8. Cathode throat region of typical CW ion laser silica discharge tube after 100 hours of operation. Discoloration is silicon or silicon monoxide from ion-decomposed SiO$_2$.

of the uniform bore region. After a few hundred hours of operation, the same sputtering damage will be seen along the entire bore, and the entrance region may be eroded completely through the capillary wall (typically 1-mm thick). The sputtering process not only limits tube life by physical damage, but also cleans up the gas in the discharge by burying it in the walls in exactly the same manner as the noble gases are pumped by an ion vacuum pump. Typical design limits for CW quartz bore ion laser tubes are 100 to 150 watts of input power per centimeter of length, 200 to 500 hours to ultimate end of life, and 10 $\mu \cdot$ l/h gas cleanup rate. Within these limits, compact lasers can be built with a few watts output, provided a gas replenishing system or a very large gas reservoir is supplied.

Several different methods have been tried in efforts to improve the characteristics of quartz bore tubes. Among the different dielectric bore materials that have been tried are boron nitride, alumina, and beryllia. Of these, beryllia (BeO) has proven to be the most successful. Standard metal–ceramic brazing techniques have been used to make beryllia bore discharge tubes which exhibit very low sputtering and gas cleanup rates; a dissipation capability of over 300 W/cm has been demonstrated to date. Fig. 9 shows a small BeO bore (25-cm active length) discharge tube which produces over an 8-watt output. This tube

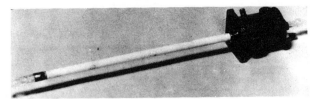

Fig. 9. Typical metal–ceramic BeO bore Ar ion laser discharge tube. 11-inch active region of this tube produces over 5 watts of output.

Fig. 10. Stacked tungsten disks used to form radiation-cooled bore of a high-power ultraviolet ion laser.

utilizes the all metal–ceramic construction techniques and impregnated tungsten cathode so familiar in modern microwave tubes. A gas return path, to equalize the gas pressure differential between cathode and anode generated by discharge processes, is incorporated as an integral part of the BeO structure. A recent life test in a similar BeO tube ran over 800 hours[1] (cycled in 6- to 9-hour operating intervals) with no measurable change in gas pressure. The estimated gas cleanup rate was less than 0.01 $\mu \cdot$ l/h [75].

It is also possible to make the bore walls out of electrically conducting materials. If the material is distributed along the bore in short segments, each insulated from the next, the discharge is substantially unaffected by the local short circuit [76]. Low-sputtering refractory materials such as tungsten [73], molybdenum [76], tantalum, and graphite [77] have been used in radiation-cooled arrangements; copper [78] and aluminum [79] have also been used in water-cooled assemblies. Tungsten has the lowest sputtering yield of the refractory metals, and has also been used as a bore material in high-performance ion lasers. Disk bore structures have been used at input power levels of 500 W/cm [80].

Fig. 10 shows a number of tungsten disks assembled in a stacked structure which is then slipped inside the vacuum envelope. The discharge and laser characteristics of this structure of relatively thin conducting disks separated by relatively large gaps are also substantially the same as those of a continuous dielectric wall. The high emissivity of the material plus the stacked-disk structure allows very high power-per-unit-length operation with radiation cooling to free space or a surrounding water jacket. Graphite can also be used in this fashion, but it has the disadvantage of low mechanical strength with a resultant tendency to form powder, a very undesirable property for devices containing optical surfaces. The low sputtering yield of carbon and its low cost make it an attractive material for laboratory-type tubes in which carbon dust can lie undisturbed in the bottom of the envelope, but it is not suitable for tubes which must be moved, operated in a vertical position, or pass a shake test.

Clearly, there will be continued developments in ion laser technology leading to higher output power, more compact packaging, and less expensive units. New wavelengths, particularly in the ultraviolet [80], will become increasingly more practical, although they will undoubtedly be used more in scientific instrumentation rather than communica-

tion. It is not likely that ion laser efficiency will increase significantly over the 0.1 to 0.2 percent maximum demonstrated to date.

C. CO_2 Lasers

The CO_2 laser is the youngest of the three types listed in Table I. Although laser oscillation was first obtained [81], [82] on wavelengths in the $00^01 \rightarrow 10^00$ band of CO_2 at about the same time as the Ar ion laser was demonstrated, it was approximately a year before the high-power high-efficiency aspects were fully realized [83] and serious device development began. The nature of the active medium of a CO_2 laser is similar in some respects to that of the He–Ne laser. A glow discharge is used with about the same input power per unit length. The electric fields in the plasma tends to be somewhat higher for a given diameter discharge (by a factor of 3 or so) because of the higher operating pressure and the electronegative character of the gas mixture. Several methods of excitation have been used successfully, including hot- and cold-cathode dc discharges, RF discharges, and microwave discharges; however, the most common method now in use is the same as that used in the He–Ne laser, i.e., the cold-cathode dc discharges. For 1- to 100-watt output CO_2 lasers, glass or quartz envelope tubes very much like those used in the He–Ne laser are usually employed, with the main differences being the addition of a water cooling jacket; NaCl, KCl, GaAs, or Ge windows instead of fused silica; and a somewhat larger diameter (1 to 2 cm instead of 2 to 4 mm).

Despite the apparent similarity of construction, the technology problems are quite different for the CO_2 and He–Ne lasers. Higher power CO_2 lasers (> 100 watts) which employ continuous gas flow are commonly made of bolt-together glass pipe of the type used for food or chemical plumbing. A laser oscillator with a CW output in excess of 9 kW has been made using this construction technique [58]; laser power amplifiers with average power outputs > 1 kW when driven by single frequency, TEM_{00} oscillators have also been made in this way by Smith and Forster [84], as shown in Fig. 11.

---

[1] The life test was terminated by a need for the test equipment and space (for a new tube), not by any change in operating characteristics.

Fig. 11. Early breadboard version of 1.5-kW average power,
15-kW peak power MOPA CO_2 laser system.

Specific output powers >1 W/cm^3 have been obtained in small tubes, although larger diameter tubes usually produce no more than 0.1 W/cm^3. A typical 10-watt output laser might have a discharge tube 1 cm in diameter by 50 cm in length and require 100 watts input power (5 kV at 20 mA). The high efficiency is primarily a consequence of the fact that both upper and lower laser levels lie very close to the ground state. Because of this low-lying position, almost all of the higher energy levels excited by the discharge funnel down to the upper laser level by cascade, and very little energy is wasted in relaxing the lower level to the ground state. Because the CO_2 laser exhibits relatively higher gains in larger diameter tubes than either He–Ne or Ar II lasers, it is relatively simple to obtain higher power by increasing the tube volume. The 1-kW value listed in Table I was obtained in about 20 meters of discharge and the 9 kW-value in about 200 meters (folded into a more reasonable length, of course).

More recently, a basically new type of CO_2 laser configuration was reported, one which uses a fast-flowing gas mixture to obtain high output power per unit volume [85]. An output in excess of 1 kW was reported for a 1-meter long CO_2 oscillator. The specific output from this device is substantially >1 W/cm^3; however, a better figure of merit for this class of fast-flow lasers is the power output per volumetric flow rate. Outputs of 1 W/ft^3/min have already been demonstrated. This number is in good agreement with the theory for fast-flow systems.

In all CO_2 lasers, mixtures of CO_2 with several other gases are employed. It is possible to make a "pure" CO_2 laser, but in reality the "pure" CO_2 is quickly dissociated by the discharge into an equilibrium mixture of CO_2, CO, and O_2 (or O, O–). The CO then populates the upper laser level by resonant collision in the same manner that helium atoms pump the Ne upper laser level in the He–Ne laser. More efficient pumping is possible if nitrogen is added to the discharge, since the net transfer of excitation from the N_2 vibrational states to the CO_2 upper laser level is better than from the CO vibrational states. The N_2 or CO molecules present are excited by collisions with the discharge electrons; at high current densities the excited N_2 and CO also can be destroyed by electron collision, with the resultant formation of a current-independent equilibrium pop-

ulation. However, this does not occur in the usual operating range of sealed-off or slow-flow CO_2 lasers. Within this range, the rate of excitation of upper levels continues to increase with increasing current. The sharp optimum which occurs in the output power versus input power characteristic results instead from the population of the lower laser level by increased thermal excitation as the input power is raised. The [10^{0}0] state lies only about 0.17 eV above the ground state and thus has a significant relative population (given by the Boltzmann factor exp $[-0.17$ eV/kT]) at room temperature and above. As a direct consequence, CO_2 lasers usually use helium as a gas additive to remove heat from the center of the discharge to the walls by improving the thermal conductivity of the gas mixture. The discharge tubes usually use water-cooled or refrigerated walls to reduce the gas temperature further. Other gas additives are used to depopulate the lower laser level directly by collisions. Water vapor and hydrogen (which probably forms water vapor in the discharge) have both been used [86]. Xenon has also been used as an additive to sealed-off CO_2 lasers to extend life, but the exact mechanism responsible for this result is not known [87].

The fast-flow CO_2 laser, on the other hand, eliminates the thermal repopulation of the lower laser level by physically transporting the "spent" mixture (i.e., the mixture with a significant lower level population) out of the optical cavity by flow, and replacing it with "fresh" mixture. In such a laser, the input power per unit volume may be greatly increased without reaching an optimum as long as the flow rate is also increased. The full potential of this fast-flow laser technique has not yet been realized.

It is easy to appreciate that the usual CO_2 laser mixture of CO_2, CO, N_2, He, H_2O, $\cdots$ can produce some very interesting chemistry in a discharge tube. Chemical reactions in the gas phase result in the formation of CN, C_2, and probably other species, so that it is not surprising that a complete understanding of the discharge processes is not yet realized. Chemical reactions between the gases and the components of the tube envelope present an even more serious problem. It is difficult to find a suitable hot cathode for this environment, thus cold cathodes are usually used. Even cold cathodes have complex chemical interactions with the discharge, resulting in the cleanup of one or more gases. The best life results to date have been obtained with small tubes (output power <50 W). Sealed-off operating lifetimes in excess of 9000 hours have been demonstrated to date [88]. These tubes usually use mixtures of CO_2–N_2–He–H_2O–Xe, although the role of all the additives is not understood. The key problem area in the development of long-life sealed-off CO_2 lasers is the understanding of the discharge chemistry. A typical life-test CO_2 laser is shown in Fig. 12. This particular glass tube has one internal mirror and one GaAs Brewster's angle window.

It is possible to circumvent the problem of discharge chemistry by flowing the gas mixture slowly through the discharge tube. Dissociation products are pumped out and fresh gas of the correct mixture is continually supplied. This is the approach presently taken with all CO_2 lasers of 100-

Fig. 12. Small glass sealed-off CO_2 laser used in life test studies.

watt or greater output. In such continuous-flow lasers, envelope cleanliness and vacuum integrity are less important than in sealed-off tubes. As a consequence, high power CO_2 lasers are generally more analogous to demountable vacuum systems than to electron tubes (as is easily seen from a close inspection of Fig. 11).

The requirement that the CO_2 laser optics must pass 10.6-μm radiation, often at very high flux densities also presents some fabrication problems [58]. The lowest loss materials at this wavelength are the alkali halides, with NaCl and KCl being the best. Unfortunately, these materials are hygroscopic, mechanically weak, and not easily vacuum sealed. Semiconductors such as germanium and GaAs have better physical properties, but they also have higher loss and suffer from thermal runaway (i.e., a positive loss temperature coefficient). The high index of refraction of germanium and GaAs makes them difficult to use for Brewster's angle windows; they are usually used as the substrate material for the output mirror in an internal mirror configuration. The development of new optical materials and sealing techniques for the CO_2 laser is a challenging area. Recently, loss figures at least a factor of 2 lower than the best GaAs were reported for single-crystal CdTe [89].

Although sealed-off high-reliability CO_2 lasers are not yet readily available for optical communications sources, several development programs are now underway. Metal-ceramic discharge tubes and versions for space application are currently being built. The ease of obtaining single-frequency output, coupled with the possibility of using a power amplifier to obtain still more power without loss of coherence makes the CO_2 laser most attractive for general communications applications. The high efficiency (20 percent) obtained in even smaller lasers and the lower quantum noise ($h\nu B$) encountered at 10.6 μm make the CO_2 laser particularly attractive for space communications.

V. SEMICONDUCTOR LASERS

The very high population inversion obtainable in semiconductors results in a large gain. In fact, the gain can be so large that the path length for the stimulated radiation need not be long and the ends of the cavity need not be highly reflecting.

Thus, with cavity lengths of only a few hundred microns, semiconductor lasers are extremely small.

In contrast to other types of lasers where the transitions occur between discrete states of excited atoms, in semiconductor lasers, the transitions involve sets of banded states.

The banding of states results from the close packing of activated atoms. Hence a high density of population inversion is required to reach threshold conditions.

A. Pumping of Semiconductor Lasers

Population inversion in semiconductor lasers can be obtained either optically or electrically. Optical excitation or photoluminescence is achieved by irradiating the semiconductor with photons that have sufficient energy to produce electron–hole pairs across the energy gap [90]. However, a high intensity of lower energy photons can also be used to generate electron–hole pairs in which case two photons [91] are absorbed to complete the transition across the gap. The main advantage of optical pumping is that it can be used with materials in which p-n junctions cannot be made, for example, in CdS or CdSe.

Electron beam excitation is also suitable for pumping wide gap semiconductors in which p-n junctions cannot be fabricated. The electron beam can be easily moved, focused, and modulated.

The most practical mode of exciting a semiconductor laser is by injection at the p-n junction. In injection lasers, the electrical energy is converted directly into coherent radiation making the injection laser the most efficient type of laser. Injection is obtained by forward biasing the p-n junction; in this case the input impedance of the diode is very low, thus making the use of stripline circuitry a convenient technique.

Another means of excitation of semiconductor lasers is by impact ionization of electron–hole pairs of avalanche breakdown [92]. This breakdown occurs when a field ($>2 \times 10^5$ V/cm) is applied across a homogeneous semiconductor comprising no p-n junction. The breakdown mode of operation is equivalent to that of gas lasers but results in a much denser plasma. Because of power dissipation, bulk excitation is obtainable in low duty cycle pulses and does not appear at present to be useful for optical communication. However, the advantage of low beam dispersion that may be anticipated in the bulk laser may spur further effort in this mode of pumping.

As we shall see later, the most practical modes of pumping semiconductor lasers are by injection, either directly into the material or by pumping optically one semiconductor with the output of an injection laser made of a different semiconductor.

B. Performance of Semiconductor Lasers

As shown in Table II the range of wavelengths achieved by semiconductor lasers extends from 0.33 to 31 μ. In certain alloys between compounds, which are miscible in all proportions, the laser wavelength is adjustable by the composition of the alloy.

The radiative recombination usually occurs in a very thin layer of the semiconductor, approximately equal to a minority carrier diffusion length if no special confinement methods are used. Thus, in injection lasers, the active region is about 1–2 μ thick at the junction. In lasers pumped

TABLE II
SPECTRAL RANGE COVERED BY SEMICONDUCTOR LASERS

	$\lambda\,(\mu)$	$h\nu$(eV)	O	E	I	A
Zns	0.33	3.8	O	E		
ZnO	0.37	3.4		E		
$Zn_{1-x}Cd_xS$	0.49–0.32	2.5–3.82	O			
ZnSe	0.46	2.7		E		
‡CdS	0.49	2.5	O	E		
ZnTe	0.53	2.3		E		
GaSe	0.59	2.1		E		
$CdSe_{1-x}S_x$	0.49–0.68	2.5–1.8	O	E†		
$CdSe_{0.95}S_{0.05}$	0.675	1.8		E		
CdSe	0.675	1.8	O	E		
‡$Al_{1-x}Ga_xAs$	0.63–0.90	2.0–1.4			I	
‡$GaAs_{1-x}P_x$	0.61–0.90	2.0–1.4		E	I	
CdTe	0.785	1.6		E		
‡GaAs	0.83–0.91**	1.50–1.38	O	E	I	A
InP	0.91	1.36			I	A
$GaAs_{1-x}Sb_x$	0.9–1.5	1.4–0.83			I	
$CdSnP_2$	1.01	1.25		E		
$InAs_{1-x}P_x$	0.9–3.2	1.4–3.9			I	
$InAs_{0.94}P_{0.06}$	0.942	1.32			I	
$InAs_{0.51}P_{0.49}$	1.6	0.78			I	
GaSb	1.55	0.80		E	I	
$In_{1-x}Ga_xAs$	0.85–3.1	1.45–3.1			I	
$In_{0.65}Ga_{0.35}As$	1.77	0.70			I	
$In_{0.75}Ga_{0.25}As$	2.07	0.60			I	
Cd_3P_2	2.1	0.58	O			
InAs	3.1	0.39	O	E	I	
$InAs_{1-x}Sb_x$	3.1–5.4	0.39–0.23			I	
$InAs_{0.98}Sb_{0.02}$	3.19	0.39			I	
$Cd_{1-x}Hg_xTe$	3–15	0.41–0.08	O	E		
$Cd_{0.32}Hg_{0.68}Te$	3.8	0.33	O			
Te	3.72	0.334		E		
PbS	4.3	0.29		E		
InSb	5.2	0.236	O	E	I	A
PbTe	6.5	0.19		E	I	
$PbS_{1-x}Se_x$	3.9–8.5	0.32–0.146		E	I	
PbSe	8.5	0.146		E	I	
$Pb_{1-x}Sn_xTe$	6–28	0.209–0.045			I	
PbSnSe	8–31.2	0.155–0.040			I	

*A Avalanche breakdown.
 O Optical pumping.
 E Electron beam pumping.
 I Injection
† Boldface indicates possible mode of excitation.
** Depending on temperature and doping.
‡ Pulsed operation at room temperature.

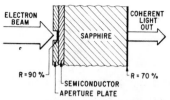

Fig. 13. Structure of electron beam excited laser in the "radiating mirror" configuration (after Basov et al. [91].

alternate p-n junction which is forward biased sees the fringing electromagnetic fields of the adjacent emitting junctions [94]. This fringing field induces a common coherence or phase locking and the radiation now can escape through an effectively larger aperture.

Although special structures can be devised to minimize beam divergence, the semiconductor laser suffers from more dispersion than other types of lasers—a limitation in some applications of optical communication.

The power levels achievable with semiconductor lasers depend on permissible operating conditions. At low temperature, several watts of CW coherent radiation can be obtained [95]. But at room temperature, because of large internal power dissipation (the threshold current increases exponentially with temperature [96]), the laser must be energized in pulses. Peak powers of 50 watts can be readily obtained. With a conservative repetition rate the average power is of the order of tens of milliwatts.

The optical system can be simplified with the use of the maximum possible optical density at the laser facet. However, the maximum peak power obtainable from a laser of a given cavity width is limited by the onset of catastrophic damage [97]. The flux density at which damage occurs depends on the pulsewidth, the degree of optical confinement in the laser active region, and to some extent on the temperature. At room temperature, the flux density at which castastrophic damage occurs is typically between 400 and 800 W/cm,² depending on the details of the fabrication process for a pulsewidth of about 100 to 200 ns. Higher peak powers are attainable with shorter pulses [97]. Recently, the use of a surface coating has raised the threshold or catastrophic damage to higher values than previously possible, thus progress in this area continues to be made [98]. A second degradation mechanism, denoted "gradual," is a function of the current density [99]. The degradation rate for a given current density depends critically on the method of laser fabrication. In particular, it appears that imperfections in the active region increase the degradation rate [99]. With current densities of 10^4 A/cm² or less, the gradual degradation has been shown to be negligible in diffused diodes operated continuously [100].

At high power levels, the output of semiconductor lasers is multimode and can occur at several wavelengths simultaneously. Yet, pure modes can also be obtained at lower power, by resorting to special narrow structures [101]. A pure mode at high power can be produced by coupling a semiconductor laser oscillator operating as a pure mode generator to another semiconductor diode operating as a

optically with nonpenetrating radiation the active region is also one diffusion length thick. With penetrating radiation, a larger volume of material is excited. The electron beam whose penetration is voltage dependent can also be used to excite a greater volume of material than feasible with injection lasers.

Where the active region is narrow, the stimulated radiation suffers considerable diffraction resulting in a divergent coherent beam. The divergence of the beam can be reduced considerably by special structures providing a large aperture to the emerging radiation. Thus, with the radiating mirror structure of Fig. 13 where an electron beam pumps a circular area of a GaAs wafer coupled to an external cavity, the beam divergence is about 20 minutes [93]. Another structure consists of a multilayered arrangement of serially connected p-n-p-n $\cdots$ regions where the spacing between adjacent junctions is so small ($\sim 1\,\mu$) that every

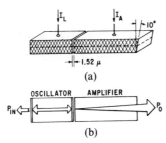

Fig. 14.   Low-noise oscillator–amplifier pair. The double lines in diagram (b) represent Fabry–Perot facets (after Kosonocky and Cornely [102]).

single-pass light amplifier, Fig. 14 [102]. When several modes can propagate simultaneously, they can interact to produce a lower beat frequency. The most remarkable mode interaction is that resulting from second-order beats [103]; because the index of refraction of semiconductors is not constant with λ (dispersion), adjacent modes are not equally spaced. Hence, frequency mixing between two pairs of adjacent modes results in two different beat frequencies; these in turn can be mixed to generate a second-order beat which occurs at about 1 GHz [104].

C. Modulation and Tuning

We have previously seen that the operating frequency of a semiconductor laser can be selected by the choice of material and in many cases adjusted by employing a suitable composition of alloy between two miscible semiconductors. Further fine tuning of the operating frequency can be obtained by adjusting the temperature (the energy gap of most semiconductors and therefore the energy of emitted photons varies at the rate of approximately -5×10^{-4} eV/$^{\circ}$K). Hydrostatic [105] or uniaxial pressure [106] can also be used to obtain frequency changes. Note that for injection lasers made with lead-salt semiconductors the emission energy shifts to lower values with increasing pressure [107]. A magnetic field breaks the bands into Landau levels which drift further apart with increasing field, thus shifting the emitted frequency as well [108]. A considerable tuning range is also obtainable by controlling the Q of the cavity. Thus, the Fabry–Perot facets can be slightly etched or coated with a lower reflectivity layer. This requires a higher threshold or higher population inversion which pushes the quasi-Fermi levels deeper into the bands with the consequence that photons of higher energy are generated [109]. The Q of the cavity can also be modulated by breaking up the active region into two separately biased sections. One of these sections can be made lossy by a reduced forward bias and even more effectively by a reverse bias [110].

The electronic control of losses in a two-section laser is an effective means of modulating the laser. The injection laser can be readily amplitude modulated up to extremely high frequencies by direct modulation of the current through the junction (up to 46 GHz have been demonstrated [111]). When the modulation is done at high frequencies, care must be taken to design the cavity such that the mode spacing corresponds to the desired modulation frequency (the modulation is equivalent to alternating the generation of coherent radiation between two allowed cavity modes).

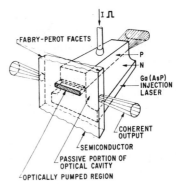

Fig. 15.   Optical pumping of one semiconductor by injection laser (after Johnson and Holonyak, Jr. [90]).

Under pulsed conditions (this is how injection lasers are operated at room temperature), modulation by varying the pulse position or the pulse length are very practical methods [112]. Note that an injection laser can be used to pump optically another semiconductor of lower energy gap, Fig. 15 [90]. Then all the modulation schemes of the injection laser affect correspondingly the output of the optically pumped material.

D. Future Prospects

Recent breakthroughs in material technology have permitted great improvement in the performance of injection lasers. The ability to make heterotransitions between the wider gap $Ga_{1-x}Al_xAs$ and GaAs has led to the close-confinement structure [113]–[115] which constrains both the inverted population (by virtue of internal barriers to carriers) and the radiation (by producing a discontinuity in refractive index at the edge of the active region). Carrier confinement results in higher gain; radiation confinement results in lower losses since the wider gap material is less absorbing than the p^+ region of earlier injection lasers. The close-confinement structure has resulted in injection lasers with substantially lower threshold current densities. Threshold current densities of 10 000 A/cm^2 are routinely obtained, as compared to typical values of 40 000–60 000 in previous GaAs lasers. This is the trend needed to eventually realize a room temperature CW injection laser. Further progress in understanding and controlling diode degradation will result in improved reliability.

Progress in doping II–VI semiconductors may succeed in generating p-n junctions in these compounds and thus broaden the spectral range accessible to injection lasers.

Finally, it must be pointed out that optical communication is also feasible with incoherent light sources, provided that they lend themselves readily to modulation. Thus, electroluminescent diodes, which are the most efficient and the fastest light sources, are well indicated for optical communication

VI. SOLID-STATE ION LASERS

Optically pumped crystalline lasers have been operated utilizing transition metal ions (Cr^{3+}, Ni^{2+}, and Co^{2+}), rare earth ions (Nd^{3+}, Pr^{3+}, Er^{3+}, Ho^{3+}, Er^{3+}, Tm^{3+}, Yb^{3+}, Yb^{2+}, Sm^{2+}, and Dy^{2+}), and the actinide ion

(U^{3+}) in various host lattices. Laser action in these solid-state ion systems has been reviewed by Johnson [7b], Kiss and Pressley [7a], and Goodwin and Heavens [5c]. Room temperature operation of a continuously pumped solid-state laser was first reported in 1962 by Johnson *et al.* [116] using Nd^{3+} in $CaWO_4$. Since then CW room temperature operation has been reported for Cr^{3+} in Al_2O_3 (ruby) [117], [118] and Nd^{3+} in glass [119], $Y_3Al_5O_{12}$ (YAlG) [120], $CaMO_4$ [121], $Ca_5(PO_4)_3F$ [122], $YAlO_3$ [123], and La_2O_2S [124]. Of these, the most highly developed continuous solid-state ion laser is Nd:YAlG. Because of its wide range of operational capabilities, it is competitive with the three prominent gas lasers mentioned in Section IV. Our discussion of CW solid-state ion systems will be divided into three subsections, i.e., ruby, Nd:YAlG, and other Nd^{3+} systems.

A. Ruby

Historically, Cr^{3+} in Al_2O_3 (ruby) is of great significance because it was employed by Maiman [2] in the first laser. The ruby laser is a three-level system in which the ground state is the terminal level. Room temperature laser oscillations were reported in ruby almost simultaneously by Evtuhov and Neeland [117] and by Röss [118]. In each case high pressure mercury arc lamps were employed for pumping. Evtuhov and Neeland employed an elliptical cylinder pump cavity; Röss employed an elliptical exfocal cavity. Detailed studies of CW ruby oscillators are found in [125]–[127]. Using a 2-mm diameter 7.5-cm long ruby rod (0.5-percent Cr^{3+} concentration), the maximum CW output was 2.37 watts at an input of 5.08 kW [126]. Substantial modulation of the output was observed, and preliminary efforts to eliminate it have been unsuccessful. A second serious liability of the CW ruby laser is the short pump lamp life (~ 25 hours) at input power levels required to produce outputs of ~ 1 watt. Although the visible output of ruby and the high peak powers obtainable in the repetitively Q-switched mode makes it attractive for certain applications, its present liabilities exceed its assets for general communication applications.

B. Nd:YAlG

Some of the distinctive characteristics of $Y_3Al_5O_{12}$ as a laser host, particularly for Nd^{3+}, are a result of its relatively low photoelastic constants, its good mechanical and thermal properties, and the fact that trivalent ions such as Nd^{3+} can be incorporated substitutionally. The commercial availability of high-quality Nd:YAlG laser rods in lengths as long as 15 cm and diameters as large as 1 cm has provided sufficient latitude to permit the operation of lasers with CW outputs as high as 750 watts. Thus, the few watt average powers which might be required in most communication systems are well within the capabilities of the Nd:YAlG laser. In Nd:YAlG, oscillation can be achieved on a number of infrared transitions, the most prominent room temperature transition occurring at 1.064 μ. When nonlinear crystals are added internally to the optical cavity, efficient

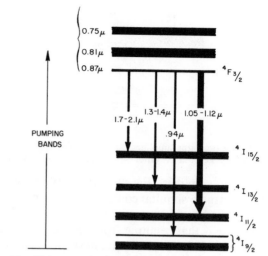

Fig. 16. Main pump bands and fluorescent transitions in Nd:YAlG. The main laser transition is at 1.064 μ.

TABLE III
MAIN ROOM TEMPERATURE TRANSITIONS IN Nd:YAlG

	Wavelength (μ)	Peak effective room temperature cross section σ_l^* (10^{-19} cm^2)†	Measured relative room temperature CW laser threshold
$^4F_{3/2} \rightarrow {}^4I_{9/2}$	0.939	0.81	
	0.946	1.34	
$^4F_{3/2} \rightarrow {}^4I_{11/2}$	1.0520	3.1	2.08
	1.0551	0.20	
	1.0615	6.65	1.15
	1.0641	8.80	1.00
	1.0682	1.10	
	1.0738	4.00	1.22
	1.0779	1.55	
	1.1055	0.32	
	1.1122	0.79	2.17
	1.1161	0.77	2.26
	1.1225	0.72	2.36
$^4F_{3/2} \rightarrow {}^4I_{13/2}$	1.319	1.50	1.60
	1.335	0.92	
	1.338	1.50	2.17
	1.342	0.63	
	1.353	0.35	
	1.357	0.88	

† $\sigma_l^* = \sigma_l(n_l/n_u)$ and $\sigma_u^* = \sigma_u$ where n_u and n_l are the upper and lower $^4F_{3/2}$ stark level populations at room temperature and the σ are the true cross sections. The peak fluorescent intensity of an individual transition at room temperature is proportion to $\sigma^* n_u$.

second-harmonic conversion of a number of the laser transitions to the visible can be achieved. The availability of CW outputs at several wavelengths in the visible and infrared and the possibility of operating it continuously pumped, repetitively pulsed (mode locked, cavity dumped, or Q-switched) make it a versatile source for communications.

1) Spectral Properties: The Nd:YAlG laser is a four-level system as depicted in Fig. 16. Of the main pump bands shown, the 0.81 μ and 0.75 μ are the strongest. The upper laser level $4F_{3/2}$, has a fluorescent efficiency greater than

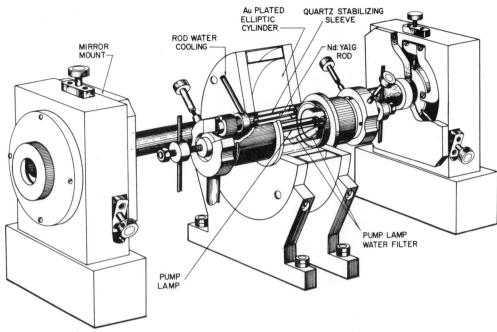

Fig. 17.   Diagram of W–I pumped Nd:YAlG laser.

99.5 percent [128] and a radiative lifetime of 230 μs [120]. The branching ratio of emission from $4F_{3/2}$ is as follows [129]:

$$^4F_{3/2} \rightarrow {}^4I_{9/2} \quad 0.25$$
$$^4F_{3/2} \rightarrow {}^4I_{11/2} \quad 0.60$$
$$^4F_{3/2} \rightarrow {}^4I_{13/2} \quad 0.14$$
$$^4F_{3/2} \rightarrow {}^4I_{15/2} \quad <0.01.$$

The effective cross section [129], [130] at room temperature of the prominent transitions and the measured relative thresholds of those transitions on which room temperature CW oscillation has been observed [120], [131] are given in Table III. At room temperature the main 1.06-μ line in Nd:YAlG is primarily homogeneously broadened by thermally activated lattice vibrations [120b]. The room temperature linewidth is 6.0 cm^{-1} or 180 GHz. The residual width at very low temperatures ($T<4°$K) is a measure of the inhomogeneous width, which is caused by variations in the crystalline electric field from ion to ion in the lattice. The inhomogeneous width varies from crystal to crystal and is highly dependent on material quality. The narrowest width measured was 0.2 cm^{-1}; strained crystals have linewidths as large as 1.0 cm^{-1} [120b].

2) Pumping Lamps and Cavities: A number of different cavities and pump lamps have been used in the operation of the Nd:YAlG laser. The original Nd:YAlG laser utilized a W–I pump lamp in an elliptical cylinder. This combination is the most highly developed, and a design utilized by one of the authors and his colleagues at Bell Telephone Laboratories is shown in Fig. 17. A packaged commercial version of a 10-watt tungsten-pumped YAlG laser system is shown in Fig. 18. W–I lamps have been developed commercially for photography and are available in a variety of

Fig. 18.   Commercial, W–I pumped Nd:YAlG laser which has rated output of ~10 watts. Lamp power supply and closed cycle water cooling systems are shown on the right.

sizes. For lamps operating at a filament temperature of 3400°K, lamp life is typically 200 hours. Lamps designed for 3000°K operation have a rated life of typically 2000 hours. In an operational test [132] on a 1-watt laser using a 3000°K lamp, a lamp life of between 3000–5000 hours was found typical. The operational life of the Nd:YAlG laser is generally determined by lamp life and the life of ancillary equipment such as the lamp power supply, etc. For moderate power (1 to 25 watts) Nd:YAlG laser applications, the W–I lamp has proven to be the most convenient and long lived of all presently available lamps.

In addition to the elliptical cylinder, spherical [133] and ellipsoidal [134] pump cavities have been used with the Nd:YAlG laser. These cavities are more costly and inconvenient to use than the elliptical cylinder. For high power lasers which use large diameter rods (>5 mm), two pump lamps are often employed in a double-elliptical cylinder configuration.

High pressure rare gas arc lamps [135]–[137] have been

TABLE IV
CHARACTERISTICS OF ROOM TEMPERATURE Nd: YAlG LASERS

Mode of operation	1.064 μ			0.532 μ
High power CW multimode	(W–I) 150 watts, Efficiency 1.8 percent [140]			
	(Kr) 100 watts, Efficiency 2.9 percent [141], [142]			
	(Kr) 250 watts, Efficiency 2.1 percent [143]			
	(Kr) 750 watts, Efficiency 1.7 percent [139]			
Low power multimode	(W–I) 10–15 watts, Efficiency 0.66–1.0 percent [138], [149]			
Single transverse mode	4–6 watts	[149]		4 watts [161]
Single axial mode	2–3 watts	[153], [149]		
Repetitively Q-switched			[52], [158]	
Peak power	$10^3 P_{\mathrm{CW}}$	$3 \times 10^2 P_{\mathrm{CW}}$	$10^1 P_{\mathrm{CW}}$	
Average power	$0.25 P_{\mathrm{CW}}$	$0.73 P_{\mathrm{CW}}$	P_{CW}	
Repetition rate (pps)	10^3	4×10^3	5×10^4	$\sim$same as 1.064 μ
Typical pulsewidth (μs)	0.25	0.61	2.0	
Repetitive cavity dumped			[159]	
Peak power	$1–2 \times 10^2 P_{\mathrm{CW}}$	$1–2 \times 10^1 P_{\mathrm{CW}}$	$1–2 P_{\mathrm{CW}}$	
Average power	P_{CW}	P_{CW}	P_{CW}	
Repetition rate (pps)	10^5	10^6	10^7	
Typical pulsewidth (ns)	(50–100)	(50–100)	(50–100)	
Mode locked			[40], [154]	
Average power	P_{CW}	P_{CW}		
Peak power	$\sim 3 \times 10^2 P_{\mathrm{CW}}$	$\sim 3 \times 10^1 P_{\mathrm{CW}}$		$\sim$same as 1.064 μ
Repetition rate (pps)	10^8	10^9		
Pulsewidth (ps)	30	30		

used to pump Nd:YAlG. The results of comparisons of the efficiency of the rare gas lamps relative to W–I are listed below:

Kr	Xe	Ar	
1.2	0.77	0.58	[135]
3.1	1.6		[136]
0.95	0.73		[137].

The measurements reported in [135] and [137] are in relatively good agreement and were carefully conducted with lamps with almost identical illuminated lengths and diameters, and under identical lamp loading conditions. The advantage of the Kr-arc lamp from a pumping point of view is the fact that it can be operated at a higher loading level than W–I; its disadvantage is its generally shorter life. When lamp life is not a major consideration, Kr is the choice for high power operation.

K–Hg discharges have also been utilized and compared with the W–I lamp [138]. When similar size K–Hg and W–I lamps were used in the same laser and operated at their maximum input powers the laser output power was almost identical; however, the efficiency with K–Hg was 2.4 times higher than with W–I. It is necessary to point out that in these experiments the K–Hg lamps could not be run with dc excitation because of cataphoresis; this, the fact that the lamp is very expensive, and the fact that it is short lived make it impractical for most applications at this time.

3) CW Performance Characteristics: The very high power capabilities of Nd:YAlG have been convincingly demonstrated by a recent announcement [139] of the achievement of 750 watts of CW output at 1.7-percent efficiency from a multiple-head Kr-pumped system. Previous work [140]–[143] on lasers operating with outputs above 100 watts is summarized in Table IV. In all cases the output is highly multitransverse moded. At these high pump power levels the effects of thermally induced lens and birefringence effects [144]–[147] make efficient operation in the TEM_{00} mode difficult. Methods to compensate for these thermal effects have been proposed [148]; however, their effective implementation has not been carried out.

Since, in most communication systems several watts of average power is probably adequate, our further discussion will be devoted to the performance of Nd:YAlG lasers with outputs of 1–15 watts. The most efficient [138], [149] performance in this power range using W–I is 1 percent. Using the pumping geometry shown in Fig. 17 with a 3.8-mm diameter 6-cm long Nd:YAlG rod (1 atomic percent Nd^{3+}) of the highest quality and a commercial (1500 W$-Q$/1500/T4/14CL) W$-$I lamp, an output of 15 watts at an overall efficiency of 1 percent was achieved [149]. With more typical high-quality Nd:YAlG rods the output powers observed were 10 watts at an efficiency of 0.66 percent. The lamp operates at 3200°K and is rated by the manufacturer (GE or Sylvania) at 500 hours. When used in the laser, typical lamp life is also 500 hours; several lamps have operated as long as 1500 hours. There are two important design features incorporated into the pumping cavity in Fig. 17 which permit good operational performance. The first is the use of the water filter around the lamp which reduces the heat load to the rod cooling system. Typically, the laser rod is cooled by water flowing around it. Because of the convective nature of this type of cooling, the rod is cooled unevenly in time and space. The quartz sleeve [150], in Fig. 17, dampens these thermal variations, and its use is particularly important for stable operation. In multitransverse mode operation amplitude fluctuations are less than 0.1 percent. Using a cavity which restricts operation to the TEM_{00} mode, a single transverse mode power was obtained which was 40 percent of the maximum multimode power. The amplitude fluctuations in TEM_{00} mode operation are less than 1 percent.

4) Single Frequency Operation: The width of the frequency spectrum of a typical multiaxial-mode Nd:YAlG lasers is 15–30 GHz. Single frequency operation has been attained [35], [151] by using an intracavity etalon, as described in Section III-B. The single frequency output of a Nd:YAlG laser can be tuned over ~100 GHz by adjusting the tilt angle or temperature of the etalon [152]. Employing a birefringent etalon to provide both axial-mode selection and a frequency discriminant, a stabilized single frequency laser has been operated with frequency fluctuations less than 2 percent [35]. In this experiment the single frequency power was 40 percent of the multimode power. Single frequency operation at the full multimode power has been demonstrated by eliminating spatial hole burning [153].

5) Continuously Pumped, Repetitively Pulsed Operation: All of the techniques discussed in Section II-C have been applied to the Nd:YAlG laser.

Mode locking has been accomplished by using both loss and phase modulation [40], [154]. Pulses with widths of 30–50 ps have been generated at rates as high as 5×10^8 pps. Pulse envelope phase instabilities have been encountered in both He–Ne [155], [156] and Nd:YAlG [156] using a phase modulator for mode locking. A solution to these problems has been reported [157], and a stabilized mode locked Nd:YAlG has been operated which is suitable for use in an optical PCM system.

The relatively long lifetime (230 μs) of the upper laser level permits the generation of high peak powers by repetitive Q switching [52]. At repetition frequencies above the reciprocal of the lifetime (>4 kHz), the average power of the repetitively Q-switched laser approaches the output power obtainable in normal CW operation. At repetition rates ≤ 2 kHz the pulses generated have a peak power ~10^3 times the laser's normal CW power. At higher repetition frequencies the peak power is approximately proportional to the reciprocal of the square of the repetition frequency. Stable operation has been achieved at repetition frequencies up to 50 kHz [55], [158]. The finite buildup time of the field inside the laser cavity and the time required to repump the inversion sets an upper limit to the repetition rates at which good pulse-to-pulse amplitude stability is obtainable. A detailed experimental and theoretical study of repetitive Q switching in Nd:YAlG is found in [158].

Pulsed operation at repetition frequencies from approximately 100 kHz to 30 MHz is best accomplished in Nd:YAlG by cavity dumping [49]. Operation over the repetition frequency range 125 kHz to 5 MHz has recently been reported [159]. At 125 kHz, for example, 200-watt pulses of 80 ns were achieved, with an average power which was equal to the normal CW power. The low-frequency limit of

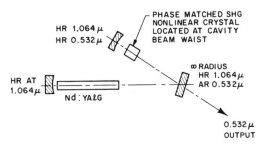

Fig. 19. Schematic diagram of a doubled Nd:YAlG laser.

stable dumping in Nd:YAlG has been discussed theoretically [49]. In Nd:YAlG cavity dumping provides efficient pulsed operation in the repetition frequency range where mode locking or stable Q switching is either impractical or impossible.

6) Doubled Nd:YAlG: Visible outputs from the Nd:YAlG laser have been obtained by incorporating into the optical cavity a nonlinear material which can be phase matched for second-harmonic generation; this is schematically shown in Fig. 19. Functionally, the second harmonic crystal acts as an output coupler in a manner analogous to the transmitting mirror of a normal laser. In the normal laser the transmitting mirror couples out power at the laser frequency, whereas the nonlinear crystal inside the laser couples power at twice the frequency. It has been shown both experimentally and theoretically [27], [29] that, by properly employing a crystal of sufficient nonlinearity in the Nd:YAlG laser, it is possible to convert completely the available output at the fundamental to the harmonic. Complete conversion of the 1.064-μ output has been reported [27] using barium sodium niobate [160]. A TEM_{00} mode output of 4-watts at 0.532 μ has been generated [161] using a W–I pumped laser of the type described in Section VI-B 3. With barium sodium niobate it is possible to efficiently double many of the other infrared outputs in Table III. In particular >70 percent CW conversion of the 1.32-μ line to 0.66-μ has been obtained recently [162].

In a doubled Nd:YAlG source having many axial modes oscillating, amplitude fluctuations at the harmonic will occur as a result of axial mode phase fluctuations at the fundamental. To achieve a stable harmonic amplitude, some form of phase stabilization such as mode locking or FM phase locking is necessary. Also, a stable harmonic output can be achieved by operating the laser in a single axial mode.

7) Diode Pumping: A compact and simple laser can be realized by utilizing efficient incoherent electroluminescent diodes to pump Nd:YAlG. For example, lasers have been operated [163], [164] utilizing an array of $GaAs_{0.87}P_{0.13}$ diodes. The diodes which were employed were cooled to 77°K in order to shift the spectral output into the desired 0.81-μ pump band region. With the diodes and rod at 77°K, laser threshold was achieved with 300 mW of electrical power into the diodes [163]. An input power of 6 watts was required to reach threshold when the rod was at room temperature [164]. A schematic diagram and a working model [164] of a diode-pumped laser are shown in Figs. 20 and 21.

Recently room temperature GaAsP diodes which emit

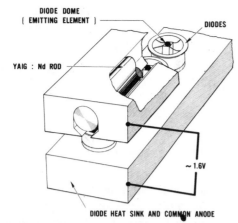

Fig. 20. Diagram of a diode-pumped Nd:YAlG laser. Diodes depicted are incoherent domed emitters such as GaAsP.

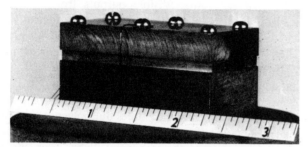

Fig. 21. Working model of diode-pumped Nd:YAlG laser. This model uses diodes which have to be cooled to 77°K in order to shift their spectral output into the desired 0.81-μ pump band region.

Fig. 22. Doubly resonant CW parametric oscillator.

at 0.81 μ have been built with typical efficiencies of 4 percent and a few as high as 8 percent [165]. With such diodes a completely room temperature diode-pumped Nd:YAlG laser should be realized in the very near future. This type of compact source could be extremely useful in the planning of a practical optical communication system.

C. Other Nd Systems

Of the other Nd^{3+} systems which have operated continuously at room temperature $Ca_5(PO_4)_3F$ [122], $YAlO_3$ [123], and La_2O_2S [124] are most noteworthy. On the basis of spectral data they should be comparable to Nd:YAlG in terms of intrinsic laser performance. Thus far, however, none of the systems has actually matched the performance of Nd:YAlG, and further development will be required before they are practically competitive. The poor thermal conductivity of $Ca_5(PO_4)_3F$ and La_2O_2S precludes the possibility of their utilization in high-power laser systems. $YAlO_3$ is mechanically and thermally similar to $Y_3Al_5O_{12}$ and, if twinning during the growth of $YAlO_3$ can be overcome, it will be useful.

VII. Parametric Oscillators

The observation of continuous parametric oscillation in 1968 [166], [167] demonstrated the feasibility of a coherent tunable continuous source for communications. The oscillator reported by Byer et al. [167] used $LiNbO_3$ as the nonlinear medium. Using an Ar laser they observed a threshold of 410 mW, an output power of 1.5 mW, and a tuning range of 0.68 μ to 0.705 μ for the signal and 2.1 μ to 1.9 μ for the

idler. In the oscillator of Smith et al. [166], the nonlinear material was $Ba_2NaNb_5O_{15}$ [160] and the pump was a 0.532-μ doubled Nd:YAlG source. A threshold of 45 mW, an output power of 3 mW, and a tuning range from 0.98 to 1.16 μ were observed. The outputs from these early continuous parametric oscillators had severe amplitude fluctuations (100 percent,) as well as oscillating bandwidths of 1–10 cm^{-1}.

In recent parametric oscillator studies (using $Ba_2NaNb_5O_{15}$) by Smith [168], the linewidth and amplitude stability have been improved by several orders of magnitude by the use of a frequency stabilized pump source which is isolated from the parametric oscillator. Using such a stabilized Ar laser 0.514-μ pump, the recent results include a reduction of the threshold power to 2.8 mW, an increase in the power output and efficiency to 45 mW and 30 percent, respectively, a reduction of amplitude fluctuations to ~20 percent, and single longitudinal mode operation of the oscillator. A typical doubly resonant continuous optical parametric oscillator used by Smith [168] in his studies is shown in Fig. 22.

Optical parametric oscillators are still in an embryonic stage of development and additional effort on amplitude and frequency stabilization will be required before they can be used in communications. The development of new nonlinear materials with even higher nonlinearities than $Ba_2NaNb_5O_{15}$ could permit the operation of continuous singly resonant oscillators in which either the signal or idler but not both are resonated. Although a singly resonant oscillator has a higher pump threshold power than an equivalent doubly resonant oscillator, the singly resonant oscillator can be stabilized and tuned more easily. At present, only pulse pumped operation of singly resonant oscillators has been possible [169]–[172]. For a detailed discussion of parametric oscillators the reader is again referred to the review by Harris [24].

VIII. SUMMARY AND CONCLUSIONS

The favored candidates for high capacity optical communications systems are the He–Ne, argon ion, Nd:YAG (plain or doubled), and CO_2 lasers. The choice of one or another will depend on the particular system requirements, such as location (terrestial or space), efficiency, life, etc. The existing state of the art in modulation [173] and detection [174] will also influence the choice. Of these sources, He–Ne is by far the most highly developed and at the present the most suitable for applications where long life and average powers less than 100 mW are required, and where large size and low efficiencies are not major considerations. Future He–Ne development is expected to result primarily in cost reduction.

In systems using average powers >100 mW, the Ar ion doubled Nd:YAlG, and CO_2 lasers are the main contenders. There has been a greater commercial development of the Ar laser, and thus it tends to be more readily available at present. Its low efficiency (0.2 percent or less) is not expected to improve and major advances are expected to be in the area of extending its present life (<1000 hours). The doubled Nd:YAlG laser is more efficient (>0.2 percent), smaller in size, and longer lived (pump lamp life is >1000 hours and that of the basic laser is much greater) than the Ar laser. The problem associated with obtaining the required high-quality nonlinear $Ba_2NaNb_5O_{15}$ crystals appears to be improving, since such crystals are now being supplied routinely by suppliers such as Linde Company. Another feature of doubled Nd:YAlG is its shorter mode locked pulses which allow for higher capacity time-multiplexed PCM systems [30]. The major area of development of this source will be centered around attempts to realize a practical diode pumped Nd:YAlG laser.

At present, the CO_2 laser is the most efficient (20 percent), and it has virtually unlimited power capabilities. In addition, near-ideal performance is obtained in high quantum efficiency detection, and an order of magnitude lower quantum noise (hvB) is realized at 10.6 μm. The major drawbacks in the use of CO_2 lasers in some types of communications systems are the requirement of cooled detectors at 10.6 μm and the somewhat higher modulation drive power per unit bandwidth. Modulator and detector advances could greatly enhance the variety of systems using CO_2 lasers in the future, although systems with bandwidths greater than 500 MHz are possible now.

Although most of the presently developed sources are adequate for communications research, none can yet compete with respect to small size and high efficiency with microwave or baseband solid-state devices used in present day low-frequency communication systems. From the results of the work to date it appears that in the near future a continuous room temperature junction laser and/or the diode-pumped Nd:YAlG laser will change this situation. It is clear at this juncture that additional optical source development is required before an optical communications system will be economically competitive with the lower frequency systems presently being used.

Continuous room temperature operation of a double heterostructure, GaAs-$Al_xGa_{1-x}As$, injection laser has been reported by I. Hayashi, M. B. Pannish, P. W. Foy, and S. Sumski in the August 1, 1970 issue of *Applied Physics Letters*.

Recent experiments on diode pumping of the Nd:YAlG laser have demonstrated the feasibility of obtaining room temperature CW operation with presently available $GaAs_{1-x}P_x$ diodes. Using 19, 4 percent efficient, incoherent, room temperature diodes with a total power input of 7.6 watts, threshold has been obtained with the laser rod cooled to $-2.5°C$. From these measurements it is concluded that 28 room temperature diodes of the same 4-percent efficiency are needed to reach threshold when the Nd:YAlG rod is at room temperature (20°C). The number of diodes in the above experimental array can be increased by a factor of 3 to 4. Using 60 diodes the Nd:YAlG laser would be 2.2 times above threshold and with 80 diodes 2.9 times above threshold. This work is described in a letter which has been submitted to *Applied Physics Letters* by F. W. Ostermayer, Jr.

ACKNOWLEDGMENT

The authors wish to thank their colleagues at Bell Telephone Laboratories, Hughes Research Laboratories, and RCA Laboratories, whose work and constructive suggestions they have used liberally. The authors also wish to thank Hughes Electron Dynamics Division for Figs. 6 and 7; Hughes Research Laboratories for Figs. 8–12; Quantronix Corporation for Fig. 18; and F.W. Ostermayer and R. G. Smith, Bell Telephone Laboratories, for Figs. 21 and 22, respectively.

REFERENCES

[1] A. L. Schawlow and C. H. Townes, "Infrared and optical masers," *Phys. Rev.*, vol. 112, pp. 1940–1949, December 1958.
[2] T. H. Maiman, "Stimulated optical radiation in ruby masers," *Nature*, vol. 187, pp. 493–494, August 1960.
[3] A. Javan, W. R. Bennett, Jr., and D. R. Herriott, "Population inversion and continuous optical maser oscillation in a gas discharge containing a helium-neon mixture," *Phys. Rev. Lett.*, vol. 6, pp. 106–110, February 1961.
[4] A. Yariv, *Quantum Electronics.* New York: Wiley, 1967, ch. 15, pp. 242–243.
[5] a) A. Yariv and J. P. Gordon, "The laser," *Proc. IEEE*, vol. 51, pp. 4–29, January 1963.
b) J. E. Geusic and H. E. D. Scovil, "Microwave and optical masers," *Rep. Prog. Phys.*, vol. 28, pp. 241–327, 1964.
c) D. W. Goodwin and O. S. Heavens, "Doped crystal and gas lasers," *Rep. Prog. Phys.*, vol. 31, pt. 2, pp. 777–859, 1968.
[6] a) A. L. Bloom, "Gas lasers," *Proc. IEEE*, vol. 54, pp. 1262–1276, October 1966.
b) C. K. N. Patel, "Gas lasers," in *Lasers*, vol. 2, A. K. Levine, Ed. New York: Dekker, 1968, pp. 1–190.
[7] a) Z. J. Kiss and R. J. Pressley, "Crystalline solid lasers," *Proc. IEEE*, vol. 54, pp. 1236–1248, October 1966.
b) L. F. Johnson, "Optically pumped pulsed crystal lasers other than ruby," in *Lasers*, vol. 1, A. K. Levine, Ed. New York: Dekker, 1966, pp. 137–180.
[8] a) M. I. Nathan, "Semiconductor lasers," *Proc. IEEE*, vol. 54, pp. 1276–1290, October 1966.
b) W. P. Dumke, "The injection laser," in *Lasers*, vol. 1, A. K. Levine, Ed. New York: Dekker, 1968, pp. 257–293.

19

[9] a) G. Birnbaum, *Optical Masers*. New York: Academic Press, 1964.
  b) B. A. Lengyel, *Introduction to Laser Physics*. New York: Wiley, 1966.
  c) W. V. Smith and P. P. Sorokin, *The Laser*. New York: McGraw-Hill, 1966.
  d) D. Röss, *Lasers, Light Amplifiers and Oscillators*. New York: Academic Press, 1969.

[10] H. Kogelnik, "Modes in optical resonators," in *Lasers*, vol. 1, A. K. Levine, Ed. New York: Dekker, 1966, pp. 295–347.

[11] H. Kogelnik and T. Li, "Laser beams and resonators," *Proc. IEEE*, vol. 54, pp. 1312–1329, October 1966.

[12] a) A. G. Fox and T. Li, "Resonant modes in a maser interferometer," *Bell Sys. Tech. J.*, vol. 40, pp. 453–488, March 1961.
  b) G. D. Boyd and J. P. Gordon, "Confocal multimode resonator for millimeter through optical wavelength lasers," *Bell Sys. Tech. J.*, vol. 40, pp. 489–508, March 1961.
  c) G. D. Boyd and H. Kogelnik, "Generalized confocal resonator theory," *Bell Sys. Tech. J.*, vol. 41, pp. 1347–1369, July 1962.

[13] M. Scully, "The quantum theory of a laser, a problem in nonequilibrium statistical mechanics," in *Quantum Optics*, R. J. Glauber, Ed. New York: Academic Press, 1969, pp. 586–629.

[14] H. Haken and W. Weidlich, "Quantum theory of the laser," in *Quantum Optics*, R. J. Glauber, Ed. New York: Academic Press, 1969, pp. 630–679.

[15] J. P. Gordon, "Quantum theory of a simple laser," in *Quantum Optics*, R. J. Glauber, Ed. New York: Academic Press, 1969, pp. 741–789.

[16] R. J. Glauber, "Coherence and quantum detection," in *Quantum Optics*, R. J. Glauber, Ed. New York: Academic Press, 1969, pp. 15–56.

[17] F. T. Arecchi, "Photocount distribution and field statistics," in *Quantum Optics*, R. J. Glauber, Ed. New York: Academic Press, 1969, pp. 57–110.

[18] Lord Rayleigh, *Phil. Mag.*, vol. S5, 1883.

[19] W. H. Louisell, *Coupled Mode and Parametric Electronics*. New York: Wiley, 1960.

[20] J. Ducuing, "Nonlinear optical processes," in *Quantum Optics*, R. J. Glauber, Ed. New York: Academic Press, 1969, pp. 421–472.

[21] Y. R. Shen, "Quantum theory of nonlinear optics," in *Quantum Optics*, R. J. Glauber, Ed. New York: Academic Press, 1969, pp. 473–492.

[22] J. A. Giordmaine, "Parametric optics," in *Quantum Optics*, R. J. Glauber, Ed. New York: Academic Press, 1969, pp. 493–519.

[23] R. W. Minck, R. W. Terhune, and C. C. Wang, "Nonlinear optics," *Proc. IEEE*, vol. 54, pp. 1357–1374, October 1966.

[24] S. E. Harris, "Tunable optical parametric oscillators," *Proc. IEEE*, vol. 57, pp. 2096–2113, December 1969.

[25] R. W. Terhune and P. D. Maker, "Nonlinear optics," in *Lasers*, vol. 2, A. K. Levine, Ed. New York: Dekker, 1968, pp. 295–372.

[26] G. D. Boyd and D. A. Kleinman, "Parametric interaction of focused Gaussian light beams," *J. Appl. Phys.*, vol. 39, pp. 3597–3639, July 1968.

[27] J. E. Geusic, H. J. Levinstein, S. Singh, R. G. Smith, and L. G. Van Uitert, "Continuous 0.532μ solid state sources using $Ba_2NaNb_5O_{15}$," *Appl. Phys. Lett.*, vol. 12, pp. 306–308, May 1, 1968.

[28] M. W. Dowley, "Efficient CW second-harmonic generation to 2573 Å," *Appl. Phys. Lett.*, vol. 13, pp. 395–397, December 1, 1968.

[29] R. G. Smith, "Theory of intracavity optical second-harmonic generation, *IEEE J. Quantum Electron.*, vol. QE-6, pp. 215–223, April 1970.

[30] a) T. S. Kinsel, "Wide-band optical communication system, pt. I: Time division multiplexing," *Proc. IEEE*, this issue, pp. 1666–1683.
  b) O. E. DeLange, "Wide-band optical communication system, pt. II: Frequency-division multiplexing," *Proc. IEEE*, this issue, pp. 1683–1690.

[31] H. Kogelnik, "Modes in optical resonators," in *Lasers*, vol. 1, A. K. Levine, Ed. New York: Dekker, 1966, pp. 330–334.

[32] A. L. Bloom, *Gas Lasers*. New York: Wiley, 1968, pp. 82–85.

[33] A. G. Fox and T. Li, "Effect of gain saturation on the oscillating modes of optical masers," *IEEE J. Quantum Electron.*, vol. QE-2, pp. 774–783, December 1966.

[34] A. L. Bloom, *Gas Lasers*. New York: Wiley, 1968, pp. 86–100, 122–137.

[35] H. G. Danielmeyer, "Stabilized efficient single-frequency Nd:YAG laser," *IEEE J. Quantum Electron.*, vol. QE-6, pp. 101–104, February 1970.

[36] T. G. Polanyi and I. Tobias, "The frequency stabilization of gas lasers," in *Lasers*, vol. 2, A. K. Levine, Ed. New York: Dekker, 1968, pp. 373–423.

[37] S. E. Harris, "Stabilization and modulation of laser oscillators by internal time-varying perturbations," *Proc. IEEE*, vol. 54, pp. 1401–1413, October 1966.

[38] L. E. Hargrove, R. L. Fork, and M. A. Pollack, "Locking of He–Ne modes induced by synchronous intracavity modulation," *Appl. Phys. Lett.*, vol. 5, pp. 4–5, July 1964.

[39] M. H. Crowell, "Characteristics of mode-coupled lasers," *IEEE J. Quantum Electron.*, vol. QE-1, pp. 12–20, April 1965.

[40] M. DiDomenico, J. E. Geusic, H. M. Marcos, and R. G. Smith, "Generation of ultrashort optical pulses by mode locking the YAlG:Nd lasers," *Appl. Phys. Lett.*, vol. 8, pp. 180–182, April 1, 1966.

[41] M. DiDomenico, "Small signal analysis of internal (coupling type) modulation of lasers," *J. Appl. Phys.*, vol. 35, pp. 2870–2876, October 1964.

[42] S. E. Harris and O. P. McDuff, "Theory of FM laser oscillation," *IEEE J. Quantum Electron.*, vol. QE-1, pp. 245–262, September 1965.

[43] P. W. Smith, "Mode-locking of lasers," *Proc. IEEE*, vol. 58, pp. 1342–1357, September 1970.

[44] A. Yariv, "Internal modulation in multimode laser oscillators," *J. Appl. Phys.*, vol. 36, pp. 388–391, February 1965.

[45] P. W. Smith, "Phase locking of laser modes by continuous cavity length variations," *Appl. Phys. Lett.*, vol. 10, pp. 51–53, January 15, 1967.

[46] ——, "The self-pulsing laser oscillator," *IEEE J. Quantum Electron.*, vol. QE-3, pp. 627–635, November 1967.

[47] A. A. Vuylsteke, "Theory of laser regeneration switching," *J. Appl. Phys.*, vol. 34, pp. 1615–1622, June 1963.

[48] W. H. Steier, "Coupling of high peak power pulses from He–Ne lasers," *Proc. IEEE* (Letters), vol. 54, pp. 1604–1606, November 1966.

[49] R. B. Chesler and D. Maydan, "A calculation of Nd:YAlG cavity dumping" (to be published).

[50] R. W. Hellwarth, "Control of fluorescent pulsations," in *Advances in Quantum Electronics*, J. R. Singer, Ed. New York: Columbia University Press, 1961, pp. 334–341.

[51] F. J. McClung and R. W. Hellwarth, "Giant optical pulsations from ruby," *J. Appl. Phys.*, vol. 33, pp. 828–829, March 1962.

[52] J. E. Geusic, M. L. Hensel, and R. G. Smith, "A repetitively Q-switched, continuously pumped YAlG:Nd laser," *Appl. Phys. Lett.*, vol. 6, pp. 175–177, May 1965.

[53] E. J. Woodbury, "Five kilohertz repetition rate pulses YAlG:Nd laser," *IEEE J. Quantum Electron.*, vol. QE-3, pp. 509–516, November 1967.

[54] V. Evtuhov and J. K. Neeland, "A continuously pumped repetitively Q-switched ruby laser and applications to frequency conversion experiments," *IEEE J. Quantum Electron.*, vol. QE-5, pp. 207–209, April 1969.

[55] R. B. Chesler, M. A. Karr, and J. E. Geusic, "A practical, high-repetition rate Q-switched Nd:YAlG laser," *IEEE J. Quantum Electron.*, vol. QE-5, p. 345, June 1969.

[56] P. K. Cheo and H. G. Cooper, "Ultraviolet ion laser transitions between 2300 Å," *J. Appl. Phys.*, vol. 36, pp. 1862–1865, January 1965.

[57] H. Steffen, J. Steffen, J. F. Moser, and F. K. Kneubühl, "Comments on a new laser emission at 0.774 mm wavelength from ICN," *Phys. Lett.*, vol. 21, pp. 425–426, 1966.

[58] F. Horrigan, R. Rudko, and D. Wilson, "High-power 10.6 μ windows," presented at the 1968 Internatl. Quantum Electron. Conf., Miami, Fla.

[59] For example, the Spectra-Physics Model 119 He–Ne laser.

[60] D. A. Leonard, "Saturation of the molecular nitrogen second positive laser transition," *Appl. Phys. Lett.*, vol. 7, pp. 4–6, July 1965.

[61] A. J. DeMaria, "Electrically excited CO_2 lasers," *Bull. Amer. Phys. Soc.*, vol. 15, p. 563, April 1970.

[62] P. Rabinowitz, S. Jacobs, and G. Goulds, "Continuous optically pumped Cs laser," *Appl. Optics*, vol. 1, pp. 513–516, 1962.

[63] I. Wieder, "Flame pumping and infrared maser action in CO_2," *Phys. Lett.*, vol. 13, pp. 759–760, June 19, 1967.

20

[9] a) G. Birnbaum, *Optical Masers.* New York: Academic Press, 1964.
b) B. A. Lengyel, *Introduction to Laser Physics.* New York: Wiley, 1966.
c) W. V. Smith and P. P. Sorokin, *The Laser.* New York: McGraw-Hill, 1966.
d) D. Röss, *Lasers, Light Amplifiers and Oscillators.* New York: Academic Press, 1969.

[10] H. Kogelnik, "Modes in optical resonators," in *Lasers*, vol. 1, A. K. Levine, Ed. New York: Dekker, 1966, pp. 295–347.

[11] H. Kogelnik and T. Li, "Laser beams and resonators," *Proc. IEEE*, vol. 54, pp. 1312–1329, October 1966.

[12] a) A. G. Fox and T. Li, "Resonant modes in a maser interferometer," *Bell Sys. Tech. J.*, vol. 40, pp. 453–488, March 1961.
b) G. D. Boyd and J. P. Gordon, "Confocal multimode resonator for millimeter through optical wavelength lasers," *Bell Sys. Tech. J.*, vol. 40, pp. 489–508, March 1961.
c) G. D. Boyd and H. Kogelnik, "Generalized confocal resonator theory," *Bell Sys. Tech. J.*, vol. 41, pp. 1347–1369, July 1962.

[13] M. Scully, "The quantum theory of a laser, a problem in nonequilibrium statistical mechanics," in *Quantum Optics*, R. J. Glauber, Ed. New York: Academic Press, 1969, pp. 586–629.

[14] H. Haken and W. Weidlich, "Quantum theory of the laser," in *Quantum Optics*, R. J. Glauber, Ed. New York: Academic Press, 1969, pp. 630–679.

[15] J. P. Gordon, "Quantum theory of a simple laser," in *Quantum Optics*, R. J. Glauber, Ed. New York: Academic Press, 1969, pp. 741–789.

[16] R. J. Glauber, "Coherence and quantum detection," in *Quantum Optics*, R. J. Glauber, Ed. New York: Academic Press, 1969, pp. 15–56.

[17] F. T. Arecchi, "Photocount distribution and field statistics," in *Quantum Optics*, R. J. Glauber, Ed. New York: Academic Press, 1969, pp. 57–110.

[18] Lord Rayleigh, *Phil. Mag.*, vol. S5, 1883.

[19] W. H. Louisell, *Coupled Mode and Parametric Electronics.* New York: Wiley, 1960.

[20] J. Ducuing, "Nonlinear optical processes," in *Quantum Optics*, R. J. Glauber, Ed. New York: Academic Press, 1969, pp. 421–472.

[21] Y. R. Shen, "Quantum theory of nonlinear optics," in *Quantum Optics*, R. J. Glauber, Ed. New York: Academic Press, 1969, pp. 473–492.

[22] J. A. Giordmaine, "Parametric optics," in *Quantum Optics*, R. J. Glauber, Ed. New York: Academic Press, 1969, pp. 493–519.

[23] R. W. Minck, R. W. Terhune, and C. C. Wang, "Nonlinear optics," *Proc. IEEE*, vol. 54, pp. 1357–1374, October 1966.

[24] S. E. Harris, "Tunable optical parametric oscillators," *Proc. IEEE*, vol. 57, pp. 2096–2113, December 1969.

[25] R. W. Terhune and P. D. Maker, "Nonlinear optics," in *Lasers*, vol. 2, A. K. Levine, Ed. New York: Dekker, 1968, pp. 295–372.

[26] G. D. Boyd and D. A. Kleinman, "Parametric interaction of focused Gaussian light beams," *J. Appl. Phys.*, vol. 39, pp. 3597–3639, July 1968.

[27] J. E. Geusic, H. J. Levinstein, S. Singh, R. G. Smith, and L. G. Van Uitert, "Continuous 0.532μ solid state sources using $Ba_2NaNb_5O_{15}$," *Appl. Phys. Lett.*, vol. 12, pp. 306–308, May 1, 1968.

[28] M. W. Dowley, "Efficient CW second-harmonic generation to 2573 Å," *Appl. Phys. Lett.*, vol. 13, pp. 395–397, December 1, 1968.

[29] R. G. Smith, "Theory of intracavity optical second-harmonic generation, *IEEE J. Quantum Electron.*, vol. QE-6, pp. 215–223, April 1970.

[30] a) T. S. Kinsel, "Wide-band optical communication system, pt. I: Time division multiplexing," *Proc. IEEE*, this issue, pp. 1666–1683.
b) O. E. DeLange, "Wide-band optical communication system, pt. II: Frequency-division multiplexing," *Proc. IEEE*, this issue, pp. 1683–1690.

[31] H. Kogelnik, "Modes in optical resonators," in *Lasers*, vol. 1, A. K. Levine, Ed. New York: Dekker, 1966, pp. 330–334.

[32] A. L. Bloom, *Gas Lasers.* New York: Wiley, 1968, pp. 82–85.

[33] A. G. Fox and T. Li, "Effect of gain saturation on the oscillating modes of optical masers," *IEEE J. Quantum Electron.*, vol. QE-2, pp. 774–783, December 1966.

[34] A. L. Bloom, *Gas Lasers.* New York: Wiley, 1968, pp. 86–100, 122–137.

[35] H. G. Danielmeyer, "Stabilized efficient single-frequency Nd:YAG laser," *IEEE J. Quantum Electron.*, vol. QE-6, pp. 101–104, February 1970.

[36] T. G. Polanyi and I. Tobias, "The frequency stabilization of gas lasers," in *Lasers*, vol. 2, A. K. Levine, Ed. New York: Dekker, 1968, pp. 373–423.

[37] S. E. Harris, "Stabilization and modulation of laser oscillators by internal time-varying perturbations," *Proc. IEEE*, vol. 54, pp. 1401–1413, October 1966.

[38] L. E. Hargrove, R. L. Fork, and M. A. Pollack, "Locking of He–Ne modes induced by synchronous intracavity modulation," *Appl. Phys. Lett.*, vol. 5, pp. 4–5, July 1964.

[39] M. H. Crowell, "Characteristics of mode-coupled lasers," *IEEE J. Quantum Electron.*, vol. QE-1, pp. 12–20, April 1965.

[40] M. DiDomenico, J. E. Geusic, H. M. Marcos, and R. G. Smith, "Generation of ultrashort optical pulses by mode locking the YAlG:Nd lasers," *Appl. Phys. Lett.*, vol. 8, pp. 180–182, April 1, 1966.

[41] M. DiDomenico, "Small signal analysis of internal (coupling type) modulation of lasers," *J. Appl. Phys.*, vol. 35, pp. 2870–2876, October 1964.

[42] S. E. Harris and O. P. McDuff, "Theory of FM laser oscillation," *IEEE J. Quantum Electron.*, vol. QE-1, pp. 245–262, September 1965.

[43] P. W. Smith, "Mode-locking of lasers," *Proc. IEEE*, vol. 58, pp. 1342–1357, September 1970.

[44] A. Yariv, "Internal modulation in multimode laser oscillators," *J. Appl. Phys.*, vol. 36, pp. 388–391, February 1965.

[45] P. W. Smith, "Phase locking of laser modes by continuous cavity length variations," *Appl. Phys. Lett.*, vol. 10, pp. 51–53, January 15, 1967.

[46] ——, "The self-pulsing laser oscillator," *IEEE J. Quantum Electron.*, vol. QE-3, pp. 627–635, November 1967.

[47] A. A. Vuylsteke, "Theory of laser regeneration switching," *J. Appl. Phys.*, vol. 34, pp. 1615–1622, June 1963.

[48] W. H. Steier, "Coupling of high peak power pulses from He–Ne lasers," *Proc. IEEE* (Letters), vol. 54, pp. 1604–1606, November 1966.

[49] R. B. Chesler and D. Maydan, "A calculation of Nd:YAlG cavity dumping" (to be published).

[50] R. W. Hellwarth, "Control of fluorescent pulsations," in *Advances in Quantum Electronics*, J. R. Singer, Ed. New York: Columbia University Press, 1961, pp. 334–341.

[51] F. J. McClung and R. W. Hellwarth, "Giant optical pulsations from ruby," *J. Appl. Phys.*, vol. 33, pp. 828–829, March 1962.

[52] J. E. Geusic, M. L. Hensel, and R. G. Smith, "A repetitively Q-switched, continuously pumped YAlG:Nd laser," *Appl. Phys. Lett.*, vol. 6, pp. 175–177, May 1965.

[53] E. J. Woodbury, "Five kilohertz repetition rate pulses YAlG:Nd laser," *IEEE J. Quantum Electron.*, vol. QE-3, pp. 509–516, November 1967.

[54] V. Evtuhov and J. K. Neeland, "A continuously pumped repetitively Q-switched ruby laser and applications to frequency conversion experiments," *IEEE J. Quantum Electron.*, vol. QE-5, pp. 207–209, April 1969.

[55] R. B. Chesler, M. A. Karr, and J. E. Geusic, "A practical, high-repetition rate Q-switched Nd:YAlG laser," *IEEE J. Quantum Electron.*, vol. QE-5, p. 345, June 1969.

[56] P. K. Cheo and H. G. Cooper, "Ultraviolet ion laser transitions between 2300 Å," *J. Appl. Phys.*, vol. 36, pp. 1862–1865, January 1965.

[57] H. Steffen, J. Steffen, J. F. Moser, and F. K. Kneubuhl, "Comments on a new laser emission at 0.774 mm wavelength from ICN," *Phys. Lett.*, vol. 21, pp. 425–426, 1966.

[58] F. Horrigan, R. Rudko, and D. Wilson, "High-power 10.6μ windows," presented at the 1968 Internatl. Quantum Electron. Conf., Miami, Fla.

[59] For example, the Spectra-Physics Model 119 He–Ne laser.

[60] D. A. Leonard, "Saturation of the molecular nitrogen second positive laser transition," *Appl. Phys. Lett.*, vol. 7, pp. 4–6, July 1965.

[61] A. J. DeMaria, "Electrically excited CO_2 lasers," *Bull. Amer. Phys. Soc.*, vol. 15, p. 563, April 1970.

[62] P. Rabinowitz, S. Jacobs, and G. Goulds, "Continuous optically pumped Cs laser," *Appl. Optics*, vol. 1, pp. 513–516, 1962.

[63] I. Wieder, "Flame pumping and infrared maser action in CO_2," *Phys. Lett.*, vol. 13, pp. 759–760, June 19, 1967.

oscillations in Nd-doped yttrium aluminum, yttrium gallium and gadolinium garnets," *Appl. Phys. Lett.*, vol. 4, pp. 182–184, May 15, 1964.
b) J. E. Geusic, "The YAlG:Nd laser," Solid-state maser research (optical) Final Rept., U. S. Army Electron. Material Agency, contract DA-36-039AMC-02333E, August 30, 1965.

[121] R. C. Duncan, Jr., "Continuous room temperature $Nd^{3+}:CaMoO_4$ laser," *J. Appl. Phys.*, vol. 36, pp. 874–875, March 1965.

[122] R. C. Ohlmann, K. B. Steinbruegge, and R. Mazelsky, "Spectroscopic and laser characteristics of neodymium-doped calcium fluoroapatite," *Appl. Opt.*, vol. 7, pp. 905–914, May 1968.

[123] a) Kh. S. Bagdasarov and A. A. Kaminskii, "YAlO₃ with TR^{3+} ion impurity as an active laser medium," *JETP Lett.*, vol. 9, p. 303, May 5, 1969.
b) M. J. Weber, M. Bass, K. Andringa, R. R. Montchamp, and E. Comperchio, "Czochralski growth and properties of YAlO₃ crystals," *Appl. Phys. Lett.*, vol. 15, pp. 342–345, November 15, 1969.

[124] R. V. Alves, K. A. Wickersheim, and R. A. Buchanan, "Optical characteristics and laser performance of neodymium activated lanthanum oxysulfide," *Bull. Amer. Phys. Soc.*, March 1970.

[125] W. A. Specht, Jr., J. K. Neeland, and V. Evtuhov, "Output spectra of Nd:YAG and ruby lasers and implications for laser linewidth determining mechanisms," *IEEE J. Quantum Electron.*, vol. QE-2, pp. 537–541, September 1966.

[126] V. Evtuhov and J. K. Neeland, "Power output and efficiency of continuous ruby lasers," *J. Appl. Phys.*, vol. 38, p. 4051, 1967.

[127] D. Röss, "Analysis of room temperature CW ruby lasers," *IEEE J. Quantum Electron.*, vol. QE-2, pp. 208–214, August 1966.

[128] T. Kushida and J. E. Geusic, "Optical refrigeration in Nd-doped yttrium aluminum garnet," *Phys. Rev. Lett.*, vol. 21, pp. 1172–1175, October 14, 1968.

[129] T. Kushida, H. M. Marcos, and J. E. Geusic, "Laser transition cross section and fluorescence branching ratio for Nd^{3+} in yttrium aluminum garnet," *Phys. Rev.*, vol. 167, pp. 289–291, March 10, 1968.

[130] ——, "Laser transition crossections of lines other than 1.064μ in Nd:YAlG" (unpublished).

[131] R. G. Smith, "New room temperature CW laser transitions in YAlG:Nd," *IEEE J. Quantum Electron.* (Correspondence), vol. QE-4, pp. 505–506, August 1968.

[132] W. W. Benson and J. E. Geusic (unpublished).

[133] C. H. Church and I. Liberman, "The spherical reflector for use in the optical pumping of lasers," *Appl. Opt.*, vol. 6, pp. 1966–1968, November 1967.

[134] D. Röss, "CW and quasi-CW solid state lasers," *Siemens Z.*, vol. 41, pp. 3–11, 1967.

[135] W. W. Benson, H. M. Marcos, and J. E. Geusic (unpublished).

[136] T. B. Read, "The CW pumping of YAlG:Nd^{3+} by water-cooled krypton arcs," *Appl. Phys. Lett.*, vol. 9, pp. 342–344, November 1, 1966.

[137] I. Liberman and R. L. Grassel, "A comparison of lamps for use in high continuous power Nd:YAlG lasers," *Appl. Opt.*, vol. 8, pp. 1875–1878, September 1969.

[138] I. Liberman, D. A. Larson, and C. H. Church, "Efficient Nd:YAG CW lasers using alkali additive lamps," *IEEE J. Quantum Electron.*, vol. QE-5, pp. 238–241, May 1969.

[139] E. G. Erickson, "Holobeam reports 760 watts CW from a segmented Nd:YAlG system," *Laser Focus*, p. 16, April 1970.

[140] W. W. Benson and J. E. Geusic (unpublished).

[141] I. Liberman, "High-power Nd:YAG continuous laser" (Abstract), *IEEE J. Quantum Electron.*, vol. QE-5, p. 345, June 1969.

[142] L. M. Osterink and J. D. Foster, "Efficient, high-power Nd:YAG laser characteristics" (Abstract), *IEEE J. Quantum Electron.*, vol. QE-5, pp. 344–345, June 1969.

[143] W. Koechner, "YAlG challenges carbon dioxide in high CW power," *Laser Focus*, pp. 29–34, September 1969.

[144] J. D. Foster and L. M. Osterink, "Birefringence in continuously pumped Nd:YAlG rods" (to be published).

[145] J. D. Foster and L. M. Osterink, "Index of refraction and expansion coefficients of Nd:YAlG," *Appl. Opt.*, vol. 7, pp. 2428–2429,

December 1968.

[146] R. T. Daly and R. A. Kaplan, "Thermally induced birefringence in a pumped laser rod," Quantronix Corporation (unpublished).

[147] W. Koechner and D. K. Rice, "Effect of birefringence on the performance of linearly polarized Nd:YAlG lasers" (to be published).

[148] L. M. Osterink and J. D. Foster, "Experimental compensation of thermally induced stress birefringence in solid state laser rods," *1969 Proc. Internatl. Electron Device Meeting*, p. 70.

[149] J. E. Geusic and H. M. Marcos (unpublished).

[150] R. B. Chesler, "A stabilizing sleeve for the YAlG laser" (to be published).

[151] J. E. Geusic, "Advances in CW solid state lasers," *1966 NEREM Rec.*, p. 192.

[152] J. E. Geusic, "The continuous Nd:YAG laser" (Abstract), *IEEE J. Quantum Electron.*, vol. QE-2, p. xvii, April 1966.

[153] H. G. Danielmeyer and W. G. Nilsen, "Spontaneous single frequency output from a spatially homogeneous Nd:YAlG laser," *Appl. Phys. Lett.*, vol. 16, pp. 124–126, February 1970.

[154] L. M. Osterink and J. D. Foster, "A mode-locked Nd:YAlG laser," *J. Appl. Phys.*, vol. 39, pp. 4163–4165, August 1968.

[155] G. W. Hong and J. R. Whinnery, "Switching of phase-locked states in the intracavity phase-modulated He–Ne laser," *IEEE J. Quantum Electron.*, vol. QE-5, pp. 367–376, July 1969.

[156] D. L. Lyon and T. S. Kinsel, "Transitions between mode-locked states in intracavity phase-modulated lasers," *Appl. Phys. Lett.*, vol. 16, pp. 89–91, February 1970.

[157] T. S. Kinsel, J. E. Geusic, H. Seidel, and R. G. Smith, "A stabilized mode-locked Nd:YAlG laser source" (Abstract), *IEEE J. Quantum Electron.*, vol. QE-5, p. 326, June 1969.

[158] R. B. Chesler, M. A. Karr, and J. E. Geusic, "A practical high-repetition rate Q-switched Nd:YAlG laser" (to be published).

[159] D. Maydan and R. B. Chesler, "Q-switching and cavity dumping of Nd:YAlG lasers" (to be published).

[160] a) J. E. Geusic, H. J. Levinstein, J. J. Rubin, S. Singh, and L. G. Van Uitert, "The nonlinear optical properties of $Ba_2NaNb_5O_{15}$," *Appl. Phys. Lett.*, vol. 11, pp. 269–271, November 1967.
b) S. Singh, D. A. Draegert, and J. E. Geusic, "Optical and ferroelectric properties of barium sodium niobate," *Phys. Rev.* (to be published).

[161] J. E. Geusic and H. M. Marcos (unpublished).

[162] M. A. Karr, Bell Telephone Laboratories, Inc., private communication.

[163] R. B. Allen and S. J. Scalise, "Continuous operation of a YAlG:Nd laser by injection pumping," *Appl. Phys. Lett.*, vol. 14, pp. 49–51, March 1969.

[164] F. W. Ostermayer, Bell Telephone Laboratories, Inc., private communication.

[165] B. Reed, Texas Instruments, Inc., private communication.

[166] R. G. Smith, J. E. Geusic, H. J. Levinstein, J. J. Rubin, S. Singh, and L. G. Van Uitert, "Continuous optical parametric oscillation in $Ba_2NaNb_5O_{15}$," *Appl. Phys. Lett.*, vol. 9, p. 308, May 1968.

[167] R. L. Byer, M. K. Oshman, J. F. Young, and S. E. Harris, "Visible CW parametric oscillator," *Appl. Phys. Lett.*, vol. 13, p. 109, August 1968.

[168] R. G. Smith, Bell Telephone Laboratories, Inc., private communication.

[169] J. E. Bjorkholm, "Efficient optical parametric oscillation using doubly and singly resonant cavities," *Appl. Phys. Lett.*, vol. 13, pp. 53–56, July 1968.

[170] J. E. Bjorkholm, "Some spectral properties of doubly and singly resonant pulsed optical parametric oscillators," *Appl. Phys. Lett.*, vol. 13, pp. 399–401, December 1960.

[171] L. B. Kreuzer, "Single mode oscillation of a pulsed singly resonant optical parametric oscillator," *Appl. Phys. Lett.*, vol. 15, pp. 263–265, October 1969.

[172] J. Falk and J. E. Murray, "Single cavity noncollinear optical parametric oscillation," *Appl. Phys. Lett.*, vol. 14, pp. 245–247, April 1969.

[173] F. S. Chen, "Modulators for optical communication," *Proc. IEEE*, this issue, pp. 1440–1457.

[174] H. Melchior, M. B. Fisher, and F. Arams, "Photodetectors for optical communication systems," *Proc. IEEE*, this issue, pp. 1466–1486.

Part II: Solid, Liquid, and Semiconductor Lasers

The first three successful lasers were all solid-state lasers. Today, however, only the original ruby laser and the Nd:YAG laser among crystalline solid-state lasers, and the Nd glass and semiconductor lasers among other forms of solid-state lasers, have found substantial practical application. The 1966 paper by Kiss and Pressley, although not especially focused on ruby or Nd:YAG lasers, does give a useful survey of these and most other transitions available in crystalline solid-state lasers. The capabilities of glass lasers are well reviewed in the 1969 paper by Young.

The semiconductor injection laser forms a very different and important class of solid-state laser devices. The fundamental concepts of this type of laser are very well covered in the paper by Nathan. More recent advances not covered by this paper, primarily the use of heterojunction wave-guiding techniques and the achievement of room-temperature CW operation, will be found in the current literature,[1] and perhaps in a future PROCEEDINGS invited review paper.

The important recent developments in liquid organic dye lasers are given an excellent introduction in the paper by Snavely, although this paper antedates the still more recent achievement of CW dye lasers.

---

[1] See M. B. Panish and I. Hayashi, "Heterostructure junction lasers," in *Applied Solid State Science*, vol. 4, R. Wolfe, Ed. New York: Academic, 1973.

H. Kressel, "Semiconductor lasers," in *Lasers: A Series of Advances*, A. K. Levine and A. J. DeMaria, Eds. New York: Marcel Dekker, 1971.

Crystalline Solid Lasers

Z. J. KISS AND R. J. PRESSLEY

Abstract—A survey of crystalline solid lasers is presented. Crystalline host materials are described, pointing out their characteristics pertinent for laser systems. Rare earth and transition metal impurities operated as lasers are tabulated and the role of sensitization in increasing the overall efficiency of laser systems is described. Characteristics of operating CW lasers are given, and some applications and research directions are suggested.

I. INTRODUCTION

THE CONDITIONS for maser action at optical frequencies were first described by Schawlow and Townes [1] in 1958. The first demonstration of laser action by Maiman [2] two years later was achieved using ruby (Al_2O_3:Cr^{3+}), a crystalline solid system. In the following years, much effort was expended on the search for new laser transitions in various media: crystalline solids, gases, liquids, glasses, plastics, and semiconductors. The progress was indeed rapid, and today we have thousand of frequencies covering the spectrum from the far IR to near UV.

The next step in the development of solid-state lasers after the pulsed three level ruby system was the operation of a pulsed four level system, CaF_2:Sm^{2+} [3] by Sorokin and Stevenson. The first continuously operating crystal laser was constructed by Johnson et al. [4] using $CaWO_4$: Nd^{3+}. In the following years, a systematic search was begun for new laser systems, using trivalent rare earths, divalent rare earths, and transition metals as the impurities in a great variety of host crystals [5]. An all important spur to the development of crystalline lasers was the generation of giant laser pulses by Q spoiling techniques [6]. Later transitions were soon found covering the spectrum from 0.55 μ (CaF_2:$Ho^{3+(7)}$) to 2.6 μ (CaF_2: $U^{3+(3)}$). The effort then shifted towards increasing the overall efficiency of crystalline lasers, finding CW systems in the visible region of the spectrum, and optimizing the parameters and techniques for Q switched laser pulses.

At the present state-of-the-art of laser technology solid crystalline lasers have certain advantages over other laser media. Since the longest lived excited metastable energy levels exist in these impurity doped solids ($\sim 10^{-3}$ seconds compared to $\sim 10^{-6}$ seconds in gases and $\sim 10^{-9}$ seconds in injection lasers) these lasers can be best utilized for energy storage, and hence for Q switching and the generation of high peak powers. The density of active impurity ions in crystal lasers is $\sim 10^{17}$–10^{20} ions/cm^3 as compared

to 10^{15}–10^{17} atoms/cm^3 in gases and 10^{22} electron hole pairs/cm^3 in injection lasers. Thus, the active ion density of crystalline solid laser medium is a good compromise for high CW powers; on one hand, it is dilute enough so that the power is not limited by cooling problems as in the case of injection lasers, and at the same time for a given power smaller active volumes are needed than for gas lasers. So far, the capabilities of crystalline lasers as frequency standards and atomic clocks have not been utilized, but it should be pointed out that in some systems the fluorescent linewidth of the "no-phonon" laser line [8] is narrower than the Doppler width in gases, and these lasers should have long term and short term stability as good as the presently used gas lasers.

There have been a number of review articles written on crystalline lasers [5], [9] and several books include chapters on such systems [10]. A recent publication by Kaminsky and Osiko [11] list all the operating crystalline laser systems and their characteristics. Since these data are available, we will not attempt to give a complete list of all the lasers, but rather concentrate on the "useful" systems and materials which represent various aspects of the state-of-the-art. First, we will discuss the materials, considering both the impurities and the hosts. In the second part, the characteristics of operating lasers will be surveyed and some considerations for future research directions will be suggested.

II. LASER MATERIALS

The laser consists of a resonant cavity, containing an amplifying medium. For crystalline lasers the cavity is usually obtained by fabricating the material into a Fabry-Perot configuration, and the amplifying medium is the excited impurity atoms in the crystal. The threshold condition for oscillation is reached when the losses of the cavity and the bulk are equaled by the gain in the amplifying medium. The excess number of atoms/cm^3 in the excited state ΔN at threshold [1], [9] is given by

$$\Delta N = \frac{3\Delta\nu}{8\pi^2\nu} \frac{hc(1 - R)}{\mu^2 L} \tag{1}$$

where ν is the frequency of the laser transition, $\Delta\nu$ is the fluorescent linewidth, and μ is the dipole moment matrix element of the transition. L is the length of the cavity and R is the product of the reflectivities of the end mirrors. The above formula assumes a Lorenzian line shape and does not take diffraction and bulk losses into account.

Manuscript received June 23, 1966; revised July 15, 1966.
The authors are with RCA Laboratories, Princeton, N. J.

Reprinted from *Proc. IEEE*, vol. 54, pp. 1236–1248, Oct. 1966.

The minimum input power required to maintain this excess number of atoms in the excited state is

$$P = \frac{h\nu \Delta N}{t}, \qquad (2)$$

where t is the lifetime of the excited state.

To select laser materials with low threshold, the fluorescent linewidth $\Delta\nu$ should be narrow. Losses at the reflective ends and bulk losses at the laser frequency originating either from impurities or from absorption by excited active ions should be minimized. For efficient utilization of the pump radiation, broad intense absorption bands which match the output of available lamps are needed. For high internal efficiency, the nonradiative relaxation time from the emitting level of the laser transition should be long compared with the spontaneous radiative lifetime of the atom. In addition, the crystal host must have good mechanical and thermal properties to withstand the severe operating conditions of useful lasers.

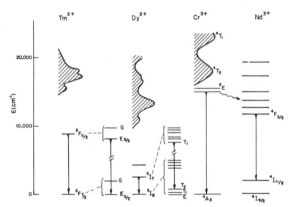

Fig. 1.   Energy level diagram of four different laser systems, illustrating the different types of crystalline lasers.

Figure 1 shows the energy level diagram of four different laser systems, illustrating the different types of crystalline lasers. $CaF_2:Tm^{2+}$ and $Al_2O_3:Cr^{3+}$ are examples of three level laser systems. For a three level laser the ground state is the terminal state, and to reach population inversion, more than half of the atoms have to be excited to the upper state. $CaF_2:Dy^{2+}$ at $4.2°K$ and YAG:Nd^{3+} at $300°K$ are examples of four level laser systems. The terminal states of these laser transitions are a certain energy ΔE above ground state, and if $\Delta E \gg kT$, the terminal level is not populated thermally. The threshold condition is thus easier to obtain than for a three level laser. The relaxation rate from the terminal level to the ground state should be fast compared to the pumping rate, but it should not be so fast as to cause line-broadening. To illustrate the threshold requirements, consider the case of ruby [1]. The required excess number of ions in the excited state at threshold is $\Delta N = 10^{17}$ ions/cm³ (see e.g., [9]).

Since the pump bands of ruby are optically thin at these concentrations, the pump threshold for obtaining inversion is almost independent of concentration. The lowest experimental thresholds are obtained with the relatively high concentration of 10^{19} ions/cm³, where the higher gain compensates for residual losses in the laser rod. Therefore, in the required three level operation, at least 5×10^{18} Cr^{3+} ions/cm³ have to be excited, 50 times more than the excess required from the threshold condition.

A. Crystalline Hosts

The crystalline host material determines the behavior of impurity ions and laser characteristics in a number of ways. The static crystalline Stark effect removes most of the degeneracy of the free-ion states, and is a major factor in determining the energy levels of the impurities. The dynamic interaction of the impurity ions with the phonons of the lattice can be responsible for the linewidth and the nonradiative relaxation process, and hence for the internal efficiency of the system.

A summary of the host crystals used for laser systems is given in Table I, together with some of the pertinent parameters. It is interesting to note that the first laser material to be discovered, ruby, in many ways is still one of the most important systems. The Al_2O_3 (sapphire) host is hard, with high thermal conductivity, and transition metals can readily be incorporated substitutionally for the Al. The Al site is too small for rare earths and it was not possible to incorporate appreciable concentration of these impurities into sapphire. Al_2O_3 can be grown by flame fusion (Verneuil technique) from the flux [12], but the best quality ruby has been obtained by pulling from the melt [13].

A number of lasers have been operated using the cubic fluoride hosts: CaF_2, SrF_2 and BaF_2. These crystals can be grown with excellent optical quality by the Stockbarger technique [14]. Rare earth doped CaF_2 crystals have been the most widely studied. A trivalent rare earth ion is incorporated substitutionally for the divalent Ca ion. Charge compensation can take place either by an interstitial fluorine or by oxygen substituting for a nearest neighbor fluorine. Charge compensation in most cases destroys the cubic symmetry of the rare earth. There are some trivalent rare earth ions, however, that remain in a cubic site with a charge compensator many lattice sites away. The fluoride hosts are particularly suitable for divalent rare earths. If the impurity is in the divalent state and substitutes for Ca^{2+}, charge compensation is not needed and the environment remains cubic. Laser action in the BaF_2 and SrF_2 hosts is more difficult to achieve than in CaF_2, but the crystal growth and purification of these hosts have not been perfected as well as CaF_2.

There is a class of tetragonal materials, the alkaline earth tungstates and molybdates, which can easily be

TABLE I
LASER HOST MATERIALS

	Symmetry	Lattice Constants Å	Melting Point °C	Refractive Index n	Hardness (Moh)	Thermal Conductivity at Room Temp. cal/cm °C	Thermal Expansion Coefficients 10^{-6}
Al_2O_3	D_{3d}^5 R3C	5.12	2040	1.765	9	0.11	5.8
CaF_2	O_h^5 Fm3m	5.451	1360	1.4335	4		19.5
SrF_2	O_h^5 Fm3m	5.78	1400	1.438			
BaF_2	O_h^5 Fm3m	6.19	1280	1.475			
$SrCl_2$	O_h^5 Fm3m	7.00	873	1.6			
LaF_3	D_{6h}^3 C6/mcm	$a_0 = 4.148$ $c_0 = 7.354$	1493				
CeF_3	D_{6h}^3 C6/mcm	$a_0 = 4.115$ $c_0 = 7.288$	1324				
$CaWO_4$	C_{4h}^6 I4/a	5.24 11.38	1570	1.918 1.934	4.5		
$SrWO_4$	C_{4h}^6 I4/a		1566				
$CaMoO_4$	C_{4h}^6 I4/a	5.23 11.44	1430	1.967 1 978	6	0.0095	25.5 c axis 19 4 a axis
$PbMoO_4$	C_{4h}^6 I4/a	5.41 12.08	1070				
Y_2O_3	T_h^7 Ia3	10.6	2450				
Gd_2O_3	T_h^7 Ia3	10.79	2330				
Er_2O_3	T_h^7 Ia3	10.54					
$Y_3Al_5O_{12}$	O_h^{10} I_a3d	12.00	1970	1.83	8.5	0.030	9.3
$Y_3Ga_5O_{12}$	O_h^{10} I_a3d	12.27		1.93	7.5		
$Gd_3Ga_5O_{12}$	O_h^{10} I_a3d		1825				
MgF_2	D_{4h}^{14} P4/mnm	$a_0 = 4.6213$ $c_0 = 3.0529$	1255	1.38			
ZnF_2	D_{4h}^{14} P4/mnm	$a_0 = 4.715$ $c_0 = 3.131$	872				
$Ca(NbO_3)_2$			1560	2.07–2.20			

grown in ambient atmosphere [15] as best exemplified by $CaWO_4$. The rare earth goes in substitutionally for Ca, but only in the trivalent oxidation state, and hence charge compensation is needed. For optimum laser performance compensation by Na^+, replacing a Ca^{++} was found to be best. Other similar hosts where laser action was observed with Nd^{3+} as the impurity ions are [11] $SrWO_4$, $Na_{0.5}Gd_{0.5}WO_4$, $CaMoO_4$, $SrMoO_4$, and $PbMoO_4$.

To incorporate trivalent rare earth into fluoride hosts without having the charge-compensation problem, the growth of LaF_3 was perfected. A similar function is served by use of the oxides as hosts, Y_2O_3, Gd_2O_3, Er_2O_3, and La_2O_3. These materials have very high melting points and good mechanical characteristics.

Some of the most useful laser hosts are the synthetic garnets, $Y_3Al_5O_{12}$, $Y_3Gd_5O_{12}$, and $Gd_3Ga_5O_{12}$ [16], [17]. These hosts provide a substitutional site both for rare earth and transition metal impurities and also have high thermal conductivities.

Many factors contribute to the linewidths in solids and a theoretical understanding of all the linewidth broadening mechanisms is still incomplete. The simplest mechanism is inhomogeneous broadening, due to random variation in the static crystalline Stark field at the position of the impurity ion. This is the mechanism that determines the width of the R lines in ruby below 78°K [18]. The temperature dependence of the R lines of Cr^{3+} in Al_2O_3 and MgO has been described in terms of Raman scattering of phonons via an intermediate electronic state [19]. In general, the width of levels 10–500 cm^{-1} above the ground state is determined at low temperatures by the spontaneous phonon emission rate [20]. These mechanisms are expected for sharp line transitions as, e.g., 4f-4f transition of a rare earth. The width of broadbands (as, e.g., the 4f-5d transitions) comes from overlapping transitions from the ground state to a number of vibrational levels of the excited electronic state.

The problem of nonradiative decay between electronic states that are separated by energies several times the Debye energy still has not been solved. Kiel [21] estimates the ratio of the probability of spontaneous phonon emission of $(n+1)/n$ phonons to be ~ 0.1. But unquestionably the nonradiative relaxation rate between levels separated by $\Delta E \gtrsim 2000$ cm^{-1} is increased by the presence of other impurities [22]. Qualitatively, for a given electronic energy separation a host with low Debye energy will lead to slower nonradiative relaxation rates. This effect is observed in hosts like $LaCl_3$, where the rare earths have many more emitting levels than in hosts like CaF_2.

B. Trivalent Rare Earth Impurities

The greatest number of crystalline lasers use trivalent rare earth as the active ions. All the laser transitions are summarized in Fig. 2, adapted from an energy level dia-

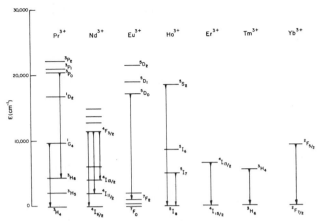

Fig. 2.   Laser transitions of trivalent rare earth.

gram by Dieke [23]. All existing trivalent rare earth lasers are four level lasers. The terminal state is either a crystal field split component of the ground level (Yb^{3+}, Tm^{3+}, Er^{3+}, Ho^{3+}, Pr^{3+}); or is a spin-orbit component above the ground level (Ho^{3+}, Eu^{3+}, Nd^{3+} and Pr^{3+}). Table II gives a summary of these laser systems.

Rare earths have a partially filled 4f shell [24], and the sharp lines observed correspond to the parity forbidden 4f-4f transitions. The oscillator strength of these transitions is $f \sim 10^{-7}$–10^{-5}, corresponding to a radiative lifetime of $\tau \approx 10^{-3}$–10^{-5} seconds in the visible region of the spectrum. The lowest energy allowed 4f-5d electric dipole transitions occur in the UV [25] and are not suitable for pumping. The intensities of the various 4f-4f transitions are determined by crystal symmetry [26], where the odd terms in the crystal potential admix other configurations (e.g., 5d) into the 4f states. These induced electric dipole transitions are at least an order of magnitude more intense than the allowed magnetic dipole transitions ($f_{MD} \sim 10^{-8}$). In all the existing trivalent rare earth systems the relatively weak 4f-4f transitions are used for pumping, and hence one has to go to high concentrations of impurities to obtain efficient coupling to the lamps. The concentration cannot be increased indefinitely because quenching of the fluorescence sets in. The inefficient pumping also limits severely the possibility of obtaining laser action at high frequencies in these systems.

Nd^{3+} is the impurity which is easiest to operate as a laser and is used in most crystal hosts. The transition is usually from the lowest crystal field level of the $^4F_{3/2}$ state to various crystal field components of the $^4I_{11/2}$ state, though transitions to the $^4I_{9/2}$ and $^4I_{13/2}$ states have also been operated as lasers. The 1.06 μ transition originates from a four level laser system with the terminal $^4I_{11/2}$ state ~ 2000 cm^{-1} above ground state. The visible trivalent rare earth lasers are CaF_2:Ho^{3+} [7] at 5512 Å, Y_2O_3:Eu^{3+} [27] at 6613 Å and LaF_3:Pr^{3+} [28] at 5985 Å. However, all these systems are marginal even in pulsed operation.

TABLE II
CRYSTALLINE LASER SYSTEM

Host	Dopant	Frequency of Laser Å	Transition	Energy of Terminal State (cm^{-1})	Mode and Highest Temperature of Operation °K	Reference
Al_2O_3	0.05% Cr^{3+}	6934	$^2E(\bar{E}) \rightarrow {}^4A_2$	0	CW, p. 350	[2], [41]
		6929	$^2E(\bar{A}) \rightarrow {}^4A_2$	0	p. 300	[42]
Al_2O_3	0.5% Cr^{3+}	7009			p. 77	[43]
		7041	pair lines	100	p. 77	[44]
		7670			p. 300	[45]
MgF_2	1% Ni^{2+}	16220	$^3T_2 \rightarrow$	340	p. 77	[35]
MgF_2	1% Co^{2+}	17500	$^4T_2$-$^4T_1$	1087	p. 77	[36]
		18030		1256	p. 77	
ZnF_2	1% Co^{2+}	26113			p. 77	[36]
$CaWO_4$	1% Nd^{3+}	10580	$^4F_{3/2} \rightarrow {}^4I_{11/2}$	2000	CW 300	[4]
		9145	$^4F_{3/2} \rightarrow {}^4I_{9/2}$	471	p. 77	[46]
		13392	$^4F_{3/2} \rightarrow {}^4I_{13/2}$	4004	p. 300	[46]
CaF_2	1% Nd^{3+}	10460	$^4F_{3/2} \rightarrow {}^4I_{11/2}$	~2000	p. 77	[47]
$CaMoO_4$	1.8% Nd^{3+}	10610	$^4F_{3/2} \rightarrow {}^4I_{11/2}$	~2000	CW 300	[48]
$Y_3Al_5O_{12}$	Nd^{3+}	10648	$^4F_{3/2} \rightarrow {}^4I_{11/2}$	2111	CW 360 p. 440	[49], [62]
LaF_3	1% Nd^{3+}	10633	$^4F_{3/2} \rightarrow {}^4I_{11/2}$	2187	p. 300	[50]
LaF_3	1% Pr^{3+}	5985	$^3P_0 \rightarrow {}^3H_6$	~4200	p. 77	[28]
$CaWO_4$	0.5% Pr^{3+}	10468	$^1G_4 \rightarrow {}^3H_4$	377	p. 77	[51]
Y_2O_3	5% Eu^{3+}	6113	$^5D_0 \rightarrow {}^7F_2$	859	p. 220	[27]
CaF_2	Ho^{3+}	5512	$^5S_2 \rightarrow {}^5I_8$	~370	p. 77	[7]
$CaWO_4$	0.5% Ho^{3+}	20460	$^5I_7 \rightarrow {}^5I_8$	250	p. 77	[50]
$Y_3Al_5O_{12}$	Ho^{3+}	20975	$^5I_7 \rightarrow {}^5I_8$	518	CW 77 p. 300	[52a, b]
$CaWO_4$	1% Er^{3+}	16120	$^4I_{13/2} \rightarrow {}^4I_{15/2}$	375	p. 77	[50]
$Ca(NbO_3)_2$	Er^{3+}	16100	$^4I_{13/2} \rightarrow {}^4I_{15/2}$		p. 77	[54]
$Y_3Al_5O_{12}$	Er^{3+}	16602	$^4I_{13/2} \rightarrow {}^4I_{15/2}$	525	p. 77	[52a]
$CaWO_4$	Tm^{3+}	19110	$^3H_4 \rightarrow {}^3H_6$	325	p. 77	[50]
$Y_3Al_5O_{12}$	Tm^{3+}	20132	$^3H_4 \rightarrow {}^3H_6$	582	CW 77 p. 300	[52a]
Er_2O_3	Tm^{3+}	19340	$^3H_4 \rightarrow {}^3H_6$		CW 77	[40]
$Y_3Al_5O_{12}$	Yb^{3+}	10296	$^2F_{5/2} \rightarrow {}^2F_{7/2}$	623	p. 77	[52a]
CaF_2	0.05% U^{3+}	26130	$^4I_{11/2} \rightarrow {}^4I_{9/2}$	609	p. 300 CW 77	[55]
SrF_2	U^{3+}	24070	$^4I_{11/2} \rightarrow {}^4I_{9/2}$	334	p. 90	[56]
CaF_2	.01% Sm^{2+}	7083	$^5D_0 \rightarrow {}^7F_1$	263	p. 20	[57]
SrF_2	.01% Sm^{2+}	6969	$^5D_0 \rightarrow {}^7F_1$	270	p. 4.2	[58]
CaF_2	.01% Dy^{2+}	23588	$^5I_7 \rightarrow {}^5I_8$	30	CW 77 p. 145	[56], [63] [22]
CaF_2	0.01% Tm^{2+}	11160	$^2F_{5/2} \rightarrow {}^2F_{7/2}$	0	p. 27 CW 4.2	[60] [61]

C. Divalent Rare Earth Impurities

Divalent rare earths have an additional electron on the 4f shell, which lowers the energy of the 5d configuration. Consequently, the allowed 4f-5d absorption bands fall in the visible region of the spectrum [29], [30]. These bands are particularly suitable for pumping the laser systems. However, for most of the rare earth ions the bands begin close to 1 μ, and only the 4f levels lying below them in the infrared can be utilized for lasers. Sm^{2+} and Eu^{2+} are the visible emitting divalent rare earths. The divalent rare earth laser systems are also shown in Table II. With the exception of the $CaF_2:Sm^{2+}$ system, all the other divalent rare earth laser systems utilize 4f-4f magnetic dipole transitions.

Eu^{2+} and Yb^{2+} have been incorporated into crystal hosts which contain oxygen (e.g., Eu^{2+} in YAG or EuO), but the other rare earth ions have only been reduced to the divalent state in crystalline hosts in which oxygen is not present. In CaF_2, trivalent rare earths can be converted to the divalent state either by high energy irradiation (X-ray, gamma-ray, or electron bombardment), or by techniques that remove interstitial fluorine charge-compensators either by electrolysis or baking the crystals in Ca vapor [32]. This latter reduction technique is preferable, since the charge-compensator of the trivalent rare earth is removed from the lattice and the divalent rare earth is stable. There is evidence [29], [33] that four divalent rare earths: La, Ce, Gd, and Tb have a 5d ground state in CaF_2. Divalent rare earths in a non-cubic site would be particularly interesting, since in addition to having convenient 4f-5d pump-bands, the 4f-4f transitions would be electric dipole ones. A class of such hosts are the mixed halofluorides, e.g., $BaClF:Sm^{2+}$ [34].

Tm^{2+}, Dy^{2+}, and Sm^{2+} have been operated as lasers, all in the CaF_2 host. In addition, Sm^{2+} was also operated as a laser in SrF_2. $CaF_2:Tm^{2+}$ is a three level laser (Fig. 1) which was also operated continuously. Sm^{2+}, even though it is a four level system, could only be operated in pulsed fashion, since there is a bottleneck in the relaxation of the F_2 terminal state to the F_0 ground state. This bottleneck would be eliminated for Sm^{2+} in a noncubic host. The $CaF_2:Dy^{2+}$ laser system is the lowest threshold CW laser at low temperature. The terminal state is ~30 cm^{-1} above ground state [22] and it becomes a four level laser only at liquid helium temperature. At room temperature this laser does not operate, since the nonradiative decay rate of $^5I_7 \rightarrow {}^5I_8$ becomes too fast compared to the radiative rate, and the fluorescence is quenched. Both the $CaF_2:Tm^{2+}$ and $CaF_2:Dy^{2+}$ laser transitions at 4.2°K have fluorescent linewidths (of the order of 0.01 cm^{-1}) smaller than the longitudinal cavity mode spacing of a 2 cm cavity. These lasers operate in a single longitudinal mode. These systems appear particularly suitable for the utilization of solid lasers as frequency standards.

D. Transition Metals

When trivalent chromium is incorporated into a octahedral site, one obtains the familiar energy-level diagram of ruby (Fig. 1). There are a number of hosts where this occurs, e.g., $Y_3Al_5O_{12}$, $LaAlO_3$, etc., but so far only $Al_2O_3:Cr^{3+}$ has been utilized as a laser. Various isotopes of Cr^{3+} in this host were also operated as lasers, as were both of the R lines and some of the lines of Cr^{3+} pairs [11]. In the other hosts mentioned, bulk losses due to excited state re-absorption and poor crystal quality so far have prevented laser action. The $LaAlO_3$ host is of particular interest for Q switching, since in the cubic site the Cr^{3+} has particularly long lifetime ($\sim10^{-2}$–10^{-1} seconds).

The isoelectronic systems of Cr^{3+}, namely V^{2+} and Mn^{4+}, have both been incorporated into a number of hosts, in particular into Al_2O_3 and MgO, but could not be made to oscillate. The other transition metal ion which emits in the visible is Mn^{2+}, often used in phosphors, but not utilized yet as a laser.

Lasers utilizing sharp vibronic transitions have been operated in the infrared. These are $MgF_2:Ni^{2+}$ [35], $MgF_2:Co^{2+}$ [36] and $ZnF_2:Co^{2+}$ [36]. These systems operate at 77°K or lower temperatures, where the vibronic lines are sufficiently narrow.

E. Sensitized Laser Systems

To improve the overall efficiency of laser systems, a second impurity can be incorporated whose absorption will match the output of optical pump sources better than the active impurity alone. Cr^{3+}–Nd^{3+} doped $Y_3Al_5O_{12}$ is such a laser material [37] and their energy level diagrams are shown on Fig. 1. Cr^{3+} absorbs the energy in the broad $^4A \rightarrow {}^4T$ bands, the Cr^{3+} atoms relax to the metastable ^2E levels, and then the excitation is transferred to some excited levels of Nd^{3+}. Finally the Nd^{3+} atom returns to its ground state via the usual $^4F_{3/2} \rightarrow {}^4I$ Nd^{3+} emission.

In this particular case, dipole-dipole interaction is responsible for the sensitization [38] and the cross transfer rates range from 10^{-5} to 10^{-3} seconds, depending on the density and separation of the Cr^{3+}–Nd^{3+} ions. Similar sensitization among rare earths is utilized in the $Y_3Al_5O_{12}$ "alphabet" [39] system, where Ho^{3+} emission at 2.1 μ is sensitized by a number of rare earths including Er^{3+} and Tm^{3+}. Sensitization of transitions in the IR is easier, since the competing spontaneous emission rates within the sensitizers decrease as $1/\lambda^3$. Sensitization of visible transitions of trivalent rare earth ions such as Sm^{3+}, Eu^{3+}, etc., still remains an important problem.

Other techniques of sensitization include host-sensitization, where pump energy absorbed by the host is transferred to the action ion, as in the case of $Er_2O_3:Tm^{3+}$ [40]. Sensitization by radiative energy transfer is a further possibility, where the emission of the sensitizer is re-absorbed by the active ion and is re-emitted in the laser transition.

III. LASER SYSTEMS

A. Experimental Laser Thresholds

The technique of optically pumping lasers places certain limitations on their short wavelength operation. These arise from the wavelength dependence of the radiative lifetime for a given dipole moment matrix element combined with the finite intensity available from present optical pump surces. Rearranging (1) we obtain the following functional form for the necessary inversion density for continuous laser threshold [1]

$$\frac{\Delta N}{V} = S\alpha\pi^2 c \frac{1}{\lambda^3} \frac{\Delta\nu}{\nu} \frac{\tau_{rad}}{\tau} \qquad (3)$$

where S is a constant of the order of one, depending upon the line shape; α is the gain necessary to overcome losses, both internal and due to the reflectors; $\Delta\nu/\nu$ is the fractional linewidth of the laser metastable level; τ_{rad}/τ is the ratio of the radiative lifetime to the observed lifetime.

From Table II it can be seen that many more materials have been operated pulsed than continuously. This follows from one or more of the following reasons. Available pulsed optical sources can provide higher pump intensities for times comparable to the typical 1 ms fluorescent lifetime over which these lasers integrate population in their upper state. The laser material may not be able to dissipate the heat caused by its internal losses under continuous illumination, or it may be impossible to transfer this heat to a coolant medium without losing thermal contact. If the system under consideration is four-level, the terminal state may not be able to depopulate at a sufficiently fast rate. In this section, the emphasis will be on numerical analysis of continuous operation in the best existing systems, with mention of pulse operation only where it has experimentally resulted in significantly higher efficiency, power, or shorter wavelength.

An evaluation of Nd^{3+}:YAG, the best room temperature continuous system, gives an indication of the present state-of-the-art. Nd^{3+}:YAG is a 1.06 μ four-level laser [49] having about a 10 Å linewidth, and an observed lifetime of 200 μs which is assumed here to be radiative. Assume also that all ions relax through the metastable level. A typical laser rod of high quality host material of 2 mm diameter by 5 cm long might require an α of 0.01/cm to overcome internal losses and lack of perfect reflectivity on the mirrors. From these values, a threshold of 0.1 watt absorbed is obtained if all of the absorbed photons are of 1.3 eV energy, i.e., are absorbed in the lowest energy pump levels.

A typical optical pump source, as mentioned, does not emit in a line but rather is a broadband emitter. A tungsten lamp, for instance, emits only 30 percent of its energy at wavelengths shorter than 1 μ. A typical Nd: YAG rod can only absorb 1/10 of this due to its narrow 4f-4f absorption bands. The average energy of the absorbed photons is also greater than 1.3 eV and the optical coupling between the lamp and crystal cannot be perfect, so another loss of a factor of two is typical. These raise the practical threshold for room temperature ND^{3+}:YAG to a value of an electrical input of 6 watts to a 3400°K tungsten optical pump.

Experimentally, thresholds close to this have been achieved [64] for Nd^{3+}:YAG. Similar evaluations have been made for other lasers and optical pumps although

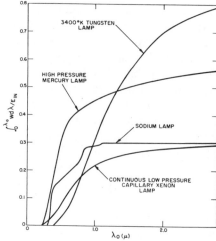

Fig. 3. The fractional pump lamp output above a given wavelength for four characteristic optical sources.

few systems have achieved the experimental state of Nd^{3+}:YAG.

Assuming that this requirement of 0.1 watt absorbed for laser action in Nd^{3+}:YAG is a realistic evaluation of the crystalline quality, fluorescent efficiency, fractional linewidths, and the dielectric reflector quality, etc., that might be obtained in a new laser system, a threshold can easily be extrapolated for an optical maser at half the wavelength, i.e., in the center of the visible at 5300 Å.

The $1/\lambda^3$ dependence of necessary inversion population gives a $1/\lambda^4$ dependence upon power due to the increase in the average energy per pump photon. This gives an immediate increase in the absorbed power to 1.6 watts. This requirement is not prohibitive, but the fractional pump lamp output above this wavelength for existing sources is very low, as shown in Fig. 3. For example, only 2.5 percent of the input to a tungsten lamp is emitted shorter than 5300 Å. If this hypothetical laser crystal has the same fractional absorption efficiency in its pump region as Nd^{3+}:YAG has (0.1), the electrical threshold is 1280 watts, or twice the power that can be dissipated in a tungsten filament the size of the laser rod.

A considerable improvement can be gained by using a Hg lamp pump, both in fractional and total pump output in the region shorter than 5300 Å, but whatever parameters one chooses, the combination of the theoretical threshold power absorbed increasing as $1/\lambda^4$ and available pump power decreasing rapidly in the UV presents serious limits on the upper frequency of any continuous solid-state laser. Unless new pump sources in the UV become available, at some wavelength the only available continuous coherent solid-state source will be the result of frequency multiplying some longer wavelength laser source. Frequency doubling may even be the lowest energy method of obtaining coherent output in regions where direct laser operation is possible. New nonlinear materials such as lithium meta-niobate can convert up to 10 percent of a continuous laser output at 1.06 μ to twice the frequency [65].

As the laser output wavelength moves into the infrared, this $1/\lambda^4$ reduction in threshold, combined with the fact that pump sources have the majority of their spectral output at energies higher than the laser, makes low level operation possible. Laser thresholds as low as a few watts [4] into a tungsten lamp or milliwatts into light emitting diodes [66] have been observed at an output wavelength of $2.36\,\mu$ in $Dy^{2+}:CaF_2$ at cryogenic temperatures.

Further into the infrared a lamp problem appears again, but in a different manner, with the overall efficiency dropping off because the lamp has too much high energy output, and the crystal is heated by the internal nonradiative decay to the metastable level. This decay may also involve other metastable levels, as in $Dy^{2+}:CaF_2$ where these contribute to excited state absorption and limit laser performance [63].

B. Efficiency of Operation Above Threshold

Optical pumped laser systems have an inherent loss due to the necessity that the pump photons be of higher energy than the emitted radiation. This loss could in principle be minimized by using a monochromatic optical pump at an energy only marginally greater than the laser. Such pumps are at present relatively inefficient, of limited intensity, and available at only a few wavelengths The usual situation is a broadband emitter and a laser material having an absorption spectrum that can utilize only a fraction of the energy emitted by the pump.

An estimate of the overall efficiency limit set by using such a conventional broadband pump is shown in Fig. 2. This plots the fraction of the electric energy input that is emitted shorter than any given wavelength. For long wavelength laser systems, a tungsten lamp is the most efficient pump as its integrated optical output is typically 90 percent of electrical input. This is considerably higher than the 60 percent obtained with Hg lamps, or the 30 percent available with continuous xenon and sodium lamps. The Hg and Na sources, however, have considerably greater output in the shorter wavelength region and are correspondingly better to pump systems with outputs in the near infrared and visible. The high pressure mercury lamp is presently the most efficient pump lamp for absorption over the entire visible region.

In principle, since the crystal and optical pump are coupled in a cavity, all of the lamp output that is not absorbed in the crystal would be reflected back into the lamp and not show up as a loss in efficiency. Practically, little gain has been achieved yet by these techniques, but there is no reason why they will not considerably improve the efficiency, although they cannot increase the power.

It can be seen from this that the maximum efficiency of optically pumped lasers drops rapidly from the optical pump lamp consideration alone as the emission wavelength moves towards the blue. Again, at some point it becomes more efficient to generate coherent energy at twice the desired wavelength, and then frequency double in a nonlinear element.

There are of course other possible losses in the laser operation. In order to achieve the limiting efficiencies in Fig. 2, it is necessary to assume that 1) all of the pump energy is imaged on the crystal; 2) all of this energy is absorbed and passes through the upper metastable laser level; 3) operation is sufficiently above threshold that stimulated emission dominates spontaneous (again much harder at shorter wavelengths as spontaneous emission is proportional to $1/\lambda^3$); and 4) that the output coupling is large enough so that internal scattering and absorption losses are negligible.

1), 3), and 4) can be approached since linear ellipses or other optical systems for coupling the lamp and crystal image more than 50 percent of the lamp output on the crystal, depending on the relative dimensions of the lamp and crystal and the complexity and size [67] that one is willing to design into them. Internal efficiencies approaching 100 percent have been measured, and thresholds in the infrared are low enough so that operation at many times threshold is attainable. Gains much higher than internal losses are also attainable, allowing high output coupling.

In order to increase 2) and approach these theoretical efficiencies, the absorption bands of the laser must overlap the entire spectral output of the lamp from the emission line to the ultaviolet lamp cutoff. As discussed in Section II, this is approached in several different ways in real crystals.

First, there are active ions with narrow emission lines and broad absorption bands such as the transition metals (Section II-D) or the divalent rare earths (Section II-C). These all have limitations in one respect or another. The efficient divalent rear earths operate in the infrared and also require cooling [29], [30]. Alone among the transition metal ions, Cr^{3+} is a three level system which has operated continuously [41], [68], [69], but it has a high threshold and its internal efficiency falls rapidly at elevated temperature, limiting high level continuous operation. Representative 77°K results using $Dy^{2+}:CaF_2$ are given in Table III, as well as results for room temperature ruby operation.

The system requirements are considerably different for three and four level systems, since in four level operation, the absorption from the pump lamp can be increased by increasing N, the total active ion concentrations, without increasing the requirement on ΔN, the necessary excited state population, at least up to concentrations where pair effects enter. In three level systems the necessary number of excited atoms varies essentially as $(N/2)+\Delta N$. Since the pump band absorption also increases linearly in N, the threshold for three level operation is almost independent of concentration, with the main concentration dependence due to the spatial distribution of excited atoms changing with concentration [70]. In efficiency of operation well above threshold, there is no real advantage to a three or four level system with the total efficiency being set mainly by the efficiency of absorption of the pump light.

TABLE III

CONTINUOUS SOLID STATE LASER POWERS AND EFFICIENCIES

Material Active System	Sensitizer	Optical Pump	λ (μ)	Eff. (%)	Power (Watts)	Operating Temp. °K	Reference
$Dy^{2+}CaF_2$	—	W	2.36	0.06	1.2	77	[63]
$Cr^{3+}Al_2O_3$	—	Hg	0.69	0.1	1.0	300	[68], [69]
$Nd^{3+}Y_3Al_5O_{12}$	—	W	1.06	0.2	2	300	[65]
			1.06	0.6	15	300	[94]
$Nd^{3+}Y_3Al_5O_{12}$	—	Plasma Arc	1.06	0.2	200	300	[73]
$Nd^{3+}Y_3Al_5O_{12}$	—	Na Doped Hg	1.06	0.2	0.5	300	[72]
$Nd^{3+}Y_3Al_5O_{12}$	Cr^{3+}	Hg	1.06	0.4	10	300	[71]
$Ho^{3+}Y_3Al_5O_{12}$	[Er^{3+}, Yb^{3+}, Tm^{3+}]	W	2.12	5.0	15	77	[52]

Secondly, as discussed in Section II-E, there are multiply doped systems where the active ion has the desired narrow levels and the activation (sensitization, or energy transfer) ion has broadbands tailored to match the lamp emission and from which the energy can transfer to the active ion. Examples of these are Nd^{3+}:Cr^{3+}:YAG with Nd^{3+} the active ion and Cr^{3+} the sensitizer [37], [71], and Ho^{3+} [ ] YAG [39] where [ ] represents a combination of Tm^{3+}, Yb^{3+}, and Er^{3+} that absorbs up to 80 percent the lamp emission above the 2.1 μ laser transition of the Ho^{3+} and transfers it reasonably efficiently to the Ho^{3+}. These techniques have resulted in the most efficient solid state lasers yet operated as shown in Table III.

A third approach uses the fortuitous coincidence of several Na vapor emission lines to Nd:YAG absorption lines to optimize the overall system [72]. Present efficiencies are, however, less than the best results with W pumping, being about 0.2 percent overall.

These multidopant approaches are excellent in theory, but must in practice be evaluated against the simpler single ion system that may well be easier to grow with extremely good quality such that its overall performance is quite close to its theoretical limit, while the theoretically more efficient multiply doped system may not as closely approach its potential. An example of this is the singly doped Nd^{3+}:YAG where the experimental efficiency pumped with a W source is comparable with [65], [94], that obtained from doubly doped material which should have at least five times higher overall efficiency.

C. Continuous Laser Output Power

The maximum continuous power output is directly related to the efficiency as well as the maximum energy densities obtainable with various pump sources. The highest power output thus far from an insulating solid laser is obtained using a vortex stabilized arc as an optical pump. This can produce the highest known continuous intensity, being able to dissipate an electrical input of 100 kw/cm². It can radiate as much as 30 percent of this in the wavelength range shorter than 1 μ. This pump and Nd^{3+}:YAG have given an output of 200 watts [73], despite the low absorption efficiency of the singly doped YAG. The necessary electrode configuration and gas containment tubes present some particular problems in

transferring this pump energy to the laser crystal efficiently but considerable improvement should still be possible with this system.

Among conventional pump sources in the few kilowatt range, the high pressure mercury lamp has not only the most short wavelength output, but also has the advantage that it has the highest surface power density. Sodium lamps can only radiate an electrical input of equivalent to 13 watts/cm². Tungsten and xenon operate as high as 350 watts/cm². A type A Hg lamp can however operate at 2000 watts/cm². This gives the potential of obtaining both high power and high efficiency in a small package. A combination of the Nd:Cr:YAG system pumped with a 2 kW Hg lamp should achieve 4 percent overall efficiency resulting in an 80 watt output. Experimentally 10 watts have been achieved [62], [71]. Another cross pumped system is the so-called alphabet YAG emitting at 2.1 μ using tungsten pumping. A theoretical efficiency of 30 percent is predicted for it and an experimental efficiency of 5 percent has been realized, giving 15 watts output [52b] with the crystal cooled to 77°K. The power limitations on this system will probably be set by cooling rather than by available pump power as nitrogen is not nearly as good a heat transfer liquid as water. Other cryogenic lasers (Dy^{2+}:CaF_2 and U^{3+}:CaF_2) that have produced continuous powers in the watt range have had these same heat transfer limitations.

D. Time Dependence of Laser Operation

A further characteristic of insulating solid lasers is the long lifetime of the upper laser level. Typical values are 40 milliseconds for Dy^{2+}:CaF_2 and almost 0.1 second for Cr^{3+} in $LaAlO_3$. This energy storage time allows considerable flexibility in the mode of laser operation, both in the optical pump necessary to populate this state and in the stimulated emission that depopulates it. This lifetime is advantageous for continuous low level operation, where populations settle down to equilibrium, or at least to a repetitive oscillatory nature, but advantage can be taken of it in other ways.

For investigating new materials and for operation at very high levels, pulsed laser operation has the primary advantage of enabling optical pump levels of higher intensity than are available continuously. A typical xenon flash lamp can sustain over 10^5 watts/cm² input for pulses of a few milliseconds duration. This is of the same order of intensity as the more complicated vortex arc. The spectral output of the pulsed xenon source is also more useful than that of the continuous xenon having a color temperature near 9000°K. The pulsed xenon lamp also has a measured efficiency of converting over 45 percent of its electrical input to spectral output of shorter than 7000 Å.

Pulsed operation as mentioned before allows use of crystals that cannot be run continuously due to any one of a variety of reasons, such as low thermal conductivity, pileup in the lower level, or strong interactions between the host and active ion broadening the levels, and raising the threshold. The latter may in fact be an advantage since above threshold the similarly broadened absorption bands should absorb a greater fraction of the pump light and lead to higher efficiencies. This is in fact true for Nd^{3+} glass where the broad absorption bands have helped to achieve an efficiency of 5 percent in pulsed operation. Ruby also fits in the category of a material that has a high continuous threshold, but is excellent in pulsed operation. Its efficiency is somewhat less than Nd^{3+} glass due in part to its emission being of shorter wavelength which allows less of the lamp spectrum to be useful, and in small part to its being limited in size as compared to the Nd^{3+} glasses.

Relatively few pulse systems have been optimized as to efficiency and power. The $Dy^{2+}:CaF_2$ system [63] is an example of a material with very low threshold, narrow emission lines, and broad absorption bands that was, however, limited in high power operation. It was found to have an internal bottleneck that delayed the energy getting from the absorption bands to the metastable laser level. This both limited the inversion obtainable and provided an excited state absorption that limited any high level operation. One requirement for efficient pulse operation is a fast decay to the metastable level.

Ruby has such a fast decay to the ^2E levels and in pulse operation has demonstrated efficiencies over 2.0 percent. The upper limit on total pulsed power output appears to be set by the quality of the host material since operation above some level causes irreversible damage to the lattice. Large rods have given hundreds of joules output, but the life of these units is limited and unpredictable. The majority of very high power work is done with Nd^{3+} glass due to its lack of size limitations.

A most important variation of pulsed laser operation is obtained by Q switching [6]. Here energy is stored by populating the upper state to as high a level as can be obtained. The limit in practice is set by superradiant fluorescent losses or pre-laser action in some undesired mode [74], but in principle most of the active ions could be pumped into the excited state and laser action prevented by complete absence of feedback. If the feedback is suddenly restored the system can emit all of this stored energy in one pulse, the time duration of which is set by the degree of inversion and the optical cavity parameters and is typically 10^{-8} to 10^{-7} seconds. The limit on the single

pulse peak power presently obtained is set by material quality, with any imperfection or inhomogeneity in the material causing enough localized loss to damage the laser rod itself. The majority of the high peak power work has been done with ruby, as it can be made in large sizes, of extremely good quality, and is a stable, strong material. The highest value claimed for commercial units that will survive many pulses is 10^9 watts in 12 ns. This system consists of an oscillator rod delivering 0.1 to 0.2 gigawatt followed by an amplifier giving a gain of 5 to 10 [75]. Higher output values can be obtained, but the present quality of ruby is such that the rods show irreversible damage after each pulse.

This high peak power also places stringent requirements on the reflectors used. Typical multiple dielectric reflectors will not survive, forcing recourse to multiple sapphire plates in a Fabry-Perot type structure [76] or to techniques utilizing total internal reflection [77]. All elements inserted in the cavity to produce the necessary change in gain are also subjected to the same fields and must have sufficiently low loss during the laser pulse to survive. Typical Q-switch techniques have consisted of mounting one reflector on a rotating shaft [6], inserting an electrooptic element in the cavity [78], or inserting a saturable absorber in the cavity [79], [80]. The highest powers and shortest pulses obtained experimentally have utilized an electrooptic element inside the cavity.

The requirement for a high repetition rate pulsed or Q switched laser are much the same as for continuous action and in fact repetitive Q switching has been demonstrated with a rotating reflector for $Dy^{2+}:CaF_2$ [63], $Nd^{3+}:YAG$ [64], and with a saturable absorber for $Cr^{3+}:Al_2O_3$ [81]. $Dy^{2+}:CaF_2$ has also been repetitively Q switched utilizing the large [63] Zeeman splitting of the laser levels in the proper orientation. Magnetic fields of a few gauss give strong Q switching at low modulation frequencies, while at near megahertz rates only fractional amplitude modulation is obtained.

The temporal characteristics of laser output under constant Q conditions vary from a smooth continuous output for $Nd^{3+}:YAG$ to the repetitive limit cycle pulses observed in most $Cr^{3+}:Al_2O_3$ lasers. The damped oscillatory behavior common to four level lasers can be quite well explained by a coupled rate equation approach, but the cause or causes of limit cycle behavior is still not understood completely, as one of two ruby lasers that are macroscopically identical will show limit cycle behavior, while the other may damp into continuous output. Experimentally the resonator configuration also contributes to the time dependence as crystals polished with flat ends tend toward limit cycle operation while spherical reflectors are most likely to give continuous output. This is particularly true if there is a bottleneck in the excitation process as in $Dy^{2+}:CaF_2$, or in the transfer from Cr^{3+} to Nd^{3+} in sensitized $Nd:Cr:YAG$.

The arc discharge lamps which give the greatest fractional emission at the shorter wavelengths, are characteristically unstable in operation. They can be externally stabilized, but their residual fluctuations generate spiking behavior in a continuous laser that may take several spikes to damp out. This noise places limitations on using these systems for communications. The most stable optical pump is of course a tungsten lamp (or outside the atmosphere the sun).

While there have been many analyses of the time dependence of laser output, a recent paper by McCumber [83] using linearized rate equations gives an excellent review of previous work and provides a more generalized theory that agrees very well with the measurements on the Nd^{3+}:YAG [82] laser. This theory predicts a peak in the noise spectrum that is observed in the continuous Nd^{3+}:YAG. The strong spiking in the Nd^{3+}:Cr^{3+}:YAG and Dy^{2+}:CaF_2 also has a repetition rate at this same characteristic frequency. The behavior of laser systems that have no saturable loss mechanism in general tends toward a truly continuous output as the crystal quality improves and overall losses are reduced.

If the internal loss or optical length of a laser cavity is modulated at a frequency equal to an integral multiple of the axial mode separation, the phenomenon of phase locking occurs [84]–[87]. The axial modes are phase locked in frequency and interfere in time giving a series of output pulses with a repetition rate at the modulation frequency. The width of each pulse is the period of the modulation signal divided by the number of modes that are locked. The peak power is directly proportional to the number of modes that are locked together. This phenomenon was first observed in gas lasers, but the broader fluorescent lines (more axial modes) available in solids have allowed the generation of a continuous train of pulses with duration as short as 8×10^{-11} seconds using Nd^{3+}:YAG operating continuously [88]. This is close to the theoretical lower limit on the length given by [88] the inverse bandwidth of the laser emission. To obtain shorter pulses would require mode locking over the greater operating frequency width of a laser with broader fluorescent transitions. The duration of these pulses is well below the present resolution of optical detectors. The most accurate method of measurement of the peak intensity (and resultant duration) of these pulses is to measure the average efficiency of a $LiNbO_3$ crystal in frequency doubling the output, and compare it with the same crystal's efficiency at the identical continuous laser power [88]. For a given laser emission linewidth, the theoretical pulse duration is independent of the length of the resonant cavity, which only determines the necessary modulating frequency and resultant pulse repetition rate.

E. Spectral Characteristics of Laser Output

The same broad fluorescent lines that enable phase locking to obtain the short pulses give rise to multimoding in a typical Fabry-Perot laser structure. The many axial modes that fit within this fluorescent linewidth of the crystal will emit simultaneously during operation. There are a few systems whose fluorescent linewidth at low temperature is less than the spacing between typical axial modes such as Dy^{2+}:CaF_2 [63] and Tm^{2+}:CaF_2 [8]. In these, normal operation is in a single axial mode,

with the corollary result that the coincidence or lack of coincidence of a resonant mode on the fluorescent peak determine the laser threshold. Using materials with broader emission lines, single mode operation has been attained by multiple interferometer [89] techniques or operating very near threshold. Single mode operation at higher level using crystals with a broad line can be achieved if a running wave cavity is used rather than a standing wave one. This has been demonstrated both by using a ring configuration of mirrors [90], [92] and laser rods and by using $\lambda/4$ plates internal to a standard Fabry-Perot cavity [91] so that there is a 90° phase difference between the oppositely running waves in the laser crystal. In a homogeneously broadened line this allows one mode to dominate and prevent oscillations of other axial modes which have slightly less gain.

The frequency or frequencies of the output in any of these configurations are then set by the mechanical dimensions of the resonator and as such will vary both with temperature and vibration [92]. The absolute stability will be quite poor (1×10^{-6}/C° for a typical ruby laser) [92], unless one of the sharp line systems is used and the emission locked to the laser line, in which case a truly atomic standard is possible. The short term stability of a multimode output can be observed by beating the longitudinal modes in a nonlinear detector. This has been shown to be of the order of a few kilocycles or less [63]. The true experimental stability of a narrow line solid-state laser at low temperatures has not yet been investigated.

IV. Applications and Research Directions [93]

The first application of lasers as a scientific tool was utilized in the study of nonlinear effects in the interaction of light with matter. For most of the nonlinear studies crystalline solid lasers are used as the source. Studies of Raman, Rayleigh, and Brillouin scattering are being carried out on a number of systems. Q switched laser pulses can be used as an intense short pumping source to study fluorescent lifetimes. Spectroscopy using a tunable laser source can improve the resolution of present optical techniques. Experiments using lasers are already in progress to improve the measurement of the two fundamental dimensions of length and time.

The most often utilized practical application of solid lasers has been for ranging and radar. Using Q switched pulses with typical pulse-width of 10^{-8} seconds, range determination to a few centimeters can be obtained. Pulse-Doppler radar using lasers can accurately determine velocities. An example where this accuracy is needed is the initial tracking of missiles at take off. The high carrier frequency of lasers in principle opens up enormous bandwidth for communications. While the atmospheric attenuation of the laser beam is a very serious difficulty for terrestrial communications, its use in space or in light pipes in the ground for special applications is most likely. The use of lasers in information handling, data storage, and display systems will be forthcoming if suitable storage medium (e.g. photochromic materials) can be developed to complement the laser.

There is considerable interest in the application of high power CW lasers for micromachining, cutting, and welding. Furthermore, these operations can be performed in vacuum behind a transparent window. In many of these applications one is interested in a spot size of the order of 1μ, and because of diffraction limitation visible lasers are preferable.

Some of the above applications indicate the direction of the search for new crystalline laser materials. A four-level crystalline CW laser operating at room temperature in the visible region of the spectrum is still missing. To extend these lasers towards higher frequencies is a difficult problem, as outlined above, and efficient frequency multiplication using nonlinear optical elements may be the solution. Another pertinent direction of effort will be to increase the efficiency and power of presently existing lasers; by sensitization, and by more efficient matching of the optical pump sources to the lasers.

It might well be that the most fruitful advances in this field will come, as has happened so often in the past, from a totally new approach. Possibly one can combine the advantages of crystalline lasers with the efficiency of injection lasers in conducting materials doped with impurities like rare earths (e.g., rare earth doped II–VI compounds, or systems like CdF_2). More probably, the advances will be in areas now unsuspected and uninvestigated.

REFERENCES

[1] A. L. Schawlow and C. J. Townes, "Infrared and optical masers," *Phys. Rev.*, vol. 112, p. 1940, December 1958.

[2] T. H. Maiman, "Stimulated optical radiation in ruby masers," *Nature*, vol. 187, pp. 493–494, August 1960.

[3] P. P. Sorokin and M. J. Stevenson, "Stimulated infrared emission from trivalent uranium," *Phys. Rev. Lett.*, vol. 5, pp. 557–559, December 1960.

[4] L. F. Johnson, G. D. Boyd, K. Nassau, and R. R. Soden, "Continuous operation of the $CaWO_4$:Nd^{3+} optical maser," *Proc. IRE (Correspondence)*, vol. 50, p. 213, February 1962.

[5] A summary of the materials work up to 1963 can be found in *Proceedings of the Third International Congress of Quantum Electronics*, P. Grivet and N. Bloombergen, Eds. New York: Columbia University Press, 1964.

[6] F. J. McClung and R. W. Hellwarth, "Giant optical pulsations from ruby," *J. Appl. Phys.*, vol. 33, p. 828, 1962.

[7] U. K. Voronko, A. A. Kamynsky, V. V. Osiko, and A. N. Prokhorov, *Jour. of Exp. and Theor. Phys. USSR (Letters to the Editor)*, vol. 1, no. 1, p. 5, 1965.

[8] See e.g., Z. J. Kiss, "Zeeman tuning of the CaF_2:Tm^{2+} optical maser," *Appl. Phys. Lett.*, vol. 2, pp. 61–62, February 1963.

[9] A. Yariv, and J. P. Gordon, "The laser," *Proc. IEEE*, vol. 51, pp. 4–29, January 1963.

[10] See, e.g., G. Birnbaum, "Optical masers," in *Advances in Electronics and Electron Physics*, (supplement 2). New York: Academic, 1964.

[11] A. A. Kaminsky and B. B. Osiko, "Inorganic ionic crystalline laser materials," *Isv. Akad. Nauk USSR Inorganic Materials*, vol. 1, p. 2049, December 1965.

[12] R. C. Linares, "Properties and growth of flux ruby," *J. Phys. Chem. Solids*, vol. 26, p. 1817, 1965.

[13] A. E. Paladino, and R. D. Reiter, "Chochralski growth of sapphire," *J. Am. Ceram. Soc.*, vol. 47, p. 465, 1964.

[14] See, e.g., H. Guggenheim, "Growth of highly perfect fluoride single crystals for optical masers," *J. Appl. Phys.*, vol. 34, p. 2482, 1963.

[15] K. Napan and A. M. Broger, "Calcium tungstate: Czochralski growth, perfection and substitution," *J. Appl. Phys.*, vol. 33, p. 3064, 1962.

[16] See, e.g., L. G. Van Uitert, W. J. Grodkiewicz, and E. F. Dearborn, "Growth of large optical quality yttrium and rare-earth aluminum garnets," *J. Am. Ceram. Soc.*, vol. 48, p. 105, 1965.

[17] F. R. Charvat and R. M. Youmans, "Characteristics of melt pulled rare-earth garnets," *American Ceramic Soc. Bull.*, p. 409, April 1965.

[18] D. E. McCumber and M. D. Sturge, "Linewidth and temperature shift of the R lines in ruby," *J. Appl. Phys.*, vol. 34, p. 1682, June 1963.

[19] See, e.g., G. F. Imbush, W. M. Yeu, A. L. Schawlow, D. E. McCumber, and M. D. Sturge, "Temperature dependence of the width and position of the $^2E \to {}^4A_2$ fluorescence lines of Cr^{3+} and V^{2+} in MgO," *Phys. Rev.*, vol. 133, p. A 1029, February 1964.

[20] See, e.g., S. Yatsiv, "Spontaneous emission of phonons and spectral linewidths of some rare-earth ions in crystals," *Physica*, vol. 28, p. 521, 1962.

[21] A. Kiel, "Multi-phonon spontaneous emission in paramagnetic crystals," [5, p. 765].

[22] See, e.g., Z. J. Kiss, "Energy levels of Dy^{2+} in CaF_2, SrF_2 and BaF_2," *Phys. Rev.*, vol. 137, pp. A 1749–1760, March 1965.

[23] See, e.g., G. H. Dieke and B. Pandey, "Spectroscopy of trivalent rare-earths," in *Proc. of the Symposium on Optical Masers*. Brooklyn, N. Y.: Polytechnic Press, 1963, pp. 327–346.

[24] See, e.g., B. R. Judd, *Operator Techniques in Atomic Spectroscopy*. New York: McGraw-Hill, 1963.

[25] E. Loh, "Lowest 4f–5d transition of trivalent rare-earth ions in CaF_2," *Phys. Rev.*, to be published.

[26] B. R. Judd, "Optical absorption intensities of rare-earth ions," *Phys. Rev.*, vol. 127, pp. 750–761, August 1962.

[27] N. C. Change, "Fluorescence and stimulated emission from trivalent europium in yttrium oxide," *J. Appl. Phys.*, vol. 34, 3500, 1963.

[28] R. Solomon and L. Mueller, "Stimulated emission at 5983 Å from Pr^{+3} in LaF_3," *Appl. Phys. Lett.*, vol. 3, p. 135, 1963.

[29] D. S. McClure and Z. J. Kiss, "Survey of the spectra of the divalent rare-earth ions in cubic crystals," *J. Chem. Phys.*, vol. 39, p. 3251, 1963.

[30] See, e.g., P. P. Feofilov, "Monocristaux du type Fluorite Activés comme Mileux pour produire une Emission Stimulée," [5, p. 1079].

[31] See, e.g., P. P. Sorokin, "Transitions of Re^{+2} ions in alkaline earth halide lattices," [5, p. 985].

[32] Z. J. Kiss and P. N. Yocom, "Stable divalent rare-earth alkaline earth halide systems," *J. Chem. Phys.*, vol. 41, p. 1511, 1964.

[33] W. Hayes and J. W. Twidell, "Paramagnetic resonance of divalent lanthanum in irradiated CaF_2," *Proc. Phys. Soc.*, vol. 82, p. 330, 1963.

[34] See, e.g., Z. J. Kiss and H. A. Weakliem, "Stark effect of 4f state and linear crystal field in BaClF:Sm^{2+}," *Phys. Rev. Letts.*, vol. 15, p. 457, 1965.

[35] L. F. Johnson, R. E. Dietz, and H. S. Guggenheim, "Optical maser oscillation ion of Ni^{2+} in MgF_2 involving simultaneous emission of phonons," *Phys. Rev. Lett.*, vol. 11, p. 318, 1963.

[36] ——, "Spontaneous and stimulated emission from Co^{2+} ions in MgF_2 & ZnF_2," *Appl. Phys. Lett.*, vol. 5, p. 21, 1963.

[37] Z. J. Kiss and R. C. Duncan, Jr., "Cross-pumped Cr^{3+}:Nd^{3+}: YAG laser system," *Appl. Phys. Lett.*, vol. 5, p. 200, 1964.

[38] D. L. Dexter, "A theory of sensitized luminescence in solids," *J. Chem. Phys.*, vol. 21, p. 836, 1953.

[39] L. F. Johnson, J. E. Geusic, and L. G. Van Uitert, "Coherent oscillations from Tm^{3+}, Ho^{3+}, Yb^{3+}, and Er^{3+} ions in yttrium aluminum garnet," *Appl. Phys. Lett.*, vol. 7, pp. 127–129, September 1965.

[40] M. B. Soffer and R. H. Hoskins, "Energy transfer and CW laser action in Tm^{+3}:Er_2O_3," *Appl. Phys. Lett.*, vol. 6, p. 200, 1965.

[41] R. J. Collins, D. F. Nelson, A. L. Schawlow, W. Bond, C. G. B. Garrett, and W. Kaiser, "Coherence, narrowing, directionality, and relaxation oscillations in the light emission from ruby," *Phys. Rev. Lett.*, vol. 5, pp. 303–305, October 1960.

[42] F. J. McCluny, S. E. Schwartz, and F. J. Meyers, "R_2 line optical maser action in ruby," *J. Appl. Phys.*, vol. 33, pp. 3139–3140, October 1962.

[43] I. Wieder and L. R. Sarles, "Stimulated optical emission from exchange-coupled ions of Cr^{+++} in Al_2O_3," *Phys. Rev. Lett.*, vol. 6, p. 95, 1961.

[44] A. L. Schawlow and G. E. Devlin, "Simultaneous optical maser action in two ruby satellite lines," *Phys. Rev. Lett.*, vol. 6, p. 96, 1961.

[45] E. G. Woodbury and W. K. Nag, "Ruby laser operation in the near IR," *Proc. IRE (Correspondence)*, vol. 50, p. 2367, November 1962.

[46] L. F. Johnson and R. A. Thomas, "Maser oscillations at 0.9 and 1.35 microns in CaWO$_4$:Nd^{3+}," *Phys. Rev.*, vol. 131, p. 2038, 1963.

[47] L. F. Johnson, "Optical maser characteristics of Nd^{3+} in CaF$_2$," *J. Appl. Phys.*, vol. 33, p. 756, 1962.

[48] R. C. Duncan, Jr., "Continuous room-temperature Nd^{3+}: CaMoO$_4$ laser," *J. Appl. Phys.*, vol. 36, p. 874, 1965.

[49] J. E. Geusic, H. W. Marcos, and L. G. Van Uitert, "Laser oscillations in Nd-doped yttrium aluminum, yttrium gallium, and gadolinium garnets," *Appl. Phys. Lett.*, vol. 4, p. 182, 1964.

[50] L. F. Johnson, "Optical maser characteristics of rare-earth ions in crystals," *J. Appl. Phys.*, vol. 34, p. 897, 1963.

[51] A. Yariv, S. P. S. Porto, and K. Nassau, "Optical maser emission from trivalent praseodymium in calcium tungstate," *J. Appl. Phys.*, vol. 33, p. 2519, 1962.

[52a] C. F. Johnson, J. E. Geusic, and C. G. Van Uitert, "Coherent oscillations from Tm^{3+}, Ho^{3+} and Er^{3+} ions in yttrium aluminum garnet," *Appl. Phys. Lett.*, vol. 7, p. 127, 1965.

[52b] ——, "Efficient high-power coherent emission from Ho^{3+} ions in yttrium aluminum garnet, assisted by energy transfer," *Appl. Phys. Lett.*, vol. 8, p. 200, 1966.

[53] Z. J. Kiss and R. C. Duncan, Jr., "Pulsed and continuous optical maser action in CaF$_2$:Dy^{2+}," *Proc. IRE (Correspondence)*, vol. 50, pp. 1531–1532, June 1962.

[54] A. A. Ballman, S. P. S. Porto, and A. Yariv, "Calcium niobate Ca(NbO$_3$)$_2$—a new laser host material," *J. Appl. Phys.*, vol. 34, p. 3155, 1963.

[55] G. D. Boyd, R. J. Collins, S. P. S. Porto, A. Yariv, and W. A. Hargreves, "Excitation, relaxation, and continuous maser action in the 2.613 micron transition of CaF$_2$:U^{3+} masers," *Phys. Rev. Lett.*, vol. 8, pp. 269–272, April 1962.

[56] S. P. S. Porto and A. Yariv, "Excitation, relaxation and optical maser action at 2.407 μ in SrF$_2$:U^{3+}," *Proc. IRE*, vol. 50, p. 153, 1962.

[57] P. P. Sorokin and M. J. Stevenon, "Stimulated emission from CaF$_2$:U^{3+}+CaF$_2$:Sm^{2+}," in *Advances in Quantum Electronics*, J. R. Singer, Ed. New York: Columbia University Press, N. Y., and London, England: pp. 65–77, 1961.

[58] P. P. Sorokin, M. J. Stevenson, J. R. Lankard, and D. C. Petht, "Spectroscopy and optical maser action in SrF$_2$:Sm^{2+}," *Phys. Rev.*, vol. 127, p. 503, 1962.

[59] Z. J. Kiss and R. C. Duncan, Jr., "Pulsed and continuous optical maser action in CaF$_2$:Dy^{2+}," *Proc. IRE (Correspondence)*, vol. 50, pp. 1531–1532, June 1962.

[60] ——, "Optical maser action in CaF$_2$:Tm^{2+}," *Proc. IRE (Correspondence)*, vol. 50, pp. 1532–1533, June 1962.

[61] R. C. Duncan, Jr., and Z. J. Kiss, "Continuously operating CaF$_2$:Tm^{2+} optical maser," *Appl. Phys. Lett.*, vol. 3, p. 23, 1963.

[62] R. J. Pressley, J. R. Collard, P. V. Goedertier, F. Sterzer, and W. Zernik, "Solid state laser explorations." RCA Laboratories, Princeton, N. J., Tech. Rept. AFAL-TR-66-129, Contract AF33(615)2645, June 1966.

[63] J. P. Wittke, J. R. Collard, R. C. Duncan, Jr., P. V. Goedertier, Z. J. Kiss, R. J. Pressley, F. Sterzer, and T. Walsh, "Solid state laser explorations." RCA Laboratories, Princeton, N. J., Tech. Rept. AFAL-TR-64-334, Contract AF33(615)1096, Final Rept., January 1965 (also R. J. Pressley and J. P. Wittke, to be published.)

[64] J. E. Geusic, "A repetitively Q switched continuously pumped YAG:Nd laser," *Appl. Phys. Lett.*, vol. 6, p. 175, 1965.

[65] ——, "Materials for solid state optical devices," *NEREM Rec.*, p. 78, 1966.

[66] S. A. Ochs and J. I. Pankove, "Injection-luminescence pumping of a CaF$_2$:Dy^{2+} laser," *Proc. IEEE*, vol. 52, pp. 713–714, June 1964.

[67] D. Roess, "Exfocal pumping of optical masers in elliptical mirrors," *Appl. Opt.*, vol. 3, p. 259, 1964.

[68] ——, "Room temperature CW ruby laser," *Microwaves*, p. 5, April 1966.

[69] V. Evtuhov, "Continuous operation of a ruby laser at room temperature," *Appl. Phys. Lett.*, vol. 6, p. 75, 1965.

[70] I. J. D'Haenens and V. Evtuhov, "Temperature and concentration effects in a ruby laser," [5, p. 1131].

[71] R. J. Pressley and P. V. Goedertier, "A high efficiency, high power solid state laser," presented at the 1965 Internat'l Electron Device Meeting, Washington, D. C.

[72] R. G. Schlecht, C. H. Church, and D. A. Larson, "High efficiency NaI pumping of a continuous Nd^{3+}:YAG laser," *IEEE J. of Quantum Electronics*, vol. QE-2, p. xviii, April 1966; also presented at 1966 Internat'l Quantum Electronics Conference, Phoenix, Ariz.

[73] J. E. Jackson and D. M. Yenni, "High power laser operation," Union Carbide Corp., Linde Div., Indianapolis, Ind. Second Interim Tech. Rept., Contract SRCR-66-4, May 1966.

[74] W. R. Sooy, R. S. Congleton, B. E. Dobratz, and W. K. Ng, "Dynamic limitation on the attainable inversion in ruby lasers," [5, p. 1103].

[75] Korad Corporation, Santa Monica, Calif., advertised data on K-1500 ruby system.

[76] H. Pawel, J. R. Sanford, J. H. Wenzel, and G. J. Wolga, "Use of dielectric etalon as a reflector for Q-switched laser operation," *Proc. IEEE (Correspondence)*, vol. 52, pp. 1048–1049, September 1964.

[77] R. Daly and S. D. Sims, "An improved method of mechanical Q switching using total internal reflection," *Appl. Opt.*, vol. 3, p. 1063, 1964.

[78] J. L. Wentz, "Novel laser Q-switching mechanism," *Proc. IEEE (Correspondence)*, vol. 52, pp. 716–717, June 1964.

[79] P. Kalafas, J. I. Masters, and E. M. E. Murray, "Photosensitive liquid used as a nondestructive passive Q switch," *J. Appl. Phys.*, vol. 35, p. 2349, 1964.

[80] B. H. Soffer, "Giant pulse laser operation by a passive reversibly bleachable absorber," *J. Appl. Phys.*, vol. 35, p. 2551, 1964.

[81] D. Roess and G. Zeitler, "Quasicontinuous ruby giant pulse laser using a saturable absorber as a Q switch," *Appl. Phys. Lett.*, vol. 8, p. 10, 1966.

[82] J. E. Geusic, H. M. Marcos, and L. G. Van Uitert, "A study of the YAG:Nd oscillator," in *Proceedings of the 1965 Physics of Quantum Electronics Conference in San Juan, Puerto Rico*, P. Kelley, B. Lax, and P. Tannenwald, Eds. New York: McGraw-Hill, 1966, p. 725.

[83] D. E. McCumber, "Intensity fluctuations in the output of CW laser oscillators I," *Phys. Rev.*, vol. 141, p. 306, 1966.

[84] L. E. Hargrove, R. F. Fork, and M. A. Pollack, "Locking of He-Ne laser modes induced by synchronous intra-cavity modulation," *Appl. Phys. Lett.*, vol. 5, p. 4, 1964.

[85] M. H. Crowell, "Characteristics of mode-coupled lasers," *IEEE J. Quantum Electronics*, vol. QE-1, pp. 12–20, April 1965.

[86] M. Didomenico, Jr., "Small signal analysis of internal (coupling type) modulation of lasers," *J. Appl. Phys.*, vol. 35, p. 2870, 1964.

[87] A. Yariv, "Internal modulation in multimode laser oscillators," *J. Appl. Phys.*, vol. 36, p. 388, 1965.

[88] M. Didomenico, Jr., J. E. Geusic, H. M. Marcos, and R. G. Smith, "Generation of ultrashort optical pulses by mode locking in YAG:Nd laser," *Appl. Phys. Lett.*, vol. 8, p. 180, 1966.

[89] D. Roess, "Single mode operation of a room-temperature CW ruby laser," *Appl. Phys. Lett.*, vol. 8, p. 109, 1966.

[90] W. W. Rigrod, "The optical ring resonator," *Bell Sys. Tech. J.*, vol. 44, ρ. 907, 1965.

[91] V. Evtuhov and A. E. Siegman, "A twisted mode technique for obtaining axially uniform energy density in a laser cavity," *Appl. Opt.*, vol. 4, p. 142, 1965.

[92] C. L. Tang, H. Statz, G. A. DeMars, and D. T. Wilson, "Spectral properties of a single mode ruby laser: evidence of homogeneous broadening of the zero-phonon lines in solids," *Phys. Rev.*, vol. 136, p. A1, 1964.

[93] See, e.g., the collection of articles on applications from the Laser Physics and Applications Symposium, Bern, Switzerland, 1964, published in *Zeitschrift für Angewandte Mathematik & Physik*, vol. 16, 1965, as well as the Digest of Technical Papers from the 1966 International Quantum Electronics Conference in Phoenix, Ariz., April 12, 1966, *IEEE J. of Quantum Electronics*, vol. QE-2, p. lix, April 1966.

[94] An indication of the continuing rapid progress in this field was the announcement just before publication of this improvement in the operation of a Nd^{3+}:YAG laser pumped with a 2500 watt tungsten lamp to obtain 15 watts output. J. E. Geusic, private communication.

Glass Lasers

C. GILBERT YOUNG, MEMBER, IEEE

Invited Paper

Abstract—In the eight years since the first publication of glass laser operation, considerable progress has been made in this technology with 5 active laser ions operating at 7 emission wavelengths having been confirmed so far. Laser action has been obtained in time regimes from continuous wave to pulsed operation with pulse widths down to 2.5×10^{-13} second. In addition, glass lasers currently provide the highest energy, the highest peak power, and the highest radiance laser sources. An efficiency of energy conversion to the optical second harmonic at 530 nm exceeding 50 percent has been obtained. Recent developments show promise of equaling gas laser performance in coherence length and average power. Present energy conversion efficiencies of over 8 percent and the development of a fiber laser receiver preamplifier are making glass laser systems applications increasingly attractive. This paper reviews the evolution of glass lasers, describes some recent developments, and discusses a number of applications in this fast-growing field.

I. INTRODUCTION

SINCE THE published descriptions by Schawlow and Townes [1] of an optical maser, or laser, a number of solid materials have been made to lase. Maiman was the first to obtain laser action, using chromium-doped aluminum oxide [2], followed shortly by Sorokin and Stevenson who used samarium-doped calcium fluoride [3], and Johnson and Nassau with neodymium-doped calcium tungstate [4]. The next condensed phase laser was neodymium-doped glass reported by Snitzer in 1961 [5]. Subsequent to this time a large number of new lasers have been found including a variety of glass lasers. Although other glass and crystalline lasers will be included in the discussion, this paper will concentrate on three glass lasers: neodymium and neodymium-ytterbium operating at 1.06 microns, and neodymium-ytterbium-erbium operating at 1.54 microns.

There are a number of characteristics which distinguish glass from other solid laser host materials. Its properties are isotropic. It can be doped at very high concentrations with excellent uniformity. It is a material which affords considerable flexibility in size and shape, and may be obtained in large homogeneous pieces of diffraction-limited optical quality. It can also be relatively cheap in large volume production and can be fabricated by a number of processes, such as drilling, drawing, fusion and cladding, which are generally alien to crystalline materials. Good laser glasses can be chosen with indices of refraction ranging from 1.5 to nearly 2.0, or which will set the peak emission wavelength for neodymium at any one of a number of wavelengths between 1.047 and 1.063 microns ([6], Fig. 1). Of

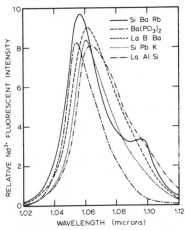

Fig. 1. The relative fluorescence of Nd^{3+} in various laser glasses at 300 K.

even greater importance is the flexibility afforded in relation to the physical constants in the ability to adjust the temperature coefficient of the index of refraction and the strain-optic coefficients so as to produce a thermally stable cavity [7], [8]. The major disadvantage of glass is its low thermal conductivity. This imposes limitations on the thickness of pieces which can be used at high average power, necessitating fairly radical configurational modifications for such operation.

The inherent nature of the glass host produces inhomogeneously broadened lines [6] which are wider than would be found for the same ion in crystals. This raises threshold, but reduces the amplified spontaneous emission losses for the same inversion in amplifier and Q-switched applications. Additionally, since the pulse duration and line width are Fourier transforms of one another, shorter pulses can be obtained in mode-locked operation using glass due to its broader fluorescent line [9].

There is thus a complementarity between glass and crystalline lasers. For continuous or very high repetition-rate operation, crystalline lasers provide desirable higher gain and greater thermal conductivity. In small systems operating at modest repetition rates, glass offers the advantage of lower cost in quantity production since the material is mixed and cast, rather than being grown or pulled from the melt as are crystals. This factor also allows the designer to use longer laser rods for simplifications in fixturing without increasing the rod price remarkably. For larger systems, glass provides greater uniformity and flexibility in physical

Manuscript received May 1, 1969. *This invited paper is one of a series planned on topics of general interest—The Editor.*
The author is with the American Optical Central Research Laboratory, Southbridge, Mass. 01550.

Reprinted from *Proc. IEEE*, vol. 57, pp. 1267–1289, July 1969.

37

parameters. Finally, the 60-times higher ratio of energy storage to optical gain of Nd:glass compared to Nd:YAG makes glass lasers particularly attractive for use in Q-switched oscillators of more than about 200 mJ output, and in power amplifiers.

This papers proceeds with a review of the basic glass laser physics, a section on the various modes of operation with examples of each, followed by descriptions of significant recent glass laser system developments, a discussion of laser glass breakdown and finally, a review of a few possible applications.

II. REVIEW OF GLASS LASERS

A. Basic Principles

In order for a fluorescent material to exhibit laser operation, the round-trip optical gain resulting from the pumping must exceed the round-trip losses within the cavity. Considering a laser of length L, end reflectivities R_1 and R_2, gain coefficient β and absorption coefficient α, the condition for oscillation is

$$R_1 R_2 \exp (\beta - \alpha) 2L \geq 1. \tag{1}$$

This equation, with the equality, expresses a steady-state condition on the inversion. The gain coefficient is related to the inversion N, which is the difference in population of the upper and lower laser levels, the line width $\Delta \nu$ of the transition, the spontaneous transition probability A_{21}, and the index of refraction n by [1] [112]

$$\beta = \frac{\lambda^2 A_{21}}{8\pi n^2 \Delta \nu} N. \tag{2}$$

Good laser design requires an accurate knowledge of β_s, the specific gain coefficient, or the optical gain available per joule of usable inversion. This can be expressed as $\beta_s = \sigma N_1$, where σ is the gain cross section and N_1 is an inversion corresponding to 1 J/cm³ of inversion. Small Q-switched systems, for example, operate most efficiently with a relatively high value of β_s because of the relatively high losses within the cavity and the requirement of only modest energy storage. Large power amplifiers can be made to perform most efficiently with a relatively low value of β_s because high energy extraction is required. Good laser glasses emitting at 1.06 μ have been made which exhibit β_s values differing by two orders of magnitude [30], [245].

Equation (2) yields β_s, but A_{21} is difficult to determine accurately, particularly in neodymium-doped glass, because of uncertainties in the multiplicities of the various levels involved. A number of techniques have been employed to determine β_s. By heating a neodymium-doped glass sample, a population of the terminal laser level has been obtained sufficient to measure either by using a very highly doped, relatively thin sample in a spectrophotometer [10], or a 20 cm long normally doped sample in a two-pass experiment employing a separate laser oscillator of the same glass as a source [11]. This method requires knowledge of the energy level multiplicities although that of the terminal laser level does not appear in the calculations. Other

methods use calculations starting from the resonance line at 880 nm and assumptions about multiplicities [11]–[13]. In addition, with assumptions concerning the quantum efficiency q, the specific gain coefficient may be calculated from the fluorescence spectrum and lifetime alone [11] by

$$\beta_s = \frac{q\eta_{1.06} K N_1 \lambda^4}{\tau_m \Sigma \eta 8\pi n^2 c \Delta \lambda} \tag{3}$$

where η refers to the number of photons emitted in fluorescence, τ_m is the measured lifetime and K is a line shape factor. A more direct method relates energy removal to change in gain in active laser elements [11], [14]–[18]. Finally, by sending through a uniformly pumped laser amplifier, a pulse of sufficient intensity to partially saturate the amplifier, and measuring the resultant pulse distortion [19], measurements of β_s have been made on a variety of laser glasses [20]. If the energy density in the amplifier equals $(\beta_s)^{-1}$, 3 dB of pulse sharpening (defined as the ratio of leading to trailing edge pulse heights) will occur. An energy density of $(4\beta_s)^{-1}$ yields 1 dB of sharpening.

The efficiency of a laser rod of given doping and diameter is a function of the length, end reflectivities, and the loss coefficient both through (1) and the conflicting requirements on the output reflectivity that it be low for maximum output coupling and high for minimum threshold. Experimentally, a length-to-diameter ratio for neodymium-doped glass of about 40 to 1 appears optimum [239]. Equation (1) can be rewritten as

$$\beta = \alpha + \frac{1}{2L} \ln \frac{1}{R_1 R_2}, \tag{4}$$

where α contains all the losses in the cavity, on a per unit length basis. If the second term of (4) is considered as a loss of energy through output, a cavity efficiency ϵ_c can be defined as the ratio of output to total loss or

$$\epsilon_c = \frac{\dfrac{1}{2L} \ln \dfrac{1}{R_1 R_2}}{\alpha + \dfrac{1}{2L} \ln \dfrac{1}{R_1 R_2}} = \frac{\beta - \alpha}{\beta}. \tag{5}$$

The cavity efficiency for a 1-meter laser oscillator with $R_1 R_2 = 0.04$ and $\alpha = 0.2$ percent/cm, would be 90 percent; for $\alpha = 0.1$ percent/cm, $\epsilon_c = 95$ percent. Such long lengths of optically homogeneous material are difficult to get in crystals, but can easily be drawn as glass rods or fibers. One of the advantages of glass becomes evident from this equation for cavity efficiency. The absorption coefficient can be made smaller than 10^{-3} (values of 3×10^{-4} have been measured in some 1-meter samples, nearly equal to the calculated absorption due to the thermal population of the terminal neodymium level at room temperature [21]) by the use of pure materials in high optical quality glass and special melting techniques. The materials must be free of absorbing contaminants. All the glass laser ions emit in the near infrared where the transition metal ions Ni, Co, Cu, Fe and V can provide absorption [8]. The most serious contaminant is the ubiquitous Fe^{2+} ion. It gives an absorption of 0.1

TABLE I
TABLE I

LASER IONS IN GLASS

Ion	Host	Transition	Wavelength	$E_1 - E_0$*	Inv. for 1% gain/cm	Ref.
Nd^{3+}	K-Ba-Si	$^4F_{3/2}$–$^4I_{11/2}$	1.06μ	1950 cm^{-1}	0.7×10^{18} cm^{-3}	[5]
	La-Ba-Th-B	$^4F_{3/2}$–$^4I_{13/2}$	1.37	4070	—	[26]
	Na-Ca-Si	$^4F_{3/2}$–$^4F_{9/2}$	0.92	470	3.5×10^{18}	[27]
Nd^{3+}	YAG†	$^4F_{3/2}$–$^4I_{11/2}$	1.065	2111	1.1×10^{16}	[28]
Yb^{3+}	Li-Mg-Al-Si	$^2F_{5/2}$–$^2F_{7/2}$	1.015	400	2.8×10^{18}	[29]
	K-Ba-Si	$^2F_{5/2}$–$^2F_{7/2}$	1.06	830	11.0×10^{18}	[30]
Ho^{3+}	Li-Mg-Al-Si	$^5I_7$–$^5I_8$	2.1	230	—	[31]
Er^{3+}	Yb-Na-K-Ba-Si	$^4I_{13/2}$–$^4I_{15/2}$	1.543	0	1.8×10^{18}	[32]
	Li-Mg-Al-Si	$^4I_{13/2}$–$^4I_{15/2}$	1.55	111	—	[33]
	Yb-Al-Zn-P_2O_5	$^4I_{13/2}$–$^4I_{15/2}$	1.536	0	9×10^{17}	[34]
	Yb-Fluorophosphate	$^4I_{13/2}$–$^4I_{15/2}$	1.54	0	—	[61]
Tm^{3+}	Li-Mg-Al-Si	$^3H_4$–$^3H_6$	1.85	—	—	[35]
	Yb-Li-Mg-Al-Si	$^3H_4$–$^3H_6$	2.015	—	—	[35]

* The energy separation between the terminal state for laser action and the ground state is given by $E_1 - E_0$.
† Neodymium-doped YAG is included for comparison.

percent/cm at 1.06μ for a concentration of 5×10^{16} ions/cm^3 [8].

An additional source of loss is possible through excited-state absorption. This comes about because any such absorption must be saturated to achieve a net gain. However, this absorption is usually followed by rapid nonradiative decay. This is probably why the uranyl ion has not been made to lase in glass [6]. In the case of neodymium in glass, experiments have been done in an attempt to measure the active loss by varying the end reflectivity on the laser rod. Using (4) for two sets of end reflectivities R_1R_2 and R_1R_3, and the two corresponding gain coefficients β_2 and β_3 a single equation for the loss coefficient α can be obtained

$$\alpha = \frac{\ln R_1R_2}{2L} + \frac{\beta_2}{\beta_3 - \beta_2}\left[\frac{\ln R_1R_2 - \ln R_1R_3}{2L}\right]. \quad (6)$$

By assuming that β is proportional to the pumping, a value of α of 10^{-3} cm^{-1} was obtained in this experiment for a glass exhibiting a 10^{-3} cm^{-1} passive absorption at 1.06μ [194]. In other similar work on a different glass, an active loss of 10^{-2} cm^{-1} was measured in a glass having a 6×10^{-3} cm^{-1} passive absorption at 1.06μ [22]. In this glass, there is evidence that a significant excited state absorption is present, which would contribute to the loss at 1.06μ [132]. Direct excited-state absorption measurements have been reported in uranium and erbium-doped silicate glasses [240].

An estimate can be made of the pump power required to reach threshold by assuming a blackbody emitter, which is a reasonable assumption for most flash-pumped solid-state lasers, and a single absorption band. The relationship can be derived to equating processes which lead to population of the metastable level to those which empty it. The result [6], [112] is

$$\exp\left(-E_p/kT_p\right) \geq \frac{N}{A_p}\frac{1}{\tau_m N_0} \quad (7)$$

where E_p is the energy of the pump band, T_p the blackbody temperature of the pump, A_p is the Einstein A coefficient for the pumping transition, τ_m is the measured lifetime, and N_0 is the concentration. In a given glass base, N and A_p are approximately independent of concentration [6]. However, τ_m decreases due to an increase in nonradiative transitions associated with concentration quenching. Therefore, from (7), minimum threshold should occur for maximum value of $\tau_m N_0$. When using thin rods, however, a higher overall efficiency may be obtained using a somewhat higher concentration due to the better coupling to the lamp.

Simple calculations from (7), assuming pumping only in the 880 nm resonance line, lead to a required blackbody pump temperature of about 3×10^3 K [52]. Summing of the pump contributions from the other absorption bands would reduce this figure. Experimentally, using a photographic flashbulb having a measured near blackbody radiation characteristic of about 4000 K, laser threshold was reached using one quarter of an imaging sphere for pump coupling [52], [60]. This would imply a minimum required T_p of less than 3×10^3 K. Experiments using the sun and a carbon-arc solar simulator as the pumping sources are in general agreement with these conclusions [62].

B. Laser Ions in Glass

Unfortunately, not much can be said about the structure of glass as its affects ions placed in it. Although some determinations have been made for transition-metal ions, for cerium [241] in a metaphosphate glass and for ytterbium in a silicate glass [242], [23]–[25], at the present time study of the behavior of ions in glass is largely an empirical science, particularly so for the rare earth ions.

Table I lists the rare earth ions that have been made to

TABLE II

Ion	Host	Sensitizer	Reference
Nd^{3+}	K BaSi	UO_2^{2+}	[38]
	Phosphate	Mn^{2+}	[39][40]
	K BaSi	Ag	[41]
	LiMgAlSi	Ce^{3+}	[40][42]
	Borosilicate	Tb^{3+}	[43]
	Borosilicate	Eu^{3+}	[43]
	—	Cr^{3+}	[44]
Yb^{3+}	LiCaB	Nd^{3+}	[45]
	LiMgAlSi	Ce^{3+}	[46]
	Alkali Si	Cr^{3+}	[47]
Ho^{3+}	LiMgAlSi	Yb^{3+}	[48]
	LiMgAlSi	Er^{3+}, Yb^{3+}	[205]
Er^{3+}	Na K BaSi	Yb^{3+}	[32]
	Alkali Si	Mo^{3+}	[47]
	Al Zn P_2O_5	Yb^{3+}	[34]
	Fluorophosphate	Yb^{3+}	[61]
Tm^{3+}	LiMgAlSi	Yb^{3+}	[35]
	LiMgAlSi	Er^{3+}	[35]

Fig. 2. Portions of the absorption and fluorescent spectra of Nd^{3+}, Yb^{3+}, and Er^{3+} in a silicate glass relevant to the energy transfer processes.

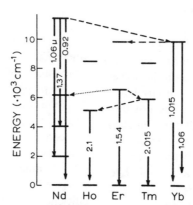

Fig. 3. Energy level diagram for the lasing ions in glass, showing the important energy transfer processes.

emit laser oscillations in glass, along with data of neodymium-doped YAG for comparison of β_s. Although laser action from Nd^{3+} is usually at 1.06 μ [5], laser action can also be achieved at 1.37 μ alone by use of a frequency-selective reflector [26], or at 0.92 μ alone by use of both a frequency selective reflector and a reduced temperature [27] as well, since the terminal laser level is separated by only 470 cm^{-1} from the ground state. Simultaneous oscillations in all three wavelengths have been seen in a calibo glass without the use of selective reflectors [36]. The energy of the terminal state for Ho^{3+} has been taken from the data on this ion in calcium tungstate [37].

Sensitization of fluorescence by the absorption of light by one ion and the subsequent transfer of energy to the fluorescent ion is of interest both in regard to increasing the pumping and for the study of energy transfer processes in glass.

For efficient radiant transfer to occur there must be overlap of the emission spectrum of the sensitizer with the absorption spectrum of the activator. In a nonradiative trans-

fer the energy is transported without intermediate radiation and absorption, but overlap is still required. The transfer rate depends on the interaction mechanism, and has an r^{-6} dependence on the interionic distance for a dipole-dipole interaction, r^{-8} for dipole-quadrupole and $\exp(-\alpha r)$ for an exchange interaction [195]. In addition to transfer between different ionic species, excitation can migrate between like ions if the absorption and fluorescent spectra overlap. This can explain concentration quenching, where the excitation wanders from ion to ion until it migrates to a site that is particularly favorable for a nonradiative transition. Fig. 2 summarizes the pertinent spectra for neodymium, ytterbium and erbium in glass. Table II gives the systems of sensitized fluorescence that have been reported. Double sensitized fluorescence of Yb^{3+} has been reported, both by direct transfer from Ce^{3+} and Nd^{3+} and by sequential transfer from UO_2^{2+} to Nd^{3+} and then to Yb^{3+} [50]. Fig. 3 summarizes the emission wavelengths and the important energy transfer processes. Detailed luminescent data may be found, for example, in [172], [231]–[234].

At the present time, the 1.06 μ emission of neodymium and of ytterbium, and the 1.54 μ emission of erbium are receiving the most attention because of their wavelengths and the fact that they operate at room temperature. The remainder of this paper will concentrate on these ions.

C. Properties of Nd³⁺ in Glass

Although neodymium has been made to lase in a large variety of glasses since 1961, it appears that the combined requirements of a relatively long fluoerscent lifetime, high fluorescence efficiency and high durability are best provided by a glass with an alkali-alkaline earth silicate base [8].

The 1.06 μ emission from neodymium-doped glass terminates on a level elevated about 2000 cm⁻¹ from the ground state. Because of this, little change in efficiency is seen by this laser operating at temperatures up to 100°C [208]. In addition, because the $^4I_{9/2}$ ground state is split by approximately 450 cm⁻¹, there is a temperature dependence to the absorption bands, which gives them an additional long-wavelength shoulder at high temperatures [6]. These provide a small additional pumping in spectral regions where the room-temperature glass is relatively transparent.

Direct quantum efficiency measurements [51], [206] and the quantum efficiency inferred from specific gain coefficient calculations compared to direct measurements of the specific gain coefficient indicate a quantum efficiency for the alkali-alkaline earth silicate laser glasses of 5 weight percent doping ranging from 50 to 75 percent. Pumping experiments, using various spectral sources and filters, [52] and excitation spectrum measurements [52], [207], indicate that the quantum efficiency is essentially constant over all the neodymium absorption bands. The fluorescent lifetime for neodymium in various laser glass bases varies from 0.04 to over 0.9 ms, depending on the concentration [8].

Three basically different time traces can be seen from a neodymium-doped glass rod, depending on the conditions of the experiment; the usually seen random spiking, limit cycles [53], [54] consisting of a series of equally spaced, nearly equally intense pulses within the overall laser pulse, and damped oscillations. Snitzer [6] has modified the rate equations of Statz and deMars [54] to explain these phenomena in glass by including the influence on the damped oscillations caused by cladding the laser rod and by high Nd³⁺ concentration effects, such as concentration quenching and cross relaxation.

The spectral output is quite different from rods showing random spiking than from clad rods that give damped oscillation. The latter produce continuous bands, whereas random spiking gives sharp lines. From an examination of the time-resolved spectra, it is found that emission can occur in several lines simultaneously in one spike [6]. By study of the time resolved emission characteristics, it is possible to conclude that the 180 Å line which occurs in spontaneous emission has an inhomogeneous broadening and ligand field splitting that are both approximately 70 Å wide. The homogeneous thermal broadening is 20 Å at 300 K and varies to about 5 Å at 77 K [6]. In [6] and [209] more complete descriptions of these phenomena are noted.

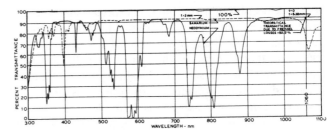

Fig. 4. Absorption spectrum of neodymium in a silicate glass base. Superimposed is the spectrum of samarium in the same base glass to show its effect when used as a cladding glass.

TABLE III

A COMPARISON OF THE USE OF Nd³⁺ AND Nd³⁺–Yb³⁺ GLASS LASERS AS POWER AMPLIFIERS AT 1.06 μ

	Nd³⁺ Glass	Nd³⁺–Yb³⁺ Glass
Fluorescent lifetime, μs	600	2200
Pumping	Nd³⁺ absorptions	Nd³⁺ and Yb³⁺ absorptions
Relative gain per unit of pumping input	4	1
Specific gain, $\dfrac{J/cm^3}{dB/cm} = \dfrac{J}{dB\text{-}cm^2}$	3	48
Energy density (J/cm²) for (a) 1 dB pulse sharpening (b) 3 dB pulse sharpening	3.3 13	53 210

The absorption properties of neodymium in glass are shown in Fig. 4.

D. Properties of Nd³⁺ — Yb³⁺ Glass

Energy transfer between neodymium and ytterbium has been reported in a number of glasses [45], [50], [55]–[57]. Interest in Nd-Yb laser glass stems from the fact that the ytterbium emits at 1.06 μ at room temperature, making it compatible for use with 1.06 μ neodymium-doped laser systems. This laser operates through absoprtion in both neodymium and ytterbium bands and nonradiative energy transfer from the neodymium to the ytterbium. By proper choice of the concentration of Nd and Yb, essentially all available flashtube excitation can be put into ytterbium inversion. The primary reason for interest in the neodymium-ytterbium laser is that the energy storage for equal gain in the Nd-Yb glass is about 16 times higher than that of the Nd glass (Table I). This is particularly important in power amplifiers or where a minimum of pulse sharpening or target feedback instability is desired. In addition, the longer fluorescent lifetime exhibited by the ytterbium, 2.2 ms, and the increased absorption provided by the ytterbium, should provide more efficient Q-switched operation and afford an improved match to the flashtubes. These factors account for the experimentally observed four times higher inversion for equal pumping in the Nd-Yb glass as compared to Nd glass. Table III summarizes the characteristics of Nd³⁺ and Yb³⁺ operating at 1.06 μ.

A significant thermal population of the terminal level of the 1.06 μ transition in ytterbium is expected at room temperature. For a fairly typical 5 percent Nd, 4 percent Yb glass this accounts for about a 0.5 percent/cm absorption at 1.06 μ and therefore makes the system partially three-level in character. However, some of this absorption is eliminated under active lasing conditions. In using this glass, full-length pumping of the laser rod is desirable.

A potentially more troublesome aspect of the Nd-Yb glass laser is related to this thermal population. As seen from Table I, the gain coefficient of the 1.015 μ transition is about four times that of the 1.06 μ transition. On the other hand, the thermal population of the terminal level of the 1.015 μ transition at room temperature is about 18 times greater than that of the 1.06 μ transition. Depending on the pumping level, temperature and dopings, these two opposing factors will cause the Nd-Yb glass laser to exhibit higher net gain at either 1.015 μ or 1.06 μ [196]. The gain of the laser rod due to each of the two transitions can be written

$$G_{1.015} = (N_u - N_A)\beta_{1.015} L \quad (8)$$

$$G_{1.06} = (N_u - N_B)\beta_{1.06} L \quad (9)$$

where N_u is the $^4F_{5/2}$ metastable population, N_A and N_B are the populations of the terminal 1.015 and 1.06 μ transitions. respectively, L is the length and the β's are the specific gain coefficients. These can be written as

$$\beta_{1.015} = 154 \times 10^{-22} \text{ dB} \cdot \text{cm}^2/\text{ion}$$

$$\beta_{1.06} = 39 \times 10^{-22} \text{ dB} \cdot \text{cm}^2/\text{ion}.$$

Assuming adequate control of off-axis spontaneous emission by surface treatment of the rod, a laser rod will self-saturate when the amplified spontaneous emission from one end, in the cone angle $d\Omega$ subtended by the opposite end and the time t of emission, is approximately equal to the energy density necessary to seriously deplete the inversion. The required energy density can conveniently be set equal to the inverse of the specific gain coefficient [68], [197]

$$\int_0^L \frac{N}{\tau} e^{\beta x} \frac{d\Omega}{4\pi} t = \frac{1}{\beta_s} \quad (10)$$

or

$$e^{\beta L} = \frac{4\pi}{d\Omega}\left(\frac{\tau}{t}\right) \quad (11)$$

where N is the inversion, τ the effective lifetime (taking competing fluorescence transitions into account) and t is the pulse width. For a clad rod 1 cm in diameter and 1 meter long, the critical gain at 1.015 μ is about 60 dB, which is about 6 dB greater than for neodymium because of the longer Yb^{3+} lifetime. Equation (8) can then be solved for N_u

$$N_u = \frac{60 + N_A\beta_{1.015}L}{\beta_{1.015}L} \quad (12)$$

and (9) becomes

$$G_{1.06} = \frac{60\beta_{1.06}}{\beta_{1.015}} + (K_A - K_B)\beta_{1.06}N_gL \quad (13)$$

where K_A and K_B are the factors giving the fractional thermal populations of levels A and B, and N_g is the ground state population. The terminal level of the 1.015 μ transition is assumed to be a pair of Kramer's doublets separated by 380 cm^{-1} from the ground Kramer's doublet. The terminal level of the 1.06 μ transition is assumed to be a Kramer's doublet elevated by 830 cm^{-1}. These assumptions are consistent with the room temperature [57] and low temperature [242] fluorescent data. Recognizing that $N=(1+K_A+K_B)N_g+N_u$ is the total ytterbium doping level, (12) becomes

$$N_g = \frac{N\beta_{1.015}L - 60}{(1 + 2K_A + K_B)\beta_{1.015}L} \quad (14)$$

and finally

$$G_{1.06} = \frac{60\beta_{1.06}}{\beta_{1.015}} - \frac{60(K_A - K_B)\beta_{1.06}}{(1 + 2K_A + K_B)\beta_{1.015}} + \frac{(K_A - K_B)\beta_{1.06}L}{(1 + 2K_A + K_B)} N. \quad (15)$$

Assuming that a gain of 60 dB at 1.015 μ will begin to seriously deplete the inversion through amplified spontaneous emission, and using the data in Table I, the room temperature gain obtainable at 1.06 μ is given by

$$G_{1.06} = [12 + 87 \times 10^{-21} N]\text{dB}$$

or

$$G_{1.06} = [12 \times 8(\text{weight-percent Yb})]\text{dB} \quad (16)$$

where N is the ytterbium concentration. For a typical 1-meter rod of 5 percent Nd, 4 percent Yb, a gain of 43 dB should therefore be obtainable at 1.06 μ before serious depletion occurs at 1.015 μ. Other dopings, sizes and temperatures impose other limits on the maximum 1.06 μ gain. Experimentally, over 20 dB has been measured in such a rod with no excess depletion being observed.

A measure of the quantum efficiency of the ytterbium ions in a 2 weight-percent Yb$_2$O$_3$ glass has been obtained by comparihg the calculated to the measured fluorescent lifetime. The calculated lifetime is obtained from the resonance transition at 0.97 μ, assuming that it arises between two Kramer's doublets [57], and measurements of the fluoresence spectra. The result is $\tau = 2.7$ ms, which, when compared to the measured value of 2.2 ms, would imply a quantum efficiency of about 80 percent.

One disadvantage in the use of the Nd-Yb laser glass as an oscillator is that its threshold is higher than that for a comparable neodymium rod because of the lower gain per unit of pumping energy obtained with the Nd-Yb glass. Long-pulse efficiency above threshold, however, is equal to that of a neodymium-doped rod.

A more detailed discussion of the spectroscopic properties of the Nd^{3+}−Yb^{3+} system may be found in [6]. The absorption properties of this system are shown in Fig. 5.

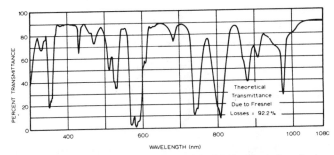

Fig. 5. Absorption spectrum of a $Nd^{3+} - Yb^{3+}$ silicate-based glass having 5 weight-percent Nd_2O_3 and 4 weight-percent of Yb_2O_3. Sample thickness was 5 mm.

E. Properties of $Nd^{3+} - Yb^{3+} - Er^{3+}$ in Glass

As shown in Tables I and II, Er^{3+} has been made to lase in both a silicate and a phosphate base. Due to the 3-level behavior of erbium and the paucity of erbium absorptions, multiple doping with neodymium and ytterbium is necessary to obtain satisfactory system efficiency. However, at high concentrations the neodymium not only sensitizes the Er^{3+} fluorescence but also quenches it through the neodymium $^4I_{15/2} \rightarrow {}^4I_{9/2}$ transition. Recently this quenching has been avoided by having only Nd^{3+} and Yb^{3+} in the cladding and Yb^{3+} and Er^{3+} alone in the active core of a laser rod. Energy is then transferred radiatively from the cladding to the core, circumventing the quenching of the Nd^{3+} [58]. Using this technique, a factor of 2 reduction in threshold has so far been obtained over triple doping in the same piece of glass. While the concentration of the erbium is kept low because it is a 3-level laser, the concentration of the Yb^{3+} can be quite high without ill effect, limited by pumping considerations, thus providing better coupling to the lamp.

Safety is one important reason for interest in the Nd-Yb-Er glass laser [59]. The transmission through the human eye to the retina is many orders of magnitude lower at 1.5 μ than at the emission wavelengths of other common solid-state lasers (Fig. 6). The eye can have an approximately 10^4 magnification of energy density to the retina. It should therefore be possible to use the Nd-Yb-Er glass laser with considerably improved eye safety, especially because at 1.5 μ the absorption is distributed throughout the eye, reducing the possibility of damage to the intervening media. Atmospheric transmission also appears to be quite good at this wavelength.

This laser can emit at either 1.543 or 1.536 μ, depending on the base composition (see Table I). Interest in the phosphate-base composition with the shorter emission wavelength is fourfold: this wavelength may be a better match to germanium avalanche photo-detectors, thus far its efficiency is better than that of the silicate, phosphate glass appears much more resistant to flashtube-induced absorptions than silicate glass, and finally, the phosphate glass is free of the temperature-dependent wavelength instability in the output of the silicate-based erbium laser which causes it to switch to emission at 1.6 μ when heated [8]. Because of its 3-level operation (see Fig. 1) this laser is not now as efficient as neodymium lasers. However, typical performance for an unclad rod is 120 mJ in a 25 ns pulse

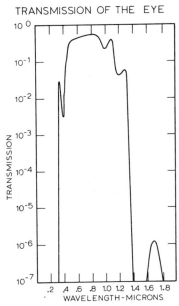

Fig. 6. Total transmission of the human eye to the retina. Data taken from [59].

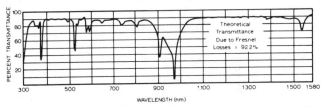

Fig. 7. Absorption spectra of a $Nd^{3+} - Yb^{3+} - Er^{3+}$ silicate-based glass having 0.25 weight-percent Nd_2O_3, 10 weight-percent Yb_2O_3 abd 1.0 weight-percent Er_2O_3. Sample thickness was 5 mm.

for 130 J input, numbers not too unlike those seen for ruby [34]. The fluorescent lifetime is about 14 ms in the silicate base and 8.2 ms in the phosphate base. A more detailed description of the spectroscopic properties of this laser is presented in [6]. The absorption and fluorescence properties of this system are shown in Fig. 7.

III. MODES OF OPERATION OF GLASS LASERS

In this section the various modes of operation of glass lasers will be briefly reviewed with representative devices described in some cases. Because glass lasers can emit continuously or with output pulses of any length down to the subpicosecond range, the discussion will be divided according to time domain.

A. Continuous Operation

Continuous laser action at 1.06 μ has been obtained at room temperature in a barium crown glass containing 6.25 weight-percent of Nd_2O_3 [60]. In this work, a 0.1-mm-diam. active core in a 1 mm-diam. inert cladding glass was used. The 3 cm-long rod and PEK type A high-pressure mercury arc lamp were side by side at the object-image conjugates inside a 23-diam. imaging sphere. Threshold was obtained at 1370 watts into the 2 kW lamp. Calculations indicate that about a watt of CW output could probably be obtained by use of longer lamps along with an appropriate rod geometry. Work has also been done on phosphate glasses for CW operation [210]. At the present time, however, the overall performance of CW neodymium YAG lasers is much superior due to the higher thermal conductivity and narrower line width.

B. Pulsed Operation in 10^{-2} to 10^{-3} second

In this time domain laser emission lasts approximately as long as the flashlamps are on. Due in part to the simplicity of this long-pulse operation, large systems of high efficiency may be built. In one such device [21], [198] an output of 5000 joules was obtained from a 30 mm by 1 meter laser rod of 3 weight-percent doping. This is the highest energy laser reported so far. The pulse duration was about 3 ms. The rod was clad with an absorbing cladding of 38 mm outside diameter. The absorbing cladding [11], which had a one part in a thousand higher index of refraction at 1.06 μ than the core glass, served to remove off-axis laser light, thus confining the laser output to a 10 milliradian cone angle. A 3 percent efficiency above threshold was measured with this system. The rod was pumped by four flashlamps in a close-wrap configuration shown schematically in Fig. 8.

In other work, an efficiency above threshold of over 8 percent was measured [21] using a 12.6 mm diam. by 55 cm long laser rod. This higher efficiency was obtained by a much closer packing of the flashtubes and coupling reflectors than was possible with the 5000 joule laser because of cooling requirements. With careful selection of flashtube gas pressure and operating parameters, good laser cavity design, and use of very low loss glass, an efficiency above threshold in excess of 12 percent should be obtainable.

In other experiments [62], [65], laser operation was obtained using the sun as the optical pump. Here, the sun's image was reduced to 3 mm diam. by a 60 cm aperture telescope system. The laser rod contained a thin active core and was end pumped by having the light reflect, either by total internal reflection [62] or by a silver reflector [65] back and forth through the core as it worked its way along the length of the fiber (see Fig. 9). One and a quarter watts of laser output were obtained in this way, although the output lasted for only 7 ms due to heating of the uncooled rod.

Several other special systems will be mentioned. In one, a long glass fiber was pumped without resonators, and a series of laser spikes separated by the round trip transit time was observed [6], [211]. In other work, a glass laser was put inside a resonant cavity with a YAG crystal to narrow the spectral emission width and reduce the threshold [212], [213]. Other approaches to spectral narrowing have been the introduction of a thin plate [214], [224], or a dye [246], or a diffraction grating [215] within the cavity. A glass laser has been pumped in the $^{4}F_{9/2}$ absorption band, resulting in enhanced output, by the output of a ruby laser [216]. Ultrasonic modulation of a glass laser rod has been used to modulate the output at 100 kHz [220]. Finally a number of traveling-wave ring lasers have been built, producing spike-free, spectrally narrowed output [217], [218].

C. Pulsed Operation in $10^{-4} - 10^{-7}$ second

Generating high-energy laser pulses of 10^{-4} to 10^{-7} second duration is considerably more difficult than generating pulses an order of magnitude longer or shorter. For longer

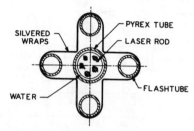

Fig. 8. "Close wrap" laser module cross section of the 5000 joule laser.

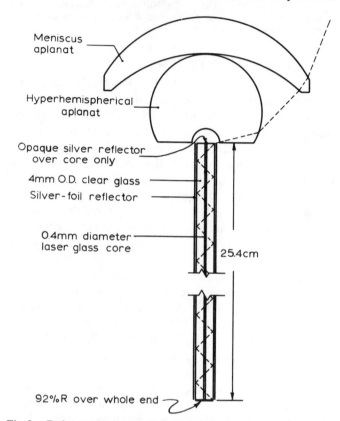

Fig. 9. End-pumped glass laser. Sunlight collected by a 60 cm parabola is imaged onto the end of the glass laser rod by the unity numerical aperture refractive elements.

pulses, the pulse duration can be set by the flashlamp pulse length. This technique becomes inefficient for pulses very much shorter than 1 ms due to the spectral shift toward shorter wavelengths of the lamps. On the other hand, it is difficult to obtain single Q-switched pulses much longer than 100 ns.

A useful technique for obtaining laser pulses in this 10^{-4} to 10^{-7} time domain is to use amplified spontaneous emission in a nonresonant system [69], [49], [219]. One such system uses a series of 5 glass laser amplifiers in series to generate amplified spontaneous emission which is then shuttered by Kerr cells to produce a square-wave pulse adjustable in length from 10^{-4} to 10^{-6} second [66]. Six Faraday rotator optical isolators [67] are interspersed through the system to provide stability in the face of the 155 dB one-way optical gain. The laser pulses have a near and far field which is smooth across the center and drops quickly to zero at the edges. The output is also spectrally smooth and gain narrowed to about 30 Å width [11]. These characteristics are consequences of the fact that the system is single-pass, rather than oscillatory, so that no modes are created, and the output beam angle is defined by the overall aspect ratio of the system. Following the 5 rods which generate the pulse are two 75 mm diameter by 1 m long glass laser power amplifier rods in parallel to provide an output of about 1000 joules. In order to prevent pulse distortion through depletion of the inversion near the output end of the power amplifier array, one of the Kerr cells is programmed to provide an increasing transmission during the pulse duration, so the resulting output is again flattopped. This device is shown schematically in Fig. 10.

For shorter pulses, in the 10^{-7} second time region, a somewhat different amplified spontaneous emission device has been built [68], [197]. This laser operates by having 12 laser amplifiers separately optically pumped and then suddenly optically connected in series. Amplified spontaneous emission from the region near either end is then amplified sufficiently after passing through all the amplifiers that it will sweep out the inversion in the last few rods in an optical avalanche. The pulse width is set by the optical transit time through the system, for an instantaneous switch. In this system, a set of four rotating prisms on a common shaft served to provide the switching between the rods in a folded, two-pass array. The measured output was about 70 joules in 70 ns in a 1 mrad beam.

One difference between amplified spontaneous emission systems and oscillatory systems is that for diffraction-limited operation with a given aperture size, the beamspread can be 30 percent less for a diffraction-limited nonoscillatory system than for the lowest order HE_{11} transverse mode due to the diffraction conditions at the aperture.

D. Pulsed Operation in 10^{-8} second

Laser pulses of about 10^{-8} second duration are obtained by Q-switching the laser cavity [70]. In this technique, one of the cavity mirrors is effectively removed during pumping and then suddenly put into place. The build-up time is determined by the switching speed and the initial gain. The pulse decay rate is set by the cavity losses. The pulse width is therefore a function of the inversion, the cavity parameters and the switching speed. Through adjustment of these variables, glass lasers have been made to emit pulses of from 10 to 120 ns. Through use of succeeding amplifier chains, energy outputs of hundreds of joules have been reported [63], [64], [221].

Laser glass can also be made to exhibit self Q-switching by double doping with the UO_2^{2+} ion [73], [74] or by suitable color centers in Nd^{3+} glass [52], [75]. Both the glasses with the uranyl ion and those with color centers require the ultraviolet light from the flashlamp. Since the uranyl ion is an excited state absorber, the ultraviolet light is needed to populate the first excited electronic state.

In some systems, nearly the same energy output is obtained in the Q-switched mode as in the long-pulse mode of operation. Direct measurements of the excited-state lifetime under moderately high inversion levels was accomplished by measuring the gain of a 1 cm core-diameter by 20 cm long clad amplifier rod as a function of time after the lamps were suddenly cut off. The lifetime was found to be

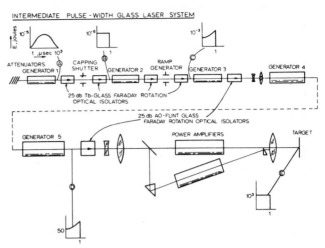

Fig. 10. Generators 1 and 2: 1 cm core, 15 mm O.D., 1 meter long, 50 dB; Generator 3: 18 mm dia., 1 meter long, 20 dB; Generator 4: 38 mm dia., 60 cm long, 11 dB; Generator 5: 38 mm dia., 1 meter long, 20 dB; Power Amplifiers: 75 mm diameter, 1 meter long, 18 dB. The ramp generator creates a rising pulse (c) which, with the pulse sharpening which occurs in the power amplifiers, results in a square-wave output pulse (E).

400 μs, in a glass exhibiting a 500 μs fluorescent lifetime in a powdered sample, for an inversion of about 1.5 J/cm³ [194]. A maximum inversion, under optimum pumping conditions and at the flashtube failure point for this rod, of 4 J/cm³ was obtained [194].

E. Pulsed Operation in 10^{-9} second

To obtain pulses shorter than the usual Q-switched pulses a number of techniques may be used. The laser may be pumped with high-reflectivity mirrors at both ends of the cavity, one of which is then suddenly removed [71]. In this mode of operation, the pulse width is set by the round-trip pulse transit time in the cavity. Pulses of a few nanoseconds have been reported through use of this technique [72].

A second technique is an electrooptic shutter within the cavity, in addition to the Q-switch, to truncate both edges of the pulse [222]. Another approach is to operate an external shutter outside the laser cavity by means of a laser-triggered spark gap in order to pass only a portion of the Q-switched pulse [77]. A different method of obtaining shortened pulses is to use saturable absorbers to produce pulse shortening of the Q-switched pulse in an amplifier [78]. Operating an amplifier in the saturation gain region will not produce pulse narrowing for the usual Q-switched laser pulse shape because the pulse has an exponentially rising leading edge which gets amplified while preserving the pulse shape and width [19]. With a saturable absorber present, the leading edge of the pulse is truncated until the absorber is saturated, at which time it quickly opens, allowing the steep-edged pulse to deplete the inversion and undergo pulse sharpening.

One reason nominal 1-ns pulses are of interest is because they represent the limitation on range resolution of a laser radar imposed by the requirement of a receiver which is both sensitive and operates in real time, such as a counter or an oscilloscope. Within this limitation, a shorter pulse could be used, but material durability may be somewhat higher for equal energy density for the longer pulse.

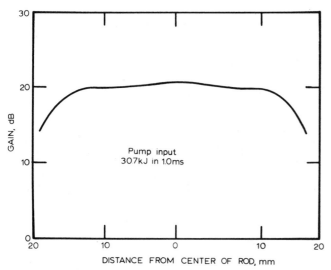

Fig. 11. Measured gain as a function of radial position in a 38 mm diameter, 1 meter long Nd^{3+}-glass laser rod of 1.1 weight percent doping. Azimuthal variations were less than 1 dB in this rod.

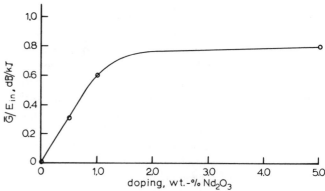

Fig. 12. Area-weighted average gain, $\overline{G}$, per unit of pumping energy versus doping level in a series of 38 mm-diameter by 1 meter long glass laser rods. Optimum doping for gain uniformity is slightly over 1 weight-percent.

F. Pulsed Operation in 10^{-10} to 10^{-11} second

Saturable absorbers were used early to Q-switch glass lasers both internally [52], [73], [76] and externally with a reversibly bleachable dye [79]. Mode locking of a gas laser was also accomplished at about the same time using an internal modulator driven by an external signal [80]. Mode locking in a solid-state laser was first reported in ruby [81], [82], with synchronous locking reported somewhat later [223]. Subsequently, there has been considerable work in this field. A comprehensive review of this area of work may be found in the PROCEEDINGS [9].

Mode-locked glass laser pulses are generally in the few picosecond range for a cavity relatively free from dispersion. One technique used to get pulses in the 10 to 100 picosecond region is to use frequency selective elements, such as a diffraction grating, inside the cavity in order to restrict the spectral bandwidth of oscillation [84]. Because the pulse length is the Fourier transform of the spectral width, reducing the spectral width sufficiently will result in a lengthened pulse.

One of the larger systems of this type so far reported yielded a pulse of 20 joules in 10^{-11} second [85] or a power of 2 TW.

G. Pulsed Operation in 10^{-12} second

Pulses as short as 2.5×10^{-13} second have been reported [238]. This corresponds to about 70 wavelengths of the 1.06 μ light. To directly achieve these short pulsewidths, a wide spectral width is needed in the laser material. The fluorescent linewidth of 200 Å in glass (to the 3 dB points) would imply that a pulsewidth of about 0.2 picosecond might be possible, if the entire linewidth is effectively available. Hole-burning experiments would indicate that an effective width of 70 Å corresponding to a pulsewidth of about a half a picosecond might be more appropriate [6]. Use of the Nd-Yb glass, with its much broader emission linewidth, might permit substantially shorter pulses to be realized. Compression of a chirped picosecond pulse [86]

can even further reduce the pulsewidth. Pulses as short as a few cycles may ultimately be obtainable by external chirping of a subpicosecond pulse, followed by subsequent compression [199].

The highest energy reported in the few picosecond time domain, is 51 joules in a pulse width of about 3 ps [87]. This represents a peak power of 17 TW and the highest peak power laser reported to date.

IV. GLASS LASER DEVICES

In this section a number of glass laser device developments will be described. The choice of subjects for discussion was predicated more on covering the broad range of glass laser systems activity than on making the treatment all inclusive. The work described will cover glass lasers of high spatial and temporal coherence, efficient second harmonic generation, special small laser fabrications, a laser preamplifier for use before a detector, and high repetition-rate operation.

A. High Radiance Glass Lasers

For many applications it is desirable to operate a high power laser with the minimum beamspread and a single forward lobe in the radiation pattern. One such diffraction-limited, lowest order mode laser has produced an output radiance of 2×10^{17} W/cm²·sr [88], [89]. The oscillator in this system is pinhole mode selected and Pockels-cell Q-switched [90], [91]. The final amplifier laser rod, preceded by one preamplifier, is doped at 1.1 weight-percent of Nd_2O_3 and has a roughened surface [226], [227], so that when pumped by four linear flashtubes in a close wrap configuration (Fig. 8), both the azimuthal and radial gain variations (Fig. 11) are less than about ± 1 dB [21]. Reducing the doping to achieve radial pumping uniformity does not seriously reduce the pumping efficiency for this size rod, as shown in Fig. 12. This laser rod is 38 mm outside diameter by 1 meter in length. The beam is stopped down so that only 30 mm of the rod is used and the output is in a 40 μrad beam of 90 joules. A 75 mm by 1 meter additional amplifier has so far increased the energy output to several hundred joules, but at no increase in beam radiance.

When the output of this laser is focused in air by a long

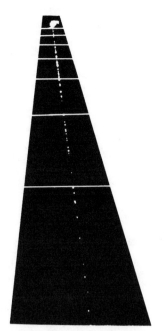

Fig. 13. Long-path air breakdown produced by a diffraction-limited glass laser. The laser output was focused by a 28 meter focal length lens at the fourth 3-meter fiducial line from the bottom. The 25-meter breakdown path was terminated by the wall of the building at the top of the figure.

focal length lens, a string of air breakdown points is created. Using a 28-meter focal length lens, a line of air breakdown points greater than 25 meters in length is obtained [89], with substantial energy being retained in the beam at the end of the line, Fig. 13. Since the first published report of this phenomenon [92], a number of explanations have been offered to explain the effect. The power density at the focal point of the lens is about 10^{11} watts/cm², which is the reported threshold value in air breakdown experiments [93], [94]. Since the breakdown extends at least half a focal length on either side of the focal point, the power densities at the extremes of the breakdown must be reduced by at least three orders of magnitude below this threshold [121]. This is suggestive of a self-focusing mechanism. Also, it has been reported [92], [95] that the breakdown is initiated at the focal point and proceeds rapidly in both directions. One suggested explanation is that the laser rod, lens combination acts as a dynamic lens system during the laser pulse, thus moving the focal point along the optic axis [92]. In addition, the air breakdown points appear apparently at random along the track, uninfluenced by the presence of smoke, for example. One explanation put forth for this randomness involves the calculation of the electric field maxima near the focal region of a lens transmitting a spatially coherent beam [96]. The data also suggest a self-focusing [97], [98] explanation alone could be involved. The electrostrictive self-focusing threshold for N_2 and O_2 has been calculated to be about 10^7 watts, and the second-order Kerr effect self-focusing threshold in these gases to be about 10^9 watts [99], the area canceling out in the calculations. Probably a combination of these explanations is required.

In other work, a laser of very uniform output energy distribution has been reported [225]. A number of other approaches have been reported for obtaining a high-power

diffraction-limited beam from a glass laser. In one system, the low-energy, diffraction-limited output from a mode selected oscillator is passed many times through a 10 cm ×10 cm×1 cm rectangular block of pumped laser glass [200], [201]. The block is face pumped by rectangular spiral flashtubes, and the 1.2 mm by 10 mm beam is made to pass through adjacent slices of the block by an elaborate optical system. In this way, unused laser material is traversed on successive passes and pulse sharpening as well as instabilities common to long amplifier chains might be avoided. A peak power of 100 MW has been reported from this system.

A second system employs large (12.5 cm) glass disks set at Brewster's angle in air, with alternate disks nearly touching each other on opposite edges [119]. The disks are essentially face pumped by two banks of flashtubes, one on either side of the array. Although the gain of this system is quite low, multiple passes through a bank of 5 disks builds the beam intensity to usable proportions.

One other high-radiance glass disk laser scheme has been reported. Here, the flashlamps are very close to the flat surface of the large-diameter disks, providing face pumping [120]. The laser beam comes into the other face of the disk at an angle, refracts into the disk and is reflected from the back of the disk. This arrangement requires a mirror on the pumped side of the disk which will transmit the pump wavelengths, but reflect the laser light.

Both of these disk systems have the advantage of providing uniform heating of the disk due to pumping, of providing a large aperture for the laser light, and of providing a relatively short heat removal path which is nearly along the beam direction. It should therefore be possible to achieve a very low beam divergence and a quite high energy output. The low pumping efficiency of these approaches, however, makes them attractive only in special applications.

B. Optical Second Harmonic Generation

Using the laser described in Section IV-A, experiments were performed to obtain efficient second harmonic generation (SHG) [100]. Using a 2.5 cm cube of DKP, an energy conversion efficiency of polarized 1.06 μ light to 0.53 μ light of 51 percent was measured. The fundamental beam was unfocused. Factors contributing to this conversion efficiency were the narrow beamspread of the laser, the high power density in the crystal, and the use of a crystal holder which altered the angular orientation of the crystal to correct for small changes in the ambient temperature. An output of 15 joules in 15 ns was measured at 0.53 μ at a radiance of about 5×10^{16} watts/cm²·sr. In earlier work, the generation of 7 joules has been reported at 5300 Å [230].

A nonlinear solarization effect was found in the flint glass lens used to focus this high-intensity. 5300 Å beam [101]. For the glasses tested, the absorption coefficient increased quadratically with the light intensity, and the glasses turned brown (solarized) along paths of maximum field strength within the sample, thus delineating the beam profile. It is thought that a double-photon absorption process is operative rather than two consecutive single-photon absorptions.

C. Integral, Compact Glass Lasers

A solid glass laser module may be constructed by incorporating a 2- or 3-mm-diam. laser glass rod and a similar-sized flashtube in a single piece of cladding glass [102], [202]. This is done by drilling two closely-spaced, parallel holes in a piece of samarium-doped cladding glass. In one hole is fused a piece of neodymium-doped laser glass; in the other, end electrodes are fixed and the tube filled with xenon gas. The outer portion of the assembly is ground and polished to produce a minimum volume, and the surface silver plated to provide the flashtube-coupling mirror. The silver is copper plated to provide a heat sink. A variation of this assembly is to use a pyrex-walled flashtube which fits snugly into a half cylindric recess in a partial unit of the type described above. The advantage of doing this is that the flashtube can be sealed more reliably when it is a separate unit, and the pyrex walls are more resistant to deterioration from the discharge. Some units have exhibited a 10 joule threshold and a 5 percent slope efficiency. Other units, when optimized for long life, have yielded at least 1 joule for 10^5 shots.

In these units heat is extracted via an all-solid conductive path. This can lead to beam bending, which is about half due to a physical deformation and about half due to an index gradient, and loss of efficiency. Two techniques which have been used to overcome this effect are to either use two flashtubes rather than one to provide thermal symmetry, or a partial cladding on the side of the laser rod core nearest the flashtube. This partial cladding has an index of refraction about one part in a thousand lower than that of the laser glass, while the surrounding matrix glass has an index of refraction about one part in a thousand higher. On-axis light which is bent toward the flashtube is therefore totally reflected back into the core, preventing its loss (Fig. 14). So far, an improvement has been seen by this technique, but further work is needed.

D. Fiber Lasers

A few of the properties of fiber lasers that are peculiar to this geometry will be described. In one configuration, a 1.24 meter fiber amplifier with a core diameter of 10 μ and a cladding diameter of 1.5 mm was wrapped into a coil about a linear flashtube inside a reflecting cylinder. The core index of refraction was 1.54 and the cladding was 1.52. The large cladding thickness was used for mechanical strength [103]. In another configuration, with a linear flashtube imaged onto a fiber amplifier, a gain of 3 dB/cm was measured [104], representing an inversion of 9 J/cm³.

A solid laser rod may be made of a fused bundle of individual fiber lasers. If care is taken to prevent cross talk by introducing emission-absorbing fibers in the bundle, the fibers may independently oscillate. Such a rod, consisting of 10 μ fibers, has given a time trace which is smooth, since it consists of the random superposition of the randomly-spiking output of many individual fibers [6]. The beamspread, set by the individual fiber core-cladding index of refraction difference, was about 2° full angle at half intensity.

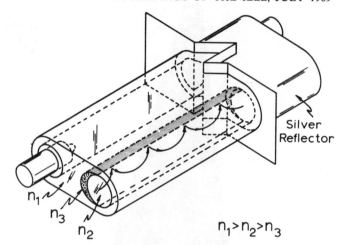

Fig. 14. One design of a compact, integral glass laser unit. Laser light which is bent off-axis by thermal-optical effects is totally reflected back into the active laser portion of the device.

E. Detection with a Fiber Laser Preamplifier at 1.06 μ

In many applications, such as laser ranging or illumination, the problem of detecting weak laser signals arises. A substantial improvement over straight detection is possible through use of a laser preamplifier (Fig. 15). Using a single-mode fiber laser [105] preamplifier and a silicon detector, a minimum detectable signal (MDS) of 4×10^3 photons has been measured for a laser pulse of $\tau = 40$ ns, $\Delta\lambda = 5$ nm and a power signal-to-noise ratio (SNR) of 64 [106], [235]. This agrees with the theoretical limiting noise due to fluctuations in spontaneous emission for an optical amplifier with gain G [107]. The number of photons generated per second as spontaneous emission in bandwidth $\Delta\nu$ is $2\Delta\nu N'(G-1)$ where N' is the number of propagating modes in the laser preamplifier. The fiber laser preamplifier had a core diameter and core-cladding index combination chosen so that only the lowest order HE_{11} mode could propagate. For a large enough number of photons so that Gaussian statistics may be used, and sufficiently high gain so that the dominant noise is that due to fluctuations in the spontaneous emission of the fiber preamplifier,

$$MDS = \sqrt{\frac{4c\Delta\lambda N'\tau(SNR)}{\lambda^2}}. \qquad (17)$$

Practically, it should be possible to reduce $\Delta\lambda$ to about 1 Å by the use of filters on the amplifier, and τ to about 3 ns which is about the practical limit of direct detection. The MDS per diffraction-limited resolution cell would then be about 100 photons for a SNR of 64. This represents an approximate 10^2 improvement over the use of a silicon diode alone, assuming it is thermal noise limited. Table IV summarizes the comparison between the performance of various detectors and laser emission wavelengths.

F. Long Coherence Length Laser System

Glass lasers have been built which provide high intensity in the lowest order spatial mode, but the temporal coherence is degraded by the large optical bandwidth. By combining the high temporal coherence of a gas laser with the high

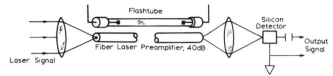

Fig. 15. Schematic of the fiber laser detector preamplifier arrangement. A reduction in the minimum detectable signal level of about 10^2 is possible, compared to a silicon detector.

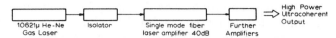

Fig. 16. Schematic of the setup used to amplify the output of a gas laser by a glass laser amplifier chain.

TABLE IV

SUMMARY OF DETECTOR PERFORMANCE FOR A $\tau = 10^{-8}$s PULSE*

Photomultipliers (24 electrons required)		MDS (Photons/Pulse)
λ	Surface	
0.69 μm	S-20	1 200
1.06	S-1	96 000
0.55	S-17	120
Semiconductor Diodes (quantum eff. = 50 percent)		
Conventional (nonmultiplying)		106 000
Avalanche (multiplying)		7 400
Fiber Laser Preamplifier		
$\Delta\lambda = 50$ Å (1 mode, $\tau = 10^{-8}$ second)		2 400
$\Delta\lambda = 1$ Å		400

* In each case, the same probability for false alarm, 10^{-3}, probability of detection of 99.99 percent, and a gating interval of $10^2\ \tau$ were used [106].

power capabilities of a glass laser, a laser system having a measured coherence length in excess of 12 meters and an output of 3.5 watts in one ms has been built [108], [192]. The gas laser used the 1.0621 μ $2s_2$ to $2p_7$ line of neon in a He-Ne mixture [109], [110]. A glass composition was used which had a fluorescence maximum at 1.062 microns. Also, since the equivalent noise input to the glass laser amplifier due to spontaneous emission is $2h\nu\Delta\nu N$ watts, where N is the number of waveguide modes [105] and $\Delta\nu$ is the linewidth, a single-mode fiber was used as the first stage of amplification (see Fig. 16). The spontaneous noise under these conditions is about 1.2 μ watts in a 100 Å bandwidth. The output of the gas laser was greater than 200 μ watts; with the insertion losses of the coupling optics and the optical isolator, 60 μ watts were incident on the fiber, providing an acceptable signal-to-noise ratio. The HE_{11} single-mode fiber had a 15 μ diameter Nd^{3+} glass laser core with an undoped outer cladding of slightly lower index of refraction. Surrounding this was a samarium-doped cladding to absorb scattered off-axis laser light and spontaneous emission out the sides. The fiber was 45 cm long and yielded an optical gain of 40 dB. A second stage of amplification provided an additional gain of 10 dB. Feedback to the fiber by the exit reflector of the gas laser was prevented by the use of an optical isolator consisting of a quarter-wave plate and polarizer. The coherence length was measured by use of a two-beam Michelson interferometer. Photographs were taken of the resultant fringes through an image converter. Good fringes were obtained using several path-length differences up through 12.2 meters.

G. High Average Power Operation—Small Systems

Under repetitive operation, a laser rod will exhibit a temperature difference between the center and the surface.

This is because heat is continually being deposited throughout the volume of the rod but can only be removed from the surface. The temperature difference depends on the thickness of the rod and the thermal conductivity of the material. Since the thermal conductivity of most glasses is essentially the same, and ten to one hundred times less than that of crystalline laser materials, a large radial temperature gradient is seen in glass laser rods when operated at high average power. This can lead to an induced lens power and ultimate fracture of the rod. For most laser glasses, the induced lens power inhibits laser action well before reaching the point where rod fracture might be a problem, although a very small amount of induced positive lens power can be used to improve the laser performance [228]. The induced lens power is usually positive, causing a focusing of the laser energy on the end mirrors or within the rod, with a consequent decrease in damage threshold of components, loss in efficiency and increase in beamspread. Of much greater importance is the fact that even a small amount of thermally-induced positive lens power will greatly reduce the threshold for self-focusing type [97] damage. Typically, a factor of ten reduction in the self-focusing damage threshold is seen with only a moderate amount of thermally induced positive lens power present. The best results reported to date are 400 mJ output in 50 ns pulses at 10 pulses per second for a 6.4 mm by 23 cm glass laser rod [116], but very rapid end-mirror deterioration occurred due to the induced positive lens power in the rod. There are three ways in which these effects may be minimized or eliminated in a laser rod: the glass may be altered to reduce the index difference resulting from a given temperature difference, the rod may have an internal structure designed to minimize the effect of the thermal lensing, or the rod may be made thin to reduce the temperature gradient. Each of these approaches will be discussed below.

There are four factors which may cause an index gradient in a laser rod. First, the inverted population produces an index change because the energy separation of the strong absorption bands in the UV, which are responsible for an increase in index or refraction, is less when the ions are in the excited $^4F_{3/2}$ state. In alkali-alkaline earth silicates the index of refraction increases by 2.1×10^{-3} for 1 weight-percent Nd_2O_3, and up to 5 percent of the ions can be raised to the $^4F_{3/2}$ state. The amount by which the index of refraction increases due to excited-state population depends upon the position of the strong UV absorption bands, and is of the order of 10^{-5} for laser glass [111]. This effect can be negated by proper choice of doping-diameter product, which should be about 50 weight-percent-mm for barium crown silicate for the most uniform inversion across the diameter of the rod.

The remaining three effects on the index of refraction are thermal: the change in index with temperature at constant density is positive because the strong UV absorption edge moves toward longer wavelengths with an increase in temperature. For the usual laser glasses the change in index with pressure at constant temperature is positive, and the change of index with expansion is negative because the ions are moving further apart. Since the center of a rod is generally higher in temperature than the surface, and therefore in compression, the first two of these three effects may be balanced by the third. The expressions which describe the change in path lengths for radially, $P_\theta(r)$ and tangentially, $P_\theta(r)$ polarized light between the center and a point r from the center of a cylindric rod are [7], [8]

$$\Delta P_r(r) = nLT \left\{ \alpha_n - \frac{\alpha}{1-s} \left[\frac{R}{T}(1+s)\left(\frac{p}{v} - \frac{q}{v}\right) \right. \right.$$
$$\left. \left. - 2(1-s)\frac{p}{v} + 2s\frac{q}{v} \right] \right\} \tag{18}$$

$$\Delta P_0(r) = nLT \left\{ \alpha_n - \frac{\alpha}{1-s} - \frac{R}{T}(1+s)\left(\frac{p}{v} - \frac{q}{v}\right) \right.$$
$$\left. - (1-3s)\frac{p}{v} - (1-s)\frac{q}{v} \right] \right\}. \tag{19}$$

In these equations, n is the index of refraction, L the rod length, T the temperature difference between the center and points at a distance r from the center, α_n is the thermal coefficient of the index, s is Poisson's ratio, R is defined by

$$R = \frac{1}{r^2} \int_0^r Tr \, dr \tag{20}$$

and p/v and q/v are the strain optic coefficients which relate the change in index of refraction to the strains in directions parallel and perpendicular, respectively, to the plane of polarization of light. An average may be taken of the Δp's

$$\frac{\Delta P_r(r) + \Delta P_\theta(r)}{2}$$
$$= nLT \left\{ \alpha_n - \frac{\alpha}{1-s} \left[-\frac{3-5s}{2}\frac{p}{v} - \frac{1-3s}{2}\frac{q}{v} \right] \right\}. \tag{21}$$

For the usual laser glasses, both ΔP_r and ΔP_θ are greater than zero, producing a positive lens effect in the laser rod under repetitive operation. This can be countered by inserting lens elements in the cavity or figuring the ends of the rod. However, the approach works for only one set of conditions, and may introduce increased losses in the cavity.

In principle, it should be possible to make both ΔP_r and ΔP_θ equal to zero if p/v can be made equal to q/v. Such a glass, a Pockels glass [113], is not a particularly good laser glass. In addition, it must be operated at reduced temperatures for ΔP_r and ΔP_θ to equal zero at 1.06 μ

A second approach is to average the two ΔP's as shown in (21) by insertion of an eighth-wave Faraday rotator inside the cavity. In this way, radially polarized light is switched to tangentially polarized light on alternate passes

and vice-versa [8]. Again, extra elements must be inserted in the cavity. Also, in Q-switched operation where the intensity increases exponentially on each pass, such a self-canceling scheme is much less effective.

A third approach is to make $\Delta P = 0$, or athermalize the glass for only one plane of polarization. This either requires the introduction of mode-selective elements inside the cavity [8], or athermalizing exactly for one plane of polarization and over athermalizing, so that the thermally-induced lens power is negative, for the other plane of polarization. In this way, light in the negative lens plane is inhibited from growth, while light in the plane of athermalization sees no induced lens power. This latter scheme seems most promising at the moment, but represents a rather delicate balance of the quantities in (18) and (19).

Putting typical values for silicate laser glasses into (18) and (19), the value of the linear expansion coefficient required to athermalize the glass is about $120 \times 10^{-7}/°C$. Silicate laser glasses of high expansion coefficient should therefore create less difficulty under high average power operation. Experimentally, for a 6.4-mm by 23-cm glass laser rod of expansion $110 \times 10^{-7}/°C$, thermal lensing becomes serious at a level of operation about one third that which produces thermal fracture.

Turning now to the second general approach to the problem of thermal lensing, a number of rod structures are possible. For example, a permanent negative lens power may be put into the rod, either over its full length or over only the part which is pumped [114]. In one rod, -8 diopters was put into the 6.3-mm-diam. by 40-cm-long rod. A thermally-induced $+8$ diopters was then introduced, and the beam-spread and output energy approached the single-shot values. A number of shortcomings to this approach are evident: it will work for only one set of operating conditions, the pumping must be quite uniform to avoid astigmatism, and the rod cannot easily be aligned with its end mirrors while cold.

A second possibility is to fabricate the cylindric laser rod of a number of fused, concentric annuli (Fig. 17). If the innermore annuli are of slightly lower index of refraction, then light initially parallel to the axis which is bent due to the thermal lensing toward the center of the rod will be totally reflected at the interface between annuli [115]. The light will therefore be contained by the combined action of thermal lensing and reflection to the various annular regions. The central core can be made lossy to eliminate off-axis light, and the whole rod clad with an absorbing, slightly lower index cladding to further eliminate large-angle off-axis light. This configuration has the advantage of operating independent of the repetition rate, and of allowing alignment of the end mirrors while cold.

The third general approach is to make the pieces thin to reduce the temperature gradient and hence the lens power in the piece. This may be done by using a single thin rod [229] or by sectioning the usually cylindric laser rod longitudinally into a bundle or fagot [114] of rods, or by sectioning the rod transversely into a stack of disks, as described below. Alternatively, a rectangular cross section or slab laser may be used [116], [117]. The slab laser has the ad-

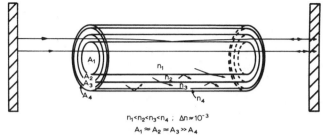

Fig. 17. Schematic diagram of a composite laser rod. Light rays which would be deviated toward the center of the rod through thermal lensing are totally reflected, thus maintaining greater uniformity of lasing across the diameter of the rod under repetitive operating conditions.

vantage that it may be possible to switch it efficiently with a single plane of polarization Q-switch, rather than dual plane switches required in many cases by circular-cross-section rods when high thermal stresses are present [118]. Slab lasers have been operated up to 100 pulses per second at good efficiency [117].

H. High Average Power Operation-Disk Lasers

When the rod diameter in a high average power system exceeds about a centimeter, a disk laser configuration is usually called for [203], [204]. There are a number of advantages to a disk laser. The temperature and index of refraction gradients can be made predominantly axial, rather than radial, so that little thermal lensing occurs. The pieces can be thin enough so that self focusing is effectively circumvented. A large laser aperture is possible. If there should be damage to a disk it can be replaced. The disks could be figured to cancel any small residual lens power left in the system. There need be no damage-prone rod edges in the laser beam, especially when using clad disks.

There are two basic approaches to a disk laser: disks at Brewster's angle for the glass-fluid combination used, and disks at normal incidence with either an index-matching fluid or antireflection coated disks for the glass-fluid combination used. All these approaches have been pursued, and at present slope efficiencies of over 2 percent are regularly measured, using disks of 18 mm to 25 mm in diameter and from 3 mm to 10 mm in thickness. Although high repetition rate work is just beginning with these units, preliminary results are encouraging. Measurements indicate that diffraction-limited beams may be passed through an active disk laser without distortion. Figs. 18 and 19 show two disk laser designs.

V. LASER-INDUCED BREAKDOWN IN GLASS

Maximum glass laser performance is generally obtained by using the smallest diameter laser rod, limited by the ability of the glass to survive the passage of the laser beam. For long-pulse operation with a pulse length of about 1 ms present laser glasses withstand over 1000 J/cm² without fracture. In the Q-switched time domain, however, passage of several tens of J/cm² may lead to fracture. Most of the work on elucidating the damage mechanisms has therefore been done in the shorter pulse regime. There appear to be three types of damage to consider: microinhomogeneities

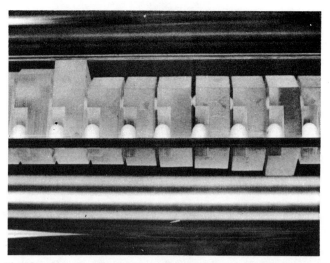

Fig. 18. A section of a disk laser. The 1 cm thick glass disks have a 18 mm diameter active core fused into a 24 mm square cladding. Fluid flow is between the disks and around two opposite corners. The disks are held apart by teflon balls and the assembly is contained in a precision bore square pyrex tube. Four water-cooled flashtubes in a close-wrap configuration provide pumping for the 45 cm array.

Fig. 19. A disk laser having the 3 mm thick elements at Brewster's angle. The 25 mm projected-diameter active laser glass is fused into a rectangular cladding piece which is supported by metal fingers at it corners. Half of the close-wrap coupling reflector is removed in this view to show the internal construction.

within the glass which serve as absorption centers causing damage, self-focusing type of damage, and surface damage. Each of these will be discussed in the sections which follow.

A. Damage Caused by Microinhomogeneities

High optical quality glass has traditionally been manufactured in platinum containers. Early laser glasses were likewise produced in platinum crucibles. It soon became evident, however, that microscopic particles of platinum appear in all glass made in a platinum environment, and that these particles serve as damage sites to the passage of intense laser beams [122]–[129]. This is because the particle is highly absorbing to the laser beam, is heated and expands, and produces a discoid fracture in the glass. Subsequently, it was established that a large part of the platinum contamination of the glass arose through condensation of platinum oxide from the atmosphere over the glass, and that this could be greatly reduced by the introduction of dry nitrogen gas instead of air over the molten glass. This led to improved thresholds for damage, but incomplete elimination of the platinum-contamination problem. For this reason, the best laser glasses are now manufactured in a

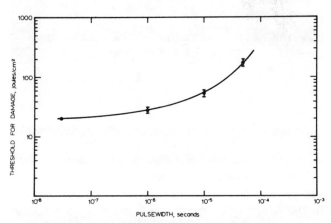

Fig. 20. Damage threshold for SF-4 glass. Although invisible to microscopic examination, it is presumed that submicroscopic platinum particles are dispersed through the glass and give rise to the fracture.

platinum free environment in ceramic crucibles. This makes the production of high optical quality glass more difficult because of the possibility of index of refraction gradients and increased absorption at 1.06 μ in the glass caused by partial dissolution of the crucible.

Experimentally, platinum particles which are too small to see under a high-power microscope under intense side illumination (and therefore of less than 1 micron dimensions) and yet are inferred to be present because the particular glass is made in a platinum crucible, can cause damage typical of that created by larger, visible platinum particles. In one series of experiments, a piece of glass containing such invisible platinum particles was measured for damage threshold with single laser pulses of 3×10^{-8}, 10^{-6}, 10^{-5}, 10^{-4} and 10^{-3} seconds duration [130]. The results are given in Fig. 20.

Because of the very small size of particle which can cause damage, even glass made in an all-ceramic environment can exhibit inclusion-type damage due to isolated damage sites. One objective in good laser glass design therefore is to make the glass as forgiving as possible to such an intrusion of foreign matter [83].

Some laser glasses are now pretested to ensure freedom from microinhomogeneities. This is a difficult procedure because damage can occur due to self focusing or surface pluming if the sample is tested with a Q-switched laser. One technique employed here is to use a laser emitting a spike-free pulse of about 10 μs duration for the damage tests. The advantages of this are that no self focusing occurs (see Section V-B); the threshold for surface damage is about 10 times higher than in the Q-switched time domain; the 10 μs pulse, because it is generated through amplified spontaneous emission, is very uniform throughout the sample volume (see Section III-C); the threshold for internal damage due to the inclusion is only about twice that in the Q-switched time domain, Fig. 20; and the size of the resulting damage, for a given small inclusion size, appears to be maximum for a 10 μs pulse width [130]. Samples are therefore tested at equivalent Q-switched energy densities of about 100 J/cm². An additional effect related to microinhomogeneities is sometimes seen. This is a fatigue phe-

nomenon whereby a piece of glass will survive tens of laser shots without damage, but fail catastrophically on the next shot with inclusion-type damage. It is possible that incompletely dissolved foreign matter can grow due to the heat generated on successive shots, until the aggregate is large enough to create a fracture center, or that progressive localized devitrification can occur.

There have been additional suggestions of damage mechanisms in laser glass. In one, it was suggested that the pump light focused at the center of a laser rod in an imaging ellipse can create the conditions necessary for devitrification [131]. Another proposal is that excited-state absorption from the $^4F_{3/2}$ level of the neodymium ion is large enough to cause a significant conversion of laser energy to heat causing damage. This mechanism would make the damage threshold dependent on the neodymium concentration [132].

B. Self Focusing in Laser Glass

If a portion of laser glass which is free of microinhomogeneities is irradiated by a 30 ns pulse through a short focal length lens, it will have a damage threshold of about 500 J/cm² [21], [64] (see Fig. 21). The damaged area will be in the form of a crushed volume of several mm dimensions. If, however, the sample is irradiated by a parallel or only slightly focused beam, the threshold for damage is an order of magnitude lower, and the form of damage is a thin (about 1 μ diameter) fossil track in the glass (see Fig. 22). A number of suggestions have been put forth as to the cause of this self-focusing behavior: electrostrictive effects, thermal effects, and dynamic focusing effects. A separate mechanism must be invoked to account for the actual absorption and fracture of the glass once self focusing has occurred, since the appearance of the fracture frequently suggests thermal shock [6], and measurements of the incandescent emission during fracture indicate the glass is heated to about 1000 K [133].

Index of refraction variations have been observed in glass just prior to fracture [134], but time resolved examination of the complete process indicates that the fracture occurs after passage of the laser beam [135]. It is therefore possible that the plasma formation due to multiphoton absorption at the output face of the rod is instrumental in the creation of the self focusing [135]. A definite focusing length is seen before the track is formed. Creation of a fossil track in the glass usually is accompanied by chipping of the output face, even when it is several centimeters distant, and the track appears always to be accompanied by a plume at the output face. A great body of literature applicable to glass has appeared on self focusing [97]–[99], [136]–[147]. A good review paper on the general subject of self focusing is found in [148]. In many cases a group of tracks is created in the glass. No one track vignettes a significant portion of the beam for subsequent shots, but the aggregate set of tracks and the concomitant damage to the output face, ultimately reduces the laser performance. There is also considerable evidence that the self focusing observed in glass is connected with stimulated Brillouin scattering [97], [149]–[153].

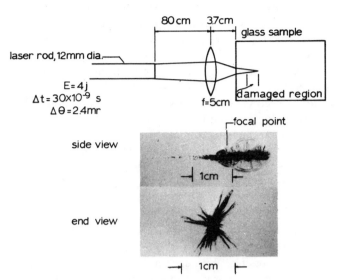

Fig. 21. Arrangement used to measure durability of laser glass without complication of self focusing. A typical damaged area is shown.

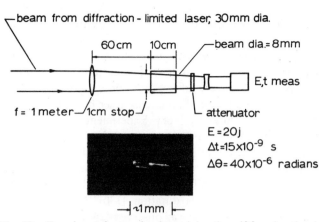

Fig. 22. Experimental setup for determining the self-focusing threshold of damage in a variety of laser glasses. For a given beam, all laser glasses appear to have about the same threshold for breakdown track formation.

TABLE V

DATA ON INTERNAL DAMAGE THRESHOLD FOR VARIOUS INCLUSIONS IN LASER GLASS, AND FOR INCLUSION-FREE LASER GLASS

Type of Inclusion, if any	Damage Threshold, J/cm² 30 ns pulse
Platinum particles (~1 micron diam.)	~10
Bubble (~1.5 mm diam.)	>500
Dielectric Inclusion (~1 micron diam.)	>50
Homogeneous glass (due to self focusing)	50 to 100, depending on laser beam profile
Homogeneous glass (no self focusing)	~500

One other aspect of this type of damage to laser glass is of interest: the threshold for appearance of the fossil track versus the pulse duration [155], [156]. It can be argued that a minimum threshold for this damage will occur in some characteristic time, because for shorter times the electrostrictive effect, since it would be expected to propagate at approximately the velocity of sound, cannot propagate over as much of the rod cross section [154]. The threshold for electrostrictive self focusing is independent of the power density [97], but does depend on the power. Experimentally, no self-focusing damage has been seen at several hundred J/cm² from a 1 μs pulse [130]. This is not to say that no self focusing occurred, since self focusing might occur without a damage mechanism to create a fossil record, and the 1 μs pulse had a very uniform energy distribution. The threshold for self-focusing damage for a 30 ns pulse appears to be about 50 to 100 J/cm² [89], [63], [236] depending very much on the beam uniformity.

Self focusing might also be explained due to a linear absorption by the glass [158]–[161]. The first portion of the laser pulse heats the glass and creates a tiny dielectric waveguide which focuses the succeeding part of the pulse. Even if the dn/dT is sufficiently negative to prevent a steady state focusing, the short time duration of the Q-switched laser pulse precludes expansion of the glass in the time of the pulse. The result is that $(\partial n/\partial T))_\rho$ should be used, and this quantity is always positive [6]. In one set of experiments on glass samples having an impurity absorption ranging from 3×10^{-4} cm^{-1} to 9×10^{-3} cm^{-1}, however, no difference in threshold for self-focusing damage threshold could be seen [64].

One other explanation can be invoked for the observed damage tracks: a dynamic lens effect in the medium, coupled with on-axis coherent addition of the light to produce a track of damage centers in a straight line [92], [147]. There is some evidence, however, that the damage tracks appear most frequently at regions of greatest optical electric field gradient, which would favor the electrostrictive self-focus-

ing explanation. In addition, chipping of the output face in line with the damage track, even when the fossil record does not appear over all the intervening distance, also favors the electrostrictive self-focusing explanation. It appears that if the beam uniformity is great enough, the self-focusing length can be made longer than desirable rod lengths. In this way, damage due to self focusing may be avoided.

Table V summarizes the data on internal damage to laser glass.

C. Surface Damage to Laser Glass

Experimentally, in the Q-switched time domain, the threshold for damage to the output face of an untreated glass laser rod is always lower by a factor of 2 to 5 than the threshold for internal damage [64], [135]. Neither immersion of the rod end in water nor optically contacting an undoped section of glass to the rod end improves the situation [64], [162]. However, treating the glass either by a hydrofluoric acid etch [163] or by washing it in dimethyldichlorosilane [164], [165] has produced a considerable increase in the resistance to damage, but the effect lasts for only a few minutes. In the case of the acid etch, the belief is that small surface irregularities and cracks are smoothed over, thus making the surface more resistant to damage. In the second case, the belief is that OH groups are substituted by CH₃ groups, thus providing stronger binding. In other work, the successive steps in the surface breakdown are suggested to be optical absorption, fluorescence, chemical reaction by quenching the fluorescence, and breakdown [169].

The coupling of the laser energy to the origin of fracture appears to be due to multiphoton excitation. A number of experiments have confirmed the production of charged particles upon passage of an intense laser beam through the end of a laser rod [166]–[168]. Experiments on creation of the surface plasma using pulses of 3×10^{-8}, 10^{-6}, 10^{-5} and 10^{-4} seconds have shown that the threshold for plasma formation depends very strongly on the pulsewidth, being at least ten times that for a 3×10^{-8} second pulsewidth for a pulse of 10^{-5} seconds duration [130]. In similar work with ruby, a ratio of 10^2 was observed [243].

The energy threshold for damage of unspecified type for a 5 ns pulse was reported to be a factor of two lower than that for a 30 ns pulse [63]. The threshold for damage for a 1 ns pulse appears to be only slightly lower than that for a 5 ns pulse [63] and that for a 10 ps pulse is evidently higher [85], [87]. It is expected theoretically that self trapping can occur in glasses for picosecond pulses [157].

In summary, much remains to be done in order to fully explain all aspects of damage to laser glass. There has been a considerable amount of theoretical work done, and experimental work over a wide enough range of parameters is now providing some clues to aid the choice between competing explanations. It appears likely that techniques will be evolved in the near future to repeatedly allow passage of about 100 J/cm² in a pulse of about 10^{-8} second, but that much greater resistance to damage will not be seen for some time.

VI. Applications for Glass Lasers

The glass laser was invented about eight years ago. Most new technologies take approximately this length of time to emerge from a laboratory demonstration to large scale use [170]. In this section speculations are made on what some of these applications might be.

In medicine, the fact that neodymium-doped glass fibers can be made quite flexible should find applications in situations where the laser light must be put into inaccessible places [171]. Also, the fact that the fourth harmonic of 1.06 μ at 2650 Å is highly absorbed by living tissue should find some medical applications.

The existence of efficient glass disk lasers should make their use in laser welding and cutting applications attractive. By using erbium-doped glass the requirement of laser eye protection on the production line might be eliminated. In microwelding, a number of schemes have been demonstrated for creating many welds simultaneously with a single laser shot [173].

Applications for high radiance lasers are many-fold. They are being used for plasma creation, and in particular for experiments in controlled thermonuclear reactions [175]–[177], [237]. The creation of neutrons by a focused laser beam has already been reported [174]. Calculations indicate that approximately 10^5 joules in about 2 ns incident on a solid deuterium particle would produce a half-percent

energy conversion to energetic neutrons [244]. This requires the use of known laser and optical isolator technology to build large laser amplifier arrays.

Plasmas can also be used to generate intense, short duration X-ray fields [178]. The long path air breakdown described in Section IV-A might be used to form a low resistance electrical path for discharging clouds in areas where it is desirable to avoid lightning flashes.

One interesting use for a diffraction-limited glass laser with frequency doubled output at 5300 Å is in lunar ranging [180]–[182]. Useful data can be obtained with 20 joules output from a 15 cm transmitting telescope and a 150 cm receiving telescope. Using a disk laser, a pulse repetition rate of 1 pulse per second or faster should be practical, and if a pulse width of 1 ns is used, which is about the limit of direct electrical detection, a range resolution of a few centimeters should be possible. With such a device, or several such devices ranging on corner cubes of the type placed on the moon by Apollo 11, precise information on lunar libration, librations of the earth, continental drift, earth swells, lunar swells, and experiments in relativity [184] can be obtained.

Such a laser can also be used for precise satellite ranging and illumination for observation with a vidicon tube. Such information is not only useful for geodesy [183], [185] but for close observation of the various earth satellites. By using the gas laser—glass laser system described in Section IV-F, it should be possible to defeat the effect of the earth's turbulent atmosphere by taking holograms of satellites by an earth-bound laser source [186]. This occurs because both the reference and the signal beams experience the same atmospheric distortions, so that if the coherence length of the pulse is at least as great as the radial size of the satellite, and the pulse is short enough to sufficiently freeze the satellite motion, a higher resolution photograph should be obtainable from the reconstructed hologram than direct photography could provide. The use of an array of fiber laser preamplifiers could allow real-time holograms to be taken at 1.06 microns with somewhat better overall system efficiency than with a vidicon tube and a 5300 Å transmitter.

In addition, detailed meteorological data can be obtained by using a high-power narrow-beamspread laser on the atmosphere [187]–[191].

Finally, since sea water is most transparent at about 5300 Å [179], a high-power diffraction-limited 5300 Å glass laser may find application there. Using range gating techniques [193], the maximum range of observation should be increased. Accurate ranging and the creation of useful holograms should also be possible, as with satellites.

In the picosecond domain interesting fundamental work can be done because the energy deposition is too fast to allow target-material transport and diffusion during the laser pulse. Stop-action photography can also be done in very short exposure times. Moreover, as technology develops, it should be possible to utilize these short pulses for accurate laser ranging, but this possibility awaits the development of an appropriate detector. Finally, picosecond pulses could ultimately be used in pulse code modulation schemes for computer and communication applications with data rates in the range of 10^{12} bits/s.

ACKNOWLEDGMENT

The author gratefully acknowledges helpful discussions n various aspects covered in this paper, particularly with . Snitzer, J. Kantorski, W. Hagen, J. Segre, P. Magnante,). LaMarre, E. Deeg, and R. Woodcock.

REFERENCES

[1] A. L. Schawlow and C. H. Townes, "Infrared and optical masers," *Phys. Rev.*, vol. 112, pp. 1940–1949, December 1958.

[2] T. H. Maiman, "Stimulated optical radiation in ruby," *Nature*, vol. 187, pp. 493–494, August 1960.

[3] P. P. Sorokin and M. J. Stevenson, "Stimulated infrared emission from trivalent uranium," *Phys. Rev. Lett.*, vol. 5, pp. 557–559, December 1960.

[4] L. F. Johnson and K. Nassau, "Infrared fluorescence and stimulated emission of Nd^{3+} in $CaWO_4$." *Proc. IRE* (Correspondence), vol. 49, pp. 1704–1706, November 1961.

[5] E. Snitzer, "Optical maser action of Nd^{3+} in a barium crown glass," *Phys. Rev. Lett.*, vol. 7, pp. 444–446. December 1961.

[6] E. Snitzer and C. G. Young, "Glass Lasers," in *Advances in Lasers*, vol. 2, A. Levine, Ed. New York: Dekker, 1968, pp. 191–256.

[7] F. W. Quelle, "Thermal distortion of diffraction-limited optical elements," *Appl. Opt.*, vol. 5, pp. 633–637, April 1966.

[8] E. Snitzer, "Glass Lasers," *Appl. Opt.*, vol. 5, pp. 1487–1499, October 1966.

[9] A. J. DeMaria, W. H. Glenn, M. J. Brienza, and M. E. Mack, "Picosecond laser pulses," *Proc. IEEE*, vol. 57, pp. 2–25, January 1969.

[10] P. Mauer, "Amplification coefficient of neodymium-doped glass at 1.06 microns," *Appl. Opt.*, vol. 3, pp. 433–434, March 1964.

[11] C. G. Young and J. W. Kantorski, "Saturation operation and gain coefficient of a neodymium-glass amplifier," *Appl. Opt.*, vol. 4, pp. 1675–1677, December 1965.

[12] J. Pantoflicek, "Calculation of the amplification coefficient of stimulated emission from fluorescence measurements," *Czech. J. Phys. B*, vol. 17, no. 1, pp. 27–33, 1967.

[13] J. K. Neeland and V. Evtuhov, "Measurement of the laser transition cross section for Nd^{3+} in yttrium aluminum garnet," *Phys. Rev.*, vol. 156, pp. 244–246, April 1967.

[14] D. E. Burlankov and A. V. Reznov, "Measurement of the amplification coefficient of an excited medium by the intensity of luminescence," *Opt. Spectry.*, vol. 20, pp. 262–263, March 1966.

[15] J. G. Edwards, "Measurement of the cross-section for stimulated emission in neodymium glass," *Nature*, vol. 212, pp. 752–753. November 1966.

[16] J.-M. Jego, "Measurement of the cross section for stimulated emission of neodymium at 1.06μ," *Compt. Rend. Acad. Sci. Paris*, vol. 264, B, pp. 1496–1498, May 1967.

[17] V. R. Belan, V. V. Grigor'yants, and M. E. Zhabotinskii, "Use of laser to measure the cross section of stimulated emission of matter," *ZhETF Pis'ma*, vol. 6, pp. 721–724, October 1967.

[18] J. G. Edwards, "Measurements of the cross section for stimulated emission in neodymium-doped glass from the output of a free-running laser oscillator," *Brit. J. Appl. Phys.*, vol. 1, pp. 449–456, April 1968.

[19] E. O. Schultz-DuBois, "Pulse sharpening and gain saturation in traveling-wave masers," *Bell Sys. Tech. J.*, vol. 43, pp. 625–658. March 1964.

[20] C. G. Young, J. W. Kantorski, C. F. Padula, and D. A. LaMarre, "Device for direct measurement of specific gain coefficient in laser glasses," to be published in *IEEE J. Quantum Electronics*.

[21] ——, "Report on glass lasers," *Microwaves*, vol. 7, pp. 69–78, July 1968.

[22] M. E. Vance, "Measured internal losses and output energies of neodymium glass lasers," *Appl. Opt.* vol. 6, pp. 775–776, April 1967.

[23] T. Bates, "Ligand-field theory and absorption spectra of the transition metal ions," in *Modern Aspects of the Vitreous State*, vol. 2, J. D. Mackensie, Ed. London: Butterworths, 1962, p. 195.

[24] A. Bishay and A. Kinawi, "Absorption spectra of iron in phosphate glasses and ligand-field theory," in *Physics of Non-Crystalline Solids*, J. A. Prins, Ed. Amsterdam: North Holland, 1965, p. 589.

[25] R. J. Landry, J. T. Fournier, and C. G. Young, "Electron spin resonance and optical absorption studies of Cr^{3+} in a phosphate glass," *J. Chem. Phys.*, vol. 46, pp. 1285–1290, February 1967.

[26] P. B. Mauer, "Laser action in neodymium-doped glass at 1.37 microns," *Appl. Opt.*, vol. 3, p. 153, January 1963.

[27] R. D. Maurer, "Operation of a Nd^{3+} glass optical maser at 9180 Å," *Appl. Opt.*, vol. 2, pp. 87–88, January 1963.

[28] T. Kushida, H. M. Marcos, and J. E. Geusic, "Laser transition cross section and fluorescence branching ratio for Nd^{3+} in yttrium aluminum garnet," *Phys. Rev.*, vol. 167, pp. 289–291, March 1968.

[29] H. W. Etzel, H. W. Gandy, and R. J. Ginther, "Stimulated emission of infrared radiation from ytterbium activated silicate glass," *Appl. Opt.*, vol. 1, pp. 534–536, July 1962.

[30] E. Snitzer, "Laser emission of Yb^{3+} at 1.06μ in Nd-, Yb-doped glass," *J. Opt. Soc. Am.*, vol. 55, p. 1574, November 1965.

[31] H. W. Gandy and R. J. Ginther, "Stimulated emission from holmium activated silicate glass," *Proc. IRE* (Correspondence), vol. 50, pp. 2113–2114, October 1962.

[32] E. Snitzer and R. Woodcock, "Yb^{3+}-Er^{3+} glass laser," *Appl. Phys. Lett.*, vol. 6, pp. 45–46, February 1965.

[33] H. W. Gandy, R. J. Ginther, and J. F. Weller, "Laser oscillations in erbium activated silicate glass," *Phys. Lett.*, vol. 16, pp. 266–267, June 1965.

[34] E. Snitzer, R. F. Woodcock, and J. Segre, "Phosphate glass Er^{3+} laser," *IEEE J. Quantum Electronics* (Digest of Technical Papers), vol. QE-4, p. 360, May 1968.

[35] H. W. Gandy, R. J. Ginther, and J. W. Weller, "Stimulated emission of TM^{3+} radiation in silicate glass," *J. Appl. Phys.*, vol. 38, pp. 3030–3031, June 1967.

[36] A. D. Pearson, S. P. S. Porto, and W. R. Northover, "Laser oscillations at 0.918, 1.057 and 1.401 microns in Nd^{3+}-doped borate glasses," *J. Appl. Phys.*, vol. 35, pp. 1704–1706, June 1964.

[37] L. F. Johnson, G. D. Boyd, and K. Nassau, "Optical maser characteristics of HO^{+3} in $CaWO_4$," *Proc. IRE*, vol. 50, pp. 87–88, January 1962.

[38] H. W. Gandy, R. J. Ginther, and J. F. Weller, "Radiationless resonance transfer from UO_2^{2+} to Nd^{3+} in coactivated barium crown glass," *Appl. Phys. Lett.*, vol. 4, pp. 188–190, June 1964.

[39] M. T. Malamed, C. Hirayama, and E. K. David, "Laser action in neodymium-doped glass produced through energy transfer," *Appl. Phys. Lett.*, vol. 7, pp. 170–172, September 1965.

[40] S. Shionoya and E. Nakazawa, "Sensitization of Nd^{3+} luminescence by Mn^{2+} and Ce^{3+} in glasses," *Appl. Phys. Lett.*, vol. 6, pp. 117–118, March 1965.

[41] S. Shionoya and E. Nakazawa, 1963 meeting, Phys. Soc. of Japan, Tokyo.

[42] H. W. Gandy, R. J. Ginther, and J. F. Weller, "Energy transfer in silicate glass coactivated with cerium and neodymium," *Phys. Lett.*, vol. 11, pp. 213–214, August 1964.

[43] A. Y. Cabezas and L. G. DeShazer, "Radiative transfer of energy between rare earth ions in glass," *Appl. Phys. Lett.*, vol. 4, pp. 37–39, January 1964.

[44] G. O. Karapetyan, V. P. Kovalyov, and S. G. Lunter, "Chromium sensitization of the neodymium luminescence in glass," *Opt. Spectry.*, vol. 19, pp. 529–531, December 1965.

[45] A. D. Pearson and S. P. S. Porto, "Nonradiative energy exchange and laser oscillation in Yb^{3+}-, Nd^{3+}-doped borate glass," *Appl. Phys. Lett.*, vol. 4, pp. 202–204, June 1964.

[46] H. W. Gandy, R. J. Ginther, and J. F. Weller, "Energy transfer in silicate glass coactivated with cerium and ytterbium," *Appl. Phys. Lett.*, vol. 5, pp. 220–222, December 1964.

[47] G. Dauge, "Nonradiative energy transfer in silicate glass," *IEEE J. Quantum Electronics* (Digest of Technical Papers), vol. QE-2, pp. lviii–lix, April 1966.

[48] H. W. Gandy, R. J. Ginther, and J. F. Weller, "Energy transfer and HO^{3+} laser action in silicate glass coactivated with Yb^{3+} and Ho^{3+}," *Appl. Phys. Lett.*, vol. 6, pp. 237–239, June 1965.

[49] V. S. Zuev, V. S. Letokhov, and Yu. U. Senatskii, "Giant superluminescent pulses," *JETP Lett.*, vol. 4, pp. 125–127, September 1966.

[50] H. W. Gandy, R. J. Ginther, and J. F. Weller, "Energy transfer in triply activated glasses," *Appl. Phys. Lett.*, vol. 6, pp. 46–49, February 1965.

[51] L. G. DeShazer and L. G. Komai, "Fluorescence conversion efficiency of neodymium glass," *J. Opt. Soc. Am.*, vol. 55, pp. 940–944, August 1965.

[52] C. G. Young, "Threshold pumping characteristics of the neodymium glass laser," *J. Opt. Soc. Am.*, vol. 52, p. 1318, November 1962.

[53] D. M. Sinnett, "An analysis of the maser equations," *J. Appl. Phys.*, vol. 33, pp. 1578–1581, April 1962.

[54] H. Statz and G. A. deMars, "Transients and oscillation pulses in masers," in *Quantum Electronics*, C. H. Townes, Ed. New York: Columbia University Press, 1960, pp. 530–537.

[55] G. E. Peterson, A. D. Pearson, and P. M. Bridenbough, "Energy exchange from Nd^{3+} to Yb^{3+} in calibo glass," *J. Appl. Phys.*, vol. 36, pp. 1962–1966, June 1965.

[56] V. P. Kovalev and G. O. Karopetzen, "Sensitizing of the luminescence of trivalent ytterbium by neodymium in silicate glasses," *Opt., Spectry.*, vol. 18, pp. 102–103, January 1965.

[57] E. Snitzer, "Laser emission at 1.06μ from Nd^{3+}–Yb^{3+} glass," *IEEE J. Quantum Electronics*, vol. 2, pp. 562–566, September 1965.

[58] R. F. Woodcock and F. Snitzer, to be published.

[59] F. Quelle, "Alternatives to Q-spoiled ruby rangefinders," *Proc. Conf. on Laser Range Instr.*, SPIE (Redondo Beach, Calif., 1968), pp. 3–10.

[60] C. G. Young, "Continuous glass laser," *Appl. Phys. Lett.*, vol. 2, pp. 151–152, April 1963.

[61] Francois Auzel, "Stimulated emission of Er^{3+} in a fluorphosphate glass," *Compt. Rend. Acad. Sci. Paris*, Ser. B, vol. 263, pp. 765–766, September 1966.

[62] C. G. Young, "A sun-pumped CW one watt laser," *Appl. Opt.*, vol. 5, pp. 993–997, June 1966.

[63] J. Robieux, J. Riffard, J. Ernest, and B. Starel, "High brightness glass lasers," *IEEE J. Quantum Electronics* (Digest of Technical Papers), vol. QE-4, p. 360, May 1968.

[64] C. G. Young and E. Snitzer, "Glass lasers," *IEEE J. Quantum Electronics* (Digest of Technical Papers), vol. QE-4, p. 360, May 1968.

[65] G. R. Simpson, "Continuous sun-pumped room temperature glass laser operation," *Appl. Opt.*, vol. 3, pp. 783–784, June 1964.

[66] C. G. Young, J. W. Kantorski, and A. D. Battista, "A high-power, intermediate pulse-width glass laser," (Abstract only), *IEEE J. Quantum Electronics*, vol. QE-3, p. 238, June 1967.

[67] C. F. Padula and C. G. Young, "Optical isolators for high power 1.06μ glass laser systems," *IEEE J. Quantum Electronics*, vol. QE-3, pp. 493–498, November 1967.

[68] C. G. Young, J. W. Kantorski, and E. O. Dixon, "Optical avalanche laser," *J. Appl. Phys.*, vol. 37, pp. 4319–4324, November 1966.

[69] R. H. Dicke, "Coherence in spontaneous radiation processes," *Phys. Rev.*, vol. 93, pp. 99–110, January 1954.

[70] R. W. Hellwarth, "Control of fluorescent pulsations," in *Advances in Quantum Electronics*, J. R. Singer, Ed. New York: Columbia University Press, 1961, pp. 334–341.

[71] A. A. Vuylsteke, "Theory of laser regeneration switching," *J. Appl. Phys.*, vol. 34, pp. 1615–1622, June 1963.

[72] W. R. Hook, R. H. Dishington, and R. P. Hilberg, "Laser cavity dumping using time variable reflection," *Appl. Phys. Lett.*, vol. 9, pp. 125–127, August 1966.

[73] N. T. Melamed, C. Hiroyama, and P. W. French, "Laser action in uranyl-sensitized Nd-doped glass," *Appl. Phys. Lett.*, vol. 6, pp. 43–45, February 1965.

[74] H. W. Gandy, R. J. Ginther, and J. F. Weller, "Laser oscillations and self Q-switching in triply-activated glass," *Appl. Phys. Lett.*, vol. 7, pp. 233–236, November 1965.

[75] W. Shiner, E. Snitzer, and R. Woodcock, "Self Q-switched Nd^{3+} glass laser," *Phys. Lett.*, vol. 21, pp. 412–413, June 1966.

[76] E. Snitzer and R. Woodcock, "Saturable absorption of color centers in Nd^{3+} and Nd^{3+}–Yb^{3+} laser glass," *IEEE J. Quantum Electronics*, vol. 2, pp. 627–632, September 1966.

[77] P. C. Magnante, to be published.

[78] R. V. Ambartsumyan, N. G. Basov, V. S. Zuev, P. G. Kryukov, and V. S. Letokhov, "Propagation of a light pulse in a nonlinearly amplifying and absorbing medium," *ZhETF Pis'ma*, vol. 4, pp. 19–22, July 1966.

[79] B. H. Soffer and R. H. Hoskins, "Generation of giant pulses from a neodymium laser by a reversibly bleachable absorber," *Nature*, vol. 204, p. 276, October 1964.

[80] L. E. Hargrove, R. L. Fork, and M. A. Pollack, "Locking of He-Ne laser modes induced by synchronous intracavity modulation," *Appl. Phys., Lett.*, vol. 5, pp. 4–5, July 1964.

[81] H. W. Mocker and R. J. Collins, "Mode competition and self-locking effects in a Q-switched ruby laser," *Appl. Phys. Lett.*, vol. 7, pp. 270–273, November 1965.

[82] T. Deutsch, "Mode locking effects in an internally modulated ruby laser," *Appl. Phys. Lett.*, vol. 7, pp. 80–82, August 1965.

[83] C. G. Young and R. F. Woodcock, "Laser-induced damage in glass," (NBS, Boulder, Colo., June 1969).

[84] P. C. Magnante, "A high spectral-radiance neodymium glass laser," *IEEE J. Quantum Electronics* (Digest of Technical Papers), vol. QE-4, pp. 363–364, May 1968.

[85] N. G. Basov, P. G. Kriukov, V. S. Letokhov, and Yu. V. Senatskii, "4–12—generation and amplification of ultra-short optical pulses," *IEEE J. Quantum Electronics*, vol. QE-4, pp. 606–609, October 1968.

[86] E. B. Treacy, "Compression of picosecond light pulses," *Phys. Lett.*, vol. 28A, pp. 34–35, October 1968.

[87] G. Gobeli, "Powerful pulsed laser bows," *Electronic News*, vol. 14, p. 72, March 17, 1969.

[88] W. F. Hagen, "Diffraction-limited high radiance Nd-glass laser system," *IEEE J. Quantum Electronics* (Digest of Technical Papers), vo. QE-4, p. 361, May 1968.

[89] ——, "Diffraction-limited high radiance Nd-glass laser system," *J. Appl. Phys.*, vol. 40, pp. 511–516, February 1969.

[90] J. G. Skinner and J. E. Geusic, "Diffraction-limited ruby oscillator," *J. Opt. Soc. Am.*, vol. 52, p. 1319, November 1962.

[91] C. G. Young and J. W. Kantorski, "Single-mode glass laser," *J. Opt. Soc. Am.*, vol. 53, p. 1339, November 1963.

[92] N. G. Basov, V. A. Boiko, O. N. Kroklin, and G. V. Sklizkov, "Formation of a long spark in air by weakly focused laser radiation," *Sov. Phys.—Doklady*, vol. 12, pp. 248–251, September 1967.

[93] P. D. Maker, R. W. Terhune, and C. M. Savage, "Optical third harmonic generation," in *Quantum Electronics III*, P. Grivet and N. Bloembergen, Eds., New York: Columbia University Press, 1964, pp. 1559–1576.

[94] Yu. P. Raizer, "Breakdown and heating of gases under the influence of a laser beam," *Sov. Phys. Uspekhi*, vol. 8, pp. 650–673, March–April 1966.

[95] V. V. Korobkin and A. J. Alcock, "Self-focusing effects associated with laser-induced air breakdown," *Phys. Rev. Lett.*, vol. 21, pp. 1433–1436, November 1968.

[96] L. R. Evans and C. G. Morgan, "Multiple collinear laser-produced sparks in gases," *Nature*, vol. 219, pp. 712–713, August 1968.

[97] R. Y. Chiao, E. Garmire, and C. H. Townes, "Self-trapping of optical beams," *Phys. Rev. Lett.*, vol. 13, pp. 479–482, October 1964.

[98] P. L. Kelley, "Self-focusing of optical beams," *Phys. Rev. Lett.*, vol. 15, pp. 1005–1008, December 1965.

[99] F. W. Quelle, "High brightness raman lasers," *Proc. 3rd Conf. Laser Technology* (Pensacola), vol. 2, C. M. Steckley and T. B. Dowd, Eds. Boston: ONR, pp. 135–137.

[100] W. F. Hagen and P. C. Magnante, "Efficient second harmonic generation with diffraction-limited and high spectral radiance Nd-glass lasers," *J. Appl. Phys.*, vol. 40, pp. 219–224, January 1969.

[101] W. F. Hagen and E. Snitzer, "Nonlinear solarization in flint glasses by intense 0.53-μm light," *IEEE J. Quantum Electronics* (Digest of Technical Papers), vol. QE-4, p. 361, May 1968.

[102] C. G. Young, D. W. Cuff, and W. P. Bazinet, "Integrated glass laser module," *J. Opt. Soc. Am.*, vol. 56, p. 1443, November 1966.

[103] C. J. Koester and E. Snitzer, "Amplification in a fiber laser," *Appl. Opt.*, vol. 3, pp. 1182–1186, October 1964.

[104] C. J. Koester, "Laser action by enhanced total internal reflection," *IEEE J. Quantum Electronics*, vol. QE-2, pp. 580–584, September 1966.

[105] E. Snitzer and H. Osterberg, "Observed dielectric wave guide modes in the visible spectrum," *J. Opt. Soc. Am.*, vol. 51, pp. 499–505, May 1961.

[106] G. C. Holst and E. Snitzer, "Detection with a fiber laser preamplifier at 1.06μ," *J. Opt. Soc. Am.*, vol. 59, p. 506, April 1969.

[107] R. J. Glauber, "Photon counting and field correlations," in *Physics of Quantum Electronics*, P. L. Kelley, B. Lax and P. E. Tannenwald, Eds. New York: McGraw-Hill, 1966, pp. 788–811.

[108] G. C. Holst, E. Snitzer, and R. Wallace, "High coherence, high power laser system at 1.0621μ," *J. Opt. Soc. Am.*, vol. 59, p. 505, April 1969.

[109] R. Zitter, "$2s-2p$ and $3p-2s$ transitions of neon in a laser ten meters long," *J. Appl. Phys.*, vol. 35, pp. 3070–3071, October 1964.

[110] I. Itzkan and G. Pincus, "1.0621μ He-Ne gas laser," *Appl. Opt.*, vol. 5, p. 349, February 1966.

[111] G. D. Baldwin and E. P. Riedel, "Measurements of dynamic optical distortion in Nd-doped glass laser rods," *J. Appl. Phys.*, vol. 38, pp. 2726–2738, June 1967.

[112] T. H. Maiman, "Stimulated optical emission in fluorescent solids, I. Theoretical considerations," *Phys. Rev.*, vol. 123, pp. 1145–1150, August 1961.

[113] F. Pockels, *Ann. Physik*, vol. 9, p. 220, 1902; vol. 11, p. 651, 1903.

[114] J. W. Kantorski, D. A. LaMarre, and C. G. Young, "Fagot lasers," (Abstract only), *IEEE J. Quantum Electronics*, vol. QE-3, p. 238, June 1967.

[115] C. G. Young and J. P. Segre, "Composite laser rod," *J. Opt. Soc. Am.*, vol. 59, p. 505, April 1969.

[116] J. P. Segre, "High repetition rate operation of Nd^{3+} glass lasers," *IEEE J. Quantum Electronics* (Digest of Technical Papers), vol. QE-4, p. 362, May 1968.

[117] J. P. Segre, "High repetition rate operation of Nd^{3+}-doped glass slab laser," to be published in *IEEE J. Quantum Electronics*.

[118] R. H. Dishington, W. R. Hook, and R. P. Hilberg, "A polarized laser cavity which is insensitive to birefringence in the laser rod," *Proc. IEEE* (Letters), vol. 55, pp. 2038–2039, November 1967.

[119] R. E. Kidder, J. E. Swain, K. Pettipiece, F. Rainer, E. D. Baird, and B. Loth, "A large-aperture disk laser system," *IEEE J. Quantum Electronics* (Digest of Technical Papers), vol. QE-4-4, pp. 377–378, May 1968.

[120] J. C. Almasi, J. P. Chernoch, W. S. Martini, and K. Tomiyasu, "Face-pumped laser," GE rept. to ONR, May 1966.

[121] C. G. Young, "Recent advances in glass lasers," *Proc. Conf. Lasers and Opto-Electronics* (University of Southampton, England, March 1969), London: Brit. Inst. Electronics and Radio Engineers, pp. 123–131.

[122] V. S. Doladugina and N. V. Korolev, "Occlusions in neodymium glass," *Sov. J. of Opt. Tech.*, vol. 35, pp. 109–111, January–February 1968.

[123] J. P. Budin and J. Raffy, "On the dynamics of laser-induced damage in glasses," *Appl. Phys. Lett.*, vol. 9, pp. 291–293, October 1966.

[124] J. H. Cullom and R. W. Waynant, "Determination of laser damage threshold for various glasses," *Appl. Opt.*, vol. 3, pp. 989–990, August 1964.

[125] J. Davit and M. Soulie, "Fractures produites dans les verres par un faisceau laser," *C. R. Acad. Sc. Paris*, vol. 261, pp. 3567–3570, November 1965.

[126] B. M. Ashkinadze, V. I. Vladimirov, V. A. Likhachev, S. M. Ryvkin, V. M. Salmanov, and I. D. Yaroshetskii, "Damage produced by a laser beam in a transparent dielectric," *Sov. Phys. Doklady*, vol. 11, pp. 703–705, February 1967.

[127] E. Snitzer, "Glass lasers," *Glass Industry*, vol. 48, pp. 11–19, September/October 1967.

[128] J. Martinelli, "Laser-induced damage thresholds for various glasses," *J. Appl. Phys.*, vol. 37, pp. 1939–1940, March 1966.

[129] P. V. Avizonis and T. Farrington, "Internal self-damage of ruby and Nd-glass lasers," *Appl. Phys. Lett.*, 7, pp. 205–208, October 1965.

[130] J. W. Kantorski, unpublished.

[131] L. I. vanTorne, "Pumping induced imperfections in glass: Nd^{3+} lasers," *Phys. Status. Solidis*, vol. 16, pp. 171–181, July 1966.

[132] M. E. Vance and R. D. Maurer, "Internal fracture of glass by high-intensity laser light," *Proc. 8th Congr. on Glass* (London, July 1968), p. 105.

[133] R. A. Miller and N. F. Borrelli, "Damage in glass induced by linear absorption of laser radiation," *Appl. Opt.*, vol. 6, pp. 164–165, January 1967.

[134] J. C. Buges, J. M. Jego, A. Terneaud, and P. Veyrie, "Variation de l'indice de refraction d'un verre soumis à un faisceau laser focalisé," *Compt. Rend. Acad. Sci. Paris*, vol. 264, pp. 871–874, March 1967.

[135] H. Dupont, A. Donzel, and J. Ernest, "On laser-induced breakdown and fracture in glasses," *Appl. Phys. Lett.*, vol. 11, pp. 271–272, November 1967.

[136] V. V. Korobkin and R. V. Seron, "Investigation of self focusing of neodymium-laser radiation," *JETP Lett.*, vol. 6, pp. 135–137, September 1967.

[137] A. H. Piekara, "Phenomenological treatment of small scale light trapping," *Appl. Phys. Lett.*, vol. 13, pp. 225–227, October 1968.

[138] C. S. Wang, "Propagation of an intense light beam in a non-linear medium," *Phys. Rev.*, vol. 173, pp. 909–917, September 1968.

[139] R. Landauer, "Sign of slow non-linearities in non-absorbing optical media," *Phys. Lett.*, vol. 25A, pp. 416–417, September 1966.

[140] S. A. Akhmanov, A. P. Sukhorukov, and R. V. Khokhlov, "Self-focusing and self trapping of intense light beams in a nonlinear medium," *Sov. Phys.—JETP*, vol. 23, pp. 1025–1033, December 1966.

[141] A. A. Chabon, "Self-focusing of light in solids via the electrostriction mechanism," *JETP Lett.*, vol. 6, pp. 20–23, July 1967.

[142] A. A. Chabon, "Ferroelectric effect in a laser beam," *JETP Lett.*, vol. 5, pp. 14–17, January 1967.

[143] R. G. Brewer, J. R. Lifsitz, E. Garmire, R. Y. Chao, and C. H. Townes, "Small-scale trapped filaments in intense laser beams," *Phys. Rev.*, vol. 166, pp. 326–331, February 1968.

[144] M. O. Degeuerce and J. C. Vienot, "Propagation de l'energie dans un faisceau laser: étude des trajectories," *Compt. Rend. Acad. Sci. Paris*, vol. 264, pp. 458–461, February 1967.

[145] Yu. P. Raizer, "Stratification of light beams in a nonlinear medium and the real threshold for self focusing," *JETP Lett.*, vol. 4, pp. 1–4, July 1966.

[146] A. L. Dyshko, V. N. Lugovoi, and A. M. Prokhorov, "Self-focusing and intense light beams," *JETP Lett.*, vol. 6, pp. 146–148, September 1967.

[147] V. N. Lugovoi and A. M. Prokhorov, "A possible explanation of the small-scale self-focusing filaments," *JETP Lett.*, vol. 7, pp. 117–119, March 1968.

[148] S. A. Akmanov, A. P. Sukhorukov, and R. V. Khokhlov, "Self focusing and diffraction of light in a nonlinear medium," *Sov. Phys.—Uspekhi*, vol. 93, pp. 609–636, March–April 1968.

[149] C. R. Giuliano, "Laser-induced damage to transparent dielectric materials," *Appl. Phys. Lett.*, vol. 5, pp. 137–139, October 1964.

[150] J. P. Budin, A. Donzel, J. Ernest, and J. Roffy, "Stimulated Brillouin scattering in glasses," *Electronics Lett.*, vol. 3, pp. 1–2, January 1967.

[151] N. M. Kroll, "Excitation of hypersonic vibrations by means of photoelastic coupling of high-intensity light waves to elastic waves," *J. Appl. Phys.*, vol. 36, pp. 34–43, January 1965.

[152] L. D. Khazov and A. N. Shestov, "Induced Mendelstam-Brillouin scattering (IMBS) in glass in the case of self focusing of the monopulse laser beam," *Opt. Spectry.* vol. 25, pp. 248–249, September 1968.

[153] B. M. Ashkinadze, V. I. Vladimirov, V. A. Likhochev, S. M. Ryvkin, V. M. Salmanov, and I. D. Yaroshetskii, "Breakdown in transparent dielectrics caused by intense laser radiation," *Sov. Phys.—JETP*, vol. 23, pp. 788–797, November 1966.

[154] F. Quelle, unpublished.

[155] D. Olness, "Laser-induced breakdown in transparent dielectrics," *J. Appl. Phys.*, vol. 39, pp. 6–8, January 1968.

[156] S. A. Akhmanov and A. D. Sukhorukov, "Nonstationary self-focusing of laser pulses in a dissipative medium," *JETP Lett.*, vol. 5, pp. 87–91, February 1967.

[157] R. G. Brewer and C. H. Lee, "Self-trapping with picosecond light pulses," *Phys. Rev. Lett.*, vol. 21, pp. 267–270, July 1968.

[158] A. G. Litvak, "Self-focusing of powerful light beams by thermal effects," *JETP Lett.*, vol. 4, pp. 230–233, November 1966.

[159] S. A. Akhmanov and A. P. Sukhorukov, "Nonstationary self-focusing of laser beams in a dissipative medium," *JETP Lett.*, vol. 5, pp. 87–91, February 1967.

[160] Yu. P. Raizer, "Suppression of self focusing of light beams and stabilization of a plane wave in a weakly absorbing medium," *JETP Lett.*, vol. 4, pp. 193–195, October 1966.

[161] ——, "Self-focusing of a homogeneous light beam in a transparent medium, due to weak absorption," *JETP Lett.*, vol. 4, pp. 85–88, August 1966.

[162] J. Ernest, unpublished.

[163] J. E. Swain, "Major improvement of surface damage threshold of laser glass by chemical polishing," *IEEE J. Quantum Electronics* (Digest of Technical Papers), vol. QE-4, pp. 362–363, May 1968.

[164] J. Davit, J. Decoux, J. Gautier, and M. Soulie, "Impression de surface des verres par un faisceau laser de grade brillance," *Compt. Rend. Acad. Sci. Paris*, vol. 266, pp. 1236–1238, April 1968.

[165] J. Davit, J. Decoux, J. Gautier, and M. Soulie, "Fractures de surface des verres sous l'action d'un faisceau laser," *Rev. Phys. Opt.*, vol. 3, pp. 118–120, June 1968.

[166] B. S. Sharma and K. E. Rieckhoff, "Laser-induced photo-conductivity in silicate glasses by multiphoton excitation, a precursor of dielectric breakdown and mechanical damage," *Can. J. Phys.*, vol. 45, pp. 3781–3791, December 1967.

[167] D. L. Rousseau, G. E. Leroi, and W. E. Falconer, "Charged particle emission after ruby laser irradiation of transparent dielectric materials," Res. Rept. AD-655-105, June 1967.

[168] L. D. Khasov and A. N. Shestov, "Surface discharge at the end of an active laser rod," *JETP Lett.*, vol. 23, pp. 261–262, September 1967.

[169] J. Davit, "Mechanism for laser surface damage of glasses," *J. Appl. Phys.*, vol. 39, pp. 6052–6056, December 1968.

[170] J. R. Bright, Ed. *Technological forecasting for industry and government*, Englewood Cliffs, N.J.: Prentice-Hall, 1968.

[171] R. E. Innis, "Flexible fiber laser probe," *Symp. on Recent Developments in Research Methods and Instrumentation* (National Institure of Health, Bethesda, Md., October 1965).

[172] E. Snitzer, "Neodymium glass laser," in *Quantum Electronics III*, P. Grivet and N. Bloembergen, Eds. New York: Columbia University Press, 1964, pp. 999–1019.

[173] W. F. Hagen, unpublished.

[174] N. G. Basov, P. G. Kriukov, S. D. Zakhorov, Yu. V. Senatsky, and S. V. Tchekolin, "Experiments on the observation of neutron emission at a focus of high-power laser radiation on a lithium deuteride surface," *IEEE J. Quantum Electronics*, vol. QE-4, pp. 864–867, November 1968.

[175] M. C. Canto, J. D. Reuss, and P. Veyrie, "Etude théorique de l'ionisation du deuterium sous l'action d'un laser à impulsion courte," *Compt. Rend. Acad. Sci. Paris*, vol. 267, pp. 878–881, October 1968.

[176] C. Colin, Y. Durand, F. Floux, D. Guyot, P. Langor, and P. Veyrie, "Laser-produced plasma from solid deuterium targets," *J. Appl. Phys.*, vol. 39, pp. 2291–2993, June 1968.

[177] H. Opower, W. Kaiser, H. Puell, and W. Heinicke, "Energetic plasmas produced by laser light on solid targets," *Z. Naturforsch.*, vol. 22a, pp. 1392–1397, September 1967.

[178] M. J. Bernstein and G. G. Comisar, "Efficiency estimates for conversion of focused laser radiation to X-rays in a hot dense plasma," Res. Rept. AD-672-964, March 1968.

[179] S. Q. Duntley, "Optical exploration of the oceans," *J. Opt. Soc. Am.*, vol. 59, p. 472, April 1969.

[180] M. I. Dzyubenko, Yu. A. Nestrizhenko, and V. G. Parusimov, "Investigation of a laser designed for lidar probling of the moon," *Ukrain. Fiz. Zh.*, vol. 13, pp. 1233–1240, August 1968.

[181] C. O. Alley and P. L. Bender, "Information obtainable from laser range measurements to a lunar corner reflector," presented at the IAU/IUGG Symp. on Continental Drift, Secular Motion of the Pole, and Rotation of the Earth (IAU Symp. no. 32) Stresa, Italy, March 1967.

[182] C. O. Alley, P. L. Bender, D. G. Currie, R. H. Dicke, and J. E. Faller, "Some implications for physics and geophysics of laser range measurements from earth to a lunar retro-reflector," *Proc. NATO Conf. on the Application of Modern Physics to the Earth and Planetary Interiors* (University of Newcastle upon Tyne, March 1967).

[183] P. L. Bender, C. O. Alley, D. G. Currie, and J. E. Faller, "Satellite geodesy using laser range measurements only," *J. Geophys. Res.*, vol. 73, pp. 5353–5358, August 1968.

[184] R. Baierlein, "Testing general relativity with laser ranging to the moon," *Phys. Rev.*, vol. 162, pp. 1275–1287, October 1967.

[185] R. Bivas, "Laser telemetry and its geodetic applications," *Recherche Spatiale*, vol. 6, pp. 1–5, November 1967.

[186] J. W. Goodman, W. H. Huntly, Jr., D. W. Jackson, and M. Lehman, "Wavefront-reconstruction imaging through random media," *Appl. Phys. Lett.*, vol. 8, pp. 311–313, June 1966.

[187] H. F. Ludloff, "Application of laser radar return to meteorological problems," *Phys. Lett.*, vol. 28A, pp. 452–453, December 1968.

[188] R. T. H. Collis and M. G. G. Ligda, "Note on lidar observations of particulate matter in the stratosphere," *J. Atmos. Sci.*, vol. 23, pp. 255–257, March 1966.

[189] A. Orszag, R. Zaharia, and Y. deValence, "Measure de la densité atmospherique au moye d'un radar optique," *Compt. Rend. Acad. Sci. Paris*, vol. 267, pp. 237–240, July 1968.

[190] G. S. Kent, B. R. Clemesha, and R. W. Wright, "High altitude atmospheric scattering of light from a laser beam," *J. Atmos. Terrest. Phys.*, vol. 9, pp. 169–181, February 1967.

[191] C. A. Northend, R. C. Honey, and W. E. Evans, "Laser radar (lidar) for meteorological observations," *Rev. Sci. Instr.*, vol. 37, pp. 393–400, April 1966.

[192] G. C. Holst, E. Snitzer, and R. Wallace, "High-coherence, high power laser system at 1.0621 microns," to be published in *IEEE J. Quantum Electronics*.

[193] D. B. Neuman, "Precision range-gated imaging technique," *J. SMPTE*, vol. 74, pp. 313–319, April 1965.

[194] C. G. Young and J. W. Kantorski, "Optical gain and inversion in neodymium glass lasers," *Proc. 1st Conf. on Laser Technology* (San Diego), G. Adelman and T. B. Dowd, Eds. Boston: ONR, 1964, pp. 75–84.

[195] P. L. Dexter, "A theory of sensitized luminescence in solids," *J. Chem. Phys.*, vol. 21, pp. 836–850, May 1953.

[196] The author is indebted to F. Byrne for first pointing out this possibility.

[197] C. G. Young and J. W. Kantorski, "High power pulsed glass laser," *Prov. 2nd Conf. on Laser Technology* (Chicago), L. Rains and T. B. Dowd, Eds. Boston: ONR, 1965, pp. 65–76.

[198] C. G. Young, "Glass laser delivers 5000-joule output," *Laser Focus*, vol. 3, p. 36, February 1967.

[199] R. A. Fisher, P. L. Kelley, and T. K. Gustafson, "Subpicosecond pulse generation using the optical kerr effect," *Appl. Phys. Lett.*, vol. 14, pp. 140–143, February 1969.

[200] E. Archbold, J. M. Burch, and R. W. E. Cook, "Generation of directional pulses in an optically swept slab laser," *IEEE J. Quantum Electronics* (Digest of Technical Papers), vol. QE-2, p. lxix, April 1966.

[201] J. G. Edwards, "A high power diffraction-limited optically swept slab laser," *Proc. Conf. on Lasers and Opto-Electronics*, University of Southampton, England). 1969, London: Brit. Inst. Electronics and Radio Engineers, pp. 13–24.

[202] C. G. Young and D. W. Cuff, "An efficient and compact glass laser illuminator," *Proc. 3rd Conf. on Laser Technology* (Pensacola), vol. 2, C. M. Stickley and T. B. Dowd, Eds., Boston: ONR, pp. 97–104.

[203] E. Booth and W. Strouse, unpublished.

[204] E. Matovitch, "Beam divergence measurements on a ruby axial gradient laser," *Proc, 3rd Conf. on Laser Technology*, (Pensacola), vol. 2, C. M. Stickley and T. B. Dowd, Eds., Boston: ONR, pp. 131–134.

[205] T. I. Veinberg, I. A. Zhmyreva, V. P. Kolobkov, and P. I. Kudryashov, "Laser action of Ho^{3+} ions in silicate glasses coactivated by holmium, erbium and ytterbium," *Opt. Spectry.* vol. 24, pp. 441–443, May 1968.

[206] N. M. Galaktionova, V. F. Egorova, V. S. Zubkova, A. M. Mak, and D. S. Prilezhaev, "Luminescence quantum yield and energy output of stimulated radiation in neodymium glass," *Sov. Phys.—Doklady*, vol. 12, pp. 360–361, October 1967.

[207] R. A. Brandewie and C. L. Telk, "Quantum efficiency of Nd^{3+} in glass, calcium tungstate, and yttrium aluminum garnet," *J. Opt. Soc. Am.*, vol. 57, pp. 1221–1225, October 1967.

[208] W. W. Holloway, Jr., M. Kestigian, F. F. Y. Wang, and G. F. Sullivan, "Temperature-dependent Nd fluorescence parameters and laser thresholds," *J. Opt. Soc. Am.*, vol. 56, pp. 1409–1419, October 1966.

[209] N. S. Belokirinitskii, A. D. Manuel'skii, and M. S. Soskin, "Studying the structure of inhomogeneously broadened luminescence bands of active media by means of induced radiation spectra," *Ukrain J. of Phys.*, vol. 12, pp. 1720–1730, 1967.

[210] O. K. Deutschbein, "The phosphate glasses, new laser materials," *Rev. Phys. Appl.*, vol. 2, pp. 29–37, August 1967.

[211] Kh. I. Gaprindashvili, V. V. Mumladze, G. G. Mshvelidze, M. E. Perel'man, and V. A. Khanevichev, "Fiber laser without a resonator," *Akad. Nauk. Gruzinskoi—SSSR, Soobishcheniya*, vol. 45, pp. 57–64, January 1967.

[212] L. G. DeShazer, "Spectral narrowing of neodymium-glass laser with a composite oscillator configuration," *J. Opt. Soc. Am.*, vol. 56, p. 1443, October 1966.

[213] A. A. Kaminskii, "Laser with combined active medium," *JETP Lett.*, vol. 7, pp. 201–203, April 1968.

[214] E. Snitzer, "Frequency control of a Nd^{3+} glass laser," *Appl. Opt.*, vol. 5, pp. 121–125, January 1966.

[215] P. C. Magnante, "A high spectral-radiance neodymium glass laser," *IEEE J. Quantum Electronics* (Digest of Technical Papers). vol. Q-4, pp. 363–364, May 1968.

[216] H. Inaba, Y. Isawa, and N. Suda, "Radiative coupling between

two different solid-state lasers," *Phys. Lett.*, vol. 22, pp. 293–295. August 1966.

[217] A. M. Bonch-Bruevich, V. Yu. Petrun'kin, V. N. Aryumanov, N. A. Esepkina, Yu. A. Imas, S. V. Kruzhalov, L. N. Pakhonov, and V. A. Chernov, "Investigation of a neodymium-glass laser with external feedback," *Sov. Phys.—Tech. Phys.*, vol. 11, pp. 1621–1623, June 1967.

[218] A. M. Bonch-Bruevich, V. Yu. Petrun'kin, N. A. Espkina, S. V. Kruzhalov, L. N. Pakhonov, V. A. Chernov, and S. L. Gulkin, "Traveling-wave laser with neodymium glass," *Sov. Phys.—Tech. Phys.*, vol. 12, pp. 1495–1499, May 1968.

[219] D. Roess, "Ruby superradiation," *Proc. IEEE* (Correspondence) vol. 52, p. 853, July 1964.

[220] H. Inaba and T. Koboyashi, "Ultrasonic frequency modulation of laser oscillation from Nd^{3+} glass rod," *J. Appl. Math. and Phys.* (ZAMP), vol. 16, 66–67, 1965.

[221] M. P. Vanyukov, V. A. Venchikov, V. Ya. Zhulai, V. I. Isaenko, and V. V. Lyubimov, "Neodymium-glass two-channel optical quantum generator with an energy of 180 J," *Instr. Exptl. Tech.*, vol. 3, pp. 634–636, May–June 1967.

[222] M. Michon, "Single nanosecond pulse generation by combining Q-switching, mode-locking by internal loss modulation, and PTM operation of a Nd^{3+} glass laser," *IEEE J. Quantum Electronics* (Digest of Technical Papers), vol. QE-3, p. 248, June 1967.

[223] M. Michon, J. Ernest, and R. Auffret, "Mode locking of a Q-spoiled Nd^{3+} doped glass laser by intracavity phase modulation," *Phys. Lett.*, vol. 23, pp. 457–458, November 1966.

[224] M. P. Vanyukov, V. I. Isaenko, L. A. Luizova, and O. A. Sharokhov, "On the cavity losses on mode selection in the stimulated emission of Nd^{3+} in glass," *Opt. Spectry.*, vol. 20, pp. 535–538, June 1966.

[225] J. deMetz, A. Terneaud, and P. Veyrie, "Optical study of the beam emitted by a high power laser," *Appl. Opt.*, vol. 5, pp. 819–822, May 1966.

[226] L. J. Aplet, E. B. Jay, and W. R. Sooy, "Effect of surface finish on temperature distributions in optically pumped Nd^{3+}: glass laser rods," *Appl. Phys. Lett.*, vol. 8, pp. 71–73, February 1966.

[227] H. Welling and C. J. Bickart, "Spatial and temporal variation of the optical path length of flash-pumped laser rods," *J. Opt. Soc. Am.*, vol. 56, pp. 611–618, May 1966.

[228] O. N. Voron'ko, N. A. Kozlov, A. A. Mak, B. G. Malinin, and A. I. Stepanov, "One possibility of increasing the Q of the resonator of a neodymium laser," *Sov. Phys.—Doklady*, vol. 12, pp. 252–253, September 1967.

[229] T. Kamogawa, H. Kotera, and H. Hazami, "High-repetition pulsed Nd^{3+} glass laser," *Japan. J. Appl. Phys.*, vol. 5, p. 449, May 1966.

[230] M. Leblanc, J. Hanus, and B. Sturel, "Generation of very intense impulses at 5300 A with a very high power laser. Saturation and possibilities in the ultraviolet," *Compt. Rend. Acad. Sci. Paris*,

vol. 263, pp. 701–704, September 1966.

[231] G. O. Karopetyan and A. L. Reishakhrit, "Luminescent glasses as laser materials," *Izv. Akad. Nauk SSSR, Inorganic Materials*, vol. 3, pp. 217–259, February 1967.

[232] D. W. Harper, "Assessment of neodymium optical maser glass," *Phys. Chem. Glasses*, vol. 5, pp. 11–16, February 1964.

[233] A. D. Pearson and G. E. Peterson, "Energy exchange processes and laser oscillation in glasses," *Proc. 7th Internatl. Comm. on Glass* (Brussels, 1965), Chorleroi, Belgium: Inst. Natl. Du Vene; 1965, pp. 10.1–10.11.

[234] A. A. Mak, Yu. A. Anan'ev, and B. A. Errvakov, "Solid state lasers," *Sov. Phys.—Uspekhi*, vol. 92, pp. 419–452, January–February 1968.

[235] G. C. Holst and E. Snitzer, "Detection with a fiber laser preamplifier at 1.06 microns," to be published in *IEEE J. Quantum Electronics*.

[236] J. Robieux, J. Riffard, J. Ernest, and B. Sturel, "High power glass lasers," to be published in *IEEE J. Quantum Electronics*.

[237] N. G. Basov, O. N. Krokhin, and G. V. Sklizkov, "Laser application for investigations of the high-temperature and plasma phenomena," *IEEE J. Quantum Electronics*, vol. QE-4, pp. 988–991, December 1968.

[238] S. L. Shapiro and M. A. Duguay, "Observation of sub-picosecond components in the mode-locked Nd-glass laser," *Phys. Lett.*, vol. 28A, pp. 698–699, February 1969.

[239] C. G. Yong, D. A. LaMarre, and A. A. Vuylsteke, "Glass laser technology," *Laser Focus*, vol. 3, pp. 21–29, December 1967.

[240] C. C. Robinson, "Excited-state absorption in fluorescent uranium, erbium and copper-tin glasses," *J. Opt. Soc. Am.*, vol. 57, pp. 4–7, January 1967.

[241] C. C. Robinson and J. T. Fournier, "Faraday effect in cerous metaphosphate glass," *Bull. Am. Ceram. Soc.*, vol. 48, p. 451, April 1969.

[242] C. C. Robinson and J. T. Fournier, to be published.

[243] Yu. K. Danileiko, V. Ya. Khaimov-Mal'kov, A. A. Manenkov, and A. M. Prokhorov, "Surface and volume scattering, surface damage, and optical homogeneity studies of ruby crystals and their connection with laser-emission characteristics," *IEEE J. Quantum Electronics*, vol. QE-5, pp. 87–92, February 1969.

[244] K. A. Bruedener, W. Bostick, A. Haught and J. Mather, "Laser produced plasma and plasma focus," *Proc. 3rd IAEA Conf.* (Novosibirsk, August 1968), Gov. Rept. TID-24804, R. L. Hirsch Ed. Oak Ridge, Tenn.: USAEC, pp. 40–49.

[245] V. F. Egorova, V. S. Zubkova, G. O. Karapetyan, A. A. Mak, D. S. Prilezhaev, and A. L. Reichakhrit, "Influence of glass composition on the luminescence characteristics of Nd^{3+} ions," *Opt. Spectry.*, vol. 23, pp. 148–151, August 1967.

[246] B. B. McFarland, R. H. Hoskins, and B. H. Soffer, "Narrow spectral emission from a passively Q-spoiled neodymeum-glass laser," *Nature*, vol. 207, pp. 1180–1181, September 1965.

Semiconductor Lasers

MARSHALL I. NATHAN

Abstract—This paper is a review of semiconductor laser work. The principles of operation are discussed. The stress is on work since early 1964. The present state-of-the-art in GaAs junction lasers is described.

SINCE the first observations of stimulated emission in GaAs in 1962 [1]–[3], considerable progress has been made in the field of semiconductor lasers. Lasers have been made from many materials. The wavelength of coherent radiation has been extended through the visible into the ultraviolet [4], [4a] and out to the far infrared [5], [6] with promise of further extension at both ends of the spectrum. Several pumping schemes have been employed; p-n junctions [1]–[3], electron beams [7], [7a], optical pumping [8], [8a] and avalanche breakdown injection [9].

This paper will review the work in semiconductor lasers. For completeness, the principles of operation will be discussed. The early work, up to February, 1964, and many of the laser principles have been covered in an earlier paper, "*P-N* Junction Lasers" by G. Burns and the present author (referred to as BN) [10]. Therefore, in this paper we shall only touch lightly on the early work and concentrate primarily on more recent developments and the present state-of-the-art. Several other review articles have recently been written [11]–[14].

Manuscript received June 24, 1966.

The author is with the IBM Thomas J. Watson Research Center, Yorktown Heights, N. Y.

I. PRINCIPLES OF OPERATION

In a laser, stimulated emission is achieved by means of a population inversion, which occurs between two electronic levels when the upper level has a greater probability of being occupied by electrons than the lower one. The probability of a photon induced downward transition with photon with energy $h\nu$ equal to the energy separation between the states, will exceed the probability of an an upward transition leading to a net stimulated emission. Laser action can occur provided a proper resonant structure is made. These ideas have been discussed in detail in many articles and books [15]. There are, however, a few aspects which are unique to semiconductor lasers, for example, 1) the continuous spectrum of the electronic states, and 2) the variety of means for achieving inversion, i.e., pumping schemes. These ideas will be reviewed in this section.

Population Inversion and Threshold

Figure 1 shows the energy vs. density of states in a semiconductor. The highest band filled with electrons is the valence band. The next higher band, the conduction band, is separated from the valence band by a region of energy, the energy gap E_g, where there are no allowed states. Figure 1(a) shows the situation in equilibrium for an intrinsic semiconductor (no electrically active impurities added). If a light wave with photon energy $h\nu$ greater than E_g is incident on the crystal, photons will be absorbed

Reprinted from *Proc. IEEE*, vol. 54, pp. 1276–1290, Oct. 1966.

60

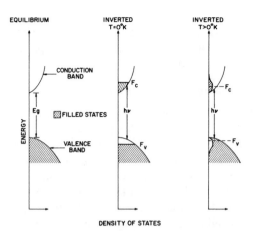

Fig. 1.   Energy vs. density of states for an intrinsic semiconductor.

by electrons making transitions from the valence band to the conduction band. Figure 1(b) shows the situation for an inverted population at 0°K. The valence band is empty of electrons (or filled with holes) down to an energy F_v, and the conduction band filled up to F_c. Now photons with energy $h\nu$ such that $E_g < h\nu < (F_c - F_v)$ will cause downward transitions and hence stimulated emission.

At finite temperature there will not be a sharp energy distinction between filled and unfilled states as shown in Fig. 1(b). Rather, the distributions of carriers will be smeared out in energy as shown in Fig. 1(c). It is still possible to give conditions for which stimulated emission will occur [16]. Overall thermal equilibrium does not exist of course, but we assume that in a given energy band, the carriers will be in thermal equilibrium with each other. The probability that a state in the conduction band is occupied is, according to Fermi-Dirac statistics,

$$f_c = \frac{1}{1 + \exp(E - F_c)/kT} \cdot \qquad (1)$$

F_c is the quasi-Fermi level for electrons in the conduction band; it is at the energy at which the probability of a state being occupied is equal to one half. A similar expression holds for the valence band. By requiring that the number of photons emitted be greater than the number absorbed, the condition for stimulated emission is found as [16]:

$$F_c - F_v > h\nu. \qquad (2)$$

The preceding discussion has been for the conduction to valence band transition in an intrinsic semiconductor. If the semiconductor contains impurities, and the impurity energy states are either the initial or final state, it is merely necessary to use the quasi-Fermi levels for the impurities with the proper degeneracy for the state [17].

In order to obtain laser action it is not only necessary that there be stimulated emission, but also that there be sufficient gain g to overcome the losses. There will be a threshold gain for this to occur. For a Fabry-Perot cavity this gain will satisfy the condition that a light wave

make a complete traversal of the cavity without attenuation, that is

$$\text{Re}^{(g_t - \alpha)l} = 1 \qquad (3)$$

where g_t is the gain per unit length at threshold; α the loss; l the length of the cavity; R the reflectivity of the ends of the cavity.[1] The primary loss mechanisms in a semiconductor are caused by penetration of the light outside the active region (the region in which the population is inverted), and by free carrier absorption of the light. The gain of a laser is proportional to the rate of stimulated emission. For a transition between two discrete energy levels, the gain is given by [15]

$$g(h\nu) = \frac{c^2[N_2 - N_1(g_2/g_1)]A(\nu)}{8\pi\nu^2 n_0^2 \tau_r} \qquad (4)$$

where $A(\nu)$ is the normalized line shape $\int A(\nu)d\nu = 1$; N_i is number/cm³ of atoms in the upper (2) or lower (1) state; g_i is the statistical weight of these states; n_0 is the index of refraction; ν is the frequency, and τ_r is the radiative lifetime. Lasing will occur at the wavelength at which g is a maximum and where $A \to A_{\max} = 1/\Delta\nu$. $\Delta\nu$ is the linewidth.

It can be seen in (4) that if $g_2 = g_1$, the gain is proportional to the number of electrons in the upper state minus the number in the lower state for a fixed $h\nu$. This excess varies linearly with the excitation rate. For a four level laser $N_1 = 0$ and the gain is proportional to the excitation rate. The shape of the curve, which depends on the line shape $A(\nu)$, is independent of excitation rate. The maximum gain occurs at a fixed frequency. In general, in a semiconductor the situation is more complex. As the excitation rate is increased, the distribution functions f_c and f_v change, i.e., F_c increases and F_v decreases. The shape of the gain curve vs. photon energy changes—its maximum shifts to higher energy. The gain is given by [17]

$$g(E)dE = \sum (4\pi^2 e^2 \hbar/(m^2 c n_0 E)) |M|^2 (f_c - f_v) \qquad (5)$$

where M is the matrix element, and the sum is over all pairs of states whose energy difference is between E and $E + dE$.

In order to illustrate explicitly some of the points just mentioned, we shall consider a simple model of a semiconductor laser. This model will also be used later in connection with the discussion of nonuniform currents in junction lasers. Its energy vs. density of states is shown in Fig. 2. The upper band, the conduction band, has a density of states ρ_c which, for example, may be given by

$$\rho_c = \rho_0 \exp E/E_0 \qquad (6)$$

where ρ_0 and E_0 are constant. The lower level is discrete, for example, an impurity level, and has a high enough density so that its population is approximately indepen-

_____
[1] If the reflectivities of the ends are different, $R = \sqrt{R_1 R_2}$.

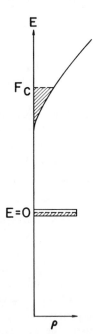

Fig. 2.   Energy vs. density of states for a simple
model of a doped semiconductor.

dent of excitation. Thus it is the same as being unoccupied. The sample is p-type, and the lower level is half filled with electrons in equilibrium. As the excitation is increased, the number of electrons in the conduction band increases, and F_c moves up. The transition probability is assumed to be independent of energy. At $T = 0°K$, the gain at energy E is proportional to the density of states in the conduction band at E.

$$g(E) \begin{array}{ll} = + K\rho_0 \exp E/E_0, & E < F_c \\ = - K\rho_0 \exp E/E_0, & E > F_c \end{array} \qquad (7)$$

where K is a constant. The transition probability has been assumed to be independent of energy. The gain at a given energy is independent of the excitation for energies less than F_c, and the maximum gain occurs at F_c. The density n of electrons in the conduction band is given by

$$n = \int_{-\infty}^{F_c} \rho_c(E)dE = \frac{E_0 g(F_c)}{K} = \frac{E_0}{K} g_{max}. \qquad (8)$$

Thus, for this model the maximum gain g_{max} is proportional to n and to the excitation rate, since the line shape is independent of F_c. However, the photon energy at which g_{max} occurs shifts to higher energy, in contrast to the behavior of a simple two level system. For an arbitrary density of states, g is not proportional to n; each case must be examined individually. For example, for a density of states which is independent of energy the gain is independent of n. For parabolic bands [17] ($\rho \propto \epsilon^{1/2}$) and for bands with Gaussian tails [18], it has been found that g proportional to n is a very good approximation at $0°K$, but at finite temperature the gain becomes superlinear especially for parabolic bands. For bands with

exponential tails g is proportional to n up to high temperatures [17]. In all these cases the transition probability is assumed to be independent of energy.

Direct and Indirect Semiconductors

The energy band structures of semiconductors are of two types; 1) those in which the lowest conduction band minimum and the highest valence band maximum are at the same wave vector in the Brillouin zone—direct semiconductors, 2) the extrema are at different wave vectors—indirect semiconductors. Since the wave vector of the light is much smaller than the electronic wave vector, first-order radiative transitions occur between states with the same electronic wave vector. Thus, the optical gain in direct semiconductors is quite high so that losses can be overcome, and laser action can occur easily. On the other hand, in indirect materials the change in wave vector must be taken up by some other agent such as a lattice vibration or an impurity. The radiative transitions are weaker, and the available gain is usually small, so that losses cannot be easily overcome.

Thus far, as will be seen in the section on materials, all well substantiated semiconductor lasers have direct band structures. These considerations have been discussed in detail by Dumke [19] and reviewed in BN [10].

Methods of Excitation

To achieve laser action it is necessary to have a method for inverting the population and a resonant structure. A variety of methods for doing this have been used.

a) Injection: The p-n junction is unique to semiconductors. Figure 3(a) shows an energy diagram of a p-n junction. Energy vs. distance is plotted; E_c is the conduction band edge; E_v the valence band edge. The semiconductor on the left is doped with donor impurities that make it degenerately n-type, that is, there are enough electrons added by the impurities to fill the conduction band up to the Fermi level F. On the right-hand side, acceptor impurities are added which deplete electrons from the valence band or, in other words, add holes down to energy F. At the junction at zero applied voltage as shown in Fig. 3(a), electrons flow from the n-side to the p-side until an electrical potential barrier V_B is built up which prevents further current flow. Now, if a voltage is applied to reduce the barrier or to raise the n-side relative to the p-side as shown in Fig. 3(b), electrons can flow over the top of barrier to the p-side where they make a transition to an empty state in the valence band and emit photons with energy approximately equal to E_g. In actual junctions, there are impurity levels in the gap near the band edges, and the photon energy can be less than E_g. It is also possible to have holes flow to the n-side where they recombine with electrons. The predominant process is determined by the relative impurity densities, the carrier lifetimes, and the carrier mobilities. In either case for high enough applied voltage, there can be a region in the vicinity of the

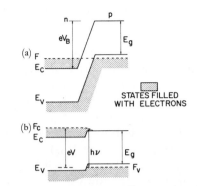

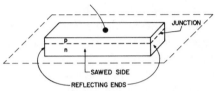

Fig. 3.   Energy vs. distance for a *p-n* junction.
(a) $V=0$, (b) $V>0$.

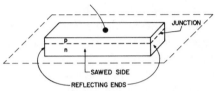

Fig. 4.   Schematic drawing for a *p-n* junction laser. The laser beam
is emitted near the junction from the reflecting ends.

junction where the population is inverted. Because the region is usually quite thin, the maximum optical gain over a reasonable distance will be in the plane of the junction. It is desirable to take advantage of this in constructing the resonant structure as shown in the Fabry-Perot structure shown in Fig. 4. At low currents there is spontaneous emission in all directions. As the current is increased, the gain increases until the threshold for laser action is reached. An increase in the light output is observed, and beams are emitted from the reflecting ends.

The current density j is related to the approximate average electron density n by the continuity equation

$$j/d = en/\tau, \qquad (9)$$

where d is the thickness of the active region and τ is the lifetime due to both radiative and nonradiative processes. We have assumed that only electrons are injected into the *p*-side (no hole injection into the *n*-side), but this is not necessary for our arguments. Using (3), (4), and (9) with $N_2 \rightarrow n$ and $N_1 = 0$, we find for the threshold current density,

$$j_t = \frac{8\pi e n_0^2 \nu^2 d \Delta\nu}{\eta c^2} \left[\frac{1}{l} \, ln(1/R) + \alpha \right] \qquad (10a)$$

$$= \frac{1}{\beta} \left[\frac{1}{l} \, ln(1/R) + \alpha \right] \qquad (10b)$$

where η is the quantum efficiency and is equal to τ_r/τ. This equation is valid only at $T=0°K$, since the lower band has been assumed to be unoccupied ($N_1=0$). The functional form of (10b) has been found experimentally in GaAs to be valid up to 300°K, but β and α are temperature dependent [21]. This will be discussed more fully in

Section III. The term outside the brackets in (10a) gives the ideal maximum value β can have.

In a typical *p-n* junction laser both the recombination and the light are confined to a narrow region near the junction. Several variations of *p-n* junction lasers have been made in which a high resistance region several microns thick is sandwiched between the *p*- and the *n*-region [22]-[25]. Depending on the particular structure, the light emission occurs at one edge of the high resistivity region [22], [24] or in a plasma throughout the high resistivity region [23], [25]. For a cavity in the junction plane, the mode confinement properties are influenced by the high resistance region. Narrower beam angles, perpendicular to the junction, have been observed than in a normal *p-n* junction, and higher order modes have been reported [24]. The cavity has also been made perpendicular to the junction for the structure in which the recombination occurs throughout the high resistivity region [25].

The *p-n* junction is a very convenient method for pumping a semiconductor laser material, but it gives rise to some difficult problems in materials preparation and structure fabrication. Junction fabrication will be discussed in detail in Section III, but we can give a few general considerations.

In most lasers, the stimulated emission occurs at photon energies close to the energy gap. In order to satisfy (2), the built in voltage V_B (see Fig. 3) must be greater than $h\nu$. This requires that the semiconductor be heavily doped with impurities on both sides of the junction so that F will be close to E_c on the *n*-side and to E_V on the *p*-side. In principle, very heavy doping is required on only one side of the junction. In fact, however, it is desirable for the semiconductor to be degenerately doped on both sides ($F>E_c$ on the *n*-side, $F<E_V$ on the *p*-side), since a good source of both electrons and holes is needed. Many semiconductors, particularly the large band gap II–VI compounds (e.g., CdS, etc.), cannot be usefully doped both *n*-type and *p*-type, so that *p-n* junctions cannot be made in them. Sometimes even if a junction can be made, inhomogeneities or other materials problems inhibit laser action.

b) Other Pumping Methods: A second pumping method, electron beam pumping [4], [7], is illustrated schematically in Fig. 5(a). A beam of high energy electrons, 20 kV or greater, is directed at a flat face of the semiconductor sample. These electrons penetrate several microns into the material (depending on the energy) and lose a large fraction of their energy by creating many low energy electron-hole pairs. It takes about two to four times the energy gap to create an electron-hole pair, so that about 10^4 electrons and holes are created per incident electron. They decay to the conduction band and valence band edges, respectively, where they form an inverted population. Coherent radiation is emitted perpendicular to the reflecting ends. The threshold condition at absolute

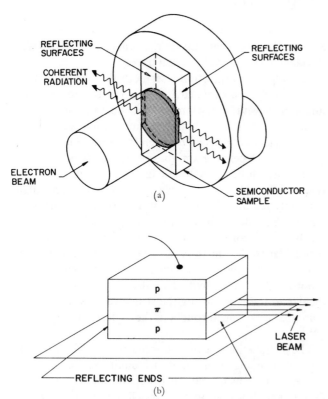

Fig. 5. (a) Electron beam pumping arrangement.
(b) Avalanche injection laser.

zero can be determined from (4) in a manner similar to (10a) provided the energy loss per electron hole pair and the penetration depth are known [26].

Optical pumping has also been used. The semiconductor is excited by radiation with photon energies greater than the band gap. The situation is then identical with electron beam pumping. A GaAs junction laser has been used to excite InSb [8] and InAs [28], for example. One problem with optical pumping is that the light is absorbed very close to the surface of the semiconductor, unless the wavelength of the exciting light is close to the energy gap. This limits the power that can be obtained compared to other optically pumped lasers because of the small volume, and makes the laser very sensitive to surface properties. In InSb and InAs diffusion of carriers away from the surface greatly reduces the seriousness of these effects. Basov et al. [8a], [29] have optically pumped GaAs with a Raman shifted ruby laser which closely matches the energy gap and thus has a relatively small absorption constant. They reported a power output of 3×10^4 watts, which is the highest for any semiconductor laser. Two-photon absorption of solid-state ion lasers has been used to pump semiconductor lasers—Nd^{+3} for GaAs [30] and ruby CdS [31]. Here, also, the absorption constant is small and large volume excitation can be obtained.

If the electrons (or holes) in a semiconductor are accelerated to sufficiently high energy by an electric field, they can ionize an electron across the energy gap and thus produce electron-hole pairs. This phenomenon, called avalanche breakdown, was suggested early as a pumping scheme for lasers [32], [33]. There are some difficulties

with the scheme, however. If a bulk crystal is used, the carriers will be very "hot," and they cannot form a degenerate distribution unless the field is turned off. Therefore, the field must be pulsed. Avalanche breakdown can also be observed in a reverse biased p-n junction. In this case the emission has a low quantum efficiency, because the field is in the direction to push the electrons to the n-side of the junction and the holes to the p-side where they are majority carriers and produce no radiation. These objections have been overcome recently in a structure discovered by Weiser and Woods [9]. The structure is essentially a p-π-p sandwich shown in Fig. 5(b). The avalanche breakdown is initiated by holes in the high resistivity π (low carrier concentration p) region where the field is high. The electrons are accelerated by the field into one of the p-regions where the field is low. They form a degenerate distribution, and laser action takes place perpendicular to the field direction.

The characteristic advantages of a p-n junction laser are that it is compact and simple, that is, it requires only a low dc voltage power supply. It can easily be modulated merely by modulating the current. It has high power conversion efficiency. Its big disadvantage is that often p-n junctions are difficult or impossible to make. The other three pumping methods eliminate this problem. Electron beam pumping and optical pumping are very convenient for searching out new laser materials. Since homogeneously doped bulk samples can be used, many uncertainties in doping of the laser in the active region are eliminated. Therefore, these methods are useful in studying the nature of the laser transition and the dependence of laser properties on doping, as for example, has been done in GaAs by Cusano [34], and InSb by Phelan [35]. However, in these methods the simplicity and compactness of the p-n junction laser system is lost. The avalanching structure eliminates the p-n junction while keeping the potential simplicity. Weiser's actual structures in GaAs are somewhat more complicated than we have indicated. They contain two high resistivity regions in series, and have a power conversion efficiency about an order of magnitude less than p-n junction lasers. However, these do not appear to be fundamental limitations. A similar type of structure has been studied in ZnTe; noncoherent high efficiency electroluminescence has been observed, but no stimulated emission has been reported [36].

II. MATERIALS

The list of semiconductor materials which have exhibited laser action has continued to grow at an undiminished rate. Table I lists the materials which have lased to date, together with the photon energy of oscillation and the method of excitation. The use of electron beam pumping and optical pumping has obviated the p-n junction, and thus has permitted laser action to be observed in several materials which cannot be made both n- and p-type.

GaAs [1]–[3], of course, was the first material to lase, and it has had the most extensive study and development.

TABLE I

SEMICONDUCTOR LASER MATERIALS. THE VALUE GIVEN FOR THE PHOTON IS THE MOST COMMON ONE REPORTED FOR LIQUID HELIUM TEMPERATURES. OBSERVED VALUES CAN VARY SLIGHTLY FROM THIS BECAUSE OF DIFFERENT IMPURITY DENSITIES OR DIFFERENT KINDS OF TRANSITIONS

Material	Photon Energy (eV)	Method of Excitation	Reference
ZnS	3.82	electron beam	[4a]
ZnO	3.30	electron beam	[4]
CdS	2.50	electron beam	[7a], [29], [57]
		optical	[29]
GaSe	2.09	electron beam	[61]
CdS_xSe_{1-x}	1.80–2.50	electron beam	[60]
CdSe	1.82	electron beam	[59]
CdTe	1.58	electron beam	[58]
$Ga(As_xP_{1-x})$	1.41–1.95	p-n junction	[37]
GaAs	1.47	p-n junction	[1]–[3]
		electron beam	[48]
		optical	[8a], [29]
		avalanche	[9]
InP	1.37	p-n junction	[39]
$In_xGa_{1-x}As$	1.5	p-n junction	[40]
GaSb	0.82	p-n junction	[44], [45]
		electron beam	[47]
InP_xAs_{1-x}	1.40	p-n junction	[41]
InAs	0.40	p-n junction	[38], [49]
		electron beam	[46]
		optical	[28], [49]
InSb	0.23	p-n junction	[42], [43]
		electron beam	[47]
		optical	[8]
Te	0.34	electron beam	[56]
PbS	0.29	p-n junction	[52]
		electron beam	[51]
PbTe	0.19	p-n junction	[50]
		electron beam	[51]
		optical	[64]
PbSe	0.145	p-n junction	[5]
		electron beam	[51]
$Hg_xCd_{1-x}Te$	0.30–0.33	optical	[64]
$Pb_xSn_{1-x}Te$	0.075–0.19	optical	[65]

In Section III we shall review the present state-of-the-art of GaAs lasers. All the other direct bandgap III–V compounds including several alloys have been reported to lase [37]–[49]. The last of this group, GaSb, was reported only two years ago [44], [45]. Apparently the difficulty was a materials problem in this semiconductor. The properties of the III–V semiconductor lasers are all quite similar. The stimulated emission usually occurs at an energy slightly less than the energy gap of the material, which means that the radiative transition involves impurity states. In InSb [35], GaSb [45], and InAs [49] lasing has also been observed at the energy gap from direct band to band recombination.

A second group of compounds to exhibit laser action is the lead salts [5], [50], [52]—PbS, PbTe, and PbSe. These are also direct gap materials. However, their extrema are not at the Brillouin zone center, but along the ⟨111⟩ directions in the zone [53]. The effect of a magnetic field on the emission from these lasers has been studied, and the results show that the laser emission is due to transitions from the conduction band to the valence band [54]. This is to be expected since the high static

dielectric constant of these materials leads to a low ionization energy for hydrogenic impurities. An interesting thing about these materials is that the p-n junctions are made by controlling the deviations from stoichiometry, rather than by doping with impurities [55].

The only elemental semiconductor yet to lase is tellurium, at a wavelength of 3.4 μ [56]. Several II–VI materials have been made to lase [4], [4a], [8a], [57]–[60], ZnS, ZnO, CdS, CdTe, CdSe, and the alloys $Cd(S_xSe_{1-x})$. As mentioned before, this has been accomplished with electron beam excitation, but not with p-n junctions, either because junctions cannot be made or, in the case of CdTe, because the doping is not heavy enough to satisfy the condition for population inversion (2). CdS, which oscillates at 4950A, is the shortest wavelength laser. Line narrowing has been observed in GaSe [61], but no further evidence of laser action has yet been reported.

Laser action has also been observed in alloys of several of the compounds, especially the III–V compounds, as can be seen from Table I. Some of these deserve special mention. $Ga(As_{1-x}P_x)$ alloys ($\lambda = 8400$–6500A, $0 < x < 0.5$) are the shortest wavelength p-n junction laser materials. They were the second semiconductors to exhibit laser action. They have threshold current densities comparable to GaAs [62], but their much lower thermal conductivity limits their potential usefulness [63]. The $(Cd_xHg_{1-x})Te$ alloys are potential long wavelength lasers [64]. Optical excitation from a GaAs diode laser was used for these materials, and spontaneous emission was observed from 3 to 15 μ and laser emission at 3.8 and 4.1 μ. $Cd(S_xSe_{1-x})$ alloys permit laser action at any wavelength in a good part of the visible spectrum. High efficiency (11 percent) and pulsed power output of 10 W have been observed in these alloys [60]. $Pb_xSn_{1-x}Te$ alloys give the longest wavelength atmospheric pressure laser to date [65].

III. JUNCTION LASERS—STATE-OF-THE-ART

In this section we survey the properties and give the present state-of-the-art in p-n junction lasers. The discussion will be primarily about GaAs, despite the fact that considerable work has now also been done in other materials.

Junction Fabrication

For lasers p-n junctions are most commonly made by diffusion of impurities. In GaAs [66], [67], an acceptor, usually Zn, is diffused into an n-type substrate having a carrier concentration in the range from 10^{17} to 6×10^{18} cm^{-3}. For typical diffusion conditions the zinc concentration rises rapidly away from the junction to about 10^{20} cm^{-3} in 5 μ, and then slowly increases to 2×10^{20} cm^{-3} at the surface. Junction depths range from 2 to 100 μ. Capacitance measurements show that the net impurity concentration (acceptors minus donors) is graded near the junction. The gradient at the junction is 2×10^{23} cm^{-4} for a typical laser with 2×10^{18} cm^{-3} electrons in the substrate.

Since the active light emitting region is only a few microns thick in the vicinity of the junction [68], it is important that the junction be flat for the laser characteristics to be good. Large deviations of the junction from planarity have been observed and found to be correlated with defects present in the substrate crystal. It has also been found that large strains are introduced during the diffusion [69].

An alternate method of preparation for lasers, epitaxial solution regrowth, has been used by Nelson [70]. In this technique a solution of GaAs plus dopants dissolved in Ga or Sn near its saturation temperature is allowed to wet an oppositely doped GaAs wafer. Cooling results in epitaxial growth on the wafer and a p-n junction at the interface. Some of the wafer is first dissolved so that a "fresh" surface is at the junction. Since the growth tends to go along (100) crystallographic planes, good planar junctions can be obtained.

Laser junctions have also been made by epitaxial regrowth from the vapor [71]. In both liquid and vapor regrown lasers, heat treatment can improve performance.

The resonant structure most often used is the Fabry-Perot structure shown in Fig. 4. The reflecting ends perpendicular to the junction are made optically flat and parallel by cleaving along (110) crystallographic planes or by polishing. The other two sides are sawed or etched to eliminate coherent feedback. Because of the high index of refraction, the reflectivity of the ends is high enough so that it is not necessary to have reflective coatings for normal operation of the laser. However, coatings can be applied. It is important that good electrical contact be made over the whole top and bottom of the laser to give a uniform current distribution, since nonuniform currents increase the threshold current. This will be discussed more fully in the following section.

Threshold Current Density

In order to compare experimental thresholds with theory it is useful to consider (10b). β and α can be determined from experiment by several techniques [10]. One that has worked well is the measurement of the dependence of j_t on laser length [72], [73]. Equation (10b) predicts a linear dependence. Figure 6 shows experimental results on a set of lasers made from the same crystal and of varying lengths. As can be seen, the linear relation holds extremely well. At 4.2°K, $\beta = 5.1 \times 10^{-2}$ cm/A and $\alpha = 13$ cm^{-1} have been found for diffused lasers with a substrate electron concentration 5×10^{18} cm^{-3} [21]. The theoretical value of β can be obtained from (10a) provided values for ν, $\Delta\nu$, d, and η are known. The photon energy $h\nu$ is 1.46 eV, and $\Delta h\nu$ is 0.02 eV [21]. The values of η and d are somewhat uncertain. The external quantum efficiency has been measured to be as high as 0.6 at 20°K [72]. For a rough comparison, we take $\eta = 1$. Direct visual observation indicates that $d < 5\ \mu$ [67], [68], and interference experiments indicate it to be greater than 1 μ [74]. We assume d is 2 μ. Then, from (9a), we find $\beta = 0.14$ cm/A. The experimental value of β is within a factor of three of

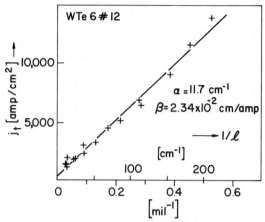

Fig. 6. Threshold current density vs. the reciprocal of the length of a set of lasers made from the same crystal. $T = 77$°K. $R = 0.25$ (After Pilkuhn and Rupprecht [72].)

theory. For less heavily doped diodes (carrier concentration of 9×10^{17} cm^{-3}) $\beta \approx 0.12$ cm/A has been found experimentally [75]. Since for typical laser lengths of 100 to 400 μ, the contribution of α to the threshold is less than 30 percent, GaAs lasers approach their ideal threshold current at 4.2°K.

In order to calculate the loss term α it is necessary to solve the electromagnetic wave problem for a thin active light emitting region sandwiched between two lossy regions. In order to get values of α small enough to agree with experiment, it is necessary to postulate some mode confinement or wave guiding of the diode, due to a higher value of the real part of the index of refraction in the active region than in the surrounding regions. Such a variation in index arises because the free carrier concentration, and the wavelength and the steepness of the absorption edge vary with position in junction. This problem has been treated by several workers and is reviewed by BN [10] and by Stern [13]. Detailed calculations by Stern [76] have shown that agreement with the experimental loss can be obtained with reasonable choices for the thickness of the active region and the index variation.

Temperature Dependence of the Threshold

Figure 7 shows the temperature dependence of j_t for diffused GaAs lasers. At high temperatures a dependence close to T^3 is observed, and at low temperatures j_t is fairly constant [77], [78]. The temperature at which the T^3 dependence begins, increases with increasing substrate carrier concentration. It has also been observed that the j_t temperature dependence is well represented by an exponential variation over the whole range [79]. Qualitatively this behavior is expected, since as the temperature is increased, the distribution of electrons becomes nondegenerate so that upward as well as downward transitions can occur at the lasing photon energy in the active region. Thus, the gain will be reduced for a given current density, and j_t will increase. At higher carrier concentrations this effect does not become important until higher temperatures.

To calculate the temperature dependence of the threshold it is necessary to have a detailed model for the laser transitions. A simple model of band to band momentum conserving transitions with constant population inversion gives $j_t \sim T$ [67]. It is well known that, for the impurity concentrations normally encountered in injection lasers, the energy bands will be seriously perturbed from their simple parabolic shape, i.e., they will have tails. The impurities will also cause a relaxation of the momentum conservation condition. Lasher and Stern [17] took partial account of these problems in threshold temperature dependence calculations. They kept the assumption of interband transitions between parabolic bands, but eliminated the momentum conservation condition, and assumed that the transition probability is the same for all pairs of states in the valence and conduction bands. They obtained closer agreement with experiment than the simple theory. Assuming that the loss is independent of temperature, they found a threshold dependence of $T^{1.8}$ to $T^{2.6}$ depending on the magnitude of the loss. A slight temperature dependence of the loss would bring the theory into very good agreement with the experimental temperature dependence of the threshold current density. However, there is a serious discrepancy with experiment; their calculations yield a nonlinear dependence of gain on current density (e.g., at 80°K $g \propto j^2$ approximately). The experimental results [21], however, indicate that $g \propto j$ holds up to 300°K. It has also been found that the loss is, in fact, not very temperature dependent and most of the temperature dependence of the threshold comes from the variation of β [78]. Recently, Stern [18] has performed calculations which take account of the energy band tails. He finds the gain much more nearly linear with current at high temperatures.

The experimental results of Fig. 7 show that curves for different carrier concentrations nearly coincide at high temperature, but differ by more than an order of magnitude at low temperature. On the other hand, on a simple basis, only a relatively small increase in threshold with carrier concentration at low temperature due to increased linewidth is expected. The curves should cross at an intermediate temperature, and the more heavily doped lasers are expected to have substantially lower threshold currents at room temperature. It is possible that the higher concentration of defects in the more heavily doped crystals is responsible for the higher thresholds observed experimentally.

Thresholds about a factor of three lower at room temperature than those of ordinary diffused lasers have been found for solution regrown lasers [80]. There is a larger temperature independent region for the threshold. A similar but less pronounced effect has been found for vapor epitaxial lasers [71]. This effect has been attributed to the presence of more pronounced energy band tails in these lasers [80].

The effect of the spatial distribution of current on threshold current has been studied using various structures [81]–[84], such as the one shown in the inset in Fig. 8

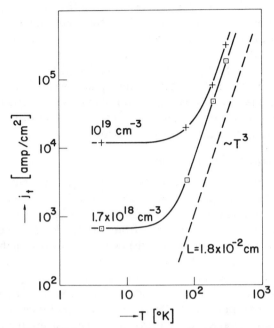

Fig. 7. Threshold current density vs. temperature. The numbers refer to the substrate carrier concentration. (After Pilkuhn and Rupprecht [18].)

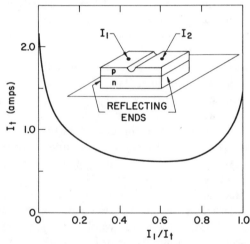

Fig. 8. Threshold current vs. the fraction of the current flowing through one of the contacts to a double unit shown in the inset. The laser beam comes out from the reflecting ends. $T = 77°K$.

[82]. The structure, called a double unit, is divided into two parts by a channel etched in the p-side of the junction parallel to the reflecting ends. The distribution of current is controlled by regulating the relative amounts of current flowing through each part of the junction. It can be seen in Fig. 8 that the threshold current I_t is much higher for nonuniform current (I_1, $I_2 \cong 0$ or 1) than for uniform ($I_1 \cong I_2 \cong 0.5$). This is true even at 2°K. At first sight this is puzzling since at $T = 0°K$, as in a four level laser, the lower level of the injection laser is unoccupied [$N_1 = 0$ in (4)]. The threshold should be independent of the distribution of the excitation, since there should be no absorption at the laser wavelength in the inactive or less excited parts of the laser. The reason this does not hold for the injection laser, lies in the fact that the continuous na-

ture of the states has been neglected. If the current is non-uniform, absorption as well as stimulated emission of photons can occur even at absolute zero. This can be seen from the simple model of the injection laser which we considered in Section I. As can be seen from Fig. 2 and (7), the gain will be negative (absorption positive) if F_c is less than $h\nu$. Therefore, the threshold will be higher for nonuniform excitation since there will be absorbing regions of the junction where F_c is low.

Operating Characteristics

Despite the fact that the internal quantum efficiency (number of photons generated divided by number of electron-hole pairs recombining) is 0.5 or higher [85], very little energy escapes from the diode laser below threshold because most of the light is totally internally reflected and absorbed in the inactive regions of the diode. Furthermore, even above threshold, external quantum efficiencies of 50 percent or less are observed at most temperatures [73], so that a good fraction of the light still remains in the laser. Dissipation of the heat produced by the absorbed light and by the series resistance is one of the major limitations on the operation of injection lasers. Because of the increase of the threshold and the decrease of the thermal conductivity of the semiconductor with increasing temperature above 30°K, the problem becomes more difficult as the temperature is raised. In order to operate an injection laser continuously, it is necessary to provide effective cooling. This can be done by placing the laser in contact with a large metallic heat sink, which is cooled by a cryogenic liquid bath. CW operation with power outputs of several watts has been obtained from GaAs injection lasers using liquid helium [11], [86], hydrogen [87], and nitrogen [88] baths. At room temperature only pulses of a few hundred nanoseconds or less can be used. The heat dissipation has been studied theoretically [89]–[93]. For CW operation, it is desirable to make the laser very narrow, that is, with a high ratio of length to width in the junction plane. Keyes [92] has shown that in order to achieve CW operation at room temperature with currently available threshold current densities, it would be necessary to make a laser with a length to width ratio of several hundred. This is beyond the present state-of-the-art. In order to summarize the present state-of-the-art, we shall describe the construction and operating characteristics of Marinace's GaAs laser heat sink package [88]. The laser diode shown in Fig. 9, is mounted between two relatively large copper bars. The junction plane is parallel to the bars, which serve as electrical contacts to the *n*- and *p*-regions. An insulating spacer is provided, and the bars are made so that it is necessary to force them open to insert the laser. Thus, the laser is under pressure and good thermal contact is then provided. Figure 10 shows a plot of optical power output vs. current density along the bottom of the figure, and current along the top for a typical continuously operating unit at 77°K. These data were taken in an integrating sphere, but because of the geometry of the

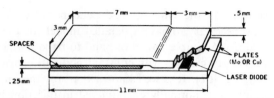

Fig. 9.   Heat sink mount for laser. (After Marinace [87].)

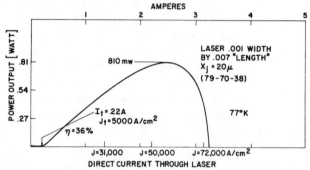

Fig. 10.   Light output vs. current for laser in a heat sink mount. A small amount of spontaneous emission is not shown, but should be visible above 3 amperes where the stimulated emission turns off due to heating. (Courtesy of Marinace.)

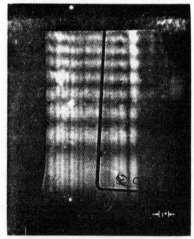

Fig. 11.   Far-field pattern of laser. The horizontal pattern is caused by diffraction of the light from the mount. The vertical pattern results from the coherence of the laser. (After Engeler and Garfinkel [87].)

mount, about 80 to 90 percent of the light comes out one end. The threshold current is about 0.2 amperes, and the differential quantum efficiency is 36 percent. The peak power output 810 mW is reached at about 2.5 amperes. Recently, peak power outputs in the 1 to 2 watt range are typical. The best values obtained in a few units thus far are 3 watts [94].

The far-field pattern of these lasers is typical of any GaAs injection lasers. A far-field pattern for a laser operating at 20°K is shown in Fig. 11 [91]. The junction plane is horizontal. It can be seen that the intensity is distributed over approximately 10° in the horizontal plane by 16° in the vertical plane. The vertical interference pattern indicates that the emission is coherent over the whole width of the junction in the horizontal plane. The fact that so many vertical bars are observed indicates that there are many modes present. Patterns with fewer vertical bars indicating fewer modes have also been observed [91].

More direct evidence on the multimode nature of a laser output can be obtained from measurement of the spectral properties. For a Fabry-Perot cavity, the wavelength of the axial modes is determined by

$$m\lambda = 2n_0 l \tag{11}$$

where m is the mode number. The spacing between successive modes is roughly 1 to 2A for typical laser lengths of 10 to 20 mils. Many axial modes are contained within the spontaneous line, which is about 100 Å wide. Ideally, the laser is expected to oscillate in one mode for which the gain is a maximum near the center of the spontaneous line, since the spontaneous line is homogeneously broadened [95]. In practice, single mode operation is quite rare. It has been observed over a range of current density from threshold to a factor of 6 above threshold at liquid helium temperature in selected small diodes, which have been etched so that the cavity has a large length to width ratio [96], [97]. However, except very close to threshold, most lasers oscillate in more than one mode. High resolution measurements on lasers near threshold have shown that the Fabry-Perot modes are actually sometimes several families of modes [98]. Oscillation is often observed in modes which are members of different families as shown in Fig. 12. Fenner [99] has observed that this multifamily behavior occurs in cleaved diodes, but polished diodes exhibit single family behavior. The different familes may result from cleavage steps or off axis modes. This multimoding not only affects the far-field pattern, but also the noise properties of the laser [96], [97]. It has been found that the noise in single mode lasers is very small and equal to the fundamental spontaneous emission noise. On the other hand, the noise in each mode in a multimode laser is quite high. However, there are correlations between the modes, so that the noise in the output integrated over wavelength is quite small again.

Figure 13 shows Konnerth's unpublished spectral results on a CW laser operating with a liquid nitrogen bath. As in Fig. 12, only close to threshold is there essentially single mode operation. The unequal spacing of the modes shows that there is more than one family of modes. There is a large shift in the wavelength of the modes with increasing current caused by heating of the diodes. This shift is brought about for the following reasons. 1) The temperature dependence of the index refraction causes a continuous shift of each mode to a longer wavelength with increasing temperature. At 77°K this shift is 0.46 Å/°K or 20 Gc/s°K [10]. 2) The spontaneous line shifts with the energy gap and causes jumping from one mode to the next toward longer wavelength with increasing temperature. This effect is about two and one-half times as big as the index effect. From the variation of the energy

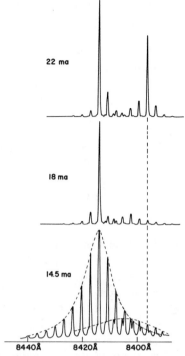

Fig. 12. Spectrum of a laser showing simultaneous oscillation in two modes at 10°K. (After Smith and Armstrong [97].)

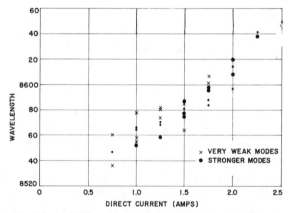

Fig. 13. Mode shifting spectrum of a CW heat-sinked laser as a function of current at 77°K. (Konnerth unpublished.)

gap with temperature [100], we can deduce that the junction temperature in Fig. 13 is 140°K at the maximum current.

Several attempts have been made to measure the linewidth [87], [101], [102] of the individual modes by interferometric techniques. For single mode operation, only an upper limit has been obtained, since measurements have been limited by instrument resolution. The lowest value reported so far is 150 kc/s at 77°K for 250 mW by Ahearn and Crowe who used a Michelson interferometer with a path imbalance of 3000 feet [102]. In multimode lasers, wider lines are measured.

At room temperature, threshold current densities are 100 000 A/cm² for ordinary diffused lasers. Heat treatment can reduce the threshold to 40 000 A/cm² [103]. Solution grown lasers have thresholds as low as 30 000

A/cm² [80]. Because of the high power required, the diodes can be operated only with pulses of about 500 ns or less duration. For a laser 5 mils wide by 15 mils long, a maximum 30 watts of light has been obtained from one end in a Marinace type heat sink with 300 amperes 30 ns input current pulses [104]. At this current level, much more power is dissipated by the series resistance, which may be about 0.05 ohms [94], than in the internally absorbed light. There are several reports of total pulse power outputs over 50 watts at room temperature [105]–[107]. The repetition rate affects the power output. At 10^5 pulses/s output as high as 3 watts with 50 amperes 70 ns pulses have been observed [108].

Modulation

One of the primary advantages of the junction laser is the ease with which it can be amplitude modulated. In efforts to determine its frequency response, a major difficulty has been in finding sensitive enough high-speed detectors. Measurements of the response of GaAs lasers to fast rise pulses have been made [109], [110]. Once the laser radiation starts to turn on, it turns on very rapidly—in less than 0.2 ns, the response time of the system [110]. However, there is a delay between the onset of the current pulse and the turn on of the stimulated emission. This delay is caused by the finite time necessary to build up the population inversion with the current through the junction, and at 77°K it is of the order of 1 ns. It decreases with increasing current above threshold and can be shown to follow the relation [110]

$$t_d = \tau ln I/(I - I_{th}). \tag{12}$$

For example, for $I=5\,I_{th}$ and $\tau=2$ ns, $t_d=0.44$ ns.

This delay does not limit the response of the laser to microwaves. The inversion will be built up when the microwaves are turned on, and it will be maintained even if the current falls below threshold, provided the period of the microwaves is less than the spontaneous lifetime.

At room temperature, delays as long as 150 ns have been observed in some lasers [111]. These delays have been shown to be due to slow traps in the diode. Vapor grown [71], solution grown [111], and heat treated diffused [103] lasers do not exhibit these delays. Amplitude modulation of a GaAs laser at 11 Gc/s has been reported [112]. The laser was operated CW at 4.2°K. The frequency limitation was imposed by the components of the system. Thus, except for the turn-on delay, no limitation has been found on the frequency response of the laser itself.

It is also possible to modulate the frequency of the laser. The index of refraction is dependent on stress applied to the semiconductor [113]–[116]. This effect has been used to frequency modulate a GaAs injection laser with ultrasonic techniques [117]. Ultrasonic waves at 2 Mc/s were applied to the laser with a quartz transducer. The sonic power was sufficient to change the frequency of the laser by at least 150 Mc/s. The maximum possible sound frequency, which is limited by the requirement that the pressure be uniform over the active

region of the diode, was estimated to be about 1 Gc/s.

A second method of frequency modulation is to vary the distribution of current in a double unit structure shown in Fig. 8 [82]–[84], [118]. One contact is used to pump the laser above threshold, and the other is used as the modulator. By varying the current in the modulator contact, continuous mode shifts of as much as 1Å have been observed [118]. It is also possible to shift the predominant mode of oscillation from one mode to the next, so that larger discrete changes in frequency can be obtained. Shifts of as much as 26Å, over several modes, have been observed in this manner [83]. A magnetic field can also be used to modulate the frequency of a laser [119].

An important point to consider about frequency modulation is that the frequency of each mode depends on temperature through the index of refraction. The difficulty is not severe at liquid helium temperature since there the index of refraction is insensitive to temperature. However, at 77°K, as discussed in the previous section, the effect produces a 0.46Å/°K shift in the modes, which is equivalent to 20 Gc/°K. Thus, extreme, if not impossible, temperature control is required to stabilize the frequency.

Applications and Devices

The field of semiconductor lasers is continuing to develop rapidly. At this stage it is difficult to say what applications will be important. In this section we discuss briefly some of the work which has been done to explore possible applications. The areas covered are opto-electronic laser devices and semiconductor laser communications.

The possibility of making an optical computer using injection laser devices has been considered, and it has been concluded that it is not a very promising approach at the present time [120]. Nevertheless, the devices which we shall discuss give a good idea of the state-of-the-art; they have interesting characteristics, and may provide a basis for the invention of more useful devices.

An optical interaction between lasers which can be used to generate the logical negation is the quenching or turning off of one laser by another laser. The output radiation of the quenching laser is passed through the active region of a second laser in a different direction from the resonant direction of the second laser. The radiation is amplified in the active region of the second laser. The power gained is subtracted from the output radiation of the second laser, and if this laser is not far above threshold, its gain can be reduced sufficiently to turn it off. Fowler [121] has observed quenching of one GaAs laser by another using the configuration shown in Fig. 14(a). The two lasers are mounted as closely as possible to each other on the same transistor header. Nevertheless, the coupling is rather inefficient, since much of the light from the quenching laser is lost because of its finite beam angle and because of scattering at the sawed surface of the quenched laser.

A much more efficient system for quenching of GaAs lasers in which both lasers are in a single block of mate-

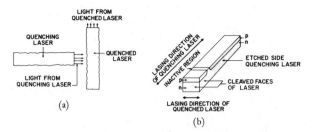

Fig. 14. Arrangements for quenching of a laser by the light from another laser. (a) Two separate lasers. (After Fowler [121].) (b) Both lasers made in a single block materials with SiO₂ masking techniques. (After Kelly [83].)

rial, has been made by Kelly [83]. With the use of SiO masking techniques, two electrically isolated junctions are made in a block of material as shown in Fig. 14 (b). All four sides perpendicular to the junctions are cleaved, but part of two of them is etched on the quenching laser so that it lases in the direction shown. The quenched laser is made narrow enough so that it can lase only perpendicular to this direction. The quenched laser is actually inside the cavity of the quenching laser so that coupling is good. A figure of merit for a quenching system is the ratio of the power extracted from the quenched laser beam to the power in the quenching beam. Kelly observed this ratio to be 0.6 to 0.75, which is an order of magnitude higher than reported by Fowler.

The double unit [82] discussed earlier and shown in Fig. 8 can be operated in an optically bistable mode [122] —two stable optical states for one electrical state. Current is put through one contact so that the unit is just below threshold. The second part of the unit is inactive. If a pulse of additional current is put into either contact, the unit will turn on and stay on after the pulse is off. The device can also be switched on with a light pulse, and a negative electrical pulse switches it off. The quenching scheme might also be used to switch it off. The explanation for the effect is thought to be that the inactive region acts as a saturable absorber which gives the laser two stable states [122].

GaAs injection lasers have also been operated as amplifiers [123]–[125]. For this purpose anti-reflecting coatings are put on the ends to reduce the feedback and inhibit oscillation. Increase of the threshold current by a factor of 10 and optical amplification by a factor of 2000 have been reported [125].

The ease of high-frequency modulation makes the GaAs injection laser attractive from a communications standpoint. In most communications systems, the laser would be operated well above threshold. Therefore, as we have seen, the output will be multimode, and heterodyne detection techniques cannot be used. The major advantages of the laser over an incoherent light emitting diode are its directionality and its high external quantum efficiency which gives high overall efficiency into an optical system. The frequency response is also better. The major disadvantage is that the laser must be refrigerated in order to take advantage of its frequency response. The big prob-

lem common to all terrestrial optical communication over long distances is atmospheric attenuation. Several optical links which use GaAs injection lasers as transmitters have been constructed [126]–[129]. One of them [129] has been demonstrated to be capable of transmitting 24 voice channels over 13 km.

ACKNOWLEDGMENT

It is a pleasure to acknowledge F. Stern, K. Weiser, and K. Konnerth for critically reading the manuscript and making many helpful comments on it. I am also indebted to J. W. Crowe, J. C. Marinace, K. E. Niebuhr, H. S. Rupprecht, and R. F. Rutz for helpful discussions. I want to thank J. C. Marinace for permission to publish Fig. 10 and K. Konnerth for permission to publish Fig. 13.

REFERENCES

[1] R. N. Hall, G. E. Fenner, J. D. Kingsley, T. J. Soltys, and R. O. Carlson, "Coherent light emission from GaAs junctions," *Phys. Rev. Lett.*, vol. 9, pp. 366–378, November 1962.
[2] M. I. Nathan, W. P. Dumke, G. Burns, F. H. Dill, Jr., and G. J. Lasher, "Stimulated emission of radiation from GaAs *p-n* junctions," *Appl. Phys. Lett.*, vol. 1, pp. 62–64, November 1962.
[3] T. M. Quist, R. H. Rediker, R. J. Keyes, W. E. Krag, B. Lax, A. L. McWhorter, and H. J. Zeiger, "Semiconductor maser of GaAs," *Appl. Phys. Lett.*, vol. 1, pp. 91–92, December 1962.
[4] F. H. Nicol, "Ultraviolet ZnO laser pumped by an electron beam," *Appl. Phys. Lett.*, vol. 9, pp. 13–15, July 1, 1966.
[4a] C. E. Hurwitz, "Efficient ultraviolet laser of ZnS by electron beam excitation," presented at the 1966 Solid-State Device Research Conference, Northwestern, Chicago, Ill., also *IEEE Trans. on Electron Devices*, to be published.
[5] J. F. Butler, A. R. Calawa, R. J. Phelan, Jr., A. J. Strauss, and R. H. Rediker, "PbSe diode laser," *Solid-State Commun.*, vol. 2, pp. 301–303, 1964.
[6] J. M. Besson, J. F. Butler, A. R. Calawa, W. Paul, and R. H. Rediker, "Pressure tuned PbSe diode laser," *Appl. Phys. Lett.*, vol. 7, pp. 200–208, October 1965, and private communication.
[7] N. G. Basov, Discussion in "*Advances in quantum electronics*," J. R. Singer, Ed., New York: Columbia University Press, 1961, p. 506.
[7a] N. G. Basov, O. V. Bogdankevich, and A. G. Devyatkov, "Exciting a semiconductor quantum generator with a fast electron beam," *Doklady Akademii Nauk SSSR*, vol. 155, p. 783, April 1964; (Translation: *Sov. Phys. Doklady*, vol. 9, p. 288, October 1964).
[8] R. J. Phelan, Jr., and R. H. Rediker, "Optically pumped semiconductor laser," *Appl. Phys. Lett.*, vol. 6, pp. 70–71, February 1965.
[8a] N. G. Basov, A. Z. Grasyuk, and V. A. Katulin, "Induced radiation in optically excited GaAs," *Doklady Akademii Nauk SSSR*, vol. 161, pp. 1306–1307, April 1965; (Translation: *Sov. Phys. Doklady*, vol. 10, pp. 343–344, October 1965).
[9] K. Weiser and J. F. Woods, "Evidence for avalanche injection laser in *p*-type GaAs," *Appl. Phys. Lett.*, vol. 7, pp. 225–228, October 1965.
[10] G. Burns and M. I. Nathan, "*P-N* junction lasers," *Proc. IEEE*, vol. 52, pp. 770–794, July 1964.
[11] Examples are given through [14]. R. H. Rediker, "Semiconductor lasers," *Phys. Today*, vol. 18, pp. 42–54, February 1965.
[12] B. Lax, "Progress in semiconductor lasers," *IEEE Spectrum*, vol. 2, pp. 62–75, July 1965.
[13] F. Stern, "Stimulated emission in semiconductors," in *Semiconductors and Semimetals, Physics of III–V Compounds*, R. K. Willardson and A. C. Beer, Eds., vol. 2. New York: Academic, 1966.
[14] W. P. Dumke, "The injection laser," in *Advances in Lasers*, A. K. Levine, Ed. New York: Dekker, to be published.
[15] See for example, B. A. Lengyel, *Lasers, Generation of Light*

by Stimulated Emission. New York: John Wiley, 1962.

[16] M. G. A. Bernard and G. Duraffourg, "Laser conditions in semiconductors," *Physica Status Solidi*, vol. 1, pp. 699–703, July 1961.

[17] G. J. Lasher and F. Stern, "Spontaneous and stimulated line shapes in semiconductor lasers," *Phys. Rev.*, vol. 133, pp. A553–563, January 1964.

[18] F. Stern, "Effect of band tails on stimulated emission of light in semiconductors," to be published.

[19] W. P. Dumke, "Interband transitions and maser action," *Phys. Rev.*, vol. 127, pp. 1559–1563, September 1962.

[20] G. J. Lasher, "Threshold relations and diffraction loss for injection lasers," *IBM J. Research and Develop.*, vol. 7, pp. 58–61, January 1963.

[21] M. H. Pilkuhn, H. Rupprecht, and S. E. Blum, "Effect of temperature on the stimulated emission from GaAs p-n junctions," *Solid-State Electronics*, vol. 7, pp. 905–909, 1964.

[22] D. K. Wilson, "Stimulated emission of exciton recombination radiation in GaAs p-n junctions," *Appl. Phys. Lett.*, vol. 3, pp. 127–129, October 1963.

[23] I. Melngailis, R. J. Phelan, Jr., and R. H. Rediker, "Luminescence and coherent emission in a large volume injection plasma in InSb," *Appl. Phys. Lett.*, vol. 5, pp. 99–100, September 1964.

[24] K. Weiser and F. Stern, "Higher order transverse modes in GaAs lasers," *Appl. Phys. Lett.*, vol. 5, pp. 115–116, September 1964.

[25] I. Melngailis, "Longitudinal injection-plasma laser in InSb," *Appl. Phys. Lett.*, vol. 6, pp. 59–60, February 1965.

[26] C. Klein, "Laser action threshold in electron-beam excited GaAs," *Appl. Phys. Lett.*, vol. 7, pp. 200–202, October 1965.

[27] C. Klein, "The excitation mechanism in electron beam pumped laser," *Physics of Quantum Electronics Conference Proceedings*, P. L. Kelley, B. Lax, and P. E. Tannenwald, Eds. New York: McGraw-Hill, 1966, pp. 424–434.

[28] I. Melngailis, "Optically pumped indium arsenide laser," *IEEE J. of Quantum Electronics (Correspondence)*, vol. QE-1, pp. 104–105, May 1965.

[29] N. G. Basov, "Quantum oscillator and amplifier investigations," in *Physics of Quantum Electronics*, P. L. Kelley, Ed. New York: McGraw-Hill, 1966, p. 411.

[30] N. G. Basov, A. Z. Grasyuk, I. G. Zabarev, and V. A. Katulyn, "Generation in GaAs under two photon optical excitation of Nd-glass laser emission," *JETP Lett.*, vol. 1, p. 118, 1965.

[31] V. K. Koniukhov, L. A. Kulevskii, and A. M. Prokhorov, "Cadmium sulfide laser with two-photon ruby excitation," *IEEE J. of Quantum Electronics*, vol. OE2, p. lxv, April 1966.

[32] N. G. Basov, O. N. Krokhin, and Y. M. Popov, "Generation, amplification, and detection of infrared and optical radiation by quantum-mechanical systems," *Usp. Fiz. Nauk.*, vol. 72, pp. 161–209, October 1960; also in *Soviet Physics Uspekhi*, vol. 3, pp. 702–728, March 1961.

[33] N. G. Basov, B. M. Vul, and Y. M. Popov, "Quantum-mechanical semiconductor generators and amplifiers of electromagnetic oscillations," *J. Exptl. Theoret. Phys. (USSR)*, vol. 37, pp. 587–588, August 1959; also in *Soviet Phys. JETP*, vol. 10, pp. 416, February 1960.

[34] D. A. Cusano, "Radiative recombination from GaAs directly excited by fast electrons," *Solid-State Commun.*, vol. 2, pp. 353–358, 1964.

[35] R. J. Phelan, Jr., "Laser emission by optical pumping of semiconductors," in *Physics of Quantum Electronics Conference Proceedings*, P. L. Kelley, B. Lax, and P. E. Tannenwald, Eds. New York: McGraw-Hill, 1966, pp. 435–441.

[36] B. L. Crowder, F. F. Morehead, and P. R. Wagner, "Efficient injection electroluminescence in ZnTe by avalanche breakdown," *Appl. Phys. Lett.*, vol. 8, pp. 148–149, March 1966.

[37] N. Holonyak, Jr., and S. F. Bevacqua, "Coherent (visible) light emission from $Ga(As_{1-x}P_x)$ junctions," *Appl. Phys. Lett.*, vol. 1, pp. 82–83, December 1962.

[38] I. Melngailis, "Maser action in InAs diodes," *Appl. Phys. Lett.*, vol. 2, pp. 176–178, May 1963.

[39] K. Weiser and R. S. Levitt, "Stimulated light emission from indium phosphide," *Appl. Phys. Lett.*, vol. 2, pp. 178–179, May 1963.

[40] I. Melngailis, A. J. Strauss, and R. H. Rediker, "Semiconductor diode masers of $(In_xGa_{1-x})As$," *Proc. IEEE (Cor-*

respondence), vol. 51, pp. 1154–1155, August 1963.

[41] F. B. Alexander et al., "Spontaneous and stimulated infrared emission from indium phosphide arsenide diodes," *Appl. Phys. Lett.*, vol. 4, pp. 13–15, January 1964.

[42] R. J. Phelan, A. R. Calawa, R. H. Rediker, R. J. Keyes, and B. Lax, "InSb diode laser," *Appl. Phys. Lett.*, vol. 3, pp. 143–145, November 1963.

[43] C. Benoit a la Guillaume and P. Lavallard, "Laser effect in indium antimonide," *Solid-State Commun.*, vol. 1, pp. 148–150, November 1963.

[44] C. Chipaux, G. Duraffourg, J. Loudette, J. P. Noblanc, and M. Bernard, "Emission stimulee dans l'antmoniure de gallium," *7th Internat'l Conference on Physics of Semiconductors, Radiative Recombination in Semiconductors* (Paris, 1964). Paris: Dunod, 1965, pp. 217–222.

[45] R. Eymard, G. Duraffourg, C. Chipaux, and M. Bernard, "Laser action in gallium antimonide diodes," in *Physics of Quantum Electronics Conference Proceedings*," P. L. Kelley, B. Lax, P. E. Tannewald, Eds. New York: McGraw-Hill, 1966, pp. 450–457.

[46] C. Benoit a la Guillaume and J. M. Debever, "Effect laser dans l'arseniure d' indium par bombardment electronique," *Solid-State Commun.*, vol. 2, pp. 145–146, 1964.

[47] ——, "Effect laser dans l'antimoniure de gallium par bombardment electronique," *Compt. Rend.*, vol. 259, pp. 2200–2206, October 1964.

[48] C. E. Hurwitz and R. J. Keyes, "Electron-beam-pumped GaAs laser," *Appl. Phys. Lett.*, vol. 5, pp. 139–141, October 1964.

[49] I. Melngailis and R. H. Rediker, "Properties of InAs lasers," *J. Appl. Phys.*, vol. 37, pp. 899–911, February 1966.

[50] J. F. Butler, A. R. Calawa, R. J. Phelan, Jr., T. C. Harman, and A. J. Strauss, "PbTe diode laser," *Appl. Phys. Lett.*, vol. 5, pp. 75–76, August 1964.

[51] C. Hurwitz, A. R. Calawa, and R. H. Rediker, "Electron beam pumped lasers of PbS, PbSe, PbTe," *IEEE J. of Quantum Electronics (Correspondence)*, vol. QE-1, pp. 102–104, May 1965.

[52] J. F. Butler and A. R. Calawa, "PbS diode laser," *J. Electrochem. Soc.*, vol. 54, pp. 1056–1057, October 1965.

[53] K. F. Cuff, M. R. Ellett, C. D. Kuglin, and L. R. Williams, "The band structure of PbTe, PbSe, and PbS," *Physics of Semiconductors Proceedings of the 7th Internat'l Conference* (Paris, 1964). Paris: Dunod, 1964, pp. 677–684.

[54] J. F. Butler and A. R. Calawa, "Magnetoemission studies of PbS, PbTe, and PbSe diode lasers," in *Physics of Quantum Electronics Conference Proceedings*, P. L. Kelley, B. Lax, and P. E. Tannenwald, Eds. New York: McGraw-Hill, 1966.

[55] J. F. Butler, "Diffused junction diodes of PbSe and PbTe," *J. Electrochem. Soc.*, vol. 111, pp. 1150–1154, October 1964.

[56] C. Benoit a la Guillaume and J. M. Debever, "Emission spontanee et stimulee du tellure par bombardment electrique," *Solid-State Communication*, vol. 3, pp. 19–20, 1965.

[57] ——, "Effet laser dans le sulfure de cadmium par bombardement electronique," *Compt. Rend.*, vol. 261, pp. 5428–5430, 1965.

[58] V. S. Vauilov and E. L. Nolle, "Cadmium telluride laser with electron excitation," *Doklady Akademii Nauk SSSR*, vol. 164, pp. 73–74, September, 1965; (Translation: *Soviet Physics—Doklady*, vol. 10, pp. 827–838, March 1966).

[59] C. E. Hurwitz, "Electron beam pumped laser of CdSe and CdS," *Appl. Phys. Lett.*, vol. 8, pp. 121–124, March 1966.

[60] C. E. Hurwitz, "Efficient visible lasers of $Cd(S_xSe_{1-x})$ by electron beam excitation," *Appl. Phys. Lett.*, vol. 8, pp. 243–245, May 15, 1966.

[61] N. G. Basov, O. V. Bogdankevich, A. N. Pechenov, G. B. Abdulaev, G. A. Akhundov, E. Yu. Salaev, "Radiation in GaSe single crystals induced by excitation with fast electrons," *Doklady Akademii Nauk SSSR*, vol. 161, p. 1059, April 1965; (Translation: *Soviet Physics-Doklady*, vol. 10, pp. 329–330, October 1965).

[62] J. J. Tietjen and S. A. Ochs, "Improved performance of $Ga(As_{1-x}P_x)$ laser diodes," *Proc. IEEE (Correspondence)*, vol. 53, pp. 180–181, February 1965.

[63] R. O. Carlson, G. A. Slack, and S. J. Silverman, "Thermal conductivity of GaAs and $GaAs_{1-x}P_x$ laser semiconductors," *J. Appl. Phys.*, vol. 36, pp. 505–507, February 1965.

[64] I. Melngailis and A. J. Strauss, "Spontaneous and coherent photoluminesce in $Cd_xHg_{1-x}Te$ alloys," *Appl. Phys. Lett.*, to be published.

[65] J. O. Dimmock, I. Melngailis, and A. J. Strauss, "Band structure and laser action in $Pb_xSn_{1-x}Te$," *Phys. Rev. Lett.*, vol. 16, pp. 1193–1196, June 27, 1966.

[66] J. C. Marinace, "Diffused junctions in GaAs injection lasers," *J. Electrochem. Soc.*, vol. 110, pp. 1153–1159, November 1963.

[67] M. H. Pilkuhn and H. S. Rupprecht, "Diffusion problems related to GaAs injection lasers," *Trans. AIME*, vol. 230, pp. 296–300, March 1964.

[68] R. N. Hall, "Coherent light emission from *p-n* junctions," *Solid-State Electronics*, vol. 6, pp. 405–416, September 1963.

[69] G. H. Schwuttke and H. S. Rupprecht," X-ray analysis of diffusion induced defects in gallium arsenide," *J. Appl. Phys.*, vol. 37, pp. 167–173, January 1966.

[70] H. Nelson, "Epitaxial growth from the liquid state and its application to the fabrication of tunnel and laser diodes," *RCA Rev.*, vol. 24, p. 603, December 1963.

[71] N. N. Winogradoff and H. K. Kessler, "Light emission and electrical characteristics of epitaxial GaAs lasers and tunnel diodes," *Solid-State Commun.*, vol. 2, pp. 119–121, 1964.

[72] M. H. Pilkuhn and H. S. Rupprecht, "A relation between the current density at threshold and the length of Fabry-Perot type GaAs lasers," *Proc. IEEE (Correspondence)*, vol. 51, pp. 1243–1244, September 1963; and "Light emission from $GaAs_xP_{1-x}$ diodes," *Trans. Met. Soc. AIME*, vol. 230, pp. 282–286, March 1964.

[73] S. V. Galginaitis, "Efficiency measurements of GaAs electroluminescent diodes," *J. Appl. Phys.*, vol. 35, pp. 295–298, February 1964.

[74] H. C. Casey, R. J. Archer, R. H. Kaiser, and J. C. Sarace, "Width of the spontaneous emission region in degenerate GaAs *p-n* junctions," *J. Appl. Phys.*, vol. 37, pp. 893–898, February 1966.

[75] M. I. Nathan, A. B. Fowler, and G. Burns, "Oscillations in GaAs spontaneous emission in Fabry-Perot cavities," *Phys. Rev. Lett.*, vol. 11, pp. 152–154, August 1963.

[76] F. Stern, "Radiation confinement in semiconductor lasers," *7th Internat'l Conference on the Physics of Semiconductors, Radiative Recombination in Semiconductors* (Paris, 1964). Paris: Dunod, 1965, pp. 165–170.

[77] G. Burns, F. H. Dill, Jr., and M. I. Nathan, "The effect of temperature on the properties of GaAs laser," *Proc. IEEE (Correspondence)*, vol. 51, pp. 947–948, June 1963.

[78] M. H. Pilkuhn and H. S. Rupprecht, "Influence of temperature on radiative recombination in GaAs *p-n* junction lasers," *7th Internat'l Conference on the Physics of Semiconductors, Radiative Recombination in Semiconductors* (Paris, 1964). Paris: Dunod, 1965, pp. 195–199.

[79] J. I. Pankove, "Temperature dependence of emission spectrum and threshold current GaAs lasers," *7th Internat'l Conference on the Physics of Semiconductors, Radiative Recombination in Semiconductors* (Paris, 1964). Paris: Dunod, 1965, pp. 201–204.

[80] G. C. Dousmanis and H. Nelson, "Temperature dependence of threshold current in GaAs lasers," *Appl. Phys. Lett.*, vol. 5, pp. 174–176, November 1964.

[81] A. Kawaji, "Some properties of junction triode laser," *J. Appl. Phys. (Japan)*, vol. 3, pp. 425–426, 1964.

[82] M. I. Nathan, J. C. Marinace, R. F. Rutz, A. E. Michel, and G. J. Lasher, "GaAs injection laser with novel mode control and switching properties," *J. Appl. Phys.*, vol. 36, pp. 473–480, February 1965.

[83] C. E. Kelly, "Interactions between closely coupled GaAs injection lasers," *IEEE Trans. on Electron Devices*, vol. ED-12, pp. 1–4, January 1965.

[84] N. G. Basov, Yu.P. Zakharov, V. V. Nikitin, and A. A. Sheronov, "GaAs *p-n* junction laser with nonuniform distribution of current," *Fizika Trerdogo Tela*, vol. 7, pp. 3128–3130, October 1965; (Translation: *Soviet Physics—Solid State*, vol. 7, pp. 2532–2533, April 1966).

[85] D. E. Hill, "Internal quantum efficiency of GaAs electroluminescent diodes," *J. Appl. Phys.*, vol. 36, pp. 3405–3409, November 1965.

[86] M. Cliftan and P. P. Debeye, "On the parameters which affect the CW output of GaAs lasers," *Appl. Phys. Lett.*, vol. 6, pp. 120–122, March 1965.

[87] W. E. Engeler and M. Garfinkel, "Characteristics of a continuous high-power GaAs junction laser," *J. Appl. Phys.*, vol. 35, pp. 1734–1741, June 1964.

[88] J. C. Marinace, "High power CW operation of GaAs injection lasers at 77°K," *IBM J. Research and Develop.*, vol. 8, pp. 543–544, November 1964.

[89] S. Mayburg, "Temperature limitation on continuous operation of GaAs lasers," *J. Appl. Phys.*, vol. 34, pp. 3417–3418, November 1963.

[90] G. J. Lasher and W. V. Smith, "Thermal limitation of the energy of single injection laser light pulse," *IBM J. Research and Develop.*, vol. 8, pp. 532–536, November 1964.

[91] W. Engeler and M. Garfinkel, "Thermal characteristics of GaAs laser junctions under high power pulsed conditions," *Solid-State Electronics*, vol. 8, pp. 585–604, 1965.

[92] R. W. Keyes, "Thermal problems of the injection laser," *IBM J. Research and Develop.*, vol. 9, pp. 303–314, March 1965.

[93] J. Vilms, L. Wandinger, and K. L. Klohn, "Optimization of the gallium arsenide injection laser for maximum CW power output," *IEEE J. of Quantum Electronics*, vol. QE-2, pp. 80–83, April 1966.

[94] J. C. Marinace, private communication.

[95] T. H. Maiman, "Stimulated optical emission in fluorescent solids I theoretical consideration," *Phys. Rev.*, vol. 123, pp. 1145–1150, August 1961.

[96] J. A. Armstrong and A. W. Smith, "Intensity fluctuations in GaAs laser emission," *Phys. Rev.*, vol. 140, pp. A155–164, October 1965.

[97] A. W. Smith and J. A. Armstrong, "Intensity noise in multimode GaAs laser emission," *IBM J. Research and Develop.*, to be published.

[98] P. P. Sorokin, J. D. Axe, and J. R. Lankard, "Spectral characteristic of GaAs lasers operating in Fabry-Perot modes," *J. Appl. Phys.*, vol. 34, pp. 2553–2556, September 1963.

[99] G. E. Fenner, private communication.

[100] M. D. Sturge, "Optical absorption of gallium arsenide between 0.6 and 2.75 ev," *Phys. Rev.*, vol. 127, pp. 768–773, August 1962.

[101] J. A. Armstrong and A. W. Smith, "Interferometric measurement of linewidth and noise in GaAs lasers," *Appl. Phys. Lett.*, vol. 4, pp. 196–198, June 1964.

[102] W. E. Ahearn and J. W. Crowe, "Linewidth measurements of CW gallium arsenide lasers at 77°K," *IEEE J. of Quantum Electronics*, vol. QE-2, p. lxvi, April 1966.

[103] R. O. Carlson and T. J. Soltys, "Low threshold room temperature GaAs laser diodes," *Bull. Am. Phys. Soc.*, vol. 10, p. 607, June 1965.

[104] J. W. Crowe, private communication.

[105] H. Nelson, J. I. Pankove, F. Hawrylo, G. C. Dousmanis, and C. Reno, "High-efficiency injection laser at room temperature," *Proc. IEEE (Correspondence)*, pp. 1360–1361, November 1964.

[106] H. J. Henkel, E. Klein, and H. Kuckuck, "Das verhalten von GaAs laser dioden bei hohen strahlungsleistangen," *Solid-State Electronics*, vol. 8, pp. 475–478, 1965.

[107] K. G. Hambleton and F. E. Birbeck, "Design of a compact 100 watt gallium arsenide laser transmitter," *SERL Tech. J.*, vol. 15, pp. 111–114, February 1965.

[108] K. Niebuhr, private communication.

[109] C. Hilsum, D. J. Oliver, and J. M. Tanner, "The speed of response of GaAs lasers," *Phys. Letters*, vol. 8, pp. 232–233, February 1964.

[110] K. Konnerth and C. Lanza, "Delay between current pulse and light emission of a gallium arsenide injection laser," *Appl. Phys. Lett.*, vol. 4, pp. 120–121, April 1964.

[111] K. Konnerth, "Turn on delay in GaAs in GaAs injection lasers operated at room temperature," to be published.

[112] B. S. Goldstein and R. M. Wiegand, "X-band modulation of GaAs lasers," *Proc. IEEE (Correspondence)*, vol. 53, p. 195, February 1965.

[113] J. Feinleib, S. Groves, W. Paul, and R. Zallen, "Effect of pressure on the spontaneous and stimulated emission from GaAs," *Phys. Rev.*, vol. 131, pp. 2070–2077, September 1963.

[114] M. J. Stevenson, J. D. Axe, and J. R. Lankard, "Line widths and pressure shifts in mode structure of stimulated emission

from GaAs junctions," *IBM J. Research and Develop.*, vol. 7, pp. 155–156, April 1963.

[115] F. M. Ryan and R. C. Miller, "The effect of uniaxial strain on the threshold current and output of GaAs laser," *Appl. Phys. Lett.*, vol. 3, pp. 162–163, November 1963.

[116] D. Mayerhofer and R. Braunstein, "Frequency tuning of GaAs laser diode by uniaxial stress," *Appl. Phys. Lett.*, vol. 3, pp. 171–172, November 1963.

[117] J. E. Ripper and G. W. Pratt, "Direct frequency modulation of semiconductor laser by ultrasonic waves," *IEEE J. of Quantum Electronics*, vol. QE-2, pp. lxvi–lxvii, April 1966.

[118] G. E. Fenner, "Internal frequency modulation of GaAs junction lasers by changing the index of refraction through electron injection," *Appl. Phys. Lett.*, vol. 5, pp. 198–199, November 1965.

[119] I. Melngailis and R. H. Rediker, "Magnetically tunable CW InAs diode maser," *Appl. Phys. Lett.*, vol. 2, pp. 202–204, June 1963.

[120] W. V. Smith, "Computer applications of lasers," this issue.

[121] A. B. Fowler, "Quenching of gallium arsenide injection lasers," *Appl. Phys. Lett.*, vol. 3, pp. 1–3, July 1963.

[122] G. J. Lasher, "Analysis of a proposed bistable injection laser," *Solid-State Electronics*, vol. 7, pp. 707–716, 1964.

[123] M. J. Coupland, K. G. Hambleton, and C. Hilsum, "Measurement of amplification in GaAs injection laser," *Phys. Rev. Lett.*, vol. 7, pp. 231–232, 1963.

[124] J. W. Crowe and R. M. Craig, Jr., "Small-signal amplification in GaAs lasers," *Appl. Phys. Lett.*, vol. 4, pp. 57–58, February 1964.

[125] J. W. Crowe and W. E. Ahearn, "Gallium arsenide laser diode amplifier," *IEEE J. of Quantum Electronics*, vol. QE-2, p. lxvii, April 1966.

[126] Examples are given in the next three references.

[127] B. A. Boershig, "A light-modulated data link," *IEEE Trans. on Broadcasting*, vol. BC-10, pp. 4–7, February 1964.

[128] E. J. Chatterton, "Semiconductor laser communication through multiple-scatter paths," *Proc. IEEE (Correspondence)*, vol. 53, pp. 2114–2115, December 1965.

[129] E. J. Schiel, E. C. Bullwinkel, and R. B. Weimer, "Pulse-code modulation multiplex transmission over an injection leser transmission system," *Proc. IEEE (Correspondence)*, vol. 353, pp. 2140–2141, December 1965.

Flashlamp-Excited Organic Dye Lasers

BENJAMIN B. SNAVELY

Invited Paper

Abstract—The flashlamp-excited dye laser is presently the only type of laser capable of tunable emission throughout most of the visible spectrum. Gain and power output of the device are comparable to solid-state systems although the laser performance is hindered by thermal effects, produced by spatially nonuniform excitation of the dye, and optical losses associated with the molecular triplet state. In most of the known laser dyes, steady-state lasing is prevented by triplet state effects.

The analysis of the gain of the dye laser is discussed in terms of the singlet-state absorption and fluorescence and triplet-state absorption spectra. The gain analysis is used to study the influence of the triplet state upon the critical inversion, and application of the analysis to a specific system is illustrated by the detailed discussion of the rhodamine 6G laser. A criterion for the maximum permissible triplet-state lifetime consistent with CW operation is given. The quenching of the triplet state of rhodamine 6G is shown to be rapid enough to allow CW operation, although thermal effects seem to be serious. The investigation of thermal effects is reviewed and the advantages of uniform excitation of the dye are pointed out. The minimum optical excitation power required for CW operation of a 7-cm-long rhodamine 6G laser of 2 mm diameter is estimated to be 850 watts neglecting thermal effects.

A catalog of dyes, with their structures, that have been used in flashlamp-excited dye lasers is given. Various methods of tuning and mode locking the dye laser are reviewed. With a single dye a tuning range of 40 nm may be obtained by substituting a diffraction grating for one of the laser mirrors. Mode locking can be produced by placing a saturable dye absorber in the cavity.

I. Introduction

A. History

THE EFFICIENT luminescence exhibited by many organic compounds makes their use as laser materials attractive and this possibility was considered early in laser research. Initial experiments produced disappointing results, however, and interest waned. Within the past three years the production of coherent, visible radiation by fluorescent organic dyes in solution has been demonstrated; the device based upon this phenomenon is called the dye laser. In this paper research on flashlamp-excited organic dye lasers will be reviewed.

The organic dye laser is to be distinguished from the inorganic liquid laser based upon the fluorescence of rare earth ions in liquid solvents, a device which is similar in many respects to the solid-state laser. The properties of the inorganic liquid laser have been extensively reviewed [1]–[3] and will not be discussed here.

One of the most attractive properties of the dye laser is tunability. In contrast to most other laser media the emission spectra of fluorescent dyes are broad, permitting the lasing wavelength to be tuned to any chosen value within a reasonably broad range. Furthermore, the number of fluorescent dyes is very large and compounds may be selected for emission in any given region of the optical spectrum. The dye laser is the first truly tunable laser which operates throughout the visible spectrum.

The dye laser combines many of the advantages of solid-state and gas lasers. The use of a liquid active medium simplifies the problem of obtaining high optical quality and facilitates cooling of the laser for operation at high pulse repetition rates. The gain of a dye solution is much greater than that obtainable from a gas and is comparable to the gain of solid-state materials. The use of a liquid active medium is not required, however; solid laser rods can also be made with dyes dispersed in plastics.

The earliest published suggestions that organic materials could be used as active media seem to be those of Brock *et al.* [4] and Rautian and Sobel'mann [5] who proposed in 1961 that triplet state phosphorescence could serve as the basis for an organic laser. In 1964 Stockman, Mallory, and Tittel [6] discussed a laser process based upon singlet-state fluorescence and Stockman [7] described early results in the experimental effort to realize a dye laser using the dye perylene excited by a fast powerful flashlamp. The first unambiguously successful effort to produce stimulated emission from organic molecules, however, was reported in 1966 by Sorokin and co-workers. These authors used a giant pulse ruby laser to excite solutions of the dyes chloro-aluminum phthalocyanine [8] (CAP) and 3,3′ diethyl-thiadicarbocyanine (DTTC) iodide [9] in an optical cavity, an arrangement referred to as the "laser-pumped laser." Similar results were obtained independently by Schäfer, Schmidt, and Volze [10], and by Spaeth and Bortfield [11] using several cyanine dyes with structures similar to DTTC.

Based upon the laser-pumped laser results Sorokin, Lankard, Hammond, and Moruzzi [12] suggested that flashlamp excitation of a dye laser should also be possible and estimated the flashlamp requirements. A suitable flashlamp was constructed and laser emission from solutions of several dyes of the xanthene class was reported by Sorokin and Lankard [13]. Shortly thereafter similar results were reported independently by Schmidt and Schäfer [14]. Flashlamp-excited dye lasers have since been studied in many laboratories.

B. Properties of Organic Dyes

To introduce the discussion of the dye laser some of the properties of organic dyes will be considered and terms defined for later use.

Manuscript received April 30, 1969. *This invited paper is one of a series planned on topics of general interest.—The Editor.*

The author is with Eastman Kodak Research Laboratories, Rochester, N. Y. 14650. He was on leave of absence at the Physikalisch-Chemisches Institut der Universität Marburg, Marburg/Lahn, Germany.

Reprinted from *Proc. IEEE*, vol. 57, pp. 1374–1390, Aug. 1969.

75

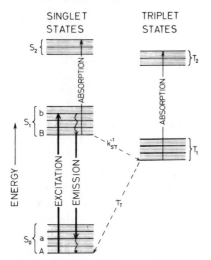

Fig. 1. Schematic representation of the energy levels of a dye molecule. The heavy horizontal lines represent vibrational states and the lighter lines represent the rotational fine structure. Excitation and laser emission are represented by the transitions $A \to b$ and $B \to a$, respectively. Other transitions represent losses in the laser process.

An energy level diagram characteristic of an organic dye molecule is shown in Fig. 1. The electronic ground state of the molecule is a singlet state, designated here as S_0, which spans a range of energies determined by the quantized vibrational and rotational excitation of the molecule. The energy between vibrational states, indicated in Fig. 1 by heavy horizontal lines, is typically of the order of 1400 to 1700 cm^{-1}. The energy spacing between rotational levels, designated by the lighter lines, is smaller than the spacing between vibrational levels by a factor of approximately 100. The rotational levels, therefore, provide a near continuum of states between the vibrational levels.

Each excited electronic state of the molecule consists of a similar broad continuum and optical transitions between these continua give rise to characteristic broad absorption and emission spectra. The first and second excited singlet states are designated as S_1 and S_2 in Fig. 1.

Transitions between singlet states are spin-allowed, giving rise to strong absorption bands. The intensity of an absorption band is conveniently specified in terms of the *molecular extinction coefficient* ε, which is defined by the relation

$$\varepsilon = (1/NL) \ln (I_0/I).$$

I_0/I is the ratio of the intensity of an incident probe light beam to the light intensity transmitted by a dye solution of path length L cm and concentration N molecules per cm^3. The extinction coefficient has the dimensions of cm^2 and is the absorption cross section for a single dye molecule.

As the first step of the laser process the molecules are excited from the lowest levels of the ground singlet S_0 state to higher vibrational–rotational levels of the S_1 state by absorbed light. This process is indicated by the transition from state A to state b ($A \to b$). The molecular energy then decays nonradiatively to level B, a low level of the S_1 state, as indicated by the wavy line $b \to B$. The laser emission

results from the stimulated transition $B \to a$. State a is a higher lying vibrational–rotational state of S_0. The laser process is terminated by the nonradiative decay $a \to A$.

The concentration of molecules in the excited singlet state S_1 must reach a certain value, the *critical inversion*, before coherent emission will be produced by the dye. The magnitude of the critical inversion depends upon the losses of the complete laser system.

The decay of radiation emitted in the spontaneous singlet-state process $B \to a$, known as *fluorescence*, is governed by the lifetime τ of state B. In the discussion which follows τ will be taken as the exponential decay lifetime for the fluorescence of a large number of excited dye molecules. For organic dye molecules τ is typically about 5×10^{-9} second. In contrast, the corresponding decay time for solid-state and inorganic laser materials is many orders of magnitude slower, about 10^{-3} second.

The photon energy for which the optical absorption of the molecule is a maximum is greater than the photon energy at the fluorescence maximum. The energy difference between absorption and emission processes is taken up by the radiationless processes $b \to B$ and $a \to A$. The separation between fluorescence and singlet absorption spectra is important for the dye laser since the unexcited dye is then transparent for the fluorescence.

Molecules in the excited singlet S_1 state may relax via a nonradiative process to a lower lying triplet state, designated as T_1 in Fig. 1, instead of decaying to the ground state. This process, known as *intersystem crossing*, proceeds at a rate governed by the *intersystem crossing rate constant* $k_{S'T}$ which has the dimensions of s^{-1}. Intersystem crossing is indicated by a dashed line in Fig. 1. Obviously intersystem crossing competes with fluorescence in the deactivation of S_1. This competition is detrimental to the operation of the laser in several ways.

The lifetime for decay of the triplet state to the ground state, denoted by τ_T, is generally much longer than τ since the triplet–singlet transition is spin-forbidden. The actual value of τ_T depends upon the experimental conditions, particularly upon the amount of oxygen present in the dye solution, and may vary from 10^{-7} second in an oxygen saturated alcohol solution to 10^{-3} second or more in a carefully degassed dye solution. The decay process $T_1 \to S_0$ may be radiative or nonradiative. The radiation accompanying the process is termed *phosphorescence*. Owing to its relatively long lifetime the triplet state acts as a trap for the excited molecules and depletes the supply of molecules available for the laser process.

The state T_1 is the lowest lying state of a manifold of excited triplet states, the first of which is indicated as T_2 in Fig. 1. Triplet–triplet transitions are spin-allowed and the optical absorption associated with these transitions is strong. Unfortunately, the corresponding absorption band generally overlaps the singlet-state fluorescence spectrum. Consequently, the accumulation of molecules in the triplet state produces a large optical loss at the wavelengths for which laser emission is most probable. The absorption associated with triplet–triplet processes can be strong enough to quench or even prevent laser emission. In order to minimize the detrimental effects of the molecular triplet state it is necessary to reach laser threshold before a significant number of molecules have accumulated in the triplet state. This requires an excitation source with an intensity which increases rapidly with time. The advantage of using a giant pulse laser for excitation, as in the first dye laser experiments, is apparent.

The triplet state of many molecules is highly reactive chemically. Thus, under the proper conditions, triplet-state molecules may react chemically and destroy the dye. The importance of this process has not yet been assessed.

Optical absorption between excited singlet states, such as the process connecting S_1 and S_2 in Fig. 1, is also a possible source of optical loss in the dye laser. The importance of this process is difficult to assess since little information has been gathered. The optical loss is probably not as serious as that associated with the triplet–triplet transitions since the population of the S_1 state is relatively small.

Competition between fluorescence and other decay processes for the excited singlet state leads to a *quantum yield* for fluorescence ϕ, defined as the ratio of the number of fluorescent photons emitted to the number of excitation photons absorbed by a large number of dye molecules, of less than unity. For most laser dyes ϕ lies between 0.5 and 1.0. The actual value of ϕ for a given dye depends upon the solvent, the temperature, and other experimental conditions.

II. *Analysis of the Dye Laser*

The dye laser process, as described in the previous section, is a four-level scheme. The analysis developed to describe four-level gas and solid-state lasers may be applied to the dye laser as well except that some modification to include triplet-state effects and singlet-state optical absorption is necessary. In this section analytical treatments of the dye laser will be reviewed.

The analysis is greatly simplified when critical inversion is produced in a time which is short compared to the inter-system crossing decay time $k_{S'T}^{-1}$, since in this case the accumulation of molecules in the triplet state is negligible. To estimate the magnitude of $k_{S'T}^{-1}$ for a typical dye it is assumed that only two decay paths exist for the state S_1 of Fig. 1: fluorescence and intersystem crossing. The quantum yield for fluorescence will then be given by [15]

$$\phi = 1/(1 + \tau k_{S'T})$$

from which

$$k_{S'T}^{-1} = \tau \phi/(1 - \phi). \tag{1}$$

Choosing values typical of laser dyes, $\tau = 5 \times 10^{-9}$ second and $\phi = 0.9$, leads to a value of $k_{S'T}^{-1} = 4.5 \times 10^{-8}$ second. Thus, neglect of the triplet state is marginally justified in the analysis of the dye laser excited by a giant pulse laser since the dye is excited to laser threshold in 10 to 50 ns. In the analysis of the flashlamp-excited laser it is not justified owing to the slower rate of rise of the excitation. The fastest flashlamps produce light pulses approximately 1 μs in duration.

An analysis of the laser-pumped dye laser has been given by Sorokin, Lankard, Hammond, and Moruzzi [12]. The coupled rate equations describing the population of the excited singlet state and the number of photons in the optical cavity were solved and used to predict the form of the laser pulse. In the calculations a Gaussian-shaped excitation pulse was assumed and the principle cavity loss was assumed to be transmission by the cavity mirrors. Absorption by molecules in the singlet and triplet states was neglected. The results described rather well the time development of the laser pulses observed by these authors. Owing to the neglect of triplet-state effects the analysis is not applicable to the flashlamp-excited laser, however.

The first analysis in which the effect of molecular accumulation in the triplet state was considered was given by Schmidt and Schäfer [14]. These authors used an analog computer to solve the coupled rate equations describing the upper laser level and triplet-state populations. An excitation flash with an intensity proportional to time was assumed. Calculations were performed for two rates of rise of the excitation, 1.7×10^{23} and 1.7×10^{22} excited molecules/cm^2 s^2, and for two quantum yields, $\phi = 0.5$ and 0.98. Under these conditions the authors showed that the excited singlet-state population reaches a maximum 20 ns or less after the beginning of the excitation and subsequently decays as molecules accumulate in the triplet state. In related experiments quenching of the stimulated emission similar to that predicted was observed although the quenching time was somewhat longer than the analytical value. The predicted quenching is entirely due to the decrease in effective dye concentration. Optical losses and thermal effects were not considered by Schmidt and Schäfer.

Dynamic effects of triplet absorption upon flashlamp-excited laser emission have also been considered by Sorokin, Lankard, Moruzzi, and Hammond [16] in an extension of the analysis developed to describe the laser-pumped laser. To the rate equations for the excited singlet-state population and the photon density in the cavity was added a third equation describing the population of the triplet state. The excitation pulse was assumed to have a Gaussian shape and the triplet state was assumed to have an infinite lifetime. The coupled rate equations were solved by using parametric values for quantum yield, optical cavity losses, excitation pulsewidth, dye concentration, and the ratio of singlet-to triplet-state extinction coefficients. Molecular concentration in the excited singlet state, photon density in the cavity, and triplet-state concentration were calculated as functions of time. The results clearly show the quenching of stimulated emission by the optical loss and the effective decrease in dye concentration associated with triplet-state accumulation.

In Fig. 2 is shown the variation in laser output efficiency as a function of the half-width of the excitation pulse calculated by Sorokin et al. The parameter α is the ratio of the total number of photons in the excitation pulse to the critical inversion as estimated from the Schawlow–Townes formula [8]. Each curve corresponds to a constant excitation energy. A value of 0.88 was chosen for the fluorescence

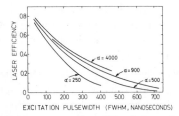

Fig. 2. Laser efficiency versus the width of the excitation pulse. The parameter α corresponds to the ratio of the total number of photons in the excitation pulse to the number of photons required to produce critical inversion as calculated from the Schawlow–Townes start-oscillation condition. Each curve represents a constant excitation energy therefore. The decrease in laser efficiency with increasing excitation pulsewidth results from the accumulation of molecules in the triplet state. The curves are taken from Sorokin et al. [16] and plotted with the assumptions $\phi = 0.88$, $\tau_T = \infty$, and $\varepsilon_{SS} = 10\varepsilon_{TT}$.

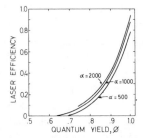

Fig. 3. Laser efficiency versus fluorescence quantum yield assuming an excitation pulsewidth of 300 ns FWHM. $\varepsilon_{SS} = 10\varepsilon_{TT}$ is assumed. The parameter α is defined in connection with Fig. 2. The decrease in laser efficiency with decreasing ϕ reflects the losses due to the molecular triplet state. The curves are taken from [16].

quantum yield and the extinction coefficient for singlet-state optical absorption is assumed to be ten times as large as the extinction coefficient for triplet state absorption. The quenching effect of the triplet state is seen in the decrease in efficiency of the laser with increasing excitation pulse width. The importance of using a short intense excitation pulse is apparent.

In Fig. 3 the calculated effect of fluorescence quantum yield upon laser efficiency is displayed for a laser excited by a Gaussian pulse of 300 ns FWHM. The triplet-state parameters are the same as those used in the calculations for Fig. 2. Under the assumed conditions laser emission from a dye with $\phi < 0.65$ appears impossible. The desirability of using dyes with high quantum yield is evident.

A short triplet-state lifetime, as the result of quenching of the state T_1 in Fig. 1, would modify the results of the analysis in such a way that the efficiency of the laser with long excitation pulses and dyes with low fluorescence quantum yields would be improved.

The time development of the gain of an organic dye has been treated by Bass, Deutsch, and Weber [17] following an analysis of phonon-terminated solid-state lasers developed by McCumber [18]. McCumber considered a four-level laser medium with relatively broad absorption and emission spectra and assumed that the energy dependence of the emission spectrum is the mirror image of the absorption spectrum but displaced toward lower energy. A *gain*

characteristic of the active medium is defined by McCumber as the difference between the probabilities for photon emission and photon absorption. The gain characteristic reflects the wavelength dependence of the absorption and fluorescence of the dye. The gain is proportional to the number of molecules in the excited singlet state and will be time dependent since the concentration of excited molecules depends upon the excitation intensity. Bass et al. [17] obtained the time-dependent population of the excited singlet state from the solution of a rate equation using an assumed shape for the excitation flash. From the excited-state population the gain curve for the dye rhodamine B at various times during the excitation was calculated. The effects of intersystem crossing and the accumulation of molecules in the triplet state were neglected.

In a further refinement of this analysis Weber and Bass [19] accounted for the triplet-state losses by introducing a triplet-state absorption term into the gain characteristic which becomes

$$G(\omega) = N_1 f(\omega) \frac{\lambda^2}{\eta^2} - N_0 \exp\left[-\hbar(\omega_0 - \omega)/kT\right] f(\omega) \frac{\lambda^2}{\eta^2} \quad (2)$$
$$- N_T \varepsilon_{TT}(\omega)$$

where N_1 is the population of the excited singlet state and N_0 is the concentration of molecules in the ground singlet state. The parameter ω_0 is taken to be the frequency corresponding to the energy difference between the lowest vibrational states of the ground singlet and first excited singlet states (see Fig. 1). In the equation η is the refractive index, and $f(\omega)$ is the spontaneous fluorescence intensity in photons per second per unit frequency per molecule normalized so that

$$\int_0^\infty f(\omega)d\omega = \phi/\tau.$$

N_T is the molecular concentration in the triplet state, $\varepsilon_{TT}(\lambda)$ is the molecular extinction coefficient for triplet-state absorption, and λ is the free-space wavelength corresponding to the frequency ω.

Rate equations for N_1, N_0, and N_T are required and are given by (3a–c) below. In these equations $P(t)$ is the intensity of the excitation flash.

$$(dN_1/dt) = N_0 P(t) - N_1/\tau \quad (3a)$$
$$(dN_T/dt) = k_{S'T} N_1 - N_T/\tau_T \quad (3b)$$
$$(dN_0/dt) = -P(t)N_0 + \left(\frac{1}{\tau} - k_{S'T}\right)N_1 + N_T/\tau_T. \quad (3c)$$

Equations (3b) and (3c) include terms describing deactivation of the triplet state through a decay lifetime τ_T.

The rate equations were solved by assuming values for τ, and $k_{S'T}$ and a form for the excitation function $P(t)$. The resulting values of $N_0(t)$, $N_1(t)$, and $N_T(t)$ were inserted into the gain characteristic (3). The time-dependent gain curves were computed for several laser dyes on the basis of this analysis. The results will be discussed in a later section of this paper.

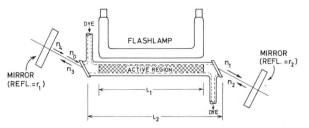

Fig. 4. Schematic diagram of the flashlamp-excited dye laser. The quantities $n_0 - n_4$ represent the density of lasing mode photons at various stages of a round trip circuit of the laser cavity starting with n_0 photons per cm^3.

The photon generation term of the gain characteristic (2) has the form of the spontaneous fluorescence lineshape. The generation of photons by stimulated emission is neglected in the above analysis. Substitution of the proper stimulated emission term into the gain characteristic introduces a wavelength dependent coefficient which modifies the form of the gain curve.

The development of a gain analysis which includes stimulated emission and triplet-state absorption has been discussed by Snavely and Peterson [20] using an approach in which the solution of the rate equations is not required. A laser system of the form shown in Fig. 4 is considered in the analysis.

The total rate per unit length at which photons are produced within the dye cell will be given by an equation analogous to the gain characteristic equation (2),

$$\left(\frac{dn}{dl}\right)_{total} = \left(\frac{dn}{dl}\right)_{stim} - \left(\frac{dn}{dl}\right)_{sing} - \left(\frac{dn}{dl}\right)_{trip}$$

where the three terms on the right correspond to the photon production by stimulated emission, photon loss by singlet-state absorption and photon loss by triplet-state absorption, respectively. To obtain the stimulated emission term a result derived by Yariv and Gordon [21] will be used. These authors show that the total rate for induced transitions from an excited state density of N_1 per cm^3 in a monochromatic radiation field at the wavelength λ produced by $n(\lambda)$ photons per cm^3 will be

$$\left(\frac{dn}{dt}\right)_{stim} = \frac{N_1 \lambda^4 E(\lambda) n(\lambda)}{8\pi\tau} \text{ cm}^{-3}\text{ s}^{-1} \qquad (4)$$

where $E(\lambda)$ is the spontaneous fluorescence lineshape function normalized so that $\int_0^\infty E(\lambda)d\lambda = \phi$. The photon production rate per unit length of the active medium dn/dl is readily obtained from the above since

$$\frac{dn}{dt} = \frac{dn}{dl}\cdot\frac{dl}{dt} = \frac{dn}{dt}\cdot\frac{c}{\eta}$$

where c is the velocity of light. Therefore

$$\left(\frac{dn}{dl}\right)_{stim} = \frac{N_1 \lambda^4 E(\lambda) n(\lambda)\eta}{8\pi\tau c}. \qquad (5)$$

The loss of photons associated with singlet- and triplet-state absorption must now be considered. The number of photons removed from the cavity by singlet-state absorption in an infinitesmal path length dl is obtained from the definition of the molecular extinction coefficient $\varepsilon_{SS}(\lambda)$ and is

$$\left(\frac{dn}{dl}\right)_{sing} = n(\lambda)N_0\varepsilon_{SS}(\lambda). \qquad (6)$$

A similar expression is obtained for the triplet-state loss in terms of the extinction coefficient for triplet-state absorption $\varepsilon_{TT}(\lambda)$:

$$\left(\frac{dn}{dl}\right)_{trip} = n(\lambda)N_T\varepsilon_{TT}(\lambda). \qquad (7)$$

Equations (5), (6), and (7) are associated with the dye alone. The oscillation threshold will be determined by the gain of the total laser system. The loss resulting from mirror transmission must also be accounted for. With reference to Fig. 4, a photon density n_0 incident upon the left-hand side of the dye cell is assumed. Using equations (5), (6), and (7) it is easily shown that the number of photons which emerges from the right-hand side of the dye cell n_1 is given by

$$n_1 = n_0 \exp\left[\frac{N_1\lambda^4 E(\lambda)\eta}{8\pi\tau c}L_1 - N_0\varepsilon_{SS}(\lambda)L_2 - N_T\varepsilon_{TT}(\lambda)L_1\right].$$

The number of photons returned to the dye cell by the right-hand mirror, with reflectance r_1, is

$$n_2 = n_1 r_1.$$

Similar arguments apply to the determination of n_3 and n_4. The photon density n_4 after the round trip through the cavity starting with the photon density n_0 is

$$n_4 = n_0 \exp\left[\frac{N_1\lambda^4 E(\lambda)\eta}{8\pi\tau c}2L_1 - N_0\varepsilon_{SS}(\lambda)2L_2\right.$$
$$\left. - N_T\varepsilon_{TT}(\lambda)2L_1 + \ln r_1 r_2\right]. \qquad (8)$$

The gain of the system $G(\lambda)$ will be defined in terms of the quantity in brackets such that

$$n_4 = n_0 e^{2G(\lambda)L_1}.$$

Triplet-state and excited singlet-state molecules are produced only in the active region L_1. In this region the molecular conservation condition yields $N_0 = N - N_1 - N_T$. N is the total molecular concentration in the unexcited end regions of the dye cell $N_0 = N$. Making these substitutions in (8) the gain equation is obtained in final form

$$G(\lambda) = N_1\left[\frac{\lambda^4 E(\lambda)\eta}{8\pi\tau c} + \varepsilon_{SS}(\lambda)\right] - N\varepsilon_{SS}(\lambda)\frac{L_2}{L_1}$$
$$- N_T\left[\varepsilon_{TT}(\lambda) - \varepsilon_{SS}(\lambda)\right] + \frac{1}{2L_1}\ln\left(r_1 r_2\right). \qquad (9)$$

At the threshold of laser oscillation $n_4 = n_0$ and $G(\lambda) = 0$. The critical inversion N_{1c}, the excited state population required for laser oscillation, is given by

$$N_{1c} = K_1 N\left(\frac{L_2}{L_1}\right) + K_2 N_T + K_3 \tag{10}$$

which is obtained by setting (9) equal to zero. The coefficients in (10) are defined by

$$K_1 = \frac{\varepsilon_{SS}(\lambda)}{A(\lambda)}; \qquad K_2 = [\varepsilon_{TT}(\lambda) - \varepsilon_{SS}(\lambda)]/A(\lambda);$$

$$K_3 = \frac{-\ln r_1 r_2}{2L_1 A(\lambda)}$$

where

$$A(\lambda) = \frac{\lambda^4 E(\lambda)\eta}{8\pi\tau c} + \varepsilon_{SS}(\lambda).$$

At the oscillation threshold not only will the gain vanish but the wavelength derivitive of $G(\lambda)$ vanishes as well since oscillation will begin at the peak of the gain curve. Differentiating $G(\lambda)$ and setting the result equal to zero yields a second equation for the critical inversion

$$N_{1c} = K_4 N\left(\frac{L_2}{L_1}\right) + K_5 N_T \tag{11}$$

where

$$K_4 = \varepsilon'_{SS}(\lambda)/A'(\lambda)$$

and

$$K_5 = [\varepsilon'_{TT}(\lambda) - \varepsilon'_{SS}(\lambda)]/A'(\lambda).$$

The primes denote differentiation with respect to wavelength. It has been assumed that the mirrors are nondispersive so that the derivitive of $\ln(r_1 r_2)$ vanishes.

Equations (10) and (11) may be solved to yield N_{1c} and N_T, as functions of the dye parameters, the laser parameters, and the laser wavelength. The results are expressed in (12) and (13).

$$N_{1c} = \frac{K_1 K_5 - K_2 K_4}{K_5 - K_2} N\left(\frac{L_2}{L_1}\right) + \frac{K_3 K_5}{K_5 - K_2}. \tag{12}$$

$$N_T = \frac{K_1 - K_4}{K_5 - K_2} N\left(\frac{L_2}{L_1}\right) + \frac{K_3}{K_5 - K_2}. \tag{13}$$

By consideration of the triplet-state rate equation, given by (3b), in connection with the gain analysis some useful relations may be obtained. If it is assumed that the excitation intensity increases linearly with time and that the excited state population is proportional to the excitation intensity the rate equation becomes

$$\frac{dN_T}{dt} = k_{S'T} K t - N_T/\tau_T$$

where K relates N_1 to the excitation intensity. Solving this equation yields the time-dependent ratio of triplet- to singlet-state populations as

$$\frac{N_T(t)}{N_1(t)} = k_{S'T}\tau_T\left[1 - \frac{\tau_T}{t}(1 - e^{-t/\tau_T})\right]. \tag{14}$$

In the limit of $t \ll \tau_T$ the ratio becomes

$$\lim_{t \ll \tau_T} \frac{N_T(t)}{N_1(t)} = \frac{k_{S'T}t}{2}. \tag{15}$$

With the use of (12) and (13) to determine the ratio of $N_T(t)/N_{1c}(t)$ in an experimental system at laser threshold, the intersystem crossing rate constant $k_{S'T}$ may be measured. The application of this technique for determining $k_{s'T}$ has been discussed by Buettner et al. [22].

Equation (15) yields a useful criterion, first derived by Sorokin et al. [16], for determining the rate of rise of excitation required to lase a given dye when triplet-state effects are important. The triplet-state losses will be given by $N_T\varepsilon_{TT}$ and the singlet-state gain, approximately, by $N_1\varepsilon_{SS}$.[1] Assuming that lasing is possible only for $N_1\varepsilon_{SS} > N_T\varepsilon_{TT}$, an upper limit for the time t_0 in which critical inversion must be reached is

$$t_0 < \frac{2\varepsilon_{SS}}{k_{S'T}\varepsilon_{TT}}. \tag{16}$$

For a dye with $k_{S'T} = 2 \times 10^7$ s^{-1} and $\varepsilon_{SS}/\varepsilon_{TT} = 10$, typical values for a laser dye, a flashlamp which will produce critical inversion in a time less than 10^{-6} second is required. This criterion, of course, applies only if no triplet quenching is present, that is, if $\tau_T \to \infty$.

If the triplet state is quenched the solution to (14) in the limit of $t \gg \tau_T$ becomes of interest and is given by

$$\lim_{t \gg \tau_T} \frac{N_T(t)}{N_1(t)} = k_{S'T}\tau_T \tag{17}$$

which is independent of time. This result suggests that long pulse laser experiments may be used to evaluate τ_T if $k_{S'T}$ can be estimated.

A maximum permissible triplet-state lifetime consistent with steady-state oscillation can be deduced from the foregoing results. In the steady state $dN_T/dt = 0$, and from (3b) $N_T = N_{1c}k_{S'T}\tau_T$. N_1 will be fixed at the critical inversion N_{1c} during lasing. Substituting this result into (11)

$$N_{1c} = N_{10} + K_5 N_{1c}k_{S'T}\tau_T$$

and

$$N_{1c} = N_{10}/(1 - K_5 k_{S'T}\tau_T)$$

where N_{10} is the critical inversion in the absence of triplet-state effects.

The denominator of this equation must be positive from which

$$\tau_T < 1/K_5 k_{S'T} \tag{18}$$

is obtained.

If τ_T is greater than the value given by (18) CW operation cannot be achieved. The evaluation of τ_T for a particular laser dye will be considered in the discussion of experimental results.

---

[1] The probabilities for stimulated emission and "stimulated" absorption at a given wavelength are assumed to be equal.

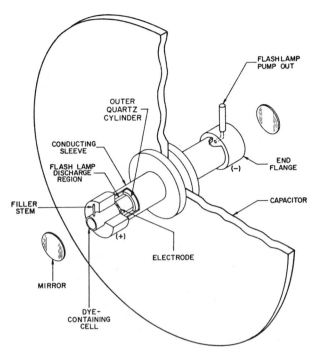

Fig. 5. Diagrammatic sketch of the low-inductance flashlamp design of Sorokin and Lankard used to produce stimulated emission from organic dyes in solution. (The sketch is reproduced from *J. Chem. Phys.*, vol. 48, p. 4735, 1968.)

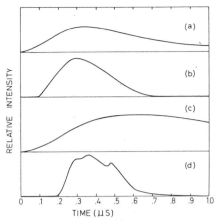

Fig. 6. Comparison of flashlamp excited dye laser results. (a) Flashlamp pulse produced by the lamp sketched in Fig. 5. (b) Laser pulse from a 10^{-4} molar solution of rhodamine 6G in ethanol as reported by Sorokin, *et al.* [16]. (c) Flashlamp pulse produced by a linear xenon flashlamp in a low inductance circuit [14]. (d) Laser pulse from a 10^{-3} molar solution of rhodamine 6G in methanol as reported by Schmidt and Schäfer [14].

III. Experimental Results

A. Excitation and Laser System Considerations

The importance of using a rapidly rising excitation flash to overcome the effects of intersystem crossing has been emphasized and the design of a suitably fast excitation source was the central problem in the early dye-laser experiments. The first flashlamp-excited dye laser, reported by Sorokin and Lankard [13], utilized a novel flashlamp which discharged 100 joules in a pulse with a risetime to peak intensity of about 0.3 μs and a duration of approximately 1 μs. A sketch of the lamp is shown in Fig. 5. In concept the lamp is similar to a design of Claesson and Lindqvist [23] in which the lamp circuit inductance is minimized by using a coaxial arrangement for the light producing discharge and the electrical connections to the flashlamp. The resistance of the lamp circuit is kept low by mounting the lamp in the center of the energy storage capacitor. The discharge takes place in the air-filled annular region between two quartz tubes, the inner of which serves as the dye cuvette. The outer quartz tube is surrounded by a divided metal sleeve which provides the electrical connections between the capacitor and the lamp and also serves as the reflector for the lamp. No advantage was found in the use of gases other than air in the lamp and the experiments of Sorokin *et al.* [16] were conducted with an air-filled lamp which was fired by reducing the pressure in the discharge region. Cooling of the lamp was facilitated by allowing the air to flow through the discharge region.

Although the risetime of such a lamp is short in comparison with that of conventional laser excitation sources it is not as short as expected. By photographing the discharge Sorokin *et al.* [16] found that the electrical discharge does not uniformly fill the annular region between quartz tubes but is filamentary. This raises the effective inductance of the lamp and slows down the discharge.

Schmidt and Schäfer [14] have obtained a similar risetime with a conventional linear xenon flashlamp driven by a low-inductance capacitor. These authors also utilized a coaxial arrangement of the lamp and its electrical connections to minimize the inductance. A spark gap in series with the lamp permitted operation with voltages which were much greater than the lamp breakdown potential.

Flashlamp and laser pulses obtained by Sorokin *et al.* [16] from a 10^{-4} molar solution of the dye rhodamine 6G in ethanol are shown in Fig. 6(a) and (b), respectively. The difference in flashlamp intensity at the onset and termination of laser emission is seen to be small, and suggests that the accumulation of molecules in the triplet state does not change significantly during this time. This result is in marked contrast to that obtained by Schmidt and Schäfer [14] for a 10^{-3} molar solution of rhodamine 6G in methanol shown in Fig. 6(d). The laser pulse is quenched when the excitation is near peak intensity as noted by comparison with the flashlamp pulse shown in Fig. 6(c). The length of the laser pulses is approximately the same in Fig. 6(b) and (d) although the excitation pulse of Fig. 6(c) is considerably longer than that of Fig. 6(a). This result was interpreted as being indicative of quenching by triplet-state losses. In view of the result shown in Fig. 6(a), and later results with rhodamine 6G, the quenching seen in Fig. 6(d) is more likely due to thermal effects than to the triplet state.

Under some conditions the risetime requirements for the flashlamp are not so severe as previously supposed. It has been shown recently by Snavely and Schäfer [24] that air-saturated methanol solutions of rhodamine 6G and rhodamine B can be maintained above laser threshold for a period greater than 50 μs while the excitation intensity remains constant, seemingly in violation of the criterion given by (16). Under the proper circumstances it is possible to use conventional flashlamp circuitry to excite a

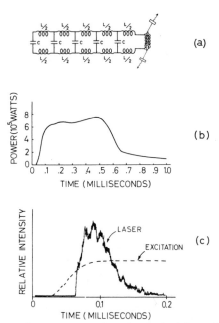

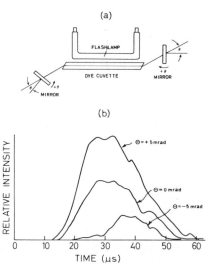

Fig. 8. Illustration of the effect of the refractive index gradient caused by nonuniform excitation of the laser dye. (a) Schematic illustration of the long-pulse dye laser excited by a linear flashlamp in a close-coupling arrangement. Rotation of the mirrors with respect to the cavity axis is indicated. (b) The effect of mirror rotation by plus or minus 5 mrad upon the laser output is shown. The results indicate that the laser beam is curved away from the flashlamp by the refractive index gradient produced during excitation of the dye.

Fig. 7. Long-pulse apparatus and results obtained from a laser experiment with an air-saturated by 5×10^{-5} molar solution of rhodamine 6G in methanol. (a) Schematic description of the spiral flashlamp-excitation source driven by a lumped-constant transmission line with $C = 10\ \mu F$, $L = 240\ \mu H$. The flashlamp used was a PEK type XE-5-3 1/2. (b) Excitation-light pulse produced by the circuit of (a). The power is obtained by normalizing the total area under the curve to the energy stored in the capacitor bank with the capacitors charged to 4 kV. (c) Laser pulse (solid curve) observed with the flash pulse (dashed curve) plotted on the same time scale.

dye laser. This is attributed to quenching of the triplet-state lifetime τ_T in a relatively short time.

The laser used in the experiments of Snavely and Schäfer consisted of an 8-cm-long quartz dye flow cuvette excited by a standard 7.5-cm spiral xenon flashlamp. The dye cuvette was provided with Brewster angle windows and the optical cavity was formed by plane dielectric mirrors. To produce relatively long flat excitation light pulses the flashlamp was driven by a lumped constant transmission line.

The circuit and excitation light pulse are shown in Fig. 7(a) and (b) and a laser pulse obtained with the apparatus is shown in (c). The laser pulse continues for a period of more than 80 μ over the flat portion of the excitation pulse before emission ceases, a period much too long to be consistent with intersystem crossing and triplet-state effects. Nonetheless, the laser emission is clearly quenched while the excitation source is above threshold. This quenching was found to be due to refractive index inhomogeneities caused by nonuniform excitation of the dye.

Optical effects produced by nonuniform heating of a liquid laser have been discussed by Winston and Gudmundscn [25]. Stockman [7] has measured the time required for optical inhomogeneity to destroy a laser cavity by inserting a flashlamp-excited benzene solution of the dye perylene in the cavity of a neodymium-glass laser. It was found that under excitation conditions similar to those used for a dye laser the optical quality of the dye solution was degraded to the point where the neodymium-glass laser would no longer oscillate in approximately 30 μs.

An experiment similar to that of Stockman was performed with a long-pulse laser of the type described above [24]. The beam from a helium–neon laser was passed through a 7-cm dye cuvette containing a 10^{-4} molar solution of rhodamine 6G in methanol. The cuvette was excited by a 7.5-cm linear flashlamp coupled to the curvette by an aluminum-foil close wrap. The flashlamp was driven from a lumped constant transmission line and produced a trapezoidal light pulse of 200 μs FWHM. The gas laser beam was detected by a photomultiplier 1.5 meters from the cuvette. It was found that the transmitted beam vanished in 20 to 30 μs, consistent with the results of Stockman [7]. This was also, approximately, the length of the laser pulse obtained with the system.

The excitation produced by the linear flashlamp in the close-coupled geometry is nonuniform and the dye closest to the flashlamp receives the most energy. As the temperature of the methanol solution increases, the refractive index decreases. The hotter dye solution near the flashlamp causes the active medium to become prismlike, bending the laser beam away from the flashlamp. The length of the laser pulse will be determined by the time during which the deformed beam stays within the optical cacity. It is advantageous, therefore, to make the cavity short, and it was found that by decreasing the mirror separation from 25 to 10 cm, otherwise maintaining the same experimental conditions, the laser pulse length and energy were increased.

The effect of mirror rotation upon laser output with this excitation geometry tends to confirm the prismlike nature of the active medium. If the beam within the cavity is deformed as suggested then the laser output should be increased by rotating the mirrors to keep them perpendicular to the deformed beam. The necessary rotation is indicated in Fig. 8, where the results of the rotation are also depicted. The maximum energy output for the system was obtained with the mirrors rotated by approximately $+7.5$ mrad. Rotating the mirrors by -5 mrad greatly decreased the laser output as shown.

The excitation power required to reach laser threshold can be estimated from the shape of the excitation flash as a function of time. It is assumed that the flash intensity is proportional to the instantaneous electrical power dissipated by the lamp. The light pulse, as obtained from an oscilloscope recording, is integrated graphically and normalized to the total stored energy of the capacitor bank. The electrical power dissipated by the lamp as a function of time is then determined and threshold is measured by noting the flashlamp power at the laser threshold. The power input to the spiral lamp, determined in this manner, is shown in Fig. 7(b).

It was found that the total energy dissipated by the flashlamp up to threshold was independent of the time required to reach threshold for the linear flashlamp-excited long pulse laser. Since these times, between 20 and 70 μs, are much longer than the excited state lifetime of the dye molecules, the result cannot be due to an integration of the excitation by the active medium. The result is interpreted as further evidence of the production of a prismlike active medium by heating effects. A particular transverse refractive index gradient was necessary to bring the cavity into resonance. The gradient produced depended upon the energy dissipated by the excitation source.

For the long pulse laser excited by a spiral flashlamp, on the other hand, the threshold excitation power was found to be nearly independent of the time required to reach threshold and was approximately 2.9×10^5 W/cm^3 of dye solution as opposed to the minimum value of 3.2×10^5 W/cm^3 for the system pumped by the linear flashtube. The optical coupling between the flashlamp and dye cuvette was considerably less efficient for the spiral lamp pumped system than for the close wrap arrangement. The actual difference in the threshold optical power density must be considerably greater than the comparison of these figures would indicate reflecting the advantage of a uniform excitation of the dye.

For the uniformly illuminated cuvette heating by the excitation flash produces a lenslike active medium. Since the hotter dye near the cuvette walls has the lower refractive index, a positive lens is produced. Considering the resonant cavity as a whole, an effective curvature of the mirrors, and thus a generalized confocal cavity, is created. The system will continue to oscillate until the effective radius of curvature of the mirrors decreases to a value for which the cavity is no longer stable. Accordingly, the longest period of stability corresponds to the shortest cavity. This is consistent with the experimental observations.

The effects of very rapid heating caused by nonuniform excitation of the dye laser have been discussed. In practice very long-term heating effects are observed as well and produce thermal schlieren which may persist for many minutes in a stagnant dye solution. These effects are easily eliminated by rapid circulation of the dye through the cuvette. Sorokin et al. [16] observed that without circu-

TABLE I
STRUCTURE, LASER WAVELENGTH, AND SOLVENTS FOR LASER DYES

Dye	Structure	Solvent	Wavelength	References
Acridine Red		EtOH	Red 600 - 630 nm	13, 16, 17, 20
Pyronin B		MeOH H$_2$O	Yellow	14
Rhodamine 6G		EtOH MeOH H$_2$O DMSO Polymethyl-methacrylate	Yellow 570 - 610 nm	13, 14, 16, 17, 19, 20, 22, 24, 44, 45, 47
Rhodamine B		EtOH MeOH Polymethyl-methacrylate	Red 605 - 635 nm	13, 14, 16, 17, 19, 20, 22, 44, 47
Na-fluorescein		EtOH H$_2$O	Green 530 - 560 nm	13, 16, 17, 20, 22, 44
2,7-Dichloro-fluorescein		EtOH	Green 530 - 560 nm	16
7-Hydroxycoumarin		H$_2$O (pH~9)	Blue 450 - 470 nm	19, 20
4-Methylumbelliferone		H$_2$O (pH~9)	Blue 450 - 470 nm	16, 26, 44
Esculin		H$_2$O (pH~9)	Blue 450 - 470 nm	16, 26
7-Diethylamino-4-Methylcoumarin		EtOH	Blue	33
Acetamidopyrene-trisulfonate		MeOH H$_2$O	Green-Yellow	14, 27
Pyrylium salt		MeOH	Green	27

lation the pulse repetition rate was limited by thermal effects to a rate of about one pulse per ten minutes. By circulating the dye through the cuvette a repetition rate of about one pulse per second was obtained, the limitation being imposed by the electrical power supply rather than by the heating effects in the dye. Dye lasers operating at repetition rates of 50 Hz or more have since been constructed in several laboratories and no limitation to the production of higher repetition rates is foreseen.

In the design of a dye circulation system care must be taken to avoid components which react with the dye solvents. This restriction is not severe as the solvents used are relatively inactive. However, some types of plastic tubing contain plasticizers which degrade the dye solutions. Another source of difficulty is lubricating material and air bubbles from the circulation pump bearings. These problems may be avoided by using a magnetically coupled Teflon pump and polypropylene tubing for the circulation system.

B. Laser Dyes

The number of dyes which have been reported to lase with flashlamp excitation is considerably smaller than the number which have been observed to emit stimulated emission when excited by a giant pulse laser, a fact which reflects the importance of triplet-state effects. Most of the reported flashlamp excited laser dyes belong to one of two structural classes, the xanthenes and the coumarins. A listing of the published laser dyes is given in Table I.

There has been little consideration in the literature of what structural properties are important for a laser dye. In order to facilitate speculation on the part of the reader as to what these might be, the structures of the laser dyes are also given.

Solvents which have been used, as well as the approximate emission wavelength, are given in Table I. The optimum dye concentration depends primarily upon the system geometry and the characteristics of the excitation light source but generally is in the range from 10^{-3} to 10^{-5} mole of dye per liter of solvent.

C. Comparison Between Laser Performance and Analytical Results

In this section the analysis previously developed will be used to interpret experimental results obtained with the dye laser. The analysis of the long-pulse dye laser, described in the last section, using the dye rhodamine 6G will be considered as an example although the techniques are applicable to any dye laser system.

The change of excitation intensity during an excited state lifetime of a dye molecule is usually small, especially for the flashlamp-excited dye laser, and the excited state population follows the excitation intensity. Therefore, the power required to reach threshold is proportional to the critical inversion. The latter, therefore, can be used as a figure of merit in comparing the effects of changes in dye and laser parameters. The basis for obtaining the critical inversion of the dye laser from a knowledge of the dye and laser parameters is contained in (10) and (11). If the wavelength dependent coefficients K_4 and K_5 are calculated, a plot of the critical inversion versus wavelength for a given triplet concentration can be obtained. The critical inversion curve can be regarded as an inverse-gain curve and will exhibit a minimum at the wavelength for which oscillation is to be expected. By choosing a series of parametric values for N_T a family of curves may be generated which displays the effect of triplet state accumulation upon critical inversion and upon lasing wavelength.

To evaluate the coefficients K_4 and K_5 of (11) careful spectrophotometric measurements of the singlet-state absorption and fluorescence spectra are required as well as the triplet-state absorption spectrum for the dye under consideration. It is also necessary to measure the fluorescence quantum yield and fluorescence decay time with reasonable accuracy.

The singlet-state absorption spectrum for rhodamine 6G, expressed in terms of the molecular extinction coefficient $\varepsilon_{SS}(\lambda)$ and the fluorescence spectrum $E(\lambda)$ obtained with a 10^{-4} molar solution of dye in ethanol, are given in Fig. 9 as measured by F. Grum of the Kodak Research

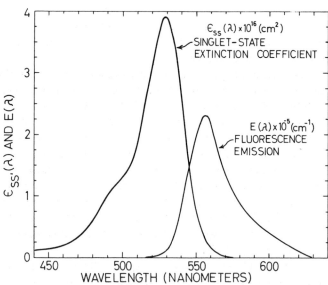

Fig. 9. Singlet-state absorption and fluorescence spectra of rhodamine 6G obtained from measurements with a 10^{-4} molar ethanol solution of the dye. The emission spectrum is normalized so that $\int_0^\infty E(\lambda)d\lambda = 0.83 = \phi$. Normalization of both curves is correct for use in the gain and critical inversion analysis presented in the text. The absorption spectrum is expressed in terms of the molecular extinction coefficient.

Laboratories. The fluorescence spectrum was obtained using a specially constructed spectrofluorimeter, which also measures the fluorescence quantum yield. A value of $\phi = 0.83$ was determined and the fluorescence spectrum shown in Fig. 9 was normalized so that $\int_0^\infty E(\lambda)d\lambda = \phi$, as required by the critical inversion analysis.

The spectral region in which dye laser emission is observed always lies at the long wavelength edge of the singlet-state absorption spectrum where the singlet extinction coefficient is small and the probability for fluorescence is large. To obtain the necessary accuracy for $\varepsilon_{SS}(\lambda)$ in this region, a long absorption path length measurement is required. The extinction coefficient for the long wavelength absorption of a 10^{-4} molar solution of rhodamine 6G in ethanol obtained using a 10-cm absorption cell is shown in Fig. 10.

The fluorescence decay time τ for rhodamine 6G in ethanol was measured [20] with a TRW fluorescence-decay-time apparatus and was 7.3×10^{-9} second. A value of 5.5×10^{-9} second has been measured by Mack [28]. In the calculations which follow the former value will be used.

The triplet-state absorption spectrum, expressed in terms of the molecular extinction coefficient $\varepsilon_{TT}(\lambda)$ for rhodamine 6G in polymethyl methacrylate is given in Fig. 11. The measurement was made by Buettner [22] using the flash-photolysis technique. Unfortunately, the triplet absorption spectrum for an alcohol solution of the dye does not seem to have been measured and it is necessary to use the spectrum of Fig. 11 in the analysis.

The differentiation of the singlet and triplet absorption spectra and of the emission spectrum necessary for the calculation of K_4 and K_5 was performed by a computer which calculated the coefficients as functions of wavelength. The results of plotting the critical inversion versus wavelength from (11) with parametric values of triplet concentration for a laser system with an active total length of 7 cm using

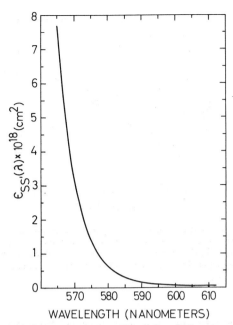

Fig. 10. Molecular extinction coefficient for singlet-state absorption versus wavelength for rhodamine 6G in the long wavelength region of the observed laser emission. The curve is obtained from a 10 cm path length absorption measurement using a 10^{-4} molar ethanol solution of the dye.

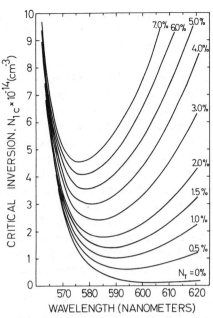

Fig. 12. Critical inversion N_{1c} versus wavelength curves with the triplet-state concentration N_T as a parameter. The triplet-state concentions are indicated at the right. The curves apply to a dye laser with $L_1 = L_2 = 7$ cm, and $r_1 = r_2 = 0.99$ (see Fig. 4) using a 5×10^{-5} molar ethanol solution of rhodamine 6G.

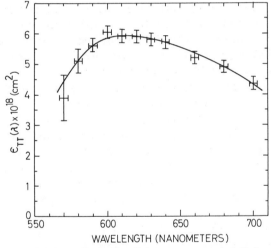

Fig. 11. Molecular extinction coefficient versus wavelength for triplet-state absorption of rhodamine 6G in polymethyl methacrylate. The measurement was made by Buettner [22] using the flash photolysis technique.

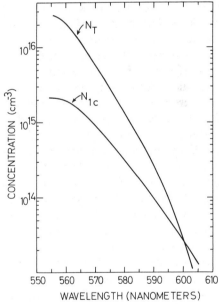

Fig. 13. Minimum values of critical inversion and triplet-state concentration corresponding to laser emission at a particular wavelength. The curves are obtained by plotting the values of N_{1c} and N_T obtained from (11) and (12). The curves apply to a laser system with the parameters given in Fig. 12. If the wavelength at the oscillation threshold of the laser is measured the N_{1c} and N_T concentrations can be determined from these curves.

5×10^{-5} molar rhodamine 6G in ethanol are shown in Fig. 12. The laser parameters correspond very nearly to the long pulse laser referred to in the previous section and displayed in Fig. 7. An important conclusion concerning the performance of this laser may be obtained from the curves. It is seen that the critical inversion minimum shifts to shorter wavelength with increasing triplet-state concentration. Since the triplet concentration increases with time during excitation of the laser it is to be expected that the rhodamine 6G laser will begin oscillation at long wavelength and sweep with time to shorter wavelengths. The breadth of the observed laser spectrum should depend upon the dynamics of the excitation, in particular on the value of the triplet-state

concentration when the critical inversion is reached. Experimental studies of the time dependence of the laser wavelength will be described in a following section.

If the wavelength of the dye laser is known the curves of Fig. 12 may be used to determine the triplet-state population and critical inversion corresponding to emission at the measured wavelength. It is only necessary to find the critical inversion curve with a minimum at the laser wavelength and read N_{1c} from the ordinate and N_T from the parametric value for that particular curve.

The determination of N_{1c} and N_T is simplified by displaying the data of Fig. 12 in a different form. N_{1c} and N_T versus wavelength, as given by (13) and (14), are plotted in Fig. 13. From these curves the concentrations of singlet- and triplet-state molecules may be read directly if the laser wavelength is known. If the time resolved spectrum for the laser is measured it is also possible to determine the singlet- and triplet-state concentrations as functions of time.

The output spectrum of the long pulse laser described in connection with Fig. 7 was measured and found to extend from 597.3 to 591.4 nm. According to the arguments presented the long wavelength limit should correspond to the laser wavelength when oscillation begins. From Fig. 13 the value for the critical inversion corresponding to oscillation at 597.3 nm is obtained as $N_{1c} = 4 \times 10^{13}$ cm^{-3} and the triplet-state concentration is $N_T = 5.8 \times 10^{13}$ cm^{-3}. That this value of N_T is consistent with a short triplet-state lifetime, as a result of quenching, is readily shown. Assume for the moment that there is no quenching. In this case N_T will increase at a rate given by (3b) with $\tau_T = \infty$. As estimated from (1), using measured values of τ and ϕ, $k_{S'T} = 2.8 \times 10^7$ s^{-1}. Using this value in the solution of (3b) it is found that the observed N_T concentration would accumulate in approximately 50 ns. The duration of the laser pulse shown in Fig. 7(c), however, is more than a thousand times greater than this. Consequently, the triplet state must be quenched to permit lasing over such a long period.

The triplet-state lifetime may be estimated with the use of (17). For laser emission at 597.3 nm the ratio $N_T/N_{1c} = 1.5$. Substituting this value into (17) and using the previously obtained value for $k_{S'T}$ a value for $\tau_T \cong 0.5 \times 10^{-7}$ second is obtained. Using the short wavelength limit of the spectrum gives $N_T/N_{1c} = 2.5$ and $\tau_T \cong 0.9 \times 10^{-7}$.

The rapid quenching of the triplet state is apparently due to oxygen dissolved in the dye solution. This conclusion is supported by experiments of Snavely and Schäfer [24], in which the oxygen concentration of the dye solution was changed by bubbling nitrogen or oxygen gas through the dye solution reservoir of a dye laser. Experiments were conducted with the previously described long pulse laser excited by a linear flashlamp. Using an air saturated methanol solution of 5×10^{-5} molar rhodamine 6G and mirrors with a reflectance of 0.99 the lasing threshold was reached with the flashlamp capacitor bank charged to 800 volts. When nitrogen gas was bubbled through the dye reservoir for 10 minutes, all other conditions remaining the same, laser threshold could not be reached with the capacitors charged to 2000 volts. By bubbling oxygen through the dye reservoir for 10 minutes the performance of the laser could be restored.

Sufficient information has been obtained to determine whether or not continuous operation of the rhodamine 6G laser is precluded by triplet-state effects. Equation (18) will be used to determine the maximum allowable value of τ_T for which CW operation is possible and this value will be

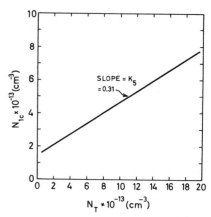

Fig. 14. Plot of the critical inversion versus triplet-state concentration obtained from Fig. 13. The slope of this curve is used to obtain the value of K_5 in (13) so that the triplet-state quenching rate required for CW operation of the rhodamine 6G laser may be estimated.

compared with the value of τ_T deduced from the laser performance. A value for the constant K_5 must first be obtained and is readily calculated from the slope of the plot of N_{1c} versus N_T. This plot, as obtained from the data of Fig. 13, is displayed in Fig. 14 and the slope at the measured value of N_{1c} yields $K_5 = 0.31$. Substituting this value and the earlier value of $k_{S'T}$ into (19) yields a maximum permissible value for τ_T of 1.2×10^{-7} second. Since this value is longer than that deduced from the laser performance it is concluded that triplet-state effects would not prevent the operation of a CW rhodamine 6G laser.

The oxygen concentration in an air saturated methanol solution at ambient pressure and temperature is approximately 2×10^{-3} mole per liter [29]. On the basis of the expected collision frequency between oxygen and dye molecules a triplet-state lifetime of less than 10^{-6} second is expected [30]. The oxygen concentration in an air-saturated water solution, on the other hand, is of the order of 2.5×10^{-4} molar [31]. Assuming that the triplet-state lifetime is inversely proportional to the oxygen concentration [30], it is to be expected that a water solution of the dye will not produce the long pulses observed with the methanol solution since a ten-fold decrease in oxygen concentration would allow the triplet-state lifetime to increase beyond the maximum value consistent with steady-state emission as calculated above.

In order to illustrate the analysis of a dye laser, an example has been chosen in which triplet-state effects are not serious. For most dyes, however, the triplet losses strongly influence the laser performance. The clearest demonstration of triplet-state quenching of stimulated emission has been given by Sorokin et al. [16] with a dye laser using diethylthiadicarbocyanine (DTDC) iodide. The DTDC solution was excited at a wavelength near its absorption maximum by a rhodamine 6G laser. The excitation conditions are thus optimized and heating effects are expected to be small.

The time development of both the excitation laser and the DTDC laser are shown in Fig. 15. The quenching of stimulated emission from the DTDC laser at the peak intensity

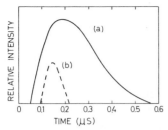

Fig. 15. Experimental demonstration of stimulated emission quenching by triplet-state absorption (from [16]). (a) Intensity of a rhodamine 6G dye laser beam used to excite a DTDC iodide solution. (b) Laser emission from the DTDC iodide solution showing quenching at the maximum excitation intensity.

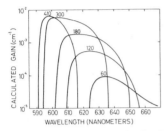

Fig. 16. Gain versus wavelength for a rhodamine B laser excited by a flashlamp as calculated by Weber and Bass [19]. Each curve corresponds to a different time during the excitation. The excitation intensity as a function of time taken to be that of an experimental flashlamp. The time in nanoseconds after initiation of the excitation flash is indicated by the numbers at the curve peaks.

of the excitation laser is evident, and is most certainly due to optical absorption associated with the triplet state. The triplet-state losses in the long chain cyanine iodide dyes, such as DTDC, are generally so great that flashlamp excitation of a laser using these dyes has not yet been achieved.

Weber and Bass [19] have calculated the time-dependent gain characteristic given by (2) for several laser dyes. The various concentrations appearing in (2) are obtained from the solution of the rate equations (3) with an assumed shape for the time dependence of the excitation. The same measurements of the dye spectral characteristics are required for the solution of (2) as for the time-independent critical inversion analysis described above. In addition, estimates of the quantities ω_0, $k_{S'T}$, and τ_T are required as well as a knowledge of the excitation function.

The results of Weber and Bass for 5×10^{-5} molar rhodamine B are shown in Fig. 16 where the gain as a function of wavelength is plotted for various times during the excitation flash. The time development of the excitation intensity was taken as that of an experimental flashlamp. Values of $k_{S'T} = 5 \times 10^7$ s^{-1} and $\tau_T = \infty$ were used. The gain curves clearly show the tendency of the laser wavelength at threshold to shift toward shorter wavelength with increasing cavity loss which delays the initiation of oscillation. With a knowledge of the system losses, the time and wavelength at which laser emission begins can be predicted from Fig. 16.

The critical inversion and lasing wavelength obtained from the gain analysis of Weber and Bass is expected to be somewhat different from those obtained from the critical inversion analysis since photon production by stimulated emission is neglected in the gain characteristic. The inclusion of stimulated emission modifies the gain characteristic by introducing an additional wavelength dependent coefficient into the photon production term of (2). This may be seen by comparing (4) with the first term of (2). A further source of difference in the results of the two analyses stems from the assumption of mirror symmetry for singlet-state absorption and emission spectra since both spectra are then normalized to the quantum yield. Using the measured singlet-state absorption spectrum in the gain analysis would eliminate the need for this assumption as well as the need for estimating ω_0.

D. Tuning of the Dye Laser

In studies of the laser-pumped laser Schäfer, Schmidt and Volze [10], Sorokin et al. [12], and others observed that the wavelength of the dye laser increases with dye concentration, other system parameters remaining fixed. With the proper choice of dye concentration any laser wavelength within a range of more than 50 nm can be obtained with a single dye.

To make use of the concentration tuning it is not necessary to actually change the concentration of the dye solution, the same effect can be obtained by altering the ratio of the active length to the total length of the dye cuvette. An arrangement for tuning the laser by changing the physical length of the dye cell has been described by Farmer et al. [32].

An elegant means of tuning the dye laser was demonstrated by Soffer and McFarland [34], who replaced one mirror of a laser-pumped rhodamine 6G laser with a diffraction grating. The grating was mounted in the Littrow arrangement and was adjusted to the angle for which the first-order reflection of the desired wavelength was reflected back upon itself along the axis of the optical cavity. In effect a mirror with a sharply peaked reflectance spectrum is provided; the wavelength of the peak is altered by rotating the grating. Since the cavity gain is very low at wavelengths other than that to which the cavity is tuned, the laser oscillates at this wavelength. In connection with the tuning experiments Soffer and McFarland observed that the output energy of the laser with the grating mirror was 60 percent as great as the output obtained with broadband dielectric mirrors and was relatively independent of wavelength. Furthermore, the spectral width of the laser output decreased from 60 to 0.6 Å as the broad-band mirror was replaced by the grating. The tuning range was found to be approximately 50 nm.

Tuning of a flashlamp-excited rhodamine 6G laser by a diffraction grating was demonstrated by Sorokin et al. [16]. The experimental arrangement is shown in Fig. 17(a). Tuning of the laser was obtained over a range of about 40 nm. The region in which the laser is tunable is shown superimposed upon the fluorescence and long wavelength absorption spectra of rhodamine 6G in Fig. 17(b). The short wavelength extreme of the tunable range is determined by

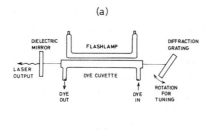

(a)

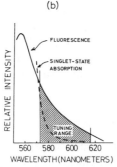

(b)

Fig. 17. Tuning of the flashlamp-excited dye laser by replacing one cavity mirror with a rotatable diffraction grating. (a) Experimental arrangement used by Sorokin et al. [16]. (b) Schematic representation of the tuning range (shaded area) superimposed on the fluorescence emission and singlet absorption spectra of rhodamine 6G.

the steeply rising loss associated with singlet-state absorption. At the long wavelength extreme the gain of the laser becomes too small for oscillation as a result of the decrease in fluorescence quantum yield.

It has been observed in several laboratories that the output spectrum of the laser with a diffraction grating in the cavity consists of a series of sharp lines of nonuniform spacing and intensity rather than of a single sharp line. The effect is well illustrated by a color photograph in an article by Sorokin [33]. This effect seems to be characteristic of the device. The wavelength of the most intense lines also varies from flash to flash even though the experimental conditions remain the same. These effects are probably caused by beam distortion within the laser cavity resulting from non-uniform excitation of the dye.

Bradley et al. [35] have described a novel tuning arrangement for the laser-pumped laser. These authors replaced one mirror of the laser cavity with an echelle grating as in the experiments of Soffer and McFarland [34]. In addition, a Fabry–Perot etalon was inserted into the cavity. Rotation of the echelle grating caused the output of the laser to change in steps corresponding to the free spectral range of the Fabry–Perot etalon. Rotation of the etalon allows tuning within the free spectral range so that continuous tuning is possible. Tuning over a range of 10 nm was reported with a line width of approximately 0.5 Å and a beam divergence of 0.5 mrad. This tuning arrangement could certainly be used with a flashlamp-excited laser as well.

Bonch–Bruyevich et al. [36] have also reported the use of a Fabry–Perot interferometer within the dye laser cavity as a means of tuning the laser output. Tuning of a laser-pumped laser over a range of 8 nm was reported.

Tuning of the laser-pumped laser by means of a Littrow mounted prism has been observed by Murakawa, Yamaguchi, and Yamanaka [37]. A laser using the dye 3,3′-dieth-yloxatricarbocyanine (DOTC) iodide in methanol was excited with a giant pulse ruby laser. A tunable range of about 40 nm was noted. The spectral width of the laser output was approximately 20 Å, considerably broader than the linewidth of the laser using tuning elements with greater spectral dispersion.

In a recent letter by Schappert et al. [38] the effects of temperature upon the spectral output of a laser-pumped laser were reported. The output wavelength of a 6×10^{-5} molar solution of DTTC was found to decrease continuously over a range of 20 nm as the temperature of the dye solution was lowered from 78 to $-117°C$. The wavelength change increased with dye concentration.

Glenn et al. [39] have shown that the output wavelength of a mixture of the dyes rhodamine 6G and rhodamine B can be varied by altering the ratio of their concentrations. A similar effect has been noted by Peterson and Snavely [40]. For tuning by this procedure, care must be taken to choose dye combinations that are compatible chemically. For many dyes the pH at which the fluorescence quantum yield is a maximum is well defined and a mixture of dyes must contain components which fluoresce at the same pH.

Another approach to the tuning of a dye laser has been demonstrated by Weber and Bass [19]. From the analysis of the time-dependent gain characteristic, as displayed in Fig. 16, it is predicted that the oscillation frequency is a function of the time required to reach threshold. Weber and Bass found that the laser could be tuned by inserting a Q switch in the laser cavity which is timed to open at a time when the gain curve has its maximum at the desired wavelength.

E. Time Dependence of the Dye Laser Emission

The results of the critical inversion analysis suggest that the output of the rhodamine 6G laser should sweep with time from longer to shorter wavelengths as the triplet-state concentration increases. For a particular dye the magnitude, and perhaps the direction, of the shift depends upon laser and excitation parameters as well as upon the spectral characteristics of the dye. It is conceivable that in some dyes the relationship between singlet and triplet spectra would be such that the output would sweep from shorter to longer wavelength. There is experimental evidence to suggest that this is the case for the cyanine dyes.

The first time-dependent spectroscopic measurements were made by Bass and Steinfeld [41], who observed that the output of the DTTC iodide laser-pumped laser swept from short to long wavelength with time. This is in opposition to the behavior predicted for rhodamine 6G. The same system was studied by Farmer, Huth, Taylor, and Kagan [42] using a streak camera and spectrograph. Their results were in agreement with the findings of Bass and Steinfeld for low dye concentrations. For the relatively high dye concentration of 5×10^{-4} molar, however, Farmer et al. found that the laser output sweeps from short to longer wavelength for approximately 10 ns of the 20-ns excitation pulse and then returns to shorter wavelengths. The total excursion of the long wavelength edge of the spectrum was approximately 25 Å. Results similar to these were reported by Gibbs and Kellock [43] for cryptocyanine and chloro-aluminum–phthalocyanine dyes excited by a ruby laser.

Measurements of the time dependence of the emission of flashlamp-excited rhodamine 6G, rhodamine B, acridine red, sodium fluorescein, and 4-methylumbelliferone dyes have been reported by Furumoto and Ceccon [44]. The behavior of these dyes was not as consistent as that of the laser-pumped laser dyes and rather complicated time-dependent patterns were found. The spectral shifts observed with the rhodamine dyes were small and only rhodamine B was reported as showing the expected shift from longer to shorter wavelength with time. A decrease in the widths of the rhodamine spectra with time was reported.

The output spectrum of the dye may be as broad as 50 Å or more. In the case of the laser-pumped laser studied by Bass and Steinfeld, time resolved spectroscopy showed that the apparent breadth is the result of a relatively narrow line sweeping with time. With very strong excitation Farmer *et al.* [42] found that the laser line could be as broad as 80 Å. According to the results of Furumoto and Ceccon the output spectrum of the flashlamp-excited laser appears to be relatively broad and does not shift with time.

F. Mode Locking of the Dye Laser

Glenn, Brienza, and de Maria [39] reported the production of a mode locked output by a dye laser excited by the second harmonic of a mode locked Nd-glass laser. The dye laser used a mixture of rhodamine B and rhodamine 6G in a dye cuvette between confocal dielectric mirrors. It was found that mode locking occurred when the optical length of the excitation laser cavity was an integer-multiple of the optical length of the dye laser cavity. The authors did not report a value for the temporal width of the mode-locked pulses.

Sorokin *et al.* [16] observed that the output of a flash-lamp-excited solution of rhodamine 6G is partially mode locked. Two or three trains of high-frequency pulses were noted which grew and decayed at different times. The time between pulses in the various pulse trains was consistent with the optical path length of the laser cavity. No nonlinear element other than the dye itself was needed in the cavity to produce the effects. The laser dye solution apparently acted as a saturable absorber.

The production of a mode locked output from a flash-lamp-excited rhodamine 6G laser with a saturable absorber in the cavity, using the arrangement shown in Fig. 18, was reported by Schmidt and Schäfer [45]. A dye-flow cuvette was used which had a silvered mirror on one end, in contact with the dye solution, and a Brewster angle window at the other end. The mode-locking behavior of the system was examined as a function of the separation between the dye cuvette and the second, dielectric, mirror. A 5×10^{-5} molar solution of 3,3'-diethyloxadicarbocyanine (DODC) iodide was used as a saturable absorber. The dye was placed in a cuvette mounted near the dielectric mirror. The spacing between mode-locked pulses was determined by the optical path length of the cavity. The pulsewidth was not given by

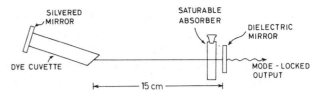

Fig. 18. Arrangement used by Schmidt and Schäfer [45] to demonstrate mode locking of the flashlamp-excited dye laser by a saturable absorber. Mode locking of a rhodamine 6G dye laser with a DODC-iodide saturable-absorber solution was demonstrated.

the authors although it was clearly less than the 4×10^{-10}-second response time of the measuring circuitry.

Soffer and Linn [46] have described a tunable mode-locked dye laser using an ethanol solution of rhodamine 6G excited by a mode locked doubled Nd-glass laser in an arrangement similar to that of Glenn *et al.* [38]. Tuning was accomplished by a rotatable grating in the arrangement described previously [33]. A two photon fluorescence technique was used to determine the mode-locked pulsewidth which the authors reported as approximately 10^{-11} second.

IV. Discussion and Conclusions

As the first truly tunable visible laser capable of reasonably high peak power, as well as high average power through high repetition rate operation, the flashlamp-excited dye laser should find useful application. These properties, in addition to the capability for mode-locked operation, make the dye laser especially attractive as a research tool. Such a device is essential for studying optical phenomena and photochemical processes on a time scale less than 10^{-9} second. Use of the laser as a monochromator in multiple photon absorption experiments also appears attractive.

Much work remains to be done, however, before the use of the dye laser in these and other applications becomes convenient. Some of the problems associated with the device have been mentioned in the foregoing discussion. One of the most serious problems seems to be the refractive index inhomogenieties in the dye solution caused by heat generated during the excitation pulse. It has been shown that nonuniform excitation of the dye cuvette can cause quenching of the stimulated emission. Even in the case of uniform excitation the detrimental effects of heating seem to be present. It seems very possible that many quenching effects previously attributed to the molecular triplet state may in fact be due to heating of the active medium. A more thorough investigation of the dynamics of the laser during oscillation is needed to understand the effects of heating in detail.

The inconsistent behavior of the flashlamp-excited laser observed in the time resolved spectroscopy experiments may very well be caused by beam distortion within the cavity resulting from refractive index inhomogeneities. The experiments reported [44] used a coaxial flashlamp which is known to produce a nonuniform excitation of the dye solution [16]. Attempts have been made to interpret the time-resolved spectroscopic experiments in terms of the broadening mechanism of the molecular states involved in the laser transitions. Triplet-state effects have been largely neglected and heating effects have been overlooked entirely. The interpretation of experimental results in terms of molecular processes is difficult for a dye laser system in which heating effects are present. It would be worthwhile to conduct time-resolved spectroscopic measurements upon a uniformly excited laser system.

The instability of the tuned laser can also very likely be traced to thermal effects. As the beam is deformed within the laser cavity by refractive index inhomogeneities it will sweep across the diffraction grating, striking the grating at an angle which varies with time. This could be expected to produce the characteristic multiline spectra observed [33]. Uniform excitation of the dye should improve the performance of the tunable laser.

Accurate measurements of the gain and efficiency of the flashlamp-excited dye laser are lacking. Data published by Sorokin and co-workers [13], [16] yields an estimated efficiency, defined as laser output divided by electrical energy stored in the capacitor, for the rhodamine 6G laser of approximately 0.4 percent. This figure is very much lower than the value obtained with the laser-pumped laser [12], although in the latter case the optical efficiency, defined as dye-laser output divided by pump-laser input, is quoted. The optical efficiency of the flashlamp-excited laser, defined in the same way, would compare more favorably with the laser-pumped laser.

It has been generally accepted that triplet-state effects will prevent the realization of a continuously operating dye laser. Results have been presented which suggest that oxygen quenching of the triplet state is, under some conditions, rapid enough to permit CW operation of a rhodamine 6G laser. The minimum power required for CW operation of a rhodamine 6G laser can be estimated on the basis of these results. The critical inversion for the long pulse laser, described in connection with Fig. 7, was found to be approximately $N_{1c} = 6 \times 10^{13}$ excited molecules per cm^3. Assuming that the dye is excited by a monochromatic light source at the wavelength of the absorption peak of rhodamine 6G, $\lambda_m = 530$ nm (see Fig. 9), the excitation power required for CW operation will be given by

$$ P = \frac{N_{1c} h c V}{\phi \tau \lambda_m} \text{ watts} $$

where V is the volume of the dye cuvette in cm^3. The dye cuvette of the long pulse laser has a length of 7 cm. A diameter of 2 mm is assumed. Using previously measured values for $\tau = 7.3 \times 10^{-9}$ second and $\phi = 0.83$ the power required is found to be approximately 850 watts. This is not an unreasonably large amount of power and suggests that experiments with continuous pumping of rhodamine 6G are worthy of consideration.

The number of known laser dyes capable of flashlamp excitation is rather small as seen by examination of Table I. A systematic search for new laser dyes among the known dyes would be very worthwhile at this stage of research on the dye laser. If structurally different dyes could be found the structural essentials for good laser dyes might be determined, thus facilitating synthesis of new and better materials. It seems unlikely that the very first laser dyes discovered, the xanthenes, are also the best which can possibly be obtained.

The development of very inexpensive lasers using plastic laser rods should be possible. It has been found that polymethylmethacrylate containing rhodamine dyes will lase [47]. In fact, some commercially available fluorescent plastics can be made to lase quite well with conventional xenon flashlamps driven by ordinary inexpensive capacitors. The production of plastic rods with a high optical quality seems to present problems however.

The invention of the laser has been a major factor in bringing the interests of electrical engineers and atomic and solid-state physicists into convergence. The dye laser, along with the saturable-absorber Q switch, is one of the first devices of a new class of electrooptical devices based upon the properties of large molecules. Research on these devices broadens interdisciplinary contact still further by involving the chemist in an important way. We hope that this contact will yield many new and useful devices.

Acknowledgment

The advice and comments of Drs. O. G. Peterson and F. P. Schäfer were of great value to the author in the preparation of this review. The author is grateful for permission to use the calculations presented in connection with the critical inversion analysis which were done by Dr. Peterson and W. C. McColgin. Finally, the author wishes to express his gratitude to Prof. H. Kuhn for the opportunity to do research at the Physikalisch–Chemisches Institute of the University of Marburg.

References

[1] A. Lempicki and H. Samelson, "Organic laser systems," *Lasers*, A. K. Levine, Ed., vol. 1, 1966, pp. 181–252.
[2] A. Heller, "Liquid lasers-design of neodymium-based inorganic ionic systems," *J. Mol. Spectry.*, vol. 28, pp. 101–117, September 1968.
[3] ——, "Liquid lasers-fluorescence, absorption and energy transfer of rare earth ion solutions in selenium oxychloride," *J. Mol. Spectry.*, pp. 208–232, October 1968.
[4] E. G. Brock, P. Czavinsky, E. Hormats, H. C. Nedderman, D. Stirpe, and F. Unterleitner, "Coherent stimulated emission from molecular crystals," *J. Chem. Phys.*, vol. 35, pp. 759–760, August 1961.
[5] S. G. Rautian and I. I. Sobel'mann, "Remarks on negative absorption," *Opt. Spectry.*, vol. 10, pp. 65–66, January 1961.
[6] D. L. Stockman, W. R. Mallory, and K. F. Tittel, "Stimulated emission in aromatic organic compounds," *Proc. IEEE* (Correspondence), vol. 52, pp. 318–319, March 1964.
[7] D. L. Stockman, "Stimulated emission considerations in fluorescent organic molecules," *Proceedings of the 1964 ONR Conf. on Organic Lasers*, available as Doc. no. AD 447468 from the Defense Documentation Center for Scientific and Technical Information, Cameron Station, Alexandria, Va.
[8] P. P. Sorokin and J. R. Lankard, "Stimulated emission observed from an organic dye, chloroaluminum phthalocyanine," *IBM J. Res. Develop.*, vol. 10, pp. 162–163, March 1966.
[9] P. P. Sorokin, W. H. Culver, E. C. Hammond, and J. R. Lankard, "End-pumped stimulated emission from a thiacarbocyanine dye," *IBM J. Res. Develop.*, vol. 10, p. 401, September 1966.
[10] F. P. Schäfer, W. Schmidt, and J. Volze, "Organic dye solution laser," *Appl. Phys. Letters*, vol. 9, pp. 306–309, October 1966.
[11] M. L. Spaeth and D. P. Bortfield, "Stimulated emission from polymethine dyes," *Appl. Phys. Letters*, vol. 9, pp. 179–181, September 1966.

[12] P. P. Sorokin, J. R. Lankard, E. C. Hammond, and V. L. Moruzzi, "Laser pumped stimulated emission from organic dyes: experimental studies and analytical comparisons," *IBM J. Res. Develop.*, vol. 11, pp. 130–148, March 1967.

[13] P. P. Sorokin and J. R. Lankard, "Flashlamp excitation of organic dye lasers: a short communication," *IBM J. Res. Develop.*, vol. 11, p. 148, March 1967.

[14] W. Schmidt and F. P. Schäfer, "Blitzlampengepumpte Farbstofflaser," *Z. Naturforsch.*, vol. 22a, pp. 1563–1566, October 1967.

[15[N. J. Turro, *Molecular Photochemistry*. New York: Benjamin, 1965, pp. 54–55.

[16] P. P. Sorokin, J. R. Lankard, V. L. Moruzzi, and E. C. Hammond, "Flashlamp pumped organic dye lasers," *J. Chem. Phys.*, vol. 48, pp. 4726–4741, May 1968.

[17] M. Bass, T. F. Deutsch, and M. J. Weber, "Frequency and time-dependent gain characteristics of laser and flashlamp-pumped dye solution lasers," *Appl. Phys. Letters*, vol. 13, pp. 120–124, August 1968.

[18] D. E. McCumber, "Theory of phonon-terminated optical masers," *Phys. Rev.*, vol. 134, pp. A299–A306, April 1964.

[19] M. J. Weber and M. Bass, "Frequency and time dependent gain characteristics of dye lasers," *IEEE J. Quantum Electronics*, vol. QE-5, pp. 175–188, April 1969.

[20] B. B. Snavely and O. G. Peterson, "Experimental measurement of the critical population inversion for the dye solution laser," *IEEE J. Quantum Electronics*, vol. QE-4, pp. 540–545, October 1968.

[21] A. Yariv and J. P. Gordon, "The laser," *Proc. IEEE*, vol. 51, pp. 4–29, January 1963.

[22] A. V. Buettner, B. B. Snavely, and O. G. Peterson, "Triplet state quenching of stimulated emission from organic dye solutions," *Proc. Internatl. Conf. on Molecular Luminescence*. New York: Benjamin, 1969, pp. 403–422.

[23] S. Claesson and L. Lindqvist, "A fast photolysis flash lamp for very high light intensities," *Arkiv Kemi*, vol. 12, pp. 1–8, January 1958.

[24] B. B. Snavely and F. P. Schäfer, "Feasibility of CW operation of dye lasers," *Phys. Letters*, vol. 28A, pp. 728–729, March 1969.

[25] H. Winston and R. A. Gudmundsen, "Refractive index effects in proposed liquid lasers," *Appl. Opt.*, vol. 3, pp. 143–146, January 1964.

[26] B. B. Snavely, O. G. Peterson, and R. F. Reithel, "Blue laser emission from a flashlamp-excited organic dye solution," *Appl. Phys. Letters*, vol. 11, pp. 275–276, November 1967.

[27] F. P. Schäfer, "Organic dye lasers," invited paper presented at the 1968 Quantum Electronics Conf., Miami, Fla., May 1968.

[28] M. E. Mack, "Measurement of nanosecond fluorescence decay times," *J. Appl. Phys.*, vol. 39, pp. 2483–2485, April 1968.

[29] Landolt-Börnstein, *Zahlenwerte und Funktionen*, vol. II b. Berlin: Springer, 1962, p. 1–75.

[30] N. J. Turro, *op. cit.* pp. 114–118.

[31] Landolt-Börnstein, *op. cit.*, pp. 1–20.

[32] G. I. Farmer, B. G. Huth, L. M. Taylor, and M. R. Kagan, "Concentration and dye length dependence of organic dye laser spectra," *Appl. Opt.*, vol. 8, pp. 363–366, February 1969.

[33] P. P. Sorokin, "Organic lasers," *Sci. Am.*, vol. 220, pp. 30–40, February 1969.

[34] B. H. Soffer and B. B. McFarland, "Continuously tunable, narrowband organic dye lasers," *Appl. Phys. Letters*, vol. 10, pp. 266–267, May 1967.

[35] D. J. Bradley, G. M. Gale, M. Moore, and P. D. Smith, "Longitudinally pumped, narrow-band continuously tunable dye laser," *Phys. Letters*, vol. 26a, pp. 378–379, March 1968.

[36] A. M. Bonch-Bruyevich, N. N. Kostin, and V. A. Khodovoi, "Selection and adjustment of generated frequencies in dye solutions," *Opt. Spectry.*, vol. 24, pp. 547–548, June 1968.

[37] S. Murakawa, G. Yamaguchi, and C. Yamanaka, "Wavelength shift of dye solution laser," *Japan J. Appl. Phys.*, vol. 7, p. 681, May 1968.

[38] G. T. Schappert, K. W. Billman, and D. C. Burnham, "Temperature tuning of an organic dye laser," *Appl. Phys. Letters*, vol. 13, pp. 124–126, August 1968.

[39] W. H. Glenn, M. J. Brienza, and A. J. de Maria, "Mode locking of an organic dye laser," *Appl. Phys. Letters*, vol. 12, pp. 54–56, January 1968.

[40] O. G. Peterson and B. B. Snavely, "Multiple-dye solution lasers," *Bull. Am. Phys. Soc.*, vol. 13, p. 397, March 1968.

[41] M. Bass and J. I. Steinfeld, "Wavelength dependent time development of the intensity of dye solution lasers," *IEEE J. Quantum Electronics*, vol. QE-4, pp. 53–58, February 1968.

[42] G. I. Farmer, B. G. Huth, L. M. Taylor, and M. R. Kagan, "Time resolved stimulated emission spectra of an organic dye laser," *Appl. Phys. Letters*, vol. 12, pp. 136–138, February 1968.

[43] W. E. K. Gibbs and H. A. Kellock, "Time-resolved spectroscopy of organic dye lasers," *IEEE J. Quantum Electronics*, vol. QE-4, pp. 293–294, May 1968.

[44] H. Furumoto and H. Ceccon, "Time dependent spectroscopy of flashlamp pumped dye lasers," *Appl. Phys. Letters*, vol. 13, pp. 335–337, November 1968.

[45] W. Schmidt and F. P. Schäfer, "Self-mode-locking of dye lasers with saturable absorbers," *Phys. Letters*, vol. 26a, pp. 558–559, April 1968.

[46] B. H. Soffer and J. W. Linn, "Continuously tunable picosecond-pulse organic-dye-laser," *J. Appl. Phys.*, vol. 39, pp. 5859–5860, December 1968.

[47] O. G. Peterson and B. B. Snavely, "Stimulated emission from flashlamp excited organic dyes in polymethyl methacrylate." *Appl. Phys. Letters*, vol. 12, pp. 238–240, April 1968.

Part III: Gas Lasers

The development of gas lasers since the first He–Ne 1.15 μ laser has proceeded along a number of significant paths, not all of which are covered here.[1] For example, the ubiquitous red He–Ne 6328 Å laser and the widely used CW 10.6 μ CO_2 laser are not treated explicitly. Among other important visible and ultraviolet lasers, however, the noble gas ion lasers are surveyed in the paper by Bridges and coauthors. Emphasis in this paper is on the plasma discharge properties, which are in many ways more significant than the purely laser properties.

Three important gas lasers for high-power applications are the transverse-excitation (TEA) type molecular laser, the gasdynamic laser, and the chemical laser. Introductory surveys of each of these three devices are given in the papers by Beaulieu, Gerry, and Chester. While further rapid advances in the state of the art of these lasers can be expected, each of the articles should continue to serve as a good introduction to their operating principles.

[1] See C. G. B. Garrett, *Gas Lasers*. New York: McGraw-Hill, 1967.

Ion Laser Plasmas

WILLIAM B. BRIDGES, FELLOW, IEEE, ARTHUR N. CHESTER, MEMBER, IEEE,
A. STEVENS HALSTED, MEMBER, IEEE, AND JERALD V. PARKER, MEMBER, IEEE

Invited Paper

Abstract—The typical noble gas ion laser plasma consists of a high-current-density glow discharge in a noble gas, in the presence of a magnetic field. Typical CW plasma conditions are current densities of 100 to 2000 A/cm^2, tube diameters of 1 to 10 mm, filling pressures of 0.1 to 1.0 torr, and an axial magnetic field of the order of 1000 G. Under these conditions the typical fractional ionization is about 2 percent and the electron temperature between 2 and 4 eV. Pulsed ion lasers typically use higher current densities and lower operating pressures.

This paper discusses the properties of ion laser plasmas, in terms of both their external discharge parameters and their internal ion and excited state densities. The effect these properties have on laser operation is explained. Many interesting plasma effects, which are important in ion lasers, are given attention. Among these are discharge nonuniformity near tube constrictions, extremely high ion radial drift velocities, wall losses intermediate between ambipolar diffusion and free fall, gas pumping effects, and radiation trapping. The current status of ion laser technology is briefly reviewed.

I. INTRODUCTION

THE ION LASER is one of the most important practical sources of visible and ultraviolet coherent light presently known. In addition, it incorporates a gaseous plasma exhibiting a number of interesting properties besides laser action. This paper briefly reviews some of the principal characteristics of ion laser plasmas.

Section I summarizes the operating principles and output characteristics of the most important ion lasers. Section II presents a basic description of the CW argon ion laser plasma as an example, covering mainly those aspects that are reasonably well understood at present. Section III discusses some of the important plasma phenomena which also occur in ion laser plasmas but which are understood only qualitatively. This discussion includes both technological problems and their underlying physical effects.

A. Characteristics of Ion Lasers

An ion laser plasma generally consists of a long narrow cylindrical high-density glow discharge. The optical radiation from the discharge is directed repeatedly through the glow region by mirrors placed at each end of the tube. The positive column of the discharge plasma constitutes the laser medium and amplifies the radiation. Many different excitation techniques have been used to produce a glow discharge of the proper characteristics. Among these are pulsed and CW dc discharges using electrodes (both hot and

Manuscript received January 16, 1971. *This invited paper is one of a series planned on topics of general interest—The Editor.*

W. B. Bridges, A. N. Chester, and J. V. Parker are with the Hughes Research Laboratories, Malibu, Calif. 90265.

A. S. Halsted is with the Electron Dynamics Division, Hughes Aircraft Company, Torrance, Calif. 90509.

cold cathode), pulsed dc discharges using electrodeless magnetic induction coupling, pulsed and CW RF discharges using induction coupling with electric fields both longitudinal and transverse to the long dimension of the plasma, microwave cyclotron resonance discharges, beam-generated plasmas, and pinch-generated plasmas (both z- and θ-pinch types). In this brief review it would be impossible to cover the detailed characteristics produced by each excitation method. Representative examples must necessarily be chosen. It is fortunate, however, that the method of producing the plasma seems to have little effect on the laser properties; CW ion lasers exhibit roughly the same overall characteristics whether produced by a hot cathode dc discharge or an RF induction excited system. Pulsed lasers are likewise relatively independent of the excitation means. In much of the discussion following, the argon ion laser [1]–[3] will be taken as the example, since the largest body of information exists for this most popular laser type.

For a typical ion laser the most important physical processes producing the laser action are as follows. In the glow discharge, approximately 1 percent of the atoms are ionized. Electrons with a mean energy of 2 to 4 eV collide with ground state ions and produce excited ions, including ionic states that constitute the upper and lower laser levels. The excitation cross sections and radiative lifetimes are such that the steady-state population density of the upper laser levels in the discharge is much larger than the density of the lower laser levels. This population inversion produces optical gain and laser action, typically on radiative transitions between several pairs of upper and lower levels. The laser output wavelengths depend on the choice of gas, the operating conditions, or the use of wavelength-selective laser mirrors.

The great practical importance of ion lasers derives from several useful characteristics. The optical gain of the ion laser plasma is very high, typically > 1 dB/m (23 percent/m) [3]–[7] and as high as 13 dB/m in argon [8]; this makes it very easy to secure laser oscillation even when mirrors and windows are not of the highest optical quality. In addition, ion lasers produce numerous laser lines that lie in or near the visible portion of the spectrum; at present, 437 different wavelengths have been observed, originating from 29 different elements, and ranging in wavelength from 0.2358 to 1.555 μm [9]. Finally, the most important laser lines (and more than half of all observed lines) employ only noble gases in the discharge tube, and the chemical inertness of these gases simplifies many aspects of tube technology.

Reprinted from *Proc. IEEE*, vol. 59, pp. 724–737, May 1971.

TABLE I
OPERATING CONDITIONS OF TYPICAL ION LASERS

Type of Laser	Tube Diameter (mm)	Tube Length (cm)	Discharge Current (A)	Gas Fill Pressure
Pulsed, noble gas	2–10	10–200	100–1000	10–50 mtorr
CW, noble gas	1–12	10–50	10–100	50–1000 mtorr
CW, He–Cd	3	150	0.04–0.10	2–10 mtorr Cd plus 2–8 torr He

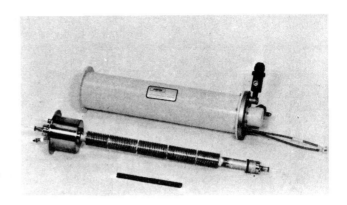

Fig. 1. A high-performance metal–ceramic argon ion laser discharge tube with its associated magnet. Metal cooling fins surround the ceramic section. (Hughes Electron Dynamics Division.)

The operating conditions of the most useful ion lasers are summarized in Table I. The simplest type of ion laser is that operated with a low-duty-cycle pulsed discharge, generally with pulses 1–100 μs in duration and repetition rates up to a few hundred pulses per second. A typical pulsed tube consists of a long glass, quartz, or ceramic capillary with an anode and a cold cathode located in side-arms at each end of the tube. An external optical resonator is usually employed, with Brewster's angle windows mounted on the discharge tube to minimize reflective loss. The laser output generally follows the current pulse. However, if the current density and the pulse length become too great, laser action may be quenched by loss of gas density in the capillary (caused by thermal driveout and gas pumping effects) or by radiation trapping of the lower laser level, as discussed later. In addition, at long pulse lengths the capillary may be thermally destroyed unless adequately cooled.

The second type of ion laser listed in Table I is the CW noble gas ion laser. The earliest CW ion lasers used a water-cooled quartz capillary to permit high average input power to the discharge. Gas pumping caused by electron and ion drift motion in the positive column led to substantial pressure gradients along the tube. To keep the discharge maintained at a convenient voltage it was found desirable to equalize the pressure at the cathode and anode by connecting them with a gas return path, which consisted of a glass tube with sufficient conductance to permit easy gas flow but long enough to prevent a discharge from striking through it. A photograph of such a tube appears in [10]. The addition of a magnetic field (about 1 kG) with field lines parallel to the optical axis increases the laser output, particularly from tubes of less than 6-mm diameter, and a magnetic field is usually used with such tubes. The higher current densities and extended lifetime currently demanded of ion lasers have led to the development of highly engineered metal–ceramic tubes of the type shown in Fig. 1.

The third major type of ion laser, typified by the CW helium–cadmium laser, operates at modest power levels and does not require special cooling. The most popular version of this laser [11] resembles in construction the familiar helium–neon lasers except for the addition of a sidearm containing cadmium metal. The vapor pressure of cadmium in the discharge is controlled by heating this sidearm. Both zinc and selenium ion lasers have been made in this form to date (see tabulation and references in [9]).

The output characteristics of the most extensively studied practical ion lasers are summarized in Table II. The argon ion laser, with blue and green output, has undergone the most advanced development to date. The newest among the sources listed is the xenon IV laser,[1] which demonstrates a high-power pulsed mode of operation that is, so far, unique among ion lasers. The high electrical efficiencies (exceeding 0.1 percent) and the availability of multiwatt output at a great variety of wavelengths (of which only a few are listed here) make ion lasers important sources of coherent radiation in the visible and near UV.

The effect of length and tube diameter on laser output power and efficiency is illustrated in Fig. 2 for the blue-green lines of the argon ion laser. In this figure the CW laser output power per unit length of discharge column is plotted as a function of the parameter JR, where J is discharge current density and R is the inner tube radius. For most curves, the externally applied magnetic field and the gas filling pressure have been adjusted to maximize the laser output at each value of current used. The curves are marked to indicate the tube diameter, which ranges from 1.2 to 12 mm. The abscissa is marked to indicate approximate input power per unit length of discharge, since the input power can be correlated with the quantity JR in a manner roughly independent of tube radius. The lines of constant electrical efficiency indicate the ratio of laser output to electrical input for the positive column only; thus the indicated efficiencies of Fig. 2, unlike those of Table II, do not include electrode potential drop, electrode heater power (if needed), or electromagnet power (if needed). The great variation that is apparent in the curves of Fig. 2 represents, in part, the gradual improvement of ion laser technology, with respect to both electrical efficiency and power output from a given length tube.

The highest reported power per unit length is 105 W/m from a 12-mm tube [12], and the highest efficiency is 0.16 percent from a 6.35-mm tube [13]. Large diameter tubes

[1] The roman numeral following the element name or symbol refers to the ionization state of the species from which the spectroscopic line arises. The numeral is one larger than the ionization. Thus "Xe^{3+}" is equivalent to "Xe IV" since the first spectrum of xenon, Xe I, arises from the neutral atom.

TABLE II
ION LASER PERFORMANCE: 1970 STATE OF THE ART

Lasing Species	Lasing Mode	Strongest Wavelengths (μm)	Output Power	Output Power per Unit Length	Overall Electrical Efficiency (%)
Ar II	CW	0.4880, 0.5145	120 W[a]	105 W/m[b]	0.16[c]
Kr II	CW	0.6471	~10 W[a]	5 W/m[a]	~0.01[a,d]
Cd II	CW	0.4416	0.2 W[e]	0.14 W/m[e]	0.09[f]
Cd II	CW	0.3250	0.02 W[e]	0.014 W/m[e]	~0.01[e,f]
Ar III	CW	0.3638, 0.3511	5 W[k]	5 W/m[g]	0.01[g]
Xe IV	CW	0.4954–0.5395 (5 lines)	0.5 W[h]	1 W/m[h]	<0.01[h]
Xe IV	pulsed	0.5395, 0.5353, 0.5260	0.4 J/pulse[i] (20 kW peak)	0.14 J/m[i]	~0.3[i,j]

[a] K. Banse, H. Boersch, G. Herziger, G. Schäfer, and W. Seelig, *Z. Angew. Phys.*, vol. 26, 1969, pp. 195–200.

[b] H. Boersch, J. Boscher, D. Hoder, and G. Schäfer, *Phys. Lett.*, vol. 31A, 1970, pp. 188–189.

[c] J. R. Fendley, Jr., *Proc. 4th DOD Conf. Laser Technology* (San Diego, Calif., Jan. 1970), vol. 1, pp. 391–398 (unpublished).

[d] W. B. Bridges and A. S. Halsted, Hughes Res. Labs., Malibu, Calif., Tech. Rep. AFAL-TR-67-89, May 1967 (unpublished); DDC accession no. AD-814-897.

[e] J. P. Goldsborough, *Appl. Phys. Lett.*, vol. 15, 1969, pp. 159–161.

[f] W. T. Silfvast, *Appl. Phys. Lett.*, vol. 13, 1968, pp. 169–171. The efficiency of 9.02 percent given in Silfvast is a misprint and should instead be 0.09 percent.

[g] W. B. Bridges and G. N. Mercer, Hughes Res. Labs., Malibu, Calif., Rep. ECOM-0229-F, Oct. 1969 (unpublished); DDC accession no. AD-861-927.

[h] ——, "CW operation of high ionization states in a xenon laser," *IEEE J. Quantum Electron.* (Corresp.), vol. QE-5, Sept. 1969, pp. 476–477.

[i] W. W. Simmons, private communication, July 31, 1970.

[j] Pulse efficiency is defined as peak laser output power, divided by the product of peak tube current and initial applied voltage.

[k] J. R. Fendley, Jr., private communication, Jan. 1970.

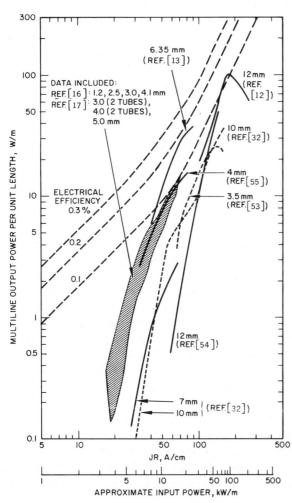

Fig. 2. Summary of reported power output characteristics of CW Ar II ion lasers from the following sources: Fig. 4 in [16]; Fig. 1 in [53]; Fig. 13 in [17]; Fig. 5 in [54]; Figs. 10 and 14 in [32]; Fig. 6 in [13]; Fig. 1 in [12]; and Table I (4-mm by 18-in tube) in [55]. Indicated electrical efficiency is calculated for positive column only, using axial electric field data from Fig. 32 in [17]; Figs. 4 and 14 in [32]; and eq. (3.1) in [26].

(~10–12 mm) have some advantages—they require less or no external magnetic field for efficient operation; their larger wall area eases the thermal load on wall materials; and the larger optical beam size reduces the flux density that must be withstood by optical components such as windows and mirrors. However, many applications require that the laser operate in a single transverse optical mode, so that the laser output is completely phase coherent across the beam front. This is very difficult to achieve in such a larger diameter tube [14] and limits the usefulness of this type of ion laser. An additional disadvantage of the larger bore tubes is that the laser gain and output power decrease with increasing tube diameter [6], [17]; thus many lines will not oscillate at all in a larger diameter tube, and those that will oscillate require higher quality windows and mirrors for efficient operation.

The relative merits of large and small diameter ion laser tubes may be summarized as follows. When maximum power per unit length is desired on the strongest laser transitions (i.e., the Ar II, Kr II, and Xe IV lines of Table II), the large diameter tubes are an appropriate choice. When minimum beam divergence (from single transverse mode operation), less output power (~1 W), or other laser transitions are required, a small diameter tube is usually the best choice. When maximum overall electrical efficiency is required, the choice of tube geometry will depend upon the wavelengths desired and other operational requirements.

Not surprisingly, the ranges of plasma characteristics that have been most fully explored correspond to the operating conditions of practical devices, as summarized in Tables I and II and in Fig. 2. We will proceed now to give a more complete description of what is known about the ion laser itself. Our discussion will center mainly on the most highly developed ion laser, the blue-green lines of Ar II.

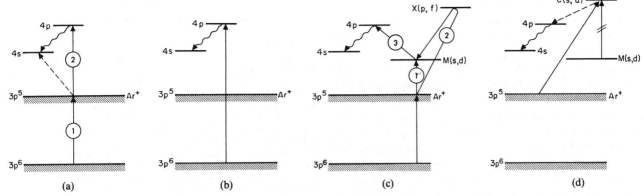

Fig. 3. Energy level diagrams for the blue and green transitions in Ar II, showing various excitation schemes (from [17]).

B. Excitation Mechanisms of Ion Lasers

The original model of laser excitation in noble gas ions was proposed by Gordon *et al.* [15], [16], and is shown schematically in Fig. 3(a). In Gordon's model, the upper laser level is excited by collisions between an electron and a ground state ion (process 2); the ion was presumably produced by the collision of an electron and a neutral atom in the ground state (process 1). Radiative cascade from higher states and de-excitation of the upper laser level by electrons are neglected. The lower laser level is rapidly depopulated by radiation (in the vacuum UV at $\lambda \approx 740$ Å for the $4s$ states of Ar II). This depopulation is assumed to occur much more rapidly than any collisional electron excitation (dotted arrow). Finally, one assumes approximate charge neutrality in the plasma and current-independent electron temperature. Then the upper laser level population N_2 turns out to vary as

$$N_2 \propto n_e n_i \approx n_e^2 \propto J^2 \qquad (1)$$

where n_e, n_i, and J are the electron density, ion density, and discharge current density, respectively. Such a quadratic current dependence has been observed in spontaneous emission measurements over a wide range of gas pressures, currents, and tube dimensions typical of CW ion lasers (see, for example, [17] and [18]). However, in pulsed operation, radiation trapping of the lower level population can upset the simple current dependence of (1). (See Section III-C for a discussion of this effect.)

An alternate excitation method involving a single electron collision was proposed by Bennett *et al.* [3] and is shown in Fig. 3(b). This excitation process apparently occurs principally in low-pressure discharges excited by short pulses. A discharge with very high E/p is required to produce an electron temperature sufficiently high to achieve a significant single step excitation rate. This mode of operation is also characterized by a unique distribution of spectral output. For example, under these conditions, 0.4765 μm is the strongest laser line in Ar II. Such a spectral distribution is never observed in CW operation, where the 0.4880-μm and 0.5145-μm lines always dominate. Moreover, the single step excitation process (assuming constant electron temperature) would lead to a spontaneous emission rate from the upper laser level which varies *linearly* with discharge current, which is not observed in CW operation.

An additional excitation process was proposed by Labuda *et al.* [19], with ionic metastables serving as the intermediate species rather than the ion ground state, as shown in Fig. 3(c). Absorption measurements [19] show that these metastable levels are highly populated. The ionic metastables are produced either by electron collision with the ionic ground state [process 1 in Fig. 3(c)] or (more probably) by cascade from higher lying states (process 2); these higher lying states exhibit populations $N_X \propto J^2$, as shown by spontaneous emission data. However, because the metastables are primarily destroyed by electron collisions rather than by spontaneous emission, it can be shown that the metastable population N_m will be proportional to J rather than J^2 (see [16] or [17]). The upper laser level is then populated by yet another electron collision (process 3), and the relation $N_2 \propto J^2$ obtains as before.

Our understanding of the upper level excitation processes remained essentially as described above until the measurements made by Rudko and Tang [20]. They found that a significant fraction of the upper laser level population was created by radiative cascade from higher lying states [see Fig. 3(d)]. By summing the spontaneous emission intensities of all spectral lines terminating on a given upper laser level (for example, $4p\ ^2D_{5/2}^0$, the upper level of the 0.4880-μm transition), and then comparing this with the sum of intensities of those lines *originating* from that level, we may determine the relative pumping rate due to cascade from the higher lying states. In a 1-mm-diameter tube operated without a magnetic field, Rudko and Tang found that approximately 50 percent of the $4p\ ^2D_{5/2}^0$ level population is created by cascade. Similar measurements were made by Bridges and Halsted [17], who obtained a value of 23 percent for the cascade contribution to this same level in a 3-mm-diameter tube operated without a magnetic field and 22 percent for the cascade contribution to the $4p\ ^4D_{5/2}^0$ (0.5145-μm) upper level. Since the populations N_C from which the cascade takes place also exhibit a quadratic dependence on J, the same quadratic dependence $N_2 \propto J^2$ results again. Note that the upper laser level population due to cascade cannot be separated from that due to two-electron-collision processes by its current dependence alone. It may be seen from the foregoing discussion that the characteristics of ion lasers are a direct consequence of the collision and radiation processes occurring within the ion laser plasma.

II. FUNDAMENTALS OF THE ION LASER PLASMA

In this section we will present a brief physical description of the basic ion laser plasma, excluding spectroscopic properties. Many interesting problems such as gas pumping, throat erosion, etc., which are associated with ion lasers will be discussed later in Section III. Although such problems are all connected with the plasma properties, they are also dependent on the overall geometry; however, in this section we will treat the local discharge properties in the laser bore without regard to geometric properties other than tube radius. On this basis the ion laser discharge can be well characterized as given by the following independent parameters.

R Tube radius, cm.
J Discharge current density, A/cm^2.
p Externally measured gas pressure, torr.
B Axial magnetic field, G.

Ignoring spectroscopic quantities, such as level populations, the only macroscopic dependent property in this description is the axial electric field E, and various derived quantities (e.g., power input/length $= IE$). In addition, there are many microscopic properties such as species number density and temperature.

The remainder of this section is organized as follows. First, we will examine a particular ion laser discharge in some detail to establish the nature of the plasma under discussion. Second, we will review the experimental results obtained when the external conditions are changed (e.g., varying J or R), including a brief discussion of the theoretical approaches that have been taken in attempting to explain these observations.

The particular example we will consider is a 2-mm-diameter smooth quartz capillary tube with a continuous dc discharge current of 5 A. This is not typical of present-day ion lasers, which commonly operate at currents of 20–30 A or more in bores constructed of smooth wall ceramics, refractory metal disks, or graphite segments. However, there are several justifications for using this particular example. Most important is the availability of a reasonable body of data taken by R. C. Miller and coworkers at Bell Telephone Laboratories [21]. The low current chosen for these measurements was dictated by the use of quartz capillaries, which fail rather quickly at high currents. Also, although 5 A is not a state-of-the-art ion laser discharge, it is well over threshold for laser action, so that the plasma conditions are not atypical of ion laser discharges. Finally, comparisons among many ion laser devices, employing a wide variety of construction techniques ranging from solid dielectric wall capillaries to stacks of thin metal disks, have always demonstrated essentially identical laser and electrical performance for the same diameter discharge.

Although the available measurements limit our discussion primarily to the argon plasma, similar plasma properties are to be expected in krypton, xenon, and other ion laser species.

Table III summarizes the information known about the 2-mm-diameter 5-A argon discharge operating at an ex-

TABLE III
SELECTED EXAMPLE OF AN ION LASER PLASMA

Laser species	argon
Discharge tube diameter	2 mm
Discharge current	5 A
External pressure	600 mtorr
Electric field	6.3 V/cm
Current density	160 A/cm^2
Power dissipation	31 W/cm
Electron temperature (double probe)	1.9 eV (22 000°K)
Ion temperature (Doppler width)	0.17 eV (2000°K)
Gas temperature (Doppler width)	0.15 eV (1700°K)
Electron density (stark broadening)	3.4×10^{13} cm^{-3}
Neutral density (X-ray absorption)	5.8×10^{15} cm^{-3}
Fractional ionization (calculated)	0.64×10^{-2}
Ion drift velocity (Doppler shift)	1.2×10^{4} cm/s
Electron drift velocity (calculated)	2.9×10^{7} cm/s
Plasma frequency	52 GHz
Debye shielding length	1.8×10^{-3} mm
Ion–atom mean free path	0.2 mm
Electron–atom mean free path	0.8 mm–12 mm (function of energy)

ternal pressure which yields maximum laser output. Several observations concerning the properties tabulated in Table III are in order.

1) The Debye length is ∼0.001 times the tube diameter, which implies that charge neutrality is closely obeyed except in the sheath region near the wall.

2) The discharge, which is often referred to as an arc, is more closely related to a glow discharge despite the high power input. The large difference between gas and electron temperature illustrates this very clearly.

3) The plasma frequency is in a difficult measurement region. It is too high for RF or microwave measurements and too low for optical measurements. As a result, electron densities have been primarily measured by Stark broadening of the H_β line using small amounts of added hydrogen.

4) The plasma is dominated by boundary phenomena. Electron and ion mean free paths are comparable to the tube diameter. Ion loss is entirely to the walls and the mechanism controlling ion motion is intermediate between collisionless free-fall motion and ambipolar diffusion.

The combination of 1) boundary effects and 2) the lack of simplifying assumptions concerning particle motion is a fundamental cause of the present inadequate theoretical understanding of ion laser plasmas. We will return to this point after concluding our discussion of the experimental data.

Having established the nature of the ion laser plasma, we can now proceed to a discussion of the effect of varying discharge diameter, gas pressure, current density, and magnetic field. The amount of experimental information on CW ion laser plasmas is very small. Measurements by Kitaeva, Osipov, and Sobelev (KOS [22]) at the Physics Institute of the USSR Academy of Sciences, and by Miller, Labuda, and Webb (MLW [21]) at Bell Telephone Laboratories, provide the bulk of the available information. The data of KOS are based on spectroscopic measurements of longitudinal and transverse linewidths for various ion and neutral lines of argon. Two different bore diameters (2.8 and 1.6 mm) were used, and measurements were made over a

TABLE IV

ION AND ATOM TEMPERATURES AS A FUNCTION OF CURRENT DENSITY

J (A/cm^2)	T_i (°K)			T_a (°K)		
	1.6 mm[a]	2 mm[b]	2.8 mm[c]	1.6 mm[a]	2 mm[b]	2.8 mm[c]
160	2550	1800	3900°K		1250	1900
200	2600		4400°K	1700	1400	2300
300	3800		5600°K	1800	1600	3300
400	3450			2350		
500	4800			3000		

[a] V. F. Kitaeva, Yu. I. Osipov, and N. N. Sobolev, "Spectroscopic studies of gas discharges used for argon ion lasers," *IEEE J. Quantum Electron.*, vol. QE-2, Sept. 1966, pp. 635–637.

[b] R. C. Miller and C. E. Webb, Bell Telephone Labs., Murray Hill, N. J., private communication.

[c] V. F. Kitaeva, Yu. I. Osipov, P. L. Rubin, and N. N. Sobolev, "On the inversion mechanism in the CW argon-ion laser," *IEEE J. Quantum Electron.*, vol. QE-5, Feb. 1969, pp. 72–77.

TABLE V

ELECTRON TEMPERATURE AND DENSITY AS A FUNCTION OF CURRENT DENSITY AND PRESSURE

J (A/cm^2)	T_e	N_e (cm^{-3}) $\times 10^{13}$			
		$p=1.0$ torr[a]	$p=0.62$ torr[b]	$p=0.37$ torr[b]	$p=0.21$ torr[b]
50	27 000[a]	6.3			
100	23 000[a]	4.3			
150	37 000[b]	5.7	2.4	1.8	1.0
200	50 000[b]	6.9	3.3	2.3	1.4
250	60 000[b]	8.9	4.1	2.8	1.6
300	70 000[b]		4.6	3.1	1.8
350	82 000[b]		4.8	3.2	1.9

[a] R. C. Miller and C. E. Webb, Bell Telephone Labs., Murray Hill, N. J., private communication. (T_e measured by double probe; N_e measured by stark broadening. Tube diameter 2 mm.) See also eq. (3.2) in [26].

[b] V. F. Kitaeva, Yu. I. Osipov, and N. N. Sobolev, *Zh. Eksp. Teor. Fiz.*, vol. 4, 1966, p. 213. (T_e deduced from T_i; N_e calculated from conductivity formula derived by V. N. Kolesnikov, doctoral dissertation, Physics Inst., USSR Academy of Sciences, 1962.

range of current densities from 100 to 500 A/cm^2 and gas pressures from 100 to 700 mtorr. Measurements by MLW have utilized a variety of techniques, including spectroscopic measurements, double probes, and X-ray absorption. However, the range of current density investigated by MLW is somewhat restricted, so that intercomparison of data between the two investigators is difficult. We will not discuss pulsed ion laser plasmas as a separate category, even though some pulsed plasma measurements have been attempted [23]–[25], since sufficient data are not yet available.

Ion and Atom Temperatures: Ion and atom temperatures are deduced directly from Doppler linewidths and as a result are rather reliable. The data in Table IV, plotted against current density, are taken from both KOS and MLW. The consistency is relatively good, and Chester [26] has suggested the empirical formula

$$T_a/300 = 1 + 0.9 \times 10^{-2} JR^{1/2} \qquad (2)$$

for the atom temperature. Since data are available for tubes of small diameter only (1.6, 2.0, 2.8, 3.0 mm) the validity of this formula has not been tested for large bore tubes. It is possible that the parameter JR suggested by Herziger and Seelig [27] might provide an equally good fit over the wider range of bore diameters employed in modern ion lasers. The pressure dependence of the longitudinal ion temperature and the neutral temperature is small, no more than 10–20 percent for pressures from 0.15–5 torr according to MLW. The transverse "temperature," or more correctly linewidth, is quite pressure dependent according to the measurements of KOS. The ion velocities transverse to the discharge are strongly affected by the electrostatic fields in the wall sheath. The observed Doppler linewidth results from a combination of random motion plus radial acceleration. This effect has been utilized by KOS to deduce electron temperatures for the ion laser plasma.

Electron Temperature: Direct electron temperature measurements utilizing a double probe have been made by MLW at low-current densities and at pressures somewhat above those commonly used in ion lasers. Their results indicate a slightly decreasing to constant electron tempera-

ture for $J < 100$ A/cm^2. Indirect measurements of electron temperature by KOS are based on the theory of Kagan and Perel' [28] which predicts the relation

$$T_i = 0.56 \, T_a + 0.13 \, T_e \qquad (3)$$

between the effective transverse temperature T_i, the neutral gas temperature T_a, and the electron temperature T_e. They find an increase of T_e proportional to J up to the highest current density measured, 350 A/cm^2. They speculate that this linear dependence will continue to 550 A/cm^2 where the temperature would be $\simeq 11$ eV. Although the ranges of measurement do not overlap, the result of these two investigations do not appear compatible. Further work on this subject is clearly needed.

It is clear that the electron temperature decreases with increasing pressure. The most rapid variation occurs in the range $0.1 < p < 1.0$. For pressures from 2 to 20 torr the electron temperature measured by MLW decreased by only 30 percent. Table V summarizes the existing measurements of T_e as a function of J at a pressure of 0.5 torr.

Electron Density: The data of Table V concerning electron density are derived from measurements of MLW and from calculations of KOS based upon an electric conductivity formula derived by Kolesnikov [29]. It is clearly established by both investigators that the electron density at constant J is approximately linearly proportional to the gas filling pressure. The calculations of KOS were checked by Stark broadening at one point ($p = 0.4$ torr, $J = 200$ A/cm^2) with nominal agreement, 2.3×10^{13} cm^{-3} calculated and 3.4×10^{13} cm^{-3} measured. The calculation technique employed by KOS and by Herziger and Seelig [27] has been criticized by Lin and Chen [30] for neglecting the effect of electron "runaway" at high current densities. A decision as to whether the more complete theory of Lin and Chen offers a significant advantage over the simple scaling laws proposed by others will be possible only if considerably more experimental work becomes available, particularly for larger diameters and higher current densities.

Neutral Atom Density: The neutral atom density is very difficult to measure in small capillary discharges. Absorption of X-rays passing along the length of the capillary was employed by MLW to deduce total atom + ion densities, but the results were reliable only at pressures greater than those employed in laser discharges. They observed a decrease in neutral atom density with increasing atom temperature as would be expected from simple gas law considerations. However, the measured densities were consistently 20–40 percent higher than those calculated from the gas law relation

$$\frac{p}{p_0} = \frac{300}{T_a}. \qquad (4)$$

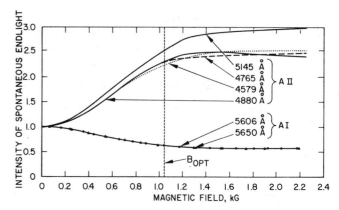

Fig. 4. Variation of the intensity of the spontaneous endlight with axial magnetic field for a number of Ar II ion laser line and neutral Ar I lines (from [17]).

This is not surprising considering the strong dynamic forces acting on the gas (see the discussion on gas pumping which follows) and the small conductance of the discharge tubes. Miller [31] has suggested that the theory of thermal transpiration might be more appropriate to this situation since the atomic mean free path is comparable to the tube radius for the smaller bore lasers. The gas law relation has not been tested experimentally in larger bore lasers where it is most likely valid. Nonetheless, it is used by most of those who advance theoretical explanations of ion laser plasmas. Both KOS and Herziger and Seelig [27] use it to obtain the neutral number density needed in the electric conductivity calculation, and Lin and Chen [30] have used a more complete form in which the external pressure is set equal to the total pressure of atoms, ions, and electrons.

Electric Field Gradient: The axial electric field measured under condition of optimum[2] laser action ($pR \simeq 0.1$ torr · cm) [32] is observed to obey the relation [17], [26], [32]

$$E_z R = 6.5 \text{ V} \qquad (5)$$

for $0.5 < R < 7.5$ mm. The dependence of E_z on J and p is small for $JR < 50$ A/cm. Herziger and Seelig [32] find $E_z \propto J$ for $JR > 200$ A/cm. The laser discharge typically exhibits a small positive resistance except at low J (usually below laser threshold) where a pronounced negative resistance is observed [26]. Again the gradient is very constant at cold filling pressures near and above optimum and begins to rise only when the fill pressure is reduced well below the optimum value.

Magnetic Field: The experimental results discussed up to this point have all been taken with no axial magnetic field; in actual fact, very few plasma data have been measured for ion laser plasmas in a magnetic field. This may seem somewhat surprising considering that almost all commercial CW ion lasers operate with a magnetic field to improve efficiency. Many investigators feel, however, that the magnetic field operates in a simple way to reduce ion loss by slowing down the electron transport to the walls, and that very little additional information will be learned about excitation mechanisms by studying the plasma in a magnetic field. This point of view is supported by measurements of

[2] Somewhat lower optimum pressures apply to the krypton and xenon ion laser plasmas.

laser output versus current which yield the same functional dependence with a magnetic field as without, except at proportionately lower discharge current. The effect of the magnetic field, however, is somewhat more complicated than this, as illustrated in Fig. 4. The spontaneous emission at constant current from ion lines is observed to increase 2.5 to 3 times with increasing magnetic field, while spontaneous emission from neutral atom lines is observed to decrease. The neutral level populations are saturated and their intensity is a function of electron temperature and not electron density (cf. [33]). The observed decrease in neutral line emission implies a decreasing electron temperature. The ion level populations are therefore determined by the interaction of ion density and electron temperature. The decrease in electron temperature is also borne out by an observed decrease in the axial electric field strength with increasing magnetic field [17].

There is an optimum value of magnetic field which depends primarily upon tube diameter. For a 1-mm-bore-diameter argon ion laser, a field of 1500 G is typical, while for 8 mm the optimum is only 400–600 G. Values for krypton and xenon ion lasers are generally lower still. The results of Herziger and Seelig [32] indicate that the improvement gained with magnetic field is not significant for bore diameters ≥ 10 mm.

Summarizing this section on plasma properties, we have seen that sufficient experimental evidence exists to establish clearly the nature of the ion laser plasma. If, however, the theory of ion laser performance is to proceed successfully to a satisfactory explanation of performance based on calculation of atomic excitation rates, from the empirical scaling law stage it occupies presently, then plasma measurements must be extended and refined to provide a more satisfactory foundation.

III. PHYSICS AND TECHNOLOGY OF PRACTICAL ION LASERS

We will now proceed to discuss a number of interesting and important effects that arise when this basic ion laser plasma is incorporated into a practical device. Some of these additional effects relate to device technology, and some represent physical processes that we have not yet discussed in sufficient detail.

A. Gas Pumping in Ion Lasers

One of the interesting and practically important characteristics of the high-current-density low-gas-pressure ion laser discharge is the rapid pumping of gas from one end of the tube to the other. In the first CW ion lasers it was found that after tens of seconds of operation the filling gas would be pumped from the cathode and bore to the anode to such a degree that laser action would cease and the discharge would extinguish. A convenient solution to this problem is to provide some form of gas return path external to the discharge which connects the anode and cathode ends and allows an external equalizing flow of neutral gas to occur [10]. A gas return path having approximately the same diameter as the laser bore, but somewhat greater length to prevent electrical breakdown, is found to be adequate to achieve maximum laser output and adequate discharge stability.

Even with a high-conductance external gas return path, very large anode to cathode pressure differentials may still occur. Fig. 5 shows the measured anode to cathode pressure differential Δp as a function of discharge current for a number of different values of gas filling pressure. These data were taken on a quartz tube filled with argon, having a 3-mm-diameter by 46-cm-long bore [17]. At 25 A and the optimum gas fill of 0.25 torr and magnetic field of 1.05 kG, this tube produced a laser output of 3 W. The notation $C_B = C_E$ indicates that the calculated gas conductance of the external gas return path was equal to that of the bore when the gas in both was at room temperature. Similar pressure differentials and current dependence were measured in tubes of different diameters, and filled with krypton or xenon [17].

The pressure differential Δp is observed to be strongly dependent on the gas pressure and discharge current. At 3 torr, the pressure differential increases linearly up to about 5 A, and then saturates. All previous experimental studies of gas pumping reported in the literature [34]–[36] were made at low discharge currents (< 1 A). Accordingly, only the initial linear increase of Δp with current predicted by theory was observed.

At the low gas pressures that are optimum for laser action (0.16 to 0.25 torr in a 3-mm tube), Δp first increases and then decreases sharply with increasing current, and actually can become negative as shown in Fig. 5. Previous to these measurements on ion lasers, no such behavior had been reported. The magnitude of Δp should be noted; in a typical laser, Δp across the bore can be as great or greater than the initial fill pressure, depending on the discharge current, pressure, and size of the gas volumes attached at anode and cathode ends.

It is extremely important that anyone performing measurements on ion lasers appreciate the strong dependence of pumping rate and Δp on discharge current, magnetic field, and filling pressure. Often measurements of line intensities, electron temperature, plasma density, etc., as a function of I, B, or p are taken at a fixed position along the bore. Obviously such measurements will be strongly affected by changes in the neutral gas density caused by gas

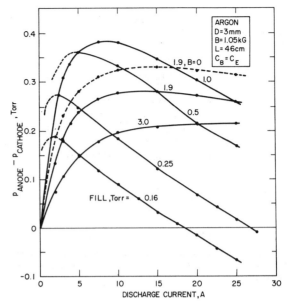

Fig. 5. Anode-to-cathode pressure differential versus discharge current in a typical argon ion laser discharge for different values of gas filling pressure (from [17]).

pumping. The characteristics measured at different positions along the bore can exhibit markedly different dependences because of gas pumping effects. The practical importance of gas pumping effects in ion lasers has prompted new theoretical analysis which has extended the early treatments of Langmuir [37] and Druyvesteyn [38]. Halsted [17] and, in a more thorough analysis, Chester [26], [39] have analyzed gas pumping in the gas pressure range of interest in ion lasers, including in their analyses provisions for an external gas return path.

A general explanation of the cause of gas pumping can be given qualitatively as follows. While ions and electrons within the positive column gain equal but opposite momentum from the axial electric field, greater momentum is delivered by the electrons to the neutral gas than by the ions, or equivalently, greater momentum is transferred to the walls by the ions than by the electrons. This difference can be analyzed by consideration of the differing longitudinal and axial drift rates, mean free paths, and recombination rates [40] of the two species. A detailed discussion of these effects is contained in [39]. Suffice it to say here that recent theory gives an accurate explanation of the pressure differences experimentally observed in ion lasers.

A basic appreciation of the factors influencing gas pumping in ion lasers may be obtained by considering the schematic representation of the laser bore and external return path shown in Fig. 6. Acting within the bore of the tube we have two pumping sources Q_{pump} and Q_{ion}. From these two sources a pressure differential Δp develops which causes a gas flow through the vacuum conductances of the external gas return tube C_E and the bore C_B. Physically, Q_{ion} results from the axial drift of ions in the positive column. Since these ions become neutral atoms at the cathode, this ion flow is equivalent to an equal amount of gas flow. The source Q_{pump}, as referred to above, is less obvious physically and results from the net transfer of mo-

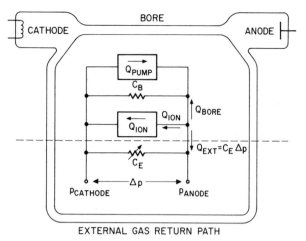

Fig. 6. Schematic representation of the laser discharge tube showing the forces acting on the gas, and the bore and external gas return path conductances.

mentum to the gas by electrons, and thus is always anode directed. Note that if Q_{pump} just equals Q_{ion}, Δp across the tube will be zero. Under these conditions, there must still be a flow of neutral gas toward the anode equal to Q_{ion}. This flow is driven by net electron collisions with neutrals even in the absence of a neutral pressure gradient. When Q_{pump} is greater than Q_{ion}, the anode pressure will increase until Δp is sufficient to cause an anode-to-cathode flow through the bore and external return path equal to $Q_{\text{pump}} - Q_{\text{ion}}$.

As reflected in the schematic diagram, Δp is found to vary inversely with the quantity $(C_E + C_B)$. This relationship is verified experimentally [17], provided one accounts for the strong dependence of the gas conductance of the bore on the temperature of the neutral gas. The high temperature of the gas in the bore (typically $> 1000°C$ at the higher discharge currents of practical interest) is the cause for the decrease in Δp with increasing discharge current shown in Fig. 5. This occurs both because of the decrease in gas number density in the bore, which reduces the pumping force, and because of an increase in the viscosity of the gas, which reduces the effectiveness of the pumping force in moving gas from cathode to anode.

The strong dependence of gas conductance on neutral gas temperature has an important role in the design of tubes having internal rather than external gas return paths. It may be shown that for viscous flow the conductance varies as the 7/4 power of the gas temperature. Thus the conductance of a path containing gas at $1000°C$ is only 1/13 of its value at room temperature. The trend in compact tube design is to try to include the gas return path within the bore structure to avoid the awkwardness and fragility of an external by pass tube such as that described in [10]. In tubes using radiation-cooled metal or graphite disks [17], [18], designers often attempt to employ off-axis holes in the segments as a gas return path. Because the gas passing through these segments is hot, typically $1000°C$, these paths are not nearly as effective in reducing the pressure differential in the bore as an external path at room temperature, with the result that very large cross-sectional areas must be devoted to gas

flow in such tubes if adequate off-axis flow is to be provided.

Ion laser tubes of advanced design (cf. Fig. 1) now use off-axis holes within a solid conduction-cooled beryllia bore to provide a gas return path. The gas in these off-axis paths is at the wall temperature of approximately $100°C$, and hence provision for adequate gas return path conductance is easily made.

B. Conditions at Laser Bore Constrictions

Physical damage to the inside of the walls confining the discharge, primarily in the vicinity of the cathode-end throat or bore constriction, often limits the maximum laser input power. In glass or quartz tubes, discoloration and erosion are observed in this region and in metal segment or disk-bore tubes, severe heating and sputtering damage is noted. Quantitative data showing an example of localized power dissipation are presented in Fig. 7. The bore of the tube shown was constructed using the common ion laser technique of electrically isolated radiation-cooled bore segments, supported in a quartz vacuum envelope, to confine the discharge. Information on conditions in the throat may be deduced from measurements of the temperature of the radiation-cooled segments. The bore segments in Fig. 7 are drawn to scale so that the temperature reading of a particular segment is plotted directly below the segment. Segment 8, the last tapered segment in the throat, operated at the highest temperature. At $1200°C$, this segment radiates (and therefore absorbs from the discharge) more than double this power of a segment at $1050°C$. From such measurements as these, one may conclude that localized power dissipation is more severe the more abrupt the throat transition, and that power dissipation is also a strong function of the position of the throat relative to the converging axial magnetic field (cf. [17]).

The conditions of localized heating occur in the throat because of the existence of a double sheath which forms to equalize the random plasma electron currents passing between the large and the small diameter regions of the bore. A general model of conditions in the vicinity of a discharge constriction has been described by Langmuir [41] and others [42]. To understand why a sheath must form, consider that the large and small diameter regions at the constriction are characterized by electron densities and temperatures n_e and T_e. The random electron current in each region is therefore proportional to $n_e T_e^{1/2}$. Since the electron density and temperature are greater in the small diameter region than in the large, the random electron currents must be greater in the small bore region. Across the constriction, the net electron current flow must be equal (ignoring for the moment the directed discharge current), and therefore a sheath of potential V_s must form at the constriction to reduce the greater outward flow of electrons from the small bore, as shown in Fig. 8. If the directed discharge current is accounted for, the effect is to increase the sheath height at the cathode-end bore constriction and to reduce the sheath height at the anode end.

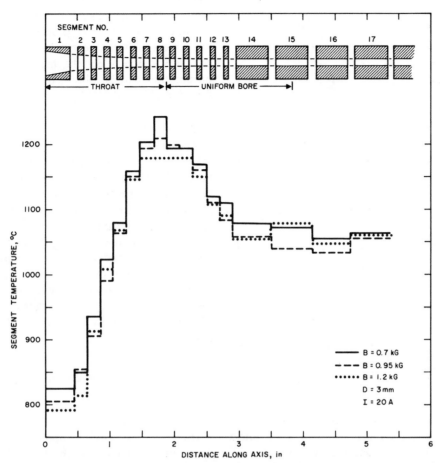

Fig. 7. Variations of the temperature along the bore of a radiation-cooled segmented graphite bore ion laser, showing the dependence on the electric field strength in the throat region. Bore segments are drawn to scale at the top of the figure (from [43]).

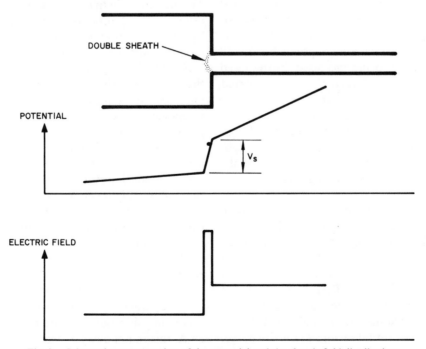

Fig. 8. Schematic representation of the potential and the electric field distributions in the vicinity of an abrupt discharge constriction.

103

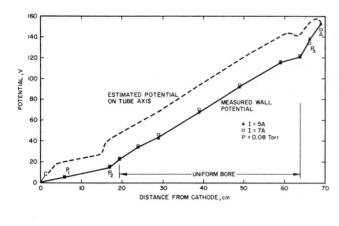

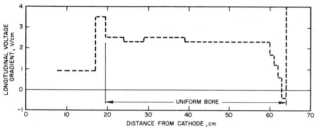

Fig. 9. Wall potential (upper curve) and longitudinal potential gradient (lower curve) as a function of distance along the discharge column of an air-cooled quartz discharge tube (from [17]).

We have directly measured the variation in wall potential and potential gradient in an air-cooled quartz bore laser using a combination of fixed and movable wall probes and a high-impedance voltmeter. The measured variation in wall potential and voltage gradient along the wall is shown in Fig. 9, and the potential on axis is sketched assuming appropriate values for electron temperature in the large and small diameter regions. The measured variation in potential and electric field is in good agreement with the behavior predicted by the simple model. The sheath heights at the cathode and anode constrictions were approximately $+10$ V and -4 V, respectively, being increased and reduced slightly by the directed discharge current, and, as expected, were independent of current to first order. Because of the presence of the double sheath, electrons will be accelerated through the sheath from the cathode region and will raise the electron energy, ionization rate, and wall sheath height in the throat region. This is believed to account for the increased heat dissipation and sputtering damage in the throat region.

In low-pressure mercury glow discharges, it has been claimed that double sheaths of the type described above produce very strong low-frequency oscillations, presumed to be ion waves [42]. Under certain discharge conditions we have observed laser modulation at discrete frequencies in the range of 150 to 220 kHz [43]. This modulation also appears as modulation of the spontaneous sidelight emission. It has been determined that this modulation is strong in the cathode and cathode throat regions of the discharge, but not in the uniform bore. This suggests that the modulation may be due to waves excited by oscillations excited in the double sheath as reported in [42]. However, this must remain as speculation pending further experiments.

The exact choice of shape for the bore constrictions at the anode and cathode has been largely determined by empirical design. Especially at high discharge current densities, and in CW krypton and xenon ion lasers which are subject to discharge instabilities, the design of the throat can very much affect the maximum current and power output that can be obtained. A fuller discussion will be found in [17] and [43].

C. Radiation Trapping

Another interesting phenomenon that plays a role in typical ion laser plasmas is the trapping or reabsorption of spontaneously emitted radiation. We mentioned earlier that the lower laser levels (the 4s levels shown in Fig. 2, in the case of the Ar II laser) are populated by electron collision, by spontaneous cascade from higher levels, and by the laser transition itself. Despite these populating processes, an inversion between upper and lower laser levels can be maintained because the lower level has a very rapid radiative decay to the ion ground state in the vacuum ultraviolet (723 Å for the Ar II $4s\ ^2P_{3/2} \to 3p^5\ ^2P_{3/2}$ transition). The ratio of A coefficients for this process to the laser A coefficients is typically large:

$$\frac{A_{723}}{A_{4880}} = 26, \qquad \frac{A_{723}}{A_{5145}} = 325 \qquad (6)$$

as calculated by Statz et al. [44, with corrections].

However, radiation trapping can reduce the effective decay rate and increase the lower level population. The 723-Å photons are absorbed by ground state ions before they can escape from the plasma, thus creating new $4s\ ^2P_{3/2}$ excited lower levels. These, in turn, reradiate, and so forth. The net effect is that the lower level is depopulated by a *diffusion* of radiation to the edge of the plasma rather than by direct radiation to the outside world. The theory of this mode of radiation transport is well known from the classical work of Holstein [45], [46]. It is difficult to apply Holstein's theory quantitatively because of uncertainties in the ion laser plasma parameters. Reasonably good qualitative agreement can be obtained for some of the more pronounced phenomena, however.

One prominent feature of pulsed ion lasers that results from radiation trapping is the occurrence of a "dead," or negative gain interval near the beginning of the light output pulse. This dead region develops gradually with increasing current, as shown in Fig. 10. The top trace shows the 0.488-μm laser output as a function of time for a 16.7-A current pulse. The corresponding current pulse is not shown in Fig. 10, but it begins at $t=0$, rises to its peak value by $t=10$ μs, remains essentially flat to $t=40$ μs, and then drops to zero at $t=55$ μs. For the small value of peak current used in the top trace, 8 μs of the 10-μs rise time is needed to reach oscillation threshold. As the peak current is increased to 25 A, this initial delay is decreased, and a slight dip in the output develops at about $t=10$ μs. With a peak current of 33.3 A (third trace from the top) the dip is more pronounced, almost reaching zero output at $t=10$ μs. At still higher currents, the output pulse has separated into two discrete parts with a zero output region in between. Transmission mea-

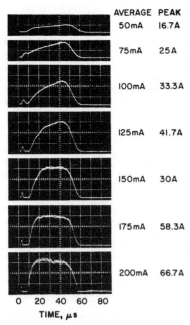

	AVERAGE	PEAK
	50mA	16.7A
	75mA	25A
	100mA	33.3A
	125mA	41.7A
	150mA	30A
	175mA	58.3A
	200mA	66.7A

TIME, μs

Fig. 10. Time variation of the 0.4880-μm Ar II output in a small pulsed ion laser (2-mm diameter by 30 cm) as the peak discharge current is increased.

surements by a number of workers [6], [23], [47], [48] have confirmed that this dead period is actually *absorbing*, that is, the lower laser level is temporarily more highly populated than the upper level.

A similar evolution of a dead period can take place if the peak discharge current is fixed and the gas pressure is increased. In this case, the width of the dead period increases in both directions in time until the entire output pulse is "swallowed up." Likewise, at appropriately adjusted current and pressure, a similar sequence of time variations can be seen as a function of radial position in the discharge. The output near the discharge wall shows no dead region, while a progressively wider dead region develops toward the tube axis. It is possible to adjust current and pressure so that the laser oscillates only near the tube walls, producing ring-like output modes as reported by Cheo and Cooper [49], provided the laser cavity will allow such high-order transverse modes.

All these characteristics were recognized early in the history of ion lasers; the temporal and spatial variations were mentioned in the initial paper on the argon ion laser [1].[3] The first explanation of these observations in terms of radiation trapping was given by E. I. Gordon (private communication) in June, 1964. This argument has been further confirmed for a pulsed laser by Klein [23], and may be summarized briefly as follows (see [17] or [23] for more detailed developments). According to Holstein, the effect of radiation trapping may be expressed by modifying the total A coefficient of the radiating level (in this case the lower laser level N_1) by a *trapping factor* γ so that A_1 is replaced by γA_1.

The gain coefficient of an ion laser can then be written

$$g_0 = K_1 J^2 \left(1 - \frac{K_2}{\gamma}\right) \qquad (7)$$

where K_1 and K_2 involve only atomic constants (Einstein coefficients, excitation cross sections, wavelength, etc.) independent of J. A two-step excitation process has been assumed in (7), giving a J^2 dependence of gain. For a cylindrical discharge configuration, Holstein gives the trapping factor as

$$\gamma = \frac{1.60}{k_0 R [\pi \ln (k_0 R)]^{1/2}} \qquad (8)$$

where k_0 is the absorption coefficient of the trapped line and R is the discharge radius. If we ignore the slowly varying logarithmic term and note that $k_0 \sim n_i / \Delta v_d$, where n_i is the ion ground state density and Δv_d is the Doppler linewidth of the absorbing vacuum UV transition, then

$$\gamma \sim \frac{\Delta v_d}{n_i R}. \qquad (9)$$

Moreover, if $n_i \sim n_e \sim pJR$ (cf. [21] quoted in [26]) for the plasma region of interest here, then

$$g_0 = K_1 J^2 \left(1 - \frac{K_3 pJR^2}{\Delta v_d}\right) \qquad (10)$$

where K_3 absorbs the remaining constants all assumed to be relatively insensitive to p, J, R, and Δv_d.

Equation (10) now indicates what happens as the pulse current builds up: the gain rises as J^2 until the second term in parentheses becomes significant. When J is large enough to make this term unity, the gain drops to zero and then becomes negative (absorption) on further increase in J. However, the Doppler linewidth also increases as the discharge heats up, a process that lags the time variation of J to some extent. When the ion temperature is sufficiently high so that the increase in Δv_d reduces this term below unity, the laser gain goes positive and oscillation returns. Increasing either the tube pressure or the tube radius makes the term larger in magnitude and lengthens the time required.

It is also clear from (10) that it is possible to increase p or R sufficiently so that no increase in Δv_d will suffice, and this also has been observed. Klein [23] has measured the time development of the vacuum UV transition and Δv_d, simultaneously with the laser output from a pulsed Ar II laser, and is able to confirm these processes qualitatively.

The argument given has said nothing explicitly about the variation of radiation trapping with magnetic field. We would expect from the discussion in Section II that the addition of a magnetic field would be equivalent to an increase in J without an increase in ion temperature. Thus the magnetic field should add to the power output when the J^2 variation dominates but detract from the laser output where trapping has become significant. Gorog and Spong [50] have investigated the output power and mode patterns as a function of both B and p. Their results are also consistent with (10). Increasing either B or p sufficiently caused the formation of ring modes and a drop in laser power, but over a reasonable range of values, an increase in B could be compensated by a decrease in p or J and vice versa.

Despite the qualitative agreement with this simple picture of radiation trapping, there is clearly much more to be learned before truly quantitative predictions can be made. The problem of ionization to higher states has been raised by Cottrell [5] and these complicate any simple picture. Dealing with the problems of highly ionized species will be necessary to explain the behavior of the strong Xe IV lines, for example. It is possible that radiation trapping of several cascading processes will have to be treated.

IV. SUMMARY

We have attempted to introduce the reader to the basic properties of ion laser plasmas and to relate these properties to the mechanisms responsible for the principal laser characteristics. Necessarily, we have had to omit many topics of interest to those engaged directly in ion laser research and development. For example, we have not gone into the details of discharge tube construction, including the choice of materials and configurations. Even though such details play important roles in obtaining reliable long-life laser operation, they do not affect the first-order plasma properties. Further details on a variety of construction techniques and technical properties are contained in [16]–[18], [43], and [51], among others. For a general review of CW argon ion lasers and a guide to the literature, we suggest the excellent paper by Kitaeva, Odintsov, and Sobolev [52].

REFERENCES

[1] W. B. Bridges, "Laser oscillation in singly ionized argon in the visible spectrum," Appl. Phys. Lett., vol. 4, April 1, 1964, pp. 128–130.

[2] G. Convert, M. Armand, and P. Martinot-Lagarde, "Effet laser dans der mélanges mercure-gaz rares," C. R. Acad. Sci., vol. 258, Mar. 23, 1964, pp. 3259–3260.

[3] W. R. Bennett, Jr., J. W. Knutson, Jr., G. N. Mercer, and J. L. Detch, "Super-radiance, excitation mechanisms, and quasi-cw oscillation in the visible Ar⁺ laser," Appl. Phys. Lett., vol. 4, May 15, 1964, pp. 180–182.

[4] V. V. Lebedeva, A. I. Odintsov, and V. M. Salimov, "Excitation conditions in argon ion lasers," Radio Eng. Electron Phys., vol. 13, Apr. 1968, pp. 655–658.

[5] T. H. E. Cottrell, "Output power characteristics of the pulsed argon ion laser," IEEE J. Quantum Electron., vol. QE-4, July 1968, pp. 435–441.

[6] S. M. Jarrett and G. C. Barker, "Direct gain measurements of pulsed argon-ion lasers," J. Appl. Phys., vol. 39, Sept. 1968, pp. 4845–4846.

[7] M. D. Sayers, "Single pass gain as a function of discharge current for the 4880 Å argon ion laser," Phys. Lett., vol. 29A, Aug. 11, 1969, pp. 591–592.

[8] W. B. Bridges and A. N. Chester, "Spectroscopy of ion lasers," IEEE J. Quantum Electron., vol. QE-1, May 1965, pp. 66–84.

[9] ——, "Ion lasers," in Laser Handbook. Cleveland, Ohio: Chemical Rubber Publishing Co., to be published. For the most complete listing published to date, see [8].

[10] E. I. Gordon and E. F. Labuda, "Gas pumping in continuously operated ion lasers," Bell Sys. Tech. J., vol. 43, July 1964, pp. 1827–1829.

[11] J. P. Goldsborough, "Stable, long life cw excitation of helium-cadmium lasers by dc cataphoresis," Appl. Phys. Lett., vol. 15, Sept. 15, 1969, pp. 159–161.

[12] H. Boersch, J. Boscher, D. Hoder, and G. Schäfer, "Saturation of laser power of cw ion laser with large bored tubes and high power cw uv," Phys. Lett., vol. 31A, Feb. 23, 1970, pp. 188–189.

[13] J. R. Fendley, Jr., "100-watt CW argon laser" (Unclassified), Proc. 4th DOD Conf. Laser Technology (San Diego, Calif., Jan. 1970), vol. 1, pp. 391–398 (unpublished).

[14] For a discussion of mode selection in lasers see for example H. Kogelnik, "Modes in optical resonators," in Lasers, vol. 1, A. K. Levine, Ed. New York: Marcel Dekker, Inc., 1966, pp. 295–347 (esp. pp. 330–333).

[15] R. C. Miller, E. F. Labuda, and E. I. Gordon, presented at the 1964 Conf. Electron Device Research, Cornell Univ., Ithaca, N. Y.

[16] E. F. Labuda, E. I. Gordon, and R. C. Miller, "Continuous-duty argon ion lasers," IEEE J. Quantum Electron., vol. QE-1, Sept. 1965, pp. 273–279.

[17] W. B. Bridges and A. S. Halsted, "Gaseous ion laser research," Hughes Res. Labs., Malibu, Calif., Tech. Rep. AFAL-TR-67-89, May 1967 (unpublished); DDC accession no. AD-814-897.

[18] W. B. Bridges and G. N. Mercer, "Ultraviolet ion lasers," Hughes Res. Labs., Malibu, Calif., Tech. Rep. ECOM-0229-F, Oct. 1969; DDC accession no. AD-861-927.

[19] E. F. Labuda, C. E. Webb, R. C. Miller, and E. I. Gordon, "A study of capillary discharges in noble gases at high current densities," presented at the 1965 18th Gaseous Electronics Conf., Minneapolis, Minn.

[20] R. I. Rudko and C. L. Tang, "Excitation mechanisms in the Ar II laser," Appl. Phys. Lett., vol. 9, July 1, 1966, pp. 41–44.

[21] R. C. Miller, E. F. Labuda, and C. E. Webb, unpublished data 1964–1967, private communication.

[22] V. F. Kitaeva, Yu. I. Osipov, and N. N. Sobolev, "Electron temperature in the electric discharge used for the argon ion laser," JETP Lett., vol. 4, Sept. 15, 1966, pp. 146–148.

[23] M. B. Klein, "Radiation trapping processes in the pulsed ion laser," Ph.D. dissertation, Univ. of California, Berkeley, Mar. 1969 (unpublished).

[24] ——, "Time-resolved temperature measurements in the pulsed argon ion laser," Appl. Phys. Lett., vol. 17, July 1, 1970, pp. 29–32.

[25] S. Hattori and T. Goto, "Excitation mechanism and plasma parameters in pulsed argon-ion lasers," IEEE J. Quantum Electron., vol. QE-5, Nov. 1969, pp. 531–538.

[26] A. N. Chester, "Experimental measurements of gas pumping in an argon discharge," Phys. Rev., vol. 169, May 5, 1968, pp. 184–193.

[27] G. Herziger and W. Seelig, "Berechnung der Besetzungsdichte und Ausgangsleistung von Ionenlasern," Z. Phys., vol. 215, 1968, pp. 437–465.

[28] Yu. M. Kagan and V. I. Perel, "On motion of ions and shape of their lines in the positive column of discharge," pt. 1, Opt. Spektrosk., vol. 2, no. 3, 1957, p. 298.
——, "Radial motion of ions in the low pressure discharge," pt. 2, ibid., vol. 4, no. 1, 1958, p. 3.

[29] V. N. Kolesnikov, "The arc discharge in inert gases," Ph.D. dissertation, 1963. Published in Proc. (Trudy) of the P. N. Lebedev Physics Institute, vol. 30, Physical Optics. New York: Consultants Bureau, 1966.

[30] S. C. Lin and C. C. Chen, "Kinetic processes in noble gas ion lasers, a review," presented at the AIAA 8th Aerospace Sciences Meeting, New York, Jan. 1970, Paper 70-82.

[31] R. C. Miller, private communication. See also discussion following (2.19) in [39].

[32] G. Herziger and W. Seelig, "Ionenlaser hoher Leistung," Z. Phys., vol. 219, pp. 5–31, 1969.

[33] C. E. Webb, "Ion laser—Part 1: The radial distribution of excited atoms and ions in a capillary discharge in argon," in Proc. Symp. Modern Optics (New York, Mar. 1967).

[34] A. Ruttenauer, "Quantitative Bestimmung der Druckdifferenzen in der positiven Säule der Edelgase Argon-Neon und Helium," Z. Phys., vol. 10, 1922, pp. 269–274.

[35] C. Kenty, "Clean-up and pressure effects in low pressure mercury vapor discharges: A reversible electrical clean-up of mercury," J. Appl. Phys., vol. 9, Dec. 1938, pp. 765–777.

[36] R. B. Cairns and K. G. Emeleus, "The longitudinal pressure gradient in discharge tubes," Proc. Phys. Soc. (London), vol. 71, Apr. 1958, pp. 694–698.

[37] I. Langmuir, "The pressure effect and other phenomena in gaseous discharges," J. Franklin Inst., vol. 196, Dec. 1923, pp. 751–762. See also The Collected Works of Irving Langmuir, vol. 4. New York: Pergamon, 1961.

[38] M. J. Druyvesteyn, "The electrophoresis in the positive column of a gas discharge," Physica, vol. 2, 1935, pp. 255–266.

[39] A. N. Chester, "Gas pumping in discharge tubes," Phys. Rev., vol. 169, May 5, 1968, pp. 172–184.

[40] C. C. Leiby, Jr., and H. J. Oskam, "Volume forces in plasmas," Phys. Fluids, vol. 10, Sept. 1967, pp. 1992–1996.

[41] I. Langmuir, op. cit., p. 751.

[42] F. W. Crawford and I. L. Freeston, "The double sheath at a discharge constriction," Stanford Univ., Stanford, Calif., Microwave Lab. Rep. 1043, June 1963 (unpublished).

[43] A. S. Halsted, W. B. Bridges, and G. N. Mercer, "Gaseous ion laser

research," Hughes Res. Lab., Malibu, Calif., Tech. Rep. AFAL-TR-68-227, July 1968; DDC accession no. AD-841-834.

[44] H. Statz, F. D. Horrigan, and S. H. Koozekanani, "Transition probabilities for some Ar II laser lines," *J. Appl. Phys.*, vol. 36, July 1965, pp. 2278–2286. The values for transitions to the ground state are underestimated by a factor of 5, as pointed out by Rubin; see [52]. A correction was published by Statz and coworkers in *J. Appl. Phys.*, vol. 39, July 1968, pp. 4045–4046.

[45] T. Holstein, "Imprisonment of resonance radiation in gases, Part I," *Phys. Rev.*, vol. 72, Dec. 15, 1947, pp. 1212–1233.

[46] ——, "Imprisonment of resonance radiation in gases, Part II, *Phys. Rev.*, vol. 83, Sept. 15, 1951, pp. 1159–1168.

[47] E. I. Gordon, E. F. Labuda, R. C. Miller, and C. E. Webb, "Excitation mechanisms of the argon-ion laser," in *Physics of Quantum Electronics*, P. L. Kelly, B. Lax, and P. E. Tannenwald, Eds. New York: McGraw-Hill, 1966.

[48] P. O. Clark, W. B. Bridges, and A. S. Halsted, 1965 (unpublished).

[49] P. K. Cheo and H. G. Cooper, "Evidence for radiation trapping as a mechanism for quenching and ring-shaped beam formation in ion lasers," *Appl. Phys. Lett.*, vol. 6, May 1, 1965, pp. 177–178.

[50] I. Gorog and F. W. Spong, "High pressure, high magnetic field effects in continuous argon ion lasers," *Appl. Phys. Lett.*, vol. 9, July 1, 1966, pp. 61–63.

[51] K. G. Hernqvist and J. R. Fendley, Jr., "Construction of long life argon lasers," *IEEE J. Quantum Electron.*, vol. QE-3, Feb. 1967, pp. 66–72.

[52] V. F. Kitaeva, A. N. Odintsov, and N. N. Sobolev, "Continuously operating argon ion lasers," *Sov. Phys.—Usp.*, vol. 12, May–June 1970, pp. 699–730.

[53] R. Paananen, "Continuously-operated ultraviolet lasers," *Appl. Phys. Lett.*, vol. 9, July 1, 1966, pp. 34–35.

[54] K. Banse, H. Boersch, G. Herziger, G. Schäffer, and W. Seelig, "Hochleistungs-Ionenlaser," *Z. Angew. Phys.*, vol. 26, Feb. 4, 1969, pp. 195–200.

[55] I. Gorog and F. W. Spong, "Experimental investigation of multiwatt argon lasers," *RCA Rev.*, vol. 28, Mar. 1967, pp. 38–57.

High Peak Power Gas Lasers

JACQUES A. BEAULIEU

Abstract—The development of transversely excited atmospheric pressure (TEA) CO_2 Lasers has led to the generation of multimegawatt laser pulses at 10.6 μm in simple laboratory prototypes. A discussion of the basic properties of these lasers is followed by the description of a number of different techniques that have been used to provide effective transverse excitation of lasers at low repetition rates. The basic technique used to increase the repetition rates to 1000 pulses per second (PPS) is described. Considering the simplicity of the technique and its low cost, it appears that in the not too distant future, pulse energies in the kilojoule region and pulse lengths of the order of one nanosecond will be practical.

INTRODUCTION

A VERY IMPORTANT characteristic of lasers is their ability to generate extremely high-power pulses of coherent radiation which can be focussed on small areas. The availability of such power concentration has led to numerous developments in nonlinear optics and in plasma physics. However, until recently only solid-state lasers such as the ruby and the neodymium lasers could achieve multimegawatt or gigawatt powers. These lasers have efficiencies less than one percent when Q switched or mode locked, to produce the desired high peak powers. The choice of wavelengths of operation is also limited and longer wavelengths than can be achieved with this type of laser are sometimes needed. Finally, higher average powers which would correspond to higher repetition rates, are also highly desirable for cases where high data rates are required.

Manuscript received October 18, 1970; revised January 18, 1971.

The author is with Defence Research Establishment Valcartier, Quebec, Canada.

Molecular lasers, specially the CO_2 laser, have a number of advantages over solid-state lasers. Their efficiency is higher (10 percent to 20 percent is common for CO_2); their average power is also much higher than for solid-state lasers; their wavelength of operation is much longer which besides being less hazardous than visible or near infrared lasers, couples better to low and medium density plasmas; finally their cost is lower since the active atoms or molecules are not in an optically perfect crystal like solid-state lasers. However, until recently, gas lasers have not been considered as possible sources of very high peak powers because of their low density of active molecules or atoms. The peak power limitations of the CO_2 laser and the most recent techniques developed to overcome these limitations will now be briefly reviewed.

PEAK POWER LIMITATIONS OF CO_2 LASERS

As for most gas lasers, the usual technique for obtaining laser action in CO_2 consists of placing the gas in a long cylindrical container inside an optical resonator and exciting the gas by means of an electrical discharge between electrodes located at each end of the long tube. Using continuous discharges, the best laser output is usually obtained for CO_2 partial pressures of the order of one torr. Such lasers can be Q switched to produce high peak power pulses but very large systems are required to produce pulse powers of the order of 100 kW. During such a Q-switched pulse, which can be approximately 100 ns long, the reexcitation of molecules by the electrical discharge can be considered to be negligible and one can expect to obtain at best one photon

Reprinted from *Proc. IEEE*, vol. 59, pp. 667–674, Apr. 1971.

per CO_2 molecule. While the volume of such lasers is considerably greater than that of solid-state lasers, the photons are less energetic due to the longer wavelength (10.6 μm for CO_2 compared to 1.06 μm for neodymium and 0.69 μm for ruby) and the density of CO_2 molecules is about three orders of magnitude smaller than that of the lasing atoms in solid-state lasers. To achieve high peak powers it appears to be necessary to operate at high pressures in large volumes.

When continuously excited, CO_2 lasers have always displayed an optimum operating pressure which varied inversely with the tube diameter. This effect was explained by Tiffany et al. [1], from thermal considerations. The same conclusions can be reached from a different approach. First, one must realize that as the gas pressure is increased in a laser of given volume and temperature, a constant fraction of the CO_2 molecules must be maintained in the upper energy level if the gain is to remain constant. This is due to the pressure broadening of the laser line which increases the frequency spread of the amplification band without increasing its amplitude. This implies that the absolute number of molecules in the upper energy state must be increased linearly with pressure. Furthermore, the deexcitation of CO_2 molecules by collisions also implies that each molecule must be excited more often as the pressure increases. Thus to maintain a constant gain, the excitation or pumping energy must increase as the square of the operating pressure. This also implies that the average power from such a laser will increase as the square of its operating pressure. The difficulty comes in trying to maintain the temperature constant. Even with an efficiency of 20 percent, most of the pump energy is dissipated as heat. By relying primarily on thermal dissipation to the laser wall to keep the temperature sufficiently low, one rapidly arrives at a pressure beyond which the temperature becomes sufficiently high to thermally populate the lower state of the laser transition and the gain drops. The larger the laser diameter, the lower this optimum pressure becomes. The use of very fast gas flows as a cooling mechanism can increase both the optimum operating pressure and the power per unit volume as was demonstrated by Tiffany et al. [1], and by Lavarini et al. [2]. However, if one is primarily interested in peak powers rather than average power, there exists another alternative for operating high pressure large volume lasers, which is the use of short excitation pulses with low repetition rates or low average powers. The first important results using pulsed excitation were obtained by Hill [3] who used 5-μs pulses at repetition rates between 20 and 100 pulses per second (PPS) in the usual long tubular discharge system. Pulse voltages of the order of a megavolt were required for the excitation of gas at pressures at least one order of magnitude higher than in previous lasers of comparable volumes. Simultaneously, the development of high pressure CO_2 lasers using transverse rather than longitudinal excitation was undertaken by a research team at the Defence Research Establishment Valcartier (DREV) [4] where the emphasis was placed on operation at atmospheric total pressure.

PROPERTIES OF TRANSVERSELY EXCITED ATMOSPHERIC PRESSURE (TEA) CO_2 LASERS

Transverse excitation, which implies a short discharge length and a large discharge area, is advantageous since the necessarily large fields required to achieve breakdown in high pressure gases can be obtained with relatively low applied voltages. The main difficulty is to achieve a large area discharge, i.e., a discharge which is more or less uniformly distributed along the entire length of the optical cavity. The next section will discuss the different electrode systems that can be used to achieve this distribution. Another important aspect of transverse excitation is that the discharge impedance is low which allows a very rapid injection of the excitation energy. This is quite important since, for good efficiencies, the excitation of the laser molecules should take place in a time which is short compared to the lifetime of the excited state of the CO_2 molecule. This lifetime at atmospheric pressure is of the order of 10 μs; hence excitation times of the order of a microsecond are desirable to achieve maximum efficiency in giant pulses. An added advantage of a short excitation time is that no external Q-switching accessories such as a spinning mirror, are needed to obtain giant-pulse action in CO_2 lasers. In TEA lasers, giant pulses are automatically obtained due to a gain-switching action which is described subsequently.

To understand the gain-switching mechanism, it is best to consider first the sequence of events illustrated in Fig. 1. Curves a and b illustrate the voltage and current behavior of the exciting pulse obtained by the discharge of a condenser into the low impedance presented by the transverse discharge system of the laser. The excitation pulse is shorter than 1 μs and the gain reaches its maximum value near the end of the discharge as illustrated in Fig. 1(c). The laser pulse illustrated in curve d occurs at a later time due to the finite time required for the laser field to grow to its maximum value. Consider for example an electrode configuration which is 1 m long in an optical cavity 1.2 m long with a 30-percent reflectivity mirror at one end. At the peak of the excitation about 10^{20} excited CO_2 molecules are in the cavity and generate photons by spontaneous emission at a rate corresponding to a few seconds of radiative lifetime. This spontaneous emission radiates in all directions and only that which lies within a solid angle of less than 1 mrad from the optical axis of the cavity is important in the build up of the laser field. Furthermore this radiation is spread over all the P and R branches of the transition from the upper to the lower laser level. Finally, it is the signal in the narrow bandwidth of a single mode of the cavity, which represents a very small fraction of the gigahertz bandwidth of the atmospheric pressure transitions that will be amplified in the laser. Thus the effective initial signal power is between 10^{-12} to 10^{-15} W. To build up to peak power levels of the order of 10^6 W requires between 36 and 42 round trips in the laser cavity which corresponds to a time of the order of 300 ns. In fact, the build up time will be longer

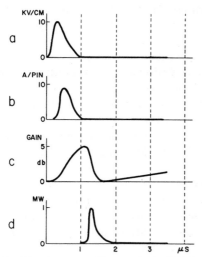

Fig. 1. Time sequence of laser action in TEA lasers.

since during the excitation the gain grows gradually and the maximum gain value is only reached at the end of the excitation pulse. Since the gain reaches a large value before the laser field becomes sufficiently strong to rapidly depopulate the upper laser level, one has the same conditions as in a Q-switching case and a giant pulse results. The theoretical and experimental studies of Gilbert [5] and Laurie [6] indicate that equal or higher peak powers are achieved by this gain-switching mechanism than by more conventional Q-switching techniques using rapidly rotating mirrors.

Another characteristic of gain-switched pulses is that they are frequently followed by a less intense and longer pulse. This phenomenon, which has been investigated by Gilbert [5], follows from the dynamics of the CO_2 molecule. The generation of the giant pulse essentially equalizes the populations in the upper and lower laser levels extracting only half the available energy. Since the lower level has a shorter lifetime than the upper laser level, a certain population inversion is rebuilt after the end of the giant pulse which gives rise to the second pulse mentioned previously if the reflectivity of the cavity mirrors is high enough. In pure CO_2 the second pulse energy is less than that of the giant pulse and can be completely absent if the laser optical resonator is made sufficiently lossy. When N_2 is present, the excited N_2 can also help to increase the upper laser level population by energy transfer and more energy can be found in the second pulse, which can last a few microseconds, than in the initial giant pulse.

At pressures substantially below atmospheric, such as used by Hill [3], the giant-pulsing effect is less pronounced because of the reduced rate of energy transfer between N_2 and CO_2 and for heavy N_2 concentrations, the laser output will take the form of a relatively small giant pulse followed by a long second pulse containing most of the energy. The peak powers are substantially smaller although the total pulse energy can be as high as 1 J/1 as reported by Hill [17].

The theoretical limit on the amount of energy that can be stored in the upper laser level of CO_2 and extracted at 10.6 μm is approximately 500 J/l of CO_2 at STP. In practice, only 10 to 20 percent of the molecules are expected to be found in the upper energy level of the laser transition. Furthermore, in atmospheric pressure lasers, mixtures of CO_2, He, and N_2 are used where the main constituent is He which is required to obtain a well diffused electrical discharge. The CO_2 content in fact varies between 5 and 20 percent. Thus more realistic figures for the energy per pulse vary from 2.5 to 20 J/l with only half that energy being released in the initial giant pulse which lasts approximately 100 ns. Thus peak powers between 25 and 200 MW/l can be achieved. Experimental values of energy per pulse for different types of TEA lasers designed at DREV vary between 2 and 10 J/l and 5 J/l have been reported by Dumanchin et al. [7], for volumes up to 30 l.

The peak power from TEA CO_2 lasers can be increased further by shortening the laser pulse. The pulsewidth associated with the gain-switched or Q-switched giant pulses is determined by the gain per unit length of the laser medium rather than the bandwidth of the CO_2 laser line. At atmospheric pressure this bandwidth is about 3 GHz which implies that subnanosecond pulses could be amplified. Since the thermalization time of the rotational levels at atmospheric pressure is less than a nanosecond, pulses with duration of a nanosecond or more could drain the energy stored in all the rotational levels. However, while it is possible to amplify very short pulses, these must be generated. Using a simple giant-pulse laser, the signal can be shortened by a combination of saturated amplifiers and leading edge sharpening by means of bleachable absorbers. Another alternative is mode locking. Spontaneous mode locking has been observed by Gilbert [8] and by Lyon et al. [9], to produce pulses with a duration of approximately 5 ns while an active mode-locking technique used by Wood et al. [10], has given mode locked pulses less than 1 ns long.

TRANSVERSE EXCITATION TECHNIQUES

Numerous techniques have been used to obtain multiple transverse discharges distributed over the length of the laser cavity. The simplest electrode system consists a long tubular anode, and a number of 1000 carbon resistors in parallel using their leads as the cathode elements which are distributed linearly along the optical cavity. Lasers using this type of electrode structures have been operated at atmospheric pressure with an efficiency of approximately 5 percent. The laser energy corresponds to a few millijoules per electrode as reported in greater details by Fortin [11]. A comprehensive study of the gain characteristics of these lasers has been undertaken by Robinson [13] and is still progressing. Some of the recent results illustrate the difficulties of interpretation with this type of measurement. Using 50 resistors distributed over a 27-cm long discharge cell and 25-mm discharge length the gain in any part of the discharge can be evaluated by observing the transmitted signal from a low power CW CO_2 laser with a beamwidth less than 1 mm over the length of the discharge structure.

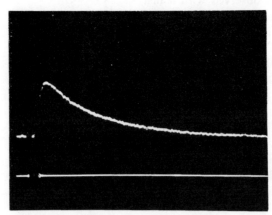

Fig. 2. Typical time behavior of gain characteristics of TEA laser.
Upper trace is amplified signal while lower trace gives input signal.

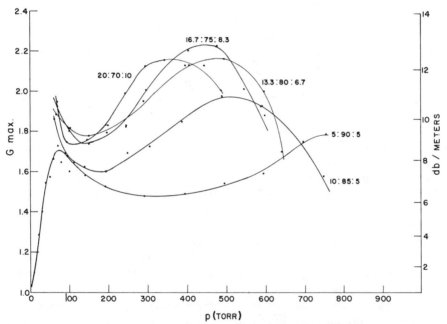

Fig. 3. Peak gain behavior for a 27-cm TEA laser with pressure. The different curves correspond to
different $CO_2:He:N_2$ concentrations.

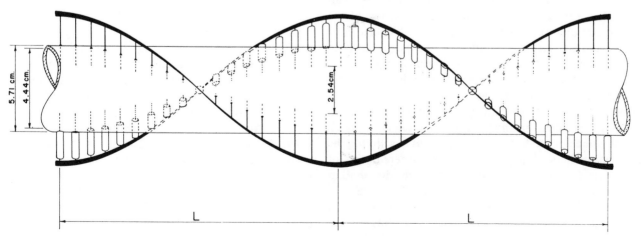

Fig. 4. Structural details of a helicoidal laser.

111

Fig. 2 illustrates the type of signal observed with a fast response Hg:Cd:Te detector. The upper trace is the transmitted signal while the lower trace is to verify the stability of the CW source during the pulse. This result was taken for a gas pressure of 300 torr, the peak gain is approximately 3 and the time of the full sweep is 50 μs. Curves of the peak gain for a given excitation energy (30 kV, in a 0.007-μF condenser) have been plotted versus pressure for the different gas mixes indicated (in Fig. 3). Gains of up to 12 dB/m indicate that a larger population inversion is achieved than in CW CO_2 lasers at low pressure. However, the interpretation of these curves is far from simple. The drop at very low pressures can be expected due to heating by the excessive pumping energy while the drop at the high pressure end in all but the highest helium concentration, is associated with the apparition of bright arcs which destroy the uniformity of excitation and create local hot spots. The variation between these extremes can be associated with the discharge characteristics. However, since the behavior of these curves vary with a number of parameters other than gas mix, more detailed explanations will have to await the detailed study of the effects of important parameters such as electrode shape, discharge energy, discharge voltage, the inductance of the discharge network, and a few others which seem to be significant.

As it was noticed that the laser output beam in a confocal mirror configuration seemed to have a section following the axial gain distribution i.e., take a rectangular shape approximately 25 by 5 mm, a helicoidal electrode distribution that would favor the TEM_{00} mode and which is illustrated in Fig. 4 was investigated. The main property of such an electrode system is that the gain would display a peak on the tube axis which intersects every discharge and would decrease smoothly away from this axis. Assuming that each discharge gives rise to a column of excited gas of radius p_0 (of the order of 2 to 4 mm as determined experimentally) the normalized gain Λ would have a radial distribution as illustrated in Fig. 5. Such a structure has been built and tested by Fortin et al. [12], which have succeeded in operating this laser reliably in a pure TEM_{00} mode, generating near one joule in a diffraction limited beam with less than 1 mrad divergence. The scattering and diffraction of the 10 μm radiation by the electrical discharges which had been feared could be a serious limitation of TEA lasers has been studied in details in the helical laser. The main effect seems to arise from the increase in temperature of the gas column through which the discharge passes, giving rise to a decrease in refractive index. In the helical laser this effect creates a diverging lens (focal length approximately 200 m per pitch of helix) which has to be taken into account in selecting the cavity parameters. Using a number of staggered rows of resistive electrodes to essentially randomize the distribution of the discharge columns the diffraction effects are substantially reduced. This method has been used to build a 6-m long laser with a nearly uniform gain distribution over a 25 by 25 mm section. As an oscillator it can generate about 8 J and as an amplifier it has a gain in excess of 20 dB, an exact value being difficult to obtain due to interaction be-

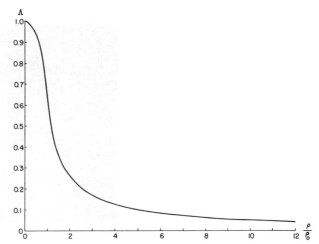

Fig. 5. Average gain distribution in helicoidal laser. The gain Λ is given in terms of the distance ρ from the laser axis, normalized with respect to the discharge radius ρ_0.

tween oscillator and amplifier. Scattering and diffraction is below the easily observable level but pulse shortening has been observed. Saturation is evident at energy densities of 200 mJ/cm^2.

As a technique to obtain discharge distribution, the resistively loaded electrodes have disadvantages such as a reduced efficiency due to the I^2R losses. Johnson [14] has investigated another technique which consists in using a small energy discharge condenser for each of the laser electrodes. This technique ensures that the energy per discharge is the same for all the discharge electrodes and the possibility of excessive discharge concentration into a single electrode is completely eliminated without resistive losses. The results obtained were similar to those published recently by Smith and DeMaria [18] although a slightly higher efficiency was obtained by Johnson using 100 pin electrodes each being connected to a 500-pF condenser, charged to 30 kV. Up to 1.2 J were obtained for a 6-percent efficiency, at atmospheric pressure but using 50 to 60 mm anode-to-cathode separation. However, the improvement in efficiency did not seem very substantial while the construction was much more complex. Simultaneously, using a resistanceless pin structure, Laurie and Hale [15] have achieved the highest efficiency to date. When higher voltages of the order of 20 kV/cm discharge are used with a low inductance discharge network, a good discharge distribution can be obtained amongst a large number of pin electrodes. A 1 m long laser with 190 pins in a single row has produced in excess of 2 J with efficiencies of at least 15 percent. This type of laser, however, has a tendency to generate bright arcs when the excitation energy is high and requires more care than the resistively loaded electrode laser to operate with good stability.

Rapid progress is presently being made in the development of "triggered" or "double-discharge" electrode systems which are most attractive due to their efficiency and the ease of construction for large volume devices. Two types of electrode structures, illustrated in Fig. 6, have been developed independently by Dumanchin et al. [7], and by

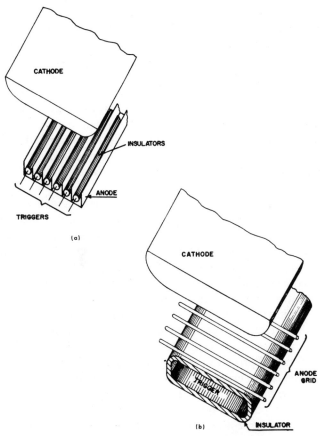

Fig. 6. Double-discharge electrode structures as developed by (a) Dumanchin *et al.* [7], (b) Laflamme [16].

Laflamme [16]. In these systems, there is a solid uniform cathode and the anode consists of either a number of parallel blades as in Fig. 6(a) or of parallel wires as in Fig. 6(b). An insulated trigger electrode structure is in close proximity to the anode such as the insulated wires interlaced with the anode blades in Fig. 6(a) or the bar next to the wire grid in Fig. 6(b), and which is electrically connected to the cathode. When the high voltage pulse is applied, it first creates a trigger to anode (low energy) discharge which is sufficient to cause an ionization cloud uniformly distributed around the anode. This preionization helps to initiate the main anode-to-cathode discharge in a very uniform fashion, reducing the effects of electrode surface irregularities and also the tendency to generate a bright arc discharge. This technique has been used by Laflamme [16] to produce very uniformly excited systems with cross sections ranging from 5 to 50 cm^2, laser energies of the order of 5 J/l and efficiencies varying from 5 to 10 percent. Similar figures were obtained by Dumanchin *et al.* [7], who for a 50-cm^2 6-m long laser have reported 135 J/pulse. All these experiments indicate that the output energy is directly proportional to the length at least up to 6 m, and also to the cross-sectional area to at least 50 cm^2. Furthermore the uniformity of the discharge is higher than in multiple pin electrode systems and the scattering by thermal gradients should be even smaller so that this seems to be the best type of structure to use in large amplifiers. The practical limits to the expansion of this

technique remains to be determined but it appears very likely that CO_2 laser pulses in excess of a kilojoule and peak powers of many gigawatts at 10.6 μm will shortly be realized.

HIGH REPETITION RATE TEA LASERS

As indicated earlier, the use of low repetition rates in TEA lasers is an adequate technique to prevent excessive heating of the gas which could destroy the laser action either by thermally populating the lower level of the laser transition thus reducing the population inversion, or by decomposing the CO_2 molecules. In imperfectly sealed TEA lasers, only a very small flow of gas is used to overcome gas losses and prevent air contamination. In such systems it is found that the practical repetition rate limit is between 10 and 30 pps before laser degradation becomes serious. Adding rapid gas recirculation and cooling in a way similar to that described by Tiffany *et al.* [1], the repetition rate of TEA lasers can be increased substantially.

Using a simple 1-m long electrode structure, either a pin structure or a multiple resistor structure, repetition rates up to 1000 pps have been achieved with pumping energies well below the maximum values used in low repetition structures. To achieve these results, the laser was placed in a larger container (50 by 80 by 110 cm) and fans were used to cause the gas to circulate rapidly transversely to the laser axis. The rapid transverse gas recirculation technique has been used in a 150-cm long laser with a more powerful discharge network that could produce laser pulses up to 0.5 J at a repetition rate up to 1200 pps. This system has operated at full power and maximum repetition rate for only a few seconds. The rapid degradation of the laser performance is associated to an alteration of the electrical discharge characteristics of the laser gas giving rise to bright arcs which not only degrade the excitation efficiency but also causes a rapid decomposition of the laser gas. The sources of this deterioration are being determined and it is expected that both higher energy per pulse and higher repetition rates will be possible with a new system presently under construction.

From the consideration of Tiffany *et al.* [1] in CW systems and those of Hill [17] on high repetition rate pulsed systems, it appears that the major factor in the achievement of compact high average power CO_2 lasers is to achieve a high mass flow through the laser system. Transverse excitation seems to be best suited for this purpose since it lends itself readily to transverse gas flows and less energy needs to be expanded to move a given gas volume through the large cross section of transverse gas flows laser than in the longitudinal gas flow considered by Hill [17] and by Lavarini *et al.* [2]. Operating at atmospheric pressure implies also that because of the high density, a greater mass flow can be achieved with subsonic gas velocities than is possible with lower pressure systems.

Although there are no sufficient data available yet to make experimental comparisons with the results of Hill [17] in terms of average power per unit mass flows, simple calculations can give an idea of the compatibility of data.

Considering a gas mixture of $1:1:4$ of $CO_2:N_2:He$, 5 J/pulse can be obtained per liter of gas at atmospheric pressure. The specific gravity of the gas mixture being approximately 10 gm/mole a complete change of gas between pulses at a mass flow rate of one pound per second means 1000 l/s. Thus a figure of 5 kW/lb mass flow/s can be obtained. The specific heat of this gas mixture being above 1 cal/gm and the efficiency of the laser action being taken as 10 percent, the temperature increase of the gas after a 50 J/l excitation pulse is only about 25°C. A lower flow rate can be considered since Hill [17] indicates that temperature of at least 150°C can be considered as efficient. Since up to 5 pulses per gas change can be assumed, 25 kW/lb s^{-1} seems reasonable which is not far from the 35 kW/lb s^{-1} given by Hill [17] for low flow rates and which contained wall cooling effects. At high flow rates values of 11 kW/lb s^{-1} were obtained which is lower than the anticipated value but comparisons are extremely difficult since the discharge systems are quite different. With TEA lasers for example, the energy per pulse is constant at low repetition rates instead of rapidly dropping to zero as in the Hill type of excitation. Because of the independence of the excitation on preionization by an earlier pulse, the behavior at high repetition rates are also quite different and complete gas changes between pulses are valid with TEA lasers but not with Hill laser. In TEA lasers with rapid gas recirculation the energy per pulse has been observed to be constant from 0.1 pps to 1000 pps instead of being very sensitive to the repetition rate as reported by Hill.

The high repetition rate capability of TEA lasers can be very useful when high data rates are required for experimental purposes. However, the possibility of compact high average power lasers is even more attractive from the industrial applications point of view since for such operations as cutting or welding the average power is often the most important characteristic. In other applications such as laser radars, the combination of high peak power short pulses and high repetition rates is a most desirable feature.

POTENTIAL AND LIMITATIONS OF TEA LASERS

A major characteristic of TEA lasers is the importance of the electrical discharge properties of the system. In most cases a lengthening of the electrical pulse leads to the formation of bright arcs which decrease the uniformity of excitation, cause excessive local heating, and give rise to scattering of the laser radiation. So far most discharge systems have been elementary, consisting of a condenser, a spark gap or thyratron, and simple electrode structures. More sophisticated electrode systems such as in the double discharge technique are beginning to bring substantial improvement in discharge distribution in large volume lasers. These could rapidly lead to energy per pulse values of 20 J/l and efficiencies of 20 percent are likely to be achieved regularly. CW operation can also be considered a possibility and with a lower laser level lifetime of 1 μs, output powers of the order of 5 MW/l could be achieved with a mass flow of 200 lb/s.

The most serious limitation at present is the degradation of optical components. With submicrosecond pulses, an energy density of 5 J/cm^2 is sufficient to bring the surface of a polished copper mirror to its evaporation temperature. In transparent window materials NaCl is the best material found so far for high peak power but is of limited value for average powers. Germanium is better for high average powers as it can be better cooled but it will damage more easily than NaCl on high peak powers. Selenium and ZnS are the best materials found for multilayer dielectric coatings although it cannot be said that the search for materials has been exhaustive.

Operation of gas lasers at pressures well above atmospheric appears promising at first sight to obtain higher powers. However, the necessity to use high pressure vessels and windows, the higher voltages required, the greater difficulty in maintaining a uniform discharge distribution, and the need for a shorter excitation time due to the shorter relaxation time are all difficulties which have to be met. Assuming that an energy density of 5 J/cm^2 can be handled by the mirrors and semitransparent windows, with an atmospheric pressure laser producing 5J/l, a maximum length of 10 m will produce the maximum energy density that the mirrors can tolerate and increasing the pressure would only decrease the maximum useful length without increasing the power or power density. To achieve higher peak powers, the laser cross section must be increased to prevent excessive power densities. Laser cross sections of one square meter are presently considered and operation at atmospheric pressure is then a great advantage. For high average power systems, it is expected that cooled metallic mirrors can be used to much higher powers than semitransparent windows and again operation at atmospheric pressure can be most important as it allows the removal of the windows without too much gas losses. The only real advantage of going to very high pressures may be for the generation of very short pulses. At atmospheric pressure, mode-locked pulses are only of the order of a nanosecond because the linewidth of the laser transitions is only of the order of 3 to 4 GHz. By operating above 10 atm the lines of the P branch would begin to overlap since the pressure broadening would be equal to or greater than the separation between lines which is of the order of 50 GHz. Thus there would be a continuous amplification band of nearly a terahertz and gain would increase with pressure above 10 atm due to this line overlap. With such a bandwidth, mode-locked pulses of the order of one picosecond could be considered.

While most of the research on TEA lasers has considered CO_2 as the lasing gas, preliminary investigations have indicated that most other lasing molecules can be used at high pressures to generate high peak powers at a large number of wavelengths varying from the near *IR* to the submillimetre region. Recent work by Kimbell *et al.* [19], has shown that the high pressure transverse excitation can also be used to initiate uniformly large volume chemical reactions leading to population inversion and short laser pulses. Chemical reactions leading to population inverted HF and CO have given rise to submicrosecond laser pulses with peak powers near 1 MW.

CONCLUSIONS

Molecular lasers, and specially the CO_2 laser, operated at atmospheric pressure and excited by means of transverse discharges constitute a new category of gas laser which is capable of peak powers comparable to those of solid-state lasers but with higher efficiencies and longer wavelength. These lasers have potential applications as research tools for nonlinear optics and high energy physics as well as industrial tools. The use of atmospheric pressure also simplifies the laser design by eliminating the need for vacuum systems and high quality sealing techniques. Relatively simple laboratory prototypes have already given results comparable to theoretical limits in both energy and efficiency. As better engineering of these lasers develops, very high peak power lasers should become available at a relatively low cost compared to solid-state lasers. Much research and development remains to be done on high peak power gas lasers but the high pressure advantages already demonstrated make the efforts worthwhile.

REFERENCES

[1] W. B. Tiffany and R. Targ, "Desk size carbon-dioxide unit delivers a kilowatt CW," *Laser Focus.* Sept. 1969, p. 48–51.
W. B. Tiffany, R. Targ, and J. D. Foster, "Kilowatt CO_2 gas transport laser," *Appl. Phys. Lett.*, vol. 15, Aug. 1, 1969, p. 91.
[2] B. Lavarini, J. P. Bettini, H. Brunet, and M. Michon, "Influence de la pression et de la vitesse du gaz sur l'émission à 10.6 μ d'un amplificateur CO_2–N_2–He," *Compt. Rend., Acad. Sc. Paris*, vol. 269, Dec. 1969, pp. 1301–1304.
[3] A. E. Hill, "Multijoule pulses from CO_2 lasers," *Appl. Phys. Lett.*, vol. 12, May 1968, pp. 324–327.
[4] A. J. Beaulieu, "Transversely excited atmospheric pressure CO_2 lasers," *Appl. Phys. Lett.*, vol. 16, June 1970, pp. 504–505; see also [5], [6], [8], [11]–[16].
[5] J. Gilbert, "Dynamics of the CO_2 laser with transverse pulsed excitation," presented at CAP-APS Meet., June 1970, see also *Bull.*

[6] K. A. Laurie and M. M. Hale, "Folded-path atmospheric-pressure CO_2 laser," *IEEE J. Quantum Electron.* (Corresp.), vol. QE-6, Aug. 1970, pp. 530–532.
[7] R. Dumanchin, J. C. Farcy, M. Michon, and J. Rocca Serra, "High-power density pulsed molecular laser," presented at the 1970 Internat. Quantum Electron. Conf. (Kyoto, Japan).
[8] J. Gilbert and J. L. Lachambre, "Self-locking of modes in a high pressure CO_2 laser with transverse pulsed excitation," to be published.
[9] D. L. Lyon, E. V. George, and H. A. Haus, "Observation of spontaneous mode locking in a high-pressure CO_2 laser," presented at the 1970 Internat. Quantum Electron. Conf. (Kyoto, Japan); see also *Appl. Phys. Lett.*, vol. 17, pp. 474–476.
[10] O. R. Wood, R. L. Abrams, and T. J. Bridges, "Mode-locking of a transversely excited atmospheric pressure CO_2 laser," *Appl. Phys. Lett.*, Nov. 1970.
[11] R. Fortin, "Preliminary measurements of a transversely excited atmospheric pressure CO_2 laser," *Can. J. Phys.*, Jan. 1970.
[12] R. Fortin, M. Gravel, J. L. Lachambre, and R. Tremblay, "Helically and transversely excited atmospheric CO_2 laser," presented at CAP-APS Meet., June 1970. *Phys. Canada*, vol. 26, p. 81; see also *Bull. Amer. Phys. Soc.*, vol. 15, sec. 11, p. 808.
[13] A. M. Robinson, "Afterglow-gain measurements in CO_2–He–N_2 mixtures at pressures up to 1 atmosphere," *Can. J. Phys.*, vol. 48, pp. 1996–2001.
——, "Effect of inductance on the small signal gain of transverse excitation atmospheric pressure discharge in carbon dioxide," *IEEE J Quantum Electron.* (Corresp.), vol. QE-7, May 1971, pp. 199–200.
[14] D. C. Johnson, "Excitation of an atmospheric pressure CO_2–N_2–He laser by capacitor discharges," to be published.
[15] K. A. Laurie, "A pin electrode TEA laser," presented at the 1970 Device Res. Conf., July 1970; to be published.
[16] A. K. Laflamme, "Double discharge excitation for atmospheric pressure CO_2 laser," *Rev. Sci. Instr.*, vol. 41, Nov. 1970, pp. 1578–1581.
[17] A. E. Hill, "Role of thermal effects and fast flow power scaling techniques," *Appl. Phys. Lett.*, vol. 16, June 1970, pp. 423–426.
[18] D. C. Smith and A. J. DaMaria, "Parametric behavior of the atmospheric pressure pulsed CO_2 laser," *J. Appl. Phys.*, vol. 41, pp. 5212–5214.
[19] T. V. Jacobson and G. H. Kimbell, "A transversely spark-initiated chemical laser with high pulse energies," *J. Appl. Phys.*, vol. 41, Dec. 1970, pp. 5210–5212. "Transverse pulse-initiated chemical laser: preliminary performance of HF system," *Chem. Phys. Lett.*, Feb. 1971.

Gasdynamic lasers

*The population inversion necessary for lasing action
in a gas can be induced by rapid expansion—without external
pumping—of a hot gas mixture to supersonic speeds*

Edward T. Gerry*   *Avco Everett Research Laboratory*

It is not enough to depend upon diffusivity to dissipate the heat from a gas laser. Under the circumstances, average power outputs are too limited. By generating flow within the gas and deriving the benefits of forced convection, performance is improved. But if flow can be used to improve power outputs, it also can be used to advance a step further and generate the conditions that are necessary for lasing action—thereby creating a gasdynamic laser. This article covers the theory behind such a device and describes an experimental unit that has proved the merit of the concept.

The most fundamental limitation on the average power output of a laser is the waste energy resulting from inefficient operation. This waste energy may appear in the form of excited metastable states or simply as heat.

Figure 1 presents an analysis of the methods by which waste energy is removed. Most laser devices, whether solid or gas, have an active medium in the form of a long, thin cylinder. In the case of a solid, waste energy simply is conducted to the wall where it is removed by a coolant; in the case of a gas laser, energy is disposed of by diffusion of metastable states or heat to the outer

*Article adapted from a paper presented by E. T. Gerry at the American Physical Society meeting, Washington, D.C., April 1970.

† In a diffusion-controlled laser, waste energy is rejected in a characteristic time approximately that of the diffusion time T_{diff}. Because the process is random walk, T_{diff} is equal to the square of the number of mean free paths $(D/\lambda)^2$ during which the energy diffuses multiplied by the mean free time between collisions $(\lambda/\overline{C})$ and, therefore, is $D^2/\lambda\overline{C}$. Here D is the characteristic dimension of the tube, λ is the mean free path, and $\overline{C}$ is the molecular speed. But if the gas is moved at a speed U in a flowing system, waste energy is rejected in a time equal to D of the tube divided by U— i.e., D/U.

Thus, for the same active volume and gas density, the ratio of the power achievable with a stagnant and with a flowing-gas laser is simply the ratio of the characteristic times, which, as shown in Fig. 1, is equal to the characteristic dimension divided by the mean free path times the flow velocity divided by the mean molecular speed—i.e., $DU/\lambda\overline{C}$. Since the speed of sound in a gas is approximately the same as the mean molecular speed, the velocity ratio essentially is the Mach number.

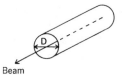
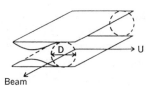

Beam   Beam

Diffusion or heat-conduction limited   Convection cooled
≈ 1 MW/km

Waste-energy rejection time is   Waste-energy rejection time is

$$\tau_{\text{diff}} = \left(\frac{D}{\lambda}\right)^2\left(\frac{\lambda}{\overline{C}}\right) = \frac{D^2}{\lambda \cdot \overline{C}}$$

$$\tau_{\text{flow}} = \frac{D}{U}$$

FIGURE 1. Comparison of characteristic times related to the removal of waste energy from cooled and uncooled high-average-power lasers.

walls of a cylindrical container.

With only diffusion or heat conduction to remove waste energy, a laser device, on an average-power basis, is limited to a maximum power of the order of 1 kW/m of solid-rod or gas-tube length—regardless of the diameter. However, by using high-speed flow to remove the waste energy more quickly, the average power capabilities of the device can be increased.†

For typical gas lasers, the factor by which the average power density can be increased through the use of high-speed flow[1] ranges from 10^3 to 10^5. In addition, since the flow time is independent of gas density, the density can be increased to provide still greater powers. Thus, high-speed flow can lead to larger devices that provide significantly greater average power densities than are achievable with diffusion-controlled lasers.

Table I indicates the application of flow to various classes of lasers. In electrically pumped lasers, flow can be used for removing waste energy. Excited by electron impact of, typically, "volts" energy, such electrically excited lasers as CO_2[2-7] operate at 10 μm (in the infrared); carbon monoxide[8,9] operates around 4 μm; the copper-vapor laser[10,11] operates in the green; and the nitrogen laser[12,13] operates in the ultraviolet. All of these, and others, are potential candidates for the application of

Reprinted from *IEEE Spectrum*, vol. 7, pp. 51–58, Nov. 1970.

I. Application of flow to various types of gasdynamic lasers

Type of Laser	Use of Gas Flow	Wavelength	Gases Used
Electrically pumped	Removal of waste energy (heat, excited states)	Widest range possible for efficient operation	CO_2, N_2, CO, Cu vapor
Chemically pumped	Removal of waste energy, temperature control, mixing, replenishing reactant	2–6 μm; longer in hybrid systems	F + HCl, F + H_2; hybrid: F + $H_2 \rightarrow$ CO_2* (where * signifies excited molecule)
Thermally pumped	Production of gas inversion from equilibrated hot gas; removal of waste	8–14 μm for efficient operation	CO_2 at 10.6 μm

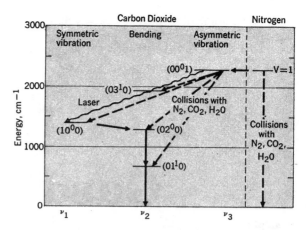

FIGURE 2. Exchanges between energy levels in a CO_2-N_2 laser. Sequential numbers relate respectively to asymmetric-stretch, bending, and symmetric-stretch mode levels. Superscript accompanying bending mode indicates the plane of vibration. V = 1 denotes first excited vibrational state.

high-speed flow cooling.

In addition to electrically excited gas lasers, there are gas lasers that are either chemically or thermally pumped. But whereas the electrically pumped gas laser benefits from the application of flow only to the extent that waste energy is carried away, the chemical and thermal types depend on the flow for their lasing action.

Flow is essential to a high-power chemical laser system to replenish reactants that are consumed in the production of population inversion. (In a chemically pumped system, the active laser species is a direct product of a chemical reaction, is produced by this reaction in an excited state, and subsequently is lased.) Also in a high-speed, flowing-gas chemical laser, active variation of flow parameters may be used to achieve temperature or reaction-rate control within the laser cavity.

For efficient operation, chemical lasers typically operate in the 2- to 6-μm region of the infrared. (The chemical energy released in a reaction generally has a value corresponding to a wavelength in this range.) Operation efficiently can be extended further into the infrared in a hybrid system—wherein the energy in the excited product molecule is transferred to another molecule, which then becomes the active laser species. Examples of such chemical laser systems include those making use of the F + HCl[14,15] exchange reaction to produce excited HF, and the F + H_2 reaction,[16,17] which also produces excited HF. This excited HF energy is transferred to CO_2 and produces laser action at 10.6 μm.[18]

Gasdynamics

The final type of laser that will be described—and to which the rest of this article will be devoted—is the thermally pumped gasdynamic laser. With it, flow is used to create an inversion from what is, initially, a completely equilibrated hot gas. A thermally pumped system starts with a hot equilibrium gas mixture in which there is no population-energy inversion. The inversion is produced "gasdynamically" by rapid expansion through a supersonic nozzle.[19]

Because a hot gas is the basic energy source, these lasers typically will operate efficiently in the 8- to 14-μm-wavelength band. The prime example of this type of laser is the nitrogen–CO_2 gasdynamic laser, operating at 10.6 μm in the standard CO_2 laser transition.

The production of vibrational nonequilibrium in CO_2 in a high-speed flow was demonstrated by Kantrowitz[20] in connection with the development of a gasdynamic method for measuring vibrational relaxation times. The inversion-production gasdynamic-method laser, in its most general form, was suggested by Basov and Oraevskii.[19] The possibility of population inversion in N_2-CO_2 mixtures by rapid expansion through a supersonic nozzle was suggested by Konyukhov and Prokhorov.[21] A similar gasdynamic approach—differential radiative relaxation in a fast expansion of an arc-heated plasma—was suggested by Hurle and Hertzberg.[22]

Figure 2 shows an energy-level diagram of the carbon dioxide and nitrogen molecules,[23] indicating the important vibrational relaxation processes that occur in such a mixture. The CO_2 gasdynamic laser typically involves a gas mixture that is mostly nitrogen, and approximately 10 percent carbon dioxide and one percent water. (Mixtures involving helium instead of water are also possible.)

Nitrogen, a simple diatomic molecule, has only one vibrational mode. Energy can be lost from this mode by collisions with nitrogen, CO_2, and water,[24] returning the excited molecule directly to the ground state. Carbon dioxide, being a linear triatomic molecule, has three basic modes of vibration: asymmetric stretch, which forms the upper laser level; symmetric stretch, which forms the lower laser level; and bending.

The energy-exchange process in the gasdynamic laser includes several transfers: (1) The very close near-resonance between nitrogen and the first asymmetric-stretch level of CO_2 causes efficient transfer of energy between these modes. Typically, the probability of this transfer at room temperature is one in every 500 collisions.[25,26] (Direct deactivation of nitrogen is relatively unimportant and energy is lost from the mixture generally through the CO_2 vibrational levels.) (2) Energy can be lost from the asymmetric-stretch mode of CO_2 by collisions with nitrogen, CO_2, and water—most probably transferring into the symmetric-stretch and bending modes of CO_2.[24,25,27-29] (However, lasing occurs only for transitions from the first asymmetric-stretch mode to the first symmetric stretch.) (3) The Fermi resonance between the bending and symmetric-stretch modes of CO_2 tightly couples these modes, Therefore, excitation energy is extracted from the lower laser level by deactivation of the bending mode—a process that can occur during collisions with all gas species, but principally water.[24]

Principle

The basic principle of the gasdynamic laser is to expand the gas rapidly through a supersonic nozzle to a high Mach number. The object is to lower the gas-mixture temperature and pressure downstream of the nozzle in a time that is short compared with the vibrational relaxation time of the upper-laser-level system (consisting of the asymmetric-stretch mode of CO_2 coupled with the nitrogen). At the same time, by addition of the catalyst (water or helium), the lower level relaxes in a time comparable to, or shorter than, the expansion time. Because of this rapid expansion, the upper-laser-level system cannot follow the rapid change in temperature and pressure and thus becomes "hung up" at a population characteristic of that in the stagnation region. Because the vibrational relaxation times associated with the upper level are long, the population will stay "hung up" for a considerable distance downstream of the nozzle. This process is known as vibrational freezing.

A basic CO_2 gasdynamic laser is presented in Fig. 3. At the top is a schematic of a supersonic nozzle. The gas flow runs from left (the stagnation region, which contains a hot, equilibrium gas mixture) to the right. A typical set of stagnation characteristics is indicated for a mixture of 7.5 percent CO_2, roughly 90 percent nitrogen, and one percent water: a temperature of 1400°K and a pressure of 17 atmospheres. (It is important to note that this is a completely equilibrated gas mixture. There is no inversion present, and since it is in an equilibrium state, this mixture can be produced in any desired way. For example, it may be produced by simply heating in a heat exchanger or nuclear reactor. Alternatively, it may be produced by combustion of a suitable fuel or by heating in a shock tube.) Typical downstream characteristics for the gasdynamic laser are an area ratio (with respect to the throat) of 14, a throat height of 0.8 mm, a Mach number of the order of 4, a pressure of about 0.1 atmosphere, and a temperature near ambient.

Figure 3(B) shows how the energy is distributed between the various degrees of freedom of the gas in a CO_2 gasdynamic laser. In the stagnation region, most of the energy is associated with the random translation and rotation of the gas molecules and 10 percent (or less) is associated with vibration. As the gas is expanded through the supersonic nozzle, the random translational and rotational energies are converted into the directed kinetic energy of flow. The vibrational energy, if it remained in equilibrium with the gas temperature, essentially would disappear downstream of the nozzle. Because of the rapid expansion, however, the vibrational energy remains "hung up" and its vibrational tempera-

ture is characteristic of that upstream of the nozzle.

Looking at this in terms of populations [Fig. 3(C)], it is apparent that the population of the lower level exceeds that of the upper level in the stagnation region, which is typical of an equilibrated gas mixture. As the gas is expanded through the nozzle, the upper-level population drops just a little bit and then remains essentially level. The lower-level population diminishes rapidly within the nozzle, continues to decrease, and virtually disappears a few centimeters downstream. Thus, downstream of the nozzle, the population of the upper level is characterized by a temperature like that of the stagnation region and the population of the lower level is characterized by a temperature like that of the downstream region. Inversion begins approximately one centimeter downstream of the nozzle throat and continues, for the gas conditions indicated in Fig. 3, for about a meter downstream of the nozzle. Because of the high gas densities involved and the high-speed flow downstream of the nozzle, an inversion capable of operation at very high powers is achieved.

General considerations

In Fig. 4, some additional considerations regarding the operation of the gasdynamic laser are presented. The flow tube has a stagnation pressure P_{stag}, stagnation tem-

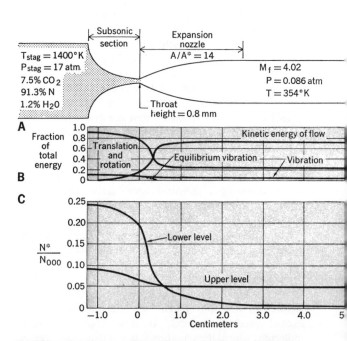

FIGURE 3. Existing conditions for a specific gas laser. N_{000} is the ground-state population. N^* is an arbitrary excited state; two such states—the 100 lower and 001 upper level—are shown.

FIGURE 4. General conditions necessary for gasdynamic lasing. Diffuser (right) is only representational.

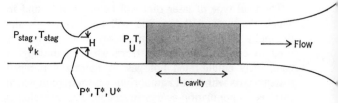

perature T_{stag}, and concentrations of species ψ_k in the stagnation region. The nozzle has a throat height H, with pressure, temperature, and velocity at the throat indicated by the starred quantities. Downstream, P, T, and U indicate pressure, temperature, and velocity in the laser cavity region—signified by the grey area—where energy is removed.

First, consider the energy that is available in the gasdynamic laser. Per unit mass, it is essentially the vibrational energy stored in the nitrogen and CO_2 asymmetric stretch upstream of the nozzle throat. For a 10 percent CO_2, 90 percent nitrogen mixture at a stagnation temperature of 1400 °K, the available laser energy (based on freezing the upper-level vibration at the stagnation temperature) is 35 J/gm of gas or 35 J/gm/s of gas flow. This maximum available energy will, of course, increase as the stagnation temperature is increased.* In any practical system, however, all of the energy available cannot be extracted because of inefficiencies and other constraints in the system. Typically only one third to one half of this available energy can be extracted from a well-designed system.

Consider what is required for efficient freezing of the upper-level vibrational energy. The maximum derivative of pressure and temperature in a supersonic flow tube occurs in the region of the throat area. Thus a typical characterizing parameter for allowable time for expansion past the throat is the ratio of the throat height to the gas velocity in the throat (which is just the speed of sound in the fluid). This flow time must be less than the effective relaxation time of the combined CO_2 and nitrogen upper laser level as it exists at the throat.† The effective relaxation time is dependent on both the temperature at the throat, and the fractional concentrations of CO_2 and water in the gas mixture; typically it is expressed in the form of a pressure–time product as it relates to gas temperature. Thus the effective time at the throat is this pressure–time product evaluated at the throat temperature divided by the static pressure (which is approximately half of the stagnation pressure) at the throat.

The effective time requirement leads to stipulations on the product of stagnation pressure and throat height in order to achieve efficient freezing of the available laser energy. For open-cycle combustion-driven systems, P_{stag} is generally fixed by conditions for diffuser-exhaust recovery to atmosphere. This relationship gives the throat

---

* $E_{max} = h\nu/[\exp(3380/T_{stag}) - 1]$ where h is Planck's constant, ν is the frequency, and T_{stag} is as in Fig. 4.

† That is, $H/U < (P\tau_{eff})_{T}*/P* \approx (P\tau_{eff})_{T}*/0.5\,P_{stag}$ or $P_{stag}H < 2(P\tau_{eff})_{T}*U$.

FIGURE 5. Shock-tube experiment to measure gain in an expanded N₂-CO₂ mixture. The gain-measuring apparatus outlined is said to be the most useful for ascertaining population inversion—distinction among gains of only a few percent being possible. Both single and array nozzles have been evaluated in the shock tunnel.

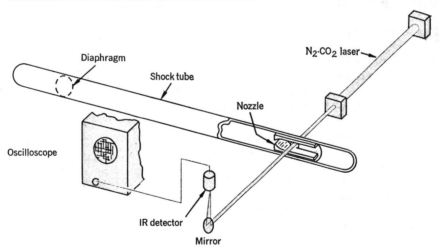

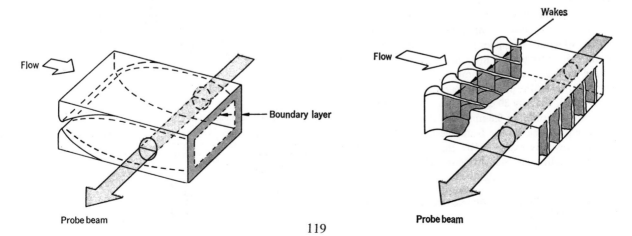

height as a function of CO_2 concentration and throat temperature. A downstream-to-throat-area ratio must be chosen sufficiently large that when the lower laser level has a population characteristic of the gas temperature downstream of the nozzle, its population is considerably less than the upper-level population.

In the cavity region, most of the available laser energy is stored in nitrogen. Since the energy is removed by laser action in CO_2, sufficient cavity length must be allowed for transfer of the energy stored in nitrogen vibration to CO_2. This length is simply the relaxation time for nitrogen's energy to transfer to CO_2 multiplied by the downstream flow velocity, and is inversely proportional to the concentration ψ of CO_2 in the gas mixture. Note that this is the minimum cavity length.‡ Generally the cavity must be longer, as energy removal also is limited by removal of lower-state energy by collisions with water or helium and, in most practical cases, by the allowable intracavity radiation flux.

In most gasdynamic lasers, laser-energy removal from the gas is flux-limited rather than kinetics-limited. It is important to note here that apart from vibrational deactivation, which occurs slowly downstream of the nozzle, energy not removed from the gas at a given point upstream is still available for removal downstream of that point and thus some flexibility in the design of optical cavities is possible.

Downstream, when an optical cavity is present, the rate at which quanta are generated in the cavity is equal to $G\varphi/h\nu$ quanta/cm³/s, where G is the local gain coefficient in cm⁻¹, φ is the optical flux in W/cm², and h is Planck's constant. To compute the power output of an optical cavity, φ is adjusted until the average gain of the cavity equals the total cavity loss.

Two basic versions tested

Several different gasdynamic laser devices have been operated at various facilities. These include various configurations based on shock-tube-generation and combustion-powered devices. The performance of both types of equipment are comparable—for comparable configurations—as expected. Figure 5 shows a shock-tube device together with instrumentation. Figure 6 is a schematic of a combustion-powered, 1.4-kg/s device, concerning which some details follow.

The burner is round in cross section with a 15-cm-diameter flow area and a length in the flow direction of the order of 45 cm. It joins a conical mixing chamber, also approximately 45 cm long, which expands the flow cross section to 30-cm diameter. Both cyanogen (C_2N_2) and carbon monoxide have been used as fuels in this device. The fuel is burned with air at the back plate of the burner and ignition is maintained by a methane pilot burner, which also supplies part of the water in the flow. Additional nitrogen is injected midway between the burner and the mixing chamber to provide the right gas mixture and the proper temperature.

An array, 3 cm by 30 cm in cross section, consisting of nozzles containing an 0.8-mm-high throat (area ratio = 14), separates the burner from the cavity. In the cavity region, power is extracted by the use of a multihole-coupled multimode stable resonator, or by a mode-

controlled unstable resonator. The former is shown in Fig. 6(A). The diffuser, which is downstream of the cavity, slows the 0.1-atmosphere Mach 4 flow to a low Mach number and raises the pressure in excess of one atmosphere so as to exhaust without pumping.

By use of the operating conditions given in Fig. 6(B), a power output of 6 kW was obtained for an operating time of 10 seconds. (The burner and nozzle row were water-cooled for steady-state operation. However, for simplicity, the cavity and diffuser, as well as the mirrors, were "heat sinked," limiting run time to the order of seconds.) Figure 7 shows the device.

Under operating conditions of the run, which produced 6 kW, the ideal laser power would have been 40 kW if complete freezing of the upper-state vibrational energy at the stagnation temperature were achieved together with 100 percent efficient energy extraction downstream. This power is a function, as discussed earlier, of the stagnation temperature and the Mach number (M). But as long as the Mach number is high enough that the lower state is well out of the way, power is relatively independent of M.

As the nozzle is not 100 percent efficient, 17.3 kW are lost to vibrational deactivation within the nozzle. This loss is a function of gas composition and the geometry of the nozzle.

FIGURE 6. A—Schematic of shock-tube-powered 1.4-kg/s gasdynamic laser and power measurements as a function of mode size. B—Exposition of combustion apparatus.

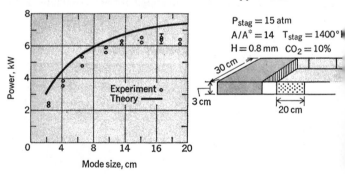

A

B

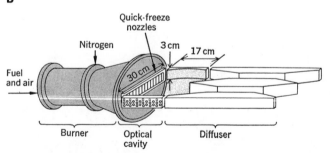

Typical Operating Conditions		Performance	
Stagnation pressure	17 atm	Power output	6 kW
Stagnation temperature	1300°K	Intracavity flux	1 kW/cm
Mach number	4.0		
Gas composition			
Carbon dioxide	8. %		
Carbon monoxide	0.2%		
Nitrogen	91. %		
Water	0.8%		
Mirror coupling	2. %		

‡ $L_{\text{cavity min}} > (P\tau_{\text{transfer}})_T \, U/P\psi_{CO_2}$ when the flux in the cavity tends to infinity.

Boundary layers on the walls of the nozzle subtract an additional 2 kW from the available laser power. (This loss would increase whereas the incomplete freezing loss would decrease if the nozzle size were smaller.)

Because the cavity was located somewhat downstream of the nozzle exit, some collisional deactivation, accounting for a loss of 1.2 kW, occurred before the gas entered the cavity.

In addition, 2 kW were lost within the cavity.

All of these losses are controlled by gas composition, Mach number, and the location of the cavity as well as—in the case of deactivation within the cavity—by the laser flux.

Mirror losses accounted for a 10.9-kW loss, largely due to the low coupling fraction used in these experiments—2.0 percent. The copper mirrors used in the tests have a combined absorption and scattering loss on reflection of the order of 1.5 to 2 percent. Mirror loss is controlled by reflectivity, the configuration of the resonator, output coupling, and the cavity flux. The mirror in these tests had an active length in the flow direction of 20 cm. Because of incomplete extraction while passing through the cavity, there was still 1.5 kW of laser energy in the gas.

These losses total 34.9 kW, leaving a net calculated laser output power of approximately 5.1 kW, which is reasonably consistent with the measured laser power of 6 kW. The largest uncertainty is in the mirror losses.

Two points should be made with regard to laser losses: (1) The aerodynamic and kinetic processes associated with gasdynamic laser operation are now relatively well understood, and fairly accurate predictions of device performance can be made. (2) Considerable improvement in the efficiency of gasdynamic lasers can be made by improving the tradeoffs between the listed losses in order to attain a higher fraction of the ideal laser power.

Mode control in gasdynamic lasers

Since the optical cavity that exists for a gasdynamic laser is basically short and fat, geometrical angles (defined by the ratio of the cavity height to the mirror spacings) are large compared with diffraction angles (defined by the ratio of the wavelength to the cavity height). The gasdynamic laser-cavity geometry, therefore, has an intrinsically high Fresnel number. With a large combustion-powered gasdynamic laser device, the ratio of geometrical angles to diffraction angles—based on the cavity size—is of the order of a thousand; i.e., the Fresnel number is of the order of a thousand.

Figure 8 shows possible schemes for obtaining near-diffraction-limited beam outputs. The stable resonator [8(A)] is the one most commonly used in ordinary gas lasers for achieving single-mode operation. ("Stable" here refers to geometrically stable, meaning that, from geometric optical considerations, off-axis rays within certain angular limits stay in the cavity. Losses from the cavity occur only by coupling through the mirrors or by diffraction.) In order to achieve diffraction-limited

FIGURE 8. Mode-control schemes.

FIGURE 7. Device diagrammed in Fig. 6(B).

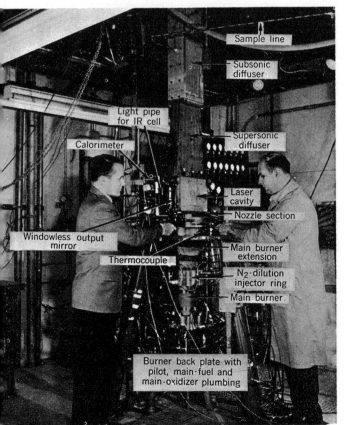

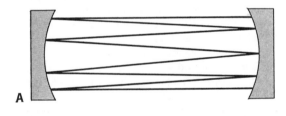

A

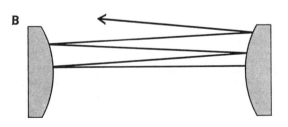

B

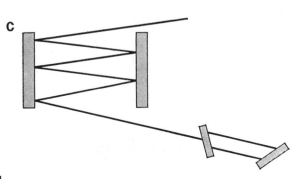

C

output (that is, minimum divergence consistent with wave optics), the stable resonator generally requires a low Fresnel number of the order of unity. Operated in this fashion, mirror-alignment requirements are less severe than for other types of resonators. In order to use this type of resonator in the high intrinsic Fresnel number medium of the gasdynamic laser, a large number of folds within the medium are required. Although this can be done, it does not appear to be the most advantageous approach to mode control in the gasdynamic laser.

A second type of resonator is the unstable resonator,[30] in which, from geometric optical considerations, off-axis rays "walk out" of the cavity. These resonators, in principle, can operate in a lowest-order mode at any Fresnel number. They are easiest to operate when high total gain exists in the cavity so that the coupling fraction over the edge of the mirror can be large. Mirror alignment requirements are severe and are similar to those for the plane–parallel resonator. Unstable resonators appear promising for use with gasdynamic lasers, as experiments with an unstable resonator in a large combustion-driven device have shown. See Fig. 8(B).

Another approach to achieving diffraction-limited output from a gasdynamic laser is to use the master oscillator–power amplifier setup [Fig.8(C)], where the output of a low-power mode-controlled CO_2 laser is amplified by folding the beam through the gasdynamic laser using a mirror system. This system can also operate with the beam-folding geometry in the gasdynamic laser having Fresnel numbers considerably greater than unity. Alignment requirements again are severe and comparable to those for the unstable resonator. The advantage of the master oscillator–power amplifier approach is that frequency control and wavelength selection can be accomplished with the gasdynamic laser by control of these parameters in the driver oscillator.

Results were obtained with an unstable resonator (see Fig. 9) used with the combustion-driven laser. The beam path of the resonator is folded several times to increase the total gain of the system. (The plane of the folding is parallel to the flow direction.) The resonator consists of three flat mirrors and one small convex coupling mirror covering 35 percent of the area of the folded

beam. The output is coupled out of the resonator over the edge of the convex coupling mirror to the outside world.

In the instance of a combustion-powered device, this output is picked up by a concentric concave mirror tilted at an angle to the system and focused out through a downstream hole so that flow disturbances introduced by entering air do not affect the optical quality of the medium. (Nor does the amount of flow exiting through the hole disturb the operation of the diffuser.) This hole location is a convenient way of circumventing the problem of suitably placing a window to handle the output

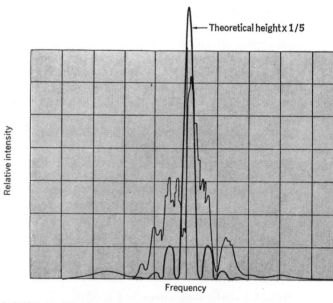

FIGURE 10. Results using an unstable resonator. Black line represents measured output power (of the order of 2 kW) whereas the colored line represents the theoretical far-field intensity distribution for an unstable resonator at 65 percent coupling operating in the lowest-order mode.

FIGURE 11. A recently constructed 14-kg/s gasdynamic device.

FIGURE 9. An unstable resonator. The output is coupled over the edge of the convex (40-meter-radius) mirror.

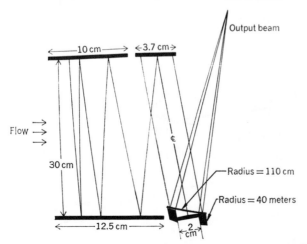

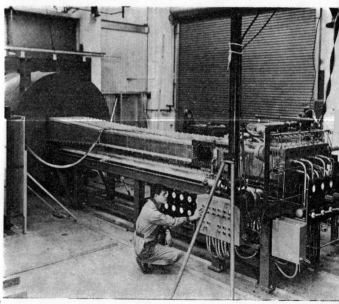

fluxes characteristic of a gasdynamic laser.

Under operating conditions similar to those for which 6 kW were produced in a multimode cavity, a power output of the order of 2 kW was achieved with the unstable resonator. Figure 10 shows IR scanner measurements. Also shown is the theoretical far-field intensity distribution for an unstable resonator of 65 percent coupling operating in the lowest-order mode. (The theoretical distribution has been multiplied by 0.2 to show it on the same scale as the output of the device.) This illustration indicates that the output radiance of the laser obtained under these operating conditions is approximately one sixth to one seventh that for pure lowest-order-mode operation, or, equivalently, about 2.5 times that for a diffraction-limited operation.

The mirrors used in these tests were copper and were operated in a heat-sink mode. Theoretical calculations indicate that significant distortion of these mirrors will take place during the run time and this distortion can lead to significant deviations from the ideal mode pattern.

In addition, in this device significant flow disturbances were present and these led to phase changes of a significant fraction of a wavelength across the beam cross section. Nozzle-array designs that eliminate or substantially reduce these flow disturbances have been developed. Also, water-cooled mirror structures that will not distort significantly under the heat loads of the gasdynamic laser have been developed. Although not conclusive, the results of these mode-control experiments on gasdynamic lasers seem to indicate that, with a uniform medium and with water-cooled mirrors, very near diffraction-limited performance of an unstable resonator can be achieved.

Finally, there are some fairly recent results to report about the large (approximately 14 kg/s) device shown in Fig. 11. This device, operated with CO as the fuel, has produced 60 kW of multimode power and 30 kW of near-diffraction-limited power in an unstable resonator, indicating the general scalability of these devices.

This work was supported in part by the Avco Corporation and in part by the Advanced Research Projects Agency and the U.S. Air Force.

REFERENCES

1. Wilson, J., "Nitrogen laser action in supersonic flow," *Appl. Phys. Letters*, vol. 8, pp. 159–161, Apr. 1966.

2. Patel, C. K. N., "Continuous wave laser action on vibrational-rotational-transitions of CO_2," *Phys. Rev.*, vol. 136, pp. A1187–A1193, Nov. 30, 1964.

3. Patel, C. K. N., "CW high power N_2-CO_2 laser," *Appl. Phys. Letters*, vol. 7, pp. 15–17, July 1, 1965.

4. Patel, C. K. N., "Selective excitation through vibrational energy transfer and optical maser action in N_2-CO_2," *Phys. Rev. Letters*, vol. 13, pp. 617–619, Nov. 23, 1964.

5. Patel, C. K. N., "CW high power CO_2-N_2-He laser," *Appl. Phys. Letters*, vol. 7, pp. 290–292, Dec. 1, 1965.

6. Sobolev, N. N., and Sokovikov, V. V., "CO_2 lasers," *Sov. Phys. Uspekhi*, vol. 10, pp. 153–170, Sept.–Oct. 1967.

7. Tiffany, W. B., Targ, R., and Foster, J. D., "Kilowatt CO_2 gas-transport laser," *Appl. Phys. Letters*, vol. 15, pp. 91–99, Aug. 1, 1969.

8. Osgood, R. M., Jr., and Eppers, W. C., Jr., "High power CO_2-H_2-He laser," *Appl. Phys. Letters*, vol. 13, pp. 409–411, Dec. 15, 1968.

9. Legay, F., "Study of a CO-N_2 laser by modulation method," *Compt. Rend. Acad. Sci. B.*, vol. 266, pp. 554–557, Feb. 26, 1968.

10. Walter, W. T., Piltch, M., Spilmene, N., and Gould, G., "Efficient pulse gas discharge laser," *IEEE J. Quantum Electronics*, vol. QE-2, pp. 474–479, Sept. 1966.

11. Leonard, D. A., "Theoretical description of 5106— a pulsed copper vapor laser," *IEEE J. Quantum Electronics*, vol. QE-3, pp. 380–381, Sept. 1967.

12. Leonard, D. A., "Saturation of the molecular nitrogen second positive laser transition," *Appl. Phys. Letters*, vol. 7, pp. 4–6, July 1, 1965.

13. Gerry, E. T., "Pulsed-molecular nitrogen laser theory," *Appl. Phys. Letters*, vol. 7, pp. 6–8, July 1, 1965.

14. Airey, J. R., "Cl + HBr pulsed chemical laser: a theoretical and experimental study," *J. Chem. Phys.*, vol. 52, pp. 156–157, Jan. 1, 1970.

15. Airey, J. R., "A supersonic mixing chemical laser," *Appl. Phys. Letters*, vol. 16, pp. 401–403, Dec. 15, 1969.

16. Spencer, D. J., Jacobs, T. A., Mirels, H., and Gross, R. W. F., "Continuous wave chemical laser," *Internat'l J. Chem. Kinetics*, vol. 1, pp. 493–494, Sept. 1969.

17. Spencer, D. J., Jacobs, T. J., Mirels, H., and Gross, R. W. F., "Preliminary performance of a CW chemical laser," *Appl. Phys. Letters*, vol. 16, pp. 235–237, Mar. 15, 1970.

18. Cool, T. A., Stephens, R. R., and Falk, T. J., "A continuous wave chemically excited CO_2 laser," *Internat'l J. Chem. Kinetics*, vol. 1, pp. 495–497, Sept. 1969.

19. Basov, N. G., and Oraevskii, A. N., "Attainment of negative temperatures by heating and cooling a system," *Sov. Phys. JETP*, vol. 17, pp. 1171–1174, Nov. 1963.

20. Kantrowitz, A. R., "Heat capacity lag in gas dynamics," *J. Chem. Phys.*, vol. 14, pp. 150–164, Mar. 1946.

21. Konyukhov, V. K., and Prokhorov, A. M., "Population inversion in adiabatic expansion of a gas mixture," *J. Exp. Theoret. Phys. Letters*, vol. 3, pp. 286–288, June 1, 1966.

22. Hurle, I. R., and Hertzberg, A., "Electronic population inversion by fluid-mechanical techniques," *Phys. Fluids*, pp. 1601–1607, Sept. 1965.

23. Herzberg, G., *Molecular Spectra and Molecular Structure*, 2nd. ed. I. *Spectra of Diatomic Molecules* (1950). II. *Infrared and Raman Spectra of Polyatomic Molecules* (1945). Princeton, N.J.: Van Nostrand.

24. Taylor, R. L., and Bitterman, S., "Survey of vibrational relaxation data for processes important in the CO_2-N_2 laser system," *Rev. Mod. Phys.*, vol. 41, pp. 26–47, Jan. 1969.

25. Rosser, W. A., Jr., Wood, A. D., and Gerry, E. T., "Deactivation of vibrationally excited carbon dioxide (v_3) by collisions with carbon dioxide or nitrogen," *J. Chem. Phys.*, vol. 50, pp. 4996–5008, June 1, 1969.

26. Sharma, R. D., and Brau, C. A., "Energy transfer in near resonance molecular collision due to long-range forces with application to transfer of vibrational energy from v_3 mode of CO_2 to N_2," *J. Chem. Phys.*, vol. 50, pp. 924–930, Jan. 1969.

27. Herzfeld, K., "Deactivation of vibrations by collision in CO_2," *Disc. Faraday Soc.*, no. 33, pp. 22–27, 1962.

28. Herzfeld, K., "Deactivation of vibrations by collision in the presence of Fermi resonance," *J. Chem. Phys.*, vol. 47, pp. 743–752, July 15, 1967.

29. Rosser, W. A., Jr., and Gerry, E. T., "De-excitation of vibrationally excited CO_2* (v_3) by collisions with He, O_2, and H_2O," *J. Chem. Phys.*, vol. 51, pp. 2286–2287, Sept. 1, 1969.

30. Siegman, A. E., and Arrathoon, R., "Modes in unstable optical resonators and lens waveguides," *IEEE J. Quantum Electronics*, vol. QE-3, pp. 156–163, Apr. 1967.

Chemical Lasers: A Survey of Current Research

ARTHUR N. CHESTER

Abstract—A brief summary is presented of recent chemical laser research. The principal emphasis is on current device performance, particularly those areas where promising new work is being done. Chemical kinetics and theoretical modeling are not discussed in detail, but extensive references are provided for the reader who wishes to pursue these subjects in greater depth. The literature citations have been updated as of November 24, 1972.

I. Introduction

A CHEMICAL laser is a laser in which a population inversion is directly produced by an elementary chemical reaction. It is common practice to regard chemical transfer lasers, e.g., DF–CO_2 [1], as chemical lasers. The ordinary CO_2 gas-dynamic laser [2] is not a chemical laser because its combustion products come essentially into thermal equilibrium before a gas expansion is used to create a population inversion. Most chemical lasers operate on a vibrational–rotational transition in a molecule (see Fig. 1), because many exothermic gas-phase reactions liberate their energy through a stretching vibration in the newly formed chemical bond; the mechanism by which this vibrational excitation occurs are discussed in the literature [3]. A wide range of wavelengths is available, presently extending from 2 μm [4] to 126.5 μm [84a].

Table I summarizes the most popular chemical laser reaction systems in use today.[1] Each of these reactions involves free atoms, since the fuels involved do not react rapidly in their molecular forms. To ensure that the chemical reaction will proceed rapidly enough to keep up with collisional relaxation processes, large numbers of free atoms must be generated to initiate the laser reaction. The techniques usually employed for laser initiation are summarized in Table II. Thus the common features of chemical lasers are the following:

1) a system for mixing together the reagent gases, shown in Table I;
2) some type of reaction initiation technique, shown in Table II, which is applied to the gases either before or after mixing;
3) a laser cavity where the excited molecules produced by the reaction undergo stimulated emission;
4) an exhaust system to remove expended gases from the laser cavity.

From Kasper and Pimentel's first work in 1964 [6] through 1969, chemical lasers were pulsed low-efficiency devices, used primarily by chemists to study chemical kinetics. The present intensive work in developing chemical lasers as practical sources dates from three important developments in 1969:

1) studies of chemical lasers in which an electric discharge

was used to set off the chemical reaction, by Tal'roze, Basov, and other Soviet scientists [7][2];
2) the continuous-wave (CW) chemical laser developed by Spencer *et al.* [8] in which the chemical reagents were rapidly mixed in a supersonic flow stream[3];
3) the "purely chemical" chemical transfer laser of Cool and Stephens [10] requiring no electrical input to start the chemical reaction, relying instead on the spontaneous reactions occurring between commercial bottled gases.

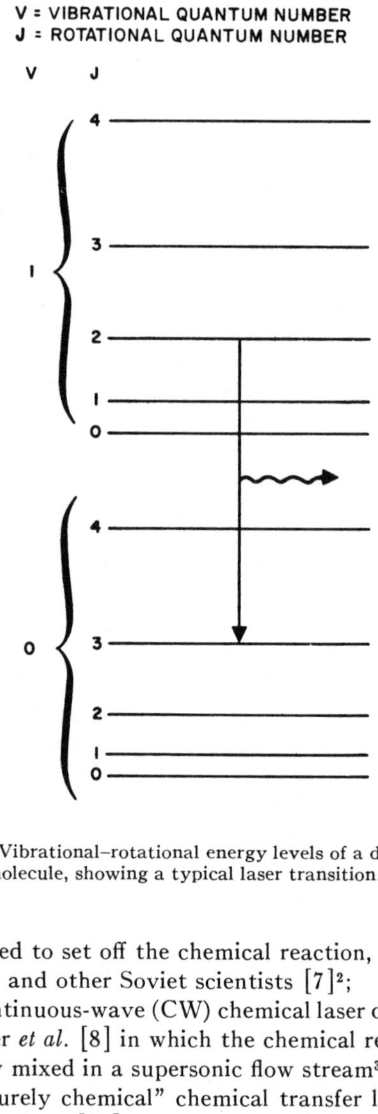

Suddenly it was possible to visualize a laser that burned bottled gases and required no electrical input; at 10-percent chemical efficiency, such a device using hydrogen and fluorine fuels would produce 1300 J of laser output for every gram of fuel consumed. Thus a period of intensive research and development ensued.

V = VIBRATIONAL QUANTUM NUMBER
J = ROTATIONAL QUANTUM NUMBER

Fig. 1. Vibrational–rotational energy levels of a diatomic molecule, showing a typical laser transition.

Manuscript received July 17, 1972; revised December 7, 1972. This paper is based on an invited talk given at the VII International Quantum Electronics Conference, Montreal, Canada, May 10, 1972.

The author is with Hughes Research Laboratories, Malibu, Calif. 90265.

[1] A more complete listing of reactions for pulsed chemical lasers is given by Basov [74] and for CW chemical lasers by Cool [83].

[2] See also the large number of papers presented at the Int. Symp. Chemical Lasers, Moscow, USSR, Sept. 2–4, 1969.
[3] CW chemical laser operation for 1.8 ms already had been achieved by Airey and McKay [9] in a shock-tube geometry, but Spencer's supersonic flow device created more interest because it could be operated in a true CW fashion for any desired length of time.

Reprinted from *Proc. IEEE*, vol. 61, pp. 414–422, Apr. 1973.

TABLE I

CHEMICAL LASER REACTION SYSTEMS AND
THEIR OUTPUT WAVELENGTHS

System	Reactions	Active Laser Molecule	Strongest Wavelengths, μm
H_2-F_2	$F + H_2 \rightarrow HF^\dagger + H$	HF	2.6 to 3.6
	$H + F_2 \rightarrow HF^\dagger + F$		
H_2-Cl_2	$H + Cl_2 \rightarrow HCl^\dagger + Cl$	HCl	3.5 to 4.1
D_2-F_2	Analogous to HF	DF	3.6 to 5.0
CS_2-O_2	$O + CS_2 \rightarrow CS + SO$	CO	4.9 to 5.7
	$SO + O_2 \rightarrow SO_2 + O$		
	$O + CS \rightarrow CO^\dagger + S$		
	$S + O_2 \rightarrow SO + O$		
DF-CO_2	$F + D_2 \rightarrow DF^\dagger + D$	CO_2	10 to 11
	$D + F_2 \rightarrow DF^\dagger + F$		
	$DF^\dagger + CO_2 \rightarrow DF + CO_2^\dagger$		

TABLE II

TECHNIQUES FOR INITIATING CHEMICAL LASER REACTIONS

Method	Example
UV Photolysis	$h\nu + F_2 \rightarrow 2F$
Electrical	$e + F_2 \rightarrow e + 2F$
Thermal	$F_2 + heat \rightarrow 2F$
Chemical	$NO + F_2 \rightarrow NOF + F$

TABLE III

SOME RECENT MEASUREMENTS OF VIBRATIONAL DISTRIBUTION
PRODUCED BY CHEMICAL REACTIONS

F Atom Reactions		H and D Atom Reactions	
Reaction	References	Reaction	References
$F + H_2$	[14], [15], [93]	$H + Br_2$	[11], [93], [102]
		$D + Br_2$	[102]
$F + D_2$	[15], [16], [17]	$H + Cl_2$	[11], [12], [88], [93]
$F + CH_4$	[16], [93]	$D + Cl_2$	[11]
$F + CD_4$	[16]	$H + F_2$	[13], [93]
$F + CHCl_3$	[18]	$H + HF$	[14]
$F + CDCl_3$	[18]	$H + ClF$	[85]
$F + HI$	[20], [21], [93]	$H + OF_2$	[87]
$F + HCl, HBr, HSiCl_3$	[93]	$H + O_3$	[92]
$F + C_2H_2$	[22]	$H + (SCl_2, S_2Cl_2,$	[88]
$F + $ Other hydrocarbons	[22], [23]	$SOCl_2, SO_2Cl_2)$	

Other Reactions	
Reaction	References
$Cl + HI$	[91]
$(Br, Cl) + (HBr, HI, DI)$	[19]
$O + CS$	[28], [29]
HF elimination	[24], [25]
HCl elimination	[26], [27]

II. CURRENT RESEARCH DIRECTIONS

Chemical laser research since 1969 may be classified into three categories: chemical kinetic studies, exploration of new chemical systems, and laser device research.

A. Chemical Kinetics Studies

This work tends to be a natural development of chemical laser work prior to 1969, in which the fundamental rates for collisional transfer, relaxation, or reaction into specified vibrational levels can be determined by measuring the laser emission. Chemical laser work and laser-motivated measurements have yielded values for many processes important to chemical laser operation, such as the following.

1) Experimental measurements of the vibrational distributions produced by various reactions, some of which are listed in Table III.

2) Experimental rates for vibrational transfer and relaxation in molecular collisions have been obtained for HCl, DCl, HBr, and HI [30], the most extensive work being that of Chen. Until recently, relatively few rates for HF and DF transfer and relaxation processes were known [31].[4] However, device development has motivated a great deal of recent work on HF and DF, both experimental [32], [33] and theoretical [34]; we now have an abundance of data involving various collision partners and temperatures, which will take some time to digest. Space does not permit a thorough discussion of these results here.

3) Spontaneous emission transition probabilities (A-coefficients) have been calculated [35] and measured [36] for the HF and DF molecules, including overtone transitions ($\Delta v > 1$). The overtone lines are beginning to find favor as a diagnostic tool [37], since the laser medium is often optically thin at those wavelengths. Calculations are also being made of optical broadening cross sections [38], which are needed when calculating optical saturation effects [39, appendix III],[5] [40].

B. Exploration of New Chemical Systems

This work is often motivated by the hope of discovering new chemical lasers with desirable wavelengths, efficiency, or operational characteristics. An amazing variety of reactions has been found to yield laser action in the hydrogen halide molecules. For HF alone, at least 19 different hydrogen sources and 29 different fluorine sources have been used successfully [41]. This escalation of reactant species is now beginning to occur for the less efficient CO chemical laser as well [42], [43]. Chain reactions [53], [89], especially those that may incorporate chain branching,[6] are preferred candidates because of their potential high efficiency.

The most interesting and promising of this work is directed toward finding radically new types of lasers. Although this

[4] For a review as of July 1971, see N. Cohen, "A review of rate coefficients for reactions in the H_2–F_2 laser system," Aerospace Corp., Tech. Rep. TR-0172 (2779)-2. A more recent compilation is currently being prepared by the same author.

[5] Experimental values of optical broadening cross sections for hydrogen halide molecules are discussed in [39, refs. [125], [126]].

[6] There were early hopes [7], [86] that chain branching in the H_2–F_2 reaction would lead to chemical lasers requiring little or no initiation energy. However, it has only recently been possible to realize sufficiently rapid reaction velocities in this chemical system to produce laser output exceeding the initiation energy [96], [104]. It now appears that the ClF_3–H_2 chemical laser system [85] exhibits chain branching and could have the potential for even more efficient operation.

TABLE IV
BEST REPORTED PERFORMANCE OF CHEMICAL LASERS, COMPARED WITH GAS-DYNAMIC AND ELECTRICAL CO_2 SYSTEMS

Type of Laser	Reaction System	Output	Volumetric Efficiency		Mass Flow Efficiency (kj/pound)	Chemical Efficiency (%)	Electrical Efficiency (%)	Gain		Operating Pressure (torr)	References
			(w/cfm)	(j/liter)				(% cm^{-1})	(db/m)		
c-w	H_2 + SF_6 or F_2	4500 w	0.5	1.0	136.0	10	∞	8	35	4 to 15	[8], [48]
pulsed	F_2, XeF_4, MoF_6 etc +H_2,CH_4 etc	10 j/pulse	36.0	80.0	140.0	12 (nonchain) 1 to 2 (chain)	>100%	10 to 73	43 to 320	10 to 1000	[7], [53], [69], [97], [105]
c-w transfer	DF-CO_2	560 w	0.26	0.54	25.0	5	∞	2 to 3	9 to 13	5 to 40	[60]
pulsed transfer	DF-CO_2	15 j/pulse	18.0	40.0	40.0	5	<1%	>0.7	>3	10 to 1000	[98], [99], [100]
free burning flame	CS_2-O_2	25 w	0.33	0.7	3.0	2.5	∞	~0.3	~1.4	10 to 40	[61]
Reported performance of nonchemical CO_2 lasers											
c-w (gas dynamic)	CO_2 (CO fuel)	60 000 w	0.7	1.5	2.0	1	∞	0.5 to 1	2 to 4	65	[2]
pulsed (electrical)	CO_2	2000 J/pulse	23.0	50.0	27	∞	24	1 to 2	4 to 9	30 to 760	[54]

work has not significantly impacted device development as yet, there have been some interesting discoveries that warrant further study. Among these are the HF/DF pure rotational laser [5], [84],[7] the HN_3–CO_2 laser system [44], the OH laser [45], and overtone chemical lasers [4]. After a lapse of some years, there is once again a commendable interest in developing a chemical laser using an electronic transition in a molecule [46]. Since the output of such a laser would be in the visible or ultraviolet part of the spectrum, where existing lasers are relatively inefficient, this class of laser could be of tremendous practical importance.[8]

C. Laser Device Research

By far the largest effort in chemical lasers today is being spent on developing improved devices. The reaction systems most studied at present are HF (from H_2, and either SF_6, F_2, or N_2F_4); DF; the DF–CO_2 transfer system; and the CO chemical laser using CS_2 and O_2 as reactants (see Table I). The largest CW output powers reported so far are 4500 W for HF/DF [48], 560 W for DF–CO_2 [49], [60],[9] and 34 W for CO [50], [51]. Maximum pulse energies, achieved with repetition rates usually well below 1 pulse per second (pps), are

10 J for HF/DF [97] and 15 J for DF–CO_2 [52]. The performance of present-day devices is summarized in Table IV and discussed more fully in the following.

1) Pulsed Chemical Lasers: In the past, pulsed chemical laser devices generally suffered from low output energy per unit electrical input. Although they may convert as much as 12 percent of their chemical energy to laser output [53], the most efficient pulsed devices require some form of electrical input to initiate the chemical reaction, either by photolysis or by means of an electrical discharge. Examples of traditional pulsed chemical lasers appear in Figs. 2 and 3. In contrast with the CO_2 electrical laser, which has an electrical efficiency exceeding 20 percent [54], most pulsed chemical lasers produce laser output energy ≤ 1 percent of their electrical input (as quoted in [48]), [55a]. Moreover, the high output energy per unit volume achieved with the DF–CO_2 system [55], [100] has been achieved using flash photolysis initiation only, which has an inherently low electrical efficiency.[10] This poor performance has not yet been significantly improved by exploratory work into such radically different reaction initiation techniques as Jensen's thermal explosion technique [56], Oraevskiy's initiation of N_2F_4–H_2 mixtures by CO_2 laser radiation [57], and Basov's proposed shock-wave and photodissociation-wave initiation [58].

However, results recently obtained promise to make pulsed chemical lasers strong competitors to pulsed electrical lasers. Parker and Stephens [96] have used a technique originated by Judd [103] in which ultraviolet radiation from a low-energy spark source is used to provide a weak background ionization density in a laser mixture (see Fig. 4). This preionization per-

---

[7] Detailed measurements have been carried out on the chemical formation rate of individual rotational states in the reaction Cl+HI; see [91].

[8] Some photodissociation lasers result in laser output on electronic transitions, but these transitions are in atoms rather than molecules, e.g., $CF_3I + h\nu \rightarrow CF_3 + I^*$ (see [6a]) and $IBr + h\nu \rightarrow I + Br^*$ (see [47]). Photodissociation lasers are not covered in this review because they are not chemical lasers within the definition proposed at the beginning of this paper; however, a good early reference with background information on these lasers is R. N. Zare and D. R. Herschbach, "Atomic and molecular fluorescence excited by photodissociation," *Appl. Opt.* (*Chem. Lasers*, suppl. 2), pp. 193–200, 1965.

[9] A multikilowatt chemical transfer laser is currently under construction as well (see [60]).

[10] This objection might not apply if more efficient ultraviolet sources were developed.

Fig. 2. Typical pulsed chemical laser using flash photolysis initiation. The stainless-steel and Teflon gas supply system provides premixed gaseous reagents to the 1-m by 1-cm quartz laser tube. Ultraviolet radiation from the ignitron-triggered flash lamp dissociates molecules within the laser tube to initiate the chemical laser reaction. (Courtesy of Hughes Res. Labs.)

Fig. 3. Single-wavelength repetitively pulsed, electrical discharge chemical laser. Using SF_6, H_2, and He gases, this laser produces 35-W, 1-μs pulses, 60 pulses per second (pps), around 3-μm wavelength. (Courtesy of Hughes Res. Labs.)

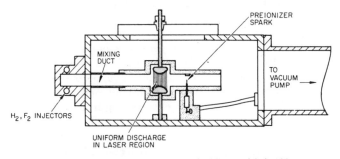

Fig. 4. Schematic view of pulsed chemical laser with >100-percent electrical efficiency. (Courtesy of Hughes Res. Labs.)

mits a uniform electrical discharge to be struck in a premixed H_2–F_2 laser mixture, initiating the chemical reaction with high efficiency. HF laser outputs greater than 17 MJ have been achieved, with a total electrical input of less than 16 MJ. Thus the "electrical efficiency" of this device exceeds 100 percent

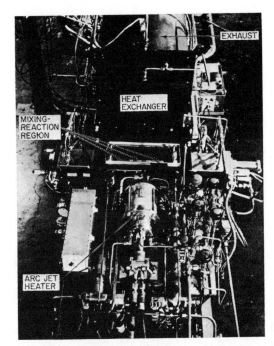

Fig. 5. CW hydrogen fluoride chemical laser. (Courtesy of Aerospace Corp.)

Fig. 6. Side view of the CW HF laser shown in Fig. 5, showing associated hardware. Since this laser is strictly a laboratory device, special precautions are taken to protect personnel from any possible hazards associated with the large flow of fluorine gas through the apparatus. The laser is remotely operated from a blockhouse, using television equipment to monitor the various pressure and flow gauges. (Courtesy of Aerospace Corp.)

(see Fig. 4). Other preionization techniques such as high-energy electron beams [54] could probably be used with similar results. Suchard and Whittier [104] have independently reported 160-percent efficiency in a conventional tube using multiple-pin electrodes, by inserting light baffles to limit the lasing volume and thereby suppress parasitic oscillations caused by wall reflections. The conversion efficiency from chemical energy to laser output in these devices is only 1–2 percent at present, but they are still extremely attractive if they prove to be scalable to higher output energies.

2) CW Chemical Lasers: CW devices in the milliwatt range can be economically constructed in any laboratory; however, chemical lasers above 1 kW in output power tend to be highly engineered, expensive systems (see Figs. 5 and 6).

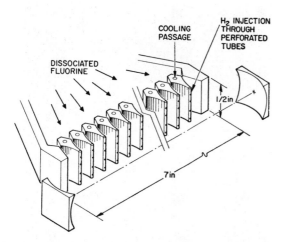

DISSOCIATED FLUORINE

COOLING PASSAGE

H₂ INJECTION THROUGH PERFORATED TUBES

1/2in

7in

Fig. 7. Schematic of mixing and reaction region in supersonic diffusion laser of Fig. 5. Redrawn from [90].

Very good HF/DF laser performance is obtained using supersonic flow and low pressures, so that optically absorbing HF molecules do not remain within the optical cavity after they are deexcited. The optical axis is normal to the gas flow direction and to the mixing interfaces. The $DF-CO_2$ system is similar, except that higher pressures can be used. Typically, an arc jet heater [8] or combustor [59] is used to provide free F atoms by thermal dissociation; their expansion through a supersonic nozzle reduces their translational temperature, which would be detrimental to laser output. These free atoms initiate the chemical laser reaction when they are rapidly mixed with other reagents after expansion (see Fig. 7). Hydrogen fluoride devices of this type can produce about 300 J of laser output per gram of total gas flow [48]; their principal practical limitation at present is their low operating pressure (<10 torr), which makes it difficult to pump the large volumes of gas required. The developing $DF-CO_2$ systems [60] will be less hampered by this limitation; their maximum operating pressures will probably be limited by the need for rapid mixing rather than by relaxation processes. The most important device research in these lasers at present involves raising the gas pressure, or pumping the gas with less expenditure of energy. Work continues on CO chemical lasers based on the $O+CS$ reaction, but power outputs and efficiencies are two orders of magnitude below values for HF devices [101].

The most novel development in CW chemical lasers is the flame laser. CS_2 and O_2 are mixed at low pressures and simply ignited; the flame is self-sustaining and produces CO laser emission. The addition of N_2O to the gas flow improves laser output, for reasons that are not well understood [61]. At present, a 75-cm² flow area produces 25 W of laser output. However, the simplicity of this device makes it very appealing, in addition to the opportunity it affords to study nonequilibrium processes in diffusion-dominated flames.

3) Laser Modeling: Although the rate equations describing chemical laser operation are well known, device analysis is very difficult because of the large numbers of competing kinetic processes and the lack of sufficiently accurate values for the critical fundamental rate constants. Analytical models have been successfully applied to the $F+H_2$ laser system [39], [62]; however, an adequate analysis of most practical devices requires digital computer solution of the rate equations, as carried out by various workers for various types of HCl lasers [63]. There is now a profusion of computer models for the

$H_2(D_2)-F_2$ and $DF-CO_2$ laser systems, both CW and pulsed [49], [64], [65]. Since the entire problem of mixing/reaction/saturation/diffraction tends to be too complicated, each model makes different assumptions and each presumably has its own strong and weak points. Only continued experimental comparisons will reveal which assumptions are most useful for understanding, optimizing, and scaling up devices.

Modeling of other systems, such as CS_2-O_2 and $N_2F_4-H_2$, has just begun because of their chemical complexity [29], [66]. Rather than waiting for two years of rate-constant measurements by numerous researchers, as was necessary to begin to understand the HF system, it would be desirable if experimenters were to make gain and saturation measurements under accurately known conditions of initiation and prereaction, over a wide range of pressures and mixtures.[11] This would enable phenomenological models to be developed, which in turn would point up which few rate constants were really crucial to the reaction system. This type of experiment would be carried out best in a flash photolysis laser, or some other system with premixed reactants and uniform measurable reaction initiation. Thus careful measurements on low-efficiency systems are favored to assist the meaningful modeling of practical high-efficiency chemical laser devices.

4) Mode Selection, Wavelength Control, and Parasitics: A number of effects constrain the optical design of chemical laser devices because of the special properties of the chemical laser medium, especially in nontransfer chemical lasers. Fundamental transverse-mode operation in CW devices is difficult to achieve because of the large Fresnel number of the gain region and the spatial inhomogeneity of the gain in the mixing region. Successful single-mode operation with an unstable resonator has often required throttling back the laser to a low power level to reduce the impact of medium inhomogeneities [67].

Wavelength control, in particular single-line operation, is inherently inefficient in most nontransfer chemical lasers. Because many vibrational levels receive inversion from the chemical reaction, maximum laser output requires utilization of cascade processes between different vibrational levels. Suppressing this cascade by using an intracavity diffraction grating results in greatly decreased laser output (see [68], for example).

Parasitics, or parasitic modes, consist of unwanted stimulated emission that competes with the intended cavity modes. They are particularly a problem in the HF/DF lasers because of their high small-signal optical gain: ~100 percent/cm in both pulsed [69] and CW [70] devices. Various types of parasitics can occur: 1) specular parasitics, or laser oscillations supported by reflections from the walls or other surfaces; 2) diffuse parasitics, nonlocalized modes supported by diffuse reflections [94]; 3) axial parasitics (usually only in high-gain laser amplifiers), caused by backscattering from windows or folding mirrors [95, fig. 2]; 4) diffractive spillage out of the main resonator mode; and 5) superradiance, which may be aided by waveguiding or other wall reflection effects, in which amplified spontaneous emission requires no feedback to compete with the intended laser mode. Parasitic thresholds can be calculated, if the degree of saturation and the mode-filling geometry of the laser oscillator or amplifier are known. Super-

---

[11] Ideally, it would be desirable to make many of these measurements with a single-transverse-mode single-axial-mode laser operating at line center, in a system free from parasitics and superradiance.

radiance has been observed in high-gain pulsed HF lasers [71], and specular parasitics have recently been invoked to help explain experimental measurements in HF devices [72].

III. REFERENCE SOURCES

Some useful review papers on chemical lasers are listed in inverse chronological order as [48] and [73] through [81]. Two of the most useful are the general review by Basov [74] and Moore's excellent general paper on the applications of lasers to chemistry [73]. A more recent bibliography is given by Chester and Hess [39, appendix IV].

In addition, Ames [82] of McDonnell–Douglas Research Laboratories is presently developing a computer data file of the chemical laser literature that will be a useful reference source in the future.

Current device research most frequently appears in *IEEE Journal of Quantum Electronics* and *Applied Physics Letters*, and chemical kinetics studies appear in *Journal of Chemical Physics* and *Chemical Physics Letters*. However, chemical laser papers appear in more than a dozen other journals as well, as indicated in the review papers mentioned.

Note Added in Proof: Three very recent publications deserve special mention. Using a beam of relativistic electrons to initiate laser action in an F_2–H_2 mixture, laser outputs of 150 to 180 percent of the deposited initiation energy have been obtained for yet another independently developed variation of the electrical laser with greater than 100-percent efficiency (V. F. Zharov, V. K. Malinovskii, Yu. S. Neganov, and G. M. Chumak, "On the effectiveness of excitation of laser generation in the F_2–H_2 mixture by a beam of relativistic electrons" (in Russian), Rep. INF 45-72, Institute of Nuclear Physics, Siberian Branch of the Academy of Sciences, Novosibirsk, USSR, 1972).

A large bibliography of chemical lasers through 1971 is now available as a University of California report (B. M. Dobratz, "Chemical lasers: An overview of the literature, 1960–1971," Rep. UCRL-51285, Lawrence Livermore Lab., Livermore, Calif., Sept. 1972), and an extensive review paper with 500 references is also soon to be available (K. Kompa, "Chemical lasers," in *Topics in Current Chemistry/Fortschritte der Chemischen Forschung*, vol. 37. New York: Springer, 1973).

ACKNOWLEDGMENT

The author wishes to thank C. Glaubach for invaluable bibliographical assistance in support of this review. The author is also grateful to the many workers in this field, only a few of whom have been referenced, for giving their advice and discussing their work in advance of publication. The reference list in this paper is not exhaustive, and apologies are offered to those whose papers did not fit into the main text discussion or were inadvertently omitted.

REFERENCES

[1] T. A. Cool, T. J. Falk, and R. R. Stephens, "DF–CO₂ and HF–CO₂ continuous-wave chemical lasers," *Appl. Phys. Lett.*, vol. 15, pp. 318–320, Nov. 15, 1969.
[2] a) E. T. Gerry, "Gasdynamic lasers," *IEEE Spectrum*, vol. 7, pp. 51–58, Nov. 1970.
   b) ——, "The gas dynamic laser," *Laser Focus*, pp. 27–31, Dec. 1970.
[3] a) J. C. Polanyi, "Vibrational–rotational population inversion," *Appl. Opt.* (Chem. Lasers, suppl. 2), pp. 109–127, 1965.
   b) D. R. Herschbach, "Reactive scattering in molecular beams," in *Molecular Beams*, vol. X, *Advances in Chemical Physics*, J. Ross, Ed. New York: Interscience, 1966, pp. 367–387.
   c) J. C. Polanyi, "Dynamics of chemical reactions," in "Molecular Dynamics of the Chemical Reactions of Gases," *Discuss. Faraday Soc.*, vol. 44, 1967, pp. 293–307, esp. pp. 297–298.
[4] a) S. N. Suchard and G. C. Pimentel, "Deuterium fluoride overtone chemical laser," *Appl. Phys. Lett.*, vol. 18, pp. 530–531, June 15, 1971.
   b) F. G. Sadie, P. A. Büger, and O. G. Malan, "Continuous-wave overtone bands in a CS₂–O₂ chemical laser," *J. Appl. Phys.*, vol. 43, pp. 2906–2907, June 1972.
[5] T. F. Deutsch, "Laser emission from HF rotational transitions," *Appl. Phys. Lett.*, vol. 11, pp. 18–20, July 1, 1967.
[6] a) J. V. V. Kasper and G. C. Pimentel, "Atomic iodine photodissociation laser," *Appl. Phys. Lett.*, vol. 5, pp. 231–233, Dec. 1, 1964.
   b) ——, "HCl chemical laser," *Phys. Rev. Lett.*, vol. 14, pp. 352–354, Mar. 8, 1965.
[7] a) O. M. Batovskii, G. K. Vasil'ev, E. F. Makarov, and V. L. Tal'roze, "Chemical laser operating on branched chain reaction of fluorine with hydrogen," *JETP Lett.*, vol. 9, pp. 200–201, June 5, 1969.
   b) N. G. Basov, L. V. Kulakov, E. P. Markin, A. I. Nikitin, and A. N. Oraevskii, "Emission spectrum of a chemical laser using an H₂–F₂ mixture," *JETP Lett.*, vol. 9, pp. 375–378, June 5, 1969.
   c) N. G. Basov, E. P. Markin, A. I. Nikitin, and A. N. Oraevskii, "Branching reactions and chemical lasers," presented at the Symp. Chemical Lasers, St. Louis, Mo., May 22–24, 1969.
[8] a) D. J. Spencer, T. A. Jacobs, H. Mirels, and R. W. F. Gross, "Continuous-wave chemical laser," *Int. J. Chem. Kinet.* (Commun.), vol. 1, pp. 493–494, Sept. 1969; Addendum, *Int. J. Chem. Kinet.*, vol. 2, p. 337, July 1970.
   b) D. J. Spencer, H. Mirels, T. A. Jacobs, and R. W. F. Gross, "Preliminary performance of a CW chemical laser," *Appl. Phys. Lett.*, vol. 16, pp. 235–237, Mar. 15, 1970.
   c) D. J. Spencer, H. Mirels, and T. A. Jacobs, "Comparison of HF and DF continuous chemical lasers—I: Power," *Appl. Phys. Lett.*, vol. 16, pp. 384–386, May 15, 1970.
   d) M. A. Kwok, R. R. Giedt, and R. W. F. Gross, "Comparison of HF and DF continuous chemical lasers—II: Spectroscopy," *Appl. Phys. Lett.*, vol. 16, pp. 386–387, May 15, 1970.
   e) D. J. Spencer, H. Mirels, and T. A. Jacobs, "Initial performance of a C.W. chemical laser," *Opto-Electron.*, vol. 2, pp. 155–160, 1970.
   f) H. Mirels and D. J. Spencer, "Power and efficiency of a continuous HF chemical laser," *IEEE J. Quantum Electron.*, vol. QE-7, pp. 501–507, Nov. 1971.
   g) D. J. Spencer, D. A. Durran, and H. A. Bixler, "Continuous-chemical-laser cavity studies," *Appl. Phys. Lett.*, vol. 20, pp. 164–167, Feb. 15, 1972.
   h) D. J. Spencer, H. Mirels, and D. A. Durran, "Performance of a cw HF chemical laser with N₂ or He diluent," *J. Appl. Phys.*, vol. 43, pp. 1151–1157, Mar. 1972.
[9] J. R. Airey and S. F. McKay, "A supersonic mixing chemical laser," *Appl. Phys. Lett.*, vol. 15, pp. 401–403, Dec. 15, 1969.
[10] a) T. A. Cool, R. R. Stephens, and T. J. Falk, "A continuous-wave chemically excited CO₂ laser," *Int. J. Chem. Kinet.*, vol. 1, pp. 495–497, Sept. 1969.
   b) T. A. Cool and R. R. Stephens, "Chemical laser by fluid mixing," *J. Chem. Phys.* (Commun.), vol. 51, pp. 5175–5176, Dec. 1969.
   c) ——, "Efficient purely chemical CW laser operation," *Appl. Phys. Lett.*, vol. 16, pp. 55–58, Jan. 15, 1970.
[11] a) P. D. Pacey and J. C. Polanyi, "Energy distribution among reaction products—III: The method of measured relaxation applied to H+Cl₂," *Appl. Opt.*, vol. 10, pp. 1725–1737, Aug. 1971.
   b) K. G. Anlauf, D. S. Horne, R. G. Macdonald, J. C. Polanyi, and K. B. Woodall, "Energy distribution among reaction products—V: H+X₂(X≡Cl, Br), D+Cl₂," *J. Chem. Phys.*, vol. 57, pp. 1561–1574, Aug. 15, 1972.
[12] P. H. Corniel, Ph.D. dissertation, Univ. of California, Berkeley, 1967.
[13] N. Jonathan, C. M. Melliar-Smith, and D. H. Slater, "Initial vibrational energy level populations resulting from the reaction H+F₂ as studied by infrared chemiluminescence," *J. Chem. Phys.* (Commun.), vol. 53, pp. 4396–4397, Dec. 1, 1970.
[14] J. C. Polanyi and D. C. Tardy, "Energy distribution in the exothermic reaction F+H₂ and the endothermic reaction HF+H," *J. Chem. Phys.* (Commun.), vol. 51, pp. 5717–5719, Dec. 15, 1969.
[15] J. C. Polanyi and K. B. Woodall, "Energy distribution among reaction products—VI: F+H₂, D₂," *J. Chem. Phys.*, vol. 57, pp. 1574–1586, Aug. 15, 1972.
[16] J. H. Parker and G. C. Pimentel, "Vibrational energy distribution through chemical laser studies—I: Fluorine atoms plus hydrogen or methane," *J. Chem. Phys.*, vol. 51, pp. 91–96, July 1, 1969.
[17] T. P. Schafer et al., "Crossed molecular beam study of F+D₂," *J. Chem. Phys.* (Commun.), vol. 53, pp. 3385–3387, Oct. 15, 1970.
[18] J. H. Parker and G. C. Pimentel, "Vibrational energy distribution through chemical laser studies—II: Fluorine atoms plus chloroform," *J. Chem. Phys.*, vol. 55, pp. 857–861, July 15, 1971.
[19] D. H. Maylotte, J. C. Polanyi, and K. B. Woodall, "Energy dis-

tribution among reaction products—IV: $X+HY(X \equiv Cl,$ Br; $Y \equiv Br,$ I), $Cl+DI$," *J. Chem. Phys.*, vol. 57, pp. 1547–1560, Aug. 15, 1972.

[20] N. Jonathan, C. M. Melliar-Smith, S. Okuda, D. H. Slater, and D. Timlin, "Initial vibrational energy level distributions determined by infra-red chemiluminescence—II: The reaction of fluorine atoms with hydrogen halides," *Mol. Phys.*, vol. 22, pp. 561–574, 1971.

[21] R. D. Coombe, G. C. Pimentel, and M. J. Berry, "Chemical laser study of the reaction $F+HI$ by the equal-gain technique," presented at the 3rd Conf. Chem. Mol. Lasers, St. Louis, Mo., May 1–3, 1972.

[22] M. J. Berry, "Energy partitioning in fluorine-atom bimolecular reactions with saturated and unsaturated hydrocarbons," presented at the 3rd Conf. Chem. Mol. Lasers, St. Louis, Mo., May 1–3, 1972.

[23] J. M. Parson and Y. T. Lee, "Crossed molecular beam study of $F+C_2H_4, C_2D_4$," *J. Chem. Phys.*, vol. 56, pp. 4658–4666, May 1, 1972.

[24] T. D. Padrick and G. C. Pimentel, "Hydrogen fluoride elimination chemical laser from N, N-difluoromethylamine," *J. Chem. Phys.*, vol. 54, pp. 720–723, Jan. 15, 1971.

[25] Papers by P. R. Poole and G. C. Pimentel; J. L. Roebber and G. C. Pimentel; and E. Cuellar-Ferreira and G. C. Pimentel, presented at the 3rd Conf. Chem. Mol. Lasers, St. Louis, Mo., May 1–3, 1972.

[26] M. J. Molina and G. C. Pimentel, "Vibrational energy distribution from chemical laser studies," presented at the 3rd Conf. Chem. Mol. Lasers, St. Louis, Mo., May 1–3, 1972.

[27] ——, "Tandem chemical laser measurements of vibrational energy distribution in the dichloroethylene photoelimination reactions," *J. Chem. Phys.*, vol. 56, pp. 3988–3993, Apr. 15, 1972.

[28] S. Tsuchiya, N. Nielsen, and S. H. Bauer, "Relative populations of vibrationally excited CO produced in pulse-discharged CS_2+O_2+He mixtures," presented at the 3rd Conf. Chem. Mol. Lasers, St. Louis, Mo., May 1–3, 1972.

[29] a) G. Hancock, C. Morley, and I. W. M. Smith, "Vibrational excitation of CO in the reaction: $O+CS \rightarrow CO+S$," *Chem. Phys. Lett.*, vol. 12, pp. 193–196, Dec. 1, 1971.
b) K. D. Foster and R. D. Suart, "Studies of CS_2/O_2 chemical lasers," presented at the 3rd Conf. Chem. Mol. Lasers, St. Louis, Mo., May 1–3, 1972.
c) K. D. Foster, "Initial distribution of CO† from the reaction $O+CS \rightarrow CO†+S$," *J. Chem. Phys.*, vol. 57, pp. 2451–2455, Sept. 15, 1972.

[30] a) H.-L. Chen, J. C. Stephenson, and C. B. Moore, "Laser-excited vibrational fluorescence of HCl and the $HCl-CO_2$ laser," *Chem. Phys. Lett.*, vol. 2, pp. 593–596, Dec. 1968.
b) A. Henry, L. Doyennette, and M. Margottin-Maclou, "Détermination des coefficients de diffusion et d'accomodation aux parois des molécules de gaz chlorhydrique excitées dans le premier état vibrationnel," *C. R. Acad. Sci.*, ser. B, vol. 271, pp. 634–637, Sept. 28, 1970.
c) H.-L. Chen and C. B. Moore, "Vibration→rotation energy transfer in hydrogen chloride," *J. Chem. Phys.*, vol. 54, pp. 4072–4080, May 1, 1971.
d) ——, "Vibration→vibration energy transfer in hydrogen chloride mixtures," *J. Chem. Phys.*, vol. 54, pp. 4080–4084, May 1, 1971.
e) M. Margottin-Maclou, L. Doyennette, and L. Henry, "Relaxation of vibrational energy in CO, HCl, CO_2, and N_2O," *Appl. Opt.*, vol. 10, pp. 1768–1780, Aug. 1971.
f) V. I. Gorshkov *et al.*, "Investigation of vibrational relaxation in the hydrogen chloride chemical laser," *Appl. Opt.*, vol. 10, pp. 1781–1785, Aug. 1971.
g) H.-L. Chen, "Vibrational relaxation of hydrogen bromide in gaseous hydrogen halide mixtures," *J. Chem. Phys.*, vol. 55, pp. 5551–5556, Dec. 15, 1971.
h) ——, "Vibration-to-vibration energy transfer in hydrogen bromide mixtures," *J. Chem. Phys.*, vol. 55, pp. 5557–5560, Dec. 15, 1971.
i) M. Y.-D. Chen and H.-L. Chen, "Vibration-to-rotation energy transfer in HBr and DBr mixtures," *J. Chem. Phys.*, vol. 56, pp. 3315–3317, Apr. 1, 1972.
j) H.-L. Chen and B. M. Hopkins, "Vibrational relaxation of HCl $(v=2)$ state," presented at the 3rd Conf. Chem. Mol. Lasers, St. Louis, Mo., May 1–3, 1972.
k) L. Henry, "Chemical laser probing of reactions," presented at the 3rd Conf. Chem. Mol. Lasers, St. Louis, Mo., May 1–3, 1972.
l) I. Burak, Y. Noter, A. M. Ronn, and A. Szöke, "Vibration-vibration energy transfer in gaseous HBr," *Chem. Phys. Lett.*, vol. 16, pp. 306–309, Oct. 1, 1972.

[31] a) W. C. Solomon, J. A. Blauer, F. C. Jaye, and J. G. Hnat, "The vibrational excitation of hydrogen fluoride behind incident shock waves," *Int. J. Chem. Kinet.*, vol. 3, pp. 215–222, May 1971.
b) J. F. Bott and N. Cohen, "Shock tube measurements of the vibrational relaxation of DF," presented at the Sept. 1971 Meeting of the Amer. Chem. Soc.
c) H. K. Shin, "Vibration–rotation–translation energy transfer in

HF–HF and DF–DF," *Chem. Phys. Lett.*, vol. 10, pp. 81–85, July 1, 1971.
d) G. K. Vasil'ev, E. F. Makarov, V. G. Papin, and V. L. Tal'roze, "Spectroscopic investigation of radiative and nonradiative relaxation of vibrationally excited HF molecules" (in Russian), *Dokl. Akad. Nauk SSSR*, vol. 191, pp. 1077–1080, May 1970.
e) J. R. Airey and S. F. Fried, "Vibrational relaxation of hydrogen fluoride," *Chem. Phys. Lett.*, vol. 8, pp. 23–26, Jan. 1, 1971.
f) G. K. Vasil'ev, E. F. Makarov, V. G. Papin, and V. L. Tal'roze, "An investigation of vibrational energy transfer from HF and DF molecules to CO_2 molecules," *Sov. Phys.—JETP*, vol. 34, pp. 51–52, Jan. 1972.

[32] a) N. G. Basov *et al.*, "Spectra of stimulated emission in the hydrogen–fluorine reaction process and energy transfer from DF to CO_2," *Appl. Opt.*, vol. 10, pp. 1814–1820, Aug. 1971.
b) J. K. Hancock and W. H. Green, "Laser-excited vibrational relaxation studies of hydrogen fluoride," *J. Chem. Phys.*, vol. 56, pp. 2474–2475, Mar. 1, 1972.
c) J. A. Blauer, W. C. Solomon, and T. W. Owens, "Vibrational excitation of deuterium fluoride and hydrogen fluoride by atomic and diatomic species," *Int. J. Chem. Kinet.*, vol. 4, pp. 293–306, May 1972.
d) J. C. Stephenson, J. Finzi, and C. B. Moore, "Vibration→vibration energy transfer in CO_2–hydrogen halide mixtures," *J. Chem. Phys.*, vol. 56, pp. 5214–5221, June 1, 1972.
e) R. R. Stephens and T. A. Cool, "Vibrational energy transfer and de-excitation in the HF, DF, $HF-CO_2$, and $DF-CO_2$ systems," *J. Chem. Phys.*, vol. 56, pp. 5863–5878, June 15, 1972.
f) R. M. Osgood, Jr., A. Javan, and P. B. Sackett, "Measurement of vibration–vibration energy transfer time in HF gas," *Appl. Phys. Lett.*, vol. 20, pp. 469–472, June 15, 1972.
g) J. F. Bott, "HF vibrational relaxation measurements using the combined shock tube–laser-induced fluorescence technique," *J. Chem. Phys.*, vol. 57, pp. 96–102, July 1, 1972.
h) J. R. Airey and I. W. M. Smith, "Quenching of infrared chemiluminescence: Rates of energy transfer from HF $(v \leq 5)$ to CO_2 and HF, and from DF $(v \leq 3)$ to CO_2 and HF," *J. Chem. Phys.*, vol. 57, pp. 1669–1676, Aug. 15, 1972.
i) J. A. Blauer, W. C. Solomon, L. H. Sentman, and T. W. Owens, "Catalytic efficiencies of H_2O, D_2O, NO, and HCl in the vibrational relaxation of HF and DF," *J. Chem. Phys.*, vol. 57, pp. 3277–3281, Oct. 15, 1972.
j) H. K. Shin, "De-excitation of $CO_2(00°1)$ by hydrogen fluoride," *J. Chem. Phys.*, vol. 57, pp. 3484–3490, Oct. 15, 1972.

[33] Papers by J. Bott; T. Just and G. Rimpel; M. A. Kwok; W. H. Green and J. K. Hancock; S. Fried; J. V. Parker; J. J. Hinchen; P. Gensel, K. L. Kompa, and J. R. MacDonald; L. D. Hess; S. Marcus and R. J. Carbone; and R. M. Osgood, Jr., A. Javan, and P. Sackett, presented at the 3rd Conf. Chem. Mol. Lasers, St. Louis, Mo., May 1–3, 1972.

[34] a) H. K. Shin, "Temperature dependence of the probability of vibrational energy transfer between HF and F," *Chem. Phys. Lett.*, vol. 14, pp. 64–69, May 1, 1972.
b) Papers by G. C. Berend and R. L. Thommarson; H. J. Kolker; and J. D. Kelley, presented at the 3rd Conf. Chem. Mol. Lasers, St. Louis, Mo., May 1–3, 1972.

[35] R. E. Meredith and F. G. Smith, "Computation of vibration–rotation matrix elements for diatomic molecules" (vol. I), Rep. 84130-39-T (I), May 1971; and "Computation of electric dipole matrix elements for hydrogen fluoride and deuterium fluoride" (vol. II), Rep. 84130-39-T (II), Nov. 1971, Willow Run Labs., Univ. of Michigan, Ann Arbor.

[36] D. I. Rosen, R. N. Sileo, and T. A. Cool, "A spectroscopic study of CW chemical lasers," presented at the 3rd Conf. Chem. Mol. Lasers, St. Louis, Mo., May 1–3, 1972.

[37] a) F. N. Mastrup and A. S. Whiteman, "Diagnostic techniques for HF/DF laser cavities," presented at the 3rd Conf. Chem. Mol. Lasers, St. Louis, Mo., May 1–3, 1972.
b) J. J. Hinchen, "Vibrational relaxation of hydrogen fluoride and deuterium fluoride," presented at the 3rd Conf. Chem. Mol. Lasers, St. Louis, Mo., May 1–3, 1972.

[38] R. E. Meredith, private communication, June 21, 1971.

[39] A. N. Chester and L. D. Hess, "Study of the HF chemical laser by pulse-delay measurements," *IEEE J. Quantum Electron.*, vol. QE-8, pp. 1–13, Jan. 1972.

[40] B. M. Shaw and R. J. Lovell, "Foreign-gas broadening of HF by CO_2," *J. Opt. Soc. Amer.*, vol. 59, pp. 1598–1601, Dec. 1969. See references.
b) G. Bachet and R. Coulon, "Spectrometre pour l'infrarouge lointain elargissement par l'argon de la premiere raie de rotation de HF gazeux a 40 cm^{-1}," *J. Quant. Spectrosc. Radiat. Transfer*, vol. 11, pp. 1827–1837, Dec. 1971.
c) R. E. Meredith, "A new method for the direct measurement of spectral line strengths and widths," *J. Quant. Spectrosc. Radiat. Transfer*, vol. 12, pp. 455–484, Apr. 1972.
d) ——, "Strengths and widths in the first overtone band of hydro-

gen fluoride," *J. Quant. Spectrosc. Radiat. Transfer*, vol. 12, pp. 485–503, Apr. 1972.

e) P. Varanasi, S. K. Sarangi, and G. D. T. Tejwani, "Line shape parameters for HCl and HF in a CO_2 atmosphere," *J. Quant. Spectrosc. Radiat. Transfer*, vol. 12, pp. 857–872, May 1972.

f) D. F. Smith, "Molecular properties of hydrogen fluoride," in *Proc. 2nd UN Geneva Conf.* London, England: Pergamon, pp. 130–138.

[41] a) D. Ames, private communication, Feb. 1972.

b) S. M. King, M. A. Kwok, D. Taylor, and S. W. Mayer, "New compounds for use in HF lasers," Aerospace Corp., El Segundo, Calif., Tech. Rep. TR-0172 (9210-02)-2, June 30, 1972.

c) M. J. Berry, "A comparison of photolytic fluorine-atom sources for chemical laser studies," *Chem. Phys. Lett.*, vol. 15, pp. 269–273, Aug. 1, 1972.

[42] M. C. Lin and K. E. Brus, "Chemical CO laser from the $O(^1D)+C_3O_2(^1\Sigma_g^+)\rightarrow 3CO(^1\Sigma^+)$ reaction," *J. Chem. Phys.* (Commun.), vol. 54, pp. 5423–5424, June 15, 1971.

[43] a) L. E. Brus and M. C. Lin, "Chemical CO laser produced from flash-initiated $O_3 + XCN$ (X = Cl, Br, I, and CN) systems," presented at the 3rd Conf. Chem. Mol. Lasers, St. Louis, Mo., May 1–3, 1972.

b) S. K. Searles and N. Djeu, "Flame laser studies (C_2H_2–O_2 combustion)," presented at the 3rd Conf. Chem. Mol. Lasers, St. Louis, Mo., May 1–3, 1972.

c) J. D. Barry, W. E. Boney, J. E. Brandelik, D. M. Mulder, and J. K. Woessner, "CO laser action by C_2H_2 oxidation," *Appl. Phys. Lett.*, vol. 20, pp. 243–244, Apr. 1, 1972.

[44] a) N. G. Basov, V. V. Gromov, Yu. L. Koshelev, Yu. P. Markin, and A. N. Oraevskii, "Stimulated emission during detonation of HN_3 in CO_2," *JETP Lett.*, vol. 10, pp. 2–4, July 5, 1969.

b) M. S. Dzhidzhoev, M. I. Pimenov, V. G. Platonenko, Yu. V. Filippov, and R. V. Khokhlov, "Creation of a population inversion in polyatomic molecules through the energy of chemical reactions," *Sov. Phys.—JETP*, vol. 30, pp. 225–229, Feb. 1970.

[45] a) E. Vietzke, H. I. Schiff, and K. H. Welge, "Stimulated infrared emission of OH* produced by a pulsed electric discharge," *Chem. Phys. Lett.*, vol. 12, pp. 429–430, Dec. 1971.

b) T. W. Ducas, L. D. Geoffrion, R. M. Osgood, Jr., and A. Javan, "Observation of laser oscillation in pure rotational transitions of OH and OD free radicals," *Appl. Phys. Lett.*, vol. 21, pp. 42–44, July 1, 1972.

[46] a) R. S. Anderson, R. A. McFarlane, and G. J. Wolga, "Laser motivated chemiluminescence studies in metal oxides and nitrides," presented at the 3rd Conf. Chem. Mol. Lasers, St. Louis, Mo., May 1–3, 1972.

b) F. N. Mastrup, private communication, Feb. 1972.

[47] C. R. Guiliano and L. D. Hess, "Reversible photodissociative laser system," *J. Appl. Phys.*, vol. 40, pp. 2428–2430, May 1969.

[48] A. N. Chester, "Chemical lasers: A status report," *Laser Focus*, pp. 25–30, Nov. 1971.

[49] T. J. Falk, "Parametric studies of the DF–CO_2 chemical transfer laser," presented at the 3rd Conf. Chem. Mol. Lasers, St. Louis, Mo., May 1–3, 1972.

[50] L. R. Boedeker, J. A. Shirley, and B. R. Bronfin, "Arc-excited flowing CO chemical laser," *Appl. Phys. Lett.*, vol. 21, pp. 247–249, Sept. 15, 1972.

[51] ——, "An arc-excited flowing CO chemical laser," *IEEE J. Quantum Electron.* (Digest of Technical Papers), vol. QE-8, pp. 582–583, June 1972.

[52] N. G. Basov, S. I. Zavorotnyi, E. P. Markin, A. I. Nikitin, and A. N. Oraevskii, "Pulsed chemical high pressure laser using the mixture $D_2 + F_2 + CO_2$," *JETP Lett.*, vol. 15, pp. 93–94, Feb. 5, 1972.

[53] L. D. Hess, "Pulsed laser emission chemically pumped by the chain reaction between hydrogen and fluorine," *J. Chem. Phys.*, vol. 55, pp. 2466–2473, Sept. 1, 1971.

[54] a) J. D. Daugherty, E. R. Pugh, and D. H. Douglas-Hamilton, "A stable, scalable high pressure gas discharge as applied to the CO_2 laser," presented at the 24th Annu. Gaseous Electronics Conf., Gainesville, Fla., Oct. 5–8, 1971.

b) *Physics Today*, vol. 25, pp. 17–19, Jan. 1972.

[55] a) >17 J/l reported by T. O. Poehler, M. Shandor, and R. E. Walker, "A high pressure pulsed CO_2 chemical transfer laser," *Appl. Phys. Lett.*, vol. 20, pp. 497–499, June 15, 1972; also presented at the 3rd Conf. Chem. Mol. Lasers, St. Louis, Mo., May 1–3, 1972.

b) 20 to 40 J/l estimated by S. N. Suchard (postdeadline paper), presented at the 3rd Conf. Chem. Mol. Lasers, St. Louis, Mo., May 1–3, 1972.

[56] a) R. J. Jensen and W. W. Rice, "Thermally initiated HF chemical laser," *Chem. Phys. Lett.*, vol. 7, pp. 214–216, Jan. 15, 1971.

b) W. W. Rice and R. J. Jensen, "Evaluation of chemical lasers based on ClN_3," presented at the 3rd Conf. Chem. Mol. Lasers, St. Louis, Mo., May 1–3, 1972.

[57] a) A. N. Oraevskii, private communication. The technique is discussed in the following:

b) N. G. Basov, E. P. Markin, A. N. Oraevskii, A. V. Pankratov, and A. N. Skachkov, "Stimulation of chemical processes by infrared laser radiation," *JETP Lett.*, vol. 14, pp. 165–167, Aug. 20, 1971.

c) J. L. Lyman and R. J. Jensen, "CO_2 laser induced explosions of mixtures of N_2F_4 and H_2," presented at the 3rd Conf. Mol. Lasers, St. Louis, Mo., May 1–3, 1972.

d) C. E. Turner, Jr., and Y. L. Pan, "Effects of intense CO_2 radiation on pulsed HF chemical laser systems," presented at the 3rd Conf. Mol. Lasers, St. Louis, Mo., May 1–3, 1972.

e) J. L. Lyman and R. J. Jensen, "Laser induced dissociation of N_2F_4," *Chem. Phys. Lett.*, vol. 13, pp. 421–424, Mar. 15, 1972.

[58] N. G. Basov, V. I. Igoshin, Yu. P. Markin, and A. N. Oraevskii, "Dynamics of chemical lasers," *Sov. J. Quantum Electron.*, vol. 1, pp. 119–134, Sept.–Oct. 1971.

[59] a) R. A. Meinzer, "A continuous-wave combustion laser," *Int. J. Chem. Kinet.* (Commun.), vol. 2, p. 335, July 1970.

b) F. N. Mastrup, J. Broadwell, and T. Jacobs, "Experimental investigation and description of an HF-combustion laser cavity," presented at the 3rd Conf. Mol. Lasers, St. Louis, Mo., May 1–3, 1972.

c) R. A. Meinzer, "HF–DF CW combustion-mixing chemical laser," presented at the 3rd Conf. Mol. Lasers, St. Louis, Mo., May 1–3, 1972.

[60] a) T. A. Cool, "Transfer chemical lasers: A review of recent research," presented at the 3rd Conf. Mol. Lasers, St. Louis, Mo., May 1–3, 1972.

b) T. J. Falk, "Parametric studies of the DF–CO_2 chemical transfer laser," Cornell Aeronautical Lab., Inc., Buffalo, N. Y., Tech. Rep. AFWL-TR-71-96, June 1972.

[61] a) H. S. Piloff, S. K. Searles, and N. Djeu, "CW CO laser from the CS_2–O_2 flame," *Appl. Phys. Lett.*, vol. 19, pp. 9–11, July 1, 1971.

b) S. K. Searles and N. Djeu, "Characteristics of a CW CO laser resulting from a CS_2–O_2 additive flame," *Chem. Phys. Lett.*, vol. 12, pp. 53–56, Dec. 1, 1971.

c) K. D. Foster and G. H. Kimbell, *Fourteenth International Symposium on Combustion*, in press.

d) M. J. Linevsky and R. A. Carabetta, "Combustion lasers," General Electric Co., Space Sciences Lab., Valley Forge, Pa., Final Rep., to be published Feb. 1973.

e) ——, "CW laser power from carbon bisulfide flames," *Appl. Phys. Lett.*, to be published.

[62] L. D. Hess, "Enhanced initiation and high pressure operation of the HF chemical laser," presented at the 3rd Conf. Mol. Lasers, St. Louis, Mo., May 1–3, 1972.

[63] a) J. R. Airey, "Cl+HBr pulsed chemical laser: A theoretical and experimental study," *J. Chem. Phys.*, vol. 52, pp. 156–167, Jan. 1, 1970.

b) N. Cohen, T. A. Jacobs, G. Emanuel, and R. L. Wilkins, "Chemical kinetics of hydrogen halide lasers—1: The H_2–Cl_2 system," *Int. J. Chem. Kinet.*, vol. 1, pp. 551–569, Nov. 1969; Erratum, *Int. J. Chem. Kinet.*, vol. 2, p. 339, July 1970.

c) V. I. Igoshin and A. N. Oraevskii, "Kinetics of HCl chemical lasers," presented at the Int. Symp. Chemical Lasers, Moscow, USSR, Sept. 2–4, 1969.

d) B. F. Gordiets, A. I. Osipov, and L. A. Shelepin, "Kinetics of vibrational exchange in molecules. Amplification of radiation in hydrogen halides by electric or chemical pumping," *Sov. Phys.—JETP*, vol. 32, pp. 334–341, Feb. 1971.

[64] a) R. Hofland and H. Mirels, "Flame-sheet analysis of C.W. diffusion-type chemical lasers—I: Uncoupled radiation," *AIAA J.*, vol. 10, pp. 420–428, Apr. 1972.

b) ——, "Flame-sheet analysis of C.W. diffusion-type chemical lasers—II: Coupled radiation," *AIAA J.*, vol. 10, pp. 1271–1280, Oct. 1972.

c) R. L. Kerber, G. Emanuel, and J. S. Whittier, "Computer modeling and parametric study for a pulsed $H_2 + F_2$ laser," *Appl. Opt.*, vol. 11, pp. 1112–1123, May 1972.

d) G. Emanuel, "Gain saturation in a laser amplifier driven by a CW oscillator," *J. Quant. Spectrosc. Radiat. Transfer*, vol. 12, pp. 913–924, May 1972.

e) H. Mirels, R. Hofland, and W. S. King, "Simplified model of CW diffusion-type chemical laser," AIAA Paper 72-145.

f) W. S. King and H. Mirels, "Numerical study of a diffusion type chemical laser," AIAA Paper 72-146.

[65] a) L. D. Hess and R. R. Stephens, "DF–CO_2 chemical transfer laser," presented at the VII Int. Quantum Electron. Conf., Montreal, Canada, May 8–11, 1972.

b) Papers by R. L. Kerber, N. Cohen, and G. Emanuel; R. L. Kerber, G. Emanuel, and J. S. Whittier; S. N. Suchard, R. L. Kerber, G. Emanuel, and J. S. Whittier; G. Emanuel, N. Cohen, and T. A. Jacobs; J. V. Parker; D. L. Bullock and I. M. Green; W. P. Curry; T. O. Poehler, M. Shandor, and R. E. Walker; J. Wilson, D. Northam, and P. Lewis, presented at the 3rd Conf. Mol. Lasers, St. Louis, Mo., May 1–3, 1972.

[66] L. S. Bender, R. Tripodi, R. J. Hall, and B. R. Bronfin, "Kinetic

theory of CO chemical lasers," presented at the 3rd Conf. Mol. Lasers, St. Louis, Mo., May 1–3, 1972.

[67] R. Chodzko, private communication.

[68] W. Q. Jeffers, "Single line operation of CO chemical lasers," presented at the 3rd Conf. Mol. Lasers, St. Louis, Mo., May 1–3, 1972.

[69] R. J. Carbone and S. Marcus, private communication. Also S. Marcus and R. J. Carbone, "Gain and relaxation studies in transversely excited HF lasers," *IEEE J. Quantum Electron.*, vol. QE-8, pp. 651–655, July 1972.

[70] F. Mastrup, private communication.

[71] J. Goldhar, R. M. Osgood, Jr., and A. Javan, "Observation of intense superradiant emission in the high-gain infrared transitions of HF and DF molecules," *Appl. Phys. Lett.*, vol. 18, pp. 167–169, Mar. 1, 1971.

[72] S. N. Suchard, R. L. Kerber, G. Emanuel, and J. S. Whittier, "Effect of H_2 pressure on pulsed H_2+F_2 laser; experiments and theory," presented at the 3rd Conf. Mol. Lasers, St. Louis, Mo., May 1–3, 1972.

[73] C. B. Moore, "Lasers in chemistry," *Ann. Rev. Phys. Chem.*, vol. 22, pp. 387–428, 1971.

[74] N. G. Basov, V. I. Igoshin, Yu. P. Markin, and A. N. Oraevskii, "Dynamics of chemical lasers," *Sov. J. Quantum Electron.*, vol. 1, pp. 119–134, Sept.–Oct. 1971.

[75] S. H. Bauer, "Recent developments in chemical lasers," in *High Frequency Generation and Amplification: Devices and Applications; Proc. Third Biennial Cornell Electrical Engineering Conf.* (Cornell Univ., Ithaca, N. Y., Aug. 17–19, 1971), pp. 45–56.

[76] J. C. Polanyi, "Nonequilibrium processes," *Appl. Opt.*, vol. 10, pp. 1717–1724, Aug. 1971.

[77] M. S. Dzhidzhoev, V. T. Plantonenko, and R. V. Khokhlov, "Chemical lasers," *Sov. Phys.—Usp.*, vol. 13, pp. 247–268, Sept.–Oct. 1970.

[78] K. L. Kompa, "Chemische laser" (in German), *Chem. Ing. Tech.*, vol. 42, pp. 573–579, May 1970.

[79] J. R. Airey, "Report on the international symposium on chemical lasers" (Moscow, USSR, Sept. 2–4, 1969), *Int. J. Chem. Kinet.*, vol. 2, pp. 65–68, 1970.

[80] D. L. Ball, "Chemical laser research," *Air Force Res. Rev.*, pp. 1–4, Nov.–Dec. 1970.

[81] G. C. Pimentel, "Infrared study of transient molecules in chemical lasers," *Pure Appl. Chem.*, vol. 18, pp. 275–284, 1969.

[82] C. E. Wiswall, D. P. Ames, and T. J. Menne, "Chemical laser device bibliography," *IEEE J. Quantum Electron.*, vol. QE-9, pp. 181–188, Jan. 1973.

[83] T. A. Cool, R. R. Stephens, and J. A. Shirley, "HCl, HF and DF partially inverted CW chemical lasers," *J. Appl. Phys.*, vol. 41, pp. 4038–4050, Sept. 1970.

[84] a) N. Skribanowitz, I. P. Herman, R. M. Osgood, Jr., M. S. Feld, and A. Javan, "Anisotropic ultrahigh gain emission observed in rotational transitions in optically pumped HF gas," *Appl. Phys. Lett.*, vol. 20, pp. 428–431, June 1, 1972.
b) J. C. Polanyi and K. B. Woodall, "Mechanism of rotational relaxation," *J. Chem. Phys.*, to be published. See references.

[85] O. D. Krogh and G. C. Pimentel, "Chemical lasers from the reactions of ClF and ClF_3 with H_2 and CH_4: A possible chain-branching chemical laser," *J. Chem. Phys.*, vol. 56, pp. 969–975, Jan. 15, 1972.

[86] a) N. G. Basov, E. P. Markin, A. I. Nikitin, and A. N. Oraevskii, "Branching reactions and chemical lasers," *IEEE J. Quantum Electron.* (Digest of Technical Papers), vol. QE-6, pp. 183–184, Mar. 1970.
b) A. N. Oraevskii, "A chemical laser based on branched reactions," *Sov. Phys.—JETP*, vol. 28, pp. 744–747, Apr. 1969.

[87] M. J. Perona, "Infrared chemiluminescence from the reaction of hydrogen atoms with oxygen difluoride," *J. Chem. Phys.*, vol. 54, pp. 4024–4028, May 1, 1971.

[88] a) R. L. Johnson, M. J. Perona, and D. W. Setser, "Hydrogen chloride vibrational population produced by the H and D atom reactions with SCl_2 and S_2Cl_2," *J. Chem. Phys.*, vol. 52, pp. 6372–6383, June 15, 1970.
b) H. Heydtmann and J. C. Polanyi, "Energy distribution among reaction products: $H+SCl_2\rightarrow HCl+SCl$," *Appl. Opt.*, vol. 10, pp. 1738–1746, Aug. 1971.
c) J. R. English III, H. C. Gardner, R. W. Mitchell, and J. A. Merritt, "HCl chemical lasers with $SOCl_2$, SO_2Cl_2, Cl_2CNCl and ClCN," *Chem. Phys. Lett.*, vol. 16, pp. 180–182, Sept. 15, 1972.

[89] G. G. Dolgov-Savel'ev, V. F. Zharov, Yu. S. Neganov, and G. M. Chumak, "Vibrational–rotational transitions in an H_2+F_2 chemical laser," *Sov. Phys.—JETP*, vol. 34, pp. 34–37, Jan. 1972.

[90] D. J. Spencer, H. Mirels, and T. A. Jacobs, "Comparison of HF and DF continuous chemical lasers—I: Power," *Appl. Phys. Lett.*, vol. 16, pp. 384–386, May 15, 1970.

[91] L. T. Cowley, D. S. Horne, and J. C. Polanyi, "Infrared chemiluminescence study of the reaction $Cl+HI\rightarrow HCl+I$ at enhanced collision energies," *Chem. Phys. Lett.*, vol. 12, pp. 144–149, Dec. 1, 1971.

[92] a) K. G. Anlauf, R. G. Macdonald, and J. C. Polanyi, "Infrared chemiluminescence from $H+O_3$ at low pressure," *Chem. Phys. Lett.*, vol. 1, pp. 619–622, Apr. 1968.
b) P. E. Charters, R. G. Macdonald, and J. C. Polanyi, "Formation of vibrationally excited OH by the reaction $H+O_3$," *Appl. Opt.*, vol. 10, pp. 1747–1754, Aug. 1971.

[93] N. Jonathan, C. M. Melliar-Smith, D. Timlin, and D. H. Slater, "Analysis of hydrogen fluoride infrared chemiluminescence from simple atom–molecule reactions," *Appl. Opt.*, vol. 10, pp. 1821–1826, Aug. 1971.

[94] A. N. Chester, "Gain thresholds for diffuse parasitic laser modes," submitted to *Appl. Opt.*

[95] P. O. Clark, "Design considerations for high power laser cavities," presented at the AIAA 5th Fluid and Plasma Dynamics Conf., Boston, Mass., June 26–28, 1972, AIAA Paper 72-708.

[96] J. V. Parker and R. R. Stephens, to be published. Preliminary results from these experiments were reported in A. N. Chester, "High power chemical laser technology," presented at the Northeast Electronics Research and Engineering Meeting, Boston, Mass., Oct. 30–Nov. 3, 1972.

[97] R. G. Wenzel and G. P. Arnold, "A high energy pulsed HF laser," presented at the Int. Electron Devices Meeting (late news paper), Washington, D. C., Dec. 4–6, 1972.

[98] S. N. Suchard, "Small signal gain and time-resolved spectral measurements in a pulsed chemical transfer laser," presented at the Int. Electron Devices Meeting (late news paper), Washington, D. C., Dec. 4–6, 1972.

[99] J. Wilson, private communication.

[100] a) A. N. Oraevskii, private communication.
b) S. N. Suchard, A. Ching, and J. S. Whittier, "Efficient pulsed chemical laser," *Appl. Phys. Lett.*, vol. 21, pp. 274–275, Sept. 15, 1972.

[101] a) B. Ahlborn, P. Gensel, and K. L. Kompa, "Transverse-flow transverse-pulsed chemical CO laser," *J. Appl. Phys.*, vol. 43, pp. 2487–2489, May 1972.
b) W. Q. Jeffers, "R-branch emission from a cw CO chemical laser," *Appl. Phys. Lett.*, vol. 21, pp. 267–269, Sept. 15, 1972.

[102] J. M. White and H. Y. Su, "Hot atom reactions in the $HBr-Br_2$, $DBr-Br_2$ systems," *J. Chem. Phys.*, vol. 57, pp. 2344–2349, Sept. 15, 1972.

[103] O. P. Judd, to be published.

[104] S. N. Suchard and J. S. Whittier, to be published.

[105] H.-L. Chen *et al.*, to be published.

Part IV: Nonlinear Optical Devices

Lasers as active optical sources are extensively supplemented by nonlinear optical devices, such as harmonic generators and tunable optical parametric oscillators, to extend their outputs to other optical, ultraviolet, and infrared wavelengths. The early review paper by Minck, Terhune, and Wang provides a good introduction to the basic physical concepts and device possibilities of nonlinear optics, while the more recent authoritative survey by Harris treats optical parametric devices.

Nonlinear Optics

R. W. MINCK, R. W. TERHUNE, AND C. C. WANG

R. W. MINCK, R. W. TERHUNE, AND C. C. WANG

Abstract—Recent advances in the field of nonlinear optical phenomena are reviewed with particular emphasis placed on such topics as parametric oscillation, self-focusing and trapping of laser beams, and stimulated Raman, Rayleigh, and Brillouin scattering. The optical frequency radiation is treated classically in terms of the amplitudes and phases of the electromagnetic fields. The interactions of light waves in a material are then formulated in terms of Maxwell's equations and the electric dipole approximation. In this method, non-linear susceptibility tensors are introduced which relate the induced dipole moment to a power series expansion in field strengths. The tensor nature and the frequency dependence of the nonlinearity coefficients are considered. The various experimental observations are described and interpreted in terms of this formalism.

I. INTRODUCTION

THE TERM "nonlinear optics" is used here to refer to those phenomena involving light waves, which must be described in terms of induced charges and currents dependent upon other than the first power of the electric and magnetic field strengths. These include effects such as Faraday, Kerr, and Raman effects which are already well understood in the usual context of "linear optics"; they also include effects such as harmonic generation, parametric amplification, and induced changes in the dielectric constants, the observation of which has been made possible only with the advent of high-intensity coherent laser beams. In the past, the latter three effects and some related phenomena have been studied extensively in the microwave region of the spectrum, as it was only in this "low" frequency region of the spectrum that intense enough fields were available to induce such nonlinear behavior in materials.

The first observation of a new nonlinear optical effect using a ruby laser was made by Franken, Hill, Peters, and Weinreich [1] in 1961 when they observed optical second harmonic generation. At about the same time Kaiser and Garret [2] reported the observation of nonlinear absorption due to a two photon process, again using a ruby laser. In these experiments the effects were very weak. However, with the much more intense Q-switched laser beams, fractional conversion efficiencies and nonlinear absorptions in the range 0.01 to 0.5 have been obtained for these effects. High conversion efficiencies for many nonlinear optical effects have opened up the possibility of effectively utilizing these effects for a wide range of new types of optical devices. The operation

Manuscript received June 13, 1966; revised July 14, 1966.
The authors are with the Ford Scientific Laboratory, Dearborn, Mich.

of a tunable optical parametric oscillator [3], for example, has already been demonstrated. On the other hand, these ever-present nonlinearities (particularly stimulated scattering, self-focusing, and two photon absorption processes) ultimately place an upper limit on the power handling capability of materials, including the materials in which laser action is obtained. Thus the study of nonlinear optics is necessary as well as rewarding.

The literature on nonlinear optical phenomena has been expanding rapidly. A complete survey of the work to date would require much more space than is available for this paper. Recognizing this, we have attempted in this paper only to present an outline of the theoretical treatment so as to catalogue the various effects; and to review in depth only the more recent experimental work. More complete reviews with references covering the earlier work are available in the literature [4]–[7].

The theoretical treatment used in this paper follows that used by Armstrong, Bloembergen, Ducuing, and Pershan [8]. In this treatment the optical frequency radiation is described classically in terms of the amplitudes and phases of the electromagnetic fields; and a perturbation method is used to handle the nonlinear coupling between various frequency and spatial components of the fields. This approach appears preferable to using second quantization for the fields, since in most of the cases of interest, the number of photons per mode is very high and the classical limit is very nearly approached. The mathematics involved in the perturbation method is conceptually straightforward, but the notation rapidly becomes quite complex because of the tensor character of the coefficients and because of their multiple frequency dependences. Consequently, many simplifying assumptions have been made to reduce this complexity, and only the essential features of the various nonlinear effects have been dealt with.

In this paper those nonlinear effects which can be associated with an induced dipole moment proportional to the square and cube of the electric field strength are emphasized. These include most of the subjects of current interest. Higher order effects will be mentioned only briefly.

We have complied with the editor's request to use MKS units. However, as electrostatic units (esu) are employed in almost all the papers in the field, all quanties quoted in this paper are written in terms of a constant times their values in esu.

Reprinted from *Proc. IEEE*, vol. 54, pp. 1357–1374, Oct. 1966.

134

II. Perturbation Approach to Nonlinear Effects on Wave Propagation

The propagation of an electromagnetic wave through a material produces changes in the spatial and temporal distribution of electrical charges as the electrons and atoms react to the fields of the wave. In linear optics, this effect is characterized by an *assumed* linear relationship between the induced dipole moment $\bar{\mathcal{P}}$ per unit volume and the electric field strength: $\bar{\mathcal{P}} = \epsilon_0 \chi \bar{\mathcal{E}}$, where ϵ_0 is free-space permittivity, and χ is the electric susceptibility. Using $\bar{\mathcal{P}}$ as a source term in Maxwell's equations one then calculates the dielectric constant $\kappa = \epsilon/\epsilon_0 = 1 + \chi$.

The perturbation approach which will be used in this paper assumes that the interactions in nonlinear optics are relatively weak, giving rise to small nonlinear contributions $\bar{\mathcal{P}}^{NL}$ to the induced dipolar field [8]. The nonlinear polarization $\bar{\mathcal{P}}^{NL}$ is assumed to be an arbitrary function of $\bar{\mathcal{E}}$ and $\bar{\mathcal{H}}$, as well as their spatial and time derivatives. $\bar{\mathcal{P}}^{NL}$ appears as an additional source term in Maxwell's equations as follows:

$$\mathbf{\nabla} \times \bar{\mathcal{H}} = \frac{\partial}{\partial t}(\epsilon \bar{\mathcal{E}}) + \frac{\partial \bar{\mathcal{P}}^{NL}}{\partial t}. \tag{1}$$

The electric field strength $\bar{\mathcal{E}}(r, t)$ associated with a plane wave will be represented in terms of its Fourier frequency components $E(\omega_l, r)$ as follows:

$$\mathcal{E}_i(r, t) = E_i(0, r)$$
$$+ \frac{1}{2} \sum_{l=-N}^{N} E_i(\omega_l, r) \cdot \exp[i(k_l r - \omega_l t)] \tag{2}$$

where i, j, k denote the coordinate variables. The $2N$ Fourier components are assumed to be plane wave solutions $\exp[i(k_l r - \omega_l t)]$ of the homogeneous wave equation, but with a complex amplitude $E_i(\omega_l, r)$ regarded as a slowly varying function of distance to account for the nonlinear effects. Since $\bar{\mathcal{E}}(r, t)$ is real in our notation, $E_i(-\omega_l, r) = E_i^*(\omega_l, r)$, $\omega_{-l} \equiv -\omega_l$, and $k_{-l} \equiv -k_l^*$. $k_l = n_l \omega_l/c$, where c is the speed of light in vacuum, and $n_l \equiv n_{-l}^* = k_l^{1/2}$ is the linear refractive index of the medium at frequency ω_l.

The nonlinear polarization $\bar{\mathcal{P}}^{NL}(r, t)$ will likewise be assumed to be expandable in terms of its harmonic components, and will thus be written as

$$\mathcal{P}_i^{NL}(r, t) = P_i^{NL}(0, r)$$
$$+ \frac{1}{2} \sum_{l=-M}^{M} P_i^{NL}(\omega_l, r) \cdot \exp(-i\omega_l t) \tag{3}$$

where $M \geq N$. Note that a factor of the form $\exp(ikr)$ is not explicitly displayed in (3). This is in conformity with the assumption that $P_i^{NL}(\omega_l, r)$ is a general function of the applied electric and magnetic fields.

As $E_i(\omega_l, r)$ is assumed to change very little per wavelength, its second spatial derivative can be neglected when (2) and (3) are substituted in (1). Assuming that the

medium is isotropic in the linear approximation, the steady-state solution for each frequency component is obtained from the following complex nonlinear differential equation

$$\frac{d}{dr} E_i(\omega_l, r) = i \frac{k_l}{2\epsilon_l} P_i^{NL}(\omega_l, r) \exp(-ik_l r). \tag{4}$$

The above equation expresses the fact that each $P^{NL}(\omega_l, r)$ radiates an electromagnetic wave 90° out of time phase with itself; it also shows that the various Fourier components $E_i(\omega_l, r)$ are coupled together through the assumed dependence of $P_i^{NL}(\omega_l, r)$ on the various field amplitudes $E_i(\omega_l, r)$.

In general many nonlinear interactions occur simultaneously. However, experimental conditions are usually such that only one process is dominant. In the following the effect of each interaction will be considered separately.

To illustrate the application of (4), consider the polarization $P_i^{NL}(\omega_l, r) = \epsilon_0 \delta \chi_1^{ij} E_j(\omega_l, r) \exp(ik_l r)$ associated with an induced change $\delta \chi_1$ in the linear susceptibility due to some constant external influence such as a dc electric or magnetic field. Letting $\delta \chi_1$ be a scalar one then finds

$$E(\omega_l, r) = E(\omega_l, 0) \exp[i(k_l/2\kappa_l)\delta \chi_1 r]. \tag{5}$$

Thus $\delta \chi_1$ leads to a change in the complex index of refraction: the real part of $\delta \chi_1$ leads to a change in phase velocity, and the imaginary part to a change in the absorption strength of the material. Phenomena such as changes in the polarization direction due to the Faraday effect can also be solved for by considering the tensorial properties of $\delta \chi_1$.

III. Nonlinear Polarization Second-Order in the Electric Field Strength

The nonlinear optical effect associated with the second-order nonlinear polarization,[1]

$$\bar{\mathcal{P}}^{NL}(r, t) = 2\epsilon_0 \chi_2 : \bar{\mathcal{E}}(r, t)\bar{\mathcal{E}}(r, t)$$
$$+ \text{(similar terms involving time}$$
$$\text{derivative operators)}, \tag{6}$$

will be considered in this section. Here χ_2 is a nonlinear susceptibility tensor of third rank, and the factor 2 is included in the definition so as to conform to common usage [9]. To begin with, only the first term on the right-hand side of (6) will be considered. Since the square of the electric field strength is involved, each frequency component of the electric field will contribute to the dc as well as the second harmonic components of $\bar{\mathcal{P}}^{NL}(r, t)$. Expressing these components in terms of the amplitudes of the Fourier components as defined in (2) and (3), one has

---

[1] In our definition, χ_2(MKS) has the dimensions M/volt, and is related to χ_2(esu) as follows: χ_2(MKS)$=[4\pi/(3\times10^4)]\chi_2$(esu). Similarly, χ_3(MKS)$=[4\pi/(3\times10^4)^2]\chi_3$(esu), and χ_n(MKS)$=[4\pi/(3\times10^4)^{n-1}]\chi_n$(esu).

$$P_i{}^{NL}(2\omega_l, r) = \epsilon_0\chi_2{}^{ijk}E_j(\omega_l, r)E_k(\omega_l, r) \exp{(i2k_lr)} \quad (7)$$

and

$$P_i{}^{NL}(0, r) = \frac{1}{2}\epsilon_0 \sum_{l=-N}^{N} \chi_2{}^{ijk}E_j(\omega_l, r)E_k{}^*(\omega_l, r) \quad (8)$$

where summation over repeated indices is implied.

With more than one frequency component present in $\bar{\mathcal{E}}(r, t)$, $\bar{\mathcal{P}}^{NL}(r, t)$ will also contain frequency components at the various sum and difference frequencies. Direct substitution in (6) shows that

$$P_i{}^{NL}(\omega_l, r) = 2\epsilon_0\chi_2{}^{ijk}E_j(\omega_m, r)E_k(\omega_n, r)$$
$$\cdot \exp{[i(k_m + k_n)r]} \quad (9)$$

or that

$$P_i{}^{NL}(\omega_l, r) = 2\epsilon_0\chi_2{}^{ijk}E_j(\omega_m, r)E_k{}^*(\omega_n, r)$$
$$\cdot \exp{[i(k_m - k_n)r]} \quad (10)$$

where $\omega_l = \omega_m \pm \omega_n$. The *linear* electrooptic effect is a special case of (9) with

$$P_i{}^{NL}(\omega_l, r) = [4\epsilon_0\chi_2{}^{ijk}E_j(0, r)]E_k(\omega_l, r) \exp{(ik_lr)}. \quad (11)$$

The bracketed term in (11) gives the induced $\delta\chi_1$ proportional to the applied dc electric field.

A. The Nonlinear Susceptibility Tensor of Third Rank

Equation (6) describes an assumed nonlinear response of a material medium. It must be invariant to those symmetry operations which transform the material into itself. These symmetry operations thus impose certain restrictions on the form of the susceptibility tensor χ_2, in particular permitting nonzero values only for materials lacking inversion symmetry. In general, χ_2 has the same form as the piezoelectric "d" tensors. These "d" tensors have been tabulated [10], and we find, for example, that for the point group $42m$, which includes ammonium dihydrogen phosphate (ADP) and potassium dihydrogen phosphate (KDP),

$$P_x{}^{NL} = 2\epsilon_0 d_{14}E_yE_z$$
$$P_y{}^{NL} = 2\epsilon_0 d_{14}E_zE_x$$
$$P_z{}^{NL} = 2\epsilon_0 d_{36}E_xE_y \quad (12)$$

where the usual contracted form of "d" has been used: Here $d_{im} = \chi_2{}^{ijk}$, $j = k$; and $d_{im} = \frac{1}{2}(\chi_2{}^{ijk} + \chi_2{}^{ikj})$, $j \neq k$. The first suffix takes values 1 through 3 for x, y, and z, whereas the second suffix runs from 1 through 6 standing, respectively, for xx, yy, zz, yz, zx, and xy.

The terms in (6) involving time derivatives lead to a frequency dependence or dispersion in the χ_2 coefficients. They can be included in the formalism by assuming that the χ_2 relating the Fourier components of $\bar{\mathcal{P}}^{NL}(r, t)$ with $\bar{\mathcal{E}}(r, t)$ are functions of the frequencies involved in a particular interaction. The χ_2 coefficients thus defined will be written as $\chi_2{}^{ijk}(-\omega_1, \omega_2, \omega_3)$, where the ordering of the frequencies indicates their association with their cor-

responding coordinates. Approximate expressions for $\chi_2{}^{ijk}(-\omega_1, \omega_2, \omega_3)$ can be found using time-dependent perturbation theory [8]. The frequency dependence of $\chi_2{}^{ijk}(-\omega_1, \omega_2, \omega_3)$ is found to be similar to that of the linear susceptibility $\chi_1{}^{ij}(-\omega_1, \omega_1)$ in that for most materials it exhibits resonant behavior in the infrared and in the ultraviolet. It has been shown [11], [11a] that Kramers-Kronig dispersion relations similar to that for the linear susceptibilities also exist for the nonlinear susceptibilities $\chi_2{}^{ijk}(-\omega_1, \omega_2, \omega_3)$.

With the above definition it follows that $\chi_2{}^{ijk}(-\omega_1, \omega_2, \omega_3)$ is symmetric in j and k

$$\chi_2{}^{ijk}(-\omega_1, \omega_2, \omega_3) = \chi_2{}^{ikj}(-\omega_1, \omega_3, \omega_2), \quad (13)$$

and is related to $\chi_2{}^*$ through the relation

$$\chi_2{}^{ijk}(-\omega_1, \omega_2, \omega_3) = \chi_2{}^{*ijk}(\omega_1, -\omega_2, -\omega_3). \quad (14)$$

If the material has negligible losses at the frequencies involved in the interaction [12], χ_2 is real and is symmetric in i and j, i.e.,

$$\chi_2{}^{ijk}(-\omega_1, \omega_2, \omega_3) = \chi_2{}^{jik}(\omega_2, -\omega_1, \omega_3). \quad (15)$$

Equation (15) may also be derived by showing that all the coefficients involved in a particular interaction are obtainable by differentiation of a common free energy term [13].

B. Second Harmonic Generation of Light in Bulk Transparent Media

The equations for second harmonic generation through the second-order nonlinear polarization will now be considered. Using (7), (9), and (4), one obtains

$$\frac{d}{dr}E_i(\omega, r)$$
$$= i\left(\frac{k_1}{2\kappa_1}\right)2\chi_2{}^{ijk}(-\omega, 2\omega, -\omega)E_j(2\omega, r)E_k{}^*(\omega, r)$$
$$\cdot \exp{(i\Delta kr)} \quad (16)$$

$$\frac{d}{dr}E_j(2\omega, r)$$
$$= i\left(\frac{k_2}{2\kappa_2}\right)\chi_2{}^{jik}(-2\omega, \omega, \omega)E_i(\omega, r)E_k(\omega, r)$$
$$\cdot \exp{(-i\Delta kr)} \quad (17)$$

where $\Delta k = k_2 - 2k_1 = (2\omega/c)(n_2 - n_1)$ is the momentum mismatch for collinear fundamental and second harmonic waves; Δk is positive for normal dispersion. Also using (14) and (15) one sees that

$$\chi_2{}^{ijk}(-\omega, 2\omega, -\omega) = \chi_2{}^{jik}(-2\omega, \omega, \omega).$$

Although solutions to (16) and (17) have been obtained in closed form [8], the essential features of second harmonic generation can be demonstrated with the aid of certain simplifying assumptions. Thus, if the amount of second harmonic generation is assumed to be small,

$E_i(\omega, r)$ can be regarded as being constant and

$$E_j(2\omega, r) = i\left(\frac{k_2}{2\kappa_2}\right)\chi_2{}^{jik}(-2\omega, \omega, \omega)E_i(\omega, 0)E_k(\omega, 0)$$

$$\cdot\left[\frac{1 - \exp(-i\Delta kr)}{i\Delta k}\right] \quad (18)$$

where the initial value $E_j(2\omega, 0)$ of the harmonic beam has been neglected. If $\Delta k = 0$, (18) predicts that $E_j(2\omega, r)$ grows linearly with distance until the fundamental wave $E_i(\omega, r)$ becomes attenuated. If $\Delta k \neq 0$, the bracketed term on the right indicates that $E_j(2\omega, r)$ undergoes periodic variation as a function of distance (Fig. 1), the period of the variation being determined by $l_{\mathrm{coh}} = \pi/\Delta k$ which is the phase coherence length [14].

For large second harmonic conversion efficiencies, $E_i(\omega, r)$ can no longer be considered constant. With $\Delta k = 0$, the exact solution of (16) and (17) shows that the conversion efficiency should increase montonically with distance to 100 percent, converting all the energy from the fundamental to the harmonic beam (Fig. 2). With $\Delta k \neq 0$, a periodic variation in the second harmonic intensity with distance is again obtained [8].

The nonlinear optical effects per wavelength are usually very small, so that unless $l_{\mathrm{coh}} \gg \lambda$ very little conversion of energy from one frequency to another can be expected. Various techniques for increasing l_{coh} have been suggested [14], [15]. The most successful to date has been the use of birefringence in an uniaxial nonlinear crystal to balance out the effect of dispersion (Fig. 3). Using this technique to obtain phase-matching, second harmonic conversion efficiency of up to 30 to 35 percent has been obtained in ADP and KDP crystals [16]. For a beam of finite cross section, the interaction length for second harmonic generation is ultimately limited by double refraction, which causes the energy of the laser and second harmonic beams to propagate in slightly different directions [17]. It has been shown that this limitation is responsible for the observed saturation of second harmonic conversion efficiencies [18]. The effect of double refraction is minimized if phase matching occurs in a direction making an angle of 90° with the crystal optic axis. At this angle, the ray and phase velocities are collinear. The condition of phase matching at 90° to the optic axis has been achieved in LiNbO₃ by changing the temperature of the crystal [19], and in KDP by applying a dc bias field combined with temperature tuning [19a].

The first experiments on second harmonic generation were performed with pulsed ruby lasers. Quantitative measurements of the χ_2 coefficients obtained with these lasers were quite uncertain because of the incompletely known characteristics of the laser beams. More accurate values of χ_2 were obtained from the observation of second harmonic generation using CW gas lasers. The relative value of χ_2 for a selected number of materials [20]–[24] are listed in Table I. Materials with high indices of

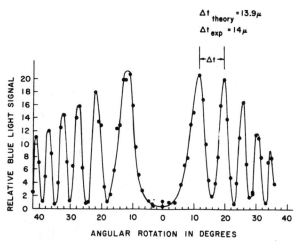

Fig. 1. Observed dependence of second harmonic generation in a quartz platelet as a function of angle between the surface normal of the platelet and the direction of the laser beam. The effective thickness of the crystal changes with rotation. The data demonstrate the oscillatory behavior of harmonic generation due to an index mismatch such as described in (18) (from Maker, Terhune, Nisenoff, and Savage [14].)

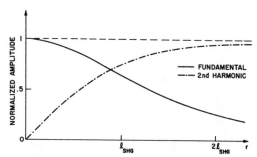

Fig. 2. The growth of the normalized second harmonic amplitude and decrease of the normalized fundamental amplitude for perfect phase matching, the second harmonic being initially zero. The interaction length $1/l_{\mathrm{SHG}} = (k_2/2\kappa_2)\chi_2|E(\omega, 0)|$ is a measure of the length necessary for appreciable second harmonic generation (from Armstrong et al. [8]).

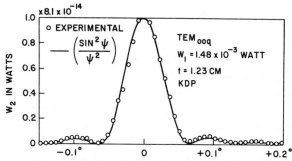

Fig. 3. Variation of second harmonic power with σ, the angle between the fundamental beam in the lowest order mode and the phase matching direction as observed by Ashkin, Boyd, and Dziedzic [25] using a gas laser. The observed results in ADP agree well with the predicted dependence of $(\sin^2\psi/\psi^2)$, where $\psi = \beta l\sigma$ with β proportional to the double refraction angle.

TABLE I
SUMMARY OF SECOND HARMONIC GENERATION EXPERIMENTS*

Crystal	Symmetry	Laser wavelength (micron)	d	
KDP	42 m	0.69[a]	$d_{36} = 1.00$	
			$d_{14} = 0.95 \pm 0.06$	
		1.06[a]	$d_{36} = 1.00$	
			$d_{14} = 1.01 \pm 0.05$	
ADP	42 m	1.06[a]	$d_{36} = 0.93 \pm 0.06$	
			$d_{14} = 0.89 \pm 0.04$	
LiNbO$_3$	3 m	1.06[b]	$d_{22} = 6.3 \pm 0.6$	
			$d_{31} = 11.9 \pm 1.7$	
CdS	6 mm	1.06[a]	$d_{15} = 35 \pm 2$	
			$d_{31} = 32 \pm 2$	
			$d_{33} = 63 \pm 4$	
GaAs	43 m	0.69[c]	$d_{14} = 165-i475$	
		1.06[d]	$d_{14} = 560 \pm 140$	
		10.6[e]	$d_{14} = 294 \pm 100$	
Se	32	10.6[e]	$d_{11} = 63 \pm 33$	
Te	32	10.6[e]	$d = 4230 \pm 670$	

[a] R. C. Miller, D. A. Kleinman, and A. Savage [20].

[b] G. D. Boyd, R. C. Miller, K. Nassau, W. L. Bond, and A. Savage [21].

[c] R. K. Chang, J. Ducuing, and N. Bloembergen [22].

[d] R. C. Miller [23].

[e] C. K. N. Patel [24].

* The absolute values of the nonzero components of the "d" tensor, as obtained from second harmonic generation experiments, are listed for a selected number of crystals. The d values are given relative to d_{36} for KDP at 6328 Å. Using a He-Ne gas laser, Ashkin et al. have measured d_{36} for KDP to be $[4\pi/(3\times10^4)](3\pm1)\times10^{-9}$ M/volt [25]; however, a somewhat lower value, $d_{36}=[4\pi/(3\times10^4)]$ $(1.36\pm0.16)\times10^{-9}$ M/volt was obtained for ADP with a single-frequency gas laser [26].

refraction have large values of χ_2 since large local field corrections are involved. Values of d_{36} of $[4\pi/(3\times10^4)]$ $\times(3\pm1)\times10^{-9}$ M/V for KDP [25] and of $[4\pi/(3\times10^4)]$ $\times(1.36\pm0.16)\times10^{-9}$ M/V for ADP [26] have been obtained with gas lasers. Some results of measurements in the infrared using the 10.6 μ CO$_2$ gas laser are also included in Table I.

Second harmonic generation is now commonly used to produce intense light beams at additional frequencies. Combined with the technique of sum and difference frequency generation (this section) and the techniques of stimulated Raman emission in liquids (Section IV), coherent radiation has been obtained at a host of frequencies ranging from the infrared to the ultraviolet. The studies of second harmonic generation in birefringent crystals with both focused and unfocused beams have received much attention, and details may be found in the literature [27]–[29].

Second harmonic generation is intimately related to optical rectification whereby radiation at a given frequency beats with itself to produce a dc or low-frequency polarization in the medium [30]. This dc effect is given from (8) by

$$P_i^{NL}(0, r) = \epsilon_0 \chi_2^{ijk}(0, \omega, -\omega) E_j(\omega_l, r) E_k^*(\omega_l, r). \quad (19)$$

The symmetry relation (15) requires that the optical rectification coefficient and the linear electrooptic coefficient be equal, i.e., $\chi_2^{ijk}(0, \omega, -\omega) = \chi_2^{kij}(-\omega, 0, \omega)$. Experiments [31] indicate that this relation is indeed satisfied.

C. Three-Wave Parametric Interactions

Interactions involving three distinct frequency components through χ_2 will now be considered. Using (4) and (9) the following set of equations are obtained:

$$\frac{d}{dr} E_i(\omega_1, r)$$
$$= i\left(\frac{k_1}{2\kappa_1}\right) 2\chi_2^{ijk}(-\omega_1, \omega_2, \omega_3) E_j(\omega_2, r) E_k(\omega_3, r)$$
$$\cdot \exp(-i\Delta kr) \quad (20)$$

$$\frac{d}{dr} E_j(\omega_2, r)$$
$$= i\left(\frac{k_2}{2\kappa_2}\right) 2\chi_2^{jik}(-\omega_2, \omega_1, -\omega_3) E_i(\omega_1, r) E_k^*(\omega_3, r)$$
$$\cdot \exp(i\Delta kr) \quad (21)$$

$$\frac{d}{dr} E_k(\omega_3, r)$$
$$= i\left(\frac{k_3}{2\kappa_3}\right) 2\chi_2^{kij}(-\omega_3, \omega_1, -\omega_2) E_i(\omega_1, r) E_j^*(\omega_2, r)$$
$$\cdot \exp(i\Delta kr) \quad (22)$$

where $\Delta k = k_1 - (k_2 + k_3)$, $\omega_1 = \omega_2 + \omega_3$, and $\chi_2^{ijk}(-\omega_1, \omega_2, \omega_3) = \chi_2^{*jik}(-\omega_2, \omega_1, -\omega_3) = \chi_2^{*kij}(-\omega_3, \omega_1, -\omega_2)$. Exact solutions of the above equations have been obtained with various amounts of radiation present initially at the three different frequencies [8].

First, the sum and difference frequency generation will be considered. Here two frequency components are initially present with comparable intensity, and a wave at the third frequency is created through the nonlinear interaction. Thus, if the waves at ω_2 and ω_3 are initially present, the initial growth of a wave at the sum frequency ω_1 will be given from (20) by

$$E_i(\omega_1, r)$$
$$= E_i(\omega_1, 0)$$
$$+ i\left(\frac{k_1}{2\kappa_1}\right) 2\chi_2^{ijk}(-\omega_1, \omega_2, \omega_3) E_j(\omega_2, 0) E_k(\omega_3, 0)$$
$$\cdot \left[\frac{1 - \exp(-i\Delta kr)}{i\Delta k}\right]. \quad (23)$$

Equation (23) is quite similar to (18) describing the growth of a second harmonic wave. Similarly, a wave at the difference frequency may be generated through (22). In this manner an infrared [32] or even a far-infrared [33] wave may be obtained by mixing two laser beams of different frequencies in a nonlinear crystal. Experimentally, conversion efficiencies in the percent range have been obtained for the mixing of two different frequency components to create a sum or difference frequency. The generation of sum and difference frequencies using incoherent sources has also been observed [34].

The three-wave interactions through χ_2 can also lead to parametric amplification and parametric oscillation. In this case, parametric amplification refers to the coupled growth of two frequency components in the presence of a strong frequency component, which will be referred to as the pump [35]. To describe these possibilities, it is convenient to write ω_p, ω_s, and ω_I as the pump, signal, and idler frequencies. The corresponding wavenumber vectors are written k_p, k_s, and k_I. Here $\omega_p = \omega_s + \omega_I$ and $\Delta k = k_p - (k_s + k_I)$. It is assumed that the percentage energy conversion is small so that the amplitude and phase of the pump radiation remain approximately constant. Under index matched conditions, solutions of (21) and (22) give

$$E_j(\omega_s, r) = E_j(\omega_s, 0) \cosh (gr) \tag{24}$$

$$E_k(\omega_I, r) = i(\omega_I/\omega_s)^{1/2} E_j(\omega_s, 0) \sinh (gr) \tag{25}$$

$$g = \left[\frac{k_I k_s}{\kappa_I \kappa_s}\right]^{1/2} \chi_2^{ijk}(-\omega_I, \omega_p, \omega_s) \, | \, E_i(\omega_p, 0) | \tag{26}$$

where the initial amplitude $E_k(\omega_I, 0)$ at the idler frequency has been neglected. From an exact solution of (20)–(22), one can show that energy and momentum are conserved. This result is equivalent to the Manley-Rowe relations in the theory of microwave parametric amplifiers. For $gr \gg 1$, (24) and (25) behave asymptotically as

$$E(\omega_s, r) = \frac{1}{2} E(\omega_s, 0)e^{gr} \tag{27}$$

$$E(\omega_I, r) = \frac{i}{2} (\omega_I/\omega_s)^{1/2}E(\omega_s, 0)e^{gr}. \tag{28}$$

Thus the signal wave at ω_s is amplified at a rate determined by the gain constant g per unit length. If $\Delta k \neq 0$, the solutions of (21) and (22) still assume the same form, but the gain constant g is modified to become

$$g \to g[1 - (\Delta k/2g)^2]^{1/2} \tag{29}$$

indicating that the gain is reduced when $\Delta k \neq 0$. Furthermore, a threshold phenomenon now occurs, namely, there will be no gain until the pump exceeds the value such that

$$g > \Delta k/2. \tag{30}$$

It is apparent that optimum gain is achieved when the phase condition $k_p = k_s + k_I$ is satisfied. These results are identical with those for microwave parametric amplifiers [35].

The above description pertains to traveling-wave parametric amplification. If a plane parallel Fabry-Perot cavity of length l filled with nonlinear material is employed, then parametric oscillation may occur when the gain exceeds the reflection and propagation losses. The minimum power required for parametric oscillation is given by

$$gl = (1 - R) \tag{31}$$

where the reflectivity $R \approx 1$ at ω_s and ω_I, but $R \approx 0$ at ω_p. It follows from (30) and (31) that oscillation will tend to occur at frequencies ω_s and ω_I for which $\Delta k = 0$.

In the first experimental attempt to demonstrate traveling-wave parametric amplification [36] in ADP, the measured gains were very small. However, soon afterwards parametric oscillations were obtained by Giordmaine and Miller [3] using a crystal of LiNbO$_3$ in a cavity configuration (Fig. 4). The nonlinear coefficient for LiNbO$_3$ is about 10 times that for ADP, and further its optical properties permit an index match at 90° to the optic axis. With a pump power of 7 kW, energy conversion efficiencies of the order of 0.2 percent were obtained. With further development efforts one can expect to achieve lower pump power threshold values and much higher energy conversion efficiencies. Calculations indicate that CW parametric oscillation in LiNbO$_3$ should be possible with a pump power of 10 mW in a single mode [36a]. Thus, with the use of parametric oscillation, one can expect to obtain in the near future both CW and high power pulsed coherent beams tunable over a wide frequency range. Recently, parametric oscillation has been observed in KDP using much higher pump powers [37].

The operation of a parametric amplifier or oscillator under degenerate condition with $\omega_s = \omega_I = \omega_p/2$ follows directly from (16) and (17) when the initial conditions $|E(2\omega, 0)| \gg |E(\omega, 0)|$ are assumed. This operation can be regarded as the inverse of second harmonic generation, in which one considers the growth of the fundamental beam in the presence of a strong second harmonic. Parametric amplification under degenerate condition has been observed in KDP [38]. It has been proposed [3] that failure to observe this subharmonic oscillation in LiNbO$_3$ is due to the increasing sensitivity of the index matching condition to small temperature gradients near the subharmonic frequency $\omega_p/2$ (Fig. 4).

Detailed analysis of parametric oscillation has been given by various authors, and may be found in the literature [35].

D. Other Second-Order Effects

Second harmonic generation at the boundary of nonlinear media [39]–[41] has been studied both theoretically and experimentally. In transparent media, this surface effect is usually negligible compared to the harmonic generation in the bulk. However, if the material is opaque at either frequency, only harmonic generation at the boundary is possible. In the latter case, the nonlinear coefficients χ_2 are complex, whereas in transparent material, they are real. Figure 5 shows the interference between the second harmonic lights produced by the same laser beam in a KDP platelet and from a GaAs mirror. From such

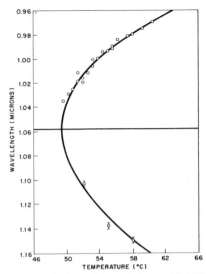

Fig. 4. Experimental data showing the wavelengths at which parametric oscillations were observed as a function of crystal temperature in LiNbO₃. The wavelengths at a given temperature are those for which $\Delta k = 0$ in a direction making an angle of 90° with the optic axis of the crystal; by scanning the crystal temperature, oscillation was scanned over the wavelength range indicated above. This parametric oscillator was pumped with the coherent radiation at 0.529 μ obtained by doubling the radiation from a Nd-doped CaWO₄ laser. The solid curve is a theoretical fit using the known temperature-dependence of birefringence (from Giordmaine and Miller [3]).

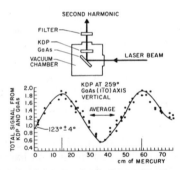

Fig. 5. Schematic of the experimental arrangement used to measure the phase difference between χ_2 in GaAs and KDP. Data show the observed interference effect between the second harmonic beams produced in the KDP platelet and from the boundary of a GaAs mirror as a function of the air pressure in the space between them. The phases of the second harmonic beams generated in both the KDP and GaAs bear a fixed relationship to the phase of the laser beam where the second harmonic is generated. Since the index mismatch in air between the laser frequency and its second harmonic varies linearly with the air pressure, the relative phases of the two second harmonic beams is scanned by varying the air pressure (from Chang, Ducuing, and Bloembergen [22]).

interference patterns, the phase shifts between the fundamental and second harmonic waves have been determined in a number of III–V and II–VI compounds. Similar techniques can be used to determine the phase of other nonlinear susceptibility coefficients.

Second harmonic generation is also possible in crystals possessing inversion symmetry. In this case the nonlinear interaction is believed due either to the Lorentz force with a second harmonic polarization proportional to $E \times \partial H / \partial t$, or to an electric quadrupole interaction of the

form $E \nabla \cdot E$. Theory and experiment [42]–[44] indicate that both terms are important in the observed light harmonics reflected from silver. Second harmonic generation due to the latter term has been observed in calcite [45] under index-matched conditions.

Second-order nonlinear effects should also occur in optically active liquids [46]. The nonlinear polarization is of the form $E(\omega_1) \times E(\omega_2)$; it leads to sum and difference frequency generation, although second harmonic generation is forbidden. This effect has recently been observed [47].

IV. NONLINEAR POLARIZATION THIRD-ORDER IN THE ELECTRIC FIELD STRENGTH

The nonlinear optical effects due to an induced polarization third order in the electric field strength can be considered in a manner analogous to that developed in Section III. The induced polarization will be assumed to have the form

$$\bar{\mathcal{P}}^{NL}(r, t) = 4\epsilon_0 \chi_3 : \bar{\mathcal{E}}(r, t) \, \bar{\mathcal{E}}(r, t) \, \bar{\mathcal{E}}(r, t)$$
$$+ \text{ (similar terms involving time}$$
$$\text{derivative operators)} \qquad (32)$$

where χ_3 is a nonlinear susceptibility tensor of fourth rank, and a factor 4 has been introduced in the definition [9]. Like χ_2, spatial symmetry operations impose certain restrictions on the form of χ_3 in a given material; but unlike χ_2, χ_3 may exist even in materials possessing inversion symmetry. The nonzero coefficients of χ_3 have been tabulated for the 32 crystallographic point groups [48]. In most of the following cases, we shall restrict our discussions to media which are isotropic in the macroscopic sense, although such discussions can be extended to media of lower symmetry.

Equation (32) will be analyzed in terms of the Fourier components of $\bar{\mathcal{P}}^{NL}(r, t)$ and $\bar{\mathcal{E}}(r, t)$ defined in (2) and (3). As before, the effect of the terms in (32) involving the time derivatives will be included in the formalism by allowing χ_3 to be complex and frequency-dependent. The components of χ_3 will be written as $\chi_3{}^{ijlk}(-\omega_1, \omega_2, \omega_3, \omega_4)$, where the ordering of the frequencies again indicates their association with the corresponding coordinates. Permutation relations similar to those exhibited in (13) and (14) also apply for χ_3 [8], [13].

In an actual material system, the nonlinear coefficients of χ_3 are complex. The imaginary part of χ_3 leads to Raman, Brillouin, and Rayleigh scattering, and to two-photon absorption of the light beam. In a lossless medium, the susceptibility coefficients of χ_3 are real. In this case, the primary nonlinear optical effects are the generation of new frequency components, and the intensity-dependent changes in the refractive index for the existing frequency components. For example, the third harmonic generated by an intense light beam at frequency ω can be described in terms of (4) with an induced polarization

$P_i{}^{NL}(3\omega, r)$

$$= 3\epsilon_0\chi_3{}^{1122}(-3\omega, \omega, \omega, \omega)E_i(\omega, r)E_j(\omega, r)E_j(\omega, r)$$

$$\cdot \exp(i3kr) \tag{33}$$

where $k = n\omega/c$. Here $\chi_3{}^{1122}(-3\omega, \omega, \omega, \omega) = (1/3)\chi_3{}^{1111}$ $(-3\omega, \omega, \omega, \omega)$ is the only nonvanishing independent coefficient of $\chi_3(-3\omega, \omega, \omega, \omega)$ in an isotropic medium, and is of the order of $[4\pi/(3\times10^4)^2]\times10^{-15}$ M^2/V^2. Conversion efficiencies of the order of 3 part in 10^6 have been reported for third harmonic generation in calcite under index matched condition [49]. Similarly, the dc electric field induced second harmonic generation [45] arises from an induced polarization of the form

$P_i{}^{NL}(2\omega, r)$

$$= 6\epsilon_0\chi_3{}^{ijkl}(-2\omega, 0, \omega, \omega)E_j(0, r)E_k(\omega, r)E_l(\omega, r)$$

$$\cdot \exp(i2kr). \tag{34}$$

In calcite, $\chi_3(-2\omega, 0, \omega, \omega)$ was observed to be of the order of $[4\pi/(3\times10^4)^2]\times10^{-14}$ M^2/V^2.

In general, χ_3 will couple together four frequency components, and lead to the generation of sum and difference frequencies. The corresponding nonlinear polarization is given by

$P_i{}^{NL}(\omega_1, r)$

$$= D\epsilon_0\chi_3{}^{ijkl}(-\omega_1, \omega_2, \omega_3, \omega_4)E_j(\omega_2, r)E_k(\omega_3, r)E_l(\omega_4, r)$$

$$\cdot \exp[i(k_2 + k_3 + k_4)r] \tag{35}$$

where $\omega_1 = \omega_2 + \omega_3 + \omega_4$ and $D = 1$, 3, or 6 depending on whether three, two, or none of the frequencies are the same.

When two components at frequencies ω_1 and ω_2 are present in an isotropic medium, the nonlinear polarization

$$P_i{}^{NL}(\omega_1, r) = 6\epsilon_0\chi_3{}^{1122}(-\omega_1, \omega_1, \omega_2, -\omega_2)E_i(\omega_1, r)$$

$$\cdot E_j(\omega_2, r)E_j{}^*(\omega_2, r)\exp(ik_1r)$$

$$+ 6\epsilon_0\chi_3{}^{1212}(-\omega_1, \omega_1, \omega_2, -\omega_2)E_j(\omega_1, r)$$

$$\cdot E_i(\omega_2, r)E_j{}^*(\omega_2, r)\exp(ik_1r)$$

$$+ 6\epsilon_0\chi_3{}^{1221}(-\omega_1, \omega_1, \omega_2, -\omega_2)E_j(\omega_1, r)$$

$$\cdot E_j(\omega_2, r)E_i{}^*(\omega_2, r)\exp(ik_1r) \tag{36}$$

describes the change in the linear susceptibility at ω_1 proportional to the intensity of a light wave at ω_2. Note that the susceptibility tensor $\chi_3(-\omega_1, \omega_1, \omega_2, -\omega_2)$ has three nonzero independent components differing only by the assignment of the frequencies to the coordinates. The above equation can be used to calculate the optical Kerr effect [50], that is, the induced birefringence at ω_1, due to $E(\omega_2, r)$ [Fig. 6(a)]. One deduces that

$$\Delta n = \delta n_\parallel - \delta n_\perp$$

$$= \frac{2\pi}{n_1}[6\chi_3{}^{1212}(-\omega_1, \omega_1, \omega_2, -\omega_2)$$

$$+ 6\chi_3{}^{1221}(-\omega_1, \omega_1, \omega_2, -\omega_2)]E_i(\omega_2, r)E_i{}^*(\omega_2, r) \tag{37}$$

where $\delta n_\parallel$ and $\delta n_\perp$ are, respectively, the change in the refractive index in a direction parallel and perpendicular to the electric field $E(\omega_2, r)$. In liquids with anisotropic molecules, the optical Kerr effect is associated primarily with molecular reorientation. For these liquids, the coefficients are large, on the order of $[4\pi/(3\times10^4)^2]\times10^{-12}$ M^2/V^2. For liquids with symmetrical molecules, the effect is expected to be primarily due to the nonlinear electronic polarizability arising from the distortion of the electron cloud in the molecules. The coefficients in this case are of the order of $[4\pi/(3\times10^4)^2]\times10^{-15}$ M^2/V^2, as they should be comparable to those describing third harmonic generation.

The quadratic dc Kerr effect is a special case of (37) with $\omega_2 = 0$ and an approrpiate change in the degeneracy factor.

With $\omega_1 = \omega_2 = \omega$, the susceptibility tensor $\chi_3(-\omega, \omega, \omega, -\omega) \equiv \chi_3[\omega]$ describes the effect of an intense light beam upon the propagation of the beam itself. For the most general case of an elliptically polarized beam with finite cross section, the self-induced effects manifest themselves as changes in the refractive index which are dependent upon both the intensity and the state of polarization of the beam. These effects lead to self-focusing of the beam [52] and a rotation of the vibrational ellipse [53] as a function of distance [see Fig. 6(b)]. When the elliptical polarization is decomposed into two orthogonal circular polarizations, $E_\pm = (1/\sqrt{2})(E_x \pm iE_y)$, the corresponding intensity-dependent changes in the refractive index are given, respectively, by

$$\delta n_+ = (2\pi/n)3\{(\chi_3{}^{1122}[\omega] + \chi_3{}^{1212}[\omega])\,|\,E_+\,|^2$$

$$+ (\chi_3{}^{1122}[\omega] + \chi_3{}^{1212}[\omega]$$

$$+ 2\chi_3{}^{1221}[\omega])\,|\,E_-\,|^2\}$$

$$\delta n_- = (2\pi/n)3\{(\chi_3{}^{1122}[\omega] + \chi_3{}^{1212}[\omega])\,|\,E_-\,|^2$$

$$+ (\chi_3{}^{1122}[\omega] + \chi_3{}^{1212}[\omega]$$

$$+ 2\chi_3{}^{1221}[\omega])\,|\,E_+\,|^2\}. \tag{38}$$

Thus the change in the refractive index is in general different for the two senses of circular polarization. In the special case of a linearly polarized beam, $\delta n_+ = \delta n_- = \delta n_\parallel$, and (38) reduces to

$$\delta n_\parallel = (2\pi/n)3\{\chi_3{}^{1122}[\omega] + \chi_3{}^{1212}[\omega]$$

$$+ \chi_3{}^{1221}[\omega]\}\,|\,E\,|^2. \tag{39}$$

This change is directly related to the critical power for self-trapping with a linearly polarized beam, as will be shown in Section IV-C.

A. Resonant Phenomena

Figure 7 is a schematic of an experimental arrangement that could be used to observe resonant two photon processes. The figure is intended to illustrate the effect of an intense laser beam on the absorption spectrum of the material. In the absence of the laser beam, this material is assumed to be transparent in the visible and to exhibit

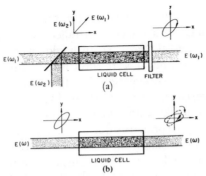

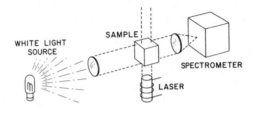

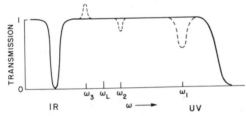

Fig. 6.   Schematic illustrating (a) the optical Kerr effect and (b) the intensity-induced rotation of the vibrational ellipse. Both of these effects are due to the intensity-dependent changes in the refractive index. In (a), the light beam at ω_2 induces birefringence at ω_1 so that $E(\omega_1)$ is changed from linear polarization to elliptical polarization. In (b), the self-induced index change with an elliptically polarized beam leads to rotation of its own vibrational ellipse; the amount of rotation is proportional to $\chi_3^{1221}(-\omega, \omega, \omega, -\omega)$, and to the distance propagated.

Fig. 7.   Schematic to indicate the additional two photon processes which one would observe in absorption spectra due to the presence of an intense laser beam at frequency ω_L. The dashed lines indicate the additional features that one might observe: simultaneous absorption at ω_1 and ω_L; absorption at ω_2 with simultaneous emission at ω_L; and the inverse, absorption at ω_L with emission at ω_3. The latter process is referred to as Raman laser action.

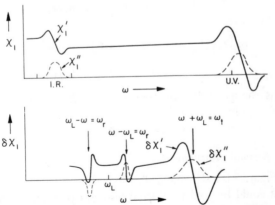

Fig. 8.   Schematic to indicate the frequency dependence of $\delta\chi_1(-\omega, \omega)$ due to the presence of the laser beam at ω_L. $\chi_1(-\omega, \omega)$ is also shown.

$$\delta\chi_1{}^{ij}(-\omega, \omega) = 6\chi_3{}^{ijkl}(-\omega, \omega, \omega_L, -\omega_L)E_k(\omega_L, r)E_l{}^*(\omega_L, r).$$

absorptions in the infrared and ultraviolet. With the laser beam at frequency ω_L present, a new set of resonances is observed. As shown, an absorption would appear at frequency ω_1 with $\omega_1 + \omega_L = \omega_t$, where ω_t is the frequency associated with an electronic transition of even parity. This process involves the simultaneous absorption of a photon from the white light source and one from the laser beam. The absorption at $\omega_2 = \omega_L + \omega_r$, corresponds to the simultaneous absorption of a photon at ω_2 and emission of one at ω_L, where ω_r is the frequency of a Raman transition. The inverse process, namely the emission of a photon at $\omega_3 = \omega_L - \omega_r$ and the simultaneous absorption of a photon at ω_L, is observed as an increase in the transmission at ω_3.

The frequency dependence of the real and imaginary parts of the *linear* susceptibility χ_1' and χ_1'' is indicated in Fig. 8 for the material whose spectrum is shown in Fig. 7. The *linear* absorption can be seen to be associated with peaks in χ_1''. The intensity dependent change in the susceptibility can be derived from (36),

$$\delta\chi_1{}^{ij}(-\omega, \omega)$$
$$= 6\chi_3{}^{ijkl}(-\omega, \omega, \omega_L, -\omega_L)E_k(\omega_L, r)E_i{}^*(\omega_L, r) \quad (40)$$

and is in general complex with a frequency dependence as indicated in Fig. 8. The peaks in the two photon absorption are seen to be associated with the peaks in χ_3''. Note that the emission at $\omega_3 = \omega_L - \omega_r$ is associated with a negative χ_3''.

Approximate expressions for the frequency dependence of χ_3 near two photon resonances can be obtained by using third-order time dependent perturbation theory in the electric dipole approximation [54]. The 24 terms thus obtained for $\chi_3{}^{ijkl}(-\omega, \omega, \omega_L - \omega_L)$ can be divided into resonant and nonresonant parts in terms of molecular parameters as follows

$$6\chi_3{}^{ijkl}(-\omega, \omega, \omega_L, -\omega_L)$$
$$= 6\chi_{3NR}{}^{ijkl}(-\omega, \omega, \omega_L, -\omega_L)$$
$$+ \chi_{3t}{}^{iljk}(-\omega, -\omega_L, \omega, \omega_L)\left[\frac{\Gamma_t}{\omega_t - (\omega + \omega_L) - i\Gamma_t}\right]$$
$$+ \chi_{3r}{}^{ikjl}(-\omega, \omega_L, \omega, -\omega_L)\left[\frac{\Gamma_r}{\omega_r - (\omega - \omega_L) - i\Gamma_r}\right.$$
$$+ \left.\frac{\Gamma_r}{\omega_r - (\omega_L - \omega) + i\Gamma_r}\right]. \quad (41)$$

Here only one resonance to a state t of width Γ_t involving the simultaneous absorption of two photons, and one resonance to a state r of width Γ_r involving the emission of one photon and the absorption of another, have been assumed. The resonant nonlinear polarizabilities χ_{3t} and χ_{3r} are of the same form, given by

$$\chi_{3s}{}^{ijkl}(\omega_1, \omega_2, \omega_3, \omega_4)$$
$$= \frac{\epsilon_0 NL}{4\hbar\Gamma_s}\langle g|\alpha_{ij}(\omega_1, \omega_2)|s\rangle\langle g|\alpha_{kl}(-\omega_3, -\omega_4)|s\rangle^* \quad (42)$$

where N is the molecular density which is assumed to be in the ground state g. L is a local field correction factor of the order of $(\kappa+2)^4/81$, which leads as with χ_2 to high values of χ_3 in high index materials. $\langle g | \alpha^{ij}(\omega_1, \omega_2) | s \rangle$ are the polarizability matrix elements for the molecular transitions involved. χ_{3r} for molecular vibrational transitions is typically of the order of $[4\pi/(3\times10^4)^2]\times10^{-12}$ M^2/V^2 with Γ_r about 10^{11} rad/s. χ_{3t} for electronic transitions has about the same value as χ_{3r} with Γ_t typically 10^{13} rad/s. Values of $6\chi_{3NR}$ are about two orders of magnitude smaller.

The two photon resonance at ω_r can also couple together four components whose frequencies satisfy the relation $\omega_1-\omega_3=\omega_2-\omega_4=\omega_r$. For example, the components at ω_1, $\omega_1-\omega_r$, ω_2 and $\omega_2-\omega_r$ are coupled through a χ_3 term of the form

$$6\chi_3{}^{ijkl}[-(\omega_1 - \omega_r), \omega_1, (\omega_2 - \omega_r), -\omega_2)]$$
$$= 6\chi_{3NR}{}^{ijkl}[-(\omega_1 - \omega_r), \omega_1, (\omega_2 - \omega_r), -\omega_2]$$
$$+ \chi_{3r}{}^{ijkl}[-(\omega_1 - \omega_r), \omega_1, (\omega_2 - \omega_r), -\omega_2]. \quad (43)$$

The two photon resonances involving the difference of two frequencies give rise to spontaneous Raman scattering, from which the frequency dependence of χ_3'' associated with these resonances can be deduced. In addition to the molecular vibrational and rotational transitions in Raman spectra, one can also observe low-frequency electronic transitions [55], the scattering of acoustic waves [56], [57], and other collective excitations [58], [59]. Excellent quantitative data on Raman-type scattering are becoming available through the use of gas lasers [60]. In particular, high resolution data [61] on scattering involving small frequency shifts is now available (Fig. 9).

A large number of experiments involving two photon absorption properties have been carried out. These include the blue fluorescence in a CaF_2 sample illuminated in the red with a regular ruby laser [2], the high percentage intensity dependent absorption in CS_2 [62], the photoionization of iodine through two photon absorption processes [63], the saturation of the two photon absorption in cesium vapor [64], and the inverse Raman effect [65], namely, the absorption at $\omega_L+\omega_r$ in Raman-active liquids.

Experiments involving the interaction of white light with a laser beam are particularly important since they provide the spectra of χ_3. Hopfield and Worlock [66] first observed appreciable induced percentage absorption in a crystal of KI, thus obtaining its two photon absorption spectrum. Recently, similar spectra have also been obtained in anthracene [67] (Fig. 10).

B. Stimulated Raman Scattering

It has been noted earlier that the emission near the Stokes frequency $\omega_s=\omega_L-\omega_r$ is associated with the negative $\chi_{3r}''(-\omega_s, \omega_L, \omega_s, -\omega_L)$ in (41). This emission can be described in terms of a material gain given from (4) by

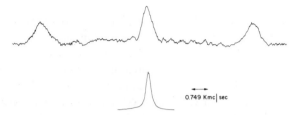

Fig. 9. High resolution spectrum of CS_2 showing the Rayleigh line and the resolved Brillouin components. The scattering was observed at 90° to the direction of the beam from a frequency-stabilized single mode He-Ne laser. The lower trace is a measurement of the scattering from large suspended particles and represents the resolution limit of the pressure scanning Fabry-Perot interferometer (from Cederquist et al. [61]).

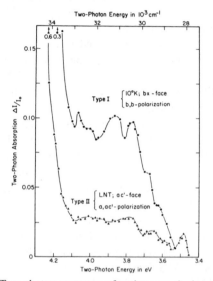

Fig. 10. Two-photon spectrum of anthracene single crystals. The relative change in transmission $\Delta I/I$ of light from a source of variable photon energy is plotted vs. the combined photon energy of both sources. For type-I spectrum, the Nd laser light and the light from the variable source were both polarized along the crystal b axis. For the type-II spectrum, the laser light was polarized along the crystal a axis and the light from the variable source was unpolarized in a plane perpendicular to the crystal b axis (from Fröhlich et al. [67]).

$$g = \frac{k_s}{2\kappa_s} \chi_{3r}{}^{1122}(-\omega_s, \omega_L, \omega_s, -\omega_L)E_j(\omega_L, r)E_j{}^*(\omega_L, r). \quad (44)$$

With $|E(\omega_L, r)|\sim2\times10^5$ V/meter, this gain is of the order of several decibels per centimeter in a typical Raman-active liquid. By placing a cell of nitrobenzene inside the cavity of a ruby giant pulse laser, Woodbury et al. [68] observed that this gain led to *Raman laser action* or stimulated Raman emission at the Stokes frequency (Fig. 11). The stimulated Raman emission has since been observed in a variety of liquids [69], solids [70], and gases [71]. The output spectrum of the stimulated emission usually consists of a series of intense lines extending toward longer wavelengths at $\omega_L-\omega_r$, $\omega_L-2\omega_r$, $\omega_L-3\omega_r$, etc., where ω_r corresponds to one of the strongest lines observed in the normal Raman spectra of the material. It is believed that the second- and higher order Stokes radiation result from the repeated processes of Raman laser action due to the Stokes lines [69].

The discovery of Raman laser action was followed by attempts to correlate the Raman gain measured with collimated laser beams with that calculated from the line strengths and widths measured in *spontaneous* Raman scattering. For liquids such as acetone and alcohol, the theory was found to be in good agreement with experimental observations [72]. However, measurements in other liquds such as CS_2 and nitrobenzene indicated that discrepancies of one to two orders of magnitude existed. It was also observed that the onset of stimulated Raman emission as a function of laser power was abrupt, contrary to the theoretical predictions (Fig. 12). It has been shown that most of these anomalies can be explained in terms of self-focusing and beam trapping which will be discussed in Section IV-D.

Experiments involving focused beams seem to be free from the effects of beam trapping, but are complicated by the introduction of geometric factors and by the possible additional interactions arising from the higher beam intensity at the focus.

With focused beams, an additional series of lines at higher frequencies (anti-Stokes) are observed emerging as concentric rings [54], as shown in Fig. 13. These anti-Stokes lines are separated by the same Raman frequency as are the Stokes lines which appeared along the axis. The major features here can be explained qualitatively as follows: first, the process is initiated by that portion of the spontaneous Stokes radiation which passes through the focal volume and is amplified there by the Raman gain in (44). Since the focal volume resembles a long thin cylinder [73], the gain will peak at nearly forward angles, thus making the stimulated Stokes emission predominantly along the axis. Secondly, with the creation of the Stokes wave at $\omega_L - \omega_r$, a nonlinear polarization is induced at the first anti-Stokes frequency $\omega_L + \omega_r$. This interaction also involves resonant values of χ_3 in (43), and will lead to large effects when the phase matching conditions are satisfied. Because the medium is dispersive, phase matching conditions can be satisfied exactly only if both the Stokes and anti-Stokes radiation emerge at different angles relative to the axis. In practice, the Stokes radiation is predominantly along the axis. The anti-Stokes radiation emerges at those angles which minimize the phase mismatch [74]. The above situation is further complicated in that these angles are affected by the intensity dependent changes of index [74]. Anti-Stokes radiation is, in general, not observed with the Raman material placed in the laser cavity, as the phase mismatch there is too great.

In the above discussion, each process has been considered separately. This is a poor approximation since many nonlinear processes are comparable in magnitude and occur simultaneously under nearly index matched conditions. For instance, if the equations for Raman laser action are considered along with those for index matched mixing of the three waves at frequencies $\omega_L - \omega_r$, ω_L, and $\omega_L + \omega_r$, one predicts a greatly reduced exponential gain factor at the Stokes frequency $\omega_L - \omega_r$ [75]. This

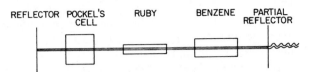

Fig. 11. Schematic of the experimental arrangement employed by Woodbury et al. [68] in the first observations of stimulated Raman effect. The mirrors are reflecting for both the laser and Stokes frequencies. The onset of Raman laser action exhibited a sharp threshold dependence on laser intensity and resulted in a large fractional conversion of energy into the Stokes lines.

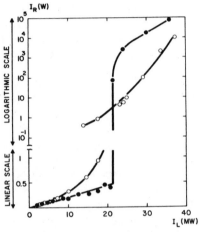

Fig. 12. Observed Stokes intensity as a function of incident laser power in several different materials. One class of materials gives qualitative agreement with the anticipated variation $I_s \sim \exp(gI_Ll)$ while the other exhibits an abrupt increase in Stokes energy at low laser power (from Bret et al. [72]).

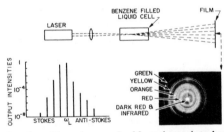

Fig. 13. Results of focusing a pulsed laser beam into benzene. The ring pattern shown was produced by placing film behind the sample at right angles to the optical axis. Each of the visible lines are separated by 992 cm^{-1}, the frequency of the strongest Raman line in benzene. The relative strengths of the lines are indicated.

effect is analogous to the lack of gain in a parametric amplifier when neither the upper nor lower sideband is suppressed.

At greatly increased laser power levels, the spectrum emitted from liquids at the focus shows that the laser frequency is broadened by tens of cm^{-1} and that the anti-Stokes lines are spread out into bands hundreds of cm^{-1} or more in width. It has been proposed that this broadening can be explained in terms of the stimulated Rayleigh scattering process associated with the alignment of molecules [76], [77].

The Raman gain in (44) is independent of the relative angle between the Stokes and laser beams polarized parallel to each other. This is consistent with the observation

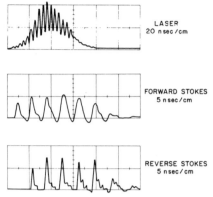

Fig. 14. Examples of transient phenomena associated with stimulated Raman scattering in hydrogen gas. The Stokes energy is observed both in the direction of the laser beam and in the reverse direction. When stimulated by a modulated laser beam, the reverse Stokes exhibits highly transient behavior, its width being instrument limited.

of Stokes gain in off-axis cavities [78], and in a cavity illuminated by a diffused laser source [79]. However, with a collimated beam, the Stokes radiation is observed predominantly in the forward direction. The backward Stokes radiation varies considerably with different experimental conditions, and can exhibit highly transient behavior [80] (Fig. 14).

Of current interest are problems associated with the details of anomalous gain, with the competition between various stimulated processes, with the variation of the *front-to-back* ratio of the Stokes emission, and with the effects of saturation of Raman transitions. Stimulated Raman emission involving pure rotational transitions [81] has recently been observed, and should also be a subject for further investigation.

C. Stimulated Brillouin Scattering

Stimulated Brillouin scattering arises from the interaction of optical and acoustic waves. During this process, the response of the medium at one point is effected by the response at neighboring points, as well as by the response in an entire region with dimensions comparable to the acoustic decay length. Whereas the stimulated Raman scattering has been described in terms of a point interaction model because of the highly localized character of Raman transitions, it is apparent that such a model is inadequate to describe the stimulated Brillouin process. The following general formalism is outlined for the Brillouin process only, but should also be applied to other scattering processes.

A proper description of Brillouin scattering can be developed from the classical Lagrangian density formalism [82]. In this approach, the interaction energy is added to the kinetic and potential energy terms for the unperturbed optical and acoustic waves. A wave equation containing coupling terms is then obtained for each component.

The linearized equations for the change of amplitudes of the Stokes and acoustic waves at constant laser inten-

sity are obtained by neglecting the second derivatives of the wave amplitudes and assuming that each wave satisfies its normal dispersion relationships. One can then write

$$\frac{\partial}{\partial t} E_s(r, t) - (c/n) \frac{\partial}{\partial r} E_s(r, t) = C_1 S^*(r, t)$$

$$\frac{\partial}{\partial t} S^*(r, t) - (1/2\tau) S^*(r, t) + v_a \frac{\partial}{\partial r} S^*(r, t)$$
$$= C_2 E_s(r, t), \quad (45)$$

where $E_s(r, t)$ and $S^*(r, t)$ are the slowly varying amplitudes of the Stokes and acoustic waves with their harmonic factors $\exp i(kr - \omega t)$ separated out; τ and v_a are, respectively, the decay time and velocity of the acoustic wave; and C_1 and C_2 are constants related to the Pockel's photoelastic coefficients and to the laser intensity. Note that the time derivatives have been retained in (45) so as to allow for the transient buildup of both the optical and the acoustic waves. Kroll [82] has treated such a system of equations, and has solved them subject to the neglect of $\partial E_s/\partial t$ and other minor approximations. For times typical of giant pulse lasers, the following solution has been obtained:

$$\frac{E_s(l, t)}{E_s(0, t)} = \exp\left[\left(\frac{g'I_L lt}{\tau}\right)^{1/2} - \frac{t}{2\tau}\right];$$
$$t < g'I_L l\tau. \quad (46)$$

Here g' is the steady-state power gain normalized per unit length per unit laser intensity, I_L is the laser intensity, and l is the interaction length. Equation (46) is valid for times t up to $g'I_L l\tau$, after which the steady-state solution $E_s(r, t) \propto \exp(g'I_L l)$ applies.

It follows from (46) that the Stokes gain is initially zero, increasing as $t^{1/2}$ for times short compared to the acoustic lifetime; thus the gain under transient conditions may be very much less than its stady-state value. Reduction of gain by several orders of magnitude is implied by measurements of the growth of the Stokes amplitude with cell length [83], and by measurements of the laser power required to achieve a large Stokes gain [82a]. If the conditions are such that stimulation occurs in a time comparable to the transit time required for an optical wave to traverse the dimensions of the interaction volume, the term $\partial E_s/\partial t$ must be retained. Extensive numerical calculations would be required to extend the treatment of transient phenomena to include effects such as modulation or depletion of the laser beam, or competition between several simultaneously stimulated processes.

The frequency ω_a of the acoustic wave is determined by the conservation of energy and momentum and is given by

$$\omega_a/\omega_L = 2(nv_a/c) \sin(\theta/2) \quad (47)$$

where θ is the angle between the laser and the scattered Stokes beam. Since v_a is much less than c/n, the acoustic

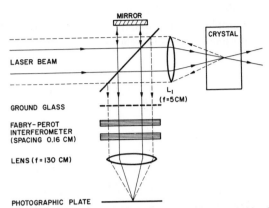

Fig. 15. Schematic of the experimental arrangement used by Chiao et al. [85] to study stimulated Brillouin scattering. The Stokes energy is radiated in the backward direction allowing it to re-enter the laser cavity, be amplified, and returned to the material where it can then cause repeated Stokes lines.

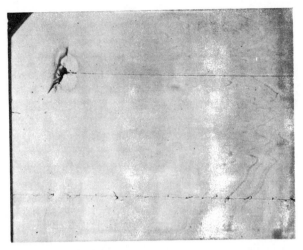

Fig. 16. Damage characteristics observed in BK-7 glass at the focus of a 46 cm lens. The ruby laser pulse used was 60 mJ in energy and 50 ns in width. The damage appears as a combination of a gross fracture and a long filament made up of very small bubbles and extending along the lens axis. In general these filaments may be has long as several centimeters and at the same time have a diameter of only a few wavelengths. In the absence of the self-focusing action of the laser beam, it would be difficult to explain the preservation of such long thin filaments over a distance longer than the order of wavelengths. (Courtesy of J. G. Atwood, Perkin-Elmer Corp.).

frequency is small, of the order 0.03 cm^{-1} for gases [84], 0.1 cm^{-1} for liquids [85], and 1 cm^{-1} for solids [86].

In most experiments stimulated Brillouin scattering is observed by passing the laser beam through the medium. The stimulated Stokes wave is usually radiated in the reverse direction (Fig. 15); thus it can re-enter the laser cavity, be amplified by the laser, and return to the medium to produce repeated Stokes lines at $\omega_L - \omega_a$, $\omega_L - 2\omega_a$, etc. Anti-Stokes waves have not been observed since the oppositely directed laser and Stokes beams produce severe phase mismatch for the induced polarization at the anti-Stokes frequency.

In anisotropic materials, the conservation of momentum and energy can be satisfied by additional combinations of Stokes and acoustic waves having frequencies and directions of propagation different from those allowable in isotropic material [87]. For example, a 75 MHz acoustic wave in crystalline quartz was observed resulting from the interaction of the laser beam and a forward Stokes wave polarized orthogonally to each other [88].

D. Self-Focusing of Laser Beams

So far in discussing nonlinear optical effects, only unbounded plane waves have been considered. In linear optics a light beam of finite cross section can be represented by a superposition of unbounded plane wave components propagating in slightly different directions. The existence of optical nonlinearities in dielectric media invalidates the principles of superposition: The plane wave components are no longer independent, but are coupled to each other through nonlinear polarization terms which bring about transfer of energy among the components. In this section the effects of intensity dependent index on the diffraction of a beam with finite cross section will be considered. At high beam intensities these effects modify the diffraction and lead to self-focusing of laser beams.

The experimental evidence for self-focusing of a laser beam has come notably from 1) studies of the laser-induced damage in glasses [89], where the damage appeared in the form of long, thin filaments along the lens axis (Fig. 16); and 2) studies of stimulated Raman emission in liquids with anisotropic molecules, where the existence of anomalous gain could be reconciled only by recognizing the formation of high-intensity filaments in the laser beam [90]. It thus appears that an intense light beam may form its own waveguide and propagate without diffraction.

The self-focusing effect of a laser beam can be readily understood in the approximations of geometrical optics. For normal dielectrics, the refractive index increases with intensity. It follows that, because of the nonuniform intensity distribution inherent with a beam of finite cross section, the intensity dependent index of refraction causes different parts of the beam to propagate with different phase velocities. A lens effect (Fig. 17) is thus produced whereby the rays move toward the region of higher intensity and increase the intensity there. This increase in intensity is accompanied by a reduction in the effective beam diameter, and continues until it is limited by other factors. A threshold exists for the onset of self-focusing as it must overcome the spreading of the beam due to diffraction. Chiao, Garmire, and Townes [52] have predicted that a light beam may be trapped at any arbitrary diameter and will thus not spread. They have further predicted that self-trapping occurs at a critical power level which is independent of the beam diameter. While it is

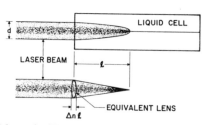

Fig. 17. Schematic illustrating the self-focusing effect and its equivalent convergent lens. The self-focusing action may be likened to a convergent lens, for which the difference in optical path length between the marginal and axial rays is compensated by the presence of lens material located on the lens axis so that both rays arrive at the focus in phase. The equivalent focal length is thus by analogy determined by the relation $\Delta nl = n_0\{l[1+(a/l)^2]^{1/2}-l\}$, where a is the beam radius at half-intensity points. Equation (48) follows after a few manipulations.

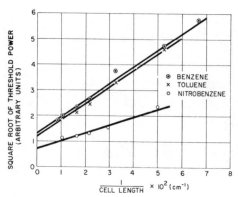

Fig. 18. Plot of the square root of the threshold laser power as a function of the inverse of the cell length for benzene, toluene, and nitrobenzene. The threshold for stimulated Raman emission was taken as the threshold for self-focusing. These plots show that the predicted dependence in (48) is well satisfied for self-focusing; the vertical intercepts of these straight lines give the values of critical power for self-trapping.

TABLE II
VALUES OF THE NONLINEAR SUSCEPTIBILITY CONSTANTS AND THE CRITICAL POWER FOR SELF-TRAPPING IN SEVERAL LIQUIDS

Liquid	$\chi_3^{1221}(-\omega, \omega, \omega, -\omega)\times 10^{14}$ $\times[(3\times 10^4)^2/4\pi](M^2/V^2)$	$[\chi_3^{1221}(-\omega, \omega, \omega, -\omega)+\chi_3^{1122}(-\omega, \omega, \omega, -\omega)]\times 10^{14}$ $\times[(3\times 10^4)^2/4\pi](M^2/V^2)$	Critical Power (kW)	
			Calculated	Measured
CS_2	34[a]	38.2[b]	16[c]	15[a] 25±5[d]
Nitrobenzene	21[a]	26.3[b]	24[c]	22[a]
Toluene	8[a]	8.9[b]	56[c]	57[a]
Benzene	6.5[a]	3.7[b]	76[c]	76[a]

[a] C. C. Wang [96].
[b] G. Mayer and F. Gires [50].
[c] Calculated with the measured values of $\chi_3^{1221}(-\omega, \omega, \omega, -\omega)$, assuming $\chi_3^{1221}(-\omega, \omega, \omega, -\omega)/[\chi_3^{1122}(-\omega, \omega, \omega, -\omega)+\chi_3^{1212}(-\omega, \omega, \omega, -\omega)]=3$.
[d] E. Garmire, R. Y. Chiao, and C. H. Townes [93].

not yet understood which factors determine the size of the observed high-intensity filaments, the distance required for the establishment of these filaments should depend very little on the terminal filament size if the beam is reduced in diameter by an appreciable factor. This distance has been referred to as the self-focusing length l, and is related to the input laser power P through the following relation [91], [92]:

$$P^{1/2} = P_{cr}^{1/2} + \frac{A}{l} \tag{48}$$

where

$$A = \frac{n}{4}\left(\frac{a^2}{f}\right)\left(\frac{c}{n_2}\right)^{1/2}. \tag{49}$$

Here n_2 is related to the index change in (39), $\delta n_{||}=(n_2/2)|E|^2$; P_{cr} is the critical power for cylindrical beam trapping [52]

$$P_{cr} = \frac{5.763\lambda^2 c}{16\pi^3 n_2}, \tag{50}$$

a is the radius of the beam; and f is a parameter introduced to account for deviations of the beam from an equiphase Gaussian intensity profile. $f=1$ for such a Gaussian beam.

The above considerations have been verified experimentally by direct observation of the evolution of beam trapping [93], and by analyzing the length-dependent threshold data [92] for stimulated Raman emission. According to (48), the critical power for self-trapping is obtained from the vertical intercept of the straight lines in Fig. 18; the results are in good agreement with the calculated values for a number of liquids (see Table II). The effect of one- and two-photon absorption has been considered [94] and found to be in good agreement with the experiments.

In addition to the molecular reorientation and nonlinear electronic polarizability, the index changes may also be associated with electrostriction which gives rise to macroscopic density changes across the beam cross section. This electrostrictive effect is characterized by a relatively long time constant, as it must propagate across the cross section of the beam at the sound velocity ($\sim 10^5$ cm/s). In liquids, the electrostrictive contribution is of the order of $[4\pi/(3\times 10^4)^2]\times 10^{-12}$ M^2/V^2 in the steady state, but its effectiveness may be reduced significantly for a Q-switched laser because of its short pulse durations,

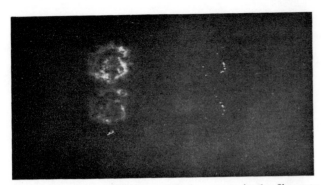

Fig. 19. Random polarization of Stokes energy in the filaments resulting from Raman laser action and self-focusing in nitrobenzene when stimulated by a circularly polarized laser beam. The radiation was analyzed for orthogonal components of linear polarization (top and bottom) and dispersed by a grating to separate the laser (left) and Stokes (right) beams for the photograph. The circularly polarized laser beam produces an identical spatial pattern having (nearly) equal intensity for the two plane components. In contrast, the ratios of intensities in the two polarizations for each small Stokes filament are uncorrelated, indicating that each filament is plane polarized in an arbitrary direction (from McClung et al. [98]).

This estimate on the order of magnitude of the coefficients is consistent with the temperature dependence of the self-focusing action [95] and with the polarization dependence of the refractive index changes [96].

The importance of the self-focusing action of a laser beam has been recognized only recently. Although this effect has now been grossly characterized and understood, many detailed features remain to be explained. In their direct observation of beam trapping, Garmire et al. [97] have observed the presence of small scale trapping (in which a multiplicity of small filaments are formed) superimposed upon large scale trapping (in which the entire beam is trapped as a unit). It is not at all understood why the beam does not focus as a unit, or which factors determine the final size of these high-intensity filaments. The state of polarization of these small filaments has been studied by McClung and co-workers [98]; with the input laser beam approximately circularly polarized, the filaments were found to be linearly polarized, but with random orientation (Fig. 19). Furthermore, it has been observed [96] that the ratio of the self-focusing threshold between circularly and linearly polarized beams disagrees with the ratio of the index changes in (38) and (39). It appears that although the optical Kerr effects associated with molecular reorientation are dominant in determining the critical power for self-trapping, other effects may also be important once the whole beam becomes self-trapped. It is possible that the small scale trapping, and thus the size of the filaments, is related to the inhomogeneity both in the laser intensity distribution, and in the liquid medium.

V. HIGHER ORDER PROCESSES

Second harmonic generation has been detected in liquids and gases composed of molecules lacking a center of inversion [99]. In this case the induced second harmonic polarization on each molecule is randomly oriented

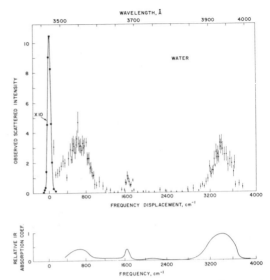

Fig. 20. Observed unpolarized nonlinear scattering spectra and infrared absorption spectra for water. The scattering intensity is given in photoelectrons per laser pulse ($\frac{1}{2}$ MW peak power) averaged over approximately 25 pulses. Standard errors are indicated. When no signal was observed, the 70 percent probability limits using Poisson statistics are shown. The spectrometer band pass was 80 cm^{-1} (from Terhune et al. [99]).

so that each molecule acts as a separate radiating source. That is, the second harmonic radiation does not emerge as a beam but rather as very weak scattered radiation analogous to Rayleigh scattering. Inelastic scattering analogous to Raman scattering also occurs. The observed spectrum for water is shown in Fig. 20.

The second harmonic scattering can be thought of as a three photon process involving the simultaneous absorption of two photons from the laser beam and the spontaneous emission of a photon near twice the laser frequency. The theory for this scattering closely parallels that for Rayleigh and Raman scattering, which only involve two photons. Much as the resonance values of χ_3 are deducible from Raman data, certain resonant values of χ_5, $(P \propto \chi_5 E^5)$, are deducible from the data of second harmonic scattering.

Multiphoton absorption processes have been postulated as an essential step preceding laser induced electrical breakdown of gases [100]. Calculations of the probability for the simultaneous absorption of up to fourteen photons by helium atoms have been carried out. The cross sections for these multiphoton processes appear sufficient to provide the few free electrons necessary to initiate breakdown.

It appears that only under very special circumstances will it be possible to measure any of the higher order coefficients such as χ_4 or χ_5. In most cases the physical effects produced by these terms are also produced by multiple application of lower order terms. For example, fourth harmonic generation can occur through a χ_4 term or through two step processes involving either the second or the third harmonic as the intermediate step. As another example, a luminescence whose intensity is proportional to the cube of the laser intensity has been observed

[101]. This luminescence could result from a three photon absorption process and thus yield values for χ_5. However, it could also be due to third harmonic generation followed by linear absorption of the harmonic radiation; or it could be due to two photon absorption followed by a linear absorption.

VI. CONCLUSIONS

Despite the progress that has been made toward understanding nonlinear optical phenomena, much remains to be done. In addition to the many problems already indicated, more theoretical and experimental work is required to predict and measure the spectral properties of the nonlinear susceptibility tensors. These studies should provide a great deal of new information about the physical properties of materials.

The view taken of nonlinear optics in this paper is really quite limited. The nonlinear interactions are also responsible for the coupling of modes within lasers [102]. Because of space limitations, we have reluctantly neglected this topic and others such as plasma interactions, breakdown phenomena, surface effects, and effects on chemical reactions. The field of nonlinear optics remains a fruitful one, for both scientists and engineers.

REFERENCES

[1] P. A. Franken, A. E. Hill, C. W. Peters, and G. Weinreich, "Generation of optical harmonics," *Phys. Rev. Lett.*, vol. 7, pp. 118–119, August 1961.
[2] W. Kaiser and C. G. B. Garrett, "Two-photon excitation in $CaF_2:Eu^{2+}$," *Phys. Rev. Lett.*, vol. 7, pp. 229–231, September 1961.
[3] J. A. Giordmaine and R. C. Miller, "Tunable coherent parametric oscillations in $LiNbO_3$ at optical frequencies," *Phys. Rev. Lett.*, vol. 14, pp. 973–976, June 1965.
[4] N. Bloembergen, *Nonlinear Optics*. New York: Benjamin, 1965.
[5] P. A. Franken and J. F. Ward, "Optical harmonics and nonlinear phenomena," *Rev. Mod. Phys.*, vol. 35, pp. 23–29, January 1963.
[6] P. S. Pershan, "Nonlinear Optics," in *Progress in Optics*, vol. 5. New York: Interscience, Amsterdam: North-Holland, 1966, in progress.
[7] L. N. Ovander, "Nonlinear optical effects in crystals," *Soviet Phys. Uspekhi*, vol. 8, pp. 337–359, November-December 1965.
[8] J. A. Armstrong, N. Bloembergen, J. Ducuing, and P. S. Pershan, "Interactions between light waves in a nonlinear dielectric," *Phys. Rev.*, vol. 127, pp. 1918–1939, September 1962.
[9] R. W. Terhune and P. D. Maker, *Advances in Lasers, II*, A. K. Levine, Ed. New York: Marcel Dekker, to be published.
[10] W. G. Cady, *Piezoelectricity*. New York: McGraw-Hill, 1946, pp. 177–199.
[11] P. J. Price, "Theory of quadratic response functions," *Phys. Rev.*, vol. 130, pp. 1792–1797, June 1963.
[11a] W. J. Caspers, "Dispersion relations for nonlinear response," *Phys. Rev.*, vol. 133, pp. A1249–A1251, March 1964.
[12] D. A. Kleinman, "Nonlinear dielectric polarization in optical media," *Phys. Rev.*, vol. 126, pp. 1977–1979, June 1962.
[13] P. S. Pershan, "Nonlinear optical properties of solids: Energy considerations," *Phys. Rev.*, vol. 130, pp. 919–928, May 1963.
[14] P. D. Maker, R. W. Terhune, M. Nisenoff, and C. M. Savage, "Effects of dispersion and focusing on the production of optical harmonics," *Phys. Rev. Lett.*, vol. 8, pp. 21–22, January 1962.
[15] J. A. Giordmaine, "Mixing of light beams in crystals," *Phys. Rev. Lett.*, vol. 8, pp. 19–20, January 1962.
[16] S. A. Akhmanov, A. I. Kovrigin, A. S. Piskarskas, and R. V. Khokhlov, "Generation of uv radiation by using cascade frequency conversion," *JETP Lett.*, vol. 2, pp. 141–143, September 1965.
[17] D. A. Kleinman, "Theory of second harmonic generation of light," *Phys. Rev.*, vol. 128, pp. 1761–1775, November 1962.
[18] C. C. Wang and G. W. Racette, "Saturation effects in second harmonic generation of light using unfocused laser beams," *J. Appl. Phys.*, vol. 36, pp. 3281–3284, October 1965.
[19] R. C. Miller, G. D. Boyd, and A. Savage, "Nonlinear optical interactions in $LiNbO_3$ without double refraction," *Appl. Phys. Lett.*, vol. 6, pp. 77–79, February 1965.
[19a] N. I. Adams, III, and J. J. Barrett, "Electric field control of 90° phase matching in KDP," *IEEE J. of Quantum Electronics*, vol. QE-2, p. xxxi, April 1966.
[20] R. C. Miller, D. A. Kleinman, and A. Savage, "Quantitative studies of optical harmonic generation in CdS, $BaTiO_3$, and KH_2PO_4 type crystals," *Phys. Rev. Lett.*, vol. 11, pp. 146–149, August 1963.
[21] G. D. Boyd, R. C. Miller, K. Nassau, W. L. Bond, and A. Savage, "$LiNbO_3$: An efficient phase matchable nonlinear optical material," *Appl. Phys. Lett.*, vol. 5, pp. 234–236, December 1964.
[22] R. K. Chang, J. Ducuing, and N. Bloembergen, "Relative phase measurement between fundamental and second-harmonic light," *Phys. Rev. Lett.*, vol. 15, pp. 6–8, July 1965.
[23] R. C. Miller, "Optical second harmonic generation in piezoelectric crystals," *Appl. Phys. Lett.*, vol. 5, pp. 17–19, July 1964.
[24] C. K. N. Patel, "Optical harmonic generation in the infrared using a CO_2 laser," *Phys. Rev. Lett.*, vol. 16, pp. 613–616, April 1966.
[25] A. Ashkin, G. D. Boyd, and J. M. Dziedzic, "Observation of continuous optical harmonic generation with gas masers," *Phys. Rev. Lett.*, vol. 11, pp. 14–17, July 1963.
[26] G. E. Francois, "CW Measurement of the optical nonlinearity of ADP," *Phys. Rev.*, vol. 143, pp. 597–600, March 1966.
[27] G. D. Boyd, A. Ashkin, J. M. Dziedzic, and D. A. Kleinman, "SHG of light with double refraction," *Phys. Rev.*, vol. 137, pp. A1305–A1320, February 1965.
[28] J. E. Bjorkholm, "Optical SHG using a focused laser beam," *Phys. Rev.*, vol. 142, pp. 126–136, February 1966.
[29] D. A. Kleinman, A. Ashkin, and G. D. Boyd, "Second harmonic generation of light by focused laser beams," *Phys. Rev.*, vol. 145, pp. 338–379, May 1966.
[30] M. Bass, P. A. Franken, J. F. Ward, and G. Weinreich, "Optical rectification," *Phys. Rev. Lett.*, vol. 9, pp. 446–448, December 1962.
[31] J. F. Ward, "Absolute measurement of an optical rectification coefficient in ADP," *Phys. Rev.*, vol. 143, pp. 569–574, March 1966.
[32] M. D. Martin and E. L. Thomas, "The generation of molecular vibrational frequencies by optical mixing," *Phys. Lett.*, vol. 19, pp. 651–652, January 1966.
[33] F. Zernike, Jr., and P. R. Berman, "Generation of far-infrared as a difference frequency," *Phys. Rev. Lett.*, vol. 15 pp. 999–1001, December 1965.
[34] D. H. McMahon and A. R. Franklin, "Detection of nonlinear optical sum spectra in ADP using incoherent light," *J. Appl. Phys.*, vol. 36, pp. 2807–2810, September 1965.
[35] W. H. Louisell, *Coupled-Mode and Parametric Electronics*. New York: Wiley, 1960.
[36] C. C. Wang and G. W. Racette, "Measurement of parametric gain accompanying optical difference frequency generation," *Appl. Phys. Lett.*, vol. 6, pp. 169–171, April 1965.
[36a] G. D. Boyd and A. Ashkin, "Theory of parametric oscillator threshold with single-mode optical masers and observation of amplification in $LiNbO_3$," *Phys. Rev.*, vol. 146, pp. 187–198, June 1966.
[37] S. A. Akhmanov, A. I. Kovrigin, V. A. Kolosov, A. S. Piskarskas, V. V. Fadeev, and R. V. Khokhlov, "Tunable parametric light generator with KDP crystal," *JETP Lett.*, vol. 3, pp. 241–245, May 1966.

[38] S. A. Akhmanov, A. I. Kovrigin, A. S. Piskarskas, V. V. Fadeev, and R. V. Khokhlov, "Observation of parametric amplification in the optical range," *JETP Lett.*, vol. 2, pp. 191–193, October 1965.

[39] N. Bloembergen and P. S. Pershan, "Light waves at the boundary of nonlinear media," *Phys. Rev.*, vol. 128, pp. 606–622, October 1962.

[40] J. Ducuing and N. Bloembergen, "Observation of reflected light harmonics at the boundary of piezoelectric crystals," *Phys. Rev. Lett.*, vol. 10, pp. 474–476, June 1963.

[41] R. K. Chang and N. Bloembergen, "Experimental verification of the laws for the reflected intensity of second harmonic light," *Phys. Rev.*, vol. 144, pp. 775–780, April 1966.

[42] S. S. Jha, "Theory of optical harmonic generation at a metal surface," *Phys. Rev.*, vol. 140, pp. A2020–A2030, December 1965

[43] F. Brown and R. E. Parks, "Magnetic-dipole contribution to optical harmonics in silver," *Phys. Rev. Lett.*, vol. 16, pp. 507–509, March 1966.

[44] N. Bloembergen, R. K. Chang, and C. H. Lee, "Second-harmonic generation of light in reflection from media with inversion symmetry," *Phys. Rev. Lett.*, vol. 16, pp. 986–989, May 1966.

[45] R. W. Terhune, P. D. Maker, and C. M. Savage, "Optical harmonic generation in Calcite," *P'ys. Rev. Lett.*, vol. 8, pp. 404–406, May 1962.

[46] J. A. Giordmaine, "Nonlinear optical properties of liquids," *Phys. Rev.*, vol. 138A, pp. A1599–A1606, June 1965.

[47] P. M. Rentzepis, J. A. Giordmaine, and K. W. Wecht, "Coherent optical mixing in optically active liquids," *Phys. Rev. Lett.*, vol. 16, pp. 792–794, May 1966.

[49] R. R. Briss, "Property tensors in magnetic crystal classes," *Proc. Phys. Soc. (London)*, vol. 79, pp. 946–953, May 1962.

[49] P. D. Maker, R. W. Terhune, and C. M. Savage, *Quantum Electronics III*. New York: Columbia University Press, 1964, p. 1559.

[50] G. Mayer and F. Gires, "Action of an intense light beam on the refractive index of liquids," *Comptes Rendus*, vol. 258, pp. 2039–2042, February 1964.

[51] L. D. Landau and E. M. Lifshitz, *Electrodynamics of Continuous Media*. New York: Pergamon, 1960, pp. 377–397.

[52] R. Y. Chiao, E. Garmire, and C. H. Townes, "Self-trapping of optical beams," *Phys. Rev. Lett.*, vol. 13, pp. 479–482, October 1964.

[53] P. D. Maker, R. W. Terhune, and C. M. Savage, "Intensity-dependent changes in the refractive index of liquids," *Phys. Rev. Lett.*, vol. 12, pp. 507–509, May 1964.

[54] P. D. Maker and R. W Terhune, "Study of optical effects due to an induced polarization third order in the electric field strength," *Phys. Rev.*, vol. 137, pp. A801–A818, February 1965.

[55] J. T. Hougen and S. Singh, "Electronic Raman effect in Pr^{3+} ions in single crystals of $PrCl_3$," *Phys. Rev. Lett.*, vol. 10, pp. 406–407, May 1963

[56] G. B. Benedek, J. B. Lastovka, K. Fritsch, and T. Greytak, "Brillouin scattering in liquids and solids using low-power lasers," *J. Opt. Soc. Am.*, vol 54, pp. 1284–1285, October 1964

[57] R. Y. Chiao and B. P. Stoicheff, "Brillouin scattering in liquids excited by the He-Ne laser," *J. Opt. Soc. Am.*, vol. 54, pp. 1286–1287, October 1964.

[58] R. J. Elliott and R. Loudon, "The possible observation of electronic Raman transitions in crystals," *Phys. Lett.*, vol. 3, pp. 189–191, January 1963.

[59] P. M. Platzman and N. Tzoar, "Nonlinear Interaction of light in a plasma," *Phys. Rev.*, vol. 136, pp. A11–A16, October 1964.

[60] T. C. Damen, R. C. C. Leite, and S. P. S. Porto, "Angular depednence of the Raman scattering from benzene excited by the He-Ne CW laser," *Phys. Rev. Lett.*, vol. 14, pp. 9–11, January 1965.

[61] A. L. Cederquist, T. Kushida, and L. Rimai, "High resolution Brillouin and Rayleigh spectra with single mode He-Ne laser and detection by photon counting," *Phys. Lett.*, to be published.

[62] J. A. Giordmaine and J. A. Howe, "Intensity-induced optical absorption cross section in CS_2," *Phys. Rev Lett.*, vol. 11, pp. 207–209, September 1963.

[63] J. L. Hall, E. J. Robinson, and L. M. Branscomb, "Laser double-quantum photodetachment of I^-," *Phys. Rev. Lett.*, vol. 14, pp. 1013–1016, June 1965.

[64] S. Yatsiv, W. G. Wagner, G. S. Picus, and F. J. McClung, "Saturation of a resonant optical double-quantum transition," *Phys. Rev. Lett.*, vol. 15, pp. 614–618, October 1965.

[65] W. J. Jones and B. P. Stoicheff, "Inverse Raman spectra: induced absorption at optical frequencies," *Phys. Rev. Lett.*, vol. 13 pp. 657–659, November 1964.

[66] J. J. Hopfield, J. M. Worlock, and K. Park, "Two quantum absorption spectrum of KI," *Phys. Rev. Lett.*, vol. 11, pp. 414–417, November 1963.

[67] D. Fröhlich and H. Mahr, "Two-photon spectroscopy in anthracene," *Phys. Rev. Lett.*, vol. 16, pp. 895–897, May 1966.

[68] E. J. Woodbury and W. K. Ng, "Ruby laser operation in the near IR," *Proc. IRE (Correspondence)*, vol. 50, p. 2367, November 1962.

[69] G. Eckhardt, R. W. Hellwarth, F. J. McClung, S. E. Schwarz, D. Weiner, and E. J. Woodbury, "Stimulated Raman scattering from organic liquids," *Phys. Rev. Lett.*, vol. 9, pp. 455–457, December 1962.

[70] G. Eckhardt, D. P. Bortfeld, and M. Geller, "Stimulated emission of Stokes and anti-Stokes Raman lines from diamond, calcite and α-sulfur single crystals," *Appl. Phys. Lett.*, vol. 3, pp. 137–138, October 1963.

[71] R. W Minck, R. W. Terhune, and W. G. Rado, "Stimulated Raman effect and resonant four photon interactions in gases," *Appl. Phys. Lett.*, vol. 3, pp. 181–184, November 1963.

[72] G. G. Bret and M. M. Denariez, "Stimulated Raman effect in acetone and acetone-carbon disulfide mixtures," *Appl. Phys. Lett.*, vol. 8, pp. 151–154, March 1966.

[73] M. Born and E. Wolf, *Principles of Optics*. New York: Pergamon, 1959, pp. 434–448.

[74] K. Shimoda, "Angular distribution of stimulated Raman radiation," *J. Appl. Phys. (Japan)*, vol. 5, pp. 86–92, January 1966.

[75] Y. R. Shen and N. Bloembergen, "Theory of stimulated Brillouin and Raman scattering," *Phys. Rev.*, vol. 137, pp. A1787–A1805, March 1965.

[76] D. I. Mash, V. V. Morozov, V. S. Starunov, and I. L. Fabelinskii, "Stimulated scattering of light of the Rayleigh-line wing," *JETP Lett.*, vol. 2, pp. 25–27, July 1965.

[77] N. Bloembergen and P. Lallemand, "Complex intensity-dependent index of refraction, frequency broadening of stimulated Raman lines, and stimulated Rayleigh scattering," *Phys. Rev. Lett.*, vol. 16, pp. 81–84, January 1966.

[78] H. Takuma and D. A. Jennings, "Coherent Raman effect in the off-axis Raman laser resonator," *Appl. Phys. Lett.*, vol. 4, pp. 185–186, June 1964.

[79] D. P. Bortfeld and W. R. Sooy, "Gain in a diffusely pumped Raman amplifier," *Appl. Phys. Lett.*, vol. 7, pp. 283–285, November 1965.

[80] E. E. Hagenlocker and R. W. Minck, "Simultaneous stimulated Brillouin and Raman scattering in gases,' *IEEE J. of Quantum Electronics*, vol. QE-2, p. liv, April 1966.

[81] R. W. Minck, E. E. Hagenlocker, and W. G. Rado, "Stimulated pure rotational Raman scattering in deuterium," *Phys. Rev. Lett.*, pp. 229–231, August 1966.

[82] N. M. Kroll, "Excitation of hypersonic vibrations by means of photo-elastic coupling of high intensity light waves to elastic waves," *J. Appl. Phys.*, vol. 36, pp. 34–43, January 1965.

[82a] E. E. Hagenlocker, R. W. Minck, and W. G. Rado, "Effects of phonon lifetime on stimulated optical scattering," to be published.

[83] E. E. Hagenlocker and W. G. Rado, "Stimulated Brillouin and Raman scattering in gases," *Appl. Phys. Lett.*, vol. 7, pp. 236–238, November 1965.

[84] E. Garmire and C. H. Townes, "Stimulated Brillouin scattering in liquids," *Appl. Phys. Lett.*, vol. 5, pp. 84–86, August 1964.

[85] R. Y. Chiao, C. H. Townes, and B. P. Stoicheff, "Stimulated Brillouin scattering and coherent generation of intense hypersonic waves," *Phys. Rev. Lett.*, vol. 12, pp. 592–595, May 1964.

[86] R. G. Brewer, "Growth of optical plane waves in stimulated

Brillouin scattering," *Phys. Rev.*, vol. 140, pp. A800–A805, November 1965.

[87] R. Loudon, "Theory of stimulated Raman scattering from lattice vibrations," *Proc. Phys. Soc. (London)*, vol. 82, pp. 393–400, September 1963.

[88] H. Hsu and W. Kavage, "Stimulated Brillouin scatterings in anisotropic media and observation of phonons," *Phys. Letters*, vol. 15, pp. 207–209, April 1965.

[89] M. Hercher, "Laser induced damage in transparent media," *J. Opt. Soc. Am.*, vol. 54, p. 563A, April 1964.

[90] G. Hauchecorne and G. Mayer, "Effects de l'anisotropie moléculaire sur la propagation d'une lumiere intense," *Comptes Rendus*, vol. 261, pp. 4014–4017, November 1965.

[91] P. L. Kelley, "Self-focusing of optical beams," *Phys. Rev. Lett.*, vol. 15, pp. 1005–1008, December 1965.

[92] C. C. Wang, "Length-dependent threshold for stimulated Raman effect and self-focusing of laser beams in liquids," *Phys. Rev. Lett.*, vol. 16, pp. 344–346, February 1966.

[93] E. Garmire, R. Y. Chiao, and C. H. Townes, "Dynamics and characteristics of the self-trapping of intense light beams," *Phys. Rev. Lett.*, vol. 16, pp. 347–349, February 1966.

[94] C. C. Wang and G. W. Racette, "Effect of linear absorption on self-focusing of laser beam in CS_2," *Appl. Phys. Lett.*, vol. 8, pp. 256–257, May 1966.

[95] Y. R. Shen and Y. J. Shaham, "Beam deterioration and stimulated Raman effect," *Phys. Rev. Lett.*, vol. 15, pp. 1008–1010, December 1965.

[96] C. C. Wang, "Nonlinear susceptibility constants and self-focusing of optical beams in liquids," *Phys. Rev.*, to be published.

[97] R. Y. Chiao and E. Garmire, "Self-trapping of optical beams," *IEEE J. of Quantum Electronics*, vol. QE-2, pp. xxxviii–xxxix, April 1966.

[98] D. L. Close, C. R. Giuliano, R. W. Hellwarth, F. J. McClung, and W. G. Wagner, "Light trapping and anomalous stimulated Raman scattering," *IEEE J. of Quantum Electronics*, vol. QE-2, p. liii, April 1966.

[99] R. W. Terhune, P. D. Maker, and C. M. Savage, "Measurement of nonlinear light scattering," *Phys. Rev. Lett.*, vol. 14, pp. 681–684, April 1965.

[100] A. Gold and H. B. Bebb, "Theory of multiphoton ionization," *Phys. Rev. Letters*, vol. 14, pp. 60–63, January 1965.

[101] S. Singh and L. T. Bradley, "Three photon absorption in naphthalene crystals by laser excitation," *Phys. Rev. Lett.*, vol. 12, pp. 612–614, June 1964.

[102] W. E. Lambe, Jr., "Theory of an optical maser," *Phys. Rev.*, vol. 134, pp. A1429–A1450, June 1964.

Tunable Optical Parametric Oscillators

STEPHEN E. HARRIS, MEMBER, IEEE

Invited Paper

Abstract—This paper reviews progress on tunable optical parametric oscillators. Topics considered include: parametric amplification of Gaussian beams; threshold; tuning techniques, spectral output, and stability; saturation and power output; spontaneous parametric emission; nonlinear materials; and far infrared generation.

Fig. 1. Schematic of optical parametric oscillator. The mirrors are highly reflecting at either the signal of idler, or both.

I. INTRODUCTION

WORK on optical parametric oscillators began in 1961 when Franken *et al.* [1] demonstrated second harmonic generation of light, and thus the existence of substantial nonlinear optical coefficients. Following a number of proposals and theoretical studies [2]–[6] Giordmaine and Miller, in 1965, constructed the first tunable optical parametric oscillator [7]. Since then, work has proceeded rapidly and it is now possible to tune through most of the visible and near infrared; to obtain greater than 50 percent conversion efficiency of the light from the laser pump; and to obtain linewidths of less than a wavenumber. With careful construction, threshold for a CW oscillator may be as low as 3 mW.

Many of the basic ideas of parametric amplification and oscillation have been extensively explored in the microwave frequency range [8]. If some upper frequency ω_p, termed as the pump, is incident on a material possessing a nonlinear reactance, then an incident signal frequency ω_s may be amplified. In the process a third frequency ω_i, termed as the idler frequency, and such that $\omega_s + \omega_i = \omega_p$ is generated. Irrespective of the phase of the incoming signal frequency, the phase of the idler may adjust such that the signal and idler are amplified, and the pump is depleted.

In the optical frequency range the nonlinear reactance is obtained via the nonlinear polarizability of noncentrosymmetric crystals [9]. This nonlinear polarizability is described by a 27 component tensor χ_{ijk} which relates the three components of the generated polarization $\mathscr{P}_i$ to the nine possible combinations of applied field $E_j E_k$. That is,

$$\mathscr{P}_i = \sum_j \sum_k \chi_{ijk} E_j E_k, \tag{1}$$

where i, j, and k may be x, y, or z. Typically over the transparency range of the crystal, the nonlinear coefficients χ_{ijk}

Manuscript received August 13, 1969; revised September 18, 1969. Preparation of this paper was sponsored jointly by the National Aeronautics and Space Administration under NASA Grant NGR-05-020-103; and by the Air Force Cambridge Research Laboratories, Office of Aerospace Research, under Contract F19628-67-C-0038. *This invited paper is one of a series planned on topics of general interest.—The Editor.*
The author is with the Department of Electrical Engineering, Microwave Laboratory, Stanford University, Stanford, Calif. 94305.

are nearly independent of frequency, and, unlike the case for a laser transition, very wide tunability is possible.

Though in principle, the χ_{ijk} allow any three optical frequencies to interact, in order to achieve significant parametric amplification it is required that at each of the three frequencies (i.e., at the signal, idler, and pump) the generated polarization travel at the same velocity as a freely propagating electromagnetic wave. This will be the case if the refractive indices of the material are such that the k vectors satisfy the momentum matching condition $\bar{k}_s + \bar{k}_i = \bar{k}_p$ [10]. For collinearly propagating waves this may be written

$$\omega_s n_s + (\omega_p - \omega_s) n_i = \omega_p n_p, \tag{2}$$

where n_s, n_i, and n_p are the refractive indices at the signal, idler, and pump. Once the pumping laser is chosen, and thus ω_p fixed, then if the refractive indices at the signal, idler, or pump frequencies are varied, the signal and idler frequencies will tune. Considerable control of the refractive indices, and very wide tuning, is possible by making use of the angular dependence of the birefringence of anisotropic crystals, and also by temperature variation. Rapid tuning over a limited range is possible by electro-optic variation of the refractive indices.

A schematic of a typical parametric oscillator is shown in Fig. 1. The oscillator consists of a nonlinear crystal and a pair of mirrors. As will be discussed later, the mirrors may be reflecting at either the signal or idler frequency, or at both frequencies. Ideally, 100 percent conversion of incident pump power to tunable signal and idler power is possible. The output of an optical parametric oscillator is very much like that of a laser. It is highly monochromatic with a spectrum consisting of one or a number of longitudinal modes. It is often a fundamental Gaussian transverse mode and may be highly collimated. To the eye, the output of a CW parametric oscillator exhibits the same sparkle effect as does a He-Ne gas laser.

One principal difference between a laser and an optical parametric oscillator is the ability of the former to collect and store wide-band uncollimated spectral energy. A laser's wavelength and linewidth are determined by the pertinent atomic transition and are not affected by the spectral or spacial distribution of the pumping radiation. However, in an optical parametric oscillator, phase coherence between

Reprinted from *Proc. IEEE*, vol. 57, pp. 2096–2113, Dec. 1969.

TABLE I
REPRESENTATIVE PARAMETRIC OSCILLATOR EXPERIMENTS

Pumping Wavelength	Nonlinear Crystal	Tuning Range	Output Power	References
$0.53\ \mu$ (doubled $CaWO_4:Nd^{3+}$)	$LiNbO_3$	$0.73\ \mu - 1.93\ \mu$ (temperature tuning)	10^3 W	[7], [23]
$0.53\ \mu$ (doubled Nd^{3+} – Glass)	KDP $LiNbO_3$	$0.96\ \mu - 1.18\ \mu$ $0.68\ \mu - 2.36\ \mu$ (angle tuning)	10^5 W 50 W	[97], [98] [25]
$0.35\ \mu$ (tripled Nd^{3+} – Glass)	KDP	$0.53\ \mu \pm 10$ percent $1.06\ \mu \pm 10$ percent (angle tuning)	$\sim 10^4$ W	[98]
$0.69\ \mu$ (ruby)	$LiNbO_3$	$1.05\ \mu - 1.20\ \mu$ $1.64\ \mu - 2.05\ \mu$ (angle, temperature, and electrooptic tuning)	$4 \ 10^5$ W	[29], [27], [34]
$0.53\ \mu$ (doubled $Nd^{3+}:YAG$)	$Ba_2NaNb_5O_{15}$	$0.98\ \mu - 1.16\ \mu$ (temperature tuning)	3 mW (CW)	[15]
$0.5145\ \mu$ (argon)	$LiNbO_3$	$0.68\ \mu - 0.71\ \mu$ $1.9\ \mu - 2.1\ \mu$ (temperature tuning)	3 mW (CW)	[19]
$0.69\ \mu$ (ruby)	$LiNbO_3$	$50\ \mu - 238\ \mu$ $0.696\ \mu - 0.704\ \mu$ (angle tuning)	~ 70 W 10^5 W	[88], [89]
$1.06\ \mu$ ($Nd^{3+}:YAG$)	$LiNbO_3$	$1.95\ \mu - 2.35\ \mu$ (temperature and angle tuning)	170 W peak 17 mW average (repetitively pulsed)	[42]

the signal, idler, and pump is very important; and either spectral or angular spread of the pump may increase its threshold or widen its linewidth.

Based on crystals presently being developed, it is likely that within a few years, narrow-band tunable sources will be available over the entire spectral region from $0.2\ \mu$ to greater than $100\ \mu$. Like fixed frequency lasers, these sources should provide at least 10^6 times as much power per bandwidth per steradian as do traditional spectroscopic sources. Such sources are likely to have significant impact on many types of excited state spectroscopy, optical pumping, semiconductor studies, and photochemistry. Table I summarizes the characteristics of a number of parametric oscillator experiments which have been performed to date.

II. PARAMETRIC AMPLIFICATION

A. Amplification of Plane Waves

Consider waves with a pumping frequency ω_p and a signal frequency ω_s to be incident on a nonlinear material having a polarizability $\mathscr{P} \sim E^2$. Mixing of these waves generates a traveling polarization wave at the difference frequency ω_i. By adjusting the birefringence of the crystal, the polarization wave may be made to travel at the same velocity as a freely propagating idler wave, thus resulting in cumulative growth. The idler wave also mixes with the pump to produce a traveling polarization wave at the signal frequency, phased such that growth of the signal field also results. The process continues with the signal and idler fields both growing, and the pumping field decaying as a function of distance

in the crystal. The equations describing this process [8], [10]–[12], in MKS units, are

$$\frac{dE_s}{dz} = -j\eta_s\omega_s dE_pE_i^* \exp -j\Delta kz \tag{3a}$$

$$\frac{dE_i}{dz} = -j\eta_i\omega_i dE_pE_s^* \exp -j\Delta kz \tag{3b}$$

$$\frac{dE_p}{dz} = -j\eta_p\omega_p dE_sE_i \exp j\Delta kz, \tag{3c}$$

where the quantities E_s, E_i, and E_p are the envelopes of the plane waves; e.g., $E_s(z,\ t) = \mathrm{Re}\ [E_s \exp j(\omega_s t - k_s z)]$. The quantities η_s, η_i, and η_p are the plane wave impedances (377/refractive index) of the three waves, and d is the effective nonlinear coefficient. In general, d depends on the direction of propagation and on the polarization of the respective waves, and will be considered further in Section VII. We allow for a $\bar{k}$ vector mismatch

$$\Delta k = k_p - k_s - k_i. \tag{4}$$

We first note that by taking the complex conjugate of (3a) and (3b) and multiplying (3a), (3b), and (3c) by $E_s/\eta_s\omega_s$, $E_i/\eta_i\omega_i$, and $E_p^*/\eta_p\omega_p$, respectively, [5] that

$$\frac{1}{\omega_s}\frac{d}{dz}\left(\frac{|E_s|^2}{2\eta_s}\right) = \frac{1}{\omega_i}\frac{d}{dz}\left(\frac{|E_i|^2}{2\eta_i}\right)$$
$$= -\frac{1}{\omega_p}\frac{d}{dz}\left(\frac{|E_p|^2}{2\eta_p}\right). \tag{5}$$

Growth at the signal implies growth at the idler. Depending on the relative phases of the three frequencies, power may flow either from the lower frequencies to the upper frequency as is the case for second harmonic or sum frequency generation; or alternately from the pump to the lower frequencies as in difference frequency generation and parametric amplification.

If we neglect depletion of the pump, then (3a) and (3b) may be solved subject to the boundary conditions that $E_s = E_s(0)$ and $E_i = E_i(0)$ at $z = 0$. For a crystal length L, we obtain

$$E_s(L) = E_s(0) \exp\left(-j \frac{\Delta k L}{2}\right)\left[\cosh sL + j \frac{\Delta k}{2s} \sinh sL\right]$$
$$- j \frac{\kappa_s}{s} E_i^*(0) \exp\left(-j \frac{\Delta k L}{2}\right)[\sinh sL], \quad (6a)$$

and

$$E_i(L) = E_i(0) \exp\left(-j \frac{\Delta k L}{2}\right)\left[\cosh sL + j \frac{\Delta k}{2s} \sinh sL\right]$$
$$- j \frac{\kappa_i}{s} E_s^*(0) \exp\left(-j \frac{\Delta k L}{2}\right)[\sinh sL] \quad (6b)$$

where

$$\kappa_s = \eta_s \omega_s d E_p$$
$$\kappa_i = \eta_i \omega_i d E_p$$
$$\Gamma^2 = \kappa_s \kappa_i^* = \omega_s \omega_i \eta_s \eta_i |d|^2 E_p|^2$$

and

$$s = (\Gamma^2 - \Delta k^2/4)^{1/2}.$$

We first examine the single pass power gain when only a signal frequency is incident, i.e., take $E_i(0) = 0$. Defining $G = |E_s(L)/E_s(0)|^2 - 1$, we find from (6a)

$$G = \Gamma^2 L^2 \frac{\sinh^2\left(\Gamma^2 - \frac{\Delta k^2}{2}\right)^{1/2} L}{\left(\Gamma^2 - \frac{\Delta k^2}{4}\right) L^2}. \quad (7)$$

For a given crystal temperature and orientation, the center of the parametric gain linewidth occurs at that signal and idler frequency where $\Delta k = 0$. At line center the gain is thus $\sinh^2 \Gamma L$, which for small gain is approximately $\Gamma^2 L^2$. Thus

$$G_{\text{small gain}} \cong \Gamma^2 L^2 = 2\omega_s \omega_i \eta_s \eta_i \eta_p |d|^2 L^2 P_p/A. \quad (8)$$

As an example, $\Gamma^2 L^2$ for a 1 cm crystal of $90°$ cut $LiNbO_3$ for $\lambda_s = \lambda_i = 1\ \mu$ is approximately 0.1 P_p/A, where P_p/A has units of MW/cm^2. From (8) it is seen that the gain of a nonlinear material is proportional to $|d|^2/n^3$, where n is the refractive index. This figure of merit, together with their transparency range, is shown for a number of nonlinear materials in Fig. 24. It is useful to note that when $\omega_s = \omega_i = \omega_p/2$ ($\omega_p/2$ is termed the degenerate frequency), that the single pass gain of an optical parametric amplifier is equal to the conversion efficiency (P_{SH}/P_F) of a second harmonic gen-

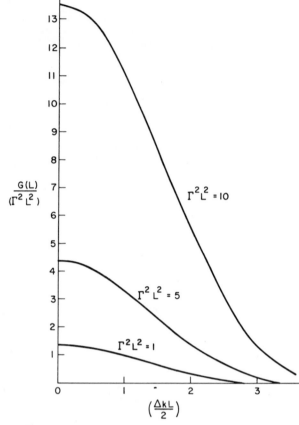

Fig. 2.   Normalized gain versus $(\Delta k L/2)$. (From Byer [96].)

erator with a fundamental frequency $\omega_p/2$. If we let $\omega_s = (\omega_p/2)(1+\delta)$ and $\omega_i = (\omega_p/2)(1-\delta)$, i.e.,

$$\delta = \frac{2\omega_s - \omega_p}{\omega_p} = \frac{2\lambda_p - \lambda_s}{\lambda_s}, \quad (9)$$

then the parametric gain off degeneracy is reduced from that on degeneracy by the factor $1 - \delta^2$.

The lineshape or dependence of the parametric gain on the optical frequency is determined by the variation of Δk with frequency. Noting that for $\Delta k^2 L^2/4 \gg \Gamma^2 L^2$ and small ΓL

$$G_{\text{small gain}} \cong \Gamma^2 L^2 \left[\frac{\sin \Delta k L/2}{(\Delta k L/2)}\right]^2,$$

it is seen that for small gain, the full half-power gain linewidth is determined by $|\Delta k L| \cong 2\pi$. For a 4 cm crystal of $LiNbO_3$ at a pump wavelength of 4880 Å and a signal wavelength of 6328 Å, the dispersion is such that the half-power linewidth is about 1.4 cm^{-1} or about 0.56 Å. However, near degeneracy, linewidths may be much larger. From (7) it is seen that the gain linewidth is also somewhat dependent on $\Gamma^2 L^2$ and thus on the strength of the pump. Fig. 2 shows the quantity $G/\Gamma^2 L^2$ versus $(\Delta k/2)$.

B. Amplification of Gaussian Beams

In determining oscillator thresholds we will be concerned with the gain experienced by a fundamental Gaussian mode of the oscillator. In general, if a beam having a Gaussian cross section is incident on a nonlinear crystal and parametrically amplified, the output beam will no longer be a simple Gaussian. This occurs as a result of Poynting vector walk-off (i.e., the fact that in an anisotropic crystal, the direction of power flow and $\bar{k}$ vector need not be the same); and also, in the absence of Poynting vector walk-off, as a result of the form of the nonlinear polarization generated in the parametric process.

The power gain experienced by an incident Gaussian mode may be found by evaluating an integral of the form $\int \bar{E}^* \cdot \bar{\mathscr{P}} dV$, where $\bar{E}$ is the electric field of the given mode, and $\bar{\mathscr{P}}$ is its driving polarization. A general analysis of this problem, allowing for arbitrarily tight focusing, and also for Poynting vector walk-off has been given by Boyd and Kleinman [13], and some of their results will be summarized in the latter part of this section. We first consider the more restricted but important case of near-field focusing and 90° phase matching (for 90° phase matching, Poynting vector walk-off is absent). An analysis of this case was first given by Boyd and Ashkin [14].

A near-field analysis keeps track of the transverse dependence of the signal, idler, and pump modes; but assumes that this transverse dependence does not change over the length of the nonlinear crystal. It thus requires that the confocal parameter of the focus (of all three beams) be as long or longer than the length of the nonlinear crystal.

To make the appropriate modification of the previous analysis we allow the signal, idler, and pump fields to have Gaussian cross-sectional dependencies of the form $E_{s0} \exp - r^2/W_s^2$, $E_{i0} \exp - r^2/W_i^2$, and $E_{p0} \exp - r^2/W_p^2$, respectively. The generated driving polarizations at the signal, idler, and pump are then also Gaussians [14] having beam waist radii $\bar{W}_s$, $\bar{W}_i$, and $\bar{W}_p$ given by

$$\frac{1}{\bar{W}_s^2} = \frac{1}{W_i^2} + \frac{1}{W_p^2} \tag{10a}$$

$$\frac{1}{\bar{W}_i^2} = \frac{1}{W_s^2} + \frac{1}{W_p^2} \tag{10b}$$

$$\frac{1}{\bar{W}_p^2} = \frac{1}{W_s^2} + \frac{1}{W_i^2}. \tag{10c}$$

For instance, the idler and pump mix to yield a polarization at the signal frequency of the form

$$\exp - \frac{r^2}{\bar{W}_s^2} = \exp - \frac{r^2}{W_i^2} \exp - \frac{r^2}{W_p^2}.$$

Note that these polarization radii are always smaller than the radius of either of the Gaussian beams which mix to produce them. We take the appropriate projections of these Gaussian polarizations by multiplying them by $\exp - r^2/W_s^2$, $\exp - r^2/W_i^2$, and $\exp - r^2/W_p^2$, respectively, and integrating over the transverse cross sections. The result of this is the set of equations

$$\frac{dE_{s0}}{dz} = - j\eta_s \omega_s d g_s E_{p0} E_{i0}^* \exp - j\Delta kz \tag{11a}$$

$$\frac{dE_{i0}}{dz} = - j\eta_i \omega_i d g_i E_{p0} E_{s0}^* \exp - j\Delta kz \tag{11b}$$

$$\frac{dE_{p0}}{dz} = - j\eta_p \omega_p d g_p E_{s0} E_{i0} \exp j\Delta kz, \tag{11c}$$

where the spatial coupling factors g_s, g_i, and g_p are

$$g_s = 2 \frac{\bar{W}_s^2}{W_s^2 + \bar{W}_s^2}; \quad g_i = 2 \frac{\bar{W}_i^2}{W_i^2 + \bar{W}_i^2}; \quad g_p = 2 \frac{\bar{W}_p^2}{W_p^2 + \bar{W}_p^2}. \tag{12}$$

These coupling factors are a measure of the failure of the driving polarizations to completely overlap the desired Gaussian modes. If (though this can never be the case) $\bar{W}_s = W_s$, $\bar{W}_i = W_i$, and $\bar{W}_p = W_p$, then $g_s = g_i = g_p = 1$. Except for these coupling factors, (11) is identical to (5); and the solutions of the previous section may be employed.

The parametric gain coefficient Γ^2 of (8) thus becomes

$$\underset{\text{Gaussian beams}}{\Gamma^2 L^2} = \omega_s \omega_i \eta_s \eta_i |d|^2 g_s g_i L^2 |E_{p0}|^2, \tag{13}$$

where, from (10) and (12) the factor $g_s g_i$ is

$$g_s g_i = 4W_p^2 \left[\frac{W_s W_i W_p}{W_s^2 W_i^2 + W_s^2 W_p^2 + W_i^2 W_p^2} \right]^2. \tag{14}$$

The power of the Gaussian pump beam is given by $(E_{p0}^2/2\eta_p)(\pi W_p^2/2)$, and thus (13) may be rewritten

$$\Gamma^2 L^2 = \frac{16}{\pi} \omega_s \omega_i \eta_s \eta_p \eta_i |d|^2 L^2 P_p M^2, \tag{15}$$

where

$$M^2 = \left[\frac{W_s W_i W_p}{W_s^2 W_i^2 + W_s^2 W_p^2 + W_i^2 W_p^2} \right]^2,$$

and P_p is the incident pump power.

Ashkin and Boyd [14] have shown that to maximize M^2 for W_s and W_i fixed, that the pumping beam size W_p^2 should be

$$\frac{1}{W_p^2} = \frac{1}{W_s^2} + \frac{1}{W_i^2}, \tag{16a}$$

in which case M^2 will be

$$M_{\max}^2 = \frac{1}{4} \frac{1}{W_s^2 + W_i^2}. \tag{16b}$$

In order to maximize the parametric gain at a given pump power, W_s and W_i should be chosen as small as possible. However, since the present analysis is restricted to the near field, the smallest allowed spot sizes are approximately those of the confocal condition, i.e., $W_s^2 = L\lambda_s/2\pi n_s$, $W_i^2 = L\lambda_i/2\pi n_i$. From (16a), if both the signal and idler are confocally focused, it is seen that the pump should also be confocally focused [14].

Combining the previous equations, and making use of the degeneracy factor of (9), the single pass parametric gain coefficient $G = \Gamma^2 L^2$ for a confocally focused signal, idler, and pump may be written

$$\Gamma^2 L^2 = 2\omega_0^2 \eta^3 |d|^2 L^2 \frac{P_p}{\pi W_0^2} (1 - \delta^2)^2, \tag{17}$$

where ω_0 is the degenerate frequency and $W_0^2 = (L\lambda_0)/(2\pi n)$

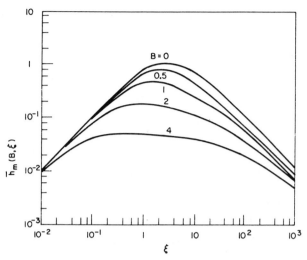

Fig. 3.   Reduction factor $\bar{h}_m(B, \xi)$ versus ξ.
(From Boyd and Kleinman [13].)

TABLE II

REPRESENTATIVE 90° PHASE MATCHABLE MATERIALS

Material	λ_p	Crystal length	P_p/A for 30 percent gain	P_p for 30 percent gain (confocal focusing)	Linewidth
ADP	2573 Å	5.0 cm	0.76 MW/cm²	39 watts	8 cm⁻¹
LiNbO₃	5300 Å	4.0 cm	0.18 MW/cm²	15 watts	1 cm⁻¹
Ba₂NaNb₅O₁₅	5300 Å	0.5 cm	1.6 MW/cm²	17 watts	8 cm⁻¹
CdSe	2.5 μ	2.0 cm	0.77 MW/cm²	141 watts	~15 cm⁻¹

is the degenerate confocal spot size. The price for moving off degeneracy is the factor $(1-\delta^2)^2$. For example, if $\lambda_p=0.473$ μ and $\lambda_s=0.54$ μ, then $\delta=0.75$ and $(1-\delta^2)^2 =0.188$; and the threshold is about five times higher than if $\lambda_s=\lambda_i=2\lambda_p$. Substituting for W_0, (17) becomes

$$\Gamma^2 L^2 = \frac{16\pi^2 d^2 L}{ce_0^3 n^2 \lambda^3} P_p (1 - \delta^2)^2. \qquad (18)$$

For a crystal of LiNbO₃ at a degenerate wavelength of 1 μ, this yields

$$\Gamma^2 L^2 \cong 0.005 \, L P_p (1 - \delta^2)^2, \qquad (19)$$

where L is in cm, and P_p in watts. Note that for optimum focusing $\Gamma^2 L^2$ increases only linearly with crystal length. Thus 1 watt of pumping power in a 4 cm LiNbO₃ crystal provides 2.0 percent single pass gain at a degenerate wavelength of 1 μ.

We now proceed with some of the results of the Boyd and Kleinman analysis [13]. Allowing for both Poynting vector walk-off and arbitrarily tight focusing they show that the effective single pass gain of interacting Gaussian modes all having the same confocal parameter is given by (18) multiplied by a reduction factor $\bar{h}_m(B, \xi)$. The parameter ξ is the ratio of the length of the nonlinear crystal to the common confocal parameter b_0; and B is a double refraction parameter [13] defined as

$$B = \frac{\rho}{2}\left(\frac{2\pi L n_0}{\lambda_0}\right)^{1/2}\left(\frac{n_3}{n_0}\right)^{1/2}, \qquad (20)$$

where ρ is the walk-off angle, L is the length of the nonlinear crystal, λ_0 is the degenerate wavelength, and n_0 and n_3 are the refractive indices at the degenerate wavelength and pump, respectively. B is approximately the ratio of the walk-off angle ρ to the far field diffraction angle of the Gaussian beam.

The reduction factor $\bar{h}_m(B, \xi)$ as a function of ξ with B as a parameter is shown in Fig. 3. In the absence of double refraction (90° phase matching), $B=0$, and for optimum focus-

ing $\xi=L/b_0$ should be 2.84. However, the increase in gain over that obtained for $L/b_0=1$, used in the previous near-field analysis, is only about 20 percent and may not be worth the increased pump power density at the tighter focus. It should also be mentioned that at the tighter focusing, Δk should be slightly different from zero. This results since, at a tight focus, the mixing of the noncollinear components of the signal and idler fields requires slightly longer $\bar{k}$ vectors than does the collinear mixing. This optimized $\bar{k}$ vector match [13] is included in the function $\bar{h}(B, \xi)$.

In many practical cases B will be sufficiently large so that, as seen in Fig. 3, the maximum value of $\bar{h}_m(B, \xi)$ will be rather independent of ξ. Boyd and Kleinman [13] have shown that for large B,

$$\bar{h}_m(B, \xi) \to \pi/4B^2 \qquad (B^2/4 > \xi > 2/B^2), \qquad (21)$$

where for the range of B and ξ shown in parenthesis, (21) is correct to within 10 percent. Since B^2 is proportional to L, the L in the numerator of (18) is cancelled, and for large birefringence the gain is nearly independent of the length of the nonlinear crystal; and subject to the criteria of (21), to the degree of focusing. Since for 90° phase matching $\bar{h}_m(B, \xi) \cong 1$, the factor $\pi/4B^2$ is approximately the gain reduction factor which is experienced as a result of walk-off.

As an example, to phase match a LiNbO₃ parametric oscillator directly pumped by a 1.06 μ Nd³⁺:YAG laser at 90° requires a crystal temperature of about 750°C [13]. Room temperature phase matching at degeneracy is accomplished by propagating at an angle of 43° with respect to the optic axis. This yields a walk-off angle $\rho=0.037$ radians and $B\cong 4.7$ $L^{1/2}$. For a 1 cm crystal the maximum gain at a given pump power is then 28 times smaller than it would have been had 90° phase matching been possible.

It should be noted that the above discussion is concerned with gain maximization at a given pump power as opposed to at a given pump power density. If maximization is with regard to power density, walk-off need not be of consequence. If the beam radii of all modes are greater than $W_0=\sqrt{2}\rho L$, then the gain reduction due to walk-off will be less than 15 percent and may be made essentially negligible for still bigger beams [14].

Table II shows the approximate pump power densities and optimized pump powers necessary to obtain 30 percent single pass gain with typically available lengths of some nonlinear crystals. Approximate gain linewidths for these crystal lengths are also shown.

III. THRESHOLD PUMPING POWER

To construct an oscillator it is necessary to resonate either the signal or the idler, or both. The latter case, where both the signal and idler are resonant, yields the lowest threshold, but poses severe stability problems and mirror requirements. Since they have the lowest threshold, oscillators of this type have thus far been the most prevalent.

To determine threshold of the oscillator we require that the single pass parametric gain (note that there is only gain when the signal and idler travel in the direction of the pump) be sufficient to offset the round-trip cavity loss. We define α_s and α_i as the round-trip $\bar{E}$ field losses at the signal and idler frequencies, e.g., $E_s(0) = (1-\alpha_s)E_s(L)$. From (6) for $\Delta k = 0$, we require

$$\frac{1}{1-\alpha_s}E_{s0}(0) = E_{s0}(0)\cosh \Gamma L - j\frac{\kappa_s}{s}E_{i0}^*(0)\sinh \Gamma L \quad (22a)$$

$$\frac{1}{1-\alpha_i}E_{i0}(0) = E_{i0}(0)\cosh \Gamma L - j\frac{\kappa_i}{s}E_{s0}^*(0)\sinh \Gamma L, \quad (22b)$$

where E_{s0} and E_{i0} are the peak amplitudes of the Gaussian modes as defined in the previous section; Γ is defined by (13); and κ_s and κ_i are given by $\eta_s \omega_s dE_p g_s$ and $\eta_i \omega_i dE_p g_i$, respectively. Taking the complex conjugate of (22b) and setting the determinant of the resulting two simultaneous equations equal to zero, we obtain

$$\left(\cosh \Gamma L - \frac{1}{1-\alpha_s}\right)\left(\cosh \Gamma L - \frac{1}{1-\alpha_i}\right) - \sinh^2 \Gamma L = 0$$

$$\cosh \Gamma L = 1 + \frac{\alpha_s \alpha_i}{2 - \alpha_s - \alpha_i}. \quad (23)$$

For low loss resonators at both the signal and idler frequencies, (23) is satisfied by

$$\Gamma^2 L^2 \underset{\substack{\text{signal and idler resonant} \\ \text{small losses}}}{\cong} \alpha_s \alpha_i. \quad (24)$$

Alternately, if only the signal is resonant and it is assumed that no idler radiation is returned to the crystal input ($\alpha_i = 1$) than for small α_s

$$\Gamma^2 L^2 \underset{\substack{\text{signal only resonant} \\ \text{small losses}}}{=} 2\alpha_s. \quad (25)$$

The ratio of threshold pump power with the signal only resonant as compared to both signal and idler resonant is $2/\alpha_i$. Thus for a 2 percent idler cavity loss, one-hundred times as much power is required for the signal-only-resonant case. It should be noted that for small losses the round-trip $\bar{E}$ field losses α_s and α_i are also the single pass power losses at the respective frequencies.

From (24) and (25) it is seen that parametric gains which are far too small to be useful for tunable amplification are sufficient to attain threshold in a parametric oscillator [14]. For single pass signal and idler power losses of 2 percent each, (24) yields $\Gamma^2 L^2 = 4 \times 10^{-4}$. Equation (19) shows that to attain this gain in a 4 cm crystal of LiNbO$_3$ at a degenerate wavelength of 1 μ requires a pump power of only

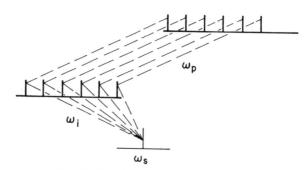

Fig. 4. Pumping with a multimode pump.

20 mW. The possibility of CW parametric oscillators with such low thresholds was first pointed out by Boyd and Ashkin [14], and first demonstrated by Smith et al. [15]. Their first oscillator employed a 5 mm crystal of Ba$_2$NaNb$_5$O$_{15}$ and was pumped by a doubled 1.06 μ Nd^{3+}:YAG laser. Threshold was observed at 45 mW of multimode power. More recently, Smith [16] has constructed a CW argon pumped oscillator with a threshold of about 2 mW.

The formulas of this section have implied that the pumping radiation consists of only a single longitudinal mode. For a parametric oscillator with both its signal and idler cavities resonant, Harris has shown [17] that if the axial mode interval of the idler frequency is set equal to that of the pumping laser, then all of the modes of the pumping laser may act in unison to produce gain at a single signal frequency mode. Though the pump modes are randomly phased, the idler modes develop compensating phases which maximize the gain of the system [18]. A schematic of this idea is shown in Fig. 4. Using it, Byer et al. [19] demonstrated an argon pumped visible CW oscillator with a tuning range of 0.68 μ–.71 μ in the visible and corresponding 1.9 μ–2.1 μ range in the IR. The oscillator used a 1.65 cm long crystal of LiNbO$_3$ and had a threshold of about 500 mW.

It has also been shown that if the dispersion of the nonlinear crystal is sufficiently small that the axial mode interval of both the signal and idler may be set equal to that of the pump, then the peak power, as opposed to the average power of the pump drives the oscillator [20]. The pump could then be phase locked and threshold obtained at a lower average power.

For the case of the signal-only resonant oscillator, the idler modes are not constrained, and thus though it has not been formally proven, the full average power of a multimode laser should be useful. Further, if the axial mode interval of the signal frequency is set equal to that of the pump, then peak power should be the pertinent quantity.

It should perhaps be mentioned that for many parametric oscillators pumped with Q-switched lasers, the problem is not one of attaining threshold, but of having sufficient gain for a sufficient time for the oscillation to build out of the noise. To deplete the pump requires about 140 dB of total gain. If we assume that the round-trip transit time of the oscillator is 1 ns, and if the length of the Q-switched pulse of the pumping laser is 20 ns; then the apparent threshold

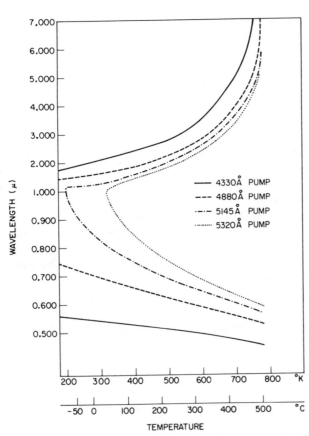

Fig. 5. Temperature tuning of LiNbO₃ for several pump wavelengths. Data were obtained from the Sellmeier equations and [21]. (From Byer [96].)

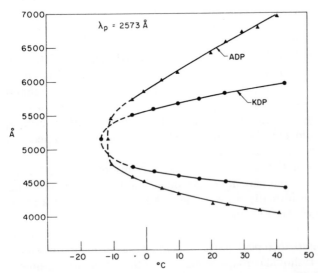

Fig. 6. Temperature tuning of ADP and KDP. These curves were obtained by Dowley [24] using the spontaneous parametric emission technique discussed in Section VI.

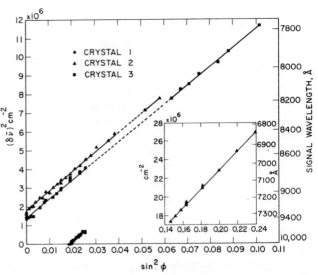

Fig. 7. Angular tuning of LiNbO₃ oscillator. The angle ϕ is the complement of the internal angle between the optic axis and the direction of propagation of the pump. The shift in wavenumbers from the degenerate frequency $\bar{\gamma} = 9434$ cm is denoted by $\delta\bar{\gamma}$. Data connected by solid lines were obtained with a single pair of mirrors. (From Miller and Nordland [25].)

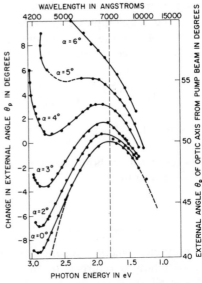

Fig. 8. Angular tuning of ADP at 3472 Å. α is the angle between the pumping beam and the direction of observation of spontaneously emitted light. Thus, $\alpha = 0°$ gives the tuning curve of a collinear oscillator. The right-hand ordinate gives the approximate angle θ_p between the optic axis and pump direction. (Magde and Mahr [46].)

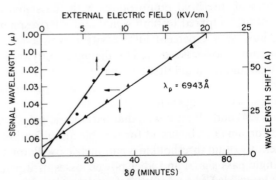

Fig. 9. Electrooptic tuning. Wavelength shift as a function of the external dc electric field; and oscillator signal wavelength as a function of $\delta\theta$, the change in angle between $\bar{k}_p$ and the crystal optic axis. (Kreuzer [27].)

of the oscillator will be a pumping power which yields somewhat less than about 7 dB gain per pass.

IV. Tuning, Spectral Output, and Stability

A. Tuning

As noted earlier, the position of the center of the parametric gain linewidth is determined by satisfaction of the $\bar{k}$ vector matching condition $\bar{k}_s + \bar{k}_i = \bar{k}_p$, and the frequency condition $\omega_s + \omega_i = \omega_p$. For collinearly propagating forward waves these are combined to yield (2). With the pump frequency fixed, any process which changes the refractive indices at the signal, idler, or pump wavelengths will tune the oscillator. Tuning methods include: temperature, angular variation of the extraordinary refractive index, electro-optic variation of the refractive indices, and, perhaps, pressure tuning via the photoelastic effect. Of these, temperature or angular tuning may be used to tune over broad ranges, and pressure or electric fields may be used for fine tuning.

Temperature tuning curves for $LiNbO_3$ for a number of different pump wavelengths are shown in Fig. 5. The curves were obtained numerically from the Sellmeier equations of Hobden and Warner [21]. They may be shifted by 30° to 100° by changes in crystal composition [22]. Giordmaine and Miller [23] have experimentally temperature tuned a $LiNbO_3$ oscillator over the range 7300 Å – 19 300 Å.

Fig. 6 shows the temperature tuning curves for ADP and KDP pumped with the doubled 5145 Å line of argon. These curves were obtained experimentally by Dowley [24] by means of the parametric spontaneous emission method which will be discussed in Section VI. Note that the full visible spectrum is tuned by a variation of only 50°C.

Angular tuning curves for a $LiNbO_3$ parametric oscillator pumped by doubled 1.06 μ [25], and for an ADP oscillator pumped by doubled ruby, are shown in Figs. 7 and 8 [26]. The first of these was obtained in an oscillator experiment, and the second was obtained via spontaneous parametric emission. The angle ϕ in Fig. 7 is the complement of the internal angle between the optic axis and the direction of propagation of the pump, i.e., $\phi = 0$ for 90° phase matching. For the ADP oscillator a change of about 8° of the angle between the optic axis and pump beam tunes most of the visible spectrum. Though the angular tuning method is mechanical and potentially fast compared to temperature tuning, its disadvantage is the reduced gain which results from Poynting vector walk-off (Section II-B).

Experimental results of electrooptic tuning are shown in Fig. 9 [27]. The oscillator was $LiNbO_3$ pumped by ruby and had a tuning rate of about 6.7 Å per kV per cm of applied electric field. The angular tuning rate of this oscillator is also shown. Electrooptic tuning has also been demonstrated by Krivoshchekov et al. [28].

Another tuning technique is shown schematically in Fig. 10 [29]. In this case, the $\bar{k}$ vector matching is not collinear [30] and the oscillator is tuned by varying the angle between the incoming pump beam and the signal cavity. This type of tuning has the advantage that the nonlinear crystal need not be rotated inside the optical cavity. Instead either the angle of the input pump is varied via a beam deflector or alter-

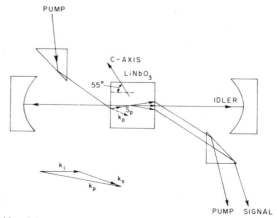

Fig. 10. Schematic of noncollinear oscillator. Tuning is accomplished by varying the angle between the pump and signal cavity. (From Falk and Murray [29].)

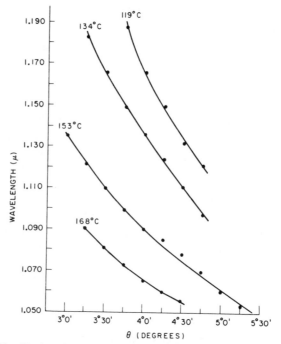

Fig. 11. Tuning of the noncollinear oscillator at different temperatures. (From Falk and Murray [29].)

nately, as was done in the experiment of Fig. 10, the entire cavity containing the nonlinear crystal is rotated. Results of this experiment are shown in Fig. 11 [29].

If we let ζ denote any variable which may be used to vary the refractive indices, and if we assume the pump frequency fixed, then for collinear phase matching the rate of signal frequency tuning with ζ is given by

$$\frac{d\omega_s}{d\zeta} = \frac{1}{b}\left[\frac{\partial k_p}{\partial \zeta} - \frac{\partial k_s}{\partial \zeta} - \frac{\partial k_i}{\partial \zeta}\right], \quad (26)$$

where b is a dispersive constant [31] given by

$$b = \frac{\partial k_i}{\partial \omega_i} - \frac{\partial k_s}{\partial \omega_s}. \quad (27)$$

In Section II-A, the half-power gain linewidth was shown to be determined approximately by the condition $|\Delta k L| = 2\pi$. Noting that $\Delta\omega_i = -\Delta\omega_s$, then from (4), $\Delta k = b\Delta\omega_s$. Thus the full half-power gain linewidth in Hz is

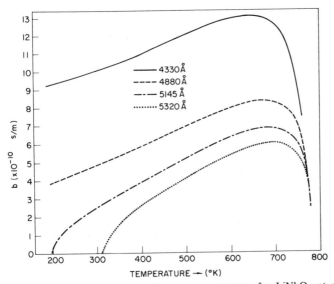

Fig. 12. Dispersion constant b versus temperature for LiNbO$_3$ at a number of pump wavelengths. Data were obtained from the Sellmeier equations of Hobden and Warner [21]. (From Byer [96].)

$$|\Delta f| \cong \frac{1}{bL}. \qquad (28)$$

From (26) and (28) it is seen that materials with small b in general have large tuning rates, but also correspondingly large linewidths. Also, near degeneracy where the linewidth of an oscillator is large, it will, in general, tune much more rapidly than when far from degeneracy. As an example, KDP has a b that is about $\frac{1}{8}$ that of LiNbO$_3$, and as a result tunes much more rapidly (see Fig. 5 and 6). On the other hand, the linewidth of a 1 cm crystal of KDP is about 40 cm^{-1}, as compared to about 5 cm^{-1} for LiNbO$_3$.

The rate of change of the center of the parametric linewidth with respect to fluctuation of the pump frequency has been examined by Kovrigin and Byer [32].

B. Spectral Properties

The spectral character of the output of an oscillator is determined by the width and saturating behavior of the gain lineshape, and by the interaction of its signal and idler modes. Even for materials with relatively large b the half-power gain width will typically be greater than 1 cm^{-1} (30 GHz); and a number of axial modes at the signal and idler frequencies will lie within the linewidth. For instance for a 4 cm crystal of LiNbO$_3$ with a pump at 4880 Å and a signal at 6328 Å, $b = 6.2 \times 10^{-10}$ s/m yielding a half-power width of 1.34 cm^{-1}. With mirrors placed on the ends of the crystal the axial mode spacing would be about 1.6 GHz and about 25 signal and idler modes would experience significant gain. Near degeneracy, typical observed linewidths are greater than 100 cm^{-1}. Fig. 12 shows the dispersive constant b for LiNbO$_3$ versus temperature at a number of pump wavelengths. Corresponding signal and idler frequencies as a function of temperature are shown in Fig. 5.

Parametric oscillators with both their signal and idler frequencies resonant pose a particularly severe problem [7], [11]. Since, as a result of dispersion, the axial mode spacing of the signal and idler frequencies are slightly different, simultaneous resonance of a signal and an idler

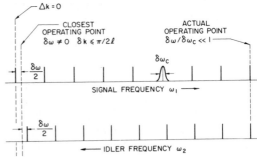

Fig. 13. Longitudinal modes of signal and idler cavities. Frequencies ω_1 and ω_2 vertically in line on the diagram satisfy $\omega_1 + \omega_2 = \omega_3$. The vertical dashed line farthest to the left indicates the frequency combination for index matching. The adjacent line represents the nearest frequency combination for which oscillation is possible. Since the frequency offset $\delta\omega$ is comparable to the cavity linewidth $\Delta\omega_c$, oscillation actually occurs at the vertical dashed line at the right where $\delta\omega = 0$. (From Giordmaine and Miller [11].)

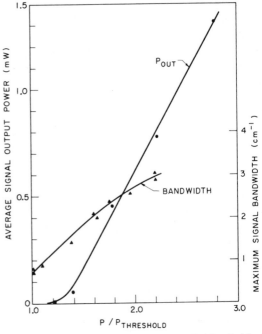

Fig. 14. Power output and linewidth versus P_p/P_p (threshold) of a CW argon pumped LiNbO$_3$ oscillator. (From Byer et al. [19].)

mode whose sum frequency is equal to the frequency of a pump mode, can only occur for certain axial modes. As shown in Fig. 13 the particular modes which happen to align may be far from the center of the gain linewidth. Furthermore, as a result of temperature changes and vibration, the modes which happen to align typically change very rapidly. This is particularly true when combined with the fact that the pump frequency is itself usually fluctuating [15]. For oscillators thus far constructed this fluctuation has had a time constant between about 10 μs and 1 ms, and has often severely reduced their average power output. Since the modes which align or nearly align are typically clustered together in groups having a spacing which is large compared to the axial mode spacing, this effect has been termed a "cluster effect" [7], [11].

The right scale of Fig. 14 shows the total spectral width of a 5145 Å argon pumped CW oscillator [19]. The data were

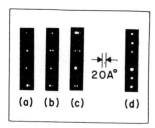

Fig. 15. Spectra of doubly and singly resonant oscillators. (a), (b), and (c) Spectra of the doubly resonant oscillator for increasing pump power. (d) Spectra of a singly resonant oscillator. (From Bjorkholm [34].)

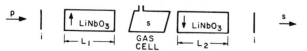

Fig. 16. Schematic of frequency locking technique. The letters s, i, and p denote the signal, idler, and pump. respectively. The gas transition is assumed to absorb at only the signal frequency; and the mirrors are assumed to have high reflectivity at only the idler frequency. The vertical arrows in the LiNbO₃ crystals denote the direction of their positive z axes. (From Harris [37].)

obtained by taking a long-term exposure with a scanning Fabry–Perot etalon. With the oscillator operated about two times above threshold, the spectral width approaches the theoretical 4 wavenumber linewidth of the 1.65 cm LiNbO₃ crystal. For weaker pump drives, only modes closer to $\Delta k = 0$ were above threshold, and the spectral width is reduced. Average signal power output is also shown.

The cluster problem may be eliminated by constructing the oscillator with only its signal or its idler cavity, but not both, resonant. Feedback at the nonresonated frequency can be prevented by careful choice of mirrors, by the use of an appropriate absorbing material in the oscillator, or by means of noncollinear $\bar{k}$ vector matching. A collinear singly resonant oscillator of this type was first demonstrated by Bjorkholm [33], [34], and some of his results are shown in Fig. 15. Spectra of a doubly resonant oscillator are shown in (a), (b), and (c) for increasing pump power. Oscillation occurred in three clusters which had a spacing of about 12 Å. Part (d) shows the spectra of his singly resonant oscillator in which, in all cases, clusters were absent. Other oscillators having only their signal or idler frequencies resonant have been constructed by Falk and Murray [29] and by Belyaev et al. [35]. The oscillator of Falk and Murray was noncollinear and is shown schematically in Fig. 10. Though no clusters were observed, an additional unexplained spectral component was often found a number of angstroms away from the primary component. Belyaev et al. also found a spectrum consisting of one or two lines spaced by a few angstroms; with the width of each individual line not exceeding 0.1 cm⁻¹.

At this time an experimental study of the competition between the longitudinal modes of a single cavity parametric oscillator has not been accomplished. A theoretical study by Kreuzer [36] has shown that if the oscillator is operated less than 4.81 times above threshold, that it should saturate uniformly and in the steady state have only a single oscillating longitudinal mode. For an oscillator operated further above threshold, the steady-state solution may consist of two or more longitudinal modes. For the latter situation, the pump is periodically depleted and restored inside the nonlinear crystal resulting in a saturated lineshape which is broadened and double peaked.

C. Locking

Many experiments require an oscillator output with far greater stability than that presently available. For instance,

optical pumping of gaseous vibrational or rotational lines will require frequency control of better than 0.03 cm⁻¹. One proposed approach which may make it possible to lock the output of an optical parametric oscillator onto a gaseous atomic absorption line is shown in Fig. 16 [37]. The usual single nonlinear crystal is replaced by two nonlinear crystals which have the direction of their +z axes reversed. Between the reversed nonlinear crystals is placed the cell containing the gas to which it is desired to lock the output frequency of the oscillator. If the absorbing transition is at the signal frequency of the oscillator, then the oscillator is made resonant at only the idler frequency. As a result of the reversed positive axes, the parametric gain of the first crystal is partially cancelled by the second crystal. That is, the relative phases of the signal, idler, and pump on entering the second crystal are such that instead of further gain the signal and idler decay to the values which they had on entering the first crystal. The pressure of an absorbing gas is then adjusted until it is nearly opaque at the pertinent atomic transition. The loss and phase shift of this gas prevents the gain cancellation in the second crystal, with a resultant sharply peaked gain function centered at the frequency of the atomic transition.

Bjorkholm has shown that it is possible to lock, at least on a transient basis, the output of a high-power pulsed optical parametric oscillator to an incident low level signal. In a recent experiment he succeeded in locking a LiNbO₃ oscillator, pumped by a ruby laser, to a stabilized CW YAG laser [38]. A minimum locking power of 1 mW, and a locking range of about 15 Å were obtained.

V. Saturation and Power Output

If the level of the pumping field is above threshold, the signal and idler fields build up and deplete the pump as it passes through the crystal. Saturation occurs when the pump intensity, appropriately averaged over the length of the nonlinear crystal, is reduced to the point where single pass gain equals single pass loss.

We first consider the case where only the signal frequency is resonant and determine the pumping power necessary to attain appreciable conversion efficiency. We assume that the round-trip cavity loss and thus the saturated single pass gain at the signal frequency are sufficiently small so that the signal field may be assumed constant over the length of the nonlinear crystal. It should be noted that this is a small gain as opposed to a small power assumption. Equation (3a) and (3c) may be solved subject to $E_i = 0$, and $E_p = E_p(0)$ at $z = 0$. Taking E_s as an undetermined constant, we find

$$E_i = -j \frac{\eta_i \omega_i d E_s^* E_p(0)}{\beta} \exp - j \frac{\Delta k L}{2} \sin \beta L$$

$$E_p = E_p(0) \exp j \frac{\Delta k L}{2} \left[\cos \beta L - j \frac{\Delta k}{2\beta} \sin \beta L \right], \tag{29}$$

where

$$\beta = \left[\omega_i \omega_p \eta_i \eta_p d^2 |E_s|^2 + \Delta k^2/4 \right]^{1/2}.$$

From (5) the generated signal power is related to the generated idler power, and thus

$$|E_s(L)|^2 - |E_s(0)|^2 = \frac{\omega_s \omega_i \eta_s \eta_i d^2 |E_s|^2 |E_p(0)|^2}{\beta^2} \sin^2 \beta L. \tag{30}$$

The single pass power gain $G = (|E_s(L)|^2 - |E_s(0)|^2)/|E_s(0)|^2$ is now set equal to the round-trip power loss $2\alpha_s$. Thus

$$\omega_s \omega_i \eta_s \eta_i d^2 |E_p(0)|^2 L^2 \frac{\sin^2 \beta L}{\beta^2 L^2} = 2\alpha_s. \tag{31}$$

With the pump level $E_p(0)$ fixed, $|E_s|$ is determined by the solution of (31). At threshold, the $|E_s| = 0$ and therefore $\beta = 0$ ($\Delta k = 0$), yielding a threshold power in agreement with (25). At line center, we see from (29), that the pump will be completely depleted when $\beta L = \pi/2$; from (31), that this will occur at a pumping power equal to $(\pi/2)^2$ times the threshold pumping power [36]. If the value of the pumping field is increased further, the pump will again begin to grow at the expense of the signal and idler field. As noted at the end of the last section, this spatially varying pump field creates an interesting type of line broadening first pointed out by Kreuzer [36].

A parametric oscillator with only its signal frequency resonant has the advantage that if the desired output of the oscillator is taken at the idler frequency, then an optimum coupling problem is avoided. That is, at any drive level the signal cavity should be made as lossless as possible. If the pump is adjusted to $(\pi/2)^2$ times its threshold value, then ω_i/ω_p of the incident pump power will be obtained at the output [33]. In a recent experiment, Falk and Murray [29] have obtained about 70 percent peak power conversion and 50 percent energy conversion from the incident ruby beam. As seen from the power-versus-time plots in Fig. 17, greater energy conversion was prevented by the build-up time of the oscillator. The schematic of this oscillator is shown in Fig. 10.

We next consider the case where both the signal and idler cavities are resonant. Here the signal and idler waves travel through the nonlinear media in both the forward and backward directions. When traveling in the backward direction, they mix to produce a pump wave traveling in the opposite direction from the incident pump wave. As first pointed out by Siegman [39], this results in a power-dependent reflection of the pump and a limiting of the transmitted pump to its threshold value. As a result, the maximum efficiency of such an oscillator is 50 percent and occurs at a pump power equal to four times the threshold pump power.

Bjorkholm has recently shown that if this backward reflection is avoided, the signal and idler resonant case may

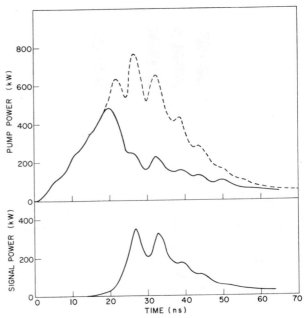

Fig. 17. Output power of noncollinear single cavity oscillator. The dashed curve shows the pump in the absence of the parametric interaction. (From Falk and Murray [29].)

also be 100 percent efficient [40]. With power-dependent pump reflections absent, the generated signal and idler powers are given by

$$\frac{\omega_p}{\omega_s} \frac{P_s}{P_p} = \frac{\omega_p}{\omega_i} \frac{P_i}{P_p} = 4 \frac{P_t}{P_p} \left[(P_p/P_t)^{1/2} - 1 \right], \tag{32}$$

where P_t is the threshold pumping power. At four times above threshold, 100 percent conversion efficiency is obtained. Fig. 18 shows the efficiency and the ratio of transmitted-to-incident pump power for the doubly resonant oscillator with and without power-dependent reflections.

Byer et al. have recently constructed a CW argon pumped ring-cavity parametric oscillator [41], shown in Fig. 19. The oscillator builds up in the direction in which the pumping wave is traveling and power-dependent reflections are avoided. Though 60 percent depletion of the incident pumping beam was observed, as a result of insufficient output coupling only a few milliwatts of output power were obtained.

Ammann et al. have recently obtained about 7 percent average power conversion in a doubly resonant $LiNbO_3$ oscillator directly pumped by a repetitively Q-switched Nd^{3+}:YAG laser [42].

Another interesting type of optical parametric oscillator is obtained when the nonlinear crystal is placed inside the cavity of the pumping laser. The mirrors for the signal and idler cavity may be coincident with those of the pumping laser or may be positioned using various types of beam splitters. Oshman and Harris [43] have shown theoretically that this type of oscillator may operate in several types of regimes. These are: an efficient regime with operating characteristics similar to those of the previously described oscillators; an inefficient regime in which the parametric coupling in effect drives the phase rather than the amplitude

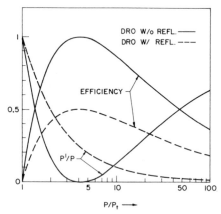

Fig. 18. Efficiency and transmitted pump power of the doubly resonant oscillator, with and without power dependent pump reflection. P'/P is the ratio of transmitted-to-incident pump power. (From Bjorkholm [40].)

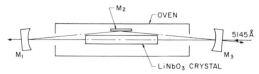

Fig. 19. Ring cavity parametric oscillator where M_1 and M_3 are 5 cm dielectric mirrors, and M_2 is a flat gold mirror. (Byer et al. [41].)

of the oscillation and where a shift of the signal, idler, and pump frequencies from their normal positions is observed; and a repetitively pulsing regime, characterized by short pulses of output power at the signal and idler. A stability analysis of these various regions shows that they are mutually exclusive and can be experimentally chosen by changing the laser gain, the oscillator output coupling, or the strength of the nonlinear interaction.

VI. SPONTANEOUS PARAMETRIC EMISSION

When light from a pumping laser is incident on a nonlinear crystal, there is spontaneous probability that pump photons will split into signal and idler photons. Without the need for optical resonators at either the signal or idler wavelengths, emission at these wavelengths may be observed. This emission has alternately been termed as spontaneous parametric emission, parametric fluorescence, parametric noise, parametric luminescence, and parametric scattering [44]–[49]. It is analogous to laser fluorescence or more exactly to spontaneous Raman and Brillouin scattering. It is important since even at pump fields which are far too low to attain oscillation it may still be observed and used to obtain temperature, angular, or electrooptic tuning curves of potential oscillator materials. The data for Figs. 6 and 8 were obtained using this technique. The fact that the spontaneously emitted power varies linearly with pump power and is independent of both the area and coherence of the pumping beam, also makes it a useful tool for the measurement of optical nonlinearities.

Spontaneous parametric emission was predicted and studied by Louisell et al. [47] and others, and was first observed at optical frequencies by Akhmanov et al. [48], Magde and Mahr [26], and Harris et al. [50].

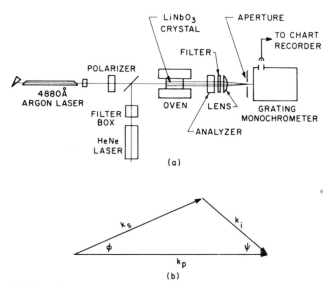

Fig. 20. (a) Apparatus for viewing spontaneous parametric emission. (b) $\bar{k}$ vector matching. (From Byer and Harris [31].)

The transition rate for spontaneous parametric emission may be calculated by forming an interaction Hamiltonian based on the nonlinear susceptibility, and then applying first order perturbation theory. Kleinman [51] and Tang [52] have shown that this approach yields the now-accepted result that the parametric emission may be considered to arise as the result of the mixing of a fictitious zero-point flux, at both the signal and idler frequency, with the incoming pump beam [31]. The effective zero-point flux is obtained by allowing one half-photon to be present in each blackbody mode of a quantizing volume. The result is a generated polarization which attempts to radiate at all frequencies and in all directions. Its ability to radiate effectively is determined by the degree of velocity synchronism with the free wave at the given frequency and in the given direction.

A typical experiment for viewing spontaneous parametric emission is shown in Fig. 20. The pump propagates along the length of a $LiNbO_3$ crystal and is polarized along its optic axis. The signal and idler waves are ordinary waves and make angles ϕ and ψ, respectively, with the pump wave. For a plane wave pump, for small ϕ and ψ, the incremental spontaneously radiated signal power [31] in a bandwidth $d\omega$ and angle $d\phi$ is given by

$$dP_s = \beta L^2 P_p f(\omega_s, \phi)\phi d\phi \, d\omega, \qquad (33)$$

where

$$B = \frac{2\omega_s^4 \omega_i d_{15}^2 \hbar n_s}{(2\pi)^2 e_0^3 c^5 n_i n_p},$$

and $f(\omega_s, \phi)$ is the velocity synchronism reduction factor given by

$$f(\omega_s, \phi) = \frac{\sin^2(|\Delta k|L/2)}{(|\Delta k|L/2)^2}, \qquad (34)$$

where $|\Delta k|$ is the length of the wave vector mismatch taken in the direction of the pump. For small angles and small dispersion Δk may be written

163

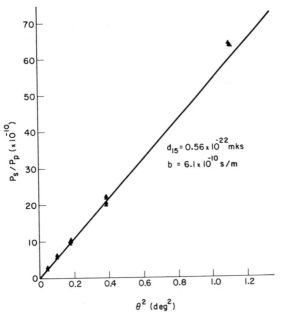

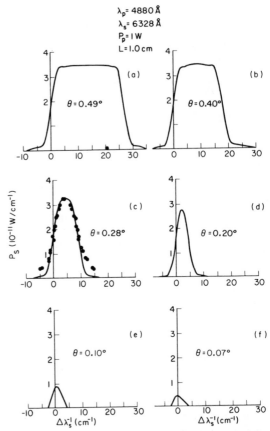

Fig. 21. Total spontaneously emitted power versus θ^2, showing theoretical and experimental results. (From Byer and Harris [31].)

Fig. 22. Spectral distribution of spontaneous power at different acceptance angles. Part (c) shows experimental points normalized to peak of the theoretical curve. (From Byer and Harris [31].)

$$\Delta k = + bd\omega_s + g\phi^2, \quad (35)$$

where

$$g = k_s k_p/2k_i,$$

and b is defined in (27). As a result of normal dispersion, b is negative, and thus higher frequencies will be obtained farther off angle.

Combining (33), (34), and (35), the total radiated power over all frequencies in a given acceptance angle θ is

$$P_s = \beta L^2 P_p \int_{-\infty}^{+\infty} \int_0^\theta \text{sinc}^2 \left[\tfrac{1}{2}(b\omega_s + g\phi^2)L\right]\phi d\phi \, d\omega_s, \quad (36)$$

which may be integrated [31] to yield

$$P_s = (\beta L P_p/b)\pi\theta^2. \quad (37)$$

The total spontaneously emitted signal power thus varies linearly with the accepted solid angle $\pi\theta^2$, pump power, and crystal length. Noting (33) it is seen to vary as the fourth power of signal frequency and the first power of the idler frequency.

The ratio of the spontaneously emitted power to the incident pumping power as a function of θ^2 is shown in Fig. 21 for a 1.1 cm crystal of LiNbO$_3$ and a CW 4880 Å pump. The triangular points are experimental and the solid line is theoretical.

The spectral distribution of the spontaneously emitted light as a function of the accepted angle θ is shown in Fig. 22. The theoretical curves were obtained by numerically integrating (36). As θ is decreased the bandwidth is at first reduced, but then approaches the limiting bandwidth $(1/bL)$ of a collinear interaction.

A detailed study of spontaneous parametric emission has been given by Kleinman [51]. He makes use of a matching surface, such as is shown in Fig. 23, which is the locus of signal and idler $\bar{k}$ vectors such that $\Delta k = 0$. For the special

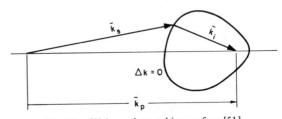

Fig. 23. Kleinman's matching surface [51].

case of emission tangent to this surface both the power and bandwidth of the spontaneously emitted signal are greatly enhanced. This effect can be considered to arise as a result of a greater number of idler modes which may contribute to the emission in the given direction and is analogous to the increased spontaneous emission which occurs when near degeneracy.

Kleinman also discusses a background spontaneous emission which is independent of crystal length and which occurs in directions where phase matching is not possible [51]. This nonphase matched emission is much smaller than the phase matched emission and has thus far not been observed.

Giallorenzi and Tang have observed and discussed spontaneous parametric emission for the case where the idler frequency lies in the infrared absorbing region of the crystal [53], and they find the intensity to be approximately the same as it would have been had there been no absorp-

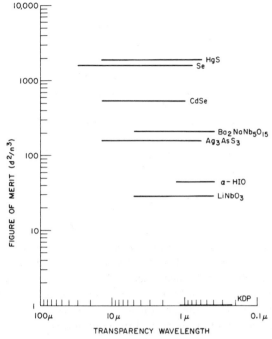

Fig. 24. d^2/n^3 and transparency of some phase matchable nonlinear optical materials.

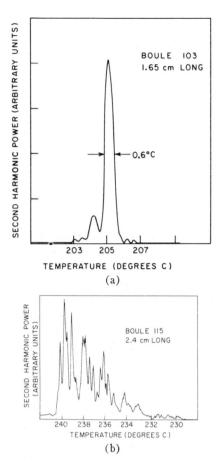

Fig. 25. Second harmonic generation versus temperature for good and bad crystals of LiNbO₃. A high quality crystal should have a half-power width of about 0.64° C/L where L is its length in centimeters.

tion. With one frequency in the absorbing region, spontaneous emission provides a convenient means to obtain the dispersion curve in the lossy region.

Recently four-photon parametric noise, corresponding to satisfaction of frequency and $\bar{k}$ vector conditions of the form $\omega_p + \omega_p = \omega_s + \omega_i$ and $\bar{k}_p + \bar{k}_p = \bar{k}_s + \bar{k}_i$, has been observed in water by Weinberg [54] and in calcite by Meadors et al. [55]. The potential advantage of a four-frequency process of this type is that the pump may be a lower frequency than the signal. Using a ruby pump, Meadors et al. tuned from 4300 Å to 5900 Å. For a four-photon process of this type, unlike the three-photon process, focusing of the pump is of consequence.

If the parametric gain is sufficient, spontaneous emission may be directly amplified without using optical resonators. One technique, demonstrated by Akmanov et al. [56] uses multiple reflections between roof top prisms. With a pump density of 70 MW/cm² of doubled 1.06 μ light, 100 kW of tunable radiation was obtained. The nonlinear crystal used was ADP, and tuning was accomplished by crystal rotation. Observed linewidths were about 1–2 Å.

VII. Nonlinear Optical Materials

From the considerations of the previous sections, a number of desirable qualities for materials to be used in optical parametric oscillators may be formulated. These are: high nonlinearity; phase matchability and, in particular, 90° phase matchability; narrow linewidth (large b); high transparency and freedom from damage; and large variation of refractive indices with temperature, angle, pressure, or electric field. Fig. 24 shows d^2/n^3 (see Section II) and the transparency range of a number of phase matchable materials. Normalization is to d^2/n^3 for KDP. Since both d and n are dependent on the particular angle and frequencies at

which phase matching occurs, there is some uncertainty to the values in this figure.

The KDP-ADP type materials are the only ones with UV transparency [57]–[59] and, as seen in Figs. 6 and 8, allow convenient visible tunability. These crystals might also be used to double a tunable visible source into the 2700 Å– 3000 Å region [57]. Unfortunately, Dowley has found that these crystals exhibit some form of UV damage which limits the average power which may be obtained when doubling the 5145 Å line of argon [58]. KDP-ADP type materials have been found to withstand very high optical power densities (> 400 MW/cm²) before exhibiting surface damage [60] and are particularly useful for high power Q-switched applications. When operated near their Curie temperature these materials should also allow wide electrooptic tuning [61].

LiNbO₃ has a very useful visible and IR transparency range, and is the most widely used oscillator material at this time [62]–[65]. Crystals of excellent optical quality are now available in 4 to 5 cm lengths. Great care must be taken to grow this material such that its refractive index does not vary with distance in the crystal [66], [67] $\bar{k}$ vector matching for a 4 cm crystal requires that the variation of refractive indices be less than 10^{-5}. It has been found that this is best achieved by growing from a melt which is about 2 percent lithium deficient. Fig. 25 shows second harmonic power versus temperature for a high quality and a low quality LiNbO₃ crystal. For the poor crystal, the refractive index varies with distance, and different parts of the crystal phase match at different temperatures. For second harmonic generation from 1.15 μ, the half-power width for a high quality crystal should be about $0.64°C/L$, where L is the length in centimeters of the nonlinear crystal. As a result of optical refractive index damage, $LiNbO_3$ must be maintained at a temperature greater than about 170°C when in the presence of intense visible radiation [68]. In a repetitively pulsed system surface damage probably occurs somewhere in the vicinity of 20 MW/cm².

$Ba_2NaNb_5O_{15}$ has a similar transparency range as $LiNbO_3$ and a d^2/n^3 which is almost ten times greater [69]–[71]. However, available crystals have lengths of 4 mm or less, and typically exhibit severe striations. At room temperature the material does not exhibit refractive index damage, but on the other hand its phase transition at about 300°, and the decrease of its nonlinearity above this transition, limit its temperature tuning range.

Two possible crystals for oscillation further in the infrared are proustite (Ag_3AsS_3) and CdSe [72]–[74]. Though proustite has been used for mixing experiments, an oscillator using it has not yet been constructed. As a result of its large birefringence, only off angle $\bar{k}$ vector matching will be possible. Also, the material is reported to damage relatively easily, i.e., at about 1 MW/cm². An absorption band with absorption of about 1 cm⁻¹ at 10.6 μ will probably prevent working with the CO_2 laser.

CdSe has a high nonlinearity [74] and should be phase matchable near 90° for a pump at about 2.5 μ.

Tellurium had attracted earlier interest when Patel used it both for doubling CO_2 [75] and to obtain a parametric gain of 3 dB at 18 μ [76]. d^2/n^3 is about $1.7\ 10^6$ on the normalized scale of Fig. 21. However, the material is hard to handle and has a relatively high loss.

Some other possible nonlinear crystals for use in oscillator applications are discussed in [77]–[84].

The conventions for specifying optical nonlinearity have caused some confusion, and may be summarized as follows [13]. If the electric field and polarization waves are written in the form

$$E_i(t) = \text{Re} \left[E_i(\omega) \exp j\omega t \right]$$

and

$$\mathscr{P}_i(t) = \text{Re} \left[\mathscr{P}_i(w) \exp j\omega t \right], \qquad (38)$$

then for three interacting frequencies with $\omega_3 = \omega_1 + \omega_2$, the generated polarization is given by

$$\mathscr{P}_i(w_3) = \sum_j \sum_k \chi_{ijk}(-\omega_3, \omega_1, \omega_2) E_j(\omega_1) E_k(\omega_2)$$

$$\mathscr{P}_i(w_2) = \sum_j \sum_k \chi_{ijk}(-\omega_2, -\omega_1, \omega_3) E_j^*(\omega_1) E_k(\omega_3) \qquad (39)$$

$$\mathscr{P}_i(w_1) = \sum_j \sum_k \chi_{ijk}(-\omega_1, -\omega_2, \omega_3) E_j^*(\omega_2) E_k(\omega_3).$$

The nonlinear susceptibility coefficients χ_{ijk} satisfy what is termed as overall permutation symmetry, which states that the subscripts and frequencies may be permuted in any order. For instance

$$\chi_{ijk}(-\omega_3, \omega_1, \omega_2) = \chi_{jik}(\omega_1, -\omega_3, \omega_2) \qquad (40)$$
$$= \chi_{kji}(\omega_2, \omega_1, -\omega_3).$$

Overall permutation symmetry in effect states that if three frequencies are involved in a lossless nonlinear process, that irrespective of which is doing the generating, or being generated, the nonlinear coefficient governing the process is the same.

The above χ_{ijk} are related to the d_{ijk} which are used to describe second harmonic generation by

$$\chi_{ijk}(-2\omega, \omega, \omega) = 2d_{ijk}(-2\omega, \omega, \omega). \qquad (41)$$

Since d_{ijk} is symmetric in the subscripts j and k, it is expressed in the usual abbreviated notation where $d_{ijk} = d_{il}$ according to $l = 1, 2, 3, 4, 5, 6$ for $jk = 11, 22, 33, 23, 13, 12$, respectively. For lossless media these coefficients may be shown to be real, and their small dispersion over the transparency range of the crystal is usually neglected. In cases where the direction of optical propagation is not along the principal crystal axes, it is necessary to take the projection of the generated polarizations in the directions of the respective optical $\bar{E}$ fields. The resulting coefficients have been termed as effective nonlinear coefficients, and their value for a number of uniaxial crystal classes have been tabulated by Boyd and Kleinman [13].

VIII. TUNABLE FAR-INFRARED GENERATION

There is great interest in extending tunable source techniques to the far IR region of the spectrum, where relatively conventional sources are at their poorest. Assuming other factors constant, the pump power necessary to achieve a given gain increases inversely as the square of the lower frequency. Also, in the far IR losses are typically somewhat greater than in the visible or near IR, and thus significantly larger nonlinearities are required. Such nonlinearities may be obtained either by making use of a very high index material, such as tellurium, or by operating with the IR frequency below the Reststrahl frequencies, and thus gaining the contribution of the lattice to the nonlinear coefficient. For instance, in $LiNbO_3$, the nonlinear coefficient governing the interaction between a microwave frequency and two optical frequencies is approximately 30 times greater than the nonlinear coefficient relating three similarly polarized optical frequencies. (This may be deduced by converting the light modulation coefficient r_{51} into an equivalent d coefficient.) Two other principal changes may occur when one of the interacting frequencies lies below the lattice absorption frequencies. First, the lattice contribution to the low-frequency dielectric constant often creates the situation where the sum of the signal and idler $\bar{k}$ vectors is greater than the pump $\bar{k}$ vector. This allows noncollinear phase matching of three waves of the same polarization. (By contrast, as a result of normal dispersion, at optical frequencies $|k_s| + |k_i| < |k_p|$.) Second, it is usually necessary to include the effect of loss on the low-frequency wave.

As the IR frequency approaches one of the vibrational modes of the lattice an increasing fraction of its energy is

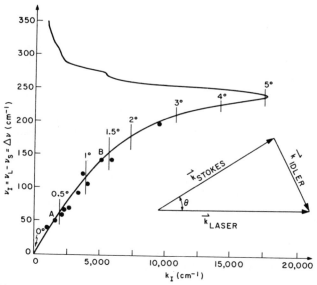

Fig. 26. Dispersion of the A_1 symmetry 248 cm^{-1} polariton mode of LiNbO$_3$. The vertical lines intersecting the dispersion curve denote the value of θ necessary for $\bar{k}$ vector matching. (From Yarborough et al. [89].)

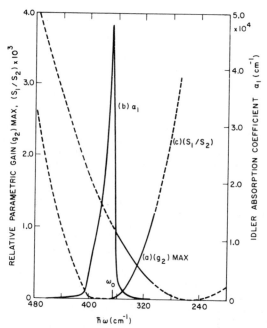

Fig. 27. Parametric gain, absorption coefficient, and ratio of IR to Stokes power, as a function of frequency near the 366 cm^{-1} lattice resonance of gallium phosphide. (a) Relative parametric gain. (b) IR absorption coefficient. (c) Ratio of IR to Stokes power densities (S_1/S_2). The solid portion of the curve denotes the region over which phase matching is possible. (From Henry and Garrett [90].)

mechanical rather than electromagnetic, and in this region it is often termed a polariton mode [85]. Gain results from the interaction with both the vibrational and electromagnetic portions of the mode; and to the extent that the vibrational portion is important, it may be considered as a tunable Raman gain [86], [87].

In two recent experiments, Gelbwachs et al. [88] and Yarborough et al. [89] have obtained tunable radiation in the vicinity of the A_1 symmetry 248 cm^{-1} polariton mode of LiNbO$_3$. The dispersion diagram for this mode is shown in Fig. 26. The vertical lines intersecting the dispersion curve denote the value of the angle θ between the pump and the Stokes beams which is necessary for $\bar{k}$ vector matching. (In analogy with the usual Raman process, the upper frequency is termed as the Stokes wave). As θ is varied, the oscillator is tuned. In the Gelbwachs et al. experiment [88], tuning was accomplished by varying the angle between the laser beam and the axis of a high-Q resonator. In the Yarborough et al. experiment [89], opposite faces of the crystal were polished flat and parallel and an external resonator was not employed. Infrared tuning from 50 μ to 238 μ was obtained with a conversion efficiency to the Stokes frequency of greater than 50 percent. Though infrared powers were not measured, it was estimated that about 70 watts were generated inside the crystal at $\lambda \cong 50\ \mu$.

As the infrared frequency approaches the lattice resonance, both the absorption coefficient and the nonlinear susceptibility become resonately large, though in such a manner as to leave the gain relatively unchanged. Fig. 27 shows theoretical results of Henry and Garrett [90] as the infrared frequency is swept through the 366 cm^{-1} lattice resonance of gallium phosphide. The absorption coefficient, parametric gain, and the ratio of generated infrared to Stokes radiation are shown. The solid portion of the curves indicate the region over which phase matching is possible.

The slow variation of the gain coefficient and its approach to zero at about 250 cm^{-1} is a result of destructive interference between the parametric and Raman type portions of the gain coefficient. This interference has been verified experimentally by Faust and Henry [91].

Though the resonant behavior of the infrared absorption does not substantially affect the gain, it may greatly influence the production of infrared radiation. Henry and Garrett [90] have shown that the ratio of infrared to Stokes powers is approximately

$$P_{\text{IR}}/P_{\text{Stokes}} = \frac{\omega_i}{\omega_s}\frac{G}{\alpha_{\text{IR}}},$$

where G is the gain and α_{IR} is the infrared loss coefficient. As seen in Fig. 27, though in the vicinity of the resonance the gain is relatively unaffected, the IR power drops sharply.

Far-infrared radiation has also been obtained by difference frequency mixing of higher frequencies. Van Tran and Patel [92] have recently reported the use of a magnetic field to achieve phase matched difference frequency generation in InSb. Phase matching was accomplished by varying the cyclotron frequency and thus the free carrier contribution to the refractive index. Discrete tuning from 95 μ to 105 μ was obtained by mixing a number of wavelengths of two synchronously Q-switched CO$_2$ lasers. Previously, nonphase matched far-infrared generation has been reported by Zernike [93].

Two other approaches to tunable generation in the far infrared are backward wave oscillation in a material with low birefringence such as LiTaO$_3$ [94], [95], and Raman down shifting of higher frequency tunable radiation.

REFERENCES

[1] P. A. Franken, A. E. Hill, C. W. Peters, and G. Weinreich, "Generation of optical harmonics," *Phys. Rev. Letters*, vol. 7, p. 118, August 1961.

[2] R. H. Kingston, "Parametric amplification and oscillation at optical frequencies," *Proc. IRE* (Correspondence), vol. 50, p. 472, April 1962.

[3] N. M. Kroll, "Parametric amplification in spatially extended media and application to the design of tunable oscillators at optical frequencies," *Phys. Rev.*, vol. 127, p. 1207, August 1962.

[4] S. A. Akhmanov and R. V. Khokhlov, "Concerning one possibility of amplification of light waves," *J. Exptl. Theoret. Phys.*, vol. 16, p. 252, January 1963.

[5] J. A. Armstrong, N. Bloembergen, J. Ducuing, and P.S. Pershan, "Interactions between light waves in a nonlinear dielectric," *Phys. Rev.*, vol. 127, p. 1918, September 1962.

[6] C. C. Wang and C. W. Racette, "Measurement of parametric gain accompanying optical difference frequency generation," *Appl. Phys. Letters*, vol. 6, p. 169, April 1965.

[7] J. A. Giordmaine and R. C. Miller, "Tunable coherent parametric oscillation in $LiNbO_3$ at optical frequencies," *Phys. Rev. Letters*, vol. 14, p. 973, June 1965.

[8] W. H. Louisell, *Coupled Mode and Parametric Electronics*. New York: Wiley, 1960.

[9] R. W. Minck, R. W. Terhune, and C. C. Wang, "Nonlinear optics," *Appl. Opt.*, vol. 5, p. 1595, October 1966.

[10] N. Bloembergen, *Nonlinear Optics*. New York: W. A. Benjamin, Inc., 1965, p. 135.

[11] J. A. Giordmaine and R. C. Miller, "Optical parametric oscillation in $LiNbO_3$," in *Physics of Quantum Electronics*, P. L. Kelley, B. Lax, and P. E. Tannenwald, Eds. New York: McGraw-Hill, 1966, pp. 31–42; also available in *Proc. Physics of Quantum Electronics Conf.* (San Juan, Puerto Rico), June 28–30, 1965.

[12] J. A. Giordmaine, "Parametric Optics," Academic Press, to be published; and presented in Quantum Optics Course XVIII, Internatl. School of Physics, Varenna, July 19–August 21, 1967.

[13] G. D. Boyd and D. A. Kleinman, "Parametric interaction of focused Gaussian light beams," *J. Appl. Phys.*, vol. 39, p. 3597, July 1968.

[14] G. D. Boyd and A. Ashkin, "Theory of parametric oscillator threshold with single-mode optical masers and observation of amplification in $LiNbO_3$," *Phys. Rev.*, vol. 146, p. 187, June 1966.

[15] R. G. Smith, J. E. Geusic, H. J. Levinstein, J. J. Rubin, S. Singh, and L. G. Van Uitert, "Continuous optical parametric oscillation in $Ba_2NaNb_5O_{15}$," *Appl. Phys. Letters*, vol. 12, p. 308, May 1968.

[16] R. G. Smith, "Nonlinear optics: recent advances," Invited paper, presented at the 1969 Conf. Laser Engrg. and Appl., Washington, D. C., May 1969.

[17] S. E. Harris, "Threshold of multimode parametric oscillators," *IEEE J. Quantum Electronics*, vol. QE-2, pp. 701–702, October 1966.

[18] H. Hsu, "Parametric interactions involving multiple elementary scattering processes," *J. Appl. Phys.*, vol. 38, p. 1787, March 1967.

[19] R. L. Byer, M. K. Oshman, J. F. Young, and S. E. Harris, "Visible CW parametric oscillator," *Appl. Phys. Letters*, vol. 13, p. 109, August 1968.

[20] S. E. Harris, "Threshold of phase-locked parametric oscillators," *IEEE J. Quantum Electronics*, vol. QE-3, pp. 205–206, May 1967.

[21] M. V. Hobden and J. Warner, "The temperature dependence of the refractive indices of pure lithium niobate," *Phys. Letters*, vol. 22, p. 243, August 1966.

[22] J. G. Bergman, A. Ashkin, A. A. Ballman, J. M. Dziedzic, H. J. Levinstein, and R. G. Smith, "Curie temperature, birefringence, and phase-matching temperature variations in $LiNbO_3$ as a function of melt stoichiometry," *Appl. Phys. Letters*, vol. 12, p. 92, February 1968.

[23] J. A. Giordmaine and R. C. Miller, "Optical parametric oscillation in the visible spectrum," *Appl. Phys. Letters*, vol. 9, p. 298, October 1966.

[24] M. W. Dowley, private communication.

[25] R. C. Miller and W. A. Nordland, "Tunable $LiNbO_3$ optical parametric oscillator with external mirrors," *Appl. Phys. Letters*, vol. 10, p. 53, January 1967.

[26] D. Magde and H. Mahr, "Study of ammonium dihydrogen phosphate of spontaneous parametric interaction tunable from 4400 to 16000 Å," *Phys. Rev. Letters*, vol. 18, p. 905, May 1967.

[27] L. B. Kreuzer, "Ruby-laser-pumped optical parametric oscillator with electro-optic effect tuning," *Appl. Phys. Letters*, vol. 10, p. 336, June 1967.

[28] G. V. Krivoshchekov, S. V. Kruglov, S. I. Marennikov, and Y. N. Polivanov, "Variation of the emission wavelength of a parametric light generator by means of an external electric field," *Zh. Eksperim. i Teor. Fiz.*, vol. 7, p. 84, February 1968.

[29] J. Falk and J. E. Murray, "Single cavity noncollinear parametric oscillation," *Appl. Phys. Letters*, vol. 14, p. 245, April 1969.

[30] S. A. Akhmanov, A. G. Ershov, V. V. Fadeev, R. V. Khokhlov, O. N. Chunaev, and E. M. Shvom, "Observation of two-dimensional parametric interaction of light waves," *J. Exptl. Theoret. Phys. Letters*, vol. 2, p. 285, November 1965.

[31] R. L. Byer and S. E. Harris, "Power and bandwidth of spontaneous parametric emission," *Phys. Rev.*, vol. 168, p. 1064, April 1968.

[32] A. I. Kovrigin and R. L. Byer, "Stability factor for optical parametric oscillators," *IEEE J. Quantum Electronics*, to be published.

[33] J. E. Bjorkholm, "Efficient optical parametric oscillation using doubly and singly resonant cavities," *Appl. Phys. Letters*, vol. 13, p. 53, July 1968.

[34] ——, "Some spectral properties of doubly and singly resonant optical parametric oscillators," *Appl. Phys. Letters*, vol. 13, p. 399, December 1968.

[35] Y. N. Belyaev, A. M. Kiselev, and J. R. Freidman, "Investigation of parametric generator with feedback in only one of the waves," *J. Expt. Theoret. Phys. Letters*, vol. 9, p. 263, April 1969.

[36] L. B. Kreuzer, "Theory of the singly resonant optical parametric oscillator," to be published.

[37] S. E. Harris, "Method to lock an optical parametric oscillator to an atomic transition," *Appl. Phys. Letters*, vol. 14, p. 335, June 1969.

[38] J. E. Bjorkholm and H. G. Danielmeyer, "Frequency control of a pulsed optical parametric oscillator by radiation injection," presented at the 1969 Conf. Laser Engrg. and Appl., Washington, D. C., May 1969.

[39] A. E. Siegman, "Nonlinear optical effects: an optical power limiter," *Appl. Optics*, vol. 1, p. 739, November 1962.

[40] J. E. Bjorkholm, "Analysis of the doubly resonant optical parametric oscillator without power dependent reflections," *IEEE J. Quantum Electronics*, vol. QE-5, pp. 293–295, June 1969.

[41] R. L. Byer, A. Kovrigin, and J. F. Young, "A CW ring cavity parametric oscillator," Post-deadline paper, 1969 Conf. Laser Engrg. and Appl., Washington, D. C., May 1969.

[42] E. O. Ammann, J. D. Foster, M. K. Oshman, and J. M. Yarborough, "Repetively pumped parametric oscillator at 2.13μ," presented at the 1969 Conf. Laser Engrg. and Appl., Washington, D. C., May 1969.

[43] M. K. Oshman and S. E. Harris, "Theory of optical parametric oscillation internal to the laser cavity," *IEEE J. Quantum Electronics*, vol. QE-4, pp. 491–502, August 1968.

[44] D. N. Klyshko, "Coherent photon decay in a nonlinear medium," *Zh. Eksperim. i Teor. Fiz.*, vol. 6, p. 490, July 1967.

[45] R. G. Smith, J. G. Skinner, J. E. Geusic, and W. G. Nilsen, "Observations of noncollinear phase matching in optical parametric noise emission," *Appl. Phys. Letters*, vol. 12, p. 97, February 1968.

[46] D. Magde and H. Mahr, "Optical parametric scattering in ammonium dihydrogen phosphate," *Phys. Rev.*, vol. 171, p. 393, July 1968.

[47] W. H. Louisell, A. Yariv, and A. E. Siegman, "Quantum fluctuations and noise in parametric processes. I.," *Phys. Rev.*, vol. 124, p. 1646, December 1961.

[48] S. A. Akhmanov, V. V. Fadeev, R. V. Khokhlov, and O. N. Chunaev, "Quantum noise in parametric light amplifiers," *J. Exptl. Theoret. Phys. Letters*, vol. 6, p. 85, August 1967.

[49] J. Budin, B. Godard, and J. Ducuing, "Noncollinear interactions in parametric luminescence," *IEEE J. Quantum Electronics*, vol. QE-4, pp. 831–837, November 1968.

[50] S. E. Harris, M. K. Oshman, and R. L. Byer, "Observation of tunable optical parametric fluorescence," *Phys. Rev. Letters*, vol. 18, p. 732, May 1967.

[51] D. A. Kleinman, "Theory of optical parametric noise," *Phys. Rev.*, vol. 174, p. 1027, October 1968.

[52] T. G. Giallorenzi and C. L. Tang, "Quantum theory of spontaneous parametric scattering of intense light," *Phys. Rev.*, vol. 166, p. 225, February 1968.

[53] ——, "CW parametric scattering in ADP with strong absorption in the idler band," *Appl. Phys. Letters*, vol. 12, p. 376, June 1968.

[54] D. L. Weinberg, "Four-photon optical parametric noise in water," *Appl. Phys. Letters*, vol. 14, p. 32, 1969.

[55] J. G. Meadors, W. T. Kavage, and E. K. Damon, "Observation of tunable four-photon parametric noise in calcite," *Appl. Phys. Letters*, vol. 14, p. 360, June 1969.

[56] A. G. Akmanov, S. A. Akhmanov, R. V. Kokhlov, A. I. Kovrigin, A. S. Piskarskas, and A. P. Sukhorukov, "Parametric interactions in optics and tunable light oscillators," *IEEE J. Quantum Electronics*, vol. QE-4, pp. 828–837, November 1968.

[57] M. W. Dowley, "Efficient CW second harmonic generation to 2573 Å," *Appl. Phys. Letters*, vol. 13, p. 395, December 1968.

[58] M. W. Dowley and E. B. Hodges, "Studies of high-power CW and quasi-CW parametric UV generation by ADP and KDP in an argon-ion laser cavity," *IEEE J. Quantum Electronics*, vol. QE-4, pp. 552–558, October 1968.

[59] W. J. Deshotels, "Ultraviolet transmission of dihydrogen arsenate and phosphate crystals," *J. Opt. Soc. Am.*, vol. 50, p. 865, September 1960.

[60] W. F. Hagen and P. C. Magnante, "Efficient second-harmonic generation with diffraction-limited and high-spectral-radiance Nd-glass lasers," *J. Appl. Phys.*, vol. 40, p. 2191, January 1969.

[61] N. I. Adams, III and J. J. Barrett, "Electric field control of 90° phasematching in KDP," *IEEE J. Quantum Electronics*, vol. QE-2, pp. 430–435, September 1966.

[62] G. D. Boyd, Robert C. Miller, K. Nassau, W. L. Bond, and A. Savage, "LiNbO$_3$: an efficient phase-matchable nonlinear optical material," *Phys. Letters*, vol. 5, p. 234, December 1964.

[63] R. C. Miller, G. D. Boyd, and A. Savage, "Nonlinear optical interactions in LiNbO$_3$ without double refraction," *Appl. Phys. Letters*, vol. 6, p. 77, February 1965.

[64] J. E. Bjorkholm, "Relative signs of the optical nonlinear coefficients d_{31} and d_{22} in LiNbO$_3$," *Appl. Phys. Letters*, vol. 13, p. 36, July 1968.

[65] ——, "Relative measurement of the optical nonlinearities of KDP, ADP, LiNbO$_3$ and α-HIO$_3$," *IEEE J. Quantum Electronics* (Correspondence), vol. QE-4, pp. 970–972, November 1968; see also correction, *IEEE J. Quantum Electronics*, vol. QE-5, p. 260, May 1969.

[66] J. E. Midwinter and J. Warner, "Up-conversion of near infrared to visible radiation in lithium-meta-niobate," *J. Appl. Phys.*, vol. 38, p. 519, February 1967.

[67] R. L. Byer and J. F. Young, "Quality testing of LiNbO$_3$ crystals," to be published.

[68] A. Ashkin, G. D. Boyd, J. M. Dziedzic, R. G. Smith, A. A. Ballman, J. J. Levinstein, and K. Nassau, "Optically induced refractive index inhomogeneities in LiNbO$_3$ and LiTaO$_3$," *Appl. Phys. Letters*, vol. 9, p. 72, July 1966.

[69] J. E. Geusic, H. J. Levinstein, J. J. Rubin, S. Singh, and L. G. Van Uitert, "The nonlinear optical properties of Ba$_2$NaNb$_5$O$_{15}$," *Appl. Phys. Letters*, vol. 11, p. 269, November 1967.

[70] R. G. Smith, J. E. Geusic, H. J. Levinstein, S. Singh, and L. G. Van Uitert, "Low threshold optical parametric oscillator using Ba$_2$NaNb$_5$O$_{15}$," *J. Appl. Phys.*, vol. 39, p. 4030, July 1968.

[71] R. L. Byer, S. E. Harris, D. J. Kuizenga, J. F. Young, and R. S. Feigelson, "Nonlinear optical properties of Ba$_2$NaNb$_5$O$_{15}$ in the tetragonal phase," *J. Appl. Phys.*, vol. 40, p. 444, January 1969.

[72] W. Bardsley, P. H. Davies, M. V. Hobden, K. F. Hulme, O. Jones, W. Pomeroy, and J. Warner, "Synthetic proustite (Ag$_3$AsS$_3$): a summary of its properties and uses," *Opto-Electronics*, vol. 1, p. 29, 1969.

[73] K. F. Hulme, O. Jones, P. H. Davies, and M. V. Hobden, "Synthetic proustite (Ag$_3$AsS$_3$): a new crystal for optical mixing," *Appl. Phys. Letters*, vol. 10, p. 133, February 1967.

[74] C. K. N. Patel, "Optical harmonic generation in the infrared using a CO$_2$ laser," *Phys. Rev. Letters*, vol. 16, p. 613, April 1966.

[75] ——, "Efficient phase-matched harmonic generation in tellurium with a CO$_2$ laser at 10.6 μ," *Phys. Rev. Letters*, vol. 15, p. 1027, December 1965.

[76] ——, "Parametric amplification in the far infrared," *Appl. Phys. Letters*, vol. 9, p. 332, November 1966.

[77] S. K. Kurtz and T. T. Perry, "A powder technique for the evaluation of nonlinear optical materials," *J. Appl. Phys.*, vol. 39, p. 3798, July 1968.

[78] F. N. H. Robinson, "Nonlinear optical coefficients," *Bell Sys. Tech. J.*, vol. 46, p. 913, May–June 1967.

[79] G. Nath and S. Haussühl, "Large nonlinear optical coefficient in phase matched second harmonic generation in LiIO$_3$," *Appl. Phys. Letters*, vol. 14, p. 154, March 1969.

[80] ——, "Strong second harmonic generation of a ruby laser in lithium iodate," *Phys. Letters*, vol. 29, p. 91, April 1969.

[81] S. K. Kurtz, T. T. Perry, and J. G. Bergman, Jr., "Alpha-iodic acid: solution-grown crystal for nonlinear optical studies and applications," *Appl. Phys. Letters*, vol. 12, p. 186, March 1968.

[82] C. K. N. Patel, R. E. Slusher, and P. A. Fleury, "Optical nonlinearities due to mobile carriers in semiconductors," *Phys. Rev. Letters*, vol. 17, p. 1011, November 1966.

[83] G. D. Boyd, T. H. Bridges, and E. G. Burkhardt, "Up-conversion of 10.6 μ radiation to the visible and second harmonic generation in HgS," *IEEE J. Quantum Electronics*, vol. QE-4, pp. 515–519, September 1968.

[84] F. R. Nash, J. G. Bergman, G. D. Boyd, and E. H. Turner, "Lithium iodate (LiIO$_3$)—a new nonlinear material," to be published.

[85] C. H. Henry and J. J. Hopfield, "Raman scattering by polaritons," *Phys. Rev. Letters*, vol. 15, p. 964, December 1965.

[86] I. P. Kaminow and W. D. Johnston, Jr., "Quantitative determination of sources of the electro-optic effect in LiNbO$_3$ and LiTaO$_3$," *Phys. Rev.*, vol. 160, p. 519, August 1967.

[87] W. D. Johnston, Jr., I. P. Kaminow, and J. G. Bergman, Jr., "Stimulated Raman gain coefficients for LiNbO$_3$, Ba$_2$NaNb$_5$O$_{15}$, and other materials," *Appl. Phys. Letters*, vol. 13, p. 190, September 1968.

[88] J. Gelbwachs, R. H. Pantell, H. E. Puthoff, and J. M. Yarborough, "A tunable stimulated Raman oscillator," *Appl. Phys. Letters*, vol. 14, p. 258, May 1969.

[89] J. M. Yarborough, S. S. Sussman, H. E. Puthoff, R. H. Pantell, and B. C. Johnson, "Efficient, tunable optical emission from LiNbO$_3$ without a resonator," *Appl. Phys. Letters*, vol. 15, p. 102, August 1969.

[90] C. H. Henry and C. G. B. Garrett, "Theory of parametric gain near a lattice resonance," *Phys. Rev.*, vol. 171, p. 1058, July 1968.

[91] W. L. Faust and C. H. Henry, "Mixing of visible and near-resonance infrared light in GaP," *Phys. Rev. Letters*, vol. 17, p. 1265, December 1966.

[92] N. Van Tran and C. K. N. Patel, "Free-carrier magneto-optical effects in far-infrared difference-frequency generation in semiconductors," *Phys. Rev. Letters*, vol. 22, p. 463, March 1969.

[93] F. Zernike, Jr. and P. R. Berman, "Generation of far-infrared as a difference frequency," *Phys. Rev. Letters*, vol. 15, p. 999, December 1965.

[94] S. E. Harris, "Proposed backward wave oscillation in the infrared," *Appl. Phys. Letters*, vol. 9, p. 114, August 1966.

[95] John G. Meadors, "Steady-state theory of backward traveling wave parametric interactions," *J. Appl. Phys.*, vol. 40, p. 2510, May 1969.

[96] R. L. Byer, "Parametric fluorescence and optical parametric oscillation," Ph.D. dissertation, Microwave Lab. Rept. 1711, Stanford University, Stanford, Calif., December 1968.

[97] S. A. Akhmanov, A. I. Kovrigin, V. A. Kolosov, A. S. Piskarskas, V. V. Fadeev, and R. V. Khokhlov, "Tunable parametric light generator with KDP crystal," *J. Exptl. Theoret. Phys. Letters*, vol. 3, p. 241, May 1966.

[98] S. A. Akhmanov, O. N. Chunaev, V. V. Fadeev, R. V. Khokhlov, D. N. Klyshko, A. I. Kovrigin, and A. S. Piskarskas, "Parametric generators of light," presented at the 1967 Symp. Mod. Optics, Polytechnic Institute of Brooklyn, Brooklyn, N. Y., March 1967.

Part V: Laser Modes and Mode-Control Techniques

Understanding the resonant modes in a laser cavity is vital to obtaining good-quality laser operation. The paper by Kogelnik and Li on transverse or spatial modes in optical beams and resonators has come to be regarded as a classic. Smith provides two extensive surveys; the first is a survey of the mode-control methods one can employ to limit a laser to single axial and transverse mode operation, and the second survey is of the mode-locking methods one can employ to couple together multiple laser modes into useful forms of laser output. DeMaria and coauthors provide extensive additional coverage of the ultrashort light pulses obtained from properly mode-locked lasers and the fascinating phenomena that can be realized using these pulses.

Laser Beams and Resonators

H. KOGELNIK AND T. LI

Abstract—This paper is a review of the theory of laser beams and resonators. It is meant to be tutorial in nature and useful in scope. No attempt is made to be exhaustive in the treatment. Rather, emphasis is placed on formulations and derivations which lead to basic understanding and on results which bear practical significance.

1. INTRODUCTION

THE COHERENT radiation generated by lasers or masers operating in the optical or infrared wavelength regions usually appears as a beam whose transverse extent is large compared to the wavelength. The resonant properties of such a beam in the resonator structure, its propagation characteristics in free space, and its interaction behavior with various optical elements and devices have been studied extensively in recent years. This paper is a review of the theory of laser beams and resonators. Emphasis is placed on formulations and derivations which lead to basic understanding and on results which are of practical value.

Historically, the subject of laser resonators had its origin when Dicke [1], Prokhorov [2], and Schawlow and Townes [3] independently proposed to use the Fabry-Perot interferometer as a laser resonator. The modes in such a structure, as determined by diffraction effects, were first calculated by Fox and Li [4]. Boyd and Gordon [5], and Boyd and Kogelnik [6] developed a theory for resonators with spherical mirrors and approximated the modes by wave beams. The concept of electromagnetic wave beams was also introduced by Goubau and Schwering [7], who investigated the properties of sequences of lenses for the guided transmission of electromagnetic waves. Another treatment of wave beams was given by Pierce [8]. The behavior of Gaussian laser beams as they interact with various optical structures has been analyzed by Goubau [9], Kogelnik [10], [11], and others.

The present paper summarizes the various theories and is divided into three parts. The first part treats the passage of paraxial rays through optical structures and is based on geometrical optics. The second part is an analysis of laser beams and resonators, taking into account the wave nature of the beams but ignoring diffraction effects due to the finite size of the apertures. The third part treats the resonator modes, taking into account aperture diffraction effects. Whenever applicable, useful results are presented in the forms of formulas, tables, charts, and graphs.

Manuscript received July 12, 1966.
H. Kogelnik is with Bell Telephone Laboratories, Inc., Murray Hill, N. J.
T. Li is with Bell Telephone Laboratories, Inc., Holmdel, N. J.

2. PARAXIAL RAY ANALYSIS

A study of the passage of paraxial rays through optical resonators, transmission lines, and similar structures can reveal many important properties of these systems. One such "geometrical" property is the stability of the structure [6], another is the loss of unstable resonators [12]. The propagation of paraxial rays through various optical structures can be described by ray transfer matrices. Knowledge of these matrices is particularly useful as they also describe the propagation of Gaussian beams through these structures; this will be discussed in Section 3. The present section describes briefly some ray concepts which are useful in understanding laser beams and resonators, and lists the ray matrices of several optical systems of interest. A more detailed treatment of ray propagation can be found in textbooks [13] and in the literature on laser resonators [14].

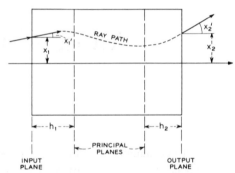

Fig. 1.   Reference planes of an optical system. A typical ray path is indicated.

2.1 Ray Transfer Matrix

A paraxial ray in a given cross section (z = const) of an optical system is characterized by its distance x from the optic (z) axis and by its angle or slope x' with respect to that axis. A typical ray path through an optical structure is shown in Fig. 1. The slope x' of paraxial rays is assumed to be small. The ray path through a given structure depends on the optical properties of the structure and on the input conditions, i.e., the position x_1 and the slope x_1' of the ray in the input plane of the system. For paraxial rays the corresponding output quantities x_2 and x_2' are linearly dependent on the input quantities. This is conveniently written in the matrix form

$$\begin{vmatrix} x_2 \\ x_2' \end{vmatrix} = \begin{vmatrix} A & B \\ C & D \end{vmatrix} \begin{vmatrix} x_1 \\ x_1' \end{vmatrix} \qquad (1)$$

Reprinted from *Proc. IEEE*, vol. 54, pp. 1312–1329, Oct. 1966.

TABLE I

RAY TRANSFER MATRICES OF SIX ELEMENTARY OPTICAL STRUCTURES

NO.	OPTICAL SYSTEM	RAY TRANSFER MATRIX
1		$\begin{vmatrix} 1 & d \\ 0 & 1 \end{vmatrix}$
2		$\begin{vmatrix} 1 & 0 \\ -\frac{1}{f} & 1 \end{vmatrix}$
3		$\begin{vmatrix} 1 & d \\ -\frac{1}{f} & 1-\frac{d}{f} \end{vmatrix}$
4		$\begin{vmatrix} 1-\frac{d_2}{f_1} & d_1+d_2-\frac{d_1 d_2}{f_1} \\ -\frac{1}{f_1}-\frac{1}{f_2}+\frac{d_2}{f_1 f_2} & 1-\frac{d_1}{f_1}-\frac{d_2}{f_2}-\frac{d_1}{f_2}+\frac{d_1 d_2}{f_1 f_2} \end{vmatrix}$
5		$\begin{vmatrix} \cos d\sqrt{\frac{n_2}{n_0}} & \frac{1}{\sqrt{n_0 n_2}}\sin d\sqrt{\frac{n_2}{n_0}} \\ -\sqrt{n_0 n_2}\,\sin d\sqrt{\frac{n_2}{n_0}} & \cos d\sqrt{\frac{n_2}{n_0}} \end{vmatrix}$
6		$\begin{vmatrix} 1 & d/n \\ 0 & 1 \end{vmatrix}$

where the slopes are measured positive as indicated in the figure. The $ABCD$ matrix is called the ray transfer matrix. Its determinant is generally unity

$$AD - BC = 1. \qquad (2)$$

The matrix elements are related to the focal length f of the system and to the location of the principal planes by

$$f = -\frac{1}{C}$$

$$h_1 = \frac{D-1}{C} \qquad (3)$$

$$h_2 = \frac{A-1}{C}$$

where h_1 and h_2 are the distances of the principal planes from the input and output planes as shown in Fig. 1.

In Table I there are listed the ray transfer matrices of six elementary optical structures. The matrix of No. 1 describes the ray transfer over a distance d. No. 2 describes the transfer of rays through a thin lens of focal length f. Here the input and output planes are immediately to the left and right of the lens. No. 3 is a combination of the first two. It governs rays passing first over a distance d and then through a thin lens. If the sequence is reversed the diagonal elements are interchanged. The matrix of No. 4 describes the rays passing through two structures of the No. 3 type. It is obtained by matrix multiplication. The ray transfer matrix for a lenslike medium of length d is given in No. 5. In this medium the refractive index varies quadratically with the distance r from the optic axis.

$$n = n_0 - \tfrac{1}{2}n_2 r^2. \qquad (4)$$

An index variation of this kind can occur in laser crystals and in gas lenses. The matrix of a dielectric material of index n and length d is given in No. 6. The matrix is referred to the surrounding medium of index 1 and is computed by means of Snell's law. Comparison with No. 1 shows that for paraxial rays the effective distance is *shortened* by the optically denser material, while, as is well known, the "optical distance" is lengthened.

2.2 Periodic Sequences

Light rays that bounce back and forth between the spherical mirrors of a laser resonator experience a periodic focusing action. The effect on the rays is the same as in a periodic sequence of lenses [15] which can be used as an optical transmission line. A periodic sequence of identical optical systems is schematically indicated in Fig. 2. A single element of the sequence is characterized by its $ABCD$ matrix. The ray transfer through n consecutive elements of the sequence is described by the nth power of this matrix. This can be evaluated by means of Sylvester's theorem

$$\begin{vmatrix} A & B \\ C & D \end{vmatrix}^n = \frac{1}{\sin \Theta} \qquad (5)$$

$$\begin{vmatrix} A\sin n\Theta - \sin(n-1)\Theta & B\sin n\Theta \\ C\sin n\Theta & D\sin n\Theta - \sin(n-1)\Theta \end{vmatrix}$$

where

$$\cos \Theta = \tfrac{1}{2}(A + D). \qquad (6)$$

Periodic sequences can be classified as either *stable* or *unstable*. Sequences are stable when the trace $(A+D)$ obeys the inequality

$$-1 < \tfrac{1}{2}(A + D) < 1. \qquad (7)$$

Inspection of (5) shows that rays passing through a stable sequence are periodically refocused. For unstable systems, the trigonometric functions in that equation become hyperbolic functions, which indicates that the rays become more and more dispersed the further they pass through the sequence.

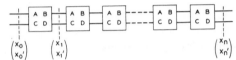

Fig. 2.   Periodic sequence of identical systems, each characterized by its *ABCD* matrix.

2.3 Stability of Laser Resonators

A laser resonator with spherical mirrors of unequal curvature is a typical example of a periodic sequence that can be either stable or unstable [6]. In Fig. 3 such a resonator is shown together with its dual, which is a sequence of lenses. The ray paths through the two structures are the same, except that the ray pattern is folded in the resonator and unfolded in the lens sequence. The focal lengths f_1 and f_2 of the lenses are the same as the focal lengths of the mirrors, i.e., they are determined by the radii of curvature R_1 and R_2 of the mirrors ($f_1 = R_1/2$, $f_2 = R_2/2$). The lens spacings are the same as the mirror spacing d. One can choose, as an element of the periodic sequence, a spacing followed by one lens plus another spacing followed by the second lens. The *ABCD* matrix of such an element is given in No. 4 of Table I. From this one can obtain the trace, and write the stability condition (7) in the form

$$0 < \left(1 - \frac{d}{R_1}\right)\left(1 - \frac{d}{R_2}\right) < 1. \qquad (8)$$

To show graphically which type of resonator is stable and which is unstable, it is useful to plot a stability diagram on which each resonator type is represented by a point. This is shown in Fig. 4 where the parameters d/R_1 and d/R_2 are drawn as the coordinate axes; unstable systems are represented by points in the shaded areas. Various resonator types, as characterized by the relative positions of the centers of curvature of the mirrors, are indicated in the appropriate regions of the diagram. Also entered as alternate coordinate axes are the parameters g_1 and g_2 which play an important role in the diffraction theory of resonators (see Section 4).

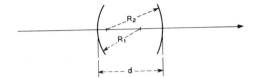

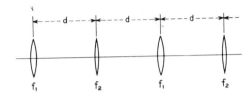

$$R_1 = 2f_1 \quad , \quad R_2 = 2f_2$$

Fig. 3.   Spherical-mirror resonator and the equivalent sequence of lenses.

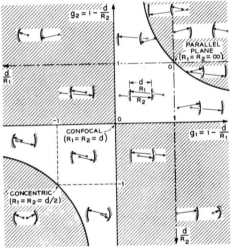

Fig. 4.   Stability diagram. Unstable resonator systems lie in shaded regions.

3. Wave Analysis of Beams and Resonators

In this section the wave nature of laser beams is taken into account, but diffraction effects due to the finite size of apertures are neglected. The latter will be discussed in Section 4. The results derived here are applicable to optical systems with "large apertures," i.e., with apertures that intercept only a negligible portion of the beam power. A theory of light beams or "beam waves" of this kind was first given by Boyd and Gordon [5] and by Goubau and Schwering [7]. The present discussion follows an analysis given in [11].

3.1 Approximate Solution of the Wave Equation

Laser beams are similar in many respects to plane waves; however, their intensity distributions are not uniform, but are concentrated near the axis of propagation and their phase fronts are slightly curved. A field component or potential u of the coherent light satisfies the scalar wave equation

$$\nabla^2 u + k^2 u = 0 \qquad (9)$$

where $k = 2\pi/\lambda$ is the propagation constant in the medium.

For light traveling in the z direction one writes

$$u = \psi(x, y, z) \exp(-jkz) \tag{10}$$

where ψ is a slowly varying complex function which represents the differences between a laser beam and a plane wave, namely: a nonuniform intensity distribution, expansion of the beam with distance of propagation, curvature of the phase front, and other differences discussed below. By inserting (10) into (9) one obtains

$$\frac{\partial^2 \psi}{\partial x^2} + \frac{\partial^2 \psi}{\partial y^2} - 2jk \frac{\partial \psi}{\partial z} = 0 \tag{11}$$

where it has been assumed that ψ varies so slowly with z that its second derivative $\partial^2\psi/\partial z^2$ can be neglected.

The differential equation (11) for ψ has a form similar to the time dependent Schrödinger equation. It is easy to see that

$$\psi = \exp\left\{-j\left(P + \frac{k}{2q}r^2\right)\right\} \tag{12}$$

is a solution of (11), where

$$r^2 = x^2 + y^2. \tag{13}$$

The parameter $P(z)$ represents a *complex* phase shift which is associated with the propagation of the light beam, and $q(z)$ is a *complex* beam parameter which describes the Gaussian variation in beam intensity with the distance r from the optic axis, as well as the curvature of the phase front which is spherical near the axis. After insertion of (12) into (11) and comparing terms of equal powers in r one obtains the relations

$$q' = 1 \tag{14}$$

and

$$P' = -\frac{j}{q} \tag{15}$$

where the prime indicates differentiation with respect to z. The integration of (14) yields

$$q_2 = q_1 + z \tag{16}$$

which relates the beam parameter q_2 in one plane (output plane) to the parameter q_1 in a second plane (input plane) separated from the first by a distance z.

3.2 Propagation Laws for the Fundamental Mode

A coherent light beam with a Gaussian intensity profile as obtained above is not the only solution of (11), but is perhaps the most important one. This beam is often called the "fundamental mode" as compared to the higher order modes to be discussed later. Because of its importance it is discussed here in greater detail.

For convenience one introduces two *real* beam parameters R and w related to the complex parameter q by

$$\frac{1}{q} = \frac{1}{R} - j\frac{\lambda}{\pi w^2}. \tag{17}$$

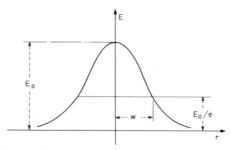

Fig. 5. Amplitude distribution of the fundamental beam.

When (17) is inserted in (12) the physical meaning of these two parameters becomes clear. One sees that $R(z)$ is the radius of curvature of the wavefront that intersects the axis at z, and $w(z)$ is a measure of the decrease of the field amplitude E with the distance from the axis. This decrease is Gaussian in form, as indicated in Fig. 5, and w is the distance at which the amplitude is $1/e$ times that on the axis. Note that the intensity distribution is Gaussian in every beam cross section, and that the width of that Gaussian intensity profile changes along the axis. The parameter w is often called the beam radius or "spot size," and $2w$, the beam diameter.

The Gaussian beam contracts to a minimum diameter $2w_0$ at the *beam waist* where the phase front is plane. If one measures z from this waist, the expansion laws for the beam assume a simple form. The complex beam parameter at the waist is purely imaginary

$$q_0 = j\frac{\pi w_0^2}{\lambda} \tag{18}$$

and a distance z away from the waist the parameter is

$$q = q_0 + z = j\frac{\pi w_0^2}{\lambda} + z. \tag{19}$$

After combining (19) and (17) one equates the real and imaginary parts to obtain

$$w^2(z) = w_0^2 \left[1 + \left(\frac{\lambda z}{\pi w_0^2}\right)^2 \right] \tag{20}$$

and

$$R(z) = z \left[1 + \left(\frac{\pi w_0^2}{\lambda z}\right)^2 \right]. \tag{21}$$

Figure 6 shows the expansion of the beam according to (20). The beam contour $w(z)$ is a hyperbola with asymptotes inclined to the axis at an angle

$$\theta = \frac{\lambda}{\pi w_0}. \tag{22}$$

This is the far-field diffraction angle of the fundamental mode.

Dividing (21) by (20), one obtains the useful relation

$$\frac{\lambda z}{\pi w_0^2} = \frac{\pi w^2}{\lambda R} \tag{23}$$

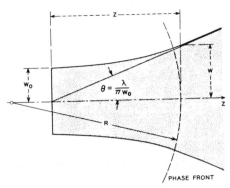

Fig. 6.   Contour of a Gaussian beam.

which can be used to express w_0 and z in terms of w and R:

$$w_0{}^2 = w^2 \Big/ \left[1 + \left(\frac{\pi w^2}{\lambda R} \right)^2 \right] \tag{24}$$

$$z = R \Big/ \left[1 + \left(\frac{\lambda R}{\pi w^2} \right)^2 \right]. \tag{25}$$

To calculate the complex phase shift a distance z away from the waist, one inserts (19) into (15) to get

$$P' = -\frac{j}{q} = -\frac{j}{z + j(\pi w_0{}^2/\lambda)} \cdot \tag{26}$$

Integration of (26) yields the result

$$jP(z) = \ln[1 - j(\lambda z/\pi w_0{}^2)]$$
$$= \ln\sqrt{1 + (\lambda z/\pi w_0{}^2)^2} - j \arctan(\lambda z/\pi w_0{}^2). \tag{27}$$

The real part of P represents a phase shift difference Φ between the Gaussian beam and an ideal plane wave, while the imaginary part produces an amplitude factor w_0/w which gives the expected intensity decrease on the axis due to the expansion of the beam. With these results for the fundamental Gaussian beam, (10) can be written in the form

$$u(r, z) = \frac{w_0}{w}$$
$$\cdot \exp\left\{ -j(kz - \Phi) - r^2 \left(\frac{1}{w^2} + \frac{jk}{2R} \right) \right\} \tag{28}$$

where

$$\Phi = \arctan(\lambda z/\pi w_0{}^2). \tag{29}$$

It will be seen in Section 3.5 that Gaussian beams of this kind are produced by many lasers that oscillate in the fundamental mode.

3.3 Higher Order Modes

In the preceding section only one solution of (11) was discussed, i.e., a light beam with the property that its intensity profile in every beam cross section is given by the same function, namely, a Gaussian. The width of this Gaussian distribution changes as the beam propagates along its axis. There are other solutions of (11) with sim-

ilar properties, and they are discussed in this section. These solutions form a complete and orthogonal set of functions and are called the "modes of propagation." Every arbitrary distribution of monochromatic light can be expanded in terms of these modes. Because of space limitations the derivation of these modes can only be sketched here.

a) Modes in Cartesian Coordinates: For a system with a rectangular (x, y, z) geometry one can try a solution for (11) of the form

$$\psi = g\left(\frac{x}{w} \right) \cdot h\left(\frac{y}{w} \right)$$
$$\cdot \exp\left\{ -j\left[P + \frac{k}{2q} (x^2 + y^2) \right] \right\} \tag{30}$$

where g is a function of x and z, and h is a function of y and z. For real g and h this postulates mode beams whose intensity patterns scale according to the width $2w(z)$ of a Gaussian beam. After inserting this trial solution into (11) one arrives at differential equations for g and h of the form

$$\frac{d^2 H_m}{dx^2} - 2x \frac{dH_m}{dx} + 2m H_m = 0. \tag{31}$$

This is the differential equation for the Hermite polynomial $H_m(x)$ of order m. Equation (11) is satisfied if

$$g \cdot h = H_m\left(\sqrt{2}\, \frac{x}{w} \right) H_n\left(\sqrt{2}\, \frac{y}{w} \right) \tag{32}$$

where m and n are the (transverse) mode numbers. Note that the same pattern scaling parameter $w(z)$ applies to modes of all orders.

Some Hermite polynomials of low order are

$$H_0(x) = 1$$
$$H_1(x) = 2x$$
$$H_2(x) = 4x^2 - 2$$
$$H_3(x) = 8x^3 - 12x. \tag{33}$$

Expression (28) can be used as a mathematical description of higher order light beams, if one inserts the product $g \cdot h$ as a factor on the right-hand side. The intensity pattern in a cross section of a higher order beam is, thus, described by the product of Hermite and Gaussian functions. Photographs of such mode patterns are shown in Fig. 7. They were produced as modes of oscillation in a gas laser oscillator [16]. Note that the number of zeros in a mode pattern is equal to the corresponding mode number, and that the area occupied by a mode increases with the mode number.

The parameter $R(z)$ in (28) is the same for all modes, implying that the phase-front curvature is the same and changes in the same way for modes of all orders. The phase shift Φ, however, is a function of the mode numbers. One obtains

$$\Phi(m, n; z) = (m + n + 1) \arctan(\lambda z/\pi w_0{}^2). \tag{34}$$

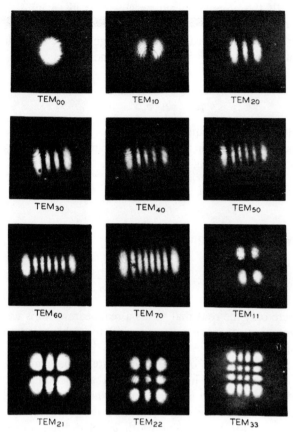

Fig. 7.   Mode patterns of a gas laser oscillator (rectangular symmetry).

This means that the phase velocity increases with increasing mode number. In resonators this leads to differences in the resonant frequencies of the various modes of oscillation.

b) Modes in Cylindrical Coordinates: For a system with a cylindrical (r, ϕ, z) geometry one uses a trial solution for (11) of the form

$$\psi = g\left(\frac{r}{w}\right) \cdot \exp\left\{-j\left(P + \frac{k}{2q}r^2 + l\phi\right)\right\}. \quad (35)$$

After some calculation one finds

$$g = \left(\sqrt{2}\,\frac{r}{w}\right)^l \cdot L_p^l\left(2\,\frac{r^2}{w^2}\right) \quad (36)$$

where L_p^l is a generalized Laguerre polynomial, and p and l are the radial and angular mode numbers. $L_p^l(x)$ obeys the differential equation

$$x\frac{d^2 L_p^l}{dx^2} + (l + 1 - x)\frac{d L_p^l}{dx} + pL_p^l = 0. \quad (37)$$

Some polynomials of low order are

$$L_0^l(x) = 1$$

$$L_1^l(x) = l + 1 - x$$

$$L_2^l(x) = \tfrac{1}{2}(l+1)(l+2) - (l+2)x + \tfrac{1}{2}x^2. \quad (38)$$

As in the case of beams with a rectangular geometry, the beam parameters $w(z)$ and $R(z)$ are the same for all cylindrical modes. The phase shift is, again, dependent on the mode numbers and is given by

$$\Phi(p, l; z) = (2p + l + 1)\,\text{arc}\,\tan(\lambda z/\pi w_0^2). \quad (39)$$

3.4 *Beam Transformation by a Lens*

A lens can be used to focus a laser beam to a small spot, or to produce a beam of suitable diameter and phase-front curvature for injection into a given optical structure. An ideal lens leaves the transverse field distribution of a beam mode unchanged, i.e., an incoming fundamental Gaussian beam will emerge from the lens as a fundamental beam, and a higher order mode remains a mode of the same order after passing through the lens. However, a lens does change the beam parameters $R(z)$ and $w(z)$. As these two parameters are the same for modes of all orders, the following discussion is valid for all orders; the relationship between the parameters of an incoming beam (labeled here with the index 1) and the parameters of the corresponding outgoing beam (index 2) is studied in detail.

An ideal thin lens of focal length f transforms an incoming spherical wave with a radius R_1 immediately to the left of the lens into a spherical wave with the radius R_2 immediately to the right of it, where

$$\frac{1}{R_2} = \frac{1}{R_1} - \frac{1}{f}. \quad (40)$$

Figure 8 illustrates this situation. The radius of curvature is taken to be positive if the wavefront is convex as viewed from $z = \infty$. The lens transforms the phase fronts of laser beams in exactly the same way as those of spherical waves. As the diameter of a beam is the same immediately to the left and to the right of a *thin* lens, the q-parameters of the incoming and outgoing beams are related by

$$\frac{1}{q_2} = \frac{1}{q_1} - \frac{1}{f}, \quad (41)$$

where the q's are measured at the lens. If q_1 and q_2 are measured at distances d_1 and d_2 from the lens as indicated in Fig. 9, the relation between them becomes

$$q_2 = \frac{(1 - d_2/f)q_1 + (d_1 + d_2 - d_1 d_2/f)}{-(q_1/f) + (1 - d_1/f)}. \quad (42)$$

This formula is derived using (16) and (41).

More complicated optical structures, such as gas lenses, combinations of lenses, or thick lenses, can be thought of as composed of a series of thin lenses at various spacings. Repeated application of (16) and (41) is, therefore, sufficient to calculate the effect of complicated structures on the propagation of laser beams. If the *ABCD* matrix for the transfer of paraxial rays through the structure is known, the q parameter of the output beam can be calculated from

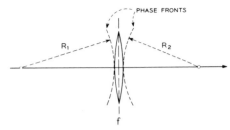

Fig. 8.   Transformation of wavefronts by a thin lens.

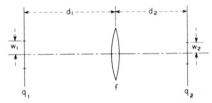

Fig. 9.   Distances and parameters for a
beam transformed by a thin lens.

$$q_2 = \frac{Aq_1 + B}{Cq_1 + D}. \qquad (43)$$

This is a generalized form of (42) and has been called the *ABCD* law [10]. The matrices of several optical structures are given in Section II. The *ABCD* law follows from the analogy between the laws for laser beams and the laws obeyed by the spherical waves in geometrical optics. The radius of the spherical waves R obeys laws of the same form as (16) and (41) for the complex beam parameter q. A more detailed discussion of this analogy is given in [11].

3.5 *Laser Resonators (Infinite Aperture)*

The most commonly used laser resonators are composed of two spherical (or flat) mirrors facing each other. The stability of such "open" resonators has been discussed in Section 2 in terms of paraxial rays. To study the *modes* of laser resonators one has to take account of their wave nature, and this is done here by studying wave beams of the kind discussed above as they propagate back and forth between the mirrors. As aperture diffraction effects are neglected throughout this section, the present discussion applies only to stable resonators with mirror apertures that are large compared to the spot size of the beams.

A mode of a resonator is defined as a self-consistent field configuragion. If a mode can be represented by a wave beam propagating back and forth between the mirrors, the beam parameters must be the same after one complete return trip of the beam. This condition is used to calculate the mode parameters. As the beam that represents a mode travels in both directions between the mirrors it forms the axial standing-wave pattern that is expected for a resonator mode.

A laser resonator with mirrors of equal curvature is shown in Fig. 10 together with the equivalent unfolded system, a sequence of lenses. For this symmetrical structure it is sufficient to postulate self-consistency for one transit of the resonator (which is equivalent to one full period of the lens sequence), instead of a complete return

trip. If the complex beam parameter is given by q_1, immediately to the right of a particular lens, the beam parameter q_2, immediately to the right of the next lens, can be calculated by means of (16) and (41) as

$$\frac{1}{q_2} = \frac{1}{q_1 + d} - \frac{1}{f}. \qquad (44)$$

Self-consistency requires that $q_1 = q_2 = q$, which leads to a quadratic equation for the beam parameter q at the lenses (or at the mirrors of the resonator):

$$\frac{1}{q^2} + \frac{1}{fq} + \frac{1}{fd} = 0. \qquad (45)$$

The roots of this equation are

$$\frac{1}{q} = -\frac{1}{2f} \, (\overset{-}{+}) \, j \sqrt{\frac{1}{fd} - \frac{1}{4f^2}} \qquad (46)$$

where only the root that yields a real beamwidth is used. (Note that one gets a real beamwidth for stable resonators only.)

From (46) one obtains immediately the real beam parameters defined in (17). One sees that R is equal to the radius of curvature of the mirrors, which means that the mirror surfaces are coincident with the phase fronts of the resonator modes. The width $2w$ of the fundamental mode is given by

$$w^2 = \left(\frac{\lambda R}{\pi}\right) \bigg/ \sqrt{2\frac{R}{d} - 1}. \qquad (47)$$

To calculate the beam radius w_0 in the center of the resonator where the phase front is plane, one uses (23) with $z = d/2$ and gets

$$w_0^2 = \frac{\lambda}{2\pi} \sqrt{d(2R - d)}. \qquad (48)$$

The beam parameters R and w describe the modes of all orders. But the phase velocities are different for the different orders, so that the resonant conditions depend on the mode numbers. Resonance occurs when the phase shift from one mirror to the other is a multiple of π. Using (28) and (34) this condition can be written as

$$kd - 2(m + n + 1) \text{ arc } \tan(\lambda d/2\pi w_0^2) = \pi(q + 1) \quad (49)$$

where q is the number of nodes of the axial standing-wave pattern (the number of half wavelengths is $q+1$),[1] and m and n are the rectangular mode numbers defined in Section 3.3. For the modes of circular geometry one obtains a similar condition where $(2p+l+1)$ replaces $(m+n+1)$.

The fundamental beat frequency ν_0, i.e., the frequency spacing between successive longitudinal resonances, is given by

$$\nu_0 = c/2d \qquad (50)$$

[1] This q is not to be confused with the complex beam parameter.

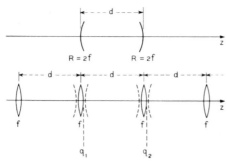

Fig. 10. Symmetrical laser resonator and the equivalent sequence of lenses. The beam parameters, q_1 and q_2, are indicated.

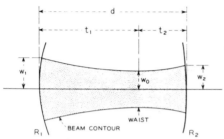

Fig. 11. Mode parameters of interest for a resonator with mirrors of unequal curvature.

where c is the velocity of light. After some algebraic manipulations one obtains from (49) the following formula for the resonant frequency ν of a mode

$$\nu/\nu_0 = (q+1) + \frac{1}{\pi}(m+n+1) \text{ arc } \cos(1-d/R). \quad (51)$$

For the special case of the confocal resonator $(d=R=b)$, the above relations become

$$w^2 = \lambda b/\pi, \qquad w_0{}^2 = \lambda b/2\pi;$$
$$\nu/\nu_0 = (q+1) + \tfrac{1}{2}(m+n+1). \quad (52)$$

The parameter b is known as the confocal parameter.

Resonators with mirrors of unequal curvature can be treated in a similar manner. The geometry of such a resonator where the radii of curvature of the mirrors are R_1 and R_2 is shown in Fig. 11. The diameters of the beam at the mirrors of a stable resonator, $2w_1$ and $2w_2$, are given by

$$w_1{}^4 = (\lambda R_1/\pi)^2 \frac{R_2 - d}{R_1 - d} \frac{d}{R_1 + R_2 - d}$$

$$w_2{}^4 = (\lambda R_2/\pi)^2 \frac{R_1 - d}{R_2 - d} \frac{d}{R_1 + R_2 - d}. \quad (53)$$

The diameter of the beam waist $2w_0$, which is formed either inside or outside the resonator, is given by

$$w_0{}^4 = \left(\frac{\lambda}{\pi}\right)^2 \frac{d(R_1 - d)(R_2 - d)(R_1 + R_2 - d)}{(R_1 + R_2 - 2d)^2} \cdot \quad (54)$$

The distances t_1 and t_2 between the waist and the mirrors, measured positive as shown in the figure, are

$$t_1 = \frac{d(R_2 - d)}{R_1 + R_2 - 2d}$$

$$t_2 = \frac{d(R_1 - d)}{R_1 + R_2 - 2d} \cdot \quad (55)$$

The resonant condition is

$$\nu/\nu_0 = (q + 1) + \frac{1}{\pi}(m + n + 1)$$
$$\text{arc } \cos\sqrt{(1 - d/R_1)(1 - d/R_2)} \quad (56)$$

where the square root should be given the sign of $(1-d/R_1)$, which is equal to the sign of $(1-d/R_2)$ for a stable resonator.

There are more complicated resonator structures than the ones discussed above. In particular, one can insert a lens or several lenses between the mirrors. But in every case, the unfolded resonator is equivalent to a periodic sequence of identical optical systems as shown in Fig. 2. The elements of the $ABCD$ matrix of this system can be used to calculate the mode parameters of the resonator. One uses the $ABCD$ law (43) and postulates self-consistency by putting $q_1=q_2=q$. The roots of the resulting quadratic equation are

$$\frac{1}{q} = \frac{D - A}{2B} \overset{-}{(+)} \frac{j}{2B} \sqrt{4 - (A + D)^2}, \quad (57)$$

which yields, for the corresponding beam radius w,

$$w^2 = (2\lambda B/\pi)/\sqrt{4 - (A + D)^2}. \quad (58)$$

3.6 Mode Matching

It was shown in the preceding section that the modes of laser resonators can be characterized by light beams with certain properties and parameters which are defined by the resonator geometry. These beams are often injected into other optical structures with different sets of beam parameters. These optical structures can assume various physical forms, such as resonators used in scanning Fabry-Perot interferometers or regenerative amplifiers, sequences of dielectric or gas lenses used as optical transmission lines, or crystals of nonlinear dielectric material employed in parametric optics experiments. To match the modes of one structure to those of another one must transform a given Gaussian beam (or higher order mode) into another beam with prescribed properties. This transformation is usually accomplished with a thin lens, but other more complex optical systems can be used. Although the present discussion is devoted to the simple case of the thin lens, it is also applicable to more complex systems, provided one measures the distances from the principal planes and uses the combined focal length f of the more complex system.

The location of the waists of the two beams to be transformed into each other and the beam diameters at the waists are usually known or can be computed. To match the beams one has to choose a lens of a focal length

f that is larger than a characteristic length f_0 defined by the two beams, and one has to adjust the distances between the lens and the two beam waists according to rules derived below.

In Fig. 9 the two beam waists are assumed to be located at distances d_1 and d_2 from the lens. The complex beam parameters at the waists are purely imaginary; they are

$$q_1 = j\pi w_1^2/\lambda, \qquad q_2 = j\pi w_2^2/\lambda \qquad (59)$$

where $2w_1$ and $2w_2$ are the diameters of the two beams at their waists. If one inserts these expressions for q_1 and q_2 into (42) and equates the imaginary parts, one obtains

$$\frac{d_1 - f}{d_2 - f} = \frac{w_1^2}{w_2^2}. \qquad (60)$$

Equating the real parts results in

$$(d_1 - f)(d_2 - f) = f^2 - f_0^2 \qquad (61)$$

where

$$f_0 = \pi w_1 w_2/\lambda. \qquad (62)$$

Note that the characteristic length f_0 is defined by the waist diameters of the beams to be matched. Except for the term f_0^2, which goes to zero for infinitely small wavelengths, (61) resembles Newton's imaging formula of geometrical optics.

Any lens with a focal length $f > f_0$ can be used to perform the matching transformation. Once f is chosen, the distances d_1 and d_2 have to be adjusted to satisfy the matching formulas [10]

$$d_1 = f \pm \frac{w_1}{w_2} \sqrt{f^2 - f_0^2},$$

$$d_2 = f \pm \frac{w_2}{w_1} \sqrt{f^2 - f_0^2}. \qquad (63)$$

These relations are derived by combining (60) and (61). In (63) one can choose either both plus signs or both minus signs for matching.

It is often useful to introduce the confocal parameters b_1 and b_2 into the matching formulas. They are defined by the waist diameters of the two systems to be matched

$$b_1 = 2\pi w_1^2/\lambda, \qquad b_2 = 2\pi w_2^2/\lambda. \qquad (64)$$

Using these parameters one gets for the characteristic length f_0

$$f_0^2 = \tfrac{1}{4}b_1 b_2, \qquad (65)$$

and for the matching distances

$$d_1 = f \pm \tfrac{1}{2}b_1 \sqrt{(f^2/f_0^2) - 1},$$

$$d_2 = f \pm \tfrac{1}{2}b_2 \sqrt{(f^2/f_0^2) - 1}. \qquad (66)$$

Note that in this form of the matching formulas, the wavelength does not appear explicitly.

Table II lists, for quick reference, formulas for the two important parameters of beams that *emerge* from various

TABLE II

FORMULAS FOR THE CONFOCAL PARAMETER AND THE LOCATION OF BEAM WAIST FOR VARIOUS OPTICAL STRUCTURES

NO	OPTICAL SYSTEM	$\tfrac{1}{2}b = \pi w_0^2/\lambda$	t
1		$\sqrt{d(R-d)}$	–
2		$\tfrac{1}{2}\sqrt{d(2R-d)}$	$\tfrac{1}{2}d$
3		$\dfrac{\sqrt{d(R_1-d)(R_2-d)(R_1+R_2-d)}}{R_1+R_2-2d}$	$\dfrac{d(R_2-d)}{R_1+R_2-2d}$
4		$\dfrac{R\sqrt{d(2R-d)}}{2R+d(n^2-1)}$	$\dfrac{ndR}{2R+d(n^2-1)}$
5		$\tfrac{1}{2}\sqrt{d(4f-d)}$	$\tfrac{1}{2}d$
6		$\tfrac{1}{2}d$	$\tfrac{1}{2}d$
7		$\dfrac{d}{2n}$	$\dfrac{d}{2n}$
8		$\dfrac{nR\sqrt{d(2R-d)}}{2n^2R-d(n^2-1)}$	$\dfrac{dR}{2n^2R-d(n^2-1)}$

optical structures commonly encountered. They are the confocal parameter b and the distance t which gives the waist location of the emerging beam. System No. 1 is a resonator formed by a flat mirror and a spherical mirror of radius R. System No. 2 is a resonator formed by two equal spherical mirrors. System No. 3 is a resonator formed by mirrors of unequal curvature. System No. 4

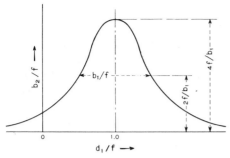

Fig. 12. The confocal parameter b_2 as a function of the lens-waist spacing d_1.

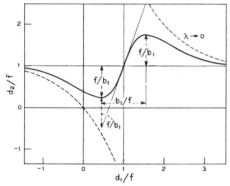

Fig. 13. The waist spacing d_2 as a function of the lens-waist spacing d_1.

is, again, a resonator formed by two equal spherical mirrors, but with the reflecting surfaces deposited on plano-concave optical plates of index n. These plates act as negative lenses and change the characteristics of the emerging beam. This lens effect is assumed not present in Systems Nos. 2 and 3. System No. 5 is a sequence of thin lenses of equal focal lengths f. System No. 6 is a system of two irises with equal apertures spaced at a distance d. Shown are the parameters of a beam that will pass through both irises with the least possible beam diameter. This is a beam which is "confocal" over the distance d. This beam will also pass through a tube of length d with the optimum clearance. (The tube is also indicated in the figure.) A similar situation is shown in System No. 7, which corresponds to a beam that is confocal over the length d of optical material of index n. System No. 8 is a spherical mirror resonator filled with material of index n, or an optical material with curved end surfaces where the beam passing through it is assumed to have phase fronts that coincide with these surfaces.

When one designs a matching system, it is useful to know the accuracy required of the distance adjustments. The discussion below indicates how the parameters b_2 and d_2 change when b_1 and f are fixed and the lens spacing d_1 to the waist of the input beam is varied. Equations (60) and (61) can be solved for b_2 with the result [9]

$$b_2/f = \frac{b_1/f}{(1 - d_1/f)^2 + (b_1/2f)^2}. \tag{67}$$

This means that the parameter b_2 of the beam emerging from the lens changes with d_1 according to a Lorentzian functional form as shown in Fig. 12. The Lorentzian is centered at $d_1 = f$ and has a width of b_1. The maximum value of b_2 is $4f^2/b_1$.

If one inserts (67) into (60) one gets

$$1 - d_2/f = \frac{1 - d_1/f}{(1 - d_1/f)^2 + (b_1/2f)^2} \tag{68}$$

which shows the change of d_2 with d_1. The change is reminiscent of a dispersion curve associated with a Lorentzian as shown in Fig. 13. The extrema of this curve occur at the halfpower points of the Lorentzian. The slope of the curve at $d_1 = f$ is $(2f/b_1)^2$. The dashed curves in the figure correspond to the geometrical optics imaging relation between d_1, d_2, and f [20].

3.7 Circle Diagrams

The propagation of Gaussian laser beams can be represented graphically on a circle diagram. On such a diagram one can follow a beam as it propagates in free space or passes through lenses, thereby affording a graphic solution of the mode matching problem. The circle diagrams for beams are similar to the impedance charts, such as the Smith chart. In fact there is a close analogy between transmission-line and laser-beam problems, and there are analog electric networks for every optical system [17].

The first circle diagram for beams was proposed by Collins [18]. A dual chart was discussed in [19]. The basis for the derivation of these charts are the beam propagation laws discussed in Section 3.2. One combines (17) and (19) and eliminates q to obtain

$$\left(\frac{\lambda}{\pi w^2} + j\frac{1}{R}\right)\left(\frac{\pi w_0^2}{\lambda} - jz\right) = 1. \tag{69}$$

This relation contains the four quantities w, R, w_0, and z which were used to describe the propagation of Gaussian beams in Section 3.2. Each pair of these quantities can be expressed in complex variables W and Z:

$$W = \frac{\lambda}{\pi w^2} + j\frac{1}{R}$$

$$Z = \frac{\pi w_0^2}{\lambda} - jz = b/2 - jz, \tag{70}$$

where b is the confocal parameter of the beam. For these variables (69) defines a conformal transformation

$$W = 1/Z. \tag{71}$$

The two dual circle diagrams are plotted in the complex planes of W and Z, respectively. The W-plane diagram [18] is shown in Fig. 14 where the variables $\lambda/\pi w^2$ and $1/R$ are plotted as axes. In this plane the lines of constant $b/2 = \pi w_0^2/\lambda$ and the lines of constant z of the Z plane appear as circles through the origin. A beam is represented by a circle of constant b, and the beam parameters w and R at a distance z from the beam waist can be easily read

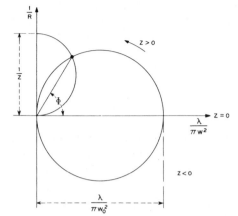

Fig. 14. Geometry for the W-plane circle diagram.

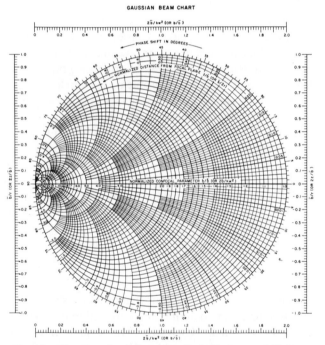

Fig. 15. The Gaussian beam chart. Both W-plane and Z-plane
circle diagram are combined into one.

such transformation makes it possible to use the Smith chart for determining complex mismatch coefficients for Gaussian beams [20]. Other circle diagrams include those for optical resonators [21] which allow the graphic determination of certain parameters of the resonator modes.

4. LASER RESONATORS (FINITE APERTURE)

4.1 General Mathematical Formulation

In this section aperture diffraction effects due to the finite size of the mirrors are taken into account; these effects were neglected in the preceding sections. There, it was mentioned that resonators used in laser oscillators usually take the form of an open structure consisting of a pair of mirrors facing each other. Such a structure with finite mirror apertures is intrinsically lossy and, unless energy is supplied to it continuously, the electromagnetic field in it will decay. In this case a mode of the resonator is a *slowly decaying* field configuration whose relative distribution does not change with time [4]. In a laser oscillator the active medium supplies enough energy to overcome the losses so that a steady-state field can exist. However, because of nonlinear gain saturation the medium will exhibit less gain in those regions where the field is high than in those where the field is low, and so the oscillating modes of an active resonator are expected to be somewhat different from the decaying modes of the passive resonator. The problem of an active resonator filled with a saturable-gain medium has been solved recently [22], [23], and the computed results show that if the gain is not too large the resonator modes are essentially unperturbed by saturation effects. This is fortunate as the results which have been obtained for the passive resonator can also be used to describe the active modes of laser oscillators.

The problem of the open resonator is a difficult one and a rigorous solution is yet to be found. However, if certain simplifying assumptions are made, the problem becomes tractable and physically meaningful results can be obtained. The simplifying assumptions involve essentially the quasi-optic nature of the problem; specifically, they are 1) that the dimensions of the resonator are large compared to the wavelength and 2) that the field in the resonator is substantially transverse electromagnetic (TEM). So long as those assumptions are valid, the Fresnel-Kirchhoff formulation of Huygens' principle can be invoked to obtain a pair of integral equations which relate the fields of the two opposing mirrors. Furthermore, if the mirror separation is large compared to mirror dimensions and if the mirrors are only slightly curved, the two orthogonal Cartesian components of the vector field are essentially uncoupled, so that separate scalar equations can be written for each component. The solutions of these scalar equations yield resonator modes which are uniformly polarized in one direction. Other polarization configurations can be constructed from the uniformly polarized modes by linear superposition.

from the diagram. When the beam passes through a lens the phase front is changed according to (40) and a new beam is formed, which implies that the incoming and outgoing beams are connected in the diagram by a vertical line of length $1/f$. The angle Φ shown in the figure is equal to the phase shift experienced by the beam as given by (29); this is easily shown using (23).

The dual diagram [19] is plotted in the Z plane. The sets of circles in both diagrams have the same form, and only the labeling of the axes and circles is different. In Fig. 15 both diagrams are unified in one chart. The labels in parentheses correspond to the Z-plane diagram, and $\bar{b}$ is a normalizing parameter which can be arbitrarily chosen for convenience.

One can plot various other circle diagrams which are related to the above by conformal transformations. One

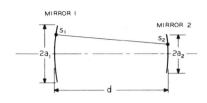

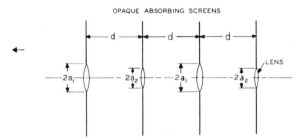

Fig. 16. Geometry of a spherical-mirror resonator with finite mirror apertures and the equivalent sequence of lenses set in opaque absorbing screens.

In deriving the integral equations, it is assumed that a traveling TEM wave is reflected back and forth between the mirrors. The resonator is thus analogous to a transmission medium consisting of apertures or lenses set in opaque absorbing screens (see Fig. 16). The fields at the two mirrors are related by the equations [24]

$$\gamma^{(1)}E^{(1)}(s_1) = \int_{S_2} K^{(2)}(s_1, s_2)E^{(2)}(s_2)dS_2$$

$$\gamma^{(2)}E^{(2)}(s_2) = \int_{S_1} K^{(1)}(s_2, s_1)E^{(1)}(s_1)dS_1 \qquad (72)$$

where the integrations are taken over the mirror surfaces S_2 and S_1, respectively. In the above equations the subscripts and superscripts one and two denote mirrors one and two; s_1 and s_2 are symbolic notations for transverse coordinates on the mirror surface, e.g., $s_1=(x_1, y_1)$ and $s_2=(x_2, y_2)$ or $s_1=(r_1, \phi_1)$ and $s_2=(r_2, \phi_2)$; $E^{(1)}$ and $E^{(2)}$ are the relative field distribution functions over the mirrors; $\gamma^{(1)}$ and $\gamma^{(2)}$ give the attenuation and phase shift suffered by the wave in transit from one mirror to the other; the kernels $K^{(1)}$ and $K^{(2)}$ are functions of the distance between s_1 and s_2 and, therefore, depend on the mirror geometry; they are equal $[K^{(1)}(s_2, s_1)=K^{(2)}(s_1, s_2)]$ but, in general, are not symmetric $[K^{(1)}(s_2, s_1)\neq K^{(1)}(s_1, s_2), K^{(2)}(s_1, s_2)\neq K^{(2)}(s_2, s_1)]$.

The integral equations given by (72) express the field at each mirror in terms of the reflected field at the other; that is, they are single-transit equations. By substituting one into the other, one obtains the double-transit or round-trip equations, which state that the field at each mirror must reproduce itself after a round trip. Since the kernel for each of the double-transit equations is symmetric [24], it follows [25] that the field distribution functions corresponding to the different mode orders are orthogonal over their respective mirror surfaces; that is

$$\int_{S_1} E_m^{(1)}(s_1)E_n^{(1)}(s_1)dS_1 = 0, \qquad m \neq n$$

$$\int_{S_2} E_m^{(2)}(s_2)E_n^{(2)}(s_2)dS_2 = 0, \qquad m \neq n \qquad (73)$$

where m and n denote different mode orders. It is to be noted that the orthogonality relation is non-Hermitian and is the one that is generally applicable to lossy systems.

4.2 Existence of Solutions

The question of the existence of solutions to the resonator integral equations has been the subject of investigation by several authors [26]–[28]. They have given rigorous proofs of the existence of eigenvalues and eigenfunctions for kernels which belong to resonator geometries commonly encountered, such as those with parallel-plane and spherically curved mirrors.

4.3 Integral Equations for Resonators with Spherical Mirrors

When the mirrors are spherical and have rectangular or circular apertures, the two-dimensional integral equations can be separated and reduced to one-dimensional equations which are amenable to solution by either analytical or numerical methods. Thus, in the case of rectangular mirrors [4]–[6], [24], [29], [30], the one-dimensional equations in Cartesian coordinates are the same as those for infinite-strip mirrors; for the x coordinate, they are

$$\gamma_x^{(1)}u^{(1)}(x_1) = \int_{-a_2}^{a_2} K(x_1, x_2)u^{(2)}(x_2)dx_2$$

$$\gamma_x^{(2)}u^{(2)}(x_2) = \int_{-a_1}^{a_1} K(x_1, x_2)u^{(1)}(x_1)dx_1 \qquad (74)$$

where the kernel K is given by

$$K(x_1, x_2) = \sqrt{\frac{j}{\lambda d}}$$
$$\cdot \exp\left\{-\frac{jk}{2d}(g_1x_1^2 + g_2x_2^2 - 2x_1x_2)\right\}. \qquad (75)$$

Similar equations can be written for the y coordinate, so that $E(x, y)=u(x)v(y)$ and $\gamma=\gamma_x\gamma_y$. In the above equation a_1 and a_2 are the half-widths of the mirrors in the x direction, d is the mirror spacing, k is $2\pi/\lambda$, and λ is the wavelength. The radii of curvature of the mirrors R_1 and R_2 are contained in the factors

$$g_1 = 1 - \frac{d}{R_1}$$

$$g_2 = 1 - \frac{d}{R_2}. \qquad (76)$$

For the case of circular mirrors [4], [31], [32] the equations are reduced to the one-dimensional form by using cylindrical coordinates and by assuming a sinusoidal azimuthal variation of the field; that is, $E(r, \phi) = R_l(r)e^{-jl\phi}$. The radial distribution functions $R_l^{(1)}$ and $R_l^{(2)}$ satisfy the one-dimensional integral equations:

$$\gamma_l^{(1)} R_l^{(1)}(r_1) \sqrt{r_1} = \int_0^{a_2} K_l(r_1, r_2) R_l^{(2)}(r_2) \sqrt{r_2}\, dr_2$$

$$\gamma_l^{(2)} R_l^{(2)}(r_2) \sqrt{r_2} = \int_0^{a_1} K_l(r_1, r_2) R_l^{(1)}(r_1) \sqrt{r_1}\, dr_1 \quad (77)$$

where the kernel K_l is given by

$$K_l(r_1, r_2) = \frac{j^{l+1} k}{d} J_l\left(k \frac{r_1 r_2}{d}\right) \sqrt{r_1 r_2}$$

$$\cdot \exp\left\{-\frac{jk}{2d}(g_1 r_1^2 + g_2 r_2^2)\right\} \quad (78)$$

and J_l is a Bessel function of the first kind and lth order. In (77), a_1 and a_2 are the radii of the mirror apertures and d is the mirror spacing; the factors g_1 and g_2 are given by (76).

Except for the special case of the confocal resonator [5] ($g_1 = g_2 = 0$), no exact analytical solution has been found for either (74) or (77), but approximate methods and numerical techniques have been employed with success for their solutions. Before presenting results, it is appropriate to discuss two important properties which apply in general to resonators with spherical mirrors; these are the properties of "equivalence" and "stability."

4.4 *Equivalent Resonator Systems*

The equivalence properties [24], [33] of spherical-mirror resonators are obtained by simple algebraic manipulations of the integral equations. First, it is obvious that the mirrors can be interchanged without affecting the results; that is, the subscripts and superscripts one and two can be interchanged. Second, the diffraction loss and the intensity pattern of the mode remain invariant if both g_1 and g_2 are reversed in sign; the eigenfunctions E and the eigenvalues γ merely take on complex conjugate values. An example of such equivalent systems is that of parallel-plane ($g_1 = g_2 = 1$) and concentric ($g_1 = g_2 = -1$) resonator systems.

The third equivalence property involves the Fresnel number N and the stability factors G_1 and G_2, where

$$N = \frac{a_1 a_2}{\lambda d}$$

$$G_1 = g_1 \frac{a_1}{a_2}$$

$$G_2 = g_2 \frac{a_2}{a_1}. \quad (79)$$

If these three parameters are the same for any two resonators, then they would have the same diffraction loss, the

same resonant frequency, and mode patterns that are scaled versions of each other. Thus, the equivalence relations reduce greatly the number of calculations which are necessary for obtaining the solutions for the various resonator geometries.

4.5 *Stability Condition and Diagram*

Stability of optical resonators has been discussed in Section 2 in terms of geometrical optics. The stability condition is given by (8). In terms of the stability factors G_1 and G_2, it is

$$0 < G_1 G_2 < 1$$

or

$$0 < g_1 g_2 < 1. \quad (80)$$

Resonators are stable if this condition is satisfied and unstable otherwise.

A stability diagram [6], [24] for the various resonator geometries is shown in Fig. 4 where g_1 and g_2 are the coordinate axes and each point on the diagram represents a particular resonator geometry. The boundaries between stable and unstable (shaded) regions are determined by (80), which is based on geometrical optics. The fields of the modes in stable resonators are more concentrated near the resonator axes than those in unstable resonators and, therefore, the diffraction losses of unstable resonators are much higher than those of stable resonators. The transition, which occurs near the boundaries, is gradual for resonators with small Fresnel numbers and more abrupt for those with large Fresnel numbers. The origin of the diagram represents the confocal system with mirrors of equal curvature ($R_1 = R_2 = d$) and is a point of lowest diffraction loss for a given Fresnel number. The fact that a system with minor deviations from the ideal confocal system may become unstable should be borne in mind when designing laser resonators.

4.6 *Modes of the Resonator*

The transverse field distributions of the resonator modes are given by the eigenfunctions of the integral equations. As yet, no exact analytical solution has been found for the general case of arbitrary G_1 and G_2, but *approximate* analytical expressions have been obtained to describe the fields in *stable* spherical-mirror resonators [5], [6]. These approximate eigenfunctions are the same as those of the optical beam modes which are discussed in Section 2; that is, the field distributions are given *approximately* by Hermite-Gaussian functions for rectangular mirrors [5], [6], [34], and by Laguerre-Gaussian functions for circular mirrors [6], [7]. The designation of the resonator modes is given in Section 3.5. (The modes are designated as TEM_{mnq} for rectangular mirrors and TEM_{plq} for circular mirrors.) Figure 7 shows photographs of some of the rectangular mode patterns of a

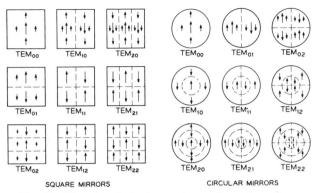

Fig. 17. Linearly polarized resonator mode configurations for square and circular mirrors.

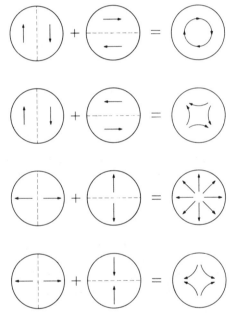

Fig. 18. Synthesis of different polarization configurations from the linearly polarized TEM_{01} mode.

laser. Linearly polarized mode configurations for square mirrors and for circular mirrors are shown in Fig. 17. By combining two orthogonally polarized modes of the same order, it is possible to synthesize other polarization configurations; this is shown in Fig. 18 for the TEM_{01} mode.

Field distributions of the resonator modes for any value of G could be obtained numerically by solving the integral equations either by the method of successive approximations [4], [24], [31] or by the method of kernel expansion [30], [32]. The former method of solution is equivalent to calculating the transient behavior of the resonator when it is excited initially with a wave of arbitrary distribution. This wave is assumed to travel back and forth between the mirrors of the resonator, undergoing changes from transit to transit and losing energy by diffraction. After many transits a quasi steady-state condition is attained where the fields for successive transits

differ only by a constant multiplicative factor. This steady-state *relative* field distribution is then an eigenfunction of the integral equations and is, in fact, the field distribution of the mode that has the lowest diffraction loss for the symmetry assumed (e.g., for even or odd symmetry in the case of infinite-strip mirrors, or for a given azimuthal mode index number l in the case of circular mirrors); the constant multiplicative factor is the eigenvalue associated with the eigenfunction and gives the diffraction loss and the phase shift of the mode. Although this simple form of the iterative method gives only the lower order solutions, it can, nevertheless, be modified to yield higher order ones [24], [35]. The method of kernel expansion, however, is capable of yielding both low-order and high-order solutions.

Figures 19 and 20 show the relative field distributions of the TEM_{00} and TEM_{01} modes for a resonator with a pair of identical, circular mirrors ($N=1$, $a_1=a_2$, $g_1=g_2$ $=g$) as obtained by the numerical iterative method. Several curves are shown for different values of g, ranging from zero (confocal) through one (parallel-plane) to 1.2 (convex, unstable). By virtue of the equivalence property discussed in Section 4.4, the curves are also applicable to resonators with their g values reversed in sign, provided the sign of the ordinate for the phase distribution is also reversed. It is seen that the field is most concentrated near the resonator axis for $g=0$ and tends to spread out as $|g|$ increases. Therefore, the diffraction loss is expected to be the least for confocal resonators.

Figure 21 shows the relative field distributions of some of the low order modes of a Fabry-Perot resonator with (parallel-plane) circular mirrors ($N=10$, $a_1=a_2$, $g_1=g_2=1$) as obtained by a modified numerical iterative method [35]. It is interesting to note that these curves are not very smooth but have small wiggles on them, the number of which are related to the Fresnel number. These wiggles are entirely absent for the confocal resonator and appear when the resonator geometry is unstable or nearly unstable. Approximate expressions for the field distributions of the Fabry-Perot resonator modes have also been obtained by various analytical techniques [36], [37]. They are represented to first order, by sine and cosine functions for infinite-strip mirrors and by Bessel functions for circular mirrors.

For the special case of the confocal resonator ($g_1=g_2$ $=0$), the eigenfunctions are self-reciprocal under the *finite* Fourier (infinite-strip mirrors) or Hankel (circular mirrors) transformation and exact analytical solutions exist [5], [38]–[40]. The eigenfunctions for infinite-strip mirrors are given by the prolate spheroidal wave functions and, for circular mirrors, by the generalized prolate spheroidal or hyperspheroidal wave functions. For large Fresnel numbers these functions can be closely approximated by Hermite-Gaussian and Laguerre-Gaussian functions which are the eigenfunctions for the beam modes.

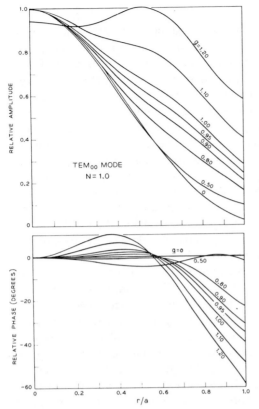

Fig. 19.   Relative field distributions of the TEM$_{00}$ mode for a resonator with circular mirrors ($N=1$).

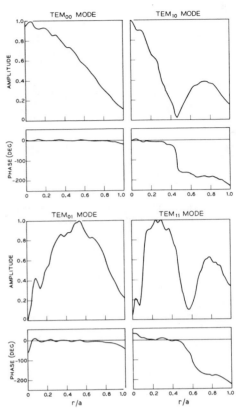

Fig. 21.   Relative field distributions of four of the low order modes of a Fabry-Perot resonator with (parallel-plane) circular mirrors ($N=10$).

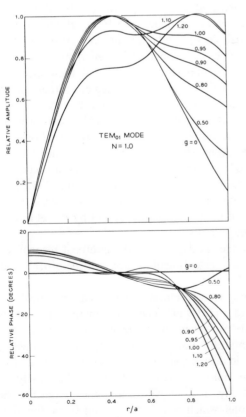

Fig. 20.   Relative field distributions of the TEM$_{01}$ mode for a resonator with circular mirrors ($N=1$).

4.7 Diffraction Losses and Phase Shifts

The diffraction loss α and the phase shift β for a particular mode are important quantities in that they determine the Q and the resonant frequency of the resonator for that mode. The diffraction loss is given by

$$\alpha = 1 - |\gamma|^2 \tag{81}$$

which is the fractional energy lost per transit due to diffraction effects at the mirrors. The phase shift is given by

$$\beta = \text{angle of } \gamma \tag{82}$$

which is the phase shift suffered (or enjoyed) by the wave in transit from one mirror to the other, in addition to the geometrical phase shift which is given by $2\pi d/\lambda$. The eigenvalue γ in (81) and (82) is the appropriate γ for the mode under consideration. If the total resonator loss is small, the Q of the resonator can be approximated by

$$Q = \frac{2\pi d}{\lambda \alpha_t} \tag{83}$$

where α_t, the total resonator loss, includes losses due to diffraction, output coupling, absorption, scattering, and other effects. The resonant frequency ν is given by

$$\nu/\nu_0 = (q + 1) + \beta/\pi \tag{84}$$

where q, the longitudinal mode order, and ν_0, the fundamental beat frequency, are defined in Section 3.5.

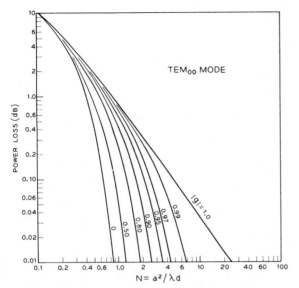

Fig. 22. Diffraction loss per transit (in decibels) for the TEM$_{00}$ mode of a stable resonator with circular mirrors.

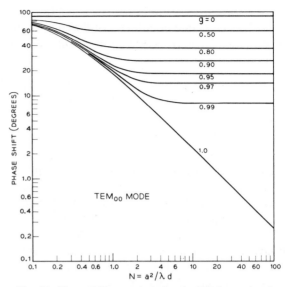

Fig. 24. Phase shift per transit for the TEM$_{00}$ mode of a stable resonator with circular mirrors.

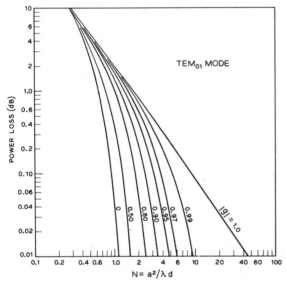

Fig. 23. Diffraction loss per transit (in decibels) for the TEM$_{01}$ mode of a stable resonator with circular mirrors.

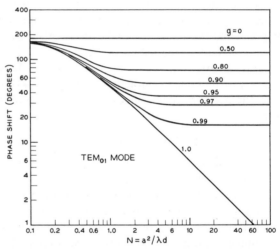

Fig. 25. Phase shift per transit for the TEM$_{01}$ mode of a stable resonator with circular mirrors.

The diffraction losses for the two lowest order (TEM$_{00}$ and TEM$_{01}$) modes of a stable resonator with a pair of identical, circular mirrors ($a_1 = a_2$, $g_1 = g_2 = g$) are given in Figs. 22 and 23 as functions of the Fresnel number N and for various values of g. The curves are obtained by solving (77) numerically using the method of successive approximations [31]. Corresponding curves for the phase shifts are shown in Figs. 24 and 25. The horizontal portions of the phase shift curves can be calculated from the formula

$$\beta = (2p + l + 1) \arccos \sqrt{g_1 g_2}$$
$$= (2p + l + 1) \arccos g, \quad \text{for } g_1 = g_2 \quad (85)$$

which is equal to the phase shift for the beam modes derived in Section 3.5. It is to be noted that the loss curves are applicable to both positive and negative values of g

while the phase-shift curves are for positive g only; the phase shift for negative g is equal to 180 degrees minus that for positive g.

Analytical expressions for the diffraction loss and the phase shift have been obtained for the special cases of parallel-plane ($g=1.0$) and confocal ($g=0$) geometries when the Fresnel number is either very large (small diffraction loss) or very small (large diffraction loss) [36], [38], [39], [41], [42]. In the case of the parallel-plane resonator with circular mirrors, the approximate expressions valid for large N, as derived by Vainshtein [36], are

$$\alpha = 8\kappa_{pl}^2 \frac{\delta(M + \delta)}{[(M + \delta)^2 + \delta^2]^2} \quad (86)$$

$$\beta = \left(\frac{M}{4\delta}\right)\alpha \quad (87)$$

where $\delta = 0.824$, $M = \sqrt{8\pi N}$, and κ_{pl} is the $(p+1)$th zero of the Bessel function of order l. For the confocal resonator with circular mirrors, the corresponding expressions are [39]

$$\alpha = \frac{2\pi(8\pi N)^{2p+l+1}e^{-4\pi N}}{p!(p+l+1)!}\left[1 + 0\left(\frac{1}{2\pi N}\right)\right] \quad (88)$$

$$\beta = (2p + l + 1)\frac{\pi}{2}. \quad (89)$$

Similar expressions exist for resonators with infinite-strip or rectangular mirrors [36], [39]. The agreement between the values obtained from the above formulas and those from numerical methods is excellent.

The loss of the lowest order (TEM$_{00}$) mode of an *unstable* resonator is, to first order, independent of the mirror size or shape. The formula for the loss, which is based on geometrical optics, is [12]

$$\alpha = 1 \pm \frac{1 - \sqrt{1 - (g_1 g_2)^{-1}}}{1 + \sqrt{1 - (g_1 g_2)^{-1}}} \quad (90)$$

where the plus sign in front of the fraction applies for g values lying in the first and third quadrants of the stability diagram, and the minus sign applies in the other two quadrants. Loss curves (plotted vs. N) obtained by solving the integral equations numerically have a ripply behavior which is attributable to diffraction effects [24], [43]. However, the average values agree well with those obtained from (90).

5. CONCLUDING REMARKS

Space limitations made it necessary to concentrate the discussion of this article on the basic aspects of laser beams and resonators. It was not possible to include such interesting topics as perturbations of resonators, resonators with tilted mirrors, or to consider in detail the effect of nonlinear, saturating host media. Also omitted was a discussion of various resonator structures other than those formed of spherical mirrors, e.g., resonators with corner cube reflectors, resonators with output holes, or fiber resonators. Another important, but omitted, field is that of mode selection where much research work is currently in progress. A brief survey of some of these topics is given in [44].

REFERENCES

[1] R. H. Dicke, "Molecular amplification and generation systems and methods," U. S. Patent 2 851 652, September 9, 1958.

[2] A. M. Prokhorov, "Molecular amplifier and generator for submillimeter waves," *JETP (USSR)*, vol. 34, pp. 1658–1659, June 1958; *Sov. Phys. JETP*, vol. 7, pp. 1140–1141, December 1958.

[3] A. L. Schawlow and C. H. Townes, "Infrared and optical masers," *Phys. Rev.*, vol. 29, pp. 1940–1949, December 1958.

[4] A. G. Fox and T. Li, "Resonant modes in an optical maser," *Proc. IRE (Correspondence)*, vol. 48, pp. 1904–1905, November 1960; "Resonant modes in a maser interferometer," *Bell Sys. Tech. J.*, vol. 40, pp. 453–488, March 1961.

[5] G. D. Boyd and J. P. Gordon, "Confocal multimode resonator for millimeter through optical wavelength masers," *Bell Sys. Tech. J.*, vol. 40, pp. 489–508, March 1961.

[6] G. D. Boyd and H. Kogelnik, "Generalized confocal resonator theory," *Bell Sys. Tech. J.*, vol. 41, pp. 1347–1369, July 1962.

[7] G. Goubau and F. Schwering, "On the guided propagation of electromagnetic wave beams," *IRE Trans. on Antennas and Propagation*, vol. AP-9, pp. 248–256, May 1961.

[8] J. R. Pierce, "Modes in sequences of lenses," *Proc. Nat'l Acad. Sci.*, vol. 47, pp. 1808–1813, November 1961.

[9] G. Goubau, "Optical relations for coherent wave beams," in *Electromagnetic Theory and Antennas*. New York: Macmillan, 1963, pp. 907–918.

[10] H. Kogelnik, "Imaging of optical mode—Resonators with internal lenses," *Bell Sys. Tech. J.*, vol. 44, pp. 455–494, March 1965.

[11] ——, "On the propagation of Gaussian beams of light through lenslike media including those with a loss or gain variation," *Appl. Opt.*, vol. 4, pp. 1562–1569, December 1965.

[12] A. E. Siegman, "Unstable optical resonators for laser applications," *Proc. IEEE*, vol. 53, pp. 277–287, March 1965.

[13] W. Brower, *Matrix Methods in Optical Instrument Design*. New York: Benjamin, 1964. E. L. O'Neill, *Introduction to Statistical Optics*. Reading, Mass.: Addison-Wesley, 1963.

[14] M. Bertolotti, "Matrix representation of geometrical properties of laser cavities," *Nuovo Cimento*, vol. 32, pp. 1242–1257, June 1964. V. P. Bykov and L. A. Vainshtein, "Geometrical optics of open resonators," *JETP (USSR)*, vol. 47, pp. 508–517, August 1964. B. Macke, "Laser cavities in geometrical optics approximation," *J. Phys. (Paris)*, vol. 26, pp. 104A–112A, March 1965. W. K. Kahn, "Geometric optical derivation of formula for the variation of the spot size in a spherical mirror resonator," *Appl. Opt.*, vol. 4, pp. 758–759, June 1965.

[15] J. R. Pierce, *Theory and Design of Electron Beams*. New York: Van Nostrand, 1954, p. 194.

[16] H. Kogelnik and W. W. Rigrod, "Visual display of isolated optical-resonator modes," *Proc. IRE (Correspondence)*, vol. 50, p. 220, February 1962.

[17] G. A. Deschamps and P. E. Mast, "Beam tracing and applications," in *Proc. Symposium on Quasi-Optics*. New York: Polytechnic Press, 1964, pp. 379–395.

[18] S. A. Collins, "Analysis of optical resonators involving focusing elements," *Appl. Opt.*, vol. 3, pp. 1263–1275, November 1964.

[19] T. Li, "Dual forms of the Gaussian beam chart," *Appl. Opt.*, vol. 3, pp. 1315–1317, November 1964.

[20] T. S. Chu, "Geometrical representation of Gaussian beam propagation," *Bell Sys. Tech. J.*, vol. 45, pp. 287–299, February 1966.

[21] J. P. Gordon, "A circle diagram for optical resonators," *Bell Sys. Tech. J.*, vol. 43, pp. 1826–1827, July 1964. M. J. Offerhaus, "Geometry of the radiation field for a laser interferometer," *Philips Res. Rept.*, vol. 19, pp. 520–523, December 1964.

[22] H. Statz and C. L. Tang, "Problem of mode deformation in optical masers," *J. Appl. Phys.*, vol. 36, pp. 1816–1819, June 1965.

[23] A. G. Fox and T. Li, "Effect of gain saturation on the oscillating modes of optical masers," *IEEE J. of Quantum Electronics*, vol. QE-2, pp. 774–783, December 1966.

[24] ——, "Modes in a maser interferometer with curved and tilted mirrors," *Proc. IEEE*, vol. 51, pp. 80–89, January 1963.

[25] F. B. Hildebrand, *Methods of Applied Mathematics*. Englewood Cliffs, N. J.: Prentice Hall, 1952, pp. 412–413.

[26] D. J. Newman and S. P. Morgan, "Existence of eigenvalues of a class of integral equations arising in laser theory," *Bell Sys. Tech. J.*, vol. 43, pp. 113–126, January 1964.

[27] J. A. Cochran, "The existence of eigenvalues for the integral equations of laser theory," *Bell Sys. Tech. J.*, vol. 44, pp. 77–88, January 1965.

[28] H. Hochstadt, "On the eigenvalue of a class of integral equations arising in laser theory," *SIAM Rev.*, vol. 8, pp. 62–65, January 1966.

[29] D. Gloge, "Calculations of Fabry-Perot laser resonators by scattering matrices," *Arch. Elect. Ubertrag.*, vol. 18, pp. 197–203, March 1964.

[30] W. Streifer, "Optical resonator modes—rectangular reflectors of spherical curvature," *J. Opt. Soc. Am.*, vol. 55, pp. 868–877, July 1965

[31] T. Li, "Diffraction loss and selection of modes in maser resonators with circular mirrors," *Bell Sys. Tech. J.*, vol. 44, pp. 917–932, May–June, 1965.

[32] J. C. Heurtley and W. Streifer, "Optical resonator modes—

circular reflectors of spherical curvature," *J. Opt. Soc. Am.*, vol. 55, pp. 1472–1479, November 1965.

[33] J. P. Gordon and H. Kogelnik, "Equivalence relations among spherical mirror optical resonators," *Bell Sys. Tech. J.*, vol. 43, pp. 2873–2886, November 1964.

[34] F. Schwering, "Reiterative wave beams of rectangular symmetry," *Arch. Elect. Übertrag.*, vol. 15, pp. 555–564, December 1961

[35] A. G. Fox and T. Li, to be published.

[36] L. A. Vainshtein, "Open resonators for lasers," *JETP (USSR)*, vol. 44, pp. 1050–1067, March 1963; *Sov. Phys. JETP*, vol. 17, pp. 709–719, September 1963.

[37] S. R. Barone, "Resonances of the Fabry-Perot laser," *J. Appl. Phys.*, vol. 34, pp. 831–843, April 1963.

[38] D. Slepian and H. O. Pollak, "Prolate spheroidal wave functions, Fourier analysis and uncertainty—I," *Bell Sys. Tech. J.*, vol. 40, pp. 43–64, January 1961.

[39] D. Slepian, "Prolate spheroidal wave functions, Fourier analysis and uncertainty—IV: Extensions to many dimensions; generalized prolate spheroidal functions," *Bell Sys. Tech. J.*, vol. 43, pp. 3009–3057, November 1964.

[40] J. C. Heurtley, "Hyperspheroidal functions—optical resonators with circular mirrors," in *Proc. Symposium on Quasi-Optics*. New York: Polytechnic Press, 1964, pp. 367–375.

[41] S. R. Barone and M. C. Newstein, "Fabry-Perot resonances at small Fresnel numbers," *Appl. Opt.*, vol. 3, p. 1194, October 1964.

[42] L. Bergstein and H. Schachter, "Resonant modes of optic cavities of small Fresnel numbers," *J. Opt. Soc. Am.*, vol. 55, pp. 1226–1233, October 1965.

[43] A. G. Fox and T. Li, "Modes in a maser interferometer with curved mirrors," in *Proc. Third International Congress on Quantum Electronics*. New York: Columbia University Press, 1964, pp. 1263–1270.

[44] H. Kogelnik, "Modes in optical resonators," in *Lasers*, A. K. Levine, Ed. New York: Dekker, 1966.

Mode Selection in Lasers

P. W. SMITH, MEMBER, IEEE

Invited Paper

Abstract—This is a tutorial review on the subject of mode selection in lasers. We begin with a historical review. After an introduction to the subject of modes in laser resonators and a brief review of the theory of laser gain saturation, the main body of the paper is devoted to a discussion of various mode-selection techniques, many which can be used to produce single-frequency laser operation. We discuss some systems for frequency stabilization of single-frequency lasers, and conclude with examples of laser applications where mode-selection techniques are required.

I. INTRODUCTION

BECAUSE THE resonators that are used for typical lasers have dimensions which are large compared to an optical wavelength, they will, in general, have a large number of closely spaced modes. If the gain medium placed in such a resonator exhibits gain at several of these mode frequencies, we might expect the laser output to consist of light at a number of closely spaced frequencies. Although Schawlow and Townes [1] in their classic paper proposing laser operation had suggested that for a sufficiently stable laser, gain nonlinearities might cause mode suppression which would result in single-mode oscillation, the early lasers were found to oscillate in a band of discrete frequencies with a bandwidth typically 10^{-4}–10^{-5} of the laser frequency. Although this is a rather monochromatic light source, there are still many applications for which greater spectral purity is required.

There has been an increasing interest in light scattering experiments in recent years as a result of the availability of high-intensity CW laser light sources. Although for most Raman scattering studies the output of a multimode laser can be used without decreasing resolution, essentially single-frequency laser output is required in order to resolve the features seen in many Brillouin scattering experiments. Another rapidly developing area where laser mode-selection techniques are finding use is that of tunable laser spectroscopy. Not only has Lamb-dip spectroscopy enabled us to study the properties of the laser transitions themselves, but the large tuning range available with tunable narrow-band dye lasers is making it possible to study a wide variety of materials with a precision never before achieved. Single-frequency lasers are important components in some proposed laser communications systems, where they generate the carrier on which information is imposed by suitable modulation techniques. Laser mode-selection techniques are also required for making holograms with appreciable depth of field. Other uses of single-frequency lasers include precision interferometry, pumps for parametric oscillators, and even, with suitable stabilization, possible standards of length and time. Some of these uses will be discussed in more detail later in this article.

Because of the need for high-power narrow-bandwidth laser sources, many techniques for reducing the number of oscillating laser modes have been developed. This article will present a

Manuscript received January 10, 1972; revised February 1, 1972.
This invited paper is one of a series planned on topics of general interest— The Editor.
The author is with Bell Telephone Laboratories, Inc., Holmdel, N. J.

tutorial review of these methods. Section II is a historical review of the subject of mode selection in lasers. Section III is a brief outline of the subject of modes in laser resonators. In many laser systems, the characteristics of the laser gain saturation will affect the output mode structure. The characteristics of laser gain saturation pertinent to mode-selection schemes are briefly covered in Section IV. In Sections V and VI we discuss specific mode-selection techniques which have been used for the selection of transverse and longitudinal resonator modes. By using one or several of these techniques, single-frequency laser operation can be obtained with almost any type of laser. For many applications it is important not just to have single-frequency operation, but to have this frequency stabilized and precisely controlled. In Section VII we discuss some representative frequency-stabilization schemes. Section VIII concludes the paper with a discussion of a number of laser applications requiring lasers with some type of mode control.

II. HISTORICAL REVIEW

Less than two years after laser action was first observed in a laboratory [2], Kleinman and Kisliuk [3], in early 1962, published the first proposal for a complex laser resonator to obtain single-frequency laser operation. Shortly thereafter, Kogelnik and Patel [4] described the successful utilization of this technique to construct a single-frequency He–Ne gas laser. In the same year Baker and Peters [5] described what appear to be the first experiments on the use of an aperture in a laser resonator to restrict oscillation to one or a few low-order transverse resonator modes.

During 1963, a number of workers began investigations of laser mode-selection techniques. Collins and White [6] reported the use of an interferometric mode selector to reduce the number of oscillating resonator modes of a ruby laser. McFarlane *et al.* [7] reported what was probably the first scientific investigation undertaken using a single-frequency laser. They observed the output as a function of length of a 1.15-μm He–Ne laser, and were able to make an estimate of the homogeneous linewidth of the 1.15-μm $2s_2 \rightarrow 2p_4$ neon laser line. Tang *et al.* [8], [9] wrote the first papers discussing the role of "spatial hole burning" in single-frequency laser operation. They showed that by constructing a laser resonator in such a way that the light interacting with the ruby laser medium was in the form of a traveling wave, single-frequency operation could be obtained.

Also in 1963, Li [10] published one of a series of papers he and Fox would write on the theory of modes in laser resonators. This paper showed how transverse mode selection could be accomplished by the use of an aperture at the center of a laser resonator. This and later papers by Fox and Li [11], [12] helped to put on a quantitative basis the subject of transverse mode selection in laser resonators.

In 1964, Lamb [52] published his now classic paper on the theory of an optical maser, and showed that under certain conditions mode competition could be strong enough to cause the suppression of oscillation on some resonator modes, with resultant

Reprinted from *Proc. IEEE*, vol. 60, pp. 422–440, Apr. 1972.

single-frequency operation. Around the year 1964, several authors reported the development of gas laser tubes sufficiently compact that they could be used in resonators short enough to produce single-frequency operation. In this way single-frequency operation was obtained with He–Ne lasers at 1.15 μm [7] and 6328 Å [13]–[16].

During 1965, a number of new techniques for longitudinal mode selection were proposed and demonstrated. Hercher [17] described a Q-switched ruby laser which oscillated in a single mode because of the use of a multiple reflector as an end mirror. This so-called "resonant-reflector" technique had been described earlier by Burch [18]. Massey *et al.* [19] described the demodulation of the output of an FM-mode-locked He–Ne laser to obtain single-frequency output. A related scheme involving an FM laser with an etalon output reflector was proposed and demonstrated by Harris and McMurtry [20]. A novel and efficient interferometric laser resonator proposed by Fox [21] was shown by Smith [22] to produce high-power single-frequency output from a 6328-Å He–Ne laser. Also during 1965, Sooy wrote a noteworthy paper explaining how saturable dyes could act to narrow the spectral output of Q-switched lasers. In the field of transverse mode selection, Li and Smith [24] described the use of a "cat's-eye" resonator to obtain high-power fundamental-transverse-mode operation of a 6328-Å He–Ne laser, and Siegman [25] wrote the first paper on the use of unstable optical resonators for transverse mode selection.

During 1966, "vernier" types of interferometric mode-selection devices were described by DiDomenico [26], [27] and White [28]. Much of the work since that time has been devoted to the improvement of existing techniques, and the application of the techniques to other laser systems. In the next few years, mode-selective and frequency-selective techniques were applied to water vapor [29], Nd:glass [30], [31], Nd:YAG [32]–[34], CO_2 [35], mercury vapor [36], and other laser systems. Mode-selective techniques were also used to produce single-frequency ring lasers [37].

In 1968 Chebotaev and his co-workers [38] and independently Lee [39] described a novel mode-selective laser system in which a resonant absorber was used inside a laser resonator. In both cases a neon discharge was used as the resonant absorber for a 6328-Å He–Ne laser. This type of laser–absorber system has also been used for laser frequency stabilization [40], [41]. Another new mode-selective technique to be introduced in 1968 was the metal film technique of Troitskii and his co-workers [43], [43]. They were able to obtain single-frequency operation of a He–Ne laser by inserting a thin metal film at a nodal plane in the laser resonator. This technique was later applied to argon ion lasers and Nd:YAG lasers by Smith *et al.* [44]. Internal modulation of single-frequency lasers was also investigated during this period [45], [46]. In recent years, many workers have been concerned with the problem of obtaining tunable single-frequency operation of dye lasers, and with the problems of laser frequency stability and reproducibility.

Before discussing specific mode-selection techniques it will be useful to review the subject of modes in laser resonators, and to discuss briefly the influence of the gain medium on the resonator modes.

III. MODES IN LASER RESONATORS

A mode of a resonator can be defined as a self-consistent field configuration. That is, the optical field distribution reproduces itself after one round trip in the resonator. The modes of a resonator formed by a pair of coaxial plane or spherical mirrors have been studied extensively (see, for example, [10]–[12] and [47]–

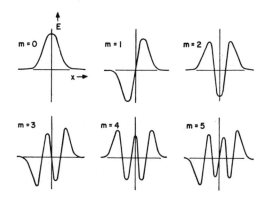

Fig. 1. Cross-sectional amplitude distribution of some low-order Gaussian beam modes of the type TEM$_{m0}$.

[49]). The reader is referred in particular to the excellent review article by Kogelnik and Li [47]. There exist sets of modes having the same spatial energy distribution transverse to the resonator axis, but having different numbers of half wavelengths of light along the axis of the resonator. These are called the longitudinal modes of the resonator and are spaced in frequency by $c/2L$, where c is the velocity of light and L is the spacing between the resonator mirrors. Corresponding to each longitudinal mode number, i.e., to a given number of half wavelengths of light along the resonator axis, there exists a set of modes which have different distributions of energy in the plane transverse to the resonator axis. These are called the transverse modes of the resonator. Fig. 1 shows the field distributions for some low-order transverse modes with rectangular symmetry. There exists a complementary set with circular symmetry. The properties of both sets are similar, however, and further discussion in this paper will refer to the rectangular-symmetry modes. The cross-sectional amplitude distribution of these modes $A(x, y)$ is given closely by [47]

$$A(x,y) = A_{m,n}\left[H_m\left(\frac{\sqrt{2}\,x}{w}\right) H_n\left(\frac{\sqrt{2}\,y}{w}\right) \right]$$
$$\cdot \exp\left[-(x^2 + y^2)/w^2\right] \quad (1)$$

where x and y are the transverse coordinates, $A_{m,n}$ is a constant whose value depends on the field strength of the mode, w is the radius of the fundamental mode ($m=0$, $n=0$) at $1/e$ maximum amplitude, $H_a(b)$ is the ath-order Hermite polynomial with argument b, and m and n are the transverse mode numbers. These modes are often called the TEM$_{mn}$ modes by analogy with modes in waveguides. Note that Fig. 1 shows the x variation of the field for different mode order numbers m. The same variations occur in the y direction for the appropriate values of n.

It can be shown that for a given laser resonator, the resonant frequency of a given mode is

$$\nu_{m,n,q} = (c/2L)\{(q+1) + [(1+m+n)/\pi]\arccos\sqrt{g_1 g_2}\} \quad (2)$$

where q, the longitudinal mode order number, is the number of nodes in the axial standing-wave pattern (the number of half wavelengths is $q+1$), m and n are the transverse mode order numbers, and $g_i = 1 - L/R_i$, where R_i is the radius of curvature of mirror i ($i = 1, 2$). Note that adjacent longitudinal modes are spaced in frequency by $c/2L$, and that different transverse modes have, in general, different frequencies.

In any real laser resonator some part of the laser beam will be lost by leaking around mirrors of finite size or by interrupting an

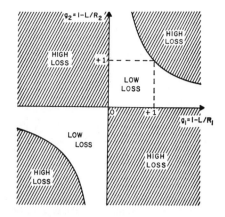

Fig. 2. Stability diagram showing stable (low-loss) and unstable (high-loss) values of mirror radius of curvature (R_1 and R_2) and mirror separation L.

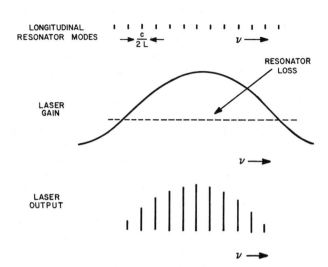

Fig. 4. Laser oscillation on a number of longitudinal resonator modes.

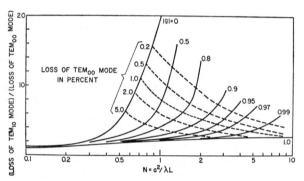

Fig. 3. Ratio of losses per pass for the TEM_{10} and TEM_{00} modes in resonators with two mirrors of equal radii of curvature, as a function of the Fresnel number N for various values of g ($=1-L/R$) (after Li [11]).

aperture within the laser resonator. These losses will depend on the diameter of the laser beam in the plane of the aperture (w) and the aperture radius. For some values of mirror curvature and spacing, the beam size at the mirrors approaches infinity, and large diffraction losses will result. These high-loss configurations are known as unstable resonator geometries. Fig. 2 shows the stable and unstable regions for a resonator with mirrors of radius R_1 and R_2, respectively, separated by a distance L. The Fresnel number $N = a^2/(\lambda L)$ is a measure of the effect of an aperture of radius a at both mirrors of a resonator of length L. λ is the wavelength of the light. For resonators with a large Fresnel number ($N \gg 1$), the transition from low- to high-loss regions in Fig. 2 is quite abrupt, but for Fresnel numbers $N \stackrel{<}{\sim} 1$, the transition is more gradual.

Because of the fact that higher order transverse modes have a larger spatial extent than the fundamental transverse mode (see Fig. 1), a given size aperture will preferentially discriminate against higher order transverse modes in a laser resonator. Fig. 3 shows how the relative diffraction losses for the fundamental and second-order laser modes depend on the resonator geometry. We shall return to this point in our discussion of transverse mode selection later in this paper.

IV. INFLUENCE OF THE LASER GAIN MEDIUM ON RESONATOR MODES: MODE COMPETITION

Let us consider the effect of placing a laser gain medium into a resonator such as we have been discussing in Section III. Light energy will begin to build up in those resonator modes for which the net gain exceeds the losses. This is illustrated in Fig. 4 where we show one possible laser output when a number of longitudinal resonator modes are above threshold for oscillation. In any real laser there are several effects which will to some extent modify this simple picture. It does illustrate, however, that although the bandwidth of a single laser mode can in theory be as narrow as a fraction of a hertz [50], the actual oscillation bandwidth of a typical laser with no mode control will be governed by the width of the gain curve, which may be several gigahertz for gas lasers in the visible region of the spectrum and much larger for solid-state lasers. The total output of the laser as a function of time will depend on the amplitudes, frequencies, and phases of the oscillating modes. Due to random phase fluctuations the output will fluctuate in a random way as a function of time. One way to control this is to fix the amplitudes and phases of the modes. This is called "mode-locking" a laser, and this technique is very useful when an output pulse train is desired. (For a recent review of mode-locking of lasers see [51]). Alternatively, mode-selection techniques may be used to obtain single-frequency CW laser output.

To some extent, the presence of the laser medium will affect the transverse mode distributions in the resonator. Fox and Li [12] have shown, however, that even for moderately high-gain gas lasers this effect is small, and to a good approximation the transverse mode intensity distribution and diffraction losses for an active laser will be the same as those calculated for a passive resonator. The gain medium will have an appreciable effect on the transverse mode distributions for such cases as the extremely high-gain gas lasers (such as the 3.5-μm He–Xe laser) and optically pumped solid-state lasers, where thermal gradients in the laser material may produce appreciable mode distortion.

The dispersive effects of the laser medium will modify somewhat the mode frequencies of the passive cavity. Once again, these effects are found to be small for low-gain lasers (see, for example [52]), but can become quite large in some of the high-gain laser systems [52a].

For most lasers the major effect of the gain medium on the oscillating laser modes is that of mode competition. Fig. 5(a) shows the multimode output of a high-power argon ion gas laser. The amplitudes of these modes are fluctuating with time. Fig. 5(b) shows the single-frequency output *at the same output power level* obtained by the use of a mode-selective device within the laser resonator. How is it possible to channel all of the power

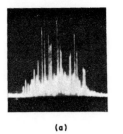
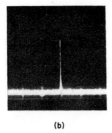

(a)   **(b)**

Fig. 5. (a) Multimode output of an argon laser. (b) Single-frequency output of the same laser. The laser output power is the same in both photographs.

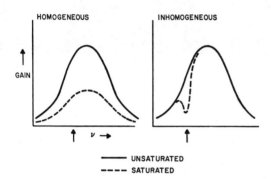

Fig. 6. The saturation behavior of homogeneously broadened and inhomogeneously broadened resonance lines under influence of intense monochromatic radiation at the frequency indicated by the arrow.

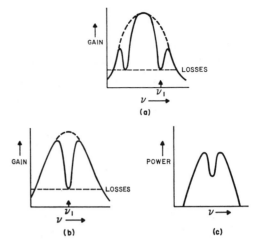

Fig. 7. Plots of laser gain as a function of frequency. (a) Single mode at frequency ν_1 away from atomic line center. (b) Single mode at frequency ν_1 at line center. (c) Plot of single-mode output power as a function of frequency.

distributed among the resonator modes in Fig. 5(a) into a single resonator mode? To answer this question we must examine the manner in which the gain medium saturates.

Fig. 6 shows the way in which homogeneously and inhomogeneously broadened resonance lines saturate under the influence of a traveling wave of intense monochromatic radiation at the frequency indicated by the arrow. For an inhomogeneous line it is possible to "burn a hole," i.e., saturate only those atoms with resonant frequencies close to that of the incident radiation.

Although a detailed treatment of the theory of laser operation is rather complicated (see, for example, [52]), we can obtain physical understanding from a "hole-burning" model due to Bennett [53]. He shows that the halfwidth of the "hole," γ_h in the population inversion-versus-frequency curve is

$$\gamma_h = \gamma(1 + I/I_s)^{1/2} \qquad (3)$$

where γ is the homogeneous linewidth (the halfwidth of the atomic response of an individual atom), I is the incident light intensity, and I_s, the saturation parameter, is a group of atomic constants. We see, then, that radiation in a single resonator mode can interact with atoms over a frequency range of roughly $2\gamma_h$. If, as is often the case, the resonator mode spacing is less than $2\gamma_h$, then several resonator modes will be interacting with the same group of atoms. This is termed mode competition, for a given atom can only contribute its energy to one of the modes, and this energy is then unavailable for the others. If some mode-suppression technique is used to favor one mode within the $2\gamma_h$ frequency region, this mode will compete more favorably for the available energy, and thus grow at the expense of the other less favored modes. In this way mode competition can aid us when we wish to obtain laser oscillation on a single resonator mode.

In a typical laser with a simple two-mirror resonator, the radiation in the resonator consists of approximately equal intensi-

ties of light traveling in each direction, and standing waves are formed in the resonator. If the laser medium is a gas whose gain curve is inhomogeneously broadened because of the Doppler effect due to the distribution of thermal velocities, then a single frequency of radiation in the resonator will, in general, interact with two groups of atoms—one group in resonance with the light traveling one direction, and the other group with light traveling in the opposite direction. This is illustrated in Fig. 7(a) where we show the two holes burned in the laser gain profile due to radiation in a single resonator mode at frequency ν_1. When ν_1 coincides with the atomic resonance frequency, both traveling waves will interact with the same group of atoms—those with zero axial velocity. This is shown in Fig. 7(b). Because at line center fewer atoms give energy to the laser mode, the output power is reduced. This effect is evident in Fig. 7(c) which shows the so-called 'Lamb dip' at the center of the output-power-versus-frequency curve. The width of the Lamb dip at low laser powers gives a measure of the homogeneous linewidth γ.

On the basis of our previous discussion it would appear that one might expect spontaneous single-frequency operation when the laser medium has a homogeneously broadened resonance line. Such operation is usually not observed, however, due to a phenomenon called spatial hole burning [8], [9]. This is illustrated in Fig. 8. We see that due to the standing-wave nature of the fields in a two-mirror laser resonator, there will be spatial regions in the gain medium which are not saturated by a given mode, and which can give gain to another mode of a different frequency. Thus even with a completely homogeneously broadened laser line, multimode oscillation will take place in a standing-wave resonator unless one utilizes one or more of the mode-selection techniques to be discussed in the next sections.

V. TRANSVERSE MODE SELECTION TECHNIQUES

From Fig. 1 it can be seen that higher and higher order transverse modes have their energy less and less concentrated along the axis of the laser resonator. For this reason a given circular aperture in a laser resonator gives progressively higher diffraction losses as one goes to higher and higher order transverse modes. This fact can conveniently be used if, as is often the case, we wish to select the lowest order (fundamental) transverse resonator mode. Then, in principle, we need only use the largest aperture for

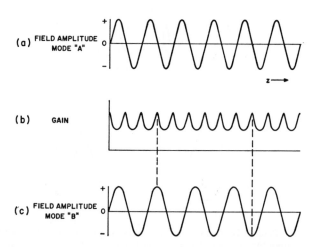

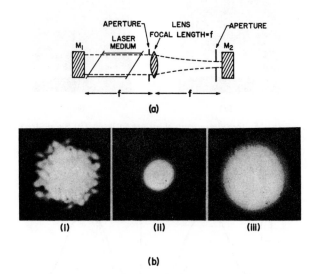

Fig. 8. Spatial hole burning in a laser medium in a standing-wave resonator. (a) Electric field of one longitudinal mode (mode *A*). (b) Gain saturation due to mode *A*. (c) Electric field of adjacent longitudinal mode (mode *B*). Note that mode *B* will experience gain as it can interact with atoms that are not saturated by mode *A*.

Fig. 9. (a) The "cat's eye" resonator. (b) Laser output from mirror M_1 for (i) both apertures open—multimode operation; (ii) aperture at lens closed—fundamental transverse mode operation; and (iii) aperture at mirror M_2 closed—fundamental transverse mode operation. The output power for (iii) is 2.5 times that of (ii) (after Li and Smith [24]).

which the losses for the second-order mode are greater than the laser gain. (The diffraction loss caused by a given aperture can be found from the curves given in [11].) In practice, due to transverse mode competition, a somewhat larger size can often be used. The transverse mode selectivity achievable with a circular aperture of radius *a* (strictly with two circular apertures—one at each resonator mirror) is illustrated in Fig. 3. Note that this selectivity is strongly dependent on the resonator geometry, and is greatest for a confocal resonator (two mirrors of equal radii of curvature separated by a distance equal to the radius of curvature) and least for the plane parallel resonator (two plane mirrors). Several authors have reported modifying the plane-parallel resonators of solid-state lasers to obtain regions with better mode-selective properties [5], [67], [68]. Evtuhov and Nieland [69] reported good mode selectivity in a ruby laser by making the resonator end reflector have high reflectivity only where the fields of the desired mode are high. A similar technique was used in [70].

Because a confocal resonator has the smallest fundamental mode volume of any of the resonator geometries, a number of authors have considered ways of increasing the mode volume and at the same time maintaining high mode selectivity [71]–[75]. For many applications, a good compromise is a resonator consisting of a plane mirror and curved mirror separated by a distance somewhat less than the mirror radius of curvature [71]. Another technique is the use of a "cat's-eye" resonator [24], [76]. Fig. 9(a) shows a schematic diagram of this type of resonator. The mode volume can be increased to fill the laser gain medium by closing the aperture at mirror M_2. Under these conditions the mode selectivity is essentially that of a confocal resonator. Fig. 9(b) shows the increases in fundamental mode volume and output power that have been achieved by Li and Smith [24] using such a resonator with a 6328 Å He–Ne gas laser.

Many authors have used resonators in the unstable (high-loss) regions of Fig. 2 to achieve fundamental mode oscillation. These resonators are used in particular with high-gain lasers. The power "lost" by diffraction around one of the laser mirrors can be utilized as the useful laser output. These unstable resonator systems have been analyzed by several authors [25], [77]–[83]. Fig. 10 shows two examples of unstable laser resonators which have been

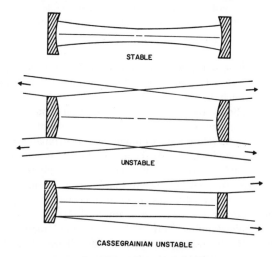

Fig. 10. Unstable (high diffraction loss) laser resonators (after Siegman and Arrathoon [77]).

used. Mode-selection experiments with unstable resonators have been reported for ruby [25], Nd:glass [30], pulsed argon [82], and CO_2 [83] laser systems.

There are a number of other fundamental-transverse-mode selection techniques which have been found useful in specific laser systems [84]–[97], [172], [173]. Experimenters have found that reduction of the angular divergence of the beam is a useful method of mode control [86]–[89]. The use of roof-top prisms with angle slightly less than 90° to make a resonator has been found to be a useful method to obtain single-transverse-mode operation of a ruby laser [89], [90]. The use of a saturable absorber dye for Q-switching a laser has been shown to be also a method of achieving transverse mode selection [91], [172], [173]. Because the lens effect due to the heating of an optically pumped laser rod by the pumping light will reduce the fundamental mode volume of the passive resonator, this effect must be compensated for if high-power fundamental-mode operation of these lasers is desired [84].

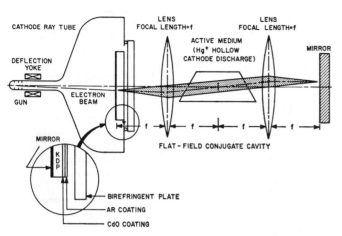

Fig. 11.   An early version of a scanlaser (after Myers [36]).

An analysis of thermal effects in a Nd:YAG laser has recently been made by Fricke [85].

Higher order transverse modes can be selected by the use of complex apertures [98], [99] or reflectors with regions of low reflectivity [100], which discriminate against (i.e., give high loss to) all but the desired mode. A simple example is the use of a single fine wire perpendicular to the resonator axis to select the second-order (TE_{01}) mode.

For some applications it may be desirable to have the ability to switch from one transverse mode to another. An example of this type of system is the "scanlaser" described by Myers [36]. This is an image display device which uses a very high-Fresnel-number laser that is capable of oscillating in a large number of transverse modes. The operation of this device is illustrated in Fig. 11. By using a mirror whose reflectivity can be varied over its surface, it is possible to have the laser oscillate with some given intensity distribution, i.e., some given combination of transverse modes. If the left-hand mirror in Fig. 11 had uniformly high reflectivity, the laser would oscillate in a large number of transverse modes. Oscillation is prevented, however, by the quartz plate which rotates the plane of polarization of the incident light sufficiently so that the loss the light experiences in passing through the Brewster-angle windows of the laser tube is larger than the laser gain. By focusing an electron beam onto a region of potassium dihydrogen phosphate (KDP) slab in front of the laser mirror, one can create areas of charge on the KDP surface, and thus fields across the KDP. The local regions of birefringence in the KDP can be made to just compensate the birefringence introduced by the quartz slab. Thus modes with radiation concentrated in these regions will experience low loss and can be made to oscillate. In this way complicated patterns of high-order laser modes can be produced and the mode patterns can be rapidly changed by "writing" a new charge distribution on the KDP slab. Resolution of more than 200 spots/cm has been obtained with devices of this type.

Johnston et al. [101] have performed "mode-switching" experiments with 6328 Å He–Ne lasers. They showed that a laser normally oscillating in the fundamental mode could be switched to second-order-mode operation by the injection of a low-power second-order-mode signal of the correct frequency. Under these conditions the two lasers will be locked together in frequency. The use of a low-power precisely mode-controlled laser to regulate a high-power laser could be a useful technique. In a manner similar

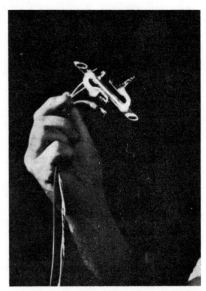

Fig. 12.   A small 6328-Å He–Ne laser tube designed for single-frequency operation.

to the scanlaser system described earlier, one could also picture the use of this mode-switching technique to force a highly multimode laser oscillator to operate with a given transverse energy distribution.

VI. LONGITUDINAL MODE-SELECTION TECHNIQUES

As we discussed in Section III, the longitudinal modes of a simple resonator are spaced in frequency by $c/2L$ where c is the velocity of light and L is the mirror spacing. We can see from Fig. 4 that it is always possible to increase the resonator losses to the point that only one or a few longitudinal modes are above threshold for oscillation. In this way we obtain narrow-bandwidth laser operation—but at the expense of output power. A somewhat more practical idea is to reduce the length of the laser resonator, and thus increase the frequency spacing between longitudinal resonator modes to the point that only one mode is above threshold for oscillation. Although this technique also sacrifices output power (for the laser gain medium must be small enough to fit within the resonator!), there are a number of laser systems for which practical single-frequency laser oscillators can be constructed in this manner. Fig. 12 shows a small 6328 Å He–Ne laser tube developed for single-frequency operation. Short single-frequency He–Ne gas lasers have been built for operation at 6328 Å [13]–[16], [102], [103] or 1.15 μm [7].

Stillman et al. [104] have shown that by reducing the effective resonator length of a cadmium selenide laser to a value of approximately 2 μm, the longitudinal mode spacing could be made sufficiently great so that the laser would oscillate on a single longitudinal mode. There are many systems, however, which would require impractically short lengths of active medium for this method to be effective. For this reason many other techniques of longitudinal mode selection have been developed, and we shall review them in this section.

Let us begin by noting that it would be possible to pass the output of a laser oscillating in a number of longitudinal modes through a filter (for example a short Fabry–Perot resonator) external to the laser cavity, and in this way obtain a narrow-band-width laser beam. For a great many laser systems, mode competi-

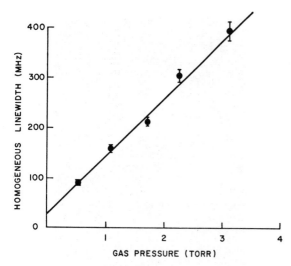

Fig. 13. Homogeneous linewidth of the 6328-Å neon laser line versus pressure for a 7:1 mixture of He³:Ne²⁰ (after Smith and Hänsch [182]).

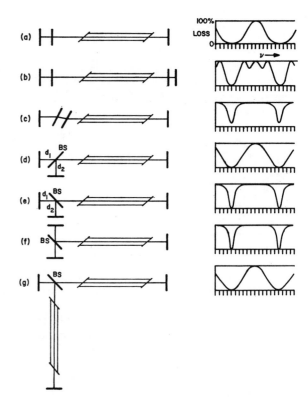

Fig. 14. Schematic representations of some interferometric schemes for laser resonators with longitudinal mode selectivity. The curves to the right of each figure show the round-trip loss for radiation in an empty resonator with perfect reflectors (reflectivity = 100 percent) and beamsplitters (reflectivity plus transmissivity = 100 percent). For comparison the frequencies of the longitudinal modes of the same laser resonator with a simple end reflector are indicated.

tion effects are such that if some modes are suppressed within the laser resonator, more energy will be available for the remaining oscillating mode or modes. Thus more power will be obtained by using mode-selection techniques within the laser resonator. A second reason for the use of internal mode-selection techniques is that one can usually obtain greater selectivity with a given device inside the laser resonator, as it is only necessary to introduce a loss greater than the gain to completely suppress an unwanted mode. For these reasons, essentially all the techniques we shall discuss are internal mode-selection techniques.

In Section IV it was pointed out that for a laser with a homogeneously broadened resonance line, the elimination of spatial hole burning may be all that is required to obtain single-frequency operation [9], [32], [33], [176]–[179]. Tang et al. [9] reported what was apparently single-frequency operation of a ruby laser with a ring resonator when one direction of oscillation was suppressed so that there was no standing wave within the resonator. Other experimenters have reported eliminating spatial hole-burning effects by physically moving the active medium [32], [176], using λ/4 plates [177], [178] or 45° rotators [179] to obtain different polarizations for the forward and backward traveling waves, and using electrooptic phase modulators within the laser resonator [33]. Using these techniques, experimenters have observed single-frequency operation of the Nd:YAG laser [32], [33], [178] and the ruby laser [9], [176], [177], [177a]. The reduction of spatial hole burning in optically pumped semiconductor lasers due to the large diffraction losses experienced by the laser radiation can result in single-frequency operation of these lasers [180].

Danielmeyer [181] has shown that if the drift or diffusion rate for the atoms is fast enough, spatial hole-burning effects will be averaged out. Thus although spatial hole burning is important for solid-state lasers, the effect is usually negligible for gas lasers because the thermal velocities of most of the atoms are sufficient for the atoms to traverse several standing-wave maxima in an excited-state lifetime. Thus a gas laser would be expected to operate spontaneously in a single frequency if the *spectral* hole burning due to the inhomogeneously broadened resonance line were eliminated. Fig. 13 shows recent measurements of the homogeneous linewidth of the 6328-Å He–Ne laser line as a function of

gas pressure [182]. It can be seen by extrapolating these data that at a pressure of about 10 torr the homogeneous linewidth becomes larger than the inhomogenous broadening (about 1600 MHz) and thus one might expect spontaneous single-frequency operation at these gas pressures. Recent experiments with waveguide gas lasers have demonstrated that He–Ne laser operation is possible in this pressure range, and Smith [183] has reported the observation of spontaneous single-frequency operation of such a high-pressure 6328-Å He–Ne laser. Similar effects are probably responsible for the spontaneous single-frequency operation of an argon ion laser reported by Yarborough and Hobart [184], Bridges and Rigrod [185], and Borisova and Pyndyk [186]. The experiments of Yarborough and Hobart are particularly noteworthy as they could obtain an output power of 2W with 90 percent of the output in a single frequency.

There are a number of mode-selection techniques that we will call interferometric techniques. In each case some type of complex laser resonator using more than two mirrors is used to provide a resonant structure for which only one high-Q (low-loss) resonator mode will be above threshold for laser oscillation. Fig. 14 shows a number of different interferometric mode-selection techniques. The first scheme shown in Fig. 14(a) was originally proposed by Kleinman and Kisliuk [3]. The two mirrors at the left-hand end of the laser resonator act as a reflector whose reflectivity varies with frequency. The reflectivity of a two-mirror interferometer of this type will have maxima and minima as functions of frequency (see, for example, [105]). The round-trip loss of the entire laser

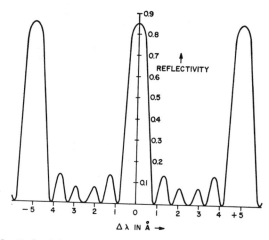

Fig. 15. Reflectivity versus wavelength for a 3-plate resonant reflector designed for use with a ruby laser (after Mahlein and Schollmeier [125]).

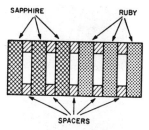

Fig. 16. CW ruby laser designed for single-frequency operation at 4.2°K and 77°K (after Birnbaum et al. [92]).

resonator as a function of frequency is shown in Fig. 14(a) together with the positions of the longitudinal resonances of the same laser resonator with a simple end reflector. The periodicity of the loss is $c/2d$, where c is the velocity of light and d is the spacing of the two left-hand mirrors. The width of the region of low loss is governed by the reflectivity of the left-hand mirrors. Clearly, modes which have frequencies near the high-loss regions will be discriminated against. By making d sufficiently small, we can arrange to have only one low-loss region occurring in the frequency band over which the laser medium has appreciable gain. We cannot, however, choose mirror reflectivities which will make the low-loss region as narrow as we wish, and therefore there may be several modes which will oscillate in this low-loss region. The Kleinman and Kisliuk technique has been analyzed in detail in [3] and [106]–[109] and experimental results have been reported for the 6328-Å He–Ne gas laser [4], [110], the argon ion laser [111], and solid-state lasers [112]–[117]. Kobayashi and Matsuo [118] have described a modification of this technique which involves placing a modulator in the Fabry–Perot interferometer. Under suitable conditions they could achieve enhanced mode selectivity.

It is also possible to achieve enhanced mode selectivity by using such interferometers at each end of the laser resonator, as shown in Fig. 14(b). By using different interferometer spacings at each end, somewhat narrower low-loss regions can be obtained [119]. There have been a number of experiments with solid-state lasers where a number of low-finesse interferometers have been used to narrow the oscillation bandwidth. When several low-finesse etalons are stacked at one end of the resonator, they form a 'resonant reflector' which can be designed to have narrow regions of high reflectivity [17], [18], [125]–[129]. Fig. 15 shows the theoretical reflectivity of a resonant reflector with three identical uncoated optical plates with equal spacing [125]. Note that the peak reflectivity is only about 85 percent. Because of this low peak reflectivity, the resonant-reflector technique is used primarily with relatively high-gain solid-state lasers. Some resonant-reflector design considerations have been discussed by Watts [128] and Magyar [129].

Birnbaum and Stocker [120], [121] and Pratesi et al. [122], [123] have made experiments with segmented ruby lasers where the ends of the ruby rod acted as low-reflectivity mirrors within the laser resonator. They observed narrow-bandwidth laser output, and later experiments by Pratesi [124] indicate single-frequency output from a 10-element pulsed ruby laser.

Fig. 16 is a schematic diagram of a ruby laser system designed for CW operation at 4.2°K and 77°K [92]. It uses a combination of the segmented-laser and resonant-reflector techniques to achieve single-frequency operation.

The technique shown in Fig. 14(c) was suggested by Collins and White [6] and Manger and Rothe [130], [131], and has been analyzed in [132]. It has been used with solid-state lasers [6], [115]–[117], [130], [131], [133]–[137], gas lasers [133], [138], [139], dye lasers [140], [141], and parametric oscillators [142]. In this scheme a tilted Fabry–Perot etalon is inserted at a small angle in the laser resonator. The reflectivity of the composite mirror consisting of the tilted etalon and adjacent end mirror corresponds closely to the transmission curve of a simple Fabry–Perot resonator. Thus a narrow region of high reflectivity can be obtained by using sufficiently high-reflectivity coatings on the etalon. The spacing between the reflectivity maxima is $c/2d$ where d is the etalon thickness. There will be some loss for the favored mode or modes, however, due to the "walk-off" associated with the tilt of the etalon. For many applications, this loss is low enough to make this technique a useful one (see, for example, [134]). Because the walk-off loss increases as the reflectivity of the etalon surfaces is increased, the technique is used primarily with lasers for which there is appreciable mode competition, and a low-finesse etalon is sufficient to produce single-frequency operation. Single-frequency operation with a low-finesse tilted etalon in the laser resonator has been reported for He–Ne lasers [138], argon ion lasers [133], and solid-state lasers [133], [134]. The use of a tilted etalon in a ring laser has also been reported [155], [159].

Fig. 14(d) shows a Michelson interferometer used as a complex end reflector [143]–[145]. The loss maxima are spaced by $c/2(d_1-d_2)$ where d_1 and d_2 are the spacings of each mirror from the beam splitter (labeled BS in the figure). The mode selectivity of this device is limited because the reflectivity of the Michelson interferometer varies sinusoidally with frequency. The ring-laser analog of this device is discussed in [157] and [159].

The device shown in Fig. 14(e) was originally described by Fox [21] and Smith [22], and has been used by a number of experimenters to obtain single-frequency operation of gas lasers [22], [146]–[153]. As before, the three left-hand mirrors can be thought of as a mirror of variable reflectivity. In this case the reflectivity peaks are spaced by $c/2(d_1+d_2)$. By increasing the reflectivity of the beamsplitter, it is possible to make the width of the low-loss region as narrow as desired, while maintaining low "on-resonance" losses for the favored mode. This versatility has made this technique one of the most used methods of achieving single-frequency gas laser operation.

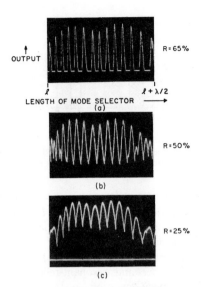

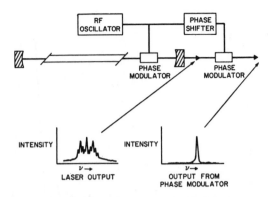

Fig. 19. Single frequency obtained by demodulating the output from an FM laser. Experimental frequency spectra after Osterink and Targ [161].

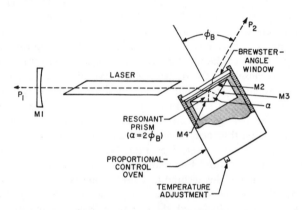

Fig. 17. Experimental displays of laser output as a function of the length of the Fox–Smith mode-selector interferometer for three values of the reflectivity R of the beamsplitter (after Nash and Smith [46]).

Fig. 18. Single-frequency argon ion laser using temperature-stabilized coated quartz prism to select a single longitudinal resonator mode (after Rigrod and Johnson [149]).

Fig. 17 shows the single-frequency output of a long He–Ne laser with a Fox–Smith mode selector as a function of the length of the mode-selector interferometer. A peak in the laser output occurs when there is a coincidence of a resonator mode with an interferometer resonance. The overall envelope shows the Lamb dip characteristic of single-frequency gas lasers. The width of an individual peak is a measure of the interferometer detuning required to introduce a certain loss, and this is a measure of the selectivity of the mode-selector interferometer. It is clear from Fig. 17 that the selectivity can be adjusted by selecting the reflectivity of the beamsplitter.

Rigrod and Johnson [149] used a coated solid quartz prism to form a mode-selector interferometer of the Fox–Smith type for use with an argon ion laser. Their laser system is shown in Fig. 18. By placing the interferometer in a temperature-controlled oven they were able to keep a resonance of the mode-selector stabilized to within ∓ 3 MHz.

Ring-laser counterparts of the Fox–Smith interferometer have been discussed by several authors [37], [155], [157], [159], [160] and experiments have been reported with He–Ne lasers [37] and argon ion lasers [160]. Sinclair [154] has described how an off-axis

confocal interferometer can function as a Fox–Smith interferometer, and has demonstrated its use with an argon ion laser.

The device shown in Fig. 14(f) has mode-selective properties identical to those of the device in Fig. 14(e) if the values of beamsplitter transmissivity and reflectivity are interchanged [155], [156]. Ring-laser counterparts of this device have also been proposed [155], [157].

The device shown in Fig. 14(g) was proposed by DiDomenico and Seidel, and the operation of this device has been demonstrated with a He–Ne laser [26], [27]. A similar device using a polarizing prism in place of the beamsplitter has been described by White [28]. This device operates on a "vernier" principle, and the favored modes are those that are resonant in both of the two coupled resonators of almost equal length. The regions of low loss are spaced by $c/2(d_1 - d_2)$ where d_1 and d_2 are the distances from the beamsplitter to the two end mirrors. There has appeared recently in the literature a comparison of the Fox–Smith interferometer and the DiDomenico–Seidel device [157]. The conclusion of [157] is that appreciably better mode selectivity can be obtained with the Fox–Smith interferometer. A general mathematical analysis of this type of complex resonator is presented in [158].

Because the amplitude of a laser using any of the previously described interferometric mode-selection schemes will depend on the relative tuning of two or more coupled resonators, a number of workers have developed electronic feedback systems to "lock" these resonators together. Representative systems are described in [22], [146], and [147].

It is possible by means of internal modulation to mode-lock a set of longitudinal modes of a laser so that the output mode amplitudes and phases correspond to an FM-modulated wave [51], [162]. We will now discuss several mode-selection schemes which can produce essentially single-frequency output from an FM-locked laser. Because the single-frequency energy is derived from all of the oscillating resonator modes, these techniques in principle should be capable of producing greater single-frequency output than most of the other mode-selection techniques discussed. In practice, however, the output powers obtained with the FM-locked laser techniques are comparable with those obtained by other methods. Fig. 19 illustrates the first of the FM-locked laser techniques, which was proposed and demonstrated by Massey et al. [19]. By passing the output of an FM-locked laser through a suitable FM demodulator, Massey et al. [19] showed that they could recover the "carrier," i.e., all of the laser output light was converted to this single output frequency. The experimentally ob-

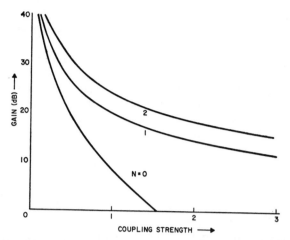

Fig. 20. Threshold gain versus coupling strength (size of periodic gain variation) for several longitudinal resonances of a distributed-feedback resonator. Modes are numbered from the center of the gain curve with $N=0$ being the highest gain mode (after Kogelnik and Shank [175]).

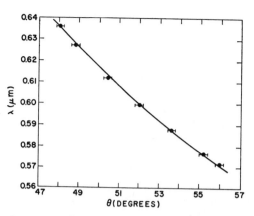

Fig. 21. Output wavelength as a function of the angle between the pumping beams for a rhodamine 6G distributed feedback dye laser (after Shank et al. [66]).

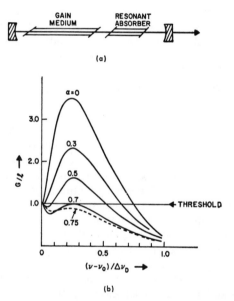

Fig. 22. (a) Use of a resonant absorber in a laser resonator to obtain single-frequency operation. (b) Saturated values of net gain for the laser in (a) normalized to the fixed (nonsaturable) resonator losses (G/l) for a single frequency oscillating at line center (ν_0). $\Delta\nu_D$ is the assumed to be the same for both the gain and the absorber media (after Bennett [166]).

served frequency spectra [161] of an FM-locked argon ion gas laser are shown in Fig. 19 before and after demodulation. It is clear that essentially single-frequency output was obtained. Weber and Mathieu [163] have proposed using a rather complex scheme to produce single-frequency output from an FM-locked laser at the second harmonic of the laser frequency. The last FM-locked laser technique that we shall discuss was proposed and demonstrated by Harris and McMurtry [20]. They used a high-finesse etalon as one of the end reflectors of the FM-locked laser resonator. This etalon was tuned to transmit primarily a single resonator mode. Because the FM modulator in the resonator couples all the modes together with a definite amplitude relationship, in theory all of the multimode power can be obtained in the single-frequency output. There are various practical difficulties, however, which limit the usefulness of this technique.

A new type of laser resonator has recently been proposed by Miller [63] and Kogelnik, and demonstrated by Kogelnik and Shank [64] and Kaminow et al. [65].They showed that either an amplitude (gain) grating or a phase grating within the (solid) laser medium can be used to form a "distributed-feedback" resonator, which will have low losses only over a narrow band of frequencies. Kogelnik and Shank [175] have made a detailed analysis of the behavior of a distributed-feedback laser considering both gain and refractive index periodicites. Fig. 20 shows the results of their calculations for the thresholds of various longitudinal resonator modes for a distributed-feedback laser in which there is a periodic variation of the gain. It can be seen that for a range of coupling strengths, the threshold for the second resonator mode is almost 20 dB higher than that for the highest gain mode (because of the way in which they defined gain, their threshold gain for the lowest order mode at high coupling strengths is actually less than zero). Thus in theory a single-frequency distributed feedback laser can be built which remains single-frequency even when pumped 100 times above threshold. In practice, the distributed-feedback lasers which have been built exhibit narrow-bandwidth, but not single-frequency output. Output frequency widths of 0.1 Å have been reported for distributed-feedback dye lasers which have gain bandwidths of several hundred angstroms. Shank et al. [66] have pumped an organic dye laser with two crossed coherent light beams. The interference of these beams produces a spatial modu-

lation of both gain and index of refraction to create a distributed-feedback grating within the laser medium. The advantage of this system is that the frequency of the resonator formed by this distributed-feedback grating can be tuned by changing the angle between the two interfering pump beams. Using a solution of rhodamine 6G in ethanol, Shank et al. were able to observe narrow-band laser oscillation over a tuning range of 640 Å as the angle between the pumping beams was varied. Fig. 21 is a plot of their results.

Chebotayev et al. [38] and Lee et al. [39] have discussed the use of a resonant absorber in the laser resonator to obtain single-frequency operation. This technique is illustrated in Fig. 22. The operation is one of mode selection and is much more selective than the coarse frequency selectivity obtained by using a gas absorption to prevent laser oscillation in the absorption band of the gas (see, for example [35], [59], [60]). Mode-selection experiments have been reported with the 6328-Å He–Ne laser using

199

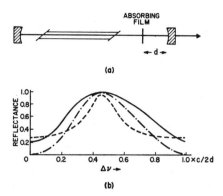

Fig. 23. (a) Absorbing film technique of longitudinal mode selection. (b) Reflectance of absorbing film–mirror combination versus tuning (after Smith *et al.* [44]).

a discharge of pure Ne as the resonant absorber [38], [39], [164] and a theoretical explanation of the results based on Lamb's theory of a gas laser has been given [165]. The experiments have been carried out, however, at power levels well above those for which Lamb's theory would be expected to be valid. Bennett [166] has given an explanation for these observations in terms of a "hole-burning" model of gain saturation. He shows that for a suitable choice of loss characteristics, it is possible to predict, on the basis of his model, behavior similar to that observed experimentally [38], [39], [164]. Fig. 22(b) shows Bennett's results for the saturated net gain as a function of frequency (when the laser is oscillating at a single frequency at line center) for various values of a parameter α, which measures the relative effect of the saturable absorber [166]. It can be seen that for values of $\alpha > 0.7$, the gain is below threshold for all modes except the favored mode.

A Lyot filter consists of a length of birefringent material with polarizers on either side. High transmission occurs only for those wavelengths for which the rotation of the polarized light passing through the birefringent material is such that it is transmitted through the second polarizer with low loss. By using such a Lyot filter in the resonator of a broad-bandwidth laser, it is possible to obtain some degree of mode selectivity. If an electrooptic crystal is used as the birefringent element, it is possible to tune the laser output frequency simply by varying the voltage applied to the crystal. Using such a filter, Walther and Hall [61] obtained a dye laser output bandwidth of 0.01 Å, and Cirkovic *et al.* [62] obtained narrow-linewidth emission from a ruby laser.

Because the standing-wave patterns of different longitudinal modes have nulls at different points along the axis of the resonator, a thin absorbing film placed in a laser resonator as shown in Fig. 23(a) will preferentially absorb those modes which have appreciable electric fields in the film. The mode with a standing-wave null at the film will experience little loss if the film is a small fraction of an optical wavelength thick. This technique was proposed and demonstrated by Troitskii and Goldina [42], [167]. They obtained single-frequency operation with 6238-Å He–Ne lasers [42] and argon ion lasers [43]. Smith *et al.* [55] presented a detailed theory of the absorbing-film mode selector. Their results for a 150-Å metal film are shown in Fig. 23(b). Smith *et al.* [44] also reported single-frequency operation of the argon ion laser and the Nd:YAG laser using thin metal films in the laser resonator. A similar theoretical treatment has been given by Troitskii [168], who also compared the absorbing-film technique with other mode-selection schemes. He concluded that the technique is most useful for relatively low-power lasers, and other schemes must be used to obtain efficient high-power operation. The reason for this is the

heating of the metal film in the high-power lasers. One method of overcoming this problem is the use of a thin nonabsorbing diffracting film [169], [170]. The use of metal films with other mode-selection schemes is discussed in [170a].

Several authors have reported mode selection caused by injection of radiation into the laser resonator. DeShazer and Maunders [188] have obtained narrow-band operation of a Nd:glass laser by injecting a beam of narrow-band light into the laser. Bondarenko *et al.* [189], [190] obtained single-frequency operation of a Q-switched ruby laser by a similar technique. Bjorkholm and Danielmeyer [191] have obtained single-frequency oscillation of a pulsed parametric oscillator by injecting a signal from a single-frequency Nd:YAG laser into the parametric oscillator resonator. Single-axial-mode oscillation was also observed by Briquet *et al.* [192] with a Nd:glass laser pumped by second harmonic light from another Nd:glass laser.

There have been a number of reports of the observation of single-frequency or narrow-bandwidth operation of Q-switched lasers when a bleachable dye is placed in the laser resonator [17], [91], [115], [171]–[173]. Sooy [23] has explained this as a transient effect where the first mode above threshold grows more rapidly than the others over a number of resonator transits and is the only mode to attain appreciable intensity within the duration of the pumping pulse.

A unique and as yet unexplained single-frequency technique was reported by Forsyth [187], who found that if the 4880-Å and 5145-Å radiation from an argon ion laser were resonated in separate resonators, single-frequency output could be obtained at 5145 Å.

Prisms [54] and diffraction gratings [29], [55]–[58] have been used as dispersive elements in laser resonators, but their frequency selectivity is usually not large enough to consider them in the category of mode-selective techniques. They are much used in dye lasers, however, to narrow the output bandwidth, and in CO_2 lasers, to restrict the oscillation to a single laser line.

VII. FREQUENCY STABILIZATION OF SINGLE-FREQUENCY LASERS

The frequency stability of an unstabilized single-frequency laser is determined by its environment. Even if the laser resonator is made of a low-thermal-expansion-coefficient material such as Invar, a change of ambient temperature of 1°C will change the frequency of a laser oscillating in the visible region of the spectrum by about 500 MHz. Mechanical fluctuations or air pressure changes which affect the resonator length also have a large effect, for a change in optical length of $\frac{1}{2}$ of an optical wavelength will change the laser frequency by $c/2L$—the longitudinal mode spacing. This is 150 MHz for a 1-m resonator.

Clearly the frequency stability of a laser can be improved by reducing these environmental effects: building mechanically stable resonators shielded from air pressure fluctuations, etc. (See, for example, [50], [193].) It has often been found more practical, however, to build electronic feedback systems to stabilize the laser. There has been a great deal of work on such feedback systems and in this section we will discuss some representative examples. For further information, the reader is referred to the review articles of Birnbaum [195], [196], Polanyi and Tobias [197], and Basov and Letokhov [198].

A. Stabilization to an External Resonator

Fig. 24 shows a simple feedback system for stabilizing a laser to an external Fabry–Perot interferometer resonance. This system is described in [22]. We have considered a case where the laser is oscillating in a number of longitudinal modes, and the Fabry–

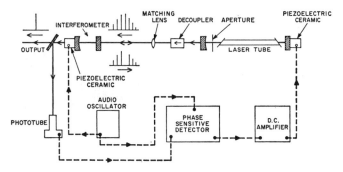

Fig. 24. Stabilization of a laser to an external Fabry–Perot interferometer which is also used to select a single frequency of the multimode laser output (after Smith [22]).

Perot interferometer serves not only as a frequency reference, but also as a filter to transmit a single laser frequency. The electronic feedback system operates as follows: The resonant frequency of the Fabry–Perot interferometer is modulated at an audio rate, and the effect of this on the output power is monitored with a detector. When the laser output drifts to one side of the interferometer response, the output will be amplitude modulated at the audio frequency of the modulation. At resonance, there will be no signal at this audio frequency (but one at twice the frequency). On the other side of the interferometer resonance, a fundamental signal of the opposite phase will be produced. Thus a phase-sensitive detector tuned to this audio frequency may be used to derive from the laser output a dc signal whose amplitude is proportional to the deviation from the center of the interferometer response, and whose sign depends on the direction of this deviation. This signal, when suitably amplified, may be used to control the frequency of the laser resonator so that the output is stabilized on the center of the Fabry–Perot interferometer response.

In a manner similar to that described in the preceding paragraph, it is possible to stabilize the frequency of a laser which is made single-frequency by using one of the interferometric mode-selection schemes shown in Fig. 14, by locking the frequency to a resonance of the mode-selector resonator [22], [146], [147]. This has the advantage that the laser is also stabilized against amplitude fluctuations due to the relative frequency displacement of the laser frequency and the mode-selector resonance. Because of the problems of long-term drift, however, all of these systems are primarily useful for frequency stabilization only when used in combination with one of the atomic-resonance systems described in the following.

B. Stabilization to an Atomic or Molecular Resonance

We will outline three representative systems.

White et al. [194], [199] have described a system which stabilizes the frequency of a 6328-Å He–Ne laser to the center of the atomic absorption profile of a pure neon discharge. The system is illustrated in Fig. 25(a). The plane-polarized beam from a single-mode Brewster-window gas laser is switched alternately to right- and left-circularly polarized light by means of a KDP electrooptic switch driven by a 400-Hz square-wave voltage. The circularly polarized light passes through the neon absorption cell which is in an axial magnetic field of approximately 350 G.

As illustrated in Fig. 25(b), this magnetic field splits the absorption profile for right- and left-circularly polarized light. The 400-Hz signal detected by the photodetector thus has an amplitude which increases as the laser frequency drifts away from the neon atomic resonance frequency (ν_0) and the phase of this signal

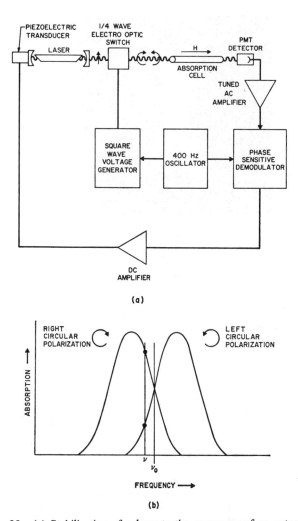

(a)

(b)

Fig. 25. (a) Stabilization of a laser to the resonance of an external absorption cell. (b) Absorber response in the presence of an axial magnetic field (after White [194]).

gives the direction of the drift. The phase-sensitive detector is then used to generate the error signal which is amplified and applied to a piezoelectric ceramic which controls the length of the laser cavity. By beating together two independently stabilized lasers, White [194] measured the frequency stability of his lasers to be one part in 10^9.

Tomlinson and Fork [200] have stabilized a He–Ne laser oscillating on the 1.52-μm neon line by utilizing the mode competition between two oppositely circularly polarized laser resonator modes. For the 1.52-μm neon line, the mode competition is a function of the applied magnetic field. For fields of a few gauss, a very sharp variation of mode intensity versus laser tuning occurs at the center of the atomic resonance line. This is demonstrated in Fig. 26(a) which shows the intensity of the right- and left-circularly polarized laser modes as a function of laser tuning. ν_0 is the atomic resonance frequency. From this mode-competition "crossover," Tomlinson and Fork derived a signal which they used to stabilize the laser frequency. Fig. 26(b) shows the beat frequency between two independent lasers stabilized in this manner. The average frequencies (10-s averages) correspond to a standard deviation of one part in 10^{10} per laser.

Recently, Barger and Hall and their colleagues have developed a laser with a very high frequency stability and resetability [40], [41]. They utilized as a reference a resonance of the methane (CH_4) molecule which coincides almost exactly with the 3.39-μm

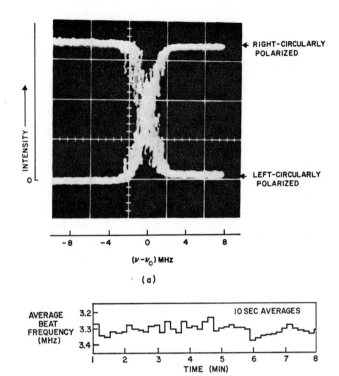

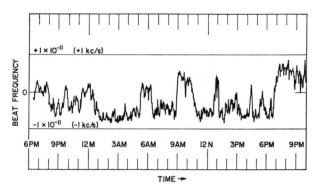

Fig. 28. Beat frequency between two independent methane-stabilized lasers (after Hall [210]).

Fig. 26. (a) Right- and left-circularly polarized output intensities as a function of laser tuning. (b) Beat frequency between two independently stabilized lasers (after Tomlinson and Fork [200]).

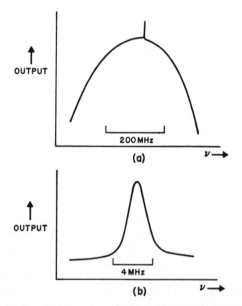

Fig. 27. (a) Output of 3.39-μm neon laser with a methane absorber in laser resonator as a function of frequency. (b) Expanded region around the methane absorption frequency (after Barger and Hall [40]).

He–Ne laser line. By placing a gas cell containing methane within the laser cavity they were able to observe the inverse "Lamb dip" —an increase in laser output power at the frequency for which both traveling waves in the laser resonator interact with the same group of methane atoms (see the explanation for a conventional Lamb dip in Section IV). Fig. 27(a) shows the output of a 3.39-μm He–Ne laser containing a methane cell as a function of tuning. The narrow feature near line center is due to the methane absorp-

tion, and has a width of about 2 MHz, as shown in the expanded-scale data in Fig. 27(b). In later experiments, linewidths as narrow as 50 kHz have been obtained [201]. By modulating the resonator length by a small amount and detecting the component of the laser output at the modulation frequency, a derivative signal can be obtained for stabilizing the laser to the center of the methane resonance.

Because the methane resonance frequency is insensitive to magnetic and electrical fields, and because the cross section is so large that appreciable absorption will occur in a low-pressure absorption cell where pressure effects are minimized and the absorption linewidth is very narrow, very precise frequency control is possible. A frequency reproducibility of 5 parts in 10^{12} and a short-term frequency stability of better than 1 part in 10^{13} have been achieved with this system [201]. Fig. 28 shows the beat frequency between two independent methane-stabilized lasers. We will describe in the next section how such lasers have been used for precision interferometry.

VIII. Uses of Lasers with Mode Selection

In this section we will briefly describe some applications where lasers using mode-selection techniques have been used to achieve results that would have been impossible to obtain with highly multimode lasers, or with conventional light sources.

A. Light-Scattering Experiments

In recent years, a large number of light-scattering experiments using lasers have been reported (for a recent review of such experiments see [202]). If the features in such experiments have a width of less than about 500 MHz in the visible region of the spectrum, then severe instrumental broadening will occur if a conventional multimode laser source is used, and some form of laser mode control is required. The high resolution which can be obtained with single-frequency lasers is necessary, in particular, for Brillouin scattering experiments. Single-frequency lasers have been used to observe Brillouin spectra of gasses [203], liquids [204], [205], and solids [206].

Yeh and Keeler [207] have recently proposed that light-scattering experiments performed with a high-power single-frequency laser could be used to study fast chemical reactions.

B. Mode Selection for Laser Holography

To obtain a good hologram using a laser light source it is desirable to have both high laser output power and a long coherence length. The high power is required in order to be able to make the hologram with a reasonably short exposure time, and the long coherence length is required if a hologram with appreciable depth of field is desired. This is because the light scattered from the back of the object of which a hologram is to be made must be coherent

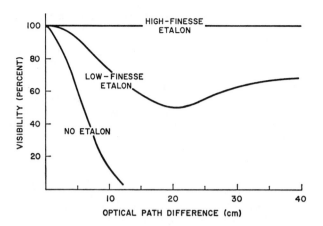

Fig. 29. Visibility of interference fringes formed by splitting the output beam of an argon ion laser with an internal Fabry–Perot etalon and recombining the two beams with a given path length difference. In all cases the output power was 5 mW (after Lin and LoBianco [208]).

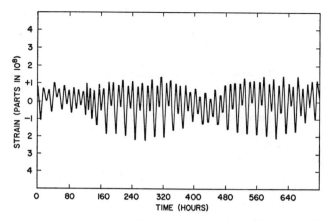

Fig. 30. Earth strain as a function of time as measured with the 30-m underground interferometer (see text) (after Levine and Hall [210]).

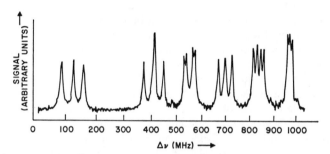

Fig. 31. Hyperfine structure of the P (117) 21-1 B←X transition of molecular iodine observed by laser saturation spectroscopy (after Hänsch et al. [222]).

with the light scattered from the front of the object in order to obtain a sharp reconstruction.

Light with a frequency width of $\Delta\nu$ has a coherence length of approximately $c/(2\Delta\nu)$. Light from an argon ion laser with no mode selection will have a coherence length of about 5 cm. Thus in order to use an argon laser to make holograms with a depth of field of greater than about 5 cm, some form of mode selection must be used to narrow the output bandwidth.

Lin and LoBianco [208] have plotted the visibility of interference fringes as a function of optical path difference between the two interfering beams for an argon ion laser with a tilted Fabry–Perot etalon as a mode selector in the laser resonator. Their results are shown in Fig. 29. It can be seen that, although with no etalon in the resonator the coherence length (path difference for which fringe visibility drops to 50 percent) is approximately 5 cm, with a sufficiently high-finesse etalon, a coherence length of much greater than 40 cm was obtained. Note that for many holographic applications it is not necessary to produce single-frequency laser operation—it is just necessary to obtain a narrow enough bandwidth to obtain the required depth of field.

Currie [209] has shown that with an essentially single-frequency laser, a hologram with a depth of field of 2.9 m could be produced.

C. Laser Interferometry

Using frequency-stabilized 3.39-μm single-frequency lasers of the type discribed in Section VII, Hall and his colleagues [210] have been able to study minute earth movements with a long-baseline interferometer.

A 30-m interferometer was set up in an unworked gold mine in Colorado. Using an electronic servo loop, a single-frequency laser was locked to a resonance of this interferometer. By beating the frequency of this laser with the output of a reference laser which was stabilized to the methane resonance, a signal proportional to the changes in interferometer length could be obtained. This system is capable of recording not only long-term changes in the earth's strain field, but also rock tides, and the relatively high frequencies associated with seismic events [210]. Fig. 30 shows 700 h of a record of rock tides obtained with this instrument. Microseismic fluctuations have been removed with a low-pass filter.

Using the same interferometer, Hall et al. [211] propose to measure the velocity of light by simultaneously measuring the frequency difference between a pair of closely spaced laser lines, and

measuring the difference in wavelength by interferometric techniques. They estimate that they will be able to measure c with a probable error of 1×10^{-8}.

The high reproducibility (5×10^{12}) which has been achieved with these methane-stabilized lasers suggests that this type of system may some day become not only the new standard of length, but also of time.

D. High-Resolution Laser Spectroscopy

In recent years, there has been a great deal of interest in the use of lasers for high-resolution absorption spectroscopy. Many workers [7], [150], [212]–[220] have used observations of the Lamb dip in the tuning curve of a single-frequency gas laser to study the homogenous lineshape of the laser transition. By placing an absorbing gas within a single-frequency laser, it is possible to observe the inverse Lamb dip as the laser is tuned over the absorption line(s) of the sample gas. In this way Hanes and Dahltrom [221], using a 6328-Å He–Ne laser, were able to observe fourteen of the hyperfine components of one rotational vibrational line of $^{127}I_2$. Similar data can be obtained from crossed-beam experiments external to the laser cavity. By crossing two beams from a tunable single-frequency laser in an external absorption cell, it is possible to obtain saturated absorption spectra without some of the constraints imposed by placing the absorbing gas within the laser cavity [182]. Hänsch et al. [222] have used this technique to investigate the complete hyperfine structure of a molecular iodine line. They used a krypton ion laser which incorporated an intraresonator etalon to obtain single-frequency operation. Fig. 31 shows the results that they obtained. It can be seen that they have achieved a resolution in excess of 10^8. Such extremely high resolutions have only been achieved by the use of narrow-bandwidth laser sources.

There is clearly a need for narrow-bandwidth laser sources with wide tuning ranges for use in laser spectroscopy. Basically, the tuning range is limited by the frequency width of the laser gain curve. Hinkley [223] has described a tunable single-frequency semiconductor diode laser which he has used to obtain the absorption spectrum of SF_6 in the 10-μm region of the spectrum. Patel *et al.* [224] have demonstrated the usefulness of the tunable spin-flip Raman laser for infrared spectroscopy by using it to observe the absorption spectrum of NH_3 in the 12-μm region of the spectrum. They demonstrated that they could obtain higher resolution than that attainable with a conventional grating spectrometer. In recent experiments [225], a spin-flip laser spectrometer used for air pollution studies has been shown capable of detecting concentrations of NO of less than 0.01 ppm in the atmosphere.

Dye lasers have very large gain bandwidths in the visible region of the spectrum. Recently, several authors [61], [226], [227] have reported the use of mode-selective techniques to obtain tunable narrow-bandwidth dye-laser output. Walther and Hall [61] obtained a dye laser spectral output of approximately 0.01 Å by using a birefringent filter (Lyot filter) in the laser cavity. Using this laser for spectroscopic studies of the sodium D_1 line, they were able to observe line broadening due to saturation. Hänsch [226] has described the use of several mode-selective techniques to obtain a dye laser spectral output of less than 0.004 Å. Using such a laser, he has observed the hyperfine structure of the sodium D_1 and D_2 lines with a resolution of better than 40 MHz [227]. By using such a laser locked to the hydrogen Balmer alpha line, Hänsch *et al.* [228] speculate that the Rydberg constant could be measured to one part in 2×10^8.

It seems clear that the future holds great promise for tunable laser spectroscopy.

IX. CONCLUDING REMARKS

We have attempted to review in a tutorial fashion the field of mode selection in lasers, and have concentrated primarily on those schemes which are capable of producing single-frequency laser operation. Our discussion of specific techniques has been of necessity brief, but we have provided a fairly complete set of references.

REFERENCES

[1] A. L. Schawlow and C. H. Townes, "Infrared and optical masers," *Phys. Rev.*, vol. 112, pp. 1940–1949, Dec. 1958.
[2] T. H. Maiman, "Stimulated optical radiation in ruby," *Nature*, vol. 187, pp. 493–494, Aug. 1960.
[3] D. A. Kleinman and P. P. Kisliuk, "Discrimination against unwanted orders in the Fabry–Perot resonator," *Bell Syst. Tech. J.*, vol. 41, pp. 453–462, Mar. 1962.
[4] H. Kogelnik and C. K. N. Patel, "Mode suppression and single frequency operation in gaseous optical masers," *Proc. IRE* (Corresp.), vol. 50, pp. 2365–2366, Nov. 1962.
[5] J. A. Baker and C. W. Peters, "Mode selection and enhancement with a ruby laser," *Appl. Opt.*, vol. 1, pp. 674–675, Sept. 1962.
[6] S. A. Collins and G. R. White, *Lasers and Applications*, W. S. C. Chang, Ed. Columbus, Ohio: Ohio State Univ. Press, 1963, pp. 96–108.
——, "Interferometer laser mode selector," *Appl. Opt.*, vol. 2, pp. 448–449, Apr. 1963.
[7] R. A. McFarlane, W. R. Bennett, and W. E. Lamb, "Single mode tuning dip in the power output of an He–Ne optical maser," *Appl. Phys. Lett.*, vol. 2, pp. 189–190, May 1963.
[8] C. L. Tang, H. Statz, and G. deMars, "Spectral output and spiking behavior of solid-state lasers," *J. Appl. Phys.*, vol. 34, pp. 2289–2295, Aug. 1963.
[9] ——, "Regular spiking and single-mode operation of ruby laser," *Appl. Phys. Lett.*, 2, pp. 222–224, June 1, 1963.
[10] T. Li, "Mode selection in an aperture-limited concentric maser interferometer," *Bell Syst. Tech. J.*, vol. 42, pp. 2609–2620, Nov. 1963.
[11] ——, "Diffraction loss and selection of modes in maser resonators with circular mirrors," *Bell Syst. Tech. J.*, vol. 44, pp. 917–932, May–June, 1965.
[12] A. G. Fox and T. Li, "Effect of gain saturation on the oscillating modes of optical masers," *IEEE J. Quantum Electron.*, vol. QE-2, pp. 774–783, Dec. 1966.
[13] E. I. Gordon and A. D. White, "Single frequency gas lasers at 6328Å," *Proc. IEEE* (Corresp.), vol. 52, pp. 206–207, Feb. 1964.
[14] K. M. Baird, D. S. Smith, G. R. Hanes, and S. Tsunekane, "Characteristics of a simple single-mode He–Ne laser," *Appl. Opt.*, vol. 4, pp. 569–571, May 1965.
[15] D. Gloge, J. Helmcke, and P. Runge, "A stable gas laser with a single emission frequency," *Frequenz*, vol. 18, pp. 367–374, Nov. 1964.
[16] J. A. Collinson, "A stable, single-frequency RF-excited gas laser at 6328Å," *Bell Syst. Tech. J.*, vol. 44, pp. 1511–1519, Sept. 1965.
[17] M. Hercher, "Single-mode operation of a Q-switched ruby laser," *Appl. Phys. Lett.*, vol. 7, pp. 39–41, July 1965.
[18] J. M. Burch, "Design of resonators," in *Quantum Electronics III*, P. Grivet and N. Bloembergen, Eds. New York: Columbia Univ. Press, 1964, pp. 1187–1202.
[19] G. A. Massey, M. K. Oshman, and R. Targ, "Generation of single-frequency light using the FM laser," *Appl. Phys. Lett.*, vol. 6, pp. 10–11, Jan. 1965.
[20] S. E. Harris and B. J. McMurtry, "Frequency selective coupling to the FM laser," *Appl. Phys. Lett.*, vol. 7, pp. 265–267, Nov. 1965.
[21] A. G. Fox, "Optical maser mode selector," U. S. Patent 3 504 299.
[22] P. W. Smith, "Stabilized, single-frequency output from a long laser cavity," *IEEE J. Quantum Electron.*, vol. QE-1, pp. 343–348, Nov. 1965.
[23] W. R. Sooy, "The natural selection of modes in a passive Q-switched laser," *Appl. Phys. Lett.*, vol. 7, pp. 36–37, July 1965.
[24] T. Li and P. W. Smith, "Mode selection and mode volume enhancement in a gas laser with internal lens," *Proc. IEEE* (Corresp.), vol. 53, pp. 399–400, Apr. 1965.
[25] A. E. Siegman, "Unstable optical resonators for laser applications," *Proc. IEEE*, vol. 53, pp. 277–287, Mar. 1965.
[26] M. DiDomenico, Jr., "A single-frequency TEM(00)-mode gas laser with high output power," *Appl. Phys. Lett.*, vol. 8, pp. 20–22, Jan. 1966.
[27] ——, "Characteristics of a single-frequency Michelson-type He–Ne gas laser," *IEEE J. Quantum Electron.*, vol. QE-2, pp. 311–322, Aug. 1966.
[28] A. D. White, "Laser cavities with increased axial mode separation," *Bell Syst. Tech. J.*, vol. 45, pp. 339–343, Feb. 1966.
[29] W. Q. Jeffers, "Single wavelength operation of a pulsed water-vapor laser," *Appl. Phys. Lett.*, vol. 11, pp. 178–180, Sept. 1967.
[30] Yu. A. Ananev, N. A. Sventsitskaya, and V. E. Sherstobitov, "Transverse mode selection in a laser with convex mirrors," *Sov. Phys.—Dokl.*, vol. 13, pp. 351–352, Oct. 1968.
[31] R. P. Flam and E. R. Schineller, "A single-mode waveguide laser," *Proc. IEEE* (Lett.), vol. 56, pp. 195–196, Feb. 1968.
[32] H. G. Danielmeyer and W. G. Nilsen, "Spontaneous single-frequency output from a spatially homogeneous Nd:YAG laser," *Appl. Phys. Lett.*, vol. 16, pp. 124–126, Feb. 1, 1970
[33] H. G. Danielmeyer and E. H. Turner, "Electro-optic elimination of spatial hole burning in lasers," *Appl. Phys. Lett.*, vol. 17, pp. 519–521, Dec. 1970.
[34] H. G. Danielmeyer, "Stabilized efficient single-frequency Nd:YAG laser," *IEEE J. Quantum Electron.*, vol. QE-6, pp. 101–104, Feb. 1970.
[35] N. V. Karlov, Yu. N. Petrov, and O. M. Stel'makh, "Control of the frequency of a CO_2 laser by a boron trichloride filter," *JETP Lett.*, vol. 8, pp. 224–226, Oct. 5, 1968.
[36] See, for example, R. A. Myers, "Fast electron beam scanlaser," *IEEE J. Quantum Electron.*, vol. QE-4, pp. 408–411, June 1968.
[37] P. W. Smith, "Stabilized single-frequency output from a long ring laser," *IEEE J. Quantum Electron.*, vol. QE-4, pp. 485–490, Aug. 1968.
[38] V. P. Chebotayev, I. M. Beterov, and V. N. Lisitsyn, "Selection and self-locking of modes in a He–Ne laser with nonlinear absorption," *IEEE J. Quantum Electron.*, vol. QE-4, pp. 788–790, Nov. 1968.
[39] P. H. Lee, P. B. Schoefer, and W. B. Barker, "Single-mode power from 6328Å laser incorporating neon absorption," *Appl. Phys. Lett.*, vol. 13, pp. 373–375, Dec. 1, 1968.
[40] R. L. Barger and J. L. Hall, "Pressure shift and broadening of the

methane line at 3.39μ studied by laser-saturated molecular absorption," *Phys. Rev. Lett.*, vol. 22, pp. 4–7, Jan. 1969.

[41] E. E. Uzgiris, J. L. Hall, and R. L. Barger, "Precision infrared Zeeman spectra of CH_4 studied by laser-saturated absorption," *Phys. Rev. Lett.*, vol. 26, pp. 289–299, Feb. 1971.

[42] Yu. V. Troitskii and N. D. Goldina, "Separation of one mode in a laser," *JETP Lett.*, vol. 7, pp. 36–38, Jan. 1968.

[43] V. I. Donin, Yu. V. Troitskii, and N. D. Goldina, "Single-mode generation and Lamb dip in an argon laser," *Opt. Spectrosc.* (USSR), vol. 26, pp. 64–65, Jan. 1969.

[44] P. W. Smith, M. V. Schneider, and H. G. Danielmeyer, "High-power single-frequency lasers using thin metal film mode-selection filters," *Bell Syst. Tech. J.*, vol. 48, pp. 1405–1420, May 1969.

[45] J. Kupka, "Frequency-modulated single-mode laser," *Electron. Lett.*, vol. 4, pp. 31–32, Jan. 1968.

[46] F. R. Nash and P. W. Smith, "Broadband optical coupling modulation," *IEEE J. Quantum Electron.*, vol. QE-4, pp. 26–34, Jan. 1968.

[47] H. Kogelnik and T. Li, "Laser beams and resonators," *Appl. Opt.*, vol. 5, pp. 1550–1567, Oct. 1966.

[48] A. G. Fox and T. Li, "Resonant modes in an optical maser," *Proc. IRE* (Corresp.), vol. 48, pp. 1904–1905, Nov. 1960.
—, "Resonant modes in a maser interferometer," *Bell Syst. Tech. J.*, vol. 40, pp. 453–488, Mar. 1961.

[49] —, "Modes in a maser interferometer with curved and tilted mirrors," *Proc. IEEE*, vol. 51, pp. 80–89, Jan. 1963.

[50] T. S. Jaseja, A. Javan, and C. H. Townes, "Frequency stability of He–Ne masers and measurements of length," *Phys. Rev. Lett.*, vol. 10, pp. 165–167, Mar. 1963.

[51] P. W. Smith, "Mode-locking of lasers," *Proc. IEEE*, vol. 58, pp. 1342–1357, Sept. 1970.

[52] W. E. Lamb, Jr., "Theory of an optical maser," *Phys. Rev.*, vol. 134, pp. A1429–A1450, June 1964.
a) A. Yariv and L. Casperson, "Longitidinal modes of a high-gain laser," *Appl. Phys. Lett.*, vol. 17, pp. 259–261, Sept. 1970.

[53] W. R. Bennett, Jr., "Hole burning effects in a He–Ne optical maser," *Phys. Rev.*, vol. 126, pp. 580–593, Apr. 1962.

[54] A. D. White, "Reflecting prisms for dispersive optical maser cavities," *Appl. Opt.*, vol. 3, pp. 431–432, Mar. 1964.

[55] B. H. Soffer and B. B. McFarland, "Continuously tunable, narrow-band organic dye lasers," *Appl. Phys. Lett.*, vol. 10, pp. 266–267, May 1967.

[56] T. J. Mauccia and G. J. Wolga, "The properties and application of diffraction gratings in frequency-selective laser resonators," *IEEE J. Quantum Electron.*, vol. QE-6, p. 185, Mar. 1970.

[57] G. J. Ernst and W. J. Witteman, "Transition selection with adjustable outcoupling for a laser device applied to CO_2," *IEEE J. Quantum Electron.*, vol. QE-7, pp. 484–488, Oct. 1971.

[58] J. E. Bjorkholm, T. C. Damen, and J. Shah, "Improved use of gratings in tunable lasers," to be published.

[59] A. L. Bloom, W. E. Bell, and R. C. Rempel, "Laser operation at 3.39μ in a helium–neon mixture," *Appl. Opt.*, vol. 2, pp. 317–318, Mar. 1963.

[60] C. B. Moore, "Gas laser frequency selection by molecular absorption," *Appl. Opt.*, vol. 4, pp. 252–253, Feb 1965.

[61] H. Walther and J. L. Hall, "Tunable dye laser with narrow spectral output," *Appl. Phys. Lett.*, vol. 17, pp. 239–242, Sept. 1970.

[62] L. Cirkovic, D. E. Evans, M. J. Forrest, and J. Katzenstein, "Use of a birefringent plate to control the spectral emission of a ruby laser," *Appl. Opt.*, vol. 7, pp. 981–982, May 1968.

[63] S. E. Miller, "Integrated optics—An introduction," *Bell Syst. Tech. J.*, vol. 48, pp. 2059–2069, Sept. 1969.

[64] H. Kogelnik and C. V. Shank, "Stimulated emission in a periodic structure," *Appl. Phys. Lett.*, vol. 18, pp. 152–154, Feb. 1971.

[65] I. P. Kaminow, H. P. Weber, and E. A. Chandross, "A poly (methyl methacrylate) dye laser with internal diffraction grating resonator," *Appl. Phys. Lett.*, vol. 18, pp. 497–499, June 1971.

[66] C. V. Shank, J. E. Bjorkholm, and H. Kogelnik, "A tunable distributed-feedback dye laser," *Appl. Phys. Lett.*, vol. 18, pp. 395–396, May 1971.

[67] J. G. Skinner and J. E. Geusic, "A diffraction limited oscillator," in *Quantum Electronics*, vol. III, P. Grivet and N. Bloembergen, Eds. New York: Columbia Univ. Press, 1964, pp. 1437–1444.

[68] J. A. Giordmaine and W. Kaiser, "Mode-selecting prism reflectors for optical masers," *J. Appl. Phys.*, vol. 35, pp. 3446–3451, Dec. 1964.

[69] V. Evtuhov and J. K. Neeland, "Study of the output spectra of ruby lasers," *IEEE J. Quantum Electron.*, vol. QE-1, pp. 7–12,

Apr. 1965.

[70] Y. Suematsu and K. Iga, "Experiment on quasi-fundamental mode oscillation of ruby laser," *Proc. IEEE* (Corresp.), vol. 52, pp. 87–88, Jan. 1964.

[71] D. C. Sinclair, "Choice of mirror curvatures for gas laser cavities," *Appl. Opt.*, vol. 3, pp. 1067–1072, Sept. 1964.

[72] A. V. Korovitsyn, L. V. Naumova, and Z. T. Lebedinskaya, "Mode selection in the semiconcentric cavity of a gaseous laser oscillator," *Radio Eng. Electron Phys.* (USSR), vol. 11, pp. 572–577, Apr. 1966.

[73] I. M. Belousova and O. B. Danilov, "An investigation of various types of resonators for producing a single mode laser with optimum parameters," *Sov. Phys.—Tech. Phys.*, vol. 12, pp. 1104–1109, Feb. 1968.

[74] D. V. Gordeev, V. M. Grimblatov, Ye. P. Ostapchenko, and V. V. Teselkin, "Feasibility of using a resonator with an aperture in the mirror of argon ion lasers," *Radiotekh. Elektron.*, vol. 14, pp. 1637–1640, Sept. 1969.

[75] A. L. Mikaelyan, A. V. Korovitsyn, and L. D. Naumova, "Single-mode operation in a CO_2 laser," *Radio Eng., Electron Phys.* (USSR), vol. 14, pp. 93–95, Jan. 1969.

[76] J. M. Burch, "Ruby masers with afocal resonators, '*J. Opt. Soc. Amer.*, vol. 52, p. 602, May 1962.

[77] A. E. Siegman and R. Arrathoon, "Modes in unstable optical resonators and lens waveguides," *IEEE J. Quantum Electron.*, vol. QE-3, pp. 156–163, Apr. 1967.

[78] S. R. Barone, "Optical resonators in the unstable region," *Appl. Opt.*, vol. 6, pp. 861–863, May 1967.

[79] W. K. Kahn, "Unstable optical resonators," *Appl. Opt.*, vol. 5, pp. 407–413, Mar. 1966.

[80] L. Bergstein, "Modes of stable and unstable optical resonators," *Appl. Opt.*, vol. 7, pp. 495–504, Mar. 1968.

[81] W. Streifer, "Unstable optical resonators and waveguides," *IEEE J. Quantum Electron.* (Corresp.), vol. QE-4, pp. 229–230, Apr. 1968.

[82] D. C. Sinclair and T. H. E. Cottrel, "Transverse mode structure in unstable optical resonators," *Appl. Opt.*, vol. 6, pp. 845–849, May 1967.

[83] W. F. Krupke and W. R. Sooy, "Properties of an unstable confocal resonator CO_2 laser system," *IEEE J. Quantum Electron.*, vol. QE-5, pp. 575–586, Dec. 1969.
E. V. Locke, R. Hella, and L. Westra, "Performance of an unstable oscillator on a 30-kW CW gas dynamic laser," *IEEE J. Quantum Electron.*, vol. QE-7, pp. 581–583, Dec. 1971.
E. B. Treacy, "Diffractive coupling from a CO_2 laser," *Appl. Opt.*, vol. 8, pp. 1107–1109, June 1969.

[84] C. M. Stickley, "Laser brightness gain and mode control by compensation for thermal distortion," *IEEE J. Quantum Electron.*, vol. QE-2, pp. 511–518, Sept. 1966.

[85] W. D. Fricke, "Fundamental mode YAG:Nd laser analysis," *Appl. Opt.*, vol. 9, pp. 2045–2052, Sept. 1970.

[86] L. G. DeShazer and E. A. Maunders, "Laser mode selection by internal reflection prisms,"' *Appl. Opt.*, vol. 6, pp. 431–435, Mar. 1967.

[87] N. G. Bondarenko, I. V. Eremina, and B. I. Talanov, "Solid state laser with mode selection within an active element," *JETP Lett.*, vol. 6, pp. 1–3, July 1967.

[88] J. A. Giordmaine and W. Kaiser, "Mode-selecting prism reflectors for optical masers," *J. Appl. Phys.*, vol. 35, pp. 3446-3451, Dec. 1964.

[89] G. Soncini and O. Svelto, "Single transverse mode pulsed ruby laser," *Appl. Phys. Lett.*, vol. 11, pp. 261–263, Oct. 1967.

[90] —, "Single mode passive Q-switched ruby laser," *IEEE J. Quantum Electron.* (Corresp.), vol. QE-4, p. 422, June 1968.

[91] J. M. McMahon, "Laser mode selection with slowly opened Q switches," *IEEE J. Quantum Electron.*, vol. QE-5, pp. 489–495, Oct. 1969.

[92] M. Birnbaum, P. H. Wendzikowski, and C. L. Fincher, "Continuous wave nonspiking single-mode ruby lasers," *Appl. Phys. Lett.*, vol. 16, pp. 436–438, June 1, 1970.

[93] A. Stein, "Mode selection for giant pulse ruby lasers," *Appl. Opt.*, vol. 6, pp. 2193–2194, Dec. 1967.

[94] D. Roess, "An optical ruby maser with high mode selection," *Frequenz*, vol. 17, pp. 61–63, Feb. 1963.

[95] D. H. Arnold and D. C. Hanna, "Transverse-mode selection of rotating-mirror Q-switched lasers," *Electron. Lett.*, vol. 5, pp. 354–355, Aug. 7, 1969.

[96] M.S. Lipsett and M. W. Strandberg, "Mode control in ruby op-

tical masers by means of elastic deformations," *Appl. Opt.*, vol. 1, pp. 343–357, May 1962.

[97] D. C. Hanna, "Increasing laser brightness by transverse mode selection. I," *Optics Tech.*, vol. 2, pp. 122–125, Aug. 1970.
——, "Increasing laser brightness by transverse mode selection. II," *ibid.*, vol. 2, pp. 175–178, Nov. 1970.

[98] A. Okaya, "Mode suppression on lasers by metal wires," *Proc. IEEE* (Corresp.), vol. 52, p. 1741, Dec. 1964.

[99] H. Kogelnik and W. W. Rigrod, "Visual display of isolated optical-resonator modes," *Proc. IRE*, vol. 50, p. 220, Feb. 1962. W. W. Rigrod, "Isolation of axi-symmetrical optical-resonator modes," *Appl. Phys. Lett.*, vol. 2, pp. 51–53, Feb. 1963.

[100] R. J. Collins and J. A. Giordmaine, "New modes of optical oscillation in closed resonators," in *Quantum Electronics*, vol. III, P. Grivet and N. Bloembergen, Eds. New York: Columbia Univ. Press, 1964, pp. 1239–1246.

[101] W. D. Johnston, Jr., T. Li, and P. W. Smith, "Competition and stimulated switching of transverse laser modes," *IEEE J. Quantum Electron.* (Corresp.), vol. QE-4, pp. 469–471, July 1968.

[102] J. V. Ramsay and K. Tanaka, "Construction of single-mode DC operated He/Ne lasers," *Japan J. Appl. Phys.*, vol. 5, pp. 918–923, Oct. 1966.

[103] C. F. Bruce, "Stable single frequency He–Ne laser," *Appl. Opt.*, vol. 10, pp. 880–883, Apr. 1971.

[104] G. E. Stillman, M. D. Stirkis, J. A. Rossi, M. R. Johnson, and N. Holonyak, "Volume excitation of an ultrathin single-mode cadmium selenide laser," *Appl. Phys. Lett.*, vol. 9, pp. 268–269, Oct. 1966.

[105] See, for example, P. Jacquinot, "New developments in interference spectroscopy," *Rep. Progr. Phys.*, vol. 23, pp. 267–312, 1960.

[106] V. I. Perel and I. V. Rogova, "On the theory of lasers with an additional mirror I," *Opt. Spectrosc.*, (USSR), vol. 25, pp. 401–403, Nov. 1968.

[107] ——, "On the theory of lasers with an additional mirror II," *Opt. Spectrosc.* (USSR), vol. 25, pp. 520–521, Dec. 1968.

[108] A. A. Bakeyev and N. V. Cheburkin, "Natural frequencies of a three-mirror resonator," *Radio Eng. Electron. Phys.* (USSR), vol. 14, pp. 1125–1130, July 1969.

[109] G. Bouwhuis, "Eingenfrequencies and quality factors of multimirror etalons," *Philips Res. Rep.*, vol. 19, pp. 422–428, Oct. 1964.

[110] G. D. Currie, "High power, single mode He–Ne laser," *Appl. Opt.*, vol. 8, pp. 1068–1069, May 1969.
——, "Single mode output with the Spectra-Physics Model 125," *Rev. Sci. Instr.*, vol. 40, pp. 1342–1343, Oct. 1969.

[111] E. Gregor, "Laser resonant cavity for increased coherence," *Appl. Opt.*, vol. 7, pp. 2138–2139, Oct. 1968.

[112] H. I. Pawel, J. R. Stanford, J. H. Wenzel, and G. J. Wolga, "Use of dielectric etalon as a reflector for Q-switched laser operation," *Proc. IEEE* (Corresp.), vol. 52, pp. 1048–1049, Sept. 1964.

[113] D. Roess, "Ruby laser with mode-selective etalon reflector," *Proc. IEEE* (Corresp.), vol. 52, pp. 196–197, Feb. 1964.

[114] W. B. Tiffany, "Repetitively pulsed, tunable ruby laser with solid etalon mode control," *Appl. Opt.*, vol. 7, pp. 67–71, Jan. 1968.

[115] F. J. McClung and D. Weiner, "Longitudinal mode control in giant pulse lasers," *IEEE J. Quantum Electron.*, vol. QE-1, pp. 94–99, May 1965.

[116] E. Snitzer, "Frequency control of a Nd^{3+} glass laser," *Appl. Opt.*, vol. 5, pp. 121–125, Jan. 1966.

[117] N. M Galaktionova, G. A. Garkavi, V. F. Egorova, A. A. Mak, and V. A. Fromzel, "Selection of a single longitudinal mode in solid-state lasers," *Opt. Spectrosc.* (USSR), vol. 28, pp. 404–408, Apr. 1970.

[118] T. Kobayashi and Y. Matsuo, "Single-frequency oscillation using two coupled cavities incorporating a Fabry–Perot electro-optic modulator," *Appl. Phys. Lett.*, vol. 16, pp. 217–218, Mar. 1970.

[119] N. Kumagai, M. Matsuhara, and H. Mori, "Design considerations for mode selective Fabry–Perot laser resonator," *IEEE J. Quantum Electron.*, vol. QE-1, pp. 85–94, May 1965.

[120] M. Birnbaum and T. L. Stocker, "Mode selection properties of segmented rod lasers," *J. Appl. Phys.*, vol. 34, pp. 3414–3415, Nov. 1963.

[121] ——, "Mode selection properties of segmented-rod giant pulse lasers," *J. Appl. Phys.*, vol. 37, pp. 531–534, Feb. 1966.

[122] R. Pratesi and G. T. diFrancia, "Many-element lasers," *Nuovo Cimento*, vol. 34, pp. 40–50, Oct. 1964.

[123] ——, "Regular emission from a many-element laser during the pumping pulse," *Proc. IEEE* (Corresp.), vol. 53, pp. 196–197, Feb. 1965.

[124] R. Pratesi, "Spiking emission from many-element lasers," *Appl. Opt.*, vol. 6, pp. 1243–1253, July 1967.

[125] H. F. Mahlein and G. Schollmeier, "Analysis and synthesis of periodic optical resonant reflectors," *Appl. Opt.*, vol. 8, pp. 1197–1202, June 1969.

[126] E. I. Nikonova, E. N. Pavlovskaya, and C. P. Startsev, "Production of one longitudinal mode in a ruby laser," *Opt. Spectrosc.* (USSR), vol. 22, pp. 535–536, June 1967.

[127] R. M. Scotland, "A mode controlled Q-switched tunable ruby laser," *Appl. Opt.*, vol. 9, pp. 1211–1213, May 1970.

[128] J. K. Watts, "Theory of multiplate resonant reflectors," *Appl. Opt.*, vol. 7, pp. 1621–1623, Aug. 1968.

[129] G. Magyar, "Simple giant pulse ruby laser of high spectral brightness," *Rev. Sci. Instr.*, vol. 38, pp. 517–519, Apr. 1967.

[130] H. Manger and H. Rothe, "Selection of axial modes in optical masers," *Phys. Lett.*, vol. 7, pp. 330–331, Dec. 1963.

[131] ——, "Single and double axial mode operation of a Nd optical maser," *Phys. Lett.*, vol. 12, pp. 182–183, Oct. 1964.

[132] N. Kumagai and M. Matsuhara, "Theory of interferometric laser mode selector," *Tech. Rep. Osaka Univ.*, vol. 16, pp. 189–198, 1966.

[133] M. Hercher, "Tunable single mode operation of gas lasers using intracavity tilted etalons," *Appl. Opt.*, vol. 8, pp. 1103–1106, June 1969.

[134] H. G. Danielmeyer," Stabilized efficient single-frequency Nd: YAG laser," *IEEE J. Quantum Electron.*, vol. QE-6, pp. 101–104, Feb. 1970.

[135] H. Manger, "Selektion Axialer Eigerschwingungen in Optischen Maser-Oscillatoren," *Z. Angew. Phys.*, vol. 18, pp. 265–270, Jan. 1965.

[136] D. Roess, "Single-mode operation of a room-temperature CW ruby laser," *Appl. Phys. Lett.*, vol. 8, pp. 109–111, Mar. 1966.

[137] M. P. Vanyukov, V. I. Isaenko, L. A. Luizova, and O. A. Shorokhov, "The cavity losses on mode selection in the stimulated emission of Nd^{3+} in glass," *Opt. Spectrosc.* (USSR), vol. 20, pp. 535–538, June 1966.

[138] H. P. Barber, "Coherence length extension of He–Ne lasers," *Appl. Opt.*, vol. 7, pp. 559–560, Mar. 1968.

[139] L. H. Lin and C. V. LoBianco, "Experimental techniques in making multicolor white light reconstructed holograms," *Appl. Opt.*, vol. 6, pp. 1255–1258, July 1967.

[140] A. J. Gibson, "A flashlamp-pumped dye laser for resonance scattering studies of the upper atmosphere," *J. Sci. Instrum.*, vol. 2, pp. 802–806, Sept. 1969.

[141] D. J. Bradley, A. J. F. Durrant, G. M. Gale, M. Moore, and P. D. Smith, "Characteristics of organic dye lasers as tunable frequency sources for nanosecond absorption spectroscopy," *IEEE J. Quantum Electron.*, vol. QE-4, pp. 707–711, Nov. 1968.

[142] L. B. Kreuzer, "Single mode oscillation of a pulsed singly resonant optical parametric oscillator," *Appl. Phys. Lett.*, vol. 15, pp. 263–265, Oct. 15, 1969.

[143] Yu. D. Kolomnikov, V. N. Lisitsyn, and V. P. Chebotaev, "Laser Michelson interferometer," *Opt. Spectrosc.* (USSR), vol. 22, pp. 449–450, May 1967.

[144] T. Uchida, U. S. Patent 3 402 365.

[145] K. Kantor, A. Kiss, and T. Salamon, "Michelson interferometer used as a tunable mirror in laser resonators," *Sov. Phys.—JETP*, vol. 25, pp. 221–222, Aug. 1967.

[146] Y. Cho, T. Tajime, and Y. Matsuo, "Stabilization of a composite-cavity single-frequency laser," *IEEE J. Quantum Electron.* (Corresp.), vol. QE-4, pp. 699–701, Oct. 1968.

[147] P. W. Smith, "On the stabilization of a high-power single-frequency laser," *IEEE J. Quantum Electron.*, vol. QE-2, pp. 666–668, Sept. 1966.

[148] P. Zory, "Measurements of argon single-frequency laser power and the 6328-Å neon isotope shift using an interferometer laser," *J. Appl. Phys.*, vol. 37, pp. 3643–3644, Aug. 1966.

[149] W. W. Rigrod and A. M. Johnson, "Resonant prism mode selector for gas lasers," *IEEE J. Quantum Electron.*, vol. QE-3, pp. 644–646, Nov. 1967.

[150] P. Zory, "Single-frequency operation of argon ion lasers," *IEEE J. Quantum Electron.*, vol. QE-3, pp. 390–398, Oct. 1967.

[151] L. Gorog and F. W. Spong, "Single-frequency argon laser," *RCA Rev.*, vol. 30, pp. 277–284, June 1969.

[152] B. Dessus and P. Laures, "Single-mode helium-neon high power

laser (Laser monomode de puissance à helium-neon)," *Ann. Telecommun.*, vol. 24, pp. 164–170, May 1969.

[153] V. P. Belyaiev, V. A. Burmakin, A. N. Evtyunin, F. A. Korolyov, V. V. Lebedeva, and A. I. Odintzov, "High-power single-frequency argon ion laser," *IEEE J. Quantum Electron.*, vol. QE-5, pp. 589–591, Dec. 1969.

[154] D. C. Sinclair, "A confocal longitudinal mode selector for single-frequency operation of gas lasers," *Appl. Phys. Lett.*, vol. 13, pp. 98–100, Aug. 1968.

[155] V. Yu. Petrun'kin, M. G. Vysotskii, and R. I. Okunev, "Selection of longitudinal modes in a ring gas laser," *Sov. Phys.—Tech. Phys.*, vol. 14, pp. 694–695, Nov. 1969.

[156] ——, "Longitudinal mode selection in a He–Ne laser with a four-mirror T-shaped resonator," *Sov. Phys.—Tech. Phys.*, vol. 13, pp. 1591–1592, May 1969.

[157] W. W. Rigrod, "Selectivity of open-ended interferometric resonators," *IEEE J. Quantum Electron.*, vol. QE-6, pp. 9–14, Jan. 1970.

[158] J. R. Fontana, "Mixed-polarization modes of Michelson-type optical resonators," *IEEE J. Quantum Electron.*, vol. QE-4, pp. 678–685, Oct. 1968.

[159] V. N. Kutin and B. I. Troshin, "Laser ring interferometer with special selective characteristics," *Opt. Spectrosc.* (USSR), vol. 29, pp. 197–198, Aug. 1970.

[160] P. J. Maloney and P. W. Smith, "On the frequency stabilization of a single-frequency argon-ion ring laser," to be published.

[161] L. M. Osterink and R. Targ, "Single-frequency light from an argon FM laser," *Appl. Phys. Lett.*, vol. 10, pp. 115–117, Feb. 1967.

[162] S. E. Harris and O. P. McDuff, "FM laser oscillation—theory," *Appl. Phys. Lett.*, vol. 5, pp. 205–206, Nov. 1964.
S. E. Harris and R. Targ, "FM oscillation of the He–Ne Laser," *Appl. Phys. Lett.*, vol. 5, pp. 202–204, Nov. 1964.

[163] H. P. Weber and E. Mathieu, "Proposal for generation of intensive single-frequency beam by second harmonic generation of an FM laser beam," *IEEE J. Quantum Electron.*, vol. QE-3, pp. 376–377, Sept. 1967.

[164] I. M. Beterov, V. N. Lisitsyn, and V. P. Chebotayev, "Oscillation mode selection and self-locking in a laser with nonlinear absorption," *Radio Eng. Electron. Phys.* (USSR), vol. 14, pp. 981–982, June 1969.

[165] M. S. Feld, A. Javan, and P. H. Lee, "Strong mode coupling induced in a gas laser by an intracavity saturable absorbing medium," *Appl. Phys. Lett.*, vol. 13, pp. 424–427, Dec. 1968.

[166] W. R. Bennett, Jr., "Hole burning in gas lasers with saturable absorbers," *Comments on Atomic and Molecular Phys.*, vol. 11, pp. 10–19, 1970.

[167] Yu. V. Troitskii, "Comparing the method of optical resonator transverse mode selection," *Zh. Prikl. Sepktrosk.* (USSR), vol. 12, pp. 425–431, Mar. 1970.

[168] ——, "Optical resonator with absorbing metal film," *Radio Eng. Electron. Phys.* (USSR), vol. 14, pp. 1423–1427, Sept. 1969.

[169] N. D. Goldina and Yu. V. Troitskii, "Experiment with a nonabsorbing diffraction selector of longitudinal modes of an optical resonator," *Opt. Spectrosc.* (USSR), vol. 28, pp. 319–320, Mar. 1970.

[170] Yu. V. Troitskii, "Thin film diffraction grating in a standing wave optical resonator," *Opt. Spectrosc.*, (USSR) vol. 27, pp. 263–265, Mar. 1969.
a) W. Culshaw and J. Kannelaud, "Two-component-mode filters for optimum single-frequency operation of Nd:YAG lasers," *IEEE J. Quantum Electron.*, vol QE-7, pp. 381–387, Aug. 1971.

[171] B. H. Soffer, "Giant pulse laser operation by a passive, reversibly bleachable absorber," *J. Appl. Phys.*, vol. 35, p. 2551, Aug. 1964.

[172] J. E. Bjorkholm and R. H. Stolen, "A simple single-mode giant pulse ruby laser," *J. Appl. Phys.*, vol. 39, pp. 4043–4044, July 1968.

[173] V. Daneu, C. A. Sacchi, and O. Svelto, "Single transverse and longitudinal mode Q-switched ruby laser," *IEEE J. Quantum Electron.*, vol. QE-2, pp. 290–293, Aug. 1966.

[174] M. Birnbaum, P. H. Wendzikowski, and C. L. Fincher, "Continuous wave nonspiking single-mode ruby lasers," *Appl. Phys. Lett.*, vol. 16, pp. 436–438, June 1, 1970.

[175] H. Kogelnik and C. V. Shank, "Coupled wave theory of stimulated emission in a periodic structure," to be published in *J. Appl. Phys.*

[176] J. Free and A. Korpel, "Laser emission from a moving ruby rod," *Proc. IEEE* (Corresp.), vol. 52, p. 90, Jan. 1964.

[177] V. Evtuhov and A. E. Siegman, "A 'twisted-mode' technique for obtaining axially uniform energy density in a laser cavity," *Appl. Opt.*, vol. 4, pp. 142–143, Jan. 1965.
a) C. L. Tang, H. Statz, G. A. Demars, and D. T. Wilson, "Spectral properties of a single-mode ruby laser—evidence of homogeneous broadening of the zero-phonon lines in solids," *Phys. Rev.*, vol. 136, p. A1–A8, Oct. 1964.

[178] D. A. Draegert, "Efficient, single-longitudinal-mode Nd:YAG laser," to be published.

[179] C. Bowness, "Single mode laser," U. S. Patent 3 409 843, Apr. 2, 1964.

[180] W. D. Johnston, Jr., "Characteristics of optically pumped platelet lasers of ZnO, CdS, CdSe, and $Cd_{.6}Se_{.4}$ between 300°K and 80°K," *J. Appl. Phys.*, vol. 42, pp. 2731–2740, June 1971.

[181] H. G. Danielmeyer, "Effects of drift and diffusion of excited states on spatial hole-burning and laser oscillation," *J. Appl. Phys.*, vol. 42, pp. 3125–3132, July 1971.

[182] P. W. Smith and T. Hänsch, "Cross-relaxation effects in the saturation of the 6328Å neon laser line," *Phys. Rev. Lett.*, vol. 26, pp. 740–743, Mar 1971.

[183] P. W. Smith, "A waveguide gas laser," *Appl. Phys. Lett.*, vol. 19, pp. 132–134, Sept. 1971.

[184] J. M. Yarborough and J. L. Hobart, "New high-power stable modes of operation of the argon laser," *Appl. Phys. Lett.*, vol. 13, pp. 305–307, Nov. 1, 1968.

[185] T. J. Bridges and W. W. Rigrod, "Output spectra of the argon ion laser," *IEEE J. Quantum Electron.*, vol. QE-1, pp. 303–308, Oct. 1965.

[186] M. S. Borisova and A. M. Pyndyk, "The frequency spectra of ion lasers," *Radio Eng. Electron. Phys.* (USSR), vol. 13, pp. 658–660, Apr. 1968.

[187] J. M. Forsyth, "Single-frequency operation of the argon ion laser at 5145Å," *Appl. Phys. Lett.*, vol. 11, pp. 391–394, Dec. 1967.

[188] L. G. DeShazer and E. A. Maunders, "Spectral control of laser oscillators by secondary light sources," *IEEE J. Quantum Electron.*, vol. QE-4, pp. 642–644, Oct. 1968.

[189] A. N. Bondarenko, G. V. Krivoshchekov, and V. A. Smirnov, "Single-frequency ruby laser with active Q-switch," *JETP Lett.*, vol 9, pp. 57–58, Jan. 20, 1969.

[190] ——, "Single frequency ruby lasers with tuned frequency under Q-switching," *Zh. Eksp. Teor. Fiz.*, vol. 56, pp. 1815–1818, June 1969.

[191] J. E. Bjorkholm and H. G. Danielmeyer, "Frequency control of a pulsed optical parametric oscillator by radiation injection," *Appl. Phys. Lett.*, vol. 15, pp. 171–173, Sept. 1969.

[192] G. Briquet, J. M. Jego, and A. Treneaud, "Generation of a monomode giant pulse laser by coherent pumping," *Appl. Phys. Lett.*, vol. 14, pp. 282–283, May 1969.

[193] U. Hochuli and P. Haldemann, "Relative frequency stability of stable He–Ne gas laser structures," *IEEE J. Quantum Electron.* (Corresp.), vol. QE-7, pp. 573–575, Dec. 1971.

[194] A. D. White, "Frequency stabilization of gas lasers," *IEEE J. Quantum Electron.*, vol. QE-1, pp. 349–357, Nov. 1965.

[195] G. Birnbaum, "Frequency stabilization of gas lasers," *Proc. IEEE*, vol. 55, pp. 1015–1026, June 1967.

[196] ——, "Frequency stabilized gas lasers. A summary," *Electron. Technol.* (Poland), vol. 2, pp. 67–69, 1969.

[197] T. G. Polanyi and I. Tobias, "Frequency stabilization of gas lasers," in *Lasers*, vol. 2, A. K. Levine, Ed. New York: Dekker, 1968.

[198] N. G. Basov and V. S. Letokhov, "Optical frequency standards," *Sov. Phys.—USP.*, vol. 11, pp. 855–880, May–June 1969.

[199] A. D. White, E. I. Gordon, and E. F. Labuda, "Frequency stabilization of single mode lasers," *Appl. Phys. Lett.*, vol. 5, p. 97, Sept. 1964.

[200] W. J. Tomlinson and R. L. Fork, "Frequency stabilization of a gas laser," *Appl. Opt.*, vol. 8, pp. 121–129, Jan. 1969.

[201] J. L. Hall, private communication.

[202] P. L. Fleury, "Light scattering as a probe of phonons and other excitations," in *Physical Acoustics*, vol. 6. New York: Academic Press, 1970.

[203] See, for example, T. J. Greytak and G. B. Benedek, "Spectrum of light scattered from thermal fluctuations in gasses," *Phys. Rev. Lett.*, vol. 17, pp. 179–182, July 1966.

[204] See, for example, P. A. Fleury and J. P. Boon, "Brillouin scattering in simple liquids: Argon and neon," *Phys. Rev.*, vol. 186, pp. 244–254, Oct. 1969.

[205] C. J. Palin, W. F. Vinen, E. R. Pike, and J. M. Vaughan, "Rayleigh and Brillouin scattering from superfluid ^{3}He–^{4}He mixtures," *J. Phys. C.*, vol. 4, pp. L225–L228, July 1971.

[206] See, for example, P. A. Fleury and P. D. Lazay, "Acoustic-soft-optic mode interactions in ferroelectric $BaTiO_3$," *Phys. Rev. Lett.*, vol. 26, pp. 1331–1334, May 1971.

[207] Y. Yeh and R. N. Keeler, "Reaction kinetics probed by dynamic light scattering," *J. Comput. Phys.*, vol. 7, pp. 566–575, June 1971.

[208] L. H. Lin and C. V. LoBianco, "Experimental techniques in making multicolor white light reconstructed holograms," *Appl. Opt.*, vol. 6, pp. 1255–1258, July 1967.

[209] G. D. Currie, "High power, single mode He–Ne laser," *Appl. Opt.*, vol. 8, pp. 1068–1069, May 1969.

[210] J. Levine and J. L. Hall, "Design and operation of a methane absorption stabilized laser strainmeter," to be published.

[211] J. L. Hall, R. L. Barger, P. L. Bender, H. S. Boyne, J. E. Faller, and J. Ward, "Precision long-path interferometry and the velocity of light," in *Electron Technology 2*, A. Smolinski and S. Hahn, Eds. Warsaw, Poland: Polish Scientific Publishers, 1969, pp. 53–66.

[212] A. Szöke and A. Javan, "Isotope shift and saturation behavior of the 1.15μ transition of Ne," *Phys. Rev. Lett.*, vol. 10, pp. 521–524, June 1963.

[213] ——, "Effects of collisions on saturation behavior of the 1.15μ transition of Ne studied with a He–Ne laser," *Phys. Rev.*, vol. 145, pp. 137–147, May 1966.

[214] P. W. Smith, "Linewidth and saturation parameters for the 6328Å transition in a He–Ne laser," *J. Appl. Phys.*, vol. 37, pp. 2090–2093, Apr. 1966.

[215] R. H. Cordover and P. A. Bonczyk, "Effect of collisions on the saturation behavior of the 6328Å transition of Ne studied with a He–Ne laser," *Phys. Rev.*, vol. 188, pp. 969–700, Dec. 1969.

[216] W. Dietel, "Nichtlineare Druckabhangingkeit des Dampfungs-parameters im 0.63 μm He–Ne single mode laser," *Phys. Lett.*, vol. 29A, pp. 268–269, May 1969.

[217] D. A. Stetser, "Research investigation of laser line profiles," *Tech. Rep.* F-920479-4, July 1967.

[218] A. I. Odintsov, V. V. Lebedeva, and G. V. Abrosimov, "Gain saturation in a single frequency argon laser," *Radio Eng. Electron. Phys.* (USSR), vol. 13, pp. 650–653, Apr. 1968.

[219] C. Bordé and L. Henry, "Study of the Lamb dip and of rotational competition in a carbon dioxide laser—Applications to the laser stabilization and to the measurement of absorption coefficients of gases," *IEEE J. Quantum Electron.*, vol. QE-4, pp. 874–880, Nov. 1968.

[220] T. Kan, H. T. Powell, and G. J. Wolga, "Observation of the central tuning dip in N_2O and CO_2 molecular lasers," *IEEE J. Quantum Electron.* (Corresp.), vol. QE-5, pp. 299–300, June 1969.

[221] G. R. Hanes and C. E. Dahlstrom, "Iodine hyperfine structure observed in saturated absorption at 633 nm," *Appl. Phys. Lett.*, vol. 14, pp. 362–364, June 1969.

[222] T. W. Hänsch, M. D. Levenson, and A. L. Schawlow, "Complete hyperfine structure of a molecular iodine line," *Phys. Rev. Lett.*, vol. 26, pp. 946–949, Apr. 1971.

[223] E. D. Hinkley, "High-resolution infrared spectroscopy with a tunable diode laser," *Appl. Phys. Lett.*, vol. 16, pp. 351–354, May 1970.

[224] C. K. N. Patel, E. D. Shaw, and R. J. Kerl, "Tunable spin-flip laser and infrared spectroscopy," *Phys. Rev. Lett.*, vol. 25, pp. 8–11, July 1970.

[225] L. B. Kreuzer and C. K. N. Patel, "Nitric oxide air pollution: Detection by optoacoustic spectroscopy," *Science*, vol. 173, pp. 45–47, July 1971.

[226] T. W. Hänsch, "Repetitively pulsed tunable dye laser for high resolution spectroscopy," *Appl. Opt.*, to be published.

[227] T. W. Hänsch, I. S. Shahin, and A. L. Schawlow, "High-resolution saturation spectroscopy of the sodium D lines with a pulsed tunable dye laser," *Phys. Rev. Lett.*, vol. 27, pp. 707–710, Sept. 1971.

[228] ——, "Optical resolution of the Lamb shift in atomic hydrogen by laser saturation spectroscopy," *Nature*, to be published.

Mode-Locking of Lasers

PETER W. SMITH, MEMBER, IEEE

Invited Paper

Abstract—This paper is a tutorial and review paper on the subject of mode-locking of lasers. It is intended as an introduction to the subject for the general reader as well as an up-to-date overview for specialists in the field. Emphasis has been placed on giving physical understanding of the phenomena and processes involved, rather than on providing details of specific theories and devices.

I. INTRODUCTION

A TYPICAL laser consists of an optical resonator formed by two coaxial mirrors and some laser gain medium within this resonator. The frequency band over which laser oscillation can occur is determined by the frequency region over which the gain of the laser medium exceeds the resonator losses. Often, there are many modes of the optical resonator which fall within this oscillation band, and the laser output consists of radiation at a number of closely spaced frequencies. The total output of such a laser as a function of time will depend on the amplitudes, frequencies, and relative phases of all of these oscillating modes. If there is nothing which fixes these parameters, random fluctuations and nonlinear effects in the laser medium will cause them to change with time, and the output will vary in an uncontrolled way. If the oscillating modes are forced to maintain equal frequency spacings with a fixed phase relationship to each other, the output as a function of time will vary in a well-defined manner. The laser is then said to be "mode-locked" or "phase-locked." The form of this output will depend on which laser modes are oscillating and what phase relationship is maintained. It is possible to obtain: an FM-modulated output, a continuous pulse train, a spacially scanning laser beam, or a "machine gun" output where pulses of light appear periodically at different spacial positions on the laser output mirror.

This paper is intended as a tutorial and review paper on the subject of mode-locking of lasers. It is hoped that it will be of use as an introduction to the subject for the general reader, as well as an up-to-date overview for the specialists in the field.

The paper is organized in the following way. Section II is a historical review of the subject. Section III discusses modes in optical resonators. Section IV treats in detail mode-locking of longitudinal modes of lasers and discusses various techniques for longitudinal mode-locking. Section V treats both transverse mode-locking and the simultaneous locking of longitudinal and transverse laser modes. The

paper concludes with a description of some recent experiments using mode-locked lasers. As a review article has recently appeared on the subject of picosecond pulses from mode-locked solid-state lasers [1], this paper will concentrate on the general principles of mode-locking and examples drawn largely from gas laser experiments, but we have attempted to provide a fairly complete set of references covering the entire field.

II. HISTORICAL REVIEW

Although there were indications of mode-locking in the work of Gürs and Müller [2], [2a] on internal modulation of ruby lasers, and the work of Statz and Tang [3] on He–Ne gas lasers, the first papers to clearly describe this effect were written in 1964 and early 1965 by DiDomencio [4], Hargrove et al. [5], and Yariv [8]. Hargrove et al. experimentally obtained a continuous train of pulses from a He–Ne laser by mode-locking with an internal acoustic loss modulator. DiDomenico, following a suggestion by E. I. Gordon, showed theoretically that mode-locking could be obtained by internal loss modulation at the resonator mode-spacing frequency. Similar theoretical predictions were made independently by Yariv [8]. Somewhat earlier, Lamb [6], in his now classic paper on the theory of laser operation, had described how the nonlinear properties of the laser medium could cause the modes of a laser to lock with equal frequency spacing. This idea was later discussed by Crowell [7] for the case of many oscillating modes, and he reported experiments with a 6328 Å He–Ne laser demonstrating the "self-locking" of laser modes due solely to the nonlinear behavior of the laser medium. Shortly thereafter, the locking of laser modes with the amplitudes and phases of an FM-modulated wave was discussed by Yariv [8] and Harris and McDuff [9], and demonstrated experimentally by Harris and Targ [10], using a 6328 Å He–Ne laser with an internal phase modulator. Harris later pointed out that internal phase modulation could also be used to produce an output pulse train [11] and this was demonstrated experimentally by Ammann et al. [12].

During 1965, experimenters were demonstrating that the techniques of mode-locking could be applied to other laser systems. The argon ion laser [13] and the ruby laser [14] were mode-locked using internal modulation techniques. A most important mode-locking technique was demonstrated by Mocker and Collins [15] who showed that the saturable dye used inside a ruby laser cavity to "Q-switch" the laser (i.e., to cause the laser to emit a short high-power burst of energy) could also be used to obtain mode-locking. Thus the laser output consisted of a number of short intense

Manuscript received April 13, 1970; revised July 9, 1970. *This invited paper is one of a series planned on topics of general interest—The Editor.*

The author is a Visiting McKay Lecturer in the Dept. of Electrical Engineering and Computer Sciences, University of California, Berkeley, Calif., on leave from the Bell Telephone Laboratories, Inc., Holmdel, N. J. 07733.

Reprinted from *Proc. IEEE*, vol. 58, pp. 1342–1357, Sept. 1970.

pulses of light. These and later experiments demonstrated that Q-switched and mode-locked solid-state lasers could produce much shorter and much higher peak power pulses than those that had been obtained from mode-locked gas lasers. Work was also continuing in 1965 on self-locking of lasers. Statz and Tang [16] published the first of a series of papers they and their collaborators would write on the theory of self-locking. They describe self-locking in terms of "combination tones" (beats) produced by pairs of modes in the nonlinear laser gain medium.

By 1966 there were many workers investigating mode-locking in various laser systems. DeMaria *et al.* reported mode-locking the Nd-doped glass laser with a loss modulator [17], and shortly thereafter, with a saturable dye [18]. The Nd-doped yttrium aluminum garnet (YAG) laser was mode-locked with an internal modulator [19]. It was now becoming clear to experimenters that there was no detector available with a fast enough response time to display the pulses from a mode-locked solid-state laser. Theory showed that the minimum duration of mode-locked pulses was of the order of the reciprocal of the oscillation bandwidth of the laser. From measurements of the bandwidth of laser oscillation it appeared that these pulses could be as short as a few picoseconds (10^{-12} seconds) in duration, yet the fastest detectors had rise times of a fraction of a nanosecond (10^{-9} seconds).

One of the first methods used to estimate these small pulsewidths was measuring the efficiency of the generation of the second harmonic of laser light in a nonlinear crystal. Theory showed that when the incident light is in the form of a pulse train, the average second-harmonic power generated is enhanced over that produced by a continuous wave of the same average power. The amount of this enhancement gives an estimate of the width of the pulses in the pulse train. This method was used by DiDomenico *et al.* [19] to estimate the width of pulses from the mode-locked Nd:YAG laser to be 80 ps (picoseconds) and by Kohn and Pantell [19a] to estimate the width of mode-locked pulses from their ruby laser to be 200 ps.

In late 1966 and early 1967, Maier *et al.* [20], Armstrong [21], and Weber [22] described a new technique for pulse-width measurement. The laser beam is split with a beam splitter and the two beams are then recombined with a relative delay in a nonlinear crystal. With proper adjustment of polarization and crystal orientation, second-harmonic radiation will be generated only when the pulses in each beam coincide in the nonlinear crystal. With this technique pulse-width measurements could be obtained in the picosecond region, and Armstrong [21] reported measuring pulse-widths of 6 ps from a Nd:glass laser mode-locked with a saturable dye. Later in the same year Giordmaine *et al.* [23] proposed and demonstrated the technique of pulsewidth measurement using two-photon fluorescence (TPF). In a manner somewhat analogous to the second-harmonic generation method, the beam is split and the two beams made to interfere in a material exhibiting two-photon fluorescence. Enhanced fluorescence is seen at points where pulses coincide in the material. This method is substantially easier to implement experimentally, and workers quickly adopted

this technique. Much of this early work had to be reexamined later, however, when several authors [24], [25], [163], [164] pointed out that free running (non-mode-locked) lasers produced TPF results very similar to those of a mode-locked laser! Only by measuring the contrast ratio (the ratio of fluorescence at the peak of the display to the fluorescence level in between the peaks) can one obtain information on the state of mode-locking.

In the meantime work was continuing on mode-locking techniques. To obtain greater pulse powers, the technique of completely coupling a single pulse out of a mode-locked laser was exploited by Michon *et al.* [26] to obtain a pulse of less than 4 ns duration with an energy of 8 mJ from a ruby laser, by Penney and Heynau [27] to obtain a pulse of less than 0.8 ns duration with peak power of 25 MW from a Nd:glass laser, and by Zitter *et al.* [28] to obtain 2 ns, 30-watt pulses from a He-Ne laser. Kachen *et al.* [28a] externally selected and amplified a single mode-locked Nd:glass laser pulse to obtain a 10 to 15 ps pulse with a peak power of 30 GW. A new mode-locking technique was described by Smith [29] who showed that mode-locking could be obtained by moving one laser mirror at a constant velocity. Self-locking was reported for the ruby [16], [26], Nd:glass [78], [78a], and argon ion [67] lasers.

During this period work was also progressing on the theory of mode-locking. Harris [30] published a good review paper on internal modulation theory and techniques. A new theory of self-locking based on the transient response of the laser medium to the incident radiation was proposed by Fox and Smith [31], and further developed by Smith [32]. Statz and his co-workers published a series of papers [33]–[35b] on self-locking, developing their theory based on the nonlinear interaction of laser modes in the gain medium. The theory of mode-locking with saturable absorbers was discussed by Garmire and Yariv [126] and Sacchi *et al.* [128].

In 1968 workers observed mode-locking with several new laser systems. Mode-locking of CO_2 gas lasers [36], [37], the CF_3I photolysis laser [37a], dye lasers [38]–[41], and semiconductor lasers [42] was reported. New techniques of mode-locking were also made available. Fox *et al.* [43] and Chebotayev *et al.* [43a] described the use of a section of laser medium excited to the lower laser state as a saturable absorber to produce mode-locking. Comly *et al.* [44] and Laussade and Yariv [45] described the use of anisotropic molecular liquids within the resonator to cause locking. Huggett [46] used the amplified beats between the laser modes to drive an internal modulator to mode-lock a He-Ne laser.

So far, we have been talking of mode-locking a set of longitudinal resonator modes, characterized by having the same spacial distribution in a plane transverse to the resonator axis, but having a different number of half-wavelengths of light along the axis of the resonator.[1] There are also sets of modes which have the same number of half-wavelengths of light along the resonator axis but have different spacial distributions of energy transverse to this axis. These modes are called transverse modes. (A more complete discussion of

---

[1] Some authors refer to these as axial modes.

modes in laser resonators will be found in Section III.) In 1968, Auston [47], [48] described how the phase-locking of a set of *transverse* modes of a laser would produce an output beam that would scan back and forth in space as a function of time. Auston [47] and Smith [49] also describe simultaneous locking of longitudinal and transverse laser modes. The light energy can be confined to a small region of space and travels a zigzag path as it bounces back and forth in the laser resonator. These first experiments were performed with a He–Ne laser, but in 1969 transverse mode-locking experiments were reported using Nd:glass [50], [51] and CO_2 [52], [53] laser systems.

Much of the work presently being done has to do with developing higher power, narrower, and better controlled pulses from mode-locked lasers, and using present techniques with new laser systems.

The availability of well-defined light outputs from mode-locked lasers has stimulated the invention of a number of devices that make use of these characteristics. In 1965, Massey *et al.* [54] described the use of an FM demodulator, external to the laser resonator, to produce a single-frequency output from an FM-locked laser. Shortly thereafter, Harris *et al.* [55] described how the single-frequency output of this laser could be stabilized. A somewhat different technique to produce single-frequency output from a mode-locked laser by coupling out only one laser mode was described by Harris and McMurtry [56].

In 1966, Duguay *et al.* [57] described a technique for frequency shifting a train of mode-locked laser pulses by passing them through an external phase modulator driven at the pulse repetition rate. 1967 brought a variety of different uses for mode-locked lasers. Trains of laser pulses were used for testing the pulse response of electrical cables [58] and for measuring the relaxation times of bleachable dyes used for laser mode-locking [59]–[59b]. In 1968, Giordmaine *et al.* [60] described a technique for shortening the duration of laser pulses by giving them a frequency sweep with an external modulator and then passing them through a dispersive structure. This technique was later demonstrated by Duguay and Hansen [61] who compressed pulses from a He–Ne laser by almost a factor of two. Fisher *et al.* [151a] proposed the use of the optical Kerr effect to give the frequency sweep and in this way Laubereau [151b] reported achieving compression of the pulses from a Nd:glass laser by a factor of 10. Optical rectification of picosecond laser pulses in a nonlinear crystal was demonstrated by Brienza *et al.* [61a]. Rentzepis [62] used mode-locked Nd:glass laser pulses to investigate radiationless molecular transitions.

Recently, Duguay and Hansen [63] have constructed a light gate with an open time of approximately 8 ps, using a mode-locked laser pulse to drive a Kerr cell. Perhaps the most dramatic of the recent uses of mode-locked laser pulses, however, are the experiments on thermonuclear neutron emission by groups in the Soviet Union [64] and the United States [159].

Some uses that are presently being investigated include mode-locked lasers as a source for pulse-code modulation communication systems, and fundamental investigations of the interaction of short pulses of radiation with resonant atomic systems.

III. MODES IN LASER RESONATORS

A mode of a resonator can be defined as a self-consistent field configuration. That is, the optical field distribution reproduces itself after one round trip in the resonator. The modes of an open resonator formed by a pair of coaxial plane or spherical mirrors have been studied in great detail [65]. One can identify a set of longitudinal (or axial) modes which all have the same form of spacial energy distribution in a transverse plane, but have different axial distributions corresponding to different numbers of half-wavelengths of light along the axis of the resonator. These longitudinal modes are spaced in frequency by $c/2L$ where c is the velocity of light and L is the optical path length between the mirrors. Fig. 1 illustrates the type of laser output one obtains when a number of longitudinal resonator modes are above threshold for laser operation. Each frequency shown in the top part of the figure corresponds to a different longitudinal mode number, i.e., a different number of half-wavelengths of light in the resonator. For a typical laser this number will be of the order of 10^6.

To obtain laser operation at a single longitudinal mode, it is usually necessary to design a laser resonator sufficiently short so that $c/2L$ is greater than the bandwidth of the gain of the laser material, or to design a complex resonator which has high loss for all modes within the oscillation bandwidth except the favored one.

It is possible to show that for each longitudinal mode number, there exists a set of solutions for the light energy inside a resonator formed by two spherical mirrors which correspond to different energy distributions in a plane transverse to the resonator axis. These solutions are called the transverse modes of the resonator. The cross-sectional amplitude distribution of these modes, for curved mirror resonators with rectangular symmetry, is given closely by [65]

$$A(x, y) = A_{m,n}\left[H_m\left(\frac{\sqrt{2}x}{w}\right) H_n\left(\frac{\sqrt{2}y}{w}\right) \right] \cdot \exp\left(-(x^2 + y^2)/w^2\right) \quad (1)$$

where x and y are the transverse coordinates, and $A_{m,n}$ is a constant whose value depends on the field strength of the mode. w is the radius of the fundamental mode ($m=0$, $n=0$) at $1/e$ maximum amplitude, $H_a(b)$ is the ath order Hermite polynominal with argument b, and m and n are called the transverse mode numbers.[2] Note that this distribution is independent of the longitudinal mode number. Fig. 2 is a plot of A along the x axis, for $m=0$ to 5. Note that as the transverse mode number increases, the energy is spread further and further from the axis of the resonator. Fig. 3 shows actual mode patterns found with a 6328 Å He–Ne gas laser. In order to obtain oscillation in a single transverse

---

[2] These modes are sometimes referred to as TEM_{mn} modes by analogy with the modes in waveguides.

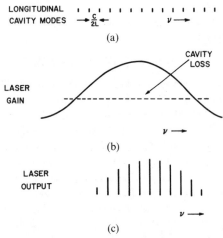

Fig. 1. Laser operation on several longitudinal resonator modes. (a) Longitudinal modes of the laser resonator. (b) Laser gain versus frequency showing the region where the gain exceeds the cavity losses. (c) Laser output: oscillation at resonator mode frequencies for which laser gain exceeds the losses.

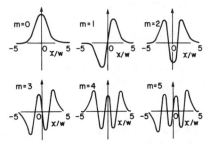

Fig. 2. Plot of field amplitude of transverse modes versus the normalized transverse coordinate x/w (w=spot size of fundamental mode), for $m=1-5$. Reversal of sign of field indicates 180° change of phase.

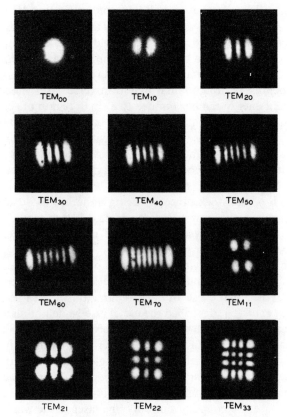

Fig. 3. Mode patterns of a gas laser oscillator with rectangular symmetry [160].

mode, it is necessary to use some device which will give high losses to all transverse modes but the desired one. Since higher order modes spread further from the resonator axis, the easiest way to accomplish single-transverse-mode operation is to insert into the laser cavity a circular aperture whose size is such that the fundamental mode experiences little diffraction loss while higher order modes suffer appreciable attenuation.

For resonators which do not have large diffraction losses, the resonant frequency of a mode can be written [65]

$$v = (c/2L)\left[(q+1) + \frac{(m+n+1)}{\pi} \\ \cdot \cos^{-1}\sqrt{(1-L/R_1)(1-L/R_2)}\right] \quad (2)$$

where $q+1$ is the number of half-wavelengths of light along the axis of the resonator, m and n are the transverse mode order numbers, and R_1 and R_2 are the radii of the two mirrors making up the laser resonator. Note the following points. 1) For a given transverse mode (given m and n), longitudinal modes with mode number differing by one are spaced in frequency by $c/2L$. 2) For a given longitudinal mode number (q), transverse modes with the sum of m and n differing by one are spaced in frequency by $(c/2\pi L)\cos^{-1}$ $\sqrt{(1-L/R_1)(1-L/R_2)}$.

Let us consider the output of a laser operating in a number of longitudinal and transverse modes. The total field will be the sum of the individual fields of each of the modes. The optical length L will not be the same for all the modes due to the dispersion of the laser material. In general, both the amplitude and the phase of these modes will vary with time due to random mechanical fluctuations of the laser resonator length and the nonlinear interaction of these modes in the laser medium. The total field will thus vary with time in some uncontrolled way with a characteristic time which is of the order of the inverse of the bandwidth of the oscillating mode frequency spectrum. In the next sections we discuss the results of fixing the frequency spacings and phases of the oscillating modes—mode-locking the laser.

IV. LONGITUDINAL MODE-LOCKING

Consider a set of oscillating longitudinal modes such as those shown in Fig. 1(c). If we somehow fix the frequency spacing, relative phases, and amplitudes of these modes, the laser output will be a well-defined function of time. Consider the nth mode to have amplitude E_n, angular frequency ω_n, and phase ϕ_n. Then the total laser output field, E_T, can be written

$$E_T = \sum_n E_n e^{i[\omega_n(t-z/c)+\phi_n]} + \text{c.c.} \quad (3)$$

where c.c. represents the complex conjugate and we have assumed the radiation is traveling in the $+z$ direction. If we have equal mode frequency spacing, $\omega_n = \omega_0 + n\Delta$ where

$\Delta = 2\pi(c/2L)$ and ω_0 is the optical frequency of the laser output. Thus, we can write

$$E_T = e^{i\omega_0(t-z/c)}\left\{\sum_n E_n e^{i[n\Delta(t-z/c)+\phi_n]}\right\} + \text{c.c.} \quad (4)$$

This corresponds to a carrier wave of frequency ω_0 whose envelope depends on the values of E_n and ϕ_n. Note, however, that

1) the envelope travels with the velocity of light;
2) the envelope is periodic with period $T = 2\pi/\Delta = 2L/c$.

For $\phi_n = $ a constant[3] independent of n, this envelope consists of a single pulse in the period T whose width is approximately the reciprocal of the frequency range over which the E_n's have an appreciable value, i.e., the ratio of the pulse spacing to the pulsewidth is approximately the number of oscillating modes. Within the laser resonator this corresponds to a pulse of light traveling back and forth between the resonator mirrors with the velocity of light. Other selections of amplitudes and phases will result in different envelopes. For any set of mode amplitudes, the narrowest pulse will always result from the phases $\phi_n = 0$, for all n. For mode amplitudes and phases corresponding to the sidebands and carrier of an FM-modulated wave, the output intensity will be constant as a function of time.

Fig. 4(a) shows the actual frequency spectrum of the output of a 6328 Å He–Ne laser. The laser is oscillating on a number of longitudinal modes and the amplitudes of these modes fluctuate with time. Fig. 4(b) shows the same laser output when the modes are mode-locked. The oscillation bandwidth has increased and the mode amplitudes are stable.

Fig. 5(a) shows the corresponding output as a function of time. The output consists of a train of narrow pulses separated by the round-trip time of the light in the optical resonator $(2L/c)$. An expanded view of one pulse in Fig. 5(b) shows the pulsewidth to be 330 ps. How can a laser be made to operate in this fashion? The remainder of this section discusses techniques for obtaining longitudinal mode-locking.

A. Self-Locking

Under certain conditions, the nonlinear effects of the laser medium itself may cause a fixed phase relationship to be maintained between the oscillating modes. The first detailed discussion of this effect appears to be that of Lamb [6], although earlier papers by Bennett [66] and Statz and Tang [3] mention phase-locking for three-mode laser operation when the central mode is tuned to the center of the atomic gain curve. The first paper to discuss self-phase-locking of a laser oscillating in many oscillating modes was that of Crowell [7], who showed that self-locking the 6328 Å He–Ne laser would result in an output train of pulses of a duration of approximately 1 ns. He found self-locking occurred when oscillation was confined to the fundamental

[3] As origin of time is arbitrary, we shall in subsequent discussion assume that this constant = 0.

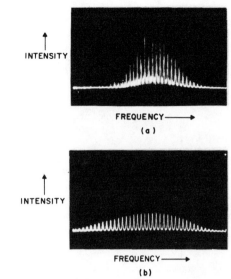

Fig. 4. (a) Output frequency spectrum from a non-mode-locked 6328 Å He–Ne laser. (b) Output frequency spectrum of the same laser when mode-locked [43].

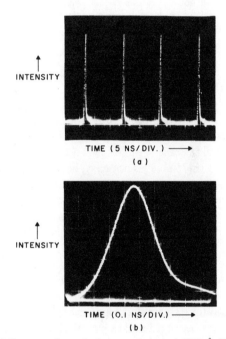

Fig. 5. (a) Output pulse train from mode-locked 6328 Å He–Ne laser. (b) Expanded view of one of the mode-locked pulses showing width of 330 ps. The "tail" is contributed by the diode detector. The laser output pulse is believed to be symmetrical [43].

transverse mode and the losses in the optical resonator were "judiciously adjusted."

There have appeared a number of papers on the theory of self-locking [16], [31]–[35b], [68]–[74]. Typically, one of two approaches is used. The first approach uses the concept of combination tones (i.e., the radiation produced by the nonlinear interaction of two or more laser modes with the laser medium). If a combination tone is close in frequency to an oscillating mode, frequency pulling effects will tend to pull the oscillating mode to the same frequency as the combination tone [75]. Fig. 6 shows a simple situation in which three modes at v_1, v_2, and v_3 are oscillating. Due to disper-

213

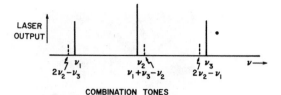

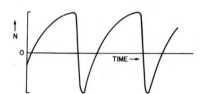

Fig. 6. Combination tones formed by three laser frequencies in a nonlinear medium.

Fig. 8. Variation of population difference as a function of time for a laser with a circulating 180° pulse.

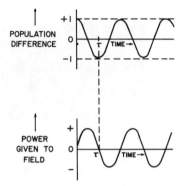

Fig. 7. Plots of normalized population difference and power given to incident field for resonant two-level atomic system for times short compared to atomic decay times.

sion in the gain medium, these modes will not, in general, be exactly equally spaced in frequency. In the nonlinear gain medium, radiation at frequencies $2v_2 - v_3$, $v_1 + v_3 - v_2$, and $2v_2 - v_1$ is generated. These frequencies will tend to pull the frequencies of the oscillating laser modes so that the oscillating modes have equal frequency spacing. Lamb [6] analyzes in detail the case of three mode oscillation at low power levels (see also [75a]). It is not easy, however, to extend this work to the case of many oscillating modes and high power levels. It has been found necessary to use restrictive assumptions and evaluate special cases [16], [33], [34], [35b], [68]–[69] or to resort to computer solutions of complex equations [71], [74].

Because of the complexity of the equations for a combination tone analysis of a large number of modes in the frequency domain, it may be easier to work in the time domain and investigate the response of the gain medium to a train of pulses of stimulating radiation. A first attempt at this type of analysis was made by Fox and Smith [31], who showed that under certain rather restrictive conditions one would expect the pulses to completely invert the populations of the laser levels. The upper portion of Fig. 7 shows the (normalized) population difference as a function of time for a two-level atomic system under the influence of radiation at the atom's resonant frequency. It is assumed that atomic lifetimes are much longer than the times being considered. +1 indicates all the atoms in the upper laser level and −1 indicates all the atoms in the lower laser level. Although complete saturation would correspond to a population difference of 0, it is seen that after the radiation has interacted with an atomic system with atoms initially in the upper state for a time τ, the population difference becomes inverted, and maximum power is given to the incident field. Thus a pulse of this "size" (called a π or 180°

pulse) [76] can interact most efficiently with the atomic gain system. In [31] and [32], it is shown that certain lasers indeed operate in such a way that the self-locked pulses in the laser resonator approximate 180° pulses for the laser medium. The population inversion recovers as a function of time with the time constant of the atomic decay time, and the population difference, as a function of time, varies as shown in Fig. 8. Similar results have also been obtained in [35a] and [74].

Experimenters have shown that most laser systems can be made to self-lock. The first self-locking experiments were performed with the 6328 Å He–Ne laser [7], [77], but soon self-locking was reported for ruby [16], [26], [77a],[4] argon ion [67], and Nd:glass [78], [78a][4] lasers. Self-locking has also been reported for Nd:YAG lasers [79][4] semiconductor lasers [42] and CO_2 gas lasers [80]. Multiple self-pulsing corresponding to several pulses traveling back and forth in the laser resonator has been observed, notably with the 6328 Å He–Ne laser [32], [77], [82]–[85] where up to six output pulses in one round-trip period have been reported. In general it is not possible to predict the exact conditions for self-pulsing, but the following conditions are usually necessary.

a) The oscillation must be confined to a single transverse mode.
b) The round-trip time for radiation in the resonator must be of the order of or greater than the atomic decay time.
c) The laser must not be operated too much above threshold.

In view of the somewhat uncertain nature of self-locking (self-locked lasers often have unstable outputs), there are some advantages in using a driven cavity perturbation to force the laser to mode-lock. Methods for accomplishing this are described in the following section.

B. Internal Modulation

Hargrove et al. [5] first used an internal modulator to produce mode-locking. They used an acoustic loss modulator, driven at the longitudinal mode-spacing frequency ($c/2L$), in the resonator of a 6328 Å He–Ne laser. Di-Domenico [4] discussed the theory of operation of a loss modulator driven at a frequency related to the mode-spac-

---

[4] Recent work has cast doubt on many of these results. Some workers believe that only partial self-locking occurs for solid-state losses (see [81], [162], [163]). There is no doubt that reliable self-locking can be obtained with gas lasers, however.

ing frequency. Shortly thereafter the use of an internal phase modulation (dielectric constant modulation) to lock laser modes with the amplitudes and phases of an FM-modulated wave was discussed by Yariv [8] and Harris and McDuff [9], and demonstrated experimentally by Harris and Targ [10]. It was later shown that internal phase modulation could also be used to produce a train of pulses similar to those produced using a loss modulator [11], [12].

Numerous internal modulation mode-locking experiments have been performed with different laser systems [5], [7], [10]–[14], [17], [19], [35a], [36], [46], [68], [86]–[100] using either acoustic [5], [7], [13], [17], [19], [37a], [86a] or electrooptic [13], [36], [92]–[94], [99], [100] loss modulators to produce output pulses, and using electrooptic phase modulators to produce FM [10]–[12], [89] or pulsing [11], [88], [89], [95] operation. It has been observed that there is a dramatic reduction of the low-frequency noise in the laser output when the laser is mode-locked [96], [99].

Pantell *et al.* [90] have discussed mode-locking a Raman laser using internal modulation techniques. Henneberger and Schulte [91] have described mode-locking a 6328 Å He–Ne laser by vibrating the resonator end mirror at the $c/2L$ rate. The resonator length modulation is equivalent to that produced by an internal phase modulator driven at $c/2L$. Several authors [86], [87], [98] have described placing the internal modulator in a resonator weakly coupled to the main laser resonator in order to reduce the effective insertion loss of the internal modulation device. Gurski [97] describes mode-locking and second-harmonic generation inside the laser resonator using the same nonlinear crystal. Huggett [46] has proposed and demonstrated the rather elegant scheme of using the amplified $c/2L$ beat frequency from the laser output to drive the internal modulator.[5]

A great deal of work has also been done on the theory of mode-locking by internal modulation [8], [30], [68], [92], [99], [101]–[113]. A good review paper on this subject is that of Harris [30]. Although formal results can be obtained, in general the resulting formulas must be solved by computer for the particular case of interest. Fig. 9 shows the results for a mode-locked pulsing operation computed from the theory of McDuff and Harris [104] for a specific case with five free-running oscillating modes. Here α_c is the mode coupling due to loss perturbation which for a small modulator near the end of the laser cavity is approximately one half the average loss per pass introduced by the loss modulator. δ is the peak single-pass phase retardation of a phase modulator.

A simple physical picture of the operation of loss or phase modulators to produce pulse train mode-locking can be given by the following.

Loss Modulation: If the loss is modulated at $c/2L$, one cycle of the modulation frequency corresponds to the time it takes the light to make a round trip in the laser resonator. Thus light incident on a modulator situated at one end of the laser resonator during a certain part of the modulation

[5] T. S. Kinsel has been able to obtain reliable locking of the Nd:YAG laser using this technique (private communication).

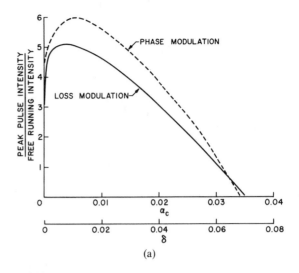

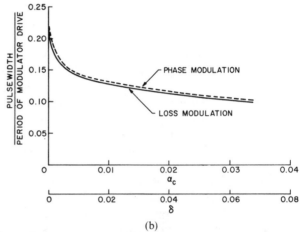

Fig. 9. (a) Pulse peak intensity versus perturbation level for constant detuning. (b) Pulsewidth versus perturbation level for constant detuning [104].

cycle will be again incident at the same point of the next cycle after one round trip in the laser resonator. Light which sees a loss at one time will again see a loss after one round in the resonator. Thus all the light in the resonator will experience loss except that light which passes through the modulator when the modulator loss is zero. Light will tend to build up in narrow pulses in these low-loss time positions. In a general way we can see that these pulses will have a width given by the reciprocal of the gain bandwidth. Pulses wider than this will experience more loss in the modulator. Pulses narrower in time will experience less gain because their frequency spectrum will be wider than the gain bandwidth.

Phase Modulation: Light passing through an electrooptic phase (dielectric constant) modulator will be up- or downshifted in frequency unless it passes through during the time of the maximum (or minimum) of the dielectric constant cycle. If the modulator is operated at the synchronous rate, this shift will increase every time the light passes through the modulator. Eventually the light frequency will fall outside the gain curve of the laser medium and the light will not experience any gain. Thus the effect of the phase modulator is similar to the loss modulator, and the previous discussion of loss modulation also applies here.

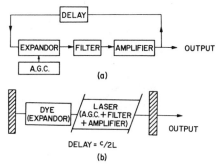

Fig. 10. (a) Block diagram of Cutler's electronic regenerative pulse generator. A.G.C. stands for automatic gain control. (b) Laser equivalent of the regenerative pulse generator.

C. Saturable Absorbers

In 1965, Mocker and Collins [15] reported simultaneously mode-locking and Q-switching a ruby laser by means of a saturable absorber in the laser resonator. (A saturable absorber is a material whose absorption coefficient for the laser light *decreases* as the light intensity is increased.) The technique of mode-locking with a saturable absorber had been described somewhat earlier in a patent application by Rigrod [15a]. DeMaria and Stetser later reported using a saturable absorber to mode-lock the Nd:glass laser [18], [114]. This technique has proved to be a very useful one for mode-locking lasers whose internal light intensity is high enough to appreciably saturate the absorber material. This requirement has confined most saturable absorber applications to liquid [40], [41], [123] and solid-state laser systems [115]–[122]. Saturable absorbers have also been used, however, with gas laser systems. Fox *et al.* [43] and Chebotayev *et al.* [43a] described the use of a pure neon discharge as a saturable absorber for the 6328 Å He–Ne laser. This was demonstrated to operate effectively to mode-lock this laser system. The use of saturable absorbers such as SF_6 to mode-lock the CO_2 gas laser is discussed in [37] and [124].

The theory of the operation of a saturable absorber to produce mode-locking can be discussed [1], [18] in terms of an electronic regenerative pulse generator described by Cutler [125]. Fig. 10(a) shows a block diagram of the system Cutler considered. The expandor is a nonlinear element whose loss decreases as the incident intensity is increased. It was shown [125] that this system would produce a pulse train with pulse separation equal to the group delay time of one round trip through the device and minimum pulsewidth, approximately given by the reciprocal of the filter bandwidth. Fig. 10(b) shows a laser with a saturable absorber in the laser resonator. This system is analogous to the electronic system with the saturable dye acting as an expandor, the laser medium acting as the amplifier and filter, and the round-trip time of light in the laser resonator as the system delay time. It is important for the saturable absorber to have a fast recovery time compared to the resonator round-trip time if it is to act as an expandor in this system. Cutler's analysis shows that if there exists a nonlinear variation of refractive index with frequency within the feedback loop, the pulsewidth will be greater than the reciprocal

of the gain bandwidth. For a Gaussian filter function and pulse shape, and assuming a square law saturable absorber, Cutler finds the pulsewidth, Δt, given by

$$\Delta t = \frac{8}{\pi \Delta \omega} \left[\frac{\left(1 + \frac{\gamma^2 \Delta \omega^4}{16} \right)}{1 + \left(1 + \frac{\gamma^2 \Delta \omega^4}{18} \right)^{1/2}} \right]^{1/2} \qquad (5)$$

where $\Delta \omega$ is the bandwidth of the filter and γ is proportional to the curvature of the plot of refractive index versus frequency. Experimental indications of this type of pulse broadening in mode-locked Nd:glass lasers have recently been observed by Treacy [1], [148]–[149].

Although an analysis of Cutler's system is helpful in understanding the operation of a saturable absorber in a mode-locked laser system, the details of the saturation of both the absorber and the laser gain must be considered in a more realistic model. Analysis of saturable absorber mode-locking has been made by several authors [126]–[131a]. In general, the results obtained are in agreement with the predictions of the pulse regenerative amplifier model.

D. Other Methods of Producing Mode-Locking

Several other methods for producing mode-locking of longitudinal modes have been reported. Smith [29] discovered that pulsing mode-locked operation of a 6328 Å He–Ne laser could be obtained by continuously translating one resonator mirror at a constant velocity. A satisfactory explanation for this behavior has not been found. Bambini and Burlamacchi [132] reported a related phenomenon where mode-locking was obtained by low-frequency resonator length modulation. Laussade and Yariv and co-workers have investigated the behavior of an anisotropic molecular liquid inside a laser resonator and have shown both theoretically [45] and experimentally [44], [133] that it may be used to obtain mode-locking. One method of obtaining population inversion and, thus, laser operation in liquid laser systems is to pump them with light from a solid-state laser. Several authors [38], [39], [123], [123a], [134] have reported mode-locked pulse-train output from laser-pumped liquid lasers by mode-locking the pumping laser and adjusting the length of the liquid laser resonator to be integrally related to the length of the pumping laser resonator.

V. TRANSVERSE MODE-LOCKING

Consider a set of transverse modes with the same longitudinal mode number but with transverse mode numbers differing by one. These modes will be approximately equally spaced in frequency (see Section III). Such a set is illustrated in Fig. 11 which shows the set TEM_{0n}, for $n = 0, 1, 2, \cdots$. What will occur if we fix the frequency spacings and relative phases of this set of simultaneously oscillating modes? This problem was first investigated by Auston [47]. He showed that a simple result could be obtained if a set of

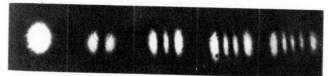

Fig. 11.   A linear set of transverse modes TEM$_{00}$, TEM$_{01}$, TEM$_{02}$, TEM$_{03}$, $\cdots$.

modes is locked with equal frequency spacings and zero phase difference with field amplitudes A_n where

$$|A_n|^2 = \frac{1}{n!}(\bar{n})^n e^{-\bar{n}}. \tag{6}$$

$\bar{n}$ is a parameter which determines the number of oscillating transverse modes. Under these conditions the intensity of the optical field of the laser output will be

$$I(\xi, t) = \frac{1}{\sqrt{\pi}} \exp -(\xi - \xi_0 \cos \Omega t)^2 \tag{7}$$

where $\xi =$ transverse coordinate in the y direction normalized with respect to the fundamental spot size, $\xi_0 = \sqrt{2\bar{n}}$, and $\Omega =$ transverse mode frequency spacing. Equation (7) represents a scanning beam of width equal to the fundamental spot size which moves back and forth in the transverse plane with maximum excursion equal to ξ_0. Fig. 12 illustrates this motion over a complete period of $2\pi/\Omega$. Fig. 13 shows how the distribution of A_n's depends on $\bar{n}$. Note that the maximum beam excursion depends on $\sqrt{\bar{n}}$. For large $\bar{n}$, the width of the distribution equals $2\sqrt{2\bar{n}}$, which is simply the total excursion (in units of number of resolvable spots). Thus one obtains the intuitively satisfying result that the maximum number of resolvable spots is roughly equal to the number of oscillating modes. This type of operation was later observed by Auston [48] using a tilting mirror acoustic modulator with a short 6328 Å He–Ne laser. Fig. 14 shows the output versus time of this laser as observed with a photodetector having an aperture in front of it. The coordinate ξ represents the displacement of the aperture from the resonator axis along the direction of beam motion.

Self-locking of transverse laser modes has also been reported by Auston [47], [48] for the 6328 Å He–Ne laser, and by Ito and Inaba [52] for the CO_2 laser.

Simultaneous locking of the modes of a laser oscillating in several longitudinal and transverse modes has also been studied. Two types of behavior have been observed. Smith [49] reports mode-locking all the longitudinal modes corresponding to each transverse mode to form a pulse inside the laser resonator. The pulses corresponding to the different transverse modes do not coincide in time, however, so that the output consists of several pulses in one resonator round-trip period, each one corresponding to a different transverse mode. This type of operation was obtained with a 6328 Å He–Ne laser using a Ne discharge as a saturable absorber [43].

A very different behavior is obtained if a set of transverse modes corresponding to each longitudinal mode number is locked together to form a scanning beam, and each of the sets of longitudinal modes is also locked with zero phase difference to form a pulse in the resonator. In this case the

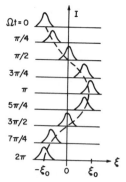

Fig. 12.   Plot of distribution of intensity for various times during the scanning period [47].

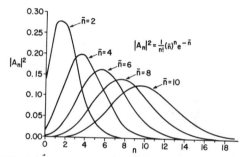

Fig. 13.   Distribution of mode amplitudes required to obtain scanning-beam operation for various numbers of oscillating modes [47].

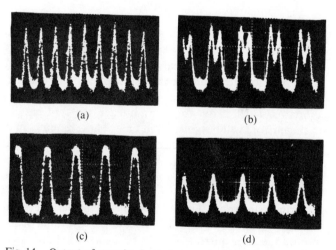

(a)   (b)

(c)   (d)

Fig. 14.   Output of scanning-beam laser versus time observed through an aperture located at various values of ξ [48]. (a) $\xi=0$. (b) $\xi=1.3$. (c) $\xi=2.6$. (d) $\xi=3.1$.

light is confined to a small region of space both in the axial and transverse directions. It can be shown [47] that this "blob" of light bounces back and forth in the laser resonator following the zigzag path to be expected from geometrical optics.

The position where this blob of light will hit the laser mirrors can be determined for a resonator consisting of two equal and coaxial mirrors using an elegant description proposed by J. R. Pierce and discussed by Herriott et al. [135]. They show that the resonator is equivalent to a series of equally spaced thin lenses and that the beam position at the nth lens can be written

$$x_n = X \sin(n\theta + \alpha) \tag{8}$$

where x_n is the displacement of the beam from the resonator

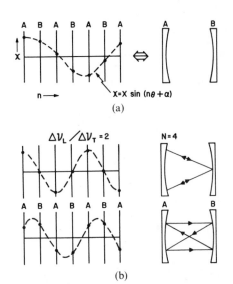

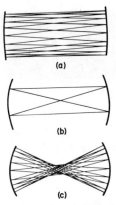

Fig. 16.   Locus of "spot" of light. (a) Close-to-plane. (b) Confocal. (c) Close-to-concentric resonators.

Fig. 15.   (a) Resonator with two mirrors, A and B, replaced by a sequence of thin lenses represented schematically by the positions marked A and B. The path of a light beam through the lenses is formed by joining with straight lines the points where the dashed sinusoid intersects the planes of the lenses. (b) Thin lens diagrams and corresponding beam paths for the case $\Delta v_L/\Delta v_T = 2$ (confocal resonator). Two possible beam paths are shown [49].

axis in the x direction at the nth lens and X and α are arbitrary constants. It can be shown that for our case

$$\theta = \pi \Delta v_T / \Delta v_L \qquad (9)$$

where Δv_T is the frequency separation between adjacent transverse modes and $\Delta v_L (= c/2L)$ is the frequency spacing between adjacent longitudinal modes. Note that if $\theta = 2\pi/N$, where N is an integer, the beam displacement will repeat itself after N lenses, i.e., after $N/2$ round trips in the laser resonator. Thus, if $\Delta v_L/\Delta v_T$ is an integer, the spot pattern on the mirrors will be constant in time. These remarks may perhaps be clarified by referring to Fig. 15. Fig. 15(a) shows a laser resonator replaced by a series of lenses represented schematically by the positions marked A and B. The "A" positions correspond to the mirror A, and the "B" positions to mirror B. The intersections of the dashed sinusoid and the marked positions indicate the locations of the spot on the laser mirrors. Fig. 15(b) shows the special case where $\Delta v_L/\Delta v_T = 2$. Positions of the spot on the laser mirror are now constant with time. Two cases are shown corresponding to two different values of α, and the corresponding beam path in the resonator is indicated. In an experimental situation, the effects of aperturing the beam will determine the value of α and thus the beam path observed. For large integral values of $\Delta v_L/\Delta v_T$, a pulse of light will appear periodically at different spacial positions on the laser output mirror. This machine gun type of operation may find use in beam scanning systems. Fig. 16 shows the light path for simultaneous locking of longitudinal and a linear set (TEM$_{0n}$) of transverse modes for (a) close-to-plane parallel, (b) confocal, and (c) close-to-concentric resonators.

Experimental observation of simultaneous mode-locking was reported by Auston [48], who observed self-locking of this type, and Smith [49], who obtained locking with a saturable absorber. Simultaneous locking has also been reported by Bambini and Burlamacchi [132] using low-frequency

resonator length modulation and, for a somewhat different situation involving two almost degenerate transverse modes, by Kohiyama et al. [136]. All of these experiments were done using the 6328 Å He–Ne laser. Simultaneous self-locking has also been reported for the Nd:glass laser [51] and the CO_2 laser [52], and saturable absorbers have also been used to obtain simultaneous locking with the Nd:glass [50] and CO_2 [53] laser systems.

VI. Current Research Interests

In this section we discuss briefly some current research topics involving mode-locked lasers. The selection of topics is somewhat arbitrary and no attempt is made to be comprehensive. The aim is to give the reader some idea of the range of current work involving mode-locked lasers.

A. Measurement of Picosecond Pulse Durations

The problem of measuring the duration of light pulses of less than several hundred picoseconds duration is not a trivial one. Sampling oscilloscopes are available with rise times of 25 ps but currently available detectors do not have rise times better than 100 to 200 ps, so that pulses less than approximately 200 ps cannot be directly displayed on an oscilloscope screen. Several techniques have been used to estimate the width of picosecond laser pulses. The efficiency of second-harmonic generation in nonlinear crystals is enhanced if the fundamental radiation is in the form of a train of narrow pulses [19], [19a], [140], [145]. The amount of this enhancement can be used to estimate the width of the pulses. DiDomenico et al. [19] used this method to estimate a pulsewidth of 80 ps for pulses from a mode-locked Nd:YAG laser. Kohn and Pantell [19a] measured 200 ps pulses from their mode-locked ruby laser.

A better technique became available when Maier et al. [20], Armstrong [21], and Weber [22] described a second-harmonic generation intensity correlation technique. In this case no comparison need be made with a non-mode-locked laser output. The technique involves splitting the laser beam with a beam splitter and recombining the beams with a relative delay in a nonlinear crystal. With a proper adjustment of polarizations and crystal orientation, second-harmonic radiation is only generated when the pulses in each beam overlap in the nonlinear crystal. Using this technique, Armstrong [21] measured pulsewidths of 6 ps

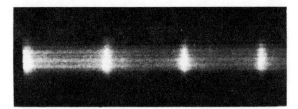

Fig. 17. Two-photon fluorescence pattern observed in rhodamine 6G dye with a mode-locked Nd:glass laser.

from a Q-switched and mode-locked Nd:glass laser, and Glenn and Brienza [141] found that pulsewidths from a Q-switched and mode-locked Nd:glass laser varied from a few ps at the beginning of the pulse train to 15 ps at the end. Krasyuk *et al.* [142] measured pulsewidths of 14 ps from a ruby ring laser. A similar technique which uses a dye exhibiting two photon fluorescence (TPF) as the nonlinear element was proposed by Giordmaine *et al.* [23]. This technique has been widely used because of its apparent simplicity. Fig. 17 shows an experimental TPF display from a mode-locked Nd:glass laser. There is a problem in the interpretation of a fluorescence pattern if the contrast ratio (the ratio of the intensity of fluorescence in the region where pulses overlap to the fluorescence intensity in between these maxima) is not the theoretical maximum [24], [25], [163], [164]. A contrast ratio of 3:1 indicates complete mode-locking with zero phases to produce the narrowest laser pulses possible with a given bandwidth. Completely random phases will give a TPF display with a contrast ratio of 1.5:1, with approximately the same peak width. An FM-locked laser output will produce no peaks (contrast ratio 1:1). There has been much recent work on the interpretation of TPF results. A detailed discussion is beyond the scope of this paper and the reader is referred to Rowe and Li [137], Weber *et al.* [138], Duguay *et al.* [162], and Picard and Schweitzer [165] for recent papers on this subject. Recently, workers have discovered that somewhat more information on pulse shape can be obtained from third-harmonic generation experiments [143]–[144].

Russian workers [139] have described a moving-image camera with an image converter which has a resolution of the order of 10 ps. They show that much more information on pulse structure can be obtained with this technique than can be obtained from TPF experiments. There seems to be no comparable instrument available in this country, however.

B. Frequency-Swept-Pulses and Pulse Compression

Duguay and co-workers showed how mode-locked pulses could be frequency shifted [57], [146], [147] or swept in frequency [60], [61] by passing them through an external phase modulator. Fig. 18 illustrates the operation of this device. Fig. 18(a) shows the pulse train incident on the modulator. Fig. 18(b) shows the variation of the refractive index of the modulator which is being driven at the pulse-repetition frequency. As can be seen from Fig. 18(c), if the pulses pass through the modulator during periods where the refractive index is not to first order changing with time, the average optical frequency will not be shifted but

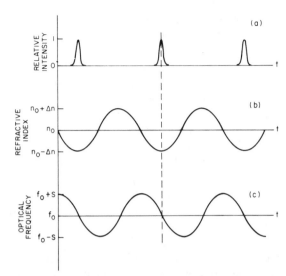

Fig. 18. (a) Laser output versus time. (b) Synchronous variation of modulator's refractive index. (c) Optical frequency of light after passing through modulator [161].

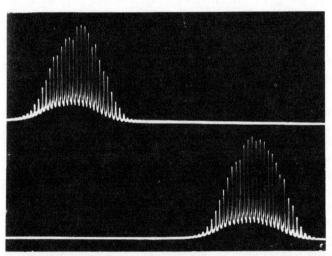

Fig. 19. Frequency spectrum of mode-locked laser pulse train. Upper trace shows unshifted spectrum and lower trace shows frequency-shifted spectrum after pulse train has passed through modulator as described in text [146].

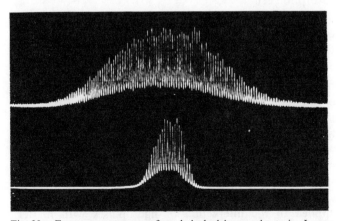

Fig. 20. Frequency spectrum of mode-locked laser pulse train. Lower trace shows spectrum of pulse train entering modulator and upper trace shows increase in width of frequency spectrum after passing through modulator giving frequency sweep to pulses as described in text [60].

pulse. If the pulse passes through the modulator $\pi/2$ later during the modulator cycle, it will be up- or down-shifted in frequency in passing through the device. Fig. 19 shows the frequency spectrum of mode-locked He–Ne laser pulses which have been frequency shifted by 4.8 GHz by passing through such a device [147]. Fig. 20 shows the change in bandwidth obtained by giving the pulses a frequency sweep [60]. These pulses may then be compressed in time to a minimum duration of the order of the reciprocal bandwidth by passing them through a suitable dispersive element [60], [61]. [161]. In this way pulse compression of about a factor of two was demonstrated with a He–Ne laser [61]. Fisher *et al.* [151a] have calculated theoretically that mode-locked pulses from a Nd:glass laser could be compressed to 0.05 ps by using a medium with an intensity-dependent index of refraction (Kerr effect) to give a frequency sweep to the pulses and then passing them through a suitable dispersive element. Laubereau [151b] has reported experimental pulsewidth compression of a factor of 10 using this technique, with CS_2 as the Kerr-effect medium. Hargrove [152] has described the reduction of pulse durations by interferometric combination of frequency shifted pulses.

Recently, Treacy [148]–[149] has reported that the pulses from a Nd:glass laser mode-locked with a saturable absorber can be compressed by passing them through a suitable dispersive element. This indicates that some kind of frequency sweep is occurring within the pulses. A theory of this effect has been given by Cubeddu and Svelto [130], on the assumption that the pulses have a linear frequency sweep caused by the dispersion of the glass matrix in which the Nd atoms are situated. A recent paper by Fisher and Fleck [150] indicates that the frequency sweep may not be a simple linear one. Duguay *et al.* [162] have discussed the effect of the nonlinear index of the glass in producing self-phase modulation of these pulses. So far the Nd:glass laser seems to be the only laser for which such a frequency sweep has been discovered. Smith [151] has shown that no frequency sweep is detectable with the 6328 Å He–Ne laser system when self-locked, or when mode-locked with a Ne absorption cell.

C. *Resonant Interaction of Radiation with Matter: Dynamic Polarization Effects*

The availability of short intense pulses of coherent radiation has stimulated work on the resonant interaction of radiation with atomic systems. In particular, it is known that pulses of radiation of duration less than the polarization memory time (T_2) of a resonant medium will interact with this medium in a fundamentally different way than will much longer pulses. These effects have been known and studied for some time in the microwave region of the spectrum and recently these ideas and techniques have been applied to the interaction of light pulses with matter. Examples of these studies are photon echo studies [153] and studies of "self-induced transparency" by McCall and Hahn [76].

Fox and Smith[31] and Smith [32] have shown that due to dynamic polarization effects a mode-locked laser pulse

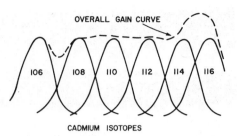

Fig. 21. Gain profile obtained using a mixture of Cd isotopes as indicated. The individual of isotope gain curves have a width of approximately 1200 MHz. Total gain curve width is approximately 8000 MHz.

may interact with gain medium in such a way that the population difference is reduced below zero after passage of the pulse through the medium, i.e., the laser medium is left in an absorbing state. This type of operation is illustrated in Figs. 7 and 8. Recent studies of the propagation of mode-locked 6328 Å He–Ne laser pulses through a resonant absorber [154], [154a] have also demonstrated the existence of dynamic polarization effects within this system. Mode-locked lasers are a convenient tool for the study of such effects and it is expected that much more work in this area will soon be reported.

D. *Communications Systems*

The ability of a mode-locked laser to emit a train of intense narrow pulses makes it attractive as the source of the carrier for an optical pulse-code modulation system. This type of system has been studied by Denton and Kinsel [155], [155a]. Kinsel [156] describes an optical pulse-code modulation system capable of handling many orders of magnitude more information than currently used microwave systems. When the need for large information capacity transmission systems stimulates their construction, some type of mode-locked laser may well be used as the source of the carrier pulse train.

E. *Mode-Locking New Laser Systems*

There is still interest in mode-locking new laser systems whose characteristics may be particularly useful for certain experiments. Faxvog *et al.* [157] have recently reported observation of self-locking of the Cd^+ laser operating at 4416 Å. Silfvast and Smith [158] have mode-locked the Cd^+ laser using an internal accoustic loss modulator at both 4416 Å and 3250 Å in the ultraviolet. As the isotope shift for these Cd^+ lines is of the order of the width of the gain curve for a single isotope, a broad gain region can be obtained by using a mixture of isotopes. A broad oscillation spectrum gives the possibility of narrow pulse output. Fig. 21 shows the gain profile obtainable using a mixture of the available Cd^+ isotopes. By using such a mixture Silfvast and Smith [158] obtained an oscillation spectrum of roughly 8000 MHz in width and were able to obtain a continuous train of pulses with pulsewidths of the order of 100 ps.

VII. CONCLUDING REMARKS

We have attempted to present a tutorial survey of the subject of mode-locking in lasers with adequate references so that the interested reader may find more detailed information if required. A minimum of mathematics has been used and physical pictures of the processes involved have been presented wherever possible.

REFERENCES

[1] A. J. DeMaria, W. H. Glenn, Jr., M. J. Brienza, and M. E. Mack, "Picosecond laser pulses," *Proc. IEEE*, vol. 57, pp. 2–25, January 1969.

[2] K. Gürs and R. Müller, "Breitband-Modulation durch Steuerung der Emission eines Optischen Masers (Auskopple-modulation)," *Phys. Lett.*, vol. 5, pp. 179–181, July 1963.

[2a] K. Gürs, "Beats and modulation in optical ruby masers," *Quantum Electronics III*, P. Grivet and N. Bloembergen, Eds. New York: Columbia University Press, 1964, pp. 1113–1119.

[3] H. Statz and C. L. Tang, "Zeeman effect and nonlinear interactions between oscillating laser modes," *Quantum Electronics III*, P. Grivet and N. Bloembergen, Eds. New York: Columbia University Press, 1964, pp. 469–498.

[4] M. DiDomenico, "Small-signal analysis of internal (coupling type) modulation of lasers," *J. Appl. Phys.*, vol. 35, pp. 2870–2876, October 1964.

[5] L. E. Hargrove, R. L. Fork, and M. A. Pollack, "Locking of He-Ne laser modes induced by synchronous intracavity modulation," *Appl. Phys. Lett.*, vol. 5, pp. 4–5, July 1964.

[6] W. E. Lamb, Jr., "Theory of an optical maser," *Phys. Rev.*, vol. 134, pp. A1429–A1450, June 1964.

[7] M. H. Crowell, "Characteristics of mode-coupled lasers," *IEEE J. Quantum Electron.*, vol. QE-1, pp. 12–20, April 1965.

[8] A. Yariv, "Internal modulation in multimode laser oscillators," *J. Appl. Phys.*, vol. 36, pp. 388–391, February 1965.

[9] S. E. Harris and O. P. McDuff, "FM laser oscillation theory," *Appl. Phys. Lett.*, vol. 5, pp. 205–206, November 1964.

[10] S. E. Harris and R. Targ, "FM oscillation of the He-Ne laser," *Appl. Phys. Lett.*, vol. 5, pp. 202–204, November 1964.

[11] S. E. Harris and O. P. McDuff, "Theory of FM laser oscillation," *IEEE J. Quantum Electron.*, vol. QE-1, pp. 245–262, September 1965.

[12] E. O. Ammann, B. J. McMurtry, and M. K. Oshman, "Detailed experiments on helium–neon FM lasers," *IEEE J. Quantum Electron.*, vol.QE-1, pp. 263–272, September 1965.

[13] A. J. DeMaria and D. A. Stetser, "Laser pulse-shaping and mode-locking with acoustic waves," *Appl. Phys. Lett.*, vol. 7, pp. 71–73, August 1, 1965.

[14] T. Deutsch, "Mode-locking effects in an internally modulated ruby laser," *Appl. Phys. Lett.*, vol. 7, pp. 80–82, August 15, 1965.

[15] H. W. Mocker and R. J. Collins, "Mode competition and self-locking effects in a Q-switched ruby laser," *Appl. Phys. Lett.*, vol. 7, pp. 270–273, November 15, 1965.

[15a] W. W. Rigrod "Mode-locked laser pulse generator," U. S. Patent 3 492 599, filed September 17, 1965.

[16] H. Statz and C. L. Tang, "Phase locking of modes in lasers," *J. Appl. Phys.*, vol. 36, pp. 3923–3927, December 1965.

[17] A. J. De Maria, C. M. Ferrar, and G. E. Danielson, "Mode locking of a Nd^{3+} doped glass laser," *Appl. Phys. Lett.*, vol. 8, pp. 22–24, January 1, 1966.

[18] A. J. DeMaria, D. A. Stetser, and H. Heynau, "Self mode-locking of lasers with saturable absorbers," *Appl. Phys. Lett.*, vol. 8, pp. 174–176, April 1, 1966.

[19] M. DiDomenico, H. M. Marcos, J. E. Geusic, and R. E. Smith, "Generation of ultrashort optical pulses by mode locking the YAG:Nd laser," *Appl. Phys. Lett.*, vol. 8, pp. 180–182, April 1, 1966.

[19a] R. L. Kohn and R. H. Pantell, "Second harmonic enhancement with an internally-modulated ruby laser," *Appl. Phys. Lett.*, vol. 8, pp. 231–233, May 1966.

[20] M. Maier, W. Kaiser, and J. A. Giordmaine, "Intense light bursts in the stimulated Raman effect," *Phys. Rev. Lett.*, vol. 17, pp. 1275–1277, December 26, 1966.

[21] J. A. Armstrong, "Measurement of picosecond laser pulse widths," *Appl. Phys. Lett.*, vol. 10, pp. 16–18, January 1, 1967.

[22] H. P. Weber, "Method for pulse width measurement of ultrashort light pulses generated by phase-locked lasers using nonlinear optics," *J. Appl. Phys.*, vol. 38, pp. 2231–2234, April 1967.

[23] J. A. Giordmaine, P. M. Rentzepis, S. L. Shapiro, and K. W. Wecht, "Two photon excitation of fluorescence by picosecond light pulses," *Appl. Phys. Lett.*, vol. 11, pp. 216–218, October 1, 1967.

[24] H. P. Weber, "Comments on the pulse width measurement with two-photon excitation of fluorescence," *Phys. Lett.*, vol. 27A, pp. 321–322, July 15, 1968.

[25] J. R. Klauder, M. A. Duguay, J. A. Giordmaine, and S. L. Shapiro, "Correlation effects in the display of picosecond pulses by two-photon techniques," *Appl. Phys. Lett.*, vol. 13, pp. 174–176, September 1968.

[26] M. Michon, J. Ernest, and R. Auffret, "Pulsed transmission mode operation in the case of a mode locking of the modes of a non Q-spoiled ruby laser," *Phys. Lett.*, vol. 21, pp. 514–515, June 15, 1966.

[27] A. W. Penney and H. A. Heynau, "PTM single-pulse selection from a mode-locked neodymium 3^+ glass laser using a bleachable dye," *Appl. Phys. Lett.*, vol. 9, pp. 257–258, October 1, 1966.

[28] R. N. Zitter, W. H. Steier, and R. Rosenberg, "Fast pulse dumping and power buildup in a mode-locked He–Ne laser," *IEEE J. Quantum Electron.*, vol. QE-3, pp. 614–617, November 1967.

[28a] G. Kachen, L. Steinmetz, and J. Kysilka, "Selection and amplification of a single mode-locked optical pulse," *Appl. Phys. Lett.*, vol. 13, pp. 229–231, October 1, 1968.

[29] P. W. Smith, "Phase locking of laser modes by continuous cavity length variation," *Appl. Phys. Lett.*, vol. 10, pp. 51–53, January 15, 1967.

[30] S. E. Harris, "Stabilization and modulation of laser oscillators by internal time-varying perturbation," *Proc. IEEE*, vol. 54, pp. 1401–1413, October 1966.

[31] A. G. Fox and P. W. Smith, "Mode-locked laser and the 180° pulse," *Phys. Rev. Lett.*, vol. 18, pp. 826–828, May 15, 1967.

[32] P. W. Smith, "The self-pulsing laser oscillator," *IEEE J. Quantum Electron.*, vol. QE-3, pp. 627–635, November 1967.

[33] H. Statz, "On the conditions for self-locking of modes in lasers," *J. Appl. Phys.*, vol. 38, pp. 4648–4655, November 1967.

[34] H. Statz, G. A. DeMars, and C. L. Tang, "Self-locking of modes in lasers," *J. Appl. Phys.*, vol. 38, pp. 2212–2222, April 1967.

[35] C. L. Tang and H. Statz, "Maximum-emission principle and phase locking in multimode lasers," *J. Appl. Phys.*, vol. 38, pp. 2963–2968, June 1967.

[35a] ——, "Large-signal effects in self-locked lasers," *J. Appl. Phys.*, vol. 39, pp. 31–35, January 1968.

[35b] H. Statz and M. Bass, "Locking in multimode solid-state lasers," *J. Appl. Phys.*, vol. 40, pp. 377–383, January 1969.

[36] D. E. Caddes, L. M. Osterink, and R. Targ, "Mode locking of the CO_2 laser," *Appl. Phys. Lett.*, vol. 12, pp. 74–76, February 1, 1968.

[37] O. R. Wood and S. E. Schwarz, "Passive mode locking of a CO_2 laser," *Appl. Phys. Lett.*, vol. 12, pp. 263–265, April 15, 1968.

[37a] C. M. Ferrar, "Q-switching and mode locking of a CF_3I photolysis laser," *Appl. Phys. Lett.*, vol. 12, pp. 381–383, June 1, 1968.

[38] W. H. Glenn, M. J. Brienza, and A. J. DeMaria, "Mode locking of an organic dye laser," *Appl. Phys. Lett.*, vol. 12, pp. 54–56, January 15, 1968.

[39] D. J. Bradley and A. J. F. Durrant, "Generation of ultrashort dye laser pulses by mode locking," *Phys. Lett.*, vol. 27A, pp. 73–74, June 3, 1968.

[40] H. Samuelson and A. Lempicki, "Q-switching and mode locking of $Nd^{3+}:SeOCl_2$ liquid laser," *J. Appl. Phys.*, vol. 39, pp. 6115–6116, December 1968.

[41] W. Schmidt and F. P. Schafer, "Self-mode-locking of dye-lasers with saturable absorbers," *Phys. Lett.*, vol. 26A, pp. 558–559, April 22, 1968.

[42] V. N. Morozov, V. V. Nikitin, and A. A. Sheronov, "Self-synchronization of modes in a GaAs semiconducting injection laser," *JETP Lett.*, vol. 7, pp. 256–258, May 1968.

[43] A. G. Fox, S. E. Schwarz, and P. W. Smith, "Use of neon as a nonlinear absorber for mode locking a He-Ne laser," *Appl. Phys. Lett.*, vol. 12, pp. 371–373, June 1, 1968.

[43a] V. P. Chebotayev, I. M. Beterov, and V. N. Lisitsyn, "Selection and self-locking of modes in a He-Ne laser with nonlinear absorption," *IEEE J. Quantum Electron.*, vol. QE-4, pp. 788–790, November 1968.

[44] J. Comly, E. Garmire, J. P. Laussade, and A. Yariv, "Observation of mode locking and ultrashort optical pulses induced by anisotropic molecular liquids," *Appl. Phys. Lett.*, vol. 13, pp. 176–178, February 1, 1968.

[45] J. P. Laussade and A. Yariv, "Mode locking and ultrashort laser pulses by anisotropic molecular liquids," *Appl. Phys. Lett.*, vol. 13, pp. 65–66, July 15, 1968.

[46] G. R. Huggett, "Mode-locking of CW lasers by regenerative RF feedback," *Appl. Phys. Lett.*, vol. 13, pp. 186–187, September 1, 1968.

[47] D. H. Auston, "Transverse mode locking," *IEEE J. Quantum Electron.* (Corresp.), vol. QE-4, pp. 420–422, June 1968.

[48] ——, "Forced and spontaneous phase locking of the transverse

modes of a He–Ne laser," *IEEE J. Quantum Electron.* (Corresp.), vol. QE-4, pp. 471–473, July 1968.

[49] P. W. Smith, "Simultaneous phase-locking of longitudinal and transverse laser modes," *Appl. Phys. Lett.*, vol. 13, pp. 235–237, October 1, 1968.

[50] M. Michon, J. Ernest, and R. Auffert, "Passive transverse mode locking in a confocal Nd^{3+} glass laser," *IEEE J. Quantum Electron.* (Corresp.), vol. QE-5, pp. 125–126, February 1969.

[51] A. A. Mak and V. A. Fromzel, "Observation of self-synchronization of transverse modes in a solid-state laser," *JETP Lett.*, vol. 10, pp. 199–201, October 5, 1969.

[52] H. Ito and H. Inaba, "Self mode locking of the transverse modes in the CO_2 laser oscillator," *Opt. Commun.*, vol. 1, pp. 61–63, May 1969.

[53] V. S. Arakelyan, N. V. Karlov, and A. M. Prokhorov, "Self synchronization of transverse modes of a CO_2 laser," *JETP Lett.*, vol. 10, pp. 178–180, September 20, 1969.

[54] G. A. Massey, M. K. Oshman, and R. Targ, "Generation of single frequency light using the FM laser," *Appl. Phys. Lett.*, vol. 6, pp. 10–11, January 1965.

[55] S. E. Harris, M. K. Oshman, B. J. McMurtry, and E. O. Ammann, "Proposed frequency stabilization of the FM laser," *Appl. Phys. Lett.*, vol. 7, pp. 184–186, October 1965.

[56] S. E. Harris and B. J. McMurtry, "Frequency selective coupling to the FM laser," *Appl. Phys. Lett.*, vol. 7, pp. 265–267, November 15, 1965.

[57] M. A. Duguay, K. B. Jefferts, and L. E. Hargrove, "Optical frequency translation of mode-locked laser pulses," *Appl. Phys. Lett.*, vol. 9, pp. 287–290, October 15, 1966.

[58] H. A. Heynau and A. W. Penney, "An application of mode-locked, laser generated picosecond pulses to the electrical measurement art," *1967 IEEE Int. Conv. Rec.*, vol. 15, pt. 7, pp. 80–86.

[59] J. W. Shelton and J. A. Armstrong, "Measurement of the relaxation time of the Eastman 9740 bleachable dye," *IEEE J. Quantum Electron.* (Corresp.), vol. QE-3, pp. 696–697, December 1967.

[59a] M. E. Mack, "Measurement of nanosecond fluorescence decay time," *J. Appl. Phys.*, vol. 39, pp. 2483–2484, April 1968.

[59b] R. I. Scarlet, J. F. Figueira, and H. Mahr, "Direct measurement of picosecond lifetimes," *Appl. Phys. Lett.*, vol. 13, pp. 71–73, July 15, 1968.

[60] J. A. Giordmaine, M. A. Duguay, and J. W. Hansen, "Compression of optical pulses," *IEEE J. Quantum Electron.*, vol. QE-4, pp. 252–255, May 1968.

[61] M. A. Duguay and J. W. Hansen, "Compression of pulses from a mode-locked He–Ne laser," *Appl. Phys. Lett.*, vol. 14, pp. 14–15, January 1, 1969.

[61a] M. J. Brienza, A. J. DeMaria, and W. H. Glenn, "Optical rectification of mode-locked laser pulses," *Phys. Lett.*, vol. 26A, pp. 390–391, March 1968.

[62] P. M. Rentzepis, "Direct measurements of radiationless transitions in liquids," *Chem. Phys. Lett.*, vol. 2, pp. 117–120, June 1968.

[63] M. A. Duguay and J. W. Hansen, "An ultrafast light gate," *Appl. Phys. Lett.*, vol. 15, pp. 192–194, September 1969.

[64] See, for example, N. G. Basov, P. G. Kriukov, S. D. Zakharov, Y. V. Senatsky, and S. V. Tchekalin, "Experiments on the observation of neutron emission at the focus of high-power laser radiation on a lithium deuteride surface," *IEEE J. Quantum Electron.*, vol. QE-4, pp. 864–867, November 1968.

[65] For an excellent review of this subject see, H. Kogelnik and T. Li, "Laser beams and resonators," *Appl. Opt.*, vol. 5, pp. 1550–1567, October 1966.

[66] W. R. Bennett, Jr., "Hole burning effects in a He–Ne laser," *Phys. Rev.*, vol. 126, pp. 580–593, April 15, 1962.

[67] O. L. Gaddy and E. M. Schaefer, "Self locking of modes in the argon ion laser," *Appl. Phys. Lett.*, vol. 9, pp. 281–282, October 15, 1966.

[68] T. Uchida, "Dynamic behavior of gas lasers," *IEEE J. Quantum Electron.*, vol. QE-3, pp. 7–16, January 1967.

[68a] S. E. Schwarz and P. L. Gordon, "Hamilton's principle and the maximum emission coincidence," *J. Appl. Phys.*, vol. 40, pp. 4441–4447, October 1968.

[69] A. Bambini and P. Burlamacchi, "Stability conditions for mode-locked gas lasers," *IEEE J. Quantum Electron.* (Corresp.), vol. QE-4, pp. 101–102, March 1968.

[70] L. N. Magdich, "Nonstationary phenomena in a laser with interacting modes," *Sov. Phys.—JETP*, vol. 26, pp. 492–494, March 1968.

[71] J. A. Fleck, "Emission of pulse trains by Q-switched lasers," *Phys. Rev. Lett.*, vol. 21, pp. 131–133, July 15, 1968.

[72] D. G. C. Jones, M. D. Sayers, and L. Allen, "Mode self-locking in gas lasers," *J. Phys. A, Proc. Phys. Soc. London (Gen)*, ser. 2, vol. 2, pp. 95–101, January 1969.

[73] D. G. C. Jones, "Self-locking of three modes in the He–Ne laser," *Appl. Phys. Lett.*, vol. 13, pp. 301–302, November 1, 1968.

[74] H. Risken and K. Nummedal, "Self-pulsing in lasers," *J. Appl. Phys.*, vol. 39, pp. 4662–4672, September 1968.

[75] For a discussion of frequency pulling of a laser see, for example, R. H. Pantell, "The laser oscillator with an external signal," *Proc. IEEE*, vol. 53, pp. 474–477, May 1965.

[75a] M. Sargent III, "Mode-locking according to the Lamb theory," *IEEE J. Quantum Electron.*, vol. QE-4, p. 346, May 1968.

[76] For a recent discussion of π and 2π pulses and their interaction with an atomic system see S. L. McCall and E. L. Hahn, "Self-induced transparency," *Phys. Rev.*, vol. 183, pp. 457–485, July 10, 1969.

[77] R. E. McClure, "Mode locking behavior of gas lasers in long cavities," *Appl. Phys. Lett.*, vol. 7, pp. 148–150, September 15, 1965.

[77a] V. V. Korbkin and M. Y. Schelev, "Self-locking of axial emission modes of a ruby laser in the free generation regime," *Sov. Phys.—JETP*, vol. 26, pp. 721–722, April 1968.

[78] M. A. Duguay, S. L. Shapiro, and P. M. Rentzepis, "Spontaneous appearance of picosecond pulses in ruby and Nd:glass lasers," *Phys. Rev. Lett.*, vol. 19, pp. 1014–1016, October 30, 1967.

[78a] S. L. Shapiro, M. A. Duguay, and L. B. Kreuzer, "Picosecond substructure of laser spikes," *Appl. Phys. Lett.*, vol. 12, pp. 36–37, January 15, 1968.

[79] M. Bass and D. Woodward, "Observation of picosecond pulses from Nd:YAG lasers," *Appl. Phys. Lett.*, vol. 12, pp. 275–277, April 15, 1968.

[80] T. J. Bridges and P. K. Cheo, "Spontaneous self-pulsing and cavity dumping in a CO_2 laser with electro-optic Q-switching," *Appl. Phys. Lett.*, vol. 14, pp. 262–264, May 1, 1969.

[81] V. I. Malyshev, A. S. Markin, A. V. Masalov, and A. A. Sychov, "On mode locking in ruby and neodymium lasers operating under free oscillation conditions," *Zh. Eksp. Teor. Fiz.*, vol. 57, pp. 834–840, 1969.

[82] T. Uchida and A. Ueki, "Self locking of gas lasers," *IEEE J. Quantum Electron.*, vol. QE-3, pp. 17–30, January 1967.

[83] F. R. Nash, "Observations of spontaneous phase locking of TEM_{00q} modes at 0.63μ," *IEEE J. Quantum Electron.*, vol. QE-3, pp. 189–196, May 1967.

[84] J. Hirano and T. Kimura, "Generation of high-repetition-rate optical pulses by a He–Ne laser," *Appl. Phys. Lett.*, vol. 12, pp. 196–198, March 1, 1968.

[85] J. Hirano and T. Kimura, "Multiple mode locking of lasers," *IEEE J. Quantum Electron.*, vol. QE-5, pp. 219–224, May 1969.

[86] M. DiDomenico and V. Czarniewski, "Locking of He–Ne laser modes by intracavity acoustic modulation in coupled interferometers," *Appl. Phys. Lett.*, vol. 6, pp. 150–152, April 15, 1965.

[86a] C. M. Ferrar, "Mode-locked flashlamp-pumped coumarin dye laser at 4600 Å," *IEEE J. Quantum Electron.* (Corresp.), vol. QE-5, pp. 550–551, November 1969.

[87] L. C. Foster, M. D. Ewy, and C. B. Crumly, "Laser mode locking by an external Doppler cell," *Appl. Phys. Lett.*, vol. 6, pp. 6–8, January 1, 1965.

[88] M. Michon, J. Ernest, and R. Auffret, "Mode locking of a Q-spoiled Nd^{3+} doped glass laser by intracavity phase modulation," *Phys. Lett.*, vol. 23, pp. 457–458, November 14, 1966.

[89] G. A. Massey, "Laser mode control by internal modulation using the transverse electrooptic effect in quartz," *Appl. Opt.*, vol. 5, pp. 999–1001, June 1966.

[90] R. H. Pantell, H. E. Puthoff, B. G. Huth, and R. L. Kohn, "Mode coupling in an external Raman resonator," *Appl. Phys. Lett.*, vol. 9, pp. 104–106, August 1, 1966.

[91] W. C. Henneberger and H. J. Schulte, "Optical pulses produced by laser length variation," *J. Appl. Phys.* (Communications), vol. 37, p. 2189, April 1966.

[92] R. H. Pantell and R. L. Kohn, "Mode coupling in a ruby laser," *IEEE J. Quantum Electron.*, vol. QE-2, pp. 306–310, August 1966.

[93] K. Gürs, "Modulation and mode locking of the continuous ruby laser," *IEEE J. Quantum Electron.*, vol. QE-3, pp. 175–180, May 1967.

[94] V. Degiorgio and B. Querzola, "Output characteristics of a mode-locked laser," *Nuovo Cimento*, vol. 55B, pp. 272–275, May 11, 1968.

[95] L. M. Osterink and J. D. Foster, "A mode-locked Nd:YAG laser," *J. Appl. Phys.*, vol. 39, pp. 4163–4165, August 1968.

[96] R. Targ and J. M. Yarborough, "Mode-locked quieting of He–Ne and argon lasers," *Appl. Phys. Lett.*, vol. 12, pp. 3–4, January 1, 1968.

[97] T. R. Gurski, "Simultaneous mode-locking and second-harmonic

generation by the same nonlinear crystal," *Appl. Phys. Lett.*, vol. 15, pp. 5–6, July 1, 1969.

[98] L. B. Allen, R. R. Rice, and R. F. Mathews, "Two cavity mode-locking of a He-Ne laser," *Appl. Phys. Lett.*, vol. 15, pp. 416–418, December 15, 1969.

[99] T. Uchida, "Direct modulation of gas lasers," *IEEE J. Quantum Electron.*, vol. QE-1, pp. 336–343, November 1965.

[100] G. W. Hong and J. R. Whinnery, "Switching of phase-locked states in the intracavity phase-modulated He-Ne laser," *IEEE J. Quantum Electron.*, vol. QE-5, pp. 367–376, July 1969.

[101] A. Yariv, "Parametric interactions of optical modes," *IEEE J. Quantum Electron.*, vol. QE-2, pp. 30–37, February 1966.

[102] L. N. Magdich, "Laser mode interaction in the course of Q-switching," *Soviet Phys.—JETP*, vol. 24, pp. 11–15, January 1967.

[103] O. P. McDuff, "Internal modulation of lasers," *1967 IEEE Region 3 Conv.*, pp. 121–132.

[104] O. P. McDuff and S. E. Harris, "Nonlinear theory of the internally loss-modulated laser," *IEEE J. Quantum Electron.*, vol. QE-3, pp. 101–111, March 1967.

[105] L. N. Magdich, "Laser mode synchronization by dielectric constant modulation," *Soviet Phys.—JETP*, vol. 25, pp. 223–226, August 1967.

[106] Y. P. Yegorov and A. S. Petrov, "Axial mode locking in internally modulated gas lasers," *Radio Eng. Electron Phys. (USSR)*, vol. 12, pp. 1365–1373, August 1967.

[107] H. Haken and M. Pauthier, "Nonlinear theory of multimode action in loss modulated lasers," *IEEE J. Quantum Electron.*, vol. QE-4, pp. 454–459, July 1968.

[108] V. S. Letokhov, "Dynamics of generation of pulsed mode-locking lasers," *Sov. Phys.—JETP*, vol. 27, pp. 746–751, November 1968.

[109] O. P. McDuff and A. L. Pardue, Jr., "Theory of laser mode coupling produced by cavity-length modulation," *IEEE J. Quantum Electron.* (Corresp.), vol. QE-4, pp. 99–101, March 1968.

[110] P. J. Titterton, "Quantum theory of an internally phase-modulated laser—I.," *IEEE J. Quantum Electron.*, vol. QE-4, pp. 85–92, March 1968.

[111] A. E. Siegman and D. J. Kuizenga, "Simple analytic expressions for AM and FM mode-locked pulses in homogeneous lasers," *Appl. Phys. Lett.*, vol. 14, pp. 181–182, March 15, 1969.

[112] J. B. Gunn, "Spectrum and width of mode-locked laser pulses," *IEEE J. Quantum Electron.*, vol. QE-5, pp. 513–516, October 1969.

[113] A. W. Smith, "Effect of host dispersion on mode-locked lasers," *Appl. Phys. Lett.*, vol. 15, pp. 194–196, September 15, 1969.

[114] D. A. Stetser and A. J. DeMaria, "Optical spectra of ultrashort optical pulses generated by mode-locked glass/neodymium lasers," *Appl. Phys. Lett.*, vol. 9, pp. 118–120, August 1, 1966.

[115] A. J. DeMaria, "Ultrashort laser pulses by simultaneously mode-locking and Q-switching Nd^{3+} glass lasers," *1967 NEREM Rec.*, pp. 34–35, 1967.

[116] A. J. DeMaria, D. A. Stetser, and W. H. Glenn, "Ultrashort light pulses," *Science*, vol. 156, pp. 1557–1568, 1967.

[117] V. I. Malyshev, A. S. Markin, and A. A. Sychev, "Mode self-synchronization in giant pulse of a ruby laser with broad spectrum," *JETP Lett.*, vol. 6, pp. 34–35, July 15, 1967.

[118] A. Schmackpfeffer and H. Weber, "Mode locking and mode competition by saturable absorbers in a ruby laser," *Phys. Lett.*, vol. 24A, pp. 190–191, January 30, 1967.

[119] R. Harrach and G. Kachen, "Pulse trains from mode-locked lasers," *J. Appl. Phys.*, vol. 39, pp. 2482–2483, April 1968.

[120] M. E. Mack, "Mode locking the ruby laser," *IEEE J. Quantum Electron.* (Corresp.), vol. QE-4, pp. 1015–1016, December 1968.

[121] A. R. Clobes and M. J. Brienza, "Passive mode locking of a pulsed Nd:YAG laser," *Appl. Phys. Lett.*, vol. 14, pp. 287–288, May 1, 1969.

[121a] P. C. Magnante, "Mode-locked and bandwidth narrowed Nd: glass laser," *J. Appl. Phys.*, vol. 40, pp. 4437–4440, October 1969.

[122] R. Cubeddu, R. Polloni, C. A. Sacchi, and O. Svelto, "Picosecond pulses, TEM_{00} mode, mode-locked ruby laser," *IEEE J. Quantum Electron.* (Corresp.), vol. QE-5, pp. 470–471, September 1969.

[123] D. J. Bradley, A. J. F. Durrant, F. O'Neill, and B. Sutherland, "Picosecond pulses from mode-locked dye lasers," *Phys. Lett.*, vol. 30A, pp. 535–536, December 29, 1969.

[123a] L. D. Derkachyova, A. I. Krymova, V. I. Malyshev, and A. S. Markin, "Mode locking in polymethylene dye lasers," *Opt. Spect.*, vol. 26, pp. 572–573, June 1969.

[124] J. H. McCoy, "Continuous passive mode locking of a CO_2 laser,"

Appl. Phys. Lett., vol. 15, pp. 357–360, December 1, 1969.

[125] C. C. Cutler, "The regenerative pulse generator," *Proc. IRE*, vol. 43, pp. 140–148, February 1955.

[126] E. M. Garmire and A. Yariv, "Laser mode-locking with saturable absorbers," *IEEE J. Quantum Electron.*, vol. QE-3, pp. 222–226, June 1967; "Correction," *IEEE J. Quantum Electron.*, vol. QE-3, p. 377, September 1967.

[126a] J. A. Fleck, "Mode-locked pulse generation in passively switched lasers," *Appl. Phys. Lett.*, vol. 12, pp. 178–181, March 1, 1968.

[126b] ——, "Origin of short-pulse emission by passively switched lasers," *J. Appl. Phys.*, vol. 39, pp. 3318–3327, June 1968.

[127] V. S. Letokhov, "Formation of ultrashort pulses of coherent light," *JETP Lett.*, vol. 7, pp. 25–28, January 15, 1968.

[128] C. A. Sacchi, G. Soncini, and O. Svelto, "Self-locking of modes in a passive Q-switched laser," *Nuovo Cimento*, vol. 48B, pp. 58–71, March 11, 1967.

[129] V. S. Letokhov, "Generation of ultrashort light pulses in a laser with a nonlinear absorber," *Sov. Phys.—JETP*, vol. 28, pp. 562–568, 1969.

[129a] T. I. Kuznetsova, V. I. Malyshev, and A. S. Markin, "Self-synchronization of axial modes of a laser with saturable filters," *Soviet Phys.—JETP*, vol. 25, pp. 286–291, August 1967.

[130] R. R. Cubeddu and O. Svelto, "Theory of laser self-locking in the presence of host dispersion," *IEEE J. Quantum Electron.*, vol. QE-5, pp. 495–502, October 1969.

[130a] ——, "Effect of dispersion on laser self-locking," *Phys. Lett.*, vol. 29A, pp. 78–79, April 7, 1969.

[131] S. E. Schwarz, "Theory of an optical pulse generator," *IEEE J. Quantum Electron.*, vol. QE-4, pp. 509–514, September 1968.

[131a] V. S. Letokhov and V. N. Marzov, "Generation of ultrashort duration coherent light pulses," *Sov. Phys.—JETP*, vol. 25, pp. 862–866, November 1967.

[132] A. Bambini and P. Burlamacchi, "Phase locking of a multimode gas laser by means of low-frequency cavity-length modulation," *J. Appl. Phys.*, vol. 39, pp. 4864–4865, September 1968.

[133] J. P. Laussade and A. Yariv, "Analysis of mode locking and ultrashort laser pulses with a nonlinear refractive index," *J. Quantum Electron.*, vol. QE-5, pp. 435–441, September 1969.

[134] B. H. Soffer and J. W. Luin, "Continuously tunable picosecond pulse organic dye laser," *J. Appl. Phys.*, vol. 39, pp. 5859–5860, December 1968.

[135] D. Herriott, H. Kogelnik, and R. Kompfner, "Off-axis paths in spherical mirror interferometers," *Appl. Opt.*, vol. 3, pp. 523–526, April 1964.

[136] K. Kohiyama, T. Fujioka, and M. Kobayashi, "Self-locking of transverse higher-order modes in a He-Ne laser," *Proc. IEEE* (Letters), vol. 56, pp. 333–335, March 1968.

[137] H. E. Rowe and T. Li, "Theory of two-photon measurement of laser output," *IEEE J. Quantum Electron.*, vol. QE-6, pp. 49–67, January 1970.

[138] H. P. Weber and R. Dändliker, "Intensity interferometry by two-photon excitation of fluorescence," *IEEE J. Quantum Electron.*, vol. QE-4, pp. 1009–1013, December 1968.

[138a] A. A. Grutter, H. P. Weber, and R. Dändliker, "Imperfectly mode-locked laser emission and its effects on nonlinear optics," *Phys. Rev.*, vol. 185, pp. 629–643, September 10, 1969.

[139] See, for example, A. A. Malyutin and M. Y. Shchelev, "Investigation of the temporal structure of neodymium-laser emission in the mode self-locking regime," *JETP Lett.*, vol. 9, pp. 266–268, April 20, 1969.

[140] M. Bass and K. Andringa, "Reproducible optical second-harmonic generation using a mode-locked laser," *IEEE J. Quantum Electron.*, vol. QE-3, pp. 621–626, November 1967.

[141] W. H. Glenn, Jr., and M. J. Brienza, "Time evolution of picosecond optical pulses," *Appl. Phys. Lett.*, vol. 10, pp. 221–224, April 15, 1967.

[142] I. K. Krasyuk, P. P. Pashkin, and A. M. Prokhorov, "Ring ruby laser for ultrashort pulses," *Soviet Phys.—JETP* (Lett.), vol. 7, pp. 89–91, February 1968.

[143] C. C. Wang and E. L. Baardsen, "Optical third harmonic generation using mode-locked and non mode-locked lasers," *Appl. Phys. Lett.*, vol. 15, pp. 396–398, November 15, 1969.

[143a] H. P. Weber and R. Dändliker, "Method for measuring the shape asymmetry of picosecond light pulses," *Phys. Lett.*, vol. 28A, pp. 77–78, November 4, 1968.

[143b] E. I. Blount and J. R. Klauder, "Recovery of laser intensity from

correlation data," *J. Appl. Phys.*, vol. 40, pp. 2874–2875, June 1969.

[144] R. C. Eckardt and C. H. Lee, "Optical third harmonic measurements of subpicosecond light pulses," *Appl. Phys. Lett.*, vol. 15, pp. 425–427, December 15, 1969.

[145] D. M. Thymian and J. A. Carruthers, "Second-harmonic enhancement using a self-locked 0.63-μ He–Ne laser," *IEEE J. Quantum Electron.*, vol. QE-5, pp. 83–86, February 1969.

[146] M. A. Duguay and J. W. Hansen, "Optical frequency shifting of a mode-locked laser beam," *IEEE J. Quantum Electron.*, vol. QE-4, pp. 477–481, August 1968.

[147] C. G. B. Garrett and M. A. Duguay, "Theory of the optical frequency translator," *Appl. Phys. Lett.*, vol. 9, p. 374, November 15, 1966.

[148] E. B. Treacy, "Compression of picosecond light pulses," *Phys. Lett.*, vol. 28A, pp. 34–35, October 1968.

[148a] ——, "Measurement of picosecond pulse substructure using compression techniques," *Appl. Phys. Lett.*, vol. 14, pp. 112–114, February 1, 1969.

[149] ——, "Optical pulse compression with diffraction gratings," *IEEE J. Quantum Electron.*, vol. QE-5, pp. 454–458, September 1969.

[150] R. A. Fisher and J. A. Fleck, Jr., "On the phase characteristics and compression of picosecond pulses," *Appl. Phys. Lett.*, vol. 15, pp. 287–290, November 1, 1969.

[151] P. W. Smith, "On the phase relationship between oscillating modes in a mode-locked 6328 Å laser," to be published.

[151a] R. A. Fisher, P. L. Kelley, and T. K. Gustafson, "Subpicosecond pulse generation using the optical Kerr effect," *Appl. Phys. Lett.*, vol. 14, pp. 140–143, February 15, 1969.

[151b] A. Laubereau, "External frequency modulation and compression of picosecond pulses," *Phys. Lett.*, vol. 29A, pp. 539–540, July 28, 1969.

[152] L. E. Hargrove, "Reduction of mode-locked-laser pulse duration by interferometric combination of frequency-shifted pulses," *J. Opt. Soc. Am.*, vol. 59, pp. 1680–1681, December 1969.

[153] I. D. Abella, N. A. Kurnit, and S. R. Hartmann, "Photon echoes," *Phys. Rev.*, vol. 141, pp. 391–406, January 1966.

[154] A. Frova, M. A. Duguay, C. G. B. Garrett, and S. L. McCall, "Pulse delay effects in the He–Ne laser mode-locked by a Ne absorption cell," *J. Appl. Phys.*, vol. 40, pp. 3969–3972, September 1969.

[154a] P. W. Smith, "Pulse velocity in a resonant absorber," *IEEE J. Quantum Electron.*, vol. QE-6, pp. 416–422, July 1970.

[155] R. T. Denton and T. S. Kinsel, "Terminals for a high-speed optical pulse code modulation communication system: I. 224-Mbit/s single channel," *Proc. IEEE*, vol. 56, pp. 140–145, February 1968.

[155a] T. S. Kinsel and R. T. Denton, "Terminals for a high-speed optical pulse code modulation communication system: II. optical multiplexing and demultiplexing," *Proc. IEEE*, vol. 56, pp. 146–154, February 1968.

[156] T. S. Kinsel, "Light wave of the future: optical PCM," *Electronics*, pp. 123–128, September 16, 1968.

[157] F. R. Faxvog, C. R. Willenbring, and J. A. Carruthers, "Self-pulsing in the He-Cd laser," *Appl. Phys. Lett.*, vol. 16, pp. 8–10, January 1, 1970.

[158] W. T. Silfvast and P. W. Smith, "Mode-locking the He-Cd laser at 4416 Å and 3250 Å," *Appl. Phys. Lett.*, vol. 17, pp. 70–73, July 15, 1970.

[159] G. W. Gobeli, J. C. Bushnell, P. S. Peercy, and E. D. Jones, "Observation of neutrons produced by laser irradiation of lithium deuteride," *Phys. Rev.*, vol. 188, pp. 300–302, December 1969.

[160] H. Kogelnik and W. W. Rigrod, "Visual display of isolated optical-resonator modes," *Proc. IRE* (Corresp.), vol. 50, p. 220, February 1962.

[161] J. W. Hansen, "Optical pulse compression," *Proc. Nat. Electron. Conf.*, vol. 24, pp. 148–151, 1968.

[162] M. A. Duguay, J. W. Hansen, and S. L. Shapiro, "Study of the Nd:glass laser radiation by means of two-photon fluorescence," to be published.

[163] V. S. Letokhov, "Ultrashort fluctuation pulses of light in a laser," *Soviet Phys.—JETP*, vol. 28, pp. 1026–1027, May 1969.

[164] T. I. Kuznetsova, "Concerning the problem of registration of ultrashort light pulses," *Soviet Phys.—JETP*, vol. 28, pp. 1303–1305, June 1969.

[165] R. H. Picard and P. Schweitzer, "Theory of intensity-correlation measurements on imperfectly mode-locked lasers," *Phys. Rev.*, vol. 1, pp. 1803–1819, June 1970.

Picosecond Laser Pulses

A. J. DeMARIA, MEMBER, IEEE, WILLIAM H. GLENN, JR., MEMBER, IEEE,
MICHAEL J. BRIENZA, AND MICHAEL E. MACK

Invited Paper

Abstract—The broad bandwidth and long storage lifetimes of Nd^{3+}: glass and ruby lasers have made possible the generation of picosecond laser pulses having peak powers in excess of one gigawatt and repetition rates in the microwave range. The numerous application areas of these pulses include research in nonlinear optics, transient response of atomic and molecular systems, optically generated plasmas, spectroscopy, ranging, optical information processing, and high-speed photography. This paper reviews several experimental techniques for generating, measuring, and utilizing these ultrashort laser pulses.

I. INTRODUCTION

IN 1961 HELLWARTH proposed an experimental technique for generating large output bursts of radiation from laser devices [1]. This experimental technique, now called laser Q-switching, was first achieved by McClung and Hellwarth in 1962 with a Kerr cell [2], by Collins and Kisliuk with a rotating disk [3], and by DeMaria, Gagosz, and Barnard with an ultrasonic-refraction shutter [4]. Rotating mirrors and prisms, Pockels cells, and saturable absorbers have also been utilized in Q-switching experiments [5]–[7]. The availability of these Q-switched optical pulses has made possible the experimental investigation of such phenomena as optically generated plasmas [8], optical harmonic generation [9]–[10], stimulated Raman, Brillouin, and Rayleigh-wing scattering [11],[1] photon echoes [12], self-induced optical transparency [13], optical self-trapping [14], and optical parametric amplification [15]. Examples of applications of these short-duration, high-peak-power Q-switched pulses include semiactive guidance, ranging, illumination, high-speed photography and holography, and material working and removal.

The minimum pulse widths obtainable with existing Q-switching techniques are limited to approximately 10^{-8} s because of the required pulse buildup time. Peak powers of approximately 5×10^8 W without any additional stages of amplification have been obtained with various straightforward, Q-switching, experimental arrangements. This paper will review the generation, measurement, and utiliza-

tion of what is believed to be the second generation of short-time-duration, high-peak-power laser pulses, i.e., laser pulses having picosecond time duration and peak power in excess of 10^9 W [16].

There are numerous reasons why researchers have become interested in the generation of picosecond light pulses with gigawatts of peak power. Picosecond light pulses can be conveniently generated in laser media having wide spectral bandwidths and long fluorescent lifetimes, such as ruby or Nd:glass. The availability of optical pulses of such high power and short time duration has aroused considerable interest among military, academic, and industrial researchers. For example, a pulse of 10^{-12} s at a wavelength of 1 micron has a length in free space of 0.03 cm and therefore offers the possibility of measuring long distances to fractions of millimeters. An event would have to move an appreciable fraction of the velocity of light in order to realize the full potentials of such pulses in high-speed photography applications. Electrical pulses having a rise time of less than 10^{-10} s, amplitudes of 60–100 V, and repetition periods as short as 1.5 ns have been obtained through the use of fast photodiode detectors. These electrical pulses were previously unattainable and should find application for determining the location and severity of internal reflections in wide-bandwidth transmission systems, studying propagation delay, transient response of wide-bandwidth systems, etc. The application of high-energy picosecond pulses to controlled thermonuclear plasma, optical radar, optical information processing, spectroscopy, nonlinear optical properties of materials, transient response of quantum systems, and ultrashort acoustic shock research appears very promising.

II. GENERATION OF PICOSECOND LASER PULSES

A. Basic Operating Principles

It is well known that a feedback loop, as illustrated by Fig. 1(a), encompassing an amplifier, a filter, a delay line, and a nonlinear element that provides less attenuation for a high-level signal than for a low-level signal, behaves as a regenerative pulse generator [17]. When the loop gain exceeds unity, a pulse recirculates indefinitely around the loop and each traversal gives rise to an output pulse at the output terminal. It is evident that such a pulse would soon be degraded unless the effects of noise and distortion can be counteracted. The nonlinear element (called an "expandor" by Cutler [17]) has the effect of 1) emphasizing the peak region of the recirculating pulse while reducing the lower amplitude regions, 2) discriminating against noise

Manuscript received June 17, 1968; revised October 28, 1968. The technical concepts and achievements discussed here have evolved over a period of two years, under the sponsorship of the U. S. Air Force Cambridge Laboratories, Office of Aerospace Research; Project Defender (under the joint sponsorship of the Advanced Research Project Agency, the Office of Naval Research, and the Department of Defense); and the U. S. Army Missile Command. *This invited paper is one of a series planned on topics of general interest.—The Editor.*

The authors are with the United Aircraft Research Laboratories, East Hartford, Conn. 06108

[1] Stimulated thermal Rayleigh scattering has also been observed with Q-switched ruby pulses. See D. H. Rank *et al.*, "Stimulated thermal Rayleigh scattering," *Phys. Rev. Lett.*, vol. 19, pp. 828–830, October 1967.

Reprinted from *Proc. IEEE*, vol. 57, pp. 2–25, Jan. 1969.

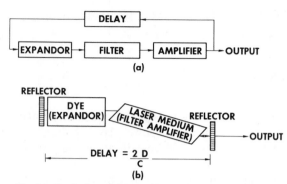

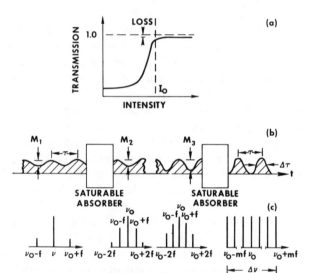

Fig. 1. Equivalence of (a) an electronic and (b) an optical regenerative pulse generator.

Fig. 2. Operation of a saturable absorber in the time and frequency domain.

and reflections, and 3) acting to shorten the pulse until the pulse width is limited by the frequency response of the circuit. The output of the regenerative oscillator has a pulse rate equal to the reciprocal of the loop delay, pulse widths equal to the reciprocal of the overall system bandwidth, and a center frequency determined by the filter frequency. Utilizing this technique, Cutler was able to generate microwave pulses having a carrier frequency of 4 GHz and pulse widths of 2 ns.

An ordinary laser possesses all the basic elements of the regenerative pulse generator operated in the microwave region with the exception of the expandor element. The laser medium serves as the amplifier, the combination of the Fabry-Perot resonances and the line width of the laser transition serve as the filter, and the time required for an optical pulse to traverse twice the distance between the reflectors serves as the loop-time-delay [see Fig. 1(b)]. The optical analog of the electronic expandor circuit element is a saturable absorber, such as the reversible bleachable dye solutions commonly used as laser Q-switches [18]. The fundamental requirements of the saturable absorber are 1) that it have an absorption line at the laser wavelength, 2)

that it have a line width equal to or greater than the laser line width, and 3) that the dye recovery time be shorter than the loop-time-delay of the laser.

A simplified explanation of the operation of the regenerative pulse laser oscillator illustrated by Fig. 1(b) can be given with the aid of Fig. 2. If an optical carrier frequency v_0 along with two sidebands at $v_0 \pm f$ are superimposed, an amplitude modulation of the light results at a frequency $f = \Delta f$ with some peak-to-peak variation M_1 and a peak intensity I_0 [see Figs. 2(b) and 2(c)]. When this beam is passed through a saturable absorber having the typical characteristics illustrated by Fig. 2(a), the initial sinusoidal amplitude fluctuation of the input beam will be found to be distorted, the peak-to-peak excursions of the fluctuation will be increased, i.e., $M_2 > M_1$, and the time duration of the fluctuation will be shorter as a result of the nonlinear transmission characteristics of the saturable absorbers. With the sharpening of the amplitude variation, additional sidebands are added to the spectrum. In an optical cavity of length $L = c/2\Delta f$ this process is repeated over and over again by reflecting the light beam back and forth between two mirrors placed on both sides of the saturable absorber cell. The fluctuation will continue to sharpen until a discrete pulse is circulating in the cavity. The laser media provide gain to compensate for the residual saturation loss of the absorber and the mirrors. The circulation rate Δf is given by $c/2L$. The pulse will eventually acquire a steady-state width $\Delta \tau$ determined by the bandwidth of the laser media. The repetitive output pulse train emitted from the laser mirrors will have discrete spectral components defined by the Fabry-Perot resonances of the cavity extending $\pm m\Delta f$ on either side of v_0 for a bandwidth Δv, as illustrated by Fig. 2(c). This result is to be expected since it is well known, from Fourier's theorem, that any repetitive pulse train can be represented by a series of discrete sinusoidal functions having integrally related frequencies and fixed phase relationships. The frequencies are all multiples of Δf, and the narrower the pulse width $\Delta \tau$, the larger the bandwidth required to reproduce the repetitive pulse [16], [19], [20], i.e., $\Delta \tau \simeq 1/\Delta v$.

The system schematically shown in Fig. 1(a) lends itself to simplified analysis by repeated application of Fourier transforms, applying the frequency or amplitude characteristics of each element, and finally equating the characteristics of the returning signal to the characteristics of the assumed initial signal. Consider a signal entering the expandor given by $S_1(t)e^{j\psi t}$. The signal leaving the expandor is $S_2(T) = K[S_1(t)]^n e^{j\psi t}$, where K is a constant indicating an amplitude change of the signal and the superscript can be taken to indicate a nonlinear operation, not necessarily a power law [17]. Notice that the nonlinear operation was performed only on the envelope portion of the signal and not on the phase portions. If this signal is now passed through a filter having a frequency function $F(\omega)$, the output from the filter has the form $F_3(\omega) = F(\omega)F_2(\omega)$, where

$$F_2(\omega) = \int_{-\infty}^{\infty} S_2(t)e^{-j\omega t}dt \quad (1)$$

is the transform of $S_2(t)$ to the frequency domain. The transformation of the signal $F_3(\omega)$ into the time domain gives

$$S_3(t) = \frac{1}{2\pi} \int_{-\infty}^{\infty} F_3(\omega) e^{j\omega t} d\omega. \tag{2}$$

The amplifier and circuit losses give a net gain G so that the output of the amplifier is $S_4(t) = GS_3(t)$. The signal $S_5(t)$ leaving the delay element is delayed by τ, and brings us back to the expandor input. We now require that $S_5(t)$ be equal to $S_1(t)$, except possibly for instantaneous phases, so that

$$S_1(t)e^{j\psi t} = \frac{GKe^{-j\theta}}{2\pi} \int_{-\infty}^{\infty} F(\omega) e^{j\omega(t-\tau)} d\omega$$
$$\cdot \int_{-\infty}^{\infty} [S_1(t-\tau)]^{\eta} e^{-j\omega(t-\tau)} e^{j(\psi t - \tau)} dt \tag{3}$$

where θ is the phase shift of the optical wave relative to the pulse time. Given a filter characteristic $F(\omega)$, an expandor nonlinear law η, and time delay τ, then (3) specifies the time function $S_1(t)e^{j\psi t}$.

In the event a function $S_5(t)$ is found to be a replica of $S_1(t)$, a solution to (3) has been found. Gaussian functions are well suited for such types of solution for (3) if a power law for η is assumed. Cutler assumed Gaussian characteristics for the pulse envelope and filter functions and a power law for η [17]. Under these assumptions, the solution for (3) gives the pulse rate Δf to be equal to the reciprocal of the group time delay $1/\tau = 2\pi/\beta = d\omega/d\phi$ around the loop, and the pulse width $\Delta\tau$ to be equal to

$$\Delta\tau = \frac{4}{\pi\Delta\omega(1 - 1/\eta)^{1/2}}$$
$$\cdot \left[\frac{2\left(1 + \frac{\gamma^2\Delta\omega^4}{16}\right)}{1 + \left(1 + \frac{\eta\gamma^2\Delta\omega^4}{4(\eta^2 + 2\eta + 1)}\right)^{1/2}} \right]^{1/2}, \tag{4}$$

where $\Delta\omega$ is the bandwidth at the 1 neper point of a Gaussian filter, and β and γ are the coefficients of linear and quadratic terms in the power series expansion of the loop's phase shift $\phi = \alpha + \beta(\omega - \omega_0) + \gamma(\omega - \omega_0)^2 + \cdots$. This power series expansion of the phase in the feedback loop can be obtained by taking a Taylor series expansion of the refractive index of the media within the feedback loop,

$$n = n_0 + \frac{dn_0}{d\nu}(\nu - \nu_0) + \frac{d^2n_0}{d\nu^2}(\nu - \nu_0)^2 + \cdots \tag{5}$$

and substituting into the phase relation $\phi = 2\pi Ln/c$, where L is the length of the media within the feedback loop. The existence of the squared frequency phase term causes the instantaneous frequency to change through the pulse as given by $\nu = \nu_0 + (bt/\pi)$, where b/π is the sweep rate. The linear frequency sweep rate is given by

$$-\frac{b}{\pi} = \frac{\pi(\eta - 1)}{\eta + 1} \frac{\gamma}{(2/\Delta\omega)^4 + \gamma^2}. \tag{6}$$

For $\gamma = 0$, the pulse width is a minimum; $\Delta\tau_{\min} \simeq 1/\Delta\nu$. A large phase curvature gives rise to a pulse whose instantaneous frequency sweeps through a bandwidth larger than the reciprocal of the pulse width; or $\Delta\tau = \Delta\tau_{\min}(\eta^2 + 1)/\eta^{1/2}$. The phase measured with respect to the pulse time, in general, is found to change from pulse to pulse in this analysis. In the observation of a series of pulses, the analysis shows that the phase would appear to move continuously through the pulse at a rate θ/τ rad/s, where θ is the phase shift change between pulses. The analysis shows that the frequency components of the repetitive pulse train are given by

$$\nu_m = N/\tau - \theta/(2\pi\tau), \tag{7}$$

where N is the number of carrier half-wavelengths in the loop. Equation (7) shows that the frequency components are not exactly harmonically related. The phase is not found to be a continuous extrapolation from previous pulses as is obtained by modulating a CW signal; nor is it directly related to the pulse time as would be obtained by shock exciting an oscillatory circuit; nor is it random and uncorrelated as is obtained from pulsed oscillators.

The frequency sweep or "chirp" described by Cutler for the microwave regenerative pulse generator case has recently been experimentally observed by E. B. Treacy [21], [22] in Nd^{3+}: glass lasers. The discovery that picosecond pulses from dye mode-locked glass: Nd^{3+} lasers are chirped explains the order of magnitude discrepancy observed between the spectral width and the pulse widths of Nd: glass regenerative pulse generators [16]. Treacy used the dispersive characteristics of two cascaded optical gratings to compress 4×10^{-12} s pulses from a dye mode-locked glass: Nd^{3+} laser down to 4×10^{-13} s [23]. These compressed pulses consisted of just over 100 optical cycles. It is not known at this time whether the laser or the dye medium is primarily responsible for the chirp characteristic of the output pulses of dye mode-locked glass: Nd^{3+} lasers.

A quantitative feeling of the magnitude of the sharpening experienced by an optical pulse in passing through a saturable absorber can be obtained by the simple calculations given below. Assume a pulse whose time duration is long compared to a saturable absorber relaxation time. The absorber will act as an intensity dependent absorber. Such a case has been treated by Hercher [24]. The intensity transport equation is given by

$$\frac{dI_{(x)}}{dt} = -\frac{\alpha_0 I_{(x)}}{1 + I_{(x)}/I_s} \tag{8}$$

where the propagation time has been neglected, α_0 is the small-signal absorption coefficient, and I_s is a saturation intensity, i.e., the intensity at which the absorption coefficient of the dye is reduced to one-half of its small-signal value. The solution to this equation is

$$I'_{(x)} e^{I_{(x)}} = TI'_{(0)} e^{I_{(0)}} \tag{9}$$

where T is the small-signal transmission of the finite length dye cell, and $I'(0)$ and $I'(x)$ are the normalized intensities

227

at the entrance and exit as a function of distance in the dye cell, respectively.

A computer experiment can be performed to illustrate the operation of Fig. 2 using (9). Fig. 3 illustrates the pulse shape normalized to unity intensity after every tenth pass of a trial Gaussian-shaped pulse initially having a peak intensity of $0.1\,I_s$ through a saturable absorber cell having a small-signal transmission of $T = 0.7$. The absorber is most effective in sharpening when the intensity is in the neighborhood of I_s. In the actual operation of a laser, the absorber concentration is much weaker and the transmission is correspondingly higher, of the order of 0.95.

It is also of interest to obtain a quantitative feeling of the magnitude of the sharpening when an optical pulse is propagated through a two-level system in which the duration of the optical pulse is less than the relaxation time of the absorber. A rate equation treatment of this case has been given by several authors [25], [26]. Their solution is applicable to the saturable gain in the laser medium as well as the saturable absorption in the absorber. When the relaxation time is assumed infinite, an exact solution of the rate equations is possible. For this case, it is more appropriate to deal with the integrated pulse intensity

$$U(t) = \int_{-\infty}^{t} I(t)\,dt \tag{10}$$

rather than the intensity. The solution for the saturable two-level medium is

$$W_{\text{out}}(t) = \ln\left[1 + G(e^{W_{\text{in}}(t)} - 1)\right], \tag{11}$$

when the normalization $W(t) = U(t)2\sigma/h\nu$ is used, σ is the cross section per active atom, and $G = \exp[\alpha_0 L]$ is the small-signal gain or attenuation of the medium.

Fig. 4 illustrates the computer simulation of a laser-medium saturable-absorber combination utilizing (11) for both the active medium and the absorber. The cross section σ_L of the laser atoms was taken to be much smaller than for the absorber molecules σ_a, i.e., $h\nu/2\sigma_L = 16$ J/cm while $h\nu/2\sigma_a = 0.016$ J/cm. The loss due to the reflectivity of the laser mirrors was assumed to be 0.2, the initial laser gain 2.0, the absorber transmission 0.8, and the starting Gaussian pulse energy 1.6×10^{-3} J/cm. Fig. 4(a) illustrates the amplitude increase of the pulse and the gain decrease of the laser medium as a function of time. The absorber was assumed to relax to its initial state between the successive passes of the pulse but the laser gain decreases as a result of all pulses that have passed through it. The amplitudes of the pulse train for the first 30 passes are small, in the range of 0.1 to 1.0 $h\nu/2\sigma_a$. When the energy finally grows to approximately the assumed absorber saturation energy (0.016 J/cm), a rapid increase in pulse energy takes place. The growth in amplitude is finally limited by the onset of saturation in the laser. The laser gain then begins to decrease. In this region, the energy of the pulses reaches a maximum and then begins to decrease. Fig. 4(b) illustrates the sharpening of the pulses as a function of the number of passes. The peak intensity of each pass has been normalized to unit energy. The most pronounced sharpening is found to occur in this

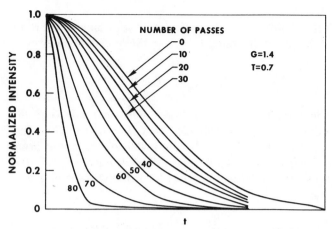

Fig. 3. Optical pulse sharpening as a function of passes through a saturable absorber having a short relaxation time with respect to the pulse duration. The assumed parameters in the calculation were intensity of the initial Gaussian pulse $0.1\,I_s$, gain between successive passes $G = 1.4$, and dye transmission $T = 0.7$.

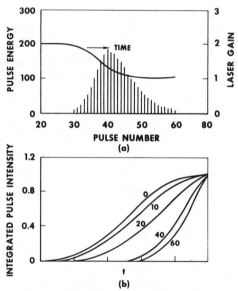

Fig. 4. Computer simulation of an optical pulse regenerative oscillator. The assumed parameters in the calculation were mirror reflectivity 0.8, starting pulse energy $= 10^{-4}$ laser saturation energy, laser saturation energy $E_L = 16$ J/cm, dye saturation energy $= 10^{-3}\,E_L$, and the dye relaxation time was assumed much, much greater than the pulse duration.

model between the twentieth and fortieth passes. Referring to Fig. 4(a), it is seen that this is just the region where the pulse energy reaches and exceeds the absorber saturation energy.

It can be concluded that much can be learned about the operation of the laser saturable-absorber combination by means of such computer simulation. Neither of the two models described here is capable of explaining the formation of the initial pulse. In addition, the absorber concentration is much weaker in the actual operation of such systems than the values assumed in these simulations. As a result, the sharpening action is much weaker. It is apparent that a more elaborate model of the dye is needed. In addition, light amplification in saturable absorbers must also be taken into account, as will be discussed in later sections of this paper.

Laser oscillators consist of a resonant system with dimensions that are large compared to the oscillating wavelengths. Consequently, mode density is high and there are $m = 2L\Delta\lambda/\lambda_0^2$ axial interferometer resonances within the line width $\Delta\lambda$ of a laser transition having a center wavelength λ_0 and reflectors separated by an optical length L. In the normal operation of a laser, these modes are to a great extent uncoupled and therefore have no fixed phase relationship between the many discrete oscillating frequencies of the laser. In the above discussion, the required coupling was supplied by a passive modulator, i.e., the saturable absorber. However, the required mode-coupling necessary to lock the phase of the axial modes, i.e., mode-locking, can also be supplied by an active modulator. This operation can best be described as follows. Assume that the Fabry-Perot mode v_0, nearest the peak of the laser gain profile, will begin to oscillate first. If an amplitude or phase modulator operating at a frequency Δf is inserted into the laser's feedback interferometer, the carrier frequency v_0 will develop sidebands at $\pm\Delta f$. If the modulating frequency Δf is chosen to be commensurate with the axial mode frequency separation $\Delta f = c/2L$, the coincidence of the upper $(v_0 + \Delta f)$ and the lower $(v_0 - \Delta f)$ sidebands with the adjacent axial mode resonances will couple the $v_0 - \Delta f$, v_0, and $v_0 + \Delta f$ modes with a well-defined amplitude and phase. As the $v_0 + \Delta f$ and $v_0 - \Delta f$ oscillations pass through the modulator, they will also become modulated and their sidebands will couple the $v_0 \pm 2\Delta f$ modes to the previous three modes. This process continues until all axial modes falling within the oscillating line width are coupled. The constructive and destructive interference of these simultaneous phase-locked modes can be described by the interference of Fourier-series components in the construction of a repetitive pulse train.

Active time-varying-loss [27]–[29] and reactive [30] modulators have been utilized for mode-locking laser oscillators. Ultrasonic standing-wave diffraction cells [27], [28], [31]–[34] and KDP Pockels cells [29] have been utilized as active-loss modulators, whereas only Pockels cells have been utilized as reactance modulators [30], [35]. Hargrove, Fork, and Pollack were the first to couple the modes of a laser for the generation of ultrashort light pulses [27]. They employed a visible He-Ne laser and an ultrasonic diffraction modulator [32]. When locking of the axial modes had been achieved; they report the virtual elimination of both random and systematic amplitude fluctuations arising from the drifting of the phases of the individual modes and a fivefold increase in peak power. Theoretically, the peak power of the pulses should be n times the average power, where n is the number of coupled modes.

Mode-locking a laser oscillator with an active modulator requires the critical adjustment of mirror spacing and modulating frequency as well as compensation for any perturbations in optical length of the feedback interferometer. Such compensation is of particular importance in the mode-locking of large solid-state lasers as a consequence of the variation in optical length of the rods during the optical pumping flash. The use of saturable absorbers as passive modulators eliminates the need for such critical adjustments.[2] In addition, the saturable absorber serves as a Q-switch and therefore makes possible the generation of pulses having peak powers in the gigawatt range.

B. Experimental Results

The relatively narrow spectral line widths of gas lasers limit pulse widths to the order of 10^{-10} s [27], [31]. The broader line widths available in solid-state lasers such as Nd^{3+}:glass [18], [28], [36], ruby [29], [37], and YAG:Nd^{3+} [38] have produced considerably narrower pulses having pulse widths in the picosecond range. The Nd^{3+}:glass is of particular interest for the generation of ultrashort pulses because of its broad oscillating line of 100 Å to 200 Å at $\lambda_0 = 1.06 \mu$. With a mirror spacing of 1.5 m. there exist approximately 6×10^4 Fabry-Perot interferometer resonances across the 200 Å line width [or 2 m = 6×10^4 spectral components in Fig. 2(c) are possible], and a pulse width in the 10^{-13} s range could theoretically be obtained.

Simultaneous Q-switching and mode-locking experiments with saturable dyes have been performed with Nd^{3+}:glass laser rods having lengths from 12.2 cm up to 100 cm [18]. One major requirement in these experiments was that the ends of the rod have Brewster's angle ends to eliminate back reflection at the end surface of the rods. The dye concentration, optical pumping intensities, and mirror reflectivities could be adjusted to obtain long pulse trains with low pulse amplitudes, short pulse trains with correspondingly higher pulse amplitudes, or single or multiple Q-switched mode-locked pulse trains. The dye cell lengths used in our experiments were typically from 1 cm to 0.1 cm. For the best results, the dye cell was placed at Brewster's angle and the dye was placed in contact with one of the mirrors. Practically all the dyes useful in the generation of ultrashort laser pulses decompose when exposed to ultraviolet light so that glass filters are useful for dye cell windows in obtaining long dye life. These comments hold equally well for both the ruby and the Nd^{3+}:glass experiments.

Eastman 9740 or 9860 saturable dyes have been used in our Nd^{3+}:glass experiments, and cryptocyanine, dicarbocyanine iodine, or dicyanine A in our ruby experiments. Eastman 9740 has an upper limit relaxation time of 25 to 35 ps [39], and 9860 an upper limit of 6 to 9 ps [40]. These dyes should therefore be well suited for obtaining pulse repetition rates up to 10^{11} or 10^{12} Hz. Of the commonly used ruby Q-switching dyes such as chloro-aluminum phthalocyanine, vanadyl phthalocyanine, and cryptocyanine, only cryptocyanine has a relaxation lifetime shorter than the recirculation time of a pulse of light between two laser Fabry-Perot mirrors separated by a rea-

---

[2] These adjustments have recently been eliminated by regenerative electronic feedback techniques. See G. R. Huggett, "Mode-locking of CW lasers by regenerative RF feedback," *Appl. Phys. Lett.*, vol. 13, pp. 186–187, September 1968.

sonable length [41]. The suitability of cryptocyanine in methanol for the generation of ultrashort laser pulses was first reported by Mocker and Collins [37]. Mode-locking of ruby lasers with cryptocyanine in nitrobenzene and ethanol has also been reported [42]–[43]. A substantial improvement in consistency of operation can be obtained by using acetone as a solvent instead of methanol. The peak of the absorption line for this dye-solvent combination coincides with the ruby line, while in the methanol solution the absorption peak shifts by 120 Å to 7060 Å. In addition, it is possible that by virtue of the lower solvent viscosity of acetone, the recovery time of the dye in the acetone solution may be shortened [44].

Two other dyes closely related to cryptocyanine also mode-lock the ruby laser effectively [45]. These dyes are dicyanine A and 1,1'-diethyl-2,2'-dicarbocyanine iodide (DDI). No solvent was found in which the absorption peak of either of these dyes coincided with the ruby line. The best solvent for dicyanine A was dimethyl sulfoxide (DMSO). The absorption peak for this dye-solvent combination occurred at 6770 Å.

DDI can be used in either water or methanol. Its absorption peak is at 7030 Å in water and at 7060 Å in methanol. The water solutions are unstable with a half-life of approximately 1 hour. Representative data of the pulse widths and peak powers obtained with the three dyes found to mode-lock the ruby laser effectively in our work appear in Table I [45]. The "nominal" power outputs are those which were found to optimize the performance of the laser. The output power level can be varied to some extent by changing the dye concentration. Several of the results listed in the table are worthy of special note. The first is that with DDI in either water or methanol, pulses nearly as short as those presently obtained with the mode-locked neodymium-glass laser can be obtained. Another is that one can choose the pulse width desired by choosing the proper dye. In all cases, a small pulse-width bandwidth product was observed, evincing the fact that a frequency chirp in the output pulses does not occur as per the Nd : glass laser.

One problem was noted with the ruby regenerative pulse oscillator. It was found that the destruction of the dielectric mirror occurred more often than for the Nd^{3+} : glass system. A Galilean telescope was included in the ruby laser cavity to double the beam diameter, thereby reducing the optical flux density and preventing the destruction of the 99^+ percent reflector. In addition, the Galilean telescope permits compensation for the thermal lensing effects within the ruby rod arising from the optical pumping by the adjustment of the separation of the telescope elements.

The Galilean telescope consisted of a 50 mm focal length plano-concave lens and a 100 mm focal length plano-convex lens. Both lenses were anti-reflection-coated. The spacing of the lenses was adjusted for a minimum beam divergence as evidenced by far field patterns. From the final spacing of the lenses, the focal length of the thermal lensing was estimated to be approximately 4 m. The beam divergence was well below 1 mrad for a 1.20 m diam, 5 cm long Verneuil rod of fair quality, or a 1 cm diam, 10 cm long Czochralski

TABLE I

PICOSECOND RUBY LASER PULSES WITH VARIOUS DYES AND SOLVENTS

Dye	Solvent	Width Pulse (ps)	Nominal Peak Power (GW)
Cryptocyanine (1,1'-diethyl-4,4'-dicarbocyanine iodide)	acetone	25	1
DDI (1,7'-diethyl-2,2'-dicarbocyanine iodine)	methanol	4	5
	water	2–4	5
dicyanine A	dimethyl sulfoxide	50	~1

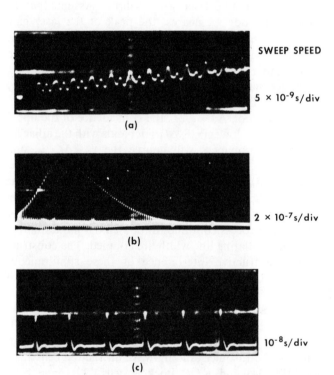

SWEEP SPEED

5×10^{-9} s/div

(a)

2×10^{-7} s/div

(b)

10^{-8} s/div

(c)

Fig. 5.   Oscillograms of the typical output of an optical regenerative pulse oscillator.

rod of excellent optical quality. The placement of the ruby rod within the cavity was not found to be critical.

Ultrashort light pulses with a Beckman & Whitley model 440 thin-film Q-switch were also unknowingly observed by Hercher in his attempts to obtain single-mode operation from a ruby laser [46]. It was not realized at that time that saturable absorbers with long relaxation times tend to narrow the spectral width while saturable absorbers with fast relaxation times tend to broaden the spectral width of lasers.

Fig. 5(a) is an oscillogram of the early portions of a slow-buildup, Q-switched Nd : glass laser pulse train. The sweep speed is 5 ns/div with the oscillogram covering 3×10^{-8} s. The oscillogram illustrates the tendency of the saturable dye to emphasize the highest amplitude fluctuation occurring at the initiation of laser oscillation and to shorten this fluctuation pulse width as it successively propagates through the saturable absorption cell. The initial periodic amplitude fluctuations are caused by the beating of two axial modes separated by a frequency $2\Delta f$. A beat frequency at $2\Delta f$ would correspond to the existence of two pulses in the feedback cavity at the same time. At the end of 3×10^{-8} s, the secondary peak of the amplitude fluctuation is practically completely eliminated. The pulse width is reduced to 0.5×10^{-9} s, and the pulse repetition period equals 2.5 ns or Δf. An oscillogram of an entire pulse train at a sweep speed of 200 ns/div is given in Fig. 5(b). An oscillogram of a pulse train at a sweep speed of 10 ns/div is given in Fig. 5(c). The recorded pulse half-widths are approximately 0.5 ns, limited by the rise time of the photodetector and traveling-wave oscilloscope combination. The fall time is larger than the rise time as a result of critical damping of a resonance at approximately 850 MHz within the ITT model F4018(S-1) biplanar photodiodes utilized in the UAC Research Laboratories Model 1240 Phototransducer.

The use of a sampling oscilloscope greatly facilitates the pulse-width measurements of CW mode-locked lasers. The relatively short overall time duration of approximately 50 to 500 ns of simultaneously Q-switched and mode-locked lasers excludes the use of sampling scopes. Unfortunately, traveling-wave scopes require large-signal inputs and have considerably narrower bandwidths than sampling scopes. For example, the Tektronix Model 519 traveling-wave scope has approximately 10 V/cm sensitivity and 1 GHz bandwidth, whereas millivolt sensitivity and bandwidths in excess of 10 GHz are available with sampling scopes. Nevertheless, an instrument-limited direct measurement of 0.15 ns rise time has been obtained [36] for a simultaneously Q-switched and mode-locked Nd^{3+}: glass laser [see Fig. 6(a)]. The measurement utilized a modified Tektronix Model 519 traveling-wave scope having a bandwidth of 3 GHz with a sensitivity of 217 V/div, and an ITT F4014 diode. If the rise time of the scope, i.e., 0.13 ns, is computed out of the measurement, a pulse width less than 90 ps is obtained out of the detector. The use of ultrashort laser pulses with fast detectors can produce electrical pulses up to 100 V in amplitude, with widths less than 90 ps and repetition times to 2 ns. Such electrical pulses should find applications in the electronics industry to measure the characteristics of wide-bandwidth systems [47]. For example, the ringing of a standard 1 GHz traveling-wave scope when excited by a F4014 photodiode is illustrated by Fig. 6(b). The reason for the grouping of three pulses illustrated by Fig. 6(a) will be given later.

A 76 cm long by 1.8 cm diam Nd^{3+}: glass rod was simultaneously mode-locked and Q-switched with a 1 cm dye cell containing 16 cm^3 of solvent and 5 cm^3 of dye. The Q-switched pulse envelope consisted of only six pulses, each having a typical pulse width of approximately 2 ns time duration. The energy output was 44 J with 72 kJ input energy for an average energy per pulse of a little over 7 J. A 47 J pulse train has been obtained with the use of a 45 cm long glass rod oscillator in conjunction with a 76 cm long glass rod amplifier. The time duration of the individual pulses of the train were typically 0.75 ns for this case. The

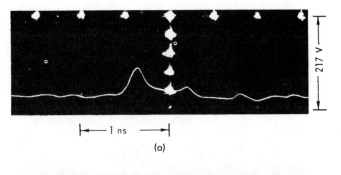

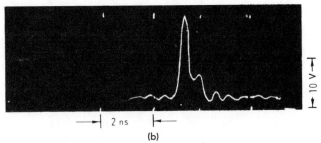

Fig. 6. Oscillogram of the output of an optical regenerative pulse oscillator taken with (a) a 0.13 ns and (b) a 0.3 ns rise time traveling-wave scope and a 90 ps detector.

use of a 16.5 cm long glass rod oscillator and a 76 cm long glass rod amplifier has resulted in the generation of pulses having a peak power between 10 GW and 100 GW and time durations between 10 ps and 2 ps, respectively. A discussion of the picosecond time duration measurement technique will be given in Section III. The use of nonlinear optical pulse measurement techniques to be described (Section III) revealed that the 2 ns and 0.75 ns pulse widths obtained with the 76 cm and 45 cm long rods were in actuality a grouping of picosecond pulses whose repetition period was much shorter than the 0.5 ns response time of the detection system. A grouping of three such pulses is illustrated by Fig. 6(a).

Energy gain versus input energy data for a 76 cm long glass rod amplifier utilized in the amplification of picosecond pulses is shown in Fig. 7. The lower curve is the actual energy gain, i.e., the ratio of measured output to input energies, of the overall amplifier system. The upper, i.e., normalized, curve is the gain of the glass laser medium without losses. Each experimental point represents an average of six to eight separate measurements. Eastman 9740 saturable absorber was utilized for optical isolation between the oscillator and the amplifier. The energy measurements were taken with ballistic thermopiles and lean heavily toward the conservative side because energy loss due to gas breakdown at the focus point within the thermopile was not taken into account.

Fig. 8(a) is a schematic diagram of a pulse regenerative laser oscillator having a low-Q resonant reflector with a thickness $d_1 = 1$ cm and surface reflectivities 5 percent and 35 percent, respectively. The distance between the 99^+ percent and 35 percent reflectivity surfaces was 120 cm. The simplified operating model of Fig. 8(a) can be given

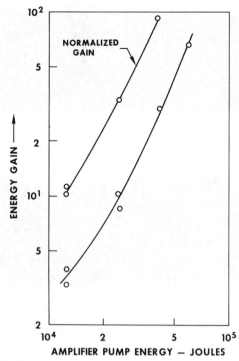

Fig. 7.   Energy gain as a function of pumping energy for a 76 cm long Nd^{3+}:glass laser amplifier.

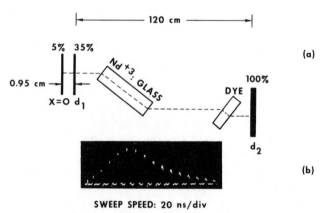

SWEEP SPEED: 20 ns/div

Fig. 8.   Oscillograms of the output of an optical regenerative pulse oscillator operated with a resonant reflector.

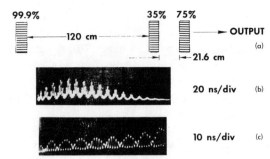

Fig. 9.   Oscillograms of the output of an optical regenerative pulse oscillator operated with two coupled cavities.

as follows. Assume that at $t=0$ two pulses exist at $x=d_1$ whose pulse width $\Delta\tau \ll d_1/c$. After $t=0$ the pulses travel in opposite directions, and the pulse traveling to the left bounces back and forth between the two surfaces of the resonant reflector for a time determined by the Q of the resonator. For the illustrated case, the delay time is approximately 50 ps. If the Q is low enough, we can assume only one extra pulse is produced by the resonant reflector. For all practical purposes, we can then assume that two pulses travel to the right and left of the resonant reflector separated by a time $t=2\,d_1/c \simeq 0.1$ ns, with the first pulse having a much larger amplitude than the second. Each time the two pulses pass through the resonant reflector the process is repeated, giving rise to $2^{(m+1)}$ pulses for m traversals in the laser cavity. If the response time of the

detection system is much slower than $2\,d_1/c$, the pulse width of the individual grouping of pulses having a fundamental repetition period of $2\,d_2/c$ will appear to increase as a result of the number of pulses within the grouping increasing with time. Fig. 8(b) is an oscillogram of the output pulse of the experiment illustrated in Fig. 8(a) which demonstrates this effect. The response time of the detection system was 0.5 ns. It is believed that this mechanism is responsible for the multiple pulsing illustrated in Fig. 6(a).

If the thickness and Q of the resonant reflector are increased to the values shown in the schematic of Fig. 9(a), the individual pulses can be resolved with currently available detection systems. Fig. 9(b) and (c) presents oscillograms of the resulting pulse trains at sweep speeds of 20 and 10 ns/div, respectively. For the case shown by Fig. 9(c), the two coupled cavities were not exact multiples of one another and phase shifts at the nodal points of the modulating envelope can be observed. The use of these techniques can be utilized to generate pulse rates extending well into the microwave region. If such multiple pulsing is to be eliminated, these data clearly reveal the necessity of placing all surfaces within a regenerative laser oscillator cavity at Brewster's angle. Similar attention must be paid to backscattering in the dye cell and other components of the laser cavity, as well as to backscattering back into the laser cavity from external objects.

If the dye cell inside a laser is moved a distance d_1 from one of the mirrors, double pulsing is obtained [16]. The time separating the two pulses corresponds to the time $t=2\,d_1/c$. The pulse-crossover points alternate between a point at a distance d_1 in front of each reflector inside the laser cavity. In general, Harrach and Kachen have found that the pulse repetition rate of the laser is $mc/2L$ when $m=L/d_1$ is an integer [48].

For some applications longer intervals between pulses are desirable. A longer pulse repetition time can be obtained by inserting an optical delay line as part of the cavity of a regenerative laser oscillator. It has been reported that such long optical delay lines can be made compact by reflecting the beam repeatedly between two spherical mirrors without interference between adjacent beams [49]. The diffraction losses of such a delay line are much lower than for an open beam because of the periodic focusing of the mirrors. Besides effectively increasing the optical distance between the cavity mirrors, this technique can yield an automatic nonmechanical digital scan of the pulsed laser output [50] if the output from one of the delay line mirrors is utilized.

Fig. 10 illustrates the arrangement used in one experiment. The radius of curvature was 1 m for mirrors R_4 and R_5. Mirrors R_3 and R_2 were used to inject the optical beam into the optical delay line through a 1.8 cm aperture in mirror R_4 in such a way that the repeated reflections traced out a circle on the surfaces of mirrors R_4 and R_5. For most cases in our experiments, the light beam slope selected by the angular adjustment of R_2 was so chosen that the light beam made six round trips between R_4 and R_5 and the trajectory traced out a six-spot circular pattern on each mirror of the delay line. After the twelfth pass in the delay line, the light beam was reflected back onto itself and through the delay line for another 12 traversals. Fig. 10(b) illustrates a pulse separation of approximately 71.5 ns.

C. Single Picosecond Pulse Generation

There are many applications where only a single ultrashort light pulse is required rather than a train. The first method for obtaining a single high-peak-power laser pulse having a pulse width narrow.. ..an those obtainable from standard Q-switched lasers was proposed by Vuylsteke [51]. This method involved Q-switching a laser with mirrors of 100 percent reflectivity on both ends of the cavity and, at the peak of the pulse, rapidly switching the output mirror from 100 percent to 0 reflectivity. In this manner, the optical energy stored within the cavity would be dumped in the time required for the round-trip transit time. This type of laser operation, called the pulse transmission mode, was reported to have resulted in pulse widths of 4 ns [52]. The experimental technique of combining the simultaneous Q-switching and mode-locking with the pulse transmission mode is, we believe, the one capable of yielding the highest peak power with the shortest pulse width [53]–[57].

A schematic diagram of an early single pulse selection experiment is given in Fig. 11. The reflectivity of both mirrors is high. In addition to the dye cell, a Glan prism polarizer and a polarizing switch such as a Pockels or Kerr cell are inserted into the feedback path of the laser. The polarizing switch was initially unenergized, and the polarizer was adjusted for maximum transmission. A pulse of radiation then "bounces" back and forth between the two reflectors. The leakage radiation from one of the mirrors is used to trigger a high-voltage pulser at a predetermined optical pulse amplitude. The high-voltage pulser energizes the polarization switch to its $\lambda/4$ voltage. After the pulse has made one traversal through the energized polarizing switch, the polarization of the pulse is rotated 45°. For a two-trip traversal of the polarizing switch, the polarization of the pulse is rotated 90°. The Glan prism prevents the propagation of the pulse with this polarization in the system, and the propagation direction of the pulse is redirected as illustrated. In effect, the high reflectivity of the cavity is suddenly changed to a low reflectivity and the pulse stored within the cavity is suddenly "dumped." The experiment can also be performed with the polarization switch and Glan polarizer outside the cavity. For this latter configura-

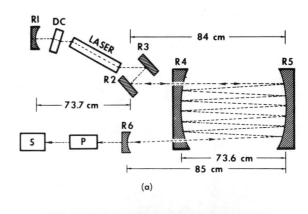

(a)

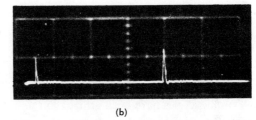

(b)

Fig. 10. Schematic diagram of an optical regenerative pulse oscillator incorporating a folded optical delay line for obtaining a long pulsating period.

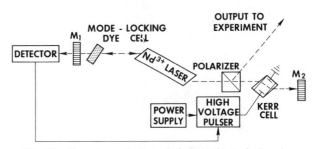

Fig. 11. Experimental schematic for selecting a single pulse from a train of ultrashort pulses.

tion, a $\lambda/2$ voltage is required across the polarizing switch [54]. This latter configuration is preferred when the minimum pulse width is desired [56].

The main problem area in performing a single ultrashort pulse selection experiment is the high-voltage pulser. The rise time of the high voltage supplied to the polarizing switch must be faster than the time required for the light pulse to make a round-trip traversal between the two mirrors. The high-voltage pulser utilized in our experiments was a Marxbank pulse generator common to many high-energy physics experiments. Avalanche transistors assembled in a Marxbank configuration were used to produce a pulse in the 1 kV range, which in turn was used to trigger a spark gap holding off a voltage in the 15 kV range. An open-ended transmission line was then used to double this voltage to 30 kV to drive the Kerr cell polarizing switch.

A single subnanosecond pulse has been obtained with the experimental arrangement shown in Fig. 11 [16], [17]. The recorded full width at the half-intensity points was measured with a 0.5 ns response time detection system and found to be 0.63 ns with an energy content of $\frac{1}{4}$ J. The use of the two-photon absorption measurement techniques to be described in Section III reveals that these pulses usually consisted of a substructure of picosecond pulses. The amplification of such single pulses with a 75 cm long amplifier has resulted in 1.8 J of energy [55]. Since the damage threshold as a function of power decreases with decreasing pulse widths, the further addition of stages of amplification should make possible the generation of picosecond pulses having peak powers well in excess of 10^{13} W. The use of these pulses in controlled thermonuclear research appears particularly promising.

III. MEASUREMENT TECHNIQUES FOR PICOSECOND LASER PULSES

A. Introduction

Direct measurement of the duration of relatively long optical pulses is most often made by displaying, by means of an oscilloscope, the output of a suitable photodetector illuminated by the optical radiation. High-speed photodiodes and oscilloscopes have been utilized in this manner to obtain direct pulse-width measurements of 1.5×10^{-10} s for a mode-locked CW YAG laser [38] and a Nd^{3+}: glass laser [16]. A power spectrum measurement of the oscillating laser bandwidth made by means of a scanning Fabry-Perot interferometer yielded an indirect measurement of 7.6×10^{-11} s for the YAG experiment, and a grating spectrometer yielded 2×10^{-13} s for the Nd^{3+}: glass laser experiment. Since it is not expected that direct electronic techniques will be capable of measuring time durations down to 10^{-13} s, new measuring techniques had to be found for the measurement of the time duration of picosecond laser pulses.

It will be shown in part **B** of this section that any linear interferometer measures the autocorrelation function of the pulse amplitude. Since the power density spectrum and the amplitude autocorrelation function are a Fourier transform pair, knowledge of one uniquely specifies the other. The power density spectrum is usually measured with a spectrometer. The equivalence arises from the fact that both instruments used to measure these two parameters are linear optical systems. The linear interferometer can also be used to measure the coherence length ΔD and the coherence time $\Delta \tau_c$; these two parameters are related to the spectral bandwidth $\Delta \omega$ and to each other by the following relationship:

$$\Delta D \simeq \frac{\pi c}{\Delta \omega} \simeq \frac{\Delta \tau_c c}{2}.$$

Since the relation $\Delta \omega \Delta \tau_c \geq 2\pi$ provides only a lower limit to a pulse duration for a given $\Delta \omega$, it can be concluded that any measurement taken with a linear optical system can provide only information establishing a lower limit to the time duration of a pulse.

It will be shown in part C that measurements taken with nonlinear optical instruments can provide information for determining the actual time duration of a pulse. Nonlinear optical instruments measure the autocorrelation of the pulse intensity.

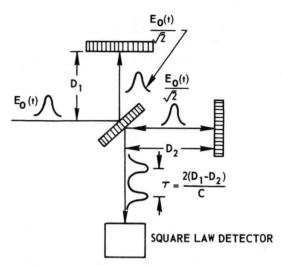

Fig. 12. A Michelson interferometer, a typical linear optical instrument.

B. Linear Optical Pulse-Width Measurement Techniques

1) *Theory*: For purposes of illustration, let us consider a linear optical instrument such as the Michelson interferometer diagramed in Fig. 12. An incident pulse having an amplitude $E(t)$ is split into two pulses, each with an amplitude $E(t)/\sqrt{2}$. Each of these pulses is made to traverse a separate orthogonal arm of the interferometer. After traversing their respective paths $2D_1$ and $2D_2$, the pulses are recombined on a square law detector such as a photographic plate or photodetector. If $D_1 \neq D_2$, the pulses can be represented by

$$E_1(t) = \frac{E_0(t)}{\sqrt{2}} e^{j\omega t}$$

and

$$E_2(t) = \frac{E_0(t-\tau)}{\sqrt{2}} e^{j\omega(t-\tau)}$$

where $\tau = 2(D_2 - D_1)/c$ and $E_0(t)$ is the slowly varying envelope of the pulse with respect to ω. The intensity incident on the detector is given by

$$I(t, \tau) = \tfrac{1}{2} |E_0(t) + E_0(t-\tau)|^2. \tag{12}$$

The response of the detector is assumed to be slow compared to the pulse duration or the delay time τ, so the output signal $S(\tau)$ of the detector is given by

$$S(\tau) = \int_{-\infty}^{\infty} I(t, \tau) dt \tag{13}$$
$$= W(1 + A(\tau))$$

where W is the pulse energy and $A(\tau)$ is the autocorrelation function of the pulse amplitude, i.e.,

$$W = \int_{-\infty}^{\infty} E_0^2(t) dt \tag{14}$$

and

$$A(\tau) = \frac{\displaystyle\int_{-\infty}^{\infty} E_0(t)E_0(t-\tau)dt}{\displaystyle\int_{-\infty}^{\infty} E_0^2(t)dt} . \qquad (15)$$

When $\tau = 0$, then $S(\tau)/W = 2$, and when τ is large enough so that no overlap between $E_1(t)$ and $E_2(t)$ exists, then $S(\tau)/W = 1$. From Fourier analysis, we find that

$$A(\tau) = \frac{1}{\sqrt{2\pi}}\int_{-\infty}^{\infty} e^{j\omega\tau}|P(\omega)|^2 d\omega \qquad (16)$$

where $|P(\omega)|^2$ is the power density spectrum of the original laser pulse [58]. Since the power density spectrum and the amplitude autocorrelation function are a Fourier transform pair, knowledge of one uniquely specifies the other. The power density spectrum is the quantity measured with a spectrometer and the amplitude autocorrelation is the quantity measured with an interferometer. The two results are essentially equivalent and the equivalence arises from the fact that both instruments are linear optical systems. Similar considerations hold for any linear optical instrument.

It is clear from Fig. 12 that an interference pattern will be observed in the plane of the detector, if the interfering pulses have a relative retardation in time equal to $\tau = 0$, $2L/c$, $4L/c$, $\cdots$, i.e., if the difference in length of the interferometer arms is $D_2 - D_1 = 0$, L, $2L$, $\cdots$, where L is the separation between the laser reflectors. The visibility of the interference pattern changes with the relative delay time of the pulses, is a maximum at $D_2 - D_1 = 0$, L, $2L$, $\cdots$, and gradually falls to zero in moving away from these points. The difference in path lengths, ΔD, over which the interference pattern is visible is called the coherence length. The coherence length and the spectral bandwidth of the pulse are related by $2\Delta D/c = 2\pi/\Delta\omega = \Delta\tau_c$, where $\Delta\tau_c$ is the coherence time of the pulse.

The coherence time $\Delta\tau_c$ will equal the duration of the pulse only in the special case where the entire spectral content of the pulse is due to the short duration of its envelope. In other words, the relation $\Delta\omega\Delta\tau \geq 2\pi$ provides only a lower limit to the pulse duration. Fig. 13 schematically summarizes the results of measurements with a linear optical instrument for pulses having a pulse-width ($\Delta\tau$) bandwidth ($\Delta\nu$) relationship given by $\Delta\tau = \Delta\tau_c = 1/\Delta\nu$ and by $\Delta\tau \gg 1/\Delta\nu$.

2) Experiments: Fig. 14 illustrates the spectral characteristics of the output from a 12.2 cm long by 0.95 cm diam Brewster-ended Nd^{3+}:glass rod operated normally, Kerr cell Q-switched, and as a regenerative pulse oscillator. Eastman 9740 saturable absorber was utilized as the optical expandor element. The spectra were obtained with a 3.4 m Jarrell-Ash spectrometer. The Kerr cell Q-switch type of operation yielded a uniform spectral width of approximately 50 Å. The overexposure of the film by overlapping the spectra of four or more Kerr cell Q-switched pulses still revealed a power spectrum of approximately 50 Å with sharply defined ends. The spectrum of the simultaneously

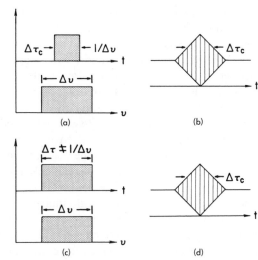

Fig. 13. For a pulse (a) having a bandwidth $\Delta\nu$ determined by the width of the pulse $\Delta\tau_c = 1/\Delta\nu$, a linear optical instrument gives a signal-to-background ratio (b) of 2 for $\tau = 0$ and 1 for $\tau \gg \Delta\tau_c$. For a pulse (c) having the characteristic $\Delta\tau > 1/\Delta\nu$, the identical result is obtained for the contrast ratio and pulse-width measurement. The instrument cannot distinguish between the two cases.

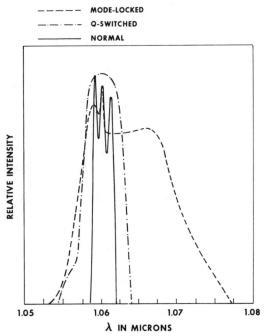

Fig. 14. Comparison of the spectral characteristics of an Nd^{3+}:glass laser in various modes of operation.

Q-switched and mode-locked Nd:glass laser with the dye cell placed at Brewster's angle revealed a uniformly distributed 180 Å wide spectrum with long leading and trailing edges. The increase in spectral width results from the tendency of the saturable absorber expandor element to distribute the energy evenly throughout the spectral line width of the laser medium by the generation of sidebands at the resonances of the Fabry-Perot interferometer. To a first approximation, the minimum pulse width obtainable with the harmonic content revealed by the spectral data is 2×10^{-13} s, with a corresponding peak power of 10^{10} W.

A close observation of the spectrum of the simultaneously Q-switched and mode-locked laser of Fig. 14 reveals a relatively intense line at 1.06 microns, i.e., the peak of the Nd^{3+} gain line. This sharp line is believed to arise from the relatively slow buildup rates of the modes within a passively Q-switched laser. For example, in a fast switched laser oscillator, i.e., switched with a Kerr cell, rotating prism, etc., the pulse buildup takes approximately 10 to 40 loop transits, whereas for a passively Q-switched laser this buildup requires typically several hundred to a thousand transits [59]. Such a long buildup time for the passively Q-switched case favors the existence of relatively few axial modes at line center for a long time before an appreciable sharpening takes place. When a streak image converter camera was used in conjunction with a 3.4 m Jarrell-Ash spectrometer, the sharp line at 1.06 microns did not appear in the time resolved spectral data taken for each individual pulse [16]. The spectral content of each pulse was found to be 100 Å to 120 Å. A channeled spectrum was obtained in any case where additional reflecting surfaces were placed in the cavity. For example, with the dye cell placed normal to the laser axis, a grouping of picosecond pulses separated in time by the optical thickness of the dye cell was obtained and it corresponded to the reciprocal of the frequency separation of the channeled spectrum. The grouping of pulses had a repetition frequency equal to the axial mode-spacing frequency of the cavity.

A still closer inspection of the spectrum of the simultaneously Q-switched and mode-locked laser of Fig. 14 reveals an unsymmetrical distribution of the spectral content about line center. The spectrum shows a tendency to expand toward the long wavelength region under mode-locked conditions. We believe this effect is caused by the fact that the peak of the laser line profile falls on the slope of the long wavelength side of the absorption line of the Eastman 9740 saturable absorber [16]. In effect, this provides less absorption, and therefore higher overall system gain toward the longer wavelength region of the oscillator spectrum. This effect is not expected to be as pronounced with Eastman 9860 dye because of the closer coincidence of the peak of the laser and dye spectral line.

In order to determine the spectral extent of the mode-locked pulses, a movable diode and slit were placed in the exit focal plane of a spectrometer [36]. Data on the temporal behavior of 1 Å and 0.5 Å selected regions of the spectra were obtained and compared with the simultaneously recorded data of the input pulse train. It was found that the oscilloscope pulse-width measurements of the radiation emitted by the limited spectral region were still instrument-limited. This is to be expected since the number of modes available in these limited apertures is still sufficient to produce pulses of the order of 3.6×10^{-11} and 1.8×10^{-11} s. Since no change in pulse shape or width was observed in random samplings of 0.6 percent and 0.3 percent portions of the total input spectral range, it can be concluded that mode-coupling is extensive over the entire spectral output.

An experiment was performed on the variation of the visibility of the fringes of pulses occurring in the middle of the pulse train recorded by an image-convertor streak camera, as a function of the length variation of one leg of the interferometer of Fig. 12 [16]. For a length variation greater than 8×10^{-3} cm, the fringes disappeared. The minimum pulse width calculated from a coherence length of 8×10^{-3} cm is 5×10^{-13} s, as expected from the discussion on theory (part B-1) in this section. These data are in excellent agreement with the spectral data of Fig. 14.

In general, it was found that when the modes of the laser were only partially locked in phase (as observed by a considerable dc level in the oscilloscope traces), a corresponding reduction in spectral content of the pulses was noted [16]. It is important to note that even under normal Q-switched or free-running operations, the spectral content of a Nd^{3+}:glass laser pulse is considerably greater than the observed pulse widths would indicate (see Fig. 14). As a result, it is to be expected that amplitude or phase fluctuations corresponding to a time duration equal to the inverse of the bandwidth of the spectrum must exist in the output. It is also expected that a finite probability exists in a series of experimental runs that a certain number of axial modes will statistically have the proper phase as to yield periodic amplitude fluctuations riding on the envelope of the Q-switched pulse. In general, the periods of these periodic fluctuations are equal to the round-trip traversal time of the cavity and can thus be easily observed on an oscilloscope. It has been recently reported that picosecond pulses normally appear in free-running and Q-switched ruby, Nd:glass, and Nd:YAG lasers even when periodic amplitude fluctuations were either not observed on an oscilloscope or, if observed, had a width longer than the response time of the detection system [60]–[62]. The conclusion of the picosecond pulsating output of free-running or Q-switched ruby, Nd:glass, or Nd:YAG lasers was based on the data obtained from two-photon absorption-fluorescence experiments. A review of the two-photon absorption-fluorescence technique for the measurement of a pulse's time duration in the picosecond range will be given in part C of this section. It will suffice to state at this point that extreme care must be taken in interpreting the fluorescence track in such pulse-width measuring experiments [63].

It is well known that the nonlinear properties of the amplifying laser medium can give rise to self-locking of the modes within gas and solid-state lasers [31], [64]–[69]. For the He-Ne laser case, it has been shown that stable mode-locking operation is most stable under conditions of low laser excitation [67]. Stable operation can be obtained under high-excitation conditions by using an auxiliary discharge tube containing pure Ne as a nonlinear loss element inside the laser cavity [70]. Self-locking of lasers is not very reproducible because it is presumably difficult, except in carefully controlled situations, to couple a large number of axial modes [31], [66], [69]. Usually self-mode-locking is an occasional occurrence and difficult to reproduce. When self-mode-locking repetitively occurs with a solid-state laser Q-switched system, it usually is peculiar to that one laser system and not necessarily reproducible with other lasers of the same type. Needless to say, self-mode-locking of lasers is not well understood at this time even though considerable effort has been expended on this topic [31], [60], [64]–[72].

C. Nonlinear Optical Pulse-Width Measurement Techniques

1) *Theory:* In part B of this section, it was noted that linear optical systems can yield only a lower limit on the pulse width of optical pulses, whereas measurements performed with nonlinear optical instruments can yield the actual pulse width. The manner in which a nonlinear optical system can perform a true measurement of a pulse can be explained as follows. Suppose one passes the two output pulses $E_1(t)$ and $E_2(t)$ from the linear interferometer of Fig. 12 through a nonlinear optical crystal. The second harmonic output from the nonlinear optical crystal will be given by

$$^{2\omega}E(t) = [E_1(t) + E_2(t - \tau)]^2 \qquad (17)$$

if one neglects the constant representing the second harmonic generation efficiency of the crystal. The output signal $S(\tau)$ from a detector having a slow response time with respect to ω and τ is given by

$$S(\tau) = \int_{-\infty}^{\infty} |^{2\omega}E(t)|^2 dt = {}^{2\omega}W(1 + 2G(\tau)) \qquad (18)$$

where $^{2\omega}W$ is the second harmonic pulse energy and $G(\tau)$ is the autocorrelation function of the pulse intensity, i.e.,

$$^{2\omega}W = \int_{-\infty}^{\infty} E^4(t)dt \qquad (19)$$

and

$$G(\tau) = \frac{\int_{-\infty}^{\infty} E^2(t)E^2(t - \tau)dt}{\int_{-\infty}^{\infty} E^4(t)dt}. \qquad (20)$$

When $\tau = 0$, $S(\tau)/^{2\omega}W = 3$, and when τ is large enough so that no overlap between $E_1(t)$ and $E_2(t)$ exists, $S(\tau)/^{2\omega}W = 1$. Measurement of the variation of $S(\tau)/^{2\omega}W$ as a function of τ gives the time duration over which the energy of the pulse is distributed. Fig. 15 schematically summarizes the results of measurements with a nonlinear optical instrument for the two basic types of pulses $\Delta\tau = \Delta\tau_c = 1/\Delta\nu$ and $\Delta\tau \gg 1/\Delta\nu$.

A nonlinear interferometer can also be constructed in which $S(\tau)/^{2\omega}W = 0$ when τ is large enough so that no overlap between $E_1(t)$ and $E_2(t)$ exists. Suppose one passes an optical pulse through a birefringent crystal of length L with the optical polarization 45° with respect to the optic axis (see Fig. 16). The crystal will resolve the single pulse into two pulses of equal amplitude and orthogonal polarization. One polarization component of the pulse propagates as an ordinary ray and the other as an extraordinary ray. For a crystal of thickness L, the delay τ so introduced is given by $\tau = L(n_e - n_o) c$, where n_e and n_o are the extraordinary and ordinary indices of refraction. For calcite, the delay is 0.59

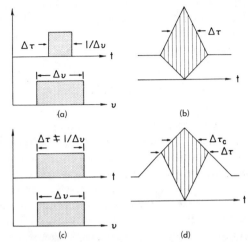

Fig. 15. For a pulse (a) having a bandwidth $\Delta\nu$ determined by the width of the pulse $\Delta\tau_c = 1/\Delta\nu$, a nonlinear optical instrument gives a signal-to-background ratio (b) of 3 for $\tau = 0$, unity for $\tau \gg \Delta\tau_c$, and a true pulse-width measurement. For a pulse (c) $\Delta\tau > 1/\Delta\nu$, a measurement of the true pulse width $\Delta\tau$ and the coherence width can simultaneously be obtained. The shaded areas of (b) and (d) represent interference fringes.

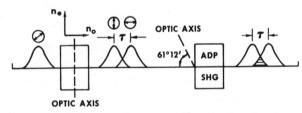

Fig. 16. Schematic experimental arrangement for measuring picosecond laser pulses with the second harmonic technique.

ps/mm at $\lambda = 1.06$. The two orthogonal polarized pulses are then caused to interact in a nonlinear optical crystal, in this case ADP, to produce second harmonic radiation at 5300 Å. The orientation of the nonlinear crystal is chosen such that no second harmonic is produced by either component of the pulse acting alone, but only when both are present. The method thus, measures the "overlap" of the pulse with a delayed replica of the pulse. When a delay is introduced such that no overlap of the two pulses occurs, the second harmonic drops to zero.

For ADP, the angle at which the second harmonic is proportional to the product of two orthogonal polarizations was reported by Weber [73] to be 61°12′. This angle produces phase matching for this combination of vertical and horizontal polarization and produces second harmonic radiation with horizontal polarization. If we let input signal E_i to the ADP crystal be represented by

$$E_i(t) = E_H(t)e^{j\omega t} + E_V(t - \tau)e^{j\omega(t - \tau)} \qquad (21)$$

the second harmonic signal $^{2\omega}E_H(t)$ is given by

$$^{2\omega}E_H(t) = E_H(t)E_V(t - \tau)e^{j\omega(2t - \tau)} \qquad (22)$$

where $E_H(t)$ and $E_V(t - \tau)$ are the slow-varying envelope of the horizontal and vertical polarized electric fields with respect to ω. The constant representing the second harmonic generating efficiency has also been neglected. If $\tau \gg \Delta\tau$, then $^{2\omega}E_H = 0$, and if $\tau = 0$, then $^{2\omega}E_H = \max$. The output signal from a detector having a slow response time with respect to 2ω is given by

$$S(\tau) = \int_{-\infty}^{\infty} E_H^2(t)E_V^2(t - \tau)dt. \tag{23}$$

Since a separate experimental run is required for each data point and since each experimental run is independent of previous runs, normalization is required in the data processing. A convenient means of normalizing is to take a pulse of any polarization and let it alone produce second harmonic energy. This reference signal can be represented by (19). The experimentally measured quantity is $S(\tau)/^{2\omega}W = I(\tau)$ as a function of τ by selecting different lengths of delay crystals. When $\tau = 0$, $S(\tau)/^{2\omega}W = 1$, and when τ is large enough so that no overlap between $E_H(t)$ and $E_V(t-\tau)$ exists, $S(\tau)/^{2\omega}W = 0$. The nonlinear optical system utilized in the measurement of picosecond pulses in [43], [74]–[76] was essentially one of the nonlinear systems described above. Fig. 17 schematically summarizes the results of measurements with the orthogonally polarized nonlinear optical instrument for the two basic types of pulses $\Delta\tau = \Delta\tau_c = 1/\Delta v$ and $\Delta\tau \gg 1/\Delta v$.

Recently, Giordmaine et al. [77], [78] reported the two-photon absorption-fluorescence technique for the measurement of picosecond laser pulses. This technique is also an intensity-correlation system. Its simplicity is its major advantage over the second harmonic systems described previously. A schematic diagram of the two-photon absorption-fluorescence measurement experimental arrangement is illustrated by Fig. 18. The correlation of one pulse with succeeding pulses (cross-correlation) illustrated by Fig. 18(a), or the correlation of a pulse with itself (autocorrelation) illustrated by Fig. 18(b), can be obtained. Both cross-correlation and autocorrelation experiments have been performed by the authors and identical results were obtained in the two measurements.

If we consider the center of the cell to be the origin ($z=0$) and denote pulses approaching the cell from the left and right as

$$E\left(t - \frac{z}{c}\right)\sin(kz - \omega t) \quad \text{and} \quad E_2\left(t + \frac{z}{c}\right)\sin(-kz - \omega t),$$

respectively, the intensity I_ω in the dye cell will then be given by

$$I_\omega = E_1^2\left(t - \frac{z}{c}\right) + 2E_1\left(t - \frac{z}{c}\right)E_2\left(t + \frac{z}{c}\right)\cos 2kz$$
$$+ E_2^2\left(t + \frac{z}{c}\right). \tag{24}$$

If we assume that the fluorescent intensity I_F is proportional to the intensity squared, $I_F \propto I_\omega^2$, then the fluorescent intensity is given by

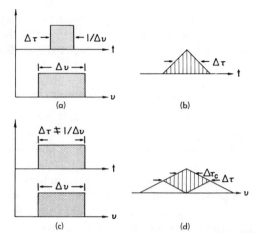

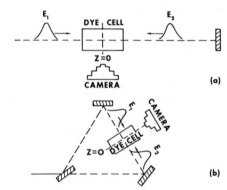

Fig. 17. For a pulse (a) $\Delta\tau = \Delta\tau_c = 1/\Delta v$, an orthogonally polarized nonlinear optical instrument gives an infinite signal-to-background ratio (b) for $\tau = 0$, zero for $\tau \gg \Delta\tau_c$, and a true pulse-width measurement. For a pulse (c) $\Delta\tau > 1/\Delta v$, a measurement of the true pulse width $\Delta\tau$ and the coherence width can simultaneously be obtained with no background signal. The shaded areas in (b) and (d) represent interference fringes.

Fig. 18. Experimental arrangement for performing intensity cross-correlations and intensity autocorrelation measurements of picosecond pulses by the two-photon absorption-fluorescence technique.

$$I_F \propto \left\{E_1^4\left(t - \frac{z}{c}\right) + E_2^4\left(t + \frac{z}{c}\right) + 2E_1^2\left(t - \frac{z}{c}\right)E_2^2\left(t + \frac{z}{c}\right)\right.$$
$$\cdot \left[1 + 2(\cos 2kz)^2\right] + 4\left[E_1^3\left(t - \frac{z}{c}\right)E_2\left(t + \frac{z}{c}\right)\right. \tag{25}$$
$$\left.\left. + E_1\left(t - \frac{z}{c}\right)E_2^3\left(t + \frac{z}{c}\right)\right]\cos 2kz\right\}.$$

The photographic film essentially records the value of I_F but it is time averaged through the photographic process and spatially averaged over several optical wavelengths due to the limited resolution of the film. The value recorded by the film is

$$S(\tau) \propto \int_{-\infty}^{\infty} E_1^4\left(t - \frac{z}{c}\right)dt + \int_{-\infty}^{\infty} E_2^4\left(t + \frac{z}{c}\right)$$
$$+ 4\int_{-\infty}^{\infty} E_1^2\left(t - \frac{z}{c}\right)E_2^2\left(t + \frac{z}{c}\right)dt. \tag{26}$$

The normalization of (26) gives

$$\frac{S(\tau)}{W} \propto 1 + 2G(\tau) \qquad (27)$$

which is identical to (18). If $E_1(t) = E_2(t)$ at $z=0$ is assumed, a bright vertical line will be recorded with a contrast ratio of 3 to 1 with respect to the background. This line will have a width $\Delta L = \Delta\tau_c c/n$, where $\Delta\tau$ is the pulse duration, c is the velocity of light, and n is the refractive index of the two-photon absorption-fluorescence dye.

It is important to realize that if a noise pulse having a time duration ΔT s and a spectral bandwidth of $\Delta\omega$ rad, in which $\Delta T \gg 2\pi/\Delta\omega$, is passed through a two-photon absorption-fluorescence cell, a bright line having a length corresponding to the coherence time $\Delta\tau_c \simeq 2\pi/\Delta\omega$ will be recorded. The evaluation of the pulse duration by this method requires caution because a very similar fluorescence structure is obtained from the radiation of an ideally mode-locked laser and from a free-running laser with the same oscillating bandwidth [63]. The proper interpretation of the data depends strongly on the contrast ratio in the photographic record. For a bandwidth-limited short pulse $(\Delta\tau \simeq 1/\Delta\nu)$, the maximum ratio of (27) is 3. For a free-running laser, the maximum ratio of (27) was calculated by Weber to be 1.5 [63]. It cannot be concluded at this point that the output of a free-running or Q-switched ruby, Nd:glass, or Nd:YAG laser *normally* consists of repetitive picosecond pulses. On the other hand, there are sufficient data available in the literature to support the conclusion that the output of regenerative pulse laser oscillators utilizing saturable absorbers consists of pulses in the 10^{-12} to 10^{-11} s range.

2) Experiments: In the analysis of the nonlinear optical system for the measurement of picosecond pulses, it has been assumed that phase matching in the nonlinear crystal is maintained over the entire bandwidth of the laser pulse. This sets an upper limit to the thickness of the crystal or a lower limit to the time resolution. ADP and KDP crystals having a thickness of 1 mm provide resolutions of approximately 1 ps. It is important to note that the group velocities of the fundamental and second harmonic frequencies must be approximately equal in order to have the envelope of the second harmonic pulse equal to the square of the fundamental pulse [79]–[82]. If the group velocities of the two harmonically related frequencies are not equal, the second harmonic pulse will have a flat top with a time duration $L_c/[v_g(2\omega)] - L_c/v_g(\omega)$, where L_c is the crystal length, and $v_g(\omega)$ and $v_g(2\omega)$ are the group velocities of the fundamental and second harmonic frequencies, respectively. Fig. 19 illustrates plots of the phase matching angles as a function of wavelength for KDP and LiNbO$_3$. Note that for KDP the curve is relatively flat around $\lambda_0 = 1.06$ microns, therefore making it possible to phase match easily over several hundred angstroms. Fig. 20 illustrates the spectrum of the second harmonic of a Nd:glass laser's picosecond pulses generated by a 1 mm thick KDP crystal. The three spectra were taken around the phase matching angle condition with a change of 0.2° between each of the three shots. It is evident

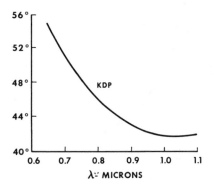
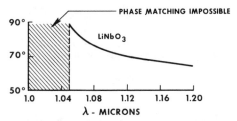

Fig. 19. A plot of the second harmonic phase matching angles as a function of wavelength for KDP and LiNbO$_3$.

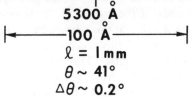

Fig. 20. The second harmonic spectrum of the picosecond pulses generated by an Nd^{3+}:glass laser in 1 mm thick KDP crystals.

that the spectral width of the second harmonic picosecond pulses can be conveniently adjusted by varying the phase matching angle.

The measurement of picosecond laser pulses by the second harmonic technique was performed by Armstrong [74], Glenn and Brienza [75], and Maier, Kaiser, and Giordmaine [76]. Glenn and Brienza also found that the early pulses in the train of pulses emitted by their laser were shorter than the later pulses in the train. Since their laser pulses had a variable pulse width, an additional parameter was needed to characterize the pulses in their measurement. This characterization was needed so that similar pulses in successive firing of the laser could be compared. The

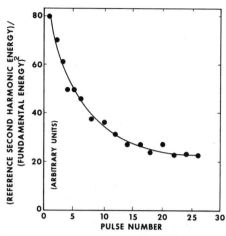

Fig. 21. The behavior of the ratio of the reference second harmonic energy to fundamental energy squared (i.e., conversion efficiency) as a function of the number of pulses of an Nd^{3+}:glass laser.

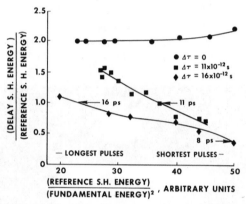

Fig. 22. The variation of the intensity correlation as a function of conversion efficiency of the picosecond pulses emitted by an Nd^{3+}:glass laser.

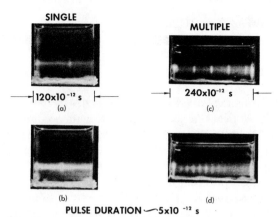

Fig. 23. Photographs of picosecond laser pulses taken with the two-photon absorption-fluorescence measurement technique.

parameter chosen was the conversion efficiency, i.e., the ratio of the energy of a reference second harmonic pulse W_r to the square of the energy of the fundamental pulse W_f^2. The highest value of this parameter occurs at the start of the pulse train and then decreases uniformly with the increasing number of pulses. This behavior is illustrated by Fig. 21 [75]. The intensity correlation ratio $S(\tau)/W_r$, plotted as a function of W_r/W_f^2 for several different delays, is illustrated by Fig. 22. The energy correlation function for a pulse of given width may be determined by drawing a vertical line on the graph and reading the value of the correlation as a function of the delay time. If the correlation function is assumed to be Lorentzian, then the shortest pulses obtained in this set of data had a full width at half-maximum of 8×10^{-12}. Treacy [23] has also noticed a periodic variation in pulse width from pulse to pulse throughout the pulse train. The periodic variation was noticed by comparing the second harmonic signal generated by the compressed picosecond pulses with the second harmonic signal generated by the uncompressed pulses.

A tentative explanation for the behavior of the pulses recorded in [75] can be summarized as follows. The action of the saturable dye in the laser cavity tends to shorten the pulse by reason of its nonlinear absorption. The theoretical

limiting width as determined from the spectral width is of the order of 10^{-13} s. The dye shortens the initial low-amplitude pulses to the same limiting widths. The pulse amplitude then begins to grow rapidly. For sufficiently high intensity, the dye will bleach completely at the beginning of each passage of the pulse. Since the dye is saturated for most of the pulse width, it will no longer be effective in shortening the pulses. Thereafter, the width of the pulse is determined by the physical properties of the laser cavity. Such factors as dispersion of the glass rod and dye cell, inhomogeneities of the optical path, transverse modes, diffraction limit, energy storage properties of mirrors, and saturation effects of the gain profile all contribute to the growth of the time duration of the pulses. The cumulative effects grow as the square root of the number of passes which could account for a near saturation in the growth of the pulse width as was experimentally observed [75]. It has not been definitely determined at the present time whether all pulse regenerative laser oscillators have variable pulse-width outputs as a function of time.

Fig. 23 presents typical data obtained by the two-photon absorption-fluorescence measurement technique. Fig. 23(a) shows the bright fluorescent line due to the overlap of a pulse with itself or with its immediate predecessor. The fact that the bright line is the overlap point of two pulses was confirmed by the observation that a movement of the reflecting mirror in (a) or the dye cell in (b), through a given distance, caused the bright line to move a corresponding distance in the dye cell. Rhodamine 6G was utilized as the dye medium. In an ethanol solution this dye has its primary absorption peak at a wavelength that is quite close to the 5300 Å second harmonic of the Nd:glass laser. In addition, it has another strong absorption line which peaks very close to the second harmonic of the ruby laser line. The two-photon absorption and subsequent fluorescence at 5500 Å of this dye is relatively much larger than for the 1,2,5,6-dibenzanthracene (DBA) dye previously used in such experiments [77]. An additional advantage of Rhodamine 6G is that the experiment can be performed by direct irradiation at either 1.06 microns or 6943 Å without requiring the conversion of these wavelengths into their second harmonic before irradiating the dye. The use of Rhodamine 6G dye in this experimental technique has made possible the measurement of a single ultrashort pulse with only one firing of the laser.

Under certain conditions multiple bright lines were obtained as shown in Fig. 23(c) and (d). The optical path length between the bright lines corresponds to one-half of the actual pulse separation in the laser output. The time scales in Fig. 23 have taken this factor into account, and therefore represent the actual pulse separation in time. These multiple pulses are caused by mode selection within the laser bandwidth as previously described in this paper (see Figs. 8 and 9). The presence of the reflecting mirror of Fig. 18(a) can act as a secondary cavity external to the main laser cavity, and gives rise to a channeled spectrum in the frequency domain or multiple pulses in the time domain [see Fig. 23(c)]. Experiments were also performed with the insertion of optical flats into the laser cavity normal to the laser axis. Multiple bright lines were again obtained with the separation of the lines in time equal to twice the optical thickness of the flat in the laser [see Fig. 23(d)]. Pulse widths as short as 2 ps and as long as 25 ps were recorded by this technique. It should be noted, however, that the photograph of Fig. 23 is an average over the entire pulse train and not the width characteristic of a single pulse in the train.

Weber [63] has noted the danger in drawing the conclusion that normally free-running or Q-switched wide-bandwidth lasers (such as ruby, Nd:glass, or YAG:Nd) consist of picosecond pulses on the basis of data obtained solely by the two-photon absorption-fluorescence measurement technique [60]–[62]. He has pointed out that a unique assignment can be made only if the contrast ratio is known with great accuracy, and has shown that the fluorescence record from N modes of equal-amplitude, equal-frequency separation and random phase relationships yields a maximum contrast ratio of 1.5. For the same case, but with fixed phase relationships between the modes, a maximum contrast ratio of 3 is obtained with this experimental technique. This second case is the mode-locked case. Similar results have also been obtained by Klauder *et al.* [83] and their measurement of the contrast ratio of the mode-locked pulses from glass:Nd^{3+} yielded 2 instead of 3. This discrepancy between theory and experiment could arise because of the simplified plane wave approximation utilized in the analysis instead of the statistical approach utilized by Ducuing and Bloembergen in their study of fluctuation in nonlinear optical processes [84].

The simplicity of the two-photon absorption-fluorescence technique is one of its major advantages if a measurement of the contrast ratio does not have to be taken. If the pulse-width measurement obtained by this technique is greater than the inverse of the spectra bandwidth of the pulse, then the measured width is the actual pulse width and a contrast ratio measurement does not have to be performed. If a periodic pulse train consisting of subnanosecond laser pulses are displayed on an oscilloscope, then the measurement of the pulse width obtained from the two-photon absorption-fluorescence can be considered to be the true pulse width with a high degree of certainty without contrast ratio measurement.

IV. RECENT EXPERIMENTS PERFORMED WITH PICOSECOND PULSES

A. Measurement of Ultrashort Decay Times

Since early in the 1930's it has been possible to measure fluorescence lifetimes shorter than 1 ns [85]. However, the fluorometers devised for these measurements are complex and cumbersome. Moreover, inasmuch as the actual decay curve is not observed, the measurements are indirect. The high intensity and short duration of mode-locked pulses make them ideal for a number of different types of lifetime measurements.

To observe a fluorescent decay time, a short-duration, high-intensity source is required to excite the fluorescence and a wide-bandwidth detection system is needed to detect it. Regenerative pulse lasers, as discussed earlier in this paper, are an ideal excitation source for such measurements. The use of the fundamental and second harmonic of ruby and neodymium lasers [86] offers promise of revolutionizing these measurement techniques [41]. With the continuously mode-locked lasers and crossed-field photomultiplier [87] used in conjunction with sampling oscilloscopes, an overall detection rise time as short as 60 ps can be obtained. With Q-switched lasers and traveling-wave oscilloscope, rise times of the order of a few tenths of a nanosecond can be obtained. In addition, the use of picosecond laser pulses has also made possible the direct measurement of population decay time in the picosecond region for the first time [39], [40]. Picosecond pulses have also been used for probing molecular orientation dynamics in liquids by the use of pulse times on the scale of the molecular orientation time [88].

The experimental method used to measure such short decay times is as follows. Absorption of a very intense light pulse of ultrashort duration by a dilute sample of dye molecules will prepare all the molecules in the light path in an excited electronic state. The subsequent decay of the population of this state is then probed by a "probe" pulse of light which can be delayed continuously to arrive before, during, or after the intense "preparing" pulse. The transmission of the sample for the probe beam is proportional at each instance to the concentration of the ground state dye molecules. As the sample is prepared in its excited state, this transmission will rise abruptly. As the ground state repopulates through decay from the excited state, transmission will decrease again. Table II illustrates the lifetimes of Eastman 9740 [39] and 9860 [40] saturable dyes directly measured by this technique. These two lifetimes are important to the use of these dyes in the generation of picosecond pulses. For example, these lifetimes indicate that pulse repetition rates as high as 10^{-11} s are possible with these dyes.

Mack investigated the fluorescence lifetime of several dyes commonly used for Q-switching ruby lasers and for liquid lasers pumped by ruby second harmonic [41]. The experimental arrangement was particularly simple. The dye cell was irradiated by the second harmonic ruby radiation for the measurements of Table III, and directly by the ruby

TABLE II
PICOSECOND ABSORPTION DECAY TIME OF EASTMAN 9740 AND 9860 SATURABLE ABSORBERS AS MEASURED BY REFERENCES [39] AND [40]

Dye	Reference	Direct Measurement (ps)	Deduced Measurement (ps)
9740	[39]	35	25
9860	[40]	9	6

TABLE III
NANOSECOND FLUORESCENT LIFETIME OF SOME LIQUID LASER DYES AS MEASURED BY REFERENCE [41]

Dye	Solvent	Fluorescent Peak (Å)	Fluorescent Lifetime (ns)
Acridine red	ethanol	5800	2.4
Acridine yellow	ethanol	5050	5.2
Sodium fluorescein	ethanol	5270	6.8
Rhodamine 6G	ethanol	5550	5.5
Rhodamine 6G	water	5550	5.5
Acridone	ethanol	4370	11.5
Anthracene	methanol	4000	4.5

TABLE IV
NANOSECOND FLUORESCENT LIFETIME OF SOME Q-SWITCHING RUBY LASER DYES AS MEASURED BY REFERENCE [41]

Dye	Solvent	Absorption Peak (A)	Fluorescent Peak (Å)	Fluorescent Lifetime (ns)
CAP	ethanol	6700	7550	10.1
CAP	methanol	6710	7550	10.3
CAP	chloronaphthalene	6970	7380	8.0
VP	nitrobenzene	6980	—	4.1
VP	chloronaphthalene	7010	—	4.2
CC	methanol	7060	7400	0.5

CAP = chloro-aluminum phthalocyanine
VP = vanadyl phthalocyanine
CC = cryptocyanine.

radiation for the measurements shown in Table IV. A copper sulfate solution and a sharp cutoff interference filter were placed in front of the photodetector to isolate the detector from the scattered ruby and second harmonic radiation, respectively. The overall response time of the detection system was 0.45 ns. The decay times were determined by fitting an exponential to each of the decaying portions in the oscillographs. The values should be accurate to ± 20 percent. The fluorescent lifetimes of Rhodamine 6G is particularly interesting since it is so useful in two-photon absorption-fluorescence measurement of picosecond pulses as discussed in Section III. In addition, it has been pumped by the second harmonic picosecond pulses of Nd lasers to generate ultrashort pulses at its fluorescent wavelength in the yellow portion of the spectrum [86].

It is now well known that the observed gain in excess of the calculated values arising in the stimulated Raman effect can be explained by self-focusing of the pumping laser light which arises primarily from the Kerr effects. The response time of the Kerr effect is given by the Debye rotation time Γ for molecules driven by an ac field:

$$\Gamma = \frac{4\pi a^3 \eta}{3kT} \qquad (28)$$

where η is the viscosity, T the temperature, and a the effective molecular radius. Values for Γ of 10^{-11} to 10^{-12} s have been measured in liquid by various indirect techniques. When an optical pulse duration $\Delta\tau$ is greater or equal to Γ, the anisotropic molecules can rotate in response to the pulse, but when $\Delta\tau < \Gamma$, the molecules cannot respond and the self-focusing threshold will increase. Shapiro, Giordmaine, and Wecht clearly showed in their experiments a definite tendency toward less Raman scattering when Γ became longer than $\Delta\tau$ [88].

B. Light Amplification in Saturable Absorbers

When power at a pump frequency and one or two sidebands is incident on a saturable power absorber, the power absorbed by the saturable absorber will vary at the difference frequency. If the magnitude of the absorption is nonlinear with respect to the power absorbed, the resulting changes in absorption will in turn alter the amount of power transmitted through the saturable absorber. If the recovery time of the saturable absorber is faster than the beat frequency between the pump and the sidebands, the absorption will vary at the difference frequency and thus pump energy into the sidebands. In principle, this process can produce more output signal at the sidebands than was present originally. This technique was first utilized to amplify millimeter waves [89].

Mack [90] has found that a saturable absorber driven into saturation by an intense light pulse from a mode-locked ruby laser can amplify a weaker light pulse simultaneously incident upon the medium. Energy gains as high as 20 times have been observed with a path length of only 1 cm in the saturable absorbing dye solution. Previous theoretical [91] and experimental [92] investigations of hole burning in saturable absorbers have indicated that the weak wave attenuation is merely reduced as compared to that for the strong wave, i.e., the hole in the absorption spectrum reaches only to the zero attenuation point. The fact that amplification of the weak wave can be achieved means that the hole in the absorption spectrum actually penetrates through to the negative absorption region. It has since been found that in addition to the gain from the saturable absorption process described above, gain from a new stimulated thermal scattering process was also taking place in these experiments [93]. This new stimulated thermal scattering process has also been observed in linear absorbers [93].

An intuitive understanding of the optical amplification process in saturable absorbers can be gained with the aid of Fig. 24. If two colinear optical waves of different frequency and of substantially different intensity are incident on such an absorber, a small amplitude modulation at the beat frequency will occur in the dye. If we assume that the intensity is sufficient to drive the dye into the nonlinear region of its transmission transfer function and that the relaxation time of the dye is sufficiently fast to follow the modulation frequency, the nonlinear transmissivity will enhance and distort the modulation as shown in Fig. 24. The distortion

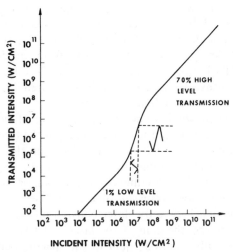

Fig. 24. Simplified explanation of optical gain in saturable absorbers by means of the transmission characteristics of cryptocyanine in methanol.

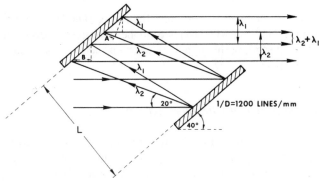

Fig. 25. Schematic diagram of two plane blazed gratings with 1200 lines/mm used to compress chirped picosecond pulses emitted by a dye mode-locked Nd^{3+}:glass laser.

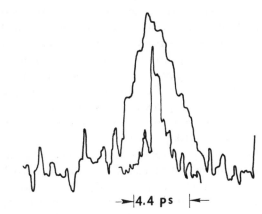

Fig. 26. Microdensitometer traces of a compressed and uncompressed pulse taken with the two-photon absorption-fluorescence technique.

indicates the generation of additional sidebands. The enhancement of the modulation indicates that the initial sideband intensity has grown relative to the intense component even though the overall intensity of the output beam is less than the input beam.

C. Compression of Picosecond Pulses

The fact that ultrashort optical pulses generated by an Nd:glass laser are typically 4 to 10×10^{-12} s long and have a spectral width of the order of 100 Å suggests that part of the observed spectral width is due to some kind of phase modulation of the carrier wave. One is led to such a conclusion from the fact that for such a bandwidth, one would expect the pulse length to be an order of magnitude smaller. Treacy has recently found that a quadratic term in the phase $\phi(t)$ exists in the picosecond pulses emitted by dye mode-locked Nd^{3+}:glass lasers [21], [23]. This phase variation $\phi(t) = \omega t + 1/2\beta t^2$ corresponds to a linear sweep of the carrier frequency $d\phi/dt = \omega + \beta t$, where the constant β is positive. This optical frequency sweep is analogous to the frequency swept wave trains emitted by chirp radar.

In chirp radar systems, the transmitted pulse is of relatively long duration $\Delta\tau$, during which time the instantaneous frequency is swept over some range Δf. The return pulse is passed through a dispersive network providing a differential delay $\Delta\tau$ over the frequency range Δf. As a result, the energy at the beginning of the pulse is delayed so as to reach the end of the network at the same time as the energy at the end of the pulse. The duration of the compressed pulse $(\Delta\tau_c)$ is of the order of $\Delta\tau_c \simeq \Delta f^{-1}$ or a time-bandwidth product of unity. The pulse normally emitted by a dye mode-locked Nd^{3+}:glass laser has a $\Delta\tau\Delta f \simeq 10$ to 20.

Fig. 25 shows the essential elements of the optical dispersion device used to compress the picosecond laser pulses. A pair of plane blazed gratings with 1200 lines per mm are arranged with their face and rulings parallel so that light of any wavelength, after diffracting twice, will exit in a beam parallel to the input beam. Since different wavelength components of a beam are diffracted through different angles, the transit time is approximately a linearly

increasing function of wavelength λ. It is easy to see that an input pulse with a positive carrier frequency sweep will be compressed by such a grating pair since the long wavelength components arrive at the input first and take a longer time to traverse the system. It is also easy to see that the time delay between the two extreme frequencies of the frequency swept pulse is proportional to the separation of the gratings.

Fig. 26 illustrates microdensitometer traces from photographs of the uncompressed and compressed pulses taken with the two-photon absorption-fluorescence technique by Treacy. The wide pulse is the pulse prior to entering the gratings, and the narrow pulse is the pulse leaving the gratings. The pulse is compressed to approximately 1 ps. Later data have resulted in pulse widths down to 4×10^{-13} s. It is not presently known whether the dye or the laser medium is primarily responsible for generating the chirp characteristic of Nd^{3+}:glass laser pulses. Treacy has also used the narrow compressed pulse to optically sample the uncompressed pulse for displaying the shape of its intensity envelope [23].[3]

Giordmaine *et al.* [97] proposed using an ultrasonic frequency translator [98], [99] for chirping the mode-locked pulses from a He-Ne or an Nd^{3+}:YAG laser. The He-Ne chirped laser would require 50 km of bromobenzene for the

---

[3] It was recently brought to our attention that picosecond pulses from a Nd:glass laser were also utilized to sample the subnanosecond pulses emitted by a He-Ne laser by M. A. Duguay and J. W. Hansen, "Optical sampling of subnanosecond light pulses," *Appl. Phys. Lett.*, vol 13, pp. 178–180, September 1968.

dispersive element to compress the pulses. The 50 ps pulses from a Nd^{3+}:YAG laser would be compressed down to 0.4 ps after going through 33 m of bromobenzene or 25 m of nitrobenzene.

D. Generation of Ultrashort Acoustic Pulses

The transient surface heating of materials by repetitive, high-power, picosecond laser pulses is capable of generating acoustic shocks of very short time duration [100]. Ready [101] has calculated that the thermal gradients produced by the absorption of a typical Q-switched ruby laser pulse could be as high as 10^6 deg/cm, with a temperature rate of change as high as 10^{16} deg/s. In a comprehensive treatment of the problem of transient surface heating, White [102] has shown that the conversion efficiency by which the sound is produced varies linearly with the incident peak power density and is inversely proportional to the first power of the sound frequency. It can therefore be expected that repetitive, high-peak-power, picosecond laser pulses can generate acoustic shocks having harmonic components well into the microwave region [100].

The output of a pulse regenerative Nd: glass laser having a repetition rate of 200 MHz, pulse widths between 10^{-12} and 10^{-11} s, and an average energy per picosecond pulse of approximately 1 mJ was used to irradiate a metal film deposited on the end of an $LiNbO_3$ bar. The entire pulse train contained from 100 to 150 individual pulses and lasted about 0.4 to 0.6 μs. The thermal stressing caused by the partial absorption of the laser pulse train by the metal film propagated acoustic shocks into the crystalline bar. The shocks traveled the length of the $LiNbO_3$ bar and were repeatedly reflected from the two ends of the bar, thus forming an echo pattern. Each time the shocks were reflected from the end of the bar opposite the thin-film end, a voltage was generated by the piezoelectric properties of the $LiNbO_3$ crystal. This voltage was detected by a radio receiver having an IF bandwidth of 8 MHz. The receiver was tuned to multiples of the 200 MHz laser pulsing rate and the radio pulse displayed on an oscilloscope. At room temperature, several echoes were observed of the fifteenth harmonic at 3 GHz. As evidenced by the sharpness of the tuning of the radio receiver, the sound was strictly confined to the harmonic frequencies of the laser pulse repetition frequency. It is felt that, besides providing a convenient method of producing discrete sets of sound frequencies well up into the microwave region, this technique demonstrates the possibility of producing ultrashort acoustic pulses with ultrashort light pulses.[4]

E. Variable Wavelength Picosecond Pulses

The availability of ultrashort optical pulses with any desired wavelength would greatly increase their applicability to studies of nonlinear transient optical effects, spectroscopy, and lifetime measurements.

It has been reported that a large number of organic dye solutions exhibit laser action when excited with a short-duration, high-intensity pump pulse [103]–[106]. Pump pulses from Q-switched lasers and from specially constructed flash tubes have been employed. The output spectra from such dye lasers are quite broad, extending in some cases over a few hundred angstroms.

It is well known that a periodic modulation of the gain of a laser medium will lead to mode-locking if the frequency of the modulation is equal to, or a multiple of, the difference frequency between longitudinal modes of the laser, and if the upper state laser lifetime is shorter than, or comparable to, the optical loop transient time of the laser cavity. If the pumping signal for a dye laser consists of a mode-locked train of pulses, the gain of the laser will have a periodic variation with a period equal to the spacing between the pumping pulses. If the length of the dye laser cavity is equal to, or a submultiple of, the length of the cavity of the laser producing the pump pulses, then the mode-locking condition will be satisfied and the output of the dye laser will consist of a series of pulses.

Initial experiments were performed with Rhodamine 6G and Rhodamine B dyes [86]. In ethanol solution these dyes have absorption peaks at 5260 Å and 5500 Å, respectively. The peaks are sufficiently broad to allow efficient pumping with the 5300 Å second harmonic of the Nd laser. The experimental arrangement for laser-pumped, mode-locked dye lasers is as follows. A 2.5 cm diam by 5 cm long dye cell was placed in a laser cavity having a 1 m radius of curvature mirrors with a reflectivity in excess of 90 percent between 5000 Å and 6000 Å. A KDP crystal was also placed in the dye laser cavity. The pulse regenerative Nd pumping laser beam was propagated through one of the mirrors and irradiated the KDP crystal, thereby producing the second harmonic radiation required for pumping the dye. The fluorescence lifetime of the Rhodamine 6G is 5.5 ns (see Table II). The Nd laser pulse period was approximately 5 ns.

Laser action in the dye was observed with its optical cavity length equal to 1, $\frac{1}{2}$, and $\frac{1}{3}$ times the length of the pumping laser cavity. The dye laser output shows very small fluorescence until the pumping signal reaches threshold, at which point the dye laser breaks into oscillation and produces a series of pulses at the repetition rate determined by its cavity length. The laser output could be varied from 5600 Å to 5900 Å by changing the composition of the solutions. The wide use of the stagger-tuning principle often utilized in the RF region can be utilized to great advantage in obtaining spectral lines of enormous widths by the proper mixing of several organic dyes [107].

F. Optical Rectification Studies with Picosecond Pulses

Optical rectification was first observed by Bass et al. in the form of an induced dc voltage across a crystal of potassium dihydrogen phosphate during the passage of an intense ruby laser pulse through the crystal [108]. Considerable difficulty is normally encountered in the unambiguous observation of the effect because of the short-duration, low-voltage signals produced which, besides requiring broad-band and thus insensitive electronic equipment, must be distinguished from electrical noise from the

---

[4] More recent work has resulted in the direction of 10 GHz signals in $LiNbO_3$ at room temperatures.

laser and spurious pyroelectric and acoustic signals. The periodic pulsating characteristics of pulse regenerative lasers allow the use of high sensitivity and selectivity of radio receivers in detecting the effect. The large signal-to-noise ratio available with such RF detection enables one to identify easily (and thus eliminate) spurious acoustic signals by their repetitive echo characteristics. In addition, signals from relatively slow effects such as the pyroelectric effect can be eliminated by detecting the microwave harmonics of the rectification signals.

Harmonically rich electrical pulses generated in crystals of $LiNbO_3$ and KDP have been observed up to approximately 10 GHz [109]. The experimental arrangement was as follows. The output of an Nd pulse regenerative oscillator having a repetition rate of 275 MHz was directed through crystals of $LiNbO_3$ or KDP either along or perpendicular to the Z-axis of the crystals. A flat-ended coaxial probe fixed against the surface normal to the Z-axis was used to sense the induced fields present at the surface. The RF signals were measured to be approximately 2 mV and were detected by a superheterodyne receiver having a 20 MHz bandwidth.

The observed signals were recorded with a signal-to-noise ratio of up to 20 to 25 dB and followed the general features of the envelope of the laser pulse train. Signals were recorded at each harmonic frequency from 275 MHz to 9.076 GHz and were limited only by the equipment available. The amplitude of the received signals followed the sin 2θ dependence as the angle θ between the plane of polarization of the light and the x-axis of the KDP crystal was rotated.

The broad-band response of the optical rectification effects, coupled with greatly increased sensitivity obtained with radio receivers and repetitive picosecond laser pulses, suggests the use of the effects for detecting picosecond light pulses and for generating millimeter wave radiation. A unique characteristic of such a microwave generating system would be its broad-band passive generating element, i.e., the nonlinear crystal, whose operating frequency can be easily changed and precisely specified by adjustment of the laser cavity length. The upper frequency of the device is limited only by the harmonic content of the available ultrashort laser pulses.

G. Optically Generated Plasma Studies

Using the focused radiation from high-powered, Q-spoiled lasers, high-density, high-temperature plasmas have been produced in gases [8], [110]–[114], from solid surfaces [115]–[118], and from single micron-sized solid particles [119]. The resulting plasmas have been employed in fundamental studies of radiation-matter interactions, to excite high-temperature gas reactions, to produce extremely thin vapor-deposited coatings, as spectral sources for microanalysis, and to study highly excited ions. The use of lasers permits the generation of plasmas with a wide range of composition, density, ionization, and temperature properties. Laser-generated gas breakdown plasmas can be produced with electron densities greater than 10^{19} cm^{-3}

and temperatures exceeding 100 eV. In the solid surface plasma plume, atoms in very high states of ionization can be obtained. A spherically symmetric expanding plasma ball can be generated by laser beam irradiation of a single solid particle. The plasma generated from the particle has a set of properties which makes it of particular interest for the study of the interaction of an expanding plasma with a magnetic field. Significant containment of the laser-irradiated single-particle plasma with low-intensity magnetic confinement has been achieved to date to indicate one use of laser-produced plasmas.

Basov and Krokhin [120] calculated that laser powers in excess of 10^9 W were needed to heat a laser-generated lithium deuteride plasma up to a temperature at which thermonuclear neutron emission may be observed. Unfortunately, optical damage to optical components resulting from such large optical intensities made it difficult to carry out such experiments at that time. It was subsequently noticed that the power damage threshold increased with decreased pulse duration. The possibility of reaching thermonuclear neutron emission temperatures with high-intensity, ultrashort laser pulses presented itself when single picosecond laser pulses became available [120]–[122]. The use of an Nd:glass single picosecond laser pulse generator in conjunction with five cascaded 60 cm long amplifiers has resulted in the generation of 20 J of energy in 10^{-11} s [123].

Preliminary results of the use of these high-energy, single ultrashort pulses in generating thermonuclear neutron emission from laser-heated lithium deuteride surfaces have been reported [123]. A total number of four coincidence neutron detections were reported in an experimental series of 14 shots. This number was calculated to be 20 times higher than the probability of obtaining accidental coincidence of a background pulse with the ultrashort laser pulse.

CONCLUSIONS

In approximately two years, a 5 order of magnitude jump has been made in our ability to generate high-peak-power, ultrashort laser pulses, i.e., from 10^{-8} s to 10^{-13} s, and a 3 order of magnitude jump in our ability to generate high peak powers, i.e., from 10^9 W to 10^{12} W. To keep up with these breakthroughs, researchers have devised techniques for measuring the time duration of ultrashort pulses which represent a 3 order of magnitude improvement over past direct measurement capability, i.e., from 10^{-10} s to 10^{-13} s. Picosecond laser pulses have proved to be valuable as scientific tools in controlled thermonuclear research, in the generation of acoustic shock waves, in nonlinear optical experiments, in the measurement of picosecond relaxation times, and in the measurement of the response time of molecular systems. It is obvious that researchers have not yet even broken the surface in finding scientific applications for picosecond laser pulses. As of this writing, no breakthrough has been made in the practical application of these "bullets of light," but the potential application of these short laser pulses in information processing, ranging, topography mapping, and high-speed photography appears extremely promising.

REFERENCES

[1] R. W. Hellwarth, "Control of fluorescent pulsations," in *Advances in Quantum Electronics*, J. R. Singer, Ed. New York: Columbia University Press, 1961, pp. 334–341.

[2] F. J. McClung and R. W. Hellwarth, "Giant optical pulses from ruby," *J. Appl. Phys.*, vol. 33, pp. 828–829, March 1962.

[3] R. J. Collins and P. Kisliuk, "Control of population inversion in pulsed optical masers by feedback modulation," *J. Appl. Phys.*, vol. 33, pp. 2009–2011, June 1962.

[4] A. J. DeMaria, R. Gagosz, and G. Barnard, "Ultrasonic-refraction shutter for optical maser oscillators," *J. Appl. Phys.*, vol. 34, pp. 453–456, March 1963.

[5] P. P. Sorokin, J. J. Luzzi, J. R. Lankard, and G. D. Pettit, "Ruby laser Q-switching elements using phthalocyanine molecules in solution," *IBM J. Research and Develop.*, vol. 8, pp. 182–194, April 1964.

[6] P. Kafalas, J. I. Masters, and E. M. E. Murray, "Photosensitive liquid used as a nondestructive passive Q-switch in a ruby laser," *J. Appl. Phys.*, vol. 35, pp. 2349–2350, August 1964.

[7] L. M. Frantz and J. S. Nodvik, "Theory of pulse propagation in a laser amplifier," *J. Appl. Phys.*, vol. 34, pp. 2346–2349, August 1963.

[8] R. G. Meyerand, Jr., and A. F. Haught, "Gas breakdown at optical frequencies," *Phys. Rev., Lett.*, vol. 11, pp. 401–402, November 1963.

[9] P. A. Franken and J. F. Ward, "Optical harmonics and nonlinear phenomena," *Rev. Mod. Phys.*, vol. 35, pp. 23–39, January 1963.

[10] N. Bloembergen, *Nonlinear Optics*. New York: W. A. Benjamin, Inc., 1965.

[11] R. Y. Chiao, E. Garmire, and G. H. Townes, "Raman and phonon masers," *Quantum Electronics and Coherent Light, Proc. Internat'l School of Physics, Enrico Fermi, Course XXXI*, P. M. Miles and C. H. Towns, Eds. New York: Academic Press, 1964, pp. 326–338.

[12] I. D. Abella, N. A. Kurnit, and S. R. Hartmann, "Photon echoes," *Phys. Rev.*, vol. 141, pp. 391–406, January 1966.

[13] S. L. McCall and E. L. Hahn, "Self-induced transparency by pulsed coherent light," *Phys. Rev. Lett.*, vol. 18, pp. 908–911, May 1967.

[14] E. Garmire, R. Y. Chiao, and C. H. Townes, "Dynamics and characteristics of the self-trapping of intense light beams," *Phys. Rev. Lett.*, vol. 16, pp. 347–349, February 1966.

[15] J. A. Giordmaine and R. C. Miller, "Tunable coherent parametric oscillation in $LiNbO_3$ at optical frequencies," *Phys. Rev. Lett.*, vol. 14, pp. 973–976, June 1965.

[16] A. J. DeMaria, D. A. Stetser, and W. H. Glenn, Jr., "Ultrashort light pulses," *Science*, vol. 156, pp. 1557–1568, June 1967.

[17] C. C. Cutler, "The regenerative pulse generator," *Proc. IRE*, vol. 43, pp. 140–148, February 1955.

[18] A. J. DeMaria, D. A. Stetser, and H. Heynau, "Self-mode-locking of laser with saturable absorbers," *Appl. Phys. Lett.*, vol. 8, pp. 174–176, April 1966.

[19] M. DiDomenico, Jr., "Small-signal analysis of internal (coupling-type) modulation of lasers," *J. Appl. Phys.*, vol. 35, pp. 3870–3876, October 1964.

[20] A. Yariv, "Internal modulation in multimode laser oscillators," *J. Appl. Phys.*, vol. 36, pp. 388–391, February 1965.

[21] E. B. Treacy, "Compression of picosecond light pulses," *Phys. Lett.*, vol. 28A, pp. 34–35, October 1968.

[22] A. J. DeMaria, "Mode-locking opens door to picosecond pulses," *Electronics*, vol. 41, pp. 112–122, September 16, 1968.

[23] E. B. Treacy, "Measurement of picosecond pulses substructure using compression techniques," *Appl. Phys. Lett.*, to be published.

[24] M. Hercher, "An analysis of saturable absorbers," *Appl. Opt.*, vol. 6, pp. 947–954, May 1967.

[25] L. M. Frantz and J. S. Nodvik, "Theory of pulse propagation in a laser amplifier," *J. Appl. Phys.*, vol. 34, pp. 2346–2349, August 1963.

[26] R. Bellman, G. Birnbaum, and W. G. Wagner, "Transmission of monochromatic radiation in a two-level material," *J. Appl. Phys.*, vol. 34, pp. 780–782, April 1963.

[27] L. E. Hargrove, R. L. Fork, and M. A. Pollack, "Locking of He-Ne laser modes induced by synchronous intracavity modulation," *Appl. Phys. Lett.*, vol. 5, pp. 4–5, July 1964.

[28] A. J. DeMaria, C. M. Ferrar, and G. E. Danielson, Jr., "Mode locking of a Nd^{3+}-doped glass laser," *Appl. Phys. Lett.*, vol. 8, pp. 22–24, January 1966.

[29] T. Deutsch, "Mode-locking effects in an internally modulated ruby laser," *Appl. Phys. Lett.*, vol. 7, pp. 80–82, August 1965.

[30] S. E. Harris and R. Targ, "FM oscillation of the He-Ne laser," *Appl. Phys. Lett.*, vol. 5, pp. 202–204, November 1964.

[31] M. H. Crowell, "Characteristics of mode-coupled lasers," *IEEE J. Quantum Electronics*, vol. QE-1, pp. 12–20, April 1965.

[32] A. J. DeMaria, "Ultrasonic-diffraction shutters for optical maser oscillators," *J. Appl. Phys.*, vol. 34, pp. 2984–2988, October 1963.

[33] L. C. Foster, M. D. Ewy, and C. B. Crumly, "Laser mode locking by an external doppler cell," *Appl. Phys. Lett.*, vol. 6, pp. 6–8, January 1965.

[34] A. J. DeMaria and D. A. Stetser, "Laser pulse-shaping and mode-locking with acoustic waves," *Appl. Phys. Lett.*, vol. 7, pp. 71–73, August 1965.

[35] G. A. Massey, M. K. Oshman, and R. Targ, "Generation of single-frequency light using the FM laser," *Appl. Phys. Lett.*, vol. 6, pp. 10–11, January 1965.

[36] D. A. Stetser and A. J. DeMaria, "Optical spectra of ultrashort optical pulses generated by mode-locked glass: Nd lasers," *Appl. Phys. Lett.*, vol. 9, pp. 118–120, August 1966.

[37] H. W. Mocker and R. J. Collins, "Mode competition and self-locking effects in a Q-switched ruby laser," *Appl. Phys. Lett.*, vol. 7, pp. 270–273, November 1965.

[38] M. DiDomenico, Jr., J. E. Geusic, H. M. Marcos, and R. G. Smith, "Generation of ultrashort optical pulses by mode-locking the YAG: Nd laser," *Appl. Phys. Lett.*, vol. 8, pp. 180–183, April 1966.

[39] J. W. Shelton and J. A. Armstrong, "Measurement of the relaxation time of the Eastman 9740 bleachable dye," *IEEE J. Quantum Electronics (Correspondence)*, vol. QE-3, pp. 696–697, December 1967.

[40] R. I. Scarlet, J. F. Figueria, and H. Mahr, "Direct measurement of picosecond lifetimes," Lab. of Atomic and Solid-State Physics, Cornell University, Ithaca, N. Y., Tech. Rept. 24, May 1968.

[41] M. E. Mack, "Measurement of nanosecond fluorescence decay time," *J. Appl. Phys.*, vol. 39, pp. 2483–2484, April 1968.

[42] V. I. Malyshev, A. S. Markin, and A. A. Sychev, "Mode self-synchronization in the giant pulse of a ruby laser with a broad spectrum," *Soviet Physics-JETP Lett.*, vol. 6, pp. 34–35, 1967.

[43] I. K. Krasyuk, P. P. Pashkin, and A. M. Prokhorov, "Ring ruby laser for ultrashort pulses," *Soviet Physics-JETP Lett.*, vol. 7, pp. 89–91, February 1968.

[44] M. L. Spaeth and W. R. Sooy, "Fluorescence and bleaching of dyes for a passive Q-switch laser," *J. Chem. Phys.*, vol. 48, pp. 2315–2319, 1968.

[45] M. E. Mack, "Mode-locking the ruby laser," *IEEE J. Quantum Electronics (Correspondence)*, vol. QE-4, pp. 1015–1016, December 1968.

[46] M. Hercher, "Single-mode operation of a Q-switched ruby laser," *Appl. Phys. Lett.*, vol. 7, pp. 39–41, July 1965.

[47] H. A. Heynau and A. W. Penney, Jr., "An application of mode-locked, laser generated picosecond pulses to the electrical measurement art," *IEEE Internat'l Conv. Rec.*, pt. 7, vol. 15, pp. 80–86, 1967.

[48] R. Harrach and G. Kachen, "Pulse trains from mode-locked lasers," *J. Appl. Phys.*, vol. 39, pp. 2482–2483, April 1968.

[49] D. R. Herriott and H. H. Schulte, "Folded optical delay lines," *Appl. Opt.*, vol. 4, pp. 883–889, August 1965.

[50] A. J. DeMaria, W. H. Glenn, Jr., and D. A. Stetser, "An extended path and digitally scanned optical pulse generator," *Appl. Opt.*, vol. 7, pp. 1405–1407, July 1968.

[51] A. A. Vuylsteke, "Theory of laser regeneration switching," *J. Appl. Phys.*, vol. 34, pp. 1615–1622, June 1963.

[52] W. R. Hook, R. H. Dishington, and R. P. Hilberg, "Laser cavity dumping using time variable reflection," *Appl. Phys., Lett.*, vol. 9, pp. 125–127, August 1966.

[53] A. W. Penney, Jr., and H. A. Heynau, "PTM single-pulse selection from a mode-locked Nd^{3+}:glass laser using a bleachable dye," *Appl. Phys. Lett.*, vol. 9, pp. 257–258, October 1966.

[54] M. Michon, J. Ernest, and R. Auffret, "Pulsed transmission mode operation in the case of a mode-locking of the modes of a non-Q-spoiled ruby laser," *Phys. Lett.*, vol. 21, pp. 514–515, June 1966.

[55] A. J. DeMaria, R. Gagosz, H. A. Heynau, A. W. Penney, Jr., and G. Wisner, "Generation and amplification of a subnanosecond laser pulse," *J. Appl. Phys.*, vol. 38, pp. 2693–2695, May 1967.

[56] N. G. Basov, P. G. Kryukov, V. S. Letokhov, and Yu. V. Senotskii, "Oscillation and amplification of ultrashort pulses of coherent light," *Internat'l Quantum Electronics Conf.* (Miami, Fla., May 1968), pp. 811–812.

[57] R. N. Lewis, E. A. Jung, G. L. Chapman, L. S. Van Loon, and T. A. Romanowski, "Some high voltage pulse techniques in use at Argonne," *IEEE Trans. Nuclear Science*, vol. NS-13, pp. 84–88, April 1966.

[58] W. B. Davenport, Jr., and W. L. Root, *An Introduction to the Theory of Random Signals and Noise*. New York: McGraw-Hill, 1958, pp. 87–99.

[59] W. R. Sooy, "The natural selection of modes in a passive Q-switched laser," *Appl. Phys. Lett.*, vol. 7, pp. 36–37, July 1965.

[60] M. A. Duguay, S. L. Shapiro, and P. M. Rentzepis, "Spontaneous appearance of picosecond pulses in ruby and Nd:glass lasers," *Phys. Rev. Lett.*, vol. 19, pp. 1014–1016, October 1967.

[61] S. L. Shapiro, M. A. Duguay, and L. B. Kreuzer, "Picosecond substructure of laser spikes," *Appl. Phys. Lett.*, vol. 12, pp. 36–37, January 1968.

[62] M. Bass and D. Woodward, "Observation of picosecond pulses from Nd:YAG lasers," *Appl. Phys. Lett.*, vol. 12, pp. 275–277, April 1968.

[63] H. P. Weber, "Comments on the pulse width measurement with two-photon excitation of fluorescence," *Phys. Lett.*, vol. 27A, pp. 321–322, 1968.

[63a] H. P. Weber and R. Dandliker, late paper presented at the International Quantum Electronics Conf., Miami, Fla., May 1968.

[64] W. E. Lamb, Jr., "Theory of an optical maser," *Phys. Rev.*, vol. 134, pp. A1429–A1450, June 1964.

[65] C. M. Ferrar, "Q-switching and mode-locking of a CF$_3$I photolysis laser," *Appl. Phys. Lett.*, vol. 12, pp. 381–383, June 1968.

[66] H. Statz and C. L. Tong, "Phase-locking of modes in lasers," *J. Appl. Phys.*, vol. 36, pp. 3923–3927, December 1965.

[67] P. W. Smith, "The self-pulsing laser oscillator," *IEEE J. Quantum Electronics*, vol. QE-3, pp. 627–635, November 1967.

[68] T. Uchida and A. Ueki, "Self locking of gas lasers," *IEEE J. Quantum Electronics*, vol. QE-3, pp. 17–30, January 1967.

[69] A. G. Fox and P. W. Smith, "Mode-locked laser and the 180° pulse," *Phys. Rev. Lett.*, vol. 18, pp. 826–828, May 1967.

[70] A. G. Fox, S. E. Schwarz, and P. W. Smith, "Use of neon as a nonlinear absorber for mode-locking a He-Ne laser," *Appl. Phys. Lett.*, vol. 12, pp. 371–373, June 1968.

[71] H. Statz, G. A. DeMars, and C. L. Tong, "Self-locking of modes in lasers," *J. Appl. Phys.*, vol. 38, pp. 2212–2222, April 1967.

[72] C. L. Tong and H. Statz, "Maximum-emission principle and phase-locking in multimode lasers," *J. Appl. Phys.*, vol. 38, pp. 2963–2968, June 1967.

[73] H. P. Weber, "Method for pulse width measurement of ultrashort light pulses generated by phase-locked lasers using nonlinear optics," *J. Appl. Phys.*, vol. 38, pp. 2231–2234, April 1967.

[74] J. A. Armstrong, "Measurement of picosecond laser pulse widths," *Appl. Phys. Lett.*, vol. 10, pp. 16–17, January 1967.

[75] W. H. Glenn and M. J. Brienza, "Time evolution of picosecond optical pulses," *Appl. Phys. Lett.*, vol. 10. pp. 221–223, April 1967.

[76] M. Maier, W. Kaiser, and J. A. Giordmaine, "Intense light bursts in the stimulated Raman effect," *Phys. Rev. Lett.*, vol. 17, pp. 1275–1277, December 1966.

[77] J. A. Giordmaine, P. M. Rentzepis, S. L. Shapiro, and K. W. Wecht, "Two-photon excitation of fluorescence by picosecond light pulses," *Appl. Phys. Lett.*, vol. 11, pp. 216–218, October 1967.

[78] P. M. Rentzepis and M. A. Duguay, "Picosecond light pulse display using two different optical frequencies," *Appl. Phys. Lett.*, vol. 11, pp. 218–220, October 1967.

[79] W. H. Glenn, Jr., "Parametric amplification of ultrashort laser pulses," *Appl. Phys. Lett.*, vol. 11, pp. 333–335, December 1967.

[80] R. C. Miller, "Second harmonic generation with broadband optical maser," *Phys. Lett.*, vol. 26A, pp. 177–178, January 1968.

[81] S. L. Shapiro, "Second harmonic generation in LiNbO$_3$ by picosecond pulses," *Appl. Phys. Lett.*, vol. 13, pp. 19–21, July 1968.

[82] J. Comly and E. Garmire, "Second harmonic generation from short pulses," *Appl. Phys. Lett.*, vol. 12, pp. 7–9, January 1968.

[83] J. R. Klauder, M. A. Duguay, J. A. Giordmaine, and S. L. Shapiro, "Correlation effects in the display of picosecond pulses by two-photon techniques," *Appl. Phys. Lett.*, vol. 13, pp. 174–176, September 1968.

[84] J. Ducuing and N. Bloembergen, "Statistical fluctuations in nonlinear optical processes," *Phys. Rev.*, vol. 133, pp. A1493–A1502, March 1964.

[85] P. Pringsheim, *Fluorescence and Phosphorescence*. New York: Interscience, 1949, pp. 10–17.

[86] W. H. Glenn, M. J. Brienza, and A. J. DeMaria, "Mode-locking of an organic dye laser," *Appl. Phys. Lett.*, vol. 12, pp. 54–56, January 1968.

[87] M. B. Fisher and R. T. McKenzie, paper presented at the Internat'l Electron Devices Meeting, Washington, D. C., October 1967.

[88] S. L. Shapiro, J. A. Giordmaine, and K. W. Wecht, "Stimulated Raman and Brillouin scattering with picosecond light pulses," *Phys. Rev. Lett.*, vol. 19, pp. 1093–1095, November 1967.

[89] B. Senityky, G. Gould, and S. Culter, "Millimeter wave amplification by resonance saturation," *Phys. Rev.*, vol. 130, pp. 1460–1465, May 15, 1963.

[90] M. E. Mack, "Light amplification in saturable absorbers," *Appl. Phys. Lett.*, vol. 12, pp. 329–330, May 1968. Presented at the Internat'l Quantum Electronics Conf., Miami, Fla., May 17, 1968, Paper 17Q-10.

[91] S. E. Schwartz and T. Y. Tan, "Wave interaction in saturable absorbers," *Appl. Phys. Lett.*, vol. 10, pp. 4–7, January 1967.

[92] B. H. Soffer and B. B. McFarland, "Frequency-locking and dye spectral hole-burning in Q-spoiled lasers," *Appl. Phys. Lett.*, vol. 8, pp. 166–169, August 1966.

[93] M. E. Mack, to be published.

[94] J. R. Klauder, A. C. Prince, S. Darlington, and W. J. Albersheim, "The theory and design of chirp radars," *Bell Sys. Tech. J.*, vol. 39, p. 745, 1960.

[95] J. R. Klauder, "The design of radar signals having both high range resolution and high velocity resolution," *Bell Sys. Tech. J.*, vol. 39, p. 809, 1960.

[96] W. B. Mims, "The detection of chirped radar signals by means of electron spin echoes," *Proc. IEEE*, vol. 51, pp. 1127–1134, August 1963.

[97] J. A. Giordmaine, M. A. Duguay, and J. W. Hansen, "Compression of optical pulses," *IEEE J. Quantum Electronics*, vol. QE-4, pp. 252–255, May 1968.

[98] C. G. B. Garrett and M. A. Duguay, "Theory of the optical frequency translator," *Appl. Phys. Lett.*, vol. 9, p. 374, 1966.

[99] M. A. Duguay and J. W. Hansen, "Optical frequency shifting of mode-locked laser beam," *IEEE J. Quantum Electronics*, vol. QE-4, pp. 477–481, August 1968.

[100] M. J. Brienza and A. J. DeMaria, "Laser-induced microwave sound by surface heating," *Appl. Phys. Lett.*, vol. 11, pp. 44–46, July 1967.

[101] J. F. Ready, "Effects due to absorption of laser radiation," *J. Appl. Phys.*, vol. 36, pp. 462–468, February 1965.

[102] R. M. White, "Generation of elastic waves by transient surface heating," *J. Appl. Phys.*, vol. 34, pp. 3559–3567, December 1963.

[103] M. Bass and T. F. Deutsch, "Broadband light amplification in organic dyes," *Appl. Phys. Lett.*, vol. 11, pp. 89–91, August 1967.

[104] B. H. Soffer and B. B. McFarland, "Continuously tunable, narrow-band organic dye lasers," *Appl. Phys. Lett.*, vol. 10, pp. 266–268, May 1967.

[105] M. L. Spaeth and D. P. Brotfeld, "Stimulated emission from polymethine dyes," *Appl. Phys. Lett.*, vol. 9, pp. 179–181, September 1966.

[106] F. P. Schafer, W. Schmidt, and J. Volze, "Organic dye solution laser," *Appl. Phys. Lett.*, vol. 9, pp. 306–309, October 1965.

[107] A. J. DeMaria and C. Miller, "Linear, wide-bandwidth optical amplifiers by staggered tuning," *J. Opt. Soc. Am.*, vol. 58, pp. 467–472, April 1968.

[108] M. Bass, P. A. Franklin, J. F. Ward, and G. Weinreich, "Optical rectification," *Phys. Rev. Lett.*, vol. 9, pp. 446–448, December 1962.

[109] M. J. Brienza, A. J. DeMaria, and W. H. Glenn, "Optical rectification of mode-locked laser pulses," *Phys. Lett.*, vol. 26A, pp. 390–391, March 1968.

[110] A. J. Alcock and S. A. Ramsden, "Two wavelength interferometry of a laser-induced spark in air," *Appl. Phys. Lett.*, vol. 8, pp. 187–188, April 1966.

[111] R. W. Minck and W. G. Rado, "Investigation of optical frequency breakdown phenomena," *Proc. Physics of Quantum Electronics Conf.*, P. L. Kelley, B. Lax, and P. E. Tannenwold, Eds. New York: McGraw-Hill, 1966, pp. 527–537.

[112] R. G. Meyerand, Jr., and A. F. Haught, "Optical energy absorption and high-density plasma production," *Phys. Rev. Lett.*, vol. 13, pp. 7–9, July 1964.

[113] A. J. Alcock, P. P. Pashinin, and S. A. Ramsden, "Temperature measurements of a laser spark from soft-X-ray emission," *Phys. Rev. Lett.*, vol. 17, pp. 528–530, September 1966.

[114] M. M. Litvak and D. F. Edwards, "Spectroscopic studies of laser-

produced hydrogen plasma," *IEEE J. Quantum Electronics*, vol. QE-2, pp. 486–492, September 1966.

[115] J. F. Ready, "Effects due to absorption of laser radiation," *J. Appl. Phys.*, vol. 36, 1965.

[116] P. Langer, G. Tonon, F. Floux, and A. Ducauze, "Laser induced emission of electrons, ions, and X rays from solid targets," *IEEE J. Quantum Electronics*, vol. QE-2, pp. 499–506, September 1966.

[117] Yu. V. Afanasyev, O. N. Krokhin, and G. V. Skilzkov, "Evaporation and heating of a substance due to laser radiation," *IEEE J. Quantum Electronics*, vol. QE-2, pp. 483–486, September 1966.

[118] B. C. Fawcett, A. H. Gabriel, F. E. Irons, N. J. Peacock, and P. A. H. Saunders, "Extreme ultraviolet spectra from laser-produced plasmas," *Proc. Phys. Soc.*, vol. 88, pp. 1051–1059, 1966.

[119] A. F. Haught and D. H. Polk, "High-temperature plasmas produced by laser beam irradiation of single solid particles," *Phys. Fluids*, vol. 9, pp. 2047–2052, 1966.

[120] N. G. Basov and O. N. Krokhin, *Soviet Physics-JETP Lett.*, vol. 46, pp. 171–172, 1964.

[121] R. V. Ambartyumian, N. G. Basov, V. S. Zuev, P. G. Kriukov, and V. S. Letokhov, *Soviet Physics-JETP Lett.*, vol. 4, pp. 19–21, July 1966.

[122] N. G. Basov, V. S. Zuev, P. G. Kriukov, V. S. Letokhov, Yu. V. Senatsky, and S. V. T. Chekalin, *Soviet Physics-JETP Lett.*, vol. 54, pp. 3–5, 1968.

[123] N. G. Basov, P. G. Kriukov, S. D. Zakharov, Yu. V. Senatsky, and S. V. Tchekalin, "Experiments on the observation of neutron emission at a focus of high-power laser radiation on lithium deuteride surfaces," *IEEE J. Quantum Electronics*, vol. QE-4, pp. **343-344, November 1968.**

Part VI: Modulation and Transmission of Laser Beams

The performance capabilities of electrooptic and acoustooptic devices are inherently limited, more by basic materials properties than by the design of these devices. The paper by Spencer, Lenzo, and Ballman gives a thorough survey of available materials and their properties for these applications.

Acoustooptic devices are widely used for modulating and deflecting laser beams. Gordon gives a review of acoustooptic devices for both of these purposes, while Kaminow and Turner, and also Chen, summarize the capabilities of electrooptic light modulators. Melchior, Fisher, and Arams survey some of the photodetection and demodulation methods employed with lasers.

Laser beams can be transmitted through lens guides and optical waveguides. Gloge summarizes many of the important innovations in optical lens waveguides. Goell and Standley review the basic properties of more closely confined optical waveguides in integrated optical circuits. Important recent advances in optical fibers for laser communications are described by Maurer.

A number of other papers covering optical communication theory principles and systems design considerations will be found in the Special Issue on Optical Communications of the PROCEEDINGS OF THE IEEE in October 1970.

Dielectric Materials for Electrooptic, Elastooptic, and Ultrasonic Device Applications

E. G. SPENCER, P. V. LENZO, AND A. A. BALLMAN

Abstract—In order to realize the advantages of using lasers for communications, means must be obtained to modulate with sufficient bandwidth and capacity, deflect, switch, frequency translate, or otherwise effect a modification of the optical beam in a predictable manner. This problem has provided the incentive for extensive materials research directed toward the development of any crystal capable of low-loss optical transmission whose properties can be modified by an electric or magnetic field or applied external stress, and whose properties, at the same time, will interact in some specified manner with the optical beam. In this paper some dielectric materials for these uses currently in various stages of development will be described.

I. INTRODUCTION

ONE OF THE expected advantages of using lasers in the field of communications is the ability to transmit a tremendous amount of information over the entire frequency range of these lasers both in the visible and the infrared regions. In order to fulfill this promise, means must be obtained to modulate with sufficient bandwidth and capacity, deflect, switch, frequency translate, or otherwise effect a modification of the optical beam in a predictable manner. This problem has provided the incentive for extensive materials research directed toward the development of any crystal capable of low-loss optical transmission whose properties can be modified by an electric or magnetic field or applied external stress, and whose properties, at the same time, will interact in some specified manner with the optical beam. In this paper some dielectric materials for these uses currently in various stages of development will be described.

The piezoelectric,[1],[2] ultrasonic,[3]–[7] electrooptic,[7]–[12] and elastooptic[13],[14] properties of single-domain ferroelectric lithium niobate,[15] $LiNbO_3$, and lithium tantalate,[1],[16],[17] $LiTaO_3$, provide an unusual combination of capabilities for useful device applications. Also, for certain optical,[18] ultrasonic,[19] and elastooptic applications, the nonferroelectric bismuth germanium oxide,[20] $Bi_{12}GeO_{20}$, and bismuth silicon oxide, $Bi_{12}SiO_{20}$, may prove to be more advantageous. A more recently synthesized[21] ferroelectric crystal system, strontium barium niobate, $Sr_xBa_{1-x}Nb_2O_6$, holds excellent promise for performance of electrooptic functions.[22] Also related to these materials, either chemically or structurally, are calcium pyroniobate,[1],[23] $Ca_2Nb_2O_7$, which has been employed for microwave modulation of light, and lead magnesium niobate,[24] $Pb_3MgNb_2O_9$, which has a large quadratic electrooptic effect and does not have the electronic conduction and compositional inhomogeneity problems of potassium tantalate niobate (KTN). For completeness, KTN,[25] $KTa_xNb_{1-x}O_3$, with its large quadratic effect, will be mentioned briefly.

For a comparison of the electrooptic properties of these materials with other crystals previously reported in the literature, a convenient tabulation has been given by Kaminow and Turner.[26]

In the following sections each of these properties will be discussed and the available measured parameters will be described and tabulated for the various crystals. In some cases a great deal of work already has been carried out, and the physical processes involved are understood at least in part. In others, however, only a sketchy amount of information is available. In all cases much more experimental data are needed in order to characterize the materials completely. These data will, in turn, lead to more detailed theoretical treatment, and will indicate, as well, possible directions for further crystal chemistry research.

It will be seen that all of these materials, even in such an early stage of development, are suitable in their present forms for actual device investigation. In fact, the lithium tantalate electrooptic pulse code modulator[27],[28] has already undergone a significant engineering development. A short description of this modulator, frequency translators, elastooptic and other devices are given which illustrate best some of the possibilities and also bring to light additional materials problems whose solutions would lead to improved performance.

II. PHYSICAL PROPERTIES AND CRYSTALLOGRAPHIC SYMMETRY

All of the crystals under discussion are congruently melting, which means that at high enough temperatures the crystals will melt and not decompose. The molten mass will have the same atomic proportions as exist in the crystals. This is a requirement for the use of the Czochralski technique in which a crystal grows on a small single crystal seed and over a period of hours is slowly *pulled from the melt*. Ideally, crystals of any desired size can be pulled from the melt in this manner.[1],[20],[21] At the same time, the axis of the boule can be chosen to be approximately in the direction of any principal crystallographic axis. Although these same crystal-growing techniques are in wide technological use, which helps to assure ultimate commercial availability, it is also true that other processes are available and may prove superior in some cases. The crystals are relatively hard and

Manuscript received September 14, 1967; revised October 23, 1967. This invited paper is one of a series planned on topics of general interest.—*The Editor*.

The authors are with Bell Telephone Laboratories, Inc., Murray Hill, N. J.

Reprinted from *Proc. IEEE*, vol. 55, pp. 2074–2108, Dec. 1967.

250

are readily processed by the usual careful optical polishing techniques. Various poling methods have been developed to insure that the ferroelectric crystals are single-domain. In general, the crystals are poled as the last step; however, they also remain single-domain during cutting and polishing.

Crystals which are noncentrosymmetric, i.e., lacking a center of symmetry, may exhibit both linear and quadratic electrooptic and elastooptic effects. In all of the crystals discussed here the linear effects are dominant. Thus, a change in optical index of refraction can be induced by either an electric field (electrooptic effect) or strain (elastooptic effect). Strain can be produced by an electric field (converse piezoelectric effect) or by a stress (elasticity).

Bismuth germanium oxide crystallizes[29] in the $I2_1 3$ crystallographic space group which is a member of the tetrahedrally coordinated cubic point group 23. These crystals are unusual in several respects. They are the first tetrahedrally coordinated cubic crystals that we know of which have been synthesized with reasonable size and good optical quality. Further, to our knowledge, no other inorganic crystals have been found that crystallize in either of the body-centered space groups belonging to the 23 point group. The crystals are moderately strong piezoelectrics,[30] weakly electrooptic, photoconductive, highly optically active,[18] photoactive,[31] have unusually good ultrasonic properties,[19] and are strongly elastooptic.[19] Bismuth germanium oxide is not a ferroelectric since it is not a member of one of the ten polar classes which can exhibit ferroelectricity. In the context of this paper ultrasonic and elastooptic uses are indicated.

Lithium niobate and lithium tantalate are isostructural trigonal crystals of point group 3m. They are ferroelectric, with the polar axis being in the direction of the x_3 or trigonal axis. $LiNbO_3$ is optically uniaxial negative (extraordinary index of refraction less than the ordinary index) and $LiTaO_3$ is optically uniaxial positive ($n_e > n_0$). Due to their high polarizabilities, the crystals are strongly piezoelectric with coupling efficiencies[2] varying from 10 to 70 percent, depending on the crystallographic orientation. For the same reasons their electrooptic effect[7]–[12] is large, being several times that of the widely used potassium dihydrogen phosphate (KDP). A further advantage is that the effect is transverse. As will be seen in the sections on electrooptics and modulators, an additional reduction of as much as 40 times in voltage requirements may be achieved by making the field path length small and the optical path length long.

The elastooptic coupling [13],[14] of $LiNbO_3$ is about ten times better than that of fused silica. The advantage is due mainly to the high indices of refraction, $n \sim 2.2$. In order to perform elastooptic experiments effectively at microwave frequencies, it is required that the ultrasonic losses be low. It was surprising at first that a trigonal crystal would have such low elastic wave propagation losses at microwave frequencies. It was subsequently found that the transverse elastic symmetry is actually much higher than is demanded by its crystallographic point group.

Calcium pyroniobate, $Ca_2Nb_2O_7$, crystallizes[1] in the monoclinic point group 2 and is optically biaxial negative (i.e., the acute bisectrix of the optic axes is parallel to the direction of the smallest index of refraction). The single twofold axis of symmetry may be taken as the crystallographic b axis. The a and c axes lie in the plane normal to the b axis. The angle β between the a and c axes is, in general, not 90 degrees. The (100) plane in calcium pyroniobate is a natural cleavage plane. The crystals grown were of good optical quality, but because of cleavage, optical polishing was not successful. The natural cleavage faces, however, were quite good enough for electrooptic measurements.[23]

A *system* of tetragonal crystals, strontium barium niobate, $Sr_xBa_{1-x}Nb_2O_6$, has been synthesized[21] with the *tungsten-bronze* structure. The values of x range from about 0.25 to 0.75. Outside this range entirely different crystals are formed with the two end products, $SrNb_2O_6$ and $BaNb_2O_6$, crystallizing in different classes from each other. The striking feature is that a *transverse, linear* electrooptic effect[22] exists which is 75 times stronger than for $LiNbO_3$ or $LiTaO_3$. The system of varying compositions has a continuous range of dielectric constants, optical indices of refraction, electrooptical constants, and Curie temperatures. The basic structural elements consist of octahedrally coordinated niobium ions, the resulting crystals being of the tungsten-bronze type. Using various cation substitutions then allows definite control over the properties just mentioned, without the more drastic substitutions for the niobium-oxygen octahedra. Thus, the material can be engineered, within limits, to fit the requirements of a specific device application.

It may be remarked at this point that each of these various crystals has some unique characteristic not possessed by the others. It is expected that a number of these will achieve technological importance, each for its own specific usefulness.

III. Optical Properties

For useful optical device applications, the optical properties of a crystal are most critical. In many cases a crystal of higher optical quality, but having inferior electrooptic or elastooptic coefficients, will actually perform better than one possessing the converse qualities. The initial stages of the materials research program are to synthesize and investigate crystals belonging to those crystallographic point groups that possess the appropriate symmetry for electrooptic and elastooptic properties. The ten polar point groups and other noncentrosymmetric groups exhibit linear electrooptic effects. Although all point groups may display quadratic electrooptic effects, large quadratic effects are most likely to appear in the centrosymmetric classes. Linear and quadratic elastooptic effects exist for all point groups. Crystals of the noncubic classes are optically uniaxial or biaxial, while those of the cubic system are optically isotropic.

Eventually the optical properties of individual ions having stronger or weaker polarizabilities must be con-

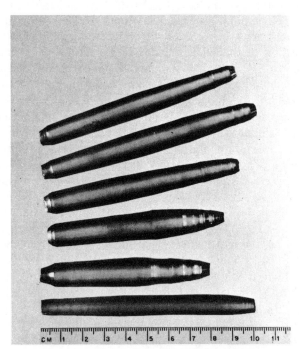

Fig. 1. Single crystals of lithium tantalate, $LiTaO_3$, grown by the Czochralski technique.

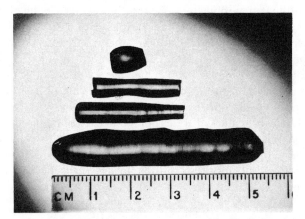

Fig. 2. Single crystals of bismuth germanium oxide, $Bi_{12}GeO_{20}$, grown by the Czochralski technique. The tendency toward growth of a six-faceted boule is apparent in the central crystal.

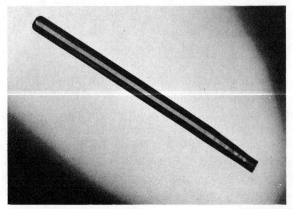

Fig. 3. A single crystal of strontium barium niobate, $Sr_{0.5}Ba_{0.5}Nb_2O_6$, ~6.3 cm long and 7 mm thick. The polar, x_3, axis is accurately aligned along the boule axis. Four major facets and 16 minor facets extend the length of the crystal.

sidered. Similarly, in the case of the strontium barium niobates, lithium niobate, lithium tantalate, and KTN, the basic unit is a Nb or Ta ion coordinated in an octahedron of oxygen ions. If the crystal is pyroelectric, the refractive index may change drastically with temperature in the neighborhood of the Curie temperature. Some of these properties will make one crystal more desirable than another. Fortunately, it appears that many new crystals differ in their performance characteristics and would be useful in different applications.

After a new crystal has been found to be promising, the problem of growing more perfect crystals becomes paramount. This is necessary both for device application and for obtaining data in order to define the intrinsic physical characteristics. The two principal problems associated with optical quality are variations of refractive index due to included impurities, residual strains, variations of growth, changes in stoichiometry or other growth faults and transient or semipermanent damage[32] to the crystal due to a high-intensity laser beam. All of these problems are, to some extent, amplified by the fact that the newer crystals generally have higher indices of refraction and smaller variations are more apparent. The most crucial test of optical quality will be the operation of a crystal inside a laser cavity for optical harmonic generation, parametric operation, electrooptic or elastooptic modulation. These are all long-range materials-development problems which will not be detailed here. They are, however, the type of problems which are amenable to steady and progressive improvements.

Finally, it should be emphasized that even though all the crystals discussed, at present, have some kind of optical difficulty that must be overcome, the same is true for all other obtainable crystals, both naturally occurring and synthetic. For example, even an examination of high-quality quartz platelets with interference optics invariably shows many unwanted strain phenomena. In any case, all of the materials being discussed here have been grown as large crystals of reasonably good optical quality for the initial stages of development. Photographs of $LiTaO_3$, $Bi_{12}GeO_{20}$, and $Sr_{0.5}Ba_{0.5}Nb_2O_6$ are shown in Figs. 1–3 to illustrate this point. The quality of the $LiTaO_3$ crystals is, perhaps, the most outstanding at this time. For example, in a cube one centimeter on an edge with the faces normal to the x_3 (optical) axis polished optically flat and parallel, the optical path length along the x_3 axis varies by less than 2λ over the area of the crystal face.[33] This would be considered excellent for some applications, such as for electrooptics or elastooptics, but unacceptable for others such as for digital beam deflectors.

IV. The Electrooptic Effect in Dielectric Crystals

The optical properties of crystals, in the visible and the near infrared regions, in many cases are determined by a single electronic transition in the ultraviolet. Similarly, the static and low-frequency dielectric properties are determined by optical lattice modes in the far infrared region. The electrooptic effect occurs whenever an applied electric field interacting directly with the combined electronic and lattice modes results in a change in the optical dielectric properties. The applied field may be a static, microwave, or another optical electric field. In some crystals electrooptic interactions have been found to be largely electronic in origin and, in some, the origin is principally due to lattice modes. Measurements and calculations on a particular crystal are needed to determine the proportions ascribable to each mode. Also, the electrooptic effect may vary linearly or quadratically with applied electric field and, in either case, the effect can be of first order in magnitude.

In the past few years, nonlinear optical properties of crystals have been investigated intensively in many laboratories, both from the theoretical and experimental standpoints. Reports have been given in several recent publications on extensions of the linear dielectric theory to nonlinear problems such as second harmonic generation[34]-[37] and electrooptic effects.[37]-[45] The results have been applied to various crystals for which measured data are available. The agreements obtained strongly suggest that the physical processes discussed are the correct ones for each crystal system evaluated.

Perhaps the simplest electrooptic crystal that can be treated in detail would be a diatomic ionic crystal such as ZnS or CuCl, having a tetrahedrally coordinated cubic structure, $\bar{4}3m$. Theoretical calculations of various aspects of the lattice dynamics have been given previously[46] and recently, calculations of the electrooptic effect have been made in detail by Kelly.[38] The dipole moment of the positive ion consists of a part due to the displacements of the ions and a part due to electronic polarization. The electronic polarization is given by the product of the polarizability, α, and an effective electric field. The nonlinear terms in polarization come in through quadrupole terms in the expansion of the effective fields. Values for the ionic charge and polarizability are obtained as phenomenological constants derived from measurements of the low-frequency dielectric constants, the optical dielectric constants, and the infrared dispersion frequencies. The fit between measured and calculated values is within a few percent and allows evaluation of the role and relative importance of the various terms involved.

In this case as in the others that follow the fundamental question is to judge how consistent the model is with the physics of the real crystal. For example, in Kelly's treatment the Szigeti relation is used to obtain a numerical value for the effective ionic charge and the Clausius-Mossotti relation is used to obtain effective polarizabilities. Thus all assumptions involved in these relations are also implicit in the electrooptic calculations. Fortunately, however, the large amount of experimental data on crystal properties which involve piezoelectric, electrooptic, and dielectric constants, optical frequency mixing, and Raman spectra combine to provide information and cross checks on the subtleties of the various theories proposed.

Other models have been proposed which are meant to apply, with reasonable accuracy, to all noncentrosymmetric crystals. In general, these are based on Bloembergen's[36]

anharmonic oscillator model and are associated with Miller's[35] rule to obtain a nonlinear constant. During an investigation of optical second harmonic generation, Miller found that by expressing energy in terms of polarization rather than applied electric fields, the second harmonic constants and the electrooptic constants had values within a factor of four of each other for all materials measured. This is now known as Miller's rule and has been found to have a much wider range of applicability than originally conceived, even to 10.6 μ in the infrared.[37] Garrett[40] has derived both a classical and a quantum mechanical model, starting from the anharmonic oscillator equations, from which Miller's constants can be estimated to within an order of magnitude.

In a similar manner, Kurtz and Robinson[39] developed a model for the electrooptic effect from the anharmonic equations for a single electron-oscillator description of optical index, the ionic terms being represented by an additive local field correction. For an applied optical field $E(\omega, t)$ and a low-frequency field $E(0)$, the displacement, x, of the electrons is given by

$$\ddot{x} + \Gamma\dot{x} + \omega_0^2 x + vx^2 = e/m[E(\omega, t) + \beta E(0)], \quad (1)$$

where Γ is the damping constant, ω_0 is the angular resonant frequency in the ultraviolet, v is the anharmonic force constant, e and m are the electronic charge and mass, and β is a local field parameter given by $\beta = (\varepsilon_0 + 2)/3$, where ε_0 is the low-frequency dielectric constant. The effect of the applied field $E(0)$ is to shift the resonant frequency from ω_0 to ω_0', where

$$(\omega_0')^2 = \omega_0^2 + 2ve\beta E(0)/m\omega_0^2. \quad (2)$$

Since the index of refraction n varies as $n \sim [(\omega_0')^2 - \omega^2]^{-1}$, the expression for the linear electrooptic coefficient r is obtained as

$$r = \frac{(n-1)^2(\varepsilon_0 + 2)v}{6\pi n^4 N_0 e\omega_0^2}, \quad (3)$$

where N_0 is the number of electrons involved. The nonlinear term v is obtained[40] from the use of Miller's rule by using $\delta = mv/2N_0 e^3$. The average value for 55 measurements[39] in 22 materials is given as $\delta = 3 \times 10^{-6}$ cm/statvolt. The value for v is then $v = 1.33 \times 10^{39}$ cm^{-1} s^{-2}. Kurtz and Robinson also derive similar expressions for ferroelectrics both above the Curie temperature (cubic, centrosymmetric, for perovskites) and below the Curie temperature (tetragonal, noncentrosymmetric). By omitting subscripts on r in (3), it is meant that the largest r_{ij} is of this magnitude. Miller's rule, however, can be applied more explicitly to specified directions in crystals of any symmetry. This refinement is not necessary for the discussion here. The electrooptic coefficients have been calculated for a number of inorganic crystals using (3) and better than an order of magnitude agreement was obtained for all cases where measured data were available.

Garrett[40] has extended the anharmonic oscillator model, based on the shell model of a one-dimensional crystal. Appropriate nonlinear terms are introduced in coupled ionic and electronic equations of motion rather than being applied separately to either the ionic or electronic motion. The solutions show that the two kinds of nonlinear contribution can partially cancel each other, at least in some crystals. It is in this manner that the results differ from those of Kurtz and Robinson.

Garrett's paper discusses many different nonlinear phenomena in crystals such as Raman effect, frequency mixing, and others. Taking his results for the electrooptic effect and casting it in a different form, his equation (29) becomes

$$r = \frac{-4\pi}{n^4 m_e(\omega_0^2 - \omega^2)}\left[3Dx_e + Cx_i\right]. \qquad (4)$$

Again, n is the optical index and ω_0 is the electronic resonant frequency in the ultraviolet. The electron mass is m_e, and x_e and x_i are electronic and ionic configurational coordinates. The anharmonicity of the electronic potential, considered separately, determines the constant D, while the constant C describes the extent to which the electronic resonant frequency is affected by ionic motion.

The Raman effect in piezoelectric crystals involves vibrational modes which are both Raman and infrared active, and as such can be excited by external electric fields. This establishes a relation between the equations for polarization in the electrooptic effect and similar equations, classically derived, for the Raman effect. Kaminow and Johnston[41] devised an experimental method, based on this relation, for evaluating the electrooptic effect in $LiNbO_3$ and $LiTaO_3$. In particular, the separate contributions due to electronic and ionic terms are obtained on an absolute basis.

Briefly, their method consists of writing expressions for the differential optical polarization for the electrooptic effect and for Raman scattering. Maintaining the full tensor notation, for the electrooptic effect the polarization is

$$P_i = \varepsilon_0 n_i^2 r_{ijk} n_j^2 E_j(\omega)E_k(0) \qquad (5)$$

where n_i and n_j are the principal refractive indices, $E_j(\omega)$ are the optical fields, and $E_k(0)$ are the applied low-frequency fields. The polarization for the Raman effect is written in terms of the differential optical polarizability α_{ijk} and the lattice displacements produced by the modulating field, $Q_k^m = \beta_k^m E_k$. The factors of proportionality β_k^m are evaluated in terms of the infrared oscillator strength, or in crystals having one active mode in terms of the dielectric constant. Thus,

$$P_i = (\alpha_{ijk}^m \beta_k^m + \xi_{ijk})E_j(\omega)E_k(0). \qquad (6)$$

The term ξ_{ijk} is the change in polarization caused by the modulation field acting directly on the electron. Combining the two equations results in

$$\varepsilon_0 n_i^2 r_{ijk} n_j^2 = \alpha_{ijk}^m \beta_k^m + \xi_{ijk}. \qquad (7)$$

The ξ_{ijk} terms involve only electronic effects and can be determined from optical second harmonic generation measurements. In order to calculate the electrooptic coefficients r_{ijk}, the final terms to be determined then are the α_{ijk}. For

$LiNbO_3$ and $LiTaO_3$ they were evaluated from measurements of the scattering efficiencies of the Raman lines.

The calculated values for the four electrooptic coefficients for $LiNbO_3$ and $LiTaO_3$ are found to agree with the measured data of Tables II and IV. It is particularly illuminating to find that the ionic or lattice contributions constitute about 90 percent of the effect and electronic contributions, about 10 percent.

V. Electrooptic Materials

The principal electrooptic effect, in the crystals discussed here, is the transverse effect and is describable in terms of the half-wave field-distance product $[E \cdot l]_{\lambda/2}$, where E is the electric field strength and l is the optical path length. This product represents the voltage required to produce half-wave retardation in a geometry $l/d=1$, where d is the crystal thickness in the direction of the applied field.

A distinction is made, at this point, between the transverse effect and the longitudinal effect of such widely used materials as potassium dihydrogen phosphate, KDP. The longitudinal effect most generally has been used and the ratio l/d is always unity. The half-wave retardation voltage required for KDP is $V_{\lambda/2} \sim 8000$ volts. For the transverse effect the half-wave field-distance product $[E \cdot l]_{\lambda/2}$ is $\sim 24\,000$ volts. Recently transverse KDP modulators have become available commercially in which, by making $l/d=250$, the half-wave retardation voltage is reduced to ~ 100 volts. In this case very long, very thin samples are required whose typical values might be $l=20$ cm and $d=0.8$ mm. All of the crystals described in this section have large transverse effects and the voltage requirements can be reduced even further by the use of favorable geometrical factors.

The optical phase retardation Γ, in radians, is given by

$$\Gamma = 2\pi l/\lambda_0\left[n_1(E) - n_2(E)\right], \qquad (8)$$

where λ_0 is the wavelength of light in vacuum, and $n_1(E)$ and $n_2(E)$ are the field-dependent indices of refraction. The form taken by $n_1(E) - n_2(E)$ depends on the crystal symmetry and is further specified by the particular direction of application of electric field and propagation and polarization directions of the optical beam. A general equation for the index ellipsoid[47] may be written in terms of the impermeability tensor B_{ij}, where by definition $B_{ij} = \partial \mathscr{E}_i / \partial D_j$, $\mathscr{E}_i$ being the optical fields. The field-dependent terms are written as $B_{ij}(E) = B_{ij}(0) + \Delta B_{ij}(E)$. The electrically deformed index ellipsoid is then given by

$$B_{ij}x_i x_j = 1, \qquad (9a)$$

subject to the conditions (for systems of symmetry higher than that of monoclinic or triclinic)

$$B_{ij} - \delta_{ij}n_i^{-2} = r_{ijk}E_k, \qquad (9b)$$

where δ_{ij} is the Kronecker delta, and the r_{ijk} are the electrooptic coefficients. The electrooptic effect then results in a change in the size, shape, or orientation of the ellipsoid.

The electrooptic properties of each of the crystals is

TABLE I

ELECTROOPTIC MATRICES r_{ij} FOR THE FOUR CRYSTALLOGRAPHIC POINT GROUPS OF INTEREST

Monoclinic (2)			Tetragonal (4mm)			Trigonal (3m)			Cubic (23)		
0	r_{12}	0	0	0	r_{13}	0	$-r_{22}$	r_{13}	0	0	0
0	r_{22}	0	0	0	r_{13}	0	r_{22}	r_{13}	0	0	0
0	r_{32}	0	0	0	r_{33}	0	0	r_{33}	0	0	0
r_{41}	0	r_{43}	0	r_{51}	0	0	r_{51}	0	r_{41}	0	0
0	r_{52}	0	r_{51}	0	0	r_{51}	0	0	0	r_{41}	0
r_{61}	0	r_{63}	0	0	0	$-r_{22}$	0	0	0	0	r_{41}

TABLE II

ELECTROOPTIC MEASUREMENTS IN LiNbO$_3$ AT $\lambda = 633$ mμ

	E applied $\|x_1$	E applied $\|x_2$	E applied $\|x_3$
light $\|x_1$ dc electrooptic coeff. $[E \cdot l]_{\lambda/2}$		$r_{22} = 2.2 \times 10^{-7}$ cm/statvolt 7230 volts	$\|0.9 r_{33} - r_{13}\| = 5.2 \times 10^{-7}$ cm/statvolt 2940 volts
light $\|x_2$ dc electrooptic coeff. $[E \cdot l]_{\lambda/2}$		$r_{22} = 1.8 \times 10^{-7}$ cm/statvolt 9000 volts	$\|0.9 r_{33} - r_{13}\| = 5.1 \times 10^{-7}$ cm/statvolt 3160 volts
light $\|x_3$ dc electrooptic coeff. $[E \cdot l]_{\lambda/2}$	$r_{22} = 2.0 \times 10^{-7}$ cm/statvolt 4000 volts	$r_{22} = 1.9 \times 10^{-7}$ cm/statvolt 4250 volts	

given in terms of the general relation of (9) leading to specific applications (8).

Lithium Niobate and Lithium Tantalate

The matrices of electrooptic coefficients in the reduced matrix form,[47] r_{ij}, for the four crystallographic point groups of interest are given in Table I. Since LiNbO$_3$ and LiTaO$_3$ are in the point group 3m, there are eight matrix elements defined by four independent constants. These data are given in Table II. One way of expressing Miller's rule is that the ratio r_{ijk}/ε_{jk} is essentially constant from one material to another, as well as being independent of temperature and frequency for a particular material. Measurements were made of the dielectric constants of LiNbO$_3$ along the x_3 axis up to 200 MHz and were found to be essentially constant. It could therefore be inferred from Miller's rule that the electrooptic effect likewise would be insensitive to frequency. The first data on the high-frequency electrooptic constants[10] were taken at 60 MHz and were found to be consistent with the low-frequency data. Recently, measurements have been extended[48] to 6 GHz with entirely similar results. From Raman scattering data[49] the infrared transverse optical mode ω_{t0} is determined to correspond to about 200 cm^{-1}. Thus, in a perfect crystal, the low-frequency electrooptic values might be expected to hold up at least through the millimeter-wave region. Also, in evaluating a series of samples it is generally much faster to study the dielectric constants than it is to measure the r_{ijk} values directly. Invoking Miller's rule then should be a good indication of the effectiveness for electrooptics.

While numerical values of the r_{ij} completely specify the crystal parameters, the use to which the crystal is put must be considered in making comparisons between differ-

ent crystals. Table III gives the equations governing amplitude modulators for crystals in the 3m point group. For these modulators incident optical polarization is set at $\pi/4$ to the principal vibration directions. In general, a $\lambda/2$ phase retardation gives rise to an elliptically polarized wave, which upon passing through a properly oriented $\lambda/4$ plate results in a plane wave rotated by $\pi/2$. This defines 100-percent modulation and the optical scheme is known as a Senarmont Compensator. Equations for phase modulation are obtained readily from the expression in Table III by assuming that the incident polarization is coincident with each principal axis separately, rather than at $\pi/4$. Referring to Table II, it is seen that the voltage required for half-wave retardation varies by as much as three times, depending on the directions of light propagation and applied field. The lower voltage requirements, 2940 volts, are for the E field along x_3 with light propagating parallel to x_1 or x_2.

In the table $[E \cdot l]_{\lambda/2}$ for E applied $\|x_2$, light $\|x_1$ should be equal to the field-distance product for E applied $\|x_2$, light $\|x_2$ and should be twice the voltage for E applied $\|x_2$, light $\|x_3$ or E applied $\|x_1$, light $\|x_3$. The deviations from these relations are attributed to residual orientation-dependent strain birefringence in the crystals.

For many applications optical phase rather than intensity modulation may be preferable or even a requirement. The appropriate electrooptic properties are obtained from Table III. Numerical comparisons can be calculated in several ways, although the most meaningful bases for comparison may differ for each specific application. A phase-modulated signal may be written in the usual manner as

$$I = I_0 \sum_{n=0}^{\infty} J_n(\Gamma) \left[\sin (v + n\omega)t + (-1)^n \sin (v - n\omega)t \right] \quad (10)$$

TABLE III

Electrooptic Properties of LiNbO$_3$ and LiTaO$_3$
(Class 3m)

	E applied $\|x_1$	E applied $\|x_2$	E applied $\|x_3$
Induced Retardation			
light $\|x_1$		$\Gamma_{(\text{ind})} = \dfrac{\pi l_1 V_2 n_0^3 r_{22}}{\lambda_0 d_2}$	$\Gamma_{(\text{ind})} = \dfrac{\pi l_1 V_3}{\lambda_0 d_3}(n_e^3 r_{33} - n_0^3 r_{13})$
light $\|x_2$		$\Gamma_{(\text{ind})} = \dfrac{\pi V_2 n_0^3 r_{22}}{\lambda_0}$	$\Gamma_{(\text{ind})} = \dfrac{\pi l_2 V_3}{\lambda_0 d_3}(n_e^3 r_{33} - n_0^3 r_{13})$
light $\|x_3$	$\Gamma_{(\text{ind})} = \dfrac{2\pi l_3 V_1 n_0^3 r_{22}}{\lambda_0 d_1}$	$\Gamma_{(\text{ind})} = \dfrac{2\pi l_3 V_2 n_0^3 r_{22}}{\lambda_0 d_2}$	

Note: In these expressions r_{51}^2 terms have been deleted since they appear as second-order effects.

l = optical path length in the crystal
d = thickness in field direction
$V = Ed$ = applied voltage
n_0 = ordinary index of refraction
n_e = extraordinary index of refraction

where I and I_0 are, respectively, the instantaneous and unmodulated amplitudes, $J_n(\Gamma)$ are Bessel functions of nth order, ν is the optical frequency, and $\omega/2\pi$ is the RF frequency. The first zero of $J_0(\Gamma)$ occurs for $\Gamma = 2.4048$ and at this value all of the optical energy will be contained in sidebands of the laser signal. Since $\Gamma = \pi$ radians represents half-wave retardation required for 100-percent amplitude modulation, the phase retardation needed for PM is comparable to that needed for AM. For the most favorable case in which the RF E field is applied parallel to x_3 with light propagating parallel to x_1 (or x_2) and polarized along x_3, the induced retardation is

$$\Gamma_{\text{ind}} = (\pi l_1 V_3/\lambda_0 d_3)n_e^3 r_{33}. \qquad (11)$$

The voltage for this half-wave retardation is computed, from values given in the tables, to be $V_3 = 1930$ volts; or for suppressed carrier phase modulation $J(2.4048)$, $V_3 = 1500$ volts.

Visual Observation

The electrooptic effect in LiNbO$_3$ is observable in a direct manner by viewing interference figures conoscopically,[9] that is, by the use of light diverging through the crystal in a polarizing microscope. Light is propagated in the x_3 direction and the interference figure is observed with static fields applied, first parallel to x_2 and then parallel to x_1. In both cases $[E \cdot l]_{\lambda/2} \sim 4000$ V. Fig. 4 is a collage of optical interference figures with field applied in the x_2 direction of a crystal ~ 1 cm on an edge. In Fig. 4(a) the slightly biaxial character of the optic axis figure with no field is attributed to residual strain. This was shown to be correct by looking at interference figures in platelets polished so thin that the strain was relieved by spontaneous distortion of the platelet. Figs. 4(b), 4(c), and 4(d) were obtained with 1000, 2000, and 3000 volts dc, respectively, and indicate the field-induced biaxiality. Here the optic plane is normal to x_1. Figs. 4(e), 4(f), and 4(g) are for the same conditions as are Figs. 4(b), 4(c), and 4(d) but with the field reversed. For this case the

optical plane is normal to x_2. Although the magnitude of the field-induced effect should be equal for opposite polarities, the birefringence apparent in the figure is not the same for both polarities since the strain birefringence noted in Fig. 4(a) aids the induced birefringence for one polarity, but opposes the induced effect for the opposite polarity. In all of these parts of Fig. 4 the principal axes of the index ellipsoid coincide with the crystallographic axes (x_1, x_2, x_3).

A similar series of interference figures was obtained for a field applied parallel to x_1, but, in contrast, the principal axes of the elliptical section normal to x_3, the direction of light propagation, were located 45 degrees from crystallographic axes x_1 and x_2 (i.e., the optic plane is at 45 degrees with respect to x_1 and x_2). These results are in agreement with the electrooptic equations.

Natural Birefringence

The optical phase retardation induced electrically is given by (8); however, in an operating modulator the change of the natural birefringence caused by a small temperature change also must be considered. The total phase retardation, for one set of conditions from Table III, may be written as

$$\Gamma = \frac{2\pi l_1}{\lambda_0}\left[n_0 - n_e + \frac{V_3}{2d_3}\left(n_e^3 r_{33} - n_0^3 r_{13}\right)\right] \qquad (12)$$

where the natural birefringence enters as the $(n_e - n_0)$ term. For LiNbO$_3$, at 6328 Å, $n_e - n_0 = -0.086$ and for LiTaO$_3$, $n_e - n_0 = +0.002$. At first it would be expected that the variation of $n_e - n_0$ with temperature might be much smaller for LiTaO$_3$ than for LiNbO$_3$. It turns out that the magnitude is approximately the same for the two cases and may be written as[28]

$$\frac{1}{l}\left.\frac{\partial l|n_e - n_0|}{\partial T}\right|_{40°C} = 4.7 \times 10^{-5} \; °C^{-1}. \qquad (13)$$

It is clear that for the crystallographic orientation for which

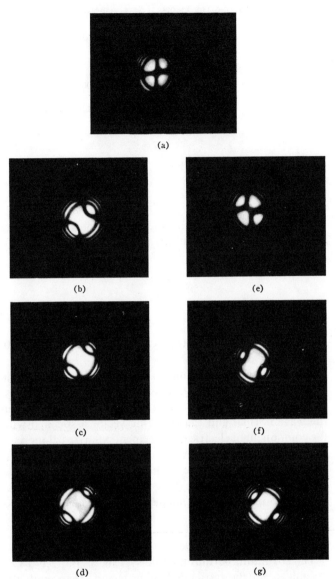

Fig. 4. Optical interference figures with light propagated in the x_3 direction and static field applied in the x_2 direction. The principal axes of the index ellipsoid coincide with the crystallographic axes (x_1, x_2, x_3).[9] (a) No field applied. (b) 1000 volts dc. (c) 2000 volts dc. (d) 3000 volts dc. (e) Polarity reversed; 1000 volts dc. (f) Polarity reversed; 2000 volts dc. (g) Polarity reversed; 3000 volts dc.

TABLE IV

ELECTROOPTIC MEASUREMENTS IN LiTaO$_3$ AT $\lambda = 632.8$ mμ*

	E applied $\| x_3$
light $\| x_1$ or x_2 electrooptic coeff. dc $[E \cdot l]_{\lambda/2}$	$\|1.007 \, r_{33} - r_{13}\| = 6.5 \times 10^{-3}$ cm/statvolt 2800 volts
electrooptic coeff. 50–86 MHz	$\|r_{33}\| = 9.1 \times 10^{-7}$; $\|r_{13}\| = 2.1 \times 10^{-7}$ $\|1.007 \, r_{33} - r_{13}\| = 7.1 \times 10^{-7}$ cm/statvolt

Note: $n_0 = 2.175$; $n_e = 2.180$; $\varepsilon_{33} = 43$ from 1.2 to 200 MHz.
* Values for r_{22} varied from sample to sample and were about two orders of magnitude smaller.

The field-distance product is increased from ~ 2900 volts to 4000 volts, a factor of 1.38 times. Under some conditions it would be preferable to supply the additional RF voltage required and avoid the necessity of a stabilized oven. In computing the additional RF power, however, the dielectric constants must also be included. For an RF E field along x_3 the power is given by $P_3 = \varepsilon_3 E_3^2 / 8\pi$ and for E along x_1 the power is $P_1 = \varepsilon_1 E_1^2 / 8\pi$. For LiNbO$_3$, $\varepsilon_1 = 98$ and $\varepsilon_3 = 51$ so that the additional power required is $P_1/P_3 = (98/51)(1.38)^2 = 3.65$ times greater.

A summary of the electrooptic properties of lithium tantalate is given in Table IV. One of the outstanding differences between LiNbO$_3$ and LiTaO$_3$ is that the ferroelectric Curie temperature is $\sim 1210°$C for the former and $\sim 650°$C for the latter.[1] The large difference in optical birefringence is at least associated with this fact. From a device point of view LiTaO$_3$ is not phase-matchable and will not be as useful for optical second harmonic generators and optical parametric devices. For electrooptic devices, the half-wave field-distance product has been found to be as low as 2700 volts for LiTaO$_3$, giving a slight advantage over lithium niobate. Although it is of no consequence here, the r_{22} measurements varied from sample to sample and were about two orders of magnitude smaller than for LiNbO$_3$. Further, it has been found that laser-induced birefringence, or damage,[27] is serious in LiNbO$_3$ and is appreciably less in LiTaO$_3$ for the powers available from the helium-neon gas lasers used.

Calcium Pyroniobate

Calcium pyroniobate, Ca$_2$Nb$_2$O$_7$, is another crystal closely related chemically and structurally to LiNbO$_3$ and LiTaO$_3$. It crystallizes[1] in the monoclinic point group 2 and is optically biaxial negative. A linear electrooptic effect was observed[23] which is sufficiently strong to be of device interest.

Measurements of the electrooptic coefficients at 3 GHz were obtained by using a simple strip-line cavity one wavelength long, shown schematically in Fig. 5. The detector was a Riesz diode followed by a microwave superheterodyne receiver. Modulation was sufficiently strong that no effort was made to optimize cavity and sample geometry.

A crystal of point group 2 has a single twofold axis of

the field-distance product is the smallest, the optical modulator must be temperature controlled. A pulse code optical modulator using a small stabilized oven to hold the temperature constant to within $\pm 0.03°$C is described in a following section.

For light propagating parallel to x_3, the index of refraction is n_0 regardless of the direction of optical polarization. Thus, there is no natural birefringence term, and an amplitude modulator would be relatively insensitive to temperature variations. This is because k_{11} equals k_{22} in the matrix of thermooptic coefficients for uniaxial crystals:

$$k_{ij} = \begin{pmatrix} k_{11} & 0 & 0 \\ 0 & k_{11} & 0 \\ 0 & 0 & k_{33} \end{pmatrix}.$$

TABLE V

ELECTROOPTIC MEASUREMENTS IN CALCIUM PYRONIOBATE WITH ELECTRIC FIELD ALONG THE TWOFOLD AXIS OF SYMMETRY

Direction of light propagation	Electrooptic coeff. (dc) (cm/statvolt)	Half-wave field-distance product $[E \cdot l]_{\lambda/2}$	
		dc (volts)	3 GHz
Normal to cleavage (100) plane, x_1	$\lvert r_{22} - 1.01 r_{32} \rvert = 4.1 \times 10^{-7}$	4550	4700 volts
Crystallographic axis, x_3	$\lvert r_{22} - 0.76 r_{12} \rvert = 3.7 \times 10^{-7}$	5080	not measured

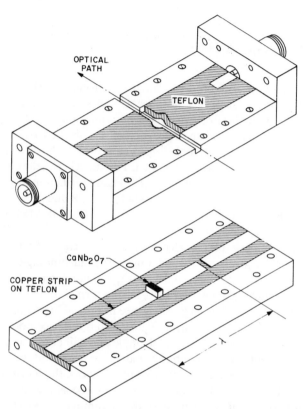

Fig. 5. Microwave strip-line cavity for modulation of light at 3 GHz using calcium pyroniobate, $Ca_2Nb_2O_7$.

symmetry which may be taken as the crystallographic b axis. The crystallographic a and c axes lie in the plane normal to the b axis. The angle β between the a and c axes is, in general, not 90 degrees. The (100) plane in calcium pyroniobate is a natural cleavage plane. The direction normal to the (100) plane is the acute bisector of the optic axes, which lie in the plane of the a and c axes. The angle between optic axes is $2v = 29$ degrees as measured on the Federov universal stage. In a right-handed Cartesian coordinate system with x_1 normal to the cleavage plane, $x_2 \Vert b$, and $x_3 \Vert c$, the equation of the index ellipsoid is

$$\left(\frac{1}{n_1^2} + r_{12}E_2\right)x_1^2 + \left(\frac{1}{n_2^2} + r_{22}E_2\right)x_2^2 + \left(\frac{1}{n_3^2} + r_{32}E_2\right)x_3^2$$
$$+ 2(r_{41}E_1 + r_{43}E_3)x_2x_3 + 2r_{52}E_2x_3x_1$$
$$+ 2(r_{61}E_1 + r_{63}E_3)x_1x_2 = 1. \qquad (14)$$

When the electric field (RF or dc) is applied along the twofold axis of symmetry, $E_1 = E_3 = 0$ and $E_2 = E$. Equation (14) becomes

$$\left(\frac{1}{n_1^2} + r_{12}E\right)x_1^2 + \left(\frac{1}{n_2^2} + r_{22}E\right)x_2^2 + \left(\frac{1}{n_3^2} + r_{32}E\right)x_3^2$$
$$+ 2r_{52}Ex_1x_3 = 1. \qquad (15)$$

For light propagating in the directions of the x_1 and x_3 axes the induced phase retardations are found to be, respectively,

$$\Gamma_1 = \frac{\pi h}{\lambda_0}(n_2^3 r_{22} - n_3^3 r_{32})E$$

and

$$\Gamma_3 = \frac{\pi l}{\lambda_0}(n_1^3 r_{12} - n_2^3 r_{22})E, \qquad (16)$$

the light path lengths being designated by h and l. For the latter the allowed polarizations of the optical electric displacements lie along the x_1 and x_2 axes. The angle between the wave normal and the Poynting vector associated with the component of electric displacement polarized along x_1 is $\alpha \approx r_{52}E$. The Poynting vector associated with the component polarized along x_2 is in the same direction as the wave normal. Measured values, given in Table V, show that the electrooptic effect is independent of frequency, at least up to 3 GHz.

Strontium Barium Niobate

The strontium barium niobate system, $Sr_xBa_{1-x}Nb_2O_6$, crystallizes in the tetragonal point group, 4mm, with the tungsten-bronze structure. The optical region over which these crystals are transparent is from about 0.4 to 8 μ. Measured data are given in Fig. 6. The crystal chemistry is given in Ballman's paper[21] along with a survey of, and references to, earlier investigations which were mainly on the ceramic phases. Some of Ballman's results which are most pertinent to the present discussion are outlined here.

The end products of the system in which the cations are all strontium or all barium do not crystallize in the tungsten-bronze structure. $BaNb_2O_6$ crystallizes in an orthorhombic point group and $SrNb_2O_6$ crystallizes in a point group which is not known at this time, and which is different from that of $BaNb_2O_6$. The range of x over which the tungsten-bronze phase occurs is $0.25 < x < 0.75$. While it has been determined that $Sr_xBa_{1-x}Nb_2O_6$ is in the 4mm point group, the detailed crystallographic analysis is currently under way.[50] The electrooptic and other dielectric measurements also are far from complete so that the physical discussion necessarily will be brief. Even so, enough is now known to indicate the possibilities of these materials for various practical devices.

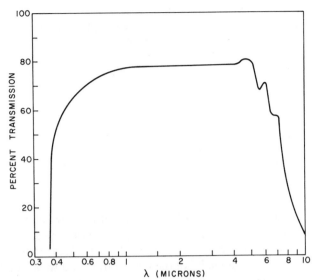

Fig. 6. Wavelength dependence of transmission for $Sr_{0.25}Ba_{0.75}Nb_2O_6$. The curve has not been corrected for surface reflection losses of ~ 25 percent at 6.33 mμ. Spectra by Miss D. M. Dodd.

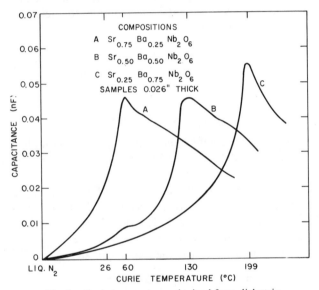

Fig. 7. Curie temperatures obtained from dielectric anomalies in $Sr_xBa_{1-x}Nb_2O_6$.[21]

The key to the strontium barium niobate system is that this is the first mixed compound investigated for electro-optic and other nonlinear optical phenomena in which the cations rather than anion groups are substituted. The niobium-oxygen octahedra remain unmodified. For the converse example, $LiNbO_3$ has a Curie temperature of 1210°C which is ~ 550°C higher than the 660°C value for $LiTaO_3$. Similarly, the Curie temperature of $KNbO_3$ is ~ 550°K higher than that of $KTaO_3$, being ~ 550°K and ~ 1°K, respectively. Mixed Nb-Ta systems, then, can have microscopic regions of widely varying dielectric properties. It is found[21] that for $Sr_{0.75}Ba_{0.25}Nb_2O_6$ the Curie temperature is ~ 60°C, for $Sr_{0.50}Ba_{0.50}Nb_2O_6$ it is ~ 130°C, and for $Sr_{0.25}Ba_{0.75}Nb_2O_6$ it is ~ 200°C. These data are shown in Fig. 7, and in Table VI the room-temperature values of the dielectric constants ε_3' taken at various frequencies are given for the same crystals. Thus it is apparent that a crystal

TABLE VI
ROOM-TEMPERATURE VALUES OF THE DIELECTRIC CONSTANTS TAKEN ALONG THE POLAR AXIS, ε_3', FOR $Sr_xBa_{1-x}Nb_2O_6$

f(MHz)	ε_3'		
	$x = 0.75$	$x = 0.50$	$x = 0.25$
0.01	6500	500	250
15	3400	450	120
100	3000	400	100

system is available in which the room-temperature polarizations, dielectric constant, electrooptic effect, piezoelectric effect, optical index, optical birefringence, and so on, all can be varied, at least within certain limits. Various other possible cation substitutions within this and related systems are also suggested.[21]

The electrooptic matrix has three independent nonzero moduli: $r_{13} = r_{23}$, $r_{42} = r_{51}$, and r_{33}. In $Sr_xBa_{1-x}Nb_2O_6$ the large electrooptic effect was observed[22] for the electric field parallel to the single tetrad symmetry axis x_3, which is also the polar axis, and with light propagation normal to the x_3 direction. The phase retardation is given by

$$\Gamma = (2\pi l/\lambda_0)/(n_2' - n_3') \quad (17)$$

where n_2' and n_3' are the principal indices of refraction normal to the direction of propagation x_1. In this case

$$n_2' = n_0 - n_0^3 r_{13}E/2 \quad \text{and} \quad n_3' = n_e - n_e^3 r_{33}E/2 \quad (18)$$

where n_0 and n_e are the ordinary and extraordinary optical indices, respectively.

For light parallel to x_3 and electric field parallel to the x_1 or a axis, Γ is written as above but for n_1' and n_2', which are

$$n_2' = n_0 \quad \text{and} \quad n_1' = n_0 - n_0^3 \frac{r_{51}^2 E_1^2}{2(n_0^{-2} - n_e^{-2})}. \quad (19)$$

Crystals in the series $Sr_xBa_{1-x}Nb_2O_6$ which have been investigated are those for which $x = 0.25$, $x = 0.50$, and $x = 0.75$. The $[E \cdot l]_{\lambda/2}$ at 632.8 mμ for the electric field along x_3, with light normal to x_3 and polarized at 45 degrees with respect to the principal axes, are shown in Table VII. In $Sr_{0.75}Ba_{0.25}Nb_2O_6$ for a one-to-one aspect ratio of electric field path to optical path length the half-wave field-distance product $[E \cdot l]_{\lambda/2}$ is 48 volts at 15 MHz. This value is comparable to the value of 28 volts[25] for the quadratic effect in KTN at dc bias fields of 2000 volts. (Also, a linear effect of the same magnitude has been reported recently for KTN.[51]) The 48 volts required is also 60 times smaller than the ~ 2800 volts obtained previously for $LiTaO_3$ and $LiNbO_3$.

The RF structure for the lithium niobate and lithium tantalate modulators which is shown in Fig. 8 used a 10-mm light path and a 0.5-mm electric field path and resulted in a reduction of the half-wave retardation voltage (by a factor of 20) to approximately 100 volts. Allowing the light beam to traverse the crystal twice further reduces the voltage requirements to ~ 50 volts. By using the same modulator structure and taking advantage of the transverse effect in

259

TABLE VII
HALF-WAVE FIELD-DISTANCE PRODUCTS FOR $Sr_xBa_{1-x}Nb_2O_6$

f	$[E \cdot l]_{\lambda/2}$		
	$x = 0.75$	$x = 0.50$	$x = 0.25$
dc	37 V dc	250 V dc	
1 MHz	80 V pp	676 V pp	1340 V pp
15 MHz	48 V pp	580 V pp	1236 V pp
100 MHz		580 V pp	

TABLE VIII
EFFECTIVE ELECTROOPTIC COEFFICIENTS IN $Sr_xBa_{1-x}Nb_2O_6$ FOR AMPLITUDE MODULATION AT 633 mμ

f	$\left\lvert \dfrac{n_e^3}{n_0^3} r_{33} - r_{13} \right\rvert$ cm/statvolt		
	$x = 0.75$	$x = 0.50$	$x = 0.25$
dc	4.15×10^{-5}	6.15×10^{-6}	
1 MHz	1.92×10^{-5}	2.28×10^{-6}	1.14×10^{-6}
15 MHz	3.2×10^{-5}	2.7×10^{-6}	1.24×10^{-6}
100 MHz		2.7×10^{-6}	

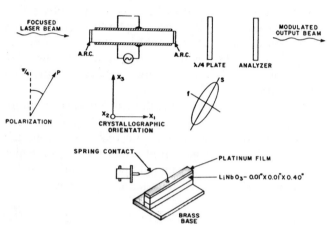

Fig. 8. RF optical modulator structure and schematic diagram using the Senarmont configuration. The ends of the crystal are antireflection-coated for 633 mμ.

strontium barium niobate in a similar manner, the half-wave retardation voltage can be reduced by an equal factor (of, say, 40 times) below the 48 volts required for the one-to-one ratio of lengths of the crystal. As an example, investigations made on a slab 7.6 mm long and 1.25 mm thick required only 8 volts to obtain half-wave retardation. It may be noted that the power required to drive a modulator is proportional to CV^2, where C is the capacitance of the unit and V is the half-wave retardation voltage. Although the capacitance increases as l/d is made larger, the voltage squared term will cause the power required to decrease by a greater amount. This results in a relative power decrease of $P \sim d/l$.

The smallest voltage and power requirements will be obtained for the composition $Sr_{0.75}Ba_{0.25}Nb_2O_6$ using the largest values of l/d. Under some conditions, however, it may be preferable to use a composition having lower dielectric constant which actually results in a higher voltage and power requirement. For example, at high microwave frequencies larger capacitances would result in low microwave impedances, making the microwave circuitry more difficult to obtain. Also, for extremely short pulse code modulation capacitance is deleterious to pulse response. On the other hand, for the modulation of higher optical power it may be advantageous to use the higher-strontium-content crystals but to increase the cross-sectional area. The point is made, then, that the actual device application will to a large extent determine the optimum chemical composition.

The normalized effective electrooptic coefficients

$|(n_e/n_0)^3 r_{33} - r_{13}|$ for amplitude modulation in which polarization of incident light is at 45 degrees with respect to the principal axes are given[22] in Table VIII. Measurements also have been made[52] of the temperature and wavelength dependence of the indices of refraction using optically polished prisms of each of the three compositions $x = 0.25$, $x = 0.50$, and $x = 0.75$. These data are given in Figs. 9 and 10, with a table of the room-temperature values being included in Fig. 9. The static electrooptic coefficients for $Sr_{0.75}Ba_{0.25}Nb_2O_6$ are

$$|r_{33}| = 4.02 \times 10^{-5} \text{ cm/statvolt}$$
$$|r_{13}| = 2.0 \times 10^{-6} \text{ cm/statvolt} \qquad (20)$$

and

$$|r_{51}| = 1.26 \times 10^{-6} \text{ cm/statvolt.}$$

Separation of the coefficients r_{33} and r_{13} was accomplished by electrooptic deflection techniques and r_{33} and r_{13} were found to be opposite in sign.

In (19), the retardation is quadratic in E and r_{51} and thus is a second-order effect. Also, since r_{13} is 20 times smaller than r_{33}, an approximate electrooptic matrix may be written in which all terms other than r_{33} are set to zero. The entire electrooptic effect is then approximately

$$\Delta B_i = 0, \quad \text{except} \quad \Delta B_3 = r_{33}E_3. \qquad (21)$$

This simplified matrix description emphasizes that the induced changes take place mainly along the ferroelectric polar axis.

The optical data are also consistent with the same point. As the compositional ratios of strontium to barium are changed, or as the temperature is changed for one particular ratio, it is found that n_e reflects these changes whereas n_0 remains relatively constant. The ordinary index refers to polarization in the x_1x_2 plane, or the plane of the a axes, and the extraordinary index refers to polarization along the x_3 or ferroelectric polar axis.

Bismuth Germanium Oxide

The electrooptic matrix for the cubic 23 point group has three equal nonzero elements, $r_{41} = r_{52} = r_{63}$. The measured value for $Bi_{12}GeO_{20}$ is $r_{41} = 0.965 \times 10^{-7}$ cm/statvolt. However, the index of refraction is large, $n = 2.55$, so that the effective modulation is 16.6 times larger, that is, $n^3 r_{41} = 1.6 \times 10^{-6}$ cm/statvolt. The half-wave field-distance prod-

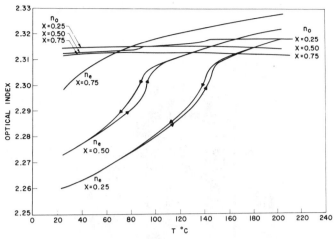

Fig. 9. Variation with temperature of the optical indices at 633 mμ for Sr$_x$Ba$_{1-x}$Nb$_2$O$_6$; $x=0.25$, $x=0.50$, and $x=0.75$.

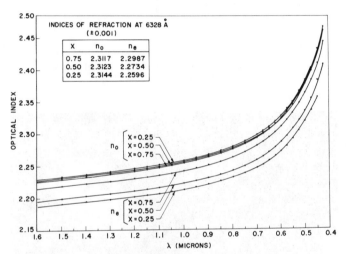

Fig. 10. Dispersion of the optical indices at room temperature for Sr$_x$Ba$_{1-x}$Nb$_2$O$_6$; $x=0.25$, $x=0.50$, and $x=0.75$.

uct is $[E \cdot l]_{\lambda/2} = 12\,000$ volts, and the effect is transverse and linear. The magnitude is approximately equal to twice that of the transverse effect in KDP, and of course is very small compared to strontium barium niobate or even lithium niobate and lithium tantalate.

The effect mentioned in Section II, which we have chosen to call *photoactivity*,[31] is also an electrooptic effect, but in a most unusual way. Briefly, it was found[18] that the optical activity of the crystals can be varied by 5 to 10 percent upon application of an electric field. If, further, a portion of the crystal is illuminated so as to excite photoconductivity, then a high electric field domain is formed across this un-illuminated region. The plane of polarization in the un-illuminated region is rotated by a large amount. For example, in an electric-field-biased crystal 1.5 mm thick the plane of polarization can be rotated by 90 degrees using only a weak monochromatic lamp. Various interesting device applications are therefore anticipated.

VI. ELECTROOPTIC HIGH-SPEED DIGITAL SYSTEM

Using a lithium tantalate modulator, Denton *et al.*[27],[28] have been able to demonstrate the feasibility of impressing the output of a high-speed digital system onto a He-Ne gas laser at 633 mμ. The approximately half-meter long laser is phase-locked by the clock from the high-speed digital terminal which results in a regular sequence of 0.6-ns light pulses occurring at a 224-MHz rate. The time duration of the light pulses is determined by the linewidth of the laser transition. The output from the laser is focused through the modulator as shown schematically in Fig. 11. Pulse information from the high-speed digital terminal controls the modulator, resulting in a coded sequence of 0.6-ns light pulses separated by 4.45 ns at the output of the modulator. Thus, coded 0.6-ns light pulses are obtained provided only that the electrical pulses applied to the modulator have rise or fall times somewhat shorter than the optical pulse separation of 4.45 ns, as shown in Fig. 12.

The electrooptic crystal and the transistor driver are shown in Fig. 13, with a cutaway diagram of the modulator crystal mounted in a temperature-controlled holder given in Fig. 14. The crystal consists of a rod of LiTaO$_3$ 1 cm long, and 0.25 mm square in cross section. The x_1 axis is along the length of the crystal and is the direction of light propagation. The x_3 or polar axis is along a crystal edge and is the direction of the electric field. The incident optical polarization is set at 45 degrees with respect to this axis. The crystal is polished optically flat and parallel on the ends and a reflective coating is deposited on the farther end so that the light beam traverses the crystal twice. The ratio of light-path length to field distance is 80-to-1, thus reducing the half-wave retardation voltage from 2600 to 32.5 volts. Only 10 mW of drive power are required for a transistor amplifier to operate the modulator crystal which acts as a capacitive load of 5 pF. The modulator rise and fall times are limited by the transistor amplifier-driver and not by the properties of the lithium tantalate crystal. Internal heating from the electrical drive is not a serious problem. However, because of changes of birefringence of LiTaO$_3$ with ambient temperature, the crystal must be maintained in a temperature-controlled oven. At an operating temperature of 40°C, the incremental change in birefringence is measured[28] to be $4.7 \times 10^{-5}/°C$. See (13).

To keep spurious signals due to changing static bi-refringence below a -20-dB level, temperature must be maintained constant and uniform to within $\pm 0.045°C$ for a one-centimeter long crystal at $\lambda = 633$ mμ.

In recently developed demonstration tests standard Bell Laboratories digital communication terminals were used. The amount of information impressed on the laser beam was equivalent to ~ 3500 voice channels, or two television signals, or various equivalent combinations of other data and signals. The modulated laser beam was transmitted a short distance, detected, and found by digital techniques to be completely error-free. The test was an evaluation of the optical modulator and detector systems only and not of the transmission medium. Fig. 15 shows the laboratory arrangement with color television and other data transmitted simultaneously.

The system described is a digital one in which the full information capacity of the optical beam is not utilized with

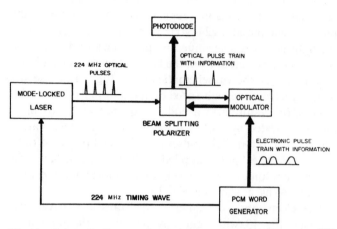

Fig. 11. Schematic diagram of optical pulse code modulation system.[27]

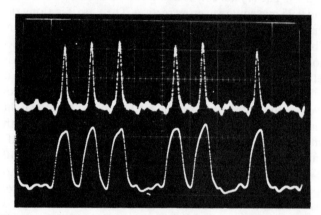

Fig. 12. Comparison of optical pulses (upper trace) and electrical pulses (lower trace) in pulse code modulation system; time scale is ∼5 ns/cm.[27]

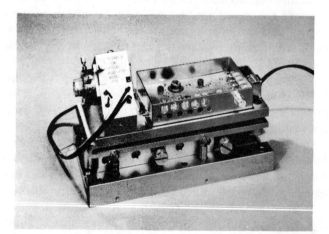

Fig. 13. Photograph of the pulse code modulation assembly, with the lithium tantalate modulator on the left and the transistor driver on the right.[28]

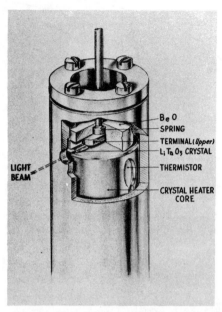

Fig. 14. Diagram of the LiTaO₃ modulator crystal holder and temperature control unit.[28]

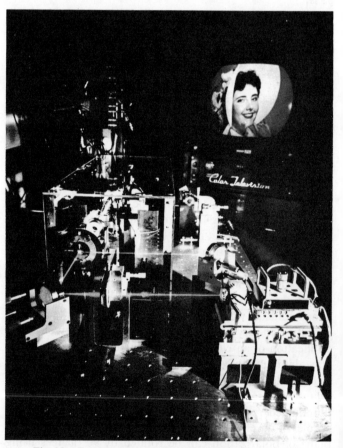

Fig. 15. Demonstration of system operation with color TV and other information on laser beam.

only one 224 megabit per second channel. If several independent channels were interleaved in time, the resulting time-multiplexed sequence of 0.6-ns pulses would be separated by a much smaller time interval. Such a system has been demonstrated by dividing the output of the phase-locked laser into two channels, each containing a modulator driven by a separate 224 megabit per second digital communication terminal. The two channels are interleaved in time and are transmitted as a single beam containing 448 megabits per second.

Details of the multiplexer and the demultiplexer, which also uses a LiTaO₃ electrooptic switch, are given by Kinsel and Denton.[53] The bit rate of the fully loaded optical system depends on the narrowness of the optical pulse which in turn depends on the width of the laser transition, i.e., $\Delta t \sim 1/\Delta v_{\rm osc}$. By using the narrower pulses obtained from a mode-locked neodymium-doped yttrium aluminum garnet laser, an information capacity of 10 000 megabits per second can be attained.

VII. MICROWAVE MODULATION

Calcium pyroniobate has proved to be effective for microwave modulation of light and an experimental strip-line cavity operating at 3 GHz was shown in Fig. 5. Kaminow and Sharpless[48] also have used lithium niobate and lithium tantalate as microwave light modulators at 4.2 GHz. In all three crystals the transverse electrooptic effect is involved. Small crystals may be used, resulting in microwave structures which are considerably different from those used previously with the longitudinal effect in KDP.

A reentrant-type microwave cavity, shown schematically in Fig. 16, was used for the LiTaO₃ and LiNbO₃ modulators. The crystal, whose dimensions are 0.25 by 0.23 by 3.88 mm, is attached to a chisel-shaped center post which represents an attempt to concentrate all of the electric field into the crystal.

The peak phase change, or phase modulation index, for light polarized along the x_3 axis is

$$\eta = \beta 2\pi l n_3^3 r_{33} V/\lambda d. \tag{22}$$

For light polarized at 45 degrees with respect to the principal axes, the peak phase retardation is

$$\Gamma = \beta 2\pi l (n_3^3 r_{33} - n_1^3 r_{13})V/\lambda d. \tag{23}$$

At microwave frequencies, the transit time for light passing through the crystal can be comparable to the modulation period. The coefficient β is the appropriate correction factor and is given by[54]

$$\beta = \tfrac{1}{2}\{(\sin u^+/u^+) + (\sin u^-/u^-)\}$$
$$u^\pm = (\pi f l/c)(\varepsilon_3^{\frac{1}{2}} \pm n) \tag{24}$$

where f is the modulation frequency.

For the LiTaO₃ modulator, the measured Γ is 0.44 rad, or 44-percent intensity modulation, over a 40-MHz bandwidth with 180 mW of modulating power. The phase modulation index, under similar conditions, is $\eta = 0.57$ rad. Thus, Kaminow and Sharpless conclude that on the order of 10 mW/MHz of bandwidth will be required to produce a unity

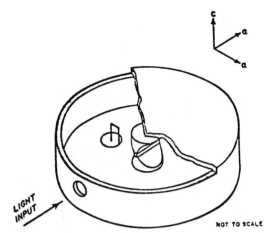

Fig. 16. Reentrant 4-GHz cavity cut away to show LiNbO₃ crystal and coupling loop. From Kaminow and Sharpless[48]

phase modulation index η or 75-percent intensity modulation in Γ for LiTaO₃ at 4 GHz. The comparable value for the baseband, 200-MHz LiTaO₃ modulation system of Denton et al.[27] is ~1 mW/MHz. Part of the difference is due to the transit time effect, β of (24), operable at the higher frequencies.

VIII. OPTICAL FREQUENCY TRANSLATOR

Lasers have been constructed with negative feedback for frequency control of one part in 10^{10} and in a single longitudinal mode. Recently, Duguay, Hargrove, and Jefferts[55] have been able to translate the frequency of a mode-locked laser continuously by as much as ± 2.4 GHz, with 100-percent conversion efficiency from incident to shifted light. This controllable frequency shift on either simple or more sophisticated lasers opens up several possible areas of new device applications and physical measurements. An exciting possibility for further materials research lies in the field of ultrahigh precision spectroscopy.

The method, as shown in Figs. 17 and 18, utilizes the 0.6-ns optical pulses from a phase-locked He-Ne laser and a synchronously modulated LiNbO₃ electrooptic modulator. The refractive index of the LiNbO₃ is then varied sinusoidally at a 56-MHz rate and the light pulses traverse the crystal during the linear portion of the decreasing index change. The 56-MHz source is derived from the output of the mode-locked laser so as to maintain the proper phase relationship. The optical signal upon passing through the crystal of length l and index n emerges with a linear increase in phase, $\Delta\Gamma$, given by

$$\Delta\Gamma = \exp 2\pi i(ft + at + c)$$

where

$$a = d/dt(nl/\lambda) \tag{25}$$

and c is a constant. The effect is equivalent to a Doppler frequency shift of $\Delta f = a$ for 100 percent of the optical energy.

The RF electric field was applied parallel to the x_3 axis, and the extraordinarily polarized laser beam was propa-

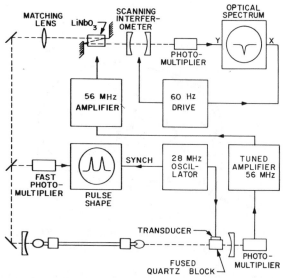

Fig. 17. Diagram of frequency translator, showing mode-locked laser, LiNbO₃ crystal, and scanning interferometer for measuring spectrum. From Duguay et al.[55]

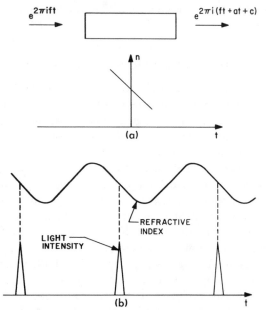

Fig. 18. The phase relation between the optical pulses and RF variation of optical index of the crystal. From Duguay et al.[55]

gated along the x_1 axis, giving rise to an index change of $\Delta n = -n_e^3 r_{33} E/2 = -1.64 \times 10^{-8} E$. In the experiment of Duguay et al., the RF field available was ~ 3 kV/cm and the beam was made to pass through the crystal three times. The frequency shift was about 800 MHz per pass or a total shift which could be varied from 0 to 2400 MHz by varying the RF power. An equivalent negative shift was obtained by reversing the phase of the RF voltage.

Since the transit time through the crystal merely changes the constant c, in the above equation, there is no essential limit on the Δf which can be obtained by using longer crystals, or several more passes through the same crystal. The practical limitations are determined by the width of the optical pulses and mode spacings. It has been predicted that using LiNbO₃ a shift of ± 30 cm^{-1} is possible. Of course,

if the full promise of some of the newer electrooptic materials is realized Δf would be even greater.

For applications other than those involving a single optical frequency it becomes important to develop the theory for arbitrary optical input signals. Garrett and Duguay[56] have analyzed the frequency translator using a series Bessel function expansion. They were able to show that a Gaussian input pulse leads to a Gaussian output pulse provided that the total frequency shift does not exceed the cube of half the number of locked modes times the mode spacings.

IX. MICROWAVE ELASTIC PROPERTIES

Within just the past few years experimental techniques have been developed which extend the frequency range over which ultrasonic, or elastic wave, propagation can be studied in crystals, to well over 100 GHz using conventional microwave equipment but operating at 4.2°K. With more specialized experimental arrangements certain elastic wave investigations have been performed in the submillimeter and far infrared regions. More important from the applied point of view, however, is that elastic wave propagation at frequencies as high as 10 GHz has been observed at room temperature in at least two materials, yttrium iron garnet[57] and lithium niobate.[5] In both cases the crystals performed as their own transducers. In a garnet rod coupling to the elastic waves is accomplished at the surface of one end by ferromagnetic resonance.[58] For a LiNbO₃ rod piezoelectric coupling[3] is used. Yttrium iron garnet is transparent in the infrared region from 1 to 10 microns while LiNbO₃ and LiTaO₃ are transparent in the visible as well as the infrared region from ~ 0.4 to 5 microns and thus are useful for elastooptic application.

Briefly, an ultrasonic wave causes a periodic variation in the index of refraction of a crystal through elastooptic coupling. A light beam incident at an appropriate angle will be partially diffracted by this index variation. This effect is explained in detail in Section X.

An ideal crystal for elastooptic applications might be one with low ultrasonic loss and large piezoelectric and elastooptic coupling efficiencies. Lithium niobate has these properties and is transparent for visible and near infrared illumination, another important criterion. For use in the near infrared region magnetic garnets also are promising in many respects; however, several other interesting magnetoelastic phenomena are involved and will not be discussed here. Furthermore, this section is not intended as a comprehensive discussion of ultrasonic phenomena. Only those properties will be described which have particular bearing on elastooptic applications.

Generation of Ultrasonic Waves

Most of the crystals we have been concerned with are ferroelectric or strongly piezoelectric and thus do not require external transducers. For frequencies up to 100 MHz several conventional transducers are available with high piezoelectric coupling efficiency and low bonding loss. From 100 MHz to above 10 GHz vacuum-deposited CdS and

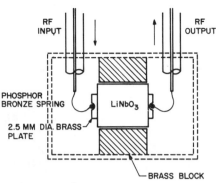

Fig. 19. Crystal holder, showing input and output coaxial coupling probes for excitation and propagation of elastic waves in piezoelectric crystal from ~1 MHz to ~3 GHz.

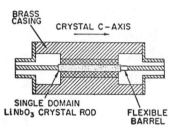

Fig. 20. Coaxial crystal structure suitable for high-frequency operation, in which the contacting element is a gold-plated stainless steel bellows. From Wen and Mayo.[4]

ZnO, as well as other thin-film transducers, can be operated at their fundamental resonance frequency with high coupling efficiency and relatively broad bandwidth.

By using a piezoelectric crystal as its own transducer, in a manner such as that shown in Fig. 19 for LiNbO₃, a large number of significant experiments can be performed readily. The details are given in a later paragraph, but it might be stated here that this rather unsophisticated design can be used for elastooptic experiments from 0.4 MHz to 2 GHz at room temperature and for elastic wave pulse echo experiments from 100 MHz to 2 GHz at temperatures from 4.2°K to 300°K. The sample holder is capable of operating at much higher frequencies; however, it was not convenient to set up additional microwave receivers at these frequencies. A different crystal holder used by Wen and Mayo[4] is shown in Fig. 20 which has a much cleaner microwave coaxial arrangement. The key to this structure is that one or both of the center conductors is terminated in a miniature gold-plated stainless steel bellows. Such a structure should have almost unlimited frquency capabilities.

Ballman,[1] Warner,[2] and Warner, Onoe, and Coquin[59],[60] have measured the piezoelectric coupling in LiNbO₃, LiTaO₃, and Bi₁₂GeO₂₀ and have found that in LiNbO₃ the coupling efficiency is as much as 30 percent for longitudinal waves propagating along the x_1 direction. For Bi₁₂GeO₂₀ the coupling is about 23 percent. These data for all crystallographic orientations are listed in Table IX.

For propagation along an x_2 direction two elastic waves can be excited piezoelectrically in LiNbO₃. These are a quasi-longitudinal mode with an effective coupling constant of $k \sim 30$ percent and a quasi-shear mode with $k \sim 60$ percent. However, for propagation at an angle ~36

TABLE IX
PIEZOELECTRIC COUPLING CONSTANTS FOR LiNbO₃ AND LiTaO₃*

Plate orientation	LiNbO₃–k(%)	LiTaO₃–k(%)
x—quasi-shear	68	44
y—quasi-shear	60	38
z—quasi-shear	0	0
y—quasi-extensional	30	20
z—quasi-extensional	18	19

* From Warner *et al.*[59]

degrees in the plane there can be a coupling of $k \sim 50$ percent to the quasi-longitudinal mode with no shear mode excitation. At an angle of ~163 degree there can be a coupling $k \sim 60$ percent to the quasi-shear mode with no longitudinal mode excitation.

For lithium tantalate the coupling for the quasi-shear mode propagating along x_2 is $k \sim 40$ percent and for the quasi-longitudinal mode $k \sim 20$ percent. Again, propagating at the angle ~43 degrees in the plane coupling to the quasi-longitudinal mode is $k \sim 30$ percent with no shear wave excitation, and for an angle of ~163 degrees coupling to the quasi-shear mode is $k \sim 40$ percent with no longitudinal wave excitation.

Propagation Losses

The losses involved in the propagation of high radio frequency or microwave elastic waves through a crystalline medium usually are expressed in terms of a damping constant Γ defined by exp $(-\Gamma z)$, where z is the path length traversed in the crystal. Through theoretical and experimental work in many laboratories significant progress has been made in relating measured elastic wave propagation losses to theoretically calculated values based on other measured fundamental constants of the material. More complete experimental data, as functions of temperature and frequency, are available for quartz, silicon, and germanium than perhaps for any other crystals. These data have been the stimuli for much of the theoretical effort. Thus, it is for these crystals that correlation between theory and experiment is most detailed.[61]

At present, materials exhibiting particularly low attenuation loss at room temperature are the magnetic and nonmagnetic garnets, Al₂O₃, MgAl₂O₄, MgO, LiNbO₃, LiTaO₃, Bi₁₂GeO₂₀, diamond, and β boron. All except the last two can be synthesized in crystals large enough to be of interest both for device applications and for detailed study of the basic damping mechanisms.

Akhiezer's[62] model for the absorption of sound has been the basis for several more recent treatments.[61],[63],[64] Summarizing these results, attenuation may be written in the form

$$\Gamma(\text{dB/cm}) = \frac{8.68 \, \Delta c}{2\rho \bar{v}^2} \, \frac{\omega^2 \tau}{1 + \omega^2 \tau^2} \qquad (26)$$

where ρ is the density, ω is the angular frequency, $\bar{v}$ is the average sound velocity in the Debye approximation, and τ is the phonon relaxation time. Essentially, the problem is to evaluate Δc, which is the difference between the adiabatic and isothermal moduli. At high temperatures, $\omega\tau < 1$, and $\Delta c = \gamma^2 C_v T$, where C_v is the specific heat and γ is an average Grüneisen constant. Taking τ to be twice the thermal rate, or $\tau = 6\kappa/C_v\bar{v}^2$, where κ is the thermal conductivity, (26) becomes

$$\Gamma(\text{dB/cm}) = \frac{8.68\,\gamma^2}{\rho\bar{v}^5}\,3\omega^2\kappa T. \qquad (27)$$

Generally, it is found that $\kappa \sim T^{-1}$ at sufficiently high temperatures, so that Γ varies as ω^2 and is independent of temperature. For lower temperatures, Mason and Bateman[61] have developed a method for evaluating the losses by using measurable quantities to obtain phonon-mode temperature variations. This brings in other relations involving the third-order as well as the second-order elastic moduli. They have been able to obtain good correlation between measured attenuation at 480 MHz in silicon, and computed attenuation values based on measured values of thermal conductivity and elastic moduli.

In a paper of special interest from the viewpoint of developing new materials for use at microwave frequencies, Oliver and Slack[65] have rewritten (27) in a form which allows them to plot the frequency-independent loss term, $\Gamma\omega^{-2}$, as a function of crystal constants, thermal conductivity, and data derived from elastic constants. Their graph, reproduced in Fig. 21, shows that the frequency-independent losses all lie within two parallel lines. A certain amount of physical order is clearly illustrated. To place a new crystal in its proper place on this chart a moderate number of measurements are needed. Even before this is done several guidelines are available. Lowest microwave elastic losses will be obtained for a crystal which has 1) the lowest density, 2) the highest Debye temperature, 3) the lowest elastic wave velocities, 4) the highest elastic isotropy, and 5) the lowest value of thermal conductivity.

The last factor represents a curious problem, but one which may be used to advantage under suitable circumstances. Intuitively, it may be expected that a more perfectly ordered crystal would have the lowest ultrasonic losses. It is true, however, that thermal conductivity increases with increasing order. The ultrasonic losses will then increase due to the stronger thermal-acoustic interaction. It has been shown experimentally[66] in some crystals that controlled disordering of the ions can result in a reduction of the measured room-temperature losses. It is interesting to speculate as to how far this process can be carried and whether the elastooptic coupling might be increased at the same time.

Lithium Niobate and Lithium Tantalate

Because of the large refractive indices of $LiNbO_3$ and $LiTaO_3$, large elastooptic effects were anticipated. It was therefore necessary to characterize the elastic properties in order to make full use of the elastooptic properties. At the same time there was no *a priori* reason to expect the room-temperature microwave elastic losses to be lower than any other trigonal crystals, such as quartz. Propagation losses

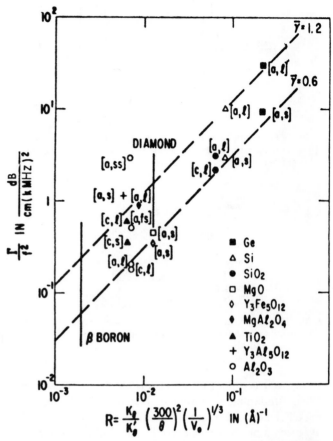

Fig. 21. Sound attenuation at room temperature normalized at 1 GHz for nine materials as a function of the thermal parameter R. Each data point is labeled with the sound propagation direction, using the axes a, b, c and with the type of sound wave propagating: longitudinal, l; shear, s; slow shear, ss; fast shear, fs. The values of R for diamond and for β-boron are drawn as solid lines. From Oliver and Slack.[65]

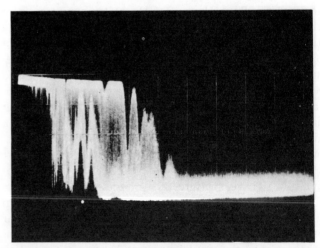

Fig. 22. A pulse echo train of longitudinal plane elastic wave pulses at 500 MHz in $LiNbO_3$; $v_l = 7.43 \times 10^5$ cm/s, pulse spacing is 1.80 μs, and $T \sim 290°$K. The scope trace is 100 μs/cm. With a few watts of peak power, the train of over 400 pulses extends to over 700 μs and completely saturates the receiver to about 500 μs.[3]

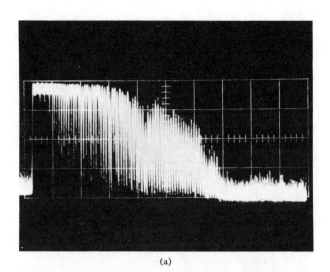

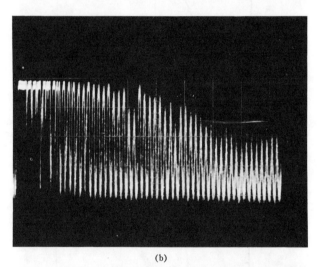

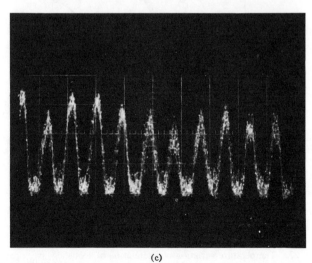

(a)

(b)

(c)

Fig. 23. Pulse echoes resulting from a 3-mW input signal. These pictures show that the crystal is remarkably free from the coupling between longitudinal and transverse waves which is usually present in anisotropic crystals.[3] (a) Trace speed is 20 μs/cm. (b) Trace speed is 10 μs/cm. (c) Trace speed is 2 μs/cm.

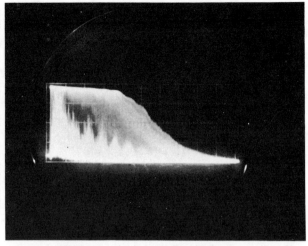

Fig. 24. A particularly smoothly decaying pulse echo train in LiNbO$_3$ at 150 MHz and $T \sim 290°$C. The oscilloscope trace speed is 200 μs/cm.

were measured[3]–[6] by the pulse echo technique. A pulse of RF or microwave energy sets up an electric field inside the lithium niobate crystal. Piezoelectric coupling converts this energy into an elastic wave pulse which then reflects back and forth between the input and output ends of the crystal, polished optically flat and parallel. The reflections from the output end piezoelectrically couple to the output line, generating a series of pulse echoes. These are detected by a superheterodyne receiver and displayed on an oscilloscope. From the path length involved and the spacing of the pulses, the velocity of sound can be determined.

With a peak power of the order of 10 watts at 500 MHz a train of about 400 pulse echoes lasting for over 700 μs can be seen at room temperature. Because of the large number of pulses involved, even the smallest misalignment of the polished end faces can cause elastic wave interferences. Fig. 22 is an oscilloscope photograph showing such a pulse echo train. The trace speed is 100 μs/cm or about 57 pulses per centimeter, unresolved in the photograph. Because of the strength of the echoes, the receiver is saturated to about 400 μs and is almost completely blocked to about 150 μs. The interferences just mentioned show up as nulls in the pulse train around 400 and 500 μs. This train of longitudinal pulses was remarkably free of extraneous transverse pulses.

A pulse echo train, of sufficient stength to be considered for device use, can be observed with RF power as low as a few microwatts. Those shown in Fig. 23 were generated with a power source of $\sim$3 mW. The spacing between pulse echoes is 1.80 μs and the pulse width of the low-power source was only about one-half of this. In Fig. 23(a) the sweep speed is 20 μs/cm. Fig. 23(b) shows a delayed portion of the trace taken at 10 μs/cm, from 80 to 100 μs from the initiating pulse, to show how clean the pulse is. Fig. 23(c) shows an expanded sweep of 2 μs/cm. A particularly smoothly decaying pulse echo train taken at $\sim$150 MHz is shown in Fig. 24.

The elastic wave losses can be determined by measuring the time interval required for the pulse echo train to decay to one-half its initial height, or by using calibrated RF

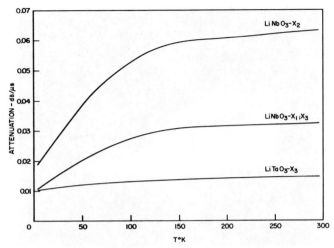

Fig. 25. Attenuation for longitudinal elastic waves at 500 MHz propagating along the x_1, x_2, and x_3 crystallographic axes in LiNbO$_3$ and along the x_3 axis in LiTaO$_3$.[6]

attenuators to measure the ratio of the RF signal in one pulse to that of a much later pulse. Data taken in this manner from 4.2°K to 296°K are shown in Fig. 25 for both LiNbO$_3$ and LiTaO$_3$. Measurements were made for longitudinal elastic waves propagating along each of the three principal crystallographic axes, x_1, x_2, and x_3, in LiNbO$_3$ and along the x_3 axis in LiTaO$_3$. These data are shown as separate curves except that for LiNbO$_3$ the data are combined as a smoothed average for the x_1 and x_3 directions. Measurements were made along the x_3 axis of two other LiNbO$_3$ crystals with entirely similar results. Although the pulse echo train contained interference nulls, it was possible to estimate the losses in all samples to within a deviation of ~ 0.005 dB/μs. At room temperature and 500 MHz for the first available samples the equivalent acoustic $Q \sim 10^5$. For the LiTaO$_3$ of Fig. 25, as well as other LiNbO$_3$ crystals, $Q \sim 1.25 \times 10^6$. Note that here Γ is given in decibels per microsecond, which represents the losses for a microsecond of information storage. For a longitudinal wave along x_3 in LiNbO$_3$ $\Gamma(\text{dB}/\mu\text{s}) = 10^{-6} v_l \Gamma(\text{dB/cm})$ $= 0.743 \Gamma(\text{dB/cm})$. Wen and Mayo[4] have measured the frequency dependence of the losses in LiNbO$_3$ at room temperature from 0.4 to 5.5 GHz, and Grace et al.[5] have measured the temperature dependence at frequencies from 2.64 to 8.96 GHz. These data are shown in Figs. 26 and 27. Below (27) it is mentioned that generally the thermal conductivity $\kappa \sim T^{-1}$ at high temperatures. In the high-temperature limit Γ should be independent of temperature and should vary as ω^2. All of the above data are consistent with these trends.

An interesting observation may be made regarding the fact that Γ at 4.2°K and at 500 MHz does not drop below the value of 10^{-2}, as compared to $\sim 10^{-4}$ for Bi$_{12}$GeO$_{20}$. Although an insufficient amount of information is available at this time, it is likely that the 10^{-2} value represents phonon scattering due to imperfections and the general disordering of ions in the crystals. This would lead to an increase in the low-temperature losses and, as mentioned earlier, a decrease in the high-temperature losses. At higher

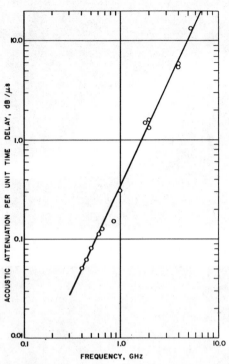

Fig. 26. Acoustic longitudinal-wave attenuation in lithium niobate at room temperature, showing an $\sim \omega^2$ dependence. From Wen and Mayo.[4]

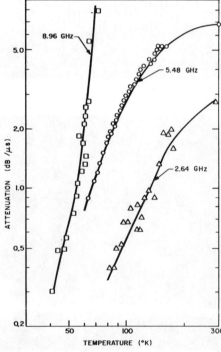

Fig. 27. Temperature variation of attenuation of acoustic waves propagating along the x_3 axis in LiNbO$_3$ at three different frequencies. The absolute attenuation at 2°K has been set equal to zero. From Grace et al.[5]

frequencies the intrinsic loss mechanisms can more easily dominate and the scattering losses remain important only at low temperatures. In most of the "low-loss" crystals it is still possible that the losses actually measured at low temperatures and reported in the literature are not intrinsic but are upper limits based on scattering losses. Also, in LiNbO$_3$ and LiTaO$_3$ the data suggest that this is the case at all temperatures for frequencies as high as 500–1000 MHz.

Velocity Surfaces and Elastic Anisotropy

In an anisotropic crystal there exist one longitudinal and two shear elastic wave velocities for propagation along any given crystallographic direction. The velocities are determined by the elastic stiffness constants, or elastic moduli c_{ijkl}, which are components of a fourth-rank tensor and are the constants of proportionality in the generalized Hooke's law: $\sigma_{ij} = c_{ijkl}\varepsilon_{kl}$, where σ_{ij} is the stress tensor and ε_{kl} is the strain tensor.

The relations between the elastic stiffness constants and the velocities are given by the Christoffel[67] equations

$$|\Gamma_{ij} - \delta_{ik}\rho v_i^2| = 0, \qquad (28)$$

where δ is the Kronecker delta and

$$\Gamma_{ik} = \Gamma_{ki} = l_j l_l c_{ijkl}. \qquad (29)$$

The direction cosines of the phase velocity vectors are represented by l_j and l_l. Wave velocity surfaces have been investigated in considerable detail for several crystals of particular interest, such as calcite, rutile, and quartz. From these data other parameters, such as the Debye temperature, and the problem of propagation of pure and mixed elastic modes have been studied. Recent papers on the Debye temperature by Robie and Edwards[68] and on the phase velocity surfaces by Farnell[69] outline the procedures and give references to the earlier theoretical work.

Using the method of resonance and antiresonance in crystallographically oriented platelets, Warner, Onoe, and Ballman[30],[59] have measured the elastic moduli and piezoelectric coupling constants for LiNbO$_3$, LiTaO$_3$, and Bi$_{12}$GeO$_{20}$. Their piezoelectric data are summarized in Table IX. Bateman and Spencer[70] have measured the velocities and elastic moduli in LiNbO$_3$ and Smith reports similar data for LiTaO$_3$.[71] These data were taken by McSkimin's high-frequency plane-wave propagation method in which all measurements are made on a single monocrystal. The values obtained are given in Tables X and XI and are in substantial agreement with those given by Warner et al.[59]

From the measured values[70] it is seen that both off-diagonal matrix elements c_{13} and c_{14} are almost vanishingly small. Since these are cross-coupling elastic terms, the result is that LiNbO$_3$ and LiTaO$_3$ are elastically more isotropic than other trigonal or hexagonal crystals for which these moduli have significant magnitudes. In this manner these crystals are similar to Al$_2$O$_3$ which also is almost completely isotropic in the transverse direction. Cross sections of the wave velocity surfaces, in the (100), (010), and (001) planes for LiNbO$_3$, are shown in Fig. 28. Elastic moduli

TABLE X

ELASTIC WAVE VELOCITIES ALONG PRINCIPAL DIRECTIONS IN LiNbO$_3$*

Mode	Propagation direction	Particle motion	Wave velocity (10^3 m · s^{-1})
L	x	x	6.54873
S	x	$\perp x$, $\approx 31°y$	4.75976
S	x	$\perp x$; $\approx 59°y$	4.03406
L	y	quasi-longit.	6.83789
S	y	quasi-shear	4.46667
S	y	x	3.94043
L	z	z	7.33059
S	z	$\perp z$; $\|x$	3.5885
S	z	$\perp z$; $\|x$	3.5885

* From Bateman and Spencer.[70]

TABLE XI

ELASTIC WAVE VELOCITIES IN PRINCIPAL DIRECTIONS IN LiTaO$_3$*

Modes†	Propagation direction	Particle displacement direction	Wave velocity (10^3 m · s^{-1})
L	x_1	x_1 plane	5.552
S	x_1	x_2x_3 plane $-55°$ to y	4.212
S	x_1	x_2x_3 plane $-35°$ to y	3.366
QL	x_2	x_2x_3 plane $+10.3°$ to y	5.690
QS	x_2	x_2x_3 plane $-10.3°$ to y	3.883
S	x_2	x_1	3.529
L	x_3	x_3	6.160
S	x_3	x_1x_3 plane $-$ arbitrary	3.604

* Computation by P. Lloyd and R. T. Smith from data of Smith.[71]
† L=longitudinal, S=shear, QL=quasi-longitudinal, QS=quasi-shear.

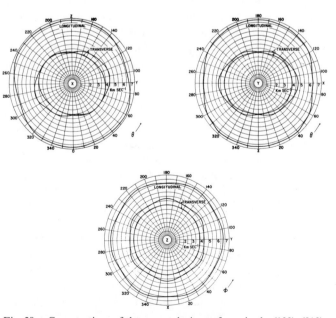

Fig. 28. Cross sections of the wave velocity surfaces, in the (100), (010), and (001) planes for LiNbO$_3$. Elastic moduli stiffened by the piezoelectric coupling were used in the computation.[70]

stiffened by the piezoelectric coupling were used for the computations. In a similar plot[75] of the unstiffened moduli, the sixfold symmetry of the lower diagram is barely observable.

Theoretical calculations of mechanisms responsible for elastic wave propagation in themselves do not reflect appreciably the lack of complete elastic isotropy. What is more important for practical considerations is the effect of elastic anisotropy on strain and other gross imperfections in the crystals. Experimentally it is generally found that smoothly decaying pulse echo trains, free of interference nulls, are more easily set up in crystals which have higher crystallographic symmetry and are most isotropic elastically. The fact that the elastic symmetry of LiNbO$_3$ and LiTaO$_3$ is higher than that demanded by their point group could be especially significant in the room-temperature measurements at 10 GHz.

Bismuth Germanium Oxide and Bismuth Silicon Oxide

Crystals belonging in the cubic point groups 43m and 23 lack a crystallographic center of symmetry and can be piezoelectric and electrooptic, at the same time being optically isotropic. Because of these promising attributes there have been many attempts in various laboratories to synthesize crystals in these point groups. Although some interesting crystals have resulted there usually have been some associated problems, such as the mixed cubic and hexagonal phases in zinc sulfide, ZnS, and chemical instability in cuprous chloride, CuCl. Bismuth germanium oxide[20],[29] and bismuth silicon oxide are the first noncentrosymmetric cubic crystals which have been synthesized as sufficiently well-grown crystals to live up to some of these promises. To our knowledge they are the only inorganic crystals presently available in the 23 point group belonging to either of the body-centered space groups, I23 or I2$_1$3. A decided practical advantage is their relatively low melting points which make the growth of fairly large, mechanically sound crystals a simpler task than is experienced in the growth of most high-temperature materials.

Bismuth germanium oxide crystals are piezoelectric with a coupling contant[30] of about 23 percent so that they can be used as their own transducers. The longitudinal wave velocity in the [110] crystallographic direction is 3.42×10^5 cm/s and one shear wave velocity is 1.77×10^5 cm/s. The velocities obtained by Onoe et al.[30] are given in Table XII. At room temperature the attenuation for 500-MHz shear waves propagating in the [110] direction is $\Gamma = 0.11$ dB/μs which is equivalent to an acoustic Q of 127 000. The crystals are thus to be considered along with LiNbO$_3$ and LiTaO$_3$ for ultra high frequency and microwave ultrasonic uses.

The low velocities of propagation allow relatively short crystals to be used for a given amount of acoustic delay. Since they are cubic, pure elastic modes of propagation can be readily obtained and in general are not as sensitive to crystallographic orientation. Further, it is found from the components of the piezoelectric matrix d_{ij} that the application of an electric field in any major crystallographic direction in bismuth germanium oxide couples to only the pure

TABLE XII

ELASTIC WAVE VELOCITIES AND COUPLING FACTORS FOR PURE MODES PIEZOELECTRICALLY EXCITED IN Bi$_{12}$GeO$_{20}$*

Mode	Propagation direction	Wave velocity (10^3 m·s^{-1})	k (%)
Shear	[100]	1.646	
Shear	[111]	1.859	
Longitudinal	[111]	3.332	15.5
Shear	[110]	1.706 (calculated)	23.5 (calculated)

* From Onoe et al.[30]

elastic shear waves. Writing out explicitly the terms of the equation, $\varepsilon_j = d_{ij}E_i$ ($i = 1$–3, $j = 1$–6), for electric fields applied in the x_1, x_2, and x_3 directions, the only strain components generated are

$$\varepsilon_4 = d_{14}E_1; \qquad \varepsilon_5 = d_{14}E_2; \qquad \varepsilon_6 = d_{14}E_3. \quad (30)$$

In tensor form these are $2\varepsilon_{23}$, $2\varepsilon_{13}$, and $2\varepsilon_{12}$, respectively. For example, ε_{23} indicates particle motion along a [001] direction and propagation along a [010] direction.

The absence of the excitation of longitudinal waves, at least in part, accounts for the particular ease with which it has been possible to excite pulse echo trains with smoothly decaying envelopes. In applications involving storage of information at high radio frequencies and microwave frequencies these are important technical considerations.

In Fig. 29 data are given for shear waves at 118 and 500 MHz propagating along a [110] direction. The temperature dependence of the attenuation is shown to consist of four peaks in the region from 50°K to 70°K. Another smaller peak occurs near 215°K and effects the room-temperature values. The origin of these peaks as yet has not been determined. Molten bismuth oxide is highly reactive and can result in the inclusion in the crystal of small amounts of platinum from the crucible during growth. It is possible that the loss maxima are caused by these or other impurities. Alternatively, they may be related to unknown loss mechanisms of the intrinsic material.

In any case, the attenuation can be separated into the sum of two sets of curves, one consisting of the loss maxima and the other consisting of the smoothly varying residual curve. The loss maxima may be interpreted as a relaxation attenuation

$$\Gamma_m = A\omega^2\tau/(1+\omega^2\tau^2) \quad (31)$$

where ω is the angular frequency and τ is the activated relaxation time given by $\tau = \tau_\infty \exp(\varepsilon/kT)$. Five such curves are present with different values of $\tau(T)$. The activation energy is $\varepsilon \sim 0.005$ eV for the low-temperature peaks and $\varepsilon \sim 0.04$ eV for the high-temperature one. It should be pointed out that the shifts in the loss maxima with frequency are more clearly seen in curves of the inverse elastic wave Q versus T. Such curves were used to evaluate the activation energies.

Stating the data in terms of acoustic Q's, we have $Q(500 \text{ MHz}) = 127\ 000$ and $Q(118 \text{ MHz}) = 430\ 000$ at

270

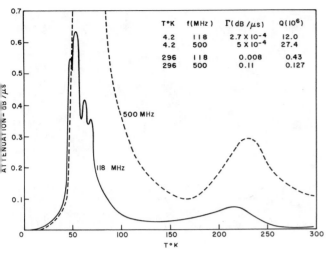

$T°K$	$f(MHz)$	$\Gamma(dB/\mu s)$	$Q(10^6)$
4.2	118	2.7×10^{-4}	12.0
4.2	500	5×10^{-4}	27.4
296	118	0.008	0.43
296	500	0.11	0.127

Fig. 29.   Temperature and frequency dependence of elastic wave propagation losses in $Bi_{12}GeO_{20}$.[19]

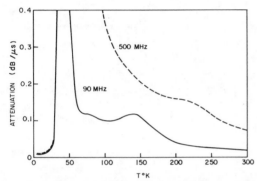

Fig. 30.   Temperature and frequency dependence of elastic wave propagation losses in $Bi_{12}SiO_{20}$.

$296°K$. It is noted again that the data also involve the tail of an absorption peak near $225°K$. Further investigation of the conditions involved in the crystal growth, the use of ultra high purity starting materials, and the elimination of platinum inclusions could determine if this high-temperature loss peak is due to a removable impurity. If so, then bismuth germanium oxide would certainly have large room-temperature acoustic Q's even to much higher microwave frequencies. Detailed investigations as to the fundamental loss mechanisms and their relation to third-order elastic moduli and thermal conductivity should await additional crystal development. At $4.2°K$ the acoutic Q's are $Q(118 \text{ MHz}) = 12 \times 10^6$ and $Q(500 \text{ MHz}) = 27 \times 10^6$. These are equivalent to $\Gamma(118 \text{ MHz}) = 2.7 \times 10^{-4}$ dB/μs and $\Gamma(500 \text{ MHz}) = 5 \times 10^{-4}$ dB/μs. The fact that $Q(500 \text{ MHz})$ is larger than $Q(118 \text{ MHz})$ is due in part to the shift of the absorption peaks to higher temperatures for higher frequencies in accordance with the activation equation above. Another possible cause is diffraction loss whereby the crystal boundaries can affect the measurements in a high-Q sample at the longer acoustic wavelengths.

Attenuation data, taken at 500 and 90 MHz, are given in Fig. 30 for bismuth silicon oxide. In general the data are quite similar to those for bismuth germanium oxide except that the attenuation at $4.2°K$ is much higher. Bismuth silicon oxide proved to be a much softer crystal, making the optical grinding and polishing more difficult. It is believed that the quality of the polished surfaces is in part responsible for the larger attenuation values.

Pulse Echo Train and Diffraction Losses

The small measured losses in $Bi_{12}GeO_{20}$ involved observation of pulse echo trains which were exceptionally long. At the same time an interesting phenomenon was observed. It was found that for certain discrete frequencies of operation the pulse echo train was modulated with almost perfect nulls appearing every 700 μs, the spacing between nulls being almost completely independent of temperature. Fig. 31 shows oscilloscope photographs taken at $4.2°K$ and at 188 MHz. The trace in the upper left was taken at 5 ms/cm, the lower left at 2 ms/cm, the upper right at 0.5 ms/cm delayed by about 20 ms, and the lower right at 0.1 ms/cm. The pulses are shear elastic waves with a spacing of 5.64 μs between pulses and with each envelope containing 110 pulses. Pulses are actually observed to 50 ms, which means 4000 to 5000 pulse echoes. Since the velocity is 1.77×10^5 cm/s, this corresponds to a path length in the crystal of ~ 88 meters. The pulse echo train decays to one half-height in about 10 ms, corresponding to a half-height path length of ~ 20 meters. Using a delayed trace and an expanded scale of 1 μs/cm sweep speed, it was found that there was no deterioration of pulse shape, the last pulse having the same shape as the first.

One possible explanation of the modulation nulls occurring every 700 μs is that two shear waves are excited simultaneously. Both propagate along a [110] direction, the direction of particle motion for one wave being different from that for the other. If, due to a slight residual strain or a slight misorientation of the crystallographic axes, the velocities differ by about one part in 10^5, then they would become out-of-phase every 700 μs and produce nulls.

It is clear that bismuth germanium oxide is an excellent acoustic Fabry-Perot interferometer. The decay to half amplitude in 20 meters corresponds to 2000 passes, one way, through a crystal one centimeter long, or to 0.025-percent loss per pass. Fox and Li[72] and Boyd and Kogelnik[73] have developed theories of the power loss and phase shift per transit in a Fabry-Perot interferometer and have plotted their results on a convenient graph. Using our values for crystal dimensions and acoustic wavelength, the diffraction losses per pass, also, are estimated to be in the range of 0.025 percent.

The measured elastic wave propagation losses in a crystal must be the sum of the losses due to intrinsic relaxation mechanisms, diffraction losses, losses due to scattering from gross defects, losses due to activated relaxation processes caused by impurity ions, and losses introduced mechanically, such as from the lack of parallelism or optical flatness of the end faces. The fact that measured losses in $Bi_{12}GeO_{20}$ are comparable to the calculated diffraction losses indicates the necessity of a much more detailed and sophisticated measurements program on even higher-purity crystals. Only then will it be possible to separate out and evaluate

271

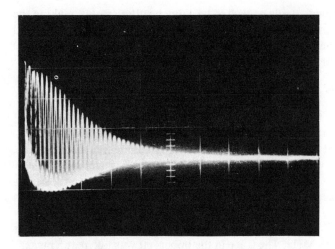
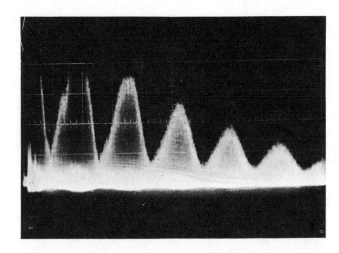
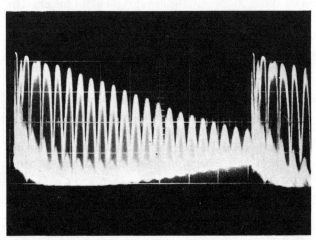
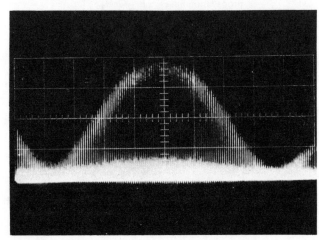

Fig. 31. Interference phenomena in pulse echo trains at 125 MHz in $Bi_{12}GeO_{20}$ at 4.2°K. Upper left trace speed is 250 ms/div showing ~ 5000 pulses spaced 5.74 μs apart. Lower left, 120 ms/div. Upper right, 0.5 ms/div. Lower right showing the individual pulses, 0.1 ms/div.

the intrinsic damping mechanisms and to understand the capabilities and limitations of the material.

X. ELASTOOPTIC PROPERTIES OF MATERIALS

Elastooptic Phenomena and Optical Diffraction

The change in index of refraction of a material with applied external stress is called the piezooptic effect and the change with respect to an internal strain is called the elastooptic effect. Exciting a plane elastic wave in a crystal creates a periodic strain pattern with spacing equal to the acoustic wavelength. This strain pattern produces an elastooptic variation in the index of refraction equivalent to a volumetric diffraction grating. Experimentally, either stationary or traveling elastic waves may be used; however, no distinction need be made at this point in the general discussion of the physical phenomena. The law of superposition also holds, for strains of moderate magnitude, so that volumetric diffraction gratings may be set up simultaneously at identical or at different frequencies. An example of this is the case in which a crystal is used as its own transducer and couples in both a longitudinal wave and one or more shear waves. A more complicated, but most interesting case is one in

which two identical diffraction gratings are set up orthogonally. This is discussed under the heading of the Schaefer-Bergmann method in Section XI.

Light incident on an elastooptic grating at an appropriate angle Θ, Fig. 32, will be partially diffracted. The angle Θ is determined from Bragg's relation commonly used for X-ray diffraction from the periodic arrangement of atoms in a crystal:

$$\sin \Theta = NK/2k \qquad (32)$$

where N is an integer indicating the order of the diffracted beam, k is the incident light wave vector $2\pi/\lambda$, and K is the acoustic wave vector $2\pi/\Lambda$.

Recently there have appeared in the literature several articles[74]–[76] in which derivations are given of the magnitude and physical characteristics of diffracted optical beams. The results of Cohen and Gordon[74] will be used for this discussion and for convenience in referencing their terminology also will be followed. For an optical beam passing through a rectangular cross-section acoustic beam of width L, and at an angle θ_0 with respect to the acoustic wave front, the amplitude $V_1(\theta_0)$ of the first-order diffracted beam is given as

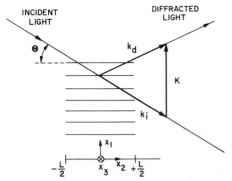

Fig. 32. Conditions for Bragg diffraction in which the elastic wave propagates in the x_1 direction and the light wave propagates approximately in the x_2 direction.

$$V_1(\theta_0) = \left[-\tfrac{1}{2} i \xi_0^* L V_0 \exp\{iK(\theta_0 - \Theta)x_2\} \right] \cdot \left[\frac{\sin \tfrac{1}{2}(\theta_0 - \Theta)L}{\tfrac{1}{2}K(\theta_0 - \Theta)L} \right] \quad (33)$$

where ξ_0^* is related to the acoustic wave amplitude, x_2 is the coordinate defined in Fig. 32, and Θ is the Bragg angle for first-order diffraction defined by (32). This expression displays a $\sin x/x$ angular dependence as the incident optical beam is varied about the Bragg angle. It is valid for acoustic frequencies such that $K^2 L > 4\pi k$. Both of these characteristics will be examined later.

At the Bragg angle the relative diffracted intensity I_1/I_0 is given by

$$\frac{I_1}{I_0} = \frac{V_1(\Theta) \cdot V_1^*(\Theta)}{V_0^2} = \tfrac{1}{4} L^2 |\xi_0|^2. \quad (34)$$

This can be rewritten in terms of the physical constants of a particular crystal, leading to a discussion of three topics of practical importance: 1) the intensity of the diffracted beam for a given acoustic power, 2) a figure of merit for elastooptic materials, and 3) the angular width of the diffracted beam as a function of acoustic frequency.

Using the geometry of Fig. 32, the acoustic wave is assumed to be uniform in the x_3 direction over a distance greater than the optical beam width. A longitudinal acoustic wave of constant cross-sectional amplitude A_0 and angular frequency Ω can be written as

$$\vec{\Psi}(x_1, t) = \hat{x}_1 A_0 \exp\left[i(Kx_1 - \Omega t) \right], \quad -\frac{L}{2} \le x_2 \le \frac{L}{2} \quad (35a)$$

$$\vec{\Psi}(x_1, t) = 0, \quad x_2 > \frac{L}{2}, x_2 < -\frac{L}{2}. \quad (35b)$$

The strain associated with this wave may be written in tensor notation as

$$\varepsilon_{ij} = \frac{1}{2}\left[\frac{\partial \Psi_i}{\partial x_j} + \frac{\partial \Psi_j}{\partial x_i} \right]. \quad (36)$$

Applying (36) to (35a), there is only one nonzero strain component:

$$\varepsilon_{11} = iKA_0 \exp\left[i(Kx_1 - \Omega t) \right]. \quad (37)$$

The optical index ellipsoid or impermeability tensor, discussed in Section V for the electrooptic effect, will be

deformed as well by a strain. This elastooptic effect is linear for experimentally produced strains, and the resulting impermeability tensor may be written as

$$B_{ij} - \delta_{ij} n_i^{-2} = \sum_1^3 p_{ijkl}\varepsilon_{kl}, \quad (38)$$

where δ_{ij} is Kronecker's delta, n_i is the appropriate index of refraction, p_{ijkl} are the elastooptic constants, and ε_{kl} are the strains.

Assuming the incident light in Fig. 32 is linearly polarized along x_3, only one component of the B_{ij} tensor is involved in the diffraction. Using reduced matrix notation, this component is

$$B_3 = \frac{1}{n_3^2} + p_{31}\varepsilon_1. \quad (39)$$

The last term is the change in B_3, $\Delta B_3 = \Delta(n_3^{-2})$. Experimentally, $\Delta n_3 \ll n_3$, so, to a good approximation,

$$\Delta n_3 \approx -\tfrac{1}{2} n_3^3 p_{31}\varepsilon_1. \quad (40)$$

The factor ξ_0^* appearing in (33) is defined as $k/\cos\theta_0 \cdot (\Delta n/n)^*$ in the derivation of that expression. Using (37) and (40), (34) can be rewritten in terms of the physical constants as

$$\frac{I_1}{I_0} = \frac{1}{16} n_3^6 p_{31}^2 \varepsilon_1^2 \omega^2 L^2/c^2 \cdot \cos^2\Theta \quad (41)$$

where c is the speed of light in vacuum and ω is the optical angular frequency.

For acoustic frequencies below 10 GHz, $\cos^2\Theta \approx 1$ and for constant strain the intensity of the diffracted light beam is essentially independent of acoustic frequency. The intensity then varies as the square of the elastooptic constant appropriate to a particular arrangement; this constant is normally about 0.1 except for unique materials. The intensity also varies with the square of the acoustic beam width and the square of the strain, $|\varepsilon_1|^2 = K^2 |A_0|^2$.

The intensity is most sensitive to the appropriate index of refraction, varying as its sixth power. The index n varies from 1.46 for fused silica to 2.2–2.3 for $LiTaO_3$ and $LiNbO_3$ to 2.55 for $Bi_{12}BeO_{20}$ and 2.6 for rutile. The advantage for $Bi_{12}GeO_{20}$ over fused silica due to the n^6 term is then ~ 25 times and for rutile is ~ 30 times.

Equation (41) indicates that $p^2 n^6$ can be a figure of merit for elastooptic materials; this would be based on an equal-strain criterion. A better figure of merit would compare two crystals on the basis of equal power in the elastic wave; this would lead to $(p^2 n^6)/(\rho v^3)$ as a figure of merit, where ρ is the density and v the appropriate sound velocity in the crystal. In practical devices the bandwidth also may be an important factor. Gordon[77] has developed a criterion including the bandwidth and arrives at a figure of merit

$$\mathscr{F} = p^2 n^7/\rho v. \quad (42)$$

The intensity profile of the diffracted beam as a function of angle is derived from (33). The width W of the diffracted beam, defined as the angular difference between the first zeros on either side of the maximum, is

$$W = 4\pi/KL. \quad (43)$$

In several applications the increase in angular width with decreasing acoustic frequencies will determine the operating frequency to be used. When the Bragg angle Θ becomes $\leq \frac{1}{2}W$, a second-order diffracted beam at frequency $\omega \pm 2\Omega$ will become observable within the main lobe of the sin x/x beam. Solving (32) and (43) yields the condition

$$K^2 L \leq 4\pi k. \quad (44)$$

This is the beginning of the Raman-Nath[74]–[76] regime, and below the frequency so defined, many diffraction grating orders may be observed. Actually, different solutions of the wave equation are then required, as (33) and (41) were derived by assuming operation in the Bragg region of a single order N.

Fig. 33 is a pictorial representation of the intensity profile of (33) for three representative frequencies. Since the diffracted beam, as in Fraunhaufer diffraction, is a far-field pattern, the angles are greatly exaggerated. The narrow intensity profile on the left would represent for, say, LiNbO$_3$, diffraction by a 1-GHz longitudinal elastic wave; the center profile would represent an $\sim$70-MHz wave, illustrating the appearance of second-order diffraction; and the profile on the right would represent a 10–20-MHz wave, illustrating several diffraction orders well within the Raman-Nath region.

Lithium Niobate, Measured Coefficients

In general, the piezoelectric coupling, the ultrasonic attenuation, and the elastooptic coefficients are all involved in the use of a dielectric crystal as the active element of an elastooptic device. As has been mentioned already, LiNbO$_3$ has unusual capabilities in each of these three respects. The photoelastic effect was determined by two methods. The initial investigation involved measuring the change in refractive indices upon application of known external stresses. Differences of pairs of piezooptical constants were obtained for certain crystallographic directions. The piezo-optical constants given in a form similar to (38) are

$$B_{ij} - \delta_{ij}n_i^{-2} = \sum_1^3 \pi_{ijkl}\sigma_{kl} \quad (45)$$

where σ_{kl} are the stresses. Differences between elastooptic constants are then obtained by using the measured values of the elastic moduli through the relation, in reduced matrix notation, $p_{mn} = \pi_{mr}c_{rn}$. The values given in Table XIII indicate that the effective elastooptic coupling for LiNbO$_3$, as given by (41), is over ten times that obtainable using fused silica.

Although this technique is informative and will continue to be used for very small initial research crystals, a more satisfactory experimental arrangement is that developed by Dixon and Cohen.[78] The method consists of scattering a laser beam at the Bragg angle with pulsed ultrasound, and comparing the intensities first as the beam traverses a fused silica rod and then as it traverses a LiNbO$_3$ crystal bonded to the fused silica. In this manner the individual elastooptic

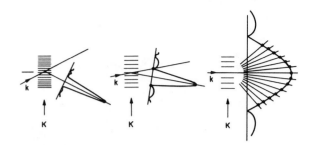

Fig. 33. Schematic representation of transition from elastooptic single-order Bragg diffraction at high frequencies to Raman-Nath multiple-order diffraction at low frequencies. As the frequency decreases from, e.g., $\sim$100 MHz for LiNbO$_3$ for profile on left to $\sim$60 MHz for profile in center to $\sim$30 MHz for profile on right, the diffraction angle decreases, (41), and the width of the intensity profile increases.

TABLE XIII
DIFFERENTIAL PIEZOOPTICAL CONSTANTS $\pi_{ij} - \pi_{jk}$ IN LiNbO$_3$*

$\pi_{12} - \pi_{11} = 1.02 \times 10^{-13}$ cm^2/dyn	σ_1, x_3
$\pi_{33} - 1.12\pi_{13} = 8.08 \times 10^{-14}$	σ_3, x_2
$\pi_{33} - 1.12\pi_{13} = 4.84 \times 10^{-14}$	σ_3, x_1
$\pi_{31} - 1.12\pi_{11} = 6.87 \times 10^{-14}$	σ_2, x_1
$\pi_{11} - \pi_{12} = 6.92 \times 10^{-14}$	σ_2, x_3

$c_{55} = 6.72 \times 10^{11}$ dyn/cm^2; $c_{44} = 6.72 \times 10^{11}$ dyn/cm^2; $c_{33} = 2.42 \times 10^{12}$ dyn/cm^2; $c_{22} = 2.04 \times 10^{12}$ dyn/cm^2.

* The crystallographic directions of applied stress and of optical propagation are designated $\sigma_{1,2,3}$ and $x_{1,2,3}$, respectively. The elastic moduli c_{ij} are from Warner *et al.*[59]

tensor components are obtained. The elastooptic constants of other crystals at 10.6 μ were measured by Carelton and Soref,[79] who also used fused silica as a reference.

The geometrical configuration is shown in Fig. 34 and typical pulse sequences detected by a photomultiplier are shown and numbered in Fig. 35. The pulses in Fig. 35(a) were obtained when the laser beam traversed the fused silica and those in Fig. 35(b) when the laser beam traversed the LiNbO$_3$. The first pulse is due to the outgoing acoustic wave; the second is due to that portion which is reflected from the LiNbO$_3$-fused silica interface; and the third pulse is due to that portion which has traveled through the LiNbO$_3$, has been reflected from the free end, and has been retransmitted through the bond into the fused silica. The pulses labeled 4 and 5 are produced by the outgoing acoustic pulse and the pulse which is reflected from the free end of the LiNbO$_3$.

The square root of the ratio of the product of the scattered light intensities of the first two pulses in the lithium niobate to the product of the intensities of the first and third pulses in the silica can be shown to yield the elastooptic constants from

$$[I_4^{LN}I_5^{LN}/I_1^{FS}I_2^{FS}]^{1/2} = (p^2 n^6 / \rho v^3)^{LN}/(p^2 n^6 / \rho v^3)^{FS}. \quad (46)$$

The value of this ratio does not depend on the acoustic loss in the material nor on the quality of the bond, provided only that the bond transmission is reciprocal. The component of the elastooptic tensor appropriate for the particular symmetry and polarization used in the experiment is denoted by p; n is the optical index, ρ the density, and v the sound velocity. The term $(\rho v^3)^{-1}$ corrects for the fact that a

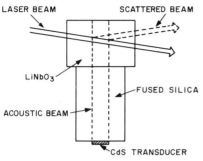

Fig. 34. Geometrical configuration of sample crystal, fused silica, and piezoelectric transducer. From Dixon and Cohen.[78]

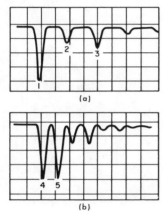

Fig. 35. Photomultiplier output when the optical beam traverses the fused silica (a) and when the beam traverses the LiNbO₃ (b). Oscilloscope sweep speed is 1 μs/cm.

TABLE XIV

ELASTOOPTIC COUPLING CONSTANTS FOR LITHIUM NIOBATE*

$p_{11} = 0.032$	$p_{13} = 0.069$	$p_{31} = 0.153$
$p_{12} = 0.063$	$p_{33} = 0.061$	$p_{41} = 0.136$
	p_{44} and p_{14} not obtained	

* From Dixon and Cohen.[78]

given acoustic power produces a larger strain in some materials than in others.

Measured values of p_{ij} obtained by Dixon and Cohen are listed in Table XIV and are consistent with our data given in the previous table. The accuracy of the former data is believed to be within ± 3 percent except for p_{41} which is estimated to be within ± 10 percent.

Using Gordon's figure of merit (42), LiNbO₃ is about 11 times better for elastooptic devices than the commonly used fused silica although the individual elastooptic terms p_{ij} are all actually smaller than that of fused silica. The advantage then comes from the larger indices of refraction. For fused silica $n = 1.46$ and for LiNbO₃ $n_e = 2.2$ and $n_0 = 2.286$ at the helium-neon wavelength of 633 mμ.

Resolvable Multiple Diffracted Beams—LiNbO₃

The intensity of the diffracted beam is a maximum when the angle between the elastic wave front and the optical beams is the Bragg angle $\Theta = \sin^{-1}(\lambda/2\Lambda)$, as in Fig. 29. The velocities in LiNbO₃ for a longitudinal wave propagating in the x_3 direction and a shear wave propagating in the x_3 direction with particle motion in the x_1 direction are

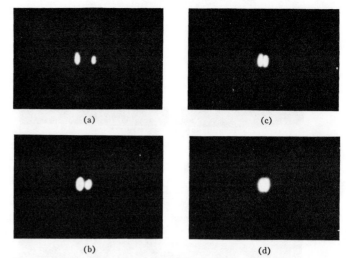

Fig. 36. Beam diffracted by two elastic waves near 500 MHz, separated by (a) 30 MHz, (b) 16 MHz, (c) 19 MHz, and (d) 5 MHz.

$v_l = 7.33 \times 10^5$ cm/s and $v_s = 3.59 \times 10^5$ cm/s, respectively. Thus at 1 GHz the two wavelengths are $\Lambda_l = 7.43\ \mu$ and $\Lambda_s = 3.71\ \mu$. The Bragg angles at this frequency for the beam from a 633-mμ helium-neon laser are $\Theta_l = 2.44$ degrees and $\Theta_s = 4.88$ degrees. The angles between the transmitted diffracted and undiffracted beams are, however, just twice this value or 4.88 and 9.76 degrees, respectively.

The diffracted beam is observable with a quite simple experimental arrangement. The output of one of several oscillators, capable of supplying 100 mW of power from about 1 MHz to 2 GHz, is fed directly to the sample probe without additional tuning, using the same coaxial structure as shown in Fig. 19. The crystal used is a cube about 6 mm on an edge, with elastic wave propagating along the trigonal, or x_3, axis. For 500 MHz the Bragg diffracted beam is displaced laterally by 8.7 cm for longitudinal waves and 17.4 cm for transverse waves when viewed at a distance of one meter from the crystal. These displacements of course increase linearly with frequency. With the coupling scheme used additional probes may be added conveniently on any of the appropriate faces. In order to investigate the resolution between diffracted beams, two CW sources each near 500 MHz were coupled into opposite ends of the crystal. Separate beams were projected onto a white surface and these beams moved closer together as the two frequencies approached each other. Photographs of these beams at different frequency intervals are given in Fig. 36.

The angular width of the beam is defined by the diffraction angle, which is the angle between the maximum and the first zero in intensity, and is given by $\delta\theta = \lambda w$, where λ is the wavelength and w the beam diameter. The beam was apertured down inside the laser to obtain a single transverse Gaussian mode of 2.5-mm spot diameter. The diffraction-limited angle is thus about 1 minute of arc (3×10^{-4} rad). It has been shown[74] that for low elastic wave losses the Bragg diffracted beams maintain the coherency of the laser beam. Then, using the Rayleigh criterion, two Bragg diffracted beams are just resolved when the angular separation is $\delta\theta$. They would be completely resolved for $2\delta\theta$. The deflection angle is $\theta = 2\Theta = 2 \sin^{-1}(\lambda/2\Lambda)$ so that the frequency

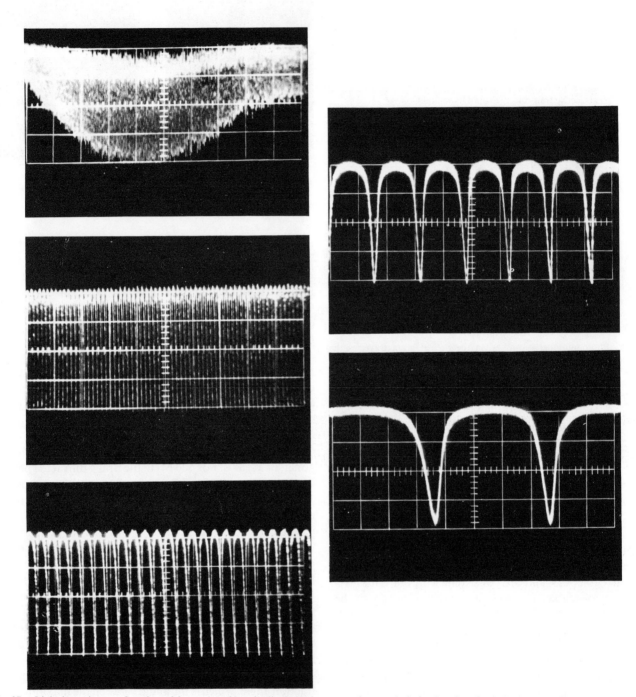

Fig. 37.   Light intensity as a function of frequency of longitudinal elastic waves; the sample is fixed at $\theta_0 = \Theta_B$ and the center frequency is 1.5 GHz. From upper left to lower right the frequency excursions are 105, 40, 14, 4, and ~ 1 MHz. The separation between maxima in all cases is 0.555 MHz.

separation corresponding to an angular separation of $\delta\theta$ is obtained by differentiation. It is $\delta F_l = (V_l \cos \Theta \lambda) \delta\theta$ for longitudinal waves in $LiNbO_3$, and for shear waves is $\delta F_s \approx \frac{1}{2} \delta F_l$. For a $\delta\theta$ of 1 minute the frequency difference is 2.5 MHz. In Fig. 36 two spots are resolved completely ($2\delta\theta$) for $\delta F_l = 5$ MHz and the above considerations apply. The photograph does not show the resolution as well as the unaided eye. The total number of resolvable spots, using the Rayleigh criterion, is given by $N = \delta F \tau_a$, where τ_a is the transit time of the elastic wave through the optical beam.[77]

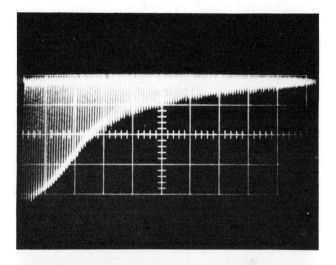

Scanning the Beam

In another experiment, a 1–2-GHz swept oscillator was used to scan the beam across the face of a photomultiplier tube, the output being presented on an oscilloscope. The crystal acts as an elastic wave Fabry-Perot interferometer with an infinite number, M, of resonant frequencies given by $F_l = MV_l/2L$ and $F_s = MV_s/2L$, where L is the length of the sample. As the frequency source is swept, the resonances as observed on the oscilloscope are as shown in Fig. 37. The general outline and the beam width are determined by the Bragg conditions, (33) and (41), and the width of the individual lines by the elastic Q. If the sample and optical beam are fixed at $\theta_0 = \Theta$ for 1.5 GHz, the frequency sweep is 120 MHz between the first zeros in intensity.

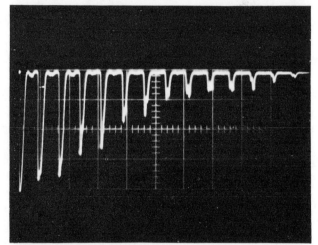

The calculated number of resolvable spots in the beam is seen to be about 24 at 1.0 GHz or 48 at 0.5 GHz, counting only those whose intensity is greater than one-half the maximum intensity. Since lithium niobate can be grown as large-size crystals, larger diameter optical beams can be used and the number of resolvable spots increased several times.

When the sweep frequency excursion is reduced so that only two resonances appear, the scope trace is calibrated in frequency by the known length of the crystal itself. The elastic Q then is measured directly from the frequency difference at half amplitude of the oscilloscope signal. A lower limit is approximately 10^5, which is consistent with the measurements using pulse echo techniques.[3]

Optical Probing of Elastic Wave Pulse Echoes

In microwave elastic propagation studies traveling waves rather than standing waves are used. A pulse echo train is set up in the crystal and is usually observed by using a microwave receiver. In $LiNbO_3$, however, a pulse echo train can be observed using the Bragg diffraction into the photomultiplier detector. Since $V_l \approx 2V_s$, the Bragg angle for shear waves is approximately twice that for longitudinal waves and the two sets of pulses are observed at separate points in space. A set of longitudinal pulse echoes is shown in Fig. 38, the upper trace being 20 μs/cm.

Although both shear and longitudinal pulse echo trains can be present in the crystal at the same time, they are observed separately. This observation can be used as a tool to determine the effectiveness of particular probes in exciting purely longitudinal or shear pulse modes.

In the middle trace of Fig. 38 the time scale is 2 μs/cm and in the lower trace it is 0.5 μs/cm. Each time an elastic

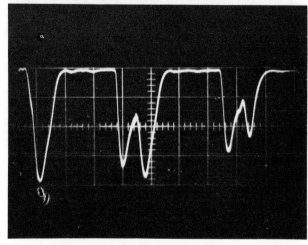

Fig. 38.   Longitudinal elastic wave pulse echo train at 0.5 GHz using a photomultiplier to detect the Bragg diffracted beam. The upper trace is 20 μs/cm, the center trace is 2 μs/cm, and the lower trace is 0.5 μs/cm. The double pulses of the lower trace show the forward and backward traveling waves separated in time.

wave pulse interrupts the optical beam a signal is deflected to the photomultiplier. The lowest curve shows the forward and backward traveling waves separated in time. The fact that the first indication is a single pulse and those following are double is caused by the optical beam being near the input rather than output probe.

Nonlinear Elastic Wave Interaction

Based on earlier work in liquids, recent investigations at Westinghouse Laboratories[80] have extended the use of elastooptic effects to measure the nonlinear elastic properties of a crystal of NaCl. An elastic wave of finite strain in a nonlinear solid will distort as it propagates so that second- and higher-order harmonics will be generated. The technique involved measuring the relative amplitudes of the positive and negative orders of the diffraction pattern at low frequencies. A crystal 15 cm long was used and it was necessary to measure the intensities as functions of distance and of strain amplitude.

In $LiNbO_3$ a pulse echo train of, say, 100 μs is equivalent to a path length of ~ 70 cm and thus the 6-mm cube can be used. As the elastic wave propagates along a crystal, the strain amplitude of the second harmonic is

$$\varepsilon^{2\omega} = \frac{Kx}{2}\left(1 + \frac{2C_{111}}{C_{11}}\right)(\varepsilon^{\omega})^2, \qquad (47)$$

where x is the distance traveled in the crystal, K is the elastic wave number, C_{11} and C_{111} are the second- and third-order elastic moduli, and ε^{ω} is the strain amplitude of the fundamental wave. The experiment then consists of observing the second harmonic optical beam diffraction at an angle which is equal to twice that of the fundamental.

The second harmonic pulse echo train is shown in Fig. 39 for a 500-MHz elastic wave. The second harmonic starts off at zero and increases with the distance the fundamental travels. Attenuation due to elastic losses deteriorates the strains at both frequencies, resulting in the bell-shaped second harmonic pulse echo train shown. Writing (47) as $\varepsilon^{2\omega} = ax(\varepsilon^{\omega})^2$ and $\varepsilon^{\omega} = b\exp(-\Gamma x)$, where Γ is the attenuation, the following holds:

$$\varepsilon^{2\omega} = Ax\exp(-2\Gamma x). \qquad (48)$$

The magnitude of $\varepsilon^{2\omega}$ is found to reach a maximum for $x = 1/2\Gamma$. Using the values of $\Delta t = 25$ μs and $v_l = 7.33 \times 10^5$ cm/s, the attenuation of 500 MHz is computed to be $\Gamma = 0.02$ cm^{-1} or 0.26 dB/cm. This value agrees with that obtained from the pulse echo decay at the fundamental frequency. Further, the third-order elastic moduli are also determined by measuring the relative amplitudes of the fundamental and second harmonic diffracted beam and applying (47).

Elastic Wave Propagation Loss Measurement Techniques

It is interesting to remark here that the elastic wave propagation losses at microwave frequencies, or the acoustic Q's, have been measured in four different ways: 1) pulse echo decay using a microwave receiver, 2) pulse echo decay using optical beam diffraction as a probe, 3) frequency scanning of optical beam diffraction, and 4)

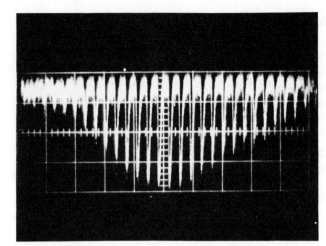

Fig. 39.   Nonlinear generation of second harmonic elastic wave pulse echo train in $LiNbO_3$ as seen by optical probing.

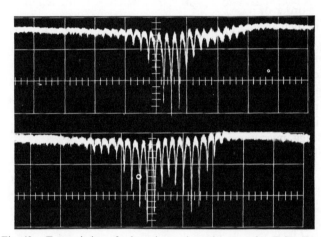

Fig. 40.   Transmission of a laser beam through a scanning Fabry-Perot interferometer, upper trace. The laser is operated in a single transverse mode but in several longitudinal modes. The Bragg diffracted beam is shown in the lower trace. Each of the longitudinal modes is upshifted and downshifted by 0.5 GHz.

second harmonic generation, again with an optical probe. The optical methods all have a quite practical importance in that optical beam diffraction occurs at different positions in space for waves of different velocities. Thus, mode conversion problems, as well as transducer efficiencies, can be readily evaluated. (Since this paper was written a paper by McMahon[81] has appeared describing other Brillouin scattering techniques for measuring microwave acoustic attenuation.)

Optical Frequency Shifts

Traveling elastic waves in opposite directions are known to diffract the beam into the same angle; however, for one direction of propagation the diffracted light is upshifted in frequency by an amount equal to the acoustic frequency. For the other direction the frequency is downshifted by the same amount. The upshifted and downshifted frequencies can be detected as a beat note of twice the acoustic frequency by using a photodiode and microwave receiver. These frequencies also can be observed as additional lines in the interferometer pattern of a Fabry-Perot etalon. Another method,[13] the one used here, is to use a scanning Fabry-Perot interferometer with an oscilloscope pattern obtained by using a photomultiplier detector. Fig. 40 shows such a pattern set up by a 0.5-GHz elastic wave using a helium-neon laser ($\lambda = 633$ mμ). The laser operates in a single transverse mode but with several equally spaced longitudinal modes (upper trace). Each of the longitudinal modes in the original laser beam is shifted upward and downward by 0.5 GHz, setting up two groups of modes (lower trace) with the separations between corresponding pairs of modes being 1 GHz apart.

Internal Laser Modulation

The optical cavity of the helium-neon laser consists of a Fabry-Perot interferometer made up of two mirrors. Experimentally and theoretically it has been determined that the maximum energy which can be coupled out of the laser is obtained when the transmittance of one mirror is of the order of one percent. The optical fields inside the laser are then about one hundred times that of the fields external to the laser. Gürs and Muller[82] first recognized that placing an electrooptic crystal inside the laser cavity and deflecting the beam so as to miss the mirror could result in an increase of the usable modulation by as much as one hundred times if no optical loss were introduced. More recently Siegmann *et al.*[83] have investigated, in a similar way, Bragg microwave elastooptic beam diffraction by fused silica placed internal to the laser.

Using lithium niobate, several experiments[84] have been carried out both in the visible region, at 633 mμ, and in the infrared region, at 1.15 μ, with the crystals inside a He-Ne gas laser. The experimental arrangement is shown in Fig. 41. The crystal holder is the same as used in the other experiments, the only difference being that a crystal approximately 1 cm square in cross section and 3 cm long used. See Fig. 42. Two opposing faces were antireflection-coated for 633 mμ and two others for 1.15 μ. The long dimension is along the optic or x_3 axis and the rectangular sides are perpendicular to the x_1 and x_2 axes. It should be noted that the sample was an early research crystal and is not representative of the optical quality more recently attained. There were serious losses due to optical imperfections. Polarizing the light along the principal optical direction, x_3, would avoid birefringence in a perfect crystal; however, any inclusions or small variations in density will give rise to microscopic birefringence and some scattering and absorption. These, in turn, also affect the normal modes of the laser and deteriorate the beam seriously. In the sample used, strains were observed in the form of striations in the beam so that it was necessary to move the crystal around in the beam until the best operating spot was found. Using an aperture to reduce the beam diameter gave an estimated minimum loss per pass of 6 percent. In any case, with the crystal inside the laser cavity it was quite easy to obtain an increase in diffracted power of about 50 times.

Siegmann *et al.*, using a rate-equation analysis, showed that the primary laser output intensity I_1 and the diffracted intensity, I_2 depend on the acoustic power input P in the form

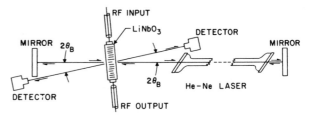

Fig. 41. Optical Bragg diffraction with the LiNbO$_3$ crystal inside the laser cavity.

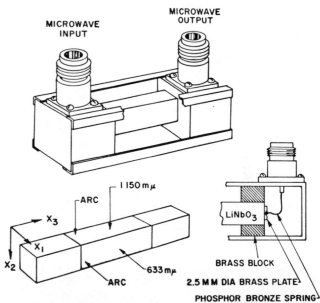

Fig. 42. Crystal holder for a LiNbO$_3$ crystal ~3 cm long. One pair of faces is antireflection-coated for 1.15 μ and another pair for 633 mμ.

$$I_1(P) = I_1(0)[(P_m - P)/(P_m + kP)] \qquad (49)$$

and

$$I_2(P) = fI_1(0)[P(P_m - P)/P_m(P_m + kP)] \qquad (50)$$

where f and k are constants depending on relative amounts of coupling to the primary and diffracted beams. The laser becomes overcoupled and ceases oscillating at a maximum acoustic drive power P_m. Similar experiments were repeated with LiNbO$_3$ at 500 MHz, with approximately one watt of RF power required to obtain the maximum power deflected. At this driving level the crystal deflects one percent of an incident beam, as measured with the crystal outside the laser cavity. No attempt was made to minimize the power required. For example, the RF elastic waves were excited along the optic axis of the crystal for which the static piezoelectric coupling is ~10 percent. In other directions the coupling can be five times greater.

The previously discussed frequency scanning experiment was also carried out, at both 633 mμ and 1.15 μ, with the sample inside the laser cavity and the diffracted beam passing to the side of the mirror. The microwave source, as before, was a 1–2-GHz frequency-swept backward-wave oscillator. The results are entirely similar and are represented as well by Fig. 37.

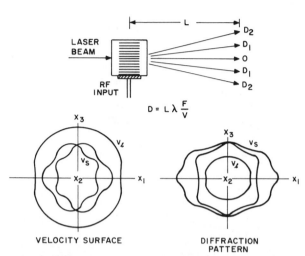

$$D = L\lambda\frac{F}{V}$$

Fig. 43. Schematic diagram for the Schaefer-Bergmann experiment, the upper diagram showing only one of the elastic strain diffraction gratings. The lower diagrams show the relations between the velocity surfaces and the Schaefer-Bergmann diffraction pattern.

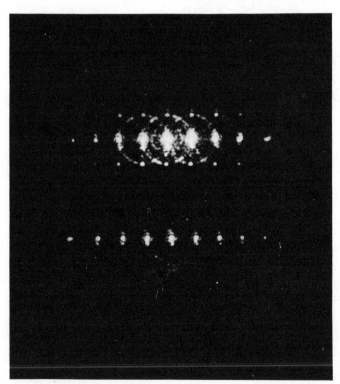

Fig. 44. Typical Schaefer-Bergmann diffraction patterns for a LiNbO₃ crystal resonant, in one direction only, for longitudinal elastic waves. The RF power is increased for the upper pattern.

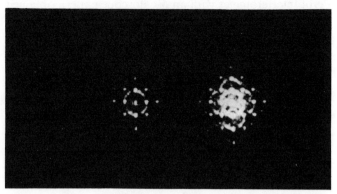

Fig. 45. Diffraction pattern similar to Fig. 44 except that the crystal is resonant simultaneously for longitudinal velocities along the x_1 and x_2 crystallographic directions. The RF power is increased for the pattern on the right.

Fig. 46. Schaefer-Bergmann pattern in which the innermost circle of diffracted laser beams represents longitudinal velocities, the outermost circle represents the slower shear velocities, and the four beams just inside this represent the faster shear velocities. The crystal is simultaneously almost resonant for the longitudinal and the slower shear velocities in both the x_1 and x_2 directions, the incident laser beam being along the x_3 direction.

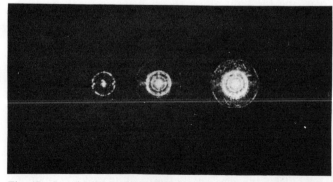

Fig. 47. Longitudinal velocity diffraction pattern for the fundamental frequency on the left. Increasing the RF power brings in the second harmonic in the center and the third harmonic on the right. The crystal was totally destroyed in attempting to photograph the fourth harmonic.

This technique makes available a single-sideband suppressed-carrier microwave modulated optical beam with the full intensity of the laser. It should be pointed out that once the system has been constructed the operation is no more critical or difficult than with the crystal outside the laser cavity. The advantages expected of $LiNbO_3$ as compared to other materials which have been used are that operation can be obtained at higher modulation frequencies with lower microwave power requirements.

XI. SCHAEFER-BERGMANN METHOD

In the 1930's Schaefer and Bergmann[85],[86] developed a method for obtaining elastic constants for a crystal from the diffraction of light by ultrasonic waves. Their method has been superseded by more accurate techniques, such as, for example, McSkimin's,[87] in which elastic moduli are determined to within one part in 10^4 or 10^5. The Schaefer-Bergmann method remains a most instructive experiment in that it represents a spectacular visualization of the elastic velocity surfaces and other related phenomena.

A crystal with two opposite faces polished optically flat and parallel is excited in one of its resonant elastic modes by using a piezoelectric transducer. For $LiNbO_3$ the crystal is used as its own transducer and again the holder shown in Fig. 19 is used without modification. An ultrasonic frequency is chosen low enough to fall within the Raman-Nath regime so that many diffracted orders are observed, yet the frequency must be high enough so that the separation between diffracted beams is well resolved. For $LiNbO_3$, frequencies from 10 to 50 MHz are appropriate. Typical diffraction patterns[88] using a helium-neon laser are shown in Figs. 44–47. As is the case for Bragg diffraction, each diffracted beam duplicates exactly all of the mode characteristics of the undiffracted beam.

If all six faces of the crystal are polished optically flat and opposing faces are made parallel, it is possible to produce index diffraction gratings in three orthogonal directions. A total of nine gratings can be established by resonating the crystal for the longitudinal or the shear waves along principal crystallographic directions. With the proper choice of crystal lengths and frequencies two or more frequencies may be made resonant simultaneously. Further, it has been found possible to excite all of the various resonances by using a probe to a single face. The resultant array of diffracted beams on a projection screen forms a pattern which is directly related to the elastic constants in a manner represented schematically by Fig. 43.

Using the Bragg relation, $\sin \Theta = K/2k$, the displacement D of the beams will vary inversely as the acoustic velocities:

$$D = L\lambda f/v \qquad (51)$$

where L is the distance to the projection screen, f is the frequency, and v is the velocity of the elastic wave for propagation in a particular direction in the crystal. For simplicity, a correction factor involving refraction at the crystal surface has not been included. As illustrated in Fig. 43 the diffracted beams will form a two-dimensional pattern defined by the inverse of the velocities. Surfaces, analogous to the velocity surfaces, but plotted in terms of the wave number vector K rather than velocity v, are called form frequency surfaces. Schaefer-Bergmann figures represent the intersection of these surfaces with planes normal to the propagation direction of the optical beam.

For $LiNbO_3$ and $LiTaO_3$, the form frequency surface is given by the determinant:

$$\begin{vmatrix} \alpha_{11} - \rho\omega^2/k^2 & \alpha_{12} & \alpha_{13} \\ \alpha_{12} & \alpha_{22} - \rho\omega^2/k^2 & \alpha_{23} \\ \alpha_{13} & \alpha_{23} & \alpha_{33} - \rho\omega^2/k^2 \end{vmatrix} = 0 \quad (52)$$

where

$$\alpha_{11} = c_{11}k_1^2 + \tfrac{1}{2}(c_{11} - c_{12})k_2^2 + c_{44}k_3^2 + c_{14}k_2k_3$$
$$\alpha_{22} = c_{66}k_1^2 + c_{11}k_2^2 + c_{44}k_3^2 - 2c_{14}k_2k_3$$
$$\alpha_{33} = c_{44}k_1^2 + c_{44}k_2^2 + c_{33}k_3^2$$
$$\alpha_{12} = 2c_{14}k_1k_3 + (c_{12} + c_{66})k_1k_2$$
$$\alpha_{13} = (c_{13} + c_{44})k_1k_3 + 2c_{14}k_1k_2$$
$$\alpha_{23} = c_{14}k_1^2 - c_{14}k_2^2 + (c_{13} + c_{44})k_2k_3.$$

The c_{ij} in this case are the elastic moduli stiffened by the piezoelectric coupling.

For example, the solution for an acoustic wave in the yz plane is resolved into a second-order and a fourth-order surface:

$$\tfrac{1}{2}(c_{11} - c_{12})k_2^2 + c_{44}k_3^2 + c_{14}k_2k_3 - \rho\omega^2 = 0 \quad (53)$$
$$(c_{11}k_2^2 + c_{44}k_3^2 - c_{14}k_2k_3 - \rho\omega^2)(c_{44}k_2^2 + c_{33}k_3^2 - \rho\omega^2)$$
$$+ [c_{14}k_2^2 - (c_{33} + c_{44})k_2k_3] = 0. \quad (54)$$

In principle, at least, the relative intensities of all diffracted beams can be calculated in a similar manner from the index ellipsoid distorted by the elastic strains. This method will be outlined to show some of the information available.

$LiNbO_3$: Elastically Deformed Ellipsoid

The undeformed index ellipsoid is defined by

$$\sum_{i=1}^{3} B_i x_i^2 = 1$$

or

$$\frac{x_1^2}{n_1^2} + \frac{x_2^2}{n_2^2} + \frac{x_3^2}{n_3^2} = 1. \quad (55)$$

Letting ΔB_i stand for the right-hand side of (38) and using the elastooptic matrix for the 3m point group, the following nonsymmetric matrix relation applies for $LiNbO_3$ and $LiTaO_3$ ($n_1 = n_2 = n_0$; $n_3 = n_e$):

$$\begin{vmatrix} \Delta B_1 \\ \Delta B_2 \\ \Delta B_3 \\ \Delta B_4 \\ \Delta B_5 \\ \Delta B_6 \end{vmatrix} = \begin{vmatrix} p_{11} & p_{12} & p_{13} & p_{14} & 0 & 0 \\ p_{12} & p_{11} & p_{13} & -p_{14} & 0 & 0 \\ p_{31} & p_{31} & p_{33} & 0 & 0 & 0 \\ p_{41} & -p_{41} & 0 & p_{44} & 0 & 0 \\ 0 & 0 & 0 & 0 & p_{44} & p_{41} \\ 0 & 0 & 0 & 0 & p_{14} & p_{66} \end{vmatrix} \begin{vmatrix} \varepsilon_1 \\ \varepsilon_2 \\ \varepsilon_3 \\ \varepsilon_4 \\ \varepsilon_5 \\ \varepsilon_6 \end{vmatrix} \quad (56)$$

where $p_{66} = \tfrac{1}{2}(p_{11} - p_{12})$. Considering separately each strain component ε_1, the deformed ellipsoid is

$$(\varepsilon_1); \left(\frac{1}{n_0^2} + p_{11}\varepsilon_1\right)x_1^2 + \left(\frac{1}{n_0^2} + p_{12}\varepsilon_1\right)x_2^2 + \left(\frac{1}{n_e^2} + p_{31}\varepsilon_1\right)x_3^2 + 2p_{41}\varepsilon_1 x_2 x_3 = 1$$

$$(\varepsilon_2); \left(\frac{1}{n_0^2} + p_{12}\varepsilon_2\right)x_1^2 + \left(\frac{1}{n_0^2} + p_{11}\varepsilon_2\right)x_2^2 + \left(\frac{1}{n_e^2} + p_{31}\varepsilon_2\right)x_3^2 - 2p_{41}\varepsilon_2 x_2 x_3 = 1$$

$$(\varepsilon_3); \left(\frac{1}{n_0^2} + p_{13}\varepsilon_3\right)x_1^2 + \left(\frac{1}{n_0^2} + p_{13}\varepsilon_3\right)x_2^2 + \left(\frac{1}{n_e^2} + p_{33}\varepsilon_3\right)x_3^2 = 1$$

$$(\varepsilon_4); \left(\frac{1}{n_0^2} + p_{14}\varepsilon_4\right)x_1^2 + \left(\frac{1}{n_0^2} - p_{14}\varepsilon_4\right)x_2^2 + \frac{1}{n_e^2} + 2p_{44}\varepsilon_4 x_2 x_3 = 1$$

$$(\varepsilon_5); \frac{x_1^2}{n_0^2} + \frac{x_2^2}{n_0^2} + \frac{x_3^2}{n_e^2} + 2p_{44}\varepsilon_5 x_1 x_3 + 2p_{14}\varepsilon_5 x_1 x_2 = 1$$

$$(\varepsilon_6); \frac{x_1^2}{n_0^2} + \frac{x_2^2}{n_0^2} + \frac{x_3^2}{n_e^2} + 2p_{41}\varepsilon_6 x_1 x_3 + 2p_{66}\varepsilon_6 x_1 x_2 = 1. \tag{57}$$

The first three of these equations are for longitudinal elastic waves and the last three are for shear waves. In particular, the last two cases are for shear waves ε_5, propagating in direction x_3 with particle displacement along x_1 (or vice versa) and for ε_6, propagating in direction x_2 with particle displacement along x_1 (or vice versa). For an optical beam propagating in the x_1 direction, the elastically deformed ellipsoid is obtained by considering the intersection of an ellipsoid with a plane at $x_1 = 0$. Thus, for the last two cases, the elastically deformed ellipse is the same as the undeformed ellipse and there is no diffraction. However, for light propagating in the x_2 direction the diffracted intensity depends only on p_{44} for shear strain, ε_5, and depends on p_{41} for shear strain, ε_6. For light propagating in the x_3 direction the related constants are p_{14} and p_{66}.

It should be pointed out that the shear wave diffraction considered above does not obey the Bragg relation, (32), in optically anisotropic crystals such as $LiNbO_3$ and $LiTaO_3$. The diffraction from a longitudinal acoustic wave which led to (41) does not change the direction of polarization of the diffracted light beam from that of the incident beam. Shear wave diffraction, however, produces a 90-degree rotation of the diffracted $\vec{E}$ vector.

Dixon[89] has shown that this rotation requires an altered Bragg relation if the optical index of the incident beam differs from that of the diffracted beam. The equations which determine the incident and diffracted beam angles, θ_i and θ_d, for optimum Bragg diffraction are

$$\sin \theta_i = \frac{\lambda_0}{2n_i\Lambda}\left[1 + \frac{\Lambda^2}{\lambda_0^2}(n_i^2 - n_d^2)\right] \tag{58}$$

$$\sin \theta_d = \frac{\lambda_0}{2n_d\Lambda}\left[1 - \frac{\Lambda^2}{\lambda_0^2}(n_i^2 - n_d^2)\right] \tag{59}$$

where λ_0 is the wavelength in vacuum of the light beam and n_i is the optical index of the incident beam n_d of the diffracted beam. These equations reduce to (32) when the diffracted beam is not rotated or when the rotation is such that the optical index remains the same. They can be derived from energy and momentum conservation conditions used to obtain (32).

Dixon also points out that the minimum acoustic frequency for anisotropic diffraction is no longer given by $K^2 L > 4\pi k$, (44), but is

$$f_{\min} = \frac{v|n_i - n_d|}{\lambda_0}. \tag{60}$$

This relation does not allow diffraction in the Raman-Nath region to be observed, since, as $f_0 \to f_{\min}$, both θ_d and θ_i approach 90 degrees.

Using the principle of anisotropic diffraction, Lean et al.[90] have shown that the number of resolvable spots obtained by frequency-scanning a laser beam with an elasto-optic device would be greatly increased over the number obtained previously with isotropic diffraction. They obtained 1000 resolvable spots for $\Delta f = 550$ MHz, using a 1600-MHz shear wave in sapphire. They predict 5000 resolvable spots under similar conditions using $LiNbO_3$, which has a larger birefringence than sapphire at room temperature.

OBSERVATIONS

Ideally the crystals should be optically perfect for the observation of all diffraction phenomena. However, all of the crystals we have investigated have been grown under research conditions and are not yet completely free of strains, imperfections, and regions of varying optical index. Extraneous scattering shows up weakly at low acoustic power and becomes most noticeable at high power levels.

With approximately one watt of 30-MHz RF power incident on the sample probe, about one-tenth of a watt of acoustic power is generated in the crystal. The frequency is adjusted for resonance of a longitudinal wave, and using a helium-neon laser, diffraction patterns such as are shown in Fig. 44 are projected on a screen. The elastic waves propagate along an x_1 direction with the laser beam along an x_3 direction being polarized along x_2. The central array of beams is typical of the Raman-Nath conditions. The diffracted beams which form circles about the central array are due to components of the elastic wave along other crystallographic directions. Each diffracted beam in all cases reproduces faithfully the mode pattern of the incident laser beam.

Experimentally, it is possible to generate this array horizontally on a projection screen by adjusting the RF frequency to the crystal resonance in the x_1 direction (horizontal). Then by shifting the frequency so that resonance is obtained for the x_2 direction (vertical), a similar array appears vertically.

If at some frequency f_0 the crystal is resonant simultaneously in both the x_1 and x_2 directions, i.e., $f_0 = Mv_l/2L$ for the x_1 direction and $f_0 = Nv_l/2L$ for the x_2 direction, then both arrays can appear together in a two-dimensional pattern as shown in Fig. 45. The RF power level is increased to show additional diffracted beams as in the diagram on the right.

Diffraction patterns represented schematically in Fig. 43 are illustrated in Fig. 46. Diffraction of longitudinal waves is observed as the smallest circle; the fast shear wave produces the four prominent spots and the slow shear wave produces the beams lying on the outer circle. The acoustic beam is along an x_1 axis and the laser beam is along the x_3 axis with polarization along x_1.

A visualization of nonlinear elastic phenomena through the generation of second and third harmonics is demonstrated in Fig. 47. The left pattern reproduces many of the characteristics of Fig. 46. The RF power is increased and the second harmonic of the slow shear wave appears in the middle pattern and the third harmonic appears in the right pattern. The sixfold elastic symmetry also is more clearly defined. It may be noted further that as the acoustic power is increased, diffraction from the imperfections shows up in a manner that indicates, at once, the degree of optical perfection.

REFERENCES

[1] A. A. Ballman, "Growth of piezoelectric and ferroelectric materials by the Czochralski technique," *J. Am. Ceramic Soc.*, vol. 48, p. 112, 1965.

[2] A. W. Warner, *Proc. 19th Annual Frequency Control Symp.* (April 1965, U. S. Army Electronics Labs., Fort Monmouth, N. J.), 1965.

[3] E. G. Spencer, P. V. Lenzo, and K. Nassau, "Elastic wave propagation in lithium niobate," *Appl. Phys. Lett.*, vol. 7, p. 67, 1965.

[4] C. P. Wen and R. F. Mayo, "Acoustic attenuation of a single-domain lithium niobate crystal at microwave frequencies," *Appl. Phys. Lett.*, vol. 9, p. 135, 1966.

[5] M. I. Grace, R. W. Kedzie, M. Kestigian, and A. B. Smith, "Elastic wave attenuation in lithium niobate," *Appl. Phys. Lett.*, vol. 9, p. 155, 1966.

[6] E. G. Spencer and P. V. Lenzo, "Temperature dependence of microwave elastic losses in LiNbO$_3$ and LiTaO$_3$," *J. Appl. Phys.*, vol. 38, p. 423, 1967.

[7] P. V. Lenzo, E. H. Turner, E. G. Spencer, and A. A. Ballman, "Electrooptic coefficients and elastic-wave propagation in single-domain ferroelectric lithium tantalate," *Appl. Phys. Lett.*, vol. 8, p. 81, 1966.

[8] G. E. Peterson, A. A. Ballman, P. V. Lenzo, and P. M. Bridenbaugh, "Electrooptic properties of LiNbO$_3$," *Appl. Phys. Lett.*, vol. 5, p. 62, 1964.

[9] P. V. Lenzo, E. G. Spencer, and K. Nassau, "Electrooptic coefficients in single-domain ferroelectric lithium niobate," *J. Opt. Soc. Am.*, vol. 56, p. 633, 1966.

[10] E. H. Turner, "High-frequency electrooptic coefficients of lithium niobate," *Appl. Phys. Lett.*, vol. 8, p. 303, 1966.

[11] E. Bernal, G. D. Chen, and T. C. Lee, "Low frequency electrooptic and dielectric constants of lithium niobate," *Phys. Lett.*, vol. 21, p. 259, 1966.

[12] P. H. Smakula and P. C. Claspy, "The electrooptic effect in LiNbO$_3$ and KTN," *Trans. Metallurgical Soc. of AIME*, vol. 239, p. 421, 1967.

[13] E. G. Spencer, P. V. Lenzo, and K. Nassau, "Optical interactions with elastic waves in lithium niobate," *IEEE J. Quantum Electronics (Correspondence)*, vol. QE-2, p. 69, March 1966.

[14] R. W. Dixon and M. G. Cohen, "A new technique for measuring magnitudes of photoelastic tensors and its application to lithium niobate," *Appl. Phys. Lett.*, vol. 8, p. 205, 1966.

[15] K. Nassau, H. J. Levinstein, and G. M. Loiacono, "Lithium niobate; growth, domain structure, dislocations and etching," *J. Phys. Chem. Solids*, vol. 27, p. 983, 1966; see also "Preparation of single domain crystals," *J. Phys. Chem. Solids*, vol. 27, pp. 989–996, 1966.

S. C. Abrahams, J. M. Reddy, and J. L. Bernstein, "Single crystal X-ray diffraction study at 24°C," *J. Phys. Chem. Solids*, vol. 27, pp. 997–1012, 1966.

S. C. Abrahams, W. C. Hamilton, and J. M. Reddy, "Single crystal neutron diffraction study at 24°C," *J. Phys. Chem. Solids*, vol. 27, pp. 1013–1018, 1966.

S. C. Abrahams, H. J. Levinstein, and J. M. Reddy, "Polycrystal X-ray diffraction study between 24° and 1200°C," *J. Phys. Chem. Solids*, vol. 27, pp. 1019–1026, 1966.

[16] H. J. Levinstein, A. A. Ballman, and C. D. Capio, "Domain structure and Curie temperatures of single crystal lithium tantalate," *J. Appl. Phys.*, vol. 37, p. 4585, 1966.

[17] A. A. Ballman, H. J. Levinstein, C. D. Capio, and H. Brown. "The Curie temperature and birefringence variation in ferroelectric lithium tantalate as a function of melt stoichiometry," to be published in *J. Am. Ceramic Soc.*

[18] P. V. Lenzo, E. G. Spencer, and A. A. Ballman, "Optical activity and electrooptic effect in bismuth germanium oxide (Bi$_{12}$GeO$_{20}$)," *Appl. Opt.*, vol. 5, p. 1688, 1966.

[19] E. G. Spencer, P. V. Lenzo, and A. A. Ballman, "Ultrasonic properties of bismuth germanium oxide," *Appl. Phys. Lett.*, vol. 8, p. 290, 1966.

[20] A. A. Ballman, "The growth and properties of piezoelectric bismuth germanium oxide, Bi$_{12}$GeO$_{20}$," *Internat'l J. Crystal Growth*, vol. 1, p. 37, 1967.

[21] A. A. Ballman and H. Brown, "The growth and properties of strontium barium metaniobate, Sr$_{1-x}$Ba$_x$Nb$_2$O$_6$, a tungsten bronze ferroelectric," to be published in *Internat'l J. Crystal Growth*.

[22] P. V. Lenzo, E. G. Spencer, and A. A. Ballman, "Electrooptic coefficients of ferroelectric strontium barium niobate," *Appl. Phys. Lett.*, vol. 11, p. 23, 1967.

[23] C. H. Holmes, E. G. Spencer, A. A. Ballman, and P. V. Lenzo, "The electrooptic effect in calcium pyroniobate," *Appl. Opt.*, vol. 4, p. 551, 1965.

[24] W. A. Bonner, E. F. Dearborn, J. E. Geusic, H. M. Marcos, and L. G. Van Uitert, "Dielectric and electrooptic properties of lead magnesium niobate," *Appl. Phys. Lett.*, vol. 10, p. 163, 1967.

[25] F. S. Chen, J. E. Geusic, S. K. Kurtz, J. G. Skinner, and S. H. Wemple, "Light modulation and beam deflection with potassium tantalate-niobate crystals," *J. Appl. Phys.*, vol. 37, p. 388, 1966.

J. E. Geusic, S. K. Kurtz, L. G. Van Uitert, and S. H. Wemple, "Electro-optic properties of some ABO$_3$ perovskites in the paraelectric phase," *Appl. Phys. Lett.*, vol. 4, p. 141, 1964.

[26] I. P. Kaminow and E. H. Turner, "Electrooptic light modulators," *Proc. IEEE*, vol. 54, pp. 1374–1390, October 1966.

[27] R. T. Denton, T. S. Kinsel, and F. S. Chen, "224 Mc/s optical pulse code modulator," *Proc. IEEE (Letters)*, vol. 54, pp. 1472–1473, October 1966.

[28] R. T. Denton, F. S. Chen, and A. A. Ballman, "Lithium tantalate light modulators," *J. Appl. Phys.*, vol. 8, p. 1611, 1967.

[29] S. C. Abrahams, P. B. Jamieson, and J. L. Bernstein, "Crystal structure of piezoelectric bismuth germanium oxide, Bi$_{12}$GeO$_{20}$" (to be published).

[30] M. Onoe, A. W. Warner, and A. A. Ballman, "Elastic and piezoelectric characteristics of bismuth germanium oxide, Bi$_{12}$GeO$_{20}$," *IEEE Trans. Sonics and Ultrasonics*, vol. SU-14, pp. 165–167, October 1967.

[31] P. V. Lenzo, E. G. Spencer, and A. A. Ballman, "Photoactivity in bismuth germanium oxide," *Phys. Rev. Lett.*, vol. 19, p. 641, 1967.

[32] A. Ashkin, G. D. Boyd, J. M. Dziedzic, R. G. Smith, A. A. Ballman, H. J. Levinstein, and K. Nassau, "Optically induced refractive index inhomogeneities in LiNbO$_3$ and LiTaO$_3$," *Appl. Phys. Lett.*, vol. 9, p. 72, 1966.

[33] J. G. Skinner (personal communication).

[34] P. A. Franken and J. F. Ward, "Optical harmonics and nonlinear

phenomena," *Rev. Modern Phys.*, vol. 35, p. 23, 1963.

[35] R. C. Miller, "Optical second harmonic generation in piezoelectric crystals," *Appl. Phys. Lett.*, vol. 5, p. 17, 1964.

[36] N. Bloembergen, *Nonlinear Optics*. New York: W. A. Benjamin, 1965.

[37] C. G. B. Garrett and F. N. H. Robinson, "Miller's phenomenological rule for computing nonlinear susceptibilities," *IEEE J. Quantum Electronics* (*Correspondence*), vol. QE-2, pp. 328–329, August 1966.

[38] R. L. Kelly, "Pockels effect in zinc-blende-structure ionic crystals," *Phys. Rev.*, vol. 151, p. 721, 1966.

[39] S. K. Kurtz and F. N. H. Robinson, "A physical model of the electrooptic effect," *Appl. Phys. Lett.*, vol. 10, p. 62, 1967.

[40] C. G. B. Garrett, "Nonlinear optics, anharmonic oscillators and pyroelectricity," to be published in *Phys. Rev.*

[41] I. P. Kaminow and W. D. Johnston, Jr., "Quantitative determination of sources of the electrooptic effect in lithium niobate and lithium tantalate," *Phys. Rev.*, vol. 160, p. 519, 1967.

[42] G. H. Heilmeier, "The dielectric and electrooptic properties of a molecular cyrstal—Hexamine," *Appl. Opt.*, vol. 3, p. 1281, 1964.

[43] M. Vassel and E. M. Conwell, "Electrooptic effect in NH_4Cl," *Phys. Rev.*, vol. 140, p. A2110, 1965.

[44] F. N. H. Robinson, "Nonlinear optical coefficients," *Bell Sys. Tech. J.*, vol. 46, p. 913, 1967.

[45] I. P. Kaminow, in *Proc. Symp. on Ferroelectricity*. New York: Elsevier, 1967, pp. 183–196.

[46] M. Born and K. Huang, *Dynamical Theory of Crystal Lattices*. London: Oxford University Press, 1954.

[47] J. F. Nye, *Physical Properties of Crystals*. London: Oxford University Press, 1960, p. 248.

[48] I. P. Kaminow and W. M. Sharpless, "Performance of $LiTaO_3$ and $LiNbO_3$ light modulators at 4 GHz," *Appl. Opt.*, vol. 6, p. 351, 1967.

[49] J. D. Axe and D. F. O'Kane, "Infrared dielectric dispersion of $LiNbO_3$," *Appl. Phys. Lett.*, vol. 9, p. 58, 1966.

[50] S. C. Abrahams (personal communication).

[51] J. A. Van Raalte, "Electrooptic effect in ferroelectric KTN," *J. Opt. Soc. Am.*, vol. 57, p. 671, 1967.

[52] E. L. Venturini, E. G. Spencer, P. V. Lenzo, and A. A. Ballman, "Refractive indices of strontium barium niobate," to be published in *J. Appl. Phys.*, January 1968.

[53] T. S. Kinsel and R. T. Denton, "Terminals for a high-speed optical pulse code modulation system—I. 224 Mbit/s single channel, II. Optical multiplexing and demultiplexing," to be published in the *Proc. IEEE*.

[54] I. P. Kaminow and J. Liu, "Propagation characteristics of partially loaded two-conductor transmission line for broadband light modulators," *Proc. IEEE*, vol. 51, pp. 132–136, January 1963.

[55] M. A. Duguay, L. E. Hargrove, and R. B. Jefferts, "Optical frequency translation of mode-locked laser pulses," *Appl. Phys. Lett.*, vol. 9, p. 287, 1966.

[56] C. G. B. Garrett and M. A. Duguay, "Theory of the optical frequency translator," *Appl. Phys. Lett.*, vol. 9, p. 274, 1966.

[57] M. Pomerantz, "Temperature dependence of microwave phonon attenuation," *Phys. Rev.*, vol. 139, p. A501, 1965.

[58] E. G. Spencer, R. T. Denton, and R. P. Chambers, "Temperature dependence of microwave acoustic losses in yttrium iron garnet," *Phys. Rev.*, vol. 125, p. 1950, 1962.

[59] A. W. Warner, M. Onoe, and G. A. Coquin, "Determination of elastic and piezoelectric constants for crystals in class (3m)," *J. Acoust. Soc. Am.*, December 1967.

[60] Onoe *et al.*[30]

[61] W. P. Mason and T. B. Bateman, "Ultrasonic-wave propagation in pure silicon and germanium," *J. Acoust. Soc. Am.*, vol. 36, p. 644, 1964.

——, "Relation between third order elastic moduli and thermal attenuation of ultrasonic waves in nonconducting and metallic crystals," *J. Acoust. Soc. Am.*, vol. 40, p. 852, 1966.

[62] A. I. Akhiezer, "On the absorption of sound in solids," *J. Phys.* (USSR), vol. 1, p. 277, 1939.

[63] H. E. Bömmel and K. Dransfeld, "Excitation and attenuation of hypersonic waves in quartz," *Phys. Rev.*, vol. 117, p. 1244, 1960.

[64] T. O. Woodruff and H. Ehrenreich, "Absorption of sound in insulators," *Phys. Rev.*, vol. 123, p. 1553, 1961.

[65] D. W. Oliver and G. A. Slack, "Ultrasonic attenuation in insulators at room temperature," *J. Appl. Phys.*, vol. 37, p. 1542, 1966.

[66] T. M. Fitzgerald, B. B. Chick, and R. Truell, "Use of the interaction of thermal and ultrasonic waves to study radiation-induced defects in quartz," *J. Appl. Phys.*, vol. 35, p. 1639, 1964.

[67] W. P. Mason, *Physical Acoustics and the Properties of Solids*. Princeton, N. J.: Van Nostrand, 1958.

[68] R. A. Robie and J. L. Edwards, "Some Debye temperatures from single crystal elastic constant data," *J. Appl. Phys.*, vol. 37, p. 2659, 1966.

[69] G. W. Farnell, "Elastic waves in trigonal crystals," *Canad. J. Phys.*, vol. 39, p. 65, 1961.

[70] T. B. Bateman and E. G. Spencer, "Zero field elastic moduli and zero strain piezoelectric constants for lithium niobate," submitted for publication in the *J. Appl. Phys.*

[71] R. T. Smith, "Elastic, piezoelectric and dielectric properties of lithium tantalate," *Appl. Phys. Lett.*, vol. 11, p. 146, 1967.

[72] A. G. Fox and T. Li, "Modes in a maser interferometer with curved and tilted mirrors," *Proc. IEEE*, vol. 51, pp. 80–89, January 1963.

[73] G. D. Boyd and H. Kogelnik, "Generalized confocal resonator theory," *Bell Sys. Tech. J.*, vol. 41, p. 1347, 1962.

[74] M. G. Cohen and E. I. Gordon, "Acoustic beam probing using optical techniques," *Bell Sys. Tech. J.*, vol. 44, p. 693, 1965.

[75] C. F. Quate, C. D. W. Wilkinson, and D. K. Winslow, "Interaction of light and microwave sound," *Proc. IEEE*, vol. 53, pp. 1604–1623, October 1965.

[76] M. Born and E. Wolf, *Principles of Optics*. New York: Pergamon Press, 1959.

[77] E. I. Gordon, "Figure of merit for acousto-optical deflection and modulation devices," *IEEE Trans. Quantum Electronics* (*Correspondence*), vol. QE-2, pp. 104–105, May 1966.

[78] R. W. Dixon and M. G. Cohen, "A new technique for measuring magnitudes of photoelastic tensors and its applications to lithium niobate," *Appl. Phys. Lett.*, vol. 8, p. 205, 1966.

[79] H. R. Carelton and R. A. Soref, "Modulation of 10.6 μ laser radiation by ultrasonic diffraction," *Appl. Phys. Lett.*, vol. 9, p. 110, 1966.

[80] J. H. Parker, Jr., F. Kelly, and D. I. Bolef, "An ultrasonic-optical determination of the third order elastic constant c_{111} for NaCl single crystals," *Appl. Phys. Lett.*, vol. 5, p. 7, 1964.

[81] D. H. McMahon, "A comparison of Brillouin scattering techniques for measuring microwave acoustic attenuation," *IEEE Trans. Sonics and Ultrasonics*, vol. SU-14, pp. 103–108, July 1967.

[82] K. Gürs and R. Muller, "Internal modulation of optical masers," *Proc. Symp. on Optical Masers*, vol. 13, Microwave Research Institute, Polytechnic Institute of Brooklyn, Brooklyn, N. Y., pp. 243–252, 1963.

[83] A. E. Siegman, C. F. Quate, J. Bjorkholm, and G. Francois, "Frequency translation of an He-Ne laser's output frequency by acoustic output coupling inside the resonant cavity," *Appl. Phys. Lett.*, vol. 5, p. 1, 1964.

[84] Unpublished work with R. N. Zitter.

[85] C. Schaefer, L. Bermann, E. Fues, and H. Ludloff, *Akad. Wiss. Berlin, Sitzberichte Phys.—Math. Kl.*, vol. 14, p. 22, 1935.

[86] L. Bergmann, *Ultrasonics and Their Scientific and Technical Applications*, H. S. Hatfield, transl. New York: Wiley, 1939.

[87] H. J. McSkimin, "Variations of the ultrasonic pulse-superposition method for increasing the sensitivity of delay-time measurements," *J. Acoust. Soc. Am.*, vol. 37, p. 864, 1965.

[88] Unpublished work with L. E. Hargrove.

[89] R. W. Dixon, "Acoustic diffraction of light in anisotropic media," *IEEE J. Quantum Electronics*, vol. QE-3, pp. 85–93, February 1967.

[90] E. G. H. Lean, C. F. Quate, and H. J. Shaw, "Continuous deflection of laser beams," *Appl. Phys. Lett.*, vol. 10, p. 48, 1967.

A Review of Acoustooptical Deflection and Modulation Devices

Abstract—A review of the principles of acoustooptical devices is given. Some very simple momentum conservation considerations indicate the optimum relationship between the optical and acoustic beam dimensions for various functions such as scanning or modulation. A calculation for the usual type of acoustic amplitude modulation is described, and serves as an example of the type of detailed considerations that are necessary and possible, as well as a verification of the validity of the simple momentum considerations. It is shown that the product of the fraction of the light that may be scattered and the bandwidth for Bragg scattering equals a materials constant times the acoustic power. This relationship is shown to be valid even to the extent of numerical constants for several configurations allowing a trade-off between these parameters. Thus, the required modulation power for any level of device performance is easily determined. The details of acoustic deflection under conditions of acoustic beam focusing or scanning are also given.

I. INTRODUCTION

A COMPREHENSIVE and informative review of the principles and history of light scattering by acoustic waves has been given recently by Quate et al. [1]. Despite considerable theoretical and experimental interest in the interaction, acoustooptic devices have experienced only limited interest as compared to electrooptic devices. However, the discovery of materials having excellent acoustical and optical properties and large photoelastic constants [2]–[4] coupled with advances in the fabrication of broadband, efficient, acoustical transducers [5] indicates increasing use of acoustic waves for systems requiring optical beam control such as flying spot memories [6]–[8], display [9], light modulation, and variable tap delay lines. The purpose of this paper is to review the principles of these devices from very simple momentum space considerations as well as a detailed scattering calculation using a Green's function solution for the scattered light beams. These considerations allow appraisal of the device parameters under conditions in which the devices would normally be operated, such as light beams with Gaussian intensity profile and acoustic beams with a rectangular profile. In particular, techniques for achieving desired parameters, i.e., a given bandwidth, by focusing the light or acoustic beam, scanning the acoustic beam, or resonating the optical path are described. Finally, efficiency×bandwidth products are derived which emphasize the optimum combination of material parameters and prescribe the required acoustic power for a given level of performance.

Manuscript received June 24, 1966; revised July 14, 1966.

The author is with Bell Telephone Laboratories, Inc., Murray Hill, N. J.

In particular, it will be shown that the proper combination of material parameters for determining optimum device performance is $n^7p^2/\rho v$ in which n is the optical index of refraction, p the appropriate component of the photoelastic tensor, ρ the mass density, and v the acoustic phase velocity. Since p, v, and ρ exhibit no extreme variations from one material to the next, the advantage usually lies with materials having large refractive index in combination with good optical and acoustical quality.

Sufficient information will be provided to establish a basis for straightforward comparison with other optical modulation and deflection devices. The interested reader should more properly make comparisons in the context of the instantaneous materials situation and the specific system requirement; hence numerical comparisons will be avoided. One numerical illustration will be given to establish relevance. However, the photoelastic properties of materials have received less attention than the electro-optic properties and considerable improvement can be expected. Acoustic transducers are also rapidly improving. In view of the ability to focus acoustic waves the interaction height can be made substantially smaller than is normally possible with bulk electrooptic devices. Numerical comparisons of the modulation power required for a given bandwidth and depth of modulation, or a given number of resolvable spots, will increasingly favor the acoustic devices. This is especially relevant for infrared modulation since the useful materials are not limited to insulators.

The following three sections are concerned with momentum considerations for several pertinent acoustooptical devices, an example of a detailed calculation for a Gaussian light beam, and efficiency-bandwidth products.

II. MOMENTUM CONSIDERATIONS

Acoustic scattering of light is associated with the change in index of refraction accompanying a progressing acoustic wave. The perturbation in optical index arises from the change in number density associated with compression, and the change in optical polarizability associated with strain of the component atoms and molecules of the scattering medium. As such, the scattering process is essentially lossless or reactive, and simple wave energy-momentum conservation principles are applicable. The frequency of the acoustic waves under discussion will be assumed to be sufficiently high so that the wavelength is very small compared to the cross-sectional dimensions of the wave. As a result, the acoustic wave behavior is pre-

Reprinted from *Proc. IEEE*, vol. 54, pp. 1391–1401, Oct. 1966.

cisely analogous to that of unguided or free-space coherent optical beams and all the familiar concepts of diffraction, focusing, near-field, far-field, beam waist, etc.,
are appropriate (see H. Kogelnik and T. Li, "Laser beams
and resonators," this issue).

The scattering process is especially simple when one
considers the interaction of plane monochromatic optical
and acoustic waves and is illustrated in Fig. 1. The incident light is denoted by a wave vector or propagation
constant $k=\omega/c'$ in which ω is the optical angular frequency and c' is the unperturbed light velocity in the
medium. The direction of the incident light (along k) is
perfectly defined since the wavefront is planar and infinitely wide. When scattering from a low intensity,
acoustic plane wave of frequency Ω and propagation constant $K=\Omega/v$ (with v the acoustic velocity) does occur,
then the energy-momentum relations

$$\omega_s = \omega \pm \Omega$$
$$k_s = k \pm K \qquad (1)$$

are appropriate. The subscript s denotes scattered light.
Since $\Omega < 2\omega v/c' \sim 10^{-5}\omega$, it follows that $k_s = k(1\pm\Omega/\omega)$
differs only trivially from k and the locus of the scattering
interaction in momentum or propagation constant space
is a circle with radius k. The angle Θ shown in Fig. 1 is
called the Bragg angle and satisfies the relation

$$\sin \Theta = \tfrac{1}{2}K/k. \qquad (2)$$

The scattering angle is 2Θ. If the angle of the incident
light differs from Θ, then the intensity of the scattered
light is identically zero. It is only for this case of extremely
wide beams that the difference between Θ defined in (2)
and the exact value [10] which differs by terms of order
v/c' and higher can be discerned.

Cohen and Gordon [10] have considered the case of
scattering with acoustic beams of finite width. As a result
of diffraction the direction of K is no longer perfectly
defined as was the case in Fig. 1, and may be depicted as
shown in Fig. 2. For this case, still assuming very wide
light beams with essentially zero diffraction angle, and
that the angular spread of acoustic energy is less than
the Bragg angle, only one component direction of K can
contribute to the scattered light, and the intensity of the
scattered light is proportional to the plane wave component of the acoustic beam moving in the appropriate
direction. As shown in [10], rocking the acoustic medium
and observing the intensity of the scattered light directly
as a function of the angle of incidence of the light θ_0
measures the angular distribution or far-field diffraction
pattern of the acoustic beam. An example of the use of
this technique is given by Korpel et al. [9], in this issue.
The analytical details are exhibited and utilized in Appendix A of this paper. In particular it is shown in Appendix
A, as might be intuitively obvious from the preceding
discussion, that the angular dependence of the scattering
interaction as measured by the technique of [10] is independent of where along the acoustic beam the measure-

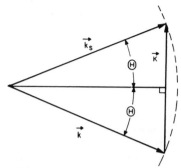

Fig. 1.   Momentum scattering for plane, monochromatic optical,
and acoustic waves. If the direction of k or K is changed, thereby
changing the angle of incidence denoted Θ, the vector sum $k+K$
no longer falls on the circle and no scattering can occur.

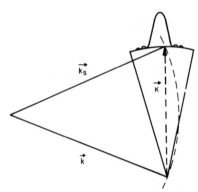

Fig. 2.   Scattering with acoustic waves of finite width corresponding to nonzero diffraction angle. The angular distribution of
acoustic energy for a rectangular, flat transducer is shown
above the cone. Only one plane wave component of the acoustic
beam is effective for scattering. The amplitude of the scattered
light as a function of the direction of the acoustic beam measures
its far-field pattern.

ment is performed. For example, Gordon and Cohen
[11], using evaporated CdS thin film transducers [12] on
cylindrical surfaces which produce cylindrical wavefronts, have obtained focused acoustic beams for which
the transducer is in the acoustic far-field, rather than at
the acoustic beam waist as is the case for flat transducers
which produce plane wavefronts. For the uniform, curved
transducer the angular distribution of acoustic energy is
approximately rectangular, and the near-field amplitude
distribution of acoustic energy is $(\sin x/x)^2$ in contrast to
the reverse situation illustrated in [10] and also [9] of
this issue.

The significance of the acoustic angular distribution can
be appreciated by noting that the intensity of the scattered
light is a function of the angle $\theta_0-\Theta$. Holding θ_0 fixed
and varying Θ by varying the acoustic frequency [see (2)],
the observed frequency dependence of the intensity of the
scattered light directly follows the angular distribution.
Focusing the acoustic beam allows considerable flexibility
in varying the frequency bandwidth of the Bragg scattering interaction. This is especially important at the higher
acoustic frequencies for which diffraction angles are small
(the diffraction angle is of the order $2\pi/KW_0$ in which W_0
is the width of the acoustic beam waist). It is worth noting, however, that focusing is merely a way of producing

a narrow beam waist or large diffraction angle which may also be produced by a correspondingly narrow, flat transducer; the choice is governed by practical considerations.

Deflectors

Any realistic appraisal of acoustooptical devices also requires consideration of light beams of finite width. Hence, as denoted in Fig. 3, the light wave vector k also has a spread $\delta\phi$ consistent with diffraction or focusing. The diffraction or focusing spread in the acoustic beam is denoted by $\delta\theta$. It can be seen from Fig. 3(a), for a given acoustic frequency for which $\Theta \approx \theta_0$ is not too large and with $\delta\theta \gg \delta\phi$, that the angular spread in the scattered light is the same as that of the incident light. Alternately, consideration of Fig. 4(a), illustrating a wide light beam and narrow acoustic beam consistent with $\delta\theta \gg \delta\phi$, illustrates that the width of the scattered beam is the same as that of the incident beam, consistent with equal diffraction angles. Hence, the spatial coherence as well as the temporal coherence of the beam is preserved. As the acoustic frequency is varied, it can be seen that the angle of the scattered beam can vary over a range $2\delta\theta$; hence, the scattered light beam may be focused to

$$N = 2\delta\theta/\delta\phi \qquad (3)$$

independent or resolvable spots. In any scanning device the significant parameter or criterion of performance is the number of resolvable spots to which the beam may be focused and this may usually be related to the ratio of scan angle to minimum angular resolution. This number is an invariant of the optical system producing the focused spot (assuming ideal optics). The required range in K is shown in Fig. 3(a) as $\Delta K = 2k\delta\theta \cos\theta_0 = \Delta\Omega/v$. Denoting $\delta\phi = 2\pi/kw_0$ in which w_0 is the width of either the incident or scattered light beam yields

$$N = (\Delta\Omega/2\pi)w_0/v \cos\theta_0 \qquad (4)$$

as the total number of resolvable spots for a frequency bandwidth $\Delta\Omega/2\pi$. Noting that $w_0/v \cos\theta_0$ is the transit time of the acoustic wave across the light beam τ, one obtains the result of [6] and [7] which is discussed also in [9] of this issue, namely $N = (\Delta\Omega/2\pi)\tau$. Note that τ^{-1} is the maximum speed of the device since the position of the diffracted spot cannot be changed randomly in a time shorter than τ. With sufficient acoustic power, virtually all of the incident light may be scattered, however, only a fraction N^{-1} of the total acoustic power is useful in scattering.

Kaminow [13] has proposed a variant of this scheme in which $\delta\phi \gg \delta\theta$ as denoted in Fig. 3(b). One notes here that only a portion of the incident light is scattered and the angular spread of the scattered light is equal to the diffraction angle of the sound; that is, the scattered light beam is considerably wider than the incident light beam as shown in Fig. 4(b). For this case it may be shown that

$$N = \tfrac{1}{2}(\Delta\Omega/2\pi)w_s/v \cos\theta_0 \qquad (5)$$

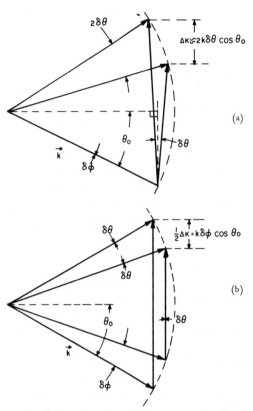

Fig. 3. Scattering with acoustic and light beams of finite width. The diffraction angle of the light is $\delta\phi$ and that of the sound is $\delta\theta$. For scanning devices there are two possible configurations: (a) $\delta\theta \gg \delta\phi$, or (b) $\delta\phi \gg \delta\theta$. For (a) the scattered light beam has a diffraction angle $\delta\phi$, for (b) the diffraction angle is $\delta\theta$.

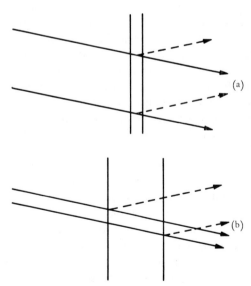

Fig. 4. Geometry of the interacting beams illustrating that the coherence width of the light beam is unchanged from that of incident light for $\delta\theta \gg \delta\phi$, but is increased for $\delta\theta \ll \delta\phi$. The diffraction angle of the scattered light always corresponds to the smaller of the two angles.

in which w_s is the width of the *scattered* beam. At best only a fraction N^{-1} of the incident light can be scattered although all of the acoustic energy may be useful in scattering.

Korpel et al. [9], [14] have shown that for scanning devices the diffraction spread in the acoustic beam is allowed to be very small compared to the desired scan angle, if the acoustic beam is made to scan or steer over an equivalent angle when the acoustic frequency is varied. Hence scanning of the acoustic beam, which was straightforwardly accomplished with an array of phased transducers, was converted into a scan of the optical beam [9], [14]. As will be seen later, this technique is better than the use of focused acoustic beams when the percentage bandwidth of the scanning device is small. In particular, all of the acoustic energy may be useful in scattering the light if the diffraction angle of the acoustic beam is about equal to that of the light beam.

Figure 5(a) illustrates the appropriate configuration for the usual type of acoustic light modulation. Here the diffraction spread in the focused light beam is made approximately equal to the diffraction spread in the acoustic beam. Scattering with amplitude modulated acoustic energy removes energy from the incident beam and produces an amplitude modulated scattered beam. In order to understand why the diffraction angles must be made comparable, consider a situation in which the acoustic wave contains two frequency components f_1 and f_2 separated by $f_2 - f_1 = \Delta f = \Delta \Omega / 2\pi$, and a diffraction spread $\delta\theta$ chosen so that each component scatters. Suppose also that the diffraction angle of the incident light $\delta\phi$ at angular frequency ω is small compared to $\delta\theta$, as shown in Fig. 5(b). The scattered light will consist of two beams, each with angular spread $\delta\phi$ but with no angular overlap, one at angular frequency $\omega + 2\pi f_1$ and the other at $\omega + 2\pi f_2$. There will be no beat frequency at $f_2 - f_1$ and no optical power modulation since the beams are not colinear. Hence, the acoustic frequency may be varied only over a range for which the scattered beams overlap, corresponding to the angular range $\delta\phi$; the modulation bandwidth is determined by $\delta\phi$ and the acoustic energy outside the angular range $\delta\phi$ is wasted.

The opposite extreme $\delta\phi \gg \delta\theta$ is likewise wasteful of optical energy. In this case the diffraction angle of the scattered light corresponds to $\delta\theta$ and there is always overlap for the two modulation frequencies. The modulation bandwidth is determined therefore by $\delta\theta$, and the light outside the angular range $\delta\theta$ is not useful.

It follows that the optimum configuration corresponds to approximate equality of $\delta\theta$ and $\delta\phi$ as illustrated in Fig. 5(a). Note that the diffraction angle of the scattered light approximates $\delta\phi$. Under these circumstances the modulation bandwidth is about half that value appropriate for Bragg scattering over the angular range $\delta\theta$; the exact value depends on the details of the angular distribution. It is interesting to note that the bandwidth for Bragg scattering Δf as follows from (2) can be written

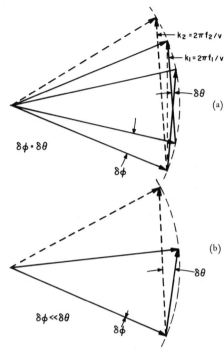

Fig. 5. Arrangement for the usual amplitude modulation device. Two configurations are shown: (a) the optimum situation for which $\delta\theta \approx \delta\phi$, and (b) the case $\delta\phi \ll \delta\theta$.

$$\frac{\pi\Delta f}{kv} = \delta\theta \cos \Theta$$

and taking

$$\delta\theta \approx \Delta\phi \approx 2\pi / k w_0$$

yields

$$\tfrac{1}{2}\Delta f \approx (v \cos \Theta)/w_0. \qquad (6)$$

Hence, the modulation bandwidth is about equal to the reciprocal of the transit time of the acoustic energy across the light beam [15], [16] in fair agreement with intuition.

In passing it should also be noted that when the diffraction angles of both beams are made equal the near-field region of the light beam exceeds the width of the acoustic beam by the factor $2f/\Delta f$. Similarly the near-field region of the acoustic beam exceeds the width of the light beam by the factor $2f/\Delta f \sin^2 \Theta$ which is even greater. Hence within the region of interaction the beams are essentially planar.

A second type of acoustic modulator has been described by Dixon and Gordon [17] and is depicted in Fig. 6. Two light beams originating from a common source intersect within an acoustic beam. The three beams are coplanar and the near-field region of all beams share the common interaction volume. The Kösters prism shown in Fig. 6 serves as a convenient means of controlling the orientation of the two light beams, and allows a symmetrical path for both light beams. The single lens produces the appropriate focus angle and focal plane. Figure 7 shows the momentum diagram for the optimum

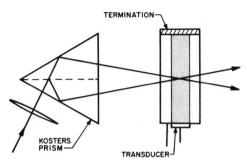

Fig. 6. Arrangement for carrier frequency modulation of the light using a Kösters prism for producing symmetrical, coherent beams.

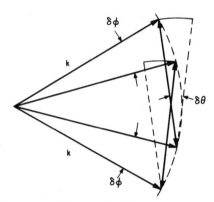

Fig. 7. Momentum diagram for the arrangement of Fig. 6.

case of equal diffraction angles of all three beams. In this case, light scattered from either light beam falls into the cone angle defined by the other. Since the scattered light is shifted in frequency by an amount equal to the acoustic frequency, each transmitted beam is intensity modulated at the acoustic frequency. For this type of modulation both the phase and amplitude of the acoustic signal is converted to variations in the light intensity, in contrast to the previous case in which variations in acoustic power are converted to variations in optical intensity.

The requirement of equal diffraction angles follows from essentially the same type of reasoning as was applied to the previous case. Efficient intensity modulation requires full overlap of the scattered and transmitted beams and full use of all of the acoustic and optical energy. As in the previous case and for the same reason, the modulation bandwidth approximates the reciprocal of the transit time of the sound across the light beam.

III. SCATTERING CALCULATION FOR ACOUSTIC MODULATOR

One approach to determining a quantitative expression for the intensity of acoustically scattered light is to formulate the amplitude ψ_s as the solution of a volume Green's function integral

$$\psi_{s,\Omega}(r) = \iiint \rho G_{k_s}(r \mid r_0) dv_0 \qquad (7)$$

in which

$$G_k(r \mid r_0) = \frac{\exp - ik \mid r - r_0 \mid}{\mid r - r_0 \mid}$$

corresponds to an outgoing wave solution and ρ is the volume polarization at frequency $\omega \pm \Omega$ induced by the acoustic wave in the presence of the incident wave. The solution is valid only in the limit of weak scattering, since rescattering is ignored. The amplitude of the scattered light is linearly related to the acoustic amplitude.

Assuming the incident light beam has a Gaussian profile, the amplitude distribution may be written

$$\psi_i(x_0, y_0, z_0)$$
$$\approx \psi_0 \exp - \left[\frac{2 \ln 2}{w_0^2} \left[(- x_0 \cos \theta_0 + y_0 \sin \theta_0)^2 + z_0^2 \right] \right.$$
$$\left. + ik(y_0 \cos \theta_0 + x_0 \sin \theta_0) \right] \qquad (8)$$

corresponding to a beam moving at angle θ_0 relative to the y_0-axis in the x_0, y_0 plane. The half-power beam diameter is w_0. The acoustic wave has the form

$$S = S_0 \exp - iKx_0 \qquad (9)$$

for $-\frac{1}{2}W_0 \leq y_0 \leq \frac{1}{2}W_0$ and $-\frac{1}{2}H \leq z_0 \leq \frac{1}{2}H$. Except for material constants

$$\rho \propto \psi_i S. \qquad (10)$$

Choosing the observation point r to lie on a large sphere of radius R (see Fig. 8) the integral for the upshifted light is straightforwardly evaluated to yield

$$\psi_{s,\Omega}(R \sin \theta, R \cos \theta \cos \phi, R \cos \theta \sin \phi)$$
$$= \frac{\psi_0 S_0}{R} (\exp - i(\omega + \Omega)R/c')(\pi W_0/\beta) \, \mathrm{Erf} \left[\frac{1}{2}\beta^{1/2}H \right]$$
$$\times \exp - \left[(\cos \theta \sin \phi)^2/4\beta + \eta^2/4\beta \cos^2 \theta_0 \right]$$
$$\times \sin \frac{1}{2}(\xi + \eta \tan \theta_0) W_0 / \frac{1}{2}(\xi + \eta \tan \theta_0) W_0 \qquad (11)$$

in which

$$\eta = - [k_s \sin \theta + k \sin \theta_0 - K]$$
$$\xi = [k_s \cos \theta \cos \phi - k \cos \theta_0]$$
$$\beta = (2 \ln 2)/w_0^2.$$

Assuming that the acoustic signal contains a multitude of frequency components the instantaneous scattered optical power is given by the surface integral

$$P_s = R^2 \int_0^{2\pi} d\phi \int_0^{\pi/2} \cos \theta d\theta \left| \sum_\Omega \psi_{s,\Omega} \exp i\Omega t \right|^2. \qquad (12)$$

For the purpose of discussion consider that the acoustic wave contains two frequency components of equal amplitude; one at frequency Ω_0 which corresponds to $\Theta = \theta_0$

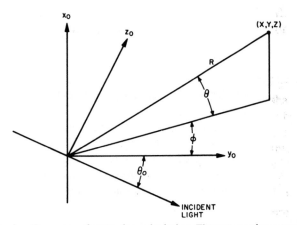

Fig. 8. Geometry of scattering calculation. The rectangular acoustic beam of height H and width W_0 moves along the x_0-axis and the incident Gaussian light beam moves at an angle θ_0. The scattering to the far-field point $X = R \sin \theta$, $Y = R \cos \theta \cos \delta$, $Z = R \cos \theta \sin \delta$ is calculated.

and one at some frequency $\Omega_0 + 2\pi f$. This corresponds to 100 percent sinusoidal modulation of the acoustic energy at frequency f. Performing the appropriate algebraic manipulations yields

$$P_s \propto [\,|S_0|^2 W_0 H][\,|\psi_0|^2/\beta]$$
$$\cdot [(\mathrm{Erf}\,[\tfrac{1}{2}\beta^{1/2}H])^2/\tfrac{1}{2}\beta^{1/2}H]\mathcal{I} \quad (13)$$

in which

$$\mathcal{I} = \pi^{-1}\int_{-\infty}^{+\infty}dx e^{-x^2}\frac{(\sin ax)^2}{ax^2}$$
$$+ \pi^{-1}\int_{-\infty}^{+\infty}dx e^{-x^2}\frac{\sin^2 a(x+b)}{a(x+b)^2}$$
$$+ 2\pi^{-1}e^{-b^2}\int_{-\infty}^{+\infty}dx e^{-x^2}\frac{\sin ax \sin a(x+b)}{ax(x+b)}$$
$$\times \cos 2\pi f(t - R/c')$$
$$a = (2\beta)^{1/2}W_0 \sin \theta_0$$
$$b = \pi f/v(2\beta)^{1/2}\cos \theta_0. \quad (14)$$

The first bracket in (13) corresponds to the power in either component of the acoustic wave; the second bracket corresponds to the optical power. The third bracket indicates that for a given acoustic power there is an optimum acoustic beam height consistent with the spot size of the optical beam. The optimum corresponds to

$$\tfrac{1}{2}\beta^{1/2}H = (\tfrac{1}{2}\ln 2)^{1/2}H/w_0 = 0.99$$

or a beam height about 70 percent larger than the optical beam half-power spot size. Note that when $\tfrac{1}{2}\beta^{1/2}H > 1$, $P_s \propto H^{-1}$.

The first term in $\mathcal{I}$ of (14) is associated with scattering by the acoustic component at Ω_0; the integral has the value

$$a^{-1}\int_0^a \mathrm{Erf}\,(a)da$$

which approaches unity monotonically for large a. Only for

$$a = (\ln 2)^{1/2}KW_0/kw_0 \lesssim 1$$

corresponding to the diffraction angle of the light less than that of the sound does the scattered intensity increase substantially with increase in the acoustic beamwidth or decrease in the acoustic diffraction angle. The scattered intensity is essentially independent of the acoustic beamwidth when the diffraction angle of the light exceeds that of the acoustic beam, corresponding to $a > 1$.

The second term in (14) corresponds to the scattering produced by the component of acoustic energy at $\Omega_0 + 2\pi f$ and the third term to the beat term. For large a, the second term approaches a value of $\exp -b^2$ indicating as expected that the falloff in scattered power is governed by the spread in the optical beam. In the same limit the third or beat term takes on the value

$$2\frac{\sin ab}{ab}(\exp - b^2)\cos 2\pi f(t - R/c')$$

indicating that the modulation bandwidth is controlled by the Bragg scattering condition since $ab \propto (fW_0/v)\tan \theta_0$.

When $fw_0/v \propto b$ is small, the two integrals have the same value as that of the first integral. The scattered light is 100 percent intensity modulated, faithfully following the acoustic intensity. Both terms vanish for large b as would be expected since this places the acoustic frequency outside the range of effective scattering. The bandwidth in the limit $a \leq 1$ is governed by $\exp -b^2$ or correspondingly by the diffraction angle of the light.

In order to give quantitative expression to the result expressed in (13) and (14) it is necessary to evaluate the integrals numerically and the results would not add much to the discussion here. Dixon and Gordon [17] have performed a similar analysis for the intersecting beam case described by Figs. 6 and 7. The results are qualitatively similar, in agreement with the conclusions of simple momentum considerations.

IV. Scattering Efficiency × Bandwidth

In a previous subsection the general considerations for acoustooptic deflection or scanning devices indicate that, ignoring transducer bandwidth, the capacity is directly proportional to the angular range of the acoustic beam. Consequently to optimize the capacity-speed product given in (4) the angular range is adjusted to allow Bragg scattering over a frequency range equal to the transducer-driving circuit bandwidth. Similarly for modulation purposes the modulation bandwidth is proportional to the angular range of the acoustic beam which, for optimum modulation efficiency, is set equal to the diffraction angle of the light. In this section the consequences of varying the angular range are examined.

For the purpose of discussion assume that the acoustic beam originates from a transducer with a flat or cylindrical face. Consequently, the acoustic beam has a waist, real or virtual, as shown in Fig. 9; with a flat transducer, the beam waist occurs at the transducer. It is shown in Appendix A that the intensity and the angle or frequency dependence of the scattered light beam is completely independent of the position of the incident light beam along the acoustic beam. It is shown also that with a rectangular amplitude distribution at the beam waist, corresponding to a beam focus angle $\Lambda/W_0 = v/f_0W_0$, the scattering bandwidth is

$$\Delta f \approx \frac{(1.8nv^2 \cos \theta_0)}{\lambda_0 f_0 W_0} \quad (15)$$

in which W_0 is the width of the acoustic beam waist and λ_0 is the free-space wavelength, and the optimum fraction of light power scattered is $\sin^2\eta^{1/2}$.[1] The parameter η is defined by

$$\eta = \frac{1}{2} \pi^2 \left(\frac{n^6p^2}{\rho v^3}\right)(\lambda_0^2H \cos^2 \theta_0)^{-1}W_0P_a \quad (16)$$

in which H is the beam height and P_a the total acoustic power. It has been assumed that the diffraction angle of the light is less than or at most equal to the diffraction angle of the sound and that the height of the acoustic beam exceeds that of the light beam. The product

$$2f_0\Delta f \cdot \eta \approx 1.8\pi^2 \left(\frac{n^7p^2}{\rho v}\right)(\lambda_0^3H \cos \theta_0)^{-1}P_a \quad (17)$$

is independent of W_0 and is essentially independent of the amplitude distribution of the acoustic beam at the beam waist; Gaussian, rectangular, half sine wave, $(\sin x)/x$, etc.[2] Varying the width of the beam waist or correspondingly the beam focus angle allows a trade-off between Δf and η keeping the product constant for constant acoustic power, center frequency, and beam height. Power loading of the transducer and acoustic medium can be avoided by working always in the far field of the acoustic beam, that is, placing the nominal beam focus or waist outside the acoustic medium by the use of cylindrical transducers.

When the acoustic transducer consists of an array with each element driven π radians out of phase with its neighbor, the acoustic energy distribution consists of two major lobes as shown in Fig. 10. The angle from the normal Ψ varies inversely with acoustic frequency. In this case the angular range of the acoustic beam is achieved by steering rather than focusing. Over a range of acoustic frequencies the angle of the light relative to one of the

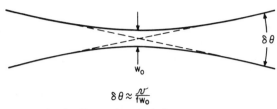

$$\delta\theta \approx \frac{\Lambda}{fW_0}$$

Fig. 9. Focused acoustic beam geometry.

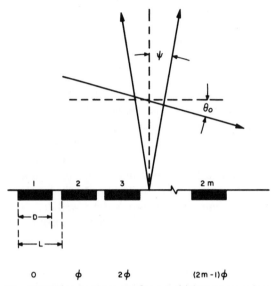

Fig. 10. Scattering arrangement for a multielement transducer. As the acoustic frequency increases, ψ decreases and the angle of incidence of the light relative to the acoustic lobe on the right increases. Thus, the angle of incidence can approximately track the Bragg angle which increases with frequency.

acoustic lobes can approximate the Bragg condition [9], [14]. As shown in Appendix B for this case the bandwidth for scanning increases inversely with the square root of the number of transducer elements. The scattered power increases proportional to the number of transducer elements. For this case the product $(\Delta f)^2 \cdot \eta$ has a value identical to that given by $2f_0\Delta f \cdot \eta$ in (17). The general considerations for beam steering, including the case for which the phase shift is not π, are given in Appendix B.

Resonating the optical path to allow multiple passes through the acoustic beam offers an alternate means of trading scattering efficiency and bandwidth. The product is still given by (17), however [18]. For beam focusing, beam steering, or optical resonance, scattering efficiency (or modulation depth) and bandwidth can be interchanged but the appropriate material parameter is $n^7p^2/\rho v$ [19].[3] Note too the strong dependence on λ_0; decreasing λ_0 from 0.633 μ to 0.488 μ increases λ_0^{-3} by a factor of two. Both n and p usually increase slightly with a decrease in λ_0. In addition, there is some evidence that near the band gap of

---

[1] When $\eta = (\pi/2)^2$ all of the incident light is scattered; for $\eta \ll 1$ the fraction of the incident light that may be scattered can be approximated by η.

[2] The numerical constant in (17) varies by less than 10 percent for the four distributions mentioned.

[3] When bandwidth is not pertinent the parameter chosen for comparing scattering efficiency is $n^6p^2/\rho v^3$, (see [1]) as follows from (16).

CdS and ZnO, p increases drastically [3]. Typical values of p range between 0.1 and 0.2.

Using this criterion, lithium niobate [4] is 11 times better than fused quartz for longitudinal waves. For this material using longitudinal waves along the x-axis with an acoustic beam height of 0.1 millimeter and the light moving at the proper angle [4] in the y-z plane.

$$\left.\begin{array}{c} 2f_0\Delta f \\ (\Delta f)^2 \end{array}\right\} \cdot \eta \approx 6 \times 10^{17} P_a (\text{watts}). \tag{18}$$

As an example, for $\Delta f = 2.3 \times 10^8$ c/s using beam steering,

$$\eta \approx 10 P_a (\text{watts})$$

and useful modulation depth is possible with modest acoustic power and large bandwidth. Acoustic transducers with conversion efficiency in the range 5 to 10 dB at microwave frequencies are now state-of-the-art [5] so that the required electrical power ($\sim$1 watt) is comparable to that required for the best bulk electrooptic modulators with similar optical aperture. The practical advantage may fall the the acoustic modulator since a well-defined interaction volume can be achieved without resorting to fabrication of crystals with small cross section. In addition, the choice of photoelastic materials is considerably broader and is not limited to high resistivity materials. This factor is especially relevant for infrared light.

V. Conclusion

Some relatively simple-minded points of view have been shown to be appropriate to describe the general principles of operation of acoustooptical deflection and modulation devices. Specific geometries have been considered and scattering-efficiency-bandwidth products determined to provide a basis for numerical comparison.

Appendix A

Angular Dependence of the Scattering Efficiency Along the Acoustic Beam

The steady-state acoustic strain wave amplitude can be written as

$$\begin{aligned} S(x, y, t) &= S_c(x, y) \cos(\Omega t - Kx) \\ &\quad + S_s(x, y) \sin(\Omega t - Kx) \\ &= \text{Re}\, \bar{S}^*(x, y) \exp i(\Omega t - Kx) \end{aligned} \tag{19}$$

in which

$$\bar{S} = S_c + iS_s \tag{20}$$

is assumed to be independent of z, Ω is the acoustic angular frequency, and $K = \Omega/v$. Assuming the wave originates outside some volume V with surface A the solution inside V satisfies the source-free Helmholtz equation. Consequently $S(x, y, t)$ can be determined from the Green's function solution

$\bar{S}^*(x', y') \exp - iKx'$

$$= \frac{1}{4\pi} \oint_A dA \left[\frac{\partial}{\partial n}(\bar{S}^* \exp - iKx) \frac{e^{-iKr}}{r} \right.$$

$$\left. - (\bar{S}^* \exp - iKx) \frac{\partial}{\partial n} \frac{e^{-iKr}}{r} \right] \tag{21}$$

in which $\partial(\ )/\partial n$ is the surface normal derivative and

$$r = [(x - x')^2 + (y - y')^2 + (z - z')^2]^{1/2}.$$

The volume V is chosen to correspond to the infinite half-space $x \geq 0$ so that $\partial(\ )/\partial n = -\partial(\ )/\partial x$. The plane $x = 0$ is placed at the beam waist. Using these definitions

$$\bar{S}^*(x', y') = \frac{e^{iKx'}}{4\pi} \int_{-\infty}^{+\infty} dy \int_{-\infty}^{+\infty} dy \bar{S}^*(0, y)$$

$$\cdot \left[iK \frac{e^{-iKr}}{r} - \frac{x'}{r} \frac{\partial}{\partial r} \frac{e^{-iKr}}{r} \right]$$

$$r = [x'^2 + (y - y')^2 + (z - z')^2]^{1/2}. \tag{22}$$

For any acoustic beam moving along the x-axis and light propagating in the x-y plane at an angle θ_0 to the y-axis, the angular dependence of the scattering interaction in the limit of weak scattering at any point along the beam can be shown to satisfy the relation [10]

$$V_1(\theta_0)\big|_{x'} = V_0 \left[-\frac{1}{2} i \exp\left(iKy \frac{\sin\theta_0 - \sin\Theta}{\cos\theta_0} \right) \right]$$

$$\times \int_{-\infty}^{+\infty} dy' \xi(x', y')^*$$

$$\cdot \exp - \left(iKy' \frac{\sin\theta_0 - \sin\Theta}{\cos\theta_0} \right). \tag{23}$$

$V_1(\theta_0)\big|_{x'}$ is the amplitude of the scattered light for an interaction taking place near the plane at x' as a function of the angle of incidence of the incident light beam, V_0 is the amplitude of the incident light, and ξ is the incremental phase retardation per unit length produced by the acoustically induced variation in the index of refraction;

$$\xi(x, y) = \frac{k_0 \Delta n(x, y)}{\cos\theta_0} \tag{24}$$

in which k_0 is the free-space propagation constant of the light and Δn is the variation in the index of refraction n. Consistent with the definition of ξ in [10]

$$\Delta n = \tfrac{1}{2} n^3 p (S_c + iS_s) = \tfrac{1}{2} n^3 p \bar{S} \tag{25}$$

in which p is the photoelastic constant. In general $\bar{S}$ and p are tensors but this notation has been suppressed. The angle Θ in (23) is defined by the Bragg condition (2)

$$\sin\Theta = \frac{\tfrac{1}{2}K}{nk_0}. \tag{26}$$

Combining (23), (24), and (25) yields

$$V_1(\theta_0)\big|_{x'} = \frac{1}{2}\frac{k_0 n^3 p V_0}{\cos\theta_0}$$

$$\cdot\left[-\frac{1}{2}i\exp\left(iKy\frac{\sin\theta_0-\sin\Theta}{\cos\theta_0}\right)\right]$$

$$\times\int_{-\infty}^{+\infty}dy'\overline{S}^*(x',y')$$

$$\cdot\exp-\left(iKy'\frac{\sin\theta_0-\sin\Theta}{\cos\theta_0}\right). \qquad (27)$$

Substituting (22) and interchanging the order of integration of y and y' and defining $y''=y'-y$ yields

$$V_1(\theta_0)\big|_{x'} = V_1(\theta_0)\big|_0 \frac{e^{iKx'}}{4\pi}\int_{-\infty}^{+\infty}dz\int_{-\infty}^{+\infty}dy''$$

$$\cdot\exp-\left(iKy''\frac{\sin\theta_0-\sin\Theta}{\cos\theta_0}\right)$$

$$\times\left[iK\frac{e^{-iK\rho}}{\rho}-\frac{x'}{\rho}\frac{\partial}{\partial\rho}\frac{e^{-iK\rho}}{\rho}\right]$$

$$\rho=[x'^2+y''^2+(z-z')^2]^{1/2}. \qquad (28)$$

To evaluate the integral one need only note the identity which follows from (21)

$$e^{-i(K_x x'+K_y y')}$$

$$\equiv\frac{1}{4\pi}\int_{-\infty}^{+\infty}dz\int_{-\infty}^{+\infty}dy[\exp-i(K_x x+K_y y)]_{x=0}$$

$$\times\left[iK_x\frac{e^{-iK\rho'}}{\rho'}-\frac{x'}{\rho'}\frac{\partial}{\partial\rho'}\frac{e^{-iK\rho'}}{\rho'}\right]$$

$$K_x^2+K_y^2=K^2$$

$$\rho'=[x'^2+(y-y')^2+(z-z')]^{1/2}. \qquad (29)$$

Comparing (28) and (29) yields

$$V_1(\theta_0)\big|_{x'} = V_1(\theta_0)\big|_0$$

$$\cdot\exp iKx'\left[1-\left[1-\left(\frac{\sin\theta_0-\sin\Theta}{\cos\theta_0}\right)^2\right]^{1/2}\right]. \qquad (30)$$

This simple result can be understood intuitively by realizing that the scattered light amplitude as a function of the angle of incidence, defined in (27), is proportional to the angular distribution of acoustic energy. This distribution, except for phase factors, cannot change as a function x.

The relative scattered energy or scattering efficiency is defined by

$$\big|V_1(\theta_0)\big|_{x'}/V_0\big|^2 = \frac{1}{16}\frac{k_0^2 p^2 n^6}{\cos^2\theta_0}\left|\int_{-\infty}^{+\infty}dy'\overline{S}(x',y')^*\right.$$

$$\left.\cdot\exp-iK\left(\frac{\sin\theta_0-\sin\Theta}{\cos\theta_0}\right)y'\right| \qquad (31)$$

which is independent of x'. Consequently one may choose the most convenient value of x' to make the evaluation. As discussed in the text the acoustic beam has a focus somewhere, real or virtual; at the focus the wavefronts are plane and the beam waist has its narrowest dimension. It is often convenient to evaluate the scattered intensity at the position of the beam waist. For example, for a beam characterized by a rectangular cross section of width W_0 at its waist and strain amplitude $\overline{S}$, (31) leads to a value for the scattering efficiency,

$$\eta = \big|V_1/V_0\big|^2$$

$$= \frac{1}{16}\frac{k_0^2 p^2 n^6 W_0^2}{\cos^2\theta_0}\left[\frac{\sin\frac{1}{2}KW_0\Delta\theta}{\frac{1}{2}KW_0\Delta\theta}\right]^2\big|\overline{S}\big|^2 \qquad (32)$$

in which $\Delta\theta$ is the excursion from the Bragg angle $\theta_0-\Theta$. The half-power points are defined by

$$\tfrac{1}{2}KW_0\Delta\theta_{1/2}\approx\pm0.45\pi \qquad (33)$$

or

$$2\Delta\theta_{1/2}\approx 1.8\pi/KW_0 \qquad (34)$$

which is just the diffraction angle of the acoustic beam $\delta\theta$. Consequently one can write for the frequency bandwidth (using $\sin\theta=\pi f/nvk_0$),

$$\delta\theta\approx\pi\Delta f/nk_0 v\cos\theta_0 \qquad (35)$$

or

$$\Delta f\approx 1.8nk_0 v^2\cos\theta_0/2\pi f_0 W_0. \qquad (36)$$

For $\theta_0=\Theta$, (32) yields

$$\eta = (\tfrac{1}{4}n^3 k_0 p W_0/\cos\theta_0)^2\big|\overline{S}\big|^2. \qquad (37)$$

The acoustic power is defined by

$$P_a = \tfrac{1}{2}\rho v^3\big|\overline{S}\big|^2 W_0 H \qquad (38)$$

in which ρ is the mass density of the medium and H is the beam height. Combining (36), (37), and (38) yields (17). Other distributions, Gaussian, half sine wave, and $\sin x/x$ yield the same result except for a slight change in the numerical constant.

APPENDIX B

ACOUSTIC BEAM STEERING USING MULTI-ELEMENT TRANSDUCERS

Consider the transducer array illustrated in Fig. 10 and assume there are $2m$ elements, of width D and center-to-center spacing L, each driven from the same source with equal amplitude and constant phase increment from one transducer to the next. Thus, the phase of the lth element counting from left to right is $-(l-1)\phi$. The strain amplitude $\overline{S}$ is assumed to be uniform across each element. Evaluating the scattered light intensity at a position immediately in front of the transducer array, (31) can be written

$$\eta = \frac{1}{16} \frac{k_0^2 p^2 n^6}{\cos^2 \theta_0}$$

$$\cdot \left| \sum_{l=0}^{2m-1} \overline{S}^* \exp il\phi \int_{-(m-l)L}^{-(m-l)L+D} \exp -i\kappa y \, dy \right|^2 \quad (39)$$

in which

$$\kappa \approx K\left[\sin\theta_0 - \tfrac{1}{2}K/k_0 n\right]/\cos\theta_0. \quad (40)$$

The approximation in (40) is associated with the substitution of K for $K\cos\psi$, since ψ is normally less than 10^{-1}. A straightforward integration and summation yields for η,

$$\eta = \frac{1}{16}\left(\frac{k_0^2 p^2 n^6 \left|\overline{S}\right|^2}{\cos^2\theta_0}\right)\left(\frac{\sin\frac{1}{2}\kappa D}{\frac{1}{2}\kappa}\right)^2$$

$$\cdot \left(\frac{\sin(\kappa L - \phi)m}{\sin\frac{1}{2}(\kappa L - \phi)}\right)^2. \quad (41)$$

In this case the interaction has an optimum whenever $\kappa L = \phi$ with half-power points at

$$\kappa L = \phi \pm 0.45\pi/m. \quad (42)$$

Under ideal circumstances the phase shift ϕ would be tailored to track the variation of κL with acoustic frequency to maintain an optimum interaction. The required variation is shown in Fig. 11. Under less than ideal circumstances ϕ can track κL approximately over some desired frequency range.

The simplest arrangement to instrument is $\phi = \pi$ at all frequencies [14]. To determine the bandwidth under these circumstances it is convenient to invert (40) and write

$$f = f_0 \pm f_0[1 - (\kappa L \cos\theta_0/\pi)\lambda_0/Ln\sin^2\theta_0]^{1/2} \quad (43)$$

in which

$$f_0 = (vn/\lambda_0)\sin\theta_0 \quad (44)$$

is the frequency at which κL has a maximum value given by

$$(\kappa L)_{max} = (\pi Ln\lambda_0)\sin^2\theta_0/\cos\theta_0. \quad (45)$$

Equation (44) may be considered to define $\sin\theta_0$ for a given center frequency f_0. From (42) the maximum allowable value of κL is $\pi(1+0.45/m)$ which defines

$$L = v^2(n/\lambda_0 f_0^2)(1 + 0.45/m)\cos\theta_0 \quad (46)$$

and makes f_0 a half-power frequency. The other half-power frequencies are defined by $\kappa L = \pi(1-0.45/m)$ which from (45) yields

$$f_{1/2} = f_0 \pm f_0\left(\frac{0.9/m}{1+0.45/m}\right)^{1/2} \quad (47)$$

and a bandwidth

$$\Delta f = 2f_0\left(\frac{0.9/m}{1+0.45/m}\right)^{1/2} \quad (48)$$

$$= v(3.6n\cos\theta_0/\lambda_0 Lm)^{1/2}. \quad (48a)$$

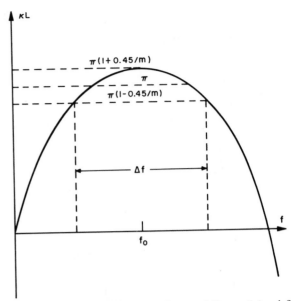

Fig. 11. A plot of the element-to-element shift $\phi = \kappa L$ [as defined by (40)] required for optimum scattering from an acoustic beam generated by a phased array of transducer elements. The construction pertains to the case $\phi \equiv \pi$.

Evaluating η at $\kappa L = \pi$ and writing for the total acoustic power

$$P_a = \tfrac{1}{2}\rho v^3 \left|\overline{S}\right|^2 2mDH \quad (49)$$

yields

$$(\Delta f)^2\eta = 14.4(n^7 p^2/\rho v)(\lambda_0^3 H\cos\theta_0)^{-1}P_a$$

$$\cdot (L/D)\sin^2(\tfrac{1}{2}\pi D/L). \quad (50)$$

Choosing $D/L = 0.742$ yields the maximum value $(L/D)\sin^2(\tfrac{1}{2}D/L) = 1.135$. With this choice

$$(\Delta f)^2\eta = 16.4(n^7 p^2/\rho v)(\lambda_0^3 H\cos\theta_0)^{-1}P_a. \quad (51)$$

The constant 16.4 is comparable to $1.8\pi^2$ and the right-hand side is identical to that of (17).

ACKNOWLEDGMENT

The author is indebted to M. G. Cohen, R. W. Dixon, and M. A. Woolf for many useful discussions.

REFERENCES

[1] C. F. Quate, C. D. W. Wilkinson, and D. K. Winslow, "Interaction of light and microwave sound," *Proc. IEEE*, vol. 53, pp. 1604–1623, October 1965.
[2] T. M. Smith and A. Korpel, "Measurement of light-sound interaction efficiencies in solids," *IEEE J. of Quantum Electronics (Correspondence)*, vol. QE-1, pp. 283–284, September 1965.
[3] B. Tell, J. M. Worlock, and R. J. Martin, "Enhancement of elasto-optic constants in the neighborhood of a band gap in ZnO and CdS," *Appl. Phys. Lett.*, vol. 6, pp. 123–124, April 1, 1965.
[4] R. W. Dixon and M. G. Cohen, "A new technique for measuring magnitudes of photoelastic tensors and its application to lithium niobate," *Appl. Phys. Lett.*, vol. 8, pp. 205–207, April 15, 1966.
E. G. Spencer, P. V. Lenzo, and K. Nassau, "Optical interactions with elastic waves in lithium niobate," *IEEE J. of Quantum Electronics (Correspondence)*, to be published.
[5] N. F. Foster and G. A. Rozgonyi, "Zinc oxide film trans-

ducers," *Appl. Phys. Lett.*, vol. 8, pp. 221–223, May 1, 1966.

[6] E. I. Gordon and M. G. Cohen, "Electro-optic gratings for light beam modulation and deflection," presented at the *1964 IEEE WESCON.*

[7] M. G. Cohen and E. I. Gordon, "Electro-optic [KTa$_x$Nb$_{1-2}$O$_3$ (KTN)] gratings for light beam modulation and deflection," *Appl. Phys. Lett.*, vol. 5, p. 181, November 1964.

[8] A. Korpel, R. Adler, P. Desmares, and T. M. Smith, "An ultrasonic light deflection system," *IEEE J. of Quantum Electronics*, vol. QE-1, pp. 60–61, April 1965.

[9] A. Korpel, R. Adler, P. Desmares, and W. Watson, "A television display using acoustic deflection and modulation of coherent light," this issue.

[10] M. G. Cohen and E. I. Gordon, "Acoustic beam probing using optical techniques," *Bell Sys. Tech. J.*, vol. XLIV, pp. 693–721, April 1965.

[11] E. I. Gordon and M. G. Cohen, "Acoustic modulation devices," presented at the 1966 Conference on Electron Device Research, California Institute of Technology, Pasadena, Calif.

[12] N. F. Foster, "Cadmium sulphide evaporated-layer transducers," *Proc. IEEE*, vol. 53, pp. 1400–1405, October 1965.

[13] I. P. Kaminow, private communication.

[14] A. Korpel, R. Adler, P. Desmares, "An improved ultrasonic light deflection system," Paper 11.5, presented at the 1965 Internat'l Electron Devices Meeting, Washington, D. C.

[15] E. I. Gordon and M. G. Cohen, "Electro-optic diffraction grating for light beam modulation and diffraction," *IEEE J. of Quantum Electronics*, vol. QE-1, pp. 191–198, August 1965.

[16] H. V. Hance and J. K. Parks, "Wide-band modulation of a laser beam using Bragg-angle diffraction by amplitude modulated ultrasonic waves," *J. Acoust. Soc. Am.*, vol. 38, pp. 14–23, July 1965.

[17] R. W. Dixon and E. I. Gordon, "Carrier frequency modulation using acoustic waves," presented at the 1966 Conference on Electron Device Research, California Institute of Technology, Pasadena, Calif.

[18] M. G. Cohen and E. I. Gordon, "Acoustic scattering of light in a Fabry-Perot resonator," *Bell Sys. Tech. J.*, vol. 45, pp. 945–966, July–August 1966.

[19] E. I. Gordon, "Figure of merit for acousto-optical deflection and modulation devices," *IEEE J. of Quantum Electronics* (*Correspondence*), vol. QE-2, pp. 104–105, May 1966.

Electrooptic Light Modulators

I. P. KAMINOW AND E. H. TURNER

Abstract—The field of electrooptic light modulation by means of the Pockels and Kerr effects in crystals is summarized with particular attention to communications applications using the optical maser. All available data on electrooptic materials are tabulated, and design considerations and operating principles for various modulator configurations are outlined.

I. INTRODUCTION

APPLICATIONS of the optical maser often require a means for modulating the amplitude, phase, frequency, or direction of a light beam at high speed. Mechanical shutters and moving mirrors have too much inertia to permit modulation at the required frequencies, which range from megahertz to gigahertz. Hence, it is necessary to rely upon optical interactions with electrical, magnetic, and acoustic fields at the modulating frequency via the nonlinearities of matter. Some of these interactions are intrinsically lossy at the optical frequency: e.g., free carrier absorption, and Franz-Keldysh effect (band edge shift by an electric field); others are intrinsically reactive (or parametric in the electrical engineering sense): e.g., magnetooptic Faraday effect, acoustooptic effect, and electrooptic Kerr and Pockels effects. This paper is concerned only with reactive electrooptic effects in solids; that is, the change in refractive index produced by an applied electric field.

Before the turn of the century Kerr observed a quadratic electrooptic effect in liquids such as carbon disulphide, and Röntgen and Kundt observed a linear effect in quartz. Pockels examined the linear effect in crystals of quartz, tourmaline, potassium chlorate, and Rochelle salt. He demonstrated the existence of a *direct* effect (independent of piezoelectrically induced strain) and characterized the linear electrooptic effect in crystals of various point symmetry using either the applied electric field or dielectric polarization as bases [1].

In 1944 Zwicker and Scherrer reported the dc electrooptic properties of KH_2PO_4 and KD_2PO_4 and related these properties to the ferroelectric behavior of these crystals. They observed that the electrooptic coefficient, based on electric fields, was proportional to the dielectric constant, exhibiting a Curie-Weiss behavior as a function of temperature. The electrooptic coefficient based on dielectric polarization is the same temperature-independent constant for both crystals [2], [3]. In 1949 Billings,

Manuscript received June 30, 1966.
The authors are with the Crawford Hill Laboratory, Bell Telephone Laboratories, Inc., Holmdel, N. J.

Reprinted from *Proc. IEEE*, vol. 54, pp. 1374–1390, Oct. 1966.

and later Carpenter, built and studied the properties of high-speed (about 1 MHz) light shutters using KH_2PO_4 and $NH_4H_2PO_4$ for use in recording sound on film and other engineering applications [4], [5]. In 1961, Holshouser, von Foerster, and Clark reported on a microwave liquid Kerr cell for use in studying high-speed photomixing [6]; and Froome and Bradsell described a microwave modulator employing the Pockels effect in $NH_4H_2PO_4$ for use in a distance measuring device [7].

Since the advent of the optical maser with its potential application to communication and switching, the electrooptic properties of a number of different materials have been studied. These measurements are summarized below, and the general properties of the materials commented on briefly. Several new modulator configurations have been proposed recently and the principles and characteristics of their operation are outlined in a later section, along with some general design considerations. Before getting into the details of modulator materials and designs, however, a description of the electrooptic effect and its application to various types of optical modulation is given.

II. Electrooptic Behavior of Crystals[1]

It is convenient to consider the change in $1/n^2$ with application of a field, where n is the refractive index, rather than a change in n directly. The quantity $1/n^2$ can be written

$$\frac{1}{n^2} = \frac{1}{n_0^2} + rE + RE^2 + \cdots, \qquad (1)$$

where r and R are the linear and quadratic electrooptic coefficients, respectively. The coefficients in (1) are those for a direct (primary) effect which is independent of crystal strain. In addition, if the crystal develops macroscopic strain under the influence of the field, there will be a change in index through the elastooptic effect. All solids exhibit an elastooptic effect and all solids are strained by an electric field, either through the converse piezoelectric effect (strain$=dE$) or through electrostriction (strain $=\gamma E^2$). The resultant secondary effect can be shown to depend on crystal symmetry in the same way as the direct effect. It is not, however, necessary that the direct and secondary effects have the same algebraic sign: the overall electrooptic effect in a crystal strained by the field can be larger or smaller than the direct effect alone [9]. If the driving fields are at a frequency corresponding to an acoustic resonance of the material, the secondary effect may be as large as, or larger than, the direct one because of large strain amplitudes. At sufficiently high frequencies the material cannot strain macroscopically and only the direct effect is important.

If the material has a center of symmetry, reversing the sense of the applied field E does not change the physical situation and, in particular, $1/n^2$ will be independent of

the sign of E. Terms of odd power in E in (1) will change sign, however, so the coefficients of these terms must vanish in centrosymmetric materials. Only noncentrosymmetric (piezoelectric) crystals can produce a linear effect. There is, of course, no restriction on the terms of even powers of E.

In order to carry the discussion further, it must be recognized that the index of refraction of a crystal depends on optical polarization relative to crystal axes, that the electric field is a vector quantity, and that the r and R coefficients must reflect the crystal symmetry. The optical properties of a crystal are frequently described in terms of the *index ellipsoid* (or *indicatrix*). The equation of this surface is

$$\frac{x_1^2}{n_1^2} + \frac{x_2^2}{n_2^2} + \frac{x_3^2}{n_3^2} = 1, \qquad (2)$$

where the coordinates x_i are parallel to the axes of the ellipsoid and n_i are the principal refractive indices. The properties of the indicatrix can be seen from a simple example. If a wavefront has its normal in the x_3 direction, then we consider the ellipse formed by the indicatrix and the $x_3=0$ plane. The wave has components polarized along x_1 and x_2 and indices given by the semi-axes of the ellipse (n_1 and n_2). In a more general case, the wave normal direction can be chosen arbitrarily and the two indices obtained as the semi-axes of the elliptical section perpendicular to the arbitrary direction.

If an electric field is applied to a crystal, the general equation of the indicatrix can be written as

$$\sum_{i,j,k,l} \left(\frac{1}{n_{ij}^2} + z_{ijk}E_k + R_{ijkl}E_kE_l \right) x_ix_j = 1 \qquad (3)$$

where the indices run from 1 to 3. The z_{ijk} and R_{ijkl} are linear and quadratic electrooptic tensor components, respectively. The indices i, j can be interchanged, as can k and l, so the usual contraction can be made: $r_{mk} \leftrightarrow z_{(ij)k}$ and $R_{mn} \leftrightarrow R_{(ij)(kl)}$, where m and n run from 1 to 6 and m is related to (ij) and n to (kl) as follows: $1 \leftrightarrow 11$, $2 \leftrightarrow 22$, $3 \leftrightarrow 33$, $4 \leftrightarrow 23$, $5 \leftrightarrow 13$, $6 \leftrightarrow 12$. The linear electrooptic matrix for a specific crystal class has the same form as the inverse of the piezoelectric coefficient matrix, but no factors of 2 appear in the electrooptic case [8], [10].

The entire formalism outlined above could have been written using dielectric polarization rather than electric field as a basis. In fact, such a description would be more appropriate since polarization is a property of the physical medium in which the effect occurs. In the case of the linear effect, however, the electric field dependence (i.e., r as defined above) is usually given, since it is most nearly the measured quantity, and this practice is followed here. In the quadratic case there is not as much precedent in the literature and, instead of terms such as $R_{mn}E_kE_l$ in (3), the polarization terms $g_{mn}P_kP_l$ are used. Since $P_k = \epsilon_0(\epsilon_{ki} - 1)E_i$, where ϵ_{ki} is the permittivity, the descriptions can be interchanged. In the situations considered here only diagonal elements, $\epsilon_{ii} \equiv \epsilon_i$, are required. For

---

[1] A clear, concise description of the properties of crystals alluded to here is given in [8].

simplicity, we denote RF and optical dielectric constants by ϵ and n^2, respectively.

It may be useful to consider an example of the use of the indicatrix in describing the electrooptic effect. In a crystal in which only the linear effect is appreciable, if a wave normal along χ_3 is chosen and the applied field is along χ_1, then the equation for the $\chi_3 = 0$ section of the indicatrix is

$$\left(\frac{1}{n_1{}^2} + r_{11}E_1\right)x_1{}^2 + \left(\frac{1}{n_2{}^2} + r_{21}E_1\right)x_2{}^2 + 2r_{61}E_1x_1x_2 = 1. \qquad (4)$$

Depending on symmetry, n_1 and n_2 may be equal and one or more r_{mi} may be equal or may vanish. If the term in r_{61} is disregarded for the moment, it is seen that the effect of $r_{11}E_1$ is to change the index of refraction for a wave polarized along x_1, so that the new index $(n_1 + \Delta n_1)$ is given by

$$\frac{1}{(n_1 + \Delta n_1)^2} = \left(\frac{1}{n_1{}^2} + r_{11}E_1\right).$$

Now, Δn_1 is a small quantity compared to n_1, so to good accuracy we have

$$\Delta n_1 = -\frac{n_1{}^3 r_{11}E_1}{2}.$$

Similarly, for the wave polarized along x_2,

$$\Delta n_2 = \frac{-n_2{}^3 r_{21}E_1}{2}.$$

The x_3-directed wave at frequency ω polarized along x_2 can be described by

$$A_2 \exp i\left(\omega t - \frac{2\pi n_2 x_3}{\lambda_0}\right),$$

where λ_0 is the free space wavelength. An x_1 polarized wave can be described similarly. Clearly the phase of the wave depends upon the value of n_2. After traversing a length $x_3 = L$, the change in phase due to the electrooptic effect is

$$-\frac{2\pi L}{\lambda_0}\left[n_2 - (n_2 + \Delta n_2)\right] = -\frac{\pi n_2{}^3 r_{21}E_1 L}{\lambda_0} \equiv \eta. \quad (5)$$

If E_1 varies sinusoidally with time, then the phase delay of the wave varies sinusoidally with peak value η, the modulation index. In this sense, the electrooptic effect leads most naturally to phase modulation.

In order to obtain amplitude modulation it is necessary to consider interference of phase modulated components. For example, in the same crystal, if the incident beam is plane polarized at 45° to x_1 and x_2 so that $A_1 = A_2 = A/\sqrt{2}$, the emergent beams are, respectively,

$$\frac{A}{\sqrt{2}} \exp i\left(\omega t - \frac{2\pi n_1 L}{\lambda_0}\right)$$

and

$$\frac{A}{\sqrt{2}} \exp i\left(\omega t - \frac{2\pi n_2 L}{\lambda_0}\right).$$

Since free space is isotropic we can choose any pair of orthogonal polarizations to describe the resultant combination. In particular, components parallel and normal to the incident polarization have amplitudes

$$A \cos \frac{\pi L}{\lambda_0}(n_1 - n_2 + \Delta n_1 - \Delta n_2) = A \cos(\Gamma/2)$$

$$\qquad (6)$$

and

$$A \sin \frac{\pi L}{\lambda_0}(n_1 - n_2 + \Delta n_1 - \Delta n_2) = A \sin(\Gamma/2)$$

where Γ is the *retardation*.

If $\pi L(n_1 - n_2)/\lambda_0$ is an even multiple of $\pi/2$, the parallel amplitude is zero except for the part contributed by the electrooptic effect. Its value is

$$\pm \cos\left[\frac{\pi L}{2\lambda_0}(n_2{}^3 r_{21} - n_1{}^3 r_{11})E_1\right]$$

$$\approx 1 - \frac{1}{2}\left[\frac{\pi L}{2\lambda_0}(n_2{}^3 r_{21} - n_1{}^3 r_{11})E_1\right]^2,$$

where the approximation is for small values of the argument. The modulation contains only even powers of E and is relatively small. If $\pi L(n_1 - n_2)/\lambda_0$ is an odd multiple of $\pi/2$, the roles of the polarizations are interchanged and the above is true for the perpendicular component. If $\pi L(n_1 - n_2)/\lambda_0$ is an odd multiple of $\pi/4$, the amplitude of either parallel or perpendicular polarization is

$$\pm \frac{1}{\sqrt{2}}\left(1 \pm \sin \frac{\pi L}{2\lambda_0}(n_2{}^3 r_{21} - n_1{}^3 r_{11})E_1\right)$$

$$\approx \pm \frac{1}{\sqrt{2}}\left(1 \pm \frac{\pi L}{2\lambda_0}(n_2{}^3 r_{21} - n_1{}^3 r_{11})E_1\right)$$

and the amplitude modulation is approximately linear in E_1 and larger than in the previous case. The condition that $\pi L(n_1 - n_2)/\lambda_0$ be an odd multiple of $\pi/4$ is sometimes referred to as *optical bias*. A fixed external compensator with axes of birefringence along x_1 and x_2 can be used to obtain this quarter wave bias if $n_1 = n_2$. If $n_1 \neq n_2$, the temperature dependence of birefringence in the crystal itself allows one to meet the condition, but at the same time it can impose stringent requirements on temperature control of the crystal in order to keep the optical bias in the proper range. Alternatively, the temperature may be fixed and a variable external compensator employed.

Referring back to (4), the third pertinent coefficient for x_3 directed waves with E_1 applied is a skew coefficient r_{61}. One effect of this coefficient is to rotate the axes of the elliptical cross section by an angle α, where

$$\tan 2\alpha = \frac{2r_{61}E_1}{\dfrac{1}{n_1{}^2} - \dfrac{1}{n_2{}^2} + (r_{11} - r_{21})E_1}. \quad (7)$$

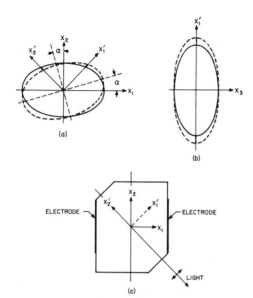

Fig. 1. (a) $x_3=0$ cross section of the indicatrix with x_1 directed field (dashed) and no field (solid). The effects of $r_{11}E_1$ and $r_{21}E_1$ have been omitted. The change in $1/x_1'^2$, where x_1' is the radius vector, is a maximum when x_1' is at 45° to x_1. The change in x_1' (and hence in index of refraction) is a maximum at angles near 45° unless the natural birefringence is very large as in the sketch. (b) $x_2'=0$ cross section of the indicatrix with x_1 directed field (dashed) and no field (solid). The change in x_1' axis is due to $r_{61}E_1$. The effect of $r_{31}E_1$ on the x_3-axis has been neglected. (c) Crystal cut for measurement or use of r_{61} coefficient. Propagation direction and polarization of the light are indicated.

If the natural birefringence (n_1-n_2) is appreciable, this angle is very small. The lengths of the axes (and hence the indices) are changed by an amount of second order in $r_{61}E_1$ that is usually negligible. However, the radius vector of the ellipse at 45° to x_1 and x_2 is changed linearly in E_1 by an amount $n'^3 r_{61}E_1/2$, where n' is an effective index intermediate between the principal indices. The situation is indicated in Fig. 1(a). Since this radius is not an axis of the elliptical section normal to x_3, it does not correspond to the polarization of a principal wave propagating along x_3. The first-order effect does occur for a wave propagating in some other direction for which the radius vector at 45° to x_1 and x_2 is a principal axis. For example, if the wave normal is at 45° to $-x_1$ and x_2 [along x_2' in Fig. 1(a)], the perpendicular section is

$$\left\{\frac{1}{2}\left(\frac{1}{n_1^2} + \frac{1}{n_2^2}\right) + r_{61}E_1 + \frac{(r_{11}+r_{21})}{2}E_1\right\}x_1'^2$$

$$+ \left\{\frac{1}{n_3^2} + r_{31}E_1\right\}x_3^2 + \sqrt{2}\,r_{51}E_1 x_1'x_3$$

$$+ \sqrt{2}\,r_{41}E_1 x_1'x_3 = 1 \qquad (8)$$

where x_1' is at 45° to x_1 and x_2. The axes of this ellipse are determined in large part by the birefringence of the crystal. The effect of r_{41} and r_{51} is a small rotation of axes, as in (7). The alteration of the $x_2'=0$ section described in (8) is indicated in Fig. 1(b). A crystal cut which allows observation of the linear change $r_{61}E_1$ is shown in Fig. 1(c). From (8) we see that r_{11} and r_{21} also cause first-order effects. If we had chosen the x_2' polarization of the x_1'

directed wave the signs of r_{11} and r_{21} terms would remain unchanged, but the r_{61} term would be reversed. Thus, for one case the effects of the coefficients add and in the other they subtract.

The fact that specific elements of the electrooptic coefficient matrix (including skew components) can be used in producing phase modulation has been shown above, and the fact that the interference of two such waves produces amplitude modulation has also been indicated. In doing this, the restrictions imposed on the matrix elements by symmetry have not been used. At this point, one effect of crystal symmetry will be mentioned specifically. In crystals with a threefold axis it is possible to rotate a birefringent section of constant retardation by rotation of the applied field, which in turn allows construction of single sideband modulators and frequency shifters [11]. If a wave normal is directed along x_3, in this case a threefold axis, then the perpendicular section is circular in the absence of an applied field; i.e., $n_1=n_2$. Moreover, symmetry requires that

$$r_{12} = r_{61} = -r_{22}$$
$$r_{21} = r_{62} = -r_{11}. \qquad (9)$$

If fields E_1 and E_2 are applied, the equation of the $x_3=0$ section is

$$\left[\frac{1}{n_1^2} + (r_{11}E_1 - r_{22}E_2)\right]x_1^2$$

$$+ \left[\frac{1}{n_1^2} + (r_{22}E_2 - r_{11}E_1)\right]x_2^2$$

$$+ 2[-r_{22}E_1 - r_{11}E_2]x_1x_2 = 1. \qquad (10)$$

If we choose new axes x_1', x_2' along the axes of the ellipse in the $x_3=0$ plane, (10) can be written

$$\left[\frac{1}{n_1^2} + (r_{11}^2 + r_{22}^2)^{1/2}E\right]x_1'^2$$

$$+ \left[\frac{1}{n_1^2} - (r_{11}^2 + r_{22}^2)^{1/2}E\right]x_2'^2 = 1, \quad (11)$$

where E is the applied field with components $E_1 = E\cos\Phi$, $E_2 = E\sin\Phi$. As shown in [11] the primed axes are rotated relative to the crystal axes through an angle

$$\theta = -\frac{1}{2}\left[\Phi + \arcsin\frac{r_{22}}{\sqrt{r_{11}^2 + r_{22}^2}}\right]. \qquad (12)$$

From (11) we note that a field of fixed magnitude causes a specific amount of birefringence (and hence a fixed magnitude of retardation Γ) in the $x_3=0$ plane, regardless of field orientation in this plane. From (12) we see that the axes of the birefringence rotate through an angle $\theta = -\frac{1}{2}\Phi$ when the field rotates through an angle Φ.

The effect of a rotating birefringent plate on an incident circularly polarized wave is merely summarized here in order to indicate how single sideband modulation can be effected [12]. The emergent wave has a component in the

TABLE I
KDP-ADP TYPE: POINT GROUP ($\bar{4}$2m) ABOVE T_c

	T_c			r_{63}	r_{41}	n_3	n_1	ϵ_3	ϵ_1	tan δ_3	tan δ_1
KH$_2$PO$_4$(KDP)	123	[2]	(T)	−10.5 [5]	+8.6 [5]	1.47 [27]	1.51 [27]	(T) 21 [28]	42 [28]	(T) see [76], [32]	
			(S)	9.7 [26]				(S) 21 [29]	44 [29]	(S) 7.5×10^{-3}[29]	4.5×10^{-3} [29]
KD$_2$PO$_4$(DKDP)	222	[16]	(T)	26.4 [16]	8.8 [24]	1.47 [27a]	1.51 [27a]	(T) 50 [16]		(T) see [33]	
								(S) 48 [14]	58 [14]	(S) 1.0×10^{-1} [14]	2.5×10^{-2} [14]
KH$_2$AsO$_4$(KDA)	97	[2]	(T)	10.9 [24]	12.5 [24]	1.52 [24]	1.57 [24]	(T) 21 [28]	54 [28]		
								(S) 19 [14]	53 [14]	(S) 8.0×10^{-3} [14]	7.5×10^{-3} [14]
RbH$_2$AsO$_4$(RDA)	110	[2]	(T)	13.0 [24]		1.52 [24]	1.56 [24]	(T) 27 [31]	41 [31]	(T) see [31]	
								(S) 24 [31]	39 [31]	(S) 5×10^{-2} [31]	3×10^{-2} [31]
NH$_4$H$_2$PO$_4$(ADP)	148*	[2]	(T)	−8.5 [5]	+24.5 [24]	1.48 [27]	1.53 [27]	(T) 15 [28]	56 [28]	(T) see [76]	
			(S)	5.5 [5]				(S) 14 [14]	58 [14]	(S) 6.0×10^{-3} [14]	7.0×10^{-3} [14]

* Antiferroelectric transition temperature; T_c in °K; r_{mi} in 10^{-12}m/V. (T) =constant stress, (S) =constant strain; refractive index at 0.546 μ; tan δ(S) at ~10^{10} c/s

original sense of circular polarization of amplitude cos $\Gamma/2$ and phase independent of θ. In addition, there is a component of opposite sense whose amplitude is sin $\Gamma/2$ and whose phase is proportional to 2θ (or $-\Phi$). Thus, if the field rotates at a rate ω_m, the phase of the wave with altered sense of circular polarization varies as $-\omega_m t$ and the instantaneous frequency of this wave is changed by ω_m. The direction of change depends on the relative direction of field rotation and incident circular polarization. The amplitude of the frequency shifted component is constant in time.

III. MATERIALS

The electrooptic materials are grouped according to their general crystallographic and physical properties as follows: 1) KDP, ADP, and their isomorphs, 2) ABO$_3$ crystals similar to perovskites, 3) AB-type semiconductors with cubic or hexagonal ZnS structure, and 4) various miscellaneous crystals. The tabulations of electrooptic coefficients may give two or more conflicting values for the same crystal. Oftentimes, poor materials are the cause. The reader should consult the original works to decide which result is more reliable.

KDP-ADP Type

Potassium dihydrogen phosphate (KDP) and ammonium dihydrogen phosphate (ADP) are the most widely known electrooptic crystals. Their general properties are reviewed elsewhere [2], [3]. They are grown at room temperature from a water solution and are free of the strains often found in crystals grown at high temperature. Excellent crystals as large as 5 cm in any dimension can be obtained commercially at nominal cost. Although the crystals are water soluble and fragile, they can be handled, cut, and polished without difficulty. The resistivity of these crystals is typically 10^{10} ohm-cm [13].

Both KDP and ADP belong to the piezoelectric point group $\bar{4}$2m at room temperature where they are normally used. Below the Curie temperature T_c, KDP type crystals become ferroelectric and, below the transition temperature, ADP type crystals become antiferroelectric. No electrooptic measurements have been made below T_c because of the experimental complexity involved.

The atoms K, H, P in KH$_2$PO$_4$ can be replaced by some of the atoms from corresponding columns in the periodic table without changing the crystal structure. A dramatic change in dielectric properties occurs when H is replaced by deuterium. For partially deuterated KDP, KD$_{2x}$H$_{2(1-x)}$PO$_4$, the Curie temperature is given [14] approximately by

$$T_c \approx (123 + 106x)°K. \qquad (13)$$

The KDP-ADP isomorphs for which electrooptic coefficients have been measured are listed in Table I along with pertinent electrical and optical properties. Additional data will often be found in the references given. Measurements at constant stress are noted by (T) and those at constant strain (high frequency) by (S). The only nonvanishing r_{mi} coefficients are $r_{41}=r_{52}$ and r_{63}. At 0.633 μ, optical loss [15] in KDP is $\frac{1}{2}$ dB/m, which is about as good as that found in the best fused quartz. The refractive indices and ultraviolet absorption, which are associated with electronic transitions in the oxygen ions, are about the same for all isomorphs. The crystals are transparent for wavelengths as short as 0.2 μ [16]–[18]. The infrared absorption is the result of hydrogen vibrations, the frequencies of which are approximately inversely proportional to the square root of the proton mass. Thus, the low-frequency absorption edge for the deuterated salt occurs at roughly $\sqrt{2}$ times the wavelength for undeuterated salts: KDP, 1.55μ [16]; ADP, 1.4 μ [19]; DKDP, 2.15 μ [19].

The quantity $r_{63}/(\epsilon_3-1)$ is roughly the same for all isomorphs and, despite the rapid increase in ϵ_3 near T_c, is independent of temperature [2], [3]. It is possible to obtain larger r_{63} and, hence, lower modulating voltage by, operating near the Curie temperature or by choosing an isomorph with T_c closer to room temperature so as to increase ϵ_3. However, the loss tangent also increases and in such a way as to make this expedient unattractive in some applications [14], [20]. The electrooptic coefficient r_{63} is practically independent of wavelength in the transparent region for KDP and ADP [5], [21], [22].

In the absence of an applied electric field, KDP-type crystals are uniaxial: light polarized parallel to or normal to the z-axis travels as a principal wave with refractive index n_3 or n_1, respectively. When a field E_3 is applied along the x_3-axis, the principal axes become x_3 and x_1' and x_2', at 45° to the x_1 and x_2 axes. The refractive index for x_3-polarization remains n_3 while,

$$n_{x_1'} \approx n_1 - \tfrac{1}{2}n_1{}^3r_{63}E_3, \quad n_{x_2'} \approx n_1 + \tfrac{1}{2}n_1{}^3r_{63}E_3.$$

Light polarized along x_1' or x_2' traveling through a crystal of length L experiences a phase modulation with index

$$\eta = \pi n_1^3 r_{63} E_3 L / \lambda. \qquad (14)$$

Light traveling along x_3 and initially polarized along x_1 or x_2 experiences a phase retardation

$$\Gamma = 2\eta. \qquad (15)$$

The voltage $E_3 L$ required to produce a retardation of π radians is called the *half-wave voltage*

$$V_{1/2} = \lambda / 2 n_1^3 r_{63}. \qquad (16)$$

For a wide area light shutter [4] or amplitude modulator both applied field and light path are along x_3. Because of natural birefringence in directions off the x_3-axis, the angular aperture is severely restricted for thick samples. Methods for increasing the aperture are described by Billings [4]. Natural birefringence is not normally a problem with a well-collimated optical maser beam. Phase or amplitude modulation with field normal to the optical path may be accomplished with electric field along x_3, optical polarization along x_1', and optical path along x_2'.

Small strains in the crystal can partially destroy the relative phase relationships required to produce amplitude modulation [119]. Normally, KDP-type crystals as grown are quite strain-free. However, despite the relatively low loss tangent, the crystal is heated throughout its volume by the modulating field but cooled only on its surface. The resultant thermal gradient produces strains that are sufficiently great when the power dissipated in the crystal is greater than a watt, independent of crystal dimensions, to make amplitude modulation ineffective [23].

A field applied along x_1 rotates the indicatrix through a small angle about x_1. The semi-axis along x_1 remains unchanged and the other axes experience changes of order $(r_{41} E_1)^2$. The situation is similar to that discussed in Section II, where the effect of the skew coefficient r_{61} was mentioned. To obtain first-order changes in refractive index [5], [24], light should be polarized in the $x_2 x_3$ plane along x_2'' or x_3'', at 45° to x_2 and x_3, and propagated in the x_3'' or x_2'' direction, respectively. The $x_2'' = 0$ section of the indicatrix is

$$\frac{x_1^2}{n_1^2} + \left[\frac{1}{2} \left(\frac{1}{n_1^2} + \frac{1}{n_3^2} \right) + r_{41} E_1 \right] x_3''^2 = 1$$

and the modulation index for x_3'' polarization

$$\eta = 2\pi\sqrt{2} \left(\frac{1}{n_1^2} + \frac{1}{n_3^2} \right)^{-3/2} r_{41} E_1 L / \lambda$$

$$\approx \pi n_1^3 r_{41} E_1 L / \lambda. \qquad (17)$$

Since r_{41} at constant stress is large in ADP, it has been suggested that a modulator with E along x_1 might be advantageous [10]. However, inasmuch as r_{41} at constant strain is not known, it is not clear that this configuration will be useful at high frequency. For intensity modulation using this configuration, there is the added complication that the ordinary and extraordinary ray paths are not parallel. A two crystal method for overcoming this difficulty is described by Ley [25], [30].

Perovskite Family

The large group of crystals with structure resembling that of the mineral perovskite, $CaTiO_3$, form the perovskite family [34]. Of particular interest are the oxides $A^{2+}B^{4+}O_3$ and $A^{1+}B^{5+}O_3$, which often exhibit ferroelectric behavior. Their properties are discussed in detail elsewhere [2], [3]. In general, they are insoluble in water and are more rugged and have larger refractive index and dielectric constant than KDP. As a rule, the oxides are transparent between 0.4 and 6 μ [35]–[37]. The infrared absorption is caused largely by vibrations of the BO_6 octahedra and the ultraviolet absorption by electronic transitions in the oxygen ions.

Perovskites may exist in several forms, with different point symmetries, that are derived from the ideal cubic perovskite structure by continuous lattice distortions. The cubic form, which is often the high temperature phase, belongs to the nonpiezoelectric, nonferroelectric point group m3m. In the ferroelectric phase, the crystals of interest here are tetragonal 4mm, with the c-axis along one of the original cube edges, or rhombohedral 3m, with the c-axis along one of the cube body diagonals. Both phases are piezoelectric. The cubic (paraelectric) and ferroelectric phases of various perovskites are discussed separately below.

Cubic: $BaTiO_3$, $SrTiO_3$, $KTaO_3$, $K(Ta, Nb)O_3$: Since the cubic phase is centrosymmetric the change in refractive index is a quadratic function of applied field or dielectric polarization. It is convenient to use the polarization P as a basis because the corresponding electrooptic coefficients g_{mn} are then insensitive to temperature and optical wavelength far from the band edges [35]. As discussed in Section II,

$$\Delta\left(\frac{1}{n_m^2} \right) = \sum g_{mn} P_j P_k \qquad (18)$$

where m, n run from 1 to 6 and j, k from 1 to 3. For m3m the only nonvanishing components are $g_{11} = g_{22} = g_{33}$, $g_{12} = g_{13} = g_{23} = g_{32} = g_{31} = g_{21}$, and $g_{44} = g_{55} = g_{66}$. To obtain a linear effect, which is both larger in magnitude and more suitable for most applications, it is customary to induce a large dc bias polarization P_{dc} along with the much smaller RF modulation polarization $(\epsilon - 1)\epsilon_0 E_i$. In effect, P_{dc} a vector quantity removes the center of symmetry. If P_{dc} is along a cube diagonal the symmetry is lowered to 3m; if P_{dc} is along a cube edge the symmetry becomes 4mm.

If P_{dc} has only an x_3-component,

$$\Delta(1/n_1^2) = \Delta(1/n_2^2) = g_{12} P_{dc}(P_{dc} + 2\epsilon\epsilon_0 E_3)$$

$$\Delta(1/n_3^2) = g_{11} P_{dc}(P_{dc} + 2\epsilon\epsilon_0 E_3)$$

$$\Delta(1/n_4^2) = g_{44} P_{dc}\epsilon\epsilon_0 E_2, \quad \Delta(1/n_5^2) = g_{44} P_{dc}\epsilon\epsilon_0 E_1$$

$$\Delta(1/n_6^2) = 0, \qquad (19)$$

where quantitites quadratic in RF polarization are neglected and it is assumed that $\epsilon \gg 1$. As the temperature is reduced toward the Curie point T_c, ϵ becomes large and large values of P_{dc} can be induced with modest fields.

Values of g_{mn} for several cubic perovskites are given in Table II. The quantity $(g_{11}-g_{12})$ appears when intensity modulation rather than phase modulation measuring techniques are employed. Only constant stress (low frequency) values have been reported. Clamping on g_{mn} (as well as ϵ) may reduce the effect by as much as 50 percent [40], [41].

Although g_{mn} is about the same for all the perovskites in Table II, $SrTiO_3$ and $KTaO_3$ are not particularly interesting because their Curie points are well below room temperature. On the other hand, $BaTiO_3$ must be heated in order to operate above T_c where the effect is large. A material with T_c near room temperature can be synthesized by forming a solid solution of two materials, one with T_c above and the other with T_c below room temperature. Such a material is $KTa_{.65}Nb_{.35}O_3(KTN)$. However, despite considerable effort, it has not been possible to grow these mixed crystals consistently with sufficient uniformity, optical quality, and electrical resistivity. Nevertheless, KTN remains interesting because of the large electrooptic effects that have been observed experimentally [43].

Several difficulties are inherent in the use of biased perovskites slightly above T_c. The properties of the material, particularly ϵ, which varies as $(T-T_c)^{-1}$, are very temperature sensitive so that T must be carefully controlled [40]. Further, even with relatively high resistivity and low photoconductivity, the dc bias field leads to space charge effects that eventually reduce the internal biasing field. A two-crystal, ac biasing scheme to overcome the latter difficulty has been proposed but not tested [42]. In connection with the former difficulty, note that if the temperature is reduced below T_c and the crystal poled into a single ferroelectric domain, then the spontaneous polarization P_s can take the part of P_{dc}. Although P_s is several times greater than the induced P_{dc}, ϵ_3 in the ferroelectric state will be much smaller than it is just above T_c. As compensation, however, ϵ_3 and the corresponding electrooptic coefficients are insensitive to T well below T_c [2]. Related single domain ferroelectrics are considered below.

Ferroelectric: $BaTiO_3$, $LiNbO_3$, $LiTaO_3$: Barium titanate is the most widely studied perovskite ferroelectric [2]. It is available in crystals with good optical and electrical properties in the form of thin platelets (typically $\frac{1}{2}$mm $\times$ 10mm $\times$ 10mm) grown by the Remeika method [47] and faceted boules (10mm $\times$ 10mm $\times$ 10mm) as grown by Linz [38]. The crystals are in a tetragonal phase (4mm) between $0°$ and $120°C$,[2] and can be poled into a single domain with the polar axis along one of the cube edges existing above $120°C$.

The only nonvanishing coefficients are $r_{13}=r_{23}$, r_{33}, and $r_{42}=r_{51}$. These coefficients are listed in Table III with

$$r_c = r_{33} - \left(\frac{n_1}{n_3}\right)^3 r_{13}, \qquad (20)$$

the quantity observed in intensity modulation measurements. Both r_{13} and r_{33} are temperature insensitive, but r_{42} increases rapidly as T approaches $0°C$, the tetragonal-to-orthorhombic transition temperature [41], [48]. At room temperature, r_{42} at constant strain is about 8×10^{-10} m/V which is about 30 times r_{33}. The temperature dependence of r_{42}, and r_{13} and r_{33} is similar to that of ϵ_1 and ϵ_3, respectively. To take advantage of the large r_{42}, it is necessary to employ a special crystal cut [49] similar to that in Fig. 1(c).

The wide disparity in clamped and free values of r_{mi} and ϵ_j indicated in Table III is due in part to the large piezoelectric coupling coefficient [28]. The large coupling coefficient and low acoustic loss mean that high-order acoustic resonances can be very troublesome in modulator design [50]. Further, large natural birefringence (n_1-n_2) implies that intensity modulators must have closely parallel end faces.

The structures of $LiNbO_3$ [51] and presumably $LiTaO_3$ [34], [51] are rhombohedral, with point group 3m below $1200°C$ and $620°C$, respectively. The structure is not strictly perovskite but it is very nearly so [34]. The electrooptic tensor has the nonvanishing components: $r_{13}=r_{23}$, r_{33}, $r_{22}=-r_{12}=-r_{61}$, and $r_{42}=r_{51}$. Crystals were first grown from a flux [52] and more recently were pulled from the melt [52a]. Techniques have now been developed that permit large crystals (typically $1 \times 1 \times 3$ cm) of good optical and electrical quality to be pulled from the melt and poled into a single domain while near T_c [51], [53]. Because of the large T_c, considerable mechanical energy would be required to depole these materials at room temperature. Hence, unlike $BaTiO_3$ with $T_c = 120°C$, these crystals may be cut, polished, and roughly handled without creating additional domains.

The properties of $LiNbO_3$, $LiTaO_3$, and $BaTiO_3$ are compared in Table II. The measurements [49], [50] indicate that $r_{13}/r_{33} > 0$ for all three materials.

Where the electrooptic properties are comparable, $LiNbO_3$ and $LiTaO_3$ are preferable to $BaTiO_3$ because of the relative ease in handling and availability of crystals. Further, it has been observed that piezoelectric resonance effects for c-axis fields are smaller in $LiNbO_3$ and $LiTaO_3$ [49], [54]. In addition the dielectric Q of $BaTiO_3$ is lower.

Recently [55] it has been observed that when exposed to intense laser beams $LiNbO_3$, $LiTaO_3$, and $BaTiO_3$ develop refractive index inhomogeneities which scatter the beam.

AB-Type Semiconductors

The group of binary compounds which crystallize in either the cubic ($\bar{4}3m$) zincblende structure or the hexagonal (6mm) wurtzite structure are similar in many re-

---

[2] T_c appears to be different for Remeika (120°C) and Linz (131°C) crystals.

TABLE II
Cubic Perovskites: m3m

	T_C	g_{11}	g_{12}	$g_{11}-g_{12}$	g_{44}	n	ϵ, tan δ
BaTiO$_3$	401	+0.12[60]	$-\lvert<0.01\rvert$[60]	+0.13 [40] 0.10 [38] (T)0.088[41] (S)0.031[41]		2.4	see [2], [38], [41]
SrTiO$_3$	low			+0.14		2.38	see [39], [44]
KTaO$_3$	4			+0.16	+0.12	2.24	see [45], [46]
KTa$_{.65}$Nb$_{.35}$O$_3$	~283	+0.136	−0.038	0.174	+0.147	2.29	see [40]

T_c, g_{mn}, n from [40] except where noted; T_c in degrees K; g_{mn} in m^4/C^2; all measurements at constant stress except where noted; (T) =constant stress, (S) =constant strain.

TABLE III
Ferroelectric Perovskites

	Sym	T_c	r_{13}	r_{33}	$r_{51}=r_{42}$	r_{22}	r_c	$n_1=n_2$	n_3	$\epsilon_1=\epsilon_2$	ϵ_3	tan δ
BaTiO$_3$	4mm	393			(T) 1640 [48] (S) 820 [41]		(T) 108 [48] (S) 23 [41]	(a) 2.44 [36]	(a) 2.37 [36]	(T) 3000 [28] (S) 2000 [41]	170 [28] 100 [41]	see [2]
			(c) (S) 8 [50]	(c) (S) 28 [50]			(S) 19 [50]	(b) 2.39 [50]	(b) 2.33 [36]			
LiNbO$_3$	3m	1470	(c) (S) 8.6 [49]	(c) (S) 30.8 [49]	(S) 28 [49]	(S) 3.4 [49]	(S) 21 [49] (T) 7 [58]	(b) 2.286 [56]	(c) 2.200 [56]	(S) 43 [59]	(S) 28 [49, 59] (T) 78 [51]	see [59]
							(T) 19 [58]				(T) 32 [51]	
LiTaO$_3$	3m	890	(c) (S) 7 [9]	(c) (S) 30.3 [9]			(S) 24 [9]				(T) 47 [9]	
			(S) 7.9 [49]	(S) 35.8 [49]	(S) 20 [49]	(S) ≈1 [49]	(S) 28 [49] (T) 22 [9]	(b) 2.176 [57]	(b) 2.180 [57]		(S) 43 [9]	

(a) at 546 μ, (b) at 633 μ, (c) $r_{13}/r_{33}>0$; T in °K, r_{mi} in 10^{-12} m/V, (T) =constant stress, (S) =constant strain.

TABLE IV
AB-Type Semiconductors

	Sym	r_{mi}	λ	n_i	λ	ϵ_i
ZnO	6mm	(S) $r_{33}=2.6$ (S) $r_{13}=1.4$ $r_{33}/r_{13}<0$	.63 [61] .63 [61] [61]	$n_3=2.123$ $n_2=n_1=2.106$ $n_3=2.015$ $n_3=n_1=1.999$	.45 [57] .45 .60 [57] .60	$\epsilon\approx8.15$ [62]
ZnS	$\bar{4}$3m	(T) $r_{41}=1.2$ 2.0 2.1	.40 [63] .546 .65	$n_0=2.471$ 2.364 2.315	.45 [57] .60 .8	(T) 16 [64] (S) 12.5 [64] 8.3 [65]
	6mm [see 66]	(S) $r_{33}=1.85$ (S) $r_{13}=.92$ $r_{33}/r_{13}<0$	.63 [66] .63 [66] [66]	$n_3=2.709$ $n_2=n_1=2.705$ $n_3=2.368$ $n_2=n_1=2.363$	.36 [67] .36 .60 [67] .60	
ZnSe	$\bar{4}$3m	(T) $r_{41}=2.0$	.546 [68]	$n_0=2.66$	.546 [69]	9.1 [65] 8.1 [70]
ZnTe	$\bar{4}$3m	(T) $r_{41}=4.55$ 3.95 (S) $r_{41}=4.3$	.59 [71] .69 [71] .63 [72]	$n_0=3.1$ 2.91	.57 [71] .70 [71]	10.1 [65]
CuCl	$\bar{4}$3m	(T) $r_{41}=6.1$ (T) $r_{41}=1.6$	[73] [74]	$n_0=1.996$ 1.933	.535 [75] .671	(T) 10 [76] (S) 8.3 [76] (S) 7.7 [77]
CuBr	$\bar{4}$3m	(T) $r_{41}=.85$	[74]	$n_0=2.16$ 2.09	.535 [75] .656	
GaP	$\bar{4}$3m	(S) $r_{41}=.5$ (S) $r_{41}=1.06$	[78] .63 [78]	$n_0=3.4595$ 3.315	.54 [57] .60 [57]	10 [83] 12 [78]
GaAs	$\bar{4}$3m	(T) $r_{41}=.27$ to 1.2 (S–T) $r_{41}=1.3$ to 1.5 (S) $r_{41}=1.2$ (T) $r_{41}=1.6$	1 to 1.8 [79] 1 to 1.8 [79] .9 to 1.08 [80] 3.39 & 10.6 [81]	$n_0=3.60$ 3.50 3.42 3.30	.90 [82] 1.02 [82] 1.25 [82] 5.0 [84]	(T) 12.5 [84] (S) 10.9 [84] (S) 11.7 [85]
CdS	6mm	(T) $r_{51}=3.7$ (T) $r_c=4$ (S) $r_{33}=2.4$ (S) $r_{13}=1.1$ $r_{33}/r_{13}<0$	.589 [86] .589 [86] .63 [72] .63 [72] [72]	$n_3=2.726$ $n_2=n_1=2.743$ $n_2=n_1=2.493$	.515 [67] .515 [67] .60 [67]	(T) $\epsilon_1=10.6$ [87] (T) $\epsilon_3=7.8$ [87] (S) $\epsilon_1=8.0$ [87] (S) $\epsilon_3=7.7$ [87]

r_{mi} in 10^{-12} m/V; λ in microns; (T) =constant stress, (S) =constant strain.

spects and are, therefore, treated together. For example, in both structures the four nearest neighbors of an $A(B)$ ion are $B(A)$ ions situated on the corners of a tetrahedron. One must look to the next nearest neighbors before the differences in the two structures become evident. Several of these compounds, including ZnS, CuCl, and CdS, can crystallize in either form. In fact, it is difficult to obtain single phase crystals of ZnS and CuCl, because the wurtzite structure is stable at high temperatures at which the crystals may be grown, and the cubic form is stable at room temperature. Some of the crystals in this group are of technical importance as semiconductors, phosphors, photodectectors, and, more recently injection laser materials. There is an extensive literature on the preparation, band structure, and electrical properties of many of these crystals, but as yet a relatively small amount of electrooptic information. All available electrooptic information is included in Table IV. With the single exception of gallium arsenide, the crystals whose electrooptic effect has been measured are transparent in at least part of the visible spectrum, and can be obtained in an insulating or semi-insulating form. Of the crystals listed in Table IV, only gallium arsenide, cadmium sulfide, and zinc telluride are readily available with dimensions of the order of a centimeter. Zinc sulphide, zinc oxide, and cuprous chloride are found in nature but are generally not of suitable quality. There is a wide variation in mechanical properties in the zincblende group even among those listed in the table. For example, the hardness (Mohs' scale) of cuprous chloride and bromide is 2 to 2.5, whereas gallium phosphide is about 7. All the cubic crystals listed have (110) cleavage, which fact is often used in preparing crystals for use although they can be cut and polished. The hexagonal crystals cleave less readily but are easily cut and polished. Cuprous chloride and bromide are attacked by moist air but the other materials can be used without protection in usual laboratory surroundings.

All the materials listed have large indices of refraction: from 2 for CuCl to 3.5 for GaAs near the band edge. As a result, the important $n^3 r$ product may be relatively large even though the coefficient r is not in itself large. The hexagonal crystals are, of course, birefringent. The maximum range of transparency for all the crystals listed in Table IV is fixed on the short wavelength end by the band edge and on the long wavelength end by reststrahl absorption. The transparency is limited, however, also by impurities and dislocations and in the long wavelength region free carriers cause appreciable absorption.

The 6mm crystals have the same r_{mi} matrix as the 4mm point group crystals: the coefficients are $r_{13} = r_{23}$, r_{33}, and $r_{42} = r_{51}$. Thus, the useful modulator configurations are the same (regarding symmetry) as those of room temperature BaTiO$_3$ described earlier.

The fact that the $\overline{4}3m$ crystals are cubic and hence optically isotropic makes them particularly attractive for modulation applications in which large acceptance angles are necessary. However, this also means that small amounts of residual birefringence, due to strain or nonuniformity, can cause an appreciable depolarizing of an

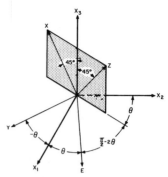

Fig. 2.   Change in orientation of indicatrix axes of a cubic crystal when applied field E lies in (001) plane. x_1, x_2, x_3 are the cubic crystal axes. The unchanged semi-axis of the indicatrix lies along y—also in (001) plane. The other semi-axes are along x and z. Maximum birefringence is obtained when light propagates along y.

optical beam. (In an anisotropic crystal, these perturbations have a relatively minor effect on the already large natural birefringence.) There is only one electrooptic coefficient ($r_{41} = r_{52} = r_{63}$) for cubic crystals, and the discussion of symmetry behavior which follows is applicable, therefore, to all cubics with a linear electrooptic effect. The zincblende crystals listed in Table IV are those for which values of r_{41} either at constant stress (T) or constant strain (S) have been reported.

Application of an electric field in a general direction relative to the crystal axes causes the initially isotropic crystals to become biaxial. There are three conditions which are of most practical importance. The first of these is a requirement of maximum induced birefringence, since this is desirable for an amplitude modulator. It can be shown that a field applied normal to one of the cube axes always produces the same (maximum) birefringence. In Fig. 2 an electric field is shown applied normal to the [001] direction (x_3) and at an angle θ to the [100] direction (x_1); i.e., the coordinate system (x_1, x_2, x_3) is parallel to the cubic crystal axes. The coordinate system (x, y, z) is used to locate the axes of the indicatrix, whose properties are altered by the electric field as follows: equal and opposite changes occur along x and z, respectively, with no change along y. In order to achieve the maximum birefringence, light is propagated in the y direction, which is in the (001) plane at an angle ($-\theta$) to the [100] axis. The change in index of refraction for light polarized along x, which is at an angle $\pi/4$ to the x_3 axis, is $+(n^3 r_{41} E)/2$ and for polarization along z, the change is $-(n^3 r_{41} E)/2$. The applied field is at right angles to the direction of propagation for $\theta = \pi/4$.

A second important crystal orientation is one which gives maximum change in index of refraction for a specific linear polarization and hence is most suitable for phase modulation. This configuration is one in which the electric field and the optical polarization are parallel and directed along a [111] crystal axis. A maximum change in index of refraction of $-n^3 r_{41} E/\sqrt{3}$ is then found. The advantage over using a field in a (100) plane is $2/\sqrt{3}$ and a given phase modulation can be obtained with $\frac{3}{4}$ of the power.

TABLE V
MISCELLANEOUS CRYSTALS

	Sym	r_{mi}	λ	n_i	λ
$Bi_4(GeO_4)_3$	$\bar{4}3m$	(T) $r_{41}=1.03$	.45 to .62 [91]	$n_0=2.07$	[91]
$C_6H_{12}N_4$—(HMT)	$\bar{4}3m$	(T) $r_{41}=4.18$	.365 to .60 [88]	$n_0=1.591$	.589 [88]
		(T) $r_{41}=.8$	.546 [90]		
		(T) $r_{41}=7.3$	.547 [92]		
Hauynite (mineral)	$\bar{4}3m$	(T) $r_{41}<.04$	[93]	$n_0=1.496$	
Langbeinites:					
$K_2Mg_2(SO_4)$	23	$r_{41}<.04$	[93]	$n_0=1.535$	[94]
$(NH_4)_2Cd_2(SO_4)_3$	23	(T) $r_{41}=.8$	.546 [102]	$n_0=1.57$	
$(NH_4)_2Mn_2(SO_4)_3$	23	(T) $r_{41}=.6$	.546 [102]	$n_0=1.57$	[94]
$NaClO_3$	23	(T) $r_{41}=.4$	.589 [75]	$n_0=1.515$	[96]
$Na_3SbS_4\cdot9H_2O$	23	(T) $r_{41}=5.66/n_0{}^3$	.42 [89]		
		(T) $r_{41}=5.62/n_0{}^3$	1.08 [89]		
Sodium Uranyl Acetate	23	(T) $r_{41}=.87$	.546 [97]	$n_0=1.507$	.546 [97]
$LiKSO_4$	6	(T) $r_c=1.6$	.546 [90]	$n_3\approx n_1=n_2=1.474$	.546 [90]
$LiNaSO_4$	3m	(T) $r_{22}<.02$	.546 [90]	$\begin{cases} n_3=1.495 \\ n_1=n_2=1.490 \end{cases}$	[96]
Tourmaline	3m	(T) $r_{22}=0.3$	.589 [75]	$\begin{cases} n_3=1.65 \\ n_1=n_2=1.63 \end{cases}$	[96]
$K_2S_2O_6$	32	(T) $r_{11}=0.26$	.546 [90]	$\begin{cases} n_3=1.1518 \\ n_1=n_2=1.456 \end{cases}$	[96] [90]
$Cs_2C_4H_4O_6$	32	(T) $r_{11}=1.0$	.546 [90]	$\begin{cases} n_3=1.546 \\ n_1=n_2=1.564 \end{cases}$	[90]
$SrS_2O_6\cdot4H_2O$	32	(T) $r_{11}=0.1$	.546 [90]	$\begin{cases} n_3=1.528 \\ n_1=n_2=1.532 \end{cases}$	[96] [96]
SiO_2—(Quartz)	32	(T) $r_{11}=-0.47$	.409 to .605 [75]	$n_3=1.555$	.546 [75]
		(T) $r_{41}=0.20$	[75]	$n_1=n_2=1.546$	.546 [75]
		(S) $r_{11}=0.23$ (calculated)	[98]		
		(S) $r_{11}=0.1$	[99]		
$(C_6H_{12}O_6)_2NaBr\cdot H_2O$	32	(T) $r_{11}=0.1$	.546 [90]	$n_3=1.560$	.546 [90]
				$n_1=n_2=1.528$	
Rochelle Salt	222	(T) $r_{41}=-2.0$	.589 [75]	$n_1=1.491$	.589 [75]
		(T) $r_{52}=-1.7$	.589 [75]	$n_2=1.493$	.589 [75]
		(T) $r_{63}=0.3$	.589 [75]	$n_3=1.497$	.589 [75]
$C(CH_2OH)_4$	2	(T) $r_{52}=1.45$	.46 to .70 [100]	$n_1=1.528$	
		(T) $\lvert r_{12}-r_{32} \rvert=0.7$	.46 to .70 [100]	$n_2\approx n_3=1.56$	[75]
$Ca_2Nb_2O_7$	2	(T) $\left\lvert r_{22}-\dfrac{n_3{}^3}{n_2{}^3}r_{32} \right\rvert=14$	.63 [101]	$n_1=1.97$	[101]
		(T) $\left\lvert r_{22}-\dfrac{n_1{}^3}{n_2{}^3}r_{12} \right\rvert=12$	.63 [101]	$n_2=2.16$	[101]
		(S) $\left\lvert r_{22}-\dfrac{n_3{}^3}{n_2{}^3}r_{32} \right\rvert=13$	.63 [101]	$n_3=2.17$	[101]

r_{mi} is 10^{-12} m/v; λ in microns; (T)=constant stress; (S)=constant strain.

The only requirement on the direction of propagation is that it be normal to the applied field, since for this configuration the crystal is uniaxial under the application of the field.

A third useful configuration is one in which light propagates along a threefold [111] axis and the electric field is applied normal to this axis. The axes of the elliptical section of the indicatrix rotate at half the rate of a rotating electric field, as was mentioned in Section II. The retardation Γ for this case is $\sqrt{\frac{2}{3}}\,n^3 r_{41}E$.

Miscellaneous Crystals

This group includes many crystals which were measured simply because they were known to be piezoelectric and were also transparent. Crystals of various symmetry are included. In many instances the other electrical and optical properties of the materials are incompletely known. The results are arranged in order of decreasing symmetry in Table V.

Among the cubic crystals, the first listed (bismuth germanate) may prove to be of technical interest since it can be obtained in crystals of good optical quality with dimensions of a few centimeters. A second $\bar{4}3m$ crystal, (HMT) hexamethylenetetramine, is of both scientific (because it is a molecular crystal) and technical interest because of the size of the electrooptic coefficient. The dielectric constant at radio frequency is close to the optical value, 2.6, and the loss tangent is small [88], [124]. The material is inexpensive and crystals can be grown with ease. However, it appears that large crystals and in particular crystals grown from solution do not exhibit a large electrooptic effect, presumably because of dislocations [88]. Additionally, practical use of the material requires protection from water vapor and elevated temperatures, and the material is soft. In the case of the mineral Hauynite only an upper limit on r_{41} has been established. The remaining cubic crystals are in point group 23, and are, in general, optically active. This fact complicates the measurement and may limit the usefulness of the materials [89].

The only representative of symmetry 6 is $LiKSO_4$. There are four coefficients and only the difference between two of them has been measured to date.

In class 3m, which has four coefficients, the value of r_{22} only has been measured in $LiNaSO_4$ and tourmaline. Since the latter is a mineral of varying composition, the measured values might be expected to vary also'.

Many of the measurements of 32 crystals were made as part of a survey of existing piezoelectric crystals [90]. There are only two coefficients for this point group. Both coefficients have been measured for quartz. The interest in quartz derives, of course, from its availability and excellent optical properties rather than a large electrooptic effect. Crystals in the 32 point group are optically active and this activity, which is manifested principally when light is propagated along the optic axis complicates the measurement of the r_{11} coefficient.

The lowest symmetry group for which coefficients have been measured is point group 2. Here there are eight independent coefficients. The coefficients or combinations of coefficients that have been measured are ones for which a field is applied along the x_2 axis. This is the axis of 2-fold symmetry and the crystal symmetry remains unchanged. The measurements on pentaerythritol yielded the coefficients that could be found with light propagating along the x_2 direction (parallel to the applied field). The calcium pyroniobate measurements were made with light propagating normal to the x_2 direction.

IV. MODULATOR DESIGN

This section contains some general design considerations as well as capsule descriptions of a few modulator configurations. Space limitations do not permit a discussion of electrooptic beam scanners such as prisms [40], electrooptic gratings [109], and digital light deflectors [110]. Some of these devices are discussed elsewhere in this issue.

Geometrical Considerations

As a rule, the light beam to be modulated passes along a straight line through the electrooptical crystal, which for efficient operation should be just large enough to contain the beam. The colinear case is considered below. Later on, we mention briefly a modulator in which the beam follows a zig-zag path through the crystal and another modulator in which the light travels as a trapped wave along a dielectric discontinuity (p-n junction).

Consider a Gaussian optical maser beam propagating

in the lowest order transverse mode and focused, by means of a lens, so as to just pass through a cylinder of length L with refractive index n. It can be shown by differentiating the Gaussian beam formulas [103] that the diameter of the cylinder will be a minimum when the lens is chosen to make L equal the confocal parameter b of the beam. Then, the beam diameter is $2w_0$ at the waist and $\sqrt{8}\ w_0$ at the ends of the cylinder, where

$$w_0^2 = \lambda L/2\pi n, \qquad (21)$$

and beam diameters are measured to points where the field is $1/e$ its value on the beam axis.

For a cylinder of diameter

$$d = S \cdot \sqrt{8}\ w_0, \qquad (22)$$

where S is a safety factor ($S \geq 1$), the minimum value of d^2/L is

$$d^2/L = S^2 \cdot 4\lambda/n\pi \qquad (23)$$

The situation is illustrated in Fig. 3.

In practice, [50], [125], beams have been passed through a rod, with some difficulty in alignment but little added loss, with $S \approx 3$ and, with easy alignment, with $S \approx 6$.

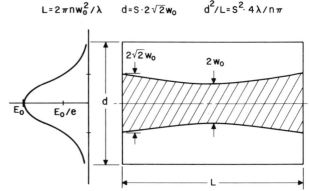

$$L = 2\pi n w_0^2/\lambda \qquad d = S \cdot 2\sqrt{2} w_0 \qquad d^2/L = S^2 \cdot 4\lambda/n\pi$$

Fig. 3.   A beam of Gaussian cross section passing through a rod of diameter d and length L.

Lumped Modulator

In Section II it was tacitly assumed that the field in the sample is uniform over the length L and constant during the time nL/c for the light to pass through the modulator, i.e.,

$$L \ll 2\pi c/2\omega_m \sqrt{\epsilon} \quad \text{and} \quad L \ll 2\pi c/2\omega_m n \qquad (24)$$

where ω_m is the highest modulating frequency and ϵ the appropriate dielectric constant at ω_m. Since $\sqrt{\epsilon} \geq n$, normally, the first restriction is dominant and when it is satisfied the modulator may be regarded as a lumped circuit element. For a rod of square cross section d^2, the lumped capacitance is

$$C = \epsilon_0 \epsilon L \qquad (25)$$

and the parallel conductance

$$G = \omega_m C \tan \delta. \qquad (26)$$

In practice, additional parallel circuit capacitance C_a and conductance G_a may be unavoidable or may be required to achieve a certain bandwidth. If the peak voltage V across the crystal is to be provided by a particular voltage generator V_g with impedance R_g, $|V/V_g|$ will be a maximum when R_g is matched to the specified load resistance by an ideal transformer.[3] The ratio $|V/V_g|^2$ will be reduced to $\frac{1}{2}$ its matched dc value at frequency

$$\Delta\omega = 2(G + G_a)/(C + C_a). \qquad (27)$$

If parallel inductance is provided to produce resonance at ω_0, the bandwidth over which $|V/V_g|^2$ is at least half its matched value at ω_0 is also $\Delta\omega$ as in (27). The resonant circuit, including an impedance matching transformer, is shown in Fig. 4.

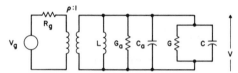

Fig. 4.   Equivalent driving circuit for a lumped modulator.

The peak voltage V is determined by the desired index η. The power dissipated is

$$P = \frac{V^2}{2}(G + G_a) = \frac{1}{2}\left(\frac{\lambda\eta d}{\pi n^3 rL}\right)^2 (G + G_a) \qquad (28)$$

or in terms of $\Delta\omega$

$$P = \frac{V^2}{4}(C + C_a)\Delta\omega = \frac{1}{4}\left(\frac{\lambda\eta d}{\pi n^3 rL}\right)^2 (C + C_a)\Delta\omega, \qquad (29)$$

where n and r are appropriate values of refractive index and electrooptic coefficient. If $C \gg C_a$, a situation which may be difficult to achieve in practice,

$$P = \left(\frac{\epsilon_0}{4\pi^2}\right)(\lambda^2)\left(\frac{\epsilon}{n^6 r^2}\right)\left(\frac{d^2}{L}\right)\cdot \eta^2 \Delta\omega. \qquad (30)$$

The maximum L is set by (24) and the minimum d^2/L by (23). For $C \gg C_a$, P is a minimum when d^2/L is a minimum. But for $C_a \gg C$, P is minimum when L is maximum *and* d^2/L is minimum.[4] Thus, for the optimum design, P is pro-

---

[3] The ratio is given by

$$\left|\frac{V}{V_g}\right| = \frac{\rho}{\rho^2 + R_g(G + G_a)},$$

where the transformer ratio ρ is supposed to be the only design parameter. Of course, $R_g(G + G_a)$ should be reduced if possible. In certain pulse applications dc coupling is required and in other cases broadband transformers are not available, then matching is not possible.

[4] This discussion assumes C_a and G_a independent of d and L.

portional to the material parameter $\epsilon/n^6 r^2$ for $C \gg C_a$ and $\epsilon^{1/2}/n^6 r^2$ for $C_a \gg C$.

When bandwidth is not important, $(G + G_a)$ should be minimized. For $G \gg G_a$, P is proportional to d^2/L and, for $G_a \gg G$, to d^2/L^2. Hence, when the added conductance G_a is dominant, as when C_a is dominant, it is advantageous to use large L.[4]

In practice, at frequencies below about 200 MHz, a capacitor may be formed by the crystal using silver paint or evaporated metal electrodes and placed into a circuit with lumped R, L, C elements or coaxial tuners [50]. At higher frequencies, the crystal must be placed in a resonant cavity [7], [104] to obtain the small parallel inductance required and to avoid series inductance in the lead wires.

Traveling Wave

As noted above, P varies as $1/L$ when C_a and/or G_a are dominant and the minimum d^2/L is employed. The transit time restriction on L given in (24) can be lifted if the modulating field is a traveling wave with phase velocity equal to the optical group velocity [105]. Then an optical disturbance will experience a constant modulating field as it passes through the crystal. The velocity match should be such that the optical disturbance does not slip more than a quarter wavelength along the modulation wave during the transit [106], i.e.,

$$\frac{\omega L}{c}(N - n) \leq \frac{\pi}{2}, \qquad (31)$$

where N is the ratio of light velocity c to the phase velocity of the modulating wave. The matching condition (31) must be satisfied over the modulating bandwidth. For finite bandwidth, then, the group velocities at optical and modulating frequencies should be equal. The matching of N and n becomes more critical for large ωL and in this sense (31) limits L.

If the modulating field travels as a plane wave completely within the crystal, then $N = \sqrt{\epsilon}$ at all modulating frequencies. In materials for which electronic, rather than lattice, polarization is dominant at the modulating frequency, $\sqrt{\epsilon} \approx n$ and, in principle, a broadband velocity match can be readily achieved. However, crystals with this property are not available in suitable size and quality at present.

When $\sqrt{\epsilon} > n$, a match can be effected over a limited band by filling a TEM line partially with the crystal and the remainder with a lower dielectric constant material, such as air [106]–[108]. In the case of an air filled, parallel plate transmission line containing a crystal of square cross section as shown in Fig. 5, the characteristic admittance at sufficiently low frequencies is

$$G_0 = \frac{N}{z_0}\left(\frac{\epsilon - 1}{N^2 - 1}\right), \qquad (32)$$

in which z_0 is 377 ohms, the impedance of free space. For

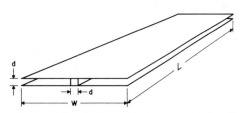

Fig. 5.   Partially filled parallel plate transmission line.

a velocity match the dimensions of the line are adjusted so that $N=n$, i.e.,

$$\frac{W}{d} = \frac{\epsilon - 1}{n^2 - 1}, \tag{33}$$

where W is the width of the plates and d a transverse dimension of the crystal. When the line is terminated in G_0, the power required is given by (28) with G_0 substituted for $(G+G_a)$. When $\epsilon=n^2=N^2$, G_0 and P have their minimum values. A wide disparity in ϵ and n^2 makes for inefficient operation because much of the input power travels outside the crystal to the load.

The line will approximate a constant phase velocity TEM line up to frequencies at which d becomes comparable with a half wavelength in the medium [106], i.e.,

$$\Delta\omega \ll \frac{2c}{d\sqrt{\epsilon}} \tag{34}$$

or, roughly,

$$\Delta\omega_{\text{max}} = \frac{c}{5d\sqrt{\epsilon}}. \tag{35}$$

Thus, d is limited by $\Delta\omega$, and L, in turn, is limited by (23) and (31). The length is further restricted by considerations of transmission line loss, optical loss, and the availability of large crystals of good quality. When transmission line loss is appreciable, it is more efficient to feed the line by a tapered coupling along its length rather than from one end [111]. For materials with $n<1.6$, optical matching liquids or cements can be found which permit a number of crystals to be joined to provide a large L.

Parallel plate modulators have been constructed by Peters [107], [108] and others [112] with $L\approx100$ cm using 20 to 25 crystals of ADP or KDP. Even with the refractive indices of crystal and cement matched within 0.001 (using a mixture of silicone oils) and the rods polished with care, the scattering at the 40 to 50 interfaces accounts for most of the observed 6 dB optical loss [112].

The broadband RF matching section, required at the input and termination to take advantage of the potential 0–2 GHz band, have not yet been designed. In addition to the impedance matching problem, operation at the low-frequency end of the band is limited by piezoelectric resonances at frequencies for which d is a small integral number of acoustic half wavelengths. However, resonances have been suppressed by clamping the crystals mechanically [113].

Standing-Wave Modulator

If broadband operation is not an important consideration, it is possible to realize the low power (large L) advantage of the traveling-wave device without the problem of impedance matching by forming a resonant cavity from a velocity matched transmission line. A resonator must be an integral number of half waves long. The resultant standing wave may be regarded as the sum of two oppositely traveling waves. The component traveling in synchronism with the optical wave produces the required modulation. The incident optical disturbance approaches the backward wave component at twice the phase velocity and, therefore, interacts with an integral number of modulating waves within the resonator. Because of the linearity of the electrooptic effect, therefore, the backward wave produces no net modulation. Standing-wave modulators have been built using a KDP rod on the axis of a TM_{01n} resonator [114].

Fabry-Perot Modulator

The effective length of time spent by an optical disturbance bouncing back and forth within a Fabry-Perot resonator is determined by the optical Q. When the optical losses are set by mirror reflectivity R, the effective distance traveled by the light before decaying to $1/e$ of its initial value is

$$L_{\text{eff}} = FL/\pi = L\sqrt{R}/(1 - R), \tag{36}$$

in which F is the *finesse* and L the length of the resonator. If the velocity matching requirements are satisfied over L_{eff}, using the standing-wave technique if necessary, it is possible to obtain a modulation corresponding to L_{eff} while dissipating power corresponding to L [115]. Only those optical carrier and sideband frequencies that fall under the Fabry-Perot transmission function will retain the proper phases on successive bounces, however. Thus, the optical carrier must fall on one mode and the sideband energy must appear within the band of that mode or any other of the modes which occur at intervals of $c/2nL$. An increase in bandwidth requires a decrease in Q and corresponding reduction L_{eff}.

A Fabry-Perot modulator has been built at 70 MHz under conditions not requiring velocity match [116]. Experiments at high frequency and/or high finesse, requiring velocity matching, appear to be difficult and have not been reported.

Zig-Zag and Phase Reversal Modulators

Up to this point, it has been assumed, unnecessarily, that the optical and modulating waves travel along the same path through a uniform medium. Electrooptic modulation is a parametric process in which the perturbed parameter is the refractive index and the interacting waves are the optical carrier (ω_0, k_0), an optical sideband (ω_s, k_s), the modulating wave (ω_m, k_m), and possibly a time-invariant periodicity in the medium $(0, k_p)$, where ω_α is an angular frequency and k_α a wave vector. The phase or velocity matching requirement is equivalent to conservation of ω and k,

$$\sum \omega_\alpha = 0, \qquad \sum k_\alpha = 0. \qquad (37)$$

When all the beams are parallel, k_p can always be designed to satisfy (37) by reversing the phase of the modulating wave at intervals [117], [118]

$$\lambda_p/2 = \pi c/\omega_m(N - n), \qquad (38)$$

or by eliminating the modulation altogether at alternate intervals. The match is effective only in the neighborhood of the specified ω_m so that the technique is narrow band.

In a uniform medium ($k_p = 0$), k conservation may be accomplished by permitting an angle γ between k_0 and k_m such that

$$\cos \gamma = n/\sqrt{\epsilon}, \qquad (39)$$

where dispersion is neglected and use is made of the fact that $\epsilon \omega_m \ll n^2 \omega_0$. In general, the component of phase velocity along the light path should equal the optical group velocity. This type of match can be achieved by having the light follow a zig-zag path at angle γ along the transmission line carrying the modulating wave [105]. Then power into the line can be used most efficiently [$N^2 = \epsilon$ in (32)]. Only relatively short lengths of crystal are required so that it may not be necessary to optically match several separate sections of crystals. Technical complexity has so far discouraged such efforts [119].

p-n Junction Modulator

A technique for confining optical and modulating fields to the same small volume, other than that using a crystal with diffraction limited dimensions, makes use of p-n junctions in GaP and GaAs. The discontinuity at the junction provides an optical guide of a few microns thickness [120], and back biasing of the junction can yield modulating field strengths approaching 10^6 volts/cm [121]. Modulating junctions have usually been made with the junction plane normal to a [111] direction in the crystal. The junction itself is 0.2 to 0.8 micron thick and the total crystal thickness a few mils. The other crystal dimensions are each typically 25 mils (0.025 inch).

The guided modes can be classified as either TE or TM, depending on whether the optical electric field or magnetic field is parallel to the plane of the junction. The guiding action is believed to depend primarily on a higher optical dielectric constant inside the depletion layer than outside, rather than on the conductivity of the p and n regions. The fractional difference in dielectric constant responsible for the guiding is found to be about 10^{-3} [121], [122]. In GaP this is larger than one can account for on the basis of a reduction in dielectric constant due to charge carriers in the n or p region. The guiding in GaAs is also not completely accounted for. Thus, this modulator is in the happy position of working better than one would predict.

The p-n junction modulator has usually been operated as an amplitude modulator. In order to do this, equal amounts of the TE and TM modes are excited by using polarization at 45° to the junction. The two modes have different phase lengths in the material and on leaving the modulator the waves are recombined to give the initial polarization plus a polarization at right angles—the amounts depending on phase difference. This difference is [121]:

$$\Delta\phi = \frac{(2\pi)^3 \sqrt{3}}{\lambda^3} n^5 t^2 L r_{41} \overline{E} \left(\Delta - \frac{r_{41} n^2 \overline{E}}{2\sqrt{3}} \right) \qquad (40)$$

where $2t$ is the junction width, L the length, and Δ the fractional change in optical dielectric constant. The average field $\overline{E}$ in the junction can be of order 10^5 volts/cm for applied voltages of 20 volts. The junction width and hence the waveguide width changes with applied field, so both $\overline{E}$ and t are functions of voltage. The experimental results [121], [122] confirm the functional form [40] which indicates that the *linear* electrooptic effect is the predominant cause of modulation *even* with the large field strengths used. The anomalously large value of dielectric discontinuity remains unexplained.

The high-frequency limit of p-n junction modulating diodes is fixed by the time constant of the series resistance and the capacitance. In order to reduce the latter, the use of a mesa structure which reduces the width of the diode has been proposed [123]. The resistance can be reduced by using more heavily doped material and by reducing the diode height as much as possible. Estimates based on properties of existing diodes indicate a phase modulation index of $\pi/4$ up to 1 GHz with 200 mW of power is within the realm of possibility [123].

CONCLUSIONS

The electrooptic properties of a large number of materials have been examined in the search for suitable modulator materials. Unfortunately, the fundamental nature of the electrooptic effect is not well understood, and the search has been guided mainly by incidental practical considerations: the availability of crystals of good optical quality transparent in the spectral region of interest, point group symmetry—piezoelectric groups for linear electrooptic effect, cubic or axial groups for simplicity of interpretation and applicability in certain devices, low dielectric loss to avoid thermal strain and reduce modulating power, high resistivity to reduce space charge effects, and heating. The particular application determines which material parameters are of greater importance, often in contradicting fashion. In all applications, the modulation produced by a given field is proportional to $n^3 r$ so that both large refractive index and electrooptic coefficient are desirable. It is often found that r is proportional to ϵ, so that large ϵ is desirable. On the other hand, it is difficult to provide optical matching from air into the material or to provide a matching cement for joining two crystals (for a long traveling-wave modulator) when n is large. When ϵ is large, the reactive power required to drive a lumped modulator is large. Further, it is difficult to provide an RF match and to make the sample small enough to avoid dielectric resonances at high frequency. Additionally, large values of ϵ generally imply a large disparity in n and $\sqrt{\epsilon}$ so that broadband velocity matching becomes inefficient for the colinear, but not necessarily the zig-zag, traveling-wave modulator.

Lumped modulators using materials like LiTaO$_3$, however, appear to be suitable for certain engineering applications [125]. The *p-n* junction modulator is also promising. For certain laboratory applications, where smaller modulation or high power are permissible, KDP and DKDP have proven useful. Nevertheless, materials with higher electrooptic coefficient, possibly also with $n \approx \sqrt{\epsilon}$ and $n \approx 1.5$, are needed for many applications.

The discovery of powerful infrared masers has created an interest in modulators that operate efficiently at wavelengths as long as 10 μ. Although many of the crystals already studied are transparent to 5 μ and some (e.g. GaAs, ZnSe, CdS) to > 10 μ, the electrooptic coefficient in the infrared would have to be proportionately larger than in the visible region to make up for the λ^{-1} dependence of modulation index. However, the electrooptic coefficient for GaAs, and probably for other known materials as well, is found to remain constant between 1 μ and 10 μ [81].

Perhaps the development of a theoretical understanding of the electrooptic effect will lead to the discovery or synthesis of the ideal substances for each application in a logical way.

A number of modulator configurations that make the most efficient use of a particular material in a given application have been described. Most of these have not been developed fully because the demand of a particular application has not warranted the effort, or because a material with the ideal properties was not available. However, as the characteristics of electrooptic modulators and modulators of other types become clearer, so that applications can be seriously considered, some of these devices may receive the full treatment. The simplest configuration is the lumped modulator. However, in order to approach the optimum efficiency, the effective path length for the electrooptic perturbation must be made very large by means of traveling-wave techniques. Where large bandwidth is important, a nonresonant structure with broadband velocity and impedance matching is required. Where narrow bandwidths are permissible, structures resonant at optical (e.g., Fabry-Perot modulator), and/or modulating (e.g., standing-wave modulator) frequencies may be employed.

REFERENCES

[1] F. Pockels, Lehrbuch der Kristalloptik. Leipzig: Teubner, 1906.
[2] F. Jona and G. Shirane, *Ferroelectric Crystals*. New York: Macmillan, 1962.
[3] W. Kanzig, *Solid State Physics*, vol. 4, F. Seitz and D. Turnbull, Eds. New York: Academic, 1957, pp. 1–197.
[4] B. H. Billings, "The electrooptic effect in uniaxial crystals of the dihydrogen phosphate (XH$_2$PO$_4$) type, Parts I, II, IV," *J. Opt. Soc. Am.*, vol. 39, pp. 797–801, and pp. 802–808, October 1949, vol. 42, pp. 12–20, January 1952.
[5] R. O'B. Carpenter, "The electrooptic effect in crystals of the dihydrogen phosphate type, Part III," Measurement of coefficients," *J. Opt. Soc. Am.*, vol. 40, pp. 225–229, April 1950.

—— "Electrooptic sound-on-film modulator," *J. Opt. Soc. Am.*, vol. 25, pp. 1145–1148, November 1953.
[6] D. F. Holshouser, H. Von Foerster, and G. L. Clark, "Microwave modulation of light using the Kerr effect," *J. Opt. Soc. Am.*, vol. 51, pp. 1360–1365, December 1961.
[7] K. D. Froome and R. H. Bradsell, "Distance measurement by means of a light ray modulated at a microwave frequency," *J. Sci. Instr.*, vol. 38, pp. 458–462, December 1961.
[8] J. F. Nye, *Physical Properties of Crystals*. Oxford, England: Oxford University Press, 1960.
[9] P. V. Lenzo, E. H. Turner, E. G. Spencer, and A. A. Ballman, "Electrooptic coefficients and elastic wave propagation in single-domain ferroelectric lithium tantalate," *Appl. Phys. Letters*, vol. 8, pp. 81–82, February 1966.
[10] Incorrect results, however, have been presented in quite recent publications such as G. N. Ramachandran and S. Ramaseshan, "Crystal Optics," in *Handbuch der Physik*, vol. 25/1. Berlin: Springer-Verlag, 1961, pp. 1–217.
[11] C. F. Buhrer, D. Baird, and E. M. Conwell, "Optical frequency shifting by electrooptic effect," *Appl. Phys. Letters*, vol. 1, pp. 46–49, October 1962.
[12] C. F. Buhrer, L. R. Bloom, and D. H. Baird, "Electrooptic light modulation with cubic crystals," *Appl. Opt.*, vol. 2, pp. 839–846, August 1963.
[13] W. P. Mason, "The elastic, piezoelectric, and dielectric constants of KDP and ADP," *Phys. Rev.*, vol. 69, pp. 173–194, March 1946.
[14] I. P. Kaminow, "Microwave dielectric properties of NH$_4$H$_2$PO$_4$, KH$_2$ASO$_4$, and partially deuterated KH$_2$PO$_4$," *Phys. Rev.*, vol. 138, pp. A1539–A1543, May 1965.
[15] F. R. Nash, "Measurements made by substitution inside laser resonator," unpublished.
[16] T. R. Sliker and S. R. Burlage, "Some dielectric and optical properties of KD$_2$PO$_4$," *J. Appl. Phys.*, vol. 34, pp. 1837–1840, July 1963.
[17] W. J. Deshotels, "Ultraviolet transmission of dihydrogen arsenate and phosphate crystals," *J. Opt. Soc. Am.*, vol. 50, p. 865, September 1960.
[18] S. F. Pellicor, "Transmittances of some optical materials for use between 0.19 and 0.34 μ," *Appl. Opt.*, vol. 3, pp. 361–366, March 1964.
[19] E. F. Kingsbury, unpublished memorandum, 1950.
[20] I. P. Kaminow, "Temperature dependence at the complex dielectric constant in KH$_2$PO$_4$-Type crystals and the design of microwave light modulators," in *Quantum Electronics III*, P. Grivet and N. Bloembergen, Eds., New York: Columbia University Press, 1964, pp. 1659–1665.
[21] O. G. Blokh, "Dispersion of r$_{63}$ for crystals of ADP and KDP," *Sov. Phys.-Cryst.*, vol. 7, pp. 509–511, January-February 1963.
[22] J. F. Ward and P. A. Franken, "Structure of nonlinear optical phenomena in KDP," *Phys. Rev.*, vol. 133, pp. A183–A190, January 1964.
[23] I. P. Kaminow, "Strain effects in electrooptic light modulators," *Appl. Opt.*, vol. 3, pp. 511–515, April 1964.
[24] J. H. Ott and T. R. Sliker, "Linear electrooptic effects in KH$_2$PO$_4$ and its isomorphs," *J. Opt. Soc. Am.*, vol. 54, pp. 1442–1444, December 1964.
[25] C. H. Clayson, "Low-voltage light-amplitude modulation," *Electronic Letters*, vol. 2, p. 138, April 1966; reply by J. M. Ley, *ibid.*, p. 139.
[26] B. H. Billings, "The electrooptic effect in crystals and its possible application to distance measure," in *Optics in Metrology*, P. Mollet, Ed. New York: Pergamon, 1960, pp. 119–135.
[27] F. Zernike, Jr., "Refractive indices of ADP and KDP between 0.2 and 1.5 μ," *J. Opt. Soc. Am.*, vol. 54, pp. 1215–1220, October 1964; V. N. Vishnevskii and I. V. Stefanski, "Temperature dependence of the dispersion of the refractivity of ADP and KDP single crystals," *Opt. and Spectr.*, vol. 20, pp. 195–196, February 1966.
[27a] R. A. Phillips, "Temperature variation of the index of refraction of ADP, KDP and deuterated KDP," *J. Opt. Soc. Am.*, vol. 56, pp. 629–632, May 1966.
[28] D. A. Berlincourt, D. R. Curran, and H. Jaffe, *Physical Acoustics*, vol. I, pt. A, W. P. Mason, Ed. New York: Academic, 1964, pp. 169–260.
[29] I. P. Kaminow and G. O. Harding, "Complex dielectric con-

stant of KH_2PO_4 at 9.2 Gc/sec," *Phys. Rev.*, vol. 129, pp. 1562–1566, February 1963.

[30] J. M. Ley, "Low voltage light-amplitude modulation," *Electronics Letters*, vol. 2, pp. 12–13, January 1966.

[31] I. S. Zheludev and T-Z Ludupov, "Complex dielectric constant of RbH_2PO_4 in the range $8 \times 10^2 - 3.86 \times 10^{10}$ cps," *Kristallografiia*, vol. 10, pp. 764–766, September–October 1965.

[32] A. von Hippel, *Dielectric Materials and Applications*. New York: Wiley, 1954.

[33] R. M. Hill and S. K. Ichiki, "Paraelectric response of KD_2PO_4," *Phys. Rev.*, vol. 130, pp. 150–151, April 1961.

[34] Helen D. Megaw, *Ferroelectricity in Crystals*. London: Methuen, 1957.

[35] J. E. Geusic, S. K. Kurtz, L. G. van Uitert, and S. H. Wemple, "Electrooptic properties of some ABO_3 perovskites in the paraelectric phase," *Appl. Phys. Letters*, vol. 4, pp. 141–143, April 1964.

[36] M. S. Schumate, "Interferometric determination of the principal refractive indices of barium titanate single crystals," *Appl. Phys. Letters*, vol. 5, pp. 178–179, November 1964.

[37] J. A. Noland, "Optical absorption of single crystal strontium titanate," *Phys. Rev.*, vol. 94, p. 724, May 1, 1954.
R. C. Casella and S. P. Keller, "Polarized light transmission of $BaTiO_3$ single crystals," *Phys. Rev.*, vol. 116, pp. 1469–1473, December 1959.
C. Hilsum, "Infrared transmission of barium titanate," *J. Opt. Soc. Am.*, vol. 45, pp. 771–772, September 1955.
J. T. Last, "Infrared-absorption studies of barium titanate and related materials," *Phys. Rev.*, vol. 105, pp. 1740–1750, March 1957.

[38] C. J. Johnson, "Some dielectric and electrooptic properties of $BaTiO_3$ single crystals," *Appl. Phys. Letters*, vol. 7, pp. 221–223, October 1965.

[39] G. Rupprecht and R. O. Bell, "Microwave losses in strontium titanate above the phase transition," *Phys. Rev.*, vol. 125, pp. 1915–1920, March 1962.

[40] F. S. Chen, J. E. Guesic, S. K. Kurtz, J. G. Skinner, and S. H. Wemple, "Light modulation and beam deflection with potassium tantalate-niobate crystals," *J. Appl. Phys.*, vol. 37, pp. 388–398, January 1966.

[41] A. R. Johnston, "The strain-free electrooptic effect in single-crystal barium titanate," *Appl. Phys. Letters*, vol. 7, pp. 195–198, October 1965.

[42] S. K. Kurtz, "Design of an electrooptic polarization switch for a high capacity high-speed digital light deflection system," *Bell Sys. Tech. J.*, to be published.

[43] F. S. Chen, J. E. Guesic, S. K. Kurtz, J. G. Skinner, and S. H. Wemple, "The use of perovskite paraelectrics in beam deflectors and light modulators," *Proc. IEEE (Correspondence)*, vol. 52, pp. 1258–1259, October 1964.

[44] A. Linz, Jr. "Some electrical properties of strontium titanate," *Phys. Rev.*, vol. 91, pp. 753–754, August 1953.

[45] J. K. Hulm, B. T. Matthias, and E. A. Long, "A ferromagnetic Curie point in $KTaO_3$ at very low temperatures," *Phys. Rev.*, vol. 79, pp. 885–886, September 1950.

[46] J. E. Guesic, S. K. Kurtz, T. J. Nelson, and S. H. Wemple, "Nonlinear dielectric properties of $KTaO_3$ near its Curie point," *Appl. Phys. Letters*, vol. 2, pp. 185–187, May 1963.

[47] J. P. Remeika, "A method for growing barium titanate single crystals," *J. Am. Chem. Soc.*, vol. 76, pp. 940–941, February 1954.

[48] A. R. Johnston and J. M. Weingart, "Determination of the low-frequency linear electrooptic effect in tetragonal $BaTiO_3$," *J. Opt. Soc. Am.*, vol. 55, pp. 828–834, July 1965.

[49] E. H. Turner, Paper 6B-2, presented at the 1966 Intnat'l Quantum Electronics Conf., Phoenix, Ariz., and "High frequency electrooptic coefficients of lithium niobate," *Appl. Phys. Letters*, vol. 8, pp. 303–304, June 1966.

[50] I. P. Kaminow, "Barium titanate light phase modulator," *Appl. Phys. Letters*, vol. 7, pp. 123–125, September 1965, "Erratum," vol. 8, p. 54, January 1966.
——, "Barium titanate light modulator II," *Appl. Phys. Letters*, vol. 8, pp. 305–306, June 1966.

[51] The structure and growth of $LiNbO_3$ are described and earlier work reviewed in a series of five papers: K. Nassau, H. J. Levinstein, and G. M. Loiacono (I and II); S. C. Abrahams, J. M. Reddy, and J. L. Bernstein (III); S. C. Abrahams, W. C. Hamilton, and J. M. Reddy (IV); S. C. Abrahams,

H. J. Levinstein, and J. M. Reddy (V); "Ferroelectric lithium niobate," *J. Phys. Chem. Solids*. vol. 27, 1966.

[52] B. T. Matthias and J. P. Remeika, "Ferroelectricity in the ilmenite structure," *Phys. Rev.*, vol. 76, pp. 1886–1887, December 1949.

[52a] A. A. Ballman, "The growth of piezoelectric and ferroelectric materials by the Czochralski technique," *J. Am. Ceram. Soc.*, vol. 48, pp. 112–113, February 1965.

[53] H. J. Levinstein, A. A. Ballman, and C. D. Capio, "The domain structure and Curie temperature of single crystal lithium tantalate," *J. Appl. Phys.*, to be published.

[54] R. T. Denton, Paper 6B-4, presented at the 1966 Internat'l Quantum Electronics Conf., Phoenix, Ariz.

[55] A. Ashkin, G. D. Boyd, J. M. Diedzic, R. G. Smith, A. A. Ballman, H. J. Levinstein, and K. Nassau, "Optically induced refractive index inhomogeneities in $LiNbO_3$ and $LiTaO_3$," *Appl. Phys. Letters*, to be published.

[56] G. D. Boyd, R. C. Miller, K. Nassau, W. L. Bond, and A. Savage, "$LiNbO_3$: An efficient phase matchable nonlinear optical material," *Appl. Phys. Letters*, vol. 5, pp. 234–236, December 1964.

[57] W. L. Bond, "Measurement of the refractive indices of several crystals," *J. Appl. Phys.*, vol. 36, pp. 1674–1677, May 1965.

[58] P. V. Lenzo, E. G. Spencer, and K. Nassau, "Electrooptic coefficients in lithium niobate," *J. Opt. Soc. Am.*, vol. 56, pp. 633–636, May 1966.

[59] F. A. Dunn, unpublished. Measurements at 9.3 GHz show that made from -180 degrees C to $+100$ degrees C ϵ_3 decreases by 15 percent and $\epsilon_1 = \epsilon_2$ by 7 percent from $+100$ to -180 degrees C. At room temperature $\epsilon_1 = 45$ and $\epsilon_3 = 27$. The dielectric loss was too low to be measured, i.e., $\tan \delta < 0.01$.

[60] E. P. Ippen, "Electrooptic deflection with $BaTiO_3$ prisms," *Proc. IEEE*, to be published.

[61] E. H. Turner, to be published. These are results of heterodyne measurements on vapor grown and hydrothermally grown crystals.

[62] R. J. Collins and D. A. Kleinman, "Infrared reflectivity of zinc oxide," *J. Phys. Chem. Solids*, vol. II, nos. 3–4, pp. 190–194, 1959.

[63] S. Namba, "Electrooptical effect of zincblende," *J. Opt. Soc. Am.*, vol. 51, pp. 76–79, January 1961.

[64] S. J. Czyzak, D. C. Reynolds et al., "On the properties of single cubic zinc sulfide crystals, *J. Opt. Soc. Am.*, vol. 44, pp. 864–867, November 1954.

[65] D. Berlincourt, H. Jaffe, and L. R. Shiozawa, "Electroelastic properties of the sulfides, selenides, and tellurides of zinc and cadmium," *Phys. Rev.*, vol. 129, pp. 1009–1017, February 1, 1963.

[66] E. H. Turner, to be published. These are results of heterodyne measurements on crystals termed "primarily hexagonal."

[67] T. M. Bieniewski and S. J. Czyzak, "Refractive indexes of single hexagonal ZnS and CdS crystals," *J. Opt. Soc. Am.*, vol. 53, pp. 496–497, April 1963.

[68] R. W. McQuaid, "Electrooptic properties of zinc selenide," *Proc. IRE (Correspondence)*, vol. 50, pp. 2484–2485, December 1962; and 'Correction to 'Electrooptic properties of zinc selenide,' " *Proc. IEEE*, vol. 51, p. 470, March 1963.

[69] D. T. F. Marple, "Refractive index of ZnSe, ZnTe, and CdTe," *J. Appl. Phys.*, vol. 35, pp. 539–542, March 1964.

[70] M. Aven, D. T. F. Marple, and B. Segall, "Some electrical and optical properties of ZnSe," *J. Appl. Phys.*, supplemental to vol. 32, pp. 2261–2265, October 1961.

[71] T. R. Sliker and J. M. Jost, "Linear electrooptic effect and refractive indices of cubic ZnTe," *J. Opt. Soc. Am.*, vol. 56, pp. 130–131, January 1966.

[72] E. H. Turner, unpublished. Heterodyne measurement.

[73] C. D. West, "Electrooptic and related properties of crystals with the zinc blend structure," *J. Opt. Soc. Am.*, vol. 43, p. 335, April 1953.

[74] L. M. Belyaev, G. F. Dobrzhanskii, Yu. U. Shaldin, "Electrooptical properties of copper chloride and bromide crystals," *Soviet Phys.—Solid State*, vol. 6, p. 2988, June 1965.

[75] Landolt-Börnstein, Zahlenwerte und Funktionen, II Band, 8 Teil, Optische Konstanten.

[76] L. M. Belyaev, G. S. Belikova, G. F. Dobrzhanskii, G. B. Netesov, and Yu. U. Shaldin, "Dielectric constant of crystals

having an electrooptical effect," *Soviet Phys.—Solid State*, vol. 6, pp. 2007–2008, February 1965.

[77] I. P. Kaminow, unpublished. This value measured at 9.3 GHz. It was also found that tan $\delta = 0.002$.

[78] K. K. Thornber, A. J. Kurtzig, and E. H. Turner, unpublished. The value $r_{41} = 0.5$ was obtained using pulse methods on relatively low resistivity samples. The 1.06 value was from a heterodyne measurement on a sample having $> 10^5$ ohm cm resistivity at 75 MHz, where ϵ was measured. More weight should be given the larger value.

[79] L. Ho and C. F. Buhrer, "Electrooptic effect of gallium arsenide," *Appl. Opt.*, vol. 2, pp. 647–648, June 1963.

[80] E. H. Turner and I. P. Kaminow, "Electrooptic effect in gallium arsenide," *J. Opt. Soc. Am.*, vol. 53, p. 523, April 1963.

[81] A. Yariv and C. A. Mead, "Semiconductors as electrooptic modulators for intrared radiation," Paper 5C-3, presented at the 1966 Internat'l Quantum Electronics Conf. Phoenix, Ariz.; also, T. E. Walsh, "Gallium-arsenide electrooptic modulators," *RCA Rev.*, to be published.

[82] D. T. F. Marple, "Refractive index of GaAs," *J. Appl. Phys.*, vol. 35, pp. 1241–1242, April 1964.

[83] C. Hilsum and A. C. Rose-Innes, *Semiconducting III–V Compounds*. New York: Pergamon, 1961.

[84] K. G. Hambleton, C. Hilsum, and B. R. Holeman, "Determination of the effective ionic change of gallium arsenide from direct measurements of the dielectric constant," *Proc. of the Physical Soc.*, vol. 77, pp. 1147–1148, June 1961.

[85] F. A. Dunn, unpublished. Measurement on semi-insulating material at 9.3 GHz, where tan $\delta < 0.01$ can be determined.

[86] D. J. A. Gainon, "Linear electrooptic effect in CdS," *J. Opt. Soc. Am.*, vol. 54, pp. 270–271, February 1964.

[87] S. J. Czyzak, H. Payne, W. M. Baker, J. E. Manthuruthil, and T. M. Bieniewski, "The study of properties of single ZnS and CdS crystals," Tech. Rept. 6, ONR Contract Nonr 1511(01)NR015218, 1960.

[88] G. H. Heilmeier, "The dielectric and electrooptical properties of a molecular crystal-hexamine," *Appl. Opt.*, vol. 3, pp. 1281–1287, November 1964.

[89] C. F. Buhrer, L. Ho, and J. Zucker, "Electrooptic effect in optically active crystals," *Appl. Opt.*, vol. 3, pp. 517–521, April 1964.

[90] T. R. Sliker, "Linear electrooptic effects in class 32, 6, 3m, and $\overline{4}3m$ crystals," *J. Opt. Soc. Am.*, vol. 54, pp. 1348–1351, November 1964.

[91] R. Nitsche, "Crystal growth and electrooptic effect of bismuth germanate, $Bi_4(GeO_4)_3$," *J. Appl. Phys.*, vol. 36, pp. 2358–2360, August 1965.

[92] R. W. McQuaid, "The Pockels effect of hexamethylenetetramine," *Appl. Opt.*, vol. 2, pp. 320–321, March 1963.

[93] K. K. Thornber and E. H. Turner, "A determination of the electrooptic coefficients of haüynite, langbeinite and gallium phosphide," unpublished.

[94] W. E. Ford, *Dana's Textbook of Mineralogy*, fourth ed. New York: Wiley, 1932. This reference gives the value $n = 1.535$ for the mineral compound $K_2Mg_2(SO_4)_3$ and is assumed approximately correct for the other two. Also $n \approx 1.572$ for $K_2Mn_2(SO_4)_3$.

[95] R. W. McQuaid, "Cubic piezoelectric crystals for electrooptic modulation," *1963 Proc. Nat'l Aerospace Electronics Conf.*, pp. 282–286.

[96] A. N. Winchell and H. Winchell, *The Microscopical Characters of Artificial Inorganic Substances*. New York: Academic, 1964.

[97] J. Warner, D. S. Robertson, and H. T. Parfit, "The electrooptic effect of sodium uranyl acetate," *Phys. Letters*, vol. 19, pp. 479–480, December 1965.

[98] W. G. Cady, *Piezoelectricity*. New York: McGraw-Hill, 1946, p. 721.

[99] D. D. Eden and G. H. Thiess, "Measurement of the direct electrooptic effect in quartz at UHF," *Appl. Opt.*, vol. 2, pp. 868–869, August 1963.

[100] O. G. Blokh, I. S. Zheludev, and U. A. Shamburov, "The electrooptic effect in crystals of pentaerythritol $C(CH_2OH)_4$," *Soviet Phys.—Cryst.*, vol. 8, pp. 37–40, July–August 1963.

[101] C. H. Holmes, E. G. Spencer, A. A. Ballman, and P. V. Lenzo, "The electrooptic effect in calcium pyroniobate," *Appl. Opt.*, vol. 4, pp. 551–553, May 1965.

[102] C. F. Buhrer and L. Ho, "Electrooptic effect in $(NH_4)_2Cd_2(SO_4)_3$ and $(NH_4)_2Mn_2(SO_4)_3$," *Appl. Opt.*, vol. 3, p. 314, February 1964.

[103] G. D. Boyd and J. P. Gordon, "Confocal multimode resonator for millimeter through optical wavelength masers," *Bell Sys. Tech. J.*, vol. 40, pp. 489–508, March 1961; H. Kogelnik and T. Li, "Laser beams and resonators," this issue.

[104] R. H. Blumenthal, "Design of a microwave-frequency light modulator," *Proc. IEEE*, vol. 50, pp. 452–456, April 1962.

[105] W. W. Rigrod and I. P. Kaminow, "Wide-band microwave light modulation," *Proc. IEEE*, vol. 51, pp. 137–140, January 1963.

[106] I. P. Kaminow and J. Liu, "Propagation characteristics of partially loaded two-conductor transmission line for broadband light modulators," *Proc. IEEE*, vol. 51, pp. 132–136, January 1963.

[107] C. J. Peters, "Gigacycle bandwidth coherent light traveling-wave phase modulator," *Proc. IEEE*, vol. 51, pp. 147–153, January 1963.

[108] ——, "Gigacycle-bandwidth coherent-light traveling-wave amplitude modulator," *Proc. IEEE*, vol. 53, pp. 455–460, May 1965.

[109] E. I. Gordon and M. G. Cohen, "Electro-Optic diffraction grating for light beam modulation and diffraction," *IEEE J. of Quantum Electronics*, vol. QE-1, pp. 191–198, August 1965.

[110] W. J. Tabor, "A high capacity digital light deflector using Wollaston prisms," *Bell Sys. Tech. J.*, to be published. R. A. Soref and D. H. McMahon, "Optical design of Wollaston-prism digital light deflectors," *Appl. Opt.*, vol. 5, pp. 425–434, March 1966.

[111] I. P. Kaminow, R. Kompfner, and W. H. Louisell, "Improvements in light modulators of the traveling-wave type," *IRE Trans. on Microwave Theory and Techniques*, vol. MTT-10, pp. 311–313, September 1962.

[112] J. A. Ernest and I. P. Kaminow, 1963, unpublished.

[113] E. A. Ohm, private communication.

[114] I. P. Kaminow, "Microwave modulation of the electrooptic effect in KH_2PO_4," *Phys. Rev. Letters*, vol. 6, pp. 528–530, May 1961. I. P. Kaminow, "Splitting of Fabry-Perot rings by microwave modulation of light," *Appl. Phys. Letters*, vol. 2, pp. 41–42, January 1963.

[115] E. I. Gordon and J. D. Rigden, "The Fabry-Perot electrooptic modulator," *Bell Sys. Tech. J.*, vol. 42, pp. 155–179, January 1963.

[116] J. T. Ruscio, "A coherent light modulator," *IEEE J. of Quantum Electronics (Correspondence)*, vol. QE-1, pp. 182–183, July 1965.

[117] S. M. Stone, "A microwave electro-optic modulator which overcomes transit time limitation," *Proc. IEEE (Correspondence)*, vol. 52, pp. 409–410, April 1964.

[118] R. A. Myers and P. S. Pershan, "Light modulation experiments at 16 Gc/sec," *J. Appl. Phys.*, vol. 36, pp. 22–28, January 1965.

[119] M. DiDomenico, Jr., and L. K. Anderson, "Broadband electrooptic traveling-wave light modulators," *Bell Sys. Tech. J.*, vol. 42, pp. 2621–2678, November 1963.

[120] A. Ashkin and M. Gershenzon, "Reflection and guiding of light at p-n junction," *J. Appl. Phys.*, vol. 34, pp. 2116–2119, July 1963.

[121] D. F. Nelson and F. K. Reinhart, "Light modulation by the electrooptic effect in reverse-biased GaP p-n junctions," *Appl. Phys. Letters*, vol. 5, pp. 148–150, October 1964.

[122] W. L. Walters, "Electrooptic effect in reverse-biased GaAs p-n junctions," *J. Appl. Phys.*, vol. 37, p. 916, February 1966.

[123] F. K. Reinhart, private communication.

[124] I. P. Kaminow, unpublished. Measurements at 9.2 GHz indicate $\epsilon = 2.6 \pm 0.2$, with no observable trend between -200 and $+100°C$, and tan $\delta < 0.005$.

[125] R. T. Denton, T. S. Kinsel, and F. S. Chen, "224 Mc/s Optical Pulse Code Modulator," *Proc. IEEE*, to be published; I. P. Kaminow, "Lithium niobate light modulator at 4 GHz (abstract)," *J. Opt. Soc. Am.*, vol. 55, November 1966, to be published.

Modulators for Optical Communications

FANG-SHANG CHEN, MEMBER, IEEE

Abstract—This paper reviews the field of high-speed small-aperture modulators for applications in optical communications, with emphasis on electrooptic modulation. The capabilities and limitations of electrooptic modulators are discussed based on a review of the physical origin of the electrooptic effect. Thermal and photoconduction phenomena, which may severely limit the operation of practical devices, are emphasized. The modulation power and bandwidth limitations using various schemes of electrooptic interaction are derived and compared. It is shown that lumped modulators are capable of efficient modulation for bandwidths up to about 1 GHz for visible wavelengths and are also attractive for their simplicity. For broader bandwidth capability the traveling wave or zigzag types of interaction become more efficient but with added complexity. Finally, acoustooptic and magnetooptic modulators are briefly discussed and compared with electrooptic modulators.

INTRODUCTION

ONE of the key links in the realization of transmitting large amounts of information over laser beams is a means of impressing the information onto the laser beam. This can be accomplished by modulation of variable reactive or absorptive elements outside the laser cavity, or by direct modulation of the laser. The reactive-type modulation includes electrooptic, acoustooptic, and magnetooptic interactions. Informative reviews of electrooptic modulators have been given by Anderson [1] and by Kaminow and Turner [2]; the latter also includes a tabulation of electrooptic crystals. Further discussions and extensive data on electrooptic crystals have been given by Rez [3]. A review of acoustooptic modulation has been given by Gordon [4].

In this paper a review of optical modulators for communications applications is given. Such modulators do not require a large aperture, which helps to reduce the modulation power. Only reactive-type modulators are discussed with emphasis placed on the electrooptic interaction. The direct modulation of lasers is included here while that of semiconductor lasers is reviewed by Paoli and Ripper [5].

Although different approaches to explaining the physical origin of electrooptic effects in solids have been proposed, their common feature is that the electrooptic effect is the result of field-induced shifts of optical transitions above the band gap. The approach convenient in describing the electrooptic effect in ferroelectrics is followed here. This analysis brings out naturally the capabilities and limitations of these crystals for device applications. Thermal and photoconduction phenomena, which may present serious problems that must be overcome in practical devices, are also discussed.

The modulation power required for various schemes of

Manuscript received in final form August 21, 1970.
The author is with Bell Telephone Laboratories, Inc., Murray Hill, N. J. 07974.

electrooptic interaction is derived and expressed as a product of parameters dependent on the material and the scheme of interaction. Comparison of modulation power and bandwidth limitations for various schemes of interaction is then made. Finally, acoustooptic and magnetooptic modulations are briefly discussed and compared with the electrooptic modulation.

THE ELECTROOPTIC EFFECT AND ELECTROOPTIC MATERIALS

Wave Propagation Inside Electrooptic Crystals

Electromagnetic wave propagation inside an anisotropic dielectric medium has two distinct properties: 1) the phase velocity depends on the propagation and light polarization directions relative to the crystal axes, and 2) the phase velocity direction may be different from the direction of energy flow. The phase velocity can be obtained by constructing an index ellipsoid [6] (or indicatrix) if the crystal does not show optical activity. The orientation of the indicatrix is related to the crystal axes (the principal axes of indicatrix are along the crystal axes except for monoclinic and triclinic systems) and the half-lengths of the principal axes are equal to the principal indices of refraction. The indicatrix has the following properties. Draw a straight line from the origin parallel to the phase velocity direction of a plane wave propagating in the crystal. A plane perpendicular to this line and passing through the origin forms an ellipse at its intersection with the ellipsoid. The semiaxes of this ellipse define two directions of light polarization which may be considered to be the directions of polarization of the normal modes. The indices of refraction of these modes are equal to the length of the semiaxes. Only an incident plane wave with its displacement vector linearly polarized along one of these two directions remains linearly polarized on emerging from the crystal. Since displacement vectors are related to electric fields by a permittivity tensor, they are not generally in the same direction. Waves of other polarizations will be decomposed into the normal modes and, on emerging from the crystal, they recombine in general with nonzero phase difference and thus the wave will be elliptically polarized.

The indicatrix can be expressed as

$$\frac{x_1^2}{n_1^2} + \frac{x_2^2}{n_2^2} + \frac{x_3^2}{n_3^2} = 1 \tag{1}$$

where $x_{1,2,3}$ are coordinate axes and $n_{1,2,3}$ are the principal refractive indices. A small change of refractive indices by application of an electric field induces a small change in the shape, size, and orientation of the indicatrix. This

Reprinted from *Proc. IEEE*, vol. 58, pp. 1440–1457, Oct. 1970.

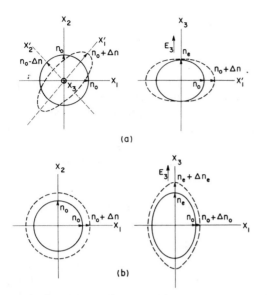

(a)

(b)

Fig. 1. Indicatrix before (solid line) and after (dashed line) a field is applied. (a) Indicatrix of KDP when a field is applied along the x_3 axis. (b) Indicatrix of LiTaO$_3$ when a field is applied along the x_3 axis.

change can be specified by giving the change in the coefficients of the indicatrix. Linear field-induced changes of the coefficients can be expressed as

$$\Delta\left(\frac{1}{n^2}\right)_{ij} = \sum_k r_{ijk}E_k \tag{2}$$

where E_k is the applied field, r_{ijk} are components of the linear electrooptic tensor, and the subscripts i and j run from 1 to 3. The tensor is expressed in contracted matrix notation, and it has been listed for all crystal classes [6], [7].

The use of the indicatrix to describe the electrooptic effect can be illustrated by considering an example of KDP (tetragonal system, $\bar{4}$2m symmetry at room temperature) which has two nonvanishing coefficients, r_{41} and r_{63}. When an electric field E_3 is applied along the crystallographic c axis, the indicatrix becomes

$$\frac{x_1^2 + x_2^2}{n_o^2} + \frac{x_3^2}{n_e^2} + 2r_{63}E_3x_1x_2 = 1, \tag{3}$$

where n_o and n_e are the ordinary and the extraordinary indices of refraction, respectively, and $x_{1,2,3}$ are along the crystallographic a, b, and c axes. By rotating the coordinates 45° around the x_3 axis, (3) can be diagonalized, and it becomes

$$\frac{x_1'^2}{(n_o + \Delta n)^2} + \frac{x_2'^2}{(n_o - \Delta n)^2} + \frac{x_3^2}{n_e^2} = 1 \tag{4}$$

$$\Delta n = \frac{n_o^3}{2}r_{63}E_3 \tag{5}$$

where $x_{1,2}'$ are the new coordinate axes, and an approximation $n_o \gg \Delta n$ is used. The indicatrices with and without E_3 are shown in Fig. 1(a).

For a plane wave with phase velocity along the x_3 axis, the ellipse defining the normal modes is obtained from (4) by setting $x_3 = 0$. The directions of polarization of these modes

are along the x_1' and x_2' axes, and their indices of refraction are $n_o \pm \Delta n$. A plane wave polarized linearly along one of these directions emerges from the crystal as a linearly polarized wave with E_3-dependent phase $(2\pi l/\lambda)\Delta n$, where l is the optical path length and λ is the vacuum wavelength.

An intensity modulation can be obtained by interference of these phase-modulated normal modes. For example, let a wave of unit intensity linearly polarized along the x_1 or x_2 axis propagate along the x_3 axis. Then the component of emergent light intensity polarized parallel to the incident polarization can be expressed as

$$I_\parallel = \frac{1}{2}\left[1 - \cos\left(\frac{4\pi l \Delta n}{\lambda}\right)\right] \tag{6}$$

and this can be spatially separated from the perpendicular component by a polarization-selective prism. The intensity of either component is not linear with E_3, however. For a small depth of modulation, this difficulty can be removed by adding a static phase difference of $\pi/2$ radians between the x_1' and x_2' axis. Then the intensity of one of the components becomes

$$I_\parallel = \frac{1}{2}\left[1 - \sin\left(\frac{4\pi l \Delta n}{\lambda}\right)\right], \tag{7}$$

which is linear with small E_3. The addition of static phase difference between the two normal modes is called optical biasing. The voltage required to change $I_\parallel$ from zero to one is usually called the half-wave voltage v_π. It is one of the parameters often used to compare materials and can be derived from (5) and (7). Since the applied voltage $v = E_3 l$ in this case,

$$v_\pi = \frac{\lambda}{2n_o^3 r_{63}}. \tag{8}$$

The electrooptic effect is called a longitudinal effect if the electric field is applied along the optical path. In this case v_π is independent of the size of the crystal as can be seen from (8).

A transverse electrooptic effect, where the applied field is normal to the light path, can be realized with KDP by choosing a light path along the x_2' axis and a field E_3 along the x_3 axis. In order to obtain an intensity modulation, the incident wave is polarized 45° to the x_1' and x_3 axes. The half-wave voltage can be shown to be

$$v_\pi = \frac{\lambda}{n_o^3 r_{63}}\left(\frac{d}{l}\right) \tag{9}$$

where d is the electrode spacing. Equation (9) shows that v_π can be reduced by choosing a small geometrical factor d/l. This is why the transverse electrooptic effect is preferred when a large index of modulation is desired.

Another well-known electrooptic material is LiTaO$_3$ which is a trigonal crystal with 3m symmetry. The indicatrix with a field E_3 along the c axis becomes

$$\frac{x_1^2}{(n_o + \Delta n_o)^2} + \frac{x_2^2}{(n_o + \Delta n_o)^2} + \frac{x_3^2}{(n_e + \Delta n_e)^2} = 1 \quad (10)$$

$$\Delta n_o = -\frac{n_o^3}{2} r_{13} E_3$$

$$\Delta n_e = -\frac{n_e^3}{2} r_{33} E_3$$

where $x_{1,2,3}$ axes are along the crystallographic a, b, and c axes, respectively. The indicatrix is shown in Fig. 1(b). Intensity modulation can be obtained by interference of the x_1 and x_3 polarized light components. The half-wave voltage becomes

$$v_\pi = \frac{\lambda}{n_e^3 r_{33} - n_o^3 r_{13}} \cdot \frac{d}{l}. \quad (11)$$

Since r_{33} and r_{13} have the same sign and $\Delta n_e > \Delta n_o$ for LiTaO$_3$ [8], the voltage required to change the phase of the x_3 polarized wave by π is less than that necessary to change the phase difference of x_3 and x_1 polarized waves by π. Thus phase modulation is more efficient than intensity modulation in LiTaO$_3$.

The indicatrix provides information about the phase velocities inside an anisotropic crystal. The direction in which a bounded part of the wavefront travels is the direction of energy flow. The directions of phase velocities and energy flow will be different unless the directions of polarization of the normal modes are parallel to the principal axes of the indicatrix [6], [9]. Thus it is possible that the two normal modes for a given wave normal propagate along different paths inside the crystal and are spatially separated on emergence. This adds inconvenience to intensity modulation, which depends on interference of the two normal modes. This inconvenience can be removed, however, by use of two pieces of crystal oriented so that the path difference is compensated [10]–[12].

Physical Origin of Electrooptic Effect

It has been shown [13], [14] that the linear electrooptic effect in piezoelectric crystals far above their acoustic resonance frequencies (i.e., clamped electrooptic effect) can be separated into two types of microscopic interaction: an applied electric field modifies the electronic polarizability directly in the absence of lattice displacements and an applied field produces a lattice displacement, which in turn modifies the electronic polarizability. The first type, which may be called a purely electronic contribution to the electrooptic effect (nonlattice contribution), is also the physical origin of optical mixing or second-harmonic generation and its magnitude can be obtained from second-harmonic measurements. The second type, the lattice contribution to the electrooptic effect, can be determined from a combination of Raman scattering and infrared absorption measurements. The electrooptic coefficients thus determined are in good agreement with the coefficients measured directly for LiNbO$_3$ and LiTaO$_3$ [13]. For LiNbO$_3$ the nonlattice contribution was found to be less than 10 percent.

There is a third contribution to the electrooptic coefficients when the frequency of the applied electric field falls below or near the acoustic resonant frequencies of the sample. The applied field strains the crystal via piezoelectric and/or electrostrictive coupling, and the indices of refraction change due to the strain-optic effect. This component of electrooptic coefficients can be neglected if the frequency of the applied field is far above the acoustic resonant frequencies, in which case the crystal is effectively clamped. The field-induced change of electronic polarizability has been shown to originate in field-dependent shifts of the frequency and strength of individual dipole oscillators representing interband optical transitions [15]–[18]. These shifts can be observed experimentally by measuring the electroreflectance above the band gap [19]–[21].

DiDomenico and Wemple described a model of electrooptic effects [18] of oxygen octahedra ferroelectrics that may be thought of as being composed of a network of oxygen octahedra (BO$_6$) with a transition metal ion (B) at its center. This class of ferroelectrics includes many of the well-known electrooptic and nonlinear crystals such as LiTaO$_3$, LiNbO$_3$, BaTiO$_3$, and Ba$_2$NaNb$_5$O$_{15}$. The change in B–O spacing as the temperature is lowered through the transition not only produces a spontaneous polarization but also shifts the energy band. Thus the energy shift ΔE, measured with respect to the energy bands in the centrosymmetric phase, can be expressed as proportional to P^2, where P is the total polarization, spontaneous plus any field-induced component. Through the Sellmeier equation, the change in index of refraction, Δn, can be linearly related to ΔE; hence $\Delta n \propto P^2$. This relation was observed earlier by Merz in BaTiO$_3$ [22]. The linear electrooptic coefficients of these ferroelectric oxides can be calculated using this relation in the following way. The change of refractive index of the perovskites (cubic system, point group m3m) can be expressed as

$$\Delta n = -\frac{n_o^3}{2} g P^2 \quad (12)$$

where n_o is the index of refraction and g is the temperature-independent quadratic-electrooptic coefficient. The linear electrooptic effect in ferroelectrics can be regarded as fundamentally a quadratic effect biased by the spontaneous polarization. Let an electric field E be applied parallel to the spontaneous polarization P_s. By substituting $P = P_s + \varepsilon_0(\kappa - 1)E$, where ε_0 is the free-space permittivity and κ is the dielectric constant of the crystal along the P_s direction, into (12), one obtains

$$\Delta n = -\frac{n_o^3}{2} g P_s^2 - n_o^3 g \varepsilon_0 (\kappa - 1) P_s E \quad (13)$$

where the term containing E^2 is neglected since $\kappa \varepsilon_0 E \ll P_s$. The first term in (13) is the spontaneous-polarization-induced index change and the second term is the linear electrooptic effect. Comparing (13) with the definition of linear electrooptic r coefficients defined by (2) and (10), and neglecting the subscripts, we find that

$$r = 2g\varepsilon_0(\kappa - 1)P_s. \tag{14}$$

Since κ follows the Curie–Weiss law, r is therefore anomalous at the transition temperature. In order to compare the g coefficients of different ferroelectrics, the volume density of BO_6 octahedra in a nonperovskite ferroelectric compared to that of a perovskite must be taken into account. This can be done [18] by defining a packing density ζ as the ratio of the BO_6 density of the nonperovskite to that of a reference perovskite with a 4-Å lattice constant. The g coefficients of nonperovskites referred to in (13) and (14) should be replaced approximately by $g\zeta^3$ where g now refers to the reference perovskite. ($\zeta = 1 \sim 1.06$ for perovskite and tungsten bronze and $\zeta = 1.2$ for $LiNbO_3$.) Using the measured values of r, κ, and P_s for various oxygen octahedra ferroelectrics and (14), the values of g are found to be essentially the same among these crystals and also the same as those observed in the paraelectric perovskites [18], [23], [24]. The existence of such universal constants in these crystals simplifies the analysis of practical modulator performances using oxygen octahedra ferroelectrics.

Temperature Dependence of Electrooptical Parameters in Oxygen Octahedra Ferroelectrics

Since the oxygen octahedra ferroelectrics include many important electrooptic materials, and since their g coefficients are not only the same both above and below their transition temperatures but also very similar among crystals having different transition temperatures, it is of interest to determine whether there is an optimum operating temperature from the device point of view.

The subscripts of the r and g coefficients neglected in (14) can be restored if the direction of polar axis of the ferroelectric with respect to the symmetry axes of the oxygen octahedron is known. It has been shown [18] that

$$r_{33} - r_{13} \approx 2\varepsilon_0(\kappa - 1)P_sG/\zeta^3 \tag{15}$$

where $G/\zeta^3 \equiv g_{11} - g_{12} \approx g_{44} \approx 0.13/\zeta^3 \ m^4/C^2$ and $g_{11,12,44}$ are the nonzero elements of the g tensor. Using (11) and (15), and defining a reduced half-wave voltage by setting $d = l$ in (11), i.e.,

$$V_\pi = v_\pi \frac{l}{d} \tag{16}$$

one obtains

$$V_\pi = \left(\frac{\lambda\zeta^3}{2n_o^3\varepsilon_0 G}\right) \cdot \left(\frac{1}{\kappa P_s}\right) \tag{17}$$

where it is assumed that $\kappa \gg 1$ and $n_o \approx n_e$. Equation (17) also applies to the paraelectric phase of the crystal if P_s is replaced by a field-induced bias polarization [24]. A parameter that often appears in the analysis of modulators is the stored-energy parameter $U_\pi = \kappa V_\pi^2$. Its value from (17) is

$$U_\pi = \left(\frac{\lambda\zeta^3}{2n_o^3\varepsilon_0 G}\right)^2 \cdot \left(\frac{1}{\kappa P_s^2}\right). \tag{18}$$

It is remarkable that (17) and (18) relate V_π and U_π of all oxygen octahedra ferroelectrics, at all temperature ranges, to only two material parameters: κ and P_s. Except for ζ, the other factors in the equations are approximately independent of materials and temperature. For a given κ and P_s, the $LiNbO_3$-type crystals have V_π higher by 1.7 due to their large $\zeta (\approx 1.2)$. Equation (17) also points out the limitation of ferroelectrics for applications in modulators, i.e., a small V_π is accompanied by a large κ. Nevertheless, V_π and U_π are still small compared to nonferroelectrics and they are still important electrooptic materials. The inverse relationship of κ and V has been verified experimentally [25] using Ba, Sr, Na niobate composite crystals that have tungsten-bronze-like structures. P_s of these crystals remains relatively constant while the Curie temperature varies between 560°C and 200°C. It has been observed that the product κV_π remains constant.

It has been shown [26] that both $V_\pi(\propto 1/\kappa P_s)$ and $U_\pi(\propto 1/\kappa P_s^2)$ decrease monotonically as the Curie temperature T_C is approached from either higher or lower temperatures. However, there are a few factors that mitigate against operating modulators very near T_C. The static birefringence $\Delta(n_e - n_o)_s$ arising from the first term in (13) is temperature sensitive via its dependence on P_s^2 and dP_s^2/dT increases as $|T - T_C|$ approaches zero.

A scheme will be discussed later that minimizes the degrading effect of fluctuations of $\Delta(n_e - n_o)_s$ on the modulator performance if $d\Delta(n_e - n_o)_s/dT$ is spatially uniform. The spatially inhomogeneous static birefringence due to internal heating of the crystal by an applied electric field is difficult to compensate, hence the temperature sensitivity of the static birefringence remains an important consideration in selecting $(T - T_C)$.

The half-wave voltage V_π is also temperature dependent. Fortunately, for the practically important case of $T < T_C$, the temperature dependence of V_π is not large enough to impose a difficulty.

The electrical Q of ferroelectrics is known to decrease as T_C is approached. Together with finite thermal conductivity and $d\Delta(n_e - n_o)_s/dT$, lower Q implies a large spatial inhomogeneity of indices of refraction which degrade the modulator performance [26]. The ferroelectrics also tend to depole near T_C unless an electric field is applied to prevent it. It can be easily seen from (13) that the completely depoled ferroelectrics show no linear electrooptic effect. Thus the choice of $(T - T_C)$ will have to be decided by balancing these opposing factors.

Other Material Properties Pertinent to Electrooptic Modulators

Practical electrooptic materials should preferably have the following properties: small V_π and U_π, small dielectric dissipation, good thermal conductivity, good optical quality in proper size, and easy polishing. In addition to these, good ohmic contacts to the crystals may be necessary in some applications, and in the case of ferroelectrics they should remain poled and free of optically induced refractive index changes (optical damage). Problems associated with electrodes and optical damage will be discussed here, since these are often critical factors in selecting the material.

When a dc electric field is applied to a solid having a finite electrical conductivity via blocking contacts (which permit transfer of charge carriers to but not from the electrode), the field distribution inside the material may no longer be uniform. In order to simplify the problem, assume that a cloud of free and trapped electrons compensated by an immobile matrix of positive charges is distributed throughout the crystal [27], [28]. When a dc field is applied, these electrons drift toward the positive electrode, in the meantime getting trapped and thermally reexcited out of the traps until eventually leaving the anode; however, they cannot reenter at the cathode due to the blocking contact. A region adjacent to the cathode will be swept clear of electrons leaving behind an immobile positive space-charge layer. This process continues until the applied voltage is completely absorbed by the space-charge layer. If the initial concentration of electrons is sufficiently small and the applied voltage sufficiently large, the electrons will be completely swept out through the anode, leaving behind only positive immobile charges in the crystal. On the other hand, for a high electron concentration or a low voltage, the final state will consist of two regions, one with positive charges near the cathode where the total voltage drops and the other a field-free neutral region. (Ferroelectric oxides are often p-type. The space-charge layer then appears at the anode.) Let V_0 be the applied voltage, ρ the electron charge density, and x_d the thickness of the space-charge layer. Then by integrating Poisson's equation, the space-charge field E becomes

$$E = \frac{\rho}{\varepsilon}(x - x_d) \qquad (19)$$

where ε is the permittivity of the crystal and $x=0$ is at the cathode. Integrating E along x and setting it equal to V_0, one obtains

$$x_d = \sqrt{\frac{2\varepsilon V_0}{\rho}}. \qquad (20)$$

The motion of the layer is regulated by a relaxation time

$$\tau = \frac{x_d D}{2\mu V_0}, \qquad (21)$$

where D is the electrode spacing, and μ is the drift mobility of the electrons (including the effect of trapping). By substituting (20) into (21), τ can be shown to be proportional to the classical dielectric relaxation time. As the solid is repeatedly charged and discharged, space-charge layers build up on both electrodes.

By applying a dc field to $BaTiO_3$ above its Curie temperature [29] and to KTN [28], [30], the building up in time of space-charge layers near the electrodes has been observed with samples viewed between crossed polarizers.[1] From a

[1] Depletion layer of 200-μ thickness has been observed in KTN of $10^8 \ \Omega \cdot cm$ resistivity [28].

practical device point of view, the formation of space-charge layers increases V_π because most of the applied field is not seen by the light beam, causes a poor extinction due to the nonuniform field distribution, and causes the optical bias to drift due to the finite relaxation times τ. The best way to avoid difficulties associated with blocking contacts is obviously to use ohmic contacts [30], although the practice of providing such contacts to most of the electrooptic crystals is not well known. If use of blocking contacts is inevitable, then (20) suggests that the resistivity of the crystal must be sufficiently large so as to extend x_d (the region of nonzero field) to a large cross section of the crystal. If the electron clouds are completely swept out of the crystal, the electric field inside the crystal will consist of a linear term expressed in (19) and a constant term. If the latter is much larger than the former, the difficulties associated with blocking contact can be expected to be minimal. Also, if the period of the applied field is much shorter than τ, there is no time for the space-charge layers to build up and the applied field distributes uniformly inside the crystal [31]–[33].

There is another anomalous behavior of electrooptic crystals often observed when a dc field is applied transverse to the illuminating light. The photoexcited carriers drifting out of the illuminated region get trapped near the periphery and remain there [34]–[36]. The local space-charge field thus created causes an inhomogeneity of the index of refraction via the electrooptic effect of the crystal. A similar mechanism is also thought to be responsible for the optical damage observed in poled ferroelectric crystals [37]–[41]. In the latter case, a spatial inhomogeneity of the refractive index near the laser beam is present without a dc applied field. The resistance to optical damage in $LiTaO_3$ has been increased from the laser intensity the order of 1 W/cm² at $\lambda = 0.488 \ \mu$ to 500 W/cm² by annealing the crystal at 700°C with an electric field along the c axis [42]. It was later shown that hydrogen was diffused into the crystal during the field annealing [43]. The mechanism by which the resistance to optical damage is enhanced by the presence of hydrogen is not understood, although it is speculated that trapping sites are modified by the presence of hydrogen in such a way that photoconduction is minimized. In addition to improving the resistance to optical damage, the presence of hydrogen in ferroelectric oxides is found to speed up the process of poling [44]. The degree to which problems associated with thermal and/or photoconduction can be overcome often dictates the choice of crystals for practical device applications.

Electrooptic Materials

Since Kaminow and Turner [2] have given an extensive list of electrooptic crystals, Table I here lists only those crystals which are in common use (DKDP, $LiTaO_3$), which represent fairly recent growth effort on oxygen octahedran ferroelectrics ($Ba_2NaNb_5O_{15}$, SBN), which are improved early ferroelectrics ($BaTiO_3$), or which are useful at 10.6 μ (GaAs, CdTe).

The Curie temperature T_C of ferroelectric oxides can be

TABLE I
ELECTROOPTIC MATERIALS PARAMETERS

Material	Sym.(a)	T_C	κ(c)	n_c	n_a	K(d)	$\dfrac{d(n_c - n_a)}{dT}$	V_π(b)	U_π(b)
		(°K)				(W/m°C)	(°C^{-1})	(kV)	(10^8 volt^2)
KD$_2$PO$_4$(DKDP)	$\bar{4}$2m	222 [45]	(T) 50 [45] (S) 48 [46]	1.47 [47]	1.51 [47]		7×10^{-6} [47]	7.5	30
LiTaO$_3$	3m	890	(T) 47 [48] (S) 43 [48]	2.180 [49]	2.176 [49]	4.4 [50]	5×10^{-5} [51]	2.8 [48]	3.7
Ba$_2$NaNb$_5$O$_{15}$	mm2	833	(T) 51 [52] (S) 33 [51]	2.22 [52]	2.32 [52] ($\approx n_b$)	3.4 [50]	4×10^{-5} [51]	1.57 [52]	1.26
Sr$_{0.25}$Ba$_{0.75}$Nb$_2$O$_6$ [53]	4mm	470	(T) 250 [51] (S) 160 [51]	2.26	2.32	1.98 [50]	1.3×10^{-4} [51]	0.48 [51]	0.58
BaTiO$_3$ [54]	4mm	405	(T) 135 [55] (S) 60 [55]	2.36	2.42		1.1×10^{-4} [54]	0.48 [56]	0.35
GaAs [57]	$\bar{4}$3m		(T) 11.5	3.34 (cubic)				91	
CdTe [58]	$\bar{4}$3m			2.3 (cubic)				44	

(a) All data listed are measured at room temperature and at $\lambda = 0.63 \mu$, except GaAs and CdTe which are for $\lambda = 10.6 \mu$.
(b) For intensity modulation; unclamped. Use r_{63} in DKDP, r_{41} in GaAs, and CdTe, r_{13}, and r_{33} for the other materials.
(c) Dielectric constant along the c axis. (T) = constant stress (unclamped). (S) = constant strain (clamped).
(d) Thermal conductivity.

varied over a wide range by mixed compounds substituting the cations, while the basic BO$_6$ oxygen octahedra remain unmodified. The values of κP_s at room temperature (especially κ) change with T_C, which in turn change V_π. It is emphasized that lower V_π alone, if accompanied by a large increase in κ, does not necessarily make the material superior. Many other composite niobates other than those listed in Table I have lower V_π by having T_C near the room temperature [25], [59], [60]. The smaller V_π is accompanied by a large dielectric constant and other difficulties associated with operating a modulator near T_C as mentioned earlier.

ELECTROOPTIC MODULATORS

In this section, various schemes of electrooptic interaction using the transverse linear electrooptic effect will be discussed. Emphasis will be placed on broad-band in-

tensity modulation using interference of the extraordinary and ordinary rays. Before discussing different schemes of interaction, considerations common to all of them will be discussed first.

General Considerations

Geometrical Factor: The half-wave voltage in materials showing a transverse electrooptic effect is proportional to a geometrical factor d/l as shown in (9) and (11). Thus the optimum geometry is to have the aperture of the crystal just large enough so that the beam passes through the sample. A Gaussian beam with the smallest cross section over the crystal length l is one that is focused so as to have a near-field length equal to l (i.e., confocal mode in the crystal [2]). In this case the geometry of the crystal is given by

$$\frac{d^2}{l} = S^2 \frac{4\lambda}{n\pi}, \qquad (22)$$

where n is the refractive index and S is a safety factor (≥ 1) for the beam to pass through the crystal. For $S=1$, the light intensity at the end faces of the crystal decreases by a factor e^{-2} at the crystal edges from its peak value (if the beam is perfectly aligned). In practice, a beam has been passed through a rod with little difficulty for $S \approx 3$ [31].

Effect of Velocity Matching and Finite Transit Time: In order to have a cumulative interaction between the modulating wave and the light wave, the component of the modulating wave group velocity in the direction of the light propagation must be equal to the group velocity of the light beam [61]. The group velocity of light is approximately the same as the phase velocity if the dispersion is small, and for broad-band modulation, the group velocity of the modulating wave must be equal to its phase velocity over the desired bandwidth. Thus the condition for matching the group velocities will be met if the phase velocities are matched.

For imperfect velocity matching, the optical phase retardation is reduced by a factor $\sin u/u$, where

$$u = \frac{\omega l}{2}\left(\frac{1}{v_0} - \frac{1}{v_m}\right), \qquad (23)$$

in which v_0 and v_m are the parallel components of the phase velocities of the light and modulating wave, respectively, and ω is the angular frequency of the modulating wave.

It will be assumed in the following that the light velocities for the extraordinary and ordinary rays are the same. If they differ, and u for both rays becomes significantly different, broad-band intensity modulation can be accomplished by phase modulating one light polarization followed by an optical discriminator [62], [63]. The entire discussion in this paper remains essentially unaltered.

For narrow-band modulation, a scheme has been proposed to overcome the problem associated with the velocity mismatch [64]. The modulating and the optical waves are allowed to propagate without velocity synchronism until a phase difference of π radians (relative to ω) is accumulated. This corresponds to $u=\pi/2$ in (23). Further interaction would reduce the accumulated modulation unless the following section of the modulator is arranged so that the induced refractive indices have opposite signs. For light propagating along the x_2 axis and the voltage along the x_3 axis of either KDP or LiTaO$_3$ this can be accomplished by inverting the x_3 axis of the second section of the modulator.

Velocity matching over a broad-band will be considered in detail later.

Compensation of Temperature Dependence of Birefringence: Temperature dependence of the static birefringence can cause fluctuation of the depth of intensity modulation and/or a poor extinction. However, this can be compensated by the schemes [65] shown in Fig. 2. In the first scheme, a half-wave plate is inserted between the two sections so that the extraordinary ray of the first section becomes the ordinary ray of the second section and vice versa.

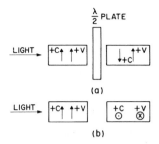

Fig. 2. Schemes of compensating temperature dependence of the static birefringence.

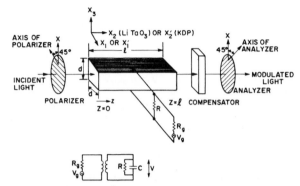

Fig. 3. Lumped modulator and its equivalent electrical circuit.

In the second scheme, the second section is rotated by 90° to achieve the same effect without a half-wave plate. In order to cancel the static birefringence and yet add the induced birefringence of the two sections, the relative sense of the $+c$ axis (or the $+x_3$ axis) and the modulating voltage must be opposite for the two sections and their optical path length must be equal.

Fluctuation of the static birefringence due to the ambient temperature can be easily stabilized with these schemes. A spatially inhomogeneous temperature distribution inside the modulator can be compensated only if the beam is essentially uncollimated by the refractive index gradient and if the temperature distribution in both sections of the modulator is the same.

Compensation of the temperature dependence of the static birefringence can also be achieved by using an interferometer with phase modulators in each arm [66]. Only the temperature difference between the two arms has to be stabilized. Decollimation of the optical beam due to the index gradient, which is caused by finite thermal conductivity, finite electrical Q, and finite temperature dependence of the refractive index, sets the maximum modulating power density that can be applied to the modulator, and this, in turn, eventually becomes a limiting factor on the bandwidth. In this paper it will be assumed that the bandwidth limitation from the temperature dependence of static birefringence can be ignored.

Various Schemes of Electrooptic Interactions

Lumped Modulators: If the optical path length of the modulator l is much shorter than one half of the modulating wavelength inside the crystal, i.e., $l \ll \pi c/\omega\sqrt{\kappa}$ where κ is the dielectric constant of the crystal, then the modulator may be regarded as a lumped capacitance C terminating the RF generator as shown in Fig. 3 and the modulating field is

TABLE II
COMPARISON OF LUMPED MODULATORS AT $\lambda = 0.63\ \mu$

Material	LiTaO$_3$ [31][b]	DKDP [68]	LiTaO$_3$ [69]	LiNbO$_3$ [70][c]
f_0 (GHz)	baseband	baseband	4.2	1.5
Δf (GHz)[a] (limited by capacitance)	1.3	0.22	0.04	1.5
Δf (GHz) (limited by transit time)	3	0.53	—	—
d (mm)	0.25	0.75	0.23	0.5
l (mm)	10	57	4.2	5
S	2.9	4.3	6.7	11.6
$P/\Delta f$[c] (mW/MHz)	1.1	60[d]	24	37

(a) Assume the load resistance $R = 50$ ohms.
(b) Based on round-trip mode.
(c) For 100 percent intensity modulation.
(d) A capacitance equalization network effective for $\Delta f = 100$ MHz is used.
(e) For LiNbO$_3$ [53], $\kappa V_\pi^2 \approx 2.9 \times 10^7$ volt^2.

uniform inside the crystal. The lossy component of the crystal can be generally ignored, and the modulator is terminated with a load resistance R that is matched to the generator impedance R_g through an ideal transformer with a bandwidth assumed to be as broad as necessary. The voltage across the modulator v decreases by a factor of $1/\sqrt{2}$ from the peak value at dc at the angular frequency

$$\Delta\omega = 2/CR, \qquad (24)$$

where $\Delta\omega$ will be defined as the bandwidth of the modulator. In (24), C includes the capacitance of the crystal and the parasitic capacitance of the crystal mount; the latter, however, will be neglected in the following analysis. The depth of modulation also decreases as the transit time of light across l becomes a significant fraction of the period of the modulating voltage. It decreases by a factor of $1/\sqrt{2}$ at the frequency which makes $u = 1.4$ in (23). For the lumped modulator, one can approximate $1/v_m$ as zero and thus the modulation bandwidth from the finite transit time effect alone is

$$\Delta\omega = \frac{2.8c}{nl}. \qquad (25)$$

The bandwidth of a lumped modulator is given either by (24) or (25) depending on which effect is dominant.

As the modulating frequency increases such that one half of its wavelength inside the medium is a significant fraction of l, the modulating field will no longer be uniformly distributed inside the crystal (i.e., $1/v_m \neq 0$). The modulating field can then be regarded as composed of two traveling waves propagating in opposite directions, each interacting independently with the light beam. The bandwidth limitation from the velocity mismatch for such modulators, which may be called standing-wave modulators, has been derived by Bicknell [67]. It will be assumed in this paper, however, that (24) determines the bandwidth of the lumped modulator. The percent modulation can be defined as [31]

$$\frac{\text{percent modulation}}{100} = 2J_1\left(\frac{\pi v}{v_\pi}\right)$$

where J_1 is the Bessel function of first order, and v is the peak modulating voltage across the crystal. A 100 percent modulation is obtained when $v = 0.383\ v_\pi \equiv v_0$. The reactive power stored in the crystal in order to obtain 100 percent modulation at low frequencies can be expressed as

$$P = \frac{\Delta\omega C}{2}\left(\frac{v_0}{\sqrt{2}}\right)^2. \qquad (26)$$

Note that P of (26) provides only 70.7 percent modulation at the band edge. P can also be regarded as a measure of efficiency of the particular type of modulation interaction. For the crystal with a square cross section as shown in Fig. 3 and using (22), (26) becomes

$$P = \frac{0.587\varepsilon_0\lambda S^2}{4n\pi}\cdot\kappa V_\pi^2\cdot\Delta\omega. \qquad (27)$$

The necessary drive power increases linearly with bandwidth because the terminating resistance R has to be decreased in order to maintain an approximately constant voltage (and a constant depth of modulation) over the increased bandwidth. If a parallel inductance is provided to resonate the capacitance at ω_0, (27) gives the power necessary to produce at least 70.7 percent modulation over the bandwidth $\Delta\omega$ centered at ω_0, provided the finite transit time of light through the crystal is not a limiting factor.

It can be easily shown that the power expressed in (27) can be reduced by one half if the light beam makes a round trip [31] inside the modulator and if the cross section of the crystal d^2 is enlarged so that S for the single-trip and the round-trip modes remains the same. There is no change in the bandwidth as long as it is not limited by transit time.

Characteristics of some lumped modulators are listed in Table II. The small $P/\Delta f$ achieved by Denton et al. [31] is mainly due to the small S and the round-trip mode used.

For all the modulators listed, the measured values of $P/\Delta f$ agree within a factor of two with the calculated values using (27) and Table I.

Other lumped or standing-wave modulators have been reported in [71]–[77].

Traveling-Wave Modulators: The bandwidth limitation resulting from the mismatch of velocities of light and modulating waves can be extended by using a substantially dispersionless structure in which both waves can propagate with equal phase velocities. One such structure is the two-parallel-plate guide shown in Fig. 4. The modulating wave propagates as a TEM wave with its phase velocity increased so as to approach the light velocity inside the crystal. This is achieved by propagating part of the energy of the modulating wave outside of the crystal. Since this part does not contribute to the interaction of light and modulating waves, the velocity synchronization is achieved at the expense of the increased modulating power in this broad-banding scheme.

The phase velocity of a modulating wave in such structures can be expressed in terms of an effective dielectric constant κ_e which is the square of the ratio of the velocity in free space to that in the structure. At the low-frequency limit, κ_e becomes [78]

$$\kappa_e \approx \frac{\kappa d}{w} \tag{28}$$

where w is the width of the plates as shown in Fig. 4. For $(n/\sqrt{\kappa}) \lesssim 0.3$, the structure is essentially dispersionless only in the frequency band where $d \lesssim 0.1\,\lambda_m$ ($\lambda_m = 2\pi c/\omega\sqrt{\kappa}$ is the modulating wavelength in the medium) is satisfied [78], which sets the maximum bandwidth achievable by a traveling-wave modulator using such structures. Within the frequency band where the phase velocity is independent of frequency, the modulator bandwidth $\Delta\omega$ depends on the mismatch between the light and modulating wave velocities and can be expressed as

$$\frac{\Delta\omega l}{2c}(\sqrt{\kappa_e} - n) = 1.4, \tag{29}$$

where $\Delta\omega$ is defined as the angular frequency at which the depth of modulation decreases to 70.7 percent.

The impedance of the transmission line is

$$Z = \frac{377}{\sqrt{\kappa_e}}\cdot\left(\frac{d}{w}\right)\text{ohms.} \tag{30}$$

In order to provide 100 percent modulation, the power applied to the transmission line (assumed lossless) is

$$P = \frac{1}{2Z}\left(0.383\,V_\pi\frac{d}{l}\right)^2. \tag{31}$$

Substituting (22), (28), and (30) into (31), one obtains

$$P = \frac{0.587\lambda S^2\varepsilon_0}{4n\pi}\cdot(\kappa V_\pi^2)\cdot\frac{2c}{l\sqrt{\kappa_e}}. \tag{32}$$

Note that P of the traveling-wave modulators is inversely

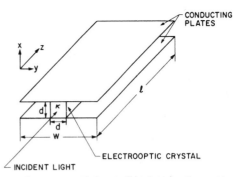

Fig. 4. Traveling-wave modulator using the two-plate guide structure.

proportional to l in contrast to the lumped modulators where P is independent of l.

By reflecting both the light beam and the modulating wave at the end of the traveling-wave structure and also by matching the velocity of the forward light wave to the forward modulating wave, the reflected light and modulating waves can also interact cumulatively [79]. This scheme is equivalent to doubling the length of the crystal. The modulation power is reduced by one half as in the round-trip lumped modulator, provided S and κ_e remain unchanged. The bandwidth for the round-trip traveling-wave modulator will be smaller by one half due to the increased light path as can be seen from (29).

Traveling-wave modulators using parallel-plate guides have been constructed [80]–[82] with $l \approx 100$ cm using many pieces of ADP or KDP. The potential modulation bandwidth of a few gigahertz can be realized if impedance matching over this bandwidth can be accomplished. Bicknell *et al.* [83] have constructed a traveling-wave modulator with $l = 16$ cm using a chain of potassium-dihydrogen-arsenate (KDA) crystals loading a coaxial transmission line. Broadband impedance matching at the input and termination of the coaxial line has been achieved over $0 \sim 3$ GHz, which is also the bandwidth as limited by the velocity matching. A traveling-wave modulator using a rod of LiTaO$_3$ of dimensions 0.25 by 0.25 by 10 mm^3 (the same as the lumped modulator of [31]) mounted in microstrip has been described by White and Chin [84]. The bandwidth is limited to 2.9 GHz by the velocity matching, in contrast to 1.3 GHz limited by the load capacitance in the case of the lumped modulator [31].

A traveling-wave modulator at 6 GHz and 10 percent bandwidth has also been constructed using KDP in the ring-plane traveling-wave circuit [85]. The phase velocity of such a structure is relatively constant only within a small percent bandwidth ($\Delta f/f_0 \approx 20$ percent), although Δf (≈ 1.2 GHz for $f_0 = 6$ GHz) is still comparable to that in baseband traveling-wave modulators using a parallel-plate guide or a coaxial line. In contrast to lumped modulators where the modulation power is proportional only to the bandwidth Δf and not to the percent bandwidth $\Delta f/f_0$, the traveling-wave modulators are not bound by such a rule.

Zigzag Modulator: The velocity matching condition of (23) states only that the parallel components of the phase velocities of optical and modulating waves be matched. In

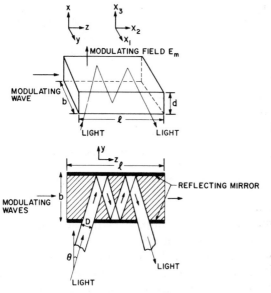

Fig. 5. Zigzag modulator.

addition to matching the velocities by increasing the modulating wave, it can also be achieved by having the light travel a zigzag path at an angle

$$\theta = \sin^{-1}\frac{n}{\sqrt{\kappa}} \qquad (33)$$

along the transmission line carrying the modulating wave [86]–[88] (see Fig. 5).

Let us assume that the light is propagating in the y–z plane of Fig. 5 and the modulating electric field is along the x axis, which is also the crystallographic x_3 axis of the crystal (assume either LiTaO₃ or KDP). The peak voltage required for 100 percent intensity modulation is

$$v_0 = 0.385\, V_\pi\!\left(\frac{d\sin\theta}{l}\right), \qquad (34)$$

and the power for 100 percent intensity modulation is

$$P = \frac{1}{2}\cdot\frac{\sqrt{\kappa}}{377}\, bd\!\left(\frac{v_0}{d}\right)^2 \qquad (35)$$

where b and d are the width and the electrode spacing, respectively, of the crystal as shown in Fig. 5. To avoid beam divergence, we impose a restriction that the light path inside the crystal ($=l/\sin\theta$) be the same as the near-field length, and modify (22) as

$$\frac{d^2\sin\theta}{l} = S^2\,\frac{4\lambda}{n\pi}. \qquad (36)$$

In order to bounce a parallel light beam of diameter $D\ (\approx d/S)$ more than once inside the crystal, it is necessary that $b \ge D/(2\sin\theta)$ from a geometrical consideration. The crystal will be swept twice by the light beam (once by zig and once by zag) when $b = D/(2\sin\theta)$ and this choice of b will result in the minimum modulation power. However, from a practical consideration of focusing the beam into the crystal with some ease, it will be assumed that

$$b = d/(2\sin\theta). \qquad (37)$$

Using (33)–(37), one obtains

$$P = \frac{0.587\lambda S^2\varepsilon_0}{4n\pi}\cdot(\kappa V_\pi^2)\cdot\frac{c}{l}. \qquad (38)$$

The bandwidth of a zigzag modulator is limited by two factors. Both the phase velocity of the modulating wave and the impedance of the circuit become frequency dependent as the width b of the modulator approaches a significant fraction of the wavelength of the modulating wave in the medium, and the projection of the beam diameter D along the direction of propagation of the modulating wave must be smaller than a half-wave of the modulating signal in the medium. Generally the latter is less restrictive than the former. For the same crystal size, the zigzag structure is less dispersive compared to the two-plate guide where the modulating crystal occupies only a small volume of the guide. Since the dispersion of these structures is known only approximately, the bandwidth of zigzag modulators will be somewhat arbitrarily defined as

$$b \approx 0.2\,\lambda_m. \qquad (39)$$

Upon substitution of (33) and (37) into (39), the bandwidth becomes

$$\Delta f = \frac{0.4c}{d}\cdot\frac{n}{\kappa}. \qquad (40)$$

In spite of a potentially efficient interaction between the optical and modulating waves over a large bandwidth, technical complexities involved in constructing zigzag modulators have made this approach rather unattractive in practice. Only recently Auth [89] reported construction of such a modulator using DKDP in X-band waveguide. Modulation was observed from 7.8 GHz to 12.4 GHz but the depth of modulation versus the modulating power was not shown. The scheme of noncollinear velocity matching has also been used to modulate a 0.63-μ laser beam with a HCN laser beam at 964 GHz using LiNbO₃ [90].

Optical Waveguide Modulators: If an optical waveguide can be built using an electrooptic material, the diffraction spread of the light beam can be eliminated and the cross section of the modulator can be reduced to the order of a wavelength, independently of the optical path length. This can reduce the required modulation power [91], [92].

One-dimensional waveguide modulators where the light is confined in only one dimension have been realized with GaP diodes [93], [94]. The reverse-biased p-n junctions of these diodes form dielectric discontinuities in the direction normal to the plane of the junction, but there are no dielectric discontinuities in the direction parallel to the junction to confine the light. Thus, guiding of the light beam is achieved only in one dimension in these diodes instead of the more desirable two-dimensional guiding. The power required for 100 percent intensity modulation in such a one-dimensional waveguide modulator can be calculated as follows. Let d be the thickness of the junction over which the modulating voltage is applied, l the optical path length (junction length), and b the junction width. It will be assumed that a light beam is focused to a diameter d to il-

TABLE III

DRIVE POWER AND BANDWIDTH FOR VARIOUS SCHEMES OF
ELECTROOPTIC INTERACTION

	Lumped	Traveling Wave	Zigzag
$\left(\dfrac{2n}{0.587\varepsilon_0\lambda S_u^2}\right)\cdot P$	Δf	$\dfrac{c}{\pi l\sqrt{\kappa_e}}$	$\dfrac{c}{2\pi l}$
Δf	$1/\pi CR^*$ $1.4c/\pi nl$	$\dfrac{1.4c}{\pi l(\sqrt{\kappa_e}-n)}$	$\dfrac{0.4c}{d}\cdot\dfrac{n}{k}$

* Limited by load capacitance. The rest are limited by transit time or velocity mismatch.

luminate the junction, and, on emerging from the junction, the beam size remains d in the direction of modulating field but becomes larger along the junction width due to diffraction spread. The dimension b must be larger than the maximum beamwidth by a safety factor S as defined in (22). Let us also assume that the diffraction spread derived for Gaussian beams [95] also applies to the one-dimensional case; then

$$b = Sd\sqrt{1+(4l\lambda/n\pi d^2)^2} \approx 4Sl\lambda/n\pi d \qquad (41)$$

where the approximation is valid for typical parameters [93], $d \approx 5$ μ, and $l=1$ mm. Substituting the junction capacitance $C = \kappa\varepsilon_0 bl/d$, $v_0 = 0.383\, V_\pi d/l$, and (41) into (26), the power for 100 percent modulation in the case of the diode modulator becomes

$$P_{\text{diode}} = \frac{0.587\varepsilon_0\lambda S^2}{4n\pi}\cdot(\kappa V_\pi^2)\cdot\frac{\Delta\omega}{S}. \qquad (42)$$

Comparing the lumped modulator power P_{lm} given by (27) with (42),

$$\frac{P_{\text{diode}}}{P_{lm}} = \frac{1}{S}. \qquad (43)$$

Thus, due to its ability to confine light beams in one dimension, the diode modulator requires only 1/3 of the power required of a lumped modulator composed of the same electrooptic crystal for a typical value of $S=3$. Since κV_π^2 of bulk GaP is approximately 1/3 of $LiTaO_3$, the modulation powers of GaP diode modulators and $LiTaO_3$ lumped modulators are approximately equal.

Comparison of the Lumped, Traveling-Wave, and Zigzag Modulators: Baseband intensity modulators using these three types of interaction will now be compared. Generally, a lumped modulator is suitable in terms of small power required for modulation bandwidths less than 1 GHz and crystal lengths of about 1 cm. As the required bandwidth or available crystal length increases, a traveling-wave or zigzag interaction becomes more efficient. As the available modulation bandwidth increases from the lumped to the traveling-wave or zig-zag modulators, technical complexities involved in building them also increase. The relationships between the modulation power, which is a measure of

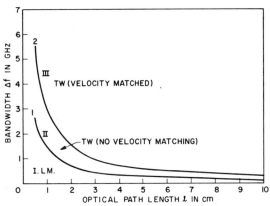

Fig. 6. Bandwidth of modulation versus optical path length for various schemes of electrooptic interaction using $LiTaO_3$. Curve 1 gives the bandwidth at which lumped modulators and traveling-wave modulators without velocity matching require equal power. Curve 2 gives the bandwidth above which traveling-wave modulators require velocity matching. In region I, the lumped modulator requires less power; in regions II and III, the traveling-wave modulator requires less power. In region II, no velocity matching is necessary; in region III, a partial velocity matching is necessary.

the efficiency of interaction, bandwidth, and crystal length, will be derived here so that the type of modulator that meets the requirements with the fewest technical complexities can be chosen. The modulation power and bandwidth are summarized in Table III. The subscripts lm, tw, and zz will be used to indicate the lumped, traveling-wave, and zigzag modulation, respectively.

From Table III, one obtains

$$\frac{P_{lm}}{P_{tw}} = \frac{\pi l\Delta f\sqrt{\kappa_e}}{c}. \qquad (44)$$

Let us assume for the moment that the effect of finite transit time for lumped modulators and velocity matching for traveling-wave modulators can be ignored; then one can set $\kappa_e = \kappa$ in (44) and P_{lm} equals P_{tw} at Δf_1, where

$$\Delta f_1 = \frac{c}{\pi\sqrt{\kappa}l}. \qquad (45)$$

Substituting (45) into (24), the terminating resistance for the lumped modulator becomes $R = 377/\sqrt{\kappa}$ ohms, which is equal to that of a traveling-wave modulator as shown in (30). Thus, at the bandwidth Δf_1 for which both types of modulator require the same power, the terminating resistances are identical. As the required bandwidth Δf becomes smaller than that given by (45), R_{lm} can be increased from $R = 377/\sqrt{\kappa}$ by a factor $\Delta f_1/\Delta f$, while R_{tw} remains unchanged. Since v_π is independent of the terminating resistance, P_{lm} becomes smaller than P_{tw} by a factor $\Delta f/\Delta f_1$. For $LiTaO_3$ (45) becomes

$$\Delta f_1 = \frac{1.4}{l(\text{cm})}\,\text{GHz} \qquad (46)$$

and it is shown as curve 1 in Fig. 6. The region below curve 1 is designated region I in Fig. 6; in this region, lumped modulators require less power than traveling-wave modulators. Modulation power for lumped modulators in region I can be further reduced by one half if a round-trip mode is used. It can be shown from (25) that the effect of finite transit time in region I is negligible and this justifies the assumption made earlier.

As the required modulation bandwidth becomes larger than Δf_1, the traveling-wave interaction becomes more efficient than lumped modulation. The bandwidth of a traveling-wave modulator without velocity matching (i.e., $\kappa_e = \kappa$) can be found from (29), and for LiTaO$_3$ it becomes

$$\Delta f_2 = \frac{3}{l(\text{cm})} \text{ GHz.} \tag{47}$$

Equation (47) is shown as curve 2 in Fig. 6. The region between curves 1 and 2 is designated region II. In region II and for a given l, the modulation power P_{tw} remains unchanged as Δf increases from Δf_1 to Δf_2, since κ_e is unchanged. Thus, a lumped modulator is not only much simpler to build but is also more efficient than a traveling-wave modulator if the required bandwidth is smaller than Δf_1 given by (46). On the other hand, for a given crystal length l and a modulation power just sufficient to drive a lumped modulator over Δf_1, a traveling-wave modulator, even without velocity matching, has a bandwidth $\Delta f_2/\Delta f_1$ ($=2.14$ for LiTaO$_3$) times larger than a lumped modulator. Thus, whether the lumped or the traveling-wave modulation is more suitable depends on the required bandwidth, power, and available crystal length. It is emphasized that the bandwidth of a lumped modulator can be increased further than that given by (46) simply by terminating the crystal with a lower resistance until the transit time effect becomes important. It is less efficient, however, than a traveling-wave modulator of the same length but has the advantage of a simpler construction.

As the bandwidth increases beyond Δf_2, a partial velocity matching ($d \neq w$ in Fig. 4) becomes necessary and P_{tw} increases correspondingly. The maximum bandwidth Δf_3 using a two-plate guide traveling-wave structure will be reached as the structure becomes dispersive, and this has been set previously to occur for $d \leq 0.1 \lambda_m$. Using this condition and (22), one obtains

$$\Delta f_3 = \frac{0.05c}{S\sqrt{\kappa}} \sqrt{\frac{n\pi}{l\lambda}}. \tag{48}$$

For LiTaO$_3$ and using $S = 3$ and $\lambda = 0.63 \ \mu$, (48) becomes

$$\Delta f_3 = \frac{24}{\sqrt{l(\text{cm})}} \text{ GHz.} \tag{49}$$

In the case of zigzag modulation, the structure becomes dispersive at Δf_4. Using (39) and (43),

$$\Delta f_4 = \frac{0.2 \, cn^2}{S\kappa\sqrt{\kappa}} \cdot \sqrt{\frac{n\pi}{l\lambda}}. \tag{50}$$

Using the same conditions as for (49), (50) becomes

$$\Delta f_4 = \frac{9.6}{\sqrt{l(\text{cm})}} \text{ GHz.} \tag{51}$$

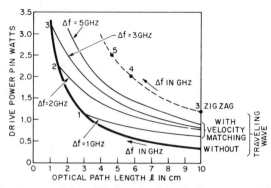

Fig. 7. Power required for 100 percent intensity modulation using LiTaO$_3$ for various interaction schemes. Bandwidths vary along the broken curves for zigzag modulators and along the heavy solid curve for traveling-wave modulators without velocity matching. With velocity matching, longer crystals permit reduction of drive power while maintaining bandwidth as shown by light solid curves. Approximately one half of the bandwidth marked on the heavy solid curve can be obtained using lumped modulators.

For a given l and S, a zigzag modulator becomes dispersive at a lower frequency than a two-plate guide traveling-wave modulator due to the large crystal width b in the zigzag case.

It is of interest to calculate the required modulation power versus the crystal length l for a given bandwidth Δf. The power required on curve 2 of Fig. 6 can be found from (32) by setting $\kappa_e = \kappa$ and, for LiTaO$_3$, $S = 3$, and $\lambda = 0.63 \ \mu$, it becomes $P = 3.5 \ W/l(\text{cm})$. This is plotted as a heavy line in Fig. 7 with Δf_2 from (47) marked on it. For the same power and l, the bandwidth will be reduced by a factor 2.14 from those marked if a lumped modulator is used instead of a traveling-wave modulator. The power required for bandwidth marked on the heavy line of Fig. 7 becomes smaller as l increases even with the expense of the necessary partial velocity matching (in region III of Fig. 6). From (29) and (32), this power can be expressed as

$$P = \frac{0.587\lambda S^2 \varepsilon_0}{2n} \cdot (\kappa V_\pi^2) \cdot \Delta fc/(nl\pi\Delta f + 1.4c). \tag{52}$$

Upon substitution of numerical values for LiTaO$_3$ and using $S = 3$ and $\lambda = 0.63 \ \mu$, (52) becomes

$$P = 0.22/\{2.2 \times 10^{-2} \times l(\text{cm}) + 0.134/\Delta f(\text{GHz})\} \text{ watts.} \tag{53}$$

Equation (53) is shown as light solid curves in Fig. 7 for various values of Δf as labeled. With velocity matching, longer crystals permit reduction of drive power while maintaining bandwidth as shown by light solid curves.

Let us now compare the modulation power of zigzag and traveling-wave modulators. From Table III, one obtains

$$\frac{P_{zz}}{P_{tw}} = \frac{\sqrt{\kappa_e}}{2}. \tag{54}$$

The bandwidth of a traveling-wave modulator without velocity matching (i.e., $\kappa_e = \kappa$) is marked on the heavy line of Fig. 7. For this bandwidth and using LiTaO$_3$, $P_{zz}/P_{tw} \approx 3.4$ from (54). On the other hand, the bandwidth of a zigzag modulator is larger than that of a traveling-wave modulator without velocity matching. Also one should note that

$P_{zz}/P_{tw} > 1$ is the result of using the electrode spacing d in (37) instead of the beam diameter D, which causes the light beam to fill only a small volume of the crystal and thus causes a poor interaction efficiency. If $D(=d/S)$ instead of d is used in (37), then the crystal is completely filled with the light beam, and (38) and (54) must be divided by S. Then for $S \approx 3$ and using LiTaO$_3$, $P_{zz}/P_{tw} \approx 1$, and the zigzag modulator still retains the advantage of a larger bandwidth. The price to be paid for such an advantage with zigzag modulators is its practical complexity in construction. The power required for 100 percent intensity modulation using the zigzag modulation [see (38)] and the limiting bandwidth [from (51)] is also shown in Fig. 7.

Optically Resonant Modulators and Coupling Modulators: Let us assume that the two end faces of a lumped modulator are mirror-coated with power reflectivity R to form an optically resonant cavity. If optical losses are set only by the mirror reflectivity, a packet of light decays to e^{-1} of its initial intensity after bouncing N times inside the cavity where

$$N \approx \frac{1}{1 - R} \quad (55)$$

for $R \lesssim 1$. It can be recognized that the right side of (55) is equal to the ratio of light intensity inside the cavity to that outside the cavity. If the transit time effect can be neglected, then it is easily seen that the depth of modulation increases by approximately a factor of N in such a modulator compared to the same modulator placed outside the optical cavity; for the same depth of modulation, the saving in modulation power is proportional to N^2. This result has been confirmed by a more sophisticated analysis [96]. Such an optical cavity has a series of passbands equally spaced in frequency by $c/2nL$, where L is the mirror spacing, and each passband has a bandwidth $c/2nNL$. The optical carrier, as well as the sidebands produced by the modulation, must fall within these passbands. Thus the factor N by which the depth of modulation is improved also limits the bandwidth available in such modulators. Another disadvantage of such modulators is that a small change in the refractive index of the crystal due to temperature has a large effect on light transmission through the modulator. Optically resonant modulators have been constructed using a cavity external to the laser cavity [97] and also within the laser cavity itself [98], [99] at the baseband with bandwidth less than 100 MHz.

The efficiency-bandwidth limitation of optically resonant modulators can be overcome by the coupling modulation scheme, in which the modulator couples out of the laser cavity an amplitude-modulated optical signal and leaves the total internal laser power essentially unperturbed [100]. The reduction in modulation power compared to the same modulator placed outside the cavity is given by the ratio of the internal to the external intensities [101], instead of the square of the ratio as for the optically resonant modulators, but the bandwidth is no longer limited by this

factor. If the laser is permitted to oscillate in only one mode, if only a small fraction of the internal intensity is coupled out, and if large attenuation can be introduced for modes outside the gain profile of the laser medium (i.e., at the cold modes), there is no bandwidth limitation except from the modulator itself [102] (i.e., from the transit time and the capacitive loading of the modulator). The experiment has been performed using a short He–Ne laser at $\lambda = 0.63 \mu$ and modulation frequency of 2 GHz [103]. In order to obtain a single-mode oscillation, the length of the laser cavity must be short enough so that only a single-cavity mode falls within the gain profile of the laser medium, which forces the output power from such lasers to be small. This difficulty has been overcome by use of a three-mirror resonator of the Fox–Smith type [101]. Efficient baseband modulation over 700 MHz has been achieved with 3 mW output optical power. The distortion due to the cold modes was suppressed to a negligible level by proper choice of reflectivity of the beam splitter in the resonator system.

Although the scheme of coupled modulation is quite promising, a practical difficulty is presented by additional optical components that must be placed inside the laser cavity. These introduce losses that reduce the available laser power, and also impose a fairly stiff requirement on the quality of crystals that can be used in such modulators due to low gain per transit of the laser medium. At present, the modulation power per megahertz of coupling modulators using DKDP, which is the most commonly used crystal for intracavity modulation because of its good optical quality, is still larger than for the LiTaO$_3$ lumped modulator external to the optical cavity [101].

ACOUSTOOPTIC AND MAGNETOOPTIC MODULATORS

There are optical interactions other than electrooptic effects which have received only limited interest for applications in broad-band optical modulators, but which may be more appealing than electrooptic modulators under some specific requirements. Only two of them, acoustooptic and magnetooptic modulators, will be discussed here.

Acoustooptic Modulators

The light beam to be modulated traverses across an acoustic beam that is amplitude-modulated by the signal to be impressed on the light with the Bragg angle θ as shown in Fig. 8. Part of the incident light will be diffracted by the periodic index variation produced by the acoustic wave via the photoelastic effect, and both the diffracted and the undiffracted light intensities follow the envelope of the acoustic wave, their variations being in-phase and out-of-phase, respectively, with the modulation of the acoustic signal. The device is essentially a baseband intensity modulator with its cutoff frequency determined by the transit time of the acoustic wave across the light beam [104], [105].

A few other factors must be considered in order to achieve successful operation of this device. The acoustic beam must be allowed to spread, either from diffraction or by focusing,

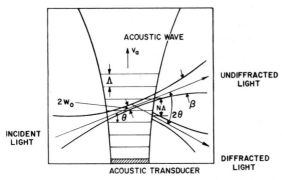

Fig. 8.   Acoustooptic modulation, where Λ is the wavelength of the acoustic wave.

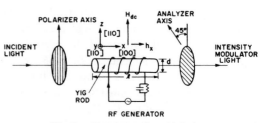

Fig. 9.   Magnetooptic modulation.

so that the Bragg scattering interaction takes place over the bandwidth of the modulator. The light beam must also have a diffraction spread of the same order of magnitude as that of the acoustic beam in order that the light diffracted by the acoustic carrier and its sidebands will overlap to produce an efficient intensity modulation [4]. To diffract nearly 100 percent of the incident light [104], the number of lines of the grating intercepted by the light beam N (see Fig. 8) must be $\gtrsim 1$. This will be satisfied when the diffraction spread of the light beam is made slightly large than that of the acoustic beam [106].

The bandwidth of the modulator due to the finite transit time of the acoustic wave across the light beam can be determined as follows. The rise time of the diffracted light intensity due to Bragg scattering by a step-function acoustic wave is shown [106] as $t_r \approx 1.3\,(w_0/v_a)$, where v_a is the velocity of the acoustic wave and $2w_0$ is the diameter of the light amplitude (assumed to be Gaussian) at its waist. Using the familiar relationship between the rise time and bandwidth of a low-pass filter $\Delta f = 0.35/t_r$, the modulation bandwidth becomes

$$\Delta f \approx 0.27\,\frac{v_a}{w_0}. \tag{56}$$

Thus the acoustic wave must traverse the light beam in approximately one half of the period of the highest modulation frequency in the band.

The diffracted light beam has the same diffraction angle β as the incident light when the diffraction spread of light and acoustic waves are about equal. In order to insure a good separation of the diffracted and undiffracted light beam, let

$$\theta = \beta. \tag{57}$$

The Bragg angle is

$$\theta = \sin^{-1}\frac{\lambda f_a}{2nv_a} \approx \frac{\lambda f_a}{2nv_a}$$

where n is the refractive index and f_a is the acoustic frequency, and $\beta \approx 2\lambda/n\pi w_0$ for a Gaussian light beam. Upon substitution into (57), one obtains

$$f_a = \frac{4v_a}{\pi w_0}. \tag{58}$$

From (56) and (58), the relation between the bandwidth Δf and the acoustic carrier frequency f_a can be found as

$$f_a \approx 5\Delta f. \tag{59}$$

Thus the bandwidth of the acoustic modulators is limited to about 20 percent of the acoustic frequency, which in turn is limited to the order of 1 GHz at present from practical considerations in making transducers [107].

Maydan [106] has built an acoustooptic pulse modulator using As_2S_3 glass. The pulse rise time was 6 ns, $f_a = 350$ MHz, and over 70 percent of the incident light at $\lambda = 0.63\,\mu$ was diffracted with an electrical power of 0.6 watt applied to the ZnO transducer. The bandwidth of this modulator is $\Delta f = 59$ MHz from (59), and the calculated pulse rise time t_r is $t_r = 0.35/\Delta f = 6$ ns, in agreement with the measurement. Using rutile and $f_a = 750$ MHz, the rise time has been decreased to 3 ns but with less efficiency (5 percent of incident light diffracted with 1 watt of electrical power at $\lambda = 0.63\,\mu$) [108].

Although the bandwidth of acoustooptic modulators is not large compared to electrooptic modulators, good acoustooptic materials with high optical quality can be found, which makes this type of modulator suitable for use inside laser cavities [109].

Magnetooptic Modulators

High-quality single crystals of yttrium iron garnet (YIG) have absorptions of less than 0.3 dB/cm at room temperature in the wavelength range of $1.15\,\mu$ to $4.5\,\mu$ [110]. Together with high-saturation optical rotation (172°/cm at $\lambda = 1.52\,\mu$) and low RF losses, this material offers an interesting possibility for efficient modulation in the near infrared [111].

Intensity modulation using the magnetic rotation can be accomplished with the arrangement shown in Fig. 9. A dc magnetic field H_{dc} (z axis) is applied to saturate the YIG rod in the direction normal to the rod axis, and a small RF magnetic field is applied along the rod axis (x axis) to wobble the magnetization within a small angle, producing a small component of RF magnetization m_x along the x axis. Since the magnetic rotation is proportional to the component of magnetization along the direction of light propagation [112], which is assumed to be along the rod axis, the emergent light beam will have its plane of polarization rotated by an amount proportional to $m_x l$ where l is the length of the rod. A linear intensity modulation is obtained after the light passes through an analyzer set to 45° with respect to the input polarizer.

The modulation power required for this type of device can be estimated as follows [113]. For a frequency much smaller than the magnetic resonance frequency, the time-varying component of magnetization along the x axis is

$$m_x \approx \frac{h_x}{H_i} M \qquad (60)$$

where h_x is the applied RF magnetic field, M is the saturation magnetization of the sample, and H_i is the dc effective internal field. For the configuration shown in Fig. 9 and if there is no magnetic anisotropy, $(H_i = (H_{dc} - 2\pi M)^{\frac{1}{2}} \cdot H_{dc}^{\frac{1}{2}}.)$ The H_i has to be such that the magnetic resonance frequency of the rod, $f_{mr}(\text{MHz}) = 2.8 \, H_i$ (Oe), is outside the modulation bandwidth in order to avoid RF losses. For baseband modulation, f_{mr} must be larger than the highest frequency in the modulation band. The RF magnetic rotation due to m_x can be expressed as

$$\phi = \phi_s l \frac{m_x}{M} \qquad (61)$$

where ϕ_s is the saturated magnetic rotation per unit length. For small intensity modulation, a modulation index $\eta (\eta = 1$ for 100 percent modulation) can be expressed as

$$\eta \approx 2\phi_{rad} \qquad (62)$$

where ϕ_{rad} is the peak value of the angle of rotation ϕ expressed in radians. Using CGS units, the time-averaged RF stored energy in the rod is

$$W = \frac{1}{8\pi} \cdot \frac{\pi d^2 l}{4} \cdot m_x h_x \text{ erg} \qquad (63)$$

where d is the diameter of the rod, and m_x and h_x are expressed in peak values. The reactive power (assuming no losses in the circuit) required for modulation bandwidth Δf is

$$P = 2\pi \Delta f \cdot W \cdot 10^{-7} \text{ watt.} \qquad (64)$$

Substituting (60)–(63) into (64) and denoting $\phi_{S \, rad}$ as ϕ_S expressed in radians, one obtains

$$P = \frac{\pi}{64} \cdot \Delta f \cdot MH_i \cdot \left(\frac{\eta}{\phi_{S \, rad}}\right)^2 \cdot \left(\frac{d^2}{l}\right) \cdot 10^{-7} \text{ watt.} \qquad (65)$$

Equation (65) shows that in order to reduce P, the saturation magnetization M must be small, the saturation optical rotation must be large, and the effective internal field H_i must be small. The dependence of P on the bandwidth and size of the crystal is the same as the lumped electrooptic modulators.

The magnetooptic modulator at $\lambda = 1.52 \, \mu$ built by LeCraw [111] used gallium-doped YIG, which has a much smaller saturation magnetization than YIG and a comparable optical rotation. The choice of crystal orientation as shown in Fig. 9, where [100] and [110] are the hard and the intermediate anisotropic axes, respectively, helps to increase the efficiency of an applied RF field in tilting the magnetization along the rod axis, and thereby increases the efficiency

of modulation. It is of interest to see how the calculation compares with LeCraw's experimental result. The following data are pertinent to the modulator: $\phi_{S \, rad} = 1.95$ rad/cm, $4\pi M = 270$ gauss, $H_i = 100$ Oe ($H_{dc} = 232$ Oe), $d = 5 \times 10^{-2}$ cm, $l = 1$ cm, $\eta = 0.2$, and $\Delta f = 200$ MHz. Substituting these into (65), one obtains $P = 53$ mW compared to the measured 86 mW—satisfactory agreement in view of the crudeness of the analysis.

The magnetooptic modulators can also be used in the round-trip mode to reduce the modulation power in the same manner as the electrooptic modulator. The power will be decreased by a factor of two if diameter of the rod d is enlarged so that the safety factor S defined in (22) remains unchanged. Using gallium-doped YIG, the round-trip mode, and $\Delta f < 200$ MHz ($H_i = 100$ Oe), the power required for 100 percent modulation per megahertz can be calculated using (65), and it becomes

$$\frac{P(\eta = 1)}{\Delta f} = 1.1 \text{ mW/MHz}$$

at $\lambda = 1.52 \, \mu$, $n = 2.2$, and $S = 3$. This is the same as the lumped LiTaO$_3$ electrooptic modulators at $\lambda = 0.63 \, \mu$. Note that $P/\Delta f$ for magnetooptic modulators increases as Δf exceeds 200 MHz due to the necessary increase in H_i. For $\lambda = 1.52 \, \mu$, LiTaO$_3$ modulators require approximately $(1.52/0.63)^3 = 14$ times larger power, keeping S unchanged. Thus, gallium-doped YIG modulators have their advantage in the near-infrared spectral range where LiTaO$_3$ modulators are less efficient. Also, the modulation power of magnetooptic modulators is supplied in the form of current rather than voltage, which makes it convenient to drive with transistors.

CONCLUSIONS

The electrooptic effects, especially these observed in ferroelectric crystals, have received much attention in recent years; their origin has been better understood and more varieties of crystals have been grown. From a device point of view, it is desirable to have materials not only with large electrooptic effects (i.e., small half-wave voltages) but also with small dielectric constants. Unfortunately, these two requirements contradict each other in the case of ferroelectrics, which is the only known class of materials with large electrooptic effects. The electric energy needed for modulation using ferroelectrics is smaller than with nonferroelectrics, and the effort in growing ferroelectrics of good optical quality in large size, with high dark resistivity and photoresistivity, and small RF dissipation should be highly rewarding. Large nonlinear optical effects observed in these crystals should also add a strong incentive to such an effort. As large crystals of good quality become available, the diameter of the incident light beam can be enlarged, which helps to increase the optical power that the modulator can handle without undue heating and optical damage. Also, fewer pieces of crystals are needed to build a long traveling-wave modulator if a very broad bandwidth is required.

It has been shown that lumped modulators are adequate

for bandwidths of about 1 GHz in the visible spectrum. Impedance matching of the crystal, which usually has a large dielectric constant, to the RF generator can be accomplished simply by a resistive termination at the crystal in such a configuration. As the required bandwidth increases further, either the traveling-wave or zigzag interaction becomes more efficient. The price to be paid for efficient modulation over a broad bandwidth is the increasing complexity in building one. Even with an efficient interaction, the required power is fairly high if the optical path is short. A long optical path can be obtained by segmenting many short crystals, which require good antireflection coatings and alignment of individual crystals. Also, each crystal must be of better optical quality so that imperfections accumulated over many pieces can still be held within a tolerable level. Impedance matching to such modulators over several gigahertz bandwidth also requires an intensive effort. When small modulation power is the dominant consideration, coupling modulators seem to be the most attractive, provided crystals become available with smaller κV_π^2 than DKDP and with much better quality than is needed for external modulation, and provided single-frequency lasers with the necessary output power are available. The bandwidth of such modulators is the same as lumped modulators.

Materials suitable for acoustooptic interactions are often of better optical quality than electrooptic crystals, which makes the acoustooptic modulation attractive if a good extinction (but of limited bandwidth) is the prime concern.

Magnetooptic modulation using gallium-doped YIG is attractive for a bandwidth of a few hundred megahertz and wavelengths in the near infrared. The optical absorption is 0.25 dB per cm at $\lambda = 1.52\ \mu$.

ACKNOWLEDGMENT

The author wishes to thank S. H. Wemple for valuable discussions on the electrooptic effect and for critically reading the manuscript, and L. K. Anderson, I. P. Kaminow, R. C. LeCraw, D. Maydan, F. R. Nash, and F. K. Reinhart for helpful discussions and comments. The constant encouragement of J. E. Geusic is deeply appreciated.

REFERENCES
[1] L. K. Anderson, "Microwave modulation of light," *Microwave J.*, vol. 4, pp. 42–53, January 1965.
[2] I. P. Kaminow and E. H. Turner, "Electrooptic light modulators," *Proc. IEEE*, vol. 54, pp. 1374–1390, October 1966.
[3] I. S. Rez, "Crystals with nonlinear polarizability," *Sov. Phys.—Usp.*, vol. 10, pp. 759–782, May–June 1968.
[4] E. I. Gordon, "A review of acoustooptical deflection and modulation devices," *Proc. IEEE*, vol. 54, pp. 1391–1401, October 1966.
[5] T. L. Paoli and J. E. Ripper, "Direct modulation of semiconductor lasers," this issue, pp. 1457–1465.
[6] J. F. Nye, *Physical Properties of Crystals*. Oxford: Oxford University Press, 1960.
[7] W. P. Mason, "Electrooptic and photoelastic effects in crystals," *Bell Syst. Tech. J.*, vol. 29, pp. 161–188, April 1950.
[8] P. V. Lenzo, E. H. Turner, E. G. Spencer, and A. A. Ballman, "Electrooptic coefficients and elastic-wave propagation in single-domain ferroelectric lithium tantalate," *Appl. Phys. Lett.*, vol. 8, pp. 81–82, February 1966.
[9] M. Born and E. Wolf, *Principles of Optics*. New York: Macmillan, 1964.
[10] J. M. Ley, "Low voltage light-amplitude modulation," *Electron. Lett.*, vol. 2, pp. 12–13, January 1966.
[11] C. H. Clayson, "Low voltage light-amplitude modulation," *Electron. Lett.*, vol. 2, p. 138, April 1966.
[12] M. Dore, "A low drive power light modulator using a readily available material ADP," *IEEE J. Quantum Electron.*, vol. QE-3, pp. 555–560, November 1967.
[13] I. P. Kaminow and W. D. Johnston, Jr., "Quantitative determination of sources of the electrooptic effect in LiNbO₃ and LiTaO₃," *Phys. Rev.*, vol. 160, pp. 519–522, August 1967.
[14] ——, "Contributions to optical nonlinearity in GaAs as determined from Raman scattering efficiencies," *Phys. Rev.*, vol. 188, pp. 1209–1211, December 1969.
[15] S. K. Kurtz and F. N. H. Robinson, "A physical model of the electrooptic effect," *Appl. Phys. Lett.*, vol. 10, pp. 62–65, January 1967.
[16] C. G. B. Garrett, "Nonlinear optics, anharmonic oscillators, and pyroelectricity," *IEEE J. Quantum Electron.*, vol. QE-4, pp. 70–84, March 1968.
[17] J. D. Zook and T. N. Casselman, "Electrooptic effects in paraelectric perovskites," *Appl. Phys. Lett.*, vol. 17, pp. 960–962, 1966.
[18] M. DiDomenico, Jr., and S. H. Wemple, "Oxygen-octahedra ferroelectrics. I. Theory of electrooptical and nonlinear optical effects," *J. Appl. Phys.*, vol. 40, pp. 720–724, February 1969.
[19] S. K. Kurtz, "Visible and ultraviolet optical properties of some ABO₃ ferroelectrics," *Proc. Int. Conf. on Ferroelectricity* (Prague, Czechoslovakia), 1966.
[20] A. Frova and P. J. Boddy, "Optical field effects and band structure of some perovskite-type ferroelectrics," *Phys. Rev.*, vol. 153, pp. 606–616, 1967.
[21] J. D. Zook, "Oscillatory electroreflectance of SrTiO₃," *Phys. Rev. Lett.*, vol. 20, pp. 848–852, 1968.
[22] W. J. Merz, "Electric and optical behavior of BaTiO₃ single-domain crystals," *Phys. Rev.*, vol. 76, pp. 1221–1225, 1949.
[23] J. D. Zook, D. Chen, and G. N. Otto, "Temperature dependence and model of the electrooptic effect in LiNbO₃," *Appl. Phys. Lett.*, vol. 11, pp. 159–161, September 1967.
[24] F. S. Chen, J. E. Geusic, S. K. Kurtz, J. G. Skinner, and S. H. Wemple, "Light modulation and beam deflection with potassium tantalate-niobate crystals," *J. Appl. Phys.*, vol. 37, pp. 388–398, January 1966.
[25] L. G. Van Uitert, J. J. Rubin, W. H. Grodkiewicz, and W. A. Bonner, "Some characteristics of Ba, Sr, Na niobates," *Mater. Res. Bull.*, vol. 4, pp. 63–74, 1969.
[26] S. H. Wemple and M. DiDomenico, Jr., "Oxygen octahedra ferroelectrics. II. Electrooptical and nonlinear-optical device applications," *J. Appl. Phys.*, vol. 40, pp. 735–752, February 1969.
[27] A. von Hippel, E. P. Gross, J. G. Jelatis, and H. Geller, "Photocurrent, space-charge buildup, and field emission in alkali halide crystals," *Phys. Rev.*, vol. 91, pp. 568–579, August 1953.
[28] S. K. Kurtz and P. J. Warter, Jr., "Space charge effects in p-type semiconducting KTa.₆₅Nb.₃₅O₃ (KTN)," *Bull. Amer. Phys. Soc.*, vol. 11, p. 34, 1966.
[29] S. Triebwasser, "Space charge fields in BaTiO₃," *Phys. Rev.*, vol. 118, pp. 100–105, April 1960.
[30] S. H. Wemple, "Electrical contact to n- and p-type ferroelectric oxides," in *Ohmic Contacts to Semi-Conductors*, B. Schwartz, Ed. New York: The Electrochemical Society, 1969, pp. 128–137.
[31] R. T. Denton, F. S. Chen, and A. A. Ballman, "Lithium tantalate light modulators," *J. Appl. Phys.*, vol. 38, pp. 1511–1617, March 1967.
[32] I. P. Kaminow, "Measurements of the electrooptic effect in CdS, ZnTe, and GaAs at 10.6 microns," *IEEE J. Quantum Electron.*, vol. QE-4, pp. 23–26, January 1968.
[33] T. Kimura and T. Yamada, "Intensity-dependent electrooptic effect of ZnTe," *IEEE J. Quantum Electron.* (Correspondence), vol. QE-6, pp. 158–159, March 1970.
[34] F. S. Chen, "A laser-induced inhomogeneity of refractive indices in KTN," *J. Appl. Phys.*, vol. 38, pp. 3418–3420, July 1967.
[35] J. P. Thaxter, "Electrical control of holographic storage in strontium-barium niobate," *Appl. Phys. Lett.*, vol. 15, pp. 210–212, October 1969.
[36] S. I. Waxman, M. Chodrow, and H. E. Panthoff, "Optical damage in KDP," *Appl. Phys. Lett.*, vol. 16, pp. 157–159, February 1970.
[37] A. Ashkin, G. D. Boyd, J. M. Dziedzic, R. G. Smith, A. A. Ballman,

H. J. Levinstein, and K. Nassau, "Optically induced refractive index inhomogeneities in LiNbO$_3$ and LiTaO$_3$," *Appl. Phys. Lett.*, vol. 9, pp. 72–74, July 1966.

[38] F. S. Chen, J. T. LaMacchia, and D. B. Fraser, "Holographic storage in LiNbO$_3$," *Appl. Phys. Lett.*, vol. 13, pp. 223–225, October 1968.

[39] F. S. Chen, "Optically induced change of refractive indices in LiNbO$_3$ and LiTaO$_3$," *J. Appl. Phys.*, vol. 40, pp. 3389–3396, July 1969.

[40] E. P. Harris and M. L. Dakss, "Optical damage to LiNbO$_3$ from GaAs laser radiation," *IBM J. Res. Develop.*, vol. 13, pp. 722–723, November 1969.

[41] W. D. Johnston, Jr., "Optical index damage in LiNbO$_3$ and other pyroelectric insulators," *J. Appl. Phys.*, vol. 41, pp. 3279–3285, July 1970.

[42] H. J. Levinstein, A. A. Ballman, R. T. Denton, A. Ashkin, and J. M. Dziedzic, "Reduction of the susceptibility to optically induced index inhomogeneities in LiTaO$_3$ and LiNbO$_3$," *J. Appl. Phys.*, vol. 38, pp. 3101–3102, July 1967.

[43] R. G. Smith, D. B. Fraser, R. T. Denton, and T. C. Rich, "Correlation of reduction in optically induced refractive-index inhomogeneity with OH content in LiTaO$_3$ and LiNbO$_3$," *J. Appl. Phys.*, vol. 39, pp. 4600–4602, September 1968.

[44] S. Singh, H. J. Levinstein, and L. G. Van Uitert, "Role of hydrogen in polarization reversal of ferroelectric Ba$_2$NaNb$_5$O$_{15}$," *Appl. Phys. Lett.*, vol. 16, pp. 176–178, February 1970.

[45] T. R. Slicker and S. R. Burlage, "Some dielectric and optical properties of KD$_2$PO$_4$," *J. Appl. Phys.*, vol. 34, pp. 1837–1840, July 1963.

[46] I. P. Kaminow, "Microwave dielectric properties of NH$_4$H$_2$PO$_4$, KH$_2$AsO$_4$, and partially deuterated KH$_2$PO$_4$," *Phys. Rev.*, vol. 138, pp. A1539–A1543, May 1965.

[47] R. A. Phillips, "Temperature variation of the index of refraction of ADP, KDP and deuterated KDP," *J. Opt. Soc. Am.*, vol. 56, pp. 629–632, May 1966.

[48] P. V. Lenzo, E. H. Turner, E. G. Spencer, and A. A. Ballman, "Electrooptic coefficients and elastic wave propagation in single-domain ferroelectric lithium tantalate," *Appl. Phys. Lett.*, vol. 8, pp. 81–82, February 1966.

[49] W. L. Bond, "Measurements of the refractive indices of several crystals," *J. Appl. Phys.*, vol. 36, pp. 1674–1677, May 1965.

[50] Measured by R. C. Bearisto of Bell Telephone Labs. and can be found in [51].

[51] F. S. Chen, "Demultipliers for a high-speed optical PCM" (to be published).

[52] J. E. Geusic, H. J. Levinstein, J. J. Rubin, S. Singh, and L. G. Van Uitert, "The nonlinear optical properties of Ba$_2$NaNb$_5$O$_{15}$," *Appl. Phys. Lett.*, vol. 11, pp. 269–271, November 1967.

[53] E. G. Spencer, P. V. Lenzo, and A. A. Ballman, "Dielectric materials for electrooptic, elastooptic, and ultrasonic device applications," *Proc. IEEE*, vol. 55, pp. 2074–2108, December 1967.

[54] S. H. Wemple, M. DiDomenico, Jr., and I. Camlibel, "Dielectric and optical properties of melt-grown BaTiO$_3$," *J. Phys. Chem. Solids*, vol. 29, pp. 1797–1803, 1968.

[55] ——, "Dielectric properties of single-domain melt-grown BaTiO$_3$," *J. Phys. Chem. Solids* (to be published).

[56] A. R. Johnston, "The strain-free electrooptic effect in single-crystal barium titanate," *Appl. Phys. Lett.*, vol. 7, pp. 195–198, October 1965.

[57] A. Yariv, C. A. Mead, and J. V. Parker, "GaAs as an electrooptic modulator at 10.6 μ," *IEEE J. Quantum Electron.*, vol. QE-2, pp. 243–245, August 1966.

[58] J. E. Kiefer and A. Yariv, "Electrooptic characteristics of CdTe at 3.39 and 10.6 μ," *Appl. Phys. Lett.*, vol. 15, pp. 26–27, July 1969.

[59] E. A. Giess, G. Burns, D. F. O'Kane, and A. W. Smith, "Ferroelectric and optical properties of KSr$_2$Nb$_5$O$_{15}$," *Appl. Phys. Lett.*, vol. 11, pp. 233–234, 1967.

[60] G. Burns, E. A. Giess, D. F. O'Kane, B. A. Scott, and A. W. Smith, "Crystal growth and ferroelectric and optical properties of K$_x$Na$_{1-x}$Ba$_2$Nb$_5$O$_{15}$," *J. Appl. Phys.*, vol. 40, pp. 901–902, 1969.

[61] W. W. Rigrod and I. P. Kaminow, "Wide-band microwave light modulation," *Proc. IEEE*, vol. 51, pp. 137–140, January 1963.

[62] S. E. Harris, "Conversion of FM light to AM light using birefringent crystals," *Appl. Phys. Lett.*, vol. 2, pp. 47–49, February 1963.

[63] I. P. Kaminow, "Balanced optical discriminator," *Appl. Opt.*, vol. 3, p. 507, April 1964.

[64] S. M. Stone, "A microwave electro-optic modulation which over-

comes transit time limitation," *Proc. IEEE* (Correspondence), vol. 2, pp. 409–410, April 1964.

[65] C. J. Peters, "Gigacycle-bandwidth coherent-light traveling-wave modulator," *Proc. IEEE*, vol. 53, pp. 455–460, May 1965.

[66] W. H. Steier, "A push-pull optical amplitude modulator," *IEEE J. Quantum Electron.*, vol. QE-3, pp. 664–667, December 1967.

[67] W. E. Bicknell, "Synchronization bandwidth of capacitive transverse-field electrooptic modulators," *IEEE J. Quantum Electron.* (Correspondence), vol. QE-4, pp. 35–37, January 1968.

[68] W. J. Rattman, W. E. Bicknell, B. K. Yap, and C. J. Peters, "Broadband, low drive power electrooptic modulator," *IEEE J. Quantum Electron.*, vol. QE-3, pp. 550–554, November 1967.

[69] I. P. Kaminow and W. M. Sharpless, "Performance of LiTaO$_3$ and LiNbO$_3$ light modulators at 4 GHz," *Appl. Opt.*, vol. 6, pp. 351–352, February 1967.

[70] K. K. Chow, R. L. Comstock, and W. B. Lonard, "1.5-GHz bandwidth light modulator," *IEEE J. Quantum Electron.* (Correspondence), vol. QE-5, pp. 618–620, December 1969.

[71] I. P. Kaminow, "Microwave modulation of the electrooptic effect in KH$_2$PO$_4$," *Phys. Rev. Lett.*, vol. 6, p. 528, May 1961.

[72] R. H. Blumenthal, "Design of a microwave-frequency light modulator," *Proc. IRE*, vol. 50, pp. 452–456, April 1962.

[73] K. M. Johnson, "Solid-state modulation and direct demodulation of gas laser light at a microwave frequency," *Proc. IRE* (Correspondence), vol. 51, pp. 1368–1369, October 1963.

[74] H. Brand, B. Hill, E. Holtz, and G. Wencker, "External light modulation with low microwave power," *Electron. Lett.*, vol. 2, pp. 317–318, August 1966.

[75] T. E. Walsh, "Gallium-arsenide electrooptical modulators," *RCA Rev.*, vol. 27, pp. 323–335, September 1966.

[76] M. Dore, "A low drive-power light modulator using a readily available material ADP," *IEEE J. Quantum Electron.*, vol. QE-3, pp. 555–560, November 1967.

[77] R. P. Riesz and M. R. Biazzo, "Gigahertz optical modulation," *Appl. Opt.*, vol. 8, pp. 1393–1395, July 1969.

[78] I. P. Kaminow and J. Liu, "Propagation characteristics of partially loaded two-conductor transmission line for broadband light modulators," *Proc. IRE*, vol. 51, pp. 132–136, January 1963.

[79] T. Sueta, K. Goto, and T. Makimoto, "A reflection-type traveling-wave light modulator," *IEEE J. Quantum Electron.*, vol. QE-5, p. 330, June 1969.

[80] C. J. Peters, "Gigacycle bandwidth coherent light traveling-wave phase modulator," *Proc. IRE*, vol. 51, pp. 147–153, January 1963.

[81] ——, "Gigacycle-bandwidth coherent-light traveling-wave modulator," *Proc. IEEE*, vol. 53, pp. 455–460, May 1965.

[82] J. A. Ernest and I. P. Kaminow (unpublished). See also [2].

[83] W. E. Bicknell, B. K. Yap, and C. J. Peters, "0 to 3 GHz traveling-wave electrooptic modulator," *Proc. IEEE* (Letters), vol. 55, pp. 225–226, February 1967.

[84] G. White and G. M. Chin, "A 1 Gbit-sec^{-1} optical PCM communications system," presented at IEEE Int. Conf. on Communications, San Francisco, Calif., June 1970.

[85] J. L. Putz, "A wide-band microwave modulator," *IEEE Trans. Electron Devices*, vol. ED-15, pp. 695–698, October 1968.

[86] W. W. Rigrod and I. P. Kaminow, "Wide-band microwave light modulation," *Proc. IEEE*, vol. 51, pp. 137–140, January 1963.

[87] M. DiDomenico, Jr., and L. K. Anderson, "Broadband electrooptic traveling-wave light modulators," *Bell Syst. Tech. J.*, vol. 42, pp. 2621–2728, November 1963.

[88] W. A. Scanga, "Traveling-wave light modulator," *Appl. Opt.*, vol. 4, pp. 1103–1106, September 1965.

[89] D. C. Auth, "Half-octave bandwidth traveling-wave X-band optical phase modulator," *IEEE J. Quantum Electron.* (Correspondence), vol. QE-5, pp. 622–623, December 1969.

[90] I. P. Kaminow, T. J. Bridges, and M. A. Pollack, "A 964 GHz traveling-wave electrooptic light modulator," *Appl. Phys. Lett.* (to be published).

[91] E. R. Schineller, "Single-mode-guide laser components," *Microwave J.*, vol. 7, pp. 77–85, January 1968.

[92] S. E. Miller, "Integrated optics: An introduction," *Bell Syst. Tech. J.*, vol. 48, pp. 2059–2069, September 1969.

[93] F. K. Reinhart, "Reverse-biased gallium phosphide diodes as high-frequency light modulators," *J. Appl. Phys.*, vol. 39, pp. 3426–3434, June 1968.

[94] F. K. Reinhart, D. F. Nelson, and J. McKenna, "Electrooptic and waveguide properties of reverse-biased gallium phosphide p-n

junctions," *Phys. Rev.*, vol. 177, pp. 1208–1221, January 1969.

[95] H. Kogelnik and T. Li, "Laser beams and resonators," *Proc. IEEE*, vol. 54, pp. 1312–1329, October 1966.

[96] E. I. Gordon and J. D. Rigden, "The Fabry–Perot electrooptic modulator," *Bell Syst. Tech. J.*, vol. 42, pp. 155–179, January 1963.

[97] J. T. Ruscio, "A coherent light modulator," *IEEE J. Quantum Electron.*, vol. QE-1, pp. 182–183, July 1965.

[98] E. A. Ohm, "A linear optical modulator with high FM sensitivity," *Appl. Opt.*, vol. 6, pp. 1233–1235, July 1967.

[99] T. Uchida, "Direct modulation of gas lasers," *IEEE J. Quantum Electron.*, vol. QE-1, pp. 336–343, November 1965.

[100] K. Gürs and R. Muller, "Internal modulation of optical masers," in *Proc. Symp. on Optical Masers.* Brooklyn, N. Y.: Polytechnic Press, 1963, pp. 243–252.

[101] F. R. Nash and P. W. Smith, "Broadband optical coupling modulation," *IEEE J. Quantum Electron.*, vol. QE-4, pp. 26–34, January 1968.

[102] I. P. Kaminow, "Internal modulation of optical masers (bandwidth limitations)," *Appl. Opt.*, vol. 4, pp. 123–127, January 1965.

[103] G. Grau and D. Rosenberger, "Low-power microwave modulation of a 0.63 μ He–Ne laser," *Phys. Lett.*, vol. 6, pp. 129–131, September 1963.

[104] H. V. Hance and J. K. Parks, "Wide-band modulation of a laser beam, using Bragg-angle diffraction by amplitude-modulated ultra-sonic wave," *J. Acoust. Soc. Am.*, vol. 38, pp. 14–23, July 1965.

[105] E. I. Gordon and M. G. Cohen, "Electrooptic diffraction grating for light beam modulation and diffraction," *IEEE J. Quantum Electron.*, vol. QE-1, pp. 191–198, August 1965.

[106] D. Maydan, "Acoustooptical pulse modulators," *IEEE J. Quantum Electron.*, vol. QE-6, pp. 15–24, January 1970.

[107] E. K. Sittig, private communication.

[108] D. Maydan, private communication.

[109] ——, "A fast modulator for extraction of internal laser power," *J. Appl. Phys.*, vol. 41, pp. 1552–1559, March 1970.

[110] R. C. LeCraw, D. L. Wood, J. F. Dillon, Jr., and J. P. Remeika, "The optical transparency of yttrium iron garnet in the near infrared," *Appl. Phys. Lett.*, vol. 7, pp. 27–28, July 1965.

[111] R. C. LeCraw, "Wideband infrared magneto-optic modulation," presented at the 1966 INTERMAG Conf., Stuttgart, Germany. *IEEE Trans. Magn.*, vol. MAG-2, p. 394, September 1966 (Abstract only).

[112] N. Bloembergen, P. S. Pershan, and L. R. Wilcox, "Microwave modulation of light in paramagnetic crystals," *Phys. Rev.*, vol. 120, pp. 2014–2023, December 1960.

[113] L. K. Anderson, private communication. Also, L. K. Anderson and R. C. LeCraw, "Practical magnetooptical modulations," presented at the 1968 New England Radio Eng. Meeting (NEREM), Boston, Mass.

Photodetectors for Optical Communication Systems

HANS MELCHIOR, MEMBER, IEEE, MAHLON B. FISHER, MEMBER, IEEE, AND FRANK R. ARAMS, FELLOW, IEEE

Abstract—The characteristics of high-sensitivity photodetectors suitable for wide bandwidth optical communication systems are summarized. Photodiodes, photomultipliers, and photoconductive detectors for wavelengths from 0.3 μm to 10.6 μm are covered. The use of internal current gain by means of avalanche and electron multiplication and by means of optical heterodyne detection to increase sensitivity of high speed photodetectors is discussed. The application to visible and infrared laser communication systems is reviewed.

I. Introduction

DEMODULATION of signals in optical communication systems requires photodetectors and detection systems which combine wide instantaneous bandwidth with high sensitivity to weak light signals. This paper will discuss the basic requirements of photodetection systems suitable for the demodulation of laser signals and present a review of the relevant characteristics of different types of high speed photodetectors.

Vacuum and Si photodiodes with response in the ultraviolet, visible, and near-infrared part of the spectrum up to about 1 μm will be described. Photomultipliers and avalanche photodiodes with wide instantaneous bandwidths that operate in the same wavelength region but provide internal gain for the photocurrent are discussed. It will be shown that internal current gain as well as optical heterodyne detection significantly improve the sensitivity of photodetection systems to weak light signals. A comparison of the characteristics of heterodyne and direct detection systems will be presented. The most important infrared photodetectors with response to a few micrometers including Ge, InAs, and InSb solid-state photodiodes with and without internal current gain will be mentioned. The extrinsic and intrinsic photoconductors and the mixed crystal photodetectors which may be utilized as detectors for 10.6-μm radiation will be discussed.

II. General Requirements for Photodetectors in Optical Communication Systems

Proper design of a demodulation system for optical signals requires specially designed photodetectors that are efficient and fast. The major requirements imposed on photodetectors and detection systems for optical communication applications thus include

1) large response to the incident optical signal,
2) sufficient instantaneous bandwidth to accommodate the information bandwidth of the incoming signal
3) minimum of noise added by the demodulation process.

Involved is the optimization of the entire detection system with respect to speed of response and sensitivity to weak light signals [1]–[4]. Attention must not only be given to the choice of the optimum detector and to the design of its load circuit and associated low noise amplifier but also to other system parameters, such as desired field-of-view [5]–[7], optical bandwidth, possible relative motion between transmitter and receiver (possibly leading to Doppler shift and/or a requirement for the generation of pointing-error information [8]), and interference due to the sun or other radiation.

Photodetectors convert the absorbed optical radiation into electrical output signals. They are square law detectors that respond to the intensity of light averaged over a number of optical cycles [5], [9]. This is because the speed of response is determined by carrier transport and relaxation processes within the photodetector. These processes do not have sufficiently short time constants to reproduce field variations which occur at optical frequencies. The general expression for the conversion of an incident photon stream of average power P_{opt} and optical frequency v into a primary photocurrent I_{ph} is

$$I_{ph} = \eta \frac{q P_{opt}}{hv} \qquad (1)$$

with P_{opt}/hv = average number of incident photons per unit time, I_{ph}/q = average number per unit time of electrons emitted from the photocathode, or electron–hole pairs collected across the junction region of photodiode, or mobile electrons and holes excited within the photoconductor, and η = conversion or quantum efficiency.

Utilization of wide information bandwidths for optical communication systems—bandwidths of several hundred MHz or even GHz are under consideration—requires demodulation systems with correspondingly wide instantaneous bandwidths. The information bandwidth of the communication system sets the longest time constants that are permissible in the photodetector and associated output amplifier and load circuit. Circuit considerations play a key role in determining the bandwidth and sensitivity of a detection system. Detector shunt capacitance, series resistance, and circuit parasitics must be considered. Because

Manuscript received June 18, 1970.

H. Melchior is with the Bell Telephone Laboratories, Inc., Murray Hill, N. J. 07922.

M. B. Fisher is with Sylvania Electronic Products, Seneca Falls, N. Y.

F. R. Arams is with AIL, a division of Cutler-Hammer, Melville, N. Y.

Reprinted from *Proc. IEEE*, vol. 58, pp. 1466–1486, Oct. 1970.

of the capacitance inherent in any photodetector and output circuit, the load or amplifier input resistance required to achieve a sufficiently large bandwidth is, with the exception of certain photoconductors and cooled long wavelength photodiodes, much smaller than the internal shunt resistance of the photodetector. As a consequence, the sensitivity to weak light signals can be quite low for large bandwidth direct detection systems. In the absence of a gain mechanism for the photosignal ahead of or within the photodetector, the thermal noise of the load resistance or the amplifier noise dominates over the quantum or shot noise of the optical signal, thus limiting the sensitivity for demodulation of weak signals.

An improvement in sensitivity is possible through the use of wide-band output transformers such as the helical coupling structure of traveling wave phototubes [10]. However, the most significant improvement in sensitivity for the demodulation of weak light signals with wide bandwidths is brought about by the introduction of gain for the photosignal before it reaches the detector output. Practically useful gain mechanisms are 1) current gain within photomultipliers, avalanche photodiodes, and photoconductors, and 2) optical mixing, that is, heterodyne detection [11]–[18]. In addition, optical preamplification [19]–[22] and parametric upconversion [23]–[28] constitute potential means for obtaining high detection sensitivity.

A. Direct Detection

The signal-to-noise ratio and the minimum detectable signal are major criteria by which the sensitivity of detection systems to weak optical signals is judged. The signal-to-noise ratio at the output of a generalized direct photodetection system that comprises a photodetector with internal current gain M and a load circuit as shown in Fig. 1 is given by [1], [29]

$$\frac{S}{N} = \frac{\frac{1}{2}(mI_{\text{ph}})^2|M(\omega)|^2}{\left[2q(I_{\text{ph}} + I_B + I_D)|M(\omega)|^2F(M) + 4k\frac{T_{\text{eff}}}{R_{\text{eq}}}\right]B} \cdot \quad (2)$$

In this equation m = modulation index of the light, $M(\omega)$ = current gain or multiplication factor within the photodetector, $2q(I_{\text{ph}} + I_B + I_D)B = \overline{i_s^2}$ = mean-square shot-noise current, I_B = background radiation induced photocurrent [3], [5], I_D = dark current component that is multiplied within photodetector, B = electrical bandwidth of detection system, $F(M)$ = factor that accounts for the increase in noise induced by the internal current gain process. The term $4k(T_{\text{eff}}/R_{\text{eq}})B$ represents the effective mean-square thermal noise current, with R_{eq} being the equivalent resistance of the photodetector and output circuit [1], and T_{eff} is an effective noise temperature that takes into account thermal noise due to the detector and load resistor, if any, and the following amplifier.

The factor $F(M)$ accounts for the increase in noise that is induced by the current gain process. While this noise factor is unity in reverse-biased solid-state, and vacuum photodiodes, it is larger than unity for photomultipliers, and

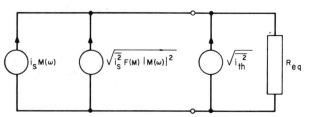

Fig. 1. Equivalent circuit of generalized photodetection system showing principal signal and noise sources of photodetector and load circuit.

especially for avalanche photodiodes. For photoconductors and unbiased photodiodes, $F = 2$ at low frequencies because, as will be shown in Section III-C, the magnitude of the radiation induced generation–recombination noise is twice as large as the shot noise.

From (2) the minimum signal that can be detected for a given bandwidth and signal-to-noise ratio can be determined.

With a direct detection system, the highest sensitivity is reached if the minimum detectable signal is limited by fluctuations in the average signal current itself. In this limit, the peak value of the minimum detectable optical power becomes

$$P_{s\,\text{min}} = mP_{\text{opt}} = 4h\nu\frac{B}{m}\left(\frac{S}{N}\right)\frac{F}{\eta} \quad (3)$$

where (3) is the quantum noise limited sensitivity of direct detection.

In practical broad-band direct photodetection systems without internal current gain, the minimum detectable signal is usually limited by the thermal noise of the detector and load resistance and by amplifier noise

$$P_{s\,\text{min}} = mP_{\text{opt}} = \frac{h\nu}{q}\left(B\left(\frac{S}{N}\right)\right)^{1/2}\left(\frac{8kT_{\text{eff}}}{\eta^2 R_{\text{eq}}}\right)^{1/2} \quad (4)$$

(thermal or amplifier noise limited sensitivity).

An appreciable improvement in sensitivity over the thermal or amplifier noise limited case (4) is possible through the use of photodetectors with internal current gain such as photomultipliers, avalanche photodiodes, and photoconductors. The ideal quantum noise limited sensitivity of (3) can, however, not be fully reached because the current gain in photomultipliers and especially in avalanche photodiodes provides excess noise ($F(M)$) and because practical quantum efficiencies are smaller than unity. In a practical photodetection system with internal current gain the highest sensitivity will be reached when the shot noise; which is induced by the average light signal, the background radiation, and the leakage currents; is multiplied to a level comparable to the thermal or amplifier noise [1], [30]. For this optimum current gain M_{opt}, the minimum detectable signal will be lower by approximately a factor of $2/M_{\text{opt}}$ as compared to the thermal noise limited case without gain (4) [1]. If more gain than this optimum value is used in photomultipliers and especially in avalanche photodiodes, the sensitivity decreases again because of the excess noise associated with carrier multiplication.

Sufficient current gain as determined by the previously mentioned noise considerations cannot be reached in all wide bandwidth detection systems because gain-bandwidth limitations ($M \cdot B$) of the current amplification mechanism set an upper limit to the maximum achievable gain M_{max}. In this case, the sensitivity increases only by a factor M_{max}.

In detection systems with moderate bandwidths, both those with and without current gain, the sensitivity is often not limited by thermal or signal induced noise but by noise generated by background radiation (I_B) [5], [6] or by dark current (I_D). Noise due to leakage currents are of importance in photoemitters and photodiodes. Noise due to background radiation often sets the sensitivity limit in direct detection systems that operate at infrared wavelengths [4]–[6]. However, by limiting the size, spectral acceptance bandwidth, and field of view of the detector and through cooling, these currents can be usually lowered considerably.

B. Optical Heterodyne Detection

In optical mixing or heterodyne detection [2]–[4], [10], [11]–[18], a coherent optical signal P_s is mixed with a laser local oscillator P_{LO} at the input of a photodetector as shown schematically in Fig. 2. The mean value of the photocurrent generated at the intermediate or difference frequency is then [11], [16], [31] for $P_{LO} \gg P_s$

$$i_{IF} = \frac{\eta q}{h\nu} \sqrt{2 P_{LO} P_s}. \tag{5}$$

For the generalized photodetection circuit of Fig. 1, the heterodyne transducer gain G_T defined [32] as the ratio of actual IF power delivered to the output amplifier with input resistance R_A, divided by the available optical signal power P_S, is

$$G_T = \frac{P_{IF}}{P_S} = 2 \left(\frac{\eta q}{h\nu} \right)^2 P_{LO} \frac{(M(\omega) R_{eq})^2}{R_A}. \tag{6}$$

This equation indicates that high conversion gain is possible using sufficient local oscillator power.

For proper mixing action, and for best response at the intermediate frequency, the polarizations of the signal and the local oscillators should be the same [33]–[35]. The signal and the local oscillator must both be normally incident and maintain parallel wavefronts over the entire sensitive area of the detector, the dimensions of which should be minimized to ease the alignment requirement. The minimum detectable signal is then as determined from (2) replacing $m I_{ph}/\sqrt{2}$ by i_{IF}, so that $m = 2(P_S/P_{LO})^{\frac{1}{2}}$

$$P_{S\,min} = \frac{S}{N} B \left\{ h\nu \frac{F}{\eta} + \left(\frac{h\nu}{g\eta} \right)^2 \frac{1}{P_{LO} M^2} \right. $$
$$\left. \cdot \left(2k \frac{T_{eff}}{R_{eq}} + q(I_B + I_D) M^2 F \right) \right\}$$

(sensitivity of heterodyne detection). (7)

For sufficient LO power, the second term representing thermal noise of detector, load and amplifier, and shot or

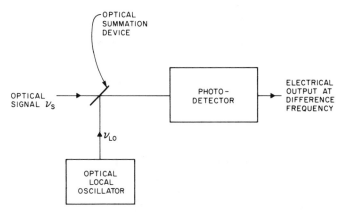

Fig. 2. Optical heterodyne detection system.

generation–recombination noise due to background and dark current, becomes small. This is the case of a well-designed heterodyne receiver in which noise is dominated by local-oscillator induced shot or generation–recombination noise.

Heterodyne detection thus allows realization of optical receivers with wide bandwidths and high sensitivity almost equal to the theoretical quantum noise limit ($h\nu B$). The availability of high conversion gain eases the design requirements for photodetectors considerably. High quantum efficiency and sufficient speed of response of the photodetector and the output circuit are necessary, but the internal current gain is not. Operation at higher temperature than direct detection may also be possible.

Practical application of heterodyne detection is most successful at infrared wavelengths because the alignment requirements are easier to maintain. For 10.6 μm, where strong and stable single line local oscillators exist, heterodyne detection systems with 1-GHz base-bandwidth and sensitivities that approach the ideal limits have been reported [17].

Heterodyne detection systems have a diffraction-limited field of view [33] and small spectral acceptance bandwidth. This makes them relatively insensitive to background radiation and interference effects. Optical mixing preserves the frequency and phase information of the input signal and can thus be used for intensity, phase, and frequency modulation.

III. REVIEW OF DETECTOR CHARACTERISTICS

A. Photoemissive Devices

The external photoelectric effect which describes the emission of electrons into a vacuum from a material which has absorbed optical radiation provides a basic mechanism for the detection of modulation on an optical carrier. Photoemission from photoelectric materials, that is photocathodes [36], [37], is used in both vacuum photodiodes and photomultipliers for the conversion of optical radiation into an electron current. The time constants involved in the photoemission process are sufficiently short [9] that the intensity modulation present on the incoming radiation is converted to a similar modulation of the electron current emitted from the photocathode for modulation frequencies

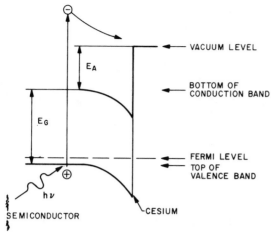

Fig. 3. Idealized energy band diagram of semiconductor photocathode. Photoexcitation of electrons from valence band over energy bandgap E_G and electron affinity E_A into vacuum is indicated.

TABLE I

BAND GAP ENERGY, ELECTRON AFFINITY, AND QUANTUM EFFICIENCY FOR SOME SEMICONDUCTORS WITH HIGH PHOTOEMISSIVE EFFICIENCY

Material	S Number	Energy-Gap E_G[eV]	Electron Affinity E_A[eV]	Quantum Efficiency η (%)	Ref.
Na_2KSB		1.0	1.0	30	[37]
$(Cs)Na_2KSB$	S-20	1.0	0.55	40	[37]
Cs_3Sb	S-17	1.6	0.45	30	[37]
$GaAs-Cs_2O$		1.4	−0.55	35 at 0.4μm 15 at 0.8μm	[38]– [40], [43]
$In_{0.16}Ga_{0.84}As-Cs_2O$		1.1		1 at 0.9μm	[42]
$InAs_{0.15}P_{0.85}-Cs_2O$		1.1	−0.25	0.8 at 1.06μm	[43]

extending into the microwave range. In a vacuum photodiode the modulated electron current is directly collected at an anode and is passed through an external load resistance to generate the output signal. In a photomultiplier, however, the electron current is first amplified in a chain of secondary emission dynodes.

The characteristics of vacuum photodiodes and photomultipliers which are of importance in the performance of optical receivers are 1) the efficiency with which the photocathode converts optical radiation to electron current, 2) time dispersion effects within the device which may limit bandwidth, 3) internal gain mechanisms which may be used to amplify the primary photocurrent, and 4) internal noise sources which may limit the sensitivity of the system. These characteristics will now be discussed with particular emphasis upon the factors which affect the performance of optical communication systems.

The net efficiency with which the incident photons are converted to emitted electrons, termed the external quantum efficiency of the photocathode, is dependent upon the incident photon energy $h\nu$, the effective diffusion length of electrons within the photosensitive material, and the work function of the photosurface [37]. Most high efficiency photocathodes are semiconductors [37] and an energy diagram of a typical semiconductor photocathode is shown in Fig. 3 where the parameters of interest are the bandgap energy E_G and the electron affinity E_A. The work function of the photosurface E_W, which defines the long wavelength threshold for photoemission, is then the sum of the bandgap energy and the electron affinity. Representative values of bandgap energy, electron affinity, and maximum quantum efficiency for several high efficiency photosurfaces are given in Table I. The newest photoemitters characterized by zero or negative electron affinity [38] such as GaAs[Cs_2O] [39]–[42] and $InAs_{0.15}P_{0.85}$[Cs_2O] [43] are of particular importance both because of high quantum efficiency and extended wavelength response. The long wavelength threshold for these surfaces is essentially determined by the bandgap of the photosensitive material rather than the work function of the surface and substantial improvements

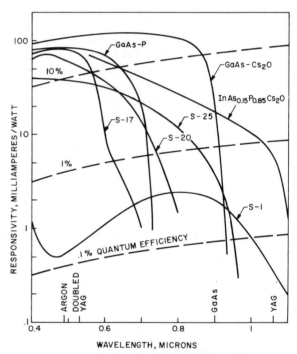

Fig. 4. Wavelength dependence of responsivity and quantum efficiency for several high efficiency photosurfaces.

in sensitivity at near infrared wavelengths, where the GaAs and the Nd:YAG lasers operate, appear possible.

The wavelength variation of the quantum efficiency is dependent upon the reflection and absorption properties of the photosurface as well as on the escape probability of the photoexcited electron in the material [37]. The typical variation of the current responsivity (A/W) and the quantum efficiency with wavelength for several photosurfaces is shown in Fig. 4 together with some of the more important laser wavelengths.

As Fig. 4 and (1) show the magnitude of the current from a photocathode is typically very small. For example, the current from an $S-20$ surface illuminated by one microwatt of argon laser power at 4880 Å is approximately 7×10^{-8} amperes. The output signal generated in a simple vacuum photodiode by passing this photocurrent directly through a load resistor is thus very small, particularly in the case of wide bandwidth devices which require small load impedances. Thus the signal-to-noise ratio of the photodiode system will be limited by noise associated with the preamplifier unless some means of amplifying the primary photocurrent prior to the load resistor is utilized. In a photomultiplier the process of secondary electron emission provides such an internal current gain with a very low noise factor. The typical photomultiplier geometry utilizes reflection dynodes in which the secondary electrons are emitted from the same side of the dynode struck by the primary beam. In most commonly used devices the electron beam is focused through the series of dynodes by a suitably shaped electrostatic field. Magnetic focusing has been used in some very wide bandwidth photomultipliers.

The gain of an electron multiplier is given by $\bar{\delta}^N$ where $\bar{\delta}$ is the average gain per stage and N is the number of stages. Commonly used secondary emission surfaces such as cesium antimonide, magnesium oxide, or beryllium oxide provide gains per stage in the range of 3 to 5 [36]. Recently developed negative electron affinity surfaces such as GaP[Cs] [44] or GaAs[Cs$_2$O] can provide gains per stage of 20 to 50. An overall gain of 10^6 to 10^8 can be achieved using any of these surfaces although obviously a larger number of stages will be required in the case of conventional low gain dynodes. The choice of dynode material must primarily be made on the basis of desired bandwidth and noise factor rather than overall gain as will now be described.

The factors which limit the bandwidth of photomultipliers are time dispersion effects within the secondary electron multiplier and capacitance associated with the anode output circuit [45], [46]. The time dispersion within the electron multiplier is caused by nonuniform electron transit times for 1) electrons emitted from different positions on the dynode surface because of spatially varying electric fields, and 2) electrons emitted from the same position on the dynode surface but with finite initial velocity distributions. The effect of nonuniform electric fields can be minimized by proper device geometry. The time dispersion due to the unavoidable initial velocity distribution can only be minimized by employing high interstage voltages and reducing the number of stages. For example, the upper half-power frequency limit of a rather idealized electron multiplier in which the electric field is assumed to have no spatial variation is given by [45]

$$f_{3\ \text{dB}} = 0.094\ \frac{2qE}{m_0 \bar{v} N^{\frac{1}{2}}} \qquad (8)$$

where E = electric field between dynodes, m_0 = electron mass, $\bar{v}$ = most probable initial velocity, and N = number of stages in the electron multiplier. Thus the bandwidth is proportional to the dynode voltage (for a given dynode spacing) and inversely proportional to the square root of the number of stages. The dynode voltage, however, also determines the secondary emission gain since most commonly used dynode materials exhibit a peak in secondary

emission gain as a function of the voltage between the stages [36]. Although this peak is rather broad it still results in a bandwidth that is not independent of the gain. Additional shot noise is introduced due to the statistics involved in the secondary emission process which amplifies both signal and primary shot noise currents [47]. The mean-square shot noise of the primary current leaving the photocathode $\overline{i_n^2} = 2q(I_{\text{ph}} + I_B + I_D)B$, with components due to signal, background, and dark current, increases through this multiplication process to

$$\overline{i^2} = 2q(I_{\text{ph}} + I_B + I_D)M_0^2 F B \qquad (9)$$

where $M_0 = \bar{\delta}^N$ = the average total multiplication, N = the number of stages, $\bar{\delta}$ = the average gain per stage, and δ is assumed to have a Poisson distribution. The noise factor F of the photomultiplier is given by [47]

$$F = \frac{\bar{\delta}^{(N+1)} - 1}{\bar{\delta}^N(\bar{\delta} - 1)} \qquad (10)$$

or, for $\bar{\delta} \gg 1$ which is the case for any practical secondary emission dynode material

$$F \approx \frac{\bar{\delta}}{\bar{\delta} - 1}. \qquad (11)$$

Thus it may be observed that a photomultiplier has a very low noise factor, a typical example being $F \approx 1.5$ for $\bar{\delta} = 3$. The new high gain dynodes result in even lower noise factors. Since the electron multiplier structure is in effect a cascaded series of amplifiers, the gain of the first stage is more important in determining excess shot noise than the gain of the remaining stages.

The optimum photomultiplier for an optical communication system operating at a specific wavelength should have the following characteristics based upon the previous discussion. The photosurface should be chosen for maximum quantum efficiency at the operating wavelength using antireflection coatings and multiple-pass techniques to enhance optical absorption whenever possible [36], [37]. High gain dynodes such as GaP should be utilized, particularly for the first stage, since excess shot noise can be minimized in this manner. These high gain dynodes typically require high dynode voltages for maximum gain and the resulting high electric field reduces transit time dispersion. In addition, their high gain reduces the number of required stages with a resultant further reduction in transit time dispersion and increase in bandwidth. Finally, the anode lead geometry or output coupling circuit must be designed to provide a bandwidth consistent with that of the communications system. Relatively narrow bandwidth photomultipliers employ simple wire connections from the anode to the output header pin, while in wide bandwidth devices, the output current must be focused into a coaxial anode and output transmission system. Factors such as photocathode area and dynamic range must be considered in addition. It is desirable to minimize the cathode area to reduce the dark current noise, consistent with optical and mechanical design requirements. The dynamic range is limited at the low end by noise and at the high end by saturation of the output current or by space charge effects that originate in the dynode chain.

The dynamic range of photomultipliers to be used in wide bandwidth systems must be carefully considered with respect to the desired S/N ratio. The S/N ratio measured at the output of the photomultiplier in a direct detection system with sufficient gain to achieve shot noise limited operation is, after (2), given by

$$\frac{S}{N} = \frac{\frac{1}{2}(mI_{ph}M)^2 R_{eq}}{2qI_{ph}M^2 FBR_{eq}}. \tag{12}$$

However, if the output current $mI_{ph}M$ becomes limited by saturation effects to I_{sat}, the S/N ratio has a limiting value of

$$\frac{S}{N} \propto \frac{I_{sat}}{M \cdot B} \tag{13}$$

which might not be sufficient to reach signal shot noise limited sensitivity.

Of the commercially available photomultipliers, the conventional electrostatically focused multipliers have bandwidths up to about 100 MHz. If a third high voltage electrode is placed between each pair of dynodes to reduce transit times, bandwidths of several hundred MHz result [49]. Improved designs that still maintain the conventional dynode chain, but use high efficiency photoemitters and the new high gain dynodes have been reported with subnanosecond response times [50]. Magnetically focused crossed-field photomultipliers can provide a bandwidth of 6 GHz by using a well-matched coaxial output line [46].

B. Solid-State Photodiodes and Avalanche Photodiodes

1) *Photodiodes:* Efficient high speed photodiodes, and avalanche photodiodes with internal current gain, have been fabricated by using specially designed p–n, p–i–n or metal–semiconductor junctions as shown in Fig. 5. These diodes are usually operated in the reverse bias region. In general, photon excited electrons and holes that are generated 1) within the high field region of the junction [51] and 2) in the bulk region and then diffuse to the junction [52], are collected as photocurrent across the high field region. However, diffusion processes are slow compared to the drift of carriers in the high field region. Therefore, in high speed photodiodes, the carriers have to be excited within the high field region of the junction or so close to the junction that diffusion times are shorter than, or at least comparable to, carrier drift times [15], [29]. Carriers are then collected across the junction at scattering limited velocities that are of the order of $v_{sat} = 10^6$ to 10^7 cm/s. Under these circumstances, the peak ac photocurrent is given by [51] (for photoexcitation of carriers at the junction edge)

$$i_{ph} = \frac{\eta q}{h\nu} m P_{opt} \frac{1 - \exp(-j\omega T_r)}{j\omega T_r} \tag{14}$$

where $T_r = $ drift transit time. As space charge layer widths (w_i) can be in the order of micrometers, carrier transit times $T_r = w_i/v_{sat}$ in the subnanosecond region are possible.

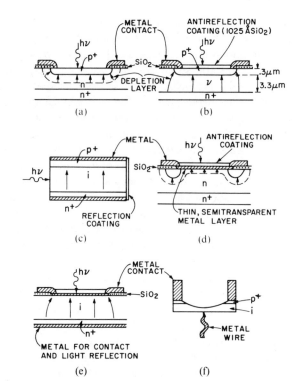

Fig. 5. Construction of different high speed photodiodes. (a) p–n diode. (b) p–i–n diode (Si optimized for 0.63 μm). (c) p–i–n diode with illumination parallel to junction. (d) Metal–semiconductor diode. (e) Metal–i–n diode. (f) Semiconductor point contact diode.

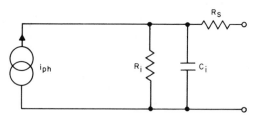

Fig. 6. Equivalent circuit of photodiode.

The ac characteristics of photodiodes can be described in terms of an equivalent circuit [15], [29], [31], [52], [54] that contains a photocurrent generator i_{ph}, diode capacitance C, series resistance R_S, and shunt resistance R_i, as shown in Fig. 6. The shunt resistance R_i is very high for diodes operating in the visible but is included to account for the relatively low leakage resistances of infrared photodiodes.

In the visible and near infrared well-designed diodes with high quantum efficiency, fast speed of response, low dark currents, and low series resistances can be obtained. High quantum efficiency requires minimization of the light reflection at the diode surface and placement of the junction in such a way that most photons are absorbed within or close to the high field region of the junction. Solid-state photodiodes are usually designed for light incidence normal to the junction plane. The quantum efficiency for a particular photodiode depends on the wavelength of operation and on the depth and width of the junction region. The spectral dependence of the quantum efficiency for a number of photodiodes is shown in Figs. 7 and 8.

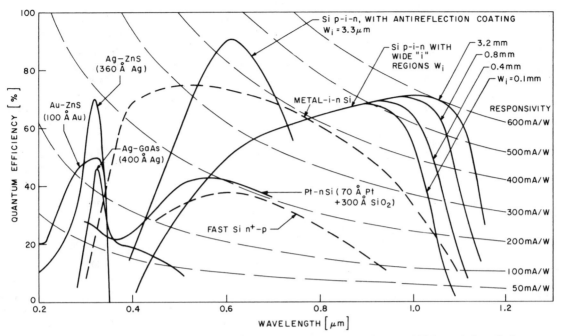

Fig. 7. Wavelength dependence of quantum efficiency and responsivity for several high speed photodiodes.

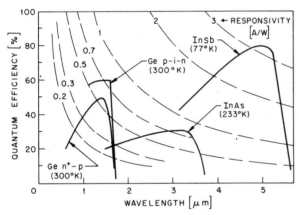

Fig. 8. Quantum efficiency and responsivity of photodiodes that operate between 1 and 6 μm.

The speed of response of photodiodes is determined by *RC* time constants [29], [54], and the already mentioned carrier diffusion or drift transit times. The characteristic cutoff frequency of a photodiode is given by $f_c = 1/2\pi R_S C(1 + (R_S/R_i))$. Practical detection systems have, however, lower cutoff frequencies because of the finite load resistance and the additional capacitance (and inductances) of the load or amplifier circuit. Fast photodiodes have usually planar junctions that are small, 50 to 1000 μm in diameter, in order to keep the diode capacitance and leakage currents small [53]. Light has to be focused onto these diodes. The highest speeds of response and the smallest capacitances might be obtained from point contact diodes [Fig. 5(f)] [55] but the focusing of the light onto the small light sensitive areas is more difficult. Low series resistances in small area photodiodes are possible if the thickness of undepleted semiconducting bulk regions is kept small, as in p–i–n or metal–i–n diodes with fully de-

pleted space charge layers or in p⁺–n diodes with an epitaxial base. For operation with baseband widths up into the GHz region, high speed photodiodes (and avalanche photodiodes) are commonly mounted into especially designed coaxial headers [54], [56].

The choice of a photodiode depends mainly on the wavelength of operation (Table II). Because of their developed technology, silicon diodes are preferentially used in the ultraviolet, visible, and near infrared part of the spectrum up to about 1 μm. With germanium diodes the response can be extended to over 1.5 μm. The geometry of both Si and Ge photodiodes can be optimized for particular applications. At short wavelengths, where light is absorbed close to the semiconductor surface it makes sense to use metal–semiconductor photodiodes with thin semitransparent metal layers [56]–[58]. Carriers are then separated in the high field region close to the surface thus yielding high quantum efficiencies and short response times. Several metal–semiconductor photodiodes [Fig. 5(d) and (e)] with response at short wavelengths have been fabricated from Si [57], GaAs [59], and ZnS [60] as indicated in Fig. 7 and Table II. For the visible range of the spectrum, where light penetrates a few micrometers into silicon, diffused p–n [61] and p–i–n junctions [53], [54], [62] are used [Fig. 5(a) and (b)]. Diffused junctions exhibit somewhat smaller reverse dark currents than metal–semiconductor junctions. For longer wavelengths close to the bandedge of the diode material, light penetrates deeply into the material. High quantum efficiency thus requires wide space charge layers. This leads to relatively long carrier transit times. For these diodes a tradeoff exists between quantum efficiency and speed of response [63]. As an example Fig. 7 shows the quantum efficiency at long wavelengths around 1 μm for Si p–i–n diodes with various space charge layer widths

TABLE II

PERFORMANCE CHARACTERISTICS OF PHOTODIODES

Diode	Wave-length Range (μm)	Peak Efficiency (%) or Respon-sivity	Sensitive Area (cm^2)	Capacitance (pF)	Series Resistance (Ω)	Response Time (seconds)	Dark Current	Operating Temperature (°K)	Comments	Ref.
Silicon n$^+$–p	0.4–1	40	2×10^{-5}	0.8 at -23 V	6	130 ps with 50-Ω load	50 pA at -10 V	300	avalanche photo-diode	[61]
Silicon p–i–n	0.6328	>90	2×10^{-5}	<1	~1	100 ps with 50-Ω load	$<10^{-9}$ A at -40 V	300	optimized for 0.6328 Å	[62]
Silicon p–i–n	0.4–1.2	>90 at 0.9 μm	5×10^{-2}	3 at -200 V	<1	7 ns	0.2 μA at -30 V	300		[65]
		>70 at 1.06 μm		3 at -200 V	<1	7 ns				
Metal–i–nSi	0.38–0.8	>70	3×10^{-2}	15 at -100 V		10 ns with 50-Ω load	2×10^{-2} A at -6 V	300		[57]
Au–nSi	0.6328	70	2			<500 ps		300	Schottky barrier, antireflection coating	[56]
PtSi-nSi	0.35–0.6	~40	2×10^{-5}	<1		120 ps		300	Schottky barrier avalanche photo-diode	[58]
Ag-GaAs	<.36	50						300		[59]
Ag-ZnS	<.35	70						300		[60]
Au-ZnS	<.35	50						300		[60]
Ge n$^+$–p	0.4–1.55	50 uncoated	2×10^{-5}	0.8 at -16 V	<10	120 ps	2×10^{-8}	300	Germainum avalanche photo-diode	[30]
Ge p–i–n	1–1.65	60	2.5×10^{-5}	3		25 ns at 500 V		77	illumination entering from side	[65] [66]
GaAs point contact	0.6328	40		0.027	30					[55]
InAs p–n	0.5–3.5	>25	3.2×10^{-4}	3 at -5 V	12	$<10^{-6}$		77		[15] [29]
InSb p–n	0.4–5.5	>25	5×10^{-4}	7.1 at -0.2 V	18	5×10^{-6}		77		[15] [29] [67]
InSb p–n	2–5.6		5×10^{-4}				1 MΩ shunt resistance	77	Reverse break down voltages 30 V	[68]
Pb$_{1-x}$Sn$_x$Te $x=0.16$	9.5 μm	45 V/W $\eta=60$	4×10^{-3}			~10^{-9}		77	shunt resistance $R_i = 10$ Ω	[72] [74]
Pb$_{1-x}$Sn$_x$Se $x=0.064$	11.4 μm	3.5 V/W $\eta=15$	7.8×10^{-3}			~10^{-9}		77	shunt resistance $R_i = 2.5$ Ω	[73] [74]
Hg$_{1-x}$Cd$_x$Te $x=0.17$	15 μm	$\eta \sim 10$–30	4×10^{-4}		8		$<3 \times 10^{-9}$	77	shunt resistance $R_i > 100\Omega$	[70] [71], [136]

[64], [65]. The corresponding carrier transit times or diode response times and bias voltages can be determined from Fig. 9 [63]. Germanium diodes [53] are used at wavelengths below 1 μm to beyond 1.5 μm. However, due to the narrower bandgap, germanium diodes exhibit larger dark currents. Reduced dark currents and thus higher sensitivity can be obtained in both Si and Ge diodes by cooling. The problems encountered in the design of Ge diodes that are fast and efficient for 1.53 (HeNe laser) and 1.54 μm (Erbium laser) are similar to those encountered with Si diodes for

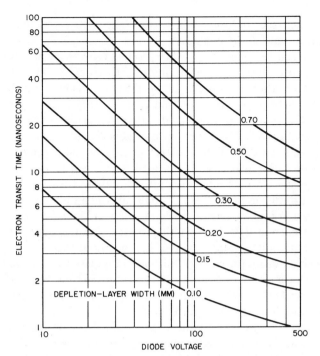

Fig. 9. Carrier transit times for Si p-i-n diodes with different depletion layer widths (after McIntyre [63], [64]).

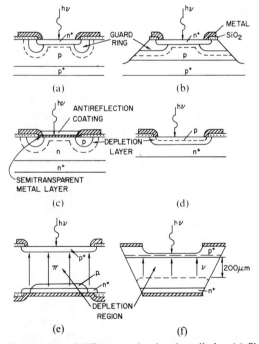

Fig. 10. Construction of different avalanche photodiodes. (a) Si guard ring structure. (b) Ge mesa structure with guard ring. (c) Metal–semiconductor structure with guard ring. (d) Planar p–n$^+$ structure with low fields at junction edges. (e) Si–n$^+$–p–π–p$^+$ structure. (f) Si structure with wide depletion region and slant surfaces.

efficiency curves of Fig. 8. For maximum sensitivity these diodes have to be operated at temperatures between 77 and 135°K.

For longer wavelengths in the 6- to 30-μm range and especially for 10.6 μm, where the CO$_2$ laser operates, mixed crystal photodiodes such as Hg$_{1-x}$Cd$_x$Te [69]–[71] and Pb$_{1-x}$Sn$_x$(Te, Se) [72]–[74] are under development. Dynamic shunt resistances in Hg$_{1-x}$Cd$_x$Te in the range 10^2 to 10^5 ohms at 77°K have been obtained [70], [71], so that photodiodes for 10.6 μm that are fast and efficient are becoming available for operation at 77°K or even 110°K.

Recently, (Hg, Cd) Te photodiodes with wider bandgaps have been investigated for use at shorter wavelengths down to 1–2 μm [69].

2) Avalanche Photodiodes: Current gain in solid-state photodiodes is possible through avalanche carrier multiplication [75], observed at high reverse bias voltages where carriers gain sufficient energy to release new electron–hole pairs through ionization. Substantial current gains have been achieved by this process even at microwave frequencies [61], [75]. Despite the fact that excess noise is introduced by the multiplication process [30], [61], [76]–[78], significant improvements in overall sensitivity are possible using silicon or germanium avalanche photodiodes for photodetection systems with wide instantaneous bandwidths.

Similar criteria apply for avalanche photodiodes, with respect to quantum efficiency and speed of response, as for conventional nonmultiplying photodiodes. Additional attention must, however, be given to the current gain and its limitations and to the noise properties of avalanche photodiodes.

In the design of avalanche photodiodes, special precautions must be taken to assure spatial uniformity of carrier multiplication over the entire light sensitive diode area. Microplasmas, i.e., small areas with lower breakdown voltages than the remainder of the junction, and excessive leakage at the junction edges can be eliminated through the use of guard ring structures [61], [79] as indicated in Fig. 10. The selection of defect-free material and cleanliness in processing allows fabrication of microplasma-free diodes. Highly uniform carrier multiplications in excess of 10^4 and 200 have been reached at room temperatures in small area silicon [61] and germanium [30] diodes, respectively. In large area diodes that are free of microplasmas, the spatial uniformity of carrier multiplication is limited either by doping inhomogeneities of the starting material or by inhomogeneities in the diffusion profile. Typical variations can be 20 to 50 percent at an average multiplication of 10^3.

The highest current gains are observed if the diodes are biased to the breakdown voltage. This is illustrated by Fig. 11. where the voltage dependence of the dark current of a Ge avalanche photodiode as well as the response to pulses from a phase locked 6328 Å laser are shown. At low reverse bias voltages, no carrier multiplication takes place. As the reverse bias voltage is increased, carrier multiplication sets in as indicated by the increase in the current pulse.

0.9 and 1.06 μm. Better results might be achieved if light is allowed to penetrate from the side [66], parallel to the junction as shown in Fig. 5(c). This may result in a very narrow sensitive area.

For wavelengths beyond 1.5 μm, up to 3.6 μm and 5.6 μm, respectively, InAs and InSb photodiodes [29], [32], [67], [68] are suitable as can be seen from the quantum

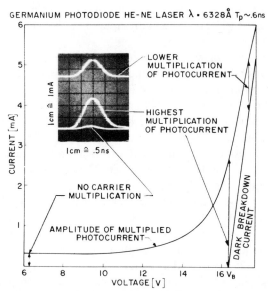

GERMANIUM PHOTODIODE HE-NE LASER λ • 6328Å T_p ~ .6ns

Fig. 11. Current voltage characteristic and pulse response of germanium avalanche photodiode.

The maximum pulse amplitude and multiplication of photocarriers is obtained at the breakdown voltage. At voltages above the breakdown voltage, a self-sustained avalanche current flows that becomes less and less sensitive to photon-excited carriers.

The maximum gain of an avalanche photodiode is limited either by current induced saturation effects [30] or by a current gain bandwidth product [80]–[83], [85]. Current induced saturation of the carrier multiplication is observed because the carriers that emerge from the multiplication region reduce the electric field within the junction and cause voltage drops across the series and load resistance of the diode [30]. This leads to a current dependent reduction of the carrier multiplication factor. This saturation manifests itself for high light intensities, at which the multiplied current will only increase as the square root of the photocurrent instead of being proportional to the photocurrent as at low light intensities. For low light intensities, the dark current will set a limit to the average value for the maximum carrier multiplication that can be achieved at low frequencies. The aforementioned maximum multiplication of 200 for germanium diodes [30] is due to the high dark current. Lowering of the temperature in diodes with sufficiently high breakdown voltages (> 20 volts for Si and Ge) so that no internal field emission takes place, decreases the dark current and leads to higher values for the maximum carrier multiplication.

At high modulation frequencies or for fast optical pulses, the current gain–bandwidth product sets a limit to the maximum achievable gain [1], [80]–[85]. For silicon [61] and germanium [30] n^+–p diodes gain–bandwidth products of 100 and 60 GHz have been reported as indicated in Table III. This current gain–bandwidth product is due to the fact that both electrons and holes excite electron–hole pairs through ionization in silicon and germanium diodes. Secondary electrons and holes thus travel back and forth

through the multiplication region long after the primary carrier has left the junction.

The current gain–bandwidth product is inversely proportional to the average transit time of the carriers through the multiplication region [80]–[83], [85] and depends on the ratio between the electron and hole ionization rates [84]. For Ge, the current gain–bandwidth product is independent of excitation because the ionization coefficients are almost equal, but for Si, higher current gain–bandwidth products result if the avalanche is initiated with electrons because they have a higher ionization rate. If one can realize carrier multiplication in solid-state diodes in which only one type of carrier ionizes, then the current gain would not be limited by a current gain–bandwidth product. The response time at high multiplications would then be about twice as long as without multiplication. Although indications exist that only electrons cause ionizations in InSb [86] and Schottky barrier GaAs [87] photodiodes, no speed of response measurements have been reported to date.

Excess noise in avalanche photodiodes is induced by fluctuations in the carrier multiplication process [76]–[79]. For practical diodes, the shot noise of the average photo and dark current has been found [30], [61], [78], [79] to increase faster than the square of the carrier multiplication M, approximately as

$$\overline{i^2} = 2q(I_{\text{ph}} + I_D)M^{2+x}B \qquad (15)$$

in reasonably close agreement with theoretical calculations based on a spatially uniform avalanche ionization region [77]. The noise factor $F = M^x$ depends on the ratio of the ionization rates and on the type of carriers that initiate the avalanche [1], [30], [77]–[79]. For germanium diodes [30], where the ionization coefficients for holes and electrons are about equal, the excess noise factor increases as the carrier multiplication ($x = 1$). This relatively high noise is due to the fact that for equal ionization rates, a relatively low number of carriers is present within the multiplication region at any one time so that fluctuations in the ionization events have a large influence on the avalanche process. Much lower noise is observed in junctions with highly unequal ionization rates if the avalanche is initiated with carriers of the higher ionization rate. Initiation of avalanche carrier multiplication by electrons leads to low excess noise in Si ($F = M^{0.4}$) [78], [79], InSb [86], and Schottky barrier GaAs [87] diodes. If only one type of carrier can ionize, theoretical calculations show that the noise factor does not follow (15), but would be equal to two at high multiplications [77]. The results for InSb and for certain GaAs diodes indicate that only electron ionization is of importance in these materials.

The different types of avalanche photodiodes that have been constructed to date are shown schematically in Fig. 10 and their performance data are listed in Table III. As Table III shows, various Si n^+–p guard ring avalanche photodiodes have been developed with diameters of the light sensitive area between 40 and 200 μm [61], [79], [89]–

TABLE III

CHARACTERISTICS OF AVALANCHE PHOTODIODES

Diode	Construction	Wavelength Range (μm)	Sensitive Area (cm^2)	Dark Current	Avalanche Breakdown Voltage (volts)	Maximum Gain	Multiplication Noise $i^2 \sim M^{2+x}$	Current Gain–Bandwidth Product (GHz)	Capacitance (pF)	Series Resistance (Ω)	Ref.
Silicon n$^+$–p	Fig. 10(a)	0.4–1	2×10^{-5}	50 pA at −10 V	23	10^4	$x \sim 0.5$	100	0.8 at −23 V	6	[61]
Silicon n$^+$–p π–p$^+$	Fig. 10 (e)	0.5–1.1	2×10^{-3}		~88	200	$x \sim 0.4$	high			[91], [92]
Silicon n$^+$–i–p$^+$	Fig. 10(f)	0.5–1.1			200 to 2 000		low	not very high			[89]
PtSi–nSi	Fig. 10(c)	0.35–0.6	4×10^{-5}	~1 nA at −10 V	50	400	$x \sim 1$ for visible illumination	40 for UV excitation	<1		[58]
Pt-GaAs	Fig. 10(c)	0.4–0.88			~60	>100	very low	>50			[87]
Germanium n$^+$–p	Fig. (10b)	0.4–1.55	2×10^{-5}	2×10^{-8} A −16 V and 300°K	16.8	250 at 300°K >10^4 at 80°K	$x \sim 1$	60	0.8 at −16 V	<10	[30]
Germanium n$^+$–p	Fig. 10(a)	for 1.54			150						[91], [92]
InAs		0.5–3.5									[88]
InSb at 77°K		0.5–5.5			a few	10	very low				[86]

[91]. These are the simplest avalanche photodiodes. They operate at relatively low voltages and are useful between about 0.4 and 0.8 μm. Similar Ge n$^+$–p diodes have been constructed [30] with a guard ring that terminates in a mesa structure in order to reduce the surface leakage current. These Ge diodes are useful as fast diodes with gain at wavelengths between 0.5 and 1.5 μm. Both the small area Si and Ge avalanche photodiodes can resolve optical pulses with widths of 130 ps between their 50-percent rise and fall time points. The current gain–bandwidth products are 100 and 60 GHz, respectively, thus indicating that current gains of 100 and 60 are possible in a detection system with 1-GHz bandwidth.

Two different structures have been developed for Si avalanche photodiodes for operation at wavelengths between 0.8 and about 1 μm. In order to achieve high quantum efficiencies these diodes have wide space charge layers. One structure, the n$^+$–ν–p$^+$ structure [89] [see Fig. 10(f)] has a wide high field region in which carrier multiplication can take place. Because the electric fields are relatively low in this case, a very low hole ionization coefficient results [84]. These diodes exhibit low excess noise [77] but will have only a low current gain–bandwidth product [81], [82]. In the n$^+$–p–π–p$^+$ structure [90], [91] [Fig. 10(c)], the multiplication is concentrated to the narrow n$^+$–p region whereas the wide π region acts as a collection region for photon excited carriers. This narrow multiplication region will result in a high gain–bandwidth product but results also in higher excess noise. The noise is higher because, at the high fields in this narrow multiplication region, the ionization

coefficients for holes are not much lower than the ionization coefficients for electrons [84].

Silicon avalanche photodiodes with high current gain and relatively low excess noise are thus available for the wavelength range between 0.4 and 0.9 μm. Germanium avalanche photodiodes have higher excess noise and higher leakage current if not cooled, but they are excellent detectors for 1.06 μm and can be used up to over 1.5 μm as well [93]. As an example, the application of a Ge avalanche diode to the demodulation of small signals at 6 GHz on a 1.15-μm laser beam is illustrated in Fig. 12. As can be seen, the signal from the diode increases as M^2 with multiplication and the noise from the avalanche diode as M^3. The best operating point (M_{opt}) with the largest S/N ratio and the highest sensitivity to weak light signals is reached if the carrier multiplication is adjusted so that the noise of the avalanche diode is about equal to the amplifier noise.

Results for GaAs and InSb diodes indicate that avalanche photodiodes with almost no excess noise (only a factor of two) and high gain that is not limited by current gain–bandwidth products will be forthcoming in the foreseeable future. The InSb avalanche diodes will extend the range of solid-state diodes with internal current gain up to 5.6 μm.

Avalanche carrier multiplication has further been observed in a number of additional materials, among them InAs [88], but no practical devices have been reported.

Extension of avalanche photodetection to wavelengths beyond the bandedge of a semiconductor material is possible in metal–semiconductor junctions as photoemission of

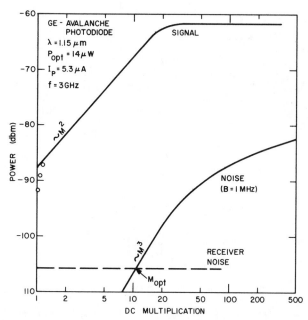

Fig. 12. Signal and noise power output of germanium photodiode in 1-MHz band at 3 GHz. Optimum operating point (M_{opt}) for best S/N ratio and sensitivity is indicated.

carriers from the metal contact can be combined with ionization of carriers within the high field region of the junction. Only small quantum efficiencies can, however, be expected. For palladium–silicon Schottky barrier diodes quantum efficiencies of 0.75 percent and avalanche multiplications >200 have been observed at 1.3 μm [94].

C. Photoconductive Detectors

Electron–hole pairs created in a bulk semiconductor by the absorption of optical radiation increase the conductivity of the material during the lifetime of the pairs. This effect, termed photoconductivity [95], provides a basic mechanism for the detection of optical signals, since upon application of a bias voltage this conductivity modulation can be translated into a modulation of the current that flows through the output circuit.

For a simple dc biased photoconductor with ohmic contacts, whose conductance is dominated by one type of excited carrier, the optically induced small signal peak ac current is [16], [31]

$$i_S = \eta\, \frac{qmP_{opt}}{h\nu}\, \frac{\tau}{T_r}\, \frac{\exp{(-j\phi)}}{(1 + \omega^2\tau^2)^{\frac{1}{2}}} \tag{16}$$

where τ = mean carrier lifetime, $T_r = L^2/\mu V$ = carrier transit time, μ = carrier mobility, L = interelectrode spacing, V = bias voltage, and $\phi = \tan^{-1}(\omega\tau)$.

The factor $M = (\tau/T_r)\exp{(-j\phi)}(1+\omega^2\tau^2)^{-1/2}$ is the photoconductive gain and indicates the number of carriers which cross the photoconductor for each absorbed photon. Practical application of photoconductors generally requires this gain to be maximized consistent with frequency response requirements by operating the photoconductor at the highest bias voltage compatible with breakdown and extraneous noise considerations. The fundamental tradeoff in photoconductors between the current gain $M = \tau/T_r$ and

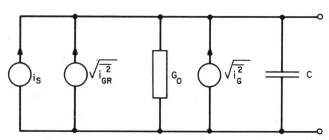

Fig. 13. Small signal equivalent circuit of photoconductive detector.

the speed of response τ is indicated by the existence of a current gain–bandwidth product [96], [97]

$$MB = \frac{1}{2\pi T_r} \leq \frac{1}{2\pi\tau_{rel}}. \tag{17}$$

The maximum value of this current gain–bandwidth product is independent of the carrier lifetime τ and limited as indicated by the dielectric relaxation time $\tau_{rel} = \rho\varepsilon$ [95], where ε = permittivity and ρ = resistivity of the photoconductor.

Evaluation of an optical detection system that utilizes photoconductors requires in addition consideration of the differential conductance G and capacitance C of the photoconductor and of the different noise sources indicated in the small signal equivalent circuit of Fig. 13. The differential conductance is given by [98]

$$G_0 = \frac{Aq\mu n}{L}\bigg|_{\text{if determined by doping}}$$
$$= \frac{\mu\tau(I_{ph} + I_B)}{L^2}\bigg|_{\text{if determined by illumination}} \tag{18}$$

where $(I_{ph} + I_B)/q \sim$ number of photons absorbed per unit time due to both signal (I_{ph}) and background (I_B) radiation.

The principal mean-square noise current generators of a photoconductor are [95]

1) generation–recombination noise [99], [100]

$$\overline{i_{GR}^2} = 4q\left(\frac{\tau}{T_r}\right)\frac{I_0 B}{1 + \omega^2\tau^2} \tag{19}$$

with $I_0 = (\tau/T_r)(I_{ph} + I_B)$ being the dc light induced output current of the photoconductor.

2) thermal noise due to the differential conductance G_0 of the photoconductor at temperature T_G

$$\overline{i_G^2} = 4kT_G G_0 B. \tag{20}$$

The S/N and sensitivity considerations of Section II can be applied to detection systems with photoconductors. The generation–recombination noise of (19) can be brought into the form $2q(I_{ph} + I_B)|M(\omega)|^2 F \cdot B$ with $|M(\omega)|^2 = M_0^2/(1 + \omega^2\tau^2)$ and $F = 2$. The thermal noise due to the output circuit with a load conductance G_L, if any, at temperature T_L and amplifier input conductance G_A and effective amplifier noise temperature T_A

$$\overline{i_L^2} = 4kT_L G_L B$$

and

$$i_A^2 = 4kT_A(G_0 + G_L)B \qquad (21)$$

can be combined with the thermal noise of the photoconductor into the single thermal noise source used in (2)–(7).

The equivalent circuit [31] of Fig. 13 takes account of the main features of the photoconductor. It should, however, be kept in mind, that in practical photoconductors material parameters often change with bias voltage, operating temperature, and optical power. For example, if the differential conductance G_0 of a photoconductor is too low, the carrier transit time T_r becomes shorter than the dielectric relaxation time τ_{rel} at relatively low bias voltages. This limits the current gain–bandwidth product [95] and leads to space charge limited currents [95]. Saturation of the current gain is often observed at high bias voltages [111]. This is because the carrier velocities cease to increase as $v = \mu E$ with electric field E but saturate at a scattering limited velocity v_{sat} thus leading to a voltage independent transit time $T_r = L/v_{sat}$ and gain $M = v_{sat}\tau/L$. In several extrinsic photoconductors an increase of the carrier lifetime τ (as shown in Fig. 14 for Ge:Cu and Ge:Hg(Sb) [102]) and of the $\mu\tau$ product [104] and impact ionization have been observed with rising bias fields. Nonohmic contacts lead to carrier depletion regions and low current gain [95]. In certain materials, such as in CdSe [105], the current gain and the carrier lifetime depend strongly on light intensity, typically decreasing like the square root of the optical power at high light intensities [16].

In many photoconductors both types of excited carriers will contribute to the photoconductivity and be able to move out through the contacts thus leading to a much lower photoconductive gain than indicated by (16). For this case and for nonohmic contacts, microwave bias has been explored as a potential means to increase the current gain and the gain–bandwidth product [106]. The effect of RF bias with a sufficiently high frequency is to effectively "trap" carriers in the bulk of the photoconductor, so that they are not swept out to the electrodes for recombination, but instead move back and forth contributing to conductivity, until volume recombination takes place. Although impressive results have been reported [101], [107], practical application of ac biasing might be somewhat limited because of complicated circuitry and because of the noise that is induced through instabilities in the microwave bias source.

In the design of photoconductive direct detection systems one strives to reach the generation–recombination noise limited sensitivity. This is obtained by the use of high gain to overcome the thermal noise of the photoconductor and output circuit. Once sensitivity is limited by background radiation induced g–r noise, a further increase in detection sensitivity is still possible through the use of aperturing, which narrows the field of view of the photoconductor, and by spectral filtering and cooling [5]–[7].

To obtain a flat frequency response and to minimize phase distortion within the bandwidth of operation B, the photoconductive lifetime should be as short as $\tau \leq 1/2\pi B$. Operation in the rolloff region ($\tau > 1/2\pi B$) is not acceptable in most communications applications. On the other hand,

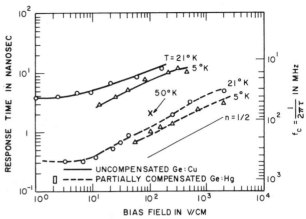

Fig. 14. Response time of photoconductive Ge:Cu and Ge:Hg(Sb) (After Buczek and Picus [102]).

the lifetime should not be shorter than that dictated by the bandwidth, in order to optimize the current gain. Once the value of τ is established the MB product can be optimized by minimizing the transit time, which is accomplished by the use of high dc bias voltages and small detector dimensions in the direction of current flow. In addition, sufficient speed of response must be provided by the output circuit to satisfy the condition $R_{eq}C \leq 1/2\pi B$. The value of C includes the capacitance of the photoconductor, its mount, and the output circuit which must thus be kept as low as possible. Given the value of B and C, a value of R_L is chosen equal to or lower than the photoconductor resistance to satisfy the previously mentioned condition. For wide signal bandwidths this loading leads to reduced sensitivity in direct detection systems since the photoconductive gain is now insufficient to overcome receiver thermal noise. However, by the use of heterodyne detection, photoconductors can yield high sensitivity and wide bandwidth in combination by using sufficient local oscillator power to overcome the thermal noise contributions.

1) Characteristics of Infrared Photoconductors: The spectral sensitivity for a number of photoconductors (and photovoltaic detectors) is shown in Fig. 15. Direct detection sensitivity is given in terms of detectivity D^*, the commonly used figure of merit for infrared detectors [5], D^* is in units of $cm \cdot Hz^{1/2} \cdot W^{-1}$, it is normalized to a sensitive area = 1 cm^2 and a postdetection bandwidth $B = 1$ Hz. D^* is related to the noise equivalent power NEP ($S/N = 1$ and $m = 1$ in (4)) by $D^* = A^{1/2}B^{1/2}/NEP$. The values for D^* given in Fig. 15 are only representative and are given mostly for a 60° field of view. The highest values of D^* are usually reached by the use of high purity materials with long time constants and high resistances.

D^* is commonly measured at audio modulation frequencies using a matched load resistance. This is satisfactory for radiation sensing applications of infrared detectors, where speed is usually secondary to sensitivity. The advent of optical communications and other laser applications gave impetus to more detailed investigations of detector speed and to the development of infrared detectors with wide bandwidth capabilities.

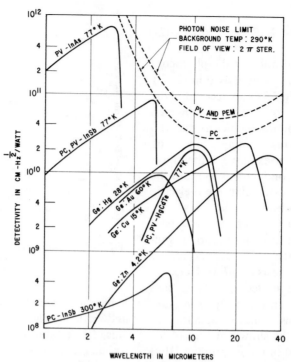

Fig. 15. Spectral dependence of detectivity for photoconductive and photovoltaic infrared detectors. Representative values are given [5], [6], [119].

TABLE IV
INFRARED PHOTOCONDUCTIVE DETECTORS

Material	Maximum Temperature for Background Limited Operation	Test Temperature (°K)	Long Wavelength Cutoff (50%) (μm)	Peak Wavelength λ_m (μm)	Absorption Coefficient (cm⁻¹)	Quantum Efficiency η	Resistance (Ω)	$D^*_{\lambda m}$ (cm·Hz$^{1/2}$/W)	Approximate Response Time (seconds)	Reference
InAs		195	3.6	3.3	$\sim 3 \times 10^3$			3×10^{11}	5×10^{-7}	[101]
InSb	110	77	5.6	5.3	$\sim 3 \times 10^3$	0.5–0.8	10^3–10^4	6×10^{10} -1×10^{11}	5×10^{-6}	[4], [108], [118], [119]
Ge:Au	60	77	9	6	~ 2	0.2–0.3	4×10^5	3×10^9–10^{10}	3×10^{-8}	[108], [110], [111], [119], [120]
Ge:Au(Sb)	60	77	9	6			10^6	6×10^9	1.6×10^{-9}	[102], [119]
Ge:Hg	35	4.2 / 27	14 / 14	11 / 10.5	~ 3 / ~ 4	0.2–0.6 / 0.62	1–4×10^4 / 1.2×10^5	7×10^9 -4×10^{10} 4×10^{10}	3×10^{-8} -10^{-9}	[108], [113]–[115], [119], [120] [115]
Ge:Hg(Sb)	35	4.2	14	11			5×10^5	1.8×10^{10}	3×10^{-10}–2×10^{-9} 3×10^{-10}–3×10^{-9}	[102], [119] [102]
Ge:Cu	17	4.2 / 20	27	23	~ 4	0.2–0.6	2×10^4	2–4×10^{10}	3×10^{-9}–10^{-8} 4×10^{-9}–1.3×10^{-7}	[108], [119], [120]
Ge:Cu(Sb)	17	4.2	27	23			2×10^5	2×10^{10}	$<2.2 \times 10^{-9}$	[119], [122]
Hg$_{1-x}$Cd$_x$Te $x=0.2$		77	14	12	$\sim 10^3$	0.05–0.3	60–400 / 20–200	10^{10} / 6×10^{10}	$<10^{-6}$ } $<4 \times 10^{-6}$ }	[4], [119], [123]–[125], [133]
Pb$_{1-x}$Sn$_x$Te $x=0.17$–0.2		77 / 4.2	11 / 15	10 / 14	$\sim 10^4$		42 / 52	3×10^8 / 1.7×10^{10}	1.5×10^{-8} } 1.2×10^{-6} }	[4], [74], [126]

Table IV lists the properties of infrared photoconductors in further detail. For wavelengths up to 3.5 to 5.7 μm, the use of cooled intrinsic InAs and InSb is indicated. For the longer wavelengths, specifically 10.6 μm, extrinsic and intrinsic photoconductors are available and will now be discussed.

a) *Extrinsic infrared photoconductors:* Extrinsic infrared photoconductors [4]–[6], [108], [109] rely on optical excitation of holes from impurity centers such as Au[110], [111], Cu [4], [112], [113], Hg [113]–[115], or Cd [116], into the valence band of a p-type semiconductor. Ge is used principally as the host crystal but Ge-Si [108] and Si [117] have also been investigated. In order to keep competing thermal excitation low, extrinsic photoconductors are cooled to temperatures between 77 and 4.2°K (see Table IV). Since optical absorption constants in the wavelength

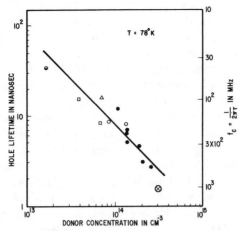

Fig. 16. Dependence of hole lifetime on antimony compensation in Ge:Au (after Bratt [128]). The bottom point is from [120].

TABLE V

SUMMARY OF RESULTS ON INFRARED HETERODYNE DETECTION

Material	Detection Mode*	Laser Wavelength (μm)	Measured Noise Equiv Power† (W/Hz)	Measured Quantum Noise Factor† (dB)	Quantum Efficiency	Calculated Noise Equiv Power (W/Hz)	Temperature (°K)	Response Time (Seconds)	Band-width	Dynamic Resistance (Ω)	Ref.
InAs	PV	3.39	1.25×10^{-19}	3.3			300				[140]
			$\sim 3 \times 10^{-19}$	7.1	0.25		300				[141]
InSb	PC	0.6328 } 3.39			0.5		77	10^{-8}		300	[142]
Ge:Au	PC	3.39			0.08		77	10^{-8}			[143]
Ge:Hg	PC	10.6	1.73×10^{-19}	9.7	0.5	7.5×10^{-20}	4.2	3.3×10^{-9}	50 MHz		[4], [144]
Ge:Hg(Sb)	PC	3.39					4.2	$< 3.7 \times 10^{-10}$	>450 MHz		[4], [102]
Ge:Cu	PC	10.6	$1.3 \times 10^{-19} S$ $7 \times 10^{-20} Q$	8.4 5.7	0.5	7.5×10^{-20}	4.2	2×10^{-9}			[4], [18], [145]
Ge:Cu(Sb)	PC	10.6	7.5×10^{-20}	6	0.56	6.7×10^{-20}	4.2	2×10^{-10}	1.5 GHz	1.2×10^3	[4], [17], [146]
$Hg_{1-x}Cd_xTe$	PC	10.6	7×10^{-20}	5.7			77	$0.5–2.5 \times 10^{-7}$	1–10 MHz	200	[133]
$Hg_{1-x}Cd_xTe$	PV	10.6	2.2×10^{-19}	10.7	0.09	2.1×10^{-19}	77		>50 MHz	190	[138]
$Pb_{1-x}Sn_xSe$	PV	10.6	$2.5 \times 10^{-19} Q$ $1.0 \times 10^{-18} S$	11.2 17.3	0.085	2.2×10^{-19}	77	2×10^{-8}		1.5	[4], [18]

* PV = photovoltaic; PC = photoconductive
† Quantum noise limit $h\nu = 1.87 \times 10^{-20}$ and 5.85×10^{-20} W/Hz at 10.6 and 3.39 μm, respectively.
Q = sensitivity calculated using LO-generated noise only.
S = measured overall receiver sensitivity.

range of extrinsic carrier excitation are generally low (typically of the order of 1 to 10 cm^{-1}), [127], the requirement for high quantum efficiency leads to detector dimensions in the direction of light incidence of several millimeters.

Although Ge:Hg and Ge:Cu have drastically different cutoff wavelengths of 14 and 27 μm, respectively, their D^* at 10.6 μm are close [113]. Time constants for uncompensated samples are typically 10^{-7} to 3×10^{-9} seconds, with

peak quantum efficiency in the 0.2 to 0.6 range [102], [113]–[115], [119], [120]. The usefulness of Ge:Au (and Ge:Au(Sb)) at 10.6 μm is limited by low quantum efficiency ($< 10^{-2}$) at this wavelength but it is used in the laboratory because it operates at liquid N_2 temperature.

For wide-band applications, short carrier lifetimes can be obtained in impurity-activated germanium by compensation with antimony donors [102], [112], [119], [128], [129]. The measured shortening of the hole lifetime with increasing compensation by donors in p-type Ge:Au is shown in Fig. 16 [128]. Compensation initially has only a small effect on D^* as the data in Table IV for Ge:Au(Sb), Ge:Hg(Sb), and Ge:Cu(Sb) show. Compensation increases the resistance of the detector [129], thereby reducing the direct detection circuit bandwidth unless a low R_L with a concomitant reduction in sensitivity is used. The usable degree of compensation has a further limit in that quantum efficiency decreases in proportion to $(N_A - N_D)$, where N_A, N_D = acceptor and donor concentrations, respectively [129]. N_A typically has a value of the order of $10^{16} cm^{-3}$ [108]. The usefulness of compensation is indicated in Table IV (and in Table V) where measured lifetime values as short as 3×10^{-10} and 2×10^{-10} seconds are given for Ge:Hg(Sb) and Ge:Cu(Sb), respectively. [4], [17], [102].

b) *Intrinsic infrared photoconductors:* Intrinsic photoconductors utilize band-to-band excitation. Consequently, the light absorption coefficient is high, typically 10^3 to 10^4 cm^{-1}, so that the dimension in the direction of light incidence need only be a few micrometers. Intrinsic infrared photoconductors are also listed in Table IV. In addition to the III–V photoconductors InAs and InSb with response to a few micrometers, narrow bandgap intrinsic detectors have been fabricated from mixed crystals $Hg_{1-x}Cd_xTe$ [4], [119], [123] and $Pb_{1-x}Sn_xTe$ [4], [74], [126]. These detectors can be tailored to the desired wavelength by varying the molar content x [4], [130]. $Hg_{1-x}Cd_xTe$ detectors have been fabricated for operation at wavelengths from 1 to 30 μm [131], [132]. Emphasis has been on the 8- to 13-μm atmospheric window. Here their technical importance comes about because they operate at higher temperatures, that is 77° to beyond 110°K [4], [125] than the extrinsic photoconductors. As Tables IV and V show, response times as short as 50 and 15 ns have been obtained in photoconductive $Hg_{1-x}Cd_xTe$ and $Pb_{1-x}Sn_xTe$, respectively [4], [131]–[133].

Background limited (BLIP) detectivity has been achieved in $Hg_{1-x}Cd_xTe$ [125]. Although the possible design of fast photoconductive $Hg_{1-x}Cd_xTe$ detectors has been considered in detail [134], [135] photovoltaic $Hg_{1-x}Cd_xTe$ and $Pb_{1-x}Sn_xTe$ detectors are demonstrating greater capability in meeting the requirements of MHz bandwidth optical communication systems [8]. Bandwidths of 30 MHz to greater than 50 MHz have been obtained in both direct detection [136], [137] and heterodyne detection [138] modes.

IV. APPLICATIONS OF PHOTODETECTORS TO VARIOUS OPTICAL COMMUNICATION SYSTEMS

The most important parameter in the choice of a photodetector is the wavelength of the optical communication system. The current state of the art in high efficiency high power lasers leads to four principal possible system wavelength regions:

1) 4880 Å—Argon, and 5300 Å—doubled YAG
2) 8500–9000 Å—GaAs
3) 1.06 μm—YAG
4) 10.6 μm—CO_2.

Although the type of the detector chosen for each system will depend upon many factors, the characteristics of well-suited detectors for each system will now be discussed.

A. Argon and Doubled YAG Laser Systems

Optical communication systems utilizing argon or doubled YAG lasers are likely to use photomultipliers in a direct detection mode of operation because of high current gain and low dark current. These wavelengths are very near to the peak response of S-20 or S-25 photosurfaces. Slight changes in the process used to form these photosurfaces can yield small changes in the wavelength of maximum response, thus allowing photomultipliers to be optimized for these systems.

Parameters affecting the type of photomultiplier to be employed are modulation bandwidth and optical signal power available at the receiver. Systems designed for modulation frequencies up to 100 MHz may employ conventional electrostatically focused photomultipliers with the number of secondary emission stages determined by the received signal power and desired (S/N) ratio. Systems with a bandwidth of several hundred MHz will require a photomultiplier in which both transit time dispersion in the electron multiplier and output capacitance have been minimized. The use of a high-voltage accelerating grid between each pair of dynodes, together with a coaxial anode structure such as employed in the RCA C70045 photomultiplier is one method for achieving the necessary modulation bandwidth. The use of high gain GaP dynode material reduces both the excess shot noise and required number of stages and it would be desirable for this application.

Very high data rate systems with modulation bandwidths of several GHz impose the most difficult requirements upon the photomultiplier. High interstage voltages must be used in the electron multiplier to minimize transit time dispersion due to initial velocity distributions and the spatial variation of electric fields must be minimized to achieve uniform interstage electron transit times. The magnetically focused crossed-field photomultiplier, such as the Sylvania 502, can provide a useful bandwidth of several GHz because of both extremely low transit time dispersion and a well-matched signal output circuit. The use of high-gain dynodes would again be very desirable to reduce excess shot noise.

Silicon avalanche photodiodes are also useful at these wavelengths. Their quantum efficiency is higher than in photomultipliers and the response and current gain mechanism are sufficiently fast. However, unless cooled, noise and dark current are larger thus limiting their sensitivity for direct detection of weak signals. The higher quantum efficiency of silicon diodes would make them useful in heterodyne detection systems where the preservation of optical frequency and phase information and rejection of interference may be of importance.

B. GaAs Systems

Optical communication systems utilizing the GaAs (8400–9000 Å) laser operate at wavelengths where both photomultipliers and solid-state photodiodes may be utilized. Small light-weight systems designed for short range communication may employ silicon photodiodes, while systems requiring maximum sensitivity will utilize photomultipliers or silicon avalanche photodiodes.

Photomultipliers with an *S*-25 or one of the new photosurfaces such as $GaAs:Cs_2O$ offer reasonable sensitivity at these wavelengths. Wide bandwidths may be achieved by incorporating these cathodes into fast photomultiplier structures. For silicon photodiodes and avalanche photodiodes, a tradeoff exists between quantum efficiency and bandwidth as mentioned in Section III-B. However, quantum efficiencies of 90 percent have been achieved for speeds of response of 10 ns [65]. Extensive development activity in many laboratories on both $GaAs:Cs_2O$ and other type III-V photoemitters and on silicon avalanche photodiodes is expected to result in practical photomultipliers and avalanche photodiodes with high sensitivity in this wavelength region.

C. YAG Systems

Photomultipliers are presently not very useful for the demodulation of 1.06-μm Nd:YAG laser radiation. This is because even the best developmental infrared photocathodes have quantum efficiencies of only 0.8 percent at 1.06 μm [43]. Silicon photodiodes which combine quantum efficiencies and response times of 75 percent and 25 ns or 45 percent and 10 ns are available [65]. Photomultipliers and Si avalanche photodiodes have recently been compared theoretically as detectors for laser pulses of 10^{-7} to 10^{-9} duration at 1.06 μm [139] and it has been found that Si avalanche diodes can detect one to two orders of magnitude lower pulse energies than the presently available photomultipliers. However, avalanche photodiodes with optimized properties for 1.06 μm are not yet available. Detection systems for 1.06 μm with signal bandwidths in excess of 100 MHz would thus use germanium avalanche photodiodes which, in order to reduce dark current, might be cooled to about 250°K.

D. CO_2 Laser Systems

Both photoconductive and photovoltaic detectors are useful for the detection of 10.6-μm radiation. Their characteristics were given in Tables II and IV. Direct detection is useful in 10.6-μm systems where simplicity but not high sensitivity are desired. Heterodyne detection is preferred because it combines high sensitivity with wide bandwidths.

Heterodyne detection is most practical at the longer wavelengths due to such factors as: sensitivity improvement due to reduced quantum noise (see (7)), easier alignment of signal and local oscillator, and greater diffraction-limited acceptance angle for a given aperture size. The CO_2 laser has proven capable of single-frequency stabilized operation as is desired for the LO [133].

Sensitivities in coherent detection approaching the theoretical quantum noise limit have been demonstrated in a variety of photodetectors. The results obtained are summarized in Table V. Results are listed for 10.6 μm (CO_2) and 3.39 μm (HeNe). Sensitivities within 3 to 10 dB of the quantum noise limit ($hv = 1.87 \times 10^{-20}$ W/Hz at 10.6 μm) have been demonstrated. Experimental sensitivity values in Table V are marked as to whether they take into account the total receiver noise (*S*), or only the LO-generated noise (*Q*). With improved heterodyne receiver design, the quantum noise limit will be more closely approached in the future. In the case of photovoltaic mixed crystal materials, particularly HgCdTe, heterodyne sensitivity is being increased as improved quantum efficiency elements become available.

As Table V shows, receiver sensitivity near 10^{-19} W/Hz simultaneously with a bandwidth exceeding 1 GHz has been obtained using highly compensated photoconductive Ge:Cu(Sb) with a time constant of 200 ps [5]. The generation–recombination noise spectrum of this mixer element measured up to 4 GHz is shown in Fig. 17, when illuminated with sufficient LO power to depress mixer resistance from 5×10^5 to 1.2×10^3 ohms. Due to the available conversion gain, heterodyne operation is obtained well beyond the 750-MHz rollover frequency. The measured variation of the heterodyne noise equivalent power ($S/N = 1$) with dc bias, and hence conversion gain, is shown in Fig. 18. The behavior shown is consistent with (7), which for a photoconductive mixer with high dark resistance and low background irradiation can be written [17]

$$P_{Smin} = \frac{S}{N} B \left[\frac{2hv}{\eta} + \frac{k(T_G + T_A)}{G} \right] \quad (22)$$

where the available conversion gain due to the local oscillator is given by

$$G = \frac{\eta q V}{2hv} \left(\frac{\tau}{T_r} \right) \frac{1}{1 + \omega^2 \tau^2}. \quad (23)$$

Note that for this photoconductive case G is independent of LO power variations since mixer conductance is directly proportional to P_{LO} (18). Expressions similar to the preceding ones but for the photodiode heterodyne case are given in [138].

For systems with bandwidths up to 100 MHz, uncompensated extrinsic photoconductors have demonstrated sensitive heterodyne performance [4], [18], [144], [145]. Photoconductive HgCdTe has given performance to 10 MHz [133]. Photovoltaic HgCdTe has emerged as a strong competitor offering in combination good frequency response, operation at moderate cryogenic temperatures, and low power dissipation [138]. As problems in mixed crystal fabrication are overcome, the use of 10.6-μm photodiode mixers will be extended to higher modulation frequencies.

Heterodyne operation requires the use of a laser local oscillator which is offset by the IF frequency, and frequency-stabilized by means of a circuit such as Fig. 19(a). The LO could be a CO_2 or possibly a PbSnTe laser [142]. In systems where significant Doppler shifts occur between transmitter and receiver, the use of the receiver configuration of Fig.

347

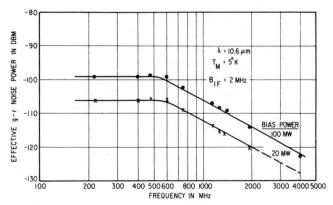

Fig. 17. Measured g–r noise spectrum of Ge:Cu(Sb) mixer [17].

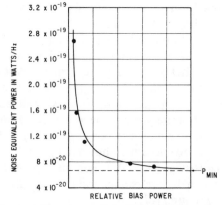

Fig. 18. Measured dependence of sensitivity of 10.6 μm heterodyne receiver on dc bias power (after [4] and [17]). (Relative bias power has linear scale.) As bias power is increased generation–recombination noise swamps out all other receiver noise contributions.

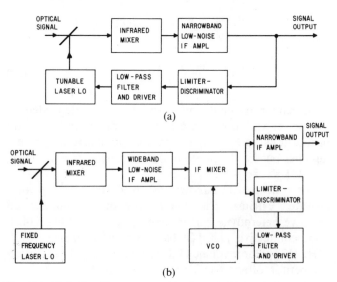

Fig. 19. Methods of frequency tracking in optical heterodyne receivers. (a) Optical tracking. (b) IF tracking.

19(b) is useful, and has been demonstrated at 10.6 μm [148] in a quantum noise limited 1.3-GHz bandwidth heterodyne receiver. Here frequency tracking is accomplished at IF and the information bandwidth established in the narrow-band second IF amplifier.

The optical heterodyne detector behaves simultaneously as a receiver and an antenna [33], [149]–[151]. This prop-

erty has been used to combine infrared heterodyne elements to form spatially coherent arrays at 10.6 μm with directional properties exactly analogous to those obtained with multi-feed microwave antennas [146]. These coherent arrays are capable of providing a multiplicity of resolution elements in the object field of the optical receiver for such purposes as spatial acquisition and tracking [8], [152].

Pyroelectric detectors with response times as short as a few nanoseconds have been reported [153]. They can be operated at room temperature, but such thermal detectors have limited sensitivity in both direct detection [4], [154], [155] and heterodyne detection [156]–[158] modes. High speed detection using metal point contacts with a time constant of 3×10^{-14} seconds has also recently been reported [159].

V. CONCLUSIONS

It is evident that there is a large number of photodetectors and detection techniques available for use in optical communication systems. The correct choice of a particular detection technique and photodetector depends mainly on the wavelength of operation, the information bandwidth of the signal, and on the sensitivity desired for the demodulation of weak signals. Direct detection is preferred for wavelengths at which photomultipliers and avalanche photodiodes with internal current gain, wide electrical bandwidth, and low excess noise are available.

Photomultipliers and silicon avalanche photodiodes are useful in the ultraviolet, visible, and near infrared to about 1 μm, germanium avalanche diodes extend this range to beyond 1.5 μm. Photomultipliers provide high-gain and low-excess noise. Specially constructed photomultipliers have achieved gains of 10^4 for baseband widths up to several GHz. Avalanche photodiodes also combine gain and high speed of response as manifested by current gain–bandwidth products of up to 100 GHz. The multiplication process in avalanche diodes, however, results in higher excess noise. But despite this excess noise, current gain in both photomultipliers and avalanche photodiodes allows realization of direct detection systems with wide bandwidths and considerably higher sensitivities to weak light signals as compared to detection systems without gain that are limited by thermal noise of the detector and output circuit. Although quite desirable, the development of avalanche photodiodes with response at longer wavelengths from materials like InAs, InSb, and even mixed crystals is still at an early stage. Photoconductors with wide bandwidths have been developed for the infrared region, but the lack of sufficient internal current gain limits their performance in direct detection systems.

Heterodyne detection in both photoconductors and photodiodes is most successfully used at infrared wavelengths where no fast detectors with sufficient internal current gain are available. For 10.6 μm, sensitivities close to the theoretical quantum noise limit ($h\nu B$) have been realized in heterodyne detection systems with bandwidths approaching 1 GHz.

ACKNOWLEDGMENT

The authors wish to thank D. Bode and P. Bratt of Santa Barbara Research Center; E. Sard, B. Peyton, and F. Pace of AIL; for supplying useful data and suggestions for this paper.

REFERENCES

[1] L. K. Anderson and B. J. McMurtry, "High-speed photodetectors," *Proc. IEEE*, vol. 54, pp. 1335–1349, October 1966.

[2] W. R. Pratt, *Laser Communication Systems*. New York: Wiley, 1969.

[3] M. Ross, *Laser Receivers*. New York: Wiley, 1967.

[4] A. Beer and R. Willardson, Eds., *Semiconductors and Semimetals*, vol. 5. *Infrared Detectors*. The following chapters are of special interest: P. Kruse, "Indium antimonide photoconductive and PEM detectors,"; D. Long and J. Schmit, "Hg$_{1-x}$Cd$_x$Te and closely related alloys as intrinsic IR detector materials,"; I. Melngailis and T. Harman, "Single crystal lead–tin chalcogenides,"; R. Keyes and T. Quist, "Low-level coherent and incoherent detection in the infrared,"; M. Teich, "Coherent detection in the infrared,"; F. Arams, E. Sard, B. Peyton, and F. Pace, "Infrared heterodyne detection with gigahertz IF response,"; E. Putley, "Pyroelectric detectors,"; H. Sommers, "Microwave biased photoconductive detector."

[5] P. W. Kruse, L. D. McGlauchlin, and R. B. McQuistan, *Elements of Infrared Technology*. New York: Wiley, 1962.

[6] R. Hudson, *Infrared System Engineering*. New York: Wiley, 1969.

[7] T. M. Quist, "Copper-doped germanium detectors," *Proc. IEEE* (Letters), vol. 56, pp. 1212–1213, July 1968.

[8] G. McElroy, McAvoy, H. Richard, T. McGunigal, and G. Schiffner, "Carbon dioxide laser communication systems for near earth applications," to be published, and H. Plotkin and J. Randall, "Systems aspects of optical space communications," to be published.

[9] A. M. Johnson, "Square law behavior of photocathodes at high light intensities and high frequencies," *IEEE J. Quantum Electron.*, vol. QE-1, pp. 99–101, May 1965.

[10] B. J. McMurtry and A. E. Siegmann, "Photomixing experiments with a ruby optical maser and a traveling-wave microwave phototube," *Appl. Opt.*, vol. 1, pp. 51–53, January 1962.

[11] B. M. Oliver, "Signal-to-noise ratios in photoelectric mixing," *Proc. IRE* (Correspondence), vol. 49, pp. 1960–1961, December 1961.

[12] H. A. Hans, C. H. Townes, and B. M. Oliver, "Comments on noise in photoelectric mixing," *Proc. IRE* (Letters), vol. 50, pp. 1544–1546, June 1962.

[13] B. M. Oliver, "Thermal and quantum noise," *Proc. IEEE*, vol. 53, pp. 436–454, May 1965.

[14] S. Jacobs and P. Rabinowitz, "Optical heterodyning with a CW gaseous laser," in *Quantum Mechanics III*, P. Grivet and N. Bloembergen Eds. New York: Columbia University Press, 1964, pp. 481–487.

[15] G. Lucovsky, M. E. Lasser, and R. B. Emmons, "Coherent light detection in solid-state photodiodes," *Proc. IEEE*, vol. 51, pp. 166–172, January 1963.

[16] O. Svelto, P. D. Coleman, M. DiDomenico, and R. H. Pantell, "Photoconductive mixing in CdSe single crystals," *J. Appl. Phys.*, vol. 34, pp. 3182–3186, November 1963.

[17] F. R. Arams, E. W. Sard, B. J. Peyton, and F. P. Pace, "Infrared 10.6 micon heterodyne detection with gigahertz IF capability," *IEEE J. Quantum Electron.*, vol. QE-3, pp. 484–492, November 1967; see also *IEEE Spectrum*, vol. 5, p. 5, June–July 1968.

[18] M. C. Teich, "Infrared heterodyne detection," *Proc. IEEE*, vol. 56, pp. 37–46, January 1968.

[19] J. E. Geusic and H. E. D. Scovil, "A unidirectional traveling-wave optical maser," *Bell Syst. Tech. J.*, vol. 41, pp. 1371–1397, July 1962.

[20] H. Kogelnik and A. Yariv, "Considerations of noise and schemes for its reduction in laser amplifiers," *Proc. IEEE*, vol. 52, pp. 165–172, February 1964.

[21] F. Arams and M. Wang, "Infrared laser preamplifier system," *Proc. IEEE* (Correspondence), vol. 53, p. 329, March 1965.

[22] G. C. Holst and E. Snitzer, "Detection with a fiber laser preamplifier at 1.06 μ," *IEEE J. Quantum Electron.*, vol. QE-5, pp. 319–320, June 1969.

[23] W. H. Louisell and A. Yariv, "Quantum fluctuations and noise in parametric processes. I," *Phys. Rev.*, vol. 124, pp. 1646–1654, December 1961.

[24] J. E. Midwinter and J. Warner, "Up-conversion of near infrared to visible radiation in lithium-metaniobate," *J. Appl. Phys.*, vol. 38, pp. 519–523, February 1967.

[25] J. Warner, "Photomultiplier detection of 10.6 μm radiation using optical up-conversion in proustite," *Appl. Phys. Lett.*, vol. 12, pp. 222–224, March 1968.

[26] G. D. Boyd, T. J. Bridges, and E. G. Burkhardt, "Up-conversion of 10.6 μ radiation to the visible and second harmonic generation in HgS," *IEEE J. Quantum Electron.*, vol. QE-4, pp. 515–519, September 1968.

[27] D. A. Kleinman and G. D. Boyd, "Infrared detection by optical mixing," *J. Appl. Phys.*, vol. 40, pp. 546–566, February 1969.

[28] Y. Klinger and F. Arams, "Infrared 10.6 micron CW up-conversion in proustite using an Nd:YAG laser pump," *Proc. IEEE*, vol. 57, pp. 1797–1798, October 1969.

[29] G. Lucovsky and R. B. Emmons, "High frequency photodiodes," *Appl. Opt.*, vol. 4, pp. 697–702, June 1965.

[30] H. Melchior and W. T. Lynch, "Signal and noise response of high speed germanium avalanche photodiodes," *IEEE Trans. Electron Devices*, vol. ED-13, pp. 829–838, December 1966.

[31] M. DiDomenico, Jr., and O. Svelto, "Solid-state photodetection: A comparison between photodiodes and photoconductors," *Proc. IEEE*, vol. 52, pp. 136–144, February 1964.

[32] G. Lucovsky, R. B. Emmons, B. Harned, and J. K. Powers, "Detection of coherent light by heterodyne techniques using solid state photodiodes," in *Quantum Electronics III*. P. Grivet and N. Bloembergen, Eds. New York: Columbia University Press, 1964, pp. 1731–1738.

[33] A. E. Siegman, "The antenna properties of optical heterodyne receivers," *Proc. IEEE*, vol. 54, pp. 1350–1356, October 1966.

[34] A. J. Bahr, "The effect of polarization selectivity on optical mixing photoelectric surfaces," *Proc. IEEE* (Correspondence), vol. 53, p. 513, May 1965.

[35] O. DeLange, "Optical heterodyne detection," *IEEE Spectrum*, vol. 5, pp. 77–85, October 1968.

[36] A. H. Sommer and W. E. Spicer, *Photoelectronic Materials and Devices*. Princeton, N. J.: Van Nostrand, 1965.

[37] A. H. Sommer, *Photoemissive Materials*. New York: Wiley, 1968.

[38] J. J. Scheer and J. van Laar, "GaAs-Cs—a new type of photoemitter," *Solid-State Commun.*, vol. 3, pp. 189–193, August 1965.

[39] A. A. Turnbill and G. B. Evans, "Photoemission from GaAs-Cs-O," *Brit. J. Appl. Phys.*, ser. 2, vol. 1, pp. 155–160, February 1968.

[40] J. J. Uebbing and R. L. Bell, "Improved photoemitters using GaAs and InGaAs," *Proc. IEEE* (Letters), vol. 56, pp. 1624–1625, September 1968.

[41] R. L. Bell and J. J. Uebbing, "Photoemission from InP-Cs-O," *Appl. Phys. Lett.*, vol. 12, pp. 76–78, February 1968.

[42] B. F. Williams, "InGaAs-CsO, a low work function (less than 1.0 eV) photoemitter," *Appl. Phys. Lett.*, vol. 14, pp. 273–275, May 1969. Also, B. F. Williams *et al.*, *Appl. Phys. Letters*, to be published.

[43] H. Sonnenberg, "InAsP-Cs$_2$O, a high efficiency infrared photocathode," *Appl. Phys. Lett.*, vol. 16, pp. 245–246, March 1970.

[44] R. E. Simon, A. H. Sommer, J. J. Tietjen, and B. F. Williams, "New high-gain dynode for photomultipliers," *Appl. Phys. Lett.*, vol. 13, pp. 355–357, November 1968.

[45] R. C. Miller and N. C. Wittwer, "Secondary emission amplification at microwave frequencies, *IEEE J. Quantum Electron.* vol. QE-1, pp. 49–59, April 1965.

[46] M. B. Fisher and R. T. McKenzie, "A traveling-wave photomultiplier," *IEEE J. Quantum Electron.*, vol. QE-2, pp. 322–327, August 1966.

[47] W. Shockley and J. R. Pierce, "A theory of noise for electron multipliers," *Proc. IRE*, vol. 26, pp. 321–332, March 1938.

[48] R. M. Matheson, "Recent photomultiplier developments at RCA," *IEEE Trans. Nucl. Sci.*, vol. NS-11, pp. 64–71, June 1964.

[49] J. R. Kerr, private communication.

[50] D. E. Persyk, "New photomultiplier detectors for laser applications," *Laser J.*, pp. 21–23, November–December 1969.

[51] W. W. Gaertner, "Depletion-layer photoeffects in semiconductors," *Phys. Rev.*, vol. 116, pp. 84–87, October 1959.

[52] D. E. Sawyer and R. H. Rediker, "Narrow base germanium photodiodes," *Proc. IRE*, vol. 46, pp. 1122–1130, June 1958.

[53] R. P. Riesz, "High-speed semiconductor photodiodes," *Rev. Sci. Instr.*, vol. 33, pp. 994–998, September 1962.

[54] L. K. Anderson, "Photodiode detection," *Proc. 1963 Symp. on Optical Masers*, Polytechnic Press, pp. 549–563.

[55] W. M. Sharpless, "Evaluation of a specially designed GaAs Schottky-barrier photodiode using 6328 Å radiation modulated at 4 GHz," *Appl. Opt.*, vol. 9, pp. 489–494, February 1970.

[56] M. V. Schneider, "Schottky barrier photodiode with antireflection coating," *Bell Syst. Tech. J.*, vol. 45, pp. 1611–1638, November 1966.

[57] United Detector Technology (manufacturers data).

[58] H. Melchior, M. P. Lepselter, and S. M. Sze, "Metal-semiconductor avalanche photodiode," presented at IEEE Solid-State Device Res. Conf., Boulder, Colo., June 1968.

[59] R. D. Baertsch and J. R. Richardson, "An Ag-GaAs Schottky barrier ultraviolet detector," *J. Appl. Phys.*, vol. 40, pp. 229–235, January 1969.

[60] J. R. Richardson and R. D. Baertsch, "Zinc sulfide Schottky barrier ultraviolet detectors," *Solid-State Electron.*, vol. 12, pp. 393–397, May 1969.

[61] L. K. Anderson, P. G. McMullin, L. A. D'Asaro, and A. Goetzberger, "Microwave photodiodes exhibiting microplasma-free carrier multiplication," *Appl. Phys. Lett.*, vol. 6, pp. 62–64, February 1965.

[62] E. Labate, private communication.

[63] R. J. McIntyre and H. C. Sprigings, "Multielement silicon photodiodes for detection at 1.06 microns," presented at the Conf. on Prep. Control of Electron. Mat., Boston, Mass., August 1966.

[64] H. C. Sprigings and R. J. McIntyre, "Improved multielement silicon photodiodes for detection of 1.06 μm," presented at the Int. Electron Devices Meet., Washington, D. C., October 1968.

[65] RCA, Montreal, Canada (manufacturers data).

[66] D. P. Mathur, R. J. McIntyre, and P. P. Webb, "A new germanium photodiode with extended long-wavelength response," presented at the Int. Electron Device Meet., Washington, D. C., October 1968.

[67] B. R. Pagel and R. L. Petritz, "Noise in InSb photodiodes," *J. Appl. Phys.*, vol. 32, pp. 1901–1904, October 1961.

[68] H. Protschka and D. C. Shang, "InSb photodiodes with high reverse breakdown voltage," presented at the Int. Electron Device Meet., Washington, D. C., October 1967.

[69] A. Kohn and J. Schlickman, "1–2 micron (HgCd)Te photodetectors," *IEEE Trans. Electron Devices*, vol. ED-16, pp. 885–890, October 1960.

[70] M. Rodot, C. Vérié, Y. Marfaing, J. Besson, and H. Lebloch, "Semiconductor lasers and fast detectors in the infrared (3 to 15 microns)," *IEEE J. Quantum Electron.*, vol. QE-2, pp. 586–593, September 1966.

[71] C. Vérié and J. Ayas, "Cd$_x$Hg$_{1-x}$Te infrared photovoltaic detectors," *Appl. Phys. Lett.*, vol. 10, pp. 241–243, May 1967.

[72] I. Melngailis and A. R. Calawa, "Photovoltaic effect in Pb$_x$Sn$_{1-x}$Te diodes," *Appl. Phys. Lett.*, vol. 9, pp. 304–306, October 1966.

[73] J. F. Butler, A. R. Calawa, I. Melngailis, T. C. Harman, and J. O. Dimmock, "Laser action and photovoltaic effect in Pb$_{1-x}$Sn$_x$Se diodes," *Bull. Am. Phys. Soc.*, vol. 12, p. 384, March 1967.

[74] I. Melngailis, "Laser action and photodetection in lead–tin chalcogenides," *J. de Physique*, vol. 29, colloque C4, supplement au no. 11–12, pp. C4–84–C4–94, November–December 1968.

[75] K. M. Johnson, "High-speed photodiode signal enhancement at avalanche breakdown voltage," *IEEE Trans. Electron Devices*, vol. ED-12, pp. 55–63, February 1965.

[76] A. S. Tager, "Current fluctuations in a semiconductor (dielectric) under conditions of impact ionization and avalanche breakdown," *Sov. Phys.—Solid State*, vol. 6, pp. 1919–1925, February 1965.

[77] R. J. McIntyre, "Avalanche multiplication noise in semiconductor junctions," *IEEE Trans. Electron Devices*, vol. ED-13, pp. 164–175, January 1966.

[78] H. Melchior and L. K. Anderson, "Noise in high speed avalanche photodiodes," presented at the Int. Electron Device Meet., Washington, D. C., October 1965.

[79] R. D. Baertsch, "Noise and ionization rate measurements in silicon photodiodes," *IEEE Trans. Electron Devices* (Correspondence), vol. ED-13, p.987, December 1966.

[80] W. T. Read, "A proposed high-frequency negative resistance diode," *Bell Syst. Tech. J.*, vol. 37, pp. 401–446, March 1958.

[81] R. B. Emmons and G. Lucovsky, "The frequency response of avalanching photodiodes," *IEEE Trans. Electron Devices*, vol. ED-13, pp. 297–305, March 1966.

[82] R. B. Emmons, "Avalanche-photodiode frequency response," *J. Appl. Phys.*, vol. 38, pp.3705–3714, August 1967.

[83] J. J. Chang, "Frequency response of PIN avalanche photodiodes," *IEEE Trans. Electron Devices*, vol. ED-14, pp. 139–145, March 1967.

[84] C. A. Lee, R. A. Logan, J. J. Kleimack, and W. Wiegmann, "Ionization rates of holes and electrons in silicon," *Phys. Rev.*, vol. 134, pp. A761–A773, May 1964.

[85] S. Donati and V. Svelto, "The statistical behavior of the avalanche photodiode," *Alta Freq.*, vol. 37, pp. 476–486, May 1968.

[86] R. D. Baertsch, "Noise and multiplication measurements in InSb avalanche photodiodes," *J. Appl. Phys.* vol. 38, pp. 4267–4274, October 1967.

[87] W. T. Lindley, R. J. Phelan, C. M. Wolfe, and A. G. Foyt, "GaAs Schottky barrier avalanche photodiodes," *Appl. Phys. Lett.*, vol. 14, pp. 197–199, March 1969.

[88] G. Lucovsky and R. B. Emmons, "Avalanche multiplication in InAs photodiodes," *Proc. IEEE* (Correspondence), vol. 53, p. 180, February 1965.

[89] R. J. Locker and G. C. Huth, "A new ionizing radiation detection concept which employs semiconductor avalanche amplification and the tunnel diode element," *Appl. Phys. Lett.*, vol. 9, pp. 227–230, September 1966.

[90] H. Ruegg, "An optimized avalanche photodiode," *IEEE Trans. Electron Devices*, vol. ED-14, pp. 239–251, May 1967.

[91] J. R. Biard and W. N. Shaunfield, "A model of the avalanche photodiode," *IEEE Trans. Electron Devices*, vol. ED-14, pp. 233–238, May 1967.

[92] W. N. Shaunfield and J. R. Biard, "A high-speed germanium photodetector for 1.54 microns," presented at the Solid-State Sensors Symp., Minneapolis, Minn., September 1968.

[93] W. T. Lynch, "Elimination of the guard ring in uniform avalanche photodiodes," *IEEE Trans. Electron Devices*, vol. ED-15, pp. 735–741, October 1968.

[94] F. D. Shepherd, Jr., A. C. Yang, and R. W. Taylor, "A 1 to 2 μm silicon avalanche photodiode," *Proc. IEEE* (Letters), vol. 58, pp. 1160–1162, July 1970.

[95] A. Rose, *Concepts in Photoconductivity and Applied Problems*. New York: Interscience, 1963.

[96] A. Rose, "Maximum performance of photoconductors," *Helv. Phys. Acta*, vol. 30, pp. 242–244, August 1957.

[97] R. W. Reddington, "Gain-bandwidth product of photoconductors," *Phys. Rev.*, vol. 115, pp. 894–896, August, 1959.

[98] P. D. Coleman, R. C. Eden, and J. N. Weaver, "Mixing and detection of coherent light in a bulk photoconductor," *IEEE Trans. Electron Devices*, vol. ED-11, pp. 488–497, November 1964.

[99] R. E. Burgess, "The statistics of charge carrier fluctuations in semiconductors," *Proc. Phys. Soc. London*, (Gen.), vol. 69, pp. 1020–1027, October 1956.

[100] K. M. Van Vliet, "Noise in semiconductors and photoconductors," *Proc. IRE*, vol. 46, pp. 1004–1018, June 1968.

[101] H. Sommer and E. K. Gatchell, "Demodulation of low-level broadband optical signals with semiconductors," *Proc. IEEE*, vol. 54, pp. 1553–1568, November 1966.

[102] C. Buczek and G. Picus, "Far infrared laser receiver investigation," Hughes Res. Labs., Malibu, Calif., A. F. Rept. AFAL-TR-68-102, May 1968.

[103] A. Yariv, C. Buczek, and G. Picus, "Recombination studies of hot holes in mercury-doped germanium," *Proc. Int. Conf. Phys. of Semicond.* (Moscow), vol. 9–10, p. 500, July 1968.

[104] R. L. Williams, "Response characteristics of extrinsic photoconductors," *J. Appl. Phys.*, vol. 40, pp. 184–192, January 1969.

[105] M. DiDomenico and L. K. Anderson, "Microwave signal-to-noise performance of CdSe bulk photoconductive detectors," *Proc. IEEE*, vol. 52, pp. 815–822, July 1964.

[106] H. Sommer and W. Teutsch, "Demodulation of low-level broadband optical signals with semiconductors; pt. II analysis of photoconductive detector," *Proc. IEEE*, vol. 52, pp. 144–153, February 1964.

[107] C. Sun and T. E. Walsh, "A solid-state microwave-biased photoconductive detector for 10.6 μm," *IEEE J. Quantum Electron.*, vol. QE-5, pp. 320–321, June 1969.

[108] H. Levinstein, "Extrinsic detectors," *Appl. Optics*, vol. 4, pp. 639–647, June 1965.

[109] E. Putley, "Far infrared photoconductivity," *Phys. Status Solidi*, vol. 6, pp. 571–614, September 1964.

[110] Johnson and H. Levinstein, "Infrared properties of gold in germanium," *Phys. Rev.*, vol. 117, pp. 1191–1203, March 1960.

[111] W. Beyer and H. Levinstein, "Cooled photoconductive infrared detectors," *J. Opt. Soc. Am.*, vol. 49, pp. 686–692, July 1959.

[112] G. S. Picus, "Carrier generation and recombination processes in copper-doped germanium photoconductors," *J. Phys. Chem. Solids*, vol. 23, pp. 1753–1761, 1962.

[113] D. Bode and H. Graham, "Comparison of performance of copper-doped germanium and mercury-doped germanium detectors," *Infrared Phys.*, vol. 3, pp. 129–137, September 1963.

[114] S. Borrello and H. Levinstein, "Preparation and Properties of mercury doped germanium," *J. Appl. Phys.*, vol. 33, pp. 2947–2950, October 1962.

[115] Y. Darviot, A. Sorrentino, B. Joly, and B. Pajot, "Metallurgy and physical properties of mercury-doped germanium related to the performance of the infrared detector," *Infrared Phys.*, vol. 7, pp. 1–10, March 1967.

[116] P. Bratt, W. Engeler, H. Levinstein, A. Mac Rae, and J. Pehek, "A status report on infrared detectors," *Infrared Phys.*, vol. 1, pp. 27–38, March 1961.

[117] R. Soref, "Extrinsic IR photoconductivity of Si doped with B, Al, Ga, P, As or Sb," *J. Appl. Phys.*, vol. 38, pp. 5201–5209, December 1967; see also, "Coherent homodyne detection at 10.6 μm with an extrinsic photoconductor," *Electron. Lett.*, vol. 2, pp. 410–411, November 1966.

[118] F. D. Morten and R. E. J. King, "Photoconductive indium antimonide detectors," *Appl. Opt.*, vol. 4, pp.659–663, June 1965.

[119] Manufacturers' Data (Santa Barbara Research Center, Texas Instruments, Honeywell, Philco-Ford, Mullard Raytheon, Société Anonyme de Télécommunications).

[120] D. Bode and P. Bratt, private communication.

[121] T. Bridges, T. Chang, and P. Cheo, "Pulse response of electrooptic modulators and photoconductive detectors at 10.6 μm," *Appl. Phys. Lett.*, vol. 12, pp. 297–300, May 1968.

[122] J. T. Yardley and C. B. Moore, "Response times of Ge:Cu infrared detectors," *Appl. Phys. Lett.*, vol. 7, pp. 311–312, December 1965.

[123] J. Schlickman, "Mercury cadmium telluride intrinsic photodetectors," *Proc. Electroopt. Syst. Conf., Industr. Sci. Conf. Mgmt, Inc.*, (Chicago, Ill.), pp. 289–309, September 1969.

[124] P. W. Kruse, "Photon effects in $Hg_{1-x}Cd_xTe$," *Appl. Opt.*, vol. 4, pp. 687–692, June 1965.

[125] B. Bartlett, D. Charlton, W. Dunn, P. Ellen, M. Jenner, and M. Jervis, "Background limited photoconductive HgCdTe detectors for use in the 8–14 micron atmospheric window," *Infrared Phys.*, vol. 9, pp. 35–36, 1969.

[126] I. Melngailis and T. Harman, "Photoconductivity in single-crystal $Pb_{1-x}Sn_xTe$," *Appl. Phys. Lett.*, vol. 13, pp. 180–183, September 1968.

[127] W. Kaiser, R. Collins, and H. Fan, "Infrared absorption in p-type germanium," *Phys. Rev.*, vol. 91, pp. 1380–1381, September 1953.

[128] P. Bratt, "Photoconductivity in impurity-activated germanium," Ph.D. dissertation, University of Syracuse, Syracuse, N. Y., January 1965.

[129] T. Vogl, J. Hansen, and M. Garbuny, "Photoconductive time constants and related characteristics of p-type gold-doped germanium," *J. Opt. Soc. Am.*, vol. 51, pp. 70–75, January 1961.

[130] J. Schmit and E. Stelzer, "Temperature and alloy compositional dependences of energy gap of $Hg_{1-x}Cd_xTe$," *J. Appl. Phys.*, vol. 40, pp. 4865–4869, November 1969.

[131] W. Saur, "Long wavelength mercury-cadmium telluride photoconductive infrared detectors," *Infrared Phys.*, vol. 8, pp. 255–258, 1968.

[132] A. Kohn and J. Schlickman, "1–2 micron (Hg, Cd)Te photodetectors," *IEEE Trans. Electron Devices*, vol. ED-16, pp. 885–890, October 1969.

[133] H. Mocker, "A 10.6 μ optical heterodyne communication system," *Appl. Opt.*, vol. 8, pp. 677–684, March 1969.

[134] D. Breitzer, E. Sard, B. Peyton, and J. McElroy, "G–R noise for auger band-to-band processes," to be published; see also: J. H. McElroy, S. C. Cohen, and H. E. Walker, "First and second summary design report ATS-F laser communication experiment infrared mixer and radiation cooler subsystem," Goddard Space Flight Center, Greenbelt, Md., Rep. X-524-69-227, 1969.

[135] D. Long, "On generation-recombination noise in infrared detection materials," *Infrared Phys.*, vol. 7, pp. 169–170, September 1967.

[136] A. Sorrentino, "IR detectors, developments at SAT," presented at Infrared Symp. Royal Radar Establishment, Malvern, England, April 1969.

[137] E. D. Hinkley and C. Fried, private communication.

[138] B. Peyton, E. Sard, R. Lange, and F. Arams, "Infrared 10.6-μm photodiode heterodyne detection," this issue, pp. 1769–1770.

[139] R. J. McIntyre, "Comparison of photomultipliers and avalanche photodiodes for laser applications," *IEEE Trans. Electron Devices*, vol. ED-17, pp. 347–352, April 1970.

[140] J. Hanlon and S. Jacobs, "Narrow-band optical heterodyne detection," *Appl. Opt.*, vol. 6, pp. 577–578, March 1967.

[141] F. E. Goodwin and M. E. Pedinoff, "Application of CCl_4 and CCl_2:CCl_2 ultrasonic modulators to infrared optical heterodyne experiments," *Appl. Phys. Lett.*, vol. 8, pp. 60–61, February 1966.

[142] E. N. Fuls, "Optical frequency mixing in photoconductive InSb," *Appl. Phys. Lett.*, vol. 4, pp. 7–8, January 1964.

[143] R. A. Wood, "Gold-doped germanium as an infrared high-frequency detector," *J. Appl. Phys.*, vol. 36, pp. 1490–1491, April 1965.

[144] C. Buczek and G. Picus, "Heterodyne performance of mercury-doped germanium," *Appl. Phys. Lett.*, vol. 11, pp. 125–126, August 1967.

[145] M. Teich, R. Keyes, and R. Kingston, "Optimum heterodyne detection at 10.6 microns in photoconductive Ge:Cu," *Appl. Phys. Lett.*, vol. 9, pp. 357–360, November 1966.

[146] F. Pace, R. Lange, B. Peyton, and F. Arams, "Infrared heterodyne receiver array with wide IF bandwidth for CO_2 laser systems," to be published.

[147] E. Hinkley, T. Harman, and C. Freed, "Optical heterodyne detection at 10.6 microns of the beat frequency between a tunable PbSnTe diode laser and a CO_2 gas laser," *Appl. Phys. Lett.*, vol. 13, pp. 49–51, July 1968.

[148] W. Chiou, T. Flattau, B. Peyton, and R. Lange, "High-sensitivity infrared heterodyne receivers with gigahertz IF bandwidth Pt. III: Packaged 10 μm heterodyne 1.3 GHz bandwidth tracking receiver," *IEEE Spectrum*, vol. 7, p. 5, January 1970 (Advertisement for AIL, Div. Cutler-Hammer, Deer Park, N. Y.).

[149] A. E. Siegman, "A maximum-signal theorem for the spatially coherent detection of scattered radiation," *IEEE Trans. Antennas Propagat.*, vol. AP-15, pp. 192–194, January 1967.

[150] V. J. Corcoran, "Directional characteristics in optical heterodyne detection process," *J. Appl. Phys.*, vol. 36, pp. 1819–1825, June 1965.

[151] W. Read and D. Fried, "Optical heterodyning and noncritical angular alignment," *Proc. IEEE* (Letters), vol. 51, p. 787, December 1963.

[152] H. A. Bostick, "A carbon dioxide laser radar system," *IEEE J. Quantum Electron.*, vol. QE-3, p. 232, June 1967.

[153] A. Glass, "Investigation of electrical properties of $Sr_{1-x}Ba_xNb_2O_6$ with special reference to pyroelectric detection," *J. Appl. Phys.*, vol. 40, pp. 4699–4713, November 1969.

[154] M. Kimmit, J. Ludlow, and E. Putley, "Use of pyroelectric detector to measure Q-switched CO_2 laser pulses," *Proc. IEEE* (Letters), vol. 56, p. 1250, 1968.

[155] A. Hadni, R. Thomas, and J. Perrin, "Response of triglycine sulfate pyroelectric detector to high frequencies (300 KHz)," *J. Appl. Phys.*, vol. 40, pp. 2740–2745, 1969.

[156] E. Leiba, "Heterodynage optique avec un detecteur pyroelectrique," *C. R. Acad. Sci.* (Paris), vol. 268, ser. B, pp. 31–33, January 1969.

[157] R. L. Abrams and A. M. Glass, "Photomixing at 10.6 μ with strontium barium niobate pyroelectric detectors," *Appl. Phys. Lett.*, vol. 15, pp. 251–253, October 1969.

[158] S. Eng and R. Gudmundsen, "Theory of optical heterodyne detection using pyroelectric effect," *Appl. Opt.*, vol. 9, pp. 161–166, January 1970.

[159] V. Danen, D. Sokoloff, A. Sanchez, and A. Javan, "Extension of laser harmonic-frequency mixing techniques into the 9μ region with an infrared metal-metal point-contract diode," *Appl. Phys. Lett.*, vol. 15, pp. 398–401, December 1969.

Optical Waveguide Transmission

DETLEF GLOGE, MEMBER, IEEE

Abstract—As optical communication systems are being studied in more detail, the need for many different types of optical waveguides becomes apparent. The applications range from miniature optical circuit connections to long-distance high-capacity transmission links. The requirements with respect to cost, attenuation, dispersion, or flexibility are vastly different. As different as the specifications are the guides evolving for various purposes. We describe the state of the art and give a survey of those guides which show the greatest future potential.

I. INTRODUCTION

AT LOWER frequencies there are two possible methods of transmission for light waves. One is the waveguide method, the other is propagation of free beams. Strictly speaking, most traditional optics use the propagation of free beams. Optical interconnections, even over distances as short as a few inches across the laboratory table, are made by light beams directed and focused by mirrors or lenses. The reason for this is that at micrometer wavelengths all the components needed for this form of transmission have convenient dimensions. True, the surface tolerances are within fractions of micrometers, but centuries of lens-making art have reduced this problem to a matter of course. More serious are the problems of ambient temperature gradients, acoustical effects and mechanical vibrations. Another reason why one looks for substitutes to the free-beam approach is the effort to miniaturize optical circuits for signal processing applications.

Another example of the preference given to the free-beam method at optical wavelengths is found in later sections of this paper. Most structures suggested there for long-distance optical communication resemble miniature radio relay systems enclosed in pipes. In these pipes, the beams propagate freely between periodic lens or mirror arrangements. Unlike the antennas in microwave radio systems, each lens or mirror collects virtually all the power radiated from the previous element. Hence, the losses are very small. It seems unlikely that any continuous optical guiding method could achieve comparable loss figures.

Nevertheless, there will be transmission systems for which attenuation may not be the most important factor. Interoffice connections within cities are an example. In urban areas the right of way is costly and environmental influences are hard to control. Continuously guiding structures may be more flexible, less sensitive, and consequently, a less expensive transmission medium in this situation.

II. HOLLOW WAVEGUIDES

Until now technological difficulties have discouraged most people from studying hollow waveguides smaller than several millimeters in diameter. This diameter corresponds

Manuscript received May 26, 1970.
The author is with Bell Telephone Laboratories, Inc., Holmdel, N. J. 07733.

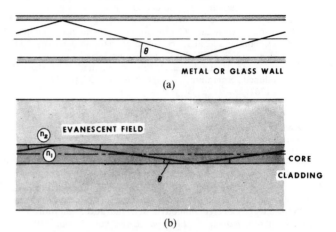

Fig. 1. Optical waveguides utilizing (a) reflecting walls and (b) total internal reflection.

to thousands of optical wavelengths, and a guide of this dimension transmits a vast number of modes. The modes of lowest loss, far from cutoff, can be conveniently described by the ray representation illustrated in Fig. 1. These rays follow zigzag paths along the guide reflected by the wall at grazing angles which are very small, though slightly different, for every mode. The grazing reflection causes relatively small loss both at metal or glass walls so that for these modes the attenuation is only a few decibels per kilometer [1]. Propagating selectively one or a few of these high-order modes requires very tight tolerances with respect to bends and irregularities in the overmoded guide [2], [3]. On the other hand, the attenuation increases only moderately if grazing angles up to 1° are tolerated [1]. Wall deformations, pipe offsets, and bends all increase the grazing angle and, in addition, transform meridional rays into skew rays [4]. Yet their effect on spread and attenuation is minor if the input beam has a divergence of 1° to begin with. In this case, for example, a bending radius of 500 meters should cause less than 1 dB/km additional loss in a pipe several centimeters in diameter [5].

Experiments have been performed with a 100-meter glass tube, 2 cm in diameter, which had a reflective coating of aluminum inside [6]. The statistical wall deformations of the unpolished glass wall were of the order of 10 μm. A loss of 57 dB/km was measured when a He–Ne laser beam at 0.63-μm wavelength was injected with a divergence of 0.2°. Data on the delay distortion introduced by this pipe are not available, but an upper limit on the group delay can be obtained by comparing the longest and shortest ray path in Fig. 1(a). This yields a time spread per unit length of

$$t = \frac{1}{v}\left(\frac{1}{\cos\theta} - 1\right) \qquad (1)$$

where v is the velocity of light in the hollow waveguide. For

Reprinted from *Proc. IEEE*, vol. 58, pp. 1513–1522, Oct. 1970.

352

a divergence angle of $1°$ one has $\theta = 0.5°$ and $t = 0.25$ ns/km. The corresponding bandwidth limitation restricts pipes of this kind to systems of fairly modest capacity.

III. SURFACE WAVEGUIDES

Rather than using the reflection at a metallic wall one can guide electromagnetic waves by total internal reflection. This is accomplished when the waves propagate in a cylindrical core made of a material of higher refractive index than the surrounding dielectric. Fig. 1(b) illustrates the propagation mechanism. The critical angle for total internal reflection is given by Snell's law

$$\cos \theta_c = \frac{n_2}{n_1} \qquad (2)$$

where n_1 and n_2 are the refractive indices in core and cladding, respectively. Only rays propagating at grazing angles smaller than the critical angle stay in the core following zigzag paths in the core material. There is an evanescent field in the cladding, but as long as the interface is perfectly cylindrical no energy is lost by radiation. Attenuation results more from the propagation loss in core and cladding than from loss at the interface. Two decades of fiber-optics research aimed at incoherent image transmission have provided a considerable body of experience in this respect [7]. For communication purposes, however, even lower attenuation and tighter tolerances are required. New thinking is necessary with respect to the influence of dispersion on modulated light waves.

At first glance, the cladding around the core seems unnecessary since free space as the outside medium exhibits both a low refractive index and low loss. Yet apart from the difficulty of supporting an unclad fiber without disturbing the outside field, there is another good reason to provide a cladding. Like hollow metallic waveguides, dielectric fibers sustain a number of modes whose propagation characteristics depend on the ratio between the core wavelength λ/n_1 and the core radius a. In the simplified picture of Fig. 1(b), modes can be understood as propagating along different zigzag paths characterized by discrete grazing angles. Since only those modes with angles smaller than the critical angle propagate, reducing the critical angle reduces the number of modes. More specifically, the term $2\pi\theta_c a/(\lambda/n_1)$ determines the cutoff. If, for example, this term is made smaller than the first root of the zeroth order Bessel function, only one mode will propagate [8]. Consequently, using (2) and the free-propagation constant $k = 2\pi/\lambda$, the single mode condition can be written in the form

$$ka(n_1^2 - n_2^2)^{1/2} < 2.4. \qquad (3)$$

At optical wavelengths, the left side of (3) is kept small through a small index difference by providing a cladding. An index difference of 1 percent is typical. For example, the core may have an index of 1.500 and the cladding 1.485. The critical angle is $8.1°$. Equation (3) is fulfilled when the core is several wavelengths in diameter. The cladding is usually hundreds of wavelengths thick so that the evanescent field, which decreases exponentially with the radius, is negligible at the external boundary of the cladding. The propagating

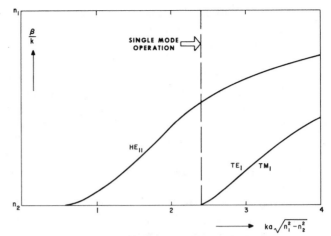

Fig. 2. Dispersion characteristic of the three lowest order modes of a cladded fiber with $n_1 - n_2 \ll n_1$.

HE_{11} mode has no cutoff. The data given in the following are for this typical single mode fiber unless specified otherwise.

Delay distortion is the reason why single mode guides are desirable for optical communication purposes. Sustaining only one desired mode or mode group in an overmoded fiber seems impossible because of unavoidable mode conversion. An estimate of the maximum possible group delay of an overmoded fiber can be obtained from (1) by introducing the relation (2) for the critical angle. From this a time delay per unit length between the longest and the shortest path length results which is

$$t_1 = \frac{1}{c} \frac{n_1}{n_2} (n_1 - n_2) \qquad (4)$$

where c is the velocity of light in vacuum. An overmoded fiber with an index difference of 1 percent could have a time spread as large as 50 ns/km. In this case, the bandwidth of a 1-km fiber is about 10 MHz.

The alternative is single mode operation. If a glass with little natural dispersion is chosen, the bandwidth is limited by the dispersion of the normal mode itself. Fig. 2 shows the dispersion characteristics of the first few modes. Plotted is the normalized propagation constant β/k versus the parameter $ka(n_1^2 - n_2^2)^{1/2}$ which is proportional to the angular frequency ω. Since optical communication systems will most likely be operated with pulses long compared to a light cycle, let us restrict our further discussion to this mode of operation. These pulses are not affected by a linear dispersion characteristic. Yet the second derivative of the β-ω function is critical. It is the change in group velocity within the modulation band that distorts the signal envelope. Gaussian pulses of an original half-width t_0 spread to

$$t_2 = \left[t_0^2 + \left(\frac{l}{t_0} \frac{d^2\beta}{d\omega^2} \right)^2 \right]^{1/2} \qquad (5)$$

in a dispersive medium of length l [9]. For the single mode fiber, the second derivative of β with respect to ω can be computed from Fig. 2. 1 km of the single mode fiber described earlier widens an 8-ps pulse to 12 ps. The corresponding bandwidth would be about 50 GHz.

Work on long-distance fibers is just beginning. For traditional fiber applications, attenuations of 1000 dB/km or more were acceptable. A reduction by almost two orders of magnitude seems necessary before experiments over distances of the order of kilometers will become feasible [10], [11]. This is primarily a material problem. Oxide glasses seem to be the most promising with respect to attenuation and fiber formation. Hopefully, reducing the impurity levels in these glasses (particularly transition metal ions) to levels below one part in a million will bring the absorption loss to something like 10 dB/km [10]. Avoiding the formation of crystallites during the cooling process should reduce the scattering loss to a few decibels per kilometer, which is the theoretical limit given by density fluctuations in the glass [10]. The best reported measurement in bulk glass samples at visible wavelengths show that the loss coefficient for absorption and for scattering are still one order of magnitude higher than the aforementioned target values [12], [13].

Another problem still unresolved is the fiber-pulling technology. In a careful measurement, the attenuation of two commercially available glass samples was compared to that of fibers pulled from these glasses [14]. The fibers were several meters long and transmitted many modes at the test wavelength of 0.63 μm. In short fiber sections, the scattering was close to that of the bulk material (equivalent loss 40 dB/km), but erratic scattering centers in the core-cladding interface increased the equivalent overall scattering loss to 250 dB/km. The absorption loss in the fibers was about equal to that of the cladding material which was 1750 dB/km compared to 1350 dB/km in the core. Theoretically, the fiber loss should have been close to that of the core. Two explanations for this discrepancy are conceivable. Either the absorption in the core increased during the pulling process or, in spite of extreme care, the energy was launched or eventually coupled into the cladding. Calculations show that if there is a roughness of the core-cladding interface as small as 0.005 μm, one meter of single mode fiber could convert 10 percent of the fundamental mode into higher mode energy which cannot propagate in the core [15]. Bends, on the other hand, cause a conversion that is six orders of magnitude smaller than this. This is because the stiffness of the cladded fiber excludes bending radii smaller than several centimeters [16].

IV. Graded Index Waveguides

The cladded fiber is just one example of a more general class of dielectric guides. All these guides have a refractive index which is constant along the guide axis z, axially symmetric, and decreases monotonically with the guide radius r. This, in fact, is a sufficient condition for electromagnetic wave guidance. Among the possible index profiles the quadratic dependence

$$n(r) = n_0\left(1 - \Delta \frac{r^2}{a^2}\right), \quad \text{for } r < a \qquad (6)$$

is of particular interest because of its lens-like characteristic. The index varies between n_0 at the axis and $n_0(1 - \Delta)$ at the circumference where $r = a$. There, $n(r)$ drops to the index

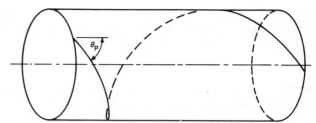

Fig. 3.   Helical path in a guide with parabolic index profile.

of the surrounding medium and (6) is no longer valid. However, for sufficiently large index gradients, modes exist whose fields are negligibly small at the boundary and which propagate entirely in the dielectric. Their attenuation is very nearly that of the dielectric. The fundamental mode is a Gaussian beam having a half-width

$$w = \frac{(a\lambda/\pi n_0)^{1/2}}{(2\Delta)^{1/4}} \qquad (7)$$

and the higher order modes have Laguerre–Gaussian field distributions [17]–[19]. The group velocities of these modes are nearly equal [20]. Provided that an absorbent wall coating eliminates modes which interfere with the wall, the inherent delay distortion can be small.

The ray representation will again help us to obtain an estimate of this delay. It is important to realize that there is no index profile which could completely eliminate delay. No index profile achieves equal transit times along all possible ray trajectories from one guide cross section to another [21]. However, the parabolic profile is close to optimal, and a comparison of the longest and the shortest transit time in the parabolic guide gives a good estimate of the time spread to be expected at optimal conditions. The transit time is longest along the helical path shown in Fig. 3. The pitch angle is given by the relation [22]

$$\cos \ \theta_p = \left(\frac{1 - 3\Delta}{1 - \Delta}\right)^{1/2}. \qquad (8)$$

The path length is by $1/\cos \theta_p$ longer than the axial path. Considering the lower refractive index at the circumference, one obtains the delay per unit guide length

$$t_3 = \frac{n_0}{c}\left(\frac{1 - \Delta}{\cos \theta_p} - 1\right). \qquad (9)$$

Inserting θ_p from (8) yields

$$t_3 \cong \frac{3}{2}\frac{n_0}{c}\Delta^2. \qquad (10)$$

If, for example, $n_0 = 1.5$ and $\Delta = 1$ percent, delays of 0.75 ns/km are possible. The corresponding bandwidth would be about 0.7 GHz.

It is a peculiarity of the parabolic profile that a Gaussian beam may follow any optical ray path in the guide, at least for a certain distance [23]. Fig. 4 shows three laser beams in a parabolic index guide, two of them on sinusoidal trajectories [24]. Beams with axial deviations comprise a number of normal guide modes whose phase relations are such that

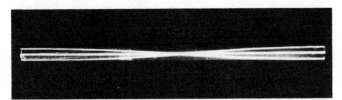

Fig. 4. Laser beams propagating in a guide with parabolic index profile [24].

over a certain distance the energy is maintained in a Gaussian beam. At longer distances this beam breaks up [25]. In a fiber with $\Delta = 1$ percent, for example, an off-axis beam of 1-μm wavelength is expected to disintegrate within a distance of 10 meters [26]. Bends cause beam deviations from the axis [27]. Because of this and the subsequent breakup, a Gaussian beam deteriorates along a guide with randomly distributed bends. If interference with the wall is to be avoided over at least 1 km, fibers with a useful diameter of 0.25 mm and $\Delta = 1$ percent should not have curvature radii smaller than 1 meter [26].

Index gradients inside glass fibers can be created by ionic exchange processes [24], [28], [29]. Potassium, for example, replaces thallium and sodium when a suitable glass is steeped in KNO_3 [30]. So far, this method has produced samples up to 10 meters long with a useful diameter of 0.25 mm and an index difference $\Delta = 1$ percent. The sources for loss are basically the same as in clad fibers with the only difference that a dielectric interface, and scattering from it, is avoided. There is, of course, a multitude of possible guides with other index profiles or combinations of the ones discussed here which might turn out to have advantages [31].

V. FILM GUIDES

A dielectric film strip provides guidance along the plane of the film. Suppose the film is a fraction of a wavelength thick and suspended in gas or vacuum. Since the energy propagates essentially outside this film there is little attenuation [32]. In order to keep the field from spreading in the plane of the film it has been suggested to twist the film or vary its thickness or refractive index along the width [33], [34]. The problem of suspending this film in a hollow tube has not yet found a practical solution at optical wavelengths.

For short optical circuit connections, delay distortion and loss is not of utmost concern and the choice of the guiding structure may be a matter of technological convenience. Sooner or later, signal processing for communication purposes will demand optical circuits integrated into common substrates. These substrates may contain or support both the optical components and their interconnections [35]. Processes may be required which permit the simultaneous formation of complicated circuits. Using some kind of masking technique, channels with an increased refractive index could be formed in a glass substrate by ion diffusion, implantation, or exchange similar to the graded index fiber formation. A protective layer of substrate glass could be sputtered on the top surface after the mask has been removed.

The diffusion would not create abrupt index steps, but the model of a rectangular dielectric guide of index n_2 em-

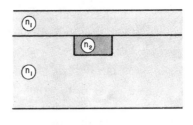

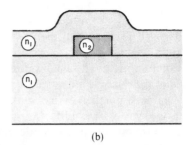

(b)

Fig. 5. High-index channels (a) induced in, and (b) deposited on, a lower-index substrate. Both are covered with a shielding layer.

TABLE I

WIDTH, INDEX DIFFERENCE, AND ESTIMATED BENDING RADIUS FOR VARIOUS DIELECTRIC GUIDING CHANNELS*

Channel Width (mm)	Index Difference	Acceptable Bending Radius (m)
1	10^{-6}	14 500
0.1	10^{-4}	14.5
0.01	10^{-2}	0.0145

* See [35].

bedded in a substrate of index n_1, as shown in Fig. 5(a), would approximate the situation reasonably well. The mode spectrum of this guide has characteristics similar to those of the cladded fiber [36]–[38]. Single mode operation with channel dimensions of the order of a wavelength requires index differences of 1 percent or less. On the other hand, the radiation loss in bends increases drastically with decreasing index difference [16], [39]. Table I illustrates this for three single mode guides of various dimensions by giving the index differences required and the bending radii acceptable for 0.25-dB loss per bend at a wavelength of 0.63 μm [35]. In miniature optical circuits, bending radii of several millimeters will be typical and, hence, index differences of at least 1 percent should be employed in bends.

Several processes for achieving guiding channels are conceivable and have been studied in preliminary experiments. Glass substrates have been irradiated with protons and it was found that channels with an increased refractive index form at a certain depth in the glass depending on the energy of the radiation [40]. These channels are completely surrounded by the substrate. Thin film strips of glass sputtered on a glass substrate have been shown to guide light around bends of about 10-mm radius [41]. These films were 0.3 μm thick and 20 μm wide and were formed by back sputtering using a glass fiber as a shadow mask. Thin-film evaporation has also been discussed [42], [43]. Both sputtered and evaporated film guides may employ a protective cover as shown in Fig. 5(b). This cover can be deposited in the same way as the film channel after the mask has been removed. A detailed description of film guides and related components can be found in [44].

VI. IRIS GUIDES

Short electromagnetic waves propagate in free space with so little divergence that, even in the millimeter-wave region, occasional interactions with the waves are sufficient to guide them along a straight pipe of practical dimensions [45]. The interaction may be in the form of irises arranged at intervals of length D, as shown in Fig. 6(a). Each iris absorbs the light outside its circular aperture of area A but transmits the remaining field. This field continues, diverging slightly because of diffraction, but is trimmed again at the next aperture. After a number of irises, the field which experiences the lowest possible diffraction loss repeats itself at every iris. For this dominant mode field, the right curve of Fig. 7 presents computer results of the diffraction loss as a function of the parameter $A/D\lambda$ (see [46], [47]). For example, a linear sequence of irises 3.4 cm in diameter and positioned at 5-meter intervals, would transmit a 1-μm laser beam with less than 1-dB loss per kilometer. Scaled experiments at 8.6 mm verified this estimate but also demonstrated that, in order to achieve this, exact alignment is crucial and bends are practically intolerable [48].

VII. GUIDES WITH SOLID LENSES OR MIRRORS

Directional changes in the iris guide could be accomplished by glass prisms or mirrors. This would introduce additional attenuation, caused mainly by the surfaces, but suitable coatings could reduce the loss per surface to a fraction of a percent. Apart from the problem of directional changes there is another good reason to have mirrors or some kind of refractive components in the guide. Diffraction not only causes the freely propagating beam to diverge but also changes its phasefront. Periodic correction of this phasefront focuses the beam and reduces the divergence [49], [50].

In the straight guide shown in Fig. 6(b), the best phase-front correction is that of ideal thin lenses with an aperture area A, whose focal length F equals half their spacing D. For this confocal case, the left curve in Fig. 7 shows the diffraction loss per lens of the dominant mode. This loss is smaller than that in the iris guide and decreases rapidly with increasing aperture area. For areas of 4 to 5 $D\lambda$, the diffraction loss is so small that it has practically no effect on the propagation characteristic and the field distribution of the dominant mode. The dominant mode field at the lenses has a Gaussian distribution which decreases to $1/e$ of its peak value at the radius

$$w_c = (D\lambda/\pi)^{1/2}. \qquad (11)$$

The attenuation of the mode is practically determined by the inherent lens losses.

Lens guides with sufficiently large apertures transmit more than one mode with negligible diffraction loss. These beam modes have Laguerre–Gaussian field distributions similar to the low-loss modes in a dielectric guide with parabolic index profile. However, in the space between the lenses, the

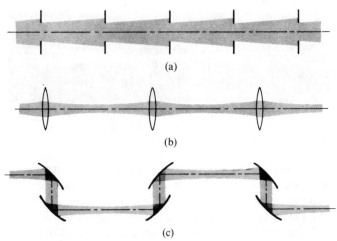

Fig. 6. Periodic guidance by (a) irises, (b) lenses, and (c) pairs of toroidal mirrors.

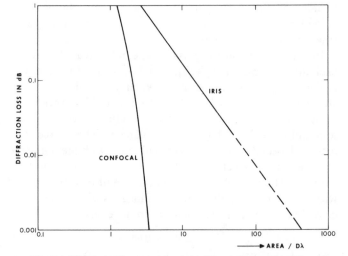

Fig. 7. Diffraction loss per iteration of the dominant mode in an iris guide and in a confocal lens guide [48].

intensity distribution of the modes is not constant along the guide axis, but shows a periodic contraction in the middle between two lenses. At this waist, the dominant Gaussian beam in a lens guide with focal length F and spacing D has a $1/e$ radius

$$w_{min} = \left(\frac{D\lambda}{2\pi}\right)^{1/2}\left(\frac{4F}{D} - 1\right)^{1/4}. \qquad (12)$$

At the lenses, the field expands to the radius

$$w = \left(\frac{2\lambda F}{\pi}\right)^{1/2}\left[\frac{4F}{D} - 1\right]^{-1/4}. \qquad (13)$$

Note that for $D=2F$ we arrive at the beam radius w_c of the confocal guide. The modes have practically all the same group velocities, and delay distortion is negligible because the influence of aberrations, as observed in parabolic dielectric guides, is negligible for typical lens guides. A detailed description of the general theory on beam modes can be found elsewhere [51], [52].

Laterally displaced lenses accomplish small directional changes of the beam direction. This fact is used for automatic realignment of lens guides [53], [54]. Larger directional changes or bends require the insertion of special deflection devices such as prisms or the mirror arrangements shown in Fig. 6(c). Every mirror is polished and coated for aberration-free reflection and focusing at 45° incidence, and directional changes are accomplished by rotating the two mirrors of a pair with respect to each other. The lens losses comprise absorption, reflection, and scattering in the material and on the surfaces. For currently available antireflection coatings, the surface losses prevail and are of the order of 1/2 percent or 0.02 dB per lens. The same can be achieved with good dielectric mirror coatings so that a mirror pair has about the same loss as a lens [55]. The spacing of the phase-transforming elements is limited to about 100 meters by the topography; thus, a loss coefficient of 0.2 dB/km represents a typical target value.

Several experimental versions very nearly verified this value when operated at a wavelength of 0.63 μm with less than 0.5-dB/km attenuation. The beam width $2w$ at 0.63 μm is roughly 1 cm. The apertures used in the experiments were several centimeters in diameter so that diffraction losses were negligible even if the alignment was not perfect. One experiment utilized 10 lenses spaced at intervals of 100 meters inside an evacuated conduit mounted on wooden poles above ground [56]. Another experiment simulated a mirror guide by shuttling a light pulse back and forth in a pipe between two mirrors 100 meters apart [57]. The 400 round trips covered a distance of 80 km. The same pipe was used to fold a laser beam between two arrays of mirrors [58]. The beam path was 6.4 km. Heterodyne and direct detection was alternately used on the signal beam returning from the guide after 64 reflections. The lack of conformity between the phasefronts of the signal and the local oscillator beam caused a signal loss of only 20 percent. Both the lens guide and the mirror arrangement were equipped with schemes for automatic realignment of the beam [58], [59].

In order to simulate actual future field conditions, an experiment was performed in an 840-meter pipe 1.50 meters below the ground level [60]. The pipe was filled with air, had a diameter of 10 cm, and consisted of six 140-meter sections with lenses. No effect of air turbulence was noticed on the beam. Surface irregularities of the lenses, though smaller than $\lambda/10$ in these experiments, were expected to have a cumulative effect in a long guide [61]–[63]. To test this, the cross-sectional profile of a short light beam was monitored which shuttled back and forth in the underground pipe [64]. The beam had a Gaussian profile when it was injected and was still approximately Gaussian when it had covered a distance of 120 km and passed 900 lenses. It seems that a gas-filled underground installation has to cope with three major instabilities [60]: 1) slowly varying thermal gradients in the ground which deflect the beam, 2) ground drifts and ground vibration, and 3) instabilities of the lens or mirror arrangements. A slight tilting of the lenses has a negligible effect on the beam though large tilt angles, which have been suggested to reduce the reflection loss, produce appreciable astigmatism [65].

Strictly speaking, the normal mode of the periodic lens guide is a Gaussian beam propagating exactly along the straight guide axis. However, Gaussian beams of the same kind can propagate along any optical ray path in the guide. A lens which is laterally displaced by δ, for example, causes an axial beam to follow a new path which, at some positions along the guide, deviates by 2δ from the axis [66]. This instability is an inherent property of lens guides in the low-loss region and disappears only when substantial additional loss is introduced [67], [68]. N random uncorrelated displacements multiply the average deviation of the beam path by the square root of N (see [69]). An angular change δ/F of the guide axis has the same effect on the beam path as a lens displacement δ. Several successive changes of this kind can be understood as a bend. Random bends, like lens displacements, have a cumulative effect on the beam location. Variations in the focal length or the spacing of the lenses act in a similarly cumulative way on the beamwidth [72]. In order to prevent the buildup of large beam excursions, automatic alignment schemes must be used [53], [54]. They measure the beam position at certain intervals and reposition the beam accordingly.

VIII. Gas Lens Guides

So far we have overlooked one potential of lens guides which is of interest when the lenses are closely spaced. By traversing these lenses off center, the beam can follow a curved path. An offset δ results in a bend with a radius $2F^2/\delta$. With $F = 50$ cm and $\delta = 1$ mm, we achieve a practical curvature radius of 500 meters. True, a spacing $D = 2F = 1$ meter results in 1000 lenses per kilometer, which is practical only if the loss per lens is at least an order of magnitude smaller than what one achieves with hard lenses. This is why gas lenses are of interest [73].

A lens effect results, for example, when a gas flowing through a heated tube warms up near the walls [74], [75]. The temperature rise decreases the density and the index of refraction and generates an approximately parabolic index profile. Unfortunately, this is a transient effect. While propagating through the heated tube the profile eventually equalizes and disappears. Yet even before this happens gravity begins to distort the density profile [76], [77]. To avoid this, the gas must be exhausted before the gravity effect becomes noticeable. This is the principle on which the gas lens operates which is shown in Fig. 8 [78]. The gas flow is in opposite directions in adjacent lens halves. The gas inlet is a porous pipe which guarantees uniform laminar flow in the end section of the path from inlet to outlet. This section is heated to about 60° above the input gas temperature. A focal length of 40 cm results. This provides a confocal lens arrangement with a normal mode radius of 0.4 mm at the He–Ne laser wavelength of 0.63 μm.

At this wavelength, a short guide of 11 lenses was tested. The transmission characteristic was found to be good as long as the beam followed a path that deviated not more than four beam radii from the axis [78]. The counterflow

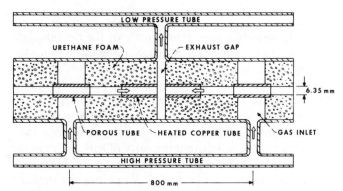

Fig. 8.   Section of a gas lens guide showing one counterflow lens [78].

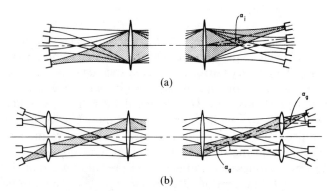

Fig. 9.   Multibeam transmission (a) by imaging, and
(b) by arranging beams in groups [91].

lens discussed in the preceding paragraph seems to have more stability and less aberration than the thermal gradient gas lens from which it has originated [74]. Yet even the latter lens proved useful in testing some experimental lens guides [79], [80]. Gentle bends with radii larger than 600 meters turned out to be feasible. The loss was unmeasurably small. A longer sequence of gas lenses was simulated by shuttling a light pulse back and forth through one of these lenses [81]. After 400 transitions, an axial beam was still well collimated, but deviations from the lens axis had a distorting effect on the beam, indicating that the thermal gas lens has considerable aberrations [25]. As mentioned earlier, beam deviations from the guide axis seem unavoidable in a lens guide in view of the cumulative effect on the beam caused by small lens displacements. The close spacing of the lenses may lead to intolerable beam deviations after relatively short distances. In this case, automatic realignment would be necessary.

Several gas lenses based on various other effects have been conceived. Among them, the quadrupole lens has received the most interest [82]–[84]. Yet none of these lenses has reached a degree of technical realization comparable to the thermal gas lens or its outgrowth, the counterflow lens.

IX. Multibeam Transmission

In principle, the periodic lens system transmits not only its natural modes or, specifically, Gaussian beams, but images of a very general nature. This is evident for one single lens with focal length F imaging a transmitter field into a detector plane [85]. Let A be the aperture area and $D = 2F$ the distance both from the transmitter to the lens and from the lens to the receiver. Then, for a coherent object field, the density of resolvable image points is given by the band of spatial frequencies $A/D^2\lambda^2$ which passes the lens pupil [86]. By multiplying with the image area A we obtain the number of the spatial degrees of freedom of this optical system which is $A^2/D^2\lambda^2$.

If, instead of the receiver, another lens with a focal length F is placed into the image plane, its exit field can be considered as the transmitter for another one-lens system. A series of these combinations results in a confocal lens guide

which images the transmitter field into the detector plane as shown in Fig. 9(a). Ideally, the sequence has the same degrees of freedom as the one-lens system. If one allows a few percent tolerance for the spacing or the focal lengths of the lenses, images will not occur exactly at every even numbered lens as in the strictly confocal guide. This is acceptable if only the detector plane at the end of the guide coincides with one of the image planes. Yet this more general guide has only half the degrees of freedom, viz. [87],

$$C_p = \frac{1}{2} A^2/D^2\lambda^2. \tag{14}$$

It is interesting to note that this is also the number of the Laguerre–Gaussian modes propagating in this guide. The capacity of the continuous dielectric guide with parabolic index profile can be calculated in a similar manner [87]. If the index profile is given by (6) this guide has

$$C_c = \Delta(\pi a n_0/\lambda)^2 \tag{15}$$

degrees of freedom.

Every degree of freedom represents an independent channel of information, but the numbers cited above must be understood as a theoretical limit achievable only with an ideal guide. Transmission of the natural orthogonal modes, for example, should ideally result in transmission without crosstalk. Practically, it is much simpler to use channels which separate naturally in the transmitter and receiver plane. Of course, this cannot be a perfect separation. Coherent sources, which produce Gaussian beams along the guide, seem to be the optimal choice. At least in an ideal guide, this method can guarantee negligible crosstalk by exploiting about 1/100 of the theoretical capacity [87].

The unavoidable imperfections of the guide present further limitations. We remember that a glass fiber with a 1-percent parabolic index profile shows aberrations at a distance of a few meters. A detailed study of this effect and its influence on the channel capacity or the crosstalk is not available. According to our previous estimates a short section of a fiber with a radius of 1/4 mm and 1-percent index difference should have 10 000 degrees of freedom and hence, exploiting 1/100 of the theoretical limit, results in a capacity of 100 beams. The transmission of a clear image through a 1-meter fiber, 0.3 mm in diameter, has been reported, but there is no information on the resolution achieved [28]. The apertures of presently available gas lenses prohibit multibeam transmission through this type of guide.

Typical lens guides made of conventional lenses or mirrors have negligible aberration even for distances of hundreds of kilometers. Mirrors can be made 20 to 30 cm in diameter without unreasonable effort. The source of crosstalk which limits the capacity and the transmission distance of these guides is scattering from the lens or mirror surfaces. This scattering seems to be somewhat smaller for mirrors than for lenses. In both cases it peaks sharply in the forward direction and decreases roughly with the third power of the deviation from this direction [88], [89]. For this reason, the crosstalk is essentially determined by the angular separation of adjacent detectors, which is called α_i in Fig. 9(a). If there are N_i channels this angle is

$$\alpha_i = \frac{1}{D}\left(\frac{A}{N_i}\right)^{1/2} \tag{16}$$

where D is the lens spacing.

The somewhat more sophisticated transmission method sketched in Fig. 9(b) alleviates this problem [87], [90]. The beams arrange in groups and open up to the normal mode radius before they enter the guide. Special collector lenses single out the groups at the end and focus the beams well separated on the detector array. In the arrangement sketched in Fig. 9(b), there are just as many groups as there are members in every group. Consequently, if N_g is the number of channels, there are $(N_g)^{1/2}$ collector lenses and $(N_g)^{1/2}$ detectors at every collector lens. The angular separation of the collector lenses as well as that of the detectors is therefore

$$a_g = \frac{1}{D}\frac{(A)^{1/2}}{(N_g)^{1/4}}. \tag{17}$$

A comparison of (16) with (17) yields $N_g = N_i^2$ when the crosstalk in both arrangements is about equal. If N_i is a relatively large number, the increase in capacity gained by the grouping method might well justify the more complicated terminals.

Calculations show that the grouping method could handle about 100 beams transmitted over a distance of 50 km with a crosstalk level of 23 dB [91]. A guide with periscopic mirror arrangements, spaced at a distance of 100 meters, is suggested for this purpose. The bundle of beams would measure about 20 cm in diameter. The estimate was based on the scattering characteristic of typical mirror samples [89]. Yet this result seems conservative in comparison with those of a laboratory experiment in which two Gaussian beams were folded side by side into a mirror cavity, 25 meters long and 15 cm in diameter [92]. The crosstalk between the two beams was only 30 dB after each beam underwent 684 reflections. In view of these results multibeam transmission seems to be a feasible means of handling large numbers of optical channels in long-distance waveguides.

X. Conclusions

Trying to meet future requirements, communication research is studying a variety of optical waveguides. There are three ranges of potential applications: 1) optical circuits, possibly integrated into small substrate chips; 2) medium range interconnections within buildings, blocks, or cities; and 3) long-distance transmission requiring low loss and high capacity. The guides evolving for these purposes are as different as the applications.

Miniature dielectric channels imbedded in a lower index substrate may some day form the basic guide for integrated optical circuits. Curvature radii of a few millimeters can be achieved. If the loss coefficient in glass fibers can be reduced by at least another order of magnitude, these fibers will represent a practical guide for medium range applications. An abrupt or graded index change of 1 percent within the fiber concentrates the propagating light in a central core region. Theoretically, a bandwidth of 100 GHz should be obtainable in a 1-km fiber if the propagation of higher order modes is avoided. The immediate problem is to make glasses with sufficiently low absorption.

The low loss which is attractive for long-range applications can only be achieved if the light propagates partly or completely in vacuum or gas. Gas lens guides avoid solid elements completely. The light beam is focused and guided by the refractive index profile in a heated gas stream. At the moment, instabilities and aberrations still pose a major problem. Using glass lenses or mirrors, one can build a guide which resembles an optical radio relay system enclosed in a pipe. The spacing of the phase-transforming elements is limited to about 100 meters by the topography. Dielectric coatings on the lenses or mirrors reduce the loss in both cases to less than 1 dB/km. If the effective cross section of the guide is 20 cm in diameter, up to 100 light beams can be transmitted simultaneously with acceptable crosstalk. How well this guide can be aligned and kept aligned over years is a problem which is currently under investigation.

References

[1] C. C. Eaglesfield, "Optical pipeline, a tentative assessment," *Proc. IEE* (London), vol. 109B, pp. 26–32, January 1962.
[2] E. A. J. Marcatili and R. A. Schmeltzer, "Hollow metallic and dielectric waveguides for long-distance optical transmission and lasers," *Bell Syst. Tech. J.*, vol. 43, pp. 1783–1809, July 1964.
[3] R. Bergeest and H. G. Unger, "Optische Wellen in Hohlleitern," *Archiv. Elek. Übertragung*, vol. 23, pp. 529–538, November 1969.
[4] N. K. Gorshkova, A. A. Dyachenko, V. A. Zyatitskiy, B. Z. Katsenelenbaum, and N. A. Kolesnikova, "Foundations of a statistical analysis of light beam propagation in a weakly deformed reflector pipe of circular cross-section," *Radio Eng. Electron. Phys.* (USSR), vol. 11, pp. 33–40, January 1966.
[5] Yu. I. Kaznacheyev, N. K. Gorshkova, and N. A. Kolesnikova, "Energy losses in flat bends of optical waveguide," *Radio Eng. Phys.* (USSR), vol. 9, pp. 883–886, June 1964.
[6] M. Procházka, J. Pachman, and J. Muzik, "Experimental investigation of a pipeline for optical communication," *Electron. Lett.*, vol. 3, pp. 73–74, February 1967.
[7] N. S. Kapany, *Fiber Optics.* New York: Academic Press, Inc., 1967.
[8] E. Snitzer, "Cylindrical dielectric waveguide modes," *J. Opt. Soc. Am.*, vol. 51, pp. 491–498, May 1961.

[9] C. G. B. Garrett and D. E. McCumber, "The propagation of a Gaussian light pulse through an anomalous dispersion medium," *Phys. Rev.*, vol. 1, pp. 305–313, February 1970.

[10] K. C. Kao and G. A. Hockham, "Dielectric-fiber surface waveguides for optical frequencies," *Proc. IEE* (London), vol. 113, pp. 1151–1158, July 1966.

[11] A. Werts, "Propagation de la lumière cohérente dans les fibres optiques," *Onde Elec.*, vol. 46, pp. 957–980, September 1966.

[12] K. C. Kao and T. W. Davies, "Spectrophotometric studies of ultra low loss optical glasses—I: single beam method," *J. Sci. Instrum.*, vol. 1, ser. 2, pp. 1063–1088, November 1968.

[13] R. D. Maurer, "Light scattering by glasses," *J. Chem. Phys.*, vol. 25, pp. 1206–1209, December 1956.

[14] A. R. Tynes, A. D. Pearson, and D. L. Bisbee, "Loss mechanisms and measurements in clad glass fibers and bulk glass" (to be published).

[15] D. Marcuse and R. M. Derosier, "Mode conversion caused by diameter changes of a round dielectric waveguide," *Bell Syst. Tech. J.*, vol. 48, pp. 3217–3232, December 1969.

[16] E. A. J. Marcatili, "Bends in optical dielectric guides," *Bell Syst. Tech. J.*, vol. 48, pp. 2103–2132, September 1969.

[17] E. A. J. Marcatili, "Modes in a sequence of thick astigmatic lens-like focusers," *Bell Syst. Tech. J.*, vol. 43, pp. 2887–2904, November 1964.

[18] H. Kogelnik, "On the propagation of Gaussian beams of light through lenslike media including those with a loss or gain variation," *Appl. Opt.*, vol. 4, pp. 1562–1569, December 1965.

[19] D. Gloge, "Bündelung kohärenter Lichtstrahlen durch ein ortsabhängiges Dielektrkum," *Archiv Elek. Übertragung*, vol. 18, pp. 451–452, July 1964.

[20] C. N. Kurtz and W. Streifer, "Guided waves in inhomogeneous focusing media—pt. I: formulation, solution for quadratic inhomogeneity," *IEEE Trans. Microwave Theory Tech.*, vol. MTT-17, pp. 11–15, January 1969.

[21] E. G. Rawson, D. R. Herriott, and J. McKenna, "Refractive index distributions in cylindrical, GRaded-INdex glass rods (GRIN rods) used as image relays," *Appl. Opt.*, vol. 9, pp. 753–759, March 1970.

[22] S. Kawakami and J. Nishizawa, "An optical waveguide with the optimum distribution of the refractive index with reference to waveform distortion," *IEEE Trans. Microwave Theory Tech.*, vol. MTT-16, pp. 814–818, October 1968.

[23] P. K. Tien, J. P. Gordon, and J. R. Whinnery, "Focusing of a light beam of Gaussian field distribution in continuous and periodic lens-like media," *Proc. IEEE*, vol. 53, pp. 129–136, February 1965.

[24] W. G. French and A. D. Pearson, "Refractive index changes produced in glass by ion exchange" (to be published).

[25] E. A. J. Marcatili, "Off-axis wave optics transmission in a lens-like medium with aberration," *Bell Syst. Tech. J.*, vol. 46, pp. 149–166, January 1967.

[26] E. A. J. Marcatili, "What kind of fiber for long-distance transmission?," presented at SPIE Seminar on Fiber Optics, Dallas, Tex., January 28–29, 1970.

[27] H. G. Unger, "Light beam propagation in curved Schlieren guides," *Archiv. Elek. Übertragung*, vol. 19, pp. 189–198, April 1965.

[28] M. Uchida, M. Furukawa, I. Kitano, K. Koizumi, and H. Matsumura, "A light-focusing fibre guide," *IEEE J. Quantum Electron.* (Digest of Technical Papers), vol. QE-5, pp. 331, June 1969.

[29] A. D. Pearson, W. G. French, and E. G. Rawson, "Preparation of a light focusing glass rod by ion-exchange techniques," *Appl. Phys. Lett.*, vol. 15, pp. 76–77, July 1969.

[30] I. Kitano, K. Koizumi, H. Matsumura, T. Uchida, and M. Furukawa, "A light-focusing fiber guide prepared by ion-exchange techniques." *J. Japan Soc. Appl. Phys.* (Supplement), vol. 39, pp. 63–70, January 1970.

[31] S. E. Miller, "Light propagation in generalized lens-like media," *Bell Syst. Tech. J.*, vol. 44, pp. 2017–2064, November 1965.

[32] A. E. Karbowiak, "New type of waveguide for light and infrared waves," *Electron. Lett.*, vol. 1, pp. 47–48, April 1965.

[33] N. Kumagai, S. Kurazono, S. Sawa, and N. Yoshikawa, "Surface waveguide consisting of inhomogeneous dielectric thin film," *Electron. Commun.* (Japan), vol. 51-B, pp. 50–54, March 1968.

[34] S. Sawa and N. Kumagai, "Surface wave along a circular *H*-bend of an inhomogeneous dielectric thin film," *Electron. Commun.* (Japan), vol. 52-B, pp. 44–50, March 1969.

[35] S. E. Miller, "Integrated optics: an introduction," *Bell Syst. Tech. J.*, vol. 48, pp. 2059–2069, September 1969.

[36] W. Schlosser and H. G. Unger, "Partially filled waveguides and surguides of rectangular cross section," in *Advances in Microwaves*. New York: Academic Press, Inc., 1966, pp. 319–387.

[37] J. E. Goell, "A circular-harmonic computer analysis of rectangular dielectric waveguides," *Bell Syst. Tech. J.*, vol. 48, pp. 2133–2160, September 1969.

[38] E. A. J. Marčatili, "Dielectric rectangular waveguide and directional coupler for integrated optics," *Bell Syst. Tech. J.*, vol. 48, pp. 2071–2102, September 1969.

[39] E. A. J. Marcatili and S. E. Miller, "Improved relations describing directional control in electromagnetic wave guidance," *Bell Syst. Tech. J.*, vol. 48, pp. 2161–2188, September 1969.

[40] E. R. Schineller, R. P. Flam, and D. W. Wilmot, "Optical waveguides formed by proton irradiation of fused silica," *J. Opt. Soc. Am.*, vol. 58, pp. 1171–1176, September 1968.

[41] J. E. Goell and R. D. Standley, "Sputtered glass waveguide for integrated optical circuits," *Bell Syst. Tech. J.*, vol. 48, pp. 3445–3448, December 1969.

[42] C. B. Shaw, Jr., and B. T. French, "Investigation of thin-film optical transmission lines," Space Research, U. S. Air Force, Washington, D. C., Contract AF49(638)-1504, Rept. C7-929.1/501, February 1968.

[43] P. K. Tien, R. Ulrich, and R. J. Martin, "Modes of propagating light waves in thin deposited semiconductor films," *Appl. Phys. Lett.*, vol. 14, pp. 291–294, May 1969.

[44] J. E. Goell and R. D. Standley, "Integrated optical circuits," this issue, pp. 1504–1512.

[45] G. Goubau and J. R. Christian, "A new waveguide for Millimeter Waves," *Proc. Army Sci. Conf. U. S. Military Acad.* (West Point, N. Y.), pp. 291–303, June 1959.

[46] A. G. Fox and T. Li, "Resonant modes in a maser interferometer," *Bell Syst. Tech. J.*, vol. 40, pp. 453–488, March 1961.

[47] L. A. Vainshtein, "Open resonators for lasers," *Soviet Phys.—JETP*, vol. 17, pp. 709–719, September 1963.

[48] J. W. Mink, "Experimental investigations with an iris beam waveguide," *IEEE Trans. Microwave Theory Tech.* (Correspondence), vol. MTT-17, pp. 48–49, January 1969.

[49] G. Goubau and F. Schwering, "On the guided propagation of electromagnetic wave beams," *IRE Trans. Antennas Propagat.*, vol. AP-9, pp. 248–256, May 1961.

[50] G. D. Boyd and J. P. Gordon, "Confocal multimode resonator for millimeter through optical wavelength masers," *Bell Syst. Tech. J.*, vol. 40, pp. 489–508, March 1961.

[51] G. Goubau, "Beam waveguides," in *Advances in Microwaves*, vol. 3. New York: Academic Press, Inc., 1968, pp. 67–126.

[52] H. Kogelnik and T. Li, "Laser beams and resonators," *Appl. Opt.*, vol. 5, pp. 1550–1567, October 1966.

[53] E. A. J. Marcatili, "Ray propagation in beam waveguides with redirectors," *Bell Syst. Tech. J.*, vol. 45, pp. 105–116, January 1966.

[54] J. R. Christian, G. Goubau, and J. Mink, "Self-aligning optical beam waveguides," *IEEE J. Quantum Electron.*, vol. QE-3, pp. 498–503, November 1967.

[55] V. R. Costich, "Coatings for 1, 2, even 3 wavelengths," *Laser Focus*, vol. 15, pp. 41–45, November 1969.

[56] G. Goubau and J. R. Christian, "Loss measurements with a beam waveguide for long-distance transmission at optical frequencies," *Proc. IEEE* (Correspondence), vol. 52, p. 1739, December 1964.

[57] O. E. DeLange, "Losses suffered by coherent light redirected and refocused many times in enclosed medium," *Bell Syst. Tech. J.*, vol. 44, pp. 283, 302, February 1965.

[58] O. E. DeLange and A. F. Dietrich, "Optical heterodyne experiments with enclosed transmission paths," *Bell Syst. Tech. J.*, vol. 47, pp. 161–178, February 1968.

[59] J. R. Christian, G. Goubau, and J. Mink, "Further investigations with an optical beam waveguide for long distance transmission," *IEEE Trans. Microwave Theory Tech.*, vol. MTT-15, pp. 216–219, April 1967.

[60] D. Gloge, "Experiments with an underground lens waveguide," *Bell Syst. Tech. J.*, vol. 46, pp. 721–735, April 1967.

[61] ——, "Mode conversion in lens guides with imperfect lenses," *Bell Syst. Tech. J.*, vol. 46, pp. 2467–2484, December 1967.

[62] V. A. Zyatitskiy, "Average conversion losses in a beam waveguide," *Radio Eng. Electron. Phys.* (USSR), vol. 12, pp. 1108–1117, July 1967.

[63] ——, "Average losses in a beam waveguide," *Radio Eng. Electron. Phys.* (USSR), vol. 13, pp. 1201–1210, August 1968.

[64] D. Gloge and W. H. Steier, "Pulse shuttling in a half-mile optical lens guide," *Bell Syst. Tech. J.*, vol. 47, pp. 767–782, May–June 1968.

[65] S. Saito, Y. Friji, and S. Shiraishi, "Low-loss laser beam transmission through lenses at the Brewster angle," *Proc. IEEE* (Letters), vol. 57, pp. 78–79, January 1969.

[66] G. Goubau and J. R. Christian, "Some aspects of beam waveguides for long-distance transmission at optical frequencies," *IEEE Trans. Microwave Theory Tech.*, vol. MTT-12, pp. 212–220, March 1964.

[67] J. R. Christian, J. W. Mink, G. Goubau, and F. K. Schwering, "Diffractional distortions in beam waveguides with off-axis beams," *Proc. IEEE*, vol. 57, pp. 829–831, May 1969.

[68] D. Gloge, "Regellose Störungen in Linsenleitungen," *Arch. Elec. Übertragung*, vol. 20, pp. 82–90, February 1966.

[69] J. Hirano and Y. Fakutsu, "Stability of a light beam in a beam waveguide," *Proc. IEEE*, vol. 52, pp. 1284–1292, November 1964.

[70] D. Marcuse, "Statistical treatment of light-ray propagation in beam waveguides," *Bell Syst. Tech. J.*, vol. 44, pp. 2065–2081, November 1965.

[71] D. Marcuse, "Probability of ray position in beam waveguides," *IEEE Trans. Microwave Theory Tech.*, vol. MTT-15, pp. 167–171, March 1967.

[72] W. H. Steier, "Statistical effects of random variations in components of a beam waveguide," *Bell Syst. Tech. J.*, vol. 45, pp. 451–471, March 1966.

[73] D. W. Berreman, "A lens or light guide using convectively distorted thermal gradient in gases," *Bell Syst. Tech. J.*, vol. 43, pp. 1469–1475, July 1964.

[74] D. Marcuse and S. E. Miller, "Analysis of a tubular gas lens," *Bell Syst. Tech. J.*, vol. 43, pp. 1759–1782, July 1964.

[75] A. C. Beck, "Thermal gas lens measurements," *Bell Syst. Tech. J.*, vol. 43, pp. 1818–1820, July 1964.

[76] W. H. Steier, "Measurements on a thermal gradient gas lens," *IEEE Trans. Microwave Theory Tech.*, vol. MTT-13, pp. 740–748, November 1965.

[77] D. Gloge, "Deformation of gas lenses by gravity," *Bell Syst. Tech. J.*, vol. 46, pp. 357–365, February 1967.

[78] P. Kaiser, "An improved thermal gas lens for optical beam waveguides," *Bell Syst. Tech. J.*, vol. 49, pp. 137–153, January 1970.

[79] A. C. Beck, "An experimental gas lens optical transmission line," *IEEE Trans. Microwave Theory Tech.* (Correspondence), vol. MTT-15, pp. 433–434, July 1967.

[80] P. Kaiser, "Measured beam deformations in a guide made of tubular gas lenses," *Bell Syst. Tech. J.*, vol. 47, pp. 179–194, February 1968.

[81] W. H. Steier, "Optical shuttle pulse measurements of gas lenses," *Appl. Opt.*, vol. 7, pp. 2295–2300, November 1968.

[82] Y. Suematsu, "Light-beam waveguide using lens-like media with periodic hyperbolic temperature distribution," *J. Inst. Electron. Commun. Japan*, vol. 49, pp. 107–113, March 1966.

[83] K. Iga and Y. Suematsu, "Experimental studies on the limitation of focusing power of hyperbolic-type gas lens," *Japan. J. Appl. Phys.*, vol. 8, pp. 255–259, February 1969.

[84] P. Marié, "Guidage de la lumière cohérente par un guide helicoidal extension a la focalisation continue des particules de haut énergie," *Ann. Télécommun.*, vol. 24, pp. 177–189, May–June 1969.

[85] N. G. Basov, A. Z. Grasyuk, and A. N. Orayevskiy, "Some properties in the transmission and reception of information using laser oscillators and amplifiers," *Radio Eng. Electron. Phys.* (USSR), vol. 9, pp. 1387–1391, September 1964.

[86] G. T. Di Francia, "Degrees of freedom of an image," vol. 59, pp. 799–804, July 1969.

[87] D. Gloge and D. Weiner, "The capacity of multiple beam waveguides and optical delay lines," *Bell Syst. Tech. J.*, vol. 47, pp. 2095–2109, December 1968.

[88] G. R. Hostetter, D. L. Patz, H. A. Hill, and C. A. Zanoni, "Measurement of scattered light from mirrors and lenses," *Appl. Opt.*, vol. 7, pp. 1383–1385, July 1968.

[89] D. Gloge, E. L. Chinnock, and H. E. Earl, "Scattering from dielectric mirrors," *Bell Syst. Tech. J.*, vol. 48, pp. 511–526, March 1969.

[90] G. Goubau and F. K. Schwering, "Diffractional crosstalk in beam waveguides for multibeam transmission," *Proc. IEEE* (Letters), vol. 56, pp. 1632–1634, September 1968.

[91] D. Gloge, "Crosstalk in multiple-beam waveguides," *Bell Syst. Tech. J.*, vol. 49, pp. 55–71, January 1970.

[92] D. Gloge and W. H. Steier, "Experimental simulation of a multiple beam optical waveguide," *Bell Syst. Tech. J.*, vol. 48, pp. 1445, 1457, May 1969.

Integrated Optical Circuits

JAMES E. GOELL, MEMBER, IEEE, AND ROBERT D. STANDLEY,
MEMBER, IEEE

Abstract—Research on optical communication systems indicates the future need for a compact, rugged, and economical medium for circuit realization. Encapsulated planar arrays of rectangular dielectric waveguides are attractive for this purpose. The individual guides would have dimensions on the order of a few micrometers while the planar arrays might cover an area of a few square centimeters. Integrated circuit technology appears adaptable to batch processing such circuits. Recent theoretical and experimental results are surveyed which support this viewpoint.

I. INTRODUCTION

THE realization of laser communication systems with enormous information capacity is envisioned. Regardless of the transmission medium, such systems may require extensive optical circuitry to process the communication signals. Toward this end consideration is being given to possible physical forms of the circuitry [1]. Natural objectives are compactness, environmental insensitivity, flexibility in performing signal processing functions, and economy.

Historically, the progression of electronic circuit development has been from a hybrid technology to a monolithic technology. The work surveyed in this paper suggests that optical circuitry may evolve along similar lines. At this juncture, a short-distance transmission medium is required which exhibits low loss over distances comparable to the circuit dimensions (a few centimeters) and lends itself to the fabrication of bends, couplers, filters, and other passive and active components required for optical signal processing.

A dielectric waveguide approach is very promising. Considerable information about such guides can be extracted from the fiber optics field where transmission losses below 0.01 dB/cm have been achieved. (The use of a metallic waveguide can be quickly eliminated since the loss of a waveguide having dimensions comparable to the wavelength would be on the order of 10^2 dB/cm.) Presently envisioned fabrication techniques make possible the realization of dielectric waveguides on dielectric substrates having approximately the cross sections of Fig. 1(a)–(c). In all cases the guided energy is concentrated in the core, i.e., the region of highest refractive index. Using such waveguides, directional couplers could be formed as shown in Fig. 1(d) and modulators as shown in Fig. 1(e). Various other active and passive circuits could also be realized [1].

II. DIELECTRIC WAVEGUIDE THEORY

In this section some of the pertinent theory of propagation in dielectric waveguide structures is given. First, we treat the slab case to give insight into general properties of

Manuscript received May 20, 1970.
The authors are with the Crawford Hill Laboratory, Bell Telephone Laboratories, Inc., Holmdel, N. J. 07733.

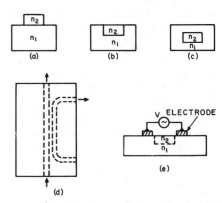

Fig. 1. Dielectric waveguide structures ($n_2 > n_1 \geq n_0$). (a) Raised guide. (b) Imbedded guide. (c) Encapsulated guide. (d) Directional coupler. (e) Modulator.

dielectric waveguides. In addition, the theory is related to some useful thin-film measurement techniques to be discussed in Section III. Following this, propagation in rectangular dielectric waveguides is treated.

A. Slab Waveguide

The theory of propagation of electromagnetic waves in dielectric slabs is well documented in [2] and [3]. We state here the pertinent results without derivation. Consider the geometry of Fig. 2, where a lossless dielectric slab of thickness b and index of refraction n_1 is bounded by regions having indices n_2 and n_3 with $n_1 > n_2 \geq n_3$. The slab is assumed to be of infinite extent in the y direction. Both transverse electric (TE) and transverse magnetic (TM) modes of propagation are possible for this geometry.

It can be shown that the electric and magnetic fields vary as

$$(E, H) \alpha \begin{cases} \sin(k_x x) \\ \cos(k_x x) \end{cases} \exp(j(\omega t - \beta z)), \qquad |x| \leq b/2 \qquad (1)$$

$$(E, H) \alpha \exp(\gamma_2(x + b/2)) \exp(j(\omega t - \beta z)), \quad x \leq -b/2 \quad (2)$$

and

$$(E, H) \alpha \exp(-\gamma_3(x - b/2)) \exp(j(\omega t - \beta z)), \quad x \geq b/2. \quad (3)$$

The dispersion relations for the propagating modes are given by

$$\tan(k_x b) = \frac{\gamma_2/k_x + \gamma_3/k_x}{1 - (\gamma_2/k_x)(\gamma_3/k_x)} \qquad (4)$$

for TE modes, and by

$$\tan(k_x b) = \frac{n_1^2(\gamma_2/n_2^2 + \gamma_3/n_3^2)}{1 - n_1^4 \gamma_2 \gamma_3/(n_2 n_3 k_x)^2} \qquad (5)$$

for the TM modes, where the n's are the refractive indices

Reprinted from *Proc. IEEE*, vol. 58, pp. 1504–1512, Oct. 1970.

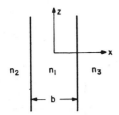

Fig. 2.   Dielectric slab waveguide.

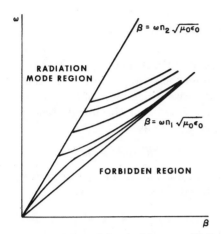

Fig. 3. ω versus β for a dielectric slab waveguide ($n_2 = n_3$).

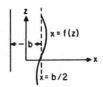

Fig. 4.   Slab guide with wall imperfection.

of the various regions shown in Fig. 2, the γ's the exponential decay constants in the outer regions, and k_x the transverse propagation constant in the central region. These constants are related to each other and the longitudinal propagation constant β by the relations

$$\beta^2 = k^2 n_2^2 + \gamma_2^2$$
$$\beta^2 = k^2 n_3^2 + \gamma_3^2$$
$$\beta^2 = k^2 n_1^2 - k_x^2 \qquad (6)$$

where

$$k = 2\pi/\lambda_0 = \omega\sqrt{\mu_0\varepsilon_0}.$$

From the preceding relations the following necessary conditions for propagation can be derived

$$bk(n_1^2 - n_2^2)^{1/2} > \tan^{-1}\left[(n_2^2 - n_3^2)/(n_1^2 - n_2^2)\right]^{1/2} \qquad (7)$$

for TE modes, and

$$bk(n_1^2 - n_2^2)^{1/2} > \tan^{-1}\left[n_1^4(n_2^2 - n_3^2)/n_3^4(n_1^2 - n_2^2)\right]^{1/2} \qquad (8)$$

for TM modes.

A typical plot of ω versus β for the case $n_2 = n_3$ is given in Fig. 3 for TE or TM propagation. Also shown are the lines $\beta = \omega n_1 \sqrt{\mu_0\varepsilon_0}$ and $\beta = \omega n_2 \sqrt{\mu_0\varepsilon_0}$. All guided propagating wave solutions lie in the region between these boundaries. Unlike the case of the hollow metallic waveguide, a principal mode exists, that is, a mode that propagates at all frequencies. Also, for all other modes, propagation ceases at a nonzero β.

In the region to the left of the upper boundary, a continuum of radiating modes exist. The slope of the upper boundary is the velocity of light in the outer medium, and the slope of the lower boundary is the velocity of light in the core. For propagation near the lower boundary, the fields are confined mainly to the core. In the limit as the upper boundary is approached, the fields extend to infinity.

The ω versus β plots for $n_2 \neq n_3$ are similar to Fig. 3, except that a principal mode does not exist.

B. Effects of Surface Imperfections

As stated previously, the rigorous solution of dielectric waveguide problems results in a discrete spectrum of guided modes plus a continuum of radiation modes. The effect of surface or bulk imperfections in a guide is to cause 1) coupling between a given guided mode and all propagating modes, and 2) coupling to the continuum of radiation modes. The latter results in increasing the waveguide loss above that due to dielectric absorption loss. Marcuse [4] has considered these effects for the dielectric slab waveguide shown in Fig. 4 with the simplifying assumption of no dimensional or other guide property variations in the y direction. Here we give only the results of his radiation loss calculations. Assuming that one wall is perfectly smooth and that for the other wall the imperfections have a sinusoidal shape of the form

$$f(z) - \frac{b}{2} = a \sin \phi z \qquad (9)$$

then for $n_1/n_2 = 1.01$, $n_2 = n_3$, $n_2 b/\lambda = 4.78$, $\lambda_0 = 1 \ \mu m$, $4\pi/b\phi = 25$, and $n_2 L/\lambda_0 = 10^4$ where L is the guide length, it is found that the fractional power loss is

$$\frac{\Delta P}{P} = 0.1 \quad \text{for} \quad n_2 a/\lambda_0 = 0.0546.$$

Thus, for $n_2 = 1.5$, a 3.19-μm-thick slab guide with wall imperfections of 364 Å would result in a radiation loss of 0.69 dB/cm for a 1-μm wavelength.

For the case of random imperfections on one wall, Marcuse has considered the case where the correlation function of $f(z)$ is of the form

$$R(u) = A^2 \exp\left(-\frac{|u|}{B}\right) \qquad (10)$$

where A is the root mean square deviation of the wall from perfect straightness, B is the correlation length, and $u = z - z'$. Fig. 5, taken from Marcuse, shows the normalized radiation loss for a guide of length L for the case $n_1 = 1.01$, while Fig. 6 gives the results for $n_1 = 1.5$. For both cases $n_2 = 1$. From these data, in a single mode waveguide, a deviation of the guide wall of about 3 percent for the worst case of correlation length results in a loss[1] of 0.5 dB/cm for an index

---

[1] It should be noted that the loss decreases with increasing correlation length for $2B/b$ sufficiently large so that a slow variation of the waveguide width is not serious. Furthermore, a few relatively large discontinuities could be tolerated.

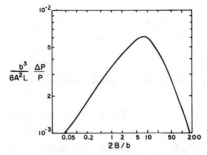

Fig. 5. Normalized radiation loss $(b^3/8A^2L)(\Delta P/P)$ as a function of the normalized correlation length $2B/b$ for $n_1/n_2 = 1.01$, $kb = 16.0$, and $n_2 = 1$. (Single guided mode operation.)

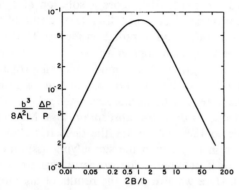

Fig. 6. Normalized radiation loss $(b^3/8A^2L)(\Delta P/P)$ as a function of the normalized correlation length $2B/b$ for $n_1/n_2 = 1.5$, $kb = 2.6$, and $n_2 = 1$. (Single guided mode operation.)

ratio of 1.01 and an outer index of 1.5. For an index ratio of 1.5 the loss is about an order of magnitude higher. For correlation functions other than (10), similar results are found for values of $2B/b$ less than that of the peak.[2]

C. Rectangular Dielectric Waveguide

The analysis of a rectangular dielectric waveguide is considerably more complex than that of the slab waveguide discussed in the preceding paragraphs. Unlike the previous case or that of the hollow metallic rectangular waveguide, the electric and magnetic fields of the rectangular dielectric waveguide modes cannot be expressed as functions exhibiting simple sinusoidal transverse variation in the core.

A number of methods have been employed to evaluate the properties of rectangular dielectric waveguide. Schlosser and Unger [5] have used a numerical approach employing sums of sine, cosine, and exponential functions. Their method is computationally best suited to cases of large aspect ratio (width to height) and to cases away from cutoff, that is, far from the radiation mode region. Marcatili [6] has obtained an approximate analytical solution in closed form also using simple sinusoidal and exponential functions. His solutions are accurate for the cases of small to moderate index difference and away from cutoff. Goell [7] has performed a numerical analysis employing cylindrical space harmonics. His solutions converge rapidly for aspect ratios between 1 and 2 and any index difference.

Fig. 7 shows the geometry used by Marcatili. In his

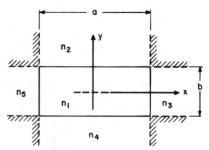

Fig. 7. Idealized rectangular waveguide geometry.

analysis the fields in the shaded regions are neglected which accounts for both the simplicity and limitations of his approach. For the case where most of the energy is confined to the core, he shows that the longitudinal propagation constant can be written as

$$\beta = (k_1^2 - k_x^2 - k_y^2)^{1/2} \quad (11)$$

where k_x and k_y are the x- and y-direction propagation constants in the core and are given by

$$k_x = \frac{\pi p}{a}\left(1 + \frac{A_3 + A_5}{\pi a}\right)^{-1} \quad (12)$$

$$k_y = \frac{\pi q}{a}\left(1 + \frac{n_2^2 A_2 + n_4^2 A_4}{\pi n_1^2 b}\right)^{-1} \quad (13)$$

for E_{pq}^y modes, and by

$$k_x = \frac{\pi p}{a}\left(1 + \frac{n_3^2 A_3 + n_5^2 A_5}{\pi n_1^2 a}\right)^{-1} \quad (14)$$

$$k_y = \frac{\pi q}{a}\left(1 + \frac{A_2 + A_4}{\pi b}\right)^{-1} \quad (15)$$

for E_{pq}^x modes, where

$$A_m = \frac{\pi}{(k_1^2 - k_m^2)^{1/2}} = \frac{\lambda}{2(n_1^2 - n_m^2)^{1/2}} \quad (16)$$

and

$$k_m = \frac{2\pi}{\lambda_m} = \omega\sqrt{\mu_0 \varepsilon_m}. \quad (17)$$

In these equations, p and q represent the number of field maxima in the x and y directions, respectively, m the region number as defined in Fig. 7, and the superscripts x and y give the polarization of the principal component of the electric field.

Fig. 8 shows some typical curves of normalized propagation constant $\mathcal{P}^2 = (k_z^2 - k_4^2)/(k_1^2 - k_4^2)$ versus normalized waveguide height $\mathcal{B} = b/\lambda_0(n_1^2 - n_4^2)^{1/2}$. Included are curves drawn using (11) and (12), curves found from more accurate transcendental relations also derived by Marcatili, and curves derived numerically by Goell. The divergence of the curves near cutoff is due to approximations imposed by Marcatili. Curves of the principal mode propagation constants for unit aspect ratio are shown in Fig. 9 for several values of $\Delta n_r = n_1/n_4 - 1$. The curves are relatively insensitive to the index difference due to the normalization employed to define the coordinates.

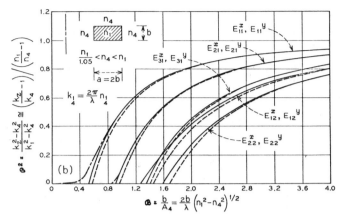

Fig. 8. Propagation constant for several modes of rectangular waveguide [6]. —— transcendental equation solution; – – – closed form solutions; – · – · – Goell's computer solutions.

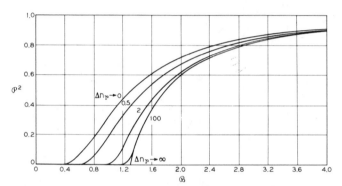

Fig. 9. Effect of index difference Δn on β [7] for unity aspect ratio.

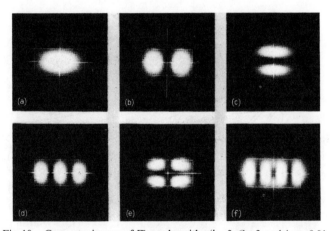

Fig. 10. Computer images of E^y_{pq} modes with $a/b = 2$, $\mathscr{B} = 2$, and $\Delta n_r = 0.01$. (a) E^y_{11}. (b) E^y_{21}. (c) E^y_{12}. (d) E^y_{31}. (e) E^y_{22}. (f) E^y_{41} [7].

Fig. 11. Power density as cutoff is approached. Computer images of the E^y_{21} mode. (a) $\mathscr{P}^2 = 0.76$. (b) $\mathscr{P}^2 = 0.31$. (c) $\mathscr{P}^2 = 0.04$ [7].

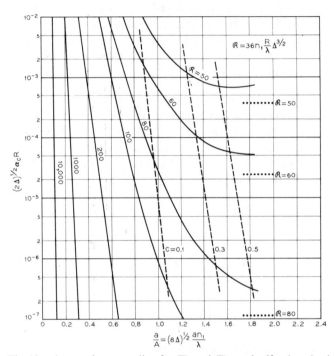

Fig. 12. Attenuation per radian for E^x_{11} and E^y_{11} modes if $n_1/n_3 = 1 + \Delta$ and $\Delta \ll 1$ [8].

Computer generated images of the first six rectangular dielectric waveguide modes are shown in Fig. 10 for a normalized guide height of 2. These images were obtained by generating dots whose area was proportional to the calculated local mode power density. The outlined rectangle represents the core boundary. Even for the E^y_{41} mode, where the normalized propagation constant is 0.12, most of the power is confined to the core and very little exists in the shaded region of Fig. 7. This accounts for the accuracy of Marcatili's analysis even for fairly small values of normalized propagation constant. Fig. 11 demonstrates the exten-

sion of the fields outside the core region as cutoff is approached. Similar results are obtained for all modes.

D. Bends

Practical circuit realization will require dielectric waveguide bends. The ability of such waveguides to negotiate bends and the radiation loss thus produced have been considered by Marcatili [8]. Fig. 12 shows normalized loss data for a typical range of parameters. Here α_c is the attenuation coefficient of the waveguide bend, and R is the radius of curvature of the bend. The parameter c is given by

$$c = \frac{a^2 \beta^2}{4 k_x R} \tag{18}$$

where β is the longitudinal propagation constant of a straight waveguide and the transverse propagation constant k_x is given by (12) for the E^y_{pq} modes or by (14) for the E^x_{pq} modes. The parameter A in the ordinate dimension is

$$A = \frac{\lambda_0}{2(n_1^2 - n_3^2)^{\frac{1}{2}}}. \tag{19}$$

The solid lines in Fig. 12 are accurate to the left of the dashed line given by $c = 0.3$. For large c the dotted curves apply.

For an example of the loss expected, consider

$$n_1 = 1.5$$

$$a = \frac{\lambda_0}{2n_1\left(1 - \frac{n_3^2}{n_1^2}\right)^{\frac{1}{2}}} \quad \text{(single mode guide)}$$

then a 1-percent attenuation (0.087 dB) resulting from radiation in a length of guide equal to R is achieved with the values shown in Table I. The smaller $n_1 - n_3$, the larger the radius of curvature. For $\lambda_0 = 0.63$ μm, if one wants to keep R below 1 mm, the difference between the internal and external refractive indices must be larger than 0.01. Table I applies to the case $b = \infty$. For finite b the value of R must be multiplied by $\left[1 - (k_y/k_3)^2\right]^{-1}$.

E. Directional Couplers

Directional couplers will form an integral part of many integrated optical circuits. Typical results of Marcatili's analysis [6] on coupling of rectangular dielectric waveguides are shown in Figs. 13 and 14. The parameters not defined in these figures are the following,

$$K = \frac{\pi}{2L}$$

$$= 2\frac{k_x^2}{k_z}\frac{\xi}{a}\frac{\exp(-c/\xi)}{1 + k_x^2\xi^2} \quad (20)$$

$$\xi = \left[\frac{1}{\left(\frac{\pi}{A_5}\right)^2 - k_x^2}\right]^{1/2} \quad (21)$$

where c is waveguide spacing and L is the length of the coupling region which results in complete power transfer from one guide to the other.

The coupled power is given by

$$P = P_0 \sin^2 Kl \quad (22)$$

where l is the length of the coupling region and P_0 the input power. We give two examples obtained from the above data. For

$$n_1 = 1.5$$

$$n_2 = 1.5/1.01$$

$$\left.\begin{array}{l} b = 1.77\,\lambda_0 \\ a = 3.54\,\lambda_0 \end{array}\right\} \quad \text{(single mode guides)}$$

then for an interguide spacing equal to the waveguide width, complete transfer of power will occur in a length of 6540 λ_0. For one fourth of this separation, the interaction length for complete coupling will be 262 λ_0.

III. Measurement of Optical Parameters

The measurement of the optical parameters of thin films has received considerable attention in the literature over a period of many years. For a thorough review of the field, the reader is referred to Heavens [9] and Abeles [10]. Our attention will be directed toward those techniques which have proven most suitable for our work. In particular, we shall treat Abeles' method of measuring the refractive index

by determining the Brewster angle of a film. The effect of a graded index on the accuracy of this method will also be discussed. A more recent method of measurement using a prism to excite propagating modes in a film will also be described as will methods of measuring attenuation.

TABLE I

$1 - (n_3/n_1)$	a/λ_0	R/λ_0
0.1	0.745	30
0.01	2.36	1060
0.001	7.45	37000

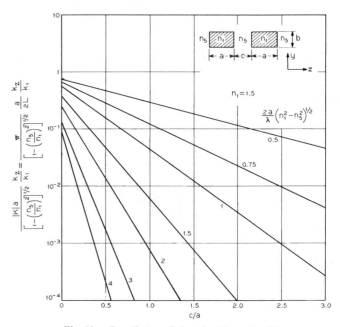

Fig. 13. Coupling coefficient for E_{1q}^y modes [6].

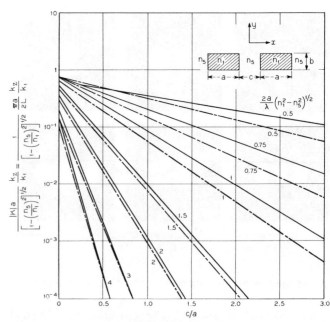

Fig. 14. Coupling coefficient for E_{1q}^x modes for $n_1/n_5 = 1.5$; — — — E_{1q}^x coupling for $n_1/n_5 = 1.1$ [6].

A. Analysis of Abeles' Method

When light polarized in the plane of incidence is incident on the air–film interface at the Brewster angle for the film, there is no reflection from the interface, i.e., the film behaves as if it did not exist. The reflected signal is then equal in magnitude to that which would be obtained from the substrate alone. More precisely, the reflection coefficient from a film of thickness d and index n_f on an infinitely thick substrate of index n_s is given by

$$R = \frac{r_f + r_s e^{-j2\delta}}{1 + r_f r_s e^{-j2\delta}} \qquad (23)$$

where r_f and r_s are Fresnel coefficients, i.e.,

$$r_f = \frac{\cos \Phi_f - n_f \cos \Phi_i}{\cos \Phi_f + n_f \cos \Phi_i} \qquad (24)$$

$$r_s = \frac{n_f \cos \Phi_s - n_s \cos \Phi_f}{n_f \cos \Phi_s + n_s \cos \Phi_f} \qquad (25)$$

and

$$\delta = \frac{2\pi n_f d}{\lambda_0} \cos \Phi_f \qquad (26)$$

where Φ_i is the angle of incidence in air; Φ_f is the angle of refraction in the film; and Φ_s is the angle of refraction in the substrate. At the Brewster angle, Φ_{iB} for the air–film boundary, $r_f = 0$, so that

$$\tan \Phi_{iB} = n_f. \qquad (27)$$

We then obtain from above

$$|R| = \frac{\cos \Phi_s - n_s \cos \Phi_i}{\cos \Phi_s + n_s \cos \Phi_i} = r_{s0} \qquad (28)$$

where r_{s0} is the Fresnel coefficient for the air-substrate boundary.

It can be shown that if $r_s = 0$, then $|r_f| = |r_{s0}|$ is also a solution. However, this root exists only in a restricted range of n_f and n_s given by

$$\frac{n_f^2 n_s^2}{n_f^2 + n_s^2} \le 1. \qquad (29)$$

Fig. 15 is a plot of the region for which this latter solution applies. In general, our work on silicate glass is in the region $n_f, n_s > 1.45$, so that this latter solution is of no concern. However, it could be significant in the evaluation of fluoride glass and other smaller index systems. In practice, the refractive index of a film can be determined to an accuracy better than ± 0.002 by this method.

B. Effect of Index Gradient

The Method described in the preceding paragraphs assumes that the film index is uniform. It is well known that for some materials, such as MgF_2, the refractive index varies with depth [11]. In order to gain insight into the effects of index gradients, consider the index profile of Fig. 16. For a transition region thickness $\Delta t = 0$, the analysis of Section III-A applies, so the true film index results from the measurement. However, for Δt finite an error will result, i.e., the mea-

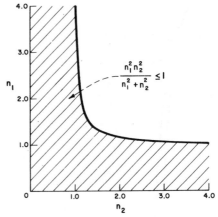

Fig. 15. Domain of substrate Brewster angle root.

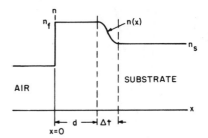

Fig. 16. Index gradient profile.

sured index, $\bar{n}$ will not be n_f. Assuming that the transition from n_f to n_s is continuous, then for Δt sufficiently large it can be shown analytically that for n_f and n_s large or for $n_f \approx n_s$, the measurement described above gives the geometric mean of the film and substrate indices.

In order to obtain further information, a numerical analysis of the problem has been performed. It was assumed that $n(x)$, the index variation in the transition region, took the form

$$n(x) = \frac{n_f + n_s}{2} - \frac{n_f - n_s}{2} \cos \left(\frac{x - d}{\Delta t} \pi \right),$$
$$\text{for } d \le x \le d + \Delta t. \quad (30)$$

The problem was solved by using a matrix formulation with $n(x)$ approximated by a series of discrete steps [12]. Fig. 17 shows the variation of the measured index $\bar{n}$ as found by Abeles' method as a function of transition region thickness for a substrate index of 1.50 and a film thickness $d = 0.1 \lambda_0$.

For a graded index, the measured index $\bar{n}$ is also a function of film thickness because the angle at which the reflection coefficient of the film on the substrate is equal to the reflection coefficient of the substrate alone is not equal to the Brewster angle for the film alone. This effect is demonstrated in Fig. 18. Clearly, even fairly short graded regions can cause significant error.

C. Film Loss Measurement

In order to determine the loss of films suitable for fabrication of waveguides, it is necessary that light be propagated in the film. Efficient injection of light by edge illumination is impractical since the films are typically about 0.3 μm thick. It has been found convenient to launch from a higher index region. One such method has been described by

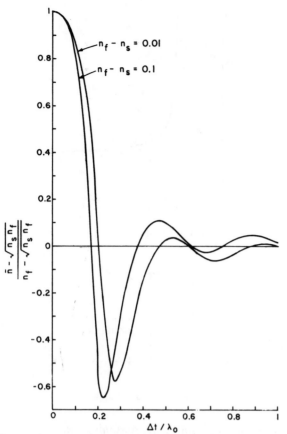

Fig. 17. Measured refractive index versus graded region thickness for $n_s = 1.5$, $d/\lambda_0 = 0.1$.

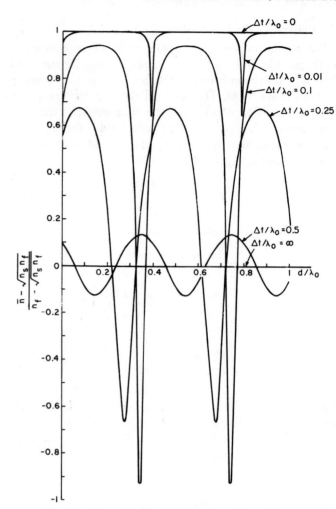

Fig. 18. Measured refractive index versus film thickness for $n_s = 1.5$, $n_f = 1.51$.

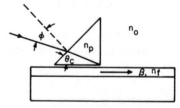

Fig. 19. Prism coupler.

Osterberg and Smith [13]. Recently Tien, Ulrich, and Martin [14] have analyzed evanescent coupling via a higher index region. These latter results have been extended by Midwinter [15] to indicate the effects of strong coupling and finite beam size.

Referring to Fig. 19, a high index prism $(n_p > n_f)$ is clamped to the film. By proper selection of the injection angle in the high index medium, the propagation constant of the incident beam can be made equal to that of the mode in the film. Assuming weak coupling, the incident angle of the input beam is related to the propagation constant of the film by

$$\phi = \sin^{-1}\left\{\frac{n_p}{n_0}\sin\left(\sin^{-1}\left(\frac{\beta}{k_0 n_p}\right) - \theta_c\right)\right\} \quad (31)$$

where n_p is the prism refractive index; β is the mode propagation constant in film [see (1)–(5)]; θ_c is the corner angle of the prism (see Fig. 19); $k_0 = 2\pi/\lambda_0$; and ϕ is the angle of incidence of light with respect to the normal to the input face of the prism.

This relationship permits calculation of several important properties. For example, knowledge of ϕ, n_p, and θ_c permits calculation of β, which in turn allows determination of n_f or d. If several modes can be launched, the correlation of the various β's with n_f and d allows determination of both parameters. Another example is that (31) permits optimization of launching efficiency of modes by selecting θ_c so that ϕ is a Brewster angle.

In our work, 45-degree isosceles prisms fabricated from SF-58 Shott glass having an index of 1.9176 were used. To

date we have not attempted to maximize launching efficiency.

Several methods have been used to measure transmission loss [16]. The simplest approach is to photograph the scattered light from the film. Taking exposures of various durations, it is possible to estimate the film attenuation. A second method is to vary the position of the launching prism and measure either the power reaching the edge of the film or the intensity of light coupled from the film by a second prism. However, our experience has been that the variability of launching efficiency makes accurate determination of film loss difficult (to an accuracy of better than ± 2 dB).

The best accuracy has been achieved by measuring the intensity of light scattered at right angles to the film using a fiber optic probe. If the scattering centers are uniformly dispersed in the film, then a plot of scattered power versus distance from the launcher permits determination of film loss. In our measurements the input beam is mechanically chopped at a 1-kHz rate. A silicon solar cell attached to the end of the fiber probe provides a signal whose strength is measured using a lock-in amplifier.

IV. FILM AND WAVEGUIDE FABRICATION

Considerable effort is being directed toward the practical realization of optical integrated circuits. This work can be divided into two areas. First, a transparent thin-film medium must be found (< 1 dB/cm loss). Second, masking and etching techniques must be found which are suitable for use with the selected medium. From Marcuse's results (Section II-B), it appears that these techniques must be capable of generating an edge whose roughness is less than 0.1 λ_0. However, occasional discontinuities of much greater magnitude are tolerable.

A. Thin-Film Media

A number of techniques for producing thin films suitable for integrated optics application have been proposed. These include ion bombardment of fused quartz, ion exchange in glass, RF and dc sputtering, thermal evaporation, pyrolytic deposition, and epitaxy. The applicability of liquids and polymers is also being considered. We briefly review the results achieved by the methods which have shown the most promise to date.

1) Film Formation by Ion Bombardment: High refractive index regions have been created in fused silica by ion bombardment. Schineller *et al.* [17] achieved a 0.01 increase in refractive index using 1.5 MeV protons at a dosage of 10^{17} protons per cm. They demonstrated mode propagation in a 4-μm-thick irradiated region. The properties of waveguides made from these surface films have not been determined.

2) Ion Exchange: Osterberg and Smith [13] appear to be the first to have investigated the propagation of surface waves in graded index films produced by a diffusion technique in glass. In their experiments Pilkington Float glass, made by floating molten glass on molten tin, was used. The refractive index of the surface layer of the glass produced by this process is higher than the bulk. However, the surface layer was too thick (0.12 mm) for a single mode waveguide.

Other authors have described the use of ion exchange with glass fibers to produce a low index cladding [18], [19]. Efforts to produce suitable guiding regions in a planar geometry are in progress.

3) Sputtered Glass Films: RF sputtering of glass has been shown to be a very promising method of producing films suitable for optical waveguide [16]. Guidance in this type media has been demonstrated with a loss below 1 dB/cm.

The best films that have been prepared by this method used ordinary laboratory slides as substrates and Corning 7059 glass as the source material for the film. The refractive index of these films is 1.62 and that of the substrates 1.515. Other materials have also shown promise. Alumina-doped SiO_2 films have been produced whose quality approached that of the 7059 glass films and, with further work, may well surpass it. This system is particularly attractive because

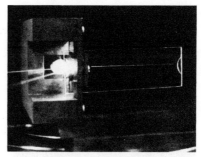

Fig. 20. Light scattered from a beam propagating in a Corning 7059 glass film [16].

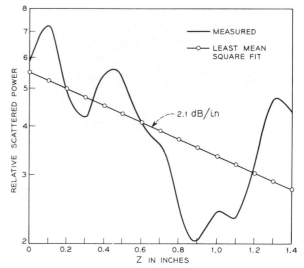

Fig. 21. Relative scattered power versus length (7059 glass film) [16].

adjustment of the doping level allows continuous variation of the film index from 1.458 to about 1.6. However, early tests indicate that the higher index films have higher scattering loss and a lower sputtering rate than the lower index films, so it is the lower end of the index range which will probably be of most interest.

Fig. 20 is a picture of the 0.6328-μm light scattered from a beam propagating in the film. The intensity of scattered light as measured by the fiber optic probe is plotted in Fig. 21. The measured loss is less than 1 dB/cm. This result is in agreement with photographic and two-prism measurements. The lack of uniformity of the scattered light intensity is due, at least in part, to inhomogeneities in the substrate. By using a higher quality substrate this source of scatter can be eliminated.

Curved sections of rectangular waveguides have been constructed from 7059 glass films by back-sputtering using quartz fibers as shadow masks. The waveguides were about 0.3 μm thick, 20 μm wide, and had a radius of curvature of about $\frac{1}{2}$ inch. A photograph of a typical section is shown in Fig. 22. Fig. 23 shows prism-launched light propagating in such a waveguide. Due to the small size of the waveguide our instrumentation will have to be improved before loss measurements can be made.

These results demonstrate the feasibility of using sputtered glass films and sputter etching in the fabrication of optical waveguides. The approach shows promise as a method of producing low-loss optical integrated circuits.

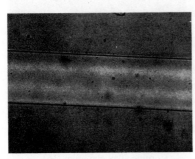

Fig. 22. Section of a rectangular waveguide ($\times$ 1000) [16].

Fig. 23. Light propagating in a curved section of rectangular waveguide [16].

B. Waveguide Fabrication

In Section II-B the edge roughness requirements were shown to be very stringent. Those computations indicate that the guide surfaces must be smooth to about 0.1 λ_0 in order that radiation loss be small. This imposes strict requirements on fabrication procedures. Conventional photolithographic techniques using optically exposed photoresist do not approach this order of edge resolution. Chemical etching of glass films usually results in clouding of the surface due to leaching of constituents, so etching by this method does not appear feasible. Hence, alternate fabrication methods are being pursued.

A promising approach is the use of a scanning electron beam to expose photoresist followed by reverse sputtering of the unwanted film. Other workers have reported exceptionally high-quality metal masks using electron-beam photoresist exposure. Gold grids consisting of 700-Å lines separated by 1400 Å have been produced [20] so that the required resolution appears achievable. Problems with this method of guide fabrication are photoresist-film compatibility, and the requirement that the resist withstand the temperature rise involved in back-sputtering. An additional problem involves the relatively slow writing times for circuits of the size we are considering using conventional scanning electron beam microscopes. By a stepping process involving mechanical movement of the substrate it should be possible to generate masks suitable for circuit fabrication.

V. Conclusions

Integrated optical circuits in the form of planar arrays of rectangular dielectric waveguides appear attractive for optical signal processing circuitry. We have summarized those aspects of the problem necessary for a basic evaluation of the approach. The problem of practical realization of such circuits appears solvable by appropriate adaptation of integrated circuit fabrication techniques. It has been shown that thin films suitable for planar dielectric waveguide applications can be produced by RF sputtering of glass. Other approaches also show promise. The problem of masking by a technique suitable for batch fabrication appears tractable, but further work is required in this area.

Acknowledgment

The authors wish to thank W. R. Sinclair for his valuable comments regarding the sputtering of glass films and the preparation of substrates, R. R. Murray who assisted in the preparation of our films and waveguides, and Miss R. E. Quinn for her aid in programming the index gradient problem.

References

[1] S. E. Miller, "Integrated optics: an introduction," *Bell Syst. Tech. J.*, vol. 48, pp. 2059–2069, September 1969.
[2] R. E. Collin, *Field Theory of Guided Waves*. New York: McGraw-Hill, 1960.
[3] D. F. Nelson and J. McKenna, "Electromagnetic modes of anisotropic dielectric waveguides at p-n junctions," *J. Appl. Phys.*, vol. 38, pp. 4057–4073, September 1957.
[4] D. Marcuse, "Mode conversion caused by surface imperfections of a dielectric slab waveguide," *Bell Syst. Tech. J.*, vol. 48, pp. 3187–3217, December 1969.
[5] W. Schlosser and H. G. Unger, "Partially filled waveguides and surface waveguides of rectangular cross section," in *Advances in Microwaves*. New York: Academic Press, Inc., pp. 319–387, 1966.
[6] E. A. J. Marcatili, "Dielectric waveguide and directional coupler for integrated optics," *Bell Syst. Tech. J.*, vol. 48, pp. 2071–2102, September 1969.
[7] J. E. Goell, "A circular-harmonic computer analysis of rectangular dielectric waveguides," *Bell Syst. Tech. J.*, vol. 48, pp. 2133–2160, September 1969.
[8] E. A. J. Marcatili, "Bends in optical dielectric guides," *Bell Syst. Tech. J.*, vol. 48, pp. 2103–2132, September 1969.
[9] O. S. Heavens, *Optical Properties of Thin Solid Films*. New York: Academic Press, Inc., 1955.
[10] F. A. Abeles, "Optics of thin films," in *Advanced Optical Techniques*. New York: Wiley, 1967, pp. 143–188.
[11] K. Nagota, "Inhomogeneity in refractive index of evaporated Mg F_2 film," *Japan J. Appl. Phys.*, vol. 7, pp. 1181–1185, October 1968.
[12] J. E. Goell and R. D. Standley, to be published.
[13] H. Osterberg and L. W. Smith, "Transmission of optical energy along surfaces: pt. II, inhomogeneous media," *J. Opt. Soc. Am.*, vol. 54, pp. 1078–1084, September 1964.
[14] P. K. Tien, R. Ulrich, and R. J. Martin, "Modes of propagating light waves in thin deposited semiconductor films," *Appl. Phys. Lett.*, vol. 14, pp. 291–294, May 1, 1969.
[15] J. E. Midwinter, "Evanscent field coupling into a thin film waveguide," to be published.
[16] J. E. Goell and R. D. Standley, "Sputtered glass waveguide for integrated optical circuits," *Bell Syst. Tech. J.*, vol. 48, pp. 3445–3448, December 1969.
[17] E. R. Schineller, R. Flam, and D. Wilmot, "Optical waveguides formed by proton irradiation of fused silica," *J. Opt. Soc. Am.*, vol. 58, pp. 1171–1176, September 1968.
[18] A. D. Pearson, W. G. French, and E. G. Rawson, "Propagation of a light focusing glass rod by ion-exchange techniques," *Appl. Phys. Lett.*, vol. 15, p. 76, 1969.
[19] T. Uchida, M. Furukawa, I. Kitano, K. Koizumi, and H. Matsumora, "A light-focusing fiber guide," presented at the 1969 IEEE Conf. on Laser Engineering and Applications, Washington, D. C., paper 7.7.
[20] A. N. Broers, "Combined electron and ion beam processes for microelectronics," *Microelectron. Rel.*, vol. 4, pp. 103–104, 1965.

Glass Fibers for Optical Communications

ROBERT D. MAURER

Invited Paper

Abstract—Glass optical waveguides with attenuations below 20 dB/km have made possible a new approach to optical communications. These glass fibers satisfy requirements for transmission over kilometer lengths with experimental systems utilizing existing devices for sources and detectors. The realization of material and fabrication advances necessary for this accomplishment are the topic of this paper. Basic theoretical principles are introduced in a review fashion. The application of these principles in choice of materials and fabrication is described. Results in fiber performance following this framework are given in a section on evaluation, which includes information capacity, attenuation, and some environmental requirements. Preliminary experiments in bundling and cabling are discussed, followed by concluding remarks.

I. INTRODUCTION

THE ATTAINMENT of glass fibers with attenuations below 20 dB/km has radically changed the outlook for optical communications [1]. Formerly, the available transmission devices were marked as complicated and prospectively expensive [2]. Most of them involved the installation of pipes filled with lensing media of various types which had to be controlled by interacting servosystems. Therefore, the prospect of flexible trouble-free transmission lines greatly enhances the practicality of future optical communication systems. Advances in solid-state detectors and light sources, both coherent and incoherent, offer additional support. With these source–detector combinations, losses of 30 to 50 dB can be tolerated with reasonable signal-to-noise ratios. Therefore,

a loss of 20 dB/km permits transmission over more than one kilometer without amplification. This attenuation factor is the approximate break-over point for widespread application and, therefore, this article will emphasize techniques for achieving such low losses. The incorporation of fibers in communication systems will be covered in a forthcoming invited paper.[1]

Signal transmission with glass waveguides presents several advantages in addition to the frequently mentioned high bandpass. Among these are low susceptibility to electromagnetic interference, small size, low weight, and dielectric isolation. These advantages may well open up specialized applications prior to widespread communication use. However, the specialized advantages are peripheral to the potential of transmission lines that are economically competitive in both high- and low-bandpass systems, thereby making optical communications superior throughout communications technology.

This review covers the progress in techniques for attaining low-loss (below 20 dB/km) glass waveguides. Cylindrical fibers with three types of radial refractive-index variation are discussed: discontinuous (step) with single-mode propagation, discontinuous with multimode propagation, and continuous (gradient) with multimode propagation. Emphasis will be placed upon demonstrations of feasibility rather than theory. The next section on theory covers only the results necessary to interpret experimental results mentioned later. There follows the bulk of the article with principles found necessary for

This invited paper is one of a series planned on topics of general interest—The Editor.

Manuscript received November 20, 1972; revised January 10, 1973.

The author is with Corning Glass Works, Research and Development Laboratories, Corning, N.Y. 14830.

[1] T. Li, E. A. J. Marcatilli, and S. E. Miller, to be published in the *Proc. IEEE.*

Reprinted from *Proc. IEEE,* vol. 61, pp. 452–462, Apr. 1973.

materials and fabrication and the emergence of these in fiber performance. Finally, some remarks are made on preliminary cable experiments and future directions.

II. Theoretical Background

A. Fiber Form

A few theoretical principles and definitions of waveguide phenomena are necessary for understanding the fabrication results discussed below. Two basically different types of cylindrical fiber offer practical advantages. One type is the step-refractive index variation in which the fiber has a core of one refractive index and a cladding of a lower refractive index. The other type is the gradient refractive index variation in which the fiber has a high refractive index at its axis which decreases continually to a lower refractive index at the surface. For theoretical purposes here, the outer boundary of both types can be neglected. That is, these refractive index variations can be considered to proceed to infinity in the radial direction.

Cylindrical waveguides with step-refractive-index variation are characterized by an important parameter [3]

$$V = \frac{2\sqrt{2}\pi r}{\lambda}(\bar{n}\Delta n)^{1/2}$$ (1)

where r is the core radius, λ the free space wavelength, $\bar{n}$ the average refractive index of core and cladding, and Δn the difference in refractive index between core and cladding. Below $V \cong 2.4$, the fibers propagate a single mode, designated HE_{11}. Because of the axial symmetry, this mode is polarization degenerate, although any core ellipticity lifts this degeneracy. The chief advantage of a single-mode waveguide is its high bandpass. Above $V \cong 2.4$, the quantity of propagating modes increases rapidly as V increases, with the approximate number given by [4], [5]

$$N \sim \frac{V^2}{2}.$$ (2)

In addition, when all modes are equally excited, an approximate value for the fractional power carried in the cladding can be given as a function of V [4]

$$\frac{\text{cladding power}}{\text{total power}} \equiv p_{cl} = \frac{8}{3V}$$ (3)

which may be used with the core power fraction to give the total attenuation [6]

$$\beta = p_{cl}\beta_{cl} + p_c\beta_c$$ (4)

where β_{cl} and β_c are the attenuation coefficients of the cladding and core, respectively.

Waveguides with continuous variation of refractive index (gradient guides) are fabricated to have a variation of the form [7], [8]

$$n = n_0 \text{ sech } \rho r$$ (5)

where n_0 is the refractive index on the fiber axis and ρ the radial variation constant related to the focussing distance. All modes corresponding to meridional rays (HE_{1m} modes) have the same group velocity, neglecting material dispersion (see below). From a macroscopic viewpoint, all meridional rays focus at successive equivalent positions on the axis of the

fiber. But skew rays do not focus with the refractive index gradient given by (5) [9]. Therefore, both experimentally and theoretically, deviations from the refractive index variation of (5) are difficult to separate from skew ray effects due to the light source.

B. Materials Properties

A major impediment to the use of glass fibers has been their attenuation. Even today this remains one of the most critical factors for system economics and manufacturing control. The attenuation can be subdivided into absorption, or conversion of light into heat, and scattering, or light escaping the bound modes. Absorption consists of three types: intrinsic, impurity, and atomic "defect" color centers. Scattering also consists of three types: intrinsic, glass inhomogeneity, and aberrations in the radial (cross-sectional) form of the refractive index. All six of these will be discussed in turn.

Intrinsic absorption, by definition, occurs when the material exists in a "perfect" state. Normally, perfect dielectric materials, like glass, are considered perfectly transparent. This is true for most applications but must be examined more closely for optical waveguides where absorption coefficients three orders of magnitude lower are needed and achieved. A knowledge of this absorption is desired not only to assess its contribution to the total absorption measured in fibers but also to assess the fundamental lower limit for any particular material. Glasses transparent in the visible have strong optical absorption bands in the ultraviolet and infrared which extend to some small degree into the visible. It is the "tail" of the ultraviolet that is considered most significant because the infrared bands, which are located beyond 4 μm, are very narrow. As a consequence, their contribution in the 600–1000-nm range is neglected. The ultraviolet absorption is due to atomic transitions involving oxygen and shifts somewhat with wavelength as the glass composition changes [10]. Little is presently known about the shape of these bands and hence their contribution to the loss. This is because fiber absorptions are so much smaller than any bulk ultraviolet absorption studied heretofore that extrapolations from the studied range are meaningless. The only recourse has been to take the smallest absorptive attenuation yet observed and conclude that the intrinsic absorption must be smaller.

Impurity absorption arises predominantly from transition-metal ions such as iron, cobalt, and chromium [11]. The absorption of these ions varies from glass to glass as does their valence state. The absorption peaks are very broad [12] so that it is difficult to identify the offending species from the spectral dependence of fiber absorption. Using iron as an example, both ferrous and ferric ions absorb in the visible range, but the ferrous ion is more troublesome with a strong absorption peak near 1000 nm. Fig. 1 shows the absorption of iron in fused silica. By doping the glass with known amounts of impurity, absorption values can be extrapolated to low concentrations, assuming the absorption is linear in concentration. When the glass has the same preparation history as the fiber, the valence state ambiguity is thus bypassed. Usually this approach shows that concentrations of impurity below a few parts per billion (ppb) are necessary if absorption below 20 dB/km is to be attained.

Another important impurity is "water" which is present as OH^- ions. This contributes sharp, easily identified, absorption bands around 950 and 725 nm [13]. These are, respectively, the third and fourth harmonic of the fundamental

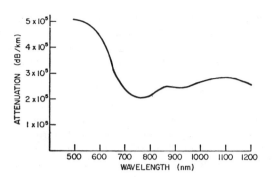

Fig. 1. Absorption spectrum of iron in fused silica for a concentration of 700 ppm atomic. The absorption in the vicinity of 1100 nm is thought due to Fe^{2+} and that at short wavelengths to Fe^{3+}.

vibrational band at about 2.8 μm. By measuring the water removed in a vacuum bakeout concurrently with the strength of the 2.8-μm band, absolute calibration can be obtained. The ratio of the 2.8-μm band to the harmonic bands can then be used for their calibration, even though these other bands are too weak to be measured except in very long paths, such as fibers. In fused silica, the peak of the 950-nm band causes 1.25 dB/km/ppm by weight [14]. The absorption per ion for water does not vary as much from glass to glass as do the transition metal ion absorptions.

The third source of absorption is atomic "defects" in glass structure including, by definition here, unwanted species of elements deliberately added to the glass composition. Oxygen defects in silica glasses have been studied by radiation [15]. These may be related to similar color centers introduced in silicate glasses by the oxidizing conditions of the melt. An example of a deliberately added element yielding an unwanted species might be titanium whose plus-three valance state has a strong absorption in the visible while titanium plus-four does not. These examples should also make it clear that oxidation states can be extremely important in glass fiber attenuation.

Some scattering attenuation arises from the intrinsic nature of glass. All transparent matter scatters light due to thermal fluctuations which, in turn, generate fluctuations in refractive index [16]. Glass differs in that these fluctuations are frozen-in when the material is cooled through the annealing range of temperature [17]. The attenuation coefficient (base e) for scattering alone is called the turbidity and is given by [16]

$$\tau \cong \frac{8\pi^3}{3\lambda^4}(n^2 - 1)^2 kT\beta \qquad (6)$$

for pure liquids. This scattering is due to compressive fluctuations, the only significant type of fluctuation in a pure liquid. Here, k is Boltzmann's constant, T the absolute temperature, and β the compressibility. Because of the frozen-liquid state of glass, high-temperature quantities are used in (6). For pure fused silica the temperature dependence of (6) has been verified and a value for the compressibility derived when the annealing temperature is used for T [18]. If other oxides are added to silica, the compressive-fluctuation scattering changes chiefly through the factors T (annealing temperature) and β. A more important change is a second type of scattering caused by fluctuations in the concentration of these

oxides [16]. This additional contribution is given by

$$\tau \cong \frac{16\pi^3 n}{3\lambda^4}\left(\frac{\partial n}{\partial c}\right)^2 \overline{\Delta c^2}\,\delta V \qquad (7)$$

where $\overline{\Delta c^2}$ is the mean-squared concentration fluctuation and δV the volume over which it occurs. Generally, if the added oxide tends to raise the bulk refractive index, fluctuations in its concentration will cause greater scattering because these fluctuations represent greater refractive index fluctuations [15]. For most high-refractive-index glasses, concentration fluctuations (7) provide by far the dominant contribution to the total scattering.

Other kinds of glass inhomogeneity can cause scattering. Some compositions tend to segregate into immiscible liquids or to precipitate crystals. Or, improper mixing during melting will result in scattering refractive-index variations. However, these other kinds of glass inhomogeneity are not intrinsic to the material and can be judiciously avoided, in contrast to the thermal fluctuations discussed in the previous paragraph.

The third source of scattering is aberration in the radial form of the refractive index. This causes conversion of light into unbound modes. For continuous refractive index variation waveguides, this scattering imperfection cannot be distinguished from the other inhomogeneities in the material previously discussed. For the discontinuous refractive index variation waveguides, this scattering imperfection is irregularity of form (roughness) in the glass boundary between core and cladding. Such irregularity could cause almost any angular and wavelength dependence of the scattering depending on the form, dimensions, etc., of the irregularity.

Glass dispersion is important through its effect in limiting the bandpass. For optical transitions in the ultraviolet approaching the visible range of wavelengths, the refractive index rises as well as the dispersion. Thus it is generally true that changing a glass composition to increase the refractive index will also increase the dispersion. The detailed results of material dispersion influence on bandpass are beyond the scope of this paper, but to a good approximation material dispersion adds linearly to the other dispersive effects [5], [19]. Often material dispersion is the major factor limiting bandpass, so the general principles influencing bandpass and their magnitude are relevant here. Dispersion causes pulses to broaden as they propagate down the waveguide, and the resultant overlap between adjacent pulses limits pulse rate and, hence, information flow. The broadening is proportional to the waveguide length. For single-mode guides the input pulse frequency breadth broadens the output to a width b,

$$b = \left[a^2 + \left(\frac{L}{a}\frac{\partial^2 h}{\partial \omega^2}\right)^2\right]^{1/2} \qquad (8)$$

where a is the Gaussian input pulsewidth, L the guide length, ω the frequency, and h the guide wave vector at center frequency [19]. Material dispersion enters through the derivative factor and is the predominant bandpass limitation for wavelengths less than about 1 μm. Table I gives some results for the material dispersion alone for single-mode waveguides with bandwidth-limited input pulses.

Equation (8) and Table I are also relevant for each mode of a multimode waveguide. However, an important practical difference is that multimode waveguides are to be used with

TABLE I

ILLUSTRATION OF THE EFFECTS OF MATERIAL DISPERSION ON BANDPASS

Glass	Single Mode—Pulse Spectral (Information) Bandwidth Limiting (GHz·km^{-1})	Multimode—Incoherent Source (Carrier) Bandwidth Alone Limiting (MHz·km^{-1})	Multimode Group Velocity Differences Alone Limiting (MHz·km^{-1})
Corning Code 7940 (fused silica)	40	418	39
Schott F-2 (lead silicate)	25	158	43
Typical soda–lime	39	408	40

Note: For the second column, the source was assumed to emit at 800 nm and to be 20 nm broad. The third column was added to compare with the second; numerical aperture was assumed to be 0.15 and the core diameter 100 μm ($V \cong 59$).

incoherent sources of broad spectral width far exceeding the information rate spectral width. Dispersion over this source emission width can influence the bandpass and is the limiting factor in some small-V waveguides, as discussed later.

Pulse broadening in multimode waveguides with step refractive index variation is mainly caused by the behavior of the different modes. When these modes are very weakly coupled, the pulsewidth is due to the different group velocities of each mode. A pulse input into all modes emerges at different times at the receiver [20]. While the mode propagation can be handled in detail [5], estimates are made using ray optics. The transit time difference between the straight through (axial) ray and the ray at the maximum (critical) angle to the axis is taken as the minimum pulsewidth. This yields

$$N_{max} \cong c/L\, \Delta n. \qquad (9)$$

Table I also includes values from (9) for comparison.

The remarks about multimode bandpass leading to (9) are not applicable when the coupling between modes is strong over the distance considered. The bandpass then becomes much higher than the estimate of (9) and material dispersion relatively more important. The most important realization for this discussion is that fabrication affects the coupling, since it is the imperfections beyond the perfect geometry which scatter light from one mode to another and cause the coupling. This is a topic of current interest, but with little information so far experimentally connected to fabrication [59].

Since, as shown, single-mode waveguide bandpass is limited primarily by material dispersion and multimode by the spread in mode group velocities, there is a point of transition. Taking the same source spectral characteristics as used in Table I, this transition can be calculated from (9) and (1). Thus a waveguide with the material dispersion of fused silica and a core of 50-μm radius will have its bandpass limited by the material below $V \cong 20$ (200 modes) and by group velocity dispersion above. As modal group velocity dispersion is decreased by mode coupling or refractive index gradients, the V for transition will increase.

Multimode waveguides with continuous refractive index variation, as previously discussed, can have less spread in group velocities for different modes. Assuming the radial refractive index variation is perfectly constructed, material dispersion can be the only limitation on bandpass. Table I again illustrates the situation. To achieve the highest poten-

tial bandpass of such waveguides, a collimated spectrally narrow source must be used with the values of column 2 increased by the proportional decrease in the spectral width over that assumed. Column 1 will then provide an upper limit.

There are many nonlinear optical effects in glass fibers which might be of device use, including Kerr effects, stimulated Brillouin emission, stimulated Raman emission, etc. These devices will not be considered here, but the limitation on power levels imposed by these effects is important. Both stimulated Brillouin emission and stimulated Raman emission dissipate power rapidly and attenuate the signal [21]. The acoustic and optical modes generated in the material absorb this lost energy. The power densities for the onset of these effects are largely independent of the glass compositions [22]. For signals (pump) of bandwidth small compared to the Raman linewidth but comparable to the Brillouin linewidth, the single-mode power is limited to a maximum of about 3×10^7 W/cm^2, and for broad spectral sources comparable in width to the Raman line, the single-mode power is limited to about 2×10^9 W/cm^2. For these numbers, a loss at the Stokes line of 4 dB/km was assumed. Therefore, some tradeoff of higher input power level can be made against the information rate limitation due to increased dispersion effects over the broader spectrum.

III. GLASS MATERIALS SYSTEMS

Suitable glass compositions should be capable of manufacture in a homogeneous state, including absence of any traces of phase separation. Since tremendous quantities are potentially needed throughout the world's communication system, no elements should be used which are not abundant in the earth's crust. All three composition systems which have been intensively investigated, the high-silica glasses, soda-lime glasses, and lead silicates, satisfy these requirements. These composition choices will be discussed in light of the previously mentioned theoretical principles.

The important wavelength region in the intrinsic absorption is the "tail," or very low absorption, part of the ultraviolet band about which little is known. Fused silica has the lowest wavelength ultraviolet absorption edge among conventional glasses. Therefore, assuming all glasses have the same shape absorption edge, the intrinsic absorption will be lowest in fused silica and generally higher the higher the refractive index. Since absorptions as low as 3 dB/km at 1.06 μm have been observed in fused silica, the intrinsic absorption is less than this [23]. Fused silica itself is unsuitable for a fiber core because there are no easily prepared glasses of lower refractive index to clad it. However, it has been recognized that doped silica could serve as a core and silica as a cladding [24]. Fibers of this type have shown absorption attenuation as low as 1 or 2 dB/km at 633 nm. In these glasses, at least, intrinsic absorption seems to be negligible. In other glasses, observed absorptions have been over 30 dB/km at 633 nm which probably means that the intrinsic absorption has not even been approached.

The choice among various compositions with respect to transition-metal ion absorption rests on the possibility of obtaining pure starting materials. There is no known way to purify glasses after melting, analogous to zone-refining silicon. The transition-metal ion content of the glass will thus com-

TABLE II

TABLE II
ATTENUATIONS DUE TO REPRESENTATIVE IMPURITIES IN SODA–LIME GLASSES

Ion	Absorption Peak (nm)	Concentration for 20 dB/km	
		Absorption at the peak (ppba)	Absorption at 800 nm (ppba)
Cu^{2+}	800	9	9
Fe^{2+}	1100	8	15
Ni^{2+}	650	4	26
V^{3+}	475	18	36
Cr^{3+}	675	8	83
Mn^{3+}	500	18	1800

TABLE III
SCATTERING LOSSES IN dB/km FOR SEVERAL REPRESENTATIVE GLASSES AS MEASURED AT 546 nm AND EXTRAPOLATED TO OTHER WAVELENGTHS

	633 nm (dB/km)	800 nm (dB/km)	1060 nm (dB/km)
Corning Code 7940 (fused silica)	4.8	1.9	0.6
Corning Code 8361 (soda–lime type)	8.5	3.3	1.1
Bausch and Lomb 517–645 (borosilicate crown)	7.7	3.0	1.0
Schott F-2 (lead silicate)	47.5	18.6	6.0

Note: These measurements on bulk glass represent only the intrinsic material loss to which must be added any other scattering losses.

prise whatever was in the starting materials plus whatever is acquired in the steps leading to fiber fabrication. These impurities have been most extensively studied in soda–lime glasses [25]. Table II shows the results for several ions taken from this literature. The numbers vary somewhat from glass to glass for a particular element because, among other things, the oxidation state is apt to be different. Changes for a given oxidation state of an element are not great and Table II is a good guide, particularly for alkali-containing glasses. It is apparent that minimizing the number of oxide components in a glass minimizes the chance for impurity pickup. Silicate glasses are preferred for most commercial applications because of their properties, and silica itself, as a constituent, can be obtained in very high purity. This is an indirect result of the work performed for semiconductor improvement, since many of the compounds used to obtain pure silicon can be oxidized to obtain silica. Synthetic quartz also can be made in high purity [26].

Glass atomic "defect" absorption, as previously defined, will depend markedly on the oxidation state of the glass. A discussion of this topic is much too complicated to be included here. Suffice it to say that high temperatures tend to cause reduction because oxygen is evolved as a gas to maximize the entropy. The free energy $U - TS$ always tends toward a minimum. Thus increasing the entropy S offsets the internal energy U. This increase becomes more important the higher the temperature because T appears as a multiplying factor. One particular case is of interest, since it has been utilized to produce low-attenuation waveguides. This is the oxidation of titanium by water in the glass [27]. Reduced titanium has a strong absorption which appears in newly made glass. However, a temperature treatment below the melting temperature apparently reverses the oxidation equilibrium with water in the glass, leading to oxidized titanium and evolved hydrogen gas. Reduced silicon also has been found in glass [28]. Perhaps one of the most fruitful areas for further research is thermal (fiber-forming) generation of light-absorbing atomic defects. Certainly, lower temperature melting glasses will reduce the severity of this problem if it exists. While both soda–lime and lead–silicate glasses are low melting, soda–lime is to be preferred in this regard, since lead is so easily reducible.

Closely allied to absorption caused by oxidation is absorption caused by radiation, since species susceptible to oxidation reduction are also apt to act as trapping sites. Precise data on radiation susceptibility have not been obtained, since observed effects often depend on impurities and data of different workers do not agree. Generally, fused silica seems the most resistant to radiation coloring. Addition of other oxides, especially alkalis or alkaline earths, tends to increase coloration. Thus soda–lime or lead–silicate glasses are much more sensitive. The addition of cerium tends to "protect" against this visible coloration by acting as a competitive trapping center that does not color in the visible range [29], [30]. The magnitude of the coloration effects that have been studied in bulk glasses is much more severe than anything tolerable in an optical waveguide. A new area of investigation will have to be opened here.

The level of intrinsic scattering in glasses is well established, as indicated by Section II. Equations (6) and (7) indicate that the scattering should vary as λ^{-4}. This has been closely verified experimentally [31]. Data on intrinsic scattering are usually measured at one wavelength and extrapolated to others with this law. Table III gives some representative glasses with their attenuation at various wavelengths extrapolated from measurements at 546 nm [17]. Fused silica, soda–lime, and alkali silicates have the lowest scattering levels. Fused silica, of which Corning Code 7940 is an example, usually does not vary more than about ±10 percent for any method of manufacture. Variations outside this precision can usually be traced to bubbles or some similar imperfection. This makes fused silica a fairly good secondary standard for checking scattering instrumentation. If any low-refractive-index glass is used for a fiber, the bulk scattering should be below ~2 dB/km at 900 nm. High-refractive-index glasses, and particularly lead-containing glasses above 1.6, do not appear suitable for low-attenuation waveguides. Glasses similar to Schott F-2 are used for the core of most commercial flexible fiber optics.

The other sources of scattering, inhomogeneity, and radial refractive index variation, offer no clear-cut choices among glasses. Inhomogeneity due to liquid–liquid phase separation is likely to arise in high silica glasses of the alkaline–earth–silica type, or, to a somewhat lesser extent, the alkali–silica type. For lower silica concentrations, borosilicates are a classic example. Such phase separation is difficult to suppress by rapid cooling because it nucleates so easily [32]. A good procedure to avoid immiscibility, which is followed in soda–lime glasses, is to adjust the composition with other oxide additives.

Dispersion choices among low-refractive-index glasses are not significant. Either high-silica or soda–lime glasses are among the best (Table I). As the index is raised, performance sacrifice occurs in some cases.

Diffusion is significant in controlling the radial refractive index profile. Fibers of continuous variation (5) are made by diffusion exchange between ions. A fiber with a discontinuous refractive index change will have the discontinuity smoothed by diffusion when the fiber is drawn. In the former case diffusion should be maximized for manufacturing speed, while in the latter it should be limited. In glass melts, mass transport is governed by convective flow as well as diffusion. Most of the processes of fiber formation occur at high viscosities so that diffusion is dominant. In this range, the mobility of ions is generally inversely proportional to their valence [33], [34]. The diffusion of trivalent and quadrivalent ions is so small that little or no data are available. Unfortunately, this does not mean that it is so small that it is unimportant here, inasmuch as very small distances are involved in single-mode waveguides. However, the valence rule alone is enough to draw some conclusions. Monovalent ions, such as alkali, thallium, and cuprous, are to be avoided if possible in discontinuous-refractive-index waveguides. Conversely, continuous variations made by diffusion should utilize these. Of the monovalent ions, thallium has the largest specific effect on the refractive index. This favors its use, since high-refractive-index gradients increase the effective numerical aperture of the waveguide.

Glass strength will be of continual interest, but it is only a secondary factor in choosing glass-composition systems. The basic primary factors, prehistory (flaw production), and water vapor, are largely independent of glass composition. Since glass is a brittle substance, any crack-like flaw will produce a high stress concentration at its tip and propagate through the specimen. Very little physical contact is necessary to produce such flaws and greatly reduce strength. The fiber surface should be coated to prevent such abuse. Water vapor acts on the tip of such flaws and diminishes the energy required to generate new surface, thereby reducing the strength still further. Fused silica has yielded the highest strengths ever obtained on bulk specimens, even though other glasses show comparable values [35].

IV. Fabrication

For discontinuous refractive index, multimode waveguide fabrication is simpler than single mode, and many of the techniques have been employed for years in the production of commercial flexible fiber optics. The two techniques utilized are referred to as the "rod in tube" and the "double crucible," and descriptions exist in the literature [36]. Fig. 2(a) illustrates the rod-in-tube method which relies on necking down the structure at high temperatures while conserving the cross-section geometry. Conservation of matter shows that

$$\left(\frac{r_b}{r_f}\right)^2 = \frac{L_b}{L_f} \tag{10}$$

where b and f refer to the rod-in-tube blank and the fiber, respectively, r is the radius, and L the length. For the cylindrical geometry, surface tension forces aid rather than distort formation of the desired shape. For more complicated geometries, such as a square cross section, the drawing through the hot zone (low viscosity) must be fast enough to prevent surface tension from greatly distorting the shape. The rounding

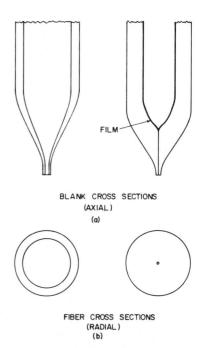

Fig. 2. Fabrication of multimode and single-mode optical waveguides. (a) Typical multimode fabrication blank sometimes called the "rod in tube." (b) A film blank technique used for making single-mode optical waveguides. This film permits simple fabrications of a very small core.

of a corner to the radius of curvature r_c is given by the crude rule of thumb

$$r_c = \frac{\Gamma t}{2\eta} \tag{11}$$

where Γ is the surface tension (about 3×10^{-9} J/m²), t the time at viscosity, and η the viscosity. The time may be altered to some extent with the traction force. Strictly speaking, it is not possible to draw down a blank with sharp corners without some rounding.

The fabrication of single-mode waveguides is much more difficult because the core is so small. Some numerical calculations can illustrate the difficulty. Examination of (1) shows that Δn must be decreased if r is to be increased, since V must be kept constant at about 2. However, very small Δn is also difficult to control—a value of about 5×10^{-3} seems a reasonable limit. (Control of absolute refractive index in conventional optical glass manufacture is about 1×10^{-3}, but such precision is not yet practical in waveguide technology.) Assuming this lowest practical value for Δn, $\bar{n} \sim 1.5$, an operating wavelength of 800 nm and $V = 2.0$, leads to $r \cong 2$ µm—a very small dimension to fabricate. Satisfactory fiber handling presently requires a radius of more than 50 µm which leads to a cladding: core ratio of 25:1. Equation (10) shows this large ratio also holds for the blank, so that a rod-in-tube blank with the rod 2 mm in diameter requires a 5-cm diameter tube. This would be extremely awkward to fabricate. The difficulty can be circumvented by several successive drawings, each followed by addition of another tube to provide cladding. A more elegant technique is shown in Fig. 2(b) [37]. A film is used, in effect, to obtain a small dimension for the core part of the blank; the hole in the center vanishes on drawing if the viscosity is low enough. Appropriate blank dimensions can be calculated, following the same principle used for calculating (10).

There is another geometrical consideration in making single-mode waveguides as previously discussed. This is the cladding thickness necessary to confine the power; if the wave reaches the cladding surface, losses occur. The radial power distribution in the waveguide depends on V, and a good rule of thumb is that the cladding radius should be about ten times the core radius ($V \sim 2$). This still leaves considerable reduction of the refractive index over that previously assumed without encountering problems of large stiff fibers. There have been no experimental demonstrations of finite cladding loss.

The double-crucible technique employs two concentric crucibles containing the core and cladding glass, respectively. Similarly, there are concentric orifices in the bottoms of the crucibles. This is a low-viscosity process with the fiber being formed directly from the melt. Relative dimensions of core and cladding are achieved by adjusting the viscosity of the glass and the orifice shape. This technique has sufficient versatility to permit the drawing of both single-mode and multi-mode fibers [38].

Different techniques are used for making continuous-index-variation fibers. One is to draw several concentric tubes of glass with step changes of refractive index between them, relying on diffusion to provide a smooth refractive index profile [39]. A more practical method is ion exchange [40], [41]. A glass fiber or rod is made from homogeneous glass and immersed in a bath of molten salts containing monovalent alkali ions. These ions exchange with ions of higher polarizability (like thallium) within the glass. The process is diffusion controlled. By stopping this process before complete exchange, a radial refractive index profile very similar to (5) is generated by diffusion laws.

These different fabrication methods offer different advantages and problems. One particular area of interest is the quality of interface between core and cladding. In conventional fiber manufacture this becomes contaminated and generates scattering loss [42] as well as, perhaps, an absorptive loss. The double-crucible method insures completely clean interfaces. If the core and cladding glasses are incompatible so that phase separation occurs on mixing at the interface, significant losses could still occur, but this can be handled by composition adjustment. The best answer to this problem is the continuous-refractive-index fiber made by ion exchange which has no interface at all.

A second area of interest is volatility of some glass component which generates inhomogeneity. The double-crucible method is most susceptible to this because of the lower viscosity. Consequently, some homogenization technique must be used within the crucible or a way found to reduce volatility.

Other important points involve the chances for contamination in the process. Whenever crucibles are used to melt glass, extreme care is necessary, because molten glass is highly corrosive and will dissolve any known material to some degree. Thus the crucible itself must be of high purity and made from elements that do not cause optical absorption. The incorporation of platinum from platinum crucibles is a frequent occurrence which causes both absorptive and scattering effects [43]. With ion-exchange processes, similar purity considerations apply to the molten salt.

High-silica glasses can be made without a crucible, using flame hydrolysis techniques. In this approach, a mixture of silicon and other compounds, depending on the oxide additive desired, are burned in a flame. The resultant small glass particles drop to the bottom of a furnace where they build up in a glass slab. This process has the advantage of maintaining high purity through a minimum of hot-glass handling.

Slow molten-salt diffusion processes are an important consideration for manufacturing speed. The approximate exchange time is given by

$$\tau \sim r^2/D \qquad (12)$$

where r is the blank radius and D the diffusion constant. In one example, fabrication of a rod blank for subsequent fiber drawing required 432 h [44]. One reason for the lengthy diffusion time in the example above was that r in (12) was artificially large. The diffusion profile best approximates (5) near the fiber axis so that only the center portion was used. As more complicated and accurate methods of heating and diffusing are devised for manufacture of the necessary gradient, ion-exchange times can be expected to drop much below the value above.

Raising the temperature to increase the diffusion constant in (12) is limited by glass deformation and by the temperature at which the molten salt begins to attack and dissolve the glass [45]. Dissolution is due to increase of alkalinity of the salt bath from decomposition, thus molten KNO_3 decomposes above about 500°C with an increase in the K_2O concentration. If the ion exchange is much faster than the dissolution, or etching, the net result can be beneficial in that a fresh clean surface is generated without hampering the primary process. The diffusion rate can also be increased by eliminating divalent ions from the glass, such as the lead which was part of the composition [45]. On the other hand, thallium volatility during melting is high and a low-temperature melting glass, like a lead silicate, is beneficial from this standpoint.

V. EVALUATION OF FIBER CHARACTERISTICS

A. Optical

The information set forth in the preceding sections can be used to evaluate fiber characteristics in relation to system performance. This section will illustrate the evolution of materials and fabrication principles alone; many other fiber characteristics, such as backscattering, bandpass, etc., which are important for system performance, are beyond the scope of the present article.

Single-mode fibers represent the most critical test of form fabrication and control. Two points are especially important —dimensional control throughout the fabrication and diffusion during fiber drawing. Fig. 3 shows titania concentration from an electron-beam microprobe scan across the core of a single-mode high-silica glass waveguide in which only the core contained titania. The resolution of the microprobe is about 2 μm so the rounding may be due to instrumental effects rather than diffusion. Additional evidence against significant diffusion is the agreement between this observed core size and that calculated from the film thickness of the blank [Fig. 2(b)]. To a first approximation, diffusion does not change the V value of a waveguide even though the core radius changes. The decrease in refractive index compensates for the increase in core radius. The refractive index change is usually linear in concentration for the important element used to increase it (titanium here). Thus $\Delta n = Kc = (KN)/(\pi r^2)$, where N is the number of ions per unit length of fiber core. From (1),

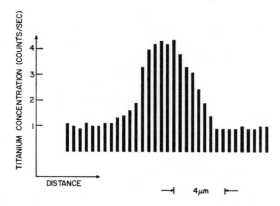

Fig. 3.   Experimental measurement of core size in a single-mode optical waveguide. Electron-beam microprobe analysis of titanium concentration perpendicular to the fiber axis with a resolution of about 2 μm. The counts far from the core center are background, since the cladding is pure silica.

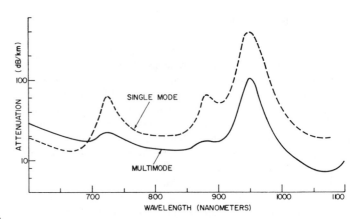

Fig. 4.   Attenuation of typical low-loss optical waveguides through the spectral region of interest for solid-state light sources. The cores are titania-doped silica and the cladding pure silica.

TABLE IV

CALCULATED MODE CUTOFF WAVELENGTH (WHERE $V = 2.4$) COMPARED WITH OBSERVED

Waveguide	Observed (nm)	Calculated (nm)
A	480	460
B	520	530

Note: These high-silica glass fibers were fabricated by the method of Fig. 2(b).

$V_1/V_2 \cong 1$. More exact numerical calculations show that the propagation vector does not greatly change for profiles as rounded as Fig. 2 when these profiles are due to diffusion [46]. On the other hand, deliberate attempts to change the core by diffusion have succeeded [47]. Table IV shows V values calculated from measurements of core size and bulk core refractive index compared with V values measured by mode cutoff patterns [48]. These experimental waveguides of Table IV thus show sufficient control for research purposes.

The single-mode designation of cylindrical waveguides is somewhat of a misnomer, since the symmetry assures two equivalent orthogonal polarizations and the mode is doubly degenerate. Any asymmetry, such as a radial stress or ellipticity in the core cross section, will destroy this degeneracy. The result is two modes of slightly different propagation characteristics. An idea of this asymmetry can be obtained from studying the relation between the state of polarization of entrance and exit beams [49]. Generally, a linearly polarized input beam will emerge circularly polarized for propagation distances greater than about a meter. On the other hand, input polarization orientations can be found which yield linearly polarized outputs. The results are thus consistent with slight asymmetry, but the origin is not yet known.

Evaluation of fabrication precision for continuous refractive index variations is extremely difficult. Equation (5) is usually handled through the expansion

$$n/n_0 = 1 - \left(\frac{\rho^2}{2}\right)r^2 + \left(\frac{5\rho^4}{24}\right)r^4 - \cdots .$$

Often only the first term is used and a "parabolic refractive index" discussed. Precision requires comparison with all the higher order terms and progress has been made utilizing the fourth order [50].

The most extensive evaluation has been on attenuation, because this was the main impediment to application. Measurement is straightforward with single-mode waveguides, because they are insensitive to the method of input as long as the input stays constant during the measurements. Unwanted light other than the single mode is easily stripped out with index fluid baths around the fiber. Multimode waveguides are entirely different. The attenuation depends markedly on the input conditions, because different modes with different attenuations are excited. Presumably, after some distance down the waveguide, coupling between the modes will lead to a uniform power distribution. However, some fibers have shown traces of the input conditions after propagation over 600 m. At least it is possible to say that the materials and fiber fabrication are consistent with the lowest attenuation observed for the modal distribution that was utilized. Physical intuition suggests that high-order modes will be more highly attenuated if the geometrical ray optic analysis is used. Rays at large angles to the fiber axis (higher order modes) travel farther and hence have higher absorption loss. However, this effect is small for small V. These large angle rays also probably suffer larger scattering loss because most scattering distributions peak in the forward direction, thus less of the scattered light is recaptured by the waveguide. So far, these expectations have been borne out [51].

Fig. 4 shows data for attenuation of two high-silica glass waveguides with titania-doped cores. One is a single mode and one a multimode in order to show that the materials principles apply equally to both. The most notable feature is the presence of sharp absorption peaks due to water, described in Section II [52]. In addition to the 950- and 725-nm bands there is another smaller band at 875 nm which is attributed to a combination of a Si–O vibration with the 950-nm harmonic. The multimode fiber was fabricated with dryer atmospheres and shows that all these bands decrease proportionately. The OH concentrations are about 320 and 80 ppm atomic, respectively.

A breakdown of the attenuation at 633 nm in the single-mode waveguide illustrates the relative importance of the factors discussed in Section II-B. A total loss measurement gives 16 dB/km and a total scattering measurement 8 dB/km. The difference of 8 dB/km is attributed to absorption. No evidence has been found for water absorption at this wavelength; the tails of the bands previously identified do not seem to contribute. This leaves intrinsic, impurity, and defect

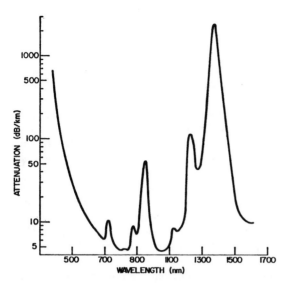

Fig. 5. Total attenuation of a very low-loss optical waveguide through a broad spectral region showing the influence of water impurity. All of the peaks are connected with light absorption due to OH⁻ present at about 100 ppm atomic.

absorption. Intrinsic absorption is probably below 1 or 2 dB/km, because this is the lowest absorption ever observed in this region, as will be mentioned later. This leaves about 6–7 dB/km due to transition-ion impurities and plus-three titanium. Plus-three titanium probably accounts for a large fraction, since it is a strong absorber (10 dB/km/ppb atomic) and would be difficult to completely oxidize by the method mentioned in Section III.

The scattering attenuation contains about 5-dB/km attenuation which can be attributed to bulk glass scattering, since measurements of doped silica indicate a slight increase of the scattering over the value for pure silica in Table III. The single-mode fiber thus contains about 3-dB/km scattering loss above the material contribution. This extra scattering has been further investigated and found to be predominantly in the forward direction [53]. It may be attributed to radial variation of the core refractive index along the guide and to glass inhomogeneity, with the relative magnitudes of each unknown.

The multimode waveguide of Fig. 4 has measurably different characteristics aside from the lower OH content already mentioned. This is notably a higher scattering attenuation of 11–12 dB/km—some 4 dB/km higher than the single-mode waveguide of the same glass composition. This leads to a higher total attenuation of about the same amount in the 633-nm region. This relative relation between single-mode and multimode scattering losses seems to be general and is perhaps connected with the fabrication. The approximate equality of the absorption between the two types suggests that it is attributable to the basic material, such as titanium plus-three.

Fig. 5 more clearly shows the importance of water in high-silica waveguides with very low loss. The series of water peaks shows decreasing absorption toward shorter wavelengths with contribution throughout the wavelength region of interest for solid-state light sources (750 to 1060 nm). This waveguide had an absorption attenuation of 1–2 dB/km at 633 nm, showing that transition-metal impurities were of small consequence.

The largest part of the attenuation at this wavelength, or 9 dB/km, was scattering. The scattering drops rapidly as the wavelength increases and the total loss reaches values as low as 4 dB/km. However, as the scattering loss drops, the curve minima show an increase from other absorption so that the overall shape is parabolic. This strongly suggests that water is the primary source of absorption attenuation, raising the loss beyond about 600 nm.

Low attenuation requires that the cladding glass also be of relatively low attenuation. From (4) it is clear that $\beta_{cl} < (\beta/p_{cl})$, so that the 4-dB/km waveguide ($V \sim 55$) from (3) had a cladding glass loss of less than 80 dB/km, assuming all modes equally excited.

Lead–silicate glass has been reduced in bulk attenuation to 50 ± 30 dB/km at 850 nm [54]. This glass has a scattering loss about 70-percent of Schott F-2 (see Table III) and hence probably is of lower refractive index. Its scattering loss at 850 nm is thus about 12 dB/km with the remainder absorption. Unfortunately, fabrication into a fiber increased the attenuation to about 400 dB/km, due to unknown causes.

Lead–silicate glass has also been used to fabricate gradient index fibers of low attenuation [55]. The glass used was similar to the Schott glasses above with the exception of thallium addition. Judging from the composition, scattering losses should be comparable or slightly higher [17]. Values of attenuation around 600 and 1000 nm were of the order of 100 dB/km. Thus at 600 nm, the absorptive loss has been reduced to the vicinity of 80 dB/km. In gradient index fibers, some types of aberration can generate loss because light reaches the fiber surface and is scattered or absorbed. Therefore, these quoted total losses are an upper limit to the material loss, which might be much lower.

A complete treatment of bandpass evaluation cannot be undertaken here. However, spectral variations in delay time illustrate the properties determining the bandpass [56]. These show that the glass dispersion is the predominant factor in single-mode fibers (Section I).

B. Environmental

Effects of environment on glass waveguide physical integrity and optical performance have not been extensively investigated because of the early stage of development.

Data on thermal changes of attenuation for commercial fibers (~1000 dB/km) have shown little or no transmission change over the range −196°C to +200°C [57]. However, the large room-temperature attenuation indicates that impurity effects were being observed. It thus seems that the impurity contribution to low-loss fibers will not introduce a temperature dependence. Titania-doped silica fibers with a room-temperature attenuation of about 40 ± 1 dB/km showed no change in attenuation when heated to 200°C. Very low attenuations have not yet been studied, but it presently seems that temperature effects will not be important.

Some preliminary observations on radiation effects in low-loss fibers are given in Table V. Aside from military applications, radiation is a unique hazard for optical fibers in conventional uses. Cosmic radiation will generate an exposure of about 1 R per year. Communication installations may thus accumulate 20–40 R over their lifetime. While this is trivial for conventional cable and glass applications, it cannot be ignored for optical waveguides which depend so critically upon

TABLE V

PRELIMINARY DATA ON THE EFFECT OF RADIATION ON ATTENUATION

Dose	633 nm (dB/km)	800 nm (dB/km)	1060 nm (dB/km)
	Neutrons		
0 n/cm²	10	5	7.5
5.5×10^{10} n/cm²	14	7.5	(9)
5.6×10^{10} n/cm²	40	9	
	Gamma Rays		
0 rad	9	4.5	4
300 rad	16	(4.5)	(4)
2300 rad	54	10	

Note: Changes at long wavelengths were too small to measure with precision.

the absence of any color centers. Assuming that a 10-percent change in attenuation is tolerable, the table indicates that the fibers tested would be satisfactory for optical frequencies above about 650 nm but not below. This neglects any beneficial bleaching effects which might occur in a long-time installation. More testing is needed in order to go beyond this very early study.

Fiber strength will undoubtedly be an area of extensive study although, again, experimentation is only beginning. Two types of strength are important—short-term strength and long-term strength, or fatigue. The former is important for the process of installation while the latter is important for installed situations, such as bends, which leave the fiber under continual stress. Requirements for these situations are not yet known and will depend upon installation design.

Present high-silica fibers have a mean strength of over 703 kg/m² (100 000 psi). Since one break in a communications link the order of a kilometer will disrupt the signal, the probability of breakage in this length fiber is needed. The data for the fibers above were obtained on samples only 20 cm long. Therefore, this type of testing will require accurate values for the distribution of strength down to very low probability of breakage in order to predict what may happen over kilometer lengths.

VI. BUNDLING AND CABLING

Bundling is a term applied to grouping many fibers together in some form of substructure of a cable. The bundle might be a group of fibers that could later be encapsulated in a plastic jacket or it might consist of some cable substructure with the individual fibers plastic coated. Both types of encapsulation have been experimentally performed with several plastics such as PVC and FEP Teflon®. At this stage the fibers could be handled and shipped if necessary.

Cabling refers to incorporating these substructures into a component including power wires, fillers, armoring, sheathing, etc., which is suitable for installation. Both near- and distant-term trends are visible. The distant trend is toward single fibers carrying an information channel analogous to wire today. This will require very low probability of fiber breakage plus advances in light sources, splicing, etc. A near-term trend is the use of a fiber bundle to carry an information channel. This gives the advantages of a large cross section to raise the capture efficiency from available light sources, retention of flexibility for installation, and redundancy to guard against fiber breakage.

The evaluation of fiber breakage is a common concern for all envisaged applications. If, under manufacture and installation, the majority of the breakage occurs in a few of the weaker fibers, one approach can be adopted. If, on the other hand, breakage of one fiber interacts in the region of breakage with other fibers in a way that breaks them as well, the situation is more difficult. For the former, more fibers remain unbroken (for a given number of breaks per unit bundle length) than if the breakage were random. In the latter, less are unbroken. For the former case, the random-breakage theory thus gives a conservative result, while for the latter, empiricism may be necessary.

The advantage of using random breakage for engineering is that it provides a single attenuation parameter characterizing broken fiber count for any length bundle when all the fibers in the bundle carry the same information. If b is the number of fibers broken per unit length, then the number of previously unbroken fibers that are broken per unit length is proportional to the unbroken fraction (random probability)

$$\frac{dN_b}{dx} = b\left(\frac{N_u}{N_0}\right)N_0$$

or

$$N_u = N_0 e^{-bx}.$$

For parallel transmission in a bundle, this has the form of an effective "attenuation factor" b. For example, the first bundle of low-loss fibers jacketed had an average attenuation of about 60 dB/km. In 305 m, the breakage represented 39 dB/km ($b = 8.9$ km^{-1}), for a total "attenuation" of 99 dB/km. Recently, breakage below 10 dB/km has been attained. This is satisfactory, without further development, for a restricted class of applications, but improvement is necessary for long distances. Protective film coating and advances in handling techniques are part of the answer. The factor b might also be used to express an additional attenuation factor due to breakage following various testing procedures. A wide variety of tests have been carried out for flexible fiber bundles to be used in the automotive industry [48]. This experience has already benefited the optical waveguide work because of the many similar aspects of the technology.

Cabling has only been carried out on an early experimental basis and breakage appeared small but the quantitative analysis has not yet been performed.

VII. CONCLUSIONS

Remarkable progress has been made in obtaining glass fibers suitable for optical communication on an experimental basis. Much work remains in developing these so that they are suitable for practical cables. Unexpected beneficial discoveries and inventions may entirely change the character of this development; but even without such assistance, the general course and technical steps toward a standard product can already be envisaged. Fiber development has spurred development of the necessary compatible components to interface these waveguides to other equipment. This includes optical couplers, splicers, etc., in addition to electronic devices. As these components take form, the final requirements for constructing system architecture will be met.

ACKNOWLEDGMENT

The author wishes to thank his colleagues at the Corning Glass Works who contributed suggestions to all parts of this paper, and to those closely involved in the technical work; namely, Dr. D. B. Keck, who assisted in guiding all phases, Dr. F. P. Kapron for theory, Dr. P. C. Schultz for glass research, and Dr. F. Zimar for fiber fabrication.

REFERENCES

[1] F. P. Kapron, D. B. Keck, and R. D. Maurer, "Radiation losses in glass optical waveguides," *Appl. Phys. Lett.*, vol. 17, pp. 423–425, Nov. 15, 1970.

[2] D. Gloge, "Optical waveguide transmission," *Proc. IEEE*, vol. 58, pp. 1513–1522, Oct. 1970.

[3] E. Snitzer, "Cylindrical dielectric waveguide modes," *J. Opt. Soc. Amer.*, vol. 51, pp. 491–498, May 1961.

[4] D. Gloge, "Weakly guiding fibers," *Appl. Opt.*, vol. 10, pp. 2252–2258, Oct. 1971.

[5] ——, "Dispersion in weakly guiding fibers," *Appl. Opt.*, vol. 10, pp. 2442–2445, Nov. 1971.

[6] A. W. Snyder, "Power loss in optical fibers," *Proc. IEEE* (Lett.), vol. 60, pp. 757–758, June 1972.

[7] A. Fletcher, T. Murphy, and A. Young, "Solution of two optical problems," *Proc. Roy. Soc.*, vol. 223, pp. 216–225, Apr. 22, 1954.

[8] F. P. Kapron, "Geometrical optics of parabolic index-gradient cylindrical lenses," *J. Opt. Soc. Amer.*, vol. 60, pp. 1433–1436, Nov. 1970.

[9] S. Kawakami and J. Nishizawa, "An optical waveguide with the optimum distribution of the refractive index with reference to waveform distortion," *IEEE Trans. Microwave Theory Tech.*, vol. MTT-16, pp. 814–818, Oct. 1968.

[10] G. H. Sigel, Jr., "Vacuum ultraviolet absorption in alkali doped fused silica and silicate glasses," *J. Phys. Chem. Solids*, vol. 32, pp. 2373–2383, Oct. 1971.

[11] D. S. McClure, "Electronic spectra of molecules and ions in crystals," in *Solid State Physics*, vol. 9, F. Seitz and D. Turnbull, Eds. New York: Academic Press, 1959, pp. 399–525.

[12] T. Bates, "Ligand field theory and absorption in spectra of transition-metal ions in glasses," in *Modern Aspects of the Vitreous State*, vol. 2, J. D. MacKenzie, Ed. Washington, D.C.: Butterworths, 1962, pp. 195–254.

[13] M. W. Jones and K. C. Kao, "Spectrophotometric studies of ultra low loss optical glasses," *J. Phys. E*, vol. 2, pp. 331–335, Apr. 1969.

[14] D. B. Keck, P. C. Schultz, and F. Zimar, "Attenuation of multimode glass optical waveguides," *Appl. Phys. Lett.*, vol. 21, pp. 215–217, Sept. 1972.

[15] E. Lell, N. J. Kreidl, and J. R. Hensler, "Radiation effects in quartz, silica, and glasses," in *Progress in Ceramic Science*, vol. 4, J. E. Burke Ed. New York: Pergamon, 1966, pp. 1–93.

[16] K. A. Stacey, *Light Scattering in Physical Chemistry*. New York: Academic Press, 1956, pp. 8–21.

[17] R. D. Maurer, "Light scattering by glasses," *J. Chem. Phys.*, vol. 25, pp. 1206–1209, Dec. 1956.

[18] D. L. Weinberg, "X-ray scattering measurements of long range thermal density fluctuations in liquids," *Phys. Lett.*, vol. 7, pp. 324–325, Dec. 15, 1963.

[19] F. P. Kapron and D. B. Keck, "Pulse transmission through a dielectric optical waveguide," *Appl. Opt.*, vol. 10, pp. 1519–1523, July 1971.

[20] M. DiDomenico, Jr., "Material dispersion in optical fiber waveguides," *Appl. Opt.*, vol. 11, pp. 652–654, Mar. 1972.

[21] N. Bloembergen, *Non-Linear Optics*. New York: Benjamin, 1965, pp. 102–120.

[22] R. H. Stolen, E. P. Ippen, and A. R. Tynes, "Raman oscillation in glass optical waveguide," *Appl. Phys. Lett.*, vol. 20, pp. 62–64, Jan. 15, 1972.

[23] T. C. Rich and D. A. Pinnow, "Total optical attenuation in bulk fused silica," *Appl. Phys. Lett.*, vol. 20, pp. 264–266, Apr. 1, 1972.

[24] R. D. Maurer and P. C. Schultz, "Fused silica optical waveguide," U. S. Patent 3 659 915 assigned to Corning Glass Works.

[25] H. L. Smith and A. J. Cohen, "Absorption spectra of cations in alkali-silicate glasses of high ultra-violet transmission," *Phys. Chem. Glasses*, vol. 4, pp. 173–187, Oct. 1963.

[26] E. D. Kolb *et al.*, "Low optical loss synthetic quartz," *Materials Res. Bull.*, vol. 7, pp. 397–406, May 1972.

[27] D. S. Carson and R. D. Maurer, "Optical attenuation in titania-silica glasses," *J. Non-Cryst. Solids*, to be published.

[28] A. J. Cohen, "The role of germanium impurity in the defect structure of silica and germanias," *Glastech. Ber.*, vol. 32K, pp. VI–53–58, 1959.

[29] J. S. Stroud, J. W. H. Schreurs, and R. F. Tucker, "Charge trapping and the electronic structure of glass," in *Seventh International Congress on Glass*. New York: Gordon and Breach, 1965, pp. 42.1–42.18.

[30] J. S. Stroud, "Color centers in cerium-containing silicate glass," *J. Chem. Phys.*, vol. 37, pp. 836–841, Aug. 15, 1962.

[31] H. N. Daglish, "Light scattering in selected optical glasses," *Glass Technol.*, vol. 11, pp. 30–35, Apr. 1970.

[32] R. D. Maurer, "Crystal nucleation in a glass containing titania," *J. Appl. Phys.*, vol. 33, pp. 2132–2139, June 1962.

[33] J. D. MacKenzie, "Semiconducting oxide glasses," in *Modern Aspects of the Vitreous State*, vol. 3, J. D. MacKenzie, Ed. Washington, D. C.: Butterworths, 1964, pp. 126–130.

[34] R. H. Doremus, "Diffusion in non-crystalline silicates," in *Modern Aspects of the Vitreous State*, vol. 2, J. D. MacKenzie, Ed. Washington, D. C.: Butterworths, 1962, pp. 1–71.

[35] W. B. Hillig, "Sources of weakness and the ultimate strength of brittle amorphous solids," in *Modern Aspects of the Vitreous State*, vol. 3, J. D. MacKenzie, Ed. Washington, D. C.: Butterworths, 1964, pp. 152–194.

[36] N. S. Kapany, *Fiber Optics*. New York: Academic Press, 1967, pp. 110–128.

[37] D. B. Keck and P. C. Schultz, "Method of producing optical waveguide fibers," U. S. Patent 3 711 262 assigned to Corning Glass Works.

[38] K. C. Kao, R. B. Dyott, and A. W. Snyder, "Design and analysis of an optical fiber waveguide for communication," in *Trunk Telecommunications by Guided Waves* (IEE Conf. Publ. 71), pp. 211–217, Sept. 1970.

[39] British Patent 1 266 521 issued to Nippon Selfoc Co., Ltd., Mar. 8, 1972.

[40] T. Uchida *et al.*, "Optical characteristics of a light-focusing fiber guide and its application," *IEEE J. Quantum Electron.*, vol. QE 6, pp. 606–612, Oct. 1970.

[41] A. D. Pearson, W. G. French, and E. G. Rawson, "Preparation of a light focussing glass rod by ion-exchange techniques," *Appl. Phys. Lett.*, vol. 15, pp. 76–77, July 15, 1969.

[42] A. R. Tynes, A. D. Pearson, and D. L. Bisbee, "Loss mechanisms and measurements in clad glass fibers and bulk glass," *J. Opt. Soc. Amer.*, vol. 61, pp. 143–153, Feb. 1971.

[43] N. Bloembergen *et al.*, "Fundamentals of damage in laser glass," National Academy of Sciences Publ. NMAB-271, Washington, D. C., pp. 29–33, July 1970.

[44] H. Kita and T. Uchida, "Light focussing fiber and rod," in *Fiber Optics*, SPIE Seminar Proc., vol. 21, pp. 117–123, Jan. 1970.

[45] H. Garfinkel, "Cation exchange properties of dry silicate membranes," in *Membranes, Microscopic Systems and Models*, vol. 1. New York: Dekker, 1972, pp. 179–247.

[46] K. B. Chan *et al.*, "Propagation characteristics of an optical waveguide with a diffused core boundary," *Electron. Lett.*, vol. 6, pp. 748–749, Nov. 12, 1970.

[47] O. Krumpholz, "Mode propagation in fibers: Discrepancies between theory and experiment," in *Trunk Telecommunications by Guided Waves* (IEE Conf. Publ. 71), pp. 56–61, Sept. 1970.

[48] E. Snitzer and H. Osterberg, "Observed dielectric waveguide modes in the visible spectrum," *J. Opt. Soc. Amer.*, vol. 5, pp. 499–505, May 1961.

[49] F. P. Kapron, N. F. Borrelli, and D. B. Keck, "Birefringence in dielectric optical waveguides," *IEEE J. Quantum Electron.*, vol. QE-8, pp. 222–225, Feb. 1972.

[50] K. B. Paxton and W. Streifer, "Aberrations and design of graded-index rods used as image relays," *Appl. Opt.*, vol. 10, pp. 2090–2096, Sept. 1971.

[51] D. Gloge, A. R. Tynes, M. A. Duguay, and J. W. Hanson, "Picosecond pulse distortion in optical fibers," *IEEE J. Quantum Electron.*, vol. QE-8, pp. 217–221, Feb. 1972.

[52] D. B. Keck and A. R. Tynes, "Spectral response of low-loss optical waveguides," *Appl. Opt.*, vol. 11, pp. 1502–1506, July 1972.

[53] E. G. Rawson, "Measurement of the angular distribution of light scattered from a glass fiber optical waveguide," *Appl. Opt.*, vol. 11, pp. 2477–2481, Nov. 1972.

[54] A. Jacobsen, N. Neuroth, and F. Reitmayer, "Absorption and scattering losses in glasses and fibers for light guidance," *J. Am. Ceramic Soc.*, vol. 54, pp. 186–187, Apr. 1971.

[55] H. Kita, I. Kitano, T. Uchida, and M. Furakawa, "Light focussing glass fibers and rods," *J. Am. Ceramic Soc.*, vol. 54, pp. 321–326, July 1971.

[56] D. Gloge and E. L. Chinnock, "Fiber-dispersion measurements using a mode-locked krypton laser," *IEEE J. Quantum Electron.* (Corresp.), vol. QE-8, pp. 852–854, Nov. 1972.

[57] R. W. Dawson, "The effect of ambient temperature on infrared transmission through a glass fiber," *Bell Syst. Tech. J.*, vol. 51, pp. 569–571, Feb. 1972.

[58] R. S. Rider, "Environmental testing of flexible jacketed fiber optic bundles," in *Fiber Optics*, SPIE Seminar Proc., vol. 21, pp. 43–48, Jan. 1970.

[59] D. Marcuse, "Higher-order loss processes and the loss penalty of multimode operation," *Bell Syst. Tech. J.*, vol. 51, pp. 1819–1836, Oct. 1972.

Part VII: Laser Applications

Laser applications in science and technology are characterized first of all by their variety and diversity. The selection of papers in this section can only provide a small sampling of the uses, both sophisticated and commonplace, that lasers have found and continue to find. With the limited number of available review articles, the applications covered in this section are neither complete nor representative.

Lasers are finding increasingly widespread applications as industrial materials processing tools. The first two papers, by Eleccion, and by Gagliano, Lumley, and Watkins, provide good introductions to this rapidly growing and commercially important area. The spatial coherence and brightness properties of steerable laser beams can also be put to use in laser displays, as reviewed by Baker, as well as in other areas, such as laser recording and laser image processing, not covered here.

Lasers, from their inception, have been thought of as potential sources for optical communications. Significant operational systems, however, are still rare. The paper by Goodwin, from the October 1970 PROCEEDINGS Special Issue on Optical Communications, surveys the operational laser communications systems studied to date. The development of low-loss optical fibers for long distance transmission described in Part VI should lead to more sophisticated systems in the near future.

Applications

Materials processing with lasers

The first major use of lasers, industrial processing, is being expanded to include bigger, tougher, tasks

Marce Eleccion   Staff Writer

Until high-power lasers came upon the scene, dynamic materials processing was the sole province of the standard high-flux sources such as gas jets, electric discharges, plasma arcs, and electron beams. The immediate realization of the laser's unique ability to generate spectrally pure high-energy radiation at enormous power densities with complete freedom from mechanical contact with a working surface resulted in a major portion of early laser applications being materials-processing oriented. This situation was abetted since materials processing exploited a primary capability of the laser—the delivery of raw energy of a certain spectral content focused to the exact power density needed at continuous, pulsed, or Q-switched operation to heat, melt, or vaporize a specific material.

As a processing tool, the laser has been called upon either to perform classic machining tasks (drilling, milling, welding, etc.) or to carry out processing that could not have been accomplished by any other method. Although the laser is restricted in size of working area because of basic design and power limitations (new high-output lasers are beginning to change this, however), whenever the laser can be used practically it usually enhances the type of materials processing desired; for example, drilling can be made faster, scribing closer, and welding of dissimilar materials better than with other devices previously used. Moreover, jobs that were once considered difficult or even impossible are now in the laser's repertory; for instance, drilling and welding through glass are now everyday realities. Aside from such "standard" materials processing as substrate scribing, hole drilling, and resistor trimming, there are many other applications for which the laser has been found to be admirably suited; thus lasers are now performing such chores as punching holes in baby-bottle nipples, dehorning and branding cattle, and tagging salmon.

Which lasers?

To match laser types to specific materials-processing tasks, the pertinent parameters must first be understood. Not only are spatial–temporal coherence and output power important in terms of obtaining large energy densities, but emitted wavelengths must be selected according to the energy-absorption characteristics of a specific material, and power–time values must be carefully chosen to meet individual processing requirements.[1] Simply put, a specific mode of operation (CW, pulsed, Q-switched) must be carefully selected to obtain the effects of heating, melting, or vaporization from a specific laser–material combination, with the character of the spatial modes in a laser output determining such processing factors as spot size and drilling depth. (See box, p. 68.)

Rough estimates of the threshold power densities needed for various types of materials processing and how they have been met over the years by available high-power commercial lasers are given in Fig. 1. It is significant that CW laser sources have only recently attained the high-power levels previously reserved for pulsed lasers. Even more significant is the fact that processing has been limited to microscopic applications because of the prevalent low overall average laser powers. Figure 2 shows the increase in average output powers for the same period as in Fig. 1. Note that not until 1971 were the multi-kilowatt industrial lasers needed for high-speed materials processing made available.[2]

Although argon-ion, chemical,[3] and liquid lasers have impressive power levels (see this writer's article on "The Family of Lasers," March SPECTRUM, pp. 26–40), at the present state of development only solid and CO_2 lasers are capable of generating extreme power outputs. Despite high powers, however, ruby and neodymium lasers suffer from low efficiency and limited repetition rates, with Nd:YAG lasers capable of higher average powers because of their lower lasing threshold and superior thermal conductivity. On the other hand, CO_2 lasers operate at high efficiencies, high repetition rates, and (despite a low gas density) even greater CW, pulsed, and Q-switched operation than solid lasers—the result of much development in recent years (see Table I). A disadvantage of the CO_2 laser in materials processing is that its 10.6-μm spectral line is reflective to metals, but this factor is overcome by higher power levels (if somewhat

Reprinted from *IEEE Spectrum*, vol. 9, pp. 62–72, Apr. 1972.

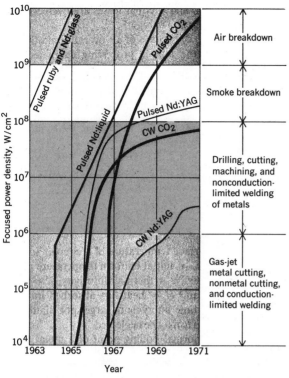

FIGURE 1. Available focused laser power densities and estimated laser application thresholds.[2]

FIGURE 2 (right). Available average laser power and power requirements of lossless applications.[2]

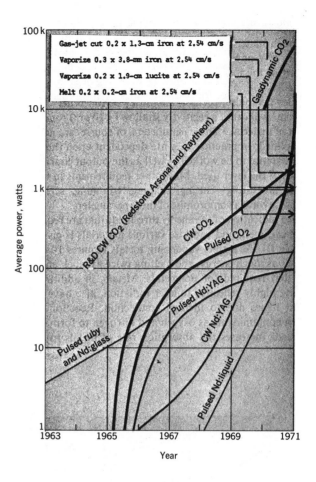

inefficiently). Table II gives an estimate of the cost of various laser devices currently available, an important consideration in any industrial processing system.[4]

In general, a laser of pulse widths greater than 1.5 ms, with a focused intensity of 10^7–10^8 W/cm², will cause the surface temperature of a material to rise above the melting point and effect *welding*. If the pulse width is decreased to less than 1.5 ms, energy absorption is increased and the

II. Price ranges for various types of available lasers[4]

Laser Type	Cost
Pulsed Nd	$8000–10 000
Doubled ruby	$12 000
Argon	$10 000
He–Ne	$100–$10 000
Ruby	$8000–20 000
Nd:glass	$8000–100 000
GaAs	$20–$1000
CO₂	$3000–15 000

I. Commercially available laser systems[1]

Laser Material	Wavelength, μm	Mode of Operation	Output, watts	Repetition Rate, pps	Pulse Length	Processing Uses
Ruby	0.6943	pulsed	1–20 (average)	1	0.3–6 ms	welding, material removal
		Q-switched	10^5 (peak)	1	0.3–2 ms	welding, material removal
			10^8 (peak)	1	5–50 ns	vaporization
Nd in glass	1.06	pulsed	1–15 (average)	1	0.5–10 ms	welding, material removal
		Q-switched	10^6 (peak)	30 ppm	0.5–1 ms	material removal
			10^9 (peak)	5 ppm	10–60 ns	vaporization
Nd in YAG	1.06	continuous	1–1100*†	—	—	heating, welding
		pulsed	1–100 (average)	1–100	0.01–5 ms	welding, material removal
		Q-switched	500–5 M* (peak)	1 k–10 k	150–300 ns	vaporization
CO₂	10.6	continuous	50–10 k*‡	—	—	welding, cutting, fracturing
		pulsed	1–250 (average)	1–1 k	5–150 μs	welding, material removal
		Q-switched	10^4 (peak)	200–500	30–300 ns	material removal, vaporization
Argon	all lines§	continuous	1–15	—	—	—
		pulsed	1–25 (peak)	1–1 k	5–100 μs	vaporization

* Revised data added by the author.
† Holobeam Model 2500 has an 1100-watt CW output.
‡ Avco Everett's CO₂ model delivers 10-kW CW power in a closed-cycle, electrically pumped continuous-flow unit.
§ 0.4579 μm, 0.4658 μm, 0.4727 μm, 0.4880 μm (strong), 0.4965 μm, 0.5017 μm, 0.5145 μm (strong).

temperature rises into the boiling range, removing just enough material to result in hole formation or *drilling*. With CW or high-repetition pulsing, the speed of movement of the material also becomes an important factor. Finally, with 5–300-ns-pulse Q-switching of up to several joules per pulse, focused intensities of the order of 10^9 W/cm² will cause a relatively shallow (<10 μm) *vaporization* of material.[1] These parameters, of course, are greatly simplified; in practice, they are dependent upon the type on material to be worked as well as the output characteristics of the laser used. Most other requirements in materials processing—cutting, shaping, trimming, scribing, etc.—involve a compromise of these parameters.

An indication of the energy thresholds that are required to heat, melt, and vaporize various materials is given in Table III. As a rule, vaporizing metals requires 10 to 20 times as much energy, and melting twice as much energy, as that required for nonmetals. Metals, in addition to being highly reflective to laser radiation, also have high heat losses due to thermal conduction. Reaction with such contaminants as oxygen and nitrogen to form compounds increases the absorption rate of metals to laser radiation, however, and thermal losses are overcome by the application of greater energy.[2]

As has been stated, because of an overall limitation in available powers, the unique processing characteristics offered by lasers have been restricted to applications on an exceedingly small scale. These capabilities—including the types of material handled and the speed, area, and depth of processing—can be expected to be enhanced during the 1970s as more versatile and higher-power lasers are introduced to the market.

Welding. Fairbanks and Adams[5] have used transient heat-transfer equations for a semiinfinite slab over which heat is uniformly transmitted in a circular spot:

$$T_s - T_0 = \frac{2H\sqrt{\alpha t}}{K}\left[\frac{1}{\sqrt{\pi}} - i \text{ erfc } \frac{a}{2\sqrt{\pi t}}\right]$$

where T_s = the surface temperature of a circular spot of radius a at time t, T_0 = initial uniform temperature of the material to be welded, α = thermal diffusivity, K = thermal conductivity, and H = heat flux absorbed into the surface. For short time intervals or large spot diameters, the error function erfc may be neglected.

In general, the normal boiling point of a metal is assumed to be the maximum temperature to which the surface can be heated without excessive vaporization or metal expulsion. As an example, the data defining welding conditions for pure aluminum are given in Fig. 3.

The maximum depth of laser penetration before surface vaporization takes place has been given by Cohen and Epperson[6]:

$$S_{max} = \frac{0.16}{\rho L H}(H^2 t_e - H^2 t_m)$$

where S_{max} = depth of penetration, ρ = density, L = latent heat of fusion, H = heat flux, t_e = time at which surface begins to vaporize, and t_m = time at which surface begins to melt. If K_m can be given as a specific constant for every material, however, this equation reduces to

$$S_{max} = \frac{K_m}{H}$$

Values of K_m for metals can be seen in Table IV.

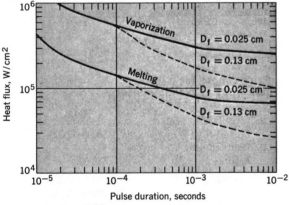

FIGURE 3. Heat flux versus laser pulse duration required for melting and vaporizing over various laser spot sizes in a semiinfinite aluminum slab.[7]

FIGURE 4. A—Two integrated circuits welded to a copper printed-circuit board. B—Close-up of welds in A.[7]

III. Energy thresholds required to heat, melt, and vaporize various materials[2]

Material	Melting Temperature, K	Vaporizing Temperature, K	Heat to Melting Temperature from 25°C, J/cm³	Melt from 25°C, J/cm³	Vaporize from 25°C, J/cm³
Gold	1 336	2 933	2 580	3 820	38 220
Silver	1 234	2 485	2 330	3 490	31 300
Copper	1 356	2 868	3 650	5 480	53 600
Aluminum	932	2 600	1 540	2 610	35 060
Iron	1 808	3 008	5 310	8 040	62 030
Cobalt	1 766	3 370	5 950	8 350	73 570
Nickel	1 728	3 110	5 640	8 370	71 150
Tungsten	3 650	6 225	8 570	12 270	114 511
Al_2O_3	2 300	decomposes	6 970	11 196	—
SiO_2	1 883	dec. 2250	2 613	2 966	dec. 3572
Acrylic plastic	dec. 230°C	—	—	—	dec. 2200
Wood	dec.	—	—	—	dec. 1800

Energy loss by reflection, diffusion into base material, or any other mechanism has not been evaluated.

Most of the systems that are commercially available for laser welding are of a pulsed-operation type, with safe and accurate positioning of the work made possible by a closed-circuit television within the system and a computer-controlled drive. Since melting can be performed at a lower heat flux by lengthening the laser pulse, some pulsed-laser welding equipments on the market employ special networks that lengthen the normal 0.5–1-ms pulses to as long as 5–10 ms.

A great variety of welding can be achieved with lasers; for example, the welding of an experimental plated-wire memory, the attaching of a thermocouple vacuum gauge to the Apollo lunar-sample return containers, the welding of IC flat packs to printed-circuit boards (Fig. 4), and wire-to-wire or wire-to-sheet welding. In most cases, the singularly attractive features of lasers in welding are its ability to weld dissimilar metals and the minimization of thermal damage to adjacent materials.[7] Welding can also be precisely located (to within ±12 μm), with weld beads as small as 25 μm in diameter.

At Varian Associates of Canada, Ltd., the inability of

IV. Values of K_m for various metals[7]

Metal	K_m*	K_m†
Cr	0.13×10^2	4.49×10^2
Cu	1.76×10^2	60.9×10^2
Au	1.89×10^2	65.4×10^2
Fe	0.31×10^2	10.7×10^2
Mo	1.51×10^2	52.2×10^2
Ni	0.34×10^2	11.8×10^2
Pt	0.61×10^2	21.1×10^2
Ag	1.60×10^2	55.3×10^2
Ta	0.42×10^2	14.5×10^2
W	1.14×10^2	39.4×10^2

* H in Btu/ft²·s and S_{max} in ft.
† H in W/cm² and S_{max} in cm.

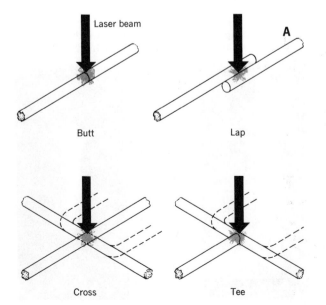

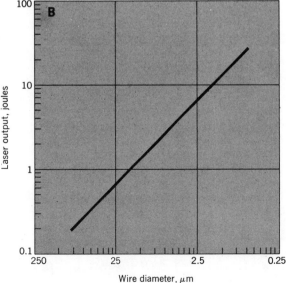

FIGURE 5. A—Joint configurations for typical wire-to-wire laser welding. B—Approximate laser output required for full-strength welds for various wire diameters.[8]

resistance welding to adhere 50-μm-thick tantalum sleeves to molybdenum collars in klystron-tube production had created a rejection rate of four in five. With laser welding (through an argon-filled glass bell jar), rejection rates dropped to zero. Varian has also been able to repair defective vacuum tubes by beaming through the glass envelope.[7]

At AiResearch Manufacturing Co., ruby lasers enabled critical welding of a stainless-steel casing in a hermetically sealed thermistor enclosure; and at Korad (Union Carbide) welds between enameled wire and electric motor terminals have been accomplished with a 4 $\times$ ⅜-inch YAG laser.

One increasingly important application of laser welding is in microelectronic packaging design of computers and

instruments.[8] Such packaging usually involves interconnection welds, circuit-board welds, and special welds. Typical interconnections are described in Fig. 5.

The various configurations used in IC–circuit-board welding are shown in Fig. 6. The board substrate preferred for laser welding is usually one of the epoxy fiberglass types, although phenolic board is also used. Nickel is considered the most versatile cladding material, but copper, Kovar, and multilayer cladding offer no real obstacles.

Some welding includes those jobs that lasers alone can perform, such as welding insulated coil wires to termination lugs and making repairs through glass. An ingenious "bending" technique associated with the latter application is shown in Fig. 7.

High-power CW YAG and CO_2 lasers are recent arrivals on the laser scene and so the literature on their use is somewhat limited. It has been determined, however, that the power required to weld with CO_2 is three times as great as with YAG, which is not surprising in view of the fact that metal absorption is much less at the longer wavelength.

As recently as a few months ago, the success of rapid, deep-penetration welding with 8–20-kW CO_2 lasers was described by Locke *et al.* of Avco Everett Research Lab. Using gasdynamic methods, the authors report 1.3-cm

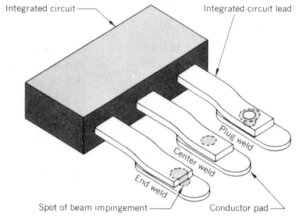

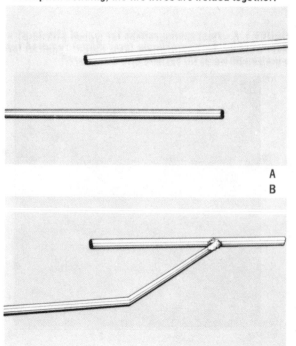

FIGURE 6. Joint designs for integrated circuit welds.[8]

FIGURE 7. A—Two 0.75-mm diameter stainless-steel wires with a 1/3-cm separation. B—After heating one wire to accomplish bending, the two wires are welded together.[8]

V. Depth of material removal for various metals[1]

	Depth, μm	
	10^7 W/cm², 600-μs Pulse	10^9 W/cm², 44-ns Q-switched Pulse
Aluminum	780	3.6
Copper	900	2.2
Nickel	580	1.2
Brass	780	2.5
Stainless steel	610	1.1

From J. F. Ready, "Effects due to absorption of laser radiation," J. Appl. Phys., vol. 36, p. 1522, 1965.

FIGURE 8. Deep welds have been made in #304 stainless steel at 1.27 m/min (A) and 2.54 m/min (B) by 20-kW lasers.[9]

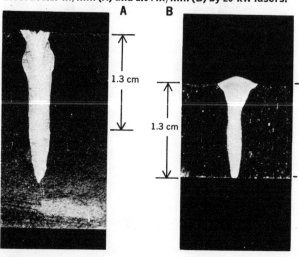

penetration at 254 cm/min rates within a 5-to-1 depth-to-average-width fusion zone in #304 stainless steel—the deepest penetration and highest speed achieved with laser welding thus far (see Fig. 8).[9]

Drilling or hole punching. In this application, bulk removal of material is accomplished at focused intensities of 10^7–10^8 W/cm². Examples of the depths attained in various metals can be seen in Table V. Normally, a 1-ms pulse can remove over 100 times as much material as a shorter Q-switched pulse.

Current applications of laser drilling include diamond piercing in order to create wire-drawing dies, drilling through extremely hard substances such as alumina ceramic (normally requiring diamond-tipped, hardened-steel drill bits), and even making holes up to a few centimeters in diameter in rubber nipples, valves, and gaskets

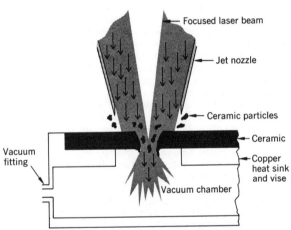

FIGURE 9. Illustration of air-jet-assisted hole drilling using a CO_2 laser.[1] The air stream and vacuum chamber prevent clogging of laser-drilled holes.

FIGURE 10. Diagram of a typical laser industrial tool used for micromachining and welding.[10]

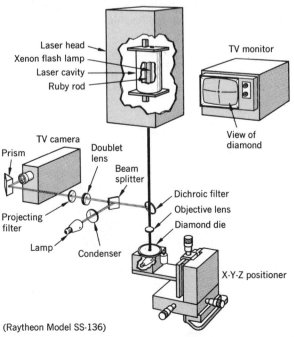

(Raytheon Model SS-136)

(Coherent Radiation). With metal bits, drilling holes less than 250 μm in diameter is extremely difficult, with drill breakage a common hazard. Lasers can drill holes as small as 10 μm across in the hardest of substances with complete safety and precision.

(One cannot completely disregard mechanical microdrilling, however. Not only have new developments by such companies as National Jet reduced the hole sizes obtainable by these older devices—an incredible 1.25 μm in some materials—but in many instances the application of both techniques produces a better job than either one could have accomplished alone. For example, although a laser is certainly faster at drilling holes than a mechanical drill, certain requirements demand that the hole be "cleaned up" by mechanical means. Such techniques as noncaptive spindling and automatic tool changing should enhance this "hybrid" relationship even further.)

An initial problem that plagued laser users attempting to drill small holes involved the buildup of resolidified material within the hole, which sometimes caused the hole to close. A solution is outlined in Fig. 9.

One commercially available laser system specializing in the drilling of holes in diamond is described in Fig. 10.[10] Consisting basically of a laser head, closed-circuit television viewing and focusing accessory, power supply, and closed-cycle refrigerated cooler, the pulsed ruby laser system illustrated (Raytheon Model SS-136) is capable of performing such tasks as microdrilling, micrometal removal, resistor trimming, wheel balancing, and welding with a total output of 50 J/pulse at 1 pulse/s.

In a novel deviation from run-of-the-mill diamond drilling, diamond imperfections are being removed by

FIGURE 11. The unique diamond-drilling capabilities of lasers may prove a boon to the precious gem industry. A—A diamond containing an imperfection. B—The flaw "disappears" after laser processing.[11] The writer has been informed that such "operations" defy detection by even a jeweler's loupe, especially when the process is accompanied by methods that fill the drilled hole back up.

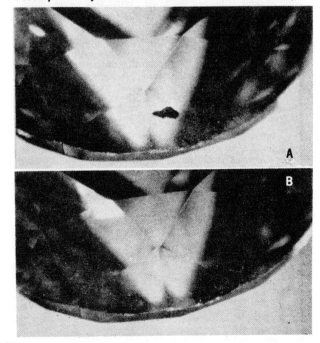

laser.[11] Natural diamonds often contain flaws such as voids, twins, and inclusions, which are even more apparent after faceting (see Fig. 11). After a hole is drilled from the surface, the imperfection is either removed or is changed in color to provide minimum contrast with the rest of the gem. Such subtle use of laser capabilities gives new meaning to the old apothegm of *caveat emptor*.

In another innovative use of laser hole drilling, Lederle Labs is streamlining the profile of the surgical needle and cord. By using a Holobeam ruby laser with a 0.6-cm aperture and 4-cm focal length, a 250-μm-diameter hole is drilled into the end of an 850-μm surgical needle to a depth of 2.5 mm (see Fig. 12). By inserting the surgical cord directly into the tapered hole and slightly crimping the sides for a secure fit, the inconvenience of the standard doubled-thread approach is totally eliminated.

Even the nuclear field has been dramatically aided by laser processing, with Holobeam perfecting a simplified

---

Getting to the point

There is more to drilling holes by laser than the minimal requirements of power density. Not only are the diameter and depth of a hole to be drilled dependent upon the "wavelengths" at which a laser beam is emitted (shorter wavelengths can be focused to smaller spots) but the dimensions of a drilled hole are also determined by the overall spatial mode of a laser output—with the lowest-order (fundamental) mode giving optimum focusing characteristics. Why this is so can easily be seen.

In laser operation, the relationship between output power P and beam energy E is dependent upon time t; thus

$$P = \frac{E}{t}$$

measured in joules per second or watts. In turn, intensity I (power density) is equal to power (energy per unit time) per unit area A transmitted normal to the direction of wave propagation:

$$I = \frac{P}{A}$$

in watts per unit area. Since even a spatially coherent laser beam has a certain divergence, the "far-field" divergence angle θ must also be considered:

$$\theta = \frac{2\lambda}{\pi D_0}$$

where λ = beam wavelength and D_0 = beam diameter as it leaves the laser cavity; this beam divergence is usually given for half-power levels at suitable aperture stops.

Since most materials processing requires that the relatively wide (a few millimeters) laser beam be focused down to the order of a micrometer or less, the focused-beam diameter D_f should also be given:

$$D_f = \frac{\lambda f}{\pi D_0}$$

where f = focal length (see Fig. A*) and D_f is wavelength- and intensity-dependent. In general, the diameter of the focused beam spot D_f determines the depth of focus d, so that to obtain a smaller focused spot one must accept a smaller depth of

focus; results usually can be varied by controlling f and D_0.

The intensity I_f at D_f can now be derived from

$$I_f = \frac{P}{\pi\left(\dfrac{D_f}{2}\right)^2}$$

Since D_f not only is proportional to laser intensity, but also to laser wavelength, it should be remembered when selecting a laser that a shorter output wavelength will give a smaller focused spot. In any case, effective spot size will ordinarily be larger than the ideal spot size because of nonuniform light transmission, propagating mode structure, and thermal properties of the material.[1]

An estimate of focused spot size as a function of f-number (f/D_0) is seen in Fig. B,† where both fundamental mode ($M^2 = 1$) and multimodes are plotted for typical 1-kW CO_2 (10.6-μm) and Nd:YAG (1.06-μm) lasers. For a given focused spot size, multimode beams have a narrower depth of focus (see Fig. C†); hence, although a multimode spot may be decreased by selecting a low enough f-number, it is impossible to increase the depth of focus. Since there is usually a minimum depth-of-focus requirement, the minimum spot size obtainable is dependent upon the available laser characteristics.

As one begins to understand the basic principles of laser processing, it becomes obvious that truly expert processing technology can only be realized as multikilowatt fundamental-mode (TEM$_{00}$) lasers become more and more available, especially since processing rate varies inversely with cut width. Since such laser systems are just beginning to be introduced, one can only guess the role that lasers will play in materials processing of the future.

* From Ref. 1.
† From Marshall, L., "Laser beam theory for industrial applications," *Laser Focus*, vol. 7, p. 26, Apr. 1971.

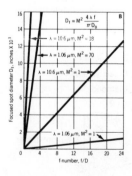

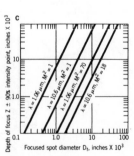

---

technique for drilling zirconium fuel rods under the rather stringent safety controls imposed (see Fig. 13). Other techniques being pursued include the welding of matrices for holding fuel rods in the reactor, and the welding of zirconium end caps to the fuel rod barrel.

Other types of processing. It has been seen that laser methods not only "take the worry out of being close," but can be practically indispensable when conventional methods are found difficult, wasteful, or time-consuming. Such is the case with *scribing*.

For as long as batch fabrication has existed in the semiconductor technology, the standard process of separating component and IC substrates has been accomplished by scribing the wafer (usually with a diamond point) and either breaking or cutting it apart. At best a risky business, this method not only required depositions that were a wasteful distance apart, but was extremely time-consuming.

With the adaptation of laser technology, component and IC devices could be spaced more closely together, and the scribing accomplished in a quicker and more efficient manner. What the laser does is to drill a "postage-stamp" pattern of tiny holes along the direction of cleavage and to about one third the substrate thickness, permitting the wafer to be snapped apart easily (see Fig. 14). Typical scribing rates are 7.5 cm/s.

In operation at such companies as Western Electric and Motorola for over a year, laser scribing systems are particularly well suited to computer control, giving higher wafer yields and hence lower cost. It should be noted that with substrates having low thermal conductivities, Q-switching is not necessarily an advantage since the short pulse spacing gives the same result as CW applications of the same power. In drilling, the required power levels tend to cause excess heating, which must be relieved by reducing the duty cycle with long millisecond-width pulsing.

It is fortunate for the laser manufacturer that thick- and thin-film resistors fall within a 5–10 percent tolerance range when fabricated, because *resistor trimming*—a major obstacle in the fast manufacturing of hybrid microelectronics—has been turned into a fast and precise art by the application of laser scribing systems. Over the

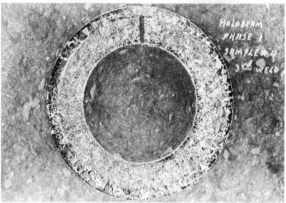

FIGURE 13. In the nuclear field, Holobeam has used a laser welder/driller system and a pressure chamber in a simplified technique for fast inert-gas purging of presealed nuclear fuel rods. After the nuclear rod has been drilled through while in a vacuum, inert gas is forced into the drilled hole under pressure and then sealed with a second laser shot through a window in the pressure chamber—all in less than 15 seconds.

FIGURE 14. A—Laser scribe in silicon produced by Motorola Central Research Labs[12] is shown at 500-X SEM magnification. B—Scribe of A at 1600-X SEM imaging.

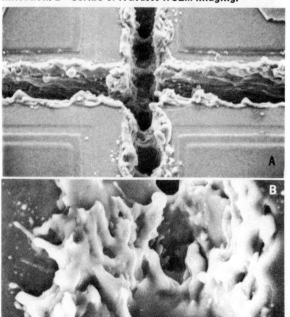

FIGURE 12. At Lederle Labs, a technique has been perfected that allows surgical cord to be inserted directly into a laser-evacuated hole in the end of a surgical needle.

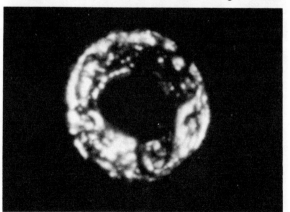

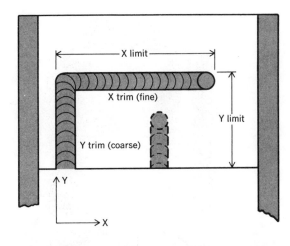

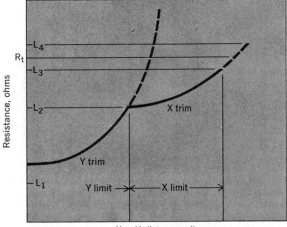

Legend:
R_t = target resistance L_3 = low tolerance limit
L_1 = blank-low limit L_4 = blank-high tolerance limit
L_2 = Y-trim proportion of range

A

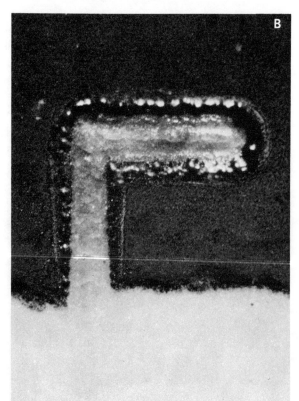

B

past several years, over ten firms have offered systems able to obtain precisions as accurate as 0.01 percent.

The method of trimming or "adjusting" a resistor is demonstrated in Fig. 15. After the laser achieves "coarse" trimming by vaporizing a channel across the length of the resistor roughly approximating the desired resistance (feedback measurements are made while the resistor is being trimmed), it then begins a "fine"-trimming maneuver at a right angle, stopping as soon as the exact resistance value is reached.

One limitation of this method is that the exposed edges of the laser cut age with time, eventually degrading the precision of the resistor. A solution has been to cover the resistor with a thin layer of glass before actual trimming, thus establishing a more stable condition at the edges.[12]

Although both YAG and CO_2 lasers are employed in resistor trimming, in general CO_2 lasers are used only for thick films, since thin-film metals do not significantly absorb the CO_2 laser's 10.6-μm wavelength. Since the wavelengths suitable for trimming both thick- and thin-films fall within 0.4–3 μm, lasers within this range are most acceptable. At present, most manufacturers are offering Nd:YAG systems for both types of film trimming, with one company (Hughes) putting its bets on a

FIGURE 15 (left). A—Criterions for measurement and control of film-resistor laser trimming.[13] B—80-X photomicrograph of laser trim cut in cermet thick-film resistor.[12]

FIGURE 16 (below). Semiautomated trimming of thick-film resistors with Korad KRT YAG-laser trimming system.

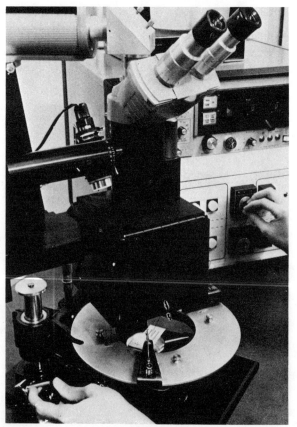

Expert opinion

The first widespread practical use of lasers was by science fiction writers. And, frankly, laboratories throughout the world are still looking for a second widespread use of these fascinating devices.

This is not to say that lasers have not found many uses in biomedicine, information handling, communications, industrial tools, and military applications. But for the most part, these have been relatively small and very specialized uses.

Because of this and because there have been so many sophisticated advances in lasers and related technologies, we tend to become impatient and forget that the first successful laser was built only 12 years ago. We also tend to forget that it took more than 15 years for the sales of transistors—devices with many obvious advantages and applications—to exceed the sales of receiving tubes. Even so, sales of lasers and related equipment last year have been estimated at $160 million, with lasers themselves accounting for $50 million of that total. The 1972 total is estimated at $187 million. . . .

Thus, lasers are a growth industry and, as this issue of IEEE SPECTRUM indicates, there is good reason to believe that their years of greatest growth lie in the future.

William M. Webster, Vice President
RCA Laboratories

There appears no question that the usefulness of lasers for micromachining has been established and will continually expand. Early attempts at using the laser for micrometallurgical reaction, such as welding and photoresist patterning, did not meet with notable success. However, with the confidence established by resistor trim micrometallurgical reaction, it appears certain that this aspect of laser materials processing will expand to encompass a wide variety of uses; this growth, however, will probably be slow.

I. Arnold Lesk, Director
Central Research Labs, Motorola Inc.

I would single out for comment three trends which should have a significant influence on the laser business over the near term. First is the continuing availability of venture capital to spawn new enterprises; second is increasing acceptance of lasers for industrial production applications; and third is the potential inhibiting influence of federal and state legislation on this neophyte industry. . . . [Regarding the second trend], it is encouraging that lasers are finding growing acceptance as true industrial production tools. In several applications, it has been demonstrated that lasers do a better and faster job, as compared to more conventional techniques. The customer can frequently recover his investment costs in a relatively expensive laser system in a period of less than one year. I would cite, for example, the extensive use of production laser systems by the microelectronics community for scribing of silicon and ceramic substrates and trimming of resistors on hybrid circuit boards. . . .

William C. Thurber
General Manager, Materials
Systems Division,
Korad Division
Union Carbide

A few years ago, the laser was sometimes described as "a solution in search of a problem." Today, more and more problems are turning up for which lasers are either the only possible solution or the cost-effective solution. Examples range from the speculative to the severely practical—fusion physics, air pollution monitoring, drilling more and better holes in jet engine parts for less money, and many others. Needless to say, this situation is a fortunate one for laser R&D: it maintains the level of support necessary for innovative developments, while present lasers make themselves useful in new scientific, military, and industrial applications. . . .

Roland W. Schmitt
R&D Manager, Physical
Science and Engineering
General Electric Company

FIGURE 17. Block diagram of a typical laser resistor-adjusting system.[13]

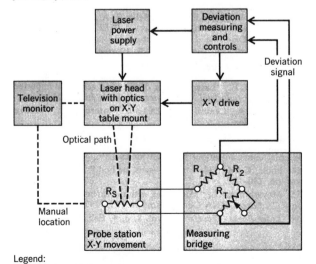

Legend:
R_1 & R_2 = fixed resistors
R_T = setting decade for target setting
R_S = substrate resistor being trimmed

FIGURE 18. The Bell Labs primary pattern generator used for creating new masks for integrated circuits.[14]

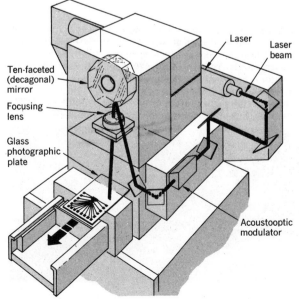

0.53-μm pulsed xenon gas laser.

An example of the type of equipment being used for resistor trimming can be seen in Fig. 16; a typical scheme for such a system is outlined in Fig. 17. The high pulse-repetition rates obtainable from equipment of this sort are capable of extremely fast trimming rates, which in turn greatly increases productivity. It is also possible to speed up trimming rates by increasing crater size; as depth of cut is increased, however, material resistance exponentially rises.

One of the most unique uses of lasers in materials processing is being carried out at Bell Labs to create new *masks for integrated circuits*.[14] This mask-making laser system—the primary pattern generator (PPG)—can quickly process the complex geometric patterns of IC's on an 8- $\times$ 10-inch photographic plate 70 times faster than machine methods still being used by industry (see Fig. 18). Using a low-power argon laser designed for high spectral purity, the PPG produces a 10-μm scan line that is directed by a computer-controlled modulator to a ten-faceted rotating mirror. Moving at right angles to the laser line, the photographic plate is eventually scanned by 32 000 lines at 26 000 positions per scan line, giving an upper limit of 832 million addressable points. Total operating time is about 10 minutes, as opposed to the more than 12 hours of machine time formerly required.

In recent years, materials-processing applications of lasers have become more and more diverse. At RCA, the fabrication of rotogravure cylinders by CO_2 lasers has proven most promising for printing systems if metal-plated plastic sleeves are used.[15] Plastic has also been used in tape form to produce optical memories by drilling holes into it at high repetition rates with lasers. At the Bulova Watch Company, the extremely precise drilling accuracy of a pulsed Nd:YAG laser developed by Sylvania has increased the precision of regulating tiny clock balance wheels by a factor of 10 while decreasing the adjustment time to 1/20 of that required by conventional methods. Ordinarily, the vibration rate of these machine-stamped balance wheels is regulated by the manual adjustment of a torsion spring from which the wheel is suspended; in Sylvania's dual-beam optical system, a laser

drills equal amounts of excess material from opposite sides of the wheel within a minute's time at 100-μs bursts.

Two major consumer areas that are certain to benefit from the unique cutting characteristics of lasers are the multi-billion-dollar garment and paper industries. At Rice-Barton, the technical feasibility of slitting paper by means of an American Optical 350-watt CO_2 laser has already been proven.[16] At Genesco, Inc., the largest apparel company in the world, a Coherent Radiation 250-watt CO_2 laser has been used by Hughes in a fabric-cutting system that is capable of cutting through over 100 layers of material, or a complete suit in three minutes (see Fig. 19). Described as the first major advance in apparel manufacturing since the invention of the sewing machine, Hughes' laser system will affect virtually every person in the United States through the $50 billion clothing industry.

The laser has already proved itself a reliable, time-saving, and relatively inexpensive tool for many types of materials processing. With the development of commercially available high-power lasers over the past year, along with the application of engineering skills in designing peripheral systems needed to solve specific problems, it can be predicted that the diversity of applications to be realized in materials processing has only just begun!

FIGURE 19. Laser-made tailoring at Genesco, Inc., is accomplished by means of Hughes' computerized CO_2-laser fabric-cutting system. Such systems will help lower industry costs, reduce the present $3 billion clothing inventory, and enable quick responses to fashion changes. Laser cutting is accurate to within the width of a single thread, thus eliminating error and waste; knife-cut frayed edges are replaced by sharp-edged unfrayed cuts.

REFERENCES

1. Gagliano, F. P., "Laser processing fundamentals: A tutorial," in *The Applications of Lasers*, Laser Industry Association, Sept. 1971, pp. 3–44.
2. Foster, J. D., "Industrial applications of high power lasers," internal report, GTE Sylvania, Electro-Optics Organization, Mountain View, Calif., 1971.
3. Chester, A. N., "Chemical lasers: a status report," *Laser Focus*, vol. 7, pp. 25–29, Nov. 1971.
4. Johnson, A. M., "Lasers in instrumentation," in *The Application of Lasers*, op. cit.
5. Fairbanks, R. H., Sr., and Adams, C. M., Jr., "Laser beam fusion welding," *Welding J.*, Welding Res. Suppl., vol. 43, pp. 97-S–102-S, Mar. 1964.
6. Cohen, M. I., and Epperson, J. P., "Application of lasers to microelectronic fabrication," in *Electron Beam and Laser Beam Technology*, 1968, pp. 139–186.
7. Thurber, W. C., "Laser welding—Theory, status and prospects," internal report, Union Carbide Corp., Korad Dept., Santa Monica, Calif., 1971.
8. Jackson, J. E., "Packaging with laser welding," Report P/B-010, Union Carbide Corp., Korad Dept.
9. Locke, E. V., Hoag, E. D., and Hella, R. A., "Deep-penetration welding with high-power CO_2 lasers," *IEEE J. Quantum Electronics*, vol. QE-8, pp. 132–135, Feb. 1972.
10. Prout, J. G., Jr., and Prifti, W. E., "Laser drilling of diamond wire drawing dies," *Laser Industrial Appl. Notes*, no. 1–70, Raytheon Co., Laser Advanced Development Center, Dec. 1970.
11. Prifti, W. E., "Laser piercing of diamond gemstone imperfections," *Laser Industrial Appl. Notes*, no. 1–71, June 1971.
12. Lesk, I. A., Central Research Labs, Motorola Inc., Personal communication.
13. Howard, R. T., and Allen, R. V., "Characterization of laser-trimmed thick-film resistors by scanning electron microscopy," internal communication, IBM-Astrionics Lab., 1971.
14. Poole, K. M., "New masks—new method," *Bell Lab. Record*, vol. pp. 130–136.
15. Meyerhofer, D., "Machining with the carbon dioxide laser," in *RCA Lasers*, RCA Corp., Camden, N.J., 1970, pp. 162–167.
16. *Laser Focus (Wavefronts)*, vol. 8, p. 4, Feb. 1972.

Lasers in Industry

FRANCIS P. GAGLIANO, ROBERT M. LUMLEY, MEMBER, IEEE,
AND LAURENCE S. WATKINS

Invited Paper

Abstract—Lasers can now be regarded as practical and economic tools with unique properties which have been utilized effectively in several applications in industry. Major applications of the laser are in thermal processes and measurements. Large amounts of concentrated energy allow microdrilling, welding, cutting, and fracturing to be simply effected in even the hardest materials. The coherence properties provide ideal sources for alignment instruments and interferometers for accurate length measurement. This paper describes these applications and their typical capabilities. Future developments and their potential are discussed, with the conclusion that the laser will be increasingly used in the manufacturing environment.

I. INTRODUCTION

IT IS more than eight years since the report of the first laser by Maiman [1] introduced a new radiation source. The properties of spatial and temporal coherence with a high intensity stimulated many ideas for new developments. Very quickly a large number of laboratories began research into applications of lasers which are now showing their value in terms of capability and economic return both in industry and elsewhere. This paper is intended to review those applications which are used in industry and describe some of the current developments.

There are many parameters which determine a laser's usefulness for an industrial manufacturing application. These include spatial and temporal coherence, power, power-time characteristics, wavelength, and cost. The spatial coherence and power allow the energy from the laser to be focused down to a small area, creating very large energy densities which can be used for various processing operations such as welding, drilling, and cutting. The wavelength is important in determining the amount of energy absorbed, and the power-time characteristics govern the rate and duration of energy input to the processing operations.

The spatial coherence for the lower powered gas lasers is essentially perfect and produces highly collimated beams. These can be used as alignment aids and as point source (Kohler) illumination for diffraction effects. The temporal coherence allows single frequency light to be generated and has brought interferometry to industrial measurements; it also introduces new tools such as holographic interferometry. Finally, the intensity and coherence properties have provided an improved source for light scattering phenomena and allow nonlinear interactions to be observed.

Many materials have been shown to be capable of stimulated emission; however, relatively few materials are used in practical laser systems. Table I is a listing of the more common types of lasers commercially available and their typical capabilities. Ruby [1] was the first material operated and remains widely used in pulse high energy processes such as welding and cutting. When Q-switched (a technique providing for extremely short pulses of very high peak powers), it is used for measurement studies as well as in processing. Other materials which are proving in many cases to be better than ruby (Cr doped Al_2O_3) are those doped with Nd [2]. The primary host materials are glass and yttrium aluminum garnet (YAG) [3]. A very efficient laser is the semiconductor junction laser [4]. However, the low output power and low coherence properties have limited its use up to now.

The first gas laser was the He–Ne [5] and it has remained a popular laser for low power (up to 100 mW) applications where coherence properties are the main requirement. These applications include interferometry, holography, and diffraction effects. Other gas lasers which have produced higher powers and shorter wavelengths are the argon ion [6] and krypton ion [7] lasers operating either pulsed or continuous. Recent developments are permitting radiation into the near ultraviolet to be produced with high coherence properties and at comparable cost to the He–Ne laser.

A longer wavelength laser [8] (10.6 μm) of much higher power (> 1 kW) and efficiency (20 percent) is the CO_2–N_2–He laser. This laser is finding a useful place in both processing and measuring systems. It can be operated with single or multimode output[1] and may be run either continuously (CW), pulsed, or Q-switched.

The material in this paper divides itself into two broad sections: processing operations, and measurement and inspection techniques. Our object has been to try to restrict the subjects to industrial manufacturing applications. Although we have attempted to give a broad view, we have not intended an exhaustive coverage.

II. PROCESSES

Introduction

The properties of intensity and spatial coherence of the laser give a source of radiant energy which can be concentrated to achieve extremely high power densities. When

Manuscript received November 14, 1968; revised December 12, 1968. *This invited paper is one of a series planned on topics of general interest.— The Editor.*

The authors are with the Engineering Research Center, Western Electric Company, Inc., Princeton, N.J. 08540

[1] When the transverse spatial configuration of a laser output is of a symmetrically uniform (i.e., Gaussian) intensity distribution, the operation is described as single mode. Conversely, multimode operation is described as a multispot pattern of varying intensities which normally has a higher total energy than the single mode. See, for example, Fox and Li [9], [10].

Reprinted from *Proc. IEEE*, vol. 57, pp. 114–147, Feb. 1969.

TABLE I

Material	Mode of Operation	Wavelength	Output	Typical Uses
Ruby (Cr^{+3} in Al_2O_3)	pulsed	0.6943 μm	1–500 J	welding, drilling, vaporization
	Q-switched		1–1000 MW	ranging, vaporization, pulsed holography, scattering spectroscopy, Raman-Brillouin
Nd in glass	pulsed	1.06 μm	1–500 J	welding, drilling, vaporization
	Q-switched		1–600 MW	ranging, vaporization, Raman-Brillouin
	mode-locked		1 MW	communications, ranging
Nd in YAG	Q-switched	1.06 μm	1–50 kW	ranging, vaporization, spectroscopy, Raman-Brillouin
	continuous		1–50 W	vaporization
GaAs	pulsed	0.900 μm	5.0 W	ranging
	pulsed (liquid N_2)	0.8490 μm	5.0 W	
CO_2-N_2-He	continuous	10.6 μm	1 kW	fracturing, vaporization, spectroscopy, communications
	Q-switched	10.6 μm	100 kW	vaporization, welding, ranging
He-Ne	continuous	0.633 μm	1–1000 mW	interferometry, diffraction, alignment, holography, scattering
Ar	continuous	0.4880 μm	0.1–5 W	holography, heating, diffraction Raman spectroscopy
	pulsed	0.4880 μm	25 W	
		0.5145 μm		
Kr	continuous	0.3507 μm		
		0.3564 μm		
		0.4762 μm		
		0.5208 μm		holography, Raman spectroscopy
		0.5682 μm		
		0.6471 μm		
N_2	pulsed	0.3371 μm	1 kW	photochemistry
Neon	continuous	0.3324 μm	50 mW	

absorbed by matter, this energy can be sufficient to overcome the binding forces associated with the atomic and molecular structures of materials, so that high temperature phase changes are easily effected.

When radiant energy is focused by a lens, the power density within the focused spot can be often represented by the following relationship:

$$P = \frac{4E}{\pi f^2 \theta^2 t} \tag{1}$$

where

P = power density at focal plane of the lens
E = energy output from the laser
f = focal length of the lens
θ = beam divergence (full angle)
t = laser pulse length.

The minimum beam divergence of a light beam is a function of its wavelength and beam diameter. The Rayleigh criterion for the minimum beam divergence of a spatially coherent light beam is

$$\theta_{\min} = \frac{1.22\lambda}{R} \tag{2}$$

where λ = wavelength, and R = radius of the beam or aperture.

With the high energy lasers available today, it is possible to approach within a factor of two or three the minimum value of beam divergence predicted by the Rayleigh criterion. In typical high powered pulsed laser systems the beam divergence will be within a range of 2 to 10 mrad (half-angle) at output levels well above threshold. Such

beam divergence is commonly stated for half-power levels using a suitable aperture stop placed at the focal plane of a long focal length lens to resolve one-half the total non-apertured energy. If it is assumed that the beam divergence of the laser is 5 mrad and that its radiation is focused with a 25 mm focal length lens, then the power density within the focal spot is 2000 times the power within the unfocused laser beam. Typically, the peak powers obtained from pulsed laser systems range from 10 000 W to approximately 1 MW. The focused beam then would have a power density of from 20 MW to 2 GW/cm^2. Assuming that only 0.1 percent of this energy were absorbed by a material, the rate of energy absorption would be 20 kW to 2 MW/cm^2.

In a typical pulsed laser system, the focused spot size S may be adequately approximated by the equation

$$S = f\theta. \tag{3}$$

Using the values of beam divergence and lens focal length from the preceding paragraph, the focal spot size is approximately 0.25 mm in diameter. With shorter focal length lenses and improved beam divergence of the laser, it is possible to reduce this focal spot size to values much less than 0.25 mm. In fact, with the proper laser system operating under the most ideal conditions, it is possible to evaporate lines in thin films where the width of the lines is essentially equal to the wavelength of the laser radiation [11].

By utilizing the various types of laser systems available today, it is possible to achieve high powered outputs over time periods which can range from nanoseconds to continuous outputs. The extremely short pulses of laser energy are achieved by Q-switching and mode-locking techniques which use electrooptic (Kerr cell), acoustooptic (ultrasonic), simple optic (rotating prism) shutters or, quite commonly for pulsed systems, bleachable organic dyes. These devices are able to spoil the stimulated emission mechanism, allowing for a large buildup of excited lasing species above the threshold level. When the Q-spoiling device is removed (i.e., the shutter between the cavity reflectors is opened), then the excess excitation very quickly produces a large amount of stimulated emission which is discharged in an extremely short time. The intensity of this short pulse greatly exceeds that obtainable from a normally pulsed laser. When the excitation mechanism can be maintained and proper cooling of the laser material can remove the associated generated heat, then CW operation (of solid state lasers) is achieved.

Because of the monochromaticity of laser light (spectral line widths of 0.1 nm or less are typical), chromatic aberrations of lenses may be ignored and thus laser radiation can be concentrated with simple optical systems. In the processing systems to be described, viewing systems have been coupled with the optics used to concentrate the laser's radiation. The use of closed-circuit television (CCTV) systems (Fig. 1) has been found to be particularly desirable in that it offers certain advantages over standard microscope and projection viewing. First, the use of CCTV systems allows the workpiece to be observed while the laser energy is actually impinging on the workpiece. Second, it is less tiring for an operator to view the workpiece on the CCTV monitor. Third, the use of CCTV offers complete safety of operation for the operator.

There are several areas of material processing and they will be discussed in the following sections:

III. Welding.
IV. Material Removal (including drilling, trimming, and evaporation).
V. Material Shaping (which covers cutting, scribing, and controlled fracturing).
VI. Thermally Induced Change (localized heat treating including annealing and surface oxidation, grain size control, diffusion, photochemical reaction, zone melting).

III. WELDING

Fusion welding is one major fabrication process where many had expected the laser to find a fertile field of application. However, the number of presently known welding applications is paradoxically small and most of these have been reported as repair or short term, low volume specialty runs.

The understandable reluctance of critical industries to set up full-scale welding production with essentially untested facilities is another major deterrent and in this respect research and development laboratories have a real challenge. Short time, high temperature reactions that one encounters with the interaction of laser energy with matter have provided the research scientist and engineer with a field where the resources of many disciplines were called upon in new experiments. Much information and time tested experi-

Fig. 1. A pulsed high power solid state laser system showing the use of closed-circuit television with the focusing optics. On the television monitor is shown a wire with a hole drilled by the laser.

mental data were required in order to provide potential applications with the necessary integrity of results. A new view of materials and their properties with respect to the fabrication force of concentrated laser light first had to be established. It is believed that this normal development phase of a new field has been the controlling factor in limiting the number of welding applications.

It is not intended in this paper to predict laser welding activities for the future but to indicate that there is now a very strong upward trend, not only in welding applications but also in areas of materials fabrication.

Materials and Their Properties

Metals are the predominant materials that have been joined by laser welding. However, it is possible that some thermoplastics will be heat sealed or welded.

The physical size and geometry of the component parts of a weldment are presently restricted to the small-scale category, from the microminiature scale [integrated circuit (IC) technology and thin films] to an upper limit related to the fusion penetration depth. This upper limit will be largely dependent on the material properties, specifically thermal and optical.

Thermal Diffusivity: One important material property that greatly affects the depth of the fusion zone is the heat diffusivity. In general, high heat diffusivity will allow for welding of thicker materials. The depth of heat penetration in a given material can be estimated from knowledge of the temperature at the material surface and its heat diffusivity. The differential equation for the linear flow of heat is

$$\frac{\partial^2 T}{\partial x^2} = \frac{1}{\alpha} \frac{\partial T}{\partial t} \tag{4}$$

where

T = temperature
x = linear diffusion depth
α = thermal or heat diffusivity = $K/\rho C$, in which K = thermal conductivity, ρ = density, and C = specific heat
t = time for diffusion to a depth x.

Carslaw and Yaeger [12] give the solution for the case of a semi-infinite solid for the boundary conditions

$$\left.\begin{array}{r} x = 0 \\ t = 0 \end{array}\right\} T = 0$$

and for $t > 0$, $T_s = $ constant.

The solution is

$$T_x = T_s \, \mathrm{erfc}\left(\frac{x}{2\sqrt{\alpha t}}\right). \tag{5}$$

The surface temperature T_s at $x = 0$ is given by

$$T_s = \frac{2F}{K}\left(\frac{\alpha t}{\pi}\right)^{\frac{1}{2}} \tag{6}$$

where $F = $ heat flux at the material surface upon absorption of laser energy, and is proportional to P of (1).

Fairbanks and Adams [13] treat the condition of a center of a hot spot on the surface of a material and use the relationship from Carslaw and Yaeger

$$T_s - T_0 = \frac{2F\sqrt{\alpha t}}{K}\left(\frac{1}{\sqrt{\pi}} - \mathrm{ierfc}\,\frac{a}{2\sqrt{\alpha t}}\right) \tag{7}$$

where

$T_s = $ surface temperature at the center of the hot spot
$T_0 = $ initial uniform temperature of the material
$a = $ radius of circular hot spot (estimated from $S = f\theta$).

If the spot size is large or the pulse length short, the second term in the brackets becomes negligibly small, which means the surface temperature quickly increases to a uniform temperature. This uniform temperature can be assumed to be constant for the pulse durations commonly used in laser welding.

Materials with high heat diffusivity accept and conduct thermal energy very quickly and experience no thermal-shock cracking. Table II lists the thermal diffusivity of several metals and alloys.

Many materials have relatively poor heat diffusivities which limit their laser weldability. McCracken [14] reports that the stainless steels and the heat-resistant alloys, such as Rene 41, have low heat diffusivities. He states that because of the poor heat diffusivity, laser welding of these materials in a single-pass process limits the fusion depths to 0.04 in (1.02 mm) because of vapor depletion on the weld surface. It is expected that increased pulse lengths or use of CW systems will significantly improve these welds. However, where there is a relatively large temperature difference between the melting and boiling points, presently available pulse lengths have provided sound welds.

Fig. 2 shows the butt weld of 0.007 in (0.178 mm) thick, 402-type stainless band steel. Defocusing the beam a small amount gave the proper power density at the joint surface to effect the weld. Schmidt *et al.* [15] have analyzed the weldability of stainless and 18 percent Ni maraging steels as related to the position of the workpiece from the focal plane of the focusing lens.

Other Material Properties: Examination of other properties of the workpiece will greatly help in understanding the manner in which the laser is used for welding and other

TABLE II

Material	Thermal Diffusivity α* ($\times 10^{-4}\,\mathrm{m^2/s}$)
Metals (commercially pure)	
Aluminum	0.91
Beryllium	0.42
Chromium	0.20
Copper	1.14
Gold	1.18
Iron	0.21
Molybdenum	0.51
Nickel	0.24
Palladium	0.24
Platinum	0.24
Silicon	0.53
Silver	1.71
Tantalum	0.23
Tin	0.38
Titanium†	0.082
Tungsten	0.62
Zinc	0.41
Alloys	
Brass (70:30)	0.38
Phosphor bronze (5% Sn)	0.21
Cupro nickel (30% Ni)	0.087
Beryllium copper (2% Be, δ phase)	0.29
Inconel (76% Ni, 16% Cr, 8% Fe)	0.039
6061, O temper aluminum alloy (1% Mg, 0.6% Si, 0.25% Cu, 0.25% Cr)	0.64
304 type stainless steel (19% Cr, 10% Ni)‡	0.041

* Calculated from the data of physical properties contained in [44].

† Values for thermal conductivity and specific heat are the average of the spread for several alloys of 99.0 percent commercially pure grades at 93°C (200°F).

‡ Wrought stainless steel in the annealed condition. Values of thermal conductivity and specific heat apply at 100°C.

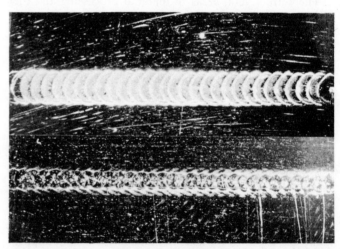

Fig. 2. An example of butt welding relatively thin materials. The photograph shows the front side to the laser beam (top), and the back of 7 mil (0.18 mm) thick stainless steel (bottom). Overlapping spots of approximately 40 mils (1 mm) in diameter gave a power density of 6×10^5 W/cm² using a 67 mm focal length lens defocused by 3.8 mm. Beam energy was 12 J/pulse. Pulse length was 5.5 ms.

thermal processes. General categories that must be considered include thermal, physical, mechanical, surface condition, metallurgical, and chemical properties. Knowledge of the thermal properties (specific heat, thermal conductivity, melting, boiling and vaporization temperatures, latent heat, heat capacity) allows for the calculation of the progress of the fusion front and of the temperature dis-

TABLE III
HEAT TRANSFER DATA FOR SEVERAL METALS*

Metal	$T_m^\dagger$ (°C)	$T_b^\dagger$ (°C)	$H^2 t_m$ ($\times 10^8$ J^2/m$^4 \cdot$s)	$H^2 t_b$ ($\times 10^8$ J^2/m$^4 \cdot$s)	S_e^1 ($\times 10^{-2}$ mm)	S_e^5 ($\times 10^{-2}$ mm)
Chromium	(1875)	(2665)	5.59	13.0	3.6	7.9
Copper	1083	2595	12.03	84.8	23.2	51.5
Gold	1063	2970	6.52	60.7	27.7	61.6
Iron	1536	(3000 ± 150)	4.79	19.2	9.1	20.3
Molybdenum	2610	5560	21.51	119.4	13.7	30.4
Nickel	1453	2730	5.76	26.0	7.6	17.2
Platinum	1769	(4530)	4.93	38.2	11.2	25.2
Silver	961	2210	6.85	45.2	27.2	60.5
Tantalum	2996 ± 50	5425 ± 100	8.58	34.4	7.4	16.2
Tungsten	3410	5930	29.24	117.0	10.4	24.4

where

H = heat flux at focal spot on the metal surface; heat flux is approximately related to the maximum depth of fusion front penetration by

$$S_{max} = \frac{0.16}{\rho L H} (H^2 t_b - H^2 t_m)$$

in which ρ and L are density and latent heat of fusion, respectively.
t_m = time at which melting first occurs.
t_b = time at which boiling first occurs.
S^1 = depth of melting when heat flux is adjusted such that surface vaporizes in 1 ms.
S_e^5 = depth of melting when heat flux is adjusted such that surface vaporizes in 5 ms.

* Taken from Cohen [16].
† Temperatures are taken from [44], and for some metals (in parentheses) are not the same as those used in Cohen's analysis.

tribution in both the molten and the solid materials for certain simplified heat transfer models [16]–[19]. Table III shows the results of some heat transfer calculations taken from Cohen's analysis [16]. The table gives the melting and boiling points of several metals and the time at which the surfaces of these metals reach the melting and boiling temperatures. The data also predict the maximum depth of stable penetration of the melt puddle for 1 ms and 5 ms pulses, respectively. Maximum penetration has occurred when the surface temperature has reached the boiling point. Materials may differ widely in their maximum achievable melt depths, and the laser welding or melting of large pieces of chromium or tantalum may be expected to be more difficult than similar processing of copper or gold. Metals with high vapor pressure have a greater tendency to vaporize when irradiated with a focused laser beam. Magnesium and similar low temperature melting alloys tend to have excessive material-depleted surfaces, particularly if their thermal diffusivity is low. Alloys with an appreciable zinc content tend to "overmelt" in attempts to weld them. Large aggregates of zinc atoms can boil out of the melt, producing uneven melting and often a resultant porous weld structure. The amount of vaporization for a particular alloy is directly related to the power density at the focal spot and the pulse shape.

The mechanical and thermomechanical properties such as strength and thermal expansion can be important when considering a particular laser welding application.

Metallurgical and Chemical: The fundamental metallurgy of alloy systems is not changed when one metal reacts with another upon absorption of laser energy. Though acceptable laser welds have been claimed for dis-

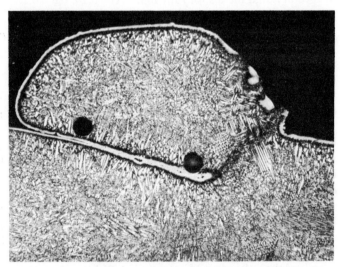

Fig. 3. Microstructure of a laser weld for beryllium copper wire. Dendritic and acicular-type structures are commonly found for many metals. Approximately 435 × mag.

similar (crystal structure and atomic radii) metals, intermetallics and other high temperature formed complex structures are by no means eliminated or avoided. The fact that welds have been made is no guarantee that they will stand up to a designed function, especially if subjected to a stress force.

In laser welding, the extremely fast heat rise and almost equally fast cooling cause a nonequilibrium type of solidification. Fig. 3 shows a fine dendritic structure fairly common for laser welds of many metal combinations. This type of structure is similar to that found for cast metals.

Consequently, for certain metal combinations it may be necessary to perform a subsequent stress-relieving or annealing operation on the weld joint. On the other hand, the swirl-like pattern resulting from the short time, high temperature reaction produces a mechanical intermixing of the metal constituents, with some amount of elemental diffusion across the interfaces. This usually contributes to the strength of the weld structure, provided the thermal expansion coefficients are compatible.

During a laser weld the chemistry involved in reactions at the surface can influence the weld integrity. Usually, normal atmospheric conditions are not detrimental to a laser weld. However, material surface contaminants and adsorbed gases can interfere with the weld results. Some knowledge of the chemistry involved will greatly aid in determining what surface preparation may be required and what welding parameters must be used.

Interaction of Photon Energy with Metal Surfaces

Reflectivity: Most metals reflect a large portion of a laser beam; therefore, correspondingly high powered beams are required for most welding processes. However, the surfaces of most materials reflect light during only a small portion of the laser pulse. As the heated surface reacts with the atmosphere in contact with the melt, reflectivity often decreases. Anderson and Jackson [20] give the absorptivity of several liquid metals for ruby light: Al=0.20, Cu=0.15, Fe=0.45, and 300-type SS=0.60. A sharp decrease in reflectivity occurs when vaporization begins. It is believed that this decrease is the result of the vaporized material scattering and "trapping" the light at the surface of the material. The refractive index of the gases immediately above the focused spot area also affects the amount of additional light that will be absorbed at the surface.

Since the reflectance of most metals increases with wavelength, more output energy will be required from a laser with a long wavelength than from one with a short wavelength. For example, as the wavelength increases from 0.69 μm up to about 1 μm, an order of magnitude increase in power is required in order to melt some materials. When a low power laser beam is pulsed on a polished metal specimen, there is no melting because the light is largely reflected. To produce some surface melting and/or vaporization, the energy of the pulse must be increased. A sudden increase in the absorption of light, however, makes it difficult to control the welding of some materials. For example, when thin ribbons of a highly reflective metal, such as gold, are welded, the intensity and duration of the laser beam must be adjusted carefully to prevent the ribbon from vaporizing [21]. This could happen because the laser beam intensity required to break down the surface reflectivity is greater than that required for stable propagation of the fusion front once the surface has melted [16]. One possible way of overcoming the problem is to decrease the beam power after applying a short breakdown pulse. A simpler approach, and often quite effective, is to intentionally coat or change the material surfaces so that they become good absorbers.

Surface finishes can affect the accepted-light to reflected-light ratio from 20 to 90 percent. A rough appearing surface is not necessarily a good absorbing surface—it can be a surface that scatters the reflected light. However, for the case of bright copper sheet, its reflectivity to 0.6943 μm laser light was reduced from approximately 95 percent to less than 20 percent by oxidizing the surface (3 to 1 ratio of CuO to Cu_2O) using a solution of a proprietary compound called Ebonol.[2]

Surface finish can also affect the absorption of light when the size of surface asperities are of the order of the wavelength of the light [22]. Fig. 4 shows the depth of penetration of the melt puddle in pieces of copper exposed to ruby radiation. The copper was prepared with surface finishes in the range of 0.025 to 100 μm; it appears that above about 2 μm the absorption process is independent of surface finish. Fig. 5 shows the appearance of the exposed copper surface for several degrees of roughness. With thin films, the optical properties of metals have an increasingly more complex role and are found to directly influence the results of thin film weldments. One example of this is illustrated in Figs. 6 through 9. Figs. 6 and 7 dramatically reveal the effects of reflectivity. The laser irradiated pattern on the nichrome metal film required about $2\frac{1}{2}$ times more energy than that required on the blued steel sample. The four-line square pattern was generated in our laboratories by a telescope lens system used for studying the laser welding of beam leaded IC devices.

Fig. 8 shows a focused line weld of a 16-beam leaded IC device. Note the good absorption of ruby light (0.6943 μm) by the beam leads, and yet very little or no absorption by the gold thin film or the ceramic. The compositional makeup of the beam lead has a titanium-platinum layer facing the incident laser light. This layer absorbs about 45 percent of the laser energy, while the gold thin film and ceramic reflect over 90 percent of the laser energy. Fig. 9 shows a transverse cross section of one of the welded beam leads, with intact thin film on either side of the well-wetted fusion joint. The fine definition of the materials shown in this photograph required a special metallographic [23] preparation process developed at the Western Electric Engineering Research Center.

Absorption of Light Energy: When laser light is directed onto a material surface, a portion of the light is absorbed and converted to thermal energy, and the rest is reflected.

It is fairly well agreed by many investigators (Mott and Jones [24] and Seitz [25]) that the absorption of light energy by materials is accomplished at a depth approximated by the mean free path of the valence electrons. The familiar equation

$$I_x = I_0 e^{-kx} \qquad (8)$$

is sufficiently accurate for describing attenuation of light intensity with depth. Here x = depth of light penetration of intensity I_x, I_0 = the incident intensity, and k = an absorp-

---

[2] Manufactured by Ethone, Inc., West Haven, Conn.

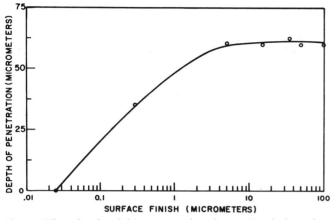

Fig. 4.   Effect of surface finish on copper for ruby wavelength absorption.

Fig. 6.   A square pattern generated by four cylindrical lenses incorporated into an inverted telescope optical system (Patent Application Pending) on a 1 μm thick nichrome thin film; 8 J and 2.2 ms pulse length resulted in melting and vaporizing a line width of approximately 2 mils (0.05 mm). 40 × mag.

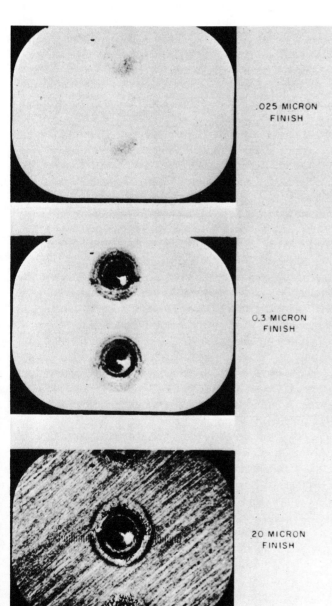

.025 MICRON FINISH

0.3 MICRON FINISH

20 MICRON FINISH

Fig. 5.   Copper surface of different surface finish after exposure to ruby radiation of equal energy. (After Cohen and Epperson [22].)

Fig. 7.   The same pattern shown in Fig. 6, developed on blued carbon steel. Beam energy was 3 J at 2.2 ms. 40 × mag.

Fig. 8. Single line welding beam leaded integrated circuits to gold thin film conductor paths on an alumina ceramic substrate. All four leads were welded with one pulse of a ruby laser at 2.2 ms pulse length and 3.5 J beam energy using a cylindrical lens. Width of a beam lead is 0.0045 in (115 μm).

Fig. 9. Transverse cross section of a beam lead weld. Note absence of a physical interface between the beam lead and thin film materials within the melted area. Approximately 435× mag.

tion coefficient. For many of the common metals the depth of maximum light penetrated ranges from about 5.0 to 50.0 nm. At this depth the converted energy is transmitted into the interior of the material by means of thermal conduction.

For sufficiently high intensities of photon energy directed at the surface, the temperature generated will become great enough for the material to melt. Subsequent absorption of light will then cause a molten puddle to propagate into the bulk material. When the power level is raised still further, gross vaporization occurs and a high velocity plume of particulate and ionized matter is ejected from the material surface [26], [27]. Several investigators [28], [29] have calculated the surface temperature of laser irradiated materials by ion emission measurements. Knecht [29] used the relationship

$$\log W = C - 0.5 \log T - \frac{B}{T}$$

where

W = rate of evaporation in g·cm^{-2} s^{-1}
T = temperature at the material surface in °K
B and C = constants of the material.

He found a calculated surface temperature for gold of 1270°K when he measured the induced ion current from a 30 ns, 45.3 kW power, Q-switched laser pulse. The ion current was converted to an evaporation rate of singly charged gold ions of 6×10^{-7} g · cm^{-2} s^{-1}.

Welding Parameters

In considering pulsed laser systems for welding applications, one major factor that must first be determined is the capacity of the system in energy or power output and in duration of lasing action. These basic parameters, along with other secondary factors, are normally used in characterizing a laser weld. These parameters or descriptive characteristics of lasing action are separated into two groups—a primary group of two and a secondary group of four.

Energy or Power: The output energy, commonly measured in joules, is a function of a charging voltage (obtained from the power supply) on a bank of capacitors (designated the energy storage unit). The stored energy is dissipated in a high intensity flash lamp which by proper coupling design excites the laser rod for stimulated emission.

Ruby and neodymium in glass are the two lasing materials used in most commercial pulsed welding systems. Ruby is generally more rugged and can be pumped harder to produce higher powers. In some respects, the shorter ruby wavelength has an advantage over that for neodymium in glass. Nd:glass has a lower lasing threshold, and while operating over a wider temperature range (up to 85°F) can give slightly higher slope efficiencies. Its lower thermal conductivity and susceptibility to damage under higher pumping levels limits its ability to achieve high average output power. However, some recent work [30], [31] in the glass technology area indicates that some of these shortcomings can be overcome.

Output Duration or Pulse Length: The parameter of pulse length, which differentiates a welding process from a drilling mechanism, is essentially determined by the amount of capacitance (or capacitance-inductance groups) in the excitation circuit. The proper matching of the flash lamp to the electronics and the optical coupling geometry in the laser head assembly both greatly aid in optimizing the laser pulse characteristics for effective welding practice. In general, the longer the pulse length, the more the workpiece or weldment will react along the fusion mechanism of energy absorption, heat transfer, melting, intermixing of constituents, and subsequent solidification. With good control of the laser beam, vaporization and possible particle expulsion can be minimized.

Present commercial systems can achieve pulse lengths up to 10 ms. Smith and Thompson [32] have compared Nd:glass to ruby for the same input parameters in the same cavity geometry. For the limited input energy covered, Nd is shown to give higher output energy at longer pulse length than ruby. With respect to pulse shape, limited studies in our laboratory and by others show that, in general, an exponentially decaying pulse shape, as opposed to a square wave shape, is favored for welding work. For example, consider a material like gold; this has a high reflectivity at the wavelength of ruby yet very good thermal properties, such as high heat diffusivity. It is desired that the first part of the laser pulse overcome the initial high reflectivity of the solid surface, then continue, but at lower intensity so as to minimize vaporization and at the same time effect sufficient heat transfer. One can see that in a decaying-type pulse shape we do have a high intensity at first and then a falloff. Fairbanks and Adams [13] have concluded that the best possible condition for energy input to a weldment is to have the energy delivered at constant surface temperature as opposed to input at constant intensity (i.e., the square wave pulse shape). Though this is highly desirable, it is not known whether anyone has demonstrated such control.

Power Density: This is not an independent variable like the primary parameters. It is commonly expressed in watts per square centimeter and is the result of concentrating the output energy, most commonly by simple lens convergence.

Wavelength: This is not a variable but can be changed by use of devices like harmonic generators. The wavelength is characteristic for each lasing material and, as noted before, it determines to a large degree how much of the beam focused on a material is absorbed and converted into heat. The wavelength is also important in determining the minimum spot to which the beam can be focused.

Focusing: This plays a major role in providing a wide latitude in determining the amount of energy a given material sees. For pulsed laser systems utilizing ruby or Nd:glass, the beam intensity at the focal plane is not truly uniform throughout the focal spot size, but actually depends on the spatial distribution of the photon intensity. In the general introduction to this section on processes, the spot diameter was simply approximated by the beam divergence and the focal length of the condensing lens used. However, since the beam intensity is normally maximum on the beam axis and decreases gradually with distance away from the axis, then the focused beam spot size or diameter should be defined with respect to some intensity level of the beam profile. Assuming single mode operation, this spot size is often defined as twice that distance from the axis at which the intensity has fallen to $1/e^2$ of the maximum value [33]. When defined in this manner, the diameter to which a laser beam will be focused by a lens just large enough to admit the entire beam is proportional to the laser wavelength and the f/number of the lens (ratio of focal length to diameter). For a laser of wavelength λ, oscillating in the lowest order mode, the diameter of the focal spot is

$$S = \frac{2\lambda f}{\pi R} \tag{9}$$

where f = focal length of the lens, and R = radius of the beam as it enters the lens. The significance of this equation as compared to the relationship $S = f\theta$ lies in the theoretical approach to the focusing of a Gaussian distribution. This equation predicts that spot sizes of the order of the output wavelength are achievable. However, the extent of the region that is melted in the target material depends, as was previously shown, upon the optical and thermal properties of the material.

Defocusing techniques (adjusting the workpiece surface at a plane other than the focal plane of the beam focusing lens unit) allow for an effective means to gain added control on the amount of heat generated in the weld area (see Fig. 10).

Other factors to be considered are means of altering the laser beam characteristics such as the use of filters, apertures, and masks. Parameters that have a bearing on laser welding, specific to production capability, are the joint geometry or design and welding speed. These process parameters are often deciding factors in determining whether a particular laser welding application can economically compete with existing methods (allowing for any improvement in the weldability).

Joint geometry has some influence on the thickness that can be welded, therefore affecting the choice of laser parameters. Since there is little time for metal flow and a relatively small amount of molten metal, the process will most always require very close-fitting joints. In the area of IC connections, intimate and positive contact of the weldments is most important. Jackson [20] has demonstrated several laser weld geometries for the more common parts.

Welding speed is limited by the pulse-repetition rate of presently available equipment. A rate of 1 pulse/s is commonplace today, and at an appreciably high energy level (>10 J). Although present laser welding equipment requires a relatively long delay between one weld spot and the next, a change in length of this delay does not affect the weld. Pulsed-laser welding is a slow process when compared to other welding techniques, averaging between 0.2 and 2 in/min (0.5 and 5 cm/min). On the other hand, "CW" welding with CO_2 (pulsed at 75–300 pulses/s) has the potential of welding at relatively higher speeds of $\frac{1}{2}$ to 6 in/min (1.3 to 15 cm/min). With higher powered CO_2, as well as with the rapidly pulsed argon ion and YAG systems, welding speeds as high as 50 in/min (130 cm/min) can be expected. Pulsing techniques of high powered CW systems hold promising expectations in large area metal melting.

At present, it can be said that the laser is quite applicable to the miniature device area of manufacture, and has a somewhat limited applicability to the standard joining operations where the mass of material in the weldment increases. As a general rule, a potential application for using the laser can be considered if the particular job cannot

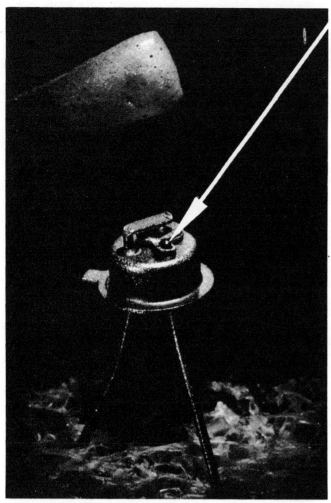

Fig. 10. A transistor unit showing the laser weld of the nickel tab containing the semiconductor material to the nickel alloy post.

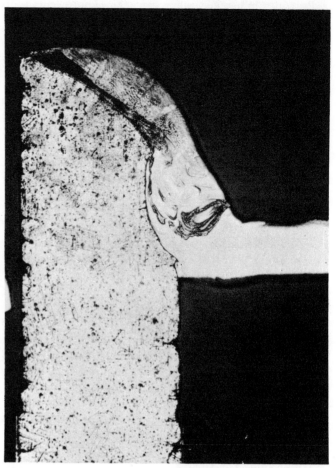

Fig. 11. Cross section representing a typical transistor unit weld. Note the flow pattern of melted material. Approximately 94 × mag.

be accomplished any other way or the economics and/or the engineering requirements are difficult to meet using conventional methods.

The laser as a welding tool will not eliminate normally encountered metallurgical difficulties. Porosity is one difficulty which is not too infrequently encountered [34]. The tendency to vaporize low boiling point (high vapor pressure) constituents of an alloy metal is apparently the cause of this difficulty. This is one of several areas that require concentrated research effort if the problem is to be effectively controlled and eliminated. Although the present pulsed laser has limitations as a welding tool, it certainly possesses several advantages which make its application as a standard manufacturing tool increasingly more likely. In addition to the monochromaticity, coherence, and intensity of the beam previously discussed, the following advantages are specific as to the laser's capability as a welding tool.

1) Because of the short time factor associated with this source of heat, welds adjacent to heat-sensitive elements can be effected. Fig. 10 shows the laser welding of transistor elements to nickel alloy parts used in the manufacture of transistor devices at the Western Electric Plant at Reading, Pa. This application is con-

sidered to be the first laser welding process for small size semiconductor devices. Tweezer welds of the Ni tab (containing the semiconductor material) to the Kovar post did not meet the vibration test requirement of 15 000 g's minimum. Centrifuge tests of laser welded samples showed bending of the tab members between 20 000 and 25 000 g's but no failure of the weld joints.

A cross section of a weld made of one of the tabs to the post is shown in Fig. 11. The tab is approximately 5 mils (0.13 mm) thick and the post about 18 mils (0.45 mm) in diameter. A laser beam of 7.5 J energy was focused with a 25 mm focal length lens at the edge of the post in line with its longitudinal axis. At this point, the beam was defocused with respect to the nickel tab by approximately 15 mils (0.4 mm). The pulse length was 3 ms. With these stated conditions, a calculated power density of 6×10^6 W/cm² was attained at the material surface area, which has been determined to be about 4×10^{-4} cm². Of course, the materials do not absorb all this energy; some losses occur in the optics and an appreciable amount is reflected. The weld area with this approach covers more than 180° around the periphery of the post.

2) Similarly, the short time factor of a laser weld allows

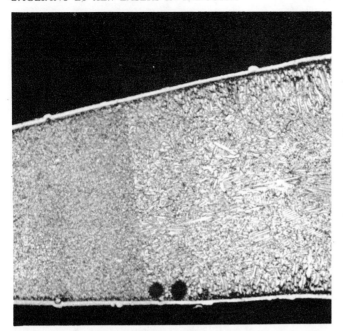

Fig. 12. Laser welding of 5 mil (0.13 mm) diameter heat treated beryllium copper wire (Fig. 3) showing extent of the heat affected zone. The weld shown in Fig. 3 would lie to the right of this photograph. The section is along the wire. Difficulties in sectioning produced the taper shown. To the left of the very pronounced transition line is the microstructure of the original heat treated condition. The light envelope about the sample is a 0.2 mil (2.5 μm) thick Permalloy electroplate. Approximately 585 × mag.

Fig. 13. A pair of 10 mil (0.25 mm) diameter precious metal wire contacts welded by one pulse from a ruby laser to a 9 mil (0.23 mm) thick phosphor bronze spring member. The weld was effected using a beam energy of 6 J at a pulse length of 3 ms. The ruby radiation was concentrated through a 43 mm focal length lens defocused from the focal plane (corresponding to the surface of the phosphor bronze spring) by −2 percent. Approximately 44 × mag.

heat treated and magnetic materials to retain their properties just outside a very small and narrow heat affected zone (Fig. 12). Rapid heating with the use of high powers and quench-like cooling are advantages for conditions as those cited in "1."

3) Welding in otherwise inaccessible areas is realized with the laser beam. Problems associated with the physical contact of weld tools, electrodes, and other media are nonexistent with a light beam.

4) Laser welding can be performed in any environment through transparent enclosures and magnetic fields.

5) Materials difficult or impossible to weld by the more conventional techniques are often quite easily and reliably joined using the laser. Fig. 13 is the laser weld of a precious metal wire contact to a phosphor bronze spring. The relatively high electrical resistance of the wire offered difficulties for the I^2R resistance heating method. The laser produced a weld equal to or greater than the strength of the 0.010 in (0.25 mm) diameter contact wire. This is an application that is currently being tooled-up at the Western Electric Plant at Oklahoma City.

6) Distortion and shrinkage of the weldment are negligible.

7) Insulated wires can be effectively welded without prior removal of the insulation (Fig. 14) [35].

8) Tooling, fixtures, setup, and inspection are in general relatively simple and easily adaptable to production procedures.

9) The optics system can be readily adaptable to CCTV viewing. Thus, on-the-spot inspection of the magnified object is realized.

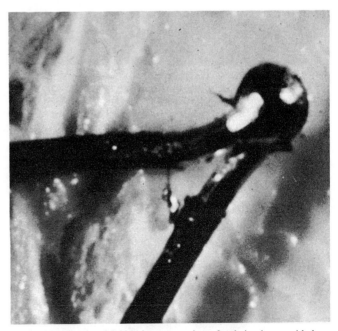

Fig. 14. Insulated 0.0056 in copper wires after being laser welded. Approximately 63 × mag.

It is also well to bear in mind that laser welding can be more economical than conventional techniques in some cases because of improved reliability and better yield.

IV. MATERIAL REMOVAL

Laser Piercing (Drilling)

The first major industrial application using the laser was in the general area of hole drilling, or more properly, piercing. The Western Electric Company's Buffalo Plant, Buffalo, N. Y., has been using the pulsed ruby laser since December, 1965, to drill holes in diamonds for the fabrication of diamond dies [36]. These dies are used in drawing intermediate 15 gauge (American Wire Gauge) 0.057 in (1.44 mm) copper wire to standard diameters (from about 20 gauge to 30 gauge) for manufacture of telephone communications cable [37].

Fig. 15 (top) shows the internal profile of a die as pierced and rough shaped by the laser at the Buffalo facility. Note that the cone sides are smooth, that the front and back cones are concentric. The size of the minimum opening is 0.018 in (0.46 mm). The center figure shows a typical die as mechanically lapped and sized, and the bottom figure shows a typical die as finished.

One can readily appreciate the advantages of using laser energy for hole piercing with consideration of the following factors.

1) Because there is no physical contact between the hole-forming tool and the material, problems such as drill-bit breakage and wear are nonexistent.
2) Precise hole location is simplified because the optics used to focus the laser beam are also used to align and locate.
3) Large aspect ratios (hole depth to hole diameter) can be achieved because of the basic characteristics of laser light.

As with the case of welding, material properties also have an influence on the laser parameters used in hole piercing. Rarely is hole piercing, when using a pulsed laser, performed with a pulse length in excess of 1.5 ms; most materials undergo vaporization (of sufficient volume for practical material removal) using pulse lengths of 300 to 700 μs. Q-switched pulse lengths (5–50 ns) produce complete vaporization over a somewhat larger area than the corresponding burst pulse area but to a much shallower depth —usually not exceeding a few micrometers. Absorption of Q-switched pulses on bulk materials conduct the thermal energy predominately in the two-dimensional plane of the surface as would be predicted by (7). Ready [38] has observed the depth of removal of material for both 600 μs and 44 ns pulse lengths in several metals (Table IV).

The number of mechanisms that have been proposed for material removal in hole drilling reveal the limited amount of systematic work performed in this area. Ready [26] and Haun [39] have suggested the generation of a type of shock wave which propagates into the material in an unidirectional manner to produce deep holes. Haun [39] gives three reasons to account for this approach.

1) To vaporize completely the amount of material removed by the laser in punching the desired hole would require much more energy than was contained in the single pulse.
2) The time required for heat to be transmitted through the entire thickness of the irradiated material is much longer than the short pulse duration.

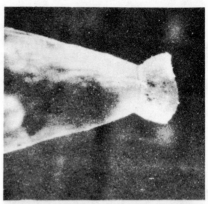
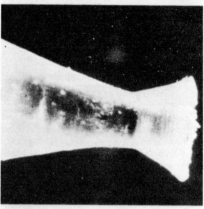
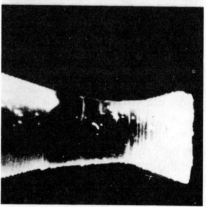

Fig. 15. Laser piercing of a diamond wire drawing die (typical for the application at the Western Electric Co., Buffalo Plant); as pierced (top), mechanically lapped and sized (center), finished die after polishing (bottom). The piercing operation (for a 100 point stone) required 250 pulses on the front side and 300 pulses for the back. Beam energy was 2–3 J and pulse length was 0.6 ms. Repetition rate was at 1 pulse/s. 30 × mag.

TABLE IV
DEPTH OF MATERIAL VAPORIZED*

| Metal | 5000 J/cm² 600 μs Laser Pulse | | 10⁹ W/cm² 44 ns Q-Switched Pulse | |
	Calculated (cm)	Observed (cm)	Calculated (μm)	Observed (μm)
Aluminum	0.084	0.078	6.2	3.6
Copper	0.083	0.090	3.0	2.2
Nickel	0.068	0.058	2.4	1.2
Brass	0.089	0.078	2.0	2.5
Stainless steel	0.076	0.061	1.8	1.1

* Taken from Ready [38].

3) The heat would tend to diffuse uniformly in all directions from the irradiated spot rather than being confined to the uniform-diameter cylinder which is punched out.

Cohen [40], on the other hand, has shown that when holes are drilled in a material, the depth of the hole increases with the energy delivered by the laser (Fig. 16). He stipulates an upper limit, however, to the depth achievable in a single laser pulse. The upper limit results from the fact that the laser plume becomes more opaque as it becomes larger and more highly excited. The latter portions of the laser pulse must pass through the plume of vaporized material, and therefore additional energy is absorbed by the plume rather than by the workpiece. Usually hole depth and desired diameters are obtained by using repetitive laser pulses at lower energy levels. Such practice produces holes that have less taper and better definition than those drilled with higher energy pulses. Exceptions are found for materials containing a large percentage of high vapor pressure elements. As was mentioned earlier, brass is difficult to laser weld with present-day pulsed units because of the high zinc content. In drilling, this condition becomes an advantage for certain applications. The temperature required to just melt the brass alloy is great enough to generate localized pockets of exceedingly high pressures so that for a drilling process, conglomerates of zinc atoms greatly enhance the vaporization mechanism.

Fig. 17 shows a 40 mil (1.0 mm) entrance and a 19 mil (0.5 mm) exit hole, in $\frac{3}{32}$ in (2.4 mm) thick 70:30 brass. One pulse from a ruby laser of approximately 75 J energy and 5 ms pulse length was used in producing the hole. A 43 mm focal length lens concentrated the energy to a power density of 10 MW/cm².

Another example of drilling through relatively thick materials is shown in Fig. 18. The sample is $\frac{1}{8}$ in (3.2 mm) thick polycrystalline, high density ceramic. The particular application is for development work in precision-insulated coaxial wire extrusion. The objective was to provide precisely spaced holes of 0.010 in (0.25 mm) to 0.012 in (0.30 mm) diameter in $\frac{1}{8}$ in (3.2 mm) alumina ceramic disks of $\frac{1}{4}$ in (6.4 mm) diameter.

Laser drilling in hard (high temperature fired) alumina ceramic is attractive because drilling ceramics by conventional means is not a simple task; it usually requires diamond tipped, hardened steel drill bits. Small size holes, less than 0.010 in (0.25 mm) in diameter, are extremely difficult for current tool technology. Breakage of the drill often results when the thickness of the ceramic is appreciably greater than the diameter of the hole to be drilled. An accepted guide in the drilling of small holes in hard brittle-like materials is the limiting ratio of hole depth (material thickness) to hole diameter. This ratio is usually designated as the aspect ratio of hole formation and is nominally 2 to 1 for conventional drilling and about 4 to 1 for ultrasonic drilling (sometimes used for ceramics and other refractory materials).

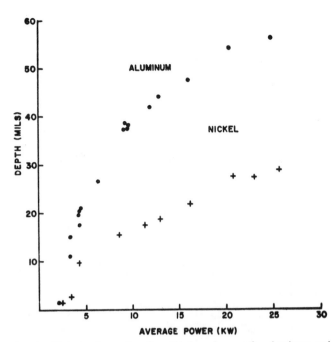

Fig. 16.   Hole depth as a function of delivered power for aluminum and nickel. Each point plotted is the average depth of several holes, using one pulse per hole. Pulse length was 0.8 ms.

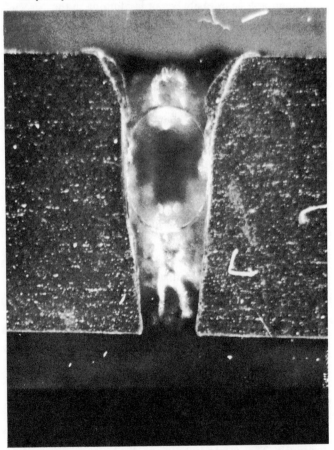

Fig. 17.   A hole made in 3/32 in (2.4 mm) 70:30 brass. One pulse of 75 J from a ruby laser produced the hole seen in cross section. Approximately 31 × mag.

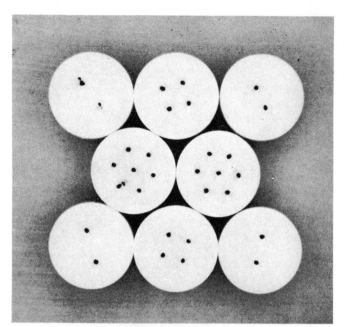

Fig. 18.   Laser drilling through high density polycrystalline alumina ceramic. Approximately 5× mag.

Though holes can easily be put in alumina in the "green" state (aspect ratio for punching techniques is approximately 3 to 1), precision of hole location is lost during the firing stage. To provide precisely located holes and maintain close tolerances (especially for small size holes) by this technique is difficult and can be expensive.

Fig. 18 shows several pattern arrays made with the ruby laser in the 99.5 percent high temperature fired Al_2O_3. The holes were drilled with a beam energy of 1.4 J, focused through a 25 mm focal length lens onto the surface of the disk. An average of 40 pulses at a repetition rate of 1 pulse every 5 seconds was used to drill through the 0.125 in (3.2 mm) thick disk. Each pulse had a time duration of 0.5 ms. A $\frac{3}{16}$ in (4.75 mm) aperture was inserted in the path of the 0.6943 μm wavelength laser beam so as to utilize the maximum intensity of the near Gaussian-type distribution of the laser output. The energy density at the focal spot at the surface of the ceramic disk was approximately 4 MW/cm^2 for the above-stated parameters.

Fig. 19 shows a longitudinal section of a laser drilled hole. It is representative of the sectional profiles of the holes shown in Fig. 18. This hole was made without the use of an aperture and at a higher energy level, 1.8 J; thus the reason for the larger size entrance diameter, which measures 0.020 in (0.50 mm). The exit hole diameter measures 0.0037 in (0.09 mm). Refocusing of 0.005 in (0.125 mm) after every third shot was also used. Note that the profile of the exit is quite square, which has been found to be the case in all our work. The entrance hole usually has some degree of taper, with tapers of a little over 1° typical for holes made under the stated conditions. Aspect ratios of better than 20 to 1 have been demonstrated in our experiments so far, as determined by using the half-depth diameter for longitudinal sections or the average diameter of the entrance and

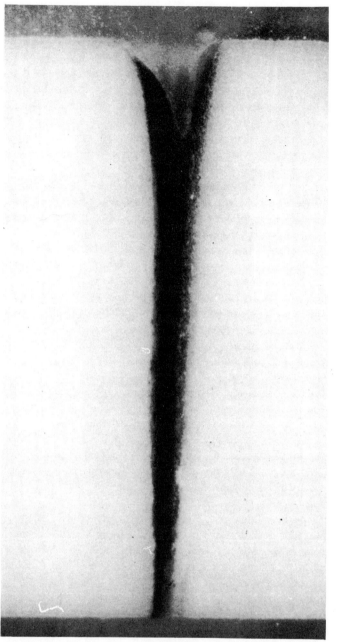

Fig. 19.   Longitudinal section of a typically drilled hole in 1/8 in (3.2 mm) thick ceramic using pulsed ruby radiation. 45× mag.

exit holes. Cohen [21] has achieved aspect ratios of 25 to 1 in similar materials.

Aperturing the emitted beam before lens convergence has a refining effect on hole formation. Fig. 20 shows the effects of increasing aperture size on hole diameter for the same lasing parameters. Ten aperture sizes were used [from $\frac{1}{32}$ in (0.8 mm) to $\frac{3}{8}$ in (9.5 mm) in increments of $\frac{1}{32}$ in (0.8 mm), except for two apertures]. The $\frac{1}{32}$ in (0.8 mm) aperture attenuated a sufficient amount of the energy so that no effect was perceived on the ceramic after several shots. Approximately 40 shots for each aperture size were used in the study. The results are summarized in the plot of Fig. 21.

Note the very pronounced change in slope at a point corresponding to an aperture size of 0.11 in (2.8 mm), close to

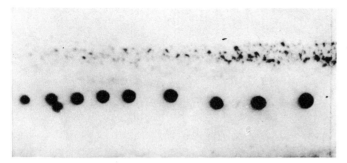

Fig. 20. The effect of using differing aperture sizes placed external to the cavity, but before beam convergence. Aperture size increases to the right. Parameters are 1.53 J beam energy and 0.5 ms pulse length, 25 mm focal length lens, 40 pulses at 1 pulse every 5 seconds. Approximately 10 × mag.

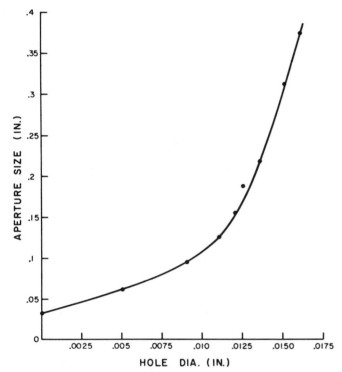

Fig. 21. Plot of aperture size versus entrance hole diameter for the same laser and optical parameters given for Fig. 20.

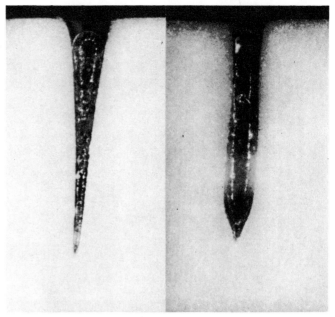

Fig. 22. The effect of refocusing versus no refocusing for multiple shot hole drilling. Approximately 26 × mag.

the $\frac{1}{8}$ in (3.2 mm) experimental aperture. Assuming that the spatial distribution of the beam intensity emitted from the face of the laser rod follows the Gaussian-type shape (actually, the multimode output might at best be represented by a quadratic form somewhat unsymmetrical or displaced from the beam axis), then the $\frac{1}{8}$ in (3.2 mm) aperture will allow for a transmittance of 90.5 percent of the total intensity of the beam. This would explain the relatively small increase in hole size with increasing aperture opening beyond the $\frac{1}{8}$ in (3.2 mm) size.

Fig. 22 shows the effect of refocusing versus no refocusing for multiple shot hole drilling in relatively thick alumina. The longitudinal section on the left shows the hole profile made by refocusing to the extent of 0.009 in (0.2 mm)/shot. The one on the right shows the profile with no refocusing but only with the initial focusing on the surface. The same

parameters were used in making both holes (3.1 J beam energy, 0.5 ms pulse length, 15 repetitive shots focused with a 39 mm focal length lens, no aperture). Contrary to what may be expected, refocusing apparently prevents straight wall holes (see Fig. 22). A "light pipe" effect is considered to be the explanation for the results obtained with ceramic, though similar results have been noted in laser drilling of some metals. On the other hand, holes of almost any taper can be achieved by the degree of refocusing.

A proposed explanation [40] for the "light pipe" mechanism is the internal reflection of the laser light as the hole progresses in depth. The light, focused at the surface, enters the hole and subsequently diverges. The diverging light is reflected from the walls of the hole because of the small incidence angle. The incidence angle at the bottom of the hole is nearly normal, however, and most of the light is absorbed there. To a first approximation, the same increase in hole depth is obtained with each of many shots. The maximum hole depth achievable is limited by the amount of energy lost due to reflections from the hole wall and by the decrease in the aperture of the hole as a result of vapor from the bottom of the hole being cooled by, and depositing on, the wall of the hole.

When laser drilling in thin [sheet stock of less than 0.020 in (0.5 mm)] materials, some uniformity of hole diameter is sacrificed for smallness of diameter. Holes of 0.5 mil (0.01 mm) to about 7 mils (0.2 mm) in diameter can be made in common metals of 15 mils (0.4 mm) and less in thickness, using relatively low powers and in three shots or less. Fig. 23 shows ruby laser drilled holes measuring 1 to 2.5 mils (0.025 to 0.06 mm) in diameter in 5 mil (0.125 mm) thick aluminum, copper, and low alloy steel.

The dynamic balancing of gyro and other rotating components has been very accurately performed with the use of the pulsed laser [41], [42]. Balancing is accomplished

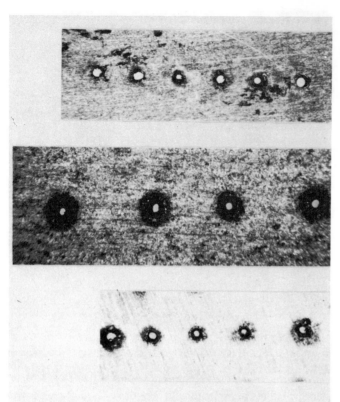

Fig. 23. Hole piercing through thin materials. (Top to bottom) 5 mil thick sections of aluminum, copper, and steel. The holes are entrance holes and measure, on the average, 2.5 mils (60 μm) for aluminum, 2 mils (50 μm) for copper, and 1.5 mils (40 μm) for steel. Approximately 36 × mag.

much in the same way as in conventional techniques, i.e., locating the point or area of imbalance by strobe light and then trimming the imbalance out. With the laser, the trimming operation is effected by material vaporization using shallow hole drilling parameters. Dynamic balancing using laser energy is simplified in the utilization of a signal from the point of imbalance to trigger the laser pulse. At a pulsing rate in phase with the point of imbalance (related to the rotating speed), material removal of milligrams per pulse is achieved.

Component Trimming

The laser is also being investigated as a means of modifying electronic components either by selective evaporation or by heating. The high power densities, small spot size, and short pulse lengths achievable with lasers make them ideal tools for this purpose. In addition, it is usually possible, when using the laser, to constantly monitor the device thus allowing it to be trimmed or adjusted to a specific value.

The determination of the exact characteristics required of a laser for material removal is done empirically because effects of parameters such as material emissivity, conduction losses, etc., cannot be calculated exactly. However, sufficient information usually exists to make preliminary calculations that will indicate which lasers are likely to be practical in a particular application. For instance, assume that it is desired to evaporate a 10 μm diameter hole in a 0.5 μm thick gold film. The volume V of gold to be evaporated is

$$V = \frac{\pi}{4} (10)^2 (0.5) \ \mu\text{m}^3 \tag{10}$$

$$V = 39.25 \ \mu\text{m}^3$$

or

$$V = 39.25 \times 10^{-12} \ \text{cm}^3$$

and the mass of material to be evaporated is

$$M = \rho V \tag{11}$$

where ρ = density of gold = 19.3 g/cm^3. Thus

$$M = (39.25 \times 10^{-12} \ \text{cm}^3) (19.3 \ \text{g/cm}^3)$$
$$= 7.58 \times 10^{-10} \ \text{g}.$$

The following equation [43] is valid for determining the energy required to raise the temperature of gold from room temperature to a temperature between 1336°K (melting point of gold) and 3000°K:

$$\Delta H = [0.0355T + 2.69] \ \text{cal/g} \tag{12}$$

where ΔH = energy required, and T = temperature in °K.

Assuming that this equation is valid to the boiling point of gold (3264°K) and that the heat of vaporization is 446 cal/g [44], the energy to vaporize a gram of gold is

$$\Delta H = [0.0355 (3264) + 2.69 + 446]$$
$$= 565 \ \text{cal/g}$$

or

$$\Delta H = 2360 \ \text{J/g}.$$

The total energy E required to evaporate the small volume of gold then becomes

$$E = \Delta H M$$
$$= (2360) (7.58 \times 10^{-10}) \ \text{J}$$
$$= 1.79 \times 10^{-6} \ \text{J}.$$

This calculation yields an estimate of the energy required to evaporate the gold, and in fact it can be seen that an estimate based on just the heat of vaporization would be lower by only 20 percent. The assumption that the vapor is at the boiling temperature of the material is probably in error; however, here again it requires an extremely large error in the assumed temperature to significantly affect the results. For instance, Kelley [43] has calculated the energy required to raise the temperature of gold vapor from the boiling temperature to twice the boiling temperature (approximately 6000°C). His calculations show a 107 cal/g requirement or a 19 percent increase. Thus it should be safe to assume that the calculated energy value is within a factor of two of the proper value.

Another contributing factor in the determination of laser energy required to evaporate a material is the absorption of the laser radiation by the material. First, the absorption of radiation by a material is a function of the material surface finish [22]. Second, the absorption is a function of the temperature and may vary radically with temperature or with the surrounding atmosphere, especially at higher temperatures. Typically, the absorption of a material increases as its temperature increases.

In this particular case it will be assumed that the radiation to be used will lie somewhere in the red portion of the visible spectrum or in the near infrared spectrum where gold will absorb approximately 1.5 percent of the incident radiation [45]. Then, assuming a constant absorption coefficient during the complete process, a minimum incident radiation of 119 μJ will be required to evaporate the 10 μm diameter hole.

Ideally, then, we need a laser that can deliver approximately 120 μJ of energy to the gold film in a period sufficiently short to reduce conduction losses to a minimum. A survey of lasers indicates that a small Q-switched YAG:Nd laser [46] with a 300ns pulse, having a peak power of approximately 500 W (150 μJ/pulse), should be adequate to accomplish the desired evaporation; this has been borne out in laboratory experiments. Similar calculations may be made for other materials and processes, and are of sufficient accuracy to determine what types of lasers are likely to be adequate to accomplish a particular operation.

Studies conducted in our laboratories resulted in the development of a laser system for adjusting glass encapsulated deposited carbon resistors to a tolerance of ± 1 percent automatically [47] (Fig. 24). Because encapsulation caused a random change in value, trimming could not be obtained by other techniques. The system used is basically a Q-switched pulsed ruby laser which will deliver $\frac{1}{2}$ J, 30 ns pulses at a rate of 1 pulse/s. The optical system for laser beam shaping consists of a mask to shape the laser beam, a telescope to control the focused beam size, and a lens to project the beam onto the resistor surface. The projection lens also serves as the objective lens of a CCTV system [36] used for observing the resistor when it is in the work station. This technique utilizes the laser's ability to perform useful operations without physical contact, its ability to vaporize high boiling point materials, and its ability to selectively vaporize materials (i.e., to evaporate the carbon while having no detrimental effect on the ceramic core and glass envelope).

In the manufacture of thin film components, the laser is a promising tool for both forming and trimming components. The techniques of trimming thin film resistors have been studied at our facilities, at Bell Telephone Laboratories [11], [48], and at other companies [49]–[51]. While there has been no reported use of the laser in mass production facilities for the purpose of trimming thin film resistors, the use of the laser appears promising, especially for resistors difficult to trim by anodization. By using the proper techniques, tolerances of 0.01 percent are achievable. In trimming (by evaporation) tantalum nitride resistors formed on high alumina ceramics with a Q-switched YAG:Nd laser, it has been found that a significant shift in resistance value occurs during the first 24 hours after trimming. This shift is a function of the material removed and appears to be predictable [48]. The magnitude of the shift for a serpentine tantalum nitride resistor of 300 ohms was 0.2 percent when

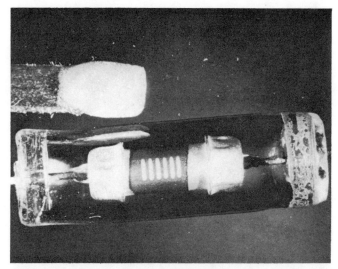

Fig. 24. Deposited carbon resistor adjusted to value by a laser after encapsulation in glass. A match head is shown (upper left) for comparison.

the resistor was trimmed 10 percent by an array of 0.001 in holes on 0.005 in centers. (The drift is primarily a function of the heat affected zone created or formed around the area where the tantalum nitride is removed.) The techniques used to trim the resistors in this series of experiments, where holes are drilled in the film, are such that they would be expected to cause the maximum amount of shift. Fig. 25 is an example of a bar-type tantalum nitride resistor trimmed by removal of circular areas (0.0005 in) of the tantalum nitride film. In this case the percentage of film area removed, and thus the total resistance change (~ 2 percent), was small. By changing the technique of removing the material (i.e., removing lines or boxes of material), the room temperature shift should be reduced considerably. Alternately, a two-step trimming process, with intermediate aging, may be used to achieve extremely close tolerances.

It has been reported that one company is near the production use of a pulsed argon ion laser for the purpose of trimming resistors of an integrated circuit [51], [52]. These resistors are chromium-silicon oxide deposited on the surface of the silicon chip. The energy of the laser heats the resistor surface, thus causing the desired change in resistance value. This technique has two striking advantages— the trimming operation can be accomplished after encapsulation of the IC, and it can be performed while the unit is electrically powered. To accomplish the desired change in resistance value, power densities of the order of 0.775×10^6 W/cm^2 in pulses having a width less than 100×10^{-6} s are utilized. It has been estimated that these pulses raise the resistor film temperature to about 1000°C [55]. Since the pulse lengths are very short, chemical processes such as diffusion and oxidation are suppressed; however, annealing effects occur and change the film resistivity.

Trimming of thin film capacitors with the laser has been attempted but success has been somewhat limited. The ideal capacitor trimming technique would not degrade the capaci-

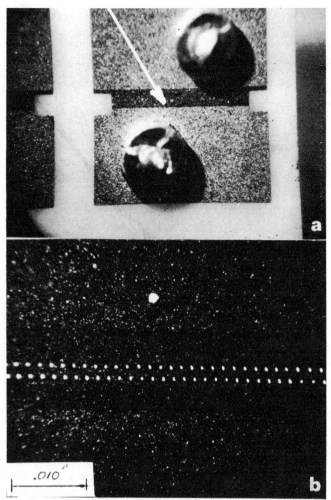

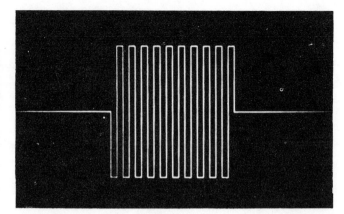

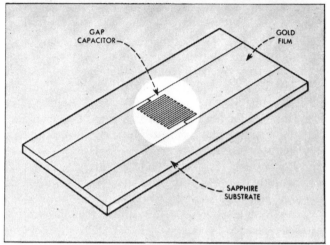

Fig. 26. Gap capacitor formed by laser machining of Au film on sapphire. (*Photograph courtesy M. I. Cohen, Bell Telephone Laboratories, Murray Hill, N.J.*).

Fig. 25. Tantalum nitride resistor trimmed by laser evaporation of resistive film. (a) Overall view of resistor. Resistive film (arrow) lies between contact electrodes which show external wire connections. (b) Magnified view of the tantalum nitride [arrow in (a)] showing individual laser evaporated areas.

tor dielectric in any way; however, currently this is virtually impossible. It is expected in the future that mode-locked lasers [53] will be available which will deliver extremely short, high peak power pulses; it is possible that these can be used to remove only the top electrode material with little if any degradation of the dielectric material.

While the process of trimming thin film capacitors is not developed at the moment, Cohen *et al.* [11] of Bell Telephone Laboratories have developed an alternate technique of making thin film capacitors with the laser. They have used a Q-switched YAG:Nd laser to vaporize extremely narrow lines in a gold conducting film to form gap capacitors and have obtained capacitance values of 4000 pF/cm² utilizing line (gap) widths as narrow as 6 μm (Fig. 26). These capacitors could easily be trimmed to value by using the laser to cut one of the meanders at the proper point.

The processes utilized in the adjustment or trimming of thick film components are much the same as those with thin film circuits, except that different lasers might be utilized. For instance, a small Q-switched YAG:Nd laser (maximum of 1.5 kW peak power in 300 ns pulses)

appears ideally suited to trimming thin film circuits. It is relatively inexpensive to build and operate, it is reliable, and it can controllably remove areas as small as a few square micrometers. With thick films, more material must normally be vaporized, thus more energy must be absorbed. This requires a large laser or a laser having an output more readily absorbed by the workpiece.

A laser system is commercially available for the purpose of trimming thick film resistors. This system has an automatic work table for locating the resistor under the focused laser beam and a Q-switched CO_2-N_2-He laser for the purpose of removing the thick film resistance material. With such a system as this, it should be possible to trim in the same manner as with thin film circuits except for two aspects. First, the CO_2-N_2-He laser wavelength is 10.6 μm whereas the YAG:Nd laser wavelength is 1.06 μm. The shorter YAG:Nd wavelength allows its energy to be focused to a spot diameter approximately one-tenth that of the CO_2-N_2-He laser [from (2) and (3)]. Second, most substrate materials do not absorb a large percentage of 1.06 μm radiation whereas they do absorb 10.6 μm radiation quite well. Thus it appears that greater substrate damage can be expected when using the CO_2-N_2-He laser.

In our studies and in studies at other companies [54], the laser has been found a useful tool in adjusting and repairing discrete electronic components—for instance, the production trimming of glass encapsulated deposited carbon resistors. In other studies it has been found that the gain of traveling wave tubes (TWT) can be adjusted by use of a pulsed ruby laser. The process, quite similar to that utilized to adjust deposited carbon resistors, uses the laser to remove a portion of a tantalum "loss pattern" from the TWT after final assembly. This process has two important advantages—it allows tubes to be used which would normally fail because of low overall gain, and it allows the adjusting of all TWT to tighter gain tolerances.

Evaporation and Deposition

The use of the laser to evaporate material, not for the purpose of removing the material but for the purpose of obtaining vapor which may then be deposited on a substrate, is a promising area. While the vaporization and deposition of materials have been accomplished by many techniques, the laser offers some unique advantages:

1) The vapor can be generated in any atmosphere transparent to the laser radiation.
2) No contaminants are introduced by the laser radiation.
3) Essentially all of the laser energy may be used for evaporation with little of it being absorbed by the substrate.
4) Small selected areas of the source material may be evaporated.
5) The evaporant source may be located very close to the substrate.
6) Compounds may be evaporated with little change in composition.

Jackson *et al.* [55] have reported a technique whereby the laser is used to vaporize material which is then deposited on a substrate. They propose that a substrate having a complete coating of some material be placed adjacent to a blank substrate such that the coating on the one substrate is very close to the surface on which it is desired to obtain a pattern. A laser is then positioned such that its energy is focused, through the first substrate, onto the thin film. As the two substrates are moved with respect to the laser beam, the material is vaporized from the one substrate and redeposited on the second substrate in a pattern which duplicates the movement of the substrates with respect to the laser beam.

This technique could be used beneficially in several different applications, such as the generation of photomasks for thin film or IC processing, the generation of conducting paths on thin film or integrated circuits, and the deposition of special thin film resistors or other types of components.

It has been shown that a laser can be used to evaporate a small volume of metal located just below a hole in a printed circuit board and have a large percentage of the vapor generated deposit on the periphery of the hole. This can be accomplished in ambient atmosphere and the deposit formed can be used for electroplating to yield a low resistivity through connect in the printed circuit board. An extension of this technique has been proposed at our laboratories as a means of creating through connects in printed circuit boards. In this process, the laser energy first drills a hole through the printed board. After the material has been removed from the hole, the laser's energy vaporizes the copper. A large percentage of the copper vapor will deposit on the edge of the hole, thus yielding a conducting surface between the two sides of the board. This conducting surface may then be electroplated to form a low resistance feedthrough.

V. MATERIAL SHAPING

The laser has been recognized as a possible tool for the purpose of shaping materials almost since its inception [56]. The accuracy of this prediction was borne out by the first announced production application of the laser [36] wherein it was used to pierce diamonds. For simplicity in presentation, we have separated the material shaping applications into three main categories: cutting (milling), scribing, and controlled fracturing. The use of the laser to shape materials by drilling or removal of surface material has been covered in previous sections.

Cutting

Laser cutting or vaporization is the technique most often thought of when the term "laser shaping" is mentioned. Most people who are aware of lasers have seen the laser burning and evaporating the workpiece, and the laser, especially the high powered CO_2-N_2-He laser, is showing great promise in this area.

As has been mentioned, a pulsed ruby laser has been quite successful in piercing diamonds for diamond wire drawing dies. From this single success it is possible to predict that the laser would be capable of shaping any material for the diamond is the hardest material to shape. Recently, the use of a CO_2-N_2-He laser for the purpose of cutting fused quartz and other materials [57] was announced. For this purpose lasers capable of generating 100–250 W continuous output were used. The laser beam is focused to spot diameters of 25–100 μm (depending on laser and conditions used) on the material, normally a flat sheet having a thickness of a few mils to approximately 100 mils thickness. The material is moved under the focused spot and is vaporized along the path the laser beam traverses.

Engineers of the British Welding Research Association [58] have developed a system where a 300 W CO_2-N_2-He laser, oxygen lance combination is used for cutting metals such as steel. In this technique, the laser is used to heat the metal over a very small area to a temperature such that the metal is burned by the oxygen jet. By moving the part in the focused beam, desired shapes have been cut from a steel plate up to $\frac{1}{8}$ in thick at rates of 40 in/min. The width of burned material is of the order of 0.020 in.

These examples show the ability of the laser to shape by "cutting," and while the examples mentioned were flat plates of materials, the processes or techniques can be extended to almost any shaping process. For instance, the cutting of such things as wire is easily accomplished with a laser. Fig. 27 shows a 0.025 in diameter tantalum wire which was cut with a 20 J, 1 ms pulsed from a ruby laser. It can be seen in this figure that the wire can be cut quite cleanly without burrs and other projections being formed. In this case the cut was circular, duplicating the shape of the beam; however, by shaping the laser beam properly, the shape of the cut can be changed.

Scribing

Scribing is another important way in which materials may be separated or shaped, especially the brittle materials such as silicon, glass, and ceramic. In the process of scribing, it is desired to remove material along a path on the surface of the material. When the material is stressed sufficiently, a fracture will occur along the scribed path, since that path is weaker than any other.

Garibotti [59] proposed that the laser be used for the purpose of scribing materials, and recent experiments have shown this to be a useful technique [60]. Our work has shown that the ideal method of scribing is by the use of a repetitively Q-switched or repetitively pulsed laser. By use of pulsed laser outputs, it is possible to use very high peak power densities which will vaporize the material with little heating and melting.

Fig. 28 shows silicon transistor wafers which were scribed with a repetitively Q-switched, YAG:Nd laser. The scribing was accomplished by using laser pulses having a peak power of approximately 300 W, pulse width of approximately 300 ns, and pulse-repetition rate of 400 pulses/s. The silicon was moved under the focused laser beam (laser beam focused with a 14 mm focal length lens) at a rate of 0.6 in/min. Using this same laser at its optimum Q-switching rate of 5000 pulses/s, it should be possible to scribe at rates of 15 in/min. With higher power outputs or wavelengths which are better absorbed by the silicon, the rates could be increased even more.

The use of a laser to scribe materials such as high alumina ceramic is also attractive. Using a CO_2-N_2-He laser, nominally 100 W continuous output, high alumina substrates 0.025 in thick, have been scribed at rates of 60 in/min. This technique is one which offers significant cost advantages over prescored substrates [61].

Techniques applicable to the scribing of ceramics and silicon are adaptable to the scribing and separation of other brittle materials such as glass; however, such parameters as power, pulse shape, and pulse rate should be optimized for each material.

Controlled Fracturing

The use of controlled fracture is a technique for separating materials which shows great promise as a means of separat-

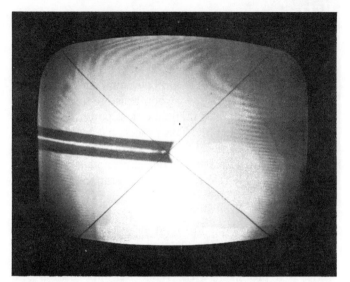

Fig. 27.   Tantalum wire cut by single pulse of energy from a pulsed ruby laser.

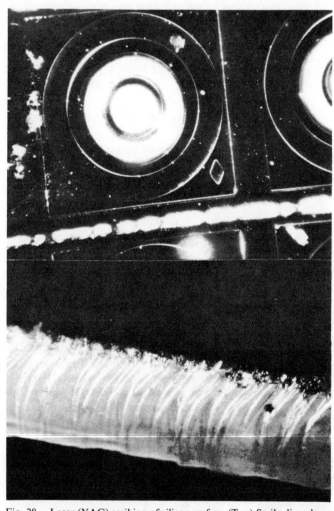

Fig. 28.   Laser (YAG) scribing of silicon wafers. (Top) Scribe line along the isolation areas between transistor units. (Bottom) Edge-on view of the wafer after fracture along the scribed line.

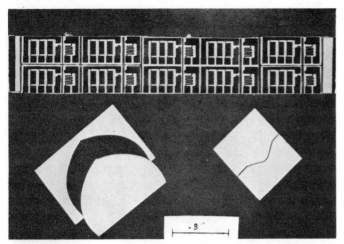

Fig. 29. Various shaped pieces of high alumina ceramic formed by controlled fracture technique, with (top) an array of thin film circuits separated by controlled fracturing.

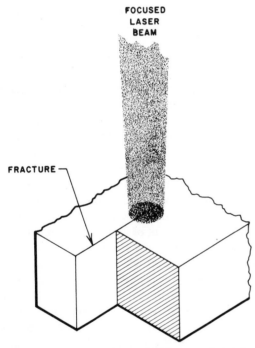

FOCUSED
LASER
BEAM

FRACTURE

Fig. 30. Schematic of controlled fracture of a brittle material by a laser beam.

ing electronic circuits or components which are batch processed, or for the generation of special shapes in brittle materials. This technique, which uses the laser to controllably fracture or break, has one inherent advantage over other separation techniques, i.e., controlled fracture allows a sample to be separated into two or more parts with no evaporation or loss of material (Fig. 29). As a result of this inherent advantage, electronic circuits may be separated with no danger of contamination as a result of the separation process, and better utilization of the raw materials may be obtained since no material need be lost.

Experiments have shown [60] that it is possible to controllably fracture high alumina ceramic 0.025 in thick at rates of 120 in/min using a CO_2-N_2-He laser capable of delivering 100 W continuously.

In the controlled fracture process, the laser's energy is applied to a small area of the surface of the material (Fig. 30). The absorption of the laser's energy creates thermal gradients which in turn create mechanical stresses sufficient to fracture the material. If the material is moved with respect to the laser beam, the fracture follows. This technique works, and works well, because the laser creates stresses sufficient to cause fracture but over a region so small that the fracture does not propagate uncontrollably.

Many brittle materials such as high alumina ceramic, ferrite sheet, quartz single crystals, sapphire, soda-lime glass, and Corning 7059 glass have been controllably fractured. Fused quartz was the only material which could not be fractured; it appears that the thermal expansion coefficient is too low to generate the necessary stresses.

In our laboratories we have also investigated a second technique of controlled fracture whereby single crystals are cleaved along certain preferred planes. When using the laser to controllably fracture single crystals, the effects of cleavage planes are usually notable because of the preference of the crystals to fracture along these planes. While this is often undesired, it can be advantageous where separation along cleavage planes is permissible or required.

Possibly the greatest advantage of cleaving crystals is that the crystal itself is the primary control of the direction of separation and, therefore, the tool or device to initiate the cleavage need not have fine control. For instance, we have found that we can cleave single crystal quartz (0.025 in thick, 0.315 in wide) by focusing the energy of a CO_2-N_2-He laser (40 W) into a line along the desired cleavage direction. Alternately, we can cleave the crystals by applying the laser's energy to a circular spot on the crystal surface. In the second case, the location of the spot determines the location of the plane along which the cleavage will occur and the crystal itself determines the path along which the fracture will occur. Some evidence indicates that focusing the laser energy into a line may not be desirable, for if the line is not perfectly aligned with the cleavage plane, it may force the fracture to deviate from the desired path.

We have experimented only with single crystal quartz, sapphire, ruby, and diamond; however, there appears to be no inherent reason why the same techniques may not be used with other crystals.

VI. THERMALLY INDUCED CHANGE

Material Surface Heating

The probable mechanism of laser light absorption has been discussed earlier. With proper power levels and pulse widths, advantage can be taken of the derived thermal effects within relatively shallow depths in many materials, especially thin film component parts. Specifically, certain electrical properties can be altered or adjusted on completed thin film integrated circuits by the heating or annealing effects from laser irradiation of discrete units. Lins and Morrison [49] have demonstrated the precision adjustment of thin film resistors using a pulsed ruby laser. The resistance change is achieved by momentarily heating the thermally alterable film above a threshold temperature without damage or melting.

They state that the extension of this method to high component densities is based on calculations which show that effective thermal energy can be localized near the optical diffraction limit of the best available lens systems. These calculations are based on consideration of single pulses of energy at low repetition rates. A minimum of about 30 pulses must be used to make a 15 percent change with a 0.1 percent tolerance.

The approximate power density required to effect laser-induced resistivity changes is 10^6 W/cm^2 for millisecond pulse lengths. For resistor widths of a size compatible with IC densities, such energy densities are achievable with continuous-wave lasers, such as the YAG system.

Levine *et al.* [62] describe experimentation whereby a Q-switched ruby laser is used to desorb neutral gases from a tungsten surface target in vacuum. Water vapor, carbon monoxide, and carbon dioxide were the gases studied on desorption under a vacuum of 10^{-8} torr. The investigators concluded upon analysis that the laser is considered to be no more than a pulsed heat source for laser energy densities up to 50 MW/cm^2.

Another laser heating technique was developed for use with an apparatus to measure relaxation times in electrolytic solutions [63]. Fast chemical reactions with relaxation times of 10^{-6} s or longer were studied in volumes of less than 0.1 cm^3 with heating to a $\Delta T \approx 0.5°$C. Both Q-switched ruby and neodymium-glass were used as the heat source. Higher Q-switched energies (up to 10^{12} W/cm^2) from a ruby laser were used to thermally etch metals and semiconductors [64]. Murphy and Ritter studied the etched surface patterns by transmission electron microscopy of replica specimens. Where grooving and striation-type etch patterns were generally observed by the more conventional etching techniques, laser thermal etching in air revealed the formation of hillocks in α-brass, iron, platinum, gallium arsenide, and germanium. Copper, however, gave the more general striation pattern.

Solid State Transformations and Diffusion

The source of heat from a defocused beam of pulsed ruby radiation has been used in our laboratory to study the recrystallization kinetics of thin sheets of 70:30 brass. Two thousandths of an inch (0.05 mm) thick, 40 percent cold worked yellow brass was pulse heated to several temperature levels, within $\pm 10°$C repeatably and at a rate of $(10^6)°$C/s. A mixture of helium gas and water vapor was used to quench the specimens at rates of $(10^5)°$C/s. This study suggests a means of attaining accurate grain size control. We are currently extending this technique of pulse heating (also CW heating using the argon laser) to study diffusion at localized areas and to measurable depths in several materials. Other metallurgical transformations such as selective heat treating of hardenable alloys is also under study. Speich *et al.* [65], [66] have successfully used the technique to study transformation mechanisms in alloy steels.

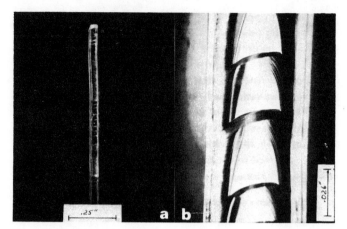

Fig. 31. Sapphire rod grown by zone melting of polycrystalline Al$_2$O$_3$ with a CO$_2$-N$_2$-He laser. (a) Overall view of sapphire rod. (b) Close-up of facets developed on rod.

Laser-Induced Chemical Reactions

Since the laser is a source of high energies, the generation of localized thermal effects has been considered for initiating certain chemical reactions. Ionization, gas breakdown, and photochromic behavior are some of the laser-induced reactions that have been looked at. Rousseau *et al.* [67] have studied the temporary ionization of transparent dielectric materials using both normal pulsed and Q-switched ruby radiation. They ascribe the dielectric breakdown of materials like glass, quartz, calcium fluoride, and sapphire —below their physical damage threshold—to thermal heating of the surface producing photons, electrons, and both positively and negatively charged ions. The emission phenomena were studied by time resolved pulses. Epstein and Sun [68] reported that when a Q-switched pulse is focused in an environment of methane gas, breakdown into hydrogen and acetylene, the main constituent for making natural rubber and polyethylene, is formed with yields of 10–20 percent. As with electric arcs, carbon dioxide was dissociated into carbon monoxide and oxygen with 200 MW pulses of 30 μs pulse length. Gas breakdown with concentrated light energy appears to offer the chemical industry a convenient and easily controllable thermal source in obtaining primary and intermediary reactants for subsequent synthesis of commercially useful products. Laser photolysis techniques [69] have been successfully used to synthesize 23 new compounds utilizing photochromic structures. Margerum [69] claims that a new group of photochromic compounds has been found; these compounds can undergo hydrogen transfer reactions with certain organic groups to form other compounds. One such reaction, that of platinum dithizonate, was found to produce an isomerized state lasting for several seconds. Solutions of this photochromic compound were found to work effectively as a passive Q-switch material.

Zone Melting

The laser has been studied in our laboratory as an energy source for purposes of zone melting materials [70]. Fig. 31 is a photograph of a single crystal sapphire grown from 96 percent alumina substrate material. Three zone passes were made to purify the material and form a "seed" crystal. A fourth and final pass was made at a rate of 1.5 in/h using 40 W of power from a CO_2-N_2-He laser. The diameter of the sapphire is approximately 0.060 in. An analysis of the material revealed that a single crystal was formed and that the impurities such as titanium and calcium were reduced by two orders of magnitude or more. While most of the studies have been confined to zone melting of Al_2O_3 to form sapphire, the technique appears to be one which can be readily adapted to the growth and purification of refractory and other materials in controlled atmospheres.

VII. Special Instrumentation Using Lasers

Spectroscopy

One of the early applications of the laser was in the spectrochemical analysis field. The laser microprobe [71] was the first such marketable unit and has been used in emission spectroscopy to increase the sensitivity of detection and the limits of trace analysis by a factor of 10. Because of this increased sensitivity, the amount of sample material needed in the excitation spark or arc can be as small as 10^{-6} grams [72]. The use of such minute amounts has qualified the microprobe as an essentially nondestructive tester, since analysis can be made directly on small, localized areas of the object without sample preparation. The unit employs the Q-switching of a ruby or Nd:glass laser to generate a plume of predominantly ionized material, which is further excited by the striking of an ac spark or dc arc across two carbon electrodes set just over the focused spot on the sample surface.

The focused beam from a Q-switched ruby laser has also been used in conjunction with a mass spectrometer to give time-of-flight analysis of solid samples in the size range of 10^{-8} to 10^{-10} grams [73].

Somewhat less success has been found in absorption spectroscopy. Since the amount of gaseous material in the laser plume is low in neutral species (i.e., nonionized material), then the ability of the analyzing beam of the absorption unit to detect a sufficient amount of neutral atoms is significantly limited. However, certain elements, specifically the high vapor pressure metals like zinc and magnesium, or those with long neutral atom life times like silver and copper, have been detected in trace amounts of 20–40 parts per million [74].

VIII. Measurement

In the earlier part of this paper the reader saw how the laser, because of its intensity and spatial coherence properties, could be used for special thermal processing purposes. In this section the exploitation of the laser as a measurement tool will be described with the emphasis on industrial applications. In many respects this is a more diverse field than processing, and effort is made to restrict the presentation to applications which use the laser's properties of spatial and temporal coherence. Lasers are becoming sufficiently low cost that in many experiments they are now used more for convenience than anything else. In many cases the phenomena involved in the measurements are well known and have been under study in the physics and optics laboratory; for example, interferometry and light scattering. Here the laser has provided a much improved source and extended the range of measurement. In the other cases the laser has allowed new measurements and techniques to be developed, such as holography, holographic interferometry, and the ring laser.

The subject of industrial measurements has been separated into the following sections:

IX. Light Scattering (which includes particle counting, contamination measurement, and velocity measurements).

X. Diffraction (which includes aligning techniques, wire size measurements using Fraunhofer diffraction, and spatial filtering).

XI. Interferometry and Holography.

XII. Ring Lasers.

XIII. Interaction Effects.

IX. Scattering

Measurements of amplitude, phase, frequency, and direction of light scattered from objects can provide information about the objects without physical contact. This has great appeal for industry where inspection and measurement techniques are needed which do not destroy or change the object itself.

The theory of light scattering is very involved and all that is pertinent here is to point out the various theories and their range of validity. (Van de Hulst [75] provides a very good treatment of light scattering.) For very small nonconducting particles the Rayleigh theory applies where the particle size is much smaller than the wavelength of the light. In this case the particle acts as an electric dipole and radiates in all directions. The intensity of the scattered energy varies as the wavelength to the inverse fourth power, the volume of the particle squared, and the number of particles. The scattered radiation as a function of angle does not depend on the shape or size of the particle.

The opposite case is where the particle is larger than the radiation wavelength and then the Mie theory is necessary. The effect of a larger scattering particle is that more radiation is scattered in the forward direction. The scattering depends on the shape and size of the particle, its refractive index and absorption coefficient, and increases rapidly with the size of the particle. These parameters also affect the polarization of the light scattered at an angle for a particular wavelength of the light, so that for Mie scattering, the size of the particle does affect the polarization of light scattered at a particular angle for a defined wavelength [76].

For totally reflecting spheres the Rayleigh scattering does not apply to small particles and the scattering is predominantly in the backward direction. As the sphere increases in size, the scattering resolves into two forms. In the forward direction the scattered light takes the form of the Fraunhofer diffraction pattern of an equivalent shaped thin conducting disk, and in the other directions the light behaves as simple reflected light.

Many authors [77]–[79] have reported work on light scattering using the laser and comparing the results for scattering with that of an incoherent source. There was some early concern [79] that scattering from laser light would be different from incoherent light scattering. Later experiments [78] have shown that the scattered light intensity is not measurably different under equivalent conditions.

The laser provides a much improved source in terms of collimation and intensity; this means that lower scattering cross sections can be measured with better angular resolution. In addition, the monochromaticity allows narrowband filtering and heterodyne techniques to be used for detection. This can be an advantage when ideal laboratory conditions are not available. One example [80]–[82] of this is the recent use of Thompson scattering of ruby laser light from a plasma to measure electron temperature and density; a 0.3 μm filter was used to minimize the photon noise from the plasma.

Contamination Measurement

Applications of light scattering to particle sizing and contamination measurement have not found extensive use as a tool in industry yet. The National Bureau of Standards [83] recently reported an instrument which was designed to measure particle contamination in liquids using laser light scattering. The system used a small He-Ne laser operating at 0.633 μm as a light source and measured the light scattered at 90° with a photomultiplier. This simple configuration was found to be very sensitive and could measure particle densities as low as a few hundred per milliliter. The measurement was empirical and needed calibration using a set of standard scattering solutions. The laser can thus be used in a quick and sensitive contamination detection system on a production line and give continuous readings of contamination levels. Neitzel [84] reports a high volume particle detector designed for contamination control in clean rooms. The instrument samples air at rates up to 50 ft³/min (1400 liters/min) and uses forward scattering of laser light to detect the dust, etc., down to a size of 0.3 μm. The instrument measures size and number of particles up to 100 000 particles/s and is quoted as being one order of magnitude faster than presently available equipment. Application of laser light scattering to clean room monitors and to perform liquid and solid inspection will no doubt continue to grow, as will the application to particle size analysis.

Doppler Effect

Doppler shifted light, scattered from moving particles, is a well known and well studied phenomena. Until recently, however, due to spectrometer limitations, only relatively large velocities could be detected. Using the laser with optical heterodyning [85], very accurate determinations of frequency shifts can be made; Cummins [85] states his system has a 10 Hz bandwidth and is able to measure velocities of 0.004 cm/s. This instrument was used to measure liquid velocities [86], [87] by scattering light from suspended particles. Velocity profiles could be determined by measuring the light scattered from various regions of the liquid.

Foreman [88] reports using a similar technique by scattering laser light from smoke particles in a wind tunnel to measure velocity profiles in addition to liquid flow measurements [89]. These would seem more research tools than industrial applications; however, the potential in industry is quite high.

An application of the Doppler effect which could have more immediate potential in industry is the use of a Doppler [90] radar instrument to measure vibration. Again this has the advantage of no contact with the object and so no interference with the object or its surrounding. The optical transducer also has no limitations due to resonances, etc., which are experienced with mechanical transducers. Laser radiation is scattered from the vibrating object and then mixed on a photodetector with the laser reference beam (local oscillator) in the usual manner (for heterodyning). The movement of the vibrating object puts a phase modulation on the scattered beam; this is analyzed with a discriminating circuit on the detector. Sensitivities to 0.1 μm in displacement are quoted for the experimental system reported.

Velocity Measurement

A more accurate velocity meter which has more immediate application uses a different effect. Stavis [91] describes this as measuring the velocity of the scattering pattern from light shone onto the surface of the object. The light scattered from the moving object falls on a diffraction grating. The scattered light forms the usual speckle pattern because of local interference effects from laser light scattering off different parts of the object. This speckle pattern moves across the grating behind which is placed a detector. Counting the light pulses caused by the movement of the speckle pattern across the grating gives a number directly proportional to the object speed. This is more applicable to measuring the velocity of objects rather than fluids. The accuracy is quoted as 0.1 percent and the instrument measures this velocity without contact with the object. The industrial application of these velocity devices is likely to be specialized initially until the economic advantages become more apparent.

X. DIFFRACTION

Diffraction effects and measurements involve the spatial coherence or collimation properties of the laser. Because of the very good spatial coherence properties, very intense equivalent point sources can be generated. This in effect means that perfect plane or spherical waves of useful intensity can be produced. Useful in an industrial sense means that instruments can be made using diffraction effects to make measurements, either automatically or using nonskilled operators.

Alignment

The first and simplest application of the spatial coherence of the laser was the use of the collimated beam as an alignment aid. This has already been used as a tunnel boring guide [92] and in geodesy; however, there is present the hazard of an open laser beam which can injure the eye. General alignment instruments have been developed [93],

[94] which are commercially available and in use, especially in the aircraft industry. They use a 1.5 mW, single transverse mode, He-Ne laser of wavelength 0.633 μm, with beam collimating optics to produce a plane parallel beam of 10 mm diameter. This is the optical line; along this line is placed a detector to determine whether objects are correctly aligned to it (Fig. 32). The detector is a circular disk divided into four quadrants, thus forming four detectors. When the beam is in the center of the detector, each quadrant receives the same light. Misalignment varies the light falling on each quadrant, and from the variation in readings an indication of the misalignment results. Accuracies are quoted for this instrument of 10 μin/ft (10^{-6} rad). This kind of alignment aid is simpler to use than the telescopes previously used and requires only one person to operate it efficiently. Because misalignment is read out by the detection system, operator judgment is not required and removes one source of error. This kind of instrument is providing a good machine shop alignment tool.

A much more sophisticated alignment system [95] is used in the Stanford Linear Accelerator Center's accelerator, measuring misalignment of 10^{-7} rad. To achieve this accuracy the laser is used as a diverging beam, being effectively a point source. This is imaged onto a detector using a Fresnel lens. The alignment line is that between the laser and the detector; the various parts of the accelerator, to which are attached the Fresnel lenses, are aligned to image the point source on the detector. The Fresnel lenses are retractable, the alignment being made with one lens at a time. To achieve this accuracy, it is necessary to enclose the laser beam path and also partially evacuate it to reduce air turbulence.

Diffraction Effects

Although diffraction effects have been studied for many years in optics, applications for measurement purposes have been very limited because the sources of monochromatic plane waves were of very low intensity. Applications such as Schlieren effect, phase contrast microscopy, and other modifications of a viewing system at its diffraction pattern plane have been restricted to visual or photographic recording techniques; also the measurements were made on the image and not on the diffraction pattern itself.

The laser as a high intensity, spatially coherent source allows much more complex diffraction phenomena to be observed with greater accuracy. The higher intensity means that detection systems can operate on the diffraction pattern. In addition, the temporal coherence allows more complicated measurements and operations to be made on the phase of the light as well as on its intensity.

Simple Diffraction Effects

An application which has received attention in our laboratories is the creation and detection of diffraction patterns to measure wire diameters. The diffraction patterns produced by bare and insulated copper wires placed in a collimated laser beam are simple Fraunhofer far field patterns [96]. For small wires, e.g, 0.5 mil (12.5 μm), the far

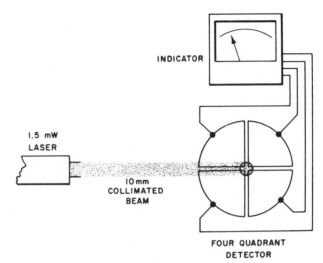

Fig. 32. Laser aligning tool showing quadrant detector to indicate misalignment from laser beam. (Laser beam is incident normally on quadrant detector plane.)

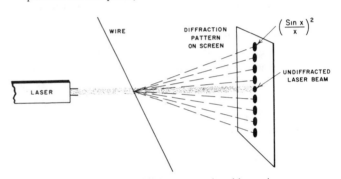

Fig. 33. Diffraction pattern produced by a wire in a spatially coherent beam.

field condition is obtained a short distance from the wire (Fig. 33). However, for larger wires it is more convenient to use a converging lens to form the diffraction pattern at its focal plane [76]. The pattern is an array of dots of decreasing intensity from the center, in which the separation of the dots is inversely proportional to the wire diameter. To use the Fraunhofer diffraction theory, the wire is assumed to be an infinitely thin conducting screen of the same width as the wire. The diffraction pattern of the wire then becomes the diffraction pattern of the original laser beam, which is the bright central spot (Fig. 33) minus (subtracting electric fields) the diffraction pattern of a slit, the array of dots. The slit diffraction pattern intensity takes the form

$$I = \left(\frac{\sin x}{x}\right)^2 \qquad (14)$$

where

$x = 2\pi\alpha r/\lambda$
α = angle subtended by diffraction pattern from lens
r = radius of wire
λ = laser wavelength.

For wire measurement processes, the central spot must be small enough not to affect the rest of the diffraction pattern, i.e., the laser beam must be several times larger than the wire. The diffraction pattern has some useful features which make it attractive for wire measurement. Movement of the wire transversely does not affect the position of the pattern; thus moving wires can be measured. Also, there is no contact with the wire, and the lower limit in terms of accuracy is governed by the wavelength of the light. In practical terms, this means measuring 0.1 mil (2.5 μm) wire to 0.5 percent accuracy or better.

The assumption made earlier must be treated with care; experiments and theory show that for conducting or absorbing objects, the Fraunhofer theory predicts very well the light pattern diffracted in the forward direction [75]. However, for dielectric nonabsorbing materials (e.g., glass fibers), Rayleigh scattering theory is required because this allows for light transmitted through the fiber [97], [98].

As previously mentioned, the laser's intensity allows electrooptic detection of the diffracted light to be used. We have tried two techniques, using either an array of photodiodes or a TV vidicon camera to detect the diffraction pattern and arrive at a measurement of wire size. Fig. 34 shows the more expensive but more accurate setup using the TV camera. The electronic signal from one line of the camera was analyzed electronically to determine the separation of the minima. A digital readout gives an indication of wire size. This system has worked very well in a laboratory setup giving 0.3 percent repeatability and is being evaluated for application on wire drawing lines.

Spatial Filtering

Another potential area of diffraction for measurement and inspection is spatial filtering where instead of analyzing the diffraction pattern, operations such as amplitude and phase multiplications are performed on the pattern. The pattern is then transformed with a second lens back to an image of the original object so that an analyzed image of the original object is seen.

Spatial filtering techniques were well known before the arrival of the laser and have been demonstrated as useful for improving noisy photographic images [99]–[101]. In some of these applications, the sharpness of images is increased by preferentially increasing the higher spatial frequency content of the picture. In others, the reverse is done to reduce high frequency noise to observe the presence of large features.

Spatial filtering is analogous to electronic signal filtering in communications systems, and O'Neil [102] and Cutrona [103] have analyzed spatial filtering networks quite thoroughly. There are two types—coherent or incoherent—depending on whether the light used is spatially coherent (point source Kohler illumination) or spatially incoherent; since the advantage of the laser is in producing the best spatially coherent light, only the former will be discussed. Spatial filtering is just part of an area of optical information processing which has had a new impetus from the laser. References [104], [105], and [106] give a good introduction to this subject.

Fig. 35 is a diagram of the basic filtering setup and uses the property of the lens in that the light pattern at one focal

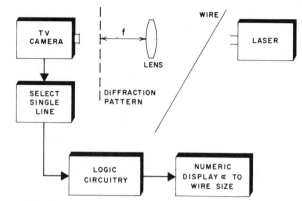

Fig. 34.   TV vidicon wire diffraction pattern detection.

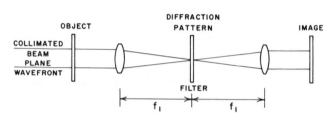

Fig. 35.   Basic spatial filtering arrangement.

plane is the Fourier transform of the pattern at the other focal plane. Spatially coherent light is passed through the object, which is a transparency, and then through the lens to form the diffraction pattern. At the diffraction pattern is placed the filter which attenuates the light at the required object spatial frequencies. The center, on the axis of the lens, is the dc component and the frequencies increase outward in both positive and negative directions in the two dimensions; the highest frequency passed is determined by the lens aperture. Also, phase multiplications can be achieved by the use of different thicknesses of high refractive index material (>1.0) on the filter. The filtered light is next passed through the second lens which makes another Fourier transform and returns the inverted image of the object, now modified by the filter. This kind of setup was used in experiments to improve photograph images [99]–[102].

One critical inspection problem we are investigating in IC manufacture is the checking of photodiffusion masks. These masks, made in either photographic emulsion or chrome on glass, can have dimensions in excess of 2 in (5.08 cm), with tolerances of ± 50 μin (1.3 μm) required over the whole field. Since the field may contain 1500 patterns, complete inspection using standard stage microscope techniques is impractical and uneconomical.

The mask is an array of the same pattern repeated at the same interval and, therefore, behaves very much like a diffraction grating. The diffraction pattern of a periodic array [76] is an interference function, which is an array of light spots, modulated in intensity by the diffraction pattern of one frame of the array. Fig. 36 shows the typical diffraction pattern of a photomask. The interference function is the frequency spectrum of a square wave grating and has components at the fundamental periodic frequency and at

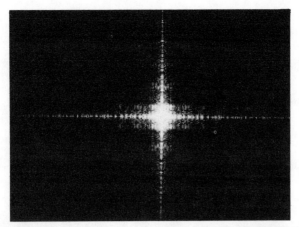

Fig. 36.   Diffraction pattern produced by semiconductor photomask.

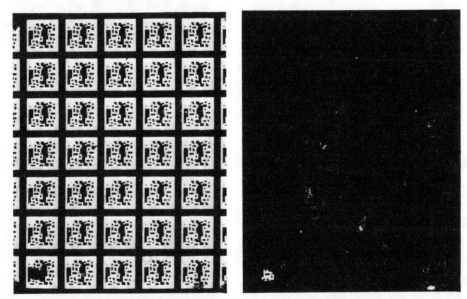

Fig. 37.   Part of a chrome on glass photomask imaged through the spatial filtering network without (left) and with (right) stop filter. Pattern repeat spacing is 62 mils (1.57 mm). Large error is displayed in lower left-hand frame; other small errors are also shown.

harmonics of this plus the dc term. The modulation represents the frequency spectrum of the frame, i.e., the individual features of the mask.

The filter we use is a stop filter corresponding to the interference function and so consists of an array of black dots. When the image is reconstructed, the result of this operation is to remove all information which is repeated at the period corresponding to the interference function. The result, as shown in Fig. 37, displays only the errors in the mask from the correct period. Thus all random errors are displayed, and the only mistakes not shown up are those which are repeated in all frames of the mask. Figs. 38 and 39 show some other results where the mask has an error in the array period in one dimension. Since the laser produces spatially coherent light, the resolution of this kind of operation is limited to the lenses used. For these results an $f/9.5$ lens was used with a He-Ne 0.633 μm laser, resulting in an error sensitivity of better than 0.4 mil (10 μm) over the 1.5 in

(3.8 cm) field. Advantages for inspection are that the whole of the mask can be examined; the alignment of the mask is critical only in rotation since the frequency spectrum is not dependent on its X, Y position and, in fact, means that the mask can be scanned over the whole area very quickly. Although the resolution at present is only 0.4 mil (10 μm), the use of a shorter wavelength laser and larger aperture lenses should bring it near the 50 μin (1.3 μm) requirement.

In [107] many other applications are given which use the Fourier transform properties of the lens for parallel analog computation, but the example described above seems to have immediate applications for measurement and inspection in industry.

The spatial coherence property with a usable intensity has thus allowed some new types of measurement and inspection techniques to be developed for industrial processes. The laser shows distinct promise for use in two-dimensional pattern measurement and inspection problems.

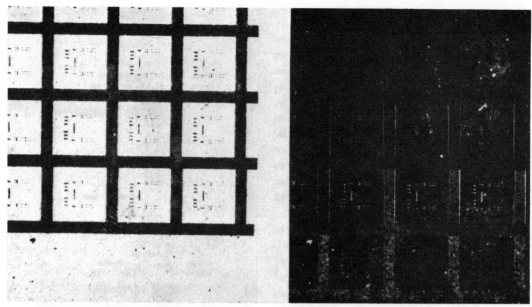

Fig. 38. Part of a chrome on glass photomask imaged through the spatial filtering network without (left) and with (right) stop filter. Pattern repeat spacing is 45 mils (1.14 mm). Step and repeat error (thin vertical stripes) of 0.8 mil is displayed in last row (error also displayed in row above). Note also error signal (less bright) produced because last row is not repeated below. Error signal occurs both in last row and in a lower position where row should occur.

Fig. 39. Part of 45 mils step and repeat emulsion photomask showing large step and repeat error in second column and smaller errors in subsequent columns. Note error signal in first column; this is due to the preceding column (not shown) which is a test pattern region and has a different format.

XI. INTERFEROMETRY AND HOLOGRAPHY

Interferometry

Predictably the laser has brought improvements to interferometry and there are now interferometer systems marketed for industrial use which incorporate a laser and are automated to the degree of giving a direct digital scale output reading. There are two principal improvements which the laser brings—increased temporal coherence length and increased intensity. These make the interferome-ter a practical tool to be considered for industrial applications where very accurate measurement for control, within a few microinches, is required over large distances.

The interferometer uses the laser wavelength as its standard and some comment is required about laser wavelength stability first. Most interferometer systems use the He-Ne [5] laser operating at 0.633 μm in the red. The neon transition has a Doppler broadened line width of about 1.5 GHz [108] and thus for an unstabilized laser there is a possible variation in frequency of about 1.0 GHz. This means an uncertainty of 1 in 10^6 in a measurement. The resonant cavity of the laser is an open-walled cavity and can operate at more than one frequency [109]. For example, a 1 m length cavity can oscillate at frequencies separated by 0.15 GHz in the fundamental TEM_{ooq} transverse mode [110]. This restricts the length over which the interferometer can operate because the interference fringes of each frequency vary in phase as the measured length increases, so destroying the visible fringe pattern. For optimum performance the laser must be a single frequency laser and must be stabilized to keep its frequency at the center of the neon laser transition. In less critical conditions of measurement accuracy and length, these requirements can be relaxed.

Fig. 40 shows the Twyman-Green interferometer used in the commercial distance measurement systems [111]–[114]. The laser beam is split into two paths—measurement and reference. In the reference arm is a mirror which returns the beam back to the beam splitter. The measurement path is parallel to the distance to be measured, and on the moving object is placed a corner cube reflector which also returns the beam to the beam splitter. The beam splitter recombines these two waves and sends them to the detector. The amplitude of the light received by the detector depends on the phase between the reference beam and the measurement beam and will go through one period as the measurement light path length changes by one wavelength, i.e., 0.6328 μm. This corresponds to a movement of the measurement arm reflector of half a wavelength. Some interferometers use a double path measurement to produce one period for every quarter-wavelength movement of the reflector [Fig. 40(b)]. The detector counts the periods, usually measuring quarter periods, and then a computer translates this to a distance measurement. To determine direction of movement, two detectors are required in some kind of quadrature phase arrangement.

Interferometer Applications

Various interferometers are available with varying accuracies which are useful for such applications as accurate machine tool calibration or control, x-y table calibration or measurement, and any other measurement where a large linear distance, up to 100 ft (30.5 m) or more, must be measured very accurately to 1 part in 10^8. The most refined system can measure to better than 2 μin (0.05 μm) over a distance of 500 in (12 m) and has compensation for atmospheric humidity, pressure, and temperature; it also allows for part temperature compensation. A simpler, less accurate instrument gives half-wavelength, 8 μin (0.3 μm) resolution over a distance of 100 ft (30.5 m).

An active area of investigation is the use of the interferometer to control x-y tables on step and repeat cameras used to make IC photomasks [115]. Work is also being done in the complimentary problem of inspecting masks using x-y tables. Two interferometers are used, one to monitor each dimension. Signals from the interferometers are fed to a computer which controls the movement of the table and the exposure circuits. Since the interferometer

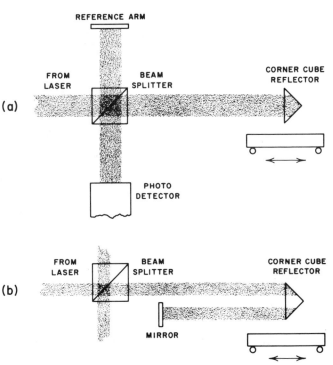

Fig. 40. Schematic of commercial interferometers, Twyman-Green type. (a) Single path. (b) Double path.

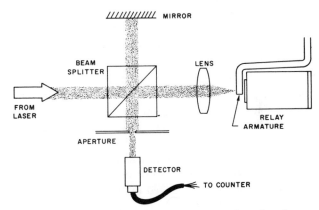

Fig. 41. Interferometer to measure movement of rough surfaces.

measures to better than 3 μin (0.08 μm) accuracies and is independent of the table moving screws, a more accurate and more reliable control results. Coupled with improved camera and table performance, the whole system produces 60 masks/h.

Motion Analysis

Techniques have been suggested to eliminate the necessity of a mirror for reflecting the measurement beam [116]. We have been working with a lens to focus the measurement beam onto a rough surface (Fig. 41). The requirement is approximately that the phase variation across the focused spot be less than a wavelength. The light reflected from the surface is collimated by the lens to a plane wave and this plane wave interfered with the reference beam in the normal manner. Using this approach, measurements were made on electromechanical relay armature and a plunge welder to

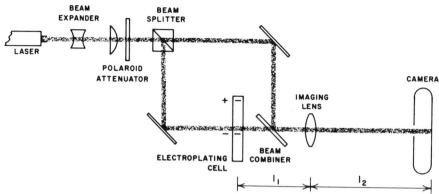

Fig. 42. Mach-Zehnder interferometer used for measuring Cu ion concentration.

determine distance and velocity with time. The advantage is that there is no influence on the motion with this optical transducer that mechanical transducers produce. The measurement distance is restricted to the focal region of the lens and cannot be used for large distances; however, this type of approach may find application on the shop floor, especially as the small He-Ne lasers used are now becoming cheaper than the associated computing electronics.

Concentration Measurement

The interferometer measures optical path length and so can measure refractive index change as well as distance change. We have used a laser Mach-Zehnder interferometer (Fig. 42) to measure ion concentration gradients in an electroplating solution [117], by measuring the associated refractive index change. Although this is not an industrial application of the interferometer on the shop floor, it indicates the kinds of concentration or density measurements which can be done easily using a very simple interferometer and a laser.

The types of interferometers and their applications are numerous; Herriott reviews some of them [118] and indicates how the laser has changed the interferometer from a sophisticated laboratory instrument into a regular tool for industry. An area already well acquainted with interferometry is the optics industry [119], and Herriott describes some interferometric measurements of large optical surfaces [118]. Although the authors have not heard of specific production applications, there is no doubt that the laser will help improve inspection of optical surfaces by simplifying and automating interferometry techniques. The laser has made a large impact on interferometry techniques and has established an area of measurement where the laser is both useful and economical.

Holography

Applications of holography in industry are still largely in the innovation stage. There are a few areas, however, which look promising and to which research and development effort is being guided. Holography, like the laser, has had a well publicized beginning but it will be some time before it is known which of the many suggested applications will become economically feasible.

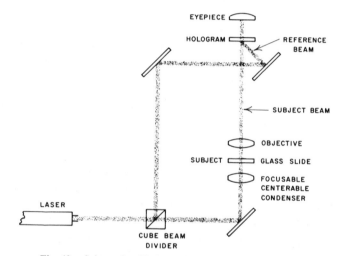

Fig. 43. Schematic of holographic microscope (from [120]).

Holography is lensless photography and has been proposed for use in a projection system in the photolithography process used for IC fabrication. Holography should be able to circumvent the limitations of a lens in projecting the large field of view required without loss of resolution and distortion due to spherical and other aberrations. There should also be an increase in resolution because higher equivalent apertures can be achieved. The main advantage, however, would be if two or more inches of field of view can be achieved keeping the 1 μm resolution and with little or no distortion. Results so far show that projection is possible over a half-inch (1.3 cm) field, with resolution of 1–2 μm, using the red He-Ne laser. Improvements in resolution should come from using blue argon ion or UV neon ion lasers.

A recently developed research tool which uses holography and could well become a manufacturing inspection aid is the holographic interference microscope [120]. This replaces one of the matched arms in the interference microscope by a hologram (Fig. 43); this hologram generates the interfering light beam and can be either the plane wave for interference microscopy or the object phase wave for differential interference effects. This produces, therefore, a much more versatile instrument since the hologram can generate any desired interference wave. The authors also state the

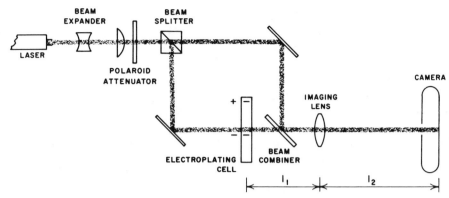

Fig. 42. Mach-Zehnder interferometer used for measuring Cu ion concentration.

determine distance and velocity with time. The advantage is that there is no influence on the motion with this optical transducer that mechanical transducers produce. The measurement distance is restricted to the focal region of the lens and cannot be used for large distances; however, this type of approach may find application on the shop floor, especially as the small He-Ne lasers used are now becoming cheaper than the associated computing electronics.

Concentration Measurement

The interferometer measures optical path length and so can measure refractive index change as well as distance change. We have used a laser Mach-Zehnder interferometer (Fig. 42) to measure ion concentration gradients in an electroplating solution [117], by measuring the associated refractive index change. Although this is not an industrial application of the interferometer on the shop floor, it indicates the kinds of concentration or density measurements which can be done easily using a very simple interferometer and a laser.

The types of interferometers and their applications are numerous; Herriott reviews some of them [118] and indicates how the laser has changed the interferometer from a sophisticated laboratory instrument into a regular tool for industry. An area already well acquainted with interferometry is the optics industry [119], and Herriott describes some interferometric measurements of large optical surfaces [118]. Although the authors have not heard of specific production applications, there is no doubt that the laser will help improve inspection of optical surfaces by simplifying and automating interferometry techniques. The laser has made a large impact on interferometry techniques and has established an area of measurement where the laser is both useful and economical.

Holography

Applications of holography in industry are still largely in the innovation stage. There are a few areas, however, which look promising and to which research and development effort is being guided. Holography, like the laser, has had a well publicized beginning but it will be some time before it is known which of the many suggested applications will become economically feasible.

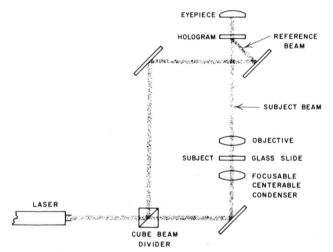

Fig. 43. Schematic of holographic microscope (from [120]).

Holography is lensless photography and has been proposed for use in a projection system in the photolithography process used for IC fabrication. Holography should be able to circumvent the limitations of a lens in projecting the large field of view required without loss of resolution and distortion due to spherical and other aberrations. There should also be an increase in resolution because higher equivalent apertures can be achieved. The main advantage, however, would be if two or more inches of field of view can be achieved keeping the 1 μm resolution and with little or no distortion. Results so far show that projection is possible over a half-inch (1.3 cm) field, with resolution of 1–2 μm, using the red He-Ne laser. Improvements in resolution should come from using blue argon ion or UV neon ion lasers.

A recently developed research tool which uses holography and could well become a manufacturing inspection aid is the holographic interference microscope [120]. This replaces one of the matched arms in the interference microscope by a hologram (Fig. 43); this hologram generates the interfering light beam and can be either the plane wave for interference microscopy or the object phase wave for differential interference effects. This produces, therefore, a much more versatile instrument since the hologram can generate any desired interference wave. The authors also state the

XIV. Conclusion

The information provided in this review has shown the breadth of applications of lasers in industry. There is no doubt that the laser has made considerable impact on the scientific community and has, in fact, become a standard research tool. We have shown here that the material progression of scientific discoveries to engineering applications is finding the laser increasingly in use in the manufacturing environment. We are confident that the economic advantages of lasers as practical tools will become more apparent as laser properties, and in particular the interaction of laser radiation with materials, are more widely understood.

Acknowledgment

The authors wish to acknowledge the contributions of several members of the Technical Staff of the Western Electric Company, Engineering Research Center. They also wish to thank M. I. Cohen of Bell Telephone Laboratories who reviewed the manuscript and offered several suggestions.

References

[1] T. H. Maiman, "Stimulated optical radiation in ruby," *Nature*, vol. 187, p. 493, 1960.

[2] E. Snitzer, "Optical maser action of Nd^{+3} in a barium crown glass," *Phys. Rev. Lett.*, vol. 7, p. 444, 1961.

[3] J. E. Geusic, H. M. Marcos, and L. G. VanUitert, "Laser oscillation in Nd doped yttrium aluminum, yttrium gallium and gadolinium garnets," *Appl. Phys. Lett.*, vol. 4, p. 182, 1964.

[4] G. E. Fenner, J. D. Kingsley, T. J. Soltys, R. N. Hall, and R. O. Carlson, "Coherent light emission from GaAs junctions," *Phys. Rev. Lett.*, vol. 9, p. 366, 1962.

[4a] M. I. Nathan, W. P. Dumke, G. Burns, F. H. Dill, Jr., and G. Lasher, "Stimulated emission of radiation from GaAs *p-n* junctions," *Appl. Phys. Lett.*, vol. 1, p. 62, 1962.

[4b] T. M. Quist, R. H. Rediker, R. J. Keyes, W. E. Krag, B. Lax, A. L. McWhorter, and H. J. Zeigler, "Semiconductor maser action of GaAs," *Appl. Phys. Lett.*, vol. 1, p. 91, 1962.

[5] A. Javan, W. R. Bennett, Jr., and D. R. Herriott, "Population inversion and continuous optical maser oscillation in a gas discharge containing a He-Ne mixture," *Phys. Rev. Lett.*, vol. 6, p. 106, 1961.

[6] W. B. Bridges and A. N. Chester, "Visible and U.V. laser oscillation at 118 wavelengths in ionized neon, argon, krypton, xenon, oxygen, and other gases," *Appl. Opt.*, vol. 4, p. 573, 1965.

[7] W. B. Bridges, "Laser action in singly ionized krypton and xenon," *Proc. IEEE* (*Correspondence*), vol. 52, pp. 843–844, July 1964.

[8] C. K. N. Patel, R. A. McFarlane, and W. L. Faust, "Optical maser action in C, N, O, S, and Br on dissociation of diatomic and polyatomic molecules," *Phys. Rev.*, vol. 133A, p. A1244, 1964.

[9] A. G. Fox and T. Li, "Resonant modes in a maser interferometer," *Bell Sys. Tech. J.*, vol. 40, p. 435, 1961.

[10] B. A. Lengyel, *Introduction to Laser Physics*. New York: Wiley, 1966.

[11] M. I. Cohen, B. A. Unger, and J. F. Milkowski, "Laser machining of thin films and integrated circuits," *Bell Sys. Tech. J.*, vol. 47, pp. 385–405, 1968.

[12] H. S. Carslaw and J. C. Yaeger, *Conduction of Heat in Solids.* London: Oxford University Press, 1959.

[13] R. H. Fairbanks and C. M. Adams, "Laser beam fusion welding," *Welding J.*, vol. 43, pp. 97s–102s, March 1964.

[14] H. S. McCracken, "Parameters effecting laser welding," presented at the 1967 Materials Engrg. Exposition and Congr., Cleveland, Ohio, ASM Tech. Rept. C7-19.1.

[15] A. O. Schmidt, I. Harn, and T. Hoski, "An evaluation of laser performance in microwelding," *Welding J.*, vol. 44, pp. 481s–489s, November 1965.

[16] M. I. Cohen, "Melting of a half-space subjected to a constant heat input," *J. Franklin Inst.*, vol. 283, pp. 271–285, April 1967.

[17] D. A. Watt, "Theory of thermal diffusivity by pulse technique," *Brit. J. Appl. Phys.*, vol. 17, pp. 231–240, 1966.

[18] N. N. Rykalin and A. A. Uglov, "Calculating heating processes in laser welding," *Welding Production*, vol. 14, June 1967.

[19] C. M. Verber and A. H. Adelman, "The interaction of laser beams with metals," *Battelle Tech. Rev.*, vol. 14, p. 5, July 1965.

[20] J. E. Anderson and J. E. Jackson, "An evaluation of pulsed laser welding," *Proc. Electron and Laser Beam Symp.* (sponsored by Pennsylvania State University and Alloyd General Corp), A. B. El-Kareh, Eds., 1965, pp. 17–50.

[21] M. I. Cohen, "Laser beams and integrated circuits," *Bell Labs. Rec.*, vol. 45, pp. 247–251, September 1967.

[22] M. I. Cohen and J. P. Epperson, "Applications of lasers to microelectronic fabrication," in *Advances in Electronics and Electron Physics*, A. B. El-Kareh, Ed. New York: Academic Press, 1968.

[23] R. H. Carter and F. P. Gagliano, "Metallographic preparation of micro-miniature devices," *Microelectronics and Reliability*, vol. 7, no. 4, 1968.

[24] N. F. Mott and H. Jones, *The Theory of the Properties of Metals and Alloys.* New York: Dover, 1958, ch. 3.

[25] F. Seitz, *The Modern Theory of Solids.* New York: McGraw-Hill, 1940, pp. 638–642.

[26] J. F. Ready, "Effects due to absorption of laser radiation," *J. Appl. Phys.*, vol. 36, p. 1522, 1965.

[27] A. W. Ehler, "Description of the plasma produced by a laser pulse striking on aluminum surface," *Bull. Am. Phys. Soc.*, ser. 10, pp. 227–Q10, February 1965.

[28] S. Namba and P. H. Kim, "The surface temperature of metals heated with laser," *Japan. J. Appl. Phys.*, vol. 4, p. 153, February 1965.

[29] W. L. Knecht, "Surface temperature of laser heated metal," *Proc. IEEE* (*Letters*), vol. 54, pp. 692–693, April 1966.

[30] A/O Corp. (Staff Members), "Glass laser technology," *Laser Focus*, vol. 3, pp. 21–28, December 1967.

[31] E. Snitzer, R. F. Woodcock, and J. P. Legre, "Phosphate glass Er^{3+} laser," presented at the Internat'l Quantum Electronics Conf., Miami, Fla., May 1968, Paper 13M-3.

[32] J. F. Smith and A. Thompson, "Metalworking lasers in engineering service applications," *Record of IEEE Annual Symp. on Electron, Ion and Laser Beam Technology*, 1967, pp. 268–277.

[33] M. I. Cohen, private communication.

[34] W. N. Platte and J. F. Smith, "Laser techniques for metals joining," *Welding J.*, vol. 42, pp. 481s–489s, November 1963.

[35] J. P. Epperson, "Laser welding in electronic circuit fabrication," *EDN Packaging and Materials Annual*, vol. 10, pp. 8–16, October 1965.

[36] J. P. Epperson, R. W. Dyer, and J. C. Grzywa, "The laser now a production tool," *Western Electric Engr.*, vol. 10, pp. 2–9, April 1966.

[37] J. C. Grzywa and A. Chesko, "Laser piercing and reworking of diamond dies," *Wire and Wire Prod.*, vol. 41, September 1966.

[38] J. F. Ready, "Interaction of high power laser radiation with absorbing surfaces," *1964 Proc. Nat'l Electronics Conf.*, vol. 22, pp. 67–71.

[39] R. D. Haun, Jr., "Laser applications," *IEEE Spectrum*, vol. 5, pp. 82–92, May 1968.

[40] M. I. Cohen, private communication.

[41] "Laser metal removal aids dynamic balancing," *Steel*, vol. 159, pp. 28–29, July 1966.

[42] "On line balancing systems," *Laser Focus*, vol. 4, p. 3, June 1968.

[43] K. K. Kelley, "Contributions to the data on theoretical metallurgy, XIII," *Bureau of Mines Bull.*, vol. 584, p. 73, 1960.

[44] *Metals Handbook*, 8th ed., vol. 1. Metals Park, Ohio: American Society for Metals, 1961, p. 1185.

[45] J. M. Bennett and E. J. Ashley, "Infrared reflectance and emittance of silver and gold evaporated in ultrahigh vacuum," *Appl. Opt.*, vol. 4, pp. 221–224, February 1965.

[46] R. G. Smith and M. L. Galvin, "Operation of the continuously pumped, repetitively Q-switched YA1G:Nd laser," *IEEE Quantum Electronics*, vol. QE-3, pp. 406–414, October 1967.

[47] "Trimming by laser," *Western Electric News Features*, p. 13, April 1968.

[48] B. A. Unger and M. I. Cohen, "Laser trimming of thin film resistors," presented at the Electronics Component Tech. Conf., Washington, D.C., 1968.

[49] S. J. Lins and R. D. Morrison, "Laser induced resistivity changes in film resistors," *WESCON Tech. Papers*, vol. 10, pt. 2, 1966, Paper 5/1.

[50] "Resistive film trimming with pulsed argon lasers," Hughes Aircraft Co., Torrance, Calif., Laser Application Note 3001.

[51] "Electronics review," *Electronics*, vol. 41, no. 3, p. 54, 1968.

[52] L. Braun and D. R. Breuer, "Laser adjustable resistors for precision monolithic circuits," *Proc. Microelectronics Symp.* (St. Louis, Mo., June 1968).

[53] M. DiDomenico, Jr., J. E. Geusic, H. M. Marcos, and R. G. Smith, "Generation of ultrashort optical pulses by mode locking the YAG:Nd laser," *Appl. Phys. Lett.*, vol. 8, no. 7, p. 180, 1966.

[54] T. A. Osial, "Industrial laser applications," *Instruments and Control Systems*, vol. 40, p. 101, October 1967.

[55] T. M. Jackson, A. D. Brisbane, and C. P. Sandbank, "Automated interconnection processes for semiconductor integrated circuit slices," *Proc. Conf. Integrated Circuits* (Eastbourne, England, May 1967).

[56] "Taming laser ray for industry," *Factory*, vol. 120, no. 4, p. 96, 1962.

[57] "Report on the '68 QEC," *Laser Focus*, vol. 4, p. 30A, June 1968.

[58] "Laser combining with oxygen jet to make a new cutting tool," *Product Engrg.*, vol. 39, p. 101, April 8, 1968.

[59] D. J. Garibotti, "Dicing of micro-semiconductors," U.S. Patent 3 112 850, December 3, 1963.

[60] R. M. Lumley, "Controlled separation of brittle materials using a laser," presented at the 70th Annual Meeting of American Ceramics Society, Chicago, Ill., April 1968, Paper 13-E-68.

[61] D. S. Paulley and D. L. Lockwood, "Method for manufacturing ceramic substrates for electrical circuits," U.S. Patent 3 324 212, June 6, 1967.

[62] L. P. Levine, J. F. Ready, and E. Bernal G., "Gas desorption produced by a giant pulse laser," *J. Appl. Phys.*, vol. 38, pp. 331–336, January 1967.

[63] H. Hoffmann, E. Yeager, and J. Striehr, "Laser temperature-jump apparatus for relaxation studies in electrolytic solutions," *Rev. Sci. Instr.*, vol. 39, May 1968.

[64] R. J. Murphy and G. J. Ritter, "Laser-induced thermal etching of metal and semi-conductor surfaces," *Nature*, vol. 210, pp. 191–192, April 1966.

[65] G. R. Speich, A. Szirmae, and R. M. Fisher, "A laser heating device for metallographic studies," *1965 ASTM Proc. Symp. Electron Metallography*, ASTM Special Tech. Publ. 396, pp. 97–114, 1966.

[66] G. R. Speich and R. M. Fisher, "Recrystallization of a rapidly heated $3\frac{1}{4}\%$ silicon steel," *Recrystallization, Grain Growth and Textures*. Metals Park, Ohio: American Society for Metals, 1966, pp. 563–598.

[67] D. L. Rousseau, G. E. Leroi, and W. E. Falconer, "Charged-particle emission upon ruby laser irradiation of transparent dielectric materials," *J. Appl. Phys.*, vol. 39, pp. 3328–3332, June 1968.

[68] L. M. Epstein and K. H. Sun, "Chemical reactions induced in gases by means of a laser," *Nature*, vol. 211, no. 5054, pp. 1173–1174, 1966.

[69] D. J. Margerum, "Acid-base characteristics of photochromism," U.S. Govt. Supported Research Rept. AD-665-426, December 1967.

[70] W. G. Pfann, "Principles of zone-melting," *Trans. AIME, J. Metals*, vol. 194, p. 747, July 1952.

[71] F. Brech and L. Cross, "Optical micromission stimulated by a ruby laser" (Abstract), *Appl. Spectroscopy*, vol. 16, no. 2, p. 59, 1962.

[72] K. G. Snetsinger and K. Keil, "Microspectro chemical analysis of minerals with the laser microprobe," *Am. Mineralogist*, vol. 52, pp. 1842–1854, November–December 1967.

[73] N. C. Fenner and N. R. Daly, "An instrument for mass analysis using a laser," *J. Materials Sci.*, vol. 3, pp. 259–261, 1968.

[74] V. G. Mossotti, K. Laqua, and W. D. Hagenah, "Laser-microanalysis by atomic absorption," *Spectrochimica Acta*, vol. 23B, pp. 197–206, 1967.

[75] H. C. Van de Hulst, *Light Scattering by Small Particles*. New York: Wiley, 1957.

[76] M. Born and E. Wolf, *Principles of Optics*. New York: Pergamon, 1965.

[77] D. H. Freeman and E. C. Kuehner, "Laser detection of small particles in liquids," presented at the Conf. on Liquid-Borne Particle Metrology, New York, February 1968.

[78] G. C. Sherman, F. S. Harris, Jr., and F. L. Morse, Jr., "Scattering of coherent and incoherent light by latex hydrosols," *Appl. Opt.*, vol. 7, p. 421, 1968. Also, "Experimental comparison of scattering of coherent and incoherent light," *IEEE Trans. Antennas and Propagation*, vol. AP-15, pp. 141–147, January 1967.

[79] L. W. Carrier and L. J. Nugent, "Comparison of some recent experimental results of coherent and incoherent light scattering with theory," *Appl. Opt.*, vol. 4, p. 1457, 1965.

[80] E. T. Gerry and D. J. Rose, "Plasma diagnostics by Thompson scattering of a laser beam," *J. Appl. Phys.*, vol. 37, p. 2715, 1966.

[81] T. S. Brown and D. J. Rose, "Plasma diagnostics using lasers: relations between scattered spectrum and electron-velocity distribution," *J. Appl. Phys.*, vol. 37, p. 2709, 1966.

[82] W. B. Johnson, "Laser interferometry and photon scattering in plasma diagnostics," *IEEE Trans. Antennas and Propagation*, vol. AP-15, pp. 152–162, January 1967.

[83] D. H. Freeman and E. C. Kuehner, "Laser detection of small particles in liquids," presented at the Conf. on Liquid-Borne Particle Metrology, New York, February 1968.

[84] W. E. Neitzel, "Development of an increased sampling rate monitoring system," presented at the Annual Conf. of American Assoc. for Contamination Control, Washington, D. C., May 1967.

[85] H. Z. Cummings, N. Knable, and Y. Yeh, "Observation of diffusion broadening of Rayleigh scattered light," *Phys. Rev. Lett.*, vol. 12, p. 150, 1964.

[86] H. Z. Cummings and Y. Yeh, "Localized fluid flow measurements with an He-Ne laser spectrometer," *Appl. Phys. Lett.*, vol. 4, p. 176, 1964.

[87] R. H. Goldstein and D. K. Kreid, "Measurement of laminar flow development in a square duct using a laser Doppler flowmeter," *J. Appl. Mech.*, vol. 34, p. 813, December 1967.

[88] J. W. Foreman, Jr., E. W. George, J. L. Jetton, R. D. Lewis, J. R. Thornton, and H. J. Watson, "Fluid flow measurements with a laser Doppler velocimeter," *IEEE J. Quantum Electronics*, vol. QE-2, pp. 260–266, August 1966.

[89] J. W. Foreman, Jr., R. D. Lewis, J. R. Thornton, and H. J. Watson, "Laser Doppler velocimeter for measurement of localized flow velocities in liquids," *Proc. IEEE (Letters)*, vol. 54, pp. 424–425, March 1966.

[90] J. T. Montonye, "Doppler optical radar and the heterodyne measurement of oscillating systems," U.S. Govt. R/D Rept. AD-651-822 May 1967. Also, G. A. Massey, "Study of vibration measurement by laser methods," NASA Contract Rept. NASA-CR-985.

[91] G. Stavis, "Optical diffraction velocimeter," *Instruments and Control Systems*, vol. 39, p. 99, 1966.

[92] D. A. Worth, "Advantages and applications of CW gas lasers," *Laser Focus*, vol. 2, p. 24, September 1966.

[93] P. A. Hickman, "Optical tooling viewed in a new light," *Laser Focus*, vol. 4, p. 23, March 1968.

[94] B. Feinberg, "Laser tooling goes to work," *Tool and Mfg. Engr.*, vol. 59, October 1967.

[95] W. B. Herrmannsfeldt, "Precision alignment using a system of large rectangular Fresnel lenses," *Appl. Opt.*, vol. 7, p. 995, 1968.

[96] M. Koedam, "Determination of small dimensions by diffraction of a laser beam," *Philips Tech. Rev.*, vol. 27, p. 208, 1966.

[97] L. A. Jeffers, "Determination of the diameter of small fibers by diffraction of a laser beam," presented at the IEEE Conf. on Laser Engrg. and Applications, Washington, D. C., June 1967, Paper THAM 11.5.

[98] W. A. Farone and M. Kerker, "Light scattering from long submicron glass cylinders at normal incidence," *J. Opt. Soc. Am.*, vol. 56, p. 481, 1966.

[99] A. Marechal and P. Croce, "Un filtre de frequences spatiales pour l'amerlioration du contraste des images optiques," *Compt. Rend. Acad. Sci.* (Paris), vol. 237, p. 607, 1953.

[100] J. Tsujiuchi, "Correction of optical images by compensation of aberrations and by spatial frequency filtering," *Progress in Optics*, vol. 2, E. Wolf, Ed. Amsterdam, Netherlands: North-Holland, 1963, ch. 4.

[101] "Restoration of atmospherically regraded images," U.S. Govt. Research Rept. AD-806-878, 1967.

[102] E. L. O'Neil, "Spatial filtering in optics," *IRE Trans. Information Theory*, vol. IT-2, pp. 56–65, June 1956.

[103] L. J. Cutrona, E. N. Leith, C. J. Palermo, and L. J. Porcello, "Optical data processing and filtering systems," *IRE Trans. Information Theory*, vol. IT-6, pp. 386–400, June 1960.

[104] D. Redman, "Processing information with light," *Science J.*, vol. 4, p. 50, 1968.

[105] D. K. Pollack, C. J. Koester, and J. T. Tippett, Eds., *Optical Processing of Information*. Washington, D. C.: Spartan Books, 1963.

[106] J. T. Tippett, D. A. Berkowitz, L. C. Clapp, C. J. Koester, and A. Vanderburgh, Jr., Eds., *Optical and Electro-Optical Information Processing*. Cambridge, Mass.: M.I.T. Press, 1965.

[107] A. Vander Lugt, "A review of optical data-processing techniques," *Optica Acata*, vol. 15, pp. 1–33, 1968.

[108] W. R. Bennett, Jr., "Inversion mechanisms in gas lasers," *Appl. Opt.* (Suppl. on Chemical Lasers), p. 3, 1965.

[109] A. Fox and T. Li, "Resonant modes in a maser interferometer," *Bell Sys. Tech. J.*, vol. 40, p. 453, 1961.

[110] G. D. Boyd and J. P. Gordon, "Confocal multimode resonator for millimeter through optical wavelength masers," *Bell Sys. Tech. J.*, vol. 25, p. 489, 1961.

[111] "Line-standard interferometer for accurate calibration of length scales," *Laser Focus*, vol. 3, p. 32, October 1967.

[112] R. W. Schede, "Interferometers for use as integral parts of machine tools," *IEEE Trans. Industry and General Applications*, vol. IGA-3, pp. 328–332, July/August 1967.

[113] W. E. Bushor and J. F. Kreidl, "Where do we stand on laser length standards," *Laser Focus*, vol. 2, p. 26, November 1966.

[114] E. A. Haley, "Lasers: multimillion dollar market," *Electronics World*, March 1968.

[115] "Laser-source masking camera promises to up production of semiconductors," *Laser Focus*, vol. 4, p. 12, March 1968.

[116] J. Kevern, "New laser profilometer offers noncontact measurement," *Product Engrg.*, vol. 38, p. 94, July 31, 1967.

[117] A. Tvarusko and L. S. Watkins, "Laser interferometric study of the diffusion layer during non-steady-state electrodeposition," to be published.

[118] D. R. Herriott, "Some applications of lasers to interferometry,"

[119] A. C. S. Van Heel and C. A. J. Simons, "Lens and surface testing with compact interferometers," *Appl. Opt.*, vol. 6, p. 803, 1967.

[120] K. Snow and R. Vandemarker, "An application of holography to interference microscopy," *Appl. Opt.*, vol. 7, p. 549, 1968.

[121] L. O. Heflinger, R. F. Wuerker, and R. E. Brooks, "Holographic interferometry," *J. Appl. Phys.*, vol. 37, p. 642, 1966.

[122] B. P. Hildebrand, K. A. Haines, and R. Larkin, "Holography as a tool in the testing of large aperture optics," *Appl. Opt.*, vol. 6, p. 1267, 1967.

[123] A. H. Rosenthal, "Regenerative circulatory multiple-beam interferometry for the study of light propagation effects," *J. Opt. Soc. Am.*, vol. 52, p. 1143, 1962.

[124] C. V. Heer, *Bull. Am. Phys. Soc.*, vol. 6, p. 58, 1961.

[125] W. M. Macek and D. I. M. Davis, Jr., "Rotation rate sensing with travelling wave ring lasers," *Appl. Phys. Lett.*, vol. 2, p. 67, February 1, 1963.

[126] P. H. Lee and J. G. Atwood, "Measurement of saturation induced optical nonreciprocity in a ring laser plasma," presented at the Internat'l Quantum Electronics Conf., Phoenix, Ariz., 1966, Paper 5B-1.

[127] P. J. Klass, "Laser unit challenges conventional gyros," *Aviation Week and Space Tech.*, p. 103, September 12, 1968.

[128] D. J. Bradley, A. J. F. Durrant, G. M. Gale, M. Moore, and P. D. Smith, "Characteristics of organic dye lasers as tunable frequency sources for nanosecond absorption spectroscopy," presented at the Internat'l Quantum Electronics Conf., Miami, Fla., May 1968, Paper 1A-3.

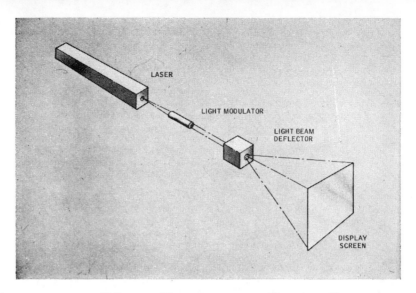

Laser display technology

Much headway has been made in developing operational experimental systems, but widespread application depends on the development of improved lasers

Charles E. Baker   Texas Instruments Incorporated

The promise of producing a cathode-ray-tube type of display with essentially unlimited screen size accounts for much of the current interest in laser displays. The first display that used a laser light source to generate a television image was operated in late 1964. Since that time, a number of experimental laser display systems have been demonstrated, including versions capable of reproducing color television images and displaying computer-generated information. Adequate light modulation and scanning techniques are presently available, but the design of broadly applicable equipment still awaits the development of more efficient multicolor lasers.

Practical, direct generation of a two-dimensional visual display by modulated light-beam scanning has been a long-sought goal. Early displays, which used this method to avoid intermediate energy-conversion steps, required complex, large-aperture optical systems and exhibited limited brightness and resolution. The laser's high brightness and directional characteristics have overcome these problems by allowing light modulation and scanning techniques to be employed that were not feasible with conventional light sources. The laser's monochromatic light, available at a variety of wavelengths, allows color displays to be readily implemented. These factors account for the present level of research and development activity in various aspects of laser display technology.[1]

Conceptually, a laser display generates an image in the same manner as a cathode-ray tube (CRT). As shown in the title illustration, the thin beam of light emerging from a continuously operating visible laser is intensity modulated by a video signal, then scanned into a two-dimensional raster. In this "open-air cathode-ray tube," the familiar electron gun, control grid, deflection coils, and

phosphor screen of the CRT have been replaced by a laser light source, electrooptic light modulator, beam deflectors, and a diffusing screen.

In comparison with the CRT, the energy beam in a laser display need not be maintained in a vacuum environment nor converted into a visible form by a special screen. This means that previous limits on screen size and brightness are removed and resolution capabilities are determined by physical optics instead of electron optics. Although there was initial concern over the lack of screen persistence in a laser display, further study[2] and direct observation have shown this to be of little

FIGURE 1. Laser display image.

Reprinted from *IEEE Spectrum*, vol. 5, pp. 39–50, Dec. 1968.

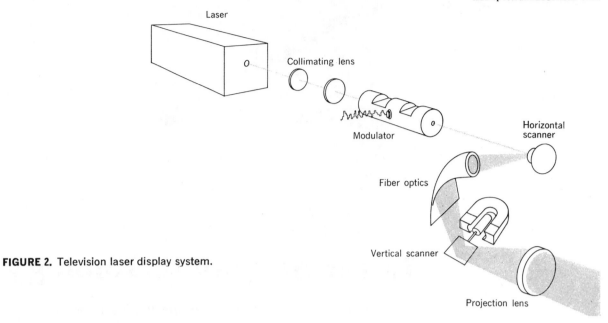

FIGURE 2. Television laser display system.

FIGURE 3. Laser display optical system.

consequence. Image quality comparable to that obtained with a CRT has been demonstrated. The image shown in Fig. 1 was produced by a television laser display system that employed an argon ion laser, an electrooptic light modulator, and moving-mirror light-beam scanners.

Experimental systems

Laser display feasibility, visual characteristics, and performance parameters have been evaluated with experimental systems developed by several different groups.

FIGURE 4. Color laser display block diagram.

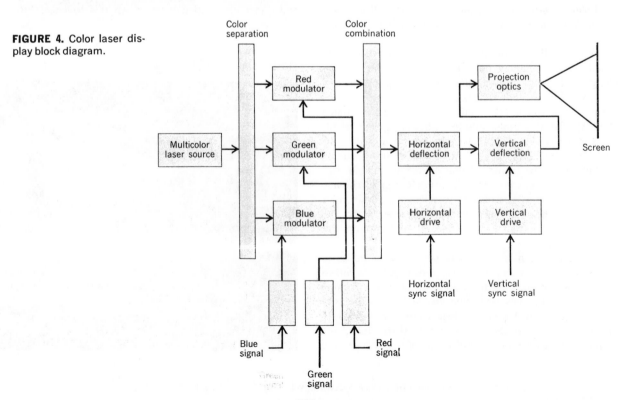

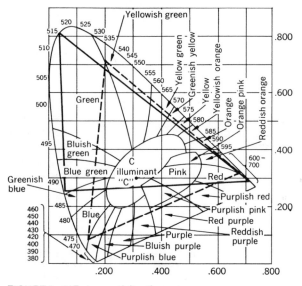

FIGURE 5. CIE chromaticity diagram.

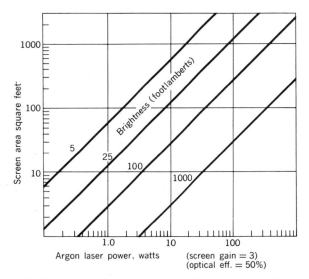

FIGURE 6. Laser display brightness. Note that a footlambert equals 3.43 candelas per square meter.

Figure 2 shows a system operated in late 1964 that used the 6328-angstrom line of a 50-mW helium-neon laser to produce a red and black image from a commercial television broadcast.[3] Laser intensity modulation was performed by a potassium dihydrogen phosphate (KDP) electrooptic light modulator. High-frequency horizontal deflection was produced by an acoustically resonant, nutating-mirror scanner. The horizontal scanner first generated a circular scan, which was then rectified into a linear scan using a fiber-optic circle-to-line converter. Vertical scanning was accomplished with a galvanometer-driven moving mirror. Video and synchronizing signals were obtained from a conventional television set. An improved version of this scanning and modulation system was subsequently constructed and a one-watt argon ion laser used as the coherent light source. This system, which is shown in Fig. 3, was demonstrated at the 1966 IEEE International Convention.

Red, green, and blue laser lines have been independently modulated and combined to form a full color display as indicated in Fig. 4. By scanning a single, variable-color, variable-intensity light beam, the convergence problems encountered in other color displays are avoided. A color laser display has been shown that has been developed by Stone.[4] This system obtains color video signals by scanning the object that is to be viewed with an unmodulated portion of the same laser beam used to generate the display. Variations in reflectivity and color are sensed by red-, green-, and blue-sensitive photomultipliers. An argon ion laser was used to provide blue and green primaries and a helium-neon laser furnished the red primary. Multiple-element, piezoelectrically driven, moving-mirror scanners were used to perform horizontal and vertical scanning. Brightness, hue, and color saturation were controlled by KDP electrooptic light modulators. A similar, commercial color television compatible laser display has also been developed by Alsabrook and Baker.[5] This system uses the scanning technique of Fig. 2.

The solid triangle on the CIE chromaticity diagram shown in Fig. 5 indicates the gamut of colors that can be generated using argon ion and helium-neon laser primaries. For reference, the color gamut of a shadow-mask color CRT is shown by the dashed lines. It can be seen that this combination of lasers gives good green coverage but the relatively long wavelength of the strong argon line results in deficient magenta coverage. Good color television displays have been produced, however, with the argon ion, helium-neon combination. This can be explained by the importance of good skin tones and lack of magenta in most scenes.

The brightness of a laser display is determined by screen size, laser power, optical system transmission efficiency, laser wavelength, and screen diffusion characteristics. Figure 6 shows screen brightness for a display with an optical transmission efficiency of 50 percent that uses an argon ion laser light source. It has been assumed that a typical rear-projection viewing screen is used, having a directional gain of three as compared with a Lambertian diffusing surface. It can be seen that a one-watt laser would produce an image of 25 fL (86 candelas per square meter) on a 1.1-square-meter screen. Typical television viewing brightness levels are between 25 and 50 fL (86 and 172 cd/m²). Much larger and brighter displays could be produced with lasers that are presently available. As much as 100 watts of visible coherent output power has been produced by experimental argon ion lasers.[6,7]

Most laser display investigators have utilized electrooptic light modulators and moving-mirror scanners; however, one system has been demonstrated that is based on quite different principles of operation. This display, which is shown in Fig. 7, points the way toward a system that does not use any moving parts. Korpel et al.[8] have used ultrasonic light-diffraction techniques to perform both video intensity control and horizontal scanning. Vertical scanning is performed with a galvanometer-driven moving mirror, but ultrasonic light-diffraction techniques have also been used for this function. This approach is reminiscent of the old Scophony[9,10]

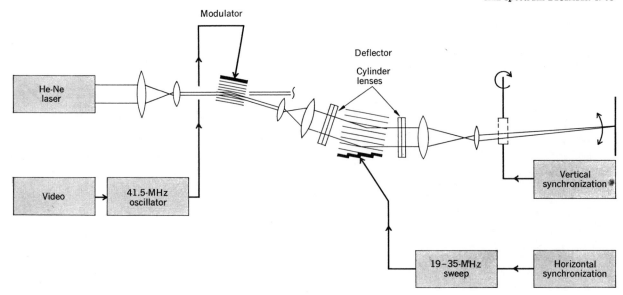

FIGURE 7. Laser display using acoustic deflection and modulation.

FIGURE 8. Stroke-writing digital laser display.

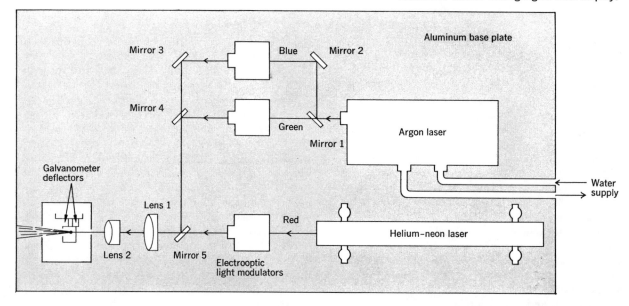

television display, which used an ultrasonic light modulator to generate simultaneously an image of an entire television line. Instead of using a rotating mirror to immobilize the line of video moving through the ultrasonic light modulator, however, the modern version uses a second ultrasonic diffraction cell. The laser version is also much more efficient, since nearly all of the light emitted by a laser can be diffracted into the proper point on the display screen. This system could be extended to full color operation, but not as readily as a display using moving-mirror scanners. The ultrasonic diffraction cells are dispersive and diffract different colors of light in different directions.

Although much laser display R&D activity has centered on television-type displays, some work has been directed toward displaying computer-generated information. Figure 8 illustrates a stroke-writing display developed by McCarthy and Lipnick.[11] High-frequency

galvanometers are used to write symbols at any desired point on the display screen. Since the laser light source is not scanned into a raster, but only directed into the areas where information is to appear, much brighter images are possible for a given laser power. Electrooptic light modulators are used to gate red, green, and blue laser lines on and off to write information in any of ten resolvable colors.

Symbols are formed by driving the orthogonal galvanometers in a programmed sequence of strokes. Galvanometer frequency-response limitations allow up to ten symbols to be generated with a 30-Hz refresh rate. Improved electrooptic or acoustic-optic deflection techniques would allow more complex displays to be generated.

Another digital laser display that has been successfully operated is shown in Fig. 9.[12] This system uses a 32-facet rotating scan mirror to perform both horizontal

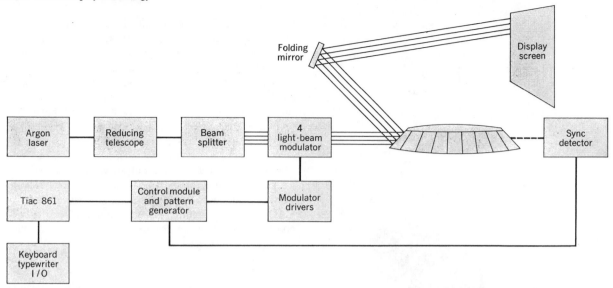

FIGURE 9. Multibeam digital laser display.

and vertical scanning. The image to be displayed is stored on a magnetic drum, which is mounted on the same shaft as the scan mirror. The rigid coupling between scanner and image storage medium ensures complete synchronization of these two functions once an image has been stored on the drum surface. The display image is regenerated once each drum revolution; thus, refresh rate is controlled by a drum rotation rate of 60 revolutions per second. Each face of the scanning mirror performs horizontal scanning as it turns with the storage drum. Vertical scanning is accomplished by having each face tilted at a slightly different angle with respect to the axis of rotation. A high-resolution display is obtained by simultaneously scanning a number of independently modulated light beams. A separate track on the magnetic drum is used to control each modulated light beam.

Lasers

The laser is noted as a source of highly directional, monochromatic light. Conventional light sources, such as a tungsten filament or arc lamp, are quite wasteful since they emit into a 360-degree solid angle from a relatively large area. When these light sources are used to illuminate the limited aperture of the usual optical system, only a small fraction of the emitted light is actually used. The additional optical complexities required to generate a color display with a conventional source tends to reduce further the amount of light available. On the other hand, all of the light emitted by a laser may be used in a display system, since the maximum optical aperture required is determined by diffraction considerations and large apertures are not required for high transmission efficiency. Light modulators and scanners have been developed over the past several years that take advantage of these unique properties of coherent light.

Several reviews[13-15] have described the wide variety of lasers that have been developed since the first ruby laser was operated in 1960. Desirable laser properties

I. Gas laser wavelengths

Laser	Wavelength, Å
Helium-neon	5939
	6046
	6118
	6294
	6328*
	6352
	6401
Argon ion	4545
	4579
	4658
	4727
	4765
	4880*
	4965
	5017
	5145*
	5289
Krypton ion	4577
	4619
	4634
	4680
	4762*
	4765
	4825
	4846
	5022
	5208*
	5309
	5682*
	6471*
	6570
	6765
	6871

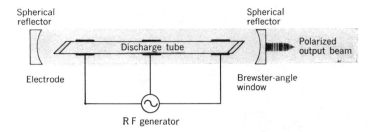

FIGURE 10. Helium-neon laser construction.

FIGURE 11. Ion laser construction.

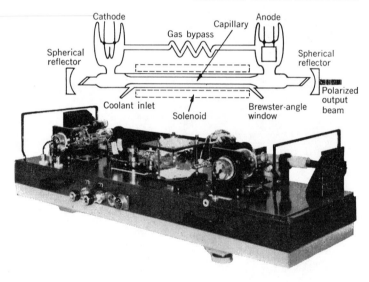

FIGURE 12. Frequency-doubled YAG:Nd laser.

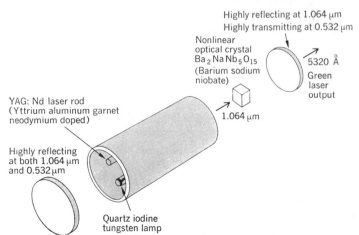

for display application include: large amounts of visible output light, continuous operation, diffraction-limited uniphase wavefront, and operation with a minimum of such ancillary equipment as a cryogenic cooler or high-energy power supply. The helium-neon and argon ion gas lasers satisfy most of these requirements. Both of these devices are commonly available and have been used in experimental displays. Use of the krypton ion laser in color displays has been discussed, but the low efficiency of this device has discouraged serious consideration. Table I lists the output wavelengths of these lasers. The dominant spectral lines that contain a major portion of the output power are marked with an asterisk. Solid-state, semiconductor, and liquid lasers will require considerably more development before display application can be considered.

Typical helium-neon and ion laser constructions are shown in Figs. 10 and 11, respectively. The helium-neon laser consists of a plasma tube that contains a low-current electric discharge. A low-loss reflector is placed at each end of the plasma tube to form an optically resonant cavity. Neon atoms that are excited by resonance collisions with helium in the electric discharge amplify light at certain wavelengths to excite this cavity. Coherent light is extracted from one of the cavity reflectors, which is partially transmissive. The maximum practical output power from a helium-neon laser is approximately 100 mW. This has proved adequate for experimental purposes, but is too low for a practical display.

The much higher power argon ion laser differs from the helium-neon laser in that it uses a more energetic electric discharge. In a typical device, a plasma is formed in a 2-mm capillary by a 20- to 30-ampere arc discharge. This creates special cooling problems, but makes it possible to generate several watts of coherent light. Argon ion lasers have produced as much as 100 watts of blue–green coherent light; however, this has been accomplished with an efficiency of only a fraction of a percent. The high output powers available from an argon ion laser presently make this the most desirable device for display system implementation.

The possibility of efficiently generating large amounts of coherent visible light was considerably improved with the development of a new nonlinear optical material—barium sodium niobate. This material has been used to convert 100 percent of the output power of a 1.064-μm infrared laser into green light at a wavelength of 5320 angstroms.[16] The configuration employed by Geusic and his co-workers is shown in Fig. 12. The $Ba_2NaNb_5O_{15}$ material is placed within the optical cavity of a conventional tungsten-lamp-pumped YAG:Nd laser. The mirrors forming the resonant cavity are coated so that power can be extracted from the laser system only at the second harmonic and not at the fundamental laser frequency. Using this technique, 1.1 watts of coherent green light have been obtained. This output represents 100 percent conversion to the green of the available infrared radiation from the basic YAG:Nd laser utilized. Tungsten-lamp pump sources have enabled as much as 50 watts to be produced with a YAG:Nd laser with an efficiency of over one percent.[17] It has been predicted that efficiencies as great as ten percent may be possible by using semiconductor light-emitting diodes as pump sources.[18]

Light-beam deflection

Scanning. A number of means of deflecting a coherent light beam have been suggested and reduced to practice.[19] These have ranged from relatively simple rotating mirrors to elegant solid-state electrooptic deflection schemes. For television-type displays, the requirement of linearly deflecting a light beam at a synchronized, 15 734-Hz rate with a short flyback time has proved particularly challenging.

One of the basic characteristics of a light-beam scanner is resolution. Using the Rayleigh criterion, the number of spot diameters N that can be resolved through a deflection angle θ is given by

$$N = \frac{A\theta}{1.27\lambda}$$

where A is the aperture of the scanner, λ is the wavelength of light, and the factor 1.27 accounts for the Gaussian intensity distribution of a laser light beam. It should be emphasized that the resolution N is more important than scan angle since θ may be increased or decreased with the use of appropriate optics. Another important characteristic is scan frequency. A useful figure of merit is the product of scan rate and the number of resolution elements. Scanners having a resolution–frequency product of over 100 million resolution elements per second have been operated.[20] This corresponds to a video bandwidth of over 50 MHz.

Mechanical scanning techniques have been used extensively for image recording and reproduction. Devices such as the Nipkow disk and various mirror and lens drum configurations have been in use since the early days of television. Generally, these devices have been regarded as being undesirable because of the problems of synchronization and the limited stability and operational life of moving parts and bearings. In the magnetic-drum system mentioned earlier, the problem of synchronization vanishes. Figure 13 displays the mirror used in this system.[12]

A d'Arsonval galvanometer can be used to drive a moving-mirror scanner, as in the unit shown in Fig. 14; this scheme has been used to perform 60-Hz vertical scanning in several experimental displays.[21] This is a conventional pen motor of the type normally used in strip-chart recorders. The upper frequency limit of a pen motor is usually assumed to be 50 to 70 Hz, although frequency components as high as 720 Hz are present in a television vertical scan sawtooth. This apparent inconsistency can be overcome and linear scans with rapid flyback obtained by placing the pen motor in a nonlinear, feedback control loop.

Figure 15 shows a moving-mirror scanner that is used to perform horizontal scanning in a television-type laser display.[3] It consists of a 1.5-mm-diameter mirror attached to a quartz fiber, which in turn is attached to a piezoelectric tube. The combination of mirror, fiber, and tube are resonant at the horizontal scan frequency. Orthogonal axes of the tube are driven by quadrature sinusoidal waveforms to generate a circular Lissajous scan pattern. The circular scan pattern is then converted to a linear scan pattern with a properly formed fiber-optic array. This technique makes possible a linear scan with essentially zero flyback time.

Another type of resonant, high-frequency moving-mirror scanner is shown in Fig. 16. Magnetostrictively

FIGURE 13. Multiple-facet scan mirror.

FIGURE 14. Galvanometer scanner.

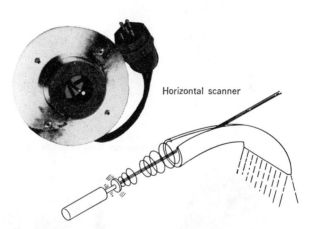

Horizontal scanner

FIGURE 15. Nutating-mirror scanner of the system appearing in Fig. 2.

FIGURE 16. Magnetostrictive torsional scanner.

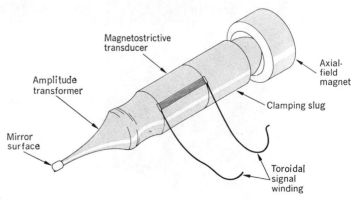

Magnetostrictive transducer

Amplitude transformer

Axial-field magnet

Clamping slug

Mirror surface

Toroidal signal winding

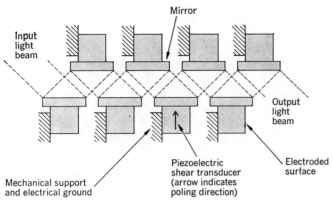

FIGURE 17. Multiple-element piezoelectric scanner.

FIGURE 18. Ultrasonic diffraction scanner principle of operation.

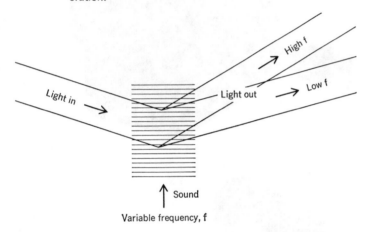

driven torsional vibrators[22] of this type have individually exhibited: (1) scan frequencies as high as 64 kHz, (2) scan amplitudes as large as 65 degrees, and (3) scanning apertures as large as 1 cm. A representative unit operates at a frequency of 28 350 Hz with a scan amplitude of 10 degrees and a scanning aperture of 4.5 mm. The number of resolution elements is about 50 MHz divided by the scan rate. Two of these devices have been operated orthogonally and driven in quadrature to produce a circular scan, which was subsequently converted into a linear scan using fiber optics. Other arrangements, with several scanners operating at various phases and frequencies, have been used to generate an approximately linear sawtooth scan without fiber optics.

The performance that can be obtained with a single, nonresonant, piezoelectrically driven moving mirror can be enhanced using the arrangement of Fig. 17. Schlafer and Fowler[19] have attached mirrors to piezoelectric shear transducers and obtained a one-degree deflection over a 17-kHz bandwidth using this technique. This general method of placing a series of devices in tandem[23] has been used quite effectively with several types of scanners to obtain deflection angles and resolutions that would not otherwise be possible.

Any scanning technique that depends on physical motion of a refracting or reflecting surface is restrained to a sequential scanning motion. This is in contrast to the random deflection capability of the inertialess electron beam used in a cathode-ray tube. The desire to overcome these limitations has motivated the development of several electrooptical and acoustooptical scanning techniques. Figure 18 depicts the operating principle of one of the most successful nonmechanical scanning tech-

FIGURE 19. Multiple-prism electrooptical light-beam scanner.

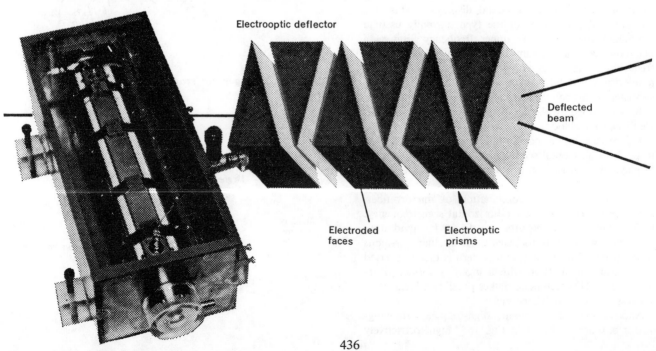

niques.[24] The ultrasonic diffraction scanner depends on the fact that a stress wave can modify the refractive index of a transparent material such as glass or water. An ultrasonic stress wave propagating through a transparent material can affect light in a manner comparable to a ruled diffraction grating. The incident light beam is deflected by an angle directly proportional to the wavelength of the light and the frequency of the ultrasonic stress wave.

Depending on the interaction medium and the deflection angle required, the ultrasonic stress-wave frequency would be between 20 and 200 MHz. A typical unit has been developed that operates at a center frequency of 60 MHz with a bandwidth of 40 MHz. The acoustooptic interaction medium of the unit is a high-index glass, a ceramic piezoelectric transducer is employed, and operation in the Bragg[25] diffraction mode allows nearly all of the incident light to be deflected in the desired direction.

Resolution of the ultrasonic diffraction scanner is directly proportional to the product of bandwidth and the time required for an ultrasonic wavefront to cross the light beam. In a television display system, the maximum allowable transit time is the horizontal blanking interval. With a 10-μs blanking interval, the diffraction scanner is capable of scanning a light beam through 400 resolvable points.

A light beam propagating through a prism is deflected by an amount depending on prism shape, orientation, and refractive index. The series arrangement of electrooptic prisms in Fig. 19 has been used[26] to deflect a light beam by varying the refractive index of each prism with an electrical field. It has proved desirable to place the prisms in an index-matching oil to minimize reflections and light loss. The limited change in refractive index that can be achieved with commonly available electrooptical materials, such as KDP, results in a maximum deflection angle of one degree. This corresponds to 100 resolution elements for a design using crystals of reasonable size. New electrooptic materials, such as barium strontium niobate, may make this type of light-beam deflector much more practical.

A digital light-beam deflector has been developed that consists of a series of binary deflection units, each of which is capable of deflecting a light beam in two different directions.[27] If N deflection units are coupled in tandem then 2^N different spot positions may be produced. The basic features of the digital deflector are shown in Fig. 20. Electrooptic crystals are used to change the direction of polarization of a transmitted light beam when a certain voltage is applied. A prism fabricated from a birefringent crystal, such as calcite, is associated with each electrooptic polarization switch. Light entering the prism will be deflected in either of two different directions depending on polarization. A typical scanner might have ten stages of X deflection and ten stages of Y deflection elements with an overall transmission efficiency of ten percent. The low efficiency results from the large number of surfaces in this device and results in serious application problems.

Another class of deflection techniques depends on internal mode selection in a highly degenerate laser cavity.[28] In this approach, light source, modulation, and deflection are all combined in a single device. In the electron-beam scan laser[29] shown in Fig. 21, mode selection is provided by a modified electrooptic light valve. A degenerate, flat-field conjugate laser resonator is formed by two mirrors and two lenses spaced such that one of the mirrors is imaged onto the other. One mirror consists of a sandwich of electrooptical and birefringent materials that serves to reflect and change the state of polarization of an incident light beam. Normally, this change in polarization is sufficient to prevent laser action from taking place. This condition may be altered by directing an electron beam to a particular point on the mode-selection sandwich. Laser action will then occur only at this point, with light intensity and direction directly controlled by the electron beam. In turn, the electron beam is controlled by conventional CRT techniques. Devices of this type have been used to deflect light in more than 10^6 different directions.

Light modulation

Lasers that are suitable for display applications can be modulated by varying their input current, but only over a limited frequency range. There has been considerable interest in developing high-frequency light-modulation techniques for laser communication systems, and much of this work is directly applicable to laser displays. The most practical light-modulation devices utilize the

FIGURE 20. Digital light-beam scanner.

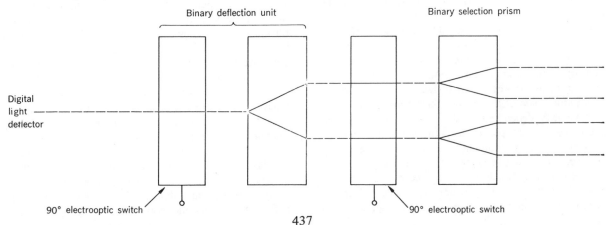

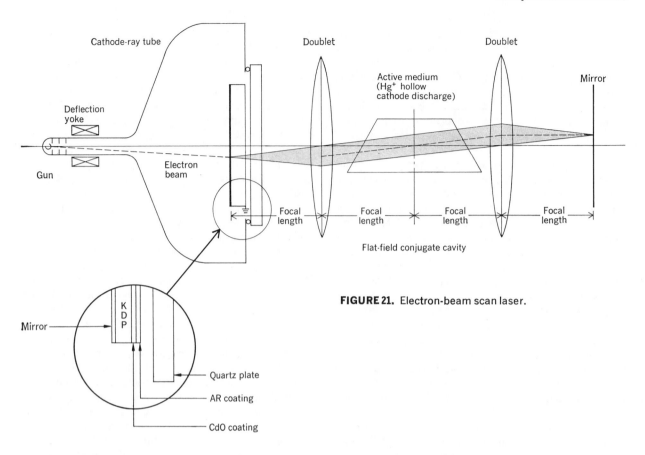

Cathode-ray tube

Doublet

Doublet

Mirror

Active medium
(Hg⁺ hollow
cathode discharge)

Deflection
yoke

Electron
beam

Gun

Focal
length

Focal
length

Focal
length

Focal
length

Flat-field conjugate cavity

Mirror

K
D
P

Quartz plate

AR coating

CdO coating

FIGURE 21. Electron-beam scan laser.

Pockels effect in an electrooptic crystal, such as KDP,[30] to control laser intensity in response to a video signal. Figure 22 shows the type of transverse field electrooptic light modulator that has been used in several laser displays. This device is constructed from two identical pieces of 45-degree Z-cut potassium dideuterium phosphate (KD*P). These crystals, which are typically 2 mm square and 35 mm long, are arranged in tandem with their optical axes orthogonal. The electrical drive field is applied parallel to the optical axis. This type of fabrication is necessary to cancel out static birefringence and minimize[31] the effect of changes in birefringence due to temperature variations. An alternate method of construction is to orient the optical axis of the two crystals in parallel with a 90-degree phase-retardation plate between the two crystals.[32]

Light modulation is performed by varying the polarization of an incident light beam. Linearly polarized light entering the crystal at the proper angle is separated into two orthogonal components, commonly called ordinary and extraordinary rays, each of which passes down the longitudinal axis of the crystal. Each component travels at a velocity depending on the index of refraction encountered in that particular path. Since the index of refraction for each component is affected by the applied voltage, a particular voltage will cause the two components to emerge exactly in phase. The result is that the transmitted light beam does not experience any change in polarization and thus cannot pass through an ortho-

gonal polarization analyzer. By applying a voltage that causes the two component velocities to differ by an amount such that they emerge 180 degrees out of phase, the output polarization is orthogonal to the input light; thus, the analyzer does not produce any attenuation. At intermediate points between these two voltages, various amounts of elliptical polarization would result so that intermediate amounts of light would emerge.

Device parameters of interest for display applications include high transmission efficiency, large contrast ratio, adequate bandwidth, and low drive power. Less than 250 volts are required to vary the transmission of a transverse modulator from minimum to maximum. The older, longitudinal type of light modulator, which operated with an applied electric field in the direction of light-beam propagation, required 20 times this driving voltage.

Typically, contrast ratios of 100:1 and transmission efficiencies as high as 90 percent are exhibited by the transverse-field light modulator.[33] Contrast ratio and transmission efficiency are primarily determined by light divergence, beam alignment with crystal axes, and scattering within the modulator. Drive voltage is determined by crystal geometry and the electrooptic coefficient. Several watts of optical power can be handled by a transverse-field modulator without taking special design precautions. Solid-state video amplifiers have been developed for use with electrooptic light modulators, which will produce the necessary voltage drive level over a 25-MHz bandwidth. Recently developed electrooptical materials such as lithium niobate and barium strontium niobate exhibit much higher electrooptical coefficients than KDP and should considerably reduce video-am-

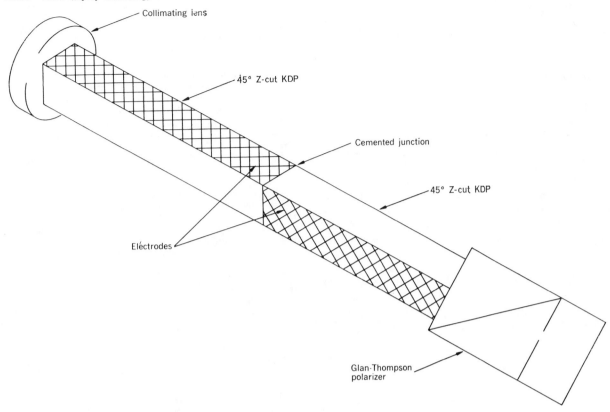

FIGURE 22. Transverse-field electrooptic modulator construction.

plifier power requirements. A 25-MHz bandwidth KDP drive amplifier dissipates over 100 watts, whereas a properly designed barium sodium niobate modulator should require less video drive power than a cathode-ray tube.

The same type of ultrasonic diffraction cell that has been used for inertialess light-beam scanning may also be used for laser intensity modulation. In this case, the video control signal is first used to modulate a carrier, which might typically be at a frequency of 40 MHz. When the diffraction cell is driven by this signal, a portion of an incident light beam will be deflected in a direction corresponding to the carrier frequency as previously described. If the light beam is sufficiently small, and the carrier amplitude is not too large, then the diffracted light intensity will follow variations in the video control signal. An appropriately placed optical stop is normally used to remove the undiffracted light. A larger incident light beam may also be used. In this case, a lens may be used to form a moving image of the video modulating signal.

Concluding remarks

Good image quality has been obtained with experimental laser display systems, and satisfactory progress is being made in overcoming various implementation problems. Resolution equivalent to a CRT television display has already been demonstrated, and much better performance appears feasible. The greater than 100-to-1 contrast ratio that can be provided by electrooptic light-

modulation techniques is a decided improvement over the 15-to-1 ratio usually realized with a CRT in such applications. This high contrast ratio results in a laser display image of sharper, more lifelike, appearance. Color displays having a satisfactory range of hues have also been produced using red, green, and blue laser lines.

Although there is a close parallel between laser displays and CRT displays, there are also substantial differences. A laser display, by its very nature, does not exhibit persistence as does a CRT display. This has been shown to have a negligible effect, however, on factors such as apparent brightness and flicker. A number of observers have commented on the characteristic "speckle"[34,35] pattern produced by coherent light. Many individuals, when seeing a laser display for the first time, do not notice this image speckle, since it has a much finer grain structure than the image being displayed. Even workers who have spent considerable amounts of time observing laser displays do not find the speckle annoying unless they concentrate on it.

Concern has been expressed over potential eye safety hazards with a laser display. In most installations, safety problems may be avoided by using a rear-projection screen with an enclosed projection path. Scanning provides a certain amount of protection by reducing average power density and some types of scanners only allow light to reach the screen when they are in operation. To keep the laser safety problem in proper perspective, it should be pointed out that a CRT envelope is a potentially lethal

439

weapon if handled improperly and cases of spontaneous implosion have been reported. Another aspect is the high voltage required for CRT operation and the resultant danger of shock and X radiation.

Fortunately, a rather simple solution to the dual problem of laser safety and image speckle is available. It was noted that an early laser display system, which utilized fiber optics in the horizontal scanner, did not exhibit image speckle due to the coherence-destroying nature of fiber-optic light transmission. Speckle may be eliminated and brightness of the projected beam greatly reduced by forming the image on an intermediate, moving, diffusing surface such as ground glass. A conventional projection lens may then be used to form the final image on either a front- or rear-projection screen. The penalty that is paid is a small reduction in transmission efficiency and resolution.

With one notable exception, the laser display systems that have been demonstrated have used helium-neon or argon ion lasers, electrooptic light modulators, and moving-mirror scanners. The use of moving parts is distasteful to many electronic engineers, but color display design is simplified by achromatic, moving-mirror scanners. The various solid-state scanners that are being developed are inherently dispersive in that they deflect different wavelengths of light by different amounts. Development of an inertialess scanner is mandatory, however, for stroke writing or randomly scanned displays. For raster-scanned displays, the inertia of a moving-mirror scanner may even be of some benefit in display stabilization.

Laser modulation and scanning techniques have reached a sufficiently advanced state of development to allow practical display system implementation; however, the low efficiency of presently available lasers prevents any serious consideration of competing with CRT displays in the immediate future. Several thousand watts from the power supply would be required by present lasers for a screen size and brightness consistent with home entertainment needs. Cathode-ray tubes will quite likely dominate the display field for a number of years to come. Technology for large-screen displays, however, is still in a state of evolution and no single approach has reached a dominant position. For many large-screen-display applications, the dynamic characteristics of a CRT-like display would be highly desirable. Since laser efficiency increases with output power level and many large-screen-display requirements cannot be satisfied with present technology, this area appears to hold more immediate interest. Further application of laser display technology rests on the development of a practical, low-cost laser with an efficiency exceeding one percent.

Revised text of a paper presented at the 1968 IEEE International Convention, New York, N.Y., March 18–21, 1968.

REFERENCES

1. "Laser display seminar," *Information Display*, vol. 5, pp. 36–65, May/June 1968.

2. Easton, R. A., Markin, J., and Sobel, A., "Subjective brightness of a very-short-persistence television display compared to one with standard persistence," *J. Opt. Soc. Am.*, vol. 57, pp. 957–962, July 1967.

3. Baker, C. E., and Rugari, A. D., "A large-screen real-time display technique," *Information Display*, vol. 3, pp. 37–46, Mar./Apr. 1966.

4. Stone, S. M., "Experimental multicolor real-time laser display system," *Proc. Society for Information Display 8th Nat'l Symp.*, pp. 161–168, 1968.

5. Baker, C. E., and Alsabrook, C. M., "A large-screen color television laser display," *Proc. 1967 Southwestern IEEE Conf. (SWIEEECO)*, pp. 5-7-1–5-7-8.

6. Boersch, H., Herziger, G., Seelig, W., and Volland, J., *Phys. Letters*, vol. 24A, p. 695, June 5, 1967.

7. Raytheon Research Division, *Laser Focus*, p. 17, Nov. 1967.

8. Korpel, A., Adler, R., Desmares, P., and Watson, W., "A television display using acoustic deflection and modulation of coherent light," *J. Appl. Opt.*, vol. 5, pp. 1667–1675, Oct. 1966.

9. Okolicsanyi, F., "The wave-slot, an optical television system," *Wireless Engr.*, vol. 14, pp. 527–536, Oct. 1937.

10. Robinson, D. M., "The supersonic light control and its application to television with special reference to the Scophony television receiver," *Proc. IRE*, vol. 27, pp. 483–486, Aug. 1939.

11. McCarthy, D., and Lipnick, R., "Large-screen multicolor laser display for trainer applications," U.S. Naval Training Devices Center, Orlando, Fla.

12. Baker, C. E., and Pipken, L. L., "A multicolor laser digital data display system," *Proc. Society for Information Display 7th Nat'l Symp.*, pp. 189–202, Oct. 1966.

13. Kiss, Z. J., and Pressley, R. J., "Crystalline solid lasers," *Proc. IEEE*, vol. 54, pp. 1236–1248, Oct. 1966.

14. Bloom, A. L., "Gas lasers," *Proc. IEEE*, vol. 54, pp. 1262–1275, Oct. 1966.

15. Nathan, M. I., "Semiconductor lasers," *Proc. IEEE*, vol. 54, pp. 1276–1289, Oct. 1966.

16. Geusic, J. F., *et al.*, "Continuous 0.532μm solid-state source using Ba$_2$NaNb$_5$O$_{15}$," *Appl. Phys. Letters*, vol. 9, p. 306, May 1, 1968.

17. Korad Dep't., Union Carbide Corp., *Laser Focus*, p. 3, Apr. 1968.

18. Anderson, L., "Laser display seminar," *Information Display*, vol. 5, p. 57, May/June 1968.

19. Fowler, V. J., and Schlafer, J., "A survey of laser beam deflection techniques," *J. Appl. Opt.*, vol. 5, pp. 1675–1681, Oct. 1966.

20. Sherman, R., "Motor-bearing breakthroughs in ultrahigh-speed laser beam scanners," *Proc. Society for Information Display 8th Nat'l Symp.*, p. 289, May 1967.

21. Fournier, J. R., and Baker, C. E., "Large angle deflection techniques for laser display," RADC-TR-66-722 (DDC No. AD651301), Rome Air Development Center, Feb. 1967.

22. Fournier, J. R., Texas Instruments Inc., Private communication.

23. Beiser, L., "Laser beam scan enhancement through periodic aperture transfer," *J. Appl. Opt.*, vol. 7, pp. 647–650, Apr. 1968.

24. Gordon, E. I., "A review of acoustooptical deflection and modulation devices," *Proc. IEEE*, vol. 54, pp. 1391–1400, Oct. 1966.

25. Adler, R., "Interaction between light and sound," *IEEE Spectrum*, vol. 4, pp. 42–54, May 1967.

26. Kiefer, J. E., Lotspeich, J. F., Brown, W. P., and Serf, H. R., "Performance characteristics of an electrooptic light-beam deflector," *Dig. 1967 IEEE Conf. on Laser Engineering and Applications*, p. 37.

27. Kulcke, W., *et al.*, "Convergent beam digital light deflector," in *Optical and Electro-Optical Information Processing*, J. T. Tippett *et al.*, eds. Cambridge, Mass.: M.I.T. Press, 1965, pp. 371–418.

28. Pole, R. V., *et al.*, "Laser deflection and scanning," in *Optical and Electro-Optical Information Processing*, J. T. Tippett *et al.*, eds. Cambridge, Mass.: M.I.T. Press, 1965, pp. 351–364.

29. Pole, R. V., and Myers, R. A., "Electron beam scan laser," *IEEE J. Quantum Electronics*, vol. QE-2, pp. 182–184, July 1966.

30. Kaminow, I. P., and Turner, E. H., "Electrooptic light modulators," *Proc. IEEE*, vol. 54, pp. 1374–1390, Oct. 1966.

31. Eden, D. D., "Solid-state techniques for modulation and demodulation of optical waves," DA 36039 AMC 03250E (DDC No. AD457311), U.S. Army Electronics Command, Feb. 1965.

32. Peters, C. J., "Gigacycle-bandwidth coherent-light traveling-wave amplitude modulator," *Proc. IEEE*, vol. 53, pp. 455–460, May 1965.

33. Bell, D. T., Texas Instruments Inc., Private communication.

34. Oliver, B. M., "Sparkling spots and random diffraction," *Proc. IEEE*, vol. 51, pp. 220–221, Jan. 1963.

35. Considine, P. S., "Effects of coherence on imaging systems," *J. Opt. Soc. Am.*, vol. 56, pp. 1001–1009, Aug. 1966.

A Review of Operational Laser Communication Systems

FRANK E. GOODWIN, MEMBER, IEEE

Abstract—Laser communication systems which have been built into serviceable units are reviewed. Although systems built prior to 1965 were more of a breadboard nature, some early experiments of historical interest are described. After 1965, techniques and component reliability were sufficiently improved to permit the development of several interesting and sophisticated systems. Performance characteristics of the more representative of these systems are listed. Recent trends show the use of infrared wavelengths, injection lasers, mode-locked/pulse-code modulation systems, optical heterodyne detection, and automatic pointing and tracking.

INTRODUCTION

THIS paper discusses the performance characteristics of laser communication systems which have actually been developed into serviceable systems. Some of the systems developed during the past five years introduce new concepts and provide important operational data. The discussion will, therefore, center around these representative systems.

The term "operational" as used here must be carefully qualified, since none of the systems described is operational in the military sense of the word. The word was chosen to be stronger than "demonstrated," to convey that the systems included are serviceable, reliable, and useful. Thus, breadboard experimental systems are omitted except for references to a few cases of particular interest.

In order to put the description of operational systems in the proper perspective and to fully appreciate the significance of the concepts demonstrated, some early experiments of historical interest will be described. These experiments generally cover the period from 1962 to 1965, prior to the availability of reliable lasers.

SOME EXPERIMENTS OF HISTORICAL INTEREST

The first communications with light beams did not use lasers, but rather reflected sunlight with a heliograph, a modern version of which is the signaling mirror. For many generations blinker signals using incandescent and arc lights provided communications between ships. Modulated arc lights provided voice communications links 30 years ago. A spectacular demonstration of transmitting television over 30 miles, using a light emitting GaAs diode, was given by M.I.T. Lincoln Labs. as early as 1962, prior to a significant laser communication demonstration.

Perhaps the first demonstration of long distance laser communication through the atmosphere was performed by a Hughes group in November 1962 over a range of 30 km

Manuscript received July 10, 1970.
The author is with Hughes Research Laboratories, Malibu, Calif. 90265.

(18 miles). This experiment was performed shortly after the 6328-Å helium–neon (He–Ne) laser was discovered. The laser was excited by an RF source high-frequency amateur radio transmitter; intensity modulation of the laser was easily achieved by modulating the intensity of the RF source. Detection was achieved with a photomultiplier. Intelligible voice communications were obtained through the use of a high-pass filter to reduce the effects of scintillation noise.

Shortly thereafter in May 1963 an all-time distance record was set by a group from Electro-Optics Systems who transmitted a voice modulated 6328-Å beam 190 km (118 miles) from Panamint Ridge near Death Valley, Calif., to a point in the San Gabriel Mountains near Pasadena. The modulation and detection schemes were similar to those of the Hughes experiment.

The first transmission of TV over a laser beam was reported by a group at North American Aviation in March 1963. Modulation was achieved through an interferometer which was driven with piezoelectric elements having a modulation bandwidth of 1.7 MHz. Later in 1963, electro-optic modulation of 5-MHz band TV signals was achieved at Hughes through the use of an eight element interdigital potassium di-hydrogen phosphate (KDP) modulator. However, both the North American and Hughes demonstrations were not suitable for transmitting signals through turbulent atmosphere. The spectral character of atmospheric turbulence interferes severely with vertical synchronization.

A significant step in avoiding the effects of atmospheric scintillation on laser modulation was demonstrated in late 1963 by a group from IBM with pulse-rate modulation of a GaAs injection laser. The pulse-rate was 8 kHz, and a pulse frequency discriminator produced a modulation bandwidth near 6 kHz. Variations in the amplitude of the received pulses were not demodulated in the output.

After the initial flurry of activity in 1962 and 1963, it was realized that amplitude modulation (AM) of laser beams with analog information was not practical in the presence of atmospheric-turbulence noise. Although experiments suggested several techniques which might be suitable to reduce the effects of atmospheric turbulence, they were as yet unproven. Some of the techniques were: 1) optical frequency modulation with optical heterodyne detection; 2) frequency modulation of a subcarrier on the laser beam; and 3) pulse code modulation.

The problem of getting useful information through a turbulent atmosphere on an optical beam became one of

Reprinted from *Proc. IEEE*, vol. 58, pp. 1746–1752, Oct. 1970.

providing enough signal margin so that a continuous flow of information was available even with deep fading in the amplitude of the received beam. If this condition were met, any of the several techniques just outlined could have been used to remove the amplitude noise. In spite of early knowledge of the necessary techniques to build a system with a channel free from atmospheric noise, the usefulness and practicality of a laser communication system was questionable for many reasons. First, existing communications systems were adequate to handle concurrent demands. Second, considerable research and development were required to improve the reliability of components to assure reliable system operation. Third, a system in the atmosphere would always be subject to interruption in the presence of heavy fog. Fourth, use of the system in space where atmospheric effects could be neglected required accurate pointing and tracking optical systems which were not then available. In view of these problems, it is not surprising that the rate of progress in the early 1960's in building operational laser communication systems was slow.

After 1965, as reliable laser sources became available, several interesting and sophisticated laser communications systems were built and evaluated. The following discussion describes the salient features of some of these systems.

OPERATIONAL SYSTEMS—DESCRIPTION AND PERFORMANCE

The following systems are described because they illustrate a particular new concept or technique. The system chosen for each illustration is considered particularly representative. It is possible that there are systems unknown to the author which may have served as a better illustration; any such oversight was unintentional.

Several trends and developments of interest occurred during the period from 1965 to 1970. First, there was a general movement toward the infrared, stimulated by the development of reliable He–Ne lasers at 1.15 μm and 3.39 μm, helium–xenon at 3.5 μm, gallium arsenide at 0.9 μm, and (most important) the Nd:YAG laser at 1.06 μm and the CO_2 laser at 10.6 μm. Second, the development of optical heterodyne techniques for detection in the infrared took place. Third, there was the development of the necessary sophisticated pointing and tracking techniques which make optical space communications feasible.

The anticipated needs of future communications systems with high quality channels having extremely large dynamic range and a high degree of linearity, as required, for example, by cable television, has introduced another trend. Electrooptic amplitude modulation, polarization modulation, and some forms of phase modulation are inherently nonlinear and cannot provide sufficient linearity to meet these requirements. The trend, therefore, has been toward optical FM, subcarrier FM, and pulse code modulation (PCM). Optical FM and subcarrier FM can provide a dynamic range up to about 55 dB. PCM, of course, limited only by available bandwidth. Dynamic ranges required in some future communication needs will exceed 55 dB, requiring the use of digital techniques. Systems which use these

TABLE I

	Hughes/U.S. Army Electronics Command SSB FM System [1]
Wavelength (μm)	0.6328
Power (watts)	0.003
Photodetector	Philco L4501 silicon photodiode
Modulator	SSB quadrature using KDP
Modulation	Subcarrier FM (FSK), 876 MHz $\pm$ 1 MHz
Coding	PCM
PCM Bit Rate	250 kbit
Base Bandwidth	1 MHz
Range Performance	Useful for 1 mile paths for PCM error count studies

later techniques are described in the paragraphs which follow.

Single Sideband FM—Optical Heterodyne Detection—1965

Single sideband (SSB) FM offers a means of avoiding the effects of amplitude fading and electrooptic modulation nonlinearities. The method is described as the single sideband modulation of the optical beam with an RF carrier, usually the output of a conventional FM transmitter. With the suppression of the original optical carrier, a new optical signal is generated which is offset in frequency from the original frequency by that of the RF carrier. Frequency modulation of the RF carrier produces true frequency modulation of the optical carrier. The receiver must be an optical hterodyne receiver in order to recover the information from the signal.

The first system of this type was a single-ended system built by Hughes for the U.S. Army Electronics Command in 1965 for PCM error rate studies [1]. The FM transmitter and receiver were elements of the AN/GRC-50, a military communication set, having a carrier frequency of 876 MHz. The system utilized a 6328-Å laser oscillator and a silicon photodiode tuned to the RF carrier frequency. Although the system was single ended, having to derive the local oscillator from the transmitter oscillator, it demonstrated techniques which were rather advanced for the time and served a useful purpose in the study of PCM error rates in the presence of atmospheric turbulence. Data for this model are given in Table I.

PCM/Pulsed Laser (PL)—1966

Wide-band PCM was first used in two laser communication systems developed in 1966. One (using the He–Ne 0.6328-μm wavelength) was built for NASA Marshall Space Flight Center (MSFC) by ITT Federal Laboratories [2], [3] and the other (using the argon ion 0.4880- or 0.5145-nm wavelengths) was built for NASA Manned Spacecraft Center (MSC) by Hughes [4], [5]. Both systems utilized orthogonal polarization modulation with balanced photomultiplier detectors to obtain symmetrical binary detection to help reduce the effects of atmospheric turbulence. Extensive measurements with the ITT system over atmospheric paths during the past two and one-half years have yielded correlation data between bit-error rates, TV picture quality, atmospheric turbulence, transmitter power, and receiver

TABLE II

	ITT/NASA MSFC System [2], [3]	Hughes/NASA MSC System [4], [5]
Wavelength (μm)	0.6328 He–Ne	0.4880/0.5154 argon ion
Power (watts)	0.005	5
Photodetector	S-20 Photomultipliers (two)	S-11 Photomultipliers (two)
Modulator	20 cm transverse field KDP	50 cm transverse field KD*P
Modulation	PCM/PL (polarization)	PCM/PL (polarization)
Coding	PCM delta	PCM
PCM Bit Rate	30 Mbit $\cdot$ s^{-1}	30 Mbit $\cdot$ s^{-1}
Base Bandwidth	10 MHz	10 MHz
Bit-Error Rate	$<10^{-6}$	$<10^{-6}$
Range Performance	Tested extensively over 5 mile path	Tested over 4.2 mile path with 50-dB system margin
	Theoretical maximum operating range = 50 000 mi	Theoretical maximum operating range = 1 500 000 mi

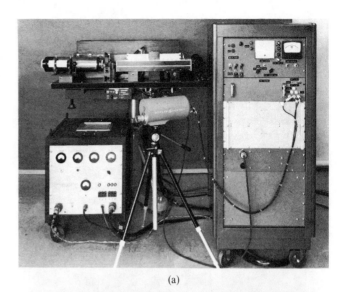

(a)

(b)

Fig. 1. (a) Hughes wide-band PCM argon laser transmitter.
(b) Hughes 5-watt argon beam at night.

aperture size. Conclusions are that TV picture quality is satisfactory with bit-error rates of 10^{-4}; the laser systems can obtain bit-error rates of less than 10^{-6} for typical atmospheric paths. The system specifications are listed in Table II.

Fig. 1(a) shows the transmitter terminal of the Hughes system. Fig. 1(b) shows the Hughes 5-watt argon laser beam in the night sky. Fig. 2 shows the receiver terminal of the ITT system (the spot of light in the background is the transmitter station).

Pulse Frequency Modulation—1965 Through 1969 (GaAs–9000 Å)

A gallium arsenide injection laser operating near room temperature ($-4°$C) was designed and built by IBM for optical communication experiments on Gemini VII [6]. The system utilized pulsed frequency modulation (PFM) for coding information; it had a pulse repetition frequency (PRF) of 8 kHz and a demodulated voice bandwidth of 6 kHz. Although the Gemini experiment was unsuccessful, the techniques employed have created a whole family of laser communication devices.

The first pulsed gallium arsenide injection lasers, operating at or near ambient $300°$K temperature, were unreliable because of metallurgical flaws in the junction and lack of understanding of the material limitations by the systems designers. Continuous efforts by RCA (Sommerville) have resulted in significant improvement in the reliability of laser diodes [7]. This improvement was achieved 1) through a close confinement construction which sharply reduces internal optical loss and 2) through better material and heterojunction diode fabrication processes.

Systems designers have now learned that the improved low-threshold diodes last longer only when they are operated at lower currents and lower temperatures. Lower design PRF rates have been a key factor in extending the life of these devices.

Recently (1968) RCA (Camden) developed a reliable field communication unit for the U.S. Army Electronics Command. Optronix, Holobeam, Saab Avionics and Santa Barbara Research Center (SBRC), now have productized versions of this type of laser communication system (see Fig. 3). The specifications listed in Table III are typical of these systems, although they refer in detail to the SBRC unit.

Wide-Band PCM Subcarrier Frequency Shift Keying—1967 (3.39 μm)

Subcarrier FM offers another means of avoiding the effects of amplitude fading and electrooptic modulation nonlinearities. The method can be simply described as the

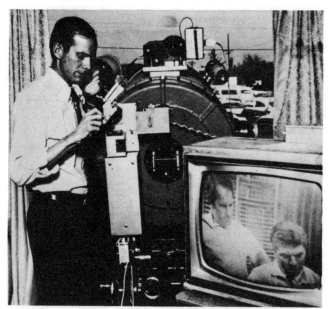

Fig. 2.   Receiver terminal of ITT system. (Courtesy of ITT Aerospace.)

Fig. 3.   Santa Barbara Research Center's gallium arsenide laser communicator.

TABLE III

	SBRC Infrared Laser Communicator
Wavelength (μm)	0.9040
Power (watts)	2 W peak, 0.003 W average, duty cycle $\cong 10^{-3}$
Photodetector	Silicon avalanche photodiode
Modulation	Direct modulation of diode current
Coding	PFM
PCM Bit Rate	6 kbit $\cdot$ s^{-1}
Base Bandwidth	2.3 kHz
Range Performance	6 miles (good seeing conditions)

amplitude modulation of the optical beam with an RF subcarrier. At the receiver, the output of an optical amplitude detector is fed into a conventional FM receiver tuned to the subcarrier frequency.

The FM subcarrier idea was applied to wide bandwidths with a 50 Mbit $\cdot$ s^{-1} frequency shift-keying (FSK) system

Fig. 4.   Hughes 3.39-μm wide-band subcarrier FSK system.

TABLE IV

	Hughes/NASA MSC 3.39-μm Wide-Band Subcarrier FSK System [8]
Wavelength (μm)	3.39 He–Ne
Power (watts)	0.002
Photodetector	Indium arsenide (300°K) photodiode
Modulator	Gallium arsenide (intracavity coupling)
Modulation	Subcarrier FSK (125 MHz$\pm$MHz)
Coding	PCM
PCM Bit Rate	50 Mbit $\cdot$ s^{-1} (nonreturn to zero)
Base Bandwidth	25 MHz
Range Performance	1 mile (good seeing conditions)

built by Hughes for the NASA MSC. Here the subcarrier center frequency was 125 ± 25 MHz. FSK, the digital form of FM, was used. Amplitude modulation of the subcarrier was provided by a gallium arsenide crystal located inside the cavity of the 3.39-μm laser. Modulating voltage applied to the crystal produces coupling of the intracavity energy off a Brewster plate [8] (see Fig. 4). The specifications are listed in Table IV.

A recent experiment by a group at Lockheed Palo Alto [9] has demonstrated the FM-subcarrier technique using a 2.8-GHz subcarrier frequency, a lithium niobate modulator, and a silicon avalanche photodetector. Sufficient linearity was obtained to transmit a band of frequencies from 54 to 216 MHz, i.e., all of the VHF TV channels simultaneously.

Optical FM—Optical Heterodyne Detection—1968 (10.6 μm)

Experiments in optical heterodyne detection have been conducted at various laboratories since 1963, although improved sensitivity in heterodyne detection was not demonstrated until 1966 [10]. Workers at Honeywell built an optical heterodyne communication system for the NASA Marshall Space Flight Center in 1968 which employed optical FM for two voice channels; the modulation was provided by mechanical motion of the laser mirror [11]. This system demonstrated the vastly improved receiver sensitivity obtainable through the use of optical heterodyne detection. Concurrent with the development of the Honeywell system the author and his coworkers at Hughes demonstrated intracavity FM modulation using an electrooptic

TABLE V

	Honeywell/NASA MSFC 10.6-μ Heterodyne System [10]	Hughes Breadboard [12]
Wavelength (μm)	10.6 CO_2	10.6 CO_2
Power (watts)	5	1
Photodetector	HgCdTe (77°K) heterodyne	Ge:Hg (21°K) heterodyne—30 MHz IF
Modulator	Piezoelectric (mechanical)	Gallium arsenide phase modulator
Modulation	Optical FM	Optical FM, intracavity
Coding	Analog	Analog
Base Bandwidth	1 MHz—2 voice channels	5 MHz—TV channel
Range Performance	Test range = 3 miles	Demonstration range = 18 miles; S/N = 50 dB

TABLE VI

	Hughes 3.39-μm Two-Way Laser Communicator (1970)
Wavelength (μm)	3.39 He–Ne
Power (watts)	0.001
Photodetector	Indium arsenide (heterodyne)–30 MHz IF
Modulator	Gallium arsenide
Modulation	Intracavity, optical FM
Coding	Analog
Base Bandwidth	to 25 MHz
Range Performance	3 miles (good seeing) / 1 mile (poor seeing) } Picture S/N greater than 40 dB.

crystal, and the transmission of TV over an 18 mile path through the atmosphere [12]. The performance characteristics are given in Table V.

The demonstration of optical heterodyne detection at 10.6 μm, along with the high power and high efficiency CO_2 laser oscillators available at this wavelength, has signaled a new era in laser communications technology. The ratio of transmitter power available to minimum detectable power in the system is already up to 170 dB and is likely to increase still more as available CO_2 laser powers increase.

Wide-Band Optical FM—Optical Heterodyne Detection—1970 (3.39 μm)

Wide-band optical FM is employed in a field-ruggedized communication system developed at Hughes (see Fig. 5). This system is ultimately capable of handling information bandwidths up to 50 Mbits · s^{-1} in either direction through the system, one direction at a time (simplex). Time sharing permits coded information to flow in two directions simultaneously. The prototype system was designed with one TV channel or four voice channels multiplexed to be available in both direction simultaneously.

This system illustrates the use of optical heterodyne detection in a ruggedized package. Heterodyne detection was (and still is) believed by many to be impractical. The frequency stability requirements for both the transmitter and receiver local oscillators are severe, but have been achieved in this system through good engineering principles. The performance specifications are given in Table VI.

Significant Achievements in Pointing and Tracking

The implementation of the laser in space communications required gimballed optical systems which will track the incoming beam and point the outgoing beam to a high degree of accuracy. Under the sponsorship of the NASA MSFC, workers at Perkin-Elmer have developed an optical system which has a dynamic tracking stability of approximately 0.3 μrad [13]. The techniques are thus available to point and track with diffraction limited beams from fairly large optical systems. Fig. 6 illustrates tracking experiment at Perkin-Elmer. Here the large tube on the upper left is the

Fig. 5. Hughes 3.39-μm optical heterodyne two-way laser communicator.

Fig. 6. Perkin-Elmer fine pointing and tracking test facility. (Courtesy of Perkin-Elmer Co.)

445

TABLE VII

	Sylvania ARL Two-Way System (1967) [14]
Wavelength (μm)	0.6328 He–Ne
Power (watts)	0.001 nominal
Photodetector	S-20 photomultipliers, (2)
Modulator	Sylvania VP2, KD*P
Modulation	Polarization
Coding	Analog
Tracking Sensor	Image dissector, photomultiplier
IMC Element	Magnetic
Base Bandwidth	5 MHz
Range Performance	1-km test range

Fig. 7. Nippon Electric's PCM/AM laser communication terminal. (Courtesy of Nippon Electric Company, Ltd.)

collimated source, simulating a signal from a distance. The telescope on the lower right is tracking the source to within less than half a microradian while the base on which the telescope is mounted is moving with a programmed motion of ±2°.

Generally speaking, laser communications equipment designed for point-to-point use on the ground does not required tracking optics. The beam is usually spread out sufficiently to provide for atmospheric image motion and diurnal refraction. An exception to this is a tracking capability included in a portable communication system built by Sylvania Applied Research Laboratory (ARL) in 1967 [14]. In this system, compensation for image motion and maintenance of fine pointing was achieved as an integral part of the system. The performance specifications are listed in Table VII.

PCM/AM PCM via Mode-Locked Lasers (1970)

The recent work of Kinsel, Denton, and Guesic [15]–[19] with amplitude PCM utilizing mode-locked laser pulses has opened the way to obtaining extremely high data rates. This work at Bell has been followed at Nippon Electric Co. (NEC) [20], by a truly monumental effort in the development of a fully operational four-leg two-way laser communication link between Yokohama and Tamagawa, a distance of 14 km. Three repeater stations are utilized in the system in addition to the two terminals. The longest leg of the link is a distance of 4.25 km. Communications occur simultaneously in both directions. To the author's knowledge, this is the first laser link built which will handle commercial traffic. Operational reliability through the atmosphere is estimated to be 99 percent for a 24 hour period (mean annual statistics) and 99.6 percent for daytime operation. Fig. 7 shows one of NEC's laser communication terminals. The performance specifications are listed in Table VIII.

New Systems Under Development

There have been many "false starts" in the development of laser communication systems during the past decade, due mainly to the lack of laser component reliability. For example, the first He–Ne lasers had lifetimes of 100 hours; they now have lifetimes exceeding 20 000 hours. Thus, operational laser communication systems are paced for the

TABLE VIII

	Nippon Electric Co. PCM/AM System [20]
Wavelength (μm)	0.6328 He–Ne
Power (watts)	0.003 in each of two redundant transmitters
Photodetector	Silicon avalanche photodiode (single)
Modulator	Lithium tantalate
Modulation	PCM/AM
Coding	PCM delta
PCM Bit Rate	123.492 MHz
Base Bandwidth	13.7 MHz per channel, 3 channels
Number of Lasers in System	16
Range Performance	Test ranges 2.5 km, 4.1 km, 4.25 km, 3.2 km. Total 14 km. 99 percent reliability 24 hour operation; 99.6 percent reliability business hours

most part by the laser technology. It would seem to make good sense to do as the group at Nippon Electric has done and base a system design around a component with proven reliability.

The first sealed-off CO_2 lasers had lifetimes of less than 100 hours, and are now approaching the 10 000 hour mark. The CO_2 laser is, therefore, now available for an operational system. The vehicle for the test is the ATS-F laser communication experiment designed to establish a two-way link between a synchronous orbit and the earth's surface. The system is being built by Aerojet General Corporation for NASA Goddard Space Flight Center (GSFC) [21]. Due to be launched in 1973, the Laser Communication Experiment (LCE) will be the first illustration of the user of lasers

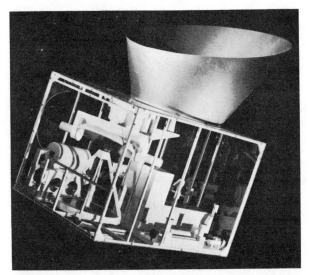

Fig. 8. Mock-up of Aerojet General's flight package for ATS-F laser communication experiment. (Courtesy of Aerojet General Corp.)

TABLE IX

	Aerojet General/NASA GFSC Laser Communication Experiment [21]
Wavelength (μm)	10.6 CO_2
Power (watts)	1.0
Photodetector	HgCdTe (100°K) optical heterodyne
Modulator	Gallium arsenide
Modulation	Intracavity FM or external AM
Coding	Analog
Base Bandwidth	5 MHz
Range Performance	3.6×10^{-7} m (23 000 mi) $S/N = 25$ dB

for space communications. To date, the configuration selected is constrained by size, weight, and power. No laser other than the CO_2 system can even approach meeting the requirements. A mock-up of the LCE is shown in Fig. 8. The large horn on top is the radiation cooler for the detector. The performance specifications are listed in Table IX.

CONCLUSIONS

Most of the techniques required to create a useful and practical laser communication system in the atmosphere or in space have been demonstrated with the model systems described in this paper. Anticipated needs for future space communication channels requiring large dynamic range and a high degree of linearity can be met with laser systems. In addition, eventual shortage of microwave channels for point communications through the atmosphere will require other modes of communication. Laser systems may be ideal to fill

the needs which cannot be met with these conventional systems, since they have the inherent advantage of not creating interference with conventional systems or with other laser systems.

Present trends indicate that laser communications systems of the future will have large dynamic range, a high degree of linearity, and will utilize PCM of extremely high data rates.

REFERENCES

[1] T. M. Strauss, "Laser communications study," Final Rep. FR 66-14-37, Contract DA28-043-AMC 00195 (E), January 1966.
[2] J. H. Ward, "A Broad bandwidth digital laser communication system," presented at the 1967 IEEE Conf. Laser Eng. Appl.
[3] J. H. Ward, "Optical communications system design and evaluation," ITT Aerospace Internal Rep., March 1970.
[4] C. V. Smith, "Wide-band laser communication systems," presented at the 1967 IEEE Conf. Laser Eng. Appl.
[5] C. V. Smith, "High data rate laser communication system," Final Rep. T.R. A6395, Contract NAS 9-4266, July 1966.
[6] D. S. Lilly, "Optical communication experiments on Gemini VII," Proc. Space Optical Tech. Conf., vol. 1, pp. 81–90, April 1966.
[7] H. Kressel and H. Nelson, "Close confinement gallium arsenide PN junction lasers with reduced optical loss at room temperature," RCA Rev., vol. 30, March 1969.
[8] C. V. Smith, "Wideband laser communication system," Final Rep. P67-113 B1480, Contract NAS 9-6420, May 1967.
[9] Aviat. Week Space Technol., pp. 50–52, June 1, 1970.
[10] F. E. Goodwin and M. E. Pedinoff, "Application of CCl_4 ultrasonic modulators to infrared optical heterodyne experiments," Appl. Phys. Lett., vol. 8, February 1, 1966.
[11] H. W. Mocker, "A 10.6 μm optical heterodyne communication system," Appl. Opt., vol. 8, p. 677, March 1969.
[12] F. E. Goodwin and T. A. Nussmeier, "Optical heterodyne communication experiments at 10.6 μ," IEEE J. Quantum Electron., vol. QE-4, pp. 612–617, October 1968.
[13] M. S. Lipsett, C. M. McIntyre, and R. C. Liu, "Space instrumentation for laser communication," presented at the Electro-Optical Sys. Design Conf., New York, N. Y., September 1969.
[14] G. Ratcliffe, "Acquisition and tracking laser communication system," Sylvania ARL Rep. SA 15, 1967.
[15] R. T. Denton and T. S. Kinsel, "Terminals for a high-speed optical pulse code modulation communication system: I. 224-M/bits single channel," Proc. IEEE, vol. 56, pp. 140–145, February 1968.
[16] T. S. Kinsel and R. T. Denton, "Terminals for a high-speed optical pulse code modulation communication system: II. Optical multiplexing and demultiplexing," Proc. IEEE, vol. 56, pp. 146–157, February 1968.
[17] R. T. Denton, "The laser and PCM," Bell Lab. Rec., vol. 46, p. 175, 1968.
[18] T. S. Kinsel, "Light wave of the future: optical PCM," Electronics, vol. 41, no. 19, p. 123, September 1968.
[19] T. S. Kinsel, J. E. Guesic, H. Seidel, and R. C. Smith, "A stabilized mode-locked Nd:YA1G laser source," IEEE J. Quantum Electron., vol. QE-5, p. 326, June 1969.
[20] T. Masuda, T. Uchida, Y. Ueno, and T. Shimamura, "An experimental PCM/AM optical communication system using mode-locked HeNe lasers," presented at 1970 Internatl. Commun. Conf., San Francisco, Calif., June 10, 1970.
[21] J. H. McElroy, "Carbon dioxide laser systems for space communications," presented at 1970, Internatl. Commun. Conf., San Francisco, Calif., June 10.

Part VIII: Holography

Holography has emerged as one of the more fascinating and potentially useful applications of the laser, on which a large amount has been published (cf. Section I of the Laser Book List in Part IX). Our coverage is limited to two outstanding papers: the 1971 Nobel Prize Lecture by Dennis Gabor published in the June 1972 issue of PROCEEDINGS OF THE IEEE, and a review of holography by Goodman that introduced a "mini Special Issue" on this topic in the September 1971 PROCEEDINGS.

Holography, 1948–1971

DENNIS GABOR

I HAVE THE ADVANTAGE in this lecture, over many of my predecessors, that I need not write down a single equation or show an abstract graph. One can of course introduce almost any amount of mathematics into holography, but the essentials can be explained and understood from physical arguments.

Holography is based on the wave nature of light, and this was demonstrated convincingly for the first time in 1801 by Thomas Young, by a wonderfully simple experiment (see Fig. 1). He let a ray of sunlight into a dark room, placed a dark screen in front of it, pierced with two small pinholes, and beyond this, at some distance, a white screen. He then saw two darkish lines at both sides of a bright line, which gave him sufficient encouragement to repeat the experiment, this time with a spirit flame as light source, with a little salt in it, to produce the bright yellow sodium light. This time he saw a number of dark lines, regularly spaced; the first clear proof that light added to light can produce darkness. This phenomenon is called interference. Thomas Young had expected it because he believed in the wave theory of light. His great contribution to Christian Huygens's original idea was the intuition that monochromatic light represents regular sinusoidal oscillations, in a medium which at that time was called "the ether." If this is so, it must be possible to produce more light by adding wavecrest to wavecrest, and darkness by adding wavecrest to wavetrough.

Light which is capable of interferences is called "coherent," and it is evident that in order to yield many interference fringes, it must be *very* monochromatic. Coherence is conveniently measured by the path difference between two rays of the same source, by which they can differ while still giving observable interference contrast. This is called the coherence length, an important quantity in the theory and practice of holography. Lord Rayleigh and Albert Michelson were the first to understand that it is a reciprocal measure of the spectroscopic line width. Michelson used it for ingenious methods of spectral analysis and for the measurement of the diameter of stars.

Let us now jump a century and a half, to 1947. At that time I was very interested in electron microscopy. This wonderful instrument had at that time produced a hundredfold improvement on the resolving power of the best light microscopes, and yet it was disappointing, because it had stopped short of resolving atomic lattices. The de Broglie wavelength of fast electrons, about 1/20 angstrom, was short enough, but the optics was imperfect. The best electron objective which one can make can be compared in optical perfection to a raindrop rather than to a microscope objective, and through the theoretical work of O. Scherzer it was known that it could never be perfected. The theoretical limit at that

Manuscript received February 21, 1972. Lecture delivered by D. Gabor on the occasion of his receiving the 1971 Nobel Prize for Physics, Stockholm, Sweden, December 13. Copyright © The Nobel Foundation 1972.

The author is Professor Emeritus and Senior Research Fellow, Imperial College of Science and Technology, University of London; and Staff Scientist, CBS Laboratories, Stamford, Conn.

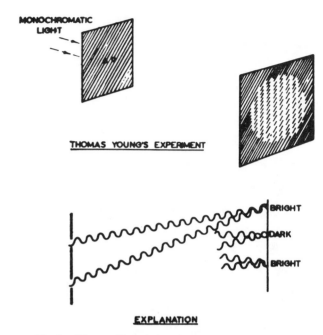

Fig. 1.   Thomas Young's interference experiments, 1801.

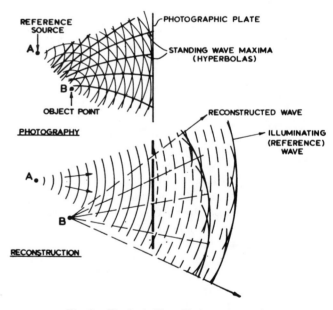

Fig. 2.   The basic idea of holography, 1947.

time was estimated at 4 Å, just about twice what was needed to resolve atomic lattices, while the practical limit stood at about 12 Å. These limits were given by the necessity of restricting the aperture of the electron lenses to a few 1/1000 radian, at which angle the spherical aberration error is about equal to the diffraction error. If one doubles this aperture so that the diffraction error is halved, the spherical aberration error is increased 8 times, and the image is hopelessly blurred.

After pondering this problem for a long time, a solution suddenly dawned on me, one fine day at Easter 1947, more or

Reprinted from *Proc. IEEE*, vol. 60, pp. 655–668, June 1972.

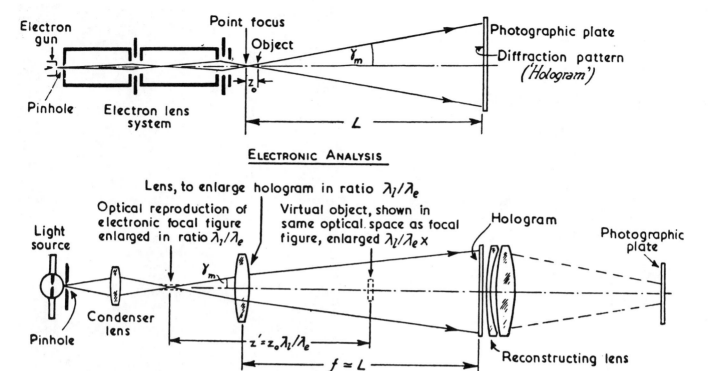

Fig. 3. The principle of electron microscopy by reconstructed wavefronts (Gabor, *Proc. Roy. Soc.*, vol. A197, p. 454, 1949 [1]).

less as shown in Fig. 2. Why not take a bad electron picture, but one which contains the *whole* information, and correct it by optical means? It was clear to me for some time that this could be done, if at all, only with coherent electron beams, with electron waves which have a definite phase. But an ordinary photograph loses the phase completely, it records only the intensities. No wonder we lose the phase, if there is nothing to compare it with! Let us see what happens if we add a standard to it, a "coherent background." My argument is illustrated in Fig. 2, for the simple case when there is only one object point. The interference of the object wave and of the coherent background or "reference wave" will then produce interference fringes. There will be maxima wherever the phases of the two waves were identical. Let us make a hard positive record, so that it transmits only at the maxima, and illuminate it with the reference source alone. Now the phases are of course right for the reference source A, but as at the slits the phases are identical, they must be right also for B; therefore, the wave of B must also appear, *reconstructed*.

A little mathematics soon showed that the principle was right, also for more than one object point, for any complicated object. Later on it turned out that in holography Nature is on the inventor's side; there is no need to take a hard positive record; one can take almost any negative. This encouraged me to complete my scheme of electron microscopy by reconstructed wavefronts, as I then called it and to propose the two-stage process shown in Fig. 3. The electron microscope was to produce the interference figure between the object beam and the coherent background, that is to say the non-diffracted part of the illuminating beam. This interference pattern I called a "hologram," from a Greek word "holos"—the whole, because it contained the whole information. The hologram was then reconstructed with light, in an optical system which corrected the aberrations of the electron optics [1].

In doing this, I stood on the shoulders of two great physi-

cists, W. L. Bragg and Fritz Zernike. Bragg had shown me, a few years earlier, his "X-ray microscope," an optical Fourier-transformer device. One puts into it a small photograph of the reciprocal lattice, and obtains a projection of the electron densities, but only in certain exceptional cases, when the phases are all real, and have the same sign. I did not know at that time, and neither did Bragg, that Mieczislav Wolfke had proposed this method in 1921, but without realising it experimentally. So the idea of a two-stage method was inspired by Bragg. The coherent background, on the other hand, was used with great success by Fritz Zernike in his beautiful investigations on lens aberrations, showing up their phase, and not just their intensity. It was only the reconstruction principle which had escaped them.

In 1947 I was working in the Research Laboratory of the British Thomson-Houston Company in Rugby, England. It was a lucky thing that the idea of holography came to me *via* electron microscopy, because if I had thought of optical holography only, the Director of Research, L. J. Davies, could have objected that the BTH company was an electrical engineering firm, and not in the optical field. But as our sister company, Metropolitan Vickers were makers of electron microscopes, I obtained the permission to carry out some optical experiments. Fig. 4 shows one of our first holographic reconstructions. The experiments were not easy. The best compromise between coherence and intensity was offered by the high pressure mercury lamp, which had a coherence length of only 0.1 mm, enough for about 200 fringes. But in order to achieve spatial coherence, we (my assistant Ivor Williams and I) had to illuminate, with one mercury line, a pinhole of 3 microns diameter. This left us with enough light to make holograms of about 1 cm diameter of objects, which were microphotographs of about 1 mm diameter, with exposures of a few minutes, on the most sensitive emulsions then available. The small coherence length forced us to arrange everything in one axis. This is now called "in line"

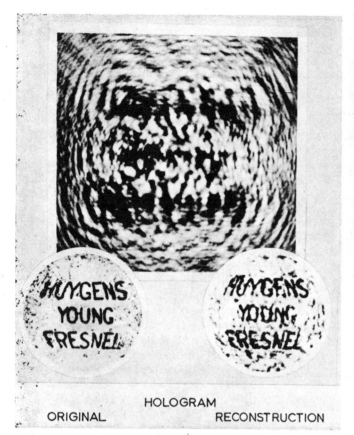

Fig. 4.   First holographic reconstruction, 1948.

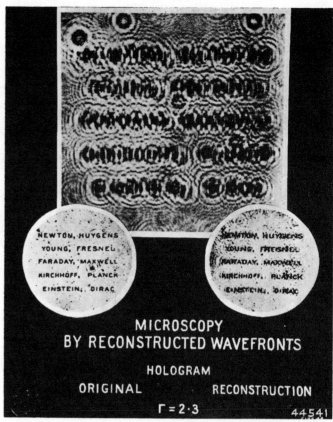

Fig. 5.   Another example of early holography, 1948 (Gabor, *Proc. Roy. Soc.*, vol. A197, p. 454, 1949 [1]).

holography, and it was the only one possible at that time. Fig. 5 shows a somewhat improved experiment, the best of our series. It was far from perfect. Apart from the *schlieren*, which cause random disturbances, there was a systematic defect in the pictures, as may be seen by the distortion of the letters. The explanation is given in Fig. 6. The disturbance arises from the fact that there is not one image but *two*. Each point of the object emits a spherical secondary wave, which interferes with the background and produces a system of circular Fresnel zones. Such a system is known after the optician who first produced it, a Soret lens. This is, at the same time, a positive and a negative lens. One of its foci is in the original position of the object point, the other in a position conjugate to it, with respect to the illuminating wavefront. If one uses "in-line holography" both images are in line, and can be separated only by focusing. But the separation is never quite perfect, because in regular coherent illumination every point leaves a "wake" behind it, which reaches to long distances.

I will tell later with what ease modern laser holography has got rid of this disturbance, by making use of the superior coherence of laser light which was not at my disposal in 1948. However, I was confident that I could eliminate the second image in the application which alone interested me at that time: seeing atoms with the electron microscope. This method, illustrated in Fig. 7, utilized the very defect of electron lenses, the spherical aberration, in order to defeat the second image. If an electron hologram is taken with a lens with spherical aberration, one can afterwards correct *one* of the two images by suitable optics, and the other has then twice the aberration, which washes it out almost completely. Fig. 7 shows that a perfectly sharp reconstruction, in which as good as nothing remains of the disturbance caused by the second image, can be obtained with a lens so bad that its definition

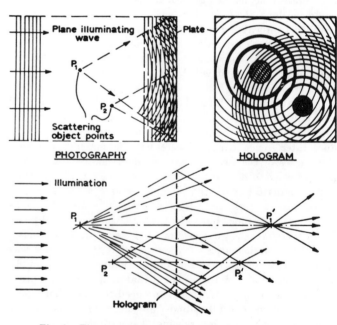

Fig. 6.   The second image. Explanation in terms of Soret lenses as holograms of single object points.

is at least 10 times worse than the resolution which one wants to obtain. Such a very bad lens was obtained using a microscope objective the wrong way round, and using it again in the reconstruction.

So it was with some confidence that two years later, in 1950, we started a programme of holographic electron microscopy in the Research Laboratory of the Associated Electrical Industries, in Aldermaston, under the direction of Dr.

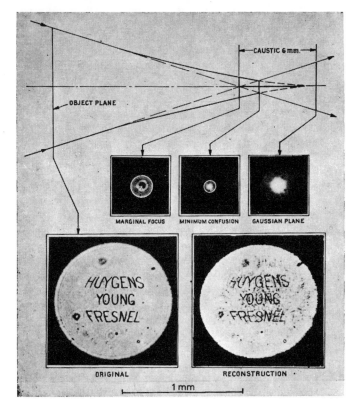

Fig. 7. Elimination of the second image by compensation of the spherical aberration in the reconstruction (Gabor, 1948; published 1951 [1]).

T. E. Allibone, with my friends and collaborators M. W. Haine, J. Dyson, and T. Mulvey,[1] By that time I had joined Imperial College, and took part in the work as a consultant. In the course of three years we succeeded in considerably improving the electron microscope, but in the end we had to give up, because we had started too early. It turned out that the electron microscope was still far from the limit imposed by optical aberrations. It suffered from vibrations, stray magnetic fields, creep of the stage, contamination of the object, all made worse by the long exposures required in the weak coherent electron beam. *Now*, 20 years later, would be the right time to start on such a programme, because in the meantime the patient work of electron microscopists has overcome all these defects. The electron microscope resolution is now right up to the limit set by the spherical aberration, about 3.5 Å, and only an improvement by a factor of 2 is needed to resolve atomic lattices. Moreover, there is no need now for such very long exposures as we had to contemplate in 1951, because by the development of the field emission cathode the coherent current has increased by a factor of 3–4 orders of magnitude. So perhaps I may yet live to see the realisation of my old ideas.

My first papers on wavefront reconstruction evoked some immediate responses. G. L. Rogers [2] in Britain made important contributions to the technique, by producing among other things the first phase holograms, and also by elucidating the theory. In California, Alberto Baez [3], Hussein El-Sum, and P. Kirkpatrick [4] made interesting forays into X-ray holography. For my part, with my collaborator W. P. Goss, I constructed a holographic interference microscope, in which

the second image was annulled in a rather complicated way by the superimposition of two holograms, "in quadrature" with one another. The response of the optical industry to this was so disappointing that we did not publish a paper on it until 11 years later, in 1966 [5]. Around 1955 holography went into a long hybernation.

The revival came suddenly and explosively in 1963, with the publication of the first successful laser[2] holograms by Emmett N. Leith and Juris Upatnieks of the University of Michigan, Ann Arbor. Their success was due not only to the laser, but to the long theoretical preparation of Emmett Leith, which started in 1955. This was unknown to me and to the world, because Leith, with his collaborators Cutrona, Palermo, Porcello, and Vivian applied his ideas first to the problem of the "side-looking radar" which at that time was classified [6]. This was in fact two-dimensional holography with electromagnetic waves, a counterpart of electron holography. The electromagnetic waves used in radar are about 100 000 times longer than light waves, while electron waves are about 100 000 times shorter. Their results were brilliant, but to my regret I cannot discuss them for lack of time.

When the laser became available, in 1962, Leith and Upatnieks could at once produce results far superior to mine, by a new, simple, and very effective method of eliminating the second image [7]. This is the method of the "skew reference wave," illustrated in Fig. 8. It was made possible by the great coherence length of the helium–neon laser, which even in 1962 exceeded that of the mercury lamp by a factor of about 3000. This made it possible to separate the reference beam from the illuminating beam; instead of going through the object, it could now go around it. The result was that the two reconstructed images were now separated not only in depth, but also angularly, by twice the incidence angle of the reference beam. Moreover, the intensity of the coherent laser light exceeded that of mercury many millionfold. This made it possible to use very fine-grain low-speed photographic emulsions and to produce large holograms, with reasonable exposure times.

Fig. 9 shows two of the earliest reconstructions made by Leith and Upatnieks, in 1963, which were already greatly superior to anything that I could produce in 1948. The special interest of these two images is, that they are reconstructions from *one* hologram, taken with different positions of the reference beam. This was the first proof of the superior storage capacity of holograms. Leith and Upatnieks could soon store 12 different pictures in one emulsion. Nowadays one can store 100 or even 300 pages of printed matter in an area which by ordinary photography would be sufficient for one.

From then on progress became very rapid. The most spectacular result of the first year was the holography of three-dimensional objects, which could be seen with the two eyes. Holography was of course three dimensional from the start, but in my early small holograms one could see this only by focusing through the field with a microscope or short-

[1] Supported by a grant of the DSIR (Direction of Scientific and Industrial Research), the first research grant ever given by that body to an industrial laboratory.

[2] I have been asked more than once why I did not invent the laser. In fact, I have thought of it. In 1950, thinking of the desirability of a strong source of coherent light, I remembered that in 1921, as a young student, in Berlin, I had heard from Einstein's own lips his wonderful derivation of Planck's law which postulated the existence of stimulated emission. I then had the idea of the pulsed laser: Take a suitable crystal, make a resonator of it by a highly reflecting coating, fill up the upper level by illuminating it through a small hole, and discharge it explosively by a ray of its own light. I offered the idea as a Ph.D. problem to my best student, but he declined it, as too risky, and I could not gainsay it, as I could not be sure that we would find a suitable crystal.

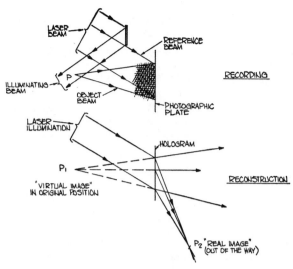

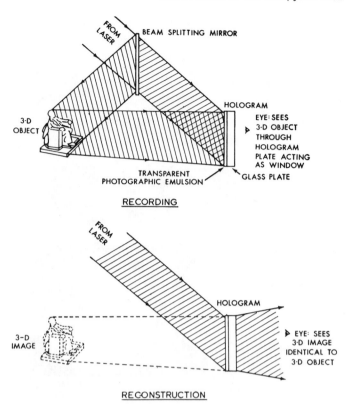

Fig. 8.   Holography with skew references beam.
(E. N. Leith and J. Upatnieks, 1963.)

Fig. 10.   Three-dimensional holography of a
diffusing object with later light.

Fig. 11.   Three-dimensional reconstruction of a small statue of Abraham Lincoln. (Courtesy of Prof. G. W. Stroke, State University of New York, Stony Brook.)

Fig. 9.   First example of multiple image storage in one hologram
(E. N. Leith and J. Upatnieks, *J. Opt. Soc. Amer.*, Nov. 1964).

focus eyepiece. But it was not enough to make the hologram large, it was also necessary that every point of the photographic plate should see every point of the object. In the early holograms, taken with regular illumination, the information was contained in a small area, in the diffraction pattern.

In the case of rough diffusing objects no special precautions are necessary. The small dimples and projections of the surface diffuse the light over a large cone. Fig. 10 shows an

example of the setup in the case of a rough object, such as a statuette of Abraham Lincoln. The reconstruction is shown in Fig. 11. With a bleached hologram ("phase hologram") one has the impression of looking through a clear window at the statuette itself. (Courtesy of Professor George W. Stroke [8].)

If the object is non-diffusing, for instance if it is a transparency, the information is spread over the whole hologram area by illuminating the object through a diffuser, such as a frosted glass plate. The appearance of such a "diffused" hologram is extraordinary; it looks like noise (Fig. 12). One can call it "ideal Shannon coding," because Claude E. Shannon has shown in his Communication Theory that the most effi-

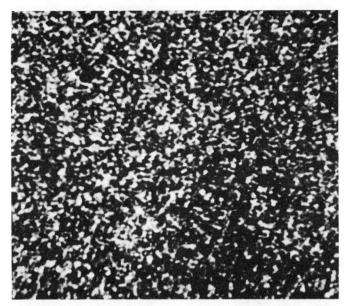

Fig. 12. Strongly magnified image of a hologram taken with diffused illumination. The information is conveyed in a "noise-like" code. (E. N. Leith and J. Upatnieks, 1964.)

Fig. 13. Reconstruction of a plane transparency, showing a restaurant, from a hologram taken with diffused illumination. (E. N. Leith and J. Upatnieks, 1964.)

Fig. 14. Equipment of a modern holographic laboratory. (Courtesy of Prof. G. W. Stroke, State University of New York, Stony Brook.)

cient coding is such that all regularities seem to have disappeared in the signal; it must be "noise-like." But where is the information in this chaos? It can be shown that it is not as irregular as it appears. It is not as if grains of sand had been scattered over the plate at random. It is rather a complicated figure, the diffraction pattern of the object, which is repeated at random intervals, but always in the same size and the same orientation.

A very interesting and important property of such diffused holograms is that any small part of it, large enough to contain the diffraction pattern, contains information on the whole object, and this can be reconstructed from the fragment, only with more noise. A diffuse hologram is therefore a *distributed memory*, and this has evoked much speculation whether human memory is not perhaps, as it were, holographic, because it is well known that a good part of the brain can be destroyed without wiping out every trace of a memory. There is no time here to discuss this very exciting question. I want only to say that in my opinion the similarity with the human memory is functional only, but certainly not structural.

It is seen that in the development of holography the hologram has become always more unlike the object, but the reconstruction always more perfect. Fig. 13 shows an excellent reconstruction by Leith and Upatnieks of a photograph, from a diffuse hologram like the one in the previous figure.

The pioneer work carried out in the University of Michigan, Ann Arbor, led also to the stabilization of holographic techniques. Today hundreds if not thousands of laboratories possess the equipment of which an example is shown in Fig. 14; the very stable granite slab or steel table, and the various optical devices for dealing with coherent light, which are now manufactured by the optical industry. The great stability is absolutely essential in all work carried out with steady-state lasers, because a movement of the order of a quarter wavelength during the exposure can completely spoil a hologram.

However, from 1965 onwards there has developed an important branch of holography where high stability is not required, because the holograms are taken in a small fraction of a microsecond, with a pulsed laser.

Imagine that you had given a physicist the problem: "Determine the size of the droplets which issue from a jet nozzle, with a velocity of 2 Mach. The sizes are probably from a few microns upwards." Certainly he would have thrown up his hands in despair! But all it takes now, is to record a simple in-line hologram of the jet, with the plate at a safe distance, with a ruby laser pulse of 20–30 nanoseconds. One then looks at the "real" image (or one reverses the illuminating beam and makes a real image of the virtual one), one dives with a microscope into the three-dimensional image of the jet and focuses the particles, one after the other. Because of the large distance, the disturbance by the second image is entirely negligible. Fig. 15 shows a fine example.

As the research workers of the TRW laboratories have shown, it is possible to record in one hologram the infusoriae in several feet of dirty water, or insects in a meter of air space. Fig. 16 shows two reconstructions of insects from one hologram, focusing on one after the other. The authors, C. Knox and R. E. Brooks have also made a cinematographic record of a holographic film, in which the flight of one mosquito is followed through a considerable depth, by refocusing in every frame [9].

Another achievement of the TRW group, Ralph Wuerker

455

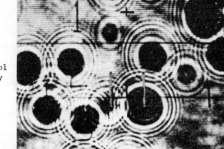

Hologram of aerosol
particles in spray

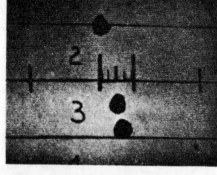

Aerosol particles
focussed

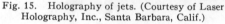

Fig. 15. Holography of jets. (Courtesy of Laser Holography, Inc., Santa Barbara, Calif.)

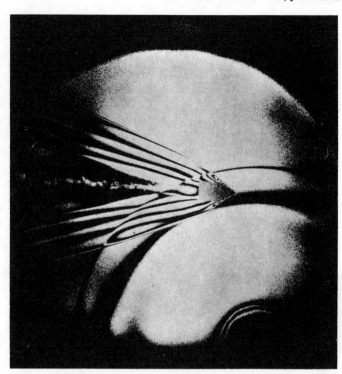

Fig. 17. Dynamic holographic interferometry. This reconstruction of a holographic interferogram shows the interaction of two air shock fronts and their associated flows. (Courtesy of Dr. R. F. Wuerker and his associates, TRW Physical Electronics Lab., Redondo Beach, Calif.)

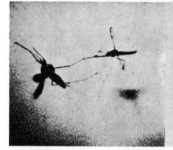

Fig. 16. Observation of mosquitos in flight. Both pictures are extracted from one hologram. (Courtesy of C. Knox and R. E. Brooks, TRW, Redondo Beach, Calif.)

and his colleagues, leads us into another branch of holography, to holographic interferometry. Fig. 17 shows a reconstruction of a bullet, with its train of shockwaves, as it meets another shockwave. But it is not just an image, it is an interferometric image. The fringes show the *loci* at which the retardation of light is by integer wavelengths, relative to the quiet air, before the event. This comparison standard is obtained by a previous exposure. This is therefore a double-exposure hologram, such as will be discussed in more detail later [10].

Fig. 18 shows another high achievement of pulse holography: a holographic three-dimensional portrait, obtained by L. Siebert in the Conductron Corporation (now merged into McDonnel-Douglas Electronics Company, St. Charles, Mo.). It is the result of outstanding work in the development of lasers. The ruby laser, as first realised by T. H. Maiman, was capable of short pulses, but its coherence length was of the order of a few centimeters only. This is no obstacle in the case of in-line holography, where the reference wave proceeds almost in step with the diffracted wavelets, but in order to take a scene of, say, one meter depth with reflecting objects one

Fig. 18. Holographic portrait. (L. Siebert, Conductron Corp., now merged into McDonnell–Douglas Electronics Co., St. Charles, Mo.)

must have a coherence length of at least one meter. Nowadays single-mode pulses of 30 nanosecond duration with 10 joule in the beam and coherence lengths of 5–8 meters are available, and have been used recently for taking my holographic portrait shown in the exhibition attached to the lecture.

In 1965, R. L. Powell and K. A. Stetson in the University of Michigan, Ann Arbor, made an interesting discovery. Holographic images taken of moving objects are washed out. But if double exposure is used, first with the object at rest, then in vibration, fringes will appear, indicating the lines where the displacement amounted to multiples of a half wavelength. Fig. 19 shows vibrational modes of a loudspeaker membrane,

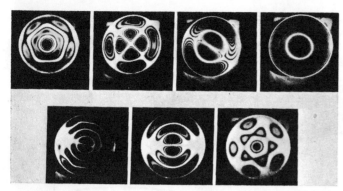

Fig. 19. Vibrational modes of a loudspeaker membrane, obtained by holographic interferometry. (R. L. Powell and K. A. Stetson, University of Michigan, Ann Arbor, 1965.)

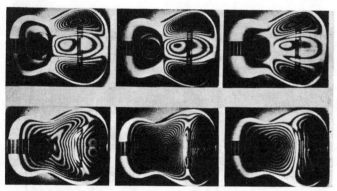

Fig. 20. Vibrational modes of a guitar, recorded by holographic interferometry. (Courtesy of Dr. K. A. Stetson and Prof. E. Ingelstam.)

Fig. 21. An early example of holographic interferometry by double exposure. (K. Haines and B. P. Hildebrand, University of Michigan, Ann Arbor, 1965.)

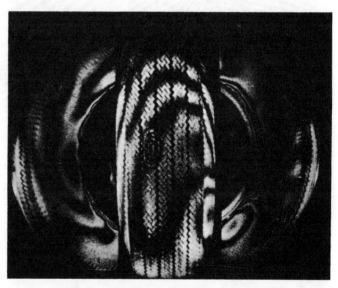

Fig. 22. Non-destructive testing by holography. Double-exposure hologram, revealing two flaws in a tyre. (Courtesy of Dr. R. Grant and GCO, Ann Arbor, Mich.)

recorded in 1965 by Powell and Stetson [11], Fig. 20 the same for a guitar, taken by K. A. Stetson in the laboratory of Professor Erik Ingelstam [12].

Curiously, both the interferograms of the TRW group and the vibrational records of Powell and Stetson preceded what is really a simpler application of the interferometrical principle, and which historically ought to have come first—if the course of science would always follow the shortest line. This is the observation of small deformations of solid bodies by double exposure holograms. A simple explanation is as follows: We take a hologram of a body in state A. This means that we freeze in the wave A by means of a reference beam. Now let us deform the body so that it assumes the state B, and take a second hologram in the same emulsion with the same reference beam. We develop the hologram, and illuminate it with the reference beam. Now the two waves A and B, frozen in at different times, and which have never seen one another, will be revived simultaneously, and they interfere with one another. The result is that Newton fringes will appear on the object, each fringe corresponding to a deformation of a half wavelength. Fig. 21 shows a fine example of such a holographic interferogram, made in 1965 by K. Haines and B. P. Hildebrand. The principle was discovered simultaneously and independently also by J. M. Burch in England.

Non-destructive testing by holographic interferometry is now by far the most important industrial application of holography. It gave rise to the first industrial firm based on holography, GCO (formerly G. C. Optronics), in Ann Arbor, Mich., and the following examples are reproduced by courtesy of GCO. Fig. 22 shows the testing of a motor car tyre. The

front of the tyre is holographed directly, the sides are seen in two mirrors, right and left. First a little time is needed for the tyre to settle down and a first hologram is taken. Then a little hot air is blown against it, and a second exposure is made, on the same plate. If the tyre is perfect, only a few, widely spaced fringes will appear, indicating almost uniform expansion. But where the cementing of the rubber sheets was imperfect, a little blister appears, as seen near the centre and near the top left corner, only a few thousandths of a millimeter high, but indicating a defect which could become serious. Alternately, the first hologram is developed, replaced exactly in the original position, and the expansion of the tyre is observed "live."

Other examples of non-destructive testing are shown in Fig. 23; all defects which are impossible or almost impossible to detect by other means, but which reveal themselves unmistakeably to the eye. A particularly impressive piece of equipment manufactured by GCO is shown in Fig. 24. It is a holographic analyser for honeycomb sandwich structures (such as shown in the middle of Fig. 23) which are used in

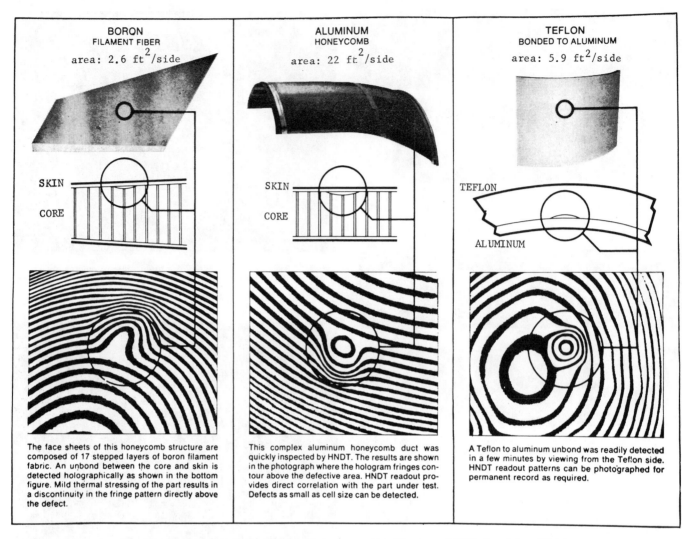

BORON
FILAMENT FIBER
area: 2.6 ft^2/side

SKIN
CORE

The face sheets of this honeycomb structure are composed of 17 stepped layers of boron filament fabric. An unbond between the core and skin is detected holographically as shown in the bottom figure. Mild thermal stressing of the part results in a discontinuity in the fringe pattern directly above the defect.

ALUMINUM
HONEYCOMB
area: 22 ft^2/side

SKIN
CORE

This complex aluminum honeycomb duct was quickly inspected by HNDT. The results are shown in the photograph where the hologram fringes contour above the defective area. HNDT readout provides direct correlation with the part under test. Defects as small as cell size can be detected.

TEFLON
BONDED TO ALUMINUM
area: 5.9 ft^2/side

TEFLON

ALUMINUM

A Teflon to aluminum unbond was readily detected in a few minutes by viewing from the Teflon side. HNDT readout patterns can be photographed for permanent record as required.

Fig. 23.   Examples of holographic non-destructive testing. (Courtesy of GCO, Ann Arbor, Mich.)

Fig. 24.   Holographic analyzer Mark II for sandwich structures. (GCO, Ann Arbor, Mich.)

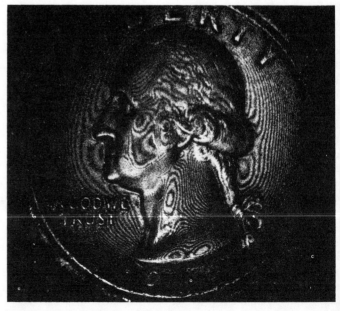

Fig. 25.   Holographically produced contour map of a medal, made by a method initiated by B. P. Hildebrand and K. A. Haines (*J. Opt. Soc. Amer.*, vol. 57, p. 155, 1967). Improved by J. Varner, University of Michigan, Ann Arbor, 1969.

aeroplane wings. The smallest welding defect between the aluminum sheets and the honeycomb is safely detected at one glance.

While holographic interferometry is perfectly suited for

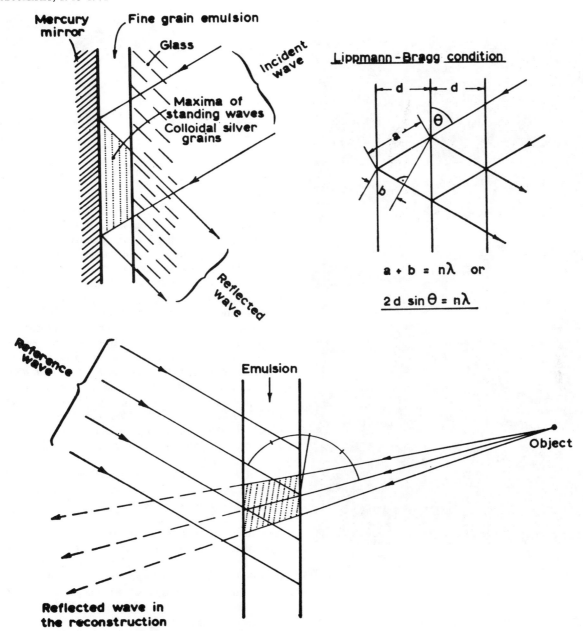

Fig. 26. Lippmann–Denisyuk reflection holography in natural colours.

the detection of very small deformations, with its fringe unit of 1/4000 mm, it is a little too fine for the checking of the accuracy of workpieces. Here another holographic technique called "contouring" is appropriate. It was first introduced by Haines and Hildebrand [13], in 1965, and has been recently much improved by J. Varner, also in Ann Arbor, Mich. Two holograms are taken of the same object, but with two wavelengths which differ by, e.g., one percent. This produces *beats* between the two-fringe system, with fringe spacings corresponding to about 1/40 mm, which is just what the workshop requires (Fig. 25).

From industrial applications I am now turning to another important development in holography. In 1962, just before the "holography explosion" the Soviet physicist Yu. N. Denisyuk [13] published an important paper in which he combined holography with the ingenious method of photography in natural colours, for which Gabriel Lippmann received the Nobel Prize in 1908. Fig. 26 illustrates Lippmann's method and Denisyuk's idea. Lippmann produced a very fine-grain

emulsion, with colloidal silver bromide, and backed the emulsion with mercury, serving as a mirror. Light falling on the emulsion was reflected at the mirror, and produced a set of standing waves. Colloidal silver grains were precipitated in the maxima of the electric vector, in layers spaced by very nearly half a wavelength. After development, the complex of layers, illuminated with white light, reflected only a narrow waveband around the original colour, because only for this colour did the wavelets scattered at the Lippmann layers add up in phase.

Denisyuk's suggestion is shown in the second diagram. The object wave and the reference wave fall in from opposite sides of the emulsion. Again standing waves are produced, and Lippmann layers, but these are no longer parallel to the emulsion surface, they bisect the angle between the two wavefronts. If now, and this is Denisyuk's principle, the developed emulsion is illuminated by the reference wave, the object will appear, in the original position and (unless the emulsion has shrunk) in the original colour.

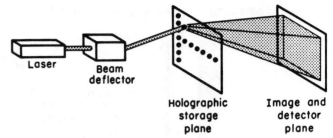

Fig. 28.   Holographic flying spot store. (L. K. Anderson and R. J. Collier, Bell Telephone Labs., 1968.)

matter uses only about 5–10% of the area, and the gradation not at all. A further factor arises from the utilization of the third dimension, the depth of the emulsion. This possibility was first pointed out in an ingenious paper by P. J. van Heerden [16], in 1963. Theoretically it appears possible to store one bit of information in about one wavelength cube. This is far from being practical, but the figure of 300, previously mentioned, is entirely realistic.

However, even without this enormous factor, holographic storage offers important advantages. A binary store, in the form of a checkerboard pattern on microfilm can be spoiled by a single grain of dust, by a hair or by a scratch, while a diffused hologram is almost insensitive to such defects. The holographic store, illustrated in Fig. 28, is according to its author L. K. Anderson [17] (1968) only a modest beginning, yet it is capable of accessing for instance any one of 64×64 printed pages in about a microsecond. Each hologram, with a diameter of 1.2 mm can contain about 10^4 bits. Reading out this information sequentially in a microsecond would of course require an impossible waveband, but powerful parallel reading means can be provided. One can confidently expect enormous extensions of these "modest beginnings" once the project of data banks will be tackled seriously.

Another application of holography which is probably only in an early stage is pattern and character recognition. I can only briefly refer to the basic work which A. Vander Lugt [18] has done in the field of pattern recognition. I will be sufficient to explain the basic principle of character recognition with the aid of Fig. 29.

Let us generalize a little the basic principle of holography. In all previous examples a complicated object beam was brought to interference with a simple plane or spherical reference beam, and the object beam was reconstructed by illuminating the hologram with the reference beam. But a little mathematics shows that this can be extended to any reference beam *which correlates sharply with itself*. The autocorrelation function is an invariant of a beam; it can be computed in any cross section. One can see at once that a spherical wave correlates sharply with itself, because it issues from a "point." But there are other beams which correlate sharply with themselves, for instance those which issue from a fingerprint, or from a Chinese ideogram, in an extreme case also those which issue from a piece of frosted glass. Hence it is quite possible for instance to translate, by means of a hologram, a Chinese ideogram into its corresponding English sentence and *vice versa*. Dr. Butters and M. Wall of Loughborough University have recently created holograms which from a portrait produce the signature of the owner, and *vice versa*.[3] In other words,

Fig. 27.   First two-colour reflecting hologram, reconstructed in white light (L. H. Lin, K. S. Pennington, G. W. Stroke, and A. E. Labeyrie [14].)

Though Denisyuk showed considerable experimental skill, lacking a laser in 1962 he could produce only an "existence proof." A colour reflecting hologram which could be illuminated with white light was first produced in 1965 by G. W. Stroke and A. Labeyrie [14a] and a two-colour version of it subsequently by L. H. Lin, K. S. Pennington, G. W. Stroke, and A. Labeyrie [14b] is shown in Fig. 27.

Since that time single-colour reflecting holograms have been developed to high perfection by new photographic processes, by K. S. Pennington [15] and others, with reflectances approaching 100 percent, but two, and even more, three-colour holograms are still far from being satisfactory. It is one of my chief preoccupations at the present to improve this situation, but it would take too long, and it would be also rather early to enlarge on this.

An application of holography which is certain to gain high importance in the next years is information storage. I have mentioned before that holography allows storing 100–300 times more printed pages in a given emulsion than ordinary micro-photography. Even without utilizing the depth dimension, the factor is better than 50. The reason is that a diffused holo-gram represents almost ideal coding, with full utilization of the area and of the gradation of the emulsion, while printed

[3] Quoted from Basil de Ferranti, "The computer, a challenge to the human brain," *Proc. Roy. Inst. of Great Britain*, p. 165, Aug. 1971.

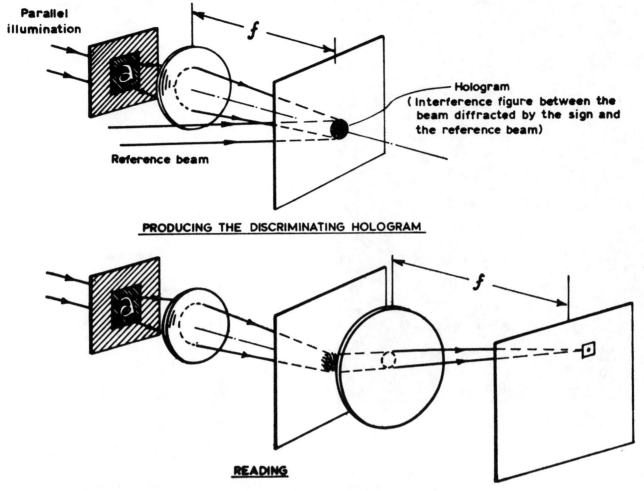

PRODUCING THE DISCRIMINATING HOLOGRAM

READING

Fig. 29.   The principle of character recognition by holography.

a hologram can be a fairly universal translator. It can for instance translate a sign which we can read to another which a machine can read.

Fig. 29 shows a fairly modest realisation of this principle. A hologram is made of a letter "a" by means of a plane reference beam. When this hologram is illuminated with the letter "a" the reference beam is reconstructed, and can activate for instance a small photocell in a certain position. This, I believe, gives an idea of the basic principle. There is of course much more needed to make a practical system of it, because there are so many ways of printing letters, but it would take me too long to explain how to deal with this and other difficulties.

With character recognition devices we have already taken half a step into the future, because these are likely to become important only in the next generation of computers or robots, to whom we must transfer a little more of human intelligence. I now want to mention briefly some other problems which are half or more than half in the future.

One, which is already very actual, is the overcoming of laser speckle. Everybody who sees laser light for the first time is surprised by the rough appearance of objects which we consider as smooth. A white sheet of paper appears as if it were crawling with ants. The crawling is put into it by the restless eye, but the roughness is real. It is called "laser speckle" and Fig. 30 shows a characteristic example of it. This is the appearance of a white sheet of paper in laser light, when viewed with a low-power optical system. It is not really noise; it is information which we do not want, information on the microscopic

unevenness of the paper in which we are not interested. What can we do against it?

In the case of rough objects the answer is, regrettably, that all we can do is to average over larger areas, thus smoothing the deviations. This means that we must throw a great part of the information away, the wanted with the unwanted. This is regrettable but we can do nothing else, and in many cases we have enough information to throw away, as can be seen by the fully satisfactory appearance of some of the reconstructions from diffuse holograms which I have shown. However, there are important areas in which we can do much more, and where an improvement is badly needed. This is the area of micro-holograms, for storing and for display. They are made as diffused holograms, in order to ensure freedom from dust and scratches, but by making them diffused, we introduce speckle, and to avoid this such holograms are made nowadays much larger than would be ideally necessary. I have shown recently [19], that the advantages of diffuse holograms can be almost completely retained, while the speckle can be completely eliminated by using, instead of a frosted glass, a special illuminating system. This, I hope, will produce a further improvement in the information density of holographic stores.

Now let us take a more radical step into the future. I want to mention briefly two of my favourite holographic brain-children. The first of these is Panoramic Holography, or one could also call it Holographic Art.

All the three-dimensional holograms made so far extend to a depth of a few meters only. Would it not be possible to

461

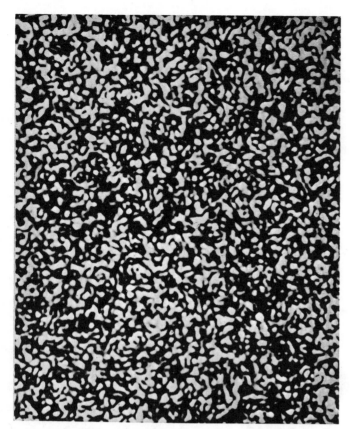

Fig. 30. Laser speckle. The appearance of, for example, a white sheet of paper, uniformly illuminated by laser light.

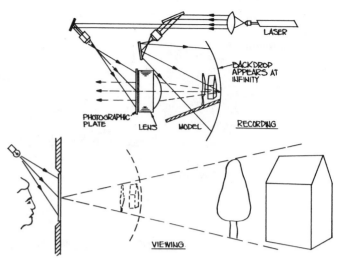

Fig. 31. Panoramic holography.

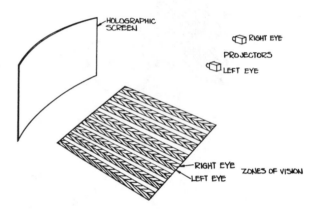

Fig. 32. Three-dimensional cinematography with holographic screen.

extend them to infinity? Could one not put a hologram on the wall, which is like a window through which one looks at a landscape, real or imaginary? I think it can be done, only it will not be a photograph but a work of art. Fig. 31 illustrates the process. The artist makes a model, distorted in such a way that it appears perspectivic, and extending to any distance when viewed through a large lens, as large as the hologram. The artist can use a smaller lens, just large enough to cover both his eyes when making the model. A reflecting hologram is made of it, and illuminated with a strong small light source. The viewer will see what the plate has seen through the lens; that is to say a scene extending to any distance, in natural colours. This scheme is under development, but considerable work will be needed to make it satisfactory, because we must first greatly improve the reflectance of three-colour holograms.

An even more ambitious scheme, probably even farther in the future, is three-dimensional cinematography, without viewing aids such as Polaroids. The problem is sketched out in Fig. 32. The audience (in one plane or two) is covered by zones of vision, with the width of the normal eye spacing, one for the right eye, one for the left, with a blank space between two pairs. The two eyes must see two different pictures; a stereoscopic pair. The viewer can move his head somewhat to the right or left. Even when he moves one eye into the blank zone, the picture will appear dimmer but not flat, because one eye gives the impression of "stereoscopy by default."

I have spent some years of work on this problem, just before holography, until I had to realise that it is strictly insolvable with the orthodox means of optics, lenticules, mirrors, prisms. One can make satisfactorily small screens for small theatres, but with large screens and large theatres one falls into a dilemma. If the lenticules or the like are large, they will

be seen from the front seats; if they are small, they will not have enough definition for the back seats.

Some years ago I realised to my surprise, that holography can solve this problem too. Use a projector as the reference source, and for instance the system of left viewing zones as the object. The screen, covered with a Lippmann emulsion, will then make itself automatically into a very complicated optical system such that when a picture is projected from the projector, it will be seen only from the left viewing zones. One then repeats the process with the right projector, and the right viewing zones. In the case of volume, (Lippmann–Denisyuk) holograms display the phenomenon of directional selectivity. If one displaces the illuminator from the original position by a certain angle, there will be no reflection. We put the two projectors at this angle (or a little more) from one another, and the effect is that the right picture will not be seen by the left eye and *vice versa*.

There remains of course one difficulty, and this is that one cannot practice holography on the scale of a theatre, and with a plate as large as a screen. But this too can be solved, by making up the screen from small pieces, not with the threatre but with a *model* of the theatre, seem through a lens, quite similar to the one used in panoramic holography.

I hope I have conveyed the feasibility of the scheme, but I feel sure that I have conveyed also its difficulties. I am not sure whether they will be overcome in this century, or in the next.

Ambitious schemes, for which I have a congenital inclina-

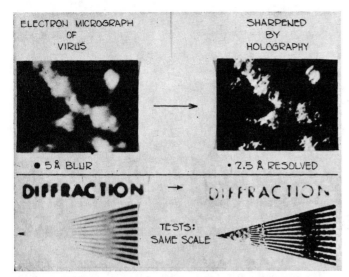

ELECTRON MICROGRAPH
OF
VIRUS

SHARPENED
BY
HOLOGRAPHY

• 5 Å BLUR

• 2.5 Å RESOLVED

DIFFRACTION → DIFFRACTION

TESTS:
SAME SCALE

Fig. 33. Scanning transmission electron micrograph (Prof. A. Crewe, University of Chicago), holographically deblurred (by Prof. G. W. Stroke, 1971). The bottom photographs prove that the effect could not be obtained by hard printing, because some spatial frequencies which appear in the original with reversed phase had to be phase-corrected.

tion, take a long time for their realisation. As I said at the beginning, I shall be lucky if I shall be able to see in my lifetime the realisation of holographic electron microscopy, on which I have started 24 years ago. But I have good hope, because I have been greatly encouraged by a remarkable achievement of G. W. Stroke [20], which is illustrated in Fig. 33. Professor Stroke has recently succeeded in deblurring micrographs taken by Professor Albert Crewe, Chicago, Ill., with his scanning transmission electron microscope, by a holographic filtering process, improving the resolution from 5 angstrom to an estimated 2.5 angstrom. This is not exactly holographic electron microscopy, because the original was not taken with coherent electrons, but the techniques used by both sides, by A. Crewe and by G. W. Stroke are so powerful, that I trust them to succeed also in the next, much greater and more important step.

Summing up, I am one of the few lucky physicists who could see an idea of theirs grow into a sizeable chapter of physics. I am deeply aware that this has been achieved by an army of young, talented, and enthusiastic researchers, of whom I could mention only a few by name. I want to express my heartfelt thanks to them, for having helped me by their work to this greatest of scientific honours.

BIBLIOGRAPHY AND REFERENCES

It is impossible to do justice to the hundreds of authors who have significantly contributed to the development of holography. The number of articles exceeds 2000, and there are more than a dozen books in several languages.

An extensive bibliography may be found for instance in, T. Kallard, Ed., *Holography*. New York: Optosonic Press, 1969 and 1970.

Books

E. S. Barrekette, W. E. Kock, T. Ose, J. Tsujiuchi, and G. W. Stroke, Eds., *Applications of Holography*. New York: Plenum, 1971.

H. J. Caulfield and S. Lu, *The Applications of Holography*. New York: Wiley–Interscience, 1970.

R. J. Collier, C. B. Burckhardt, and L. H. Lin, *Optical Holography*. New York: Academic Press, 1971.

J. B. DeVelis and G. O. Reynolds, *Theory and Applications of Holography*. Reading, Mass.: Addison Wesley, 1967.

M. Françon, *Holographie*. Paris, France: Masson et Cie, 1969.

H. Kiemle und D. Röss, *Einführung in die Technik der Holographie*. Frankfurt am Main, Germany: Akademische Verlagsgesellschaft, 1969. English translation: *Introduction to Holographic Techniques*. New York: Plenum, 1972, in print.

W. E. Kock, *Lasers and Holography (An Introduction to Coherent Optics)*. Garden City, N. Y.: Doubleday, 1969 (Russian translation, Moscow, USSR: Mir, 1967).

Yu. I. Ostrovsky, *Holography* (in Russian). Leningrad, USSR: Nauka, 1970.

E. R. Robertson and J. M. Harvey, Eds., *The Engineering Uses of Holography*. Cambridge, England: Cambridge Univ. Press, 1970.

G. W. Stroke, *An Introduction to Coherent Optics and Holography*. New York: Academic Press, 1st ed. 1966; 2nd ed. 1969.

J. C. Vienot, P. Smigielski, and H. Royer, *Holographie Optique (Developpements. Applications)*. Paris, France: Dunod, 1971.

Papers

[1] D. Gabor, "A new microscopic principle," *Nature*, vol. 161, pp. 777–778, 1948.
——, "Microscopy by reconstructed wavefronts," *Proc. Roy. Soc.*, vol. A197, pp. 454–487, 1949.
——, "Microscopy by reconstructed wavefronts: II," *Proc. Roy. Soc.* (London), vol. 64, pt. 6, pp. 449–469, 1951.

[2] G. L. Rogers, "Experiments in diffraction microscopy," *Proc. Roy. Soc.* (Edinburgh), vol. 63A, p. 193, 1952.

[3] A. Baez, "Resolving power in diffraction microscopy," *Nature*, vol. 169, pp. 963–964, 1952.

[4] H. M. A. El-Sum and P. Kirkpatrick, "Microscopy by reconstructed wavefronts," *Phys. Rev.*, vol. 85, p. 763, 1952.

[5] D. Gabor and W. P. Goss, "Interference microscope with total wavefront reconstruction," *J. Opt. Soc. Amer.*, vol. 56, pp. 849–858, 1966.

[6] L. J. Cutrona, E. N. Leith, L. J. Porcello, and W. E. Vivian, "On the application of coherent optical processing techniques to synthetic-aperture radar," *Proc. IEEE*, vol. 54, pp. 1026–1032, Aug. 1966.

[7] E. N. Leith and J. Upatnieks, "Wavefront reconstruction with continuous tone transparencies," *J. Opt. Soc. Amer.*, vol. 53, p. 522, 1963.
——, "Wavefront reconstruction with continuous-tone objects," *J. Opt. Soc. Amer.*, vol. 53, pp. 1377–1381, 1963.

[8] From D. Gabor, W. E. Kock, and G. W. Stroke, "Holography," *Science*, vol. 173, pp. 11–23, 1971. This is one of the first 3-D diffused light holograms recorded by G. W. Stroke at the University of Michigan, where he originated this work. An early reference is G. W. Stroke, "Theoretical and experimental foundations of high-resolution optical holography," presented in Rome, Italy, on Sept. 14, 1964; also in *Pubbliciazioni IV Centenario della Nascita di Galileo Galilei*. Firenze, Italy: G. Barbera, 1966, pp. 53–63.

[9] C. Knox and R. E. Brooks, "Holographic motion picture microscopy," *Proc. Roy. Soc.* (London), vol. B174, pp. 115–121, 1969.

[10] G. W. Stroke and A. E. Labeyrie, "Two-beam interferometry by successive recording of intensities in a single hologram," *Appl. Phys. Lett.*, vol. 8, pp. 42–44, Jan. 15, 1966.
L. O. Heflinger, R. F. Wuerker, and R. E. Brooks, *J. Appl. Phys.*, vol. 37, pp. 642–649, Feb. 1966.

[11] J. M. Burch, "The application of lasers in production engineering," *Production Eng.*, vol. 44, pp. 431–442, 1965.
R. L. Powell and K. A. Stetson, "Interferometric vibration analysis by wavefront reconstruction," *J. Opt. Soc. Amer.*, vol. 55, pp. 1593–1598, 1965.

[12] K. A. Stetson, thesis (under direction of E. Ingelstam), Royal Institute of Technology, Stockholm, Sweden, 1969.

[13] Yu. N. Denisyuk, "Photographic reconstruction of the optical properties of an object in its own scattered radiation," *Dokl. Akad. Nauk SSR*, vol. 144, pp. 1275–1278, 1962.

[14a] G. W. Stroke and A. E. Labeyrie, *Phys. Lett.*, vol. 20, no. 4, pp. 368–370, Mar. 1, 1966.

[14b] L. H. Lin, K. S. Pennington, G. W. Stroke, and A. E. Labeyrie, *Bell Syst. Tech. J.*, vol. 45, p. 659, 1966.

[15] K. S. Pennington and J. S. Harper, "Techniques for producing low-noise, improved-efficiency holograms," *Appl. Opt.*, vol. 9, pp. 1643–1650, 1970.

[16] P. J. Van Heerden, "A new method of storing and retrieving information," *Appl. Opt.*, vol. 2, pp. 387–392, 1963.

[17] L. K. Anderson, "Holographic optical memory for bulk data storage," *Bell Lab. Rec.*, vol. 46, p. 318, 1968.

[18] A. Vander Lugt, "Signal detection by complex spatial filtering," *IEEE Trans. Inform. Theory*, vol. IT-10, pp. 139–145, Apr. 1964.

[19] D. Gabor, "Laser speckle and its elimination," *IBM J. Res. Develop.*, vol. 14, pp. 509–514, Sept. 1970.

[20] G. W. Stroke, "Image deblurring and aperture synthesis using 'a posteriori' processing by Fourier-transform holography," *Opt. Acta*, vol. 16, pp. 401–422, 1971.
——, "Sharpening images by holography," *New Scientist*, vol. 51, pp. 671–674, 1971.
G. W. Stroke and M. Halioua, "Attainment of diffraction-limited imaging in high-resolution electron microscopy by 'a posteriori' holographic image sharpening," *Optik*, 1972, in print.

An Introduction to the Principles and Applications of Holography

JOSEPH W. GOODMAN, MEMBER, IEEE

Invited Paper

Abstract—Holography has strong historical ties with electrical engineering and potential application to many electrical engineering problems. The basic problem addressed by holography is introduced in both physical and mathematical terms. The analogy between the hologram of a point-source object and the linear FM signals of chirp radar is stressed, and the first-order imaging properties of holograms recorded in arbitrary geometries are derived.

Various types of holograms are described, including thin, thick, transmission, reflection, amplitude, and phase holograms. The important properties of each type of hologram are introduced.

A survey of various applications of holography is presented, with introductions to the use of holography in interferometry, microscopy, imaging through distorting media, optical data processing, and optical data storage. The use of simple holograms as optical elements is also described.

I. INTRODUCTION

DURING the past decade the field known as holography has undergone dramatic changes with regard to both theoretical understanding and practical application. The ideas that ten years ago were regarded primarily as an optical curiosity have developed into a branch of technology with a diversity of demonstrated applications. Why should this field, so closely tied to physical optics and seemingly so far from the more traditional areas of electrical and electronic phenomena, be of particular interest to electrical engineers? There are a variety of answers to this question.

First, there is a strong historical tie between holography and electrical engineering, for many of the most important innovative advances in the field have been made by individuals who, at the time of their contributions, were concerned with areas closely associated with the electronic sciences. Most important, of course, was the original conception of the idea behind holography by Dennis Gabor in 1948 [1] (see also [2]–[4]). Holography was envisioned by Gabor as a potential means for overcoming the spherical aberration of electron lenses in the electron microscope. A second example is the pioneering work of Leith and Upatnieks [5], who applied techniques developed in connection with synthetic aperture radars [6]–[10] to holography, thereby removing many of the practical difficulties previously encountered by Gabor. (For a discussion of the close connection between holography and synthetic aperture radars, see the paper by Leith in this issue.[1])

Above and beyond the purely historical connections, the language, concepts, and mathematical tools of electrical engineering have proved extremely useful in holography and have played an important role in its development. Notable examples are the work of Lohmann [11], based on analogies with single-sideband modulation techniques, and again the work of Leith and Upatnieks, whose

papers (e.g., [5], [12]–[15]) make liberal use of concepts from communication and information theory.

Finally, holography is of growing importance to electrical engineers because of the significant roles it can now play and will undoubtedly play in the future in the design of electronic, acoustical, and electrooptical systems of various kinds. Some of the present and future capabilities of holography in this regard are discussed in Section IV of this paper, as well as in related papers in this issue.

The goal of this paper is to present a tutorial review of the principles of holography and an introduction to a variety of its applications. An effort has been made, wherever possible, to present the discussion in terms familiar to most electrical engineers. Primary attention is devoted to optical holography, but the reader is reminded, with the help of the accompanying papers, that holographic principles can also be applied in other regions of the electromagnetic spectrum and indeed with nonelectromagnetic waves. The reader interested in pursuing the subject of holography in greater depth may wish to consult any of the several books on the subject that now exist [16]–[20]. In addition, several excellent review articles have been published [21]–[23].

II. FUNDAMENTALS

A. The Central Problem of Holography

The central problem addressed by holography is that of "wavefront reconstruction," by which we mean recording, and later reconstituting, the amplitude and phase distributions of a monochromatic (or nearly monochromatic) wave disturbance incident on a prescribed surface. The difficulties associated with this task in the optical region of the spectrum arise first in connection with the measurement of phase, for detector response is governed by the intensity (i.e., squared amplitude) of the wave and is quite independent of phase. To overcome this difficulty it is necessary to resort to an interferometric recording process, which effectively encodes the phase distribution of the incident wave as a measurable modulation (generally spatial, but in some cases temporal) of the intensity distribution. Thus the wave of interest is allowed to interfere with a mutually coherent "reference" wave, such that fringes of interference are formed and recorded. Gabor referred to such a recording as a "hologram," a term derived from the Greek word "holos" and meaning a "complete record."

Seldom is the hologram itself the desired end product. Rather the recording process is only the first step in what normally is a two-step process, the second step consisting of the construction of a new wave from the detected data. Generally the amplitude and phase distributions of the new wave are desired to be identical (up to a possible scaling of the spatial coordinates) with those of the original wavefront. In some cases, however, the reconstructed wave is intentionally made to differ from the original wave in some desired fashion.

The reader may well wonder just what is accomplished by this rather laborious two-step process, which produces in the end little

Manuscript received March 10, 1971; revised June 16, 1971. This work was supported by the Office of Naval Research. *This invited paper is one of a series planned on topics of general interest—The Editor.*

The author is with the Department of Electrical Engineering, Stanford University, Stanford, Calif. 94305.

[1] E. N. Leith, "Quasi-holographic techniques in the microwave region," this issue, pp. 1305–1318.

Reprinted from *Proc. IEEE*, vol. 59, pp. 1292–1304, Sept. 1971.

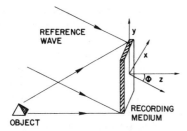

Fig. 1. Recording a hologram.

more than a duplicate of the wavefront that was already available at the start. The answers to this question will become apparent in detail only when the various applications of holography are considered. In most cases it is not the wavefront itself that is of ultimate interest, but rather some object (possibly three-dimensional) through which the wave has passed, or from which it has been reflected during propagation to the recording plane. The ability to record information about that object in the form of a hologram, rather than as a directly formed image, introduces a variety of advantages, particularly with respect to the kind and quality of information that can ultimately be recovered. For example, the image generated when the hologram is illuminated may be three-dimensional. A more detailed discussion of these advantages is best presented after the holographic process has been described in greater depth, and is therefore postponed until Section IV.

B. The Basic Mathematics of Holography

The physical quantities dealt with in this analysis are complex-valued functions defined on three-dimensional space and representing the scalar amplitude and phase distributions of a monochromatic wave. If the wave of concern is electromagnetic, the scalar amplitude may be regarded as that of a single polarization component, it being tacitly assumed that all waves are identically polarized and that their polarizations are unaffected by the various operations involved. Such an assumption must be used with some caution, for polarization effects can be important in practice [24]–[26]. As a first approximation, a more complete theory can be developed by treating the two polarization components independently.

With reference to Fig. 1, let there exist in the plane $z=0$ a detection medium, for example a photographic emulsion if the waves are optical. For the present, the physical thickness of the medium will be ignored. Let there impinge upon this medium from the left two monochromatic waves, each with wavelength λ. One wave, which we refer to here as the "object" wave, has been reflected from or transmitted by some object of interest; the complex amplitude of this incident wave is represented by $U_0(x, y)$. The second wave, which we refer to as the "reference" wave, has complex amplitude $U_r(x, y)$ at the recording medium. The physically measurable quantity at the detector is the intensity distribution of the incident wave, which can be written

$$I(x, y) = |U_r + U_0|^2 = |U_r|^2 + |U_0|^2 + U_r^* U_0 + U_r U_0^*. \quad (1)$$

Note in particular that the third term of (1) contains U_0, so there is some hope that perhaps both amplitude and phase information can be recovered.

If optical reconstruction of the wavefront U_0 is desired, the detected intensity pattern must somehow be transferred to a spatial modulator which will impress upon an incident optical wave the desired pattern of amplitude and phase. A photographic transparency provides the simplest modulator of this kind, although many other materials can and have been used, including embossed vinyl tape [27], thermoplastics [28], [29], electrooptic crystals [30],

ferromagnetic films [31], photopolymers [32], and photochromics [33].

To attempt by direct means to impress upon an incident optical wave a prescribed distribution of both amplitude and phase is an enormously difficult task, requiring simultaneous and independent control of both attenuation and phase shift through the modulator. In practice it is possible to control either attenuation or phase shift, but not both independently. Suppose, however, that it is possible to achieve an amplitude transmittance $t_A(x, y)$ (i.e., the ratio of transmitted complex field to incident complex field at each point) that is simply proportional to the intensity $I(x, y)$ of the holographic interference pattern. For example, over a limited dynamic range the amplitude transmittance of a developed photographic transparency can be approximated by

$$t_A(x, y) \cong t_b + \beta \Delta I(x, y) \quad (2)$$

where t_b and β are constants while ΔI represents the variations of intensity about its mean level. A similar proportionality can be achieved for a purely phase-shifting medium, such as a bleached photographic transparency, through the linearization

$$t_A(x, y) = \exp\{j\mu\Delta I(x, y)\} \cong 1 + j\mu\Delta I(x, y) \quad (3)$$

($\mu=$ constant) valid for small modulation depths. In practice neither a pure attenuation nor a pure phase shift can be realized, but (2) and (3) provide adequate models for thin modulation media that operate primarily through attenuation or primarily through phase shift.

To proceed with our discussion of the reconstruction process, suppose that proportionality between t_A and I is achieved. Let the modulator be illuminated by a replication of the original reference wave $U_r(x, y)$. Neglecting unimportant constants, the field $U_c(x, y)$ appearing immediately behind the modulator is

$$U_c(x, y) = U_r(x, y)I(x, y)$$
$$= U_r|U_r|^2 + U_r|U_0|^2 + |U_r|^2 U_0 + U_r^2 U_0^*. \quad (4)$$

Now if the reference wave has been chosen to have approximately constant intensity $|U_r|^2$, the third term above is clearly a duplication of the original object wavefront, and if it can be separated from the other components the wavefront reconstruction process will have succeeded.

Separation of the various reconstructed wave components represented in (4) posed the most serious obstacle to useful application of Gabor's invention until the early 1960s, when Leith and Upatnieks introduced the concept of an "offset reference" hologram [5]. The basic idea behind this innovation is best illustrated by considering a very specific reference wave, namely a plane wave with wave vector in the y-z plane inclined at an angle Φ with respect to the z axis (cf. Fig. 1). The field distribution at the detector due to this wave is of the form

$$U_r(x, y) = A \exp\{-j2\pi\alpha y\} \quad (5)$$

where A is a constant and $\alpha = \sin\Phi/\lambda$. In addition let the object wave U_0 be expressed in terms of its amplitude and phase distributions

$$U_0(x, y) = a(x, y)\exp[j\theta(x, y)]. \quad (6)$$

The pattern of interference now becomes

$$I(x, y) = A^2 + a^2(x, y) + Aa(x, y)\exp\{j[2\pi\alpha y + \theta(x, y)]\}$$
$$+ Aa(x, y)\exp\{-j[2\pi\alpha y + \theta(x, y)]\}$$
$$= A^2 + a^2(x, y) + 2Aa(x, y)\cos[2\pi\alpha y + \theta(x, y)]. \quad (7)$$

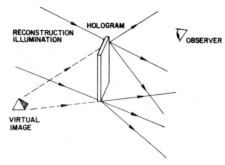

Fig. 2. Reconstructing a virtual image.

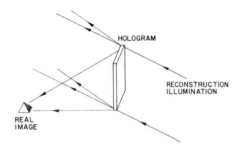

Fig. 3. Reconstructing a real image.

Clearly, the amplitude and phase distributions of the object wave have been encoded, respectively, as amplitude and phase modulations of a spatial carrier of frequency α.

Let the recorded hologram now be illuminated by a duplication of the reference wave. The transmitted field has the following components:

$$U_c(x, y) = A^3 \exp\{-j2\pi\alpha y\} + Aa^2(x, y)\exp\{-j2\pi\alpha y\} + A^2 a(x, y)$$
$$\cdot \exp\{j\theta(x, y)\} + A^2 a(x, y)\exp\{-j[4\pi\alpha y + \theta(x, y)]\}. \quad (8)$$

Up to an unimportant constant, the third term of this equation is a duplication of the original wavefront. The first two terms are multiplied by $\exp(-j2\pi\alpha y)$, which indicates they will propagate away from the modulator in the general direction of travel of the reference wave. The fourth term contains a factor $\exp\{-j4\pi\alpha y\}$, which indicates that it will propagate away at an even steeper angle. Thus a properly situated observer will intercept the reconstructed wavefront without interference, and will accordingly see behind the hologram a virtual image of the object that gave rise to U_0 (see Fig. 2).

If the hologram is illuminated not by the original reference wave but rather by its complex conjugate U_r^*, the fourth term of (7) yields a wave proportional to U_0^*. As consideration of the simple case of a point-source object shows, U_0^* corresponds to a wave converging towards a real image of the original object. One way to generate the required illumination U_r^* is to illuminate the hologram by a plane wave with direction of propagation opposite to that used during recording. Thus referring to Fig. 3, the hologram is illuminated from the right, and the wavefront U_0^* is created to the left of the transparency. The result is the formation of a real image of the object, with each image point coming to focus at the location of the object point that gave rise to it. Again the various extraneous reconstructed waves will propagate away from the image, provided the carrier frequency is chosen sufficiently high.

C. The Elementary Signals of Holography

A hologram of a complex object may be viewed as being built up as a superposition of "elementary signals" which are the holograms of individual point-source components of the object. Strictly speak-

ing, this view is not quite correct, for it neglects the signals generated by interference of each object point with all other object points, i.e., the $|U_0|^2$ term of (1). Nonetheless, under most conditions these object-object intermodulation terms do not diffract light in the direction of the desired images, and therefore there is some justification in neglecting them. As we shall see, there is a marked similarity between the hologram of a point-source object and the more familiar "chirp" signals of radar and communication theory.

For simplicity, let the reference wave again be the plane wave described by (5), and let the object wave be a simple spherical wave expanding about the source point (x_0, y_0, z_0), where z_0 is a negative number. At the recording plane, the object wave is of the form

$$U_0(x, y) = a \exp\left\{j\frac{2\pi}{\lambda}[z_0^2 + (x - x_0)^2 + (y - y_0)^2]^{1/2}\right\}. \quad (9)$$

To concentrate on the first-order properties of the imaging process, it is necessary to make a small angle or paraxial approximation, representing the square root in (9) by the first two terms of its binomial expansion. Thus the spherical wavefront is approximated by a paraboloid,

$$\exp\left\{j\frac{2\pi}{\lambda}[z_0^2 + (x - x_0)^2 + (y - y_0)^2]^{1/2}\right\}$$
$$\approx \exp\left\{-j\frac{2\pi z_0}{\lambda} - j\frac{\pi}{\lambda z_0}[(x - x_0)^2 + (y - y_0)^2]\right\} \quad (10)$$

where we have used the fact that z_0 is negative. Inclusion of higher order terms in the binomial expansion is necessary if predictions of the aberrations associated with the holographic process are to be made [34]–[37], but this rather specialized subject will not be treated here. Introducing (10) and (5) in (1), the intensity distribution at the detector is found to be

$$I(x, y) = I_r + I_0 + 2\sqrt{I_r I_0}$$
$$\cdot \cos\left\{2\pi\alpha y - \frac{\pi}{\lambda z_0}[(x - x_0)^2 + (y - y_0)^2] + \phi\right\} \quad (11)$$

where $I_r = A^2$, $I_0 = a^2$, and $\phi = 2\pi z_0/\lambda$ is a constant phase angle. Equation (11) represents what we call an "elementary signal."

Neglecting the constants I_r and I_0, the elementary signal bears a strong resemblance to the "chirp" (i.e., linear FM) signals so widely encountered in the theory of modern radar [38], [39]; it is in fact the two-dimensional analog of such signals. Let $\psi(x, y)$ represent the argument of the consinusoidal term in (11), i.e.,

$$\psi(x, y) = 2\pi\alpha y - \frac{\pi}{\lambda z_0}[(x - x_0)^2 + (y - y_0)^2] + \phi. \quad (12)$$

The "local frequency" (analogous to instantaneous frequency) of the fringe pattern is a vector quantity $\nu(x, y)$ defined by

$$\nu(x, y) \triangleq \frac{1}{2\pi}\nabla\psi(x, y) = \frac{1}{2\pi}\frac{\partial\psi}{\partial x}\hat{x} + \frac{1}{2\pi}\frac{\partial\psi}{\partial y}\hat{y} \quad (13)$$

where $\hat{x}$ and $\hat{y}$ are unit vectors. Performing the required differentiations, the vector components are found to be

$$\nu_X = \frac{1}{2\pi}\frac{\partial\psi}{\partial x} = -\frac{(x - x_0)}{\lambda z_0}$$
$$\nu_Y = \frac{1}{2\pi}\frac{\partial\psi}{\partial y} = \alpha - \frac{(y - y_0)}{\lambda z_0}. \quad (14)$$

Thus each vector component of spatial frequency sweeps linearly

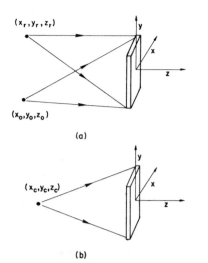

Fig. 4. Generalized geometries. (a) Recording a hologram. (b) Reconstructing images.

with its corresponding space coordinate. The length v of the local frequency vector is

$$v = \left\{ \left[\alpha - \frac{(y - y_0)}{\lambda z_0} \right]^2 + \left[\frac{x - x_0}{\lambda z} \right]^2 \right\}^{1/2}. \tag{15}$$

The encoding of a point-source object into the extended fringe pattern of (11) is significant in several respects. Most important, the highly localized point-source object has been dispersed by the propagation phenomenon to yield a detected signal occupying an extended region of the space domain. When the object wave arises from a complex source which has a large dynamic range of brightness, the redistribution of signal energy often results in a considerable relaxation of the dynamic range requirements at the detector. In addition, a spatial redundancy is introduced such that localized imperfections of the recording medium are often of little consequence.

The importance of the elementary signals represented by (11) was recognized at an early time in the history of holography by Rogers [40], who pointed out their similarity to Fresnel zone plates. The zone plate interpretation of holography provides a very physical way of explaining the holographic process [41]. The "chirp" signal interpretation likewise has many advantages, for many of the known properties of pulse compression systems can be drawn upon in discussing the properties of the reconstructed images.

D. Generalized Hologram Geometries

The assumption of a plane reference wave in the previous sections was made for analytical simplicity, but is by no means a necessary restriction in practice. Similarly, the assumption of a reconstruction wave which duplicates the original reference wave (or its conjugate) is not necessary unless the hologram thickness becomes comparable with, or greater than, the finest recorded fringe, in which case Bragg diffraction effects are important (see Section III). Several important properties of holograms as image-forming devices are discovered when a more general geometry is considered.

Let the object wave be generated by a point source at coordinates (x_0, y_0, z_0) and the reference wave be generated by a point source at coordinates (x_r, y_r, z_r), as shown in Fig. 4a. The wavelength used during the recording process is represented by λ_1. Let the resulting hologram transparency be illuminated by a spherical wave of wavelength λ_2 arising from a point source at coordinates (x_c, y_c, z_c), as shown in Fig. 4(b).

If the spherical wavefronts are again approximated by paraboloids and constant phase factors are dropped, the incident wave-

fronts are

$$U_r(x, y) \cong A \exp\left\{ -j \frac{\pi}{\lambda_1 z_r} \left[(x - x_r)^2 + (y - y_r)^2 \right] \right\} \tag{16}$$

$$U_0(x, y) \cong a \exp\left\{ -j \frac{\pi}{\lambda_1 z_0} \left[(x - x_0)^2 + (y - y_0)^2 \right] \right\}. \tag{17}$$

Similarly, the wavefront illuminating the hologram during reconstruction is

$$U_c(x, y) \cong B \exp\left\{ -j \frac{\pi}{\lambda_2 z_c} \left[(x - x_c)^2 + (y - y_c)^2 \right] \right\}. \tag{18}$$

Assuming the ideal detector characteristic described by (2), the two important terms of amplitude transmittance of the hologram are

$$t_1(x, y) \propto U_c(U_r^* U_0), \qquad t_2(x, y) \propto U_c(U_r U_0^*). \tag{19}$$

If the required multiplications are performed and the two resulting wavefronts are compared with a parabolic approximation to a spherical reconstructed wave corresponding to an image point at (x_i, y_i, z_i)

$$U_i(x, y) \cong D \exp\left\{ -j \frac{\pi}{\lambda_2 z_i} \left[(x - x_i)^2 + (y - y_i)^2 \right] \right\} \tag{20}$$

the following identifications can be made, where the upper set of signs applies to one wave and the lower set to the other:

$$x_i = \pm \frac{\lambda_2 z_i}{\lambda_1 z_0} x_0 \mp \frac{\lambda_2 z_i}{\lambda_1 z_r} x_r - \frac{z_i}{z_c} x_c$$

$$y_i = \pm \frac{\lambda_2 z_i}{\lambda_1 z_0} y_0 \mp \frac{\lambda_2 z_i}{\lambda_1 z_r} y_r - \frac{z_i}{z_c} y_c$$

$$z_i = \left(\frac{1}{z_c} \pm \frac{\lambda_2}{\lambda_1 z_r} \mp \frac{\lambda_2}{\lambda_1 z_0} \right)^{-1}. \tag{21}$$

When z_i is a negative number, the image is virtual, for the wave appears to be diverging from a point lying to the left of the hologram in Fig. 4(b). When z_i is positive, the image is real, for the light comes to a focus to the right of the hologram. Note that in general it is not necessary that one image be real and the other virtual. For example, when $\lambda_1 = \lambda_2$, $z_r = z_0$, and $z_c > 0$, both images are real, while when $\lambda_1 = \lambda_2$, $z_r = z_0$, and $z_c < 0$ both images are virtual.

From the results presented in (21) it is a simple matter to show that, for objects more complex than a single point source, the images produced by the holographic process may be magnified or demagnified with respect to the object that gives rise to them. The transverse magnification M_t is found from (21) to be

$$M_t = \left| \frac{\partial x_i}{\partial x_0} \right| = \left| \frac{\partial y_i}{\partial y_0} \right| = \left| \frac{\lambda_2 z_i}{\lambda_1 z_0} \right| = \left| 1 - \frac{z_0}{z_r} \mp \frac{\lambda_1 z_0}{\lambda_2 z_c} \right|^{-1} \tag{22}$$

while the longitudinal magnification M_l is

$$M_l = \left| \frac{\partial z_i}{\partial z_0} \right| = \frac{\lambda_1}{\lambda_2} M_t^2. \tag{23}$$

The possible use of these magnifications in microscopy, particularly X-ray microscopy, were studied by El-Sum in the 1950s [42].

An important conclusion to be drawn from the preceding equations is that, if the images are ultimately formed with radiation of different wavelength than used for recording, the longitudinal and transverse magnifications will not be the same. For a three-dimensional object, the result is an apparent distortion of the image. To emphasize this point, consider a microwave hologram recorded with

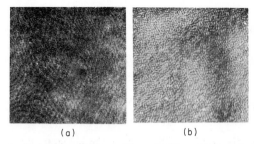

Fig. 5. Holograms of diffuse objects. (a) View with the naked eye.
(b) High magnification.

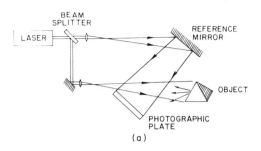

(a)

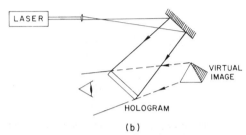

(b)

Fig. 6. Hologram of a three-dimensional object. (a) Recording.
(b) Viewing the virtual image.

(a)

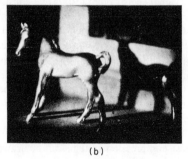

(b)

Fig. 7. Photographs of the virtual image of a three-dimensional object,
illustrating a change of both focus and perspective.

wavelength $\lambda_1 = 10$ cm. For the sake of argument, suppose that a photographic hologram of the same size as the original detecting array is produced (such a large transparency would, of course, hardly be desirable in practice). Let this transparency be illuminated by red light from an He–Ne laser ($\lambda = 632.8$ nm). The longitudinal magnification is then about 1.6×10^5 times greater than the transverse magnification.

A partial solution to this problem is afforded by scaling the linear size of the hologram by a factor m between recording and reconstruction, a step that would generally be necessary in any case to achieve a transparency of reasonable size. If $m < 1$, the size of the hologram has been reduced, while if $m > 1$ it has been increased. In this case the transverse and longitudinal magnifications become [34]

$$M_t = m\left[1 - \frac{z_0}{z_r} \mp m^2 \frac{\lambda_1}{\lambda_2}\frac{z_0}{z_c}\right]^{-1}$$

$$M_l = \frac{\lambda_1}{\lambda_2}M_t^2. \tag{24}$$

The transverse and longitudinal magnifications can be made equal, independent of position, by making M_t equal to λ_2/λ_1 and scaling the linear size of the hologram by a factor $m = \lambda_2/\lambda_1$ [35]. For the particular example outlined above, the hologram must be reduced in size by a factor 0.6×10^{-5}. If the original microwave array is 10 m long, the detected data must be transferred to a hologram of length 60 μm to achieve equal transverse and longitudinal magnifications. Unfortunately, the resulting hologram is so small that the virtual image cannot be viewed with any useful parallax. In addition, the resulting transverse demagnification may be so great as to require the use of a microscope to view the image.

E. Holograms with Diffused Illumination and Holograms of Three-Dimensional Objects

When the object of interest is a simple transparency, it is generally advantageous to illuminate that transparency with highly diffused laser light during the recording process [13]. Thus rather than simply expanding the output of a laser to fully illuminate the transparency of interest, a diffusing medium, such as opal glass, is inserted between the laser and the transparency, generally in close proximity to the transparency. The effect of the diffuser is to introduce an extremely complex phase distribution across the illuminating wave, or equivalently to randomize (in an ensemble sense) the phase of the light transmitted by each point on the object transparency. As a consequence, the elementary signals described by (11) add at the recording plane with unrelated spatial phases, producing a hologram that appears to the eye to be uniformly gray. Fig. 5(a) shows a hologram formed with diffuse object illumination, while Fig. 5(b) shows a highly magnified portion of that same hologram. The extremely complex patterns observed on a microscopic scale represent the information-bearing structure of the hologram.

Two advantages are gained by the use of diffused illumination.

First, the energy distribution across the hologram is spread out more evenly than would otherwise be the case, resulting in a reduction of the dynamic range required of the recording medium. Second, because the original phase relations between elementary signals on the hologram are randomized by the diffuser, large regions where the elementary signals would have added with destructive interference are eliminated, and the entire virtual image of the transparency can be viewed through any portion of the hologram. Thus the use of diffused illumination increases the degree of spatial redundancy present in the holographic recording. These advantages are gained only at a price, however, for the images formed from such holograms have a mottled appearance or granularity that is generally referred to as "speckle" [43], [44] and which arises because of the diffused nature of the illumination. Speckle not only reduces the esthetic qualities of the images, but in addition leads to a loss of effective resolution. Several methods for eliminating or minimizing speckle have been proposed and demonstrated [45]–[47], often through the use of specially constructed diffusing plates.

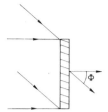

Fig. 8. Interference fringes in a thick recording medium.

A second situation in which diffused illumination arises quite naturally is in the recording of holograms of three-dimensional objects, as first successfully accomplished by Leith and Upatnieks [13]. Most three-dimensional scenes containing figurines, models, or more naturally occurring objects are composed primarily of surfaces that are rough on the scale of an optical wavelength. Thus the object itself serves to diffuse the reflected light, and the holograms formed from such scenes bear a strong resemblance to those recorded from diffusely illuminated transparencies. Fig. 6(a) illustrates a typical geometry used for recording holograms of three-dimensional objects, while Fig. 6(b) illustrates the means by which the virtual image would be viewed. Fig. 7(a) and (b) shows two photographs of the virtual image produced by such a hologram, illustrating the different views of the object obtained through different parts of the hologram. The dramatic nature of the three-dimensional images obtained from such holograms is largely responsible for the great popular interest in holography. However, as we shall see in Section IV, many of the important applications of holography are quite independent of this three-dimensional imaging capability.

III. VARIOUS TYPES OF HOLOGRAMS

In recent years, a wide variety of different types of holograms have been discussed in the literature, many of which have proved to be very useful and important from a practical point of view. Here we briefly review the properties of several different types of holograms; more detailed discussions of this material are available in the literature [15], [23], [48], [49].

A. Thin versus Thick Holograms

The importance of the third dimension, i.e., depth, of a holographic recording was recognized at an early date by Denisyuk [50] and by van Heerden [51]. Strictly speaking, a hologram may be considered a thin diffracting structure only if its optical thickness is less than a wavelength. In practice, however, it is not the relation of thickness to wavelength that influences the characteristics of the hologram, but rather the relationship between the thickness and the period of the finest fringe recorded on the hologram. If the finest fringe period is larger than the thickness, the hologram behaves essentially as a two-dimensional diffracting structure. If the fringe period is smaller than the thickness, then that fringe behaves as a three-dimensional diffracting structure. In practice, a given hologram contains many different fringe structures, some of which may be fine and some coarse, so the hologram may exhibit properties of both thin and thick structures.

Fig. 8 illustrates the formation of fringes in the volume of an emulsion for the particular case of plane object and reference waves. Regions of constructive and destructive interference move through the emulsion at an angle that bisects the angle Φ between the two interfering waves. The result is a grating structure that exists through the volume of the emulsion.

To discuss this phenomenon in greater depth, let the complex amplitudes of the object and reference waves in the emulsion be represented by

$$U_r(\boldsymbol{p}) = A \exp\{j\boldsymbol{k}_r \cdot \boldsymbol{p}\}$$

Fig. 9. Wave-vector diagram, hologram recording.

$$U_0(\boldsymbol{p}) = a \exp\{j\boldsymbol{k}_0 \cdot \boldsymbol{p}\} \qquad (25)$$

where $\boldsymbol{p}$ represents a position vector, while $\boldsymbol{k}_r$ and $\boldsymbol{k}_0$ are wave vectors of the reference and object waves. Both $\boldsymbol{k}_r$ and $\boldsymbol{k}_0$ have lengths $2\pi/\lambda$, where λ is again the wavelength of the light (in the emulsion). The intensity $I(\boldsymbol{p})$ of the light at each point $\boldsymbol{p}$ in the emulsion is given by

$$I(\boldsymbol{p}) = |U_r + U_0|^2 = A^2 + a^2 + 2Aa \cos\left[(\boldsymbol{k}_r - \boldsymbol{k}_0) \cdot \boldsymbol{p}\right]. \quad (26)$$

Thus we may specify an effective $\boldsymbol{K}$ vector for the fringes in the emulsion by the definition

$$\boldsymbol{K} \triangleq \boldsymbol{k}_r - \boldsymbol{k}_0. \qquad (27)$$

The direction of $\boldsymbol{K}$ indicates the normal to the fringes in the emulsion, while the length of $\boldsymbol{K}$ is related to the fringe period Λ (measured normally between two fringes) by

$$|\boldsymbol{K}| = \frac{2\pi}{\Lambda}. \qquad (28)$$

As shown in Fig. 9, which pictorially represents (27), the fact that $|\boldsymbol{k}_r| = |\boldsymbol{k}_0|$ implies that the direction of $\boldsymbol{K}$ is indeed such that the fringes run at an angle which bisects the angle Φ between the incident wave vectors. Simple trigonometry also shows that the fringe period Λ is given by

$$\Lambda = \frac{\lambda}{2 \sin \dfrac{\Phi}{2}}. \qquad (29)$$

A grating of thickness T may be regarded as "thin" if $\Lambda \gg T$, while it must be regarded as "thick" when $\Lambda \ll T$. For Kodak 649F plate, which is widely used in holography, the thickness ($\sim 16\,\mu$) and refractive index (~ 1.5) are such that a hologram behaves as a thin structure only for external reference-object angles of 10° or less.

For a thick hologram, the brightness of the reconstructed image is governed by Bragg diffraction, and is in general sensitive both to the angle at which the hologram is illuminated and to the wavelength of the illumination. A qualitative understanding of these dependences can be obtained from the following reasoning. A plane optical wave will be strongly reflected from a grating of *infinite* thickness only if the wave vector $\boldsymbol{k}_c$ of the illumination, the grating vector $\boldsymbol{K}$, and the wave vector $\boldsymbol{k}_i$ of the reflected light satisfy

$$\boldsymbol{k}_c - \boldsymbol{k}_i = \pm \boldsymbol{K} \qquad (30)$$

for only then will the scattered contributions from all depths of the grating add constructively. Neglecting the effects of emulsion shrinkage, the two most important ways this condition can be satisfied are those illustrated in Fig. 10(a) and (b). First, referring to Fig. 10(a), if the hologram is illuminated by a wave with wave vector $\boldsymbol{k}_c$ identically equal to the wave vector $\boldsymbol{k}_r$ of the original reference wave, then an image wave with wave vector $\boldsymbol{k}_i = \boldsymbol{k}_0$ will be generated. For a more general situation in which the object wave contains an entire family of wave-vector components, this illumination yields a virtual image of the object. On the other hand, as illustrated in Fig. 10(b), if the hologram is illuminated by a wave with wave vector $\boldsymbol{k}_c = -\boldsymbol{k}_r$, then an image wave with wave vector $\boldsymbol{k}_i = -\boldsymbol{k}_0$ will be generated. When the object wave contains a family of wave-vector components, this illumination yields a real image of the object.

469

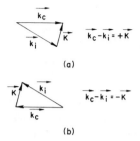

Fig. 10. Wave-vector diagrams for forming a virtual and a real image. (a) Virtual image. (b) Real image.

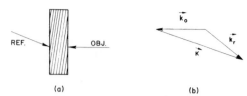

Fig. 12. A thick reflection hologram. (a) Recording geometry. (b) Wave-vector diagram.

Fig. 11. Grating vector components for a hologram of finite thickness.

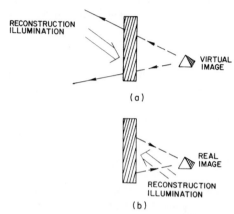

Fig. 13. Obtaining images from a thick reflection hologram. (a) Virtual image. (b) Real image.

In practice the grating is not infinitely thick, but rather has some finite thickness T. In this case we may regard the grating as containing an angular spectrum of grating-vector components, all with the same length $2\pi/\Lambda$, but distributed in direction over an approximate angle $\Delta\psi$, as shown in Fig. 11. When the bisector of the reference-object angle Φ is normal to the emulsion, we have that $\Delta\psi \cong \Lambda/T$, and using (29)

$$\Delta\psi \cong \frac{\lambda}{2T\sin\dfrac{\Phi}{2}}. \qquad (31)$$

Thus the angular spread of the grating vectors is smallest when Φ is near 180°.

As a consequence of the spread of the angular directions of grating vectors, exact Bragg alignment of the readout wave is not required. Rather it is only necessary that the wave vector k_c of the illumination be properly aligned with any one of the grating vector components. Rotational alignment of the hologram is found to be most critical when $\Phi \approx 90°$, while the wavelength selectivity is greatest when $\Phi \approx 180°$ and the reference and object waves enter from opposite sides of the plate, a condition we examine in Section III-B.

Because of the Bragg effect it is possible to utilize all three dimensions of the recording medium for information storage. Thus both angular and wavelength multiplexing of stored images are possible [33], [52], and very high information densities can be achieved. A general discussion of the information storage capabilities of thick holograms has been given by Gabor and Stroke [53].

B. Transmission versus Reflection Holograms

For the most common types of holograms, the images of interest are formed with transmitted light. It is also possible, however, to construct holograms that form images by means of reflected light. For example, if a hologram recorded on a silver halide material develops a relief pattern that is directly related to the original exposure (as often happens in practice [54]), then deposition of a metallized layer on the emulsion surface can create a highly reflective hologram from which images can be obtained [55]. Similar techniques can be applied for holograms recorded on thermoplastic materials [56].

A more complex type of reflection hologram can be produced if the recording medium is thick [57]. With reference to Fig. 12(a), let the object and reference waves be incident from opposite sides of

the emulsion, with an angular separation Φ approaching 180°. In this case the recorded fringes run nearly parallel to the emulsion surface, and from (29) we see that the fringe period Λ is approximately $\lambda/2$. The k-vector diagram appropriate for this case is illustrated in Fig. 12(b). For reconstruction of a virtual image, the Bragg condition requires that the hologram be illuminated by a duplication of the original reference wave, in which case the image appears in reflected light, as shown in Fig. 13(a). The geometry for reconstructing a real image is shown in Fig. 13(b), where it is clear that again the image is formed in reflected light.

The wavelength selectivity of a thick grating structure is highest when $\Phi \approx 180°$ [15], and as a consequence, the images formed by thick reflection holograms may be viewed with white light sources. In practice, the color of the reconstructed image is not the same as the color of the light used during recording, but rather is shifted towards shorter wavelengths by emulsion shrinkage that occurs during the fixing of the emulsion. This problem can be overcome by omitting the fixing step, or by reswelling the emulsion. Fig. 14 shows a photograph of the virtual image produced by a thick reflection hologram.

C. Amplitude Holograms

A hologram may be classified as an amplitude hologram if absorption is the primary mechanism by means of which stored information is transferred to the reconstructed optical field. The properties of such a hologram depend on whether it is thin or thick in the sense described in Section III-A.

A thin amplitude hologram is best described in terms of its amplitude transmittance, as was done in (2). Since holograms are really modulated diffraction gratings, we concentrate attention on the simple case of a sinusoidal amplitude grating, as described by

$$t_A(x, y) = \tfrac{1}{2}[1 + \cos 2\pi\alpha y]. \qquad (32)$$

Such an amplitude transmittance could in principle be obtained from a perfectly linear holographic material exposed to two equal-intensity plane waves. The field transmitted by such a hologram

Fig. 14. Photograph of the virtual image produced by a thick reflection hologram.

when it is illuminated by a unit amplitude plane wave contains three wave components,

$$U_c(x, y) = \tfrac{1}{2} + \tfrac{1}{4} \exp(j2\pi\alpha y) + \tfrac{1}{4} \exp(-j2\pi\alpha y). \qquad (33)$$

The intensity of the wave component $\tfrac{1}{4} \exp(j2\pi\alpha y)$, which leads to one of the two reconstructed images, is $\tfrac{1}{16}$. Since the hologram was illuminated by a unit intensity wave, only $\tfrac{1}{16}$ of the incident light contributes to the reconstructed image. Accordingly we say that the maximum diffraction efficiency of a thin amplitude hologram is $\tfrac{1}{16}$ or 6.25 percent. In practice, the perfect recording linearity implied by (32) cannot be achieved, and the contrast of the sinusoidal grating must be reduced to maintain linearity. Diffraction efficiencies somewhat less than 6.25 percent are therefore achieved in practice, typical numbers being in the range 1 to 2 percent.

A thick amplitude hologram formed by interference of two plane waves is described by an absorption coefficient β that varies sinusoidally with distance, i.e.,

$$\beta(x, y) = \beta_0 + \beta_1 \cos(2\pi\alpha y). \qquad (34)$$

An analysis by Kogelnik, using coupled mode theory [48], [49], has shown that the diffraction efficiency of this type of hologram depends on whether the hologram is made to be viewed with transmitted light or with reflected light. The maximum theoretical diffraction efficiency for a transmission hologram was found to be 3.7 percent, while that for a reflection hologram is 7.2 percent.

D. Phase Holograms

When the primary mechanism by which a hologram modulates an incident wave is either a change of dielectric constant or a change of physical thickness, the hologram is called a phase hologram. Because phase holograms can in principle be lossless diffracting structures, their diffraction efficiencies can be much higher than those of amplitude holograms. However, modulation of phase is intrinsically a nonlinear operation, and some care must be taken to avoid image degradations caused by nonlinearities.

For a thin phase hologram [58], [59], the most suitable description is again in terms of amplitude transmittance, as indicated previously in (3). Again treating the case of a simple sinusoidal grating, the amplitude transmittance is of the form

$$t_A(x, y) = \exp\{j\mu \sin 2\pi\alpha y\} \qquad (35)$$

where μ is the peak phase modulation amplitude. Expanding the exponential in a Fourier series, we find

$$t_A(x, y) = \sum_{q=-\infty}^{\infty} J_q(\mu) \exp(j2\pi q\alpha y) \qquad (36)$$

where J_q is a Bessel function of the first kind, order q. The $q=0$ component leads to an undiffracted component of transmitted light,

while the $q = \pm 1$ components correspond to the usual first-order images. Values of q greater than 1 and less than -1 lead to so-called "higher order images," which in general bear little resemblance to the original object, and therefore are of little interest, provided their angular separation from the primary images is sufficiently great to prevent overlap.

The diffraction efficiency of a thin phase grating is equal to $J_1^2(\mu)$, and has a maximum value of 33.9 percent, achieved when the modulation depth μ equals 1.8. However, if the object is more complicated than the simple plane wave implicitly assumed in writing (34), the choice of an rms modulation depth as large as 1.8 leads to an unacceptable degree of intermodulation distortion in the primary images. In practice, the maximum diffraction efficiency consistent with good image quality is in the range of 5 to 10 percent.

If a surface blaze occurs on a thin sinusoidal phase grating, diffraction efficiencies considerably in excess of 33.9 percent can be achieved, both in theory and in practice [60]. However, the blazed phase hologram is again intrinsically nonlinear [61], and for objects more complex than a single plane wave the achievable diffraction efficiency is severely limited by image distortions.

A thick phase hologram formed by interference of two plane waves is characterized by a refractive index ε that varies sinusoidally with space

$$\varepsilon = \varepsilon_0 + \varepsilon_1 \cos 2\pi\alpha y. \qquad (37)$$

Under the condition of Bragg alignment it is theoretically possible to achieve a diffraction efficiency of 100 percent from thick sinusoidal phase gratings of both the transmission and reflection types [48], [49]. In practice, diffraction efficiencies in the 60 to 70-percent range can typically be achieved with Kodak 649F plate. For the special case of a dichromated gelatin recording medium, diffraction efficiencies as high as 90 percent have been achieved [62], [63], but the storage mechanism is believed to involve surface cracking rather than the internal refractive index modulation of (37) [64].

When the object of interest is a complex diffuse subject rather than a simple plane wave, the theoretical results must be modified. Upatnieks and Leonard [65] have shown that, for a diffuse object with angular subtense smaller than the angle $\Delta\psi$ of (31), the speckle effect (i.e., the granularity of the diffusely reflected or diffusely transmitted coherent light) limits the theoretically achievable diffraction efficiency to 64 percent. However, if the angular subtense of the object is greater than $\Delta\psi$, this conclusion is no longer valid, and higher diffraction efficiencies are theoretically possible. In practice, diffraction efficiencies higher than 50 percent are seldom achieved with diffuse objects.

IV. Applications of Optical Holography

In this section, various applications of holography are discussed. Attention is restricted to optical holography, since microwave and acoustic holography are covered elsewhere in this issue.

A. Interferometry

In 1965 several groups of workers discovered more or less simultaneously that holography offers important new methods for testing and measurement through interferometry [66]–[72]. While a considerable variety of techniques for holographic interferometry exists, all rest on the ability of a hologram to store and regenerate both the amplitude and phase distributions of a complex optical disturbance.

Perhaps the simplest and most fundamental technique is that called "double-exposure" holographic interferometry [73]. Let a hologram be recorded with reference wave U_r, but let the exposure take place in two steps, first with an object wave U_{01} and second

Fig. 15.   Image obtained from a double-exposure hologram of a light bulb (courtesy of R. E. Brooks, L. O. Heflinger, and R. F. Wuerker).

with an object wave U_{02}. Since the exposures take place sequentially, the two interference patterns are superimposed independently on the emulsion, the effective intensity pattern being

$$I(x, y) = |U_r + U_{01}|^2 \tau_1 + |U_r + U_{02}|^2 \tau_2 \qquad (38)$$

where τ_1 and τ_2 are weighting factors representing the fractions of time devoted to each individual exposure. If the developed hologram is illuminated by a duplication of U_r, and if the usual conditions of linearity are satisfied, one component of the reconstructed field has the form

$$U'_c \propto |U_r|^2 (U_{01}\tau_1 + U_{02}\tau_2). \qquad (39)$$

Thus a linear superposition of the two object waves is generated, and these two waves will interfere.

There are several fundamental aspects of this type of interferometry that should be noted. Most important, it is a method of *differential* interferometry, for only changes of the object wave between exposures create interference fringes. If the object under study is completely stationary, and if all conditions of object illumination are the same for the two exposures, no interference fringes are seen in the reconstructed image, for U_{01} and U_{02} are identical. If, however, the object or the medium within which it is situated are perturbed between exposures, fringes are observed in the three-dimensional space of the holographic image. From this fringe structure, much information about the perturbations can be derived. The differential nature of this process allows interferometry to be performed with low quality optical elements, and even through highly inhomogeneous structures that would otherwise preclude the use of interferometric techniques. Fig. 15 shows the interference fringes generated by a hologram exposed before and after a common light bulb has been turned on. The fringes are generated by changes of the index of refraction caused by patterns of gas flow within the glass envelope.

Two additional important properties of double-exposure holographic intereometry are that it allows interference between two wave fields that existed at entirely different times, and even between two wave fields that have entirely different wavelengths.

The concept of double-exposure holographic interferometry is readily extended to multiple-exposure holographic interferometry, in which case three or more wave fields can be simultaneously reconstructed and caused to interfere. Generalization from multiple exposure to continuous exposure is also possible; continuous-exposure holography has found application particularly in the study of vibrating structures [66], [74].

Another variety of holographic interferometry is two-frequency holography, for which the object is illuminated by a source with two

separate frequencies and the image reconstructed with a single-frequency source. Each illumination frequency creates a separate and independent hologram; when the recording is properly illuminated, two separate wave fields are generated and interfere. Such techniques have been used for the generation of depth contours for profile measurement [68].

Finally, it is possible to perform "real-time" holographic interferometry by causing a holographic image to directly interfere with the object from which the hologram was made [70]. Deformations of the object can then be monitored in real time. Alternatively, real objects can be tested against an "ideal" object represented, for example, by a computer-generated hologram.

The reader interested in pursuing the topic of holographic interferometry will find a wealth of material in the recent optics literature (see, for example, [75], [76]).

B. High-Resolution Volume Imagery (Holographic Microscopy)

With a conventional imaging system, i.e., a system that uses lenses and/or mirrors as the image-forming elements, high transverse resolution is achieved only at the price of a limited depth of focus. Thus only a limited volume of object space can be recorded in a sharply focused image with a single photograph. On the other hand, with holography a single photographic recording can yield high-resolution images of a very large volume of object space. For example, if a hologram of a large object volume is properly illuminated, a real image of that entire object volume is formed behind the hologram. By sequentially examining the image at various distances from the hologram, the entire object volume can be searched with high transverse resolution.

The previously described property of holography has been usefully exploited in a number of applications, including the measurement of particle sizes in aerosols [77], the study of large volumes of living biological specimens [78], and in more general microscopy [79], [80]. This property is probably most important for the study of three-dimensional dynamic or transient phenomena with high resolution.

C. Imaging through Distorting Media

Because a hologram records information about the phase of an optical wave, several holographic methods for forming high-resolution images in the presence of wavefront distortions have proven possible. The first method of interest [81], [82] may be applied when the distorting medium is constant in time and movable in space. With reference to Fig. 16, suppose that a thin distorting medium with amplitude transmittance

$$t_A(\xi, \eta) = \exp\{jW(\xi, \eta)\} \qquad (40)$$

lies between the object and the recording plane. The wavefront immediately to the right of the distorting medium may be represented by $U_0(\xi, \eta) \exp\{jW(\xi, \eta)\}$, where $U_0(\xi, \eta)$ is the wavefront that would be present in the absence of the distorting medium. The distorted wavefront propagates to the recording plane, where a hologram is recorded. Let the hologram be illuminated by a wave traveling in a direction opposite to that of the reference wave, such that a real image is formed. Since the phase distribution in the real image is conjugate to that of a corresponding wave in the original object space, at the real image of the distorting medium there is a field distribution $U_0^*(\xi, \eta) \exp\{-jW(x, y)\}$. If the original distorting medium is now inserted to coincide with its real image, the field transmitted is

$$U_0^*(\xi, \eta) \exp\{-jW(\xi, \eta)\} \exp\{jW(\xi, \eta)\} = U_0^*(\xi, \eta). \qquad (41)$$

Thus the effect of the distorting medium has been cancelled, and an undistorted real image is recovered.

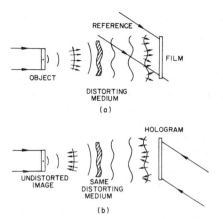

Fig. 16. Holographic imaging through a distorting medium. (a) Recording the hologram. (b) Recovering the image.

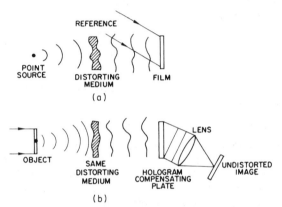

Fig. 17. Recording (a) and using (b) a hologram compensating plate.

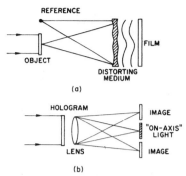

Fig. 18. Holographic imaging through a distorting medium, reference, and object waves both passing through the distorting medium. (a) Recording the hologram. (b) Obtaining images.

This method clearly has potential application to the problem of storing information in a secure fashion, such that only authorized individuals (i.e., those with the correct "decoding plate") can have access to it. However, the very stringent requirements regarding precise positioning of the decoding plate pose serious practical difficulties. A somewhat simpler problem to which the idea has been successfully applied by Loth and Collins [83] is the removal of aberrations in low quality microscope objectives, the same objective being used both in recording and in reconstruction.

A second method of interest is illustrated in Fig. 17. In this case the distorting medium may be immovable, but it must be essentially unchanging in time. A hologram of a point-source object is first recorded through the distorting medium. If the wavefront incident on this hologram is $\exp\{jW(x, y)\}$, then the portion of the hologram transmittance that normally contributes to a real image is $\exp\{-jW(x, y)\}$. If a more general object is now to be viewed through the distorting medium, each object point generates a wavefront with distortions of the form $\exp\{jW(x, y)\}$. If the object wave passes through the hologram, and if the light diffracted by the "real-image" term of the hologram transmittance is viewed, the wavefront distortions are cancelled, allowing an undistorted image to be formed by a conventional imaging system. Thus the hologram has served as a "compensating plate." This technique has been applied to the compensation of lens aberrations with some success [84].

A third and final technique, which may be applied to movable or immovable and time-invariant or time-varying distortions is illustrated in Fig. 18. In this case both the reference and object waves pass through the distorting medium [85]. If the reference and object are not too widely separated, the distortions of both the object and reference wavefronts are identical. Interference of the two identically distorted wavefronts yields a hologram that is free from distortions

and from which a high quality image can be obtained. Application of this method to the problem of imaging through atmospheric inhomogeneities has been investigated [86], [87], and some improvement of image quality has been observed with holography over long paths through the atmosphere [88].

D. Holograms as Optical Elements

Holograms have been found useful in a number of specialized applications as optical elements, playing the role of specialized lenses or gratings. A hologram of a point-source object is basically a focusing element which in many respects behaves like a lens. The obvious advantages of a hologram in this respect are that it can be more lightweight and compact than a lens. A less obvious advantage in some applications follows from the fact that holograms can be physically overlapped in the emulsion, whereas two lenses cannot physically overlap. Thus a holographic "fly's-eye" lens can have elements that are larger than the element-to-element spacing, and therefore can achieve higher resolution than might otherwise be possible. A chief disadvantage of holographic elements of this kind is their very large chromatic aberration.

Holographic lenses have found application as focusing elements in the page composer of a holographic memory [89]. Holographic gratings have been used both for spectral analysis [90] and for coupling light waves into thin films [91]. In addition, a holographic element has been used to produce line scanning in a unique scanner [92].

E. Holograms in Coherent Optical Data Processing Systems

While a complete and detailed review of the broad field of coherent optical data processing is beyond the scope of this paper, some attention will be devoted to the important role holographic techniques can play in such systems. More detailed reviews of this field can be found in the literature [6], [93]–[96].

Coherent optical data processing systems depend on the ability of simple positive (i.e., converging) lens to perform a two-dimensional Fourier transformation of a coherent field distribution impressed across its front focal plane. More specifically, if a complex amplitude distribution $U_0(x_0, y_0)$ is impressed across the front focal plane (coordinates x_0, y_0) of a positive lens of focal length f, then across the rear focal plane (coordinates x_f, y_f) there will appear a field distribution

$$U_f(x_f, y_f) = \frac{1}{\lambda f} \int\int_{-\infty}^{\infty} U_0(x_0, y_0)$$

$$\cdot \exp\left\{-j\frac{2\pi}{\lambda f}(x_0 x_f + y_0 y_f)\right\} dx_0\, dy_0. \quad (42)$$

If two such lenses are separated by a distance $2f$, as shown in Fig. 19, then the lens L_1 displays the Fourier spectrum of the input

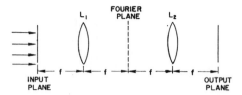

Fig. 19.   A simple coherent optical data processing system.

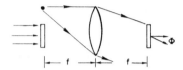

Fig. 20.   Recording an interferometrically generated spatial filter.

transparency in the Fourier plane, where the spectral components can be modified in amplitude and/or phase, and the lens L_2 performs a second transformation to yield a linearly filtered image of the original input. If a linear filtering operation with two-dimensional transfer function $H(f_X, f_Y)$ is desired, this operation can be realized by placing in the Fourier plane a transparency with amplitude transmittance proportional to H, i.e.,

$$t_A(x_f, y_f) \propto H\left(\frac{x_f}{\lambda f}, \frac{y_f}{\lambda f}\right). \qquad (43)$$

A major practical problem encountered in the realization of such filtering systems arises from the necessity to control *both* the amplitude and the phase transmission through the focal plane with complicated but related distributions. Major advances in this regard were made by Kozma and Kelly [97] by introducing the use of carrier frequency transparencies and by Vander Lugt [98] by introducing the interferometrically generated (or holographic) frequency-plane filter. If it is desired to synthesize an optical filtering system with a particular impulse response h, the required frequency plane filter may be generated with the recording system illustrated in Fig. 20. A transparency with amplitude transmittance proportional to h (which is often of simple form in practice) is introduced in the front focal plane of the transforming lens. Incident on the photographic emulsion in the rear focal plane of the lens is the sum of a plane wave inclined at angle Φ to the normal (analogous to the reference wave of holography) and a wave with amplitude distribution proportional to $H(x_f/\lambda f, y_f/\lambda f)$, where H is the desired transfer function. Assuming that the amplitude transmittance of the developed transparency can be made proportional to the intensity of the exposing light, we have

$$t_A(x_f, y_f) \propto \left| \exp\left(-j2\pi\alpha y_f\right) + H\left(\frac{x_f}{\lambda f}, \frac{y_f}{\lambda f}\right) \right|^2$$

$$= 1 + |H|^2 + H \exp\left(j2\pi\alpha y_f\right) + H^* \exp\left(-j2\pi\alpha y_f\right) \quad (44)$$

where $\alpha = \sin \Phi/\lambda$. If this transparency is inserted in the frequency plane of the coherent processing system of Fig. 19, at the output several different distributions of light appear. About the optical axis there appears an image of the input transparency which has been filtered by a transfer function $1 + |H|^2$. This output term is generally not of interest. Deflected to opposite sides of the output plane are two additional images, one filtered by transfer function H and the second by transfer function H^*. If the spatial frequency α is chosen sufficiently high, the three distributions are physically separated in space, and by choosing the proper region of the output plane the experimenter can realize a transfer function H or H^*, as he may desire.

The significance of the holographic frequency-plane filter is that, due to the spatial carrier frequency introduced, it allows effective control over both the amplitude and the phase of the transfer function by means of a purely absorbing frequency-plane transparency. True simultaneous control of amplitude and phase transmission through the frequency plane is therefore not required when this type of filter is used. This property has led to a considerable extension of the types of problems to which coherent optical data processing systems can be applied.

An important byproduct of the interferometric approach to filter generation is the presence of the term H^* in (44). Since this is precisely the transfer function of a filter "matched" to the signal h [99], matched filter detection and recognition systems can readily be synthesized optically [98], [100]. In addition, by sandwiching an interferometrically generated filter transparency with a second transparency with amplitude transmittance proportional to $|H|^{-2}$, a new filter can readily be generated [101] such that its transfer function G is given by

$$G = \frac{H^*}{|H|^2} = \frac{1}{H}. \qquad (45)$$

Thus G is an inverse filter and may be used, for example, for image deblurring [102].

F. Holographic Data Storage and Retrieval Systems

Of the various applications of holography that have been proposed, those with the greatest potential for wide-spread commercial use are undoubtedly in the data storage field. Examples include the use of holography for the storage of TV program material in vinyl tape cassettes [27], holographic read-only memories for storage of digital data [103]–[105], holographic storage of consumer credit status for credit verification purposes [106], and the use of holography in a microfiche system [107].

There are a multitude of reasons for considering holography as a means for data storage, particularly for high-density storage. Most widely mentioned, perhaps, is the highly redundant nature of a holographic recording. If a portion of the hologram is obliterated by a dust speck or a scratch, there need be no localized loss of information, for the entire image can still be recovered, albeit with some small loss of resolution. For applications requiring the storage and recovery of visual data, this type of protection can be extremely valuable. For applications involving the storage of digital data, similar protection can be provided by direct image storage using error correcting codes [108], probably with more efficient use of the resolvable spots available on the recording medium. However, the advantage of holography in this case lies with the extreme simplicity of the "decoding" operations: the stored data are recovered at the speed of light with no decoding computations required.

A second important advantage of holography, particularly in a high-density storage system, comes from the ability to obtain high magnification of the image data without a corresponding magnification of the registration errors. This property is most pronounced when the hologram is recorded using a reference point source that lies coplanar with the object data to be stored. For this so-called "lensless Fourier transform" recording geometry, the image obtained from the hologram suffers absolutely no motion as the hologram is translated in the reading beam, yet the magnification can be high in the sense that a large image can be obtained from a tiny hologram. Closely related to this insensitivity to translational movement is an insensitivity to the movement of the hologram along the optical axis, a property not shared by conventional imaging systems operating with similar magnifications.

With respect to the storage of digital data in a read-only holographic memory, a page-organized memory storing 10^8 bits with an access time of 1 μs to a single page of 10^4 bits appears feasible [105]. Efforts are underway to attempt to find suitable materials for a read-write memory [29], [31], but at this early stage it is difficult to judge the likelihood of success in this regard. Mention should also be made of a recent proposal for an associative memory based on holographic techniques [109].

Ultimately, of course, the success of holographic data storage systems will rest on economic questions. Can the data be stored holographically with less cost than offered by an alternative approach with similar performance? At present there are considerable grounds for optimism in this regard. However, the future will depend not only on advances in holography but also on advances in other competitive areas of technology.

REFERENCES

[1] D. Gabor, "A new microscope principle," *Nature*, vol. 161, May 1948, pp. 777–778.

[2] ——, "Microscopy by reconstructed wavefronts," *Proc. Roy. Soc., Ser. A*, vol. 197, July 1949, pp. 454–487.

[3] ——, "Microscopy by reconstructed wavefronts: II," *Proc. Phys. Soc. London* (Gen.), vol. 64, June 1951, pp. 449–469.

[4] D. Gabor and W. P. Goss, "Interference microscope with total wavefront reconstruction," *J. Opt. Soc. Am.*, vol. 56, July 1966, pp. 849–858.

[5] E. N. Leith and J. Upatnieks, "Reconstructed wavefronts and communication theory," *J. Opt. Soc. Am.*, vol. 52, Oct. 1962, pp. 1123–1130.

[6] L. J. Cutrona, E. N. Leith, C. J. Palermo, and L. J. Porcello, "Optical data processing and filtering systems," *IRE Trans. Inform. Theory*, vol. IT-6, June 1960, pp. 386–400.

[7] L. J. Cutrona, E. N. Leith, L. J. Porcello, and W. E. Vivian, "On the application of coherent optical processing techniques to synthetic-aperture radar," *Proc. IEEE*, vol. 54, Aug. 1966, pp. 1026–1032.

[8] E. N. Leith, "Optical processing techniques for simultaneous pulse compression and beam sharpening," *IEEE Trans. Aerosp. Electron. Syst.*, vol. AES-4, Nov. 1968, pp. 879–885.

[9] E. N. Leith and A. L. Ingalls, "Synthetic antenna data processing by wavefront reconstruction," *Appl. Opt.*, vol. 7, Mar. 1968, pp. 539–544.

[10] R. O. Harger, *Synthetic Aperture Radar Systems: Theory and Design*. New York: Academic Press, 1970.

[11] A. Lohmann, "Optical single-sideband transmission applied to the Gabor microscope," *Opt. Acta*, vol. 3, June 1956, pp. 97–99.

[12] E. N. Leith and J. Upatnieks, "Wavefront reconstruction with continuous-tone objects," *J. Opt. Soc. Am.*, vol. 53, Dec. 1963, pp. 1377–1381.

[13] ——, "Wavefront reconstruction with diffused illumination and three-dimensional objects," *J. Opt. Soc. Am.*, vol. 54, Nov. 1964, pp. 1295–1301.

[14] ——, "Photography by laser," *Sci. Amer.*, vol. 212, June 1965, pp. 24–35.

[15] E. N. Leith, A. Kozma, J. Upatnieks, J. Marks, and N. Massey, "Holographic data storage in three-dimensional media," *Appl. Opt.*, vol. 5, Aug. 1966, pp. 1303–1311.

[16] G. W. Stroke, *An Introduction to Coherent Optics and Holography*. New York: Academic Press, 1966.

[17] J. B. De Velis and G. O. Reynolds, *Theory and Applications of Holography*. Reading, Mass.: Addison-Wesley, 1967.

[18] H. M. Smith, *Principles of Holography*. New York: Wiley, 1969.

[19] H. J. Caulfield and S. Lu, *The Applications of Holography*. New York: Wiley, 1970.

[20] R. J. Collier, L. Lin, and C. Burckhardt, *Optical Holography*. New York: Academic Press, 1971.

[21] R. J. Collier, "Some current views on holography," *IEEE Spectrum*, vol. 3, July 1966, pp. 67–74.

[22] E. N. Leith and J. Upatnieks, "Recent advances in holography," in *Progress in Optics*, vol. 11, E. Wolf, Ed. Amsterdam: North-Holland, 1967.

[23] J. C. Urbach and R. W. Meir, "Properties and limitations of hologram recording materials," *Appl. Opt.*, vol. 8, Nov. 1969, pp. 2269–2281.

[24] A. W. Lohmann, "Reconstruction of vectorial wavefronts," *Appl. Opt.*, vol. 4, Dec. 1965, pp. 1667–1668.

[25] H. W. Rose, T. L. Williamson, and S. A. Collins, Jr., "Polarization effects in holography," *Appl. Opt.*, vol. 9, Oct. 1970, pp. 2394–2396.

[26] C. B. Burckhardt, "Diffraction of a plane wave at a sinusoidally stratified dielectric grating," *J. Opt. Soc. Am.*, vol. 56, Nov. 1966, pp. 1502–1509.

[27] R. Bartoline, W. Hannan, D. Karlsons, and M. Lurie, "Embossed hologram motion pictures for television playback," *Appl. Opt.*, vol. 9, Oct. 1970, pp. 2283–2290.

[28] J. C. Urbach and R. W. Meier, "Thermoplastic xerographic holography," *Appl. Opt.*, vol. 5, Apr. 1966, p. 666–667.

[29] L. H. Lin and H. L. Beauchamp, "Write-read-erase in situ optical memory using thermoplastic holograms," *Appl. Opt.*, vol. 9, Sept. 1970, pp. 2088–2092.

[30] H. R. Farrah, E. Marom, and R. K. Mueller, "An underwater viewing system using sound holography," in *Acoustical Holography*, vol. 2, A. Metherell and L. Larmore, Eds. New York: Plenum, 1970.

[31] R. S. Mezrich, "Magnetic holography," *Appl. Opt.*, vol. 9, Oct. 1970, pp. 2275–2279.

[32] J. A. Jenny, "Holographic recording with photopolymers," *J. Opt. Soc. Am.*, vol. 60, Sept. 1970, pp. 1155–1161.

[33] A. A. Friesem and J. L. Walker, "Thick absorption recording media in holography," *Appl. Opt.*, vol. 9, Jan. 1970, pp. 201–214.

[34] R. W. Meier, "Magnification and third-order aberrations in holography," *J. Opt. Soc. Am.*, vol. 55, Aug. 1965, pp. 987–992.

[35] E. N. Leith, J. Upatnieks, and K. A. Haines, "Microscopy by wavefront reconstruction," *J. Opt. Soc. Am.*, vol. 55, Aug. 1965, pp. 981–986.

[36] J. A. Armstrong, "Fresnel holograms: Their imaging properties and aberrations," *IBM J. Res. Develop.*, vol. 9, May 1965, pp. 171–178.

[37] E. B. Champagne, "Nonparaxial imaging magnification and aberration properties in holography," *J. Opt. Soc. Am.*, vol. 57, Jan. 1967, pp. 51–55.

[38] J. R. Klauder, A. C. Price, S. Darlington, and W. J. Alberscheim, "The theory and design of chirp radars," *Bell Syst. Tech. J.*, vol. 39, July 1960, pp. 745–808.

[39] A. W. Rihaczek, *Principles of High-Resolution Radar*. New York: McGraw-Hill, 1969.

[40] G. L. Rogers, "Gabor diffraction microscopy: The hologram as a generalized zone-plate," *Nature*, vol. 166, Aug. 1950, p. 237.

[41] W. Kock, *Lasers and Holography*. Garden City, N. Y.: Doubleday, 1968.

[42] H. M. A. El-Sum, "Reconstructed wavefront microscopy," Ph.D. dissertation, Dept. of Physics, Stanford University, Stanford, Calif., 1952 (available from University Microfilm, Inc., Ann Arbor, Mich.).

[43] J. D. Rigden and E. I. Gordon, "The granularity of scattered optical maser light," *Proc. IRE* (Corresp.), vol. 50, Nov. 1962, pp. 2367–2368.

[44] P. S. Consodine, "Effects of coherence on imaging systems," *J. Opt. Soc. Am.*, vol. 56, Aug. 1966, pp. 1001–1009.

[45] E. N. Leith and J. Upatnieks, "Imagery with pseudo-randomly diffused coherent illumination," *Appl. Opt.*, vol. 7, Oct. 1968, pp. 2085–2089.

[46] H. J. Gerritsen, W. J. Hannan, and E. G. Ramberg, "Elimination of speckle noise in holograms with redundancy," *Appl. Opt.*, vol. 7, Nov. 1968, pp. 2301–2311.

[47] C. B. Burckhardt, "Use of a random phase mask for the recording of Fourier transform holograms of data masks," *Appl. Opt.*, vol. 9, Mar. 1970, pp. 695–700.

[48] H. Kogelnik, "Response and efficiency of five hologram types," in *Modern Optics*, J. Fox, Ed. Brooklyn, N. Y.: Polytechnic Press, 1967.

[49] ——, "Coupled wave theory for thick hologram gratings," *Bell Syst. Tech. J.*, vol. 48, Nov. 1969, pp. 2909–2947.

[50] Y. N. Denisyuk, "Photographic reconstruction of the optical properties of an object in its own scattered radiation field," *Sov. Phys.—Dokl.*, vol. 7, Dec. 1962, pp. 543–545.

[51] P. J. van Heerden, "A new optical method of storing and retrieving information," *Appl. Opt.*, vol. 2, Apr. 1963, pp. 387–392.

[52] L. H. Lin, K. S. Pennington, G. W. Stroke, and A. E. Labeyrie, "Multicolor holographic image reconstruction with white-light illumination," *Bell Syst. Tech. J.*, vol. 45, Apr. 1966, pp. 659–660.

[53] D. Gabor and G. W. Stroke, "The theory of deep holograms," *Proc. Roy. Soc., Ser. A*, vol. 304, Apr. 1968, pp. 275–289.

[54] H. M. Smith, "Photographic relief images," *J. Opt. Soc. Am.*, vol. 58, Apr. 1968, pp. 533–539.

[55] A. K. Rigler, "Wavefront reconstruction by reflection," *J. Opt. Soc. Am.*, vol. 55, Dec. 1965, p. 1693.

[56] J. C. Urbach, private communication.

[57] G. W. Stroke and A. Labeyrie, "White-light reconstruction of holographic images using the Lippmann-Bragg diffraction effect," *Phys. Lett.*, vol. 20, Mar. 1966, pp. 368–370.

[58] W. T. Cathey, Jr., "Three-dimensional wavefront reconstruction using a phase hologram," *J. Opt. Soc. Am.*, vol. 55, Apr. 1965, p. 457.

[59] G. L. Rogers, "Experiments in diffraction microscopy," *Proc. Roy. Soc., Ser. A* (Edinburgh), vol. 63, Feb. 1952, pp. 193–221.

[60] N. K. Sheridon, "Production of blazed hologram," *Appl. Phys. Lett.*, vol. 12, May 1968, pp. 316–320.

[61] D. Kermisch, "Wavefront-reconstruction mechanism in blazed holograms," *J. Opt. Soc. Am.*, vol. 60, June 1970, pp. 782–786.

[62] T. A. Shankoff, "Phase holograms in dichromated gelatin," *Appl. Opt.*, vol. 7, Oct. 1968, pp. 2101–2105.

[63] L. H. Lin, "Hologram formation in hardened dichromated gelatin films," *Appl. Opt.*, vol. 8, May 1969, pp. 963–966.

[64] R. K. Curran and T. A. Shankoff, "The mechanism of hologram formation in dichromated gelatin," *Appl. Opt.*, vol. 9, July 1970, pp. 1651–1657.

[65] J. Upatnieks and C. Leonard, "Efficiency and image contrast of dielectric holograms," *J. Opt. Soc. Am.*, vol. 60, Mar. 1970, pp. 297–305.

[66] R. L. Powell and K. A. Stetson, "Interferometric vibration analysis by wavefront reconstruction," *J. Opt. Soc. Am.*, vol. 55, Dec. 1965, pp. 1593–1598.

[67] R. E. Brooks, L. O. Heflinger, and R. F. Wuerker, "Interferometry with a holographically reconstructed comparison beam," *Appl. Phys. Lett.*, vol. 7, Nov. 1965, pp. 248–249.

[68] B. P. Hildebrand and K. A. Haines, "Multiple-wavelength and multiple-source holography applied to contour generation," *J. Opt. Soc. Am.*, vol. 57, Feb. 1967, pp. 155–162.

[69] J. M. Burch, "The application of lasers in production engineering," *Prod. Eng.*, vol. 44, Sept. 1965, pp. 431–442.

[70] R. J. Collier, E. T. Doherty, and K. S. Pennington, "Application of moire techniques to holography," *Appl. Phys. Lett.*, vol. 7, Oct. 1965, pp. 223–225.

[71] M. H. Horman, "Application of wavefront reconstruction to interferometry," *J. Opt. Soc. Am.*, vol. 55, May 1965, p. 615.

[72] D. Gabor, G. W. Stroke, R. Restrick, A. Funkhouser, and D. Brumm, "Optical image synthesis (complex amplitude addition and subtraction) by holographic Fourier transformation," *Phys. Lett.*, vol. 18, Aug. 1965, pp. 116–118.

[73] L. O. Heflinger, R. F. Wuerker, and R. E. Brooks, "Holographic interferometry," *J. Appl. Phys.*, vol. 37, Feb. 1966, pp. 642–649.

[74] C. C. Aleksoff, "Temporally modulated holography," *Appl. Opt.*, vol. 10, June 1971, pp. 1329–1341.

[75] E. R. Robertson and J. M. Harvey, Eds., *The Engineering Uses of Holography*. Cambridge: Cambridge University Press, 1970.

[76] B. Ragent and R. M. Brown, Eds., "Holographic instrumentation applications," NASA Rep. SP-248, 1970.

[77] B. J. Thompson, J. H. Ward, and W. R. Zinky, "Application of hologram techniques for particle size analysis," *Appl. Opt.*, vol. 6, Mar. 1967, pp. 519–526.

[78] C. Knox, "Holographic microscopy as a technique for recording dynamic microscopic subjects," *Science*, vol. 153, Aug. 1966, pp. 989–990.

[79] R. F. van Ligten, "Holographic microscopy," in *Holography Seminar Proc.* Redondo Beach, Calif.: Soc. Photo-Optical Instrum. Engineers, 1968.

[80] M. E. Cox, R. G. Buckles, and D. Whitlow, "Cineholomicroscopy of small animal microcirculation," *Appl. Opt.*, vol. 10, Jan. 1971, pp. 128–131.

[81] E. N. Leith and J. Upatnieks, "Holographic imagery through diffusing media," *J. Opt. Soc. Am.*, vol. 56, Apr. 1966, p. 523.

[82] H. Kogelnik, "Holographic image projection through inhomogeneous media," *Bell Syst. Tech. J.*, vol. 44, Oct. 1965, pp. 2451–2455.

[83] L. Toth and S. A. Collins, "Reconstruction of a three-dimensional microscopic sample using holographic techniques," *Appl. Phys. Lett.*, vol. 13, July 1968, pp. 7–9.

[84] J. Upatnieks, A. Vander Lugt, and E. Leith, "Correction of lens aberrations by means of holograms," *Appl. Opt.*, vol. 5, Apr. 1966, pp. 589–593.

[85] J. W. Goodman, W. H. Huntley, Jr., D. W. Jackson, and M. Lehmann, "Wavefront-reconstruction imaging through random media," *Appl. Phys. Lett.*, vol. 8, June 1966, pp. 311–313.

[86] J. D. Gaskill, "Imaging through a randomly inhomogeneous medium by wavefront reconstruction," *J. Opt. Soc. Am.*, vol. 58, May 1968, pp. 600–608.

[87] ——, "Atmospheric degradation of holographic images," *J. Opt. Soc. Am.*, vol. 59, Mar. 1969, pp. 308–318.

[88] J. W. Goodman, D. W. Jackson, M. Lehmann, and J. Knotts, "Experiments in long-distance holographic imagery," *Appl. Opt.*, vol. 8, Aug. 1969, pp. 1581–1586.

[89] W. C. Stewart and L. S. Cosentino, "Optics for a read-write holographic memory," *Appl. Opt.*, vol. 9, Oct. 1970, pp. 2271–2275.

[90] J. Cordelle, J. Flammand, G. Pieuchard, and A. Labeyrie, "Aberration-corrected concave gratings made holographically," in *Optical Instruments and Techniques*, J. H. Dickson, Ed. Newcastle-on-Tyne, England: Oriel Press, 1970.

[91] H. Kogelnik and T. P. Sosnowski, "Thin-film coupling with holographic Bragg gratings," *J. Opt. Soc. Am.*, vol. 60, Nov. 1970, p. 1543.

[92] D. H. McMahon, A. R. Franklin, and J. B. Thaxter, "Light beam deflection using holographic scanning techniques," *Appl. Opt.*, vol. 8, Feb. 1969, pp. 399–402.

[93] E. L. O'Neill, "Spatial filtering in optics," *IRE Trans. Inform. Theory*, vol. IT-2, June 1956, pp. 56–65.

[94] A. B. Vander Lugt, "A review of optical data-processing techniques," *Opt. Acta*, vol. 15, Jan.–Feb. 1968, pp. 1–34.

[95] A. R. Shulman, *Optical Data Processing.* New York: Wiley, 1970.

[96] J. W. Goodman, *Introduction to Fourier Optics.* New York: McGraw-Hill, 1968, ch. 7.

[97] A. Kozma and D. L. Kelly, "Spatial filtering for detection of signals submerged in noise," *Appl. Opt.*, vol. 4, Apr. 1965, pp. 387–392.

[98] A. Vander Lugt, "Signal detection by complex spatial filtering," *IEEE Trans. Inform. Theory*, vol. IT-10, Apr. 1964. pp. 139–145.

[99] G. L. Turin, "An introduction to matched filters," *IRE Trans. Inform. Theory*, vol. IT-6, June 1960, pp. 311–329.

[100] A. B. Vander Lugt, F. B. Rotz, and A. Klooster, Jr., "Character reading by optical spatial filtering," in *Optical and Electro-Optical Information Processing*, J. T. Tippett, D. A. Berkowitz, L. C. Clapp, C. J. Koester, and A. Vanderburg, Jr., Eds. Cambridge, Mass.: M.I.T. Press, 1965, ch. 7.

[101] G. W. Stroke, "A new holographic method for a posteriori image-deblurring restoration of ordinary photographs using 'extended-source' lensless Fourier-transform holography compensation," *Phys. Lett.*, vol. 27a, Aug. 1968, pp. 405–406.

[102] ——, "Image deblurring and aperture synthesis using a posteriori processing by Fourier transform holography," *Opt. Acta*, vol. 16, July–Aug. 1969, pp. 401–422.

[103] J. A. Rajchman, "Promise of optical memories," *J. Appl. Phys.*, vol. 41, Mar. 1970, pp. 1376–1383.

[104] ——, "An optical read-write mass memory," *Appl. Opt.*, vol. 9, Oct. 1970, pp. 2269–2271.

[105] L. K. Anderson, "High capacity holographic optical memory," *Microwaves*, vol. 9, Mar. 1970, pp. 62–66.

[106] E. H. Chrysty and K. K. Sutherlin, "Validating credit cards using holography," *1970 IEEE Int. Conv. Dig.* (New York, March 23–26), pp. 338–339.

[107] A. A. Friesem, E. N. Tompkins, and G. E. Hoffmann, "Holographic application in high density document storage and retrieval," presented at 15th Ann. Technical Symp. Society of Photo-Optical Instrum. Engineers, Los Angeles, Calif., Sept. 1970.

[108] I. B. Oldham, R. T. Chien, and D. T. Tang, "Error detection and correction in a photo-digital memory system," *IBM J. Res. Develop.*, vol. 12, Nov. 1968, pp. 422–430.

[109] M. Sakaguchi, N. Nishida, and T. Nemoto, "A new associative memory system utilizing holography," *IEEE Trans. Comput.*, vol. C-19, Dec. 1970, pp. 1174–1181.

Part IX: Laser Book List

The comprehensive annotated book list on lasers and related topics presented in this final section is maintained in machine-readable form at Stanford University. The books are classified under the following twelve headings.

A. *Introductory-level books*, intended for high school and general nontechnical readers.

B. *Advanced-level books*, giving a general coverage of lasers.

C. *Textbooks* on lasers and quantum electronics, intended as classroom texts, with problems and exercises.

D. *Reference handbooks and review/progress series*.

E. *Reprint collections, lecture series, symposium proceedings*, other multiauthor collections.

F. Books on *Gas lasers* specifically.

G. Books on *Semiconductor injection lasers* specifically.

H. *Nonlinear optics and nonlinear optical devices*.

I. *Holography and optical data processing*.

J. *Laser applications*, other than those already listed above.

K. *Miscellaneous special topics:* masers, coherence theory, glass lasers, resonators.

L. *Foreign-language books* on all of the above topics (where "foreign" has the chauvinistic meaning of any language other than English).

Additions and corrections to this list are invited. Please address them to the attention of Prof. A. E. Siegman, Microwave Laboratory, Stanford University, Stanford, Calif. 94305.

A. INTRODUCTORY-LEVEL BOOKS, FOR HIGH-SCHOOL AND GENERAL READERS

Beesley, M. J., 'Applications of Lasers' (Wykeham Technological Series, Wykeham Publications, London, 1970). 150 pp., 25 shillings (paper). A paperback introduction to lasers and their applications.

Brotherton, M., 'Masers and Lasers: How They Work, What They Do' (McGraw-Hill, New York, 1964). 224 pp., $8.50. Nontechnical descriptions for the non-scientist layman.

Brown, Ronald, 'Lasers: Tools of Modern Technology' (Doubleday Science Series, Doubleday and Co., Inc., 1968). 192 pp., $2.45 (paper). A very well-illustrated and up-to-date description for the general reader of lasers and particularly their important applications. Strongly recommended for the lay reader, or as a supplementary book for students in laser courses.

Carroll, John M., 'The Story of the Laser' (E. P. Dutton & Co., New York, 1970). 213 pp., $5.95. Revised edition of a book originally published in 1964, providing a popularized nontechnical coverage of the laser for the layman.

Eaglesfield, C. C., 'Laser Light' (Macmillan and Co., London, and St. Martin's Press, New York, 1967). 200 pp., $6.00. An introductory account of what a laser is, how it works, and how laser light compares with ordinary light and radio waves, written in an informal chatty style with a minimum of mathematics.

Fishlock, David (ed.), 'A Guide to the Laser' (American Elsevier Publishing Co., Inc., New York, 1967). 163 pp., $8.50. A collection of 11 essays by British scientists for technically oriented laymen reviewing the historical backbround, important kinds, and (in some detail) the most promising applications of lasers. Recommended.

Heavens, O. S., 'Lasers' (Gerald Duckworth, London, 1971). 159 pp., 3.25 British pounds (hardbound), 1.25 British pounds (paper). A brief nonmathematical introduction to some of the physical and quantum concepts behind lasers, the properties of a few of the more common types of lasers, and some of the potential applications of lasers, intended for the reader "with only a limited knowledge of science".

Klein, H. A., 'Masers and Lasers' (Lippincott, Philadelphia and New York, 1963). 184 pp., $3.95 Nontechnical descriptions for the layman, including some human interest anecdotes concerning personalities in the field.

Kock, Winston, E., 'Lasers and Holography' (Doubleday Science Study Series, Doubleday and Co., Inc., New York, 1969). 120 pp., $1.25 paperback, $4.50 hardcover. An introduction to holography intended for young students or interested laymen without specialized technical backgrounds.

Leinwoll, Stanley, 'Understanding Lasers and Masers' (John F. Ryder, Publisher, Inc., New York, 1965). 88 pp., $2.75. A very elementary and now very dated paperback for semitechnical readers.

Lytel, A., 'ABC''s of Lasers and Masers' (Sams Photofact Publications, Indianapolis, 1963). 95 pp., $1.95. A few good illustrations and practical hints on ruby lasers, but otherwise very dated and limited in content.

Melia, T. P., 'An Introduction to Masers and Lasers' (Chapman and Hall London; Barnes and Noble, New York, 1967). 162 pp., $5.50. A short book intended as an elementary introduction to the subject for engineers and scientists. Not recommended.

Nehrich, R. B., Voran, G. I., and Dessel, N. F., 'Atomic Light, Lasers' (Sterling Publishing Co., 1968). 103 pp., $2.94. Nontechnical popularized description of lasers.

Patrusky, B., 'The Laser - Light That Never Was Before' (Dodd, Mead, and Co., New York, 1966). 128 pp., $3.23. Another brief nontechnical book for the layman. The text is as over-dramatic as the title, but the coverage is fairly complete and up-to-date, with a fair number of good pictures and diagrams.

Pierce, J. R., 'Quantum Electronics: The Fundamentals of Transistors and Lasers' (Doubleday Science Study Series, Garden City, 1966). 138 pp., $1.25. For the general reader, at a very introductory level.

Rodgers, G. L., 'Handbook of Gas Laser Experiments' (Iliffe Books, Ltd., London, 1970). 71 pp., 1.40 British pounds. Suggests and describes some simple gas laser experiments that can be done in an undergraduate physics laboratory.

Stehling, K. R., 'Lasers and Their Applications' (World, Cleveland, 1966). 201 pp., $6.00. Still another popularized description for the nontechnical reader.

Van Pelt, W. F., et al., 'Laser Fundamentals and Experiments' (U.S. Department of Commerce Clearinghouse, accession number PB-193 565, 1971). 117 pp., $3.00. A Public Health Service manual primarily for high school instructors which discusses laser fundamentals and safety, and laser experiments on scattering, absorption, reflection, refraction, polarization, coherency, diffraction and holography.

Wright, G., 'An Experimental Introduction to Lasers' (Wykeham Science Series, Wykeham Publications, London, 1970). 150 pp., 25 shillings (paper). An introduction to laser principles at the high school or undergraduate level by means of selected illustrative experiments demonstrating or using lasers.

B. ADVANCED-LEVEL BOOKS, GENERAL COVERAGE OF LASERS

Birnbaum, G., 'Optical Masers' (Academic Press, New York, 1964). 306 pp., $12.00. Starting as a review article for the "Advances in Electronics and Electron Physics" series, this expanded into a separate book. Fairly advanced and theoretical in tone, but also with an extensive bibliography and summaries of experimental results.

Fain, V. M. and Khanin, Ya. I., 'Quantum Electronics. Vol. 1: Basic Theory. Vol. 2: Maser Amplifiers and Oscillators' (MIT Press, Cambridge, 1969). Vol. 1: 314 pp., $16.50; Vol. 2: 312 pp., $16.50. Translation of a two-volume Soviet work originally published in 1965. Volume 1: heavily theoretical and quantum-mechanical analysis of stimulated and spontaneous emission and large-signal nonlinear effects. Volume 2: 40% on ammonia and microwave solid-state masers, 40% a rather outdated treatment of lasers, 20% a long appendix on the energy levels of paramagnetic crystals.

Haken, H., 'Licht und Materie Ic/Light and Matter Ic' (Springer-Verlag, Berlin, Heidelberg, New York). 288 pp., $8 vo. A comprehensive survey of the full quantum-mechanical, semiclassical, and rate-equation approaches to laser theory, together with theoretical treatments of optical resonators and quantum coherence.

Heavens, O. S., 'Optical Masers' (Methuen, London, 1964). 103 pp., $3.25. One of the Methuen's Monographs series. This is a brief but good introduction to fundamentals as well as a summary of important techniques. Recommended as a brief introduction to the field.

Lengyel, B. A., 'Introduction to Laser Physics' (Wiley, New York, 1966). 311 pp., $8.95. An expanded and updated version of an earlier book by Lengyel, giving a good introduction to the fundamentals of laser devices and laser physics.

Lengyel, Bela A., 'Lasers' (2nd ed.) (John Wiley & Sons, Inc., New York, 1971). 400 pp., $14.95. This latest edition of Lengyel's book (earlier versions were published in 1961 and 1966) gives an introduction to the fundamental principles of laser devices and laser physics, and a survey of most of the important types of laser devices currently known.

Maitland A. and Dunn, M. H., 'Laser Physics' (North Holland Publishing Co., Amsterdam-Holland, 1970). 425 pp., $21.00. Introduces the fundamental principles of laser physics, with emphasis on general principles more than on specific types of lasers, including the interaction of radiation and matter, optical resonators and beams, the Lamb laser theory, coherence theory, and a large number of appendices reviewing various background topics.

Mashkevitch, V. S., 'Laser Kinetics' (American Elsevier Publishing Co., New York, 1967). 235 pp., $15.00. English translation of a Russian monograph presenting a unified analysis of lasers, including semiconductor and Raman

lasers. Almost totally theoretical, with few illustrations and very little physical discussion.

Patek, Karel, 'Lasers' (The Chemical Rubber Co., Cleveland, Ohio, 1967). 288 pp., $12.50. This translation of the Czechoslovokian original is a fairly detailed and well-illustrated general account of laser theory and practice, with extensive references, intended for the specialist working with lasers or on their applications.

Ratner, A. M., 'Spectral, Spatial, and Temporal Properties of Lasers' (Plenum Press, New York, 1972). 211 pp., $19.50. A translation of a Russian book which treats the spatial and temporal mode properties of lasers in terms of a nonlinear wave equation.

Roess, Dieter, 'Lasers, Light Amplifiers and Oscillators' (Academic Press, London, 1969). 756 pp., $29.50. English translation of an extensive and detailed treatise on lasers by a well-known German expert on the subject. Recommended.

Smith, W. V. and Sorokin, P. P., 'The Laser' (McGraw-Hill, New York, 1966). 498 pp., $15.50. A general treatment of all the major types of lasers, together with a chapter on laser applications. The authors' statement in the Preface that the book "Treats the entire laser development from a unified point of view, with the depth necessary to be useful to a research worker ... and the clarity and background useful to a graduate student" is a generally accurate claim.

Thorp, J. S., 'Masers and Lasers: Physics and Design' (Macmillan, London, and St. Martin's Press, New York, 1967). 312 pp., $8.50. Intended as an account of the essentials of masers and lasers suitable for undergraduate study, as well as an introduction for post-graduate engineers and scientists. More than one-third of the text is devoted to microwave masers, the next third to a limited review of optical masers, and the final third to the preparation and evaluation of maser and laser crystals (chiefly ruby).

Vanier, Jacques, 'The Basic Theory of Lasers and Masers' (Gordon and Breach, New York, 1971). 127 pp., 4.75 British pounds (hardbound), 2.25 British pounds (paper). A monograph primarily discussing not lasers but rather the general quantum density matrix formalism and its application to various quantum problems, including the laser as one of the examples.

C. TEXTBOOKS ON LASERS AND QUANTUM ELECTRONICS

Chang, W. S. C., 'Principles of Quantum Electronics' (Addison-Wesley Publishing Co., Reading Mass., 1969). 540 pp., $17.50. A graduate text in quantum electronics which begins with a review of quantum mechanics, then discusses the theory of quantum energy levels of atoms and the interaction of radiation and matter in considerable detail, and then applies these theoretical concepts to the general analysis of lasers and laser action. Numerous appendices.

Pantell, R. H. and Puthoff, H. E., 'Fundamentals of Quantum Electronics' (John

Wiley and Sons, New York, 1969). 360 pp., $15.95. Gives a consistent quantum-mechanical theoretical treatment of quantum electronic processes (laser amplification, Raman and Brillouin effects, electrons in crystals, transition effects in semiconductors), using the density operator formalism with both semiclassical and quantized-field approaches.

Siegman, A. E., 'An Introduction to Lasers and Masers' (McGraw-Hill, New York, 1971). 530 pp., $17.50. A senior/first-year-graduate textbook in laser principles and laser devices for engineering and science students, following an electronic engineering approach with no quantum mechanics background required.

Unger, H. G., 'Introduction to Quantum Electronics' (Pergamon Press, Inc., Maxwell House, Fairview Park, Elmsford, New York, 1970) 183 pp., $6.50. Translation of a German textbook for students in physics and electronics, emphasizing the physical principles and the quantum theory of lasers and masers.

Vuylsteke, A., 'Elements of Maser Theory' (Van Nostrand, Princeton, 1960). 361 pp., $9.50. Presents practically a complete course in advanced physics, statistical mechanics, and quantum mechanics, before getting to maser theory. The notation sometimes becomes cumbersome, but the material gives a collection of much background physics in one place.

Yariv, A., 'Quantum Electronics' (Wiley, New York, 1966). 478 pp., $14.95. A graduate-level textbook giving an extensive coverage of the back-ground physics, together with device theory and some practical de- tails, for most of the major topics in lasers and masers and re-lated areas.

Yariv, A., 'Introduction to Optical Electronics' (Holt, Rinehart and Winston, New York, 1971). 342 pp., $17.00. A non-quantum-mechanical senior-level textbook giving a brief introductory coverage of topics in lasers, optical wave propa- gation and optical resonators, nonlinear optics, and light modulation and demodulation.

D. REFERENCE HANDBOOKS AND REVIEW/PROGRESS SERIES

Arecchi, F. T., and Schulz-DuBois, E. O. (eds.), 'Laser Handbook' (American Elsevier/North Holland Publishing Company, Amsterdam, 1972). Approx. 2000 pp., $160.00. A massive and carefully edited collection of 40 review articles by distinguished authors, largely European, covering nearly every aspect of lasers.

Ashburn, Edward V., 'Laser Literature - A Permutated Bibliography, 1958-1966' (Western Periodicals Co., North Hollywood, Calif., 1967). Vol. 1, 407 pp., vol. 2, 315 pp., $75.00 for both. Although the literature coverage is not fully complete through all of 1966, this massive bibliography of the laser literature up to that point has been favorably reviewed.

Goodwin, D. W. (ed.), 'Advances in Quantum Electronics, Vol. 1' (Academic Press, 111 Fifth Ave., New York 10003, 1970). 286 pp., $12.00. The first volume in this new series of detailed reviews presents articles on the carbon dioxide and YAG lasers, their development, use and potentialities; on recent advances in quantum counter action; and on interference holography.

Harvey, A. F., 'Coherent Light' (Wiley-Interscience, London, 1971). 1365 pp., $47.50. A massive intermediate-level reference book for specialists in optics and lasers, with 28 chapters ranging from Chap. 1, "Fundamentals of Electromagnetic Radiation", and Chap. 2, "Optical Properties of Media" through to Chap. 27, "Processing and Communication of Information" and Chap. 28, "Metrology and Radar Techniques". There is a bibliography of from 300 to 600 literature references for each chapter.

Heard, H. G., 'Laser Parameter Measurements Handbook' (John Wiley and Sons, Inc., New York, 1968). 489 pp., $17.50. An extensive survey, aided by many contributing authors, covering the technical and reports literature on measurement methods for determining virtually all laser parameters that might be of interest.

Levine, A. K. (ed.), 'Lasers: A Series of Advances, Vol. 1' (Marcel Dekker, New York, 1966). 365 pp., $14.50. A collection of detailed and authoritative reviews covering selected topics in laser technology (ruby lasers; solid-state lasers; organic lasers; Q-switching; and optical resonator modes), by noted experts in the field.

Levine, A. K. (ed.), 'Lasers: A Series of Advances, Vol. 2' (Marcel Dekker, Inc., New York, 1968). 450 pp., $19.75. A second volume of very good review articles by well-known experts, including Patel on gas lasers; Snitzer and Young on glass lasers; Dumke on injection diode lasers; Terhune and Maker on nonlinear optics; and Polanyi and Tobias on gas laser frequency stabilization.

Levine, Albert K., and DeMaria, Anthony J., 'Lasers: A Series of Advances, Vol. 3' (Marcel Dekker, Inc., New York, 1971). 384 pp., $22.50. A third volume in this excellent series of review papers, with articles by H. Kressel on semiconductor diode lasers, P. K. Cheo on the carbon dioxide molecular laser, and M. Bass, T. F. Deutsch and M J. Weber on laser-pumped and flashlamp-pumped dye lasers.

Pressley, Robert J. (ed.), 'Handbook of Lasers, with Selected Data on Optical Technology' (Chemical Rubber Co., Cleveland, Ohio, 1971). 655 pp., $27.50. A handbook from the publishers of the "Handbook of Chemistry and Physics", in much the same general style, containing a large amount of information but with rather widely unbalanced and spotty coverage of some important topics, and with nearly 20% of the book devoted to atmospheric transmission data alone.

Sanders, J. H., and Stevens, K. W. H. (eds.), 'Progress in Quantum Electronics' (Pergamon Press, Oxford, 1971). ??? pp., $??. Text not available for review.

Tomiyasu, K., 'The Laser Literature: An Annotated Guide' (Plenum Press, New York, 1968). 172 pp., $15.00. Lists 4000 literature references, covering the period 1963 through December 1966, indexed, and classified into 27 subject categories.

Wolf, Emil (ed.), 'Progress in Optics, Vols. V-IX' (American Elsevier Publishing Company, New York, 1966-1972). Vol. V (1966), 395 pp., $17.50; Vol. VI (1967), 404 pp., $16.50; Vol. VII (1969), 443 pp., $21.00; Vol. VIII (1970), 471 pp., $24.00; Vol. IX (1971), 437 pp., $28.50. Each of the annual volumes in this review series contains from 6 to 8 high-quality review articles on topics of current interest in optics, with increasing emphasis in recent volumes on important topics in lasers, coherent optics, and quantum electronics.

E. REPRINT COLLECTIONS, LECTURE SERIES, SYMPOSIUM PROCEEDINGS

Allen, L., 'Essentials of Lasers' (Pergamon Press, Oxford, 1969). 233 pp., $7.00. A small monograph giving a brief survey of basic laser principles in some 56 pages, followed by reprints of 12 well-chosen fundamental papers from the early laser literature.

Ambartsumyan, K. V., Basov, N. G., et al, 'Non-Resonant Feedback in Lasers' (Pergamon Press, Progress in Quantum Electronics, Vol. 1, Part 3, 1971??). 96 pp., 1.50 British pounds (paper). Paperback monograph; contents not available for review.

Barnes, Frank S. (ed.), 'Laser Theory' (IEEE Press, New York, 1972). 480 pp., $14.95 (IEEE member edition $7.50 paperbound). A reprint collection of important articles on laser theory and basic laser physics.

Chang, W. S. C. (ed.), 'Lasers and Masers' (Ohio State University Engineering Experiment Station, Columbus, 1963). 267 pp., $7.15. A symposium proceedings containing 23 papers on various laser topics presented at Ohio State in November 1962.

DeWitt, C., Blandin, A., and Cohen-Tannoudje, C. (eds.), 'Quantum Optics and Electronics: 1964 Les Houches Lectures' (Gordon and Breach, New York, 1965). 621 pp., $8.50. A set of 8 rather heavily theoretical lectures (3 in French) on quantum theory of radiation, optical maser theory and nonlinear optics, by 8 notable figures in these fields. Includes a reprint and expanded notes on Lamb's classic Phys. Rev. laser analysis paper. From the 1964 sessions of the Summer School of Theoretical Physics, University of Grenoble, held at Les Houches, France.

Fox, J. (ed.), 'Optical Masers' (Polytechnic Press, Brooklyn, 1964). 652 pp., $15.00. The proceedings of an April 1963 PIB symposium. Contains 50 papers of varying lengths, including some reasonably good survey and review papers.

Glauber, R. J. (ed.), 'Quantum Optics' (Course 42, Proceedings of the International School of Physics "Enrico Fermi", Varenna, Italy) (Academic Press, New York, 1970). 762 pp., $31.00. A fat (and expensive) collection of 29 high-quality lectures by leading experts, largely concerned with quantum and laser statistics, noise and fluctuations, parametric effects, and nonlinear optics.

Grivet, P. and Bloembergen, N. (eds.), 'Quantum Electronics III' (Columbia University Press, New York, 1964). 1923 pp., $35.00. The massive (2 volumes, 1923 pages, $35.00) proceedings of the Third International Conference on Quantum Electronics, held in Paris in February 1963. Contains several papers on every aspect of the topic by almost every worker in the laser and quantum electronics fields, giving a great deal of dross and a great deal of information.

Kaminow, Ivan P., and Siegman. A. E. (eds.), 'Laser Devices and Applications' IEEE Press, New York, 1973). 500? pp., $15.00? (IEEE member edition $7.50 paperbound). A collection of reprints of invited and review papers from the PROCEEDINGS OF THE IEEE covering laser devices and laser applications.

Kay, S. M. and Maitland, A. (eds.), 'Quantum Optics' (Academic Press, London & New York,

1970). 580 pp., 155 British shillings. Lectures presented at a NATO Advanced Study Institute and Scottish Universities' Summer School in Physics covering quantum theory of coherence, photon counting, laser theory, optical resonators, nonlinear optics, and optical pumping.

Kelley, P. L., Lax, B., and Tannenwald, P. E. (eds.), 'Physics of Quantum Electronics' (McGraw-Hill, New York, 1966). 861 pp., $24.00. Proceedings of a June 1965 conference in Puerto Rico. Concentrates on more basic physics topics, particularly nonlinear optics, Raman and Brillouin effects, spectroscopy, semiconductor lasers, gas breakdown, gas laser processes, and laser fluctuations and quantum noise theory.

Schwarz, Helmut J., and Hora, Heinrich (eds.), 'Laser Interaction and Related Plasma Phenomena, Vol 2' (Plenum Publishing Corp., New York, 19??). ??? pp., $???. An additional series of papers from a second Rensselaer Polytechnic Institute workshop, covering various types of very high power gas lasers and the gas breakdown and plasma heating effects produced by such lasers

Scientific American (ed.), 'Lasers and Light' (W. H. Freeman & Co., San Francisco, 1969). ??? pp., $5.95 (paper). A collection of Scientific American articles on lasers and optics, including the September 1968 Special Issue on Light.

Singer, J. R. (ed.), 'Advances in Quantum Electronics' (Columbia University Press, New York, 1961). 641 pp., $15.00. The one-volume proceedings of the Second Quantum Electronics Conference, held at Berkeley in March 1961. Some good early papers on lasers and other related topics are included.

Skobel'tsyn, D. V. (ed.), 'Quantum Electronics in Lasers and Masers' (Translated from Vol. 31 of the Proceedings of the P. N. Lebedev Physics Institute by Consultants Bureau, Plenum Publishing Corp., New York, 1968). 210 pp., $22.50. Articles by leading Russian scientists on gas lasers, laser action in semiconductors, regenerative laser amplifiers, and the hydrogen beam maser.

Skobel'tsyn, D. V. (ed.), 'Quantum Electronics in Lasers and Masers, Part 2' (Translated from Vol. 52 of the Proceedings of the P. N. Lebedev Physics Institute by Consultants Bureau, Plenum Publishing Corp., New York, 1972). Approx. 300 pp., $37.50 (paper). Four additional articles by Russian researchers discussing injection lasers, the effects of focused high-power laser beams on materials, and the spiking properties of laser oscillators.

Skobel'tsyn, D. V. (ed.), 'Nonlinear Optics' (Translated from the Proceedings of the P. N. Lebedev Physics Institute by Consultants Bureau, Plenum Publishing Corporation, New York, 1970). 210 pp., $22.50 (paper). A collection of entirely theoretical papers covering laser analytical models, resonator modes, laser Q-switching, laser rate equations, and a long 1965 doctoral dissertation on gas lasers by Rautian which occupies nearly half the book.

Townes, C. H. (ed.), 'Quantum Electronics' (Columbia University Press, New York, 1960). 606 pp., $15.00. The proceedings of the first

QE Conference in New York in September 1959, before lasers. Early papers may still be useful as background material.

Townes, C. H. and Miles, P. A. (eds.), 'Quantum Electronics and Coherent Light' (Academic Press, New York, 1965). 371 pp., $16.00. Collected lectures from a Varenna Summer School, giving a good overall coverage of

the quantum electronics field.

Weber, Joseph (ed.), 'Lasers' (Gordon and Breach, New York, 1968; International Science Review Series, Vols. 10a and 10b). 1400+ pp., $??. An extensive collection of reprints from the scientific literature, with limited commentary appended.

F. BOOKS ON GAS LASERS ONLY

Allen, L. and Jones, D. G. C., 'Principles of Gas Lasers' (Plenum Press, New York, and Butterworth & Co., London, 1967). 158 pp., $12.00. A brief but well-written review of the physical principles and practical details of all types of gas lasers, at a fairly advanced and detailed level, with extensive references and extensive coverage of relevant topics.

Bloom, Arnold L., 'Gas Lasers' (John Wiley and Sons, inc., New York, 1968). 172 pp., $8.95. A brief but well-written and very well-illustrated introduction to the basic principles and practical characteristics of gas lasers; important phenomena occuring in gas lasers; the properties of gas laser beams; and some of their useful applications. The author's extensive experience in both laser research and commercial laser developement make this an especially authoritative and useful book.

Garrett, G. C. B., 'Gas Lasers' (McGraw-Hill, New York, 1967). 144 pp., $10.95. A good

readable monograph on gas discharge lasers. The discussions of each topic tend to be rather brief and sketchy, but numerous literature references are given.

Sinclair, D. C. and Bell, W. E., 'Gas Laser Technology' (Holt, Rinehart and Winston, Inc., New York, 1969). 157 pp., $7.00. This book ranges over both the theoretical foundations and the technical practicalities of gas laser technology, going from rather advanced-level theory to the practical details of laser construction, but with the latter coverage limited almost completely to the He-Ne, Hg, and Argon-ion lasers only.

Stenholm, Stig, 'The Semiclassical Theory of the Gas Laser' (Pergamon Press, Progress in Quantum Electronics, Vol. 1, Part 3, 1971??). 88 pages, 1.50 British pounds (paper). A brief monograph covering the subject indicated in the title.

G. SEMICONDUCTOR AND INJECTION LASERS

Campbell, R. W., and Mims, F. W., 'Semiconductor Diode Lasers' (Howard W. Sams & Co., Indianapolis, Indiana, 1972). 192 pp., $5.95 (paperbound). An introduction to the injection laser and its applications, with heavy emphasis on the practical details of power supplies, pulsers, detectors, and transmitter and receiver optics.

Gooch, C. H. (ed.), 'Gallium Arsenide Lasers' (John Wiley and Sons, Inc., New York, 1970). 348 pp., $14.50. A series of review articles by British experts covering all aspects of GaAs injection laser theory, fabrication, laser

properties, and device applications.

Pankove, Jacques I., 'Optical Processes in Semiconductors' (Prentice-Hall, 19??). 422 pp., $21.00. Based on a series of lectures for graduate students on the physical principles of semiconductors relevant to injection lasers, electroluminescence, and other optical processes in semiconductors.

Thornton, 'The Physics of Electroluminescent Devices' (Barnes and Noble, 1967). 382 pp., $14.50. Has a large section on injection lasers and their background physics.

H. NONLINEAR OPTICS AND NONLINEAR OPTICAL DEVICES

Akhmanov, S. A., and R. V. Khokhlov, 'Problems of Nonlinear Optics' (Gordon and Breach Science Publishers, New York, 1972). 307 pp., $24.50. Translation of a very much older review monograph (references are cited only to the middle of 1963) by two well-known Russian workers covering nonlinear optical effects, optical harmonic generation, parametric effects at optical frequencies, and light modulation using nonlinear media.

Baldwin, George C., 'An Introduction to Nonlinear Optics' (Plenum Press, New York, 1969). 155 pp., $9.50. The first half of this small book reviews a number of introductory background

topics in optics, including Maxwell's equations, plane propagation, polarization, and birefringence; the second half gives brief introductory descriptions of most of the important nonlinear optical effects, including optical parametric amplification.

Bloembergen, N., 'Nonlinear Optics' (Benjamin, New York, 1965). 222 pp., $9.00 (paper). Paperback monograph covering the basic physical principles of this new area of optical physics at an advanced level, by a foremost worker in the field.

Butcher, P. N., 'Nonlinear Optical Phenomena'

(Bulletin No. 200, Engineering Experiment Station, Ohio State University, Columbus, Ohio 43210, 1964). 150 pp., $3.50 (paper). A very general and powerful quantum-density-matrix analysis of the nonlinear polarization in an assembly of molecules is derived with emphasis on symmetry properties, and this analysis is then used to discuss all the major nonlinear

optical effects.

Zernike, Fritz, and Midwinter, John E., 'Applied Nonlinear Optics' (Wiley-Interscience, New York, 1973??). 224 pp., $14.95. A non-quantum introduction to the theory and design of nonlinear optical devices, optical second-harmonic generators, and optical parametric devices.

I. HOLOGRAPHY AND OPTICAL DATA PROCESSING

Barrekette, Euval S., and Kock, Winston E. (eds.), 'Applications of Holography' (Plenum Press, New York, 1971). 404 pp., $16.50. A collection of 20 papers from a United States -Japan Seminar on Information Processing by Holography held in Washington, D.C., in October 1969.

Camatini, Ezio (ed.), 'Optical and Acoustical Holography' (Plenum Press, New York, 1972). Approx. 370 pp., $26.00. Proceedings of a NATO Advanced Study Institute on Optical and Acoustical Holography held in Milan in May 1971, with 5 general sections by 13 authors, including D. Gabor and E. N. Leith.

Caulfield, Henry J. and Sun, Lu, 'The Applications of Holography' (Wiley Series in Pure and Applied Optics, New York, 1970). 160 pp., $9.95. An introduction to holographic fundamentals for non-specialists, together with a conservative assessment of present and possible future applications, clearly indicating both the disadvantages and the advantages of holography.

Collier, Robert J., Burkhardt, Christoph D., and Lin, Lawrence H., 'Optical Holography' (Academic Press, New York, 1971). 618 pp., $22.00. A comprehensive treatment of holography, including background theory, optical techniques, properties and processing of optical recording materials, lasers useful for these purposes, and applications of holograms to imaging, spacial filtering and recognition, and three-dimensional displays.

DeVelis, John B. and Reynolds, George O., 'Theory and Applications of Holography' (Addison-Wesley Publishing Company, Reading, Mass., 1967). 196 pp., $12.95. A good brief introduction to both the theory and practice of this important new area, using the coherence theory approach.

Goodman, J. W., 'Introduction to Fourier Optics' (McGraw-Hill, New York, 1968). 288 pp., $13.50. A good elementary introduction to optical diffraction theory and its applications in image formation, spatial filtering, wavefront reconstruction, optical data processing, and holography.

Kallard, T. (ed.), 'Holography: State of the Art Review' (Optosonic Press, New York, 1969/1970).

1969 volume, 182 pp., $12.00; 1970 supplement, 240 pp., $15.00. A collection of condensed patent reprints related to holography and a bibliography on holography and related subjects, to be updated by yearly supplements.

Kiemle, Horst, and Roess, Dieter, 'Introduction to Holographic Techniques' (Plenum Publishing Corp., New York, 1971). 275 pp., $17.50. English translation of the German original, covering an introduction to holographic principles, various kinds of practical holographic techniques and their properties, and numerous examples of practical applications.

Lehman, Matt, 'Holography: Technique and Practice' (The Focal Press, London, 1971). 148 pp., $18.00. A short but very practical 'how-to-do-it' book by an experienced and skilled experimenter in photography and holography.

Robertson, E. R., and Harvey, J. M. (eds.), 'The Engineering Uses of Holography' (Cambridge University Press, New York, 1970). 620 pp., $39.50. Proceedings of a conference held at the University of Strathclyde, Glasgow, Scotland, in September 1968.

Shulman, A. R., 'Optical Data Processing' (Wiley-Interscience, John Wiley and Sons, New York, 1970). 704 pp., $32.50. Covers the general technology of optical data processing and the necessary background for understanding coherent optical data processing techniques, together with specific practical examples.

Smith, Howard, M., 'Principles of Holography' (John Wiley and Sons, Inc., New York, 1969). 239 pp., $9.95. A broad coverage of the history, the theory and practice, and the applications of holography, providing both a good introduction to and a good general reference work on this subject.

Stroke, G. W., 'An Introduction to Coherent Optics and Holography' (Academic Press, New York, 1966; Second Edition, 1969). 270 pp., $10.00. Fundamental theory and applications of diffraction, image formation, wavefront reconstruction, and spatial frequency transformation, together with reprints of three classic early papers by D. Gabor.

J. LASER APPLICATIONS

Beesley, M. J., 'Lasers and Their Applications' (Taylor & Francis, Ltd. London, and Barnes and Noble, Inc., New York, 1971). 246 pp., $16.00 (5.00 British pounds). A useful survey book at

a comparatively elementary level, divided roughly evenly between the theory and properties of lasers and the various applications of lasers, including holography.

Brown, Ronald, 'Lasers: A Survey of Their Performance and Applications' International Publications Service, ????, 1969). ??? pp., $23.75. Not available for review.

Charschan, S. S. (ed.), 'Lasers in Industry' (Van Nostrand Reinhold Co., New York, 1972). 640 pp., $24.50. An authoritative book on laser applications, especially materials processing and interferometric applications, by a group of Western Electric Company experts.

Demtroder, W., 'Laser Spectroscopy' (Springer-Verlag, New York, 1971). 95 pp., $8.10 (paper). Volume 17 of the "Topics in Current Chemistry" series. Summarizes the characteristics of lasers as spectroscopic sources, and then surveys the applications of lasers to spectroscopy, including photochemistry, laser photolysis, light scattering, absorption, fluorescence and Raman spectroscopy, plasma spectroscopy, chemical microanalysis, and high- resolution saturated-absorption spectroscopy.

Elion, H. A., 'Lasers Systems and Applications' (Pergamon Press, Inc., New York, 1967). 624 pp., $22.50. The first one-fourth of this book is devoted to a rather uneven review of potential laser applications; most of the remainder is a reprinted NASA bibliography on lasers, available at less cost directly from the Government Printing Office, and here minus the abstracts included in the NASA original.

Gibson, T. R. and Hendra, P. J., 'Laser Raman Spectroscopy' (Wiley Interscience, London, 1970). 266 pp., $15.00. A practical book written with the chemist in mind, describing general instrument design, commercial laser Raman instruments, the Raman effects observed in various kinds of materials, sample handling, and extensive references to the literature.

Goldman, Leon, M.D., 'Laser Cancer Research' (Springer-Verlag, New York, 1966). 73 pp., $4.00. An early and apparently rather hastily prepared brief introduction to a research topic whose lasting importance is still to be determined.

Goldman, Leon, M.D., 'Biomedical Aspects of the Laser' (Springer-Verlag, New York, 1967). 232 pp., $11.40. A review of available laser instrumentation relevant to biomedical applications and of current research results and future prospects for the laser in biomedicine.

Goldman, Leon, M.D., and Rockwell, R. James, Jr., 'Lasers in Medicine' (Gordon and Breach, London, 1971). 385 pp., $19.50. Text not available for review.

Goldman, Leon, M.D., and Rockwell, R. James, Jr., 'Application of the Laser' (Chemical Rubber Co., Cleveland, Ohio, 1972). Approx. 200 pp., $25.00. Applications of the laser in industry, science and technology, medicine and biology, and in the military, including safety considerations when working with lasers.

International Atomic Energy Agency (eds.), 'Laser Applications in Plasma Physics (1962-1968)' (International Atomic Energy Agency, Vienna, 1969). 228 pp., $6.00 (paper). A very brief (30 pages) introductory review of plasma diagnostic methods using lasers and of plasma production using focused laser beams in gases and solids is followed by a bibliography of 735 references with abstracts, including subject and author indices.

Loader, E. J., 'Basic Laser Raman Spectroscopy' (Heyden and Son, Limited, London, 1970). 105 pp., $4.50. This small book is intended as a practical guide for chemists who are adopting Raman spectroscopy. It explains the Raman effect in non-mathematical terms, explains how samples may be investigated by this technique, and cites calibration data for individual Raman spectra.

Marshall, Samuel L. (ed.), 'Laser Technology and Applications' (McGraw-Hill, New York, 1968). 294 pp., $14.00. A series of chapters by separate authors, aimed "for engineers and physicists on a senior or first-year graduate level," covering the major types of lasers, laser technology, and laser application.

McGuff, P. E., 'Surgical Applications of Laser' (Charles C. Thomas, Springfield, 1966). 224 pp., $10.50. An opening look at what may become an important laser application.

Pratt, W. K., 'Laser Communications Systems' (John Wiley and Sons, Inc., New York, 1969). 271 pp., $14.95. General discussions and useful factual information on laser modulation, propagation, and detection, and on laser communication systems concepts.

Ready, John F., 'Effects of High Power Laser Radiation' (Academic Press, New York, 1971). 448 pp., $17.50. A comprehensive treatment of high power laser effects and applications including heating, melting, vaporization, particle emission, plasma production, gas breakdown, breakdown of transparent materials, and biological effects, including experimental data, industrial applications, and background information on lasers useful for these purposes.

Ross, Monte, 'Laser Receivers: Devices, Techniques, Systems' (Wiley, New York, 1966). 405 pp., $14.95. A survey of laser modulation and detection methods, including signal and noise considerations and systems considerations.

Ross, Monte (ed.), 'Laser Applications, Vol. 1' (Academic Press, New York, 1972). 320 pp., $16.00. A series intended to "fill the gap between laser research...and its practical applications", with articles on holography, metrology and geodesy, the laser gyro, machining and welding, and laser communications in the first volume.

Schwarz, Helmut J. and Hora, Heinrich (eds.), 'Laser Interaction and Related Plasma Phenomena' (Plenum Publishing Corp., New York, 1971). 509 pp., $25.00. A series of papers from a workshop at Rensselaer Polytechnic Institute introducing and explaining the uses of laser technology in the field of plasma physics, including thermonuclear physics, thermophysics, and spectroscopy, as well as creating, diagnosing, and photographing plasmas.

Smith, W. V., 'Laser Applications' (Artech House, Dedham, Mass., 1971). 199 pp., $9.95. An introduction to the characteristics of lasers and their relation to laser applications, including descriptions of time and space coherence, nonlinear optics, diffraction, holography, and picosecond pulses. Includes problems and solutions.

Tobin, Marvin C., 'Laser Raman Spectroscopy' (Wiley-Interscience, New York, 1971). 180 pp., $13.50. Another review of the techniques and applications of laser Raman spectroscopy, from Vol. 35 of "Chemical Analysis: A Series of

Monographs on Analytical Chemistry and its Applications".

Wolbarsht, M. L. (ed.), 'Laser Applications in Medicine and Biology' Vol. 1 (Plenum Press, New York, 1971). 302 pp., $18.00. Nine articles on the use of laser radiation in medicine, cancer research, cell biology, dentistry and opthalmology, including laser safety and ocular damage criteria.

K. MISCELLANEOUS SPECIAL TOPICS: MICROWAVE MASERS, COHERENCE, GLASS LASERS, RESONATORS

Albers, Walter A. (ed.), 'The Physics of Opto-Electronic Materials' (Plenum Press, New York, 1971). 281 pp., $16.50. Thirteen papers on light-controlling processes and light modulation materials presented at a symposium held in 1970.

Klauder, John R. and Sudarshan, E. C. G., 'Fundamentals of Quantum Optics' (W. A. Benjamin, New York, 1968). 279 pp., $13.50. A theoretical treatment of the detailed quantum theory of optical coherence.

Lengyel, B. A., 'Lasers' (Wiley, New York, 1961). 125 pp., $6.95. A brief and now somewhat dated survey of fundamentals and early experimental results on lasers.

Marcuse, Dietrich, 'Light Transmission Optics' (Bell Laboratories Series, Van Nostrand Reinhold Co., Cincinnati, Ohio, 1972?). 444 pp., $24.50. Covers geometrical and wave optics of rays and gaussian optical beams propagating in lenses, lens waveguides, ducts, optical fibers, and optical dielectric waveguides.

Milek, J. T., and M. Neuberger, 'Linear Electrooptic Modulator Materials' (Plenum Publishing Company??, New York, 1972). 258 pp., $22.50. A review of the principles of electrooptic Pockels-cell modulators, and a comprehensive survey of the electrooptic materials properties of 13 important electrooptic crystals.

Orton, J. W., Paxton, D. H., and Walling, J. C., 'The Solid State Maser' (Pergamon Press, Oxford, 1970). 290 pp., $ not available. A brief but well-done review of the microwave solid-state maser and its applications in the opening part of the book is followed by reprints (actually, all reset in new type) of 16 major early papers of historical interest in this field.

Patek, K., 'Glass Lasers' (The Chemical Rubber Co., Cleveland, 1970). 225 pp., $32.50. Detailed coverage of the preparation and properties of glass laser materials, and the design and operation of glass lasers.

Siegman, A. E., 'Microwave Solid-State Masers' (McGraw-Hill, New York, 1964). 583 pp., $18.50. Extensive coverage of the background, theory, and techniques for the microwave maser. Many sections are relevant as introductory background or supplemental material for lasers as well.

Singer, J. R., 'Masers' (Wiley, New York, 1959). 147 pp., $6.50. The first book published on masers, pre-lasers. Good introduction to ideas, description of early microwave and ammonia masers.

Skobel'tsyn, D. V. (ed.), 'Quantum Electronics and Paramagnetic Resonance' (Translated from Vol. 49 of the Proceedings of the P. N. Lebedev Physics Institute by Consultants Bureau, Plenum Publishing Corp, New York, ????). 148 pp., $20.00. Two papers by Russian workers, covering the microwave solid-state maser as a radio receiver, plus studies in microwave magnetic resonance.

Steele, Earl L., 'Optical Lasers in Electronics' (John Wiley and Sons, Inc., New York, 1968). 267 pp., $11.95. Covers primarily topics related to ruby lasers, ruby laser amplifiers, Q-switching, and a chapter on injection lasers.

Stepin, L. D., 'Quantum Radio Frequency Physics' (MIT Press, Cambridge, Mass., 1965). 227 pp., $7.50. Primarily concerned with the physics and analysis of electron paramagnetic resonance (EPR) and nuclear magnetic resonance (NMR), with some brief coverage of radio-frequency and microwave magnetic-resonance masers. Translated from a Russian original.

Troup, G., 'Masers and Lasers' (Methuen, London, 1963; Wiley, New York, 1964; Second Edition). 192 pp., $4.50. Another brief Methuen's Monograph, originally concerned with microwave masers only, and rather sketchily up-dated to mention lasers also.

Weber, J. (ed.), 'Masers' (Gordon and Breach, New York, 1967; International Science Review Series, Vol. 9). 848 pp., $15.00. A collection of verbatim reprints of nearly 150 journal articles on microwave maser frequency standards and microwave solid-state masers.

Weinstein, Lev Albertovich, 'Open Resonators and Open Waveguides' (Golem Press, Boulder, Colorado, 1969). 439 pp., $15.00. Translation of a Russian book by a Soviet scientist (often also identified as Vainshtein) who has done important work on optical resonator theory using an analytical approach not commonly employed elsewhere.

L. FOREIGN-LANGUAGE BOOKS (I.E., OTHER THAN ENGLISH)

Akhmanov, S. A., et al., 'Quantum Electronics' (Soviet Electronics, 1969). pp. & $ not available. A review of all areas of quantum electronics, followed by an extensive alphabetical encyclopedia covering many terms and concepts used in quantum electronics (in Russian).

Bernard, M-Y., 'Masers and Lasers' (Presses Universitaires de France, Paris, 1964). 148

pp., $3.50. General review and textbook on these topics (in French).

Brand, H., 'Laser' (F. Dummlers Verlag, Bonn, 1966). pp. & $ not available. (In German). Text not available for review.

Brandli, H. D., 'Laserphysik' (Verlag Hallwag, Bern und Stuttgart, 1966). 81 pp., 18.80 Swiss francs (paper). An introductory paperback text for technicians. (In German).

Brown, Ronald, 'Laser' (Deutsche Verlags-Anstalt, Stuttgart, 1970). 124 pp., 18.40 German marks (paper). German translation of the well-recommended introductory paperback 'Lasers: Tools of Modern Technology' by R. Brown.

Carroll, John M., 'Todesstrahlen - Die Geschichte des Laser' (Verlag Ullstein GMBH, Berlin, 1965). 136 pp., 16 German marks. German translation of Carroll, 'The Story of the Laser'.

Cucurezeanu, I., 'Laseri' (Editora Academiei Republicii Socialiste Romania, Bucarest, 1966). 270 pp., 12 Rumanian lei. (In Rumanian). Reported to be a rather thorough, complete and up-to-date survey; cf. review by J. N. Howard in Applied Optics, May 1967.

Francon, M., 'Holographie' (Masson and Cie, Paris, 1969). 124 pp., 40 French francs. A well-reviewed brief introduction to the principles and applications of holography by a well-known expert in optics. (In French).

Gurs, Karl, 'Laser' (Umschau Verlag, Frankfurt am Main, 1970). 583 pp., 27 German marks. A well-illustrated paperback by a well-known industrial laser researcher, introducing the important major types of lasers, and their important areas of application in science and technology. (In German).

Ishchenko, E. F. and Klimkov, Yu. M., 'Lasers' (Soviet Radio Press, Moscow, 1968). pp. & $ not available. A general coverage of lasers and potential applications (in Russian).

Japan Electrical and Communication Society (ed.), 'Introduction to Lasers' (Japan Electrical and Communication Society, Tokyo, 1965). 256 pp., 700 Yen. A collection of reviews on lasers, including solid-state, gas and semiconductor types and their applications to communications, information, metrology, etc. (In Japanese).

Japan Physical Society (ed.), 'Quantum Electronics' (Asakura Book Co., Tokyo, 1965). 326 pp., 1800 Yen. A collection of detailed reviews covering most topics in lasers and quantum electronics by leading Japanese researchers. (In Japanese).

Kiemle, Horst and Roess, D., 'Einfuhrung in die Technik der Holographie' (Akademische Verlag gesellschaft, Frankfurt am Main, 1969). 334 pp., 67 German marks. (In German). Text not available for review.

Kleen, W. and Muller, R. (eds.), 'Laser' (Springer-Verlag, Berlin, 1969). 567 pp., $27.00. Ten sections by German authorities in the field (G. Grau, K. Gurs, W. Kleen, R. Muller, D. Rosenberger, G. H. Winstel) giving fairly advanced and detailed reviews of atomic spectroscopy, laser theory, gas, solid-state and semiconductor lasers, laser techniques, and applications, with extensive references to the literature (in German).

Klinger, H. H., 'Laser' (Franckh'sche Verlagshandlung, Stuttgart, 1964). pp. & $ not

available. (In German). Text not available for review.

Lawrence, L. G., 'Grundlagen der Lasertechnik' (Winter'sche Verlagshandlung, Prien, 1964). pp. & $ not available. (In German). Text not available for review.

Mikaelian, A. L., Ter-Mikaelian, M. L., and Turkov, Yu. G., 'Solid State Lasers' (Sovetskoe Radio, Moscow, 1967). 384 pp., $ not available. (In Russian). Text not available for review.

Mollwo, E. and Kaule, W., 'Maser und Laser' (B. I. Hochschultaschenbucher No. 79/79a, Bibliographisches Institut AG, Mannheim, 1966). 214 pp., 7.90 German marks. A paperback textbook covering the principles of maser action (including quantum concepts) and their application in microwave and optical masers. (In German).

Orszag, A., 'Les Lasers - Principes, Realisations, Applications' (Masson et Cie, Paris, 1968). 176 pp., $10.00. "Pleasant little book...excellent introduction to lasers," according to review by W. J. Condell in J. Opt. Soc. Am. 59, 1388 (October 1969). (In French).

Ostrovsky, Yu. I., 'Holography' (Nauka Press, Leningrad, 1970). 124 pp., 40 Russian kopeks. Reported to be a good brief survey of the standard techniques of holography with illustrations and references. (In Russian).

Paul, Harry, 'Lasertheorie, Vols. I and II' (Akademie-Verlag, Berlin, 1969). 184 pp., 8 German marks each volume (paper). Two small paperback monographs reviewing the theoretical aspects of lasers and quantum electronics. (In German).

Rieck, H., 'Halbleiter-Laser' (G. Braun, Karlsruhe, 1967). pp. & $ not available. (In German). Text not available for review.

Roess, D., 'Laser-Lichtverstaerker und -Oszillatoren' (Akademische Verlagsgellschaft, Frankfurt, 1966). 756 pp., 98 German marks. Extensive and detailed advanced-level coverage of lasers by a well-known German expert in the field. Recommended. (In German).

Schubert, Max and Wilhelmi, Bernd, 'Einfuhrung in die Nichtlineare Optik' (BSB B.G. Teubner Verlagsgesellschaft Leipzig, 1970). 160 pp., 18.50 German marks. (In German). Text not available for review.

Schwaiger, Egloff, 'Laser - Light von morgen' (Franz Ehrenwirth Verlag AG, Munchen, 1969). 231 pp., $ not available. A well-written popularized discussion of laser applications, particularly in holography (in German).

Sminov, V. M. et al., '----' (Soviet publication) pp. & $ not available. Book on welding and drilling with lasers (in Russian).

Stepanov, B. I. (ed.), 'Methods for Calculation of Lasers, Vols. 1 and 2' (Science and Technology Press, Minsk, 1966 and 1969). 484 pp. & 656 pp., $ not available. A collection of works of researchers at the Physics Institute of the Byelorussian S.S.R. Academy of Sciences. (In Russian).

Svelto, Orazio, 'Principi Del Laser' (Tamburini Editore, Milano, 1970). 262 pp., $ not available (paper). A good introduction to the basic physical principles and the general device characteristics of lasers, using a semiclassical approach to the theory (in Italian).

Tradowsky, K., 'Laser - kurz and bundig' (Vogel-Verlag, Wurzburg, 1968). 148 pp., 16.80 German marks (paper). Well-illustrated elementary-level paperback introduction to lasers and related optical techniques. (In German).

Unger, H-G, 'Quantenelektronik' (Hochschul-Lehrbuch, Friedr-Vieweg and Sohn, Braunschweig, 1968). 160 pp., 9.80 German marks (paper). A textbook for students in electrotechnique and physics, emphasizing the physical principles and quantum theory of laser and maser devices (in German).

Vienot, Jean-Charles, Smigielski, Paul, and Royer, Henri, 'Holographie Optique, Developpements,

Applications' (Dunod Editeur, Paris, ????). 218 pp., 68 Fr francs (paper). A well-reviewed summary of holographic theory, techniques, and applications. In French.

Volfson, N. S., and Shitova, E. I., 'Quantum Optical Generators - Lasers' (Science Publishers for the Academy of Sciences, Moscow, 1964). 176 pp., $ not available. A bibliographic index of Russian and non-Russian laser literature. (In Russian).

Yazawa, K., 'Introduction to Masers' (Ohm Book Co., Tokyo, 1964). 208 pp., 320 Yen. An introduction to the basic principles of masers, and discussion of several practical maser systems. (In Japanese).

Author Index

Subject Index

492

Editors' Biographies

Ivan P. Kaminow (A'54–M'59–SM'73) was born in Union City, N. J., on March 3, 1930. He received the B.S.E.E. degree from Union College, Schenectady, N. Y., in 1952, the M.S.E. degree from the University of California, Los Angeles, in 1954, and the A.M. and Ph.D. degrees from Harvard University, Cambridge, Mass., in 1957 and 1960, respectively. He was a Hughes Masters Fellow at UCLA from 1952 to 1954, and a Bell Laboratories Communications Development Training Program Fellow at Harvard from 1956 to 1960.

He was a member of the Technical Staff in the Microwave Laboratory, Hughes Aircraft Company, Culver City, Calif., from 1952 to 1954. He was a member of the Technical Staff in the Military Research Department, Bell Laboratories, Whippany, N. J., from 1954 to 1960, and since 1960 has been with the Research Department, Bell Laboratories, Holmdel, N. J. He has done research on microwave antennas, ferrite devices, ferrimagnetic resonance, high-pressure physics, ferroelectrics, electrooptic and nonlinear optical materials, Raman scattering, light modulation devices, integrated optics, and optical communications. He was a Visiting Lecturer at Princeton University in 1968, and has also given courses in nonlinear optics and dielectrics at U.C.L.A. and Bell Laboratories. He is the author of a forthcoming book on *Electrooptic Devices* (New York: Academic Press).

Dr. Kaminow is a member of the American Physical Society and the American Optical Society. He is a member of the Editorial Board of the PROCEEDINGS OF THE IEEE, and of the Awards Board and the Subcommittee on Ferroelectrics, and he is also Chairman of the Prize Papers Committee.

A. E. Siegman (S'54–M'57–F'66) was born in Detroit, Mich., on November 23, 1931. He received the A.B. degree summa cum laude in engineering science from Harvard University, Cambridge, Mass., in 1952. He studied at the University of California, Los Angeles, under the Hughes Cooperative Plan, receiving the M.S. degree in applied physics in 1954. From 1954 to 1957 he was a National Science Foundation Fellow and Research Assistant at Stanford University, Stanford, Calif., where he received the Ph.D. degree in electrical engineering in 1957.

Since 1957 he has been on the faculty at Stanford University, where he is now Professor of Electrical Engineering. During the spring semester of 1965 he served as Visiting Professor of Applied Physics at Harvard University. He has published numerous technical articles in the fields of microwave electronics, parametric devices and microwave solid state masers, the modulation and demodulation of light, and laser devices and their applications. He is the author of *Microwave Solid-State Masers* (New York: McGraw-Hill, 1964), and *An Introduction to Lasers* (New York: McGraw-Hill, 1970). He holds two patents on microwave-frequency photodevices. In 1969 he was awarded the John Simon Guggenheim Fellowship, and was on sabbatical leave at the IBM Research Laboratory, Zurich, Switzerland, from 1969 to 1970. He currently directs a research program in laser devices and their applications at Stanford. He also serves as a Consultant to the Sylvania Electro-Optics Organization, United Aircraft Research Laboratories, and other companies.

Dr. Siegman is a fellow of the Optical Society of America, and a member of the American Physical Society, the American Association for the Advancement of Science, the American Society for Engineering Education, the American Association of University Professors, Phi Beta Kappa, and Sigma Xi. He is a member of the Editorial Boards of the PROCEEDINGS OF THE IEEE and the IEEE PRESS.